FORMULAS FROM GEOMETRY

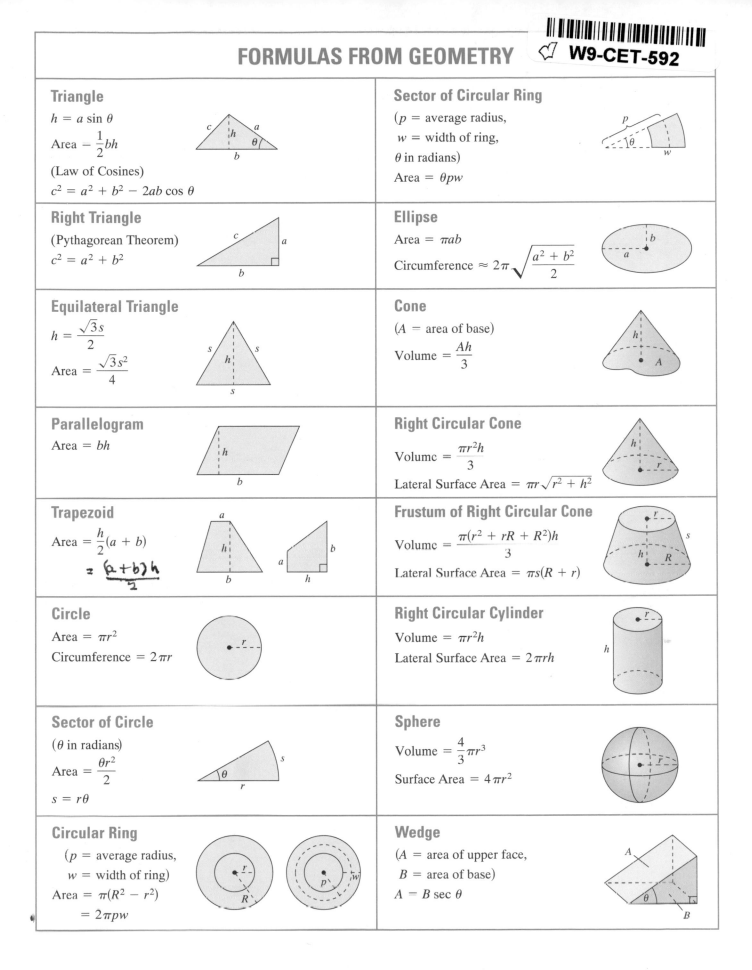

Triangle

$h = a \sin \theta$

Area $= \dfrac{1}{2}bh$

(Law of Cosines)

$c^2 = a^2 + b^2 - 2ab \cos \theta$

Right Triangle

(Pythagorean Theorem)

$c^2 = a^2 + b^2$

Equilateral Triangle

$h = \dfrac{\sqrt{3}s}{2}$

Area $= \dfrac{\sqrt{3}s^2}{4}$

Parallelogram

Area $= bh$

Trapezoid

Area $= \dfrac{h}{2}(a + b)$

$= \dfrac{(a+b)h}{2}$

Circle

Area $= \pi r^2$

Circumference $= 2\pi r$

Sector of Circle

(θ in radians)

Area $= \dfrac{\theta r^2}{2}$

$s = r\theta$

Circular Ring

(p = average radius,
w = width of ring)

Area $= \pi(R^2 - r^2)$

$= 2\pi pw$

Sector of Circular Ring

(p = average radius,
w = width of ring,
θ in radians)

Area $= \theta pw$

Ellipse

Area $= \pi ab$

Circumference $\approx 2\pi\sqrt{\dfrac{a^2 + b^2}{2}}$

Cone

(A = area of base)

Volume $= \dfrac{Ah}{3}$

Right Circular Cone

Volume $= \dfrac{\pi r^2 h}{3}$

Lateral Surface Area $= \pi r\sqrt{r^2 + h^2}$

Frustum of Right Circular Cone

Volume $= \dfrac{\pi(r^2 + rR + R^2)h}{3}$

Lateral Surface Area $= \pi s(R + r)$

Right Circular Cylinder

Volume $= \pi r^2 h$

Lateral Surface Area $= 2\pi rh$

Sphere

Volume $= \dfrac{4}{3}\pi r^3$

Surface Area $= 4\pi r^2$

Wedge

(A = area of upper face,
B = area of base)

$A = B \sec \theta$

Calculus

of a Single Variable

Seventh Edition

Calculus

of a Single Variable

Seventh Edition

Ron Larson
Robert P. Hostetler
The Pennsylvania State University
The Behrend College

Bruce H. Edwards
University of Florida

with the assistance of
David E. Heyd
The Pennsylvania State University
The Behrend College

Houghton Mifflin Company   Boston   New York

Editor in Chief, Mathematics: Jack Shira
Managing Editor: Cathy Cantin
Development Manager: Maureen Ross
Development Editor: Laura Wheel
Assistant Editor: Rosalind Horn
Supervising Editor: Karen Carter
Project Editor: Patty Bergin
Editorial Assistant: Lindsey Gulden
Production Technology Supervisor: Gary Crespo
Senior Marketing Manager: Michael Busnach
Marketing Assistant: Nicole Mollica

We have included examples and exercises that use real-life data as well as technology output from a variety of software. This would not have been possible without the help of many people and organizations. Our wholehearted thanks goes to all for their time and effort.

Trademark Acknowledgments: TI is a registered trademark of Texas Instruments, Inc. Mathcad is a registered trademark of MathSoft, Inc. Windows, Microsoft, and MS-DOS are registered trademarks of Microsoft, Inc. Mathematica is a registered trademark of Wolfram Research, Inc. DERIVE is a registered trademark of Texas Instruments, Inc. IBM is a registered trademark of International Business Machines Corporation. Maple is a registered trademark of Waterloo Maple, Inc. HMClassPrep is a trademark of Houghton Mifflin Company.

Printed in the U.S.A.

Library of Congress Control Number: 2001089302

ISBN: 0-618-14916-3

23456789–VH–05 04 03 02 01

Contents

Chapter 6	**Applications of Integration 410**

Chapter 7	**Integration Techniques, L'Hôpital's Rule, and Improper Integrals 480**

*Available in *e-solutions Calculus Learning Tools Student CD-ROM* and at the text-specific website at *college.hmco.com*.

A Word from the Authors

Welcome to *Calculus of a Single Variable*, Seventh Edition. Much has changed since we wrote the first edition—nearly 25 years ago. With each edition, we have listened to you, our users, and have tried to incorporate your suggestions for improvement.

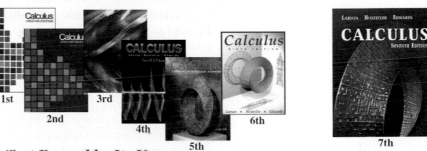

A Text Formed by Its Users

Through your support and suggestions, the text has evolved over seven editions to include these extensive enhancements:

- Expanded exercise sets containing a greater variety of tasks such as skill building, applications, explorations, writing, critical thinking, and theoretical problems
- Additional applications that more accurately represent the diverse uses of calculus in the world
- Many more open-ended activities and investigations
- Clearer, less cluttered text, full annotations and labels—carefully planned page layout
- Additional art, composed with more color, accuracy, and realism
- A more comprehensive and more mathematically rigorous text
- Increased technology use, as both a problem-solving tool and an investigative tool
- References to the history of calculus and to the mathematicians who developed it
- Updated references to current mathematical journals
- Considerably more help in the supplements package for both students and instructors
- Alternatives to the traditional print medium, particularly in the CD-ROM version
- Five different volumes from which to choose your preferred teaching approach (see page xx)

What's New and Different in the Seventh Edition

In the Seventh Edition, we continue to offer instructors and students a text that is pedagogically sound, mathematically precise, and comprehensible. There are many minor changes in the mathematics, prose, art, and design. The more significant changes are noted here.

- *New* **P.S. Problem Solving** At the end of each chapter, we have included a two-page collection of new applied and theoretical exercises. These exercises offer problems that have some unusual characteristics that set them apart from exercises in a regular exercise set.

- *New* **Getting at the Concept** Midway through each section exercise set we have added a set of problems that check a student's understanding of the basic concepts presented in the section.

- *New* **Section Objectives** Each section in the Seventh Edition begins with a list of learning objectives. These enable students to identify and focus on the key points of the section.

- *New* **Downloadable Graphs** Many exercise sets contain problems in which students are asked to draw on the graph that is provided. Because this is not feasible in the actual text, we now provide printable enlargements of these graphs on the website *www.mathgraphs.com.*

- *New* **Journal Articles on the Web** The Seventh Edition contains over 60 references to articles from mathematics journals noted in the feature *For Further Information.* In order to make the articles easily accessible to instructors and students, they are now available on the website *www.matharticles.com.*

- *Revised* **Chapter Openers** The chapter openers have been redesigned as two-page spreads in the Seventh Edition. Included in the chapter openers is a real-world application designed to motivate the calculus topics of the chapter.

- *Revised* **Review Exercises** In order to provide a more effective study tool, we have grouped the Review Exercises by text section. This reorganization allows students to target specific concepts that may require additional study and review.

- **Exercise Sets** Approximately 20 percent of the exercises in the Seventh Edition are new. The new exercises include skill, concept, applied, and theoretical problems.

- **Table of Contents** Although the organization of the table of contents is much the same as in the Sixth Edition, some notable changes are as follows. In an effort to cut back on the length of the text, we have moved Section 3.10 *Business and Economic Applications* (Appendix G in the Seventh Edition), Appendix A *Precalculus Review* (Appendix D in the Seventh Edition), Appendix E *Rotation and the General Second-Degree Equation*, Appendix F *Complex Numbers* to the text-specific website at *college.hmco.com.* We removed Appendix C *Basic Differentiation Rules for Elementary Functions* from the text; however, that material appears on the inside front cover of the text.

We have expanded coverage of differential equations topics in a new Appendix A (*Additional Topics in Differential Equations*). Topics include slope fields, Euler's Method, and first-order linear differential equations.

Although we carefully and thoroughly revised the text by enhancing the usefulness of some features and topics and by adding others, we did not change many of the things that our colleagues and the two million students who have used this book have told us work for them. We still offer comprehensive coverage of the material required by students in a three-semester or four-quarter calculus course, including carefully stated theories and proofs.

We hope you will enjoy the Seventh Edition. We are proud to have it as our first calculus book to be published in the twenty-first century.

Ron Larson   *Robert P. Hostetler*   *Bruce H. Edwards*

Features

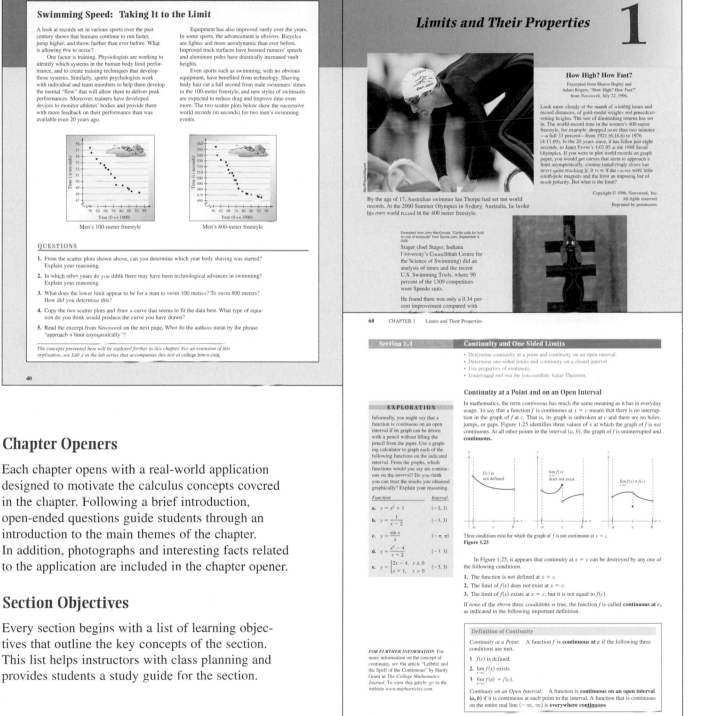

Chapter Openers

Each chapter opens with a real-world application designed to motivate the calculus concepts covered in the chapter. Following a brief introduction, open-ended questions guide students through an introduction to the main themes of the chapter. In addition, photographs and interesting facts related to the application are included in the chapter opener.

Section Objectives

Every section begins with a list of learning objectives that outline the key concepts of the section. This list helps instructors with class planning and provides students a study guide for the section.

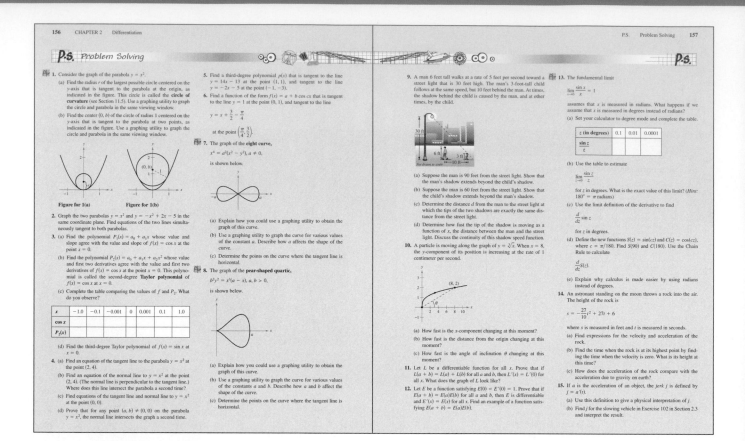

New! P.S. Problem Solving

Each chapter concludes with a collection of thought-provoking and challenging exercises that further explore and expand upon the concepts of the chapter. These exercises have unusual characteristics that set them apart from traditional calculus exercises.

Review Exercises

A set of *Review Exercises* is included at the end of each chapter. In order to provide students with a more useful study tool, these exercises are grouped by section. This organization allows students to identify specific problem types related to chapter concepts for study and review.

Getting at the Concept

These exercises contain questions that check a student's understanding of the basic concepts of the section. They are generally located midway through the section exercise sets and are boxed and titled for easy reference.

Section Projects

Appearing at the end of selected exercise sets, the *Section Projects* contain extended applications, which can be assigned as an individual or group activity.

Getting at the Concept

45. Let f be continuous on $[a, b]$ and differentiable on (a, b). If there exists c in (a, b) such that $f'(c) = 0$, does it follow that $f(a) = f(b)$? Explain.

46. Let f be continuous on the closed interval $[a, b]$ and differentiable on the open interval (a, b). Also, suppose that $f(a) = f(b)$ and that c is a real number in the interval such that $f'(c) = 0$. Find an interval for the function g over which Rolle's Theorem can be applied, and find the corresponding critical number of g (k is a constant).
    (a) $g(x) = f(x) + k$ (b) $g(x) = f(x - k)$
    (c) $g(x) = f(kx)$

47. A plane begins its takeoff at 2:00 P.M. on a 2500-mile flight. The plane arrives at its destination at 7:30 P.M. Explain why there were at least two times during the flight when the speed of the plane was 400 miles per hour.

48. When an object is removed from a furnace and placed in an environment with a constant temperature of 90°F, its core temperature is 1500°F. Five hours later the core temperature is 390°F. Explain why there must exist a time in the interval when the temperature is decreasing at a rate of 222°F per hour.

SECTION PROJECT RAINBOWS

Rainbows are formed when light strikes raindrops and is reflected and refracted, as shown in the figure. (This figure shows a cross section of a spherical raindrop.) The Law of Refraction states that $(\sin \alpha)/(\sin \beta) = k$, where $k \approx 1.33$ (for water). The angle of deflection is given by $D = \pi + 2\alpha - 4\beta$.

(a) Sketch the graph of D for $0 \le \alpha \le \pi/2$. Use a graphing utility with
$$D = \pi + 2\alpha - 4\sin^{-1}\left(\frac{1}{k}\sin\alpha\right).$$

(b) Prove that the minimum angle of deflection occurs when
$$\cos\alpha = \sqrt{\frac{k^2 - 1}{3}}.$$

For water, what is the minimum angle of deflection, D_{min}? (The angle $\pi - D_{min}$ is called the *rainbow angle*.) What value of α produces this minimum angle? (A ray of sunlight that strikes a raindrop at this angle, α, is called a *rainbow ray*.)

FOR FURTHER INFORMATION For more information about the mathematics of rainbows, see the article "Somewhere Within the Rainbow" by Steven Janke in *The UMAP Journal*. To view this article, go to the website *www.matharticles.com*.

Open Explorations

The *Interactive* CD-ROM version of this text contains open explorations, which further investigate selected examples throughout the text using computer algebra systems (*Maple*, *Mathematica*, *Derive*, and *Mathcad*). The icon 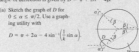 identifies an example for which an open exploration exists.

Additional Features

Additional teaching and learning resources can be found throughout the text. These resources include explorations, technology notes, historical vignettes, study tips, journal references, lab series, and notes. For a complete description of these resources, go to the text-specific website at *college.hmco.com*.

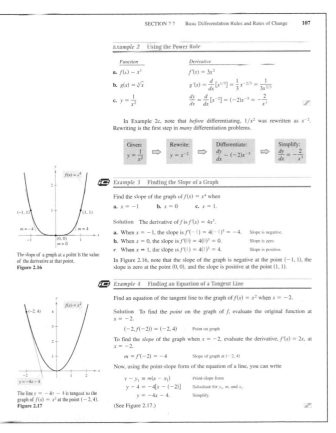

Example 2   Using the Power Rule

Function	Derivative
a. $f(x) = x^3$	$f'(x) = 3x^2$
b. $g(x) = \sqrt[3]{x}$	$g'(x) = \frac{d}{dx}[x^{1/3}] = \frac{1}{3}x^{-2/3} = \frac{1}{3x^{2/3}}$
c. $y = \frac{1}{x^2}$	$\frac{dy}{dx} = \frac{d}{dx}[x^{-2}] = (-2)x^{-3} = -\frac{2}{x^3}$

In Example 2c, note that *before* differentiating, $1/x^2$ was rewritten as x^{-2}. Rewriting is the first step in *many* differentiation problems.

Given:	Rewrite:	Differentiate:	Simplify:
$y = \dfrac{1}{x^2}$	$y = x^{-2}$	$\dfrac{dy}{dx} = (-2)x^{-3}$	$\dfrac{dy}{dx} = -\dfrac{2}{x^3}$

Example 3   Finding the Slope of a Graph

Find the slope of the graph of $f(x) = x^4$ when
a. $x = -1$ **b.** $x = 0$ **c.** $x = 1$.

Solution The derivative of f is $f'(x) = 4x^3$.

a. When $x = -1$, the slope is $f'(-1) = 4(-1)^3 = -4$.   *Slope is negative.*
b. When $x = 0$, the slope is $f'(0) = 4(0)^3 = 0$.   *Slope is zero.*
c. When $x = 1$, the slope is $f'(1) = 4(1)^3 = 4$.   *Slope is positive.*

In Figure 2.16, note that the slope of the graph is negative at the point $(-1, 1)$, the slope is zero at the point $(0, 0)$, and the slope is positive at the point $(1, 1)$.

The slope of a graph at a point is the value of the derivative at that point.
Figure 2.16

Example 4   Finding an Equation of a Tangent Line

Find an equation of the tangent line to the graph of $f(x) = x^2$ when $x = -2$.

Solution To find the *point* on the graph of f, evaluate the original function at $x = -2$.
$$(-2, f(-2)) = (-2, 4) \quad \text{Point on graph}$$
To find the *slope* of the graph when $x = -2$, evaluate the derivative, $f'(x) = 2x$, at $x = -2$.
$$m = f'(-2) = -4 \quad \text{Slope of graph at } (-2, 4)$$
Now, using the point-slope form of the equation of a line, you can write
$$y - y_1 = m(x - x_1) \quad \text{Point-slope form}$$
$$y - 4 = -4[x - (-2)] \quad \text{Substitute for } y_1, m, \text{ and } x_1.$$
$$y = -4x - 4. \quad \text{Simplify.}$$
(See Figure 2.17.)

The line $y = -4x - 4$ is tangent to the graph of $f(x) = x^2$ at the point $(-2, 4)$.
Figure 2.17

Acknowledgments

We would like to thank the many people who have helped us at various stages of this project during the past 25 years. Their encouragement, criticisms, and suggestions have been invaluable to us.

Seventh Edition Reviewers

Raymond Badalian
Los Angeles City College

Beth Long
Pellissippi State Technical College

John Santomas
Villanova University

Christopher Butler
Case Western Reserve University

Gordon Melrose
Old Dominion University

Lynn Smith
Gloucester County College

Dane R. Camp
New Trier High School, IL

Larry Norris
North Carolina State University

Anthony Thomas
University of Wisconsin–Platteville

Barbara Cortzen
DePaul University

Eleanor Palais
Belmont High School, MA

Charles Wheeler
Montgomery College

Kathy Hoke
University of Richmond

Lila Roberts
Georgia Southern University

Previous Editions' Reviewers

Dennis Alber, *Palm Beach Junior College*; James Angelos, *Central Michigan University*; Kerry D. Bailey, *Laramie County Community College*; Harry L. Baldwin, Jr., *San Diego City College*; Homer F. Bechtell, *University of New Hampshire*; Keith Bergeron, *United States Air Force Academy*; Norman Birenes, *University of Regina*; Brian Blank, *Washington University*; Andrew A. Bulleri, *Howard Community College*; Paula Castagna, *Fresno City College*; Jack Ceder, *University of California–Santa Barbara*; Charles L. Cope, *Morehouse College*; Jorge Cossio, *Miami-Dade Community College*; Jack Courtney, *Michigan State University*; James Daniels, *Palomar College*; Kathy Davis, *University of Texas*; Paul W. Davis, *Worcester Polytechnic Institute*; Luz M. DeAlba, *Drake University*; Nicolae Dinculeanu, *University of Florida*; Rosario Diprizio, *Oakton Community College*; Garret J. Etgen, *University of Houston*; Russell Euler, *Northwest Missouri State University*; Phillip A. Ferguson, *Fresno City College*; Li Fong, *Johnson County Community College*; Michael Frantz, *University of La Verne*; William R. Fuller, *Purdue University*; Dewey Furness, *Ricks College*; Javier Garza, *Tarleton State University*; K. Elayn Gay, *University of New Orleans*; Thomas M. Green, *Contra Costa College*; Ali Hajjafar, *University of Akron*; Ruth A. Hartman, *Black Hawk College*; Irvin Roy Hentzel, *Iowa State University*; Howard E. Holcomb, *Monroe Community College*; Eric R. Immel, *Georgia Institute of Technology*; Arnold J. Insel, *Illinois State University*; Elgin Johnston, *Iowa State University*; Hideaki Kaneko, *Old Dominion University*; Toni Kasper, *Borough of Manhattan Community College*; William J. Keane, *Boston College*; Timothy J. Kearns, *Boston College*;

Ronnie Khuri, *University of Florida*; Frank T. Kocher, Jr., *Pennsylvania State University*; Robert Kowalczyk, *University of Massachusetts–Dartmouth*; Joseph F. Krebs, *Boston College*; David C. Lantz, *Colgate University*; Norbert Lerner, *State University of New York at Cortland*; Maita Levine, *University of Cincinnati*; Murray Lieb, *New Jersey Institute of Technology*; Ransom Van B. Lynch, *Phillips Exeter Academy*; Bennet Manvel, *Colorado State University*; Mauricio Marroquin, *Los Angeles Valley College*; Robert L. Maynard, *Tidewater Community College*; Robert McMaster, *John Abbott College*; Darrell Minor, *Columbus State Community College*; Maurice Monahan, *South Dakota State University*; Michael Montaño, *Riverside Community College*; Philip Montgomery, *University of Kansas*; David C. Morency, *University of Vermont*; Gerald Mueller, *Columbus State Community College*; Duff A. Muir, *United States Air Force Academy*; Charlotte J. Newsom, *Tidewater Community College*; Terry J. Newton, *United States Air Force Academy*; Donna E. Nordstrom, *Pasadena City College*; Robert A. Nowlan, *Southern Connecticut State University*; Luis Ortiz-Franco, *Chapman University*; Barbara L. Osofsky, *Rutgers University*; Judith A. Palagallo, *University of Akron*; Wayne J. Peeples, *University of Texas*; Jorge A. Percz, *LaGuardia Community College*; Darrell J. Peterson, *Santa Monica College*; Donald Poulson, *Mesa Community College*; Jean L. Rubin, *Purdue University*; Barry Sarnacki, *United States Air Force Academy*; N. James Schoonmaker, *University of Vermont*; George W. Schultz, *St. Petersburg Junior College*; Richard E. Shermoen, *Washburn University*; Thomas W. Shilgalis, *Illinois State University*; J. Philip Smith, *Southern Connecticut State University*; Frank Soler, *De Anza College*; Enid Steinbart, *University of New Orleans*; Michael Steuer, *Nassau Community College*; Mark Stevenson, *Oakland Community College*; Lawrence A. Trivieri, *Mohawk Valley Community College*; John Tweed, *Old Dominion University*; Carol Urban, *College of DuPage*; Marjorie Valentine, *North Side ISD, San Antonio*; Robert J. Vojack, *Ridgewood High School, NJ*; Bert K. Waits, *Ohio State University*; Florence A. Warfel, *University of Pittsburgh*; John R. Watret, *Embry-Riddle Aeronautical University*; Carroll G. Wells, *Western Kentucky University*; Jay Wiestling, *Palomar College*; Paul D. Zahn, *Borough of Manhattan Community College*; August J. Zarcone, *College of DuPage*

During the past four years, several users of the Sixth Edition wrote to us with suggestions. We considered each and every one of them when preparing the manuscript for the Seventh Edition. We would like to extend a special thanks to Mikhail Ostrovskii of the Catholic University of America for the many thoughtful suggestions he sent to us. The time and care he invested in several correspondences was quite extraordinary.

We would like to thank the staff at Larson Texts, Inc., and the staff of Meridian Creative Group, who assisted with proofreading the manuscript, preparing and proofreading the art package, and checking and typesetting the supplements.

A special note of thanks goes to the instructors who responded to our survey and to the over 2 million students who have used earlier editions of the text.

On a personal level, we are grateful to our wives, Deanna Gilbert Larson, Eloise Hostetler, and Consuelo Edwards, for their love, patience, and support. Also, a special note of thanks goes to R. Scott O'Neil.

If you have suggestions for improving this text, please feel free to write to us. Over the past 25 years we have received many useful comments from both instructors and students, and we value these very much.

Ron Larson
Robert P. Hostetler
Bruce H. Edwards

Supplements

Resources

Website (*college.hmco.com*)

Many additional text-specific study and interactive features for students and instructors can be found at the Houghton Mifflin website.

For the Student

Study and Solutions Guide, Volume I by Bruce H. Edwards (University of Florida)

Graphing Technology Guide for Precalculus and Calculus by Benjamin N. Levy and Laurel Technical Services

Graphing Calculator Videotape by Dana Mosely

Calculus, 7E, *Videotapes* by Dana Mosely

For the Instructor

Complete Solutions Guide, Volumes I and II by Bruce H. Edwards (University of Florida)

Test Item File by Ann Rutledge Kraus (The Pennsylvania State University, The Behrend College)

Instructor's Resource Guide by Ann Rutledge Kraus (The Pennsylvania State University, The Behrend College)

Computerized Testing (WIN, Macintosh)

HMClassPrep™ (Instructor's CD-ROM)

New exciting study aids make the best supplements package for Calculus even better.

New! Text-Specific Video Series (available in VHS and DVD formats)

Tied directly to the Larson/Hostetler/Edwards *Calculus*, Seventh Edition, textbook, these videos created by Dana Mosely provide lecture-style instruction, review of key concepts, real-life data examples, and more. Ideal for students who want extra guidance or who have missed a class, the videos cover select material from Chapters P–9.

Announcing a whole new suite of electronic study tools for calculus: The Larson *eSolutions* for Calculus.

Calculus Learning Tools Student **CD-ROM**   Contains Computer Algebra System Explorations, rotatable 3-D art, printable MathGraphs and MathArticles referenced throughout the text, as well as MathBios, labs, and more.

Companion Website   Includes rotatable 3-D art and other student and instructor resources. Visit **www.college.hmco.com/mathematics.**

Interactive and *Internet Calculus 3.0* These two products are comprehensive multimedia courses in calculus. To provide you with a choice, we offer *Interactive Calculus 3.0* on CD-ROM and *Internet Calculus 3.0* online. Both contain the complete text of *Calculus*, Seventh Edition, as well as other exciting features such as solutions to odd-numbered exercises, rotatable 3-D graphs, editable 2-D graphs, Open Explorations using one of four computer algebra systems, animations, videos, simulations, Try Its for every example, and more.

CalcChat.com website   An on-line resource where students can access, discuss, *and* help each other with step-by-step solutions to all the odd-numbered exercises in the Larson *Calculus* series.

EduSpace On-Line Learning Environment   Instructors can easily assign, deliver, and grade homework and other assignments based on the even-numbered exercises in the text via Houghton Mifflin's new *EduSpace* platform.

Live, on-line tutoring from SMARTHINKING.COM

Houghton Mifflin has partnered with SMARTHINKING.com to give students the most advanced on-line tutoring possible. SMARTHINKING is a virtual learning assistance center created in conjunction with 31 schools. It provides qualified tutors (e-structors) and independent study resources for core courses and skills. Students can access tutors and resources at home, school, or anywhere else they have an Internet connection.

Instructors

To get copies of these supplements or for more information on packaging them at significant discounts with any Larson *Calculus*, Seventh Edition, textbook, please contact your Houghton Mifflin sales representative or call 1-800-733-1717. AP Instructors call McDougal Littell at 1-800-462-6595.

Students

For details on how you can order these exciting new learning aids, visit our website at *http://www.hmco.com* or email us at *college_math@hmco.com.*

Interactive Calculus 3.0 CD-ROM and Internet Calculus 3.0

To accommodate a wide variety of teaching and learning styles, *Calculus* is also available as *Interactive Calculus 3.0* on an interactive CD-ROM and *Internet Calculus 3.0*. These versions incorporate live mathematics throughout the entire program. Live mathematics helps students visualize and explore—leading to a deeper understanding of calculus concepts than has ever before been possible.

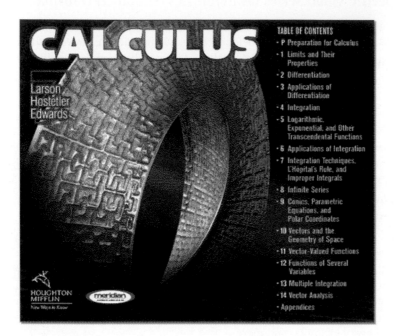

Live Mathematics Throughout

- Open Explorations give students the opportunity to explore using computer algebra systems.

- Section Quizzes require students to enter free-response answers and to click-and-drag answers into place.

- Editable two-dimensional graphs, featured throughout the entire program, provide additional opportunities to explore and investigate.

- Rotatable three-dimensional graphs allow for a whole new level of visualization.

- New and enhanced explorations, simulations, and animations make concepts come alive.

Classroom Management Tool and Syllabus Builder

All of the content of the Seventh Edition text— a wealth of applications, exercises, worked-out examples, and detailed explanations—is included in *Interactive Calculus 3.0* on CD-ROM and *Internet Calculus 3.0*. Instructors have the flexibility of customizing content and interactive features for students as desired. Instructors may simply add dates to a default syllabus or may modify the order of topics. Either way, a customized syllabus is easy to distribute electronically and update instantly. This tool is particularly useful for managing distance learning courses.

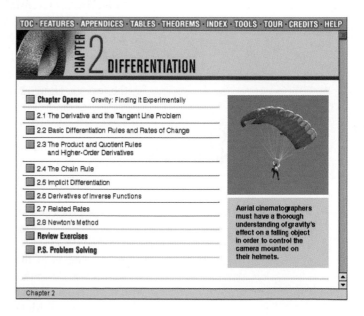

Aerial cinematographers must have a thorough understanding of gravity's effect on a falling object in order to control the camera mounted on their helmets.

Features

Exercises with solutions to all odd exercises provide immediate feedback for students.

Try Its allow students to try problems similar to the examples and to check their work using the worked-out solutions provided.

Quizzes with responses require students to enter free responses, click-and-drag answers, and choose multiple choice answers.

Editable Graphs encourage students to explore concepts by graphing "editable" graphs as well as to change the viewing window and to use *zoom* and *trace* features.

Rotatable Graphs allow students to view three-dimensional graphs as they rotate, greatly enhancing visualization.

Simulations encourage exploration and hands-on interaction with mathematical concepts.

Animations, which use motion and sound to explain concepts, can be played and replayed, or viewed one step at a time.

Complete searchable text-specific **Content, Index, Theorem Index,** and **Features Index** facilitate cross-referencing.

Video Clips engage student interest and show connections between mathematics and other disciplines.

Syllabus Builder enables instructors to save administrative time and to convey important information online.

Bookmarking capability provides fast, efficient navigation of the site.

Other special features include:

Articles • Connections • History • Look Ahead • Math Trends •
Section Projects • Technology

Calculus Options
25 Years of Success, Leadership, and Innovation

This best-selling calculus program continues to expand to offer instructors and students more flexible teaching and learning options.

Calculus with Analytic Geometry, Seventh Edition

This core text covers the entire three-semester course in fifteen chapters.

Calculus of a Single Variable, Seventh Edition

The single variable text is designed for a two-semester course in calculus, presented in ten chapters.

Multivariable Calculus, Seventh Edition

The multivariable text is designed for a third-semester course in calculus, presented in five chapters.

NEW *Calculus I with Precalculus: A One-Year Course*

This text is designed as a two-semester course that integrates the first semester of calculus with precalculus.

Calculus II, Seventh Edition

The Calculus II text is designed for a second-semester course in calculus presented in five chapters.

Calculus: Early Transcendental Functions, Third Edition

This text, which integrates coverage of transcendental functions from the beginning of the text, covers an entire three-semester course in calculus.

Calculus of a Single Variable: Early Transcendental Functions, Third Edition

This single variable text is designed for a two-semester course in calculus, presented in ten chapters, with transcendental functions integrated from the beginning of the text.

Calculus with Analytic Geometry, Alternate Sixth Edition

This text, with trigonometry introduced in Chapter 8, covers the entire three-semester course in eighteen chapters. It also offers alternative treatments of the following topics: limits, applications of integration, exponential and logarithmic functions, and vectors.

Calculus

Calculus with Precalculus

Calculus with Early Transcendental Functions

Calculus with Late Trigonometry

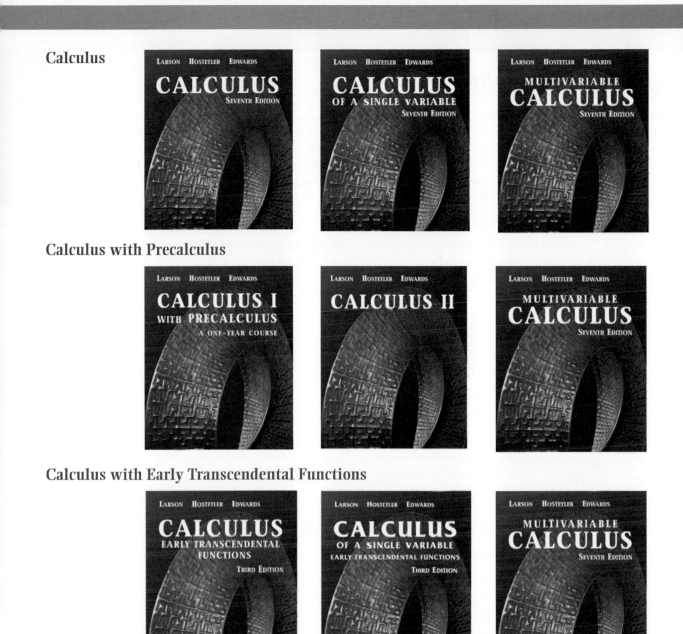

Eruptions of Old Faithful

Yellowstone National Park, located in the northwest corner of Wyoming and adjacent parts of Montana and Idaho, contains over half of all known geysers on earth. Few of these hydrothermal spectacles are regular enough to be anticipated accurately. Of the more than 400 geysers in Yellowstone National Park, predictions are posted for only seven, one of which is Old Faithful.

Old Faithful geyser was named in 1870. In 1938 geologist Harry M. Woodward noticed that there was a correlation between the duration of Old Faithful's eruption and the length of time (interval) before the next eruption.

More than 137,000 eruptions of Old Faithful have been observed and recorded. The durations have varied between 1.5 and 5.5 minutes, and the intervals have varied between 30 and 120 minutes. Since Woodward's observations, the intervals have tended to increase. It is speculated that major earthquakes in the region have shifted the circulation of hot water away from the geyser, resulting in a longer "fill time."

The data below show 35 eruptions as ordered pairs of the form (x, y), where x is the duration in minutes and y is the interval in minutes. *(Source: Yellowstone National Park)*

(1.80, 56),   (1.82, 58),   (1.88, 60),   (1.90, 62),   (1.92, 60),   (1.93, 56),

(1.98, 59),   (2.03, 60),   (2.05, 57),   (2.13, 60),   (2.30, 57),   (2.35, 57),

(2.37, 61),   (2.82, 73),   (3.13, 76),   (3.27, 77),   (3.65, 77),   (3.70, 82),

(3.78, 79),   (3.83, 85),   (3.87, 81),   (3.88, 80),   (4.10, 89),   (4.27, 90),

(4.30, 84),   (4.30, 89),   (4.43, 84),   (4.43, 89),   (4.47, 80),   (4.47, 86),

(4.53, 89),   (4.55, 86),   (4.60, 88),   (4.60, 92),   (4.63, 91)

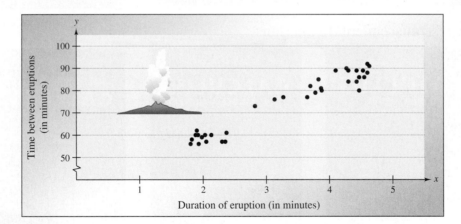

QUESTIONS

1. Describe the relationship between the durations and the intervals of Old Faithful's eruptions. Is there a correlation between the two measures?

2. Suppose that you observe an eruption that lasts for 2 minutes and 40 seconds. When would you expect the next eruption?

3. Write a model (an equation involving x and y) that can be used to predict the length of time until the next eruption. Explain how you obtained the model. Did you use technology or did you obtain the model using only hand calculations?

The concepts presented here will be explored further in this chapter. For an extension of this application, see Lab 1 in the lab series that accompanies this text at college.hmco.com.

Preparation for Calculus

P

W. Perry Conway/Tom Stack & Associates

An eruption of Old Faithful is capable of shooting up to 8400 gallons of boiling water as high as 184 feet in the air.

For a geyser to exist, three conditions must exist: a supply of water, an intense heat source, and pressure-tight plumbing.

Volcanically heated water enters a geyser's plumbing system at great depths, boiling and heating cooler water that is flowing in from the surface. Eventually the steam from the boiling water creates enough pressure to explosively force great volumes of water high into the air. Once an eruption is over, the entire filling, heating, and boiling process begins again.

National Park Services, Yellowstone National Park

Harry M. Woodward was the first to describe a relationship between the durations and intervals of Old Faithful's eruptions.

- Sketch the graph of an equation.
- Find the intercepts of a graph.
- Test a graph for symmetry with respect to an axis and the origin.
- Find the points of intersection of two graphs.
- Interpret mathematical models for real-life data.

The Graph of an Equation

In 1637 the French mathematician René Descartes revolutionized the study of mathematics by joining its two major fields—algebra and geometry. With Descartes's coordinate plane, geometric concepts could be formulated analytically and algebraic concepts could be viewed graphically. The power of this approach is such that within a century, much of calculus had been developed.

The same approach can be followed in your study of calculus. That is, by viewing calculus from multiple perspectives—*graphically*, *analytically*, and *numerically*—you will increase your understanding of core concepts.

Consider the equation $3x + y = 7$. The point $(2, 1)$ is a **solution point** of the equation because the equation is satisfied (is true) when 2 is substituted for x and 1 is substituted for y. This equation has many other solutions, such as $(1, 4)$ and $(0, 7)$. To systematically find other solutions, solve the original equation for y.

$$y = 7 - 3x \qquad \text{Analytic approach}$$

Then construct a **table of values** by substituting several values of x.

x	0	1	2	3	4
y	7	4	1	-2	-5

Numerical approach

From the table, you can see that $(0, 7)$, $(1, 4)$, $(2, 1)$, $(3, -2)$, and $(4, -5)$ are solutions of the original equation $3x + y = 7$. Like many equations, this equation has an infinite number of solutions. The set of all solution points is the **graph** of the equation, as shown in Figure P.1.

NOTE   Even though we refer to the sketch shown in Figure P.1 as the graph of $3x + y = 7$, it really represents only a *portion* of the graph. The entire graph would extend beyond the page.

In this course, you will study many sketching techniques. The simplest is point plotting—that is, you plot points until the basic shape of the graph seems apparent.

Example 1   **Sketching a Graph by Point Plotting**

Sketch the graph of $y = x^2 - 2$.

Solution   First construct a table of values. Then plot the points shown in the table.

x	-2	-1	0	1	2	3
y	2	-1	-2	-1	2	7

Finally, connect the points with a *smooth curve*, as shown in Figure P.2. This graph is a **parabola.** It is one of the conics you will study in Chapter 9.

RENÉ DESCARTES (1596–1650)

Descartes made many contributions to philosophy, science, and mathematics. The idea of representing points in the plane by pairs of real numbers and representing curves in the plane by equations was described by Descartes in his book *La Géométrie*, published in 1637.

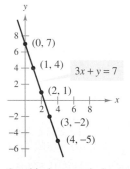

Graphical approach: $3x + y = 7$
Figure P.1

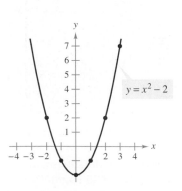

The parabola $y = x^2 - 2$
Figure P.2

One disadvantage of point plotting is that to get a good idea about the shape of a graph, you may need to plot many points. With only a few points, you could badly misrepresent the graph. For instance, suppose that to sketch the graph of

$$y = \tfrac{1}{30}x(39 - 10x^2 + x^4)$$

you plotted only five points:

$$(-3, -3), (-1, -1), (0, 0), (1, 1), \text{ and } (3, 3)$$

as shown in Figure P.3(a). From these five points, you might conclude that the graph is a line. This, however, is not correct. By plotting several more points, you can see that the graph is more complicated, as shown in Figure P.3(b).

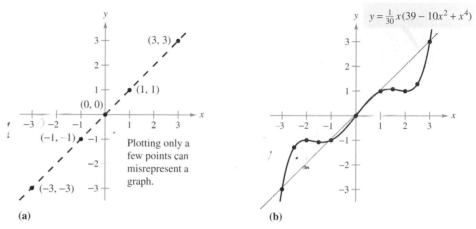

Plotting only a few points can misrepresent a graph.

(a)   (b)

Figure P.3

EXPLORATION

Comparing Graphical and Analytic Approaches   Use a graphing utility to graph each of the following. In each case, find a viewing window that shows the important characteristics of the graph.

a. $y = x^3 - 3x^2 + 2x + 5$

b. $y = x^3 - 3x^2 + 2x + 25$

c. $y = -x^3 - 3x^2 + 20x + 5$

d. $y = 3x^3 - 40x^2 + 50x - 45$

e. $y = -(x + 12)^3$

f. $y = (x - 2)(x - 4)(x - 6)$

A purely graphical approach to this problem would involve a simple "guess, check, and revise" strategy. What types of things do you think an analytic approach might involve? For instance, does the graph have symmetry? Does the graph have turns? If so, where are they?

As you proceed through Chapters 1, 2, and 3 of this text, you will study many new analytic tools that will help you analyze graphs of equations such as these.

TECHNOLOGY PITFALL   Technology has made sketching of graphs easier. Even with technology, however, it is possible to badly misrepresent a graph. For instance, each of the graphing utility screens in Figure P.4 shows a portion of the graph of

$$y = x^3 - x^2 - 25.$$

From the screen on the left, you might assume that the graph is a line. From the screen on the right, however, you can see that the graph is not a line. Thus, whether you are sketching a graph by hand or using a graphing utility, you must realize that different "viewing windows" can produce very different views of a graph. In choosing a viewing window, your goal is to show a view of the graph that fits well in the context of the problem.

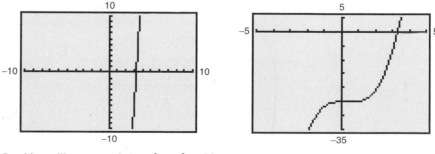

Graphing utility screens of $y = x^3 - x^2 - 25$
Figure P.4

NOTE   In this text, we use the term *graphing utility* to mean either a graphing calculator or computer graphing software such as *Maple, Mathematica, Derive, Mathcad,* or the *TI-89.*

Intercepts of a Graph

Two types of solution points that are especially useful when graphing an equation are those having zero as their *x*- or *y*-coordinate. Such points are called **intercepts** because they are the points at which the graph intersects the *x*- or *y*-axis. The point $(a, 0)$ is an **x-intercept** of the graph of an equation if it is a solution point of the equation. To find the *x*-intercepts of a graph, let *y* be zero and solve the equation for *x*. The point $(0, b)$ is a **y-intercept** of the graph of an equation if it is a solution point of the equation. To find the *y*-intercepts of a graph, let *x* be zero and solve the equation for *y*.

$y = 0$

$x = 0$

NOTE   Some texts denote the *x*-intercept as the *x*-coordinate of the point $(a, 0)$ rather than the point itself. Unless it is necessary to make a distinction, we will use the term *intercept* to mean either the point or the coordinate.

It is possible for a graph to have no intercepts, or it might have several. For instance, consider the four graphs shown in Figure P.5.

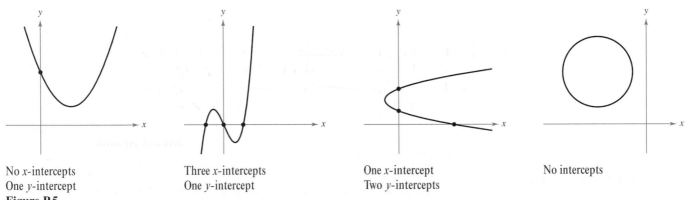

No *x*-intercepts
One *y*-intercept
Figure P.5

Three *x*-intercepts
One *y*-intercept

One *x*-intercept
Two *y*-intercepts

No intercepts

Example 2   Finding *x*- and *y*-intercepts

Find the *x*- and *y*-intercepts of the graph of $y = x^3 - 4x$.

Solution   To find the *x*-intercepts, let *y* be zero and solve for *x*.

$$x^3 - 4x = 0 \qquad \text{Let } y \text{ be zero.}$$
$$x(x - 2)(x + 2) = 0 \qquad \text{Factor.}$$
$$x = 0, 2, \text{ or } -2 \qquad \text{Solve for } x.$$

Because this equation has three solutions, you can conclude that the graph has three *x*-intercepts:

$(0, 0), (2, 0), \text{ and } (-2, 0)$.   *x*-intercepts

To find the *y*-intercepts, let *x* be zero. Doing this produces $y = 0$. So, the *y*-intercept is

$(0, 0)$.   *y*-intercept

(See Figure P.6.)

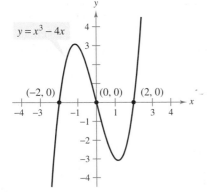

Intercepts of a graph
Figure P.6

TECHNOLOGY   Example 2 uses an analytic approach to finding intercepts. When an analytic approach is not possible, you can use a graphical approach by finding the points where the graph intersects the axes. Try using a graphing utility to approximate the intercepts.

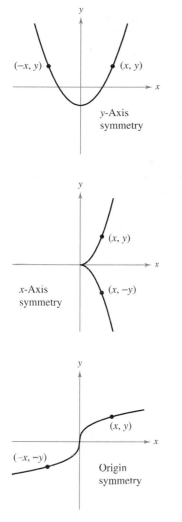

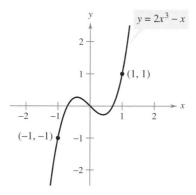

Figure P.7

Origin symmetry
Figure P.8

Symmetry of a Graph

Knowing the symmetry of a graph *before* attempting to sketch it is useful because you need only half as many points to sketch the graph. The following three types of symmetry can be used to help sketch the graph of an equation (see Figure P.7).

1. A graph is **symmetric with respect to the *y*-axis** if, whenever (x, y) is a point on the graph, $(-x, y)$ is also a point on the graph. This means that the portion of the graph to the left of the *y*-axis is a mirror image of the portion to the right of the *y*-axis.

2. A graph is **symmetric with respect to the *x*-axis** if, whenever (x, y) is a point on the graph, $(x, -y)$ is also a point on the graph. This means that the portion of the graph above the *x*-axis is a mirror image of the portion below the *x*-axis.

3. A graph is **symmetric with respect to the origin** if, whenever (x, y) is a point on the graph, $(-x, -y)$ is also a point on the graph. This means that the graph is unchanged by a rotation of 180° about the origin.

Tests for Symmetry

1. The graph of an equation in *x* and *y* is symmetric with respect to the *y*-axis if replacing *x* by $-x$ yields an equivalent equation.

2. The graph of an equation in *x* and *y* is symmetric with respect to the *x*-axis if replacing *y* by $-y$ yields an equivalent equation.

3. The graph of an equation in *x* and *y* is symmetric with respect to the origin if replacing *x* by $-x$ and *y* by $-y$ yields an equivalent equation.

The graph of a polynomial has symmetry with respect to the *y*-axis if each term has an even exponent (or is a constant). For instance, the graph of

$$y = 2x^4 - x^2 + 2 \qquad \text{\textit{y}-axis symmetry}$$

has symmetry with respect to the *y*-axis. Similarly, the graph of a polynomial has symmetry with respect to the origin if each term has an odd exponent, as illustrated in Example 3.

Example 3   **Testing for Origin Symmetry**

Show that the graph of

$$y = 2x^3 - x$$

is symmetric with respect to the origin.

Solution

$y = 2x^3 - x$	Write original equation.
$-y = 2(-x)^3 - (-x)$	Replace *x* by $-x$ and *y* by $-y$.
$-y = -2x^3 + x$	Simplify.
$y = 2x^3 - x$	Equivalent equation

Because the replacement produces an equivalent equation, you can conclude that the graph of $y = 2x^3 - x$ is symmetric with respect to the origin, as shown in Figure P.8.

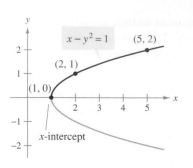

First plot the x-intercept and the points above the x-axis. Then use symmetry to complete the graph.
Figure P.9

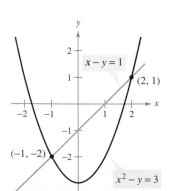

Two points of intersection
Figure P.10

Example 4   Using Intercepts and Symmetry to Sketch a Graph

Sketch the graph of $x - y^2 = 1$.

Solution   The graph is symmetric with respect to the x-axis because replacing y by $-y$ yields an equivalent equation.

$x - y^2 = 1$	Write original equation.
$x - (-y)^2 = 1$	Replace y by $-y$.
$x - y^2 = 1$	Equivalent equation

This means that the portion of the graph below the x-axis is a mirror image of the portion above the x-axis. To sketch the graph, first sketch the portion above the x-axis. Then reflect in the x-axis to obtain the entire graph, as shown in Figure P.9.

TECHNOLOGY   Graphing utilities are designed so that they most easily graph equations in which y is a function of x (see Section P.3 for a definition of **function**). To graph other types of equations, you need to split the graph into two or more parts *or* you need to use a different graphing mode. For instance, to sketch the graph of the equation in Example 4, you can split it into two parts.

$y_1 = \sqrt{x - 1}$	Top portion of graph
$y_2 = -\sqrt{x - 1}$	Bottom portion of graph

Points of Intersection

A **point of intersection** of the graphs of two equations is a point that satisfies both equations. You can find the points of intersection of two graphs by solving their equations simultaneously.

Example 5   Finding Points of Intersection

Find all points of intersection of the graphs of $x^2 - y = 3$ and $x - y = 1$.

Solution   Begin by sketching the graphs of both equations on the *same* rectangular coordinate system, as shown in Figure P.10. Having done this, it appears that the graphs have two points of intersection. To find these two points, you can use the following steps.

$y = x^2 - 3$	Solve first equation for y.
$y = x - 1$	Solve second equation for y.
$x^2 - 3 = x - 1$	Equate y-values.
$x^2 - x - 2 = 0$	Write in general form.
$(x - 2)(x + 1) = 0$	Factor.
$x = 2$ or -1	Solve for x.

The corresponding values of y are obtained by substituting $x = 2$ and $x = -1$ into either of the original equations. Doing this produces two points of intersection:

$(2, 1)$ and $(-1, -2)$.   Points of intersection

You can check these points by substituting into *both* of the original equations or by using the *intersect* feature of a graphing utility.

---

indicates that in the Interactive 3.0 *CD-ROM and* Internet 3.0 *versions of this text (available at* college.hmco.com) *you will find an Open Exploration, which further explores this example using the computer algebra systems* Maple, Mathcad, Mathematica, *and* Derive.

The Mauna Loa Observatory in Hawaii has been measuring the increasing concentration of carbon dioxide in earth's atmosphere since 1960.

Mathematical Models

Real-life applications of mathematics often use equations as **mathematical models.** In developing a mathematical model to represent actual data, you should strive for two (often conflicting) goals: accuracy and simplicity. That is, you want the model to be simple enough to be workable, yet accurate enough to produce meaningful results. Section P.4 explores these goals more completely.

Example 6   **Comparing Two Mathematical Models**

The Mauna Loa Observatory in Hawaii records the carbon dioxide concentration y (in parts per million) in earth's atmosphere. The January readings for various years are shown in Figure P.11. In the July 1990 issue of *Scientific American*, these data were used to predict the carbon dioxide level in earth's atmosphere in the year 2035, using the quadratic model

$$y = 316.2 + 0.70t + 0.018t^2 \qquad \text{Quadratic model for 1960–1990 data}$$

where $t = 0$ represents 1960, as shown in Figure P.11(a).

The data shown in Figure P.11(b) represent the years 1975 through 1998 and can be modeled by

$$y = 307.9 + 1.50t \qquad \text{Linear model for 1975–1998 data}$$

where $t = 0$ represents 1960. What was the prediction given in the *Scientific American* article in 1990? Given the new data for 1990 through 1998, does this prediction for the year 2035 seem accurate?

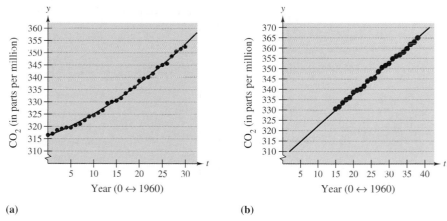

(a)   (b)

Figure P.11

Solution   To answer the first question, substitute $t = 75$ (for 2035) into the quadratic model.

$$y = 316.2 + 0.70(75) + 0.018(75)^2 = 469.95 \qquad \text{Quadratic model}$$

So, the prediction in the *Scientific American* article was that the carbon dioxide concentration in earth's atmosphere would reach about 470 parts per million in the year 2035. Using the linear model for the 1975–1998 data, the prediction for the year 2035 is

$$y = 307.9 + 1.50(75) = 420.4. \qquad \text{Linear model}$$

So, based on the linear model for 1975–1998, it appears that the 1990 prediction was too high.

NOTE   The models in Example 6 were developed using a procedure called *least squares regression* (see Section 12.9). Both the quadratic and linear models have a correlation given by $r^2 = 0.997$. The closer r^2 is to 1, the "better" the model.

EXERCISES FOR SECTION P.1

In Exercises 1–4, match the equation with its graph. [Graphs are labeled (a), (b), (c), and (d).]

(a)

(b)

(c)

(d)

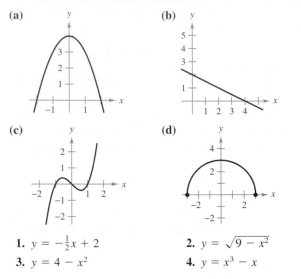

1. $y = -\frac{1}{2}x + 2$

2. $y = \sqrt{9 - x^2}$

3. $y = 4 - x^2$

4. $y = x^3 - x$

In Exercises 5–12, sketch the graph of the equation by point plotting.

5. $y = \frac{3}{2}x + 1$

6. $y = 6 - 2x$

7. $y = 4 - x^2$

8. $y = (x - 3)^2$

9. $y = |x + 2|$

10. $y = |x| - 1$

11. $y = \sqrt{x} - 4$

12. $y = \sqrt{x + 2}$

In Exercises 13 and 14, describe the viewing window that yields the figure.

13. $y = x^3 - 3x^2 + 4$

14. $y = |x| + |x - 10|$

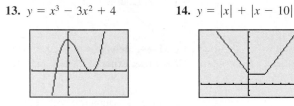

In Exercises 15 and 16, use a graphing utility to graph the equation. Move the cursor along the curve to approximate the unknown coordinate of each solution point accurate to two decimal places.

15. $y = \sqrt{5 - x}$ (a) $(2, y)$ (b) $(x, 3)$

16. $y = x^5 - 5x$ (a) $(-0.5, y)$ (b) $(x, -4)$

In Exercises 17–24, find any intercepts.

17. $y = x^2 + x - 2$

18. $y^2 = x^3 - 4x$

19. $y = x^2\sqrt{25 - x^2}$

20. $y = (x - 1)\sqrt{x^2 + 1}$

21. $y = \dfrac{3(2 - \sqrt{x})}{x}$

22. $y = \dfrac{x^2 + 3x}{(3x + 1)^2}$

23. $x^2y - x^2 + 4y = 0$

24. $y = 2x - \sqrt{x^2 + 1}$

In Exercises 25–36, test for symmetry with respect to each axis and to the origin.

25. $y = x^2 - 2$

26. $y = x^2 - x$

27. $y^2 = x^3 - 4x$

28. $y = x^3 + x$

29. $xy = 4$

30. $xy^2 = -10$

31. $y = 4 - \sqrt{x + 3}$

32. $xy - \sqrt{4 - x^2} = 0$

33. $y = \dfrac{x}{x^2 + 1}$

34. $y = \dfrac{x^2}{x^2 + 1}$

35. $y = |x^3 + x|$

36. $|y| - x = 3$

In Exercises 37–54, sketch the graph of the equation. Identify any intercepts and test for symmetry.

37. $y = -3x + 2$

38. $y = -\frac{1}{2}x + 2$

39. $y = \frac{1}{2}x - 4$

40. $y = \frac{2}{3}x + 1$

41. $y = 1 - x^2$

42. $y = x^2 + 3$

43. $y = (x + 3)^2$

44. $y = 2x^2 + x$

45. $y = x^3 + 2$

46. $y = x^3 - 4x$

47. $y = x\sqrt{x + 2}$

48. $y = \sqrt{9 - x^2}$

49. $x = y^3$

50. $x = y^2 - 4$

51. $y = \dfrac{1}{x}$

52. $y = \dfrac{10}{x^2 + 1}$

53. $y = 6 - |x|$

54. $y = |6 - x|$

In Exercises 55–58, use a graphing utility to graph the equation. (See Technology note, page 6.) Identify any intercepts and test for symmetry.

55. $y^2 - x = 9$

56. $x^2 + 4y^2 = 4$

57. $x + 3y^2 = 6$

58. $3x - 4y^2 = 8$

Getting at the Concept

In Exercises 59–62, write an equation whose graph has the indicated property. (There may be more than one correct answer.)

59. The graph has intercepts at $x = -2$, $x = 4$, and $x = 6$.

60. The graph has intercepts at $x = -\frac{5}{2}$, $x = 2$, and $x = \frac{3}{2}$.

61. The graph is symmetric with respect to the origin.

62. The graph is symmetric with respect to the x-axis.

The symbol indicates an exercise in which you are instructed to use graphing technology or a symbolic computer algebra system. The solutions of other exercises may also be facilitated by use of appropriate technology.

In Exercises 63–72, find the points of intersection of the graphs of the equations.

63. $x + y = 2$
    $2x - y = 1$

64. $2x - 3y = 13$
    $5x + 3y = 1$

65. $x + y = 7$
    $3x - 2y = 11$

66. $5x - 6y = 9$
    $-7x + 3y = -18$

67. $x^2 + y = 6$
    $x + y = 4$

68. $x = 3 - y^2$
    $y = x - 1$

69. $x^2 + y^2 = 5$
    $x - y = 1$

70. $x^2 + y^2 = 25$
    $2x + y = 10$

71. $y = x^3$
    $y = x$

72. $y = x^3 - 4x$
    $y = -(x + 2)$

In Exercises 73 and 74, use a graphing utility to find the points of intersection of the graphs. Check your results analytically.

73. $y = x^3 - 2x^2 + x - 1$
    $y = -x^2 + 3x - 1$

74. $y = x^4 - 2x^2 + 1$
    $y = 1 - x^2$

75. **Break-Even Point** Find the sales necessary to break even $(R = C)$ if the cost C of producing x units is

$$C = 5.5\sqrt{x} + 10,000 \qquad \text{Cost equation}$$

and the revenue R for selling x units is

$$R = 3.29x. \qquad \text{Revenue equation}$$

76. **Think About It** Each table shows solution points for one of the following equations.

(i) $y = kx + 5$ (ii) $y = x^2 + k$ (iii) $y = kx^{3/2}$ (iv) $xy = k$

Match each equation with the correct table and find k.

(a)

x	1	4	9
y	3	24	81

(b)

x	1	4	9
y	7	13	23

(c)

x	1	4	9
y	36	9	4

(d)

x	1	4	9
y	-9	6	71

77. **Modeling Data** The table shows the consumer price index (CPI) for selected years. *(Source: Bureau of Labor Statistics)*

Year	1970	1975	1980	1985	1990	1995	2000
CPI	38.8	53.8	82.4	107.6	130.7	152.4	168.7

(a) Use the regression capabilities of a graphing utility to find a mathematical model of the form $y = at^2 + bt + c$ for the data. In the model, y represents the consumer price index and t represents the year, with $t = 0$ corresponding to 1970.

(b) Use a graphing utility to graph the model and compare the data with the model.

(c) Use the model to predict the CPI for the year 2004.

78. **Modeling Data** The table shows the average number of acres per farm in the United States for selected years. *(Source: U.S. Department of Agriculture)*

Year	1950	1960	1970	1980	1990	1998
Acreage	213	297	374	426	460	435

(a) Use the regression capabilities of a graphing utility to find a mathematical model of the form

$$y = at^2 + bt + c$$

for the data. In the model, y represents the average acreage and t represents the year, with $t = 0$ corresponding to 1950.

(b) Use a graphing utility to graph the model and compare the data with the model.

(c) Use the model to predict the average number of acres per farm in the United States in the year 2004.

79. **Copper Wire** The resistance y in ohms of 1000 feet of solid copper wire at 77°F can be approximated by the mathematical model

$$y = \frac{10,770}{x^2} - 0.37, \qquad 5 \le x \le 100$$

where x is the diameter of the wire in mils (0.001 in.). Use a graphing utility to graph the model. If the diameter of the wire is doubled, the resistance is changed by approximately what factor?

80. (a) Prove that if a graph is symmetric with respect to the x-axis and to the y-axis, then it is symmetric with respect to the origin. Give an example to show that the converse is not true.

(b) Prove that if a graph is symmetric with respect to one axis and to the origin, then it is symmetric with respect to the other axis.

True or False? **In Exercises 81–84, determine whether the statement is true or false. If it is false, explain why or give an example that shows it is false.**

81. If $(1, -2)$ is a point on a graph that is symmetric with respect to the x-axis, then $(-1, -2)$ is also a point on the graph.

82. If $(1, -2)$ is a point on a graph that is symmetric with respect to the y-axis, then $(-1, -2)$ is also a point on the graph.

83. If $b^2 - 4ac > 0$ and $a \ne 0$, then the graph of $y = ax^2 + bx + c$ has two x-intercepts.

84. If $b^2 - 4ac = 0$ and $a \ne 0$, then the graph of $y = ax^2 + bx + c$ has only one x-intercept.

85. Find an equation of the graph that consists of all points (x, y) whose distance from the origin is K times $(K \ne 1)$ the distance from $(2, 0)$. (For a review of the Distance Formula, see Appendix D.)

- Find the slope of a line passing through two points.
- Write the equation of a line with a given point and slope.
- Interpret slope as a ratio or as a rate in a real-life application.
- Sketch the graph of a linear equation in slope-intercept form.
- Write equations of lines that are parallel or perpendicular to a given line.

The Slope of a Line

The **slope** of a nonvertical line is a measure of the number of units the line rises (or falls) vertically for each unit of horizontal change from left to right. Consider the two points (x_1, y_1) and (x_2, y_2) on the line in Figure P.12. As you move from left to right along this line, a vertical change of

$$\Delta y = y_2 - y_1 \qquad \text{Change in } y$$

units corresponds to a horizontal change of

$$\Delta x = x_2 - x_1 \qquad \text{Change in } x$$

units. (Δ is the Greek uppercase letter *delta*, and the symbols Δy and Δx are read "delta y" and "delta x.")

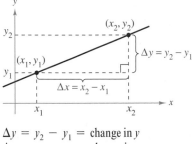

$\Delta y = y_2 - y_1 = $ change in y
$\Delta x = x_2 - x_1 = $ change in x
Figure P.12

SYMBOL FOR SLOPE

The use of the letter m to represent the slope of a line comes from *monter*, the French verb meaning to mount, to climb, or to rise.

Definition of the Slope of a Line

The **slope** m of a nonvertical line passing through the points (x_1, y_1) and (x_2, y_2) is

$$m = \frac{\Delta y}{\Delta x} = \frac{y_2 - y_1}{x_2 - x_1}, \qquad x_1 \neq x_2.$$

NOTE   When using the formula for slope, note that

$$\frac{y_2 - y_1}{x_2 - x_1} = \frac{-(y_1 - y_2)}{-(x_1 - x_2)} = \frac{y_1 - y_2}{x_1 - x_2}.$$

So, it does not matter in which order you subtract *as long as* you are consistent and both "subtracted coordinates" come from the same point.

Figure P.13 shows four lines: one has a positive slope, one has a slope of zero, one has a negative slope, and one has an "undefined slope." In general, the greater the absolute value of the slope of a line, the steeper the line is. For instance, in Figure P.13, the line with a slope of -5 is steeper than the line with a slope of $\frac{1}{5}$.

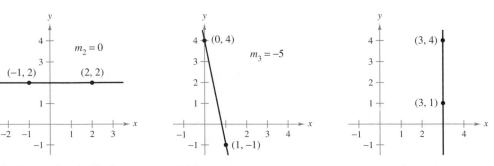
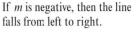

If m is positive, then the line rises from left to right.
Figure P.13

If m is zero, then the line is horizontal.

If m is negative, then the line falls from left to right.

If m is undefined, then the line is vertical.

EXPLORATION

Investigating Equations of Lines
Use a graphing utility to graph each of the linear equations. Which point is common to all seven lines? Which value in the equation determines the slope of each line?

a. $y - 4 = -2(x + 1)$

b. $y - 4 = -1(x + 1)$

c. $y - 4 = -\frac{1}{2}(x + 1)$

d. $y - 4 = 0(x + 1)$

e. $y - 4 = \frac{1}{2}(x + 1)$

f. $y - 4 = 1(x + 1)$

g. $y - 4 = 2(x + 1)$

Use your results to write an equation of a line passing through $(-1, 4)$ with a slope of m.

Equations of Lines

Any two points on a nonvertical line can be used to calculate its slope. This can be verified from the similar triangles shown in Figure P.14. (Recall that the ratios of corresponding sides of similar triangles are equal.)

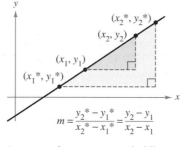

Any two points on a nonvertical line can be used to determine its slope.
Figure P.14

You can write an equation of a nonvertical line if you know the slope of the line and the coordinates of one point on the line. Suppose the slope is m and the point is (x_1, y_1). If (x, y) is any other point on the line, then

$$\frac{y - y_1}{x - x_1} = m.$$

This equation, involving the two variables x and y, can be rewritten in the form $y - y_1 = m(x - x_1)$, which is called the **point-slope equation of a line.**

> **Point-Slope Equation of a Line**
>
> An equation of the line with slope m passing through the point (x_1, y_1) is given by $y - y_1 = m(x - x_1)$.

Example 1   **Finding an Equation of a Line**

Find an equation of the line that has a slope of 3 and passes through the point $(1, -2)$.

Solution

$$\begin{aligned}
y - y_1 &= m(x - x_1) &&\text{Point-slope form}\\
y - (-2) &= 3(x - 1) &&\text{Substitute } -2 \text{ for } y_1, 1 \text{ for } x_1, \text{ and } 3 \text{ for } m.\\
y + 2 &= 3x - 3 &&\text{Simplify.}\\
y &= 3x - 5 &&\text{Solve for } y.
\end{aligned}$$

(See Figure P.15.)

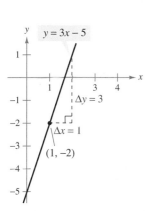

The line with a slope of 3 passing through point $(1, -2)$
Figure P.15

NOTE   Remember that only nonvertical lines have a slope. Consequently, vertical lines cannot be written in point-slope form. For instance, the equation of the vertical line passing through the point $(1, -2)$ is $x = 1$.

Ratios and Rates of Change

The slope of a line can be interpreted as either a *ratio* or a *rate*. If the *x*- and *y*-axes have the same unit of measure, the slope has no units and is a **ratio.** If the *x*- and *y*-axes have different units of measure, the slope is a rate or **rate of change.** In your study of calculus, you will encounter applications involving both interpretations of slope.

Example 2   **Population Growth and Engineering Design**

a. The population of Arizona was 2,717,000 in 1980 and 3,665,000 in 1990. Over this 10-year period, the average rate of change of the population was

$$\text{Rate of change} = \frac{\text{change in population}}{\text{change in years}}$$

$$= \frac{3,665,000 - 2,717,000}{1990 - 1980}$$

$$= 94,800 \text{ people per year.}$$

If Arizona's population had continued to increase at this same rate for the next 10 years, it would have had a 2000 population of 4,613,000. In the 2000 census, however, Arizona's population was determined to be 5,131,000, so the population's rate of change from 1990 to 2000 was greater than in the previous decade (see Figure P.16).   *(Source: U.S. Census Bureau, Population Division)*

b. In tournament water-ski jumping, the ramp rises to a height of 6 feet on a raft that is 21 feet long, as shown in Figure P.17. The slope of the ski ramp is the ratio of its height (the rise) to the length of its base (the run).

$$\text{Slope of ramp} = \frac{\text{rise}}{\text{run}} \qquad \text{Rise is vertical change, run is horizontal change.}$$

$$= \frac{6 \text{ feet}}{21 \text{ feet}}$$

$$= \frac{2}{7}$$

In this case, note that the slope is a ratio and has no units.

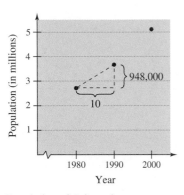

Population of Arizona in census years
Figure P.16

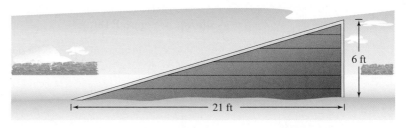

Dimensions of a water-skip ramp
Figure P.17

The rate of change found in Example 2a is an **average rate of change.** An average rate of change is always calculated over an interval. In this case, the interval is [1980, 1990]. In Chapter 2 you will study another type of rate of change called an *instantaneous rate of change.*

Graphing Linear Models

Many problems in analytic geometry can be classified in two basic categories: (1) Given a graph, what is its equation? and (2) Given an equation, what is its graph? The point-slope equation of a line can be used to solve problems in the first category. However, this form is not especially useful for solving problems in the second category. The form that is better suited to sketching the graph of a line is the **slope-intercept** form of the equation of a line.

The Slope-Intercept Equation of a Line

The graph of the linear equation

$$y = mx + b$$

is a line having a *slope* of m and a *y-intercept* at $(0, b)$.

Example 3   **Sketching Lines in the Plane**

Sketch the graph of each equation.

a. $y = 2x + 1$ **b.** $y = 2$ **c.** $3y + x - 6 = 0$

Solution

a. Because $b = 1$, the y-intercept is $(0, 1)$. Because the slope is $m = 2$, you know that the line rises two units for each unit it moves to the right, as shown in Figure P.18(a).

b. Because $b = 2$, the y-intercept is $(0, 2)$. Because the slope is $m = 0$, you know that the line is horizontal, as shown in Figure P.18(b).

c. Begin by writing the equation in slope-intercept form.

$$3y + x - 6 = 0 \qquad \text{Write original equation.}$$
$$3y = -x + 6 \qquad \text{Isolate y-term on the left.}$$
$$y = -\frac{1}{3}x + 2 \qquad \text{Slope-intercept form}$$

In this form, you can see that the y-intercept is $(0, 2)$ and the slope is $m = -\frac{1}{3}$. This means that the line falls one unit for every three units it moves to the right, as shown in Figure P.18(c).

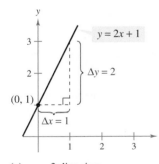

(a) $m = 2$; line rises

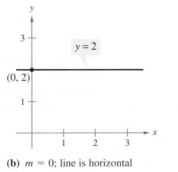

(b) $m = 0$; line is horizontal

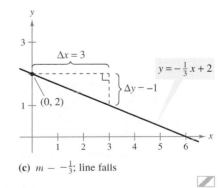

(c) $m = -\frac{1}{3}$; line falls

Figure P.18

Because the slope of a vertical line is not defined, its equation cannot be written in the slope-intercept form. However, the equation of *any* line can be written in the **general form**

$$Ax + By + C = 0$$ General form of the equation of a line

where A and B are not *both* zero. For instance, the vertical line given by $x = a$ can be represented by the general form $x - a = 0$.

Summary of Equations of Lines

1. General form: $Ax + By + C = 0$
2. Vertical line: $x = a$
3. Horizontal line: $y = b$
4. Point-slope form: $y - y_1 = m(x - x_1)$
5. Slope-intercept form: $y = mx + b$

Parallel and Perpendicular Lines

The slope of a line is a convenient tool for determining whether two lines are parallel or perpendicular, as shown in Figure P.19. Specifically, nonvertical lines with the same slope are parallel and nonvertical lines whose slopes are negative reciprocals are perpendicular.

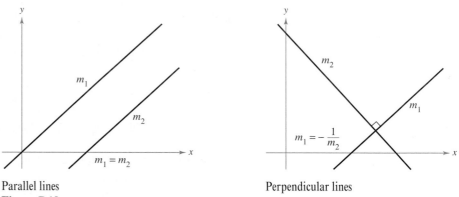

Parallel lines
Figure P.19

Perpendicular lines

STUDY TIP   In mathematics, the phrase "if and only if" is a way of stating two implications in one statement. For instance, the first statement at the right could be rewritten as the following two implications.

a. If two distinct nonvertical lines are parallel, then their slopes are equal.

b. If two distinct nonvertical lines have equal slopes, then they are parallel.

Parallel and Perpendicular Lines

1. Two distinct nonvertical lines are **parallel** if and only if their slopes are equal—that is, if and only if $m_1 = m_2$.

2. Two nonvertical lines are **perpendicular** if and only if their slopes are negative reciprocals of each other—that is, if and only if

$$m_1 = -\frac{1}{m_2}.$$

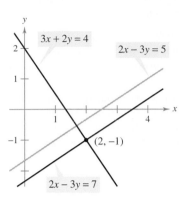

Lines parallel and perpendicular to
$2x - 3y = 5$
Figure P.20

iC *Example 4*   **Finding Parallel and Perpendicular Lines**

Find the general forms of the equations of the lines that pass through the point $(2, -1)$ and are

a. parallel to the line $2x - 3y = 5$ **b.** perpendicular to the line $2x - 3y = 5$.

(See Figure P.20.)

Solution   By writing the linear equation $2x - 3y = 5$ in slope-intercept form, $y = \frac{2}{3}x - \frac{5}{3}$, you can see that the given line has a slope of $m = \frac{2}{3}$.

a. The line through $(2, -1)$ that is parallel to the given line also has a slope of $\frac{2}{3}$.

$$y - y_1 = m(x - x_1) \qquad \text{Point-slope form}$$
$$y - (-1) = \tfrac{2}{3}(x - 2) \qquad \text{Substitute.}$$
$$3(y + 1) = 2(x - 2) \qquad \text{Simplify.}$$
$$2x - 3y - 7 = 0 \qquad \text{General form}$$

Note the similarity to the original equation.

b. Using the negative reciprocal of the slope of the given line, you can determine that the slope of a line perpendicular to the given line is $-\frac{3}{2}$. Therefore, the line through the point $(2, -1)$ that is perpendicular to the given line has the following equation.

$$y - y_1 = m(x - x_1) \qquad \text{Point-slope form}$$
$$y - (-1) = -\tfrac{3}{2}(x - 2) \qquad \text{Substitute.}$$
$$2(y + 1) = -3(x - 2) \qquad \text{Simplify.}$$
$$3x + 2y - 4 = 0 \qquad \text{General form}$$

TECHNOLOGY PITFALL   The slope of a line will appear distorted if you use different tick-mark spacing on the x- and y-axes. For instance, the graphing calculator screens in Figures P.21(a) and P.21(b) both show the lines given by $y = 2x$ and $y = -\frac{1}{2}x + 3$. Because these lines have slopes that are negative reciprocals, they must be perpendicular. In Figure P.21(a), however, the lines don't appear to be perpendicular because the tick-mark spacing on the x-axis is not the same as that on the y-axis. In Figure P.21(b) the lines appear perpendicular because the tick-mark spacing on the x-axis is the same as on the y-axis. This type of viewing window is said to have a *square setting*.

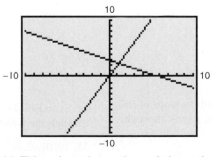

(a) Tick-mark spacing on the x-axis is not the same as tick-mark spacing on the y-axis.

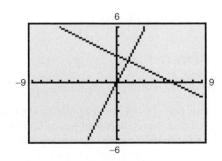

(b) Tick-mark spacing on the x-axis is the same as tick-mark spacing on the y-axis.

Figure P.21

EXERCISES FOR SECTION P.2

In Exercises 1–6, estimate the slope of the line from its graph. To print an enlarged copy of the graph, go to the website **www.mathgraphs.com.**

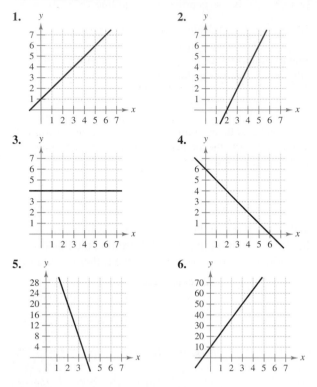

1.

2.

3.

4.

5.

6.

In Exercises 7 and 8, sketch the lines through the point with the indicated slopes. Make the sketches on the same set of coordinate axes.

Point	Slopes			
7. $(2, 3)$	(a) 1	(b) -2	(c) $-\frac{3}{2}$	(d) Undefined
8. $(-4, 1)$	(a) 3	(b) -3	(c) $\frac{1}{3}$	(d) 0

In Exercises 9–14, plot the pair of points and find the slope of the line passing through them.

9. $(3, -4), (5, 2)$ **10.** $(1, 2), (-2, 4)$

11. $(2, 1), (2, 5)$ **12.** $(3, -2), (4, -2)$

13. $\left(-\frac{1}{2}, \frac{2}{3}\right), \left(-\frac{3}{4}, \frac{1}{6}\right)$ **14.** $\left(\frac{7}{8}, \frac{3}{4}\right), \left(\frac{5}{4}, -\frac{1}{4}\right)$

In Exercises 15–18, use the point on the line and the slope of the line to find three additional points that the line passes through. (There is more than one correct answer.)

Point	Slope	Point	Slope
15. $(2, 1)$	$m = 0$	**16.** $(-3, 4)$	m undefined
17. $(1, 7)$	$m = -3$	**18.** $(-2, -2)$	$m = 2$

19. *Writing*   Write a paragraph explaining why any two points on a line can be used to calculate the slope of the line.

20. *Conveyor Design*   A moving conveyor is built to rise 1 meter for each 3 meters of horizontal change.

(a) Find the slope of the conveyor.

(b) Suppose the conveyor runs between two floors in a factory. Find the length of the conveyor if the vertical distance between floors is 10 feet.

21. *Modeling Data*   The table shows the population of the United States for 1991–1998. Time in years is represented by t, with $t = 1$ corresponding to 1991, and the population (in millions) is represented by y.   *(Source: U.S. Bureau of the Census)*

t	1	2	3	4
y	252.1	255.0	257.7	260.3

t	5	6	7	8
y	262.8	265.2	267.7	270.3

(a) Plot the data by hand and connect adjacent points with a line segment.

(b) Use the slope to determine the year when the population increased most rapidly.

22. *Rate of Change*   Each of the following is the slope of a line representing daily revenue y in terms of time x in days. Use the slope to interpret any change in daily revenue for a 1-day increase in time.

(a) $m = 400$ (b) $m = 100$ (c) $m = 0$

In Exercises 23–26, find the slope and the y-intercept (if possible) of the line.

23. $x + 5y = 20$ **24.** $6x - 5y = 15$

25. $x = 4$ **26.** $y = -1$

In Exercises 27–32, find an equation of the line that passes through the point and has the indicated slope. Sketch the line.

Point	Slope	Point	Slope
27. $(0, 3)$	$m = \frac{3}{4}$	**28.** $(-1, 2)$	m undefined
29. $(0, 0)$	$m = \frac{2}{3}$	**30.** $(0, 4)$	$m = 0$
31. $(3, -2)$	$m = 3$	**32.** $(-2, 4)$	$m = -\frac{3}{5}$

In Exercises 33–42, find an equation of the line that passes through the points, and sketch the line.

33. $(0, 0), (2, 6)$ **34.** $(0, 0), (-1, 3)$

35. $(2, 1), (0, -3)$ **36.** $(-3, -4), (1, 4)$

37. $(2, 8), (5, 0)$ **38.** $(-3, 6), (1, 2)$

39. $(5, 1), (5, 8)$ **40.** $(1, -2), (3, -2)$

41. $\left(\frac{1}{2}, \frac{7}{2}\right), \left(0, \frac{3}{4}\right)$ **42.** $\left(\frac{7}{8}, \frac{3}{4}\right), \left(\frac{5}{4}, -\frac{1}{4}\right)$

43. Find an equation of the vertical line with x-intercept at 3.

44. Show that the line with intercepts $(a, 0)$ and $(0, b)$ has the following equation.

$$\frac{x}{a} + \frac{y}{b} = 1, \qquad a \neq 0, b \neq 0$$

In Exercises 45–48, use the result of Exercise 44 to write an equation of the line.

45. x-intercept: $(2, 0)$

y-intercept: $(0, 3)$

46. x-intercept: $\left(-\frac{2}{3}, 0\right)$

y-intercept: $(0, -2)$

47. Point on line: $(1, 2)$

x-intercept: $(a, 0)$

y-intercept: $(0, a)$

$(a \neq 0)$

48. Point on line: $(-3, 4)$

x-intercept: $(a, 0)$

y-intercept: $(0, a)$

$(a \neq 0)$

In Exercises 49–56, sketch a graph of the equation.

49. $y = -3$

50. $x = 4$

51. $y = -2x + 1$

52. $y = \frac{1}{3}x - 1$

53. $y - 2 = \frac{3}{2}(x - 1)$

54. $y - 1 = 3(x + 4)$

55. $2x - y - 3 = 0$

56. $x + 2y + 6 = 0$

Square Setting In Exercises 57 and 58, use a graphing utility to graph both lines in each viewing window. Compare the graphs. Do the lines appear perpendicular? Are the lines perpendicular? Explain.

57. $y = x + 6$, $y = -x + 2$

(a)

```
Xmin = -10
Xmax = 10
Xscl = 1
Ymin = -10
Ymax = 10
Yscl = 1
```

(b)

```
Xmin = -15
Xmax = 15
Xscl = 1
Ymin = -10
Ymax = 10
Yscl = 1
```

58. $y = 2x - 3$, $y = -\frac{1}{2}x + 1$

(a)

```
Xmin = -5
Xmax = 5
Xscl = 1
Ymin = -5
Ymax = 5
Yscl = 1
```

(b)

```
Xmin = -6
Xmax = 6
Xscl = 1
Ymin = -4
Ymax = 4
Yscl = 1
```

In Exercises 59–64, write an equation of the line through the point (a) parallel to the given line and (b) perpendicular to the given line.

	Point	Line		Point	Line
59.	$(2, 1)$	$4x - 2y = 3$	**60.**	$(-3, 2)$	$x + y = 7$
61.	$\left(\frac{3}{4}, \frac{7}{8}\right)$	$5x - 3y = 0$	**62.**	$(-6, 4)$	$3x + 4y = 7$
63.	$(2, 5)$	$x = 4$	**64.**	$(-1, 0)$	$y = -3$

Rate of Change In Exercises 65–68, you are given the dollar value of a product in 2001 *and* the rate at which the value of the product is expected to change during the next 5 years. Write a linear equation that gives the dollar value V of the product in terms of the year t. (Let $t = 0$ represent 2000.)

	2001 Value	Rate
65.	$2540	$125 increase per year
66.	$156	$4.50 increase per year
67.	$20,400	$2000 decrease per year
68.	$245,000	$5600 decrease per year

In Exercises 69 and 70, use a graphing utility to graph the parabolas and find their points of intersection. Find an equation of the line through the points of intersection and sketch its graph in the same viewing window.

69. $y = x^2$

$y = 4x - x^2$

70. $y = x^2 - 4x + 3$

$y = -x^2 + 2x + 3$

In Exercises 71 and 72, determine whether the points are collinear. (Three points are *collinear* if they lie on the same line.)

71. $(-2, 1), (-1, 0), (2, -2)$ **72.** $(0, 4), (7, -6), (-5, 11)$

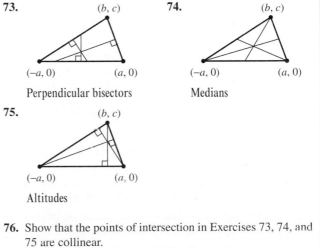

Getting at the Concept

In Exercises 73–75, find the coordinates of the point of intersection of the given segments.

73. (b, c), $(-a, 0)$, $(a, 0)$

Perpendicular bisectors

74. (b, c), $(-a, 0)$, $(a, 0)$

Medians

75. (b, c), $(-a, 0)$, $(a, 0)$

Altitudes

76. Show that the points of intersection in Exercises 73, 74, and 75 are collinear.

77. *Temperature Conversion* Find a linear equation that expresses the relationship between the temperature in degrees Celsius C and degrees Fahrenheit F. Use the fact that water freezes at $0°C$ ($32°F$) and boils at $100°C$ ($212°F$). Use the equation to convert $72°F$ to degrees Celsius.

78. Reimbursed Expenses A company reimburses its sales representatives $150 per day for lodging and meals plus 34¢ per mile driven. Write a linear equation giving the daily cost C to the company in terms of x, the number of miles driven. How much does it cost the company if a sales representative drives 137 miles on a given day?

79. Career Choice An employee has two options for positions in a large corporation. One position pays $12.50 per hour *plus* an additional unit rate of $0.75 per unit produced. The other pays $9.20 per hour *plus* a unit rate of $1.30.

(a) Find linear equations for the hourly wages W in terms of x, the number of units produced per hour, for each option.

(b) Use a graphing utility to graph the linear equations and find the point of intersection.

(c) Interpret the meaning of the point of intersection of the graphs in part (b). How would you use this information to select the correct option if the goal were to obtain the highest hourly wage?

80. Straight-Line Depreciation A small business purchases a piece of equipment for $875. After 5 years the equipment will be outdated, having no value.

(a) Write a linear equation giving the value y of the equipment in terms of the time x, $0 \le x \le 5$.

(b) Find the value of the equipment when $x = 2$.

(c) Estimate (to two-decimal-place accuracy) the time when the value of the equipment is $200.

81. Apartment Rental A real estate office handles an apartment complex with 50 units. When the rent is $580 per month, all 50 units are occupied. However, when the rent is $625, the average number of occupied units drops to 47. Assume that the relationship between the monthly rent p and the demand x is linear. (*Note:* The term *demand* refers to the number of occupied units.)

(a) Write a linear equation giving the demand x in terms of the rent p.

(b) *Linear extrapolation* Use a graphing utility to graph the demand equation and use the *trace* feature to predict the number of units occupied if the rent is raised to $655.

(c) *Linear interpolation* Predict the number of units occupied if the rent is lowered to $595. Verify graphically.

82. Modeling Data An instructor gives regular 20-point quizzes and 100-point exams in a mathematics course. Average scores for six students, given as ordered pairs (x, y) where x is the average quiz score and y is the average test score, are (18, 87), (10, 55), (19, 96), (16, 79), (13, 76), and (15, 82).

(a) Use the regression capabilities of a graphing utility to find the least squares regression line for the data.

(b) Use a graphing utility to plot the points and graph the regression line in the same viewing window.

(c) Use the regression line to predict the average exam score for a student with an average quiz score of 17.

(d) Interpret the meaning of the slope of the regression line.

(e) The instructor adds 4 points to the average test score of everyone in the class. Describe the change in the position of the plotted points and the change in the equation of the line.

Distance In Exercises 83–88, find the distance between the point and line, or between the lines, using the formula for the distance between the point (x_1, y_1) and the line $Ax + By + C = 0$.

$$\text{Distance} = \frac{|Ax_1 + By_1 + C|}{\sqrt{A^2 + B^2}}$$

83. Point: $(0, 0)$

Line: $4x + 3y = 10$

84. Point: $(2, 3)$

Line: $4x + 3y = 10$

85. Point: $(-2, 1)$

Line: $x - y - 2 = 0$

86. Point: $(6, 2)$

Line: $x = -1$

87. Line: $x + y = 1$

Line: $x + y = 5$

88. Line: $3x - 4y = 1$

Line: $3x - 4y = 10$

89. Show that the distance between the point (x_1, y_1) and the line $Ax + By + C = 0$ is

$$\text{Distance} = \frac{|Ax_1 + By_1 + C|}{\sqrt{A^2 + B^2}}.$$

90. Write the distance d between the point $(3, 1)$ and the line $y = mx + 4$ in terms of m. Use a graphing utility to graph the equation. When is the distance 0? Explain the result geometrically.

91. Prove that the diagonals of a rhombus intersect at right angles. (A rhombus is a quadrilateral with sides of equal lengths.)

92. Prove that the figure formed by connecting consecutive midpoints of the sides of any quadrilateral is a parallelogram.

93. Prove that if the points (x_1, y_1) and (x_2, y_2) lie on the same line as (x_1^*, y_1^*) and (x_2^*, y_2^*), then

$$\frac{y_2^* - y_1^*}{x_2^* - x_1^*} = \frac{y_2 - y_1}{x_2 - x_1}.$$

Assume $x_1 \ne x_2$ and $x_1^* \ne x_2^*$.

94. Prove that if the slopes of two nonvertical lines are negative reciprocals of each other, then the lines are perpendicular.

True or False? In Exercises 95 and 96, determine whether the statement is true or false. If it is false, explain why or give an example that shows it is false.

95. The lines represented by $ax + by = c_1$ and $bx - ay = c_2$ are perpendicular. Assume $a \ne 0$ and $b \ne 0$.

96. It is possible for two lines with positive slopes to be perpendicular to each other.

- Use function notation to represent and evaluate a function.
- Find the domain and range of a function.
- Sketch the graph of a function.
- Identify different types of transformations of functions.
- Classify functions and recognize combinations of functions.

Functions and Function Notation

A **relation** between two sets X and Y is a set of ordered pairs, each of the form (x, y), where x is a member of X and y is a member of Y. A **function** from X to Y is a relation between X and Y that has the property that any two ordered pairs with the same x-value also have the same y-value. The variable x is the **independent variable,** and the variable y is the **dependent variable.**

Many real-life situations can be modeled by functions. For instance, the area A of a circle is a function of the circle's radius r.

$$A = \pi r^2 \qquad \text{\small A is a function of r.}$$

In this case r is the independent variable and A is the dependent variable.

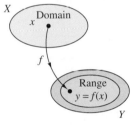

A real-valued function f of a real variable
Figure P.22

Definition of a Real-Valued Function of a Real Variable

Let X and Y be sets of real numbers. A **real-valued function f of a real variable x** from X to Y is a correspondence that assigns to each number x in X exactly one number y in Y.

The **domain** of f is the set X. The number y is the **image** of x under f and is denoted by $f(x)$. The **range** of f is a subset of Y and consists of all images of numbers in X (see Figure P.22).

Functions can be specified in a variety of ways. In this text, however, we will concentrate primarily on functions that are given by equations involving the dependent and independent variables. For instance, the equation

$$x^2 + 2y = 1 \qquad \text{\small Equation in implicit form}$$

defines y, the dependent variable, as a function of x, the independent variable. To **evaluate** this function (that is, to find the y-value that corresponds to a given x-value), it is convenient to isolate y on the left side of the equation.

$$y = \frac{1}{2}(1 - x^2) \qquad \text{\small Equation in explicit form}$$

Using f as the name of the function, you can write this equation as

$$f(x) = \frac{1}{2}(1 - x^2). \qquad \text{\small Function notation}$$

The original equation, $x^2 + 2y = 1$, **implicitly** defines y as a function of x. When you solve the equation for y, you are writing the equation in **explicit** form.

Function notation has the advantage of clearly identifying the dependent variable as $f(x)$ while at the same time telling you that x is the independent variable and that the function itself is "f." The symbol $f(x)$ is read "f of x." Function notation allows you to be less wordy. Instead of asking "What is the value of y that corresponds to $x = 3$?" you can ask "What is $f(3)$?"

FUNCTION NOTATION

The word *function* was first used by Gottfried Wilhelm Leibniz in 1694 as a term to denote any quantity connected with a curve, such as the coordinates of a point on a curve or the slope of a curve. Forty years later, Leonhard Euler used the word function to describe any expression made up of a variable and some constants. He introduced the notation

$$y = f(x).$$

In an equation that defines a function, the role of the variable x is simply that of a placeholder. For instance, the function given by

$$f(x) = 2x^2 - 4x + 1$$

can be described by the form

$$f(\,\rule{1cm}{0.4pt}\,) = 2(\,\rule{1cm}{0.4pt}\,)^2 - 4(\,\rule{1cm}{0.4pt}\,) + 1$$

where parentheses are used instead of x. To evaluate $f(-2)$, simply place -2 in each set of parentheses.

$$
\begin{aligned}
f(-2) &= 2(-2)^2 - 4(-2) + 1 && \text{Substitute } -2 \text{ for } x. \\
&= 2(4) + 8 + 1 && \text{Simplify.} \\
&= 17 && \text{Simplify.}
\end{aligned}
$$

NOTE Although f is often used as a convenient function name and x as the independent variable, you can use other symbols. For instance, the following equations all define the same function.

$$
\begin{aligned}
f(x) &= x^2 - 4x + 7 && \text{Function name is } f, \text{ independent variable is } x. \\
f(t) &= t^2 - 4t + 7 && \text{Function name is } f, \text{ independent variable is } t. \\
g(s) &= s^2 - 4s + 7 && \text{Function name is } g, \text{ independent variable is } s.
\end{aligned}
$$

Example 1   Evaluating a Function

For the function f defined by $f(x) = x^2 + 7$, evaluate each of the following.

a. $f(3a)$ **b.** $f(b - 1)$ **c.** $\dfrac{f(x + \Delta x) - f(x)}{\Delta x}, \quad \Delta x \neq 0$

Solution

$$
\begin{aligned}
\textbf{a. } f(3a) &= (3a)^2 + 7 && \text{Substitute } 3a \text{ for } x. \\
&= 9a^2 + 7 && \text{Simplify.} \\
\textbf{b. } f(b - 1) &= (b - 1)^2 + 7 && \text{Substitute } b - 1 \text{ for } x. \\
&= b^2 - 2b + 1 + 7 && \text{Expand binomial.} \\
&= b^2 - 2b + 8 && \text{Simplify.}
\end{aligned}
$$

$$
\begin{aligned}
\textbf{c. } \frac{f(x + \Delta x) - f(x)}{\Delta x} &= \frac{[(x + \Delta x)^2 + 7] - (x^2 + 7)}{\Delta x} \\
&= \frac{x^2 + 2x\Delta x + (\Delta x)^2 + 7 - x^2 - 7}{\Delta x} \\
&= \frac{2x\Delta x + (\Delta x)^2}{\Delta x} \\
&= \frac{\Delta x(2x + \Delta x)}{\Delta x} \\
&= 2x + \Delta x, \qquad \Delta x \neq 0
\end{aligned}
$$

STUDY TIP In calculus, it is important to clearly communicate the domain of a function or expression. For instance, in Example 1c the two expressions

$$\frac{f(x + \Delta x) - f(x)}{\Delta x} = 2x + \Delta x, \quad \Delta x \neq 0$$

are equivalent because $\Delta x = 0$ is excluded from the domain of each expression. Without a stated domain restriction, the two expressions would not be equivalent.

NOTE The expression in Example 1c is called a *difference quotient* and has a special significance in calculus. We will say more about this in Chapter 2.

The Domain and Range of a Function

The domain of a function can be described explicitly, or it may be described *implicitly* by an equation used to define the function. The implied domain is the set of all real numbers for which the equation is defined, whereas an explicitly defined domain is one that is given along with the function. For example, the function given by

$$f(x) = \frac{1}{x^2 - 4}, \quad 4 \le x \le 5$$

has an explicitly defined domain given by $\{x: 4 \le x \le 5\}$. On the other hand, the function given by

$$g(x) = \frac{1}{x^2 - 4}$$

has an implied domain that is the set $\{x: x \ne \pm 2\}$.

Example 2   **Finding the Domain and Range of a Function**

---

a. The domain of the function

$$f(x) = \sqrt{x - 1}$$

is the set of all x-values for which $x - 1 \ge 0$, which is the interval $[1, \infty)$. To find the range observe that $f(x) = \sqrt{x - 1}$ is never negative. So, the range is the interval $[0, \infty)$, as indicated in Figure P.23(a).

b. The domain of the tangent function, as shown in Figure P.23(b),

$$f(x) = \tan x$$

is the set of all x-values such that

$$x \ne \frac{\pi}{2} + n\pi, \quad n \text{ is an integer.} \qquad \text{Domain of tangent function}$$

The range of this function is the set of all real numbers. For a review of the characteristics of this and other trigonometric functions, see Appendix D.

Example 3   **A Function Defined by More than One Equation**

Determine the domain and range of the function.

$$f(x) = \begin{cases} 1 - x, & \text{if } x < 1 \\ \sqrt{x - 1}, & \text{if } x \ge 1 \end{cases}$$

Solution   Because f is defined for $x < 1$ and $x \ge 1$, the domain is the entire set of real numbers. On the portion of the domain for which $x \ge 1$, the function behaves as in Example 2a. For $x < 1$, the values of $1 - x$ are positive. So, the range of the function is the interval $[0, \infty)$. (See Figure P.24.)

A function from X to Y is **one-to-one** if to each y-value in the range there corresponds exactly one x-value in the domain. For instance, the function given in Example 2a is one-to-one, whereas the functions given in Examples 2b and 3 are not one-to-one. A function from X to Y is **onto** if its range consists of all of Y.

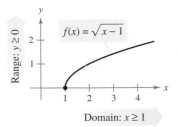

(a) The domain of f is $[1, \infty)$ and the range is $[0, \infty)$.

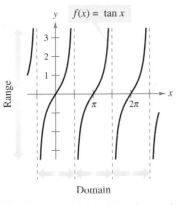

(b) The domain of f is all x-values such that $x \ne \frac{\pi}{2} + n\pi$ and the range is $(-\infty, \infty)$.

Figure P.23

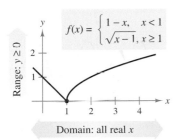
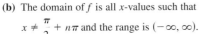

The domain of f is $(-\infty, \infty)$ and the range is $[0, \infty)$.

Figure P.24

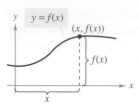

The graph of a function
Figure P.25

The Graph of a Function

The graph of the function $y = f(x)$ consists of all points $(x, f(x))$, where x is in the domain of f. In Figure P.25, note that

$x = $ the directed distance from the y-axis

$f(x) = $ the directed distance from the x-axis.

A vertical line can intersect the graph of a function of x at most *once*. This observation provides a convenient visual test, called the **vertical line test,** for functions of x. For example, in Figure P.26(a), you can see that the graph does not define y as a function of x because a vertical line intersects the graph twice, whereas in Figures P.26(b) and (c), the graphs do define y as a function of x.

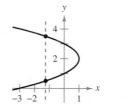

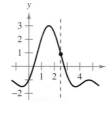

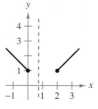

(**a**) Not a function of x (**b**) A function of x (**c**) A function of x
Figure P.26

Figure P.27 shows the graphs of eight basic functions. You should be able to recognize these graphs. (Graphs of the other four basic trigonometric functions are shown in Appendix D.)

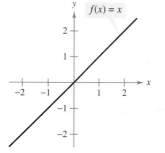

Identity function

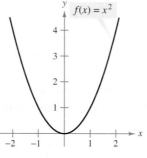

Squaring function

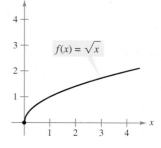

Cubing function

Square root function

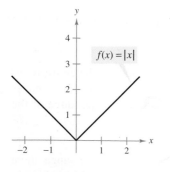

Absolute value function

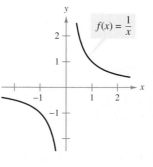

Rational function

Sine function

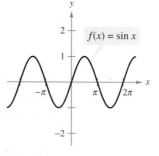

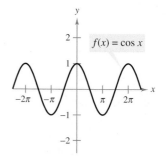

Cosine function

The graphs of eight basic functions
Figure P.27

Transformations of Functions

Some families of graphs have the same basic shape. For example, compare the graph of $y = x^2$ with the graphs of the four other quadratic functions shown in Figure P.28.

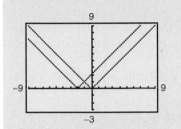
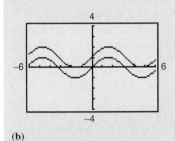
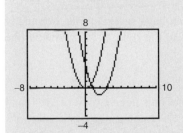
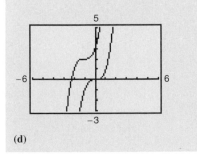
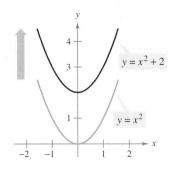

(a) Vertical shift upward

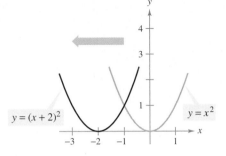

(b) Horizontal shift to the left

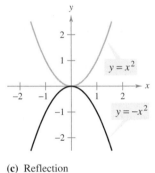

(c) Reflection

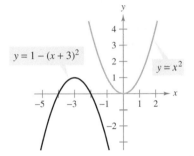

(d) Shift left, reflect, and shift upward

Figure P.28

Each of the graphs in Figure P.28 is a **transformation** of the graph of $y = x^2$. The three basic types of transformations illustrated by these graphs are vertical shifts, horizontal shifts, and reflections. Function notation lends itself well to describing transformations of graphs in the plane. For instance, if $f(x) = x^2$ is considered to be the original function in Figure P.28, the transformations shown can be represented by the following equations.

$y = f(x) + 2$ Vertical shift up 2 units

$y = f(x + 2)$ Horizontal shift to the left 2 units

$y = -f(x)$ Reflection about the x-axis

$y = -f(x + 3) + 1$ Shift left 3 units, reflect about x-axis, and shift up 1 unit

---

Basic Types of Transformations $(c > 0)$

Original graph: $\quad y = f(x)$

Horizontal shift c units to the **right**: $\quad y = f(x - c)$

Horizontal shift c units to the **left**: $\quad y = f(x + c)$

Vertical shift c units **downward**: $\quad y = f(x) - c$

Vertical shift c units **upward**: $\quad y = f(x) + c$

Reflection (about the x-axis): $\quad y = -f(x)$

Reflection (about the y-axis): $\quad y = f(-x)$

Reflection (about the origin): $\quad y = -f(-x)$

LEONHARD EULER (1707–1783)

In addition to making major contributions to almost every branch of mathematics, Euler was one of the first to apply calculus to real-life problems in physics. His extensive published writings include such topics as shipbuilding, acoustics, optics, astronomy, mechanics, and magnetism.

Bettmann/Corbis

Classifications and Combinations of Functions

The modern notion of a function is derived from the efforts of many seventeenth- and eighteenth-century mathematicians. Of particular note was Leonhard Euler, to whom we are indebted for the function notation $y = f(x)$. By the end of the eighteenth century, mathematicians and scientists had concluded that many real-world phenomena could be represented by mathematical models taken from a collection of functions called **elementary functions.** Elementary functions fall into three categories.

1. Algebraic functions (polynomial, radical, rational)

2. Trigonometric functions (sine, cosine, tangent, and so on)

3. Exponential and logarithmic functions

You can review the trigonometric functions in Appendix D. The other nonalgebraic functions, such as the inverse trigonometric functions and the exponential and logarithmic functions, are introduced in Chapter 5.

The most common type of algebraic function is a **polynomial function**

$$f(x) = a_n x^n + a_{n-1} x^{n-1} + \cdots + a_2 x^2 + a_1 x + a_0, \qquad a_n \neq 0$$

where the positive integer n is the **degree** of the polynomial function. The numbers a_i are **coefficients,** with a_n the **leading coefficient** and a_0 the **constant term** of the polynomial function. It is common practice to use subscript notation for coefficients of general polynomial functions, but for polynomial functions of low degree, the following simpler forms are often used.

Zeroth degree:	$f(x) = a$	Constant function
First degree:	$f(x) = ax + b$	Linear function
Second degree:	$f(x) = ax^2 + bx + c$	Quadratic function
Third degree:	$f(x) = ax^3 + bx^2 + cx + d$	Cubic function

Although the graph of a polynomial function can have several turns, eventually the graph will rise or fall without bound as x moves to the right or left. Whether the graph of

$$f(x) = a_n x^n + a_{n-1} x^{n-1} + \cdots + a_2 x^2 + a_1 x + a_0$$

eventually rises or falls can be determined by the function's degree (odd or even) and by the leading coefficient a_n, as indicated in Figure P.29. Note that the dashed portions of the graphs indicate that the **leading coefficient test** determines *only* the right and left behavior of the graph.

FOR FURTHER INFORMATION For more on the history of the concept of a function, see the article "Evolution of the Function Concept: A Brief Survey" by Israel Kleiner in *The College Mathematics Journal*. To view this article, go to the website *www.matharticles.com*.

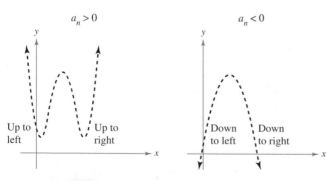

Graphs of polynomial functions of even degree

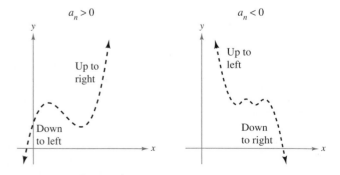

Graphs of polynomial functions of odd degree

The leading coefficient test for polynomial functions
Figure P.29

Just as a rational number can be written as the quotient of two integers, a **rational function** can be written as the quotient of two polynomials. Specifically, a function f is rational if it has the form

$$f(x) = \frac{p(x)}{q(x)}, \quad q(x) \neq 0$$

where $p(x)$ and $q(x)$ are polynomials.

Polynomial functions and rational functions are examples of **algebraic functions.** An algebraic function of x is one that can be expressed as a finite number of sums, differences, multiples, quotients, and radicals involving x^n. For example, $f(x) = \sqrt{x + 1}$ is algebraic. Functions that are not algebraic are **transcendental.** For instance, the trigonometric functions are transcendental.

Two functions can be combined in various ways to create new functions. For example, given $f(x) = 2x - 3$ and $g(x) = x^2 + 1$, you can form the following functions.

$$(f + g)(x) = f(x) + g(x) = (2x - 3) + (x^2 + 1) \qquad \text{Sum}$$
$$(f - g)(x) = f(x) - g(x) = (2x - 3) - (x^2 + 1) \qquad \text{Difference}$$
$$(fg)(x) = f(x)g(x) = (2x - 3)(x^2 + 1) \qquad \text{Product}$$
$$(f/g)(x) = \frac{f(x)}{g(x)} = \frac{2x - 3}{x^2 + 1} \qquad \text{Quotient}$$

You can combine two functions in yet another way, called **composition.** The resulting function is called a **composite function.**

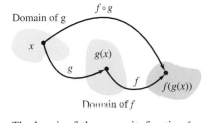

Domain of g

$f \circ g$

x

$g(x)$

g

f

$f(g(x))$

Domain of f

The domain of the composite function $f \circ g$
Figure P.30

Definition of Composite Function

Let f and g be functions. The function given by $(f \circ g)(x) = f(g(x))$ is called the **composite** of f with g. The domain of $f \circ g$ is the set of all x in the domain of g such that $g(x)$ is in the domain of f (see Figure P.30).

The composite of f with g may not be equal to the composite of g with f.

Example 4   **Finding Composites of Functions**

Given $f(x) = 2x - 3$ and $g(x) = \cos x$, find the following.

a. $f \circ g$ **b.** $g \circ f$

Solution

a. $(f \circ g)(x) = f(g(x))$ Definition of $f \circ g$

            $= f(\cos x)$ Substitute $\cos x$ for $g(x)$.

            $= 2(\cos x) - 3$ Definition of $f(x)$

            $= 2 \cos x - 3$ Simplify.

b. $(g \circ f)(x) = g(f(x))$ Definition of $g \circ f$

            $= g(2x - 3)$ Substitute $2x - 3$ for $f(x)$.

            $= \cos(2x - 3)$ Definition of $g(x)$

Note that $(f \circ g)(x) \neq (g \circ f)(x)$.

In Section P.1, an *x*-intercept of a graph was defined to be a point $(a, 0)$ at which the graph crosses the *x*-axis. If the graph represents a function f, the number a is a **zero** of f. In other words, *the zeros of a function f are the solutions of the equation* $f(x) = 0$. For example, the function $f(x) = x - 4$ has a zero at $x = 4$ because $f(4) = 0$.

In Section P.1 you also studied different types of symmetry. In the terminology of functions, a function is **even** if its graph is symmetric with respect to the *y*-axis, and is **odd** if its graph is symmetric with respect to the origin. The symmetry tests in Section P.1 yield the following test for even and odd functions.

Test for Even and Odd Functions

The function $y = f(x)$ is **even** if $f(-x) = f(x)$.
The function $y = f(x)$ is **odd** if $f(-x) = -f(x)$.

NOTE   Except for the constant function $f(x) = 0$, the graph of a function of *x* cannot have symmetry with respect to the *x*-axis because it then would fail the vertical line test for the graph of the function.

Example 5   **Even and Odd Functions and Zeros of Functions**

Determine whether each function is even, odd, or neither. Then find the zeros of the function.

a. $f(x) = x^3 - x$ **b.** $g(x) = 1 + \cos x$

Solution

a. This function is odd because

$$f(-x) = (-x)^3 - (-x) = -x^3 + x = -(x^3 - x) = -f(x).$$

The zeros of f are found as follows.

$$x^3 - x = 0 \qquad \text{Let } f(x) = 0.$$
$$x(x^2 - 1) = x(x - 1)(x + 1) = 0 \qquad \text{Factor.}$$
$$x = 0, 1, -1 \qquad \text{Zeros of } f$$

See Figure P.31(a).

b. This function is even because

$$g(-x) = 1 + \cos(-x) = 1 + \cos x = g(x). \qquad \cos(-x) = \cos(x)$$

The zeros of g are found as follows.

$$1 + \cos x = 0 \qquad \text{Let } g(x) = 0.$$
$$\cos x = -1 \qquad \text{Subtract 1 from each side.}$$
$$x = (2n + 1)\pi, \ n \text{ is an integer.} \qquad \text{Zeros of } g$$

See Figure P.31(b).

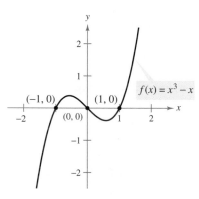

(a) Odd function

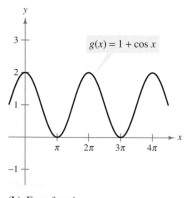

(b) Even function
Figure P.31

NOTE   Each of the functions in Example 5 is either even or odd. However, some functions, such as $f(x) = x^2 + x + 1$, are neither even nor odd.

EXERCISES FOR SECTION P.3

In Exercises 1–10, evaluate (if possible) the function at the given value(s) of the independent variable. Simplify the results.

1. $f(x) = 2x - 3$
   (a) $f(0)$
   (b) $f(-3)$
   (c) $f(b)$
   (d) $f(x - 1)$

2. $f(x) = \sqrt{x + 3}$
   (a) $f(-2)$
   (b) $f(6)$
   (c) $f(c)$
   (d) $f(x + \Delta x)$

3. $g(x) = 3 - x^2$
   (a) $g(0)$
   (b) $g(\sqrt{3})$
   (c) $g(-2)$
   (d) $g(t - 1)$

4. $g(x) = x^2(x - 4)$
   (a) $g(4)$
   (b) $g(\frac{3}{2})$
   (c) $g(c)$
   (d) $g(t + 4)$

5. $f(x) = \cos 2x$
   (a) $f(0)$
   (b) $f(-\pi/4)$
   (c) $f(\pi/3)$

6. $f(x) = \sin x$
   (a) $f(\pi)$
   (b) $f(5\pi/4)$
   (c) $f(2\pi/3)$

7. $f(x) = x^3$
   $$\frac{f(x + \Delta x) - f(x)}{\Delta x}$$

8. $f(x) = 3x - 1$
   $$\frac{f(x) - f(1)}{x - 1}$$

9. $f(x) = \dfrac{1}{\sqrt{x - 1}}$
   $$\frac{f(x) - f(2)}{x - 2}$$

10. $f(x) = x^3 - x$
    $$\frac{f(x) - f(1)}{x - 1}$$

In Exercises 11–16, find the domain and range of the function.

11. $h(x) = -\sqrt{x + 3}$

12. $g(x) = x^2 - 5$

13. $f(t) = \sec \dfrac{\pi t}{4}$

14. $h(t) = \cot t$

15. $f(x) = \dfrac{1}{x}$

16. $g(x) = \dfrac{2}{x - 1}$

In Exercises 17–20, evaluate the function as indicated. Determine its domain and range.

17. $f(x) = \begin{cases} 2x + 1, & x < 0 \\ 2x + 2, & x \geq 0 \end{cases}$
    (a) $f(-1)$ (b) $f(0)$ (c) $f(2)$ (d) $f(t^2 + 1)$

18. $f(x) = \begin{cases} x^2 + 2, & x \leq 1 \\ 2x^2 + 2, & x > 1 \end{cases}$
    (a) $f(-2)$ (b) $f(0)$ (c) $f(1)$ (d) $f(s^2 + 2)$

19. $f(x) = \begin{cases} |x| + 1, & x < 1 \\ -x + 1, & x \geq 1 \end{cases}$
    (a) $f(-3)$ (b) $f(1)$ (c) $f(3)$ (d) $f(b^2 + 1)$

20. $f(x) = \begin{cases} \sqrt{x + 4}, & x \leq 5 \\ (x - 5)^2, & x > 5 \end{cases}$
    (a) $f(-3)$ (b) $f(0)$ (c) $f(5)$ (d) $f(10)$

In Exercises 21–28, sketch a graph of the function and find its domain and range. Use a graphing utility to verify your graph.

21. $f(x) = 4 - x$

22. $g(x) = \dfrac{4}{x}$

23. $h(x) = \sqrt{x - 1}$

24. $f(x) = \frac{1}{2}x^3 + 2$

25. $f(x) = \sqrt{9 - x^2}$

26. $f(x) = x + \sqrt{4 - x^2}$

27. $g(t) = 2 \sin \pi t$

28. $h(\theta) = -5 \cos \dfrac{\theta}{2}$

In Exercises 29–32, use the vertical line test to determine whether y is a function of x. To print an enlarged copy of the graph, go to the website *www.mathgraphs.com*.

29. $x - y^2 = 0$

30. $\sqrt{x^2 - 4} - y = 0$

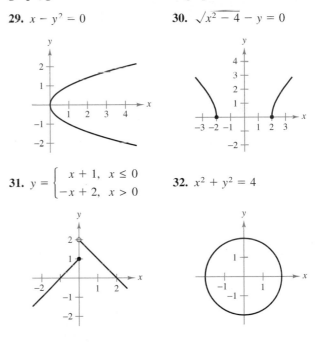

31. $y = \begin{cases} x + 1, & x \leq 0 \\ -x + 2, & x > 0 \end{cases}$

32. $x^2 + y^2 = 4$

In Exercises 33–36, determine whether y is a function of x.

33. $x^2 + y^2 = 4$

34. $x^2 + y = 4$

35. $y^2 = x^2 - 1$

36. $x^2y - x^2 + 4y = 0$

37. **Think About It** Express the function
    $$f(x) = |x| + |x - 2|$$
    without using absolute value signs. (For a review of absolute value, see Appendix D.)

38. **Writing** Use a graphing utility to graph the polynomial functions $p_1(x) = x^3 - x + 1$ and $p_2(x) = x^3 - x$. How many zeros does each function have? Is there a cubic polynomial that has no zeros? Explain.

Modeling Data In Exercises 39–42, match the data with a function from the following list.

(i) $f(x) = cx$ (ii) $g(x) = cx^2$

(iii) $h(x) = c\sqrt{|x|}$ (iv) $r(x) = c/x$

Determine the value of the constant c for each function such that the function fits the data shown in the table.

39.

x	-4	-1	0	1	4
y	-32	-2	0	-2	-32

40.

x	-4	-1	0	1	4
y	-1	$-\frac{1}{4}$	0	$\frac{1}{4}$	1

41.

x	-4	-1	0	1	4
y	-8	-32	Undef.	32	8

42.

x	-4	-1	0	1	4
y	6	3	0	3	6

Getting at the Concept

43. Water runs into a vase of height 30 centimeters at a constant rate. The vase is full after 5 seconds. Use this information and the shape of the vase shown in the figure to answer the questions if d is the depth of the water in centimeters and t is the time in seconds.

(a) Explain why d is a function of t.

(b) Determine the domain and range of the function.

(c) Sketch a possible graph of the function.

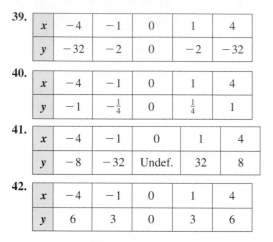

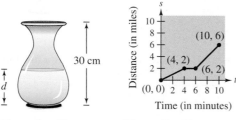

Time (in minutes)

Figure for 43   **Figure for 44**

44. The graph of the distance that a student drives in a 10-minute trip to school is shown in the figure. Give a verbal description of characteristics of the student's drive to school.

45. A student who commutes 27 miles to attend college remembers, after driving a few minutes, that a term paper that is due has been forgotten. Driving faster than usual, the student returns home, picks up the paper, and once again starts toward school. Sketch a possible graph of the student's distance from home as a function of time.

46. *Modeling Data* The table shows the average number of acres per farm in the United States for selected years. (*Source: Department of Agriculture*)

Year	1950	1960	1970	1980	1990	1998
Acreage	213	297	374	426	460	435

(a) Plot the data where A is the acreage and t is the time in years, with $t = 0$ corresponding to 1950. Sketch a freehand curve that approximates the data.

(b) Use the curve in part (a) to approximate $A(15)$.

47. Use the graph of f shown in the figure to sketch the graph of each function. To print an enlarged copy of the graph, go to the website *www.mathgraphs.com*.

(a) $f(x + 3)$ (b) $f(x - 1)$

(c) $f(x) + 2$ (d) $f(x) - 4$

(e) $3f(x)$ (f) $\frac{1}{4}f(x)$

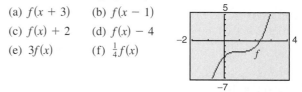

48. Use the graph of f shown in the figure to sketch the graph of each function. To print an enlarged copy of the graph, go to the website *www.mathgraphs.com*.

(a) $f(x - 4)$ (b) $f(x + 2)$

(c) $f(x) + 4$ (d) $f(x) - 1$

(e) $2f(x)$ (f) $\frac{1}{2}f(x)$

49. Use the graph of $f(x) = \sqrt{x}$ to sketch the graph of each function. In each case, describe the transformation.

(a) $y = \sqrt{x} + 2$ (b) $y = -\sqrt{x}$ (c) $y = \sqrt{x - 2}$

50. Specify a sequence of transformations that will yield each graph of h from the graph of the function $f(x) = \sin x$.

(a) $h(x) = \sin\left(x + \dfrac{\pi}{2}\right) + 1$ (b) $h(x) = -\sin(x - 1)$

51. *Graphical Reasoning* An electronically controlled thermostat is programmed to automatically lower the temperature during the night (see figure). The temperature T in degrees Celsius is given in terms of t, the time in hours on a 24-hour clock.

(a) Approximate $T(4)$ and $T(15)$.

(b) The thermostat is reprogrammed to produce a temperature $H(t) = T(t - 1)$. How does this change the temperature? Explain.

(c) The thermostat is reprogrammed to produce a temperature $H(t) = T(t) - 1$. How does this change the temperature? Explain.

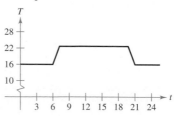

52. Given $f(x) = \sqrt{x}$ and $g(x) = x^2 - 1$, find the following.

(a) $f(g(1))$ (b) $g(f(1))$ (c) $g(f(0))$

(d) $f(g(-4))$ (e) $f(g(x))$ (f) $g(f(x))$

In Exercises 53–56, find the composite functions $(f \circ g)$ and $(g \circ f)$. What is the domain of each composite function? Are the two composite functions equal?

53. $f(x) = x^2$

$g(x) = \sqrt{x}$

54. $f(x) = x^2 - 1$

$g(x) = \cos x$

55. $f(x) = \dfrac{3}{x}$

$g(x) = x^2 - 1$

56. $f(x) = \dfrac{1}{x}$

$g(x) = \sqrt{x + 2}$

57. Ripples A pebble is dropped into a calm pond, causing ripples in the form of concentric circles. The radius (in feet) of the outer ripple is given by $r(t) = 0.6t$, where t is the time in seconds after the pebble strikes the water. The area of the circle is given by the function $A(r) = \pi r^2$. Find and interpret $(A \circ r)(t)$.

58. Automobile Aerodynamics The horsepower H required to overcome wind drag on a certain automobile is approximated by

$$H(x) = 0.002x^2 + 0.005x - 0.029, \qquad 10 \le x \le 100$$

where x is the speed of the car in miles per hour.

(a) Use a graphing utility to graph H.

(b) Rewrite the power function so that x represents the speed in kilometers per hour. [Find $H(1.6x)$.]

In Exercises 59–62, determine whether the function is even, odd, or neither. Use a graphing utility to verify your result.

59. $f(x) = x^2(4 - x^2)$

60. $f(x) = \sqrt[3]{x}$

61. $f(x) = x \cos x$

62. $f(x) = \sin^2 x$

Think About It **In Exercises 63 and 64, find the coordinates of a second point on the graph of a function f if the given point is on the graph and the function is (a) even and (b) odd.**

63. $\left(-\frac{3}{2}, 4\right)$

64. $(4, 9)$

65. Prove that the function is odd.

$$f(x) = a_{2n+1}x^{2n+1} + \cdots + a_3 x^3 + a_1 x$$

66. Prove that the function is even.

$$f(x) = a_{2n}x^{2n} + a_{2n-2}x^{2n-2} + \cdots + a_2 x^2 + a_0$$

67. Prove that the product of two even (or two odd) functions is even.

68. Prove that the product of an odd function and an even function is odd.

69. Use a graphing utility to graphically demonstrate the results of Exercises 67 and 68. Use functions of your choice.

70. What can be said about the sum or difference of (a) two even functions, (b) two odd functions, and (c) an odd function and an even function? Demonstrate your conclusions graphically.

71. Volume An open box of maximum volume is to be made from a square piece of material 24 centimeters on a side by cutting equal squares from the corners and turning up the sides (see figure).

(a) Use the *table* feature of a graphing utility to complete six rows of a table. (The first two rows are shown.) Use the result to guess the maximum volume.

Height, x	Length and Width	Volume, V
1	$24 - 2(1)$	$1[24 - 2(1)]^2 = 484$
2	$24 - 2(2)$	$2[24 - 2(2)]^2 = 800$

(b) Use a graphing utility to plot the points (x, V). Is V a function of x?

(c) If yes, write V as a function of x, and determine its domain.

(d) Use a graphing utility to graph the volume function and approximate the dimensions of the box that yield a maximum volume.

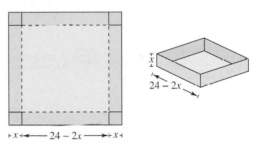

72. Length A right triangle is formed in the first quadrant by the x- and y-axes and the line through the point $(3, 2)$ (see figure). Write the length L of the hypotenuse as a function of x.

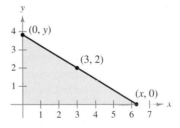

True or False? **In Exercises 73–76, determine whether the statement is true or false. If it is false, explain why or give an example that shows it is false.**

73. If $f(a) = f(b)$, then $a = b$.

74. A vertical line can intersect the graph of a function at most once.

75. If $f(x) = f(-x)$ for all x in the domain of f, then the graph of f is symmetric with respect to the y-axis.

76. If f is a function, then $f(ax) = af(x)$.

- Fit a linear model to a real-life data set.
- Fit a quadratic model to a real-life data set.
- Fit a trigonometric model to a real-life data set.

Fitting a Linear Model to Data

A basic premise of science is that much of the physical world can be described mathematically and that many physical phenomena are predictable. This scientific outlook was part of the scientific revolution that took place in Europe during the late 1500s. Two early publications that are connected with this revolution were *On the Revolutions of the Heavenly Spheres* by the Polish astronomer Nicolaus Copernicus and *On the Structure of the Human Body* by the Belgian anatomist Andreas Vesalius. Each of these books was published in 1543 and each broke with prior tradition by suggesting the use of a scientific method rather than unquestioned reliance on authority.

One characteristic of modern science is gathering data and then describing the data with a mathematical model. For instance, the data given in Example 1 are inspired by Leonardo da Vinci's famous drawing that indicates that a person's height and arm span are equal.

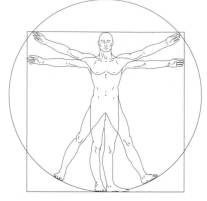

A computer graphics drawing based on the pen and ink drawing of Leonardo da Vinci's famous study of human proportions, called *Vitruvian Man*

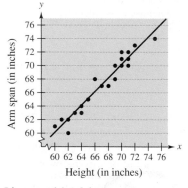

Linear model and data
Figure P.32

Example 1 Fitting a Linear Model to Data

A class of 28 people collected the following data, which represent their heights x and arm spans y (rounded to the nearest inch).

$(60, 61), (65, 65), (68, 67), (72, 73), (61, 62), (63, 63), (70, 71),$

$(75, 74), (71, 72), (62, 60), (65, 65), (66, 68), (62, 62), (72, 73),$

$(70, 70), (69, 68), (69, 70), (60, 61), (63, 63), (64, 64), (71, 71),$

$(68, 67), (69, 70), (70, 72), (65, 65), (64, 63), (71, 70), (67, 67)$

Find a linear model to represent these data.

Solution There are different ways to model these data with an equation. The simplest would be to observe that x and y are about the same and list the model as simply $y = x$. A more careful analysis would be to use a procedure from statistics called linear regression. (You will study this procedure in Section 12.9.) The least squares regression line for these data is

$$y = 1.006x - 0.225. \qquad \text{Least squares regression line}$$

The graph of the model and the data are shown in Figure P.32. From this model, you can see that a person's arm span tends to be about the same as his or her height.

TECHNOLOGY Many scientific and graphing calculators have built-in least squares regression programs. Typically, you enter the data into the calculator and then run the linear regression program. The program usually displays the slope and y-intercept of the best-fitting line and the correlation coefficient r. The closer $|r|$ is to 1, the better the model fits the data. For instance, in Example 1, the value of r is 0.97, which indicates that the model is a good fit for the data. If the r-value is positive, the variables have a positive correlation, as in Example 1. If the r-value is negative, the variables have a negative correlation.

Fitting a Quadratic Model to Data

A function that gives the height s of a falling object in terms of the time t is called a position function. If air resistance is not considered, the position of a falling object can be modeled by

$$s(t) = \tfrac{1}{2}gt^2 + v_0 t + s_0$$

where g is the acceleration due to gravity, v_0 is the initial velocity, and s_0 is the initial height. The value of g depends on where the object is dropped. On earth, g is approximately -32 feet per second per second, or -9.8 meters per second per second.

To discover the value of g experimentally, you could record the heights of a falling object at several increments, as shown in Example 2.

Example 2   **Fitting a Quadratic Model to Data**

A basketball is dropped from a height of about $5\frac{1}{4}$ feet. The height of the basketball is recorded 23 times at intervals of about 0.02 second.[*] The results are shown in the table.

Time	0.0	0.02	0.04	0.06	0.08	0.099996
Height	5.23594	5.20353	5.16031	5.0991	5.02707	4.95146

Time	0.119996	0.139992	0.159988	0.179988	0.199984	0.219984
Height	4.85062	4.74979	4.63096	4.50132	4.35728	4.19523

Time	0.23998	0.25993	0.27998	0.299976	0.319972	0.339961
Height	4.02958	3.84593	3.65507	3.44981	3.23375	3.01048

Time	0.359961	0.379951	0.399941	0.419941	0.439941
Height	2.76921	2.52074	2.25786	1.98058	1.63488

Find a model to fit these data. Then use the model to predict the time when the basketball will hit the ground.

Solution   Begin by drawing a scatter plot of the data, as shown in Figure P.33. From the scatter plot, you can see that the data do not appear to be linear. It does appear, however, that they might be quadratic. To check this, enter the data into a calculator or computer that has a quadratic regression program. You should obtain the model

$$s = -15.45t^2 - 1.30t + 5.234.$$ Least squares regression quadratic

Using this model, you can predict the time when the basketball hits the ground by substituting 0 for s and solving the resulting equation for t.

$$0 = -15.45t^2 - 1.30t + 5.234$$ Let $s = 0$.

$$t = \frac{1.30 \pm \sqrt{(-1.30)^2 - 4(-15.45)(5.234)}}{2(-15.45)}$$ Quadratic Formula

$$t \approx 0.54$$ Choose positive solution.

The solution is about 0.54 second. In other words, the basketball will continue to fall for about 0.1 second more before hitting the ground.

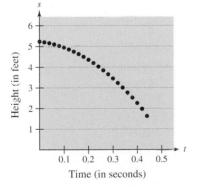

Scatter plot of data
Figure P.33

---

[*] *Data were collected with a Texas Instruments CBL (Calculator-Based Laboratory) System.*

The plane of earth's orbit about the sun and its axis of rotation are not perpendicular. Instead, earth's axis is tilted with respect to its orbit. The result is that the amount of daylight received by locations on earth varies with the time of year. That is, it varies with the position of earth in its orbit.

Fitting a Trigonometric Model to Data

What is mathematical modeling? This is one of the questions that is asked in the book *Guide to Mathematical Modeling*. Here is part of the answer.[*]

1. Mathematical modeling consists of applying your mathematical skills to obtain useful answers to real problems.

2. Learning to apply mathematical skills is very different from learning mathematics itself.

3. Models are used in a very wide range of applications, some of which do not appear initially to be mathematical in nature.

4. Models often allow quick and cheap evaluation of alternatives, leading to optimal solutions that are not otherwise obvious.

5. There are no precise rules in mathematical modeling and no "correct" answers.

6. Modeling can be learned only by *doing*.

Example 3   **Fitting a Trigonometric Model to Data**

The number of hours of daylight on earth depends on two variables: the latitude and the time of year. Here are the numbers of minutes of daylight at a location of 20° latitude on the longest and shortest days of the year: June 21 (summer solstice), 801 minutes; December 21 (winter solstice), 655 minutes. Use this data to write a model for the number of minutes of daylight d on each day of the year at a location of 20° latitude. How could you check the accuracy of your model?

Solution   Here is one way to create a model. You can hypothesize that the model is a sine function whose period is 365 days. Using the given data, you can conclude that the amplitude of the graph is $(801 - 655)/2$, or 73. Thus, one possible model is

$$d = 728 - 73 \sin\left(\frac{2\pi t}{365} + \frac{\pi}{2}\right).$$

In this model, t represents the number of each day of the year, with December 21 represented by $t = 0$. A graph of this model is shown in Figure P.34. To check the accuracy of this model, we used a weather almanac to find the numbers of minutes of daylight on different days of the year at the location of 20° latitude.

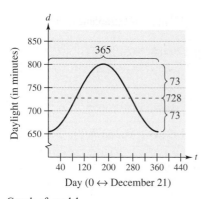

Graph of model
Figure P.34

Date	Value of t	Actual Daylight	Daylight Given by Model
Dec 21	0	655 min	655 min
Jan 1	11	657 min	656 min
Feb 1	42	676 min	673 min
Mar 1	70	705 min	702 min
Apr 1	101	740 min	740 min
May 1	131	772 min	774 min
Jun 1	162	796 min	797 min
Jun 21	182	801 min	801 min
Jul 1	192	799 min	800 min
Aug 1	223	782 min	784 min
Sep 1	254	752 min	752 min
Oct 1	284	718 min	715 min
Nov 1	315	685 min	680 min
Dec 1	345	661 min	659 min

You can see that the model is fairly accurate.

---

[*] *Text from Dilwyn Edwards and Mike Hamson,* Guide to Mathematical Modeling *(Boca Raton: CRC Press, 1990). Used by permission of the authors.*

Lab Series | LAB 1

EXERCISES FOR SECTION P.4

In Exercises 1–4, a scatter plot of data is given. Determine whether the data can be modeled by a linear function, a quadratic function, or a trigonometric function, or that there appears to be no relationship between x and y. To print an enlarged copy of the graph, go to the website *www.mathgraphs.com*.

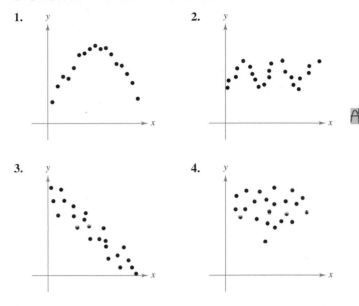

1.

2.

3.

4.

5. Carcinogens The ordered pairs give the exposure index x of a carcinogenic substance and the cancer mortality y per 100,000 people in the population.

(3.50, 150.1), (3.58, 133.1), (4.42, 132.9),

(2.26, 116.7), (2.63, 140.7), (4.85, 165.5),

(12.65, 210.7), (7.42, 181.0), (9.35, 213.4)

(a) Plot the data. From the graph, do the data appear to be approximately linear?

(b) Visually find a linear model for the data. Graph the model.

(c) Use the model to approximate y if $x = 3$.

6. Quiz Scores The ordered pairs represent the scores on two consecutive 15-point quizzes for a class of 18 students.

(7, 13), (9, 7), (14, 14), (15, 15), (10, 15), (9, 7),

(14, 11), (14, 15), (8, 10), (15, 9), (10, 11), (9, 10),

(11, 14), (7, 14), (11, 10), (14, 11), (10, 15), (9, 6)

(a) Plot the data. From the graph, does the relationship between consecutive scores appear approximately linear?

(b) If the data appear approximately linear, find a linear model for the data. If not, give some possible explanations.

7. Hooke's Law Hooke's Law states that the force F required to compress or stretch a spring (within its elastic limits) is proportional to the distance d that the spring is compressed or stretched from its original length. That is, $F = kd$, where k is a measure of the stiffness of the spring and is called the *spring constant*. The table shows the elongation d in centimeters of a spring when a force of F kilograms is applied.

F	20	40	60	80	100
d	1.4	2.5	4.0	5.3	6.6

(a) Use the regression capabilities of a graphing utility to find a linear model for the data.

(b) Use a graphing utility to plot the data and graph the model. How well does the model fit the data? Explain your reasoning.

(c) Use the model to estimate the elongation of the spring when a force of 55 kilograms is applied.

8. Falling Object In an experiment, students measured the speed s (in meters per second) of a falling object t seconds after it was released. The results are shown in the table.

t	0	1	2	3	4
s	0	11.0	19.4	29.2	39.4

(a) Use the regression capabilities of a graphing utility to find a linear model for the data.

(b) Use a graphing utility to plot the data and graph the model. How well does the model fit the data? Explain your reasoning.

(c) Use the model to estimate the speed of the object after 2.5 seconds.

9. Energy Consumption and Gross National Product The data show the per capita energy usage (in millions of Btu) and the per capita gross national product (in thousands of U.S. dollars) for a sample of countries in 1997. *(Source: International Energy Annual and the World Bank)*

Argentina	(71, 9.0)	Bangladesh	(3, 0.4)
Brazil	(48, 4.8)	Canada	(402, 19.6)
Denmark	(184, 34.9)	Finland	(232, 24.8)
France	(166, 26.3)	Greece	(112, 11.6)
India	(12, 0.4)	Italy	(131, 20.2)
Japan	(169, 38.2)	Mexico	(59, 3.7)
Pakistan	(13, 0.5)	South Korea	(162, 10.6)
Sweden	(244, 26.2)	United States	(352, 29.1)

(a) Use the regression capabilities of a graphing utility to find a linear model for the data. What is the correlation coefficient?

(b) Use a graphing utility to plot the data and graph the model.

(c) Interpret the graph in part (b). Use the graph to identify the three countries that differ most from the linear model.

(d) Delete the data for the three countries identified in part (c). Fit a linear model to the resulting data and give the correlation coefficient.

10. *Brinell Hardness* The data in the table show the Brinell hardness H of 0.35 carbon steel when hardened and tempered at temperature t (degrees Fahrenheit). *(Source: Standard Handbook for Mechanical Engineers)*

t	200	400	600	800	1000	1200
H	534	495	415	352	269	217

(a) Use the regression capabilities of a graphing utility to find a linear model for the data.

(b) Use a graphing utility to plot the data and graph the model. How well does the model fit the data? Explain your reasoning.

(c) Use the model to estimate the hardness when t is 500°F.

11. *Automobile Costs* The data in the table show the variable costs for operating an automobile in the United States for the years 1990 through 1997. The functions y_1, y_2, and y_3 represent the costs in cents per mile for gas and oil, maintenance, and tires. *(Source: American Automobile Manufacturers Association)*

Year	y_1	y_2	y_3
1990	5.40	2.10	0.90
1991	6.70	2.20	0.90
1992	6.00	2.20	0.90
1993	6.00	2.40	0.90
1994	5.60	2.50	1.10
1995	6.00	2.60	1.40
1996	5.90	2.80	1.40
1997	6.60	2.80	1.40

(a) Let t be the time in years where $t = 0$ represents 1990. Use the regression capabilities of a graphing utility to find a cubic model for y_1 and linear models for y_2 and y_3.

(b) Use a graphing utility to graph y_1, y_2, y_3, and $y_1 + y_2 + y_3$ in the same viewing window. Use the model to estimate the total variable cost per mile in 2002.

12. *Beam Strength* Students in a lab measured the breaking strength S (in pounds) of wood 2 inches thick, x inches high, and 12 inches long. The results are shown in the table.

x	4	6	8	10	12
S	2370	5460	10,310	16,250	23,860

(a) Use the regression capabilities of a graphing utility to fit a quadratic model to the data.

(b) Use a graphing utility to plot the data and graph the model.

(c) Use the model to approximate the breaking strength when $x = 2$.

13. *Health Maintenance Organizations* The bar graph shows the number of people N (in millions) receiving care in HMOs for the years 1990 through 1998. *(Source: Interstudy Publications)*

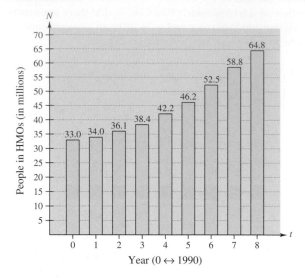

(a) Let t be the time in years, with $t = 0$ corresponding to 1990. Use the regression capabilities of a graphing utility to find linear and cubic models for the data.

(b) Use a graphing utility to graph the data and the linear and cubic models.

(c) Use the graphs in part (b) to determine which is the better model.

(d) Use a graphing utility to find and graph a quadratic model for the data.

(e) Interpret the slope of the linear model in terms of the context of the data.

(f) Use the linear and cubic models to estimate the number of people receiving care in HMOs in the year 2000.

14. *Car Performance* The time t (in seconds) required to attain a speed of s miles per hour from a standing start for a Dodge Avenger is shown in the table. *(Source: Road & Track)*

s	30	40	50	60	70	80	90
t	3.4	5.0	7.0	9.3	12.0	15.8	20.0

(a) Use the regression capabilities of a graphing utility to find a quadratic model for the data.

(b) Use a graphing utility to plot the data and graph the model.

(c) Use the graph in part (b) to state why the model is not appropriate for determining the time required to attain speeds less than 20 miles per hour.

(d) Because the test began from a standing start, add the point (0, 0) to the data. Fit a quadratic model to the revised data and graph the new model. Does it more accurately model the behavior of the car for low speeds? Explain.

15. Car Performance A V8 car engine is coupled to a dynamometer and the horsepower y is measured at different engine speeds x (in thousands of revolutions per minute). The results are shown in the table.

x	1	2	3	4	5	6
y	40	85	140	200	225	245

(a) Use the regression capabilities of a graphing utility to find a cubic model for the data.

(b) Use a graphing utility to plot the data and graph the model.

(c) Use the model to approximate the horsepower when the engine is running at 4500 revolutions per minute.

16. Boiling Temperature The table shows the temperature (°F) at which water boils at selected pressures p (pounds per square inch). *(Source: Standard Handbook for Mechanical Engineers)*

p	5	10	14.696 (1 atmosphere)	20
T	162.24°	193.21°	212.00°	227.96°

p	30	40	60	80	100
T	250.33°	267.25°	292.71°	312.03°	327.81°

(a) Use the regression capabilities of a graphing utility to find a cubic model for the data.

(b) Use a graphing utility to plot the data and graph the model.

(c) Use the graph to estimate the pressure required for the boiling point of water to exceed 300°F.

(d) Explain why the model would not be correct for pressures exceeding 100 pounds per square inch.

17. Harmonic Motion The motion of an oscillating weight suspended by a spring was measured by a motion detector. The data collected and the approximate maximum (positive and negative) displacements from equilibrium are shown in the figure. The displacement y is measured in centimeters and the time t is measured in seconds.

(a) Is y a function of t? Explain.

(b) Approximate the amplitude and period of the oscillations.

(c) Find a model for the data.

(d) Use a graphing utility to graph the model in part (c). Compare the result with the data in the figure.

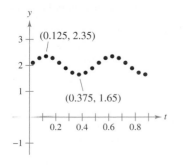

18. Temperature The table shows the normal daily high temperatures for Honolulu H and Chicago C (in degrees Fahrenheit) for month t, with $t = 1$ corresponding to January. *(Source: NOAA)*

t	1	2	3	4	5	6
H	80.1	80.5	81.6	82.8	84.7	86.5
C	29.0	33.5	45.8	58.6	70.1	79.6

t	7	8	9	10	11	12
H	87.5	88.7	88.5	86.9	84.1	81.2
C	83.7	81.8	74.8	63.3	48.4	34.0

(a) A model for Honolulu is

$$H(t) = 84.40 + 4.28 \sin\left(\frac{\pi t}{6} + 3.86\right).$$

Find a model for Chicago.

(b) Use a graphing utility to graph the data and the model for the temperatures in Honolulu. How well does the model fit?

(c) Use a graphing utility to graph the data and the model for the temperatures in Chicago. How well does the model fit?

(d) Use the models to estimate the average annual temperature in each city. What term of the model did you use? Explain.

(e) What is the period of each model? Is it what you expected? Explain.

(f) Which city has a greater variability of temperatures throughout the year? Which factor of the models determines this variability? Explain.

Getting at the Concept

19. Search for real-life data in a newspaper or magazine. Fit the data to a model. What does your model imply about the data?

20. Describe a possible real-life situation for each data set. Then describe how a model could be used in the real-life setting.

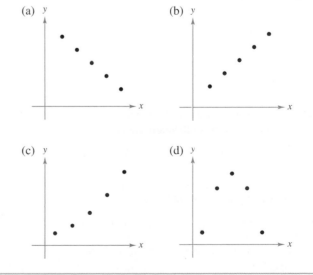

REVIEW EXERCISES FOR CHAPTER P

P.1   In Exercises 1–4, find the intercepts (if any).

1. $y = 2x - 3$

2. $y = (x - 1)(x - 3)$

3. $y = \dfrac{x - 1}{x - 2}$

4. $xy = 4$

In Exercises 5 and 6, check for symmetry with respect to both axes and to the origin.

5. $x^2 y - x^2 + 4y = 0$

6. $y = x^4 - x^2 + 3$

In Exercises 7–14, sketch the graph of the equation.

7. $y = \frac{1}{2}(-x + 3)$

8. $4x - 2y = 6$

9. $-\frac{1}{3}x + \frac{5}{6}y = 1$

10. $0.02x + 0.15y = 0.25$

11. $y = 7 - 6x - x^2$

12. $y = 6x - x^2$

13. $y = \sqrt{5 - x}$

14. $y = |x - 4| - 4$

In Exercises 15 and 16, describe the viewing window of a graphing utility that yields the figure.

15. $y = 4x^2 - 25$

16. $y = 8\sqrt[3]{x - 6}$

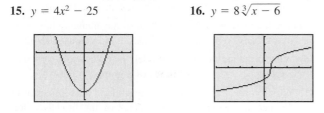

In Exercises 17 and 18, use a graphing utility to find the point(s) of intersection of the graphs of the equations.

17. $3x - 4y = 8$
    $x + y = 5$

18. $x - y + 1 = 0$
    $y - x^2 = 7$

19. **Think About It**   Write an equation whose graph has intercepts at $x = -2$ and $x = 2$ and is symmetric with respect to the origin.

20. **Think About It**   For what value of k does the graph of $y = kx^3$ pass through the point?

    (a) $(1, 4)$ (b) $(-2, 1)$

    (c) $(0, 0)$ (d) $(-1, -1)$

P.2   In Exercises 21 and 22, plot the points and find the slope of the line passing through the points.

21. $\left(\frac{3}{2}, 1\right), \left(5, \frac{5}{2}\right)$

22. $(7, -1), (7, 12)$

In Exercises 23 and 24, use the concept of slope to find t such that the three points are collinear.

23. $(-2, 5), (0, t), (1, 1)$

24. $(-3, 3), (t, -1), (8, 6)$

In Exercises 25–28, find an equation of the line that passes through the point with the indicated slope. Sketch the line.

25. $(0, -5)$, $m = \frac{3}{2}$

26. $(-2, 6)$, $m = 0$

27. $(-3, 0)$, $m = -\frac{2}{3}$

28. $(5, 4)$, m is undefined.

29. Find the equations of the lines passing through $(-2, 4)$ and having the following characteristics.

    (a) Slope of $\frac{7}{16}$

    (b) Parallel to the line $5x - 3y = 3$

    (c) Passing through the origin

    (d) Parallel to the y-axis

30. Find the equations of the lines passing through $(1, 3)$ and having the following characteristics.

    (a) Slope of $-\frac{2}{3}$

    (b) Perpendicular to the line $x + y = 0$

    (c) Passing through the point $(2, 4)$

    (d) Parallel to the x-axis

31. **Rate of Change**   The purchase price of a new machine is $12,500, and its value will decrease by $850 per year. Use this information to write a linear equation that gives the value V of the machine t years after it is purchased. Find its value at the end of 3 years.

32. **Break-Even Analysis**   A contractor purchases a piece of equipment for $36,500 that costs an average of $9.25 per hour for fuel and maintenance. The equipment operator is paid $13.50 per hour, and customers are charged $30 per hour.

    (a) Write an equation for the cost C of operating this equipment for t hours.

    (b) Write an equation for the revenue R derived from t hours of use.

    (c) Find the break-even point for this equipment by finding the time at which $R = C$.

P.3   In Exercises 33–36, sketch the graph of the equation and use the vertical line test to determine whether the equation expresses y as a function of x.

33. $x - y^2 = 0$

34. $x^2 - y = 0$

35. $y = x^2 - 2x$

36. $x = 9 - y^2$

37. Evaluate (if possible) the function $f(x) = 1/x$ at the specified values of the independent variable, and simplify the results.

    (a) $f(0)$ (b) $\dfrac{f(1 + \Delta x) - f(1)}{\Delta x}$

38. Evaluate (if possible) the function at each value of the independent variable.

    $$f(x) = \begin{cases} x^2 + 2, & x < 0 \\ |x - 2|, & x \geq 0 \end{cases}$$

    (a) $f(-4)$ (b) $f(0)$ (c) $f(1)$

39. Find the domain and range of each function.

(a) $y = \sqrt{36 - x^2}$ (b) $y = \dfrac{7}{2x - 10}$ (c) $y = \begin{cases} x^2, & x < 0 \\ 2 - x, & x \geq 0 \end{cases}$

40. Given $f(x) = 1 - x^2$ and $g(x) = 2x + 1$, find the following.

(a) $f(x) - g(x)$ (b) $f(x)g(x)$ (c) $g(f(x))$

41. Sketch (on the same set of coordinate axes) a graph of f for $c = -2, 0,$ and 2.

(a) $f(x) = x^3 + c$ (b) $f(x) = (x - c)^3$

(c) $f(x) = (x - 2)^3 + c$ (d) $f(x) = cx^3$

42. Use a graphing utility to graph $f(x) = x^3 - 3x^2$. Use the graph to write a formula for the function g shown in the figure. To print an enlarged copy of the graph, go to the website *www.mathgraphs.com*.

(a)   (b)

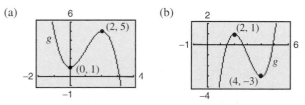

43. *Conjecture*

(a) Use a graphing utility to graph the functions f, g, and h in the same viewing window. Write a description of any similarities and differences you observe among the graphs.

Odd powers: $f(x) = x$, $g(x) = x^3$, $h(x) = x^5$

Even powers: $f(x) = x^2$, $g(x) = x^4$, $h(x) = x^6$

(b) Use the result in part (a) to make a conjecture about the graphs of the functions $y - x^7$ and $y = x^8$. Use a graphing utility to verify your conjecture.

44. *Think About It* Use the result of Exercise 43 to guess the shapes of the graphs of the functions f, g, and h. Then use a graphing utility to graph each function and compare the result with your guess.

(a) $f(x) = x^2(x - 6)^2$ (b) $g(x) = x^3(x - 6)^2$

(c) $h(x) = x^3(x - 6)^3$

45. *Area* A wire 24 inches long is to be cut into four pieces to form a rectangle whose shortest side has a length of x.

(a) Express the area A of the rectangle as a function of x.

(b) Determine the domain of the function and use a graphing utility to graph the function over that domain.

(c) Use the graph of the function to approximate the maximum area of the rectangle. Make a conjecture about the dimensions that yield a maximum area.

46. *Writing* The following graphs give the profits P for two small companies over a period p of 2 years. Create a story to describe the behavior of each profit function for some hypothetical product the company produces.

(a)   (b)

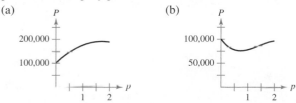

47. *Think About It* What is the minimum degree of the polynomial function whose graph approximates the given graph? What sign must the leading coefficient have?

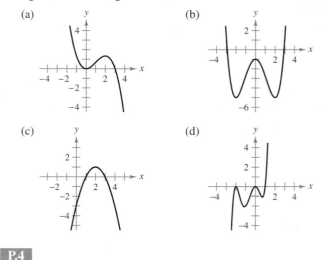

P.4

48. *Stress Test* A machine part was tested by bending it x centimeters ten times per minute until the time y (in hours) of failure. The results are recorded in the table.

x	3	6	9	12	15	18	21	24	27	30
y	61	56	53	55	48	35	36	33	44	23

(a) Use the regression capabilities of a graphing utility to find a linear model for the data.

(b) Use a graphing utility to plot the data and graph the model.

(c) Use the graph to determine whether there may have been an error made in conducting one of the tests or in recording the results. If so, eliminate the erroneous point and find the model for the revised data.

49. *Harmonic Motion* The motion of an oscillating weight suspended by a spring was measured by a motion detector. The data collected and the approximate maximum (positive and negative) displacements from equilibrium are shown in the figure. The displacement y is measured in feet and the time t is measured in seconds.

(a) Is y a function of t? Explain.

(b) Approximate the amplitude and period of the oscillations.

(c) Find a model for the data.

(d) Use a graphing utility to graph the model in part (c). Compare the result with the data in the figure.

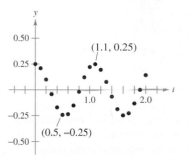

P.S. Problem Solving

1. Consider the circle
$$x^2 + y^2 - 6x - 8y = 0.$$

(a) Find the center and radius of the circle.

(b) Find an equation of the tangent line to the circle at the point $(0, 0)$.

(c) Find an equation of the tangent line to the circle at the point $(6, 0)$.

(d) Where do the two tangent lines intersect?

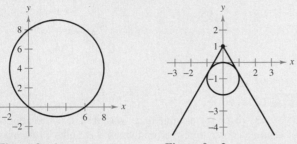

Figure for 1   **Figure for 2**

2. There are two tangent lines from the point $(0, 1)$ to the circle $x^2 + (y + 1)^2 = 1$. Find equations of these two lines by using the fact that each tangent line intersects the circle in *exactly* one point.

3. The Heaviside function $H(x)$ is widely used in engineering applications.

$$H(x) = \begin{cases} 1, & x \geq 0 \\ 0, & x < 0 \end{cases}$$

Sketch the graph of the Heaviside function and the graphs of the following functions by hand.

(a) $H(x) - 2$ (b) $H(x - 2)$ (c) $-H(x)$

(d) $H(-x)$ (e) $\frac{1}{2}H(x)$ (f) $-H(x - 2) + 2$

OLIVER HEAVISIDE (1850–1925)

Heaviside was a British mathematician and physicist who contributed to the field of applied mathematics, especially applications of mathematics to electrical engineering. The *Heaviside function* is a classic type of "on-off" function that has applications to electricity and computer science.

4. Consider the graph of the function f shown below. Use this graph to sketch the graphs of the following functions. To print an enlarged copy of the graph, go to the website *www.mathgraphs.com*.

(a) $f(x + 1)$ (b) $f(x) + 1$ (c) $2f(x)$ (d) $f(-x)$

(e) $-f(x)$ (f) $|f(x)|$ (g) $f(|x|)$

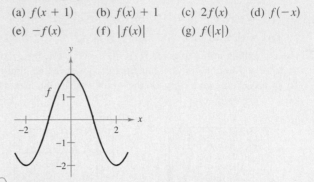

5. A rancher plans to fence a rectangular pasture adjacent to a river. The rancher has 100 meters of fence, and no fencing is needed along the river.

(a) Express the area A of the pasture as a function of x, the length of the side parallel to the river. What is the domain of A?

(b) Graph the area function $A(x)$ and estimate the dimensions that yield the maximum amount of area for the pastures.

(c) Find the dimensions that yield the maximum amount of area for the pastures by completing the square.

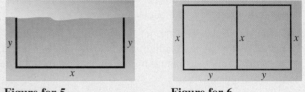

Figure for 5   **Figure for 6**

6. A rancher has 300 feet of fence to enclose two adjacent pastures.

(a) Express the total area A of the two pastures as a function of x. What is the domain of A?

(b) Graph the area function and estimate the dimensions that yield the maximum amount of area for the pastures.

(c) Find the dimensions that yield the maximum amount of area for the pastures by completing the square.

7. You are in a boat 2 miles from the nearest point on the coast. You are to go to a point Q located 3 miles down the coast and 1 mile inland (see figure). You can row at 2 miles per hour and walk at 4 miles per hour. Express the total time T of the trip as a function of x.

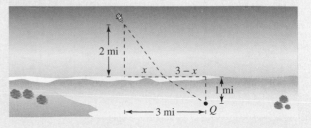

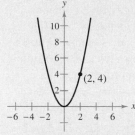

8. You drive to the beach at a rate of 120 kilometers per hour. On the return trip, you drive at a rate of 60 kilometers per hour. What is your average speed for the entire trip? Explain your reasoning.

9. One of the fundamental themes of calculus is to find the slope of the tangent line to a curve at a point. To see how this can be done, consider the point $(2, 4)$ on the graph of $f(x) = x^2$.

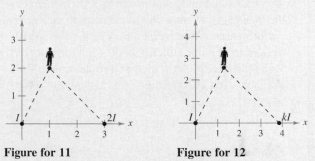

(a) Find the slope of the line joining $(2, 4)$ and $(3, 9)$. Is the slope of the tangent line at $(2, 4)$ greater than or less than this number?

(b) Find the slope of the line joining $(2, 4)$ and $(1, 1)$. Is the slope of the tangent line at $(2, 4)$ greater than or less than this number?

(c) Find the slope of the line joining $(2, 4)$ and $(2.1, 4.41)$. Is the slope of the tangent line at $(2, 4)$ greater than or less than this number?

(d) Find the slope of the line joining $(2, 4)$ and $(2 + h, f(2 + h))$ in terms of the nonzero number h. Verify that $h = 1$, -1, and 0.1 yield the solutions to parts (a)–(c) above.

(e) What is the slope of the tangent line at $(2, 4)$? Explain how you arrived at your answer.

10. Sketch the graph of the function $f(x) = \sqrt{x}$ and label the point $(4, 2)$ on the graph.

(a) Find the slope of the line joining $(4, 2)$ and $(9, 3)$. Is the slope of the tangent line at $(4, 2)$ greater than or less than this number?

(b) Find the slope of the line joining $(4, 2)$ and $(1, 1)$. Is the slope of the tangent line at $(4, 2)$ greater than or less than this number?

(c) Find the slope of the line joining $(4, 2)$ and $(4.41, 2.1)$. Is the slope of the tangent line at $(4, 2)$ greater than or less than this number?

(d) Find the slope of the line joining $(4, 2)$ and $(4 + h, f(4 + h))$ in terms of the nonzero number h.

(e) What is the slope of the tangent line at the point $(4, 2)$? Explain how you arrived at your answer.

11. A large room contains two speakers that are 3 meters apart. The sound intensity I of one speaker is twice that of the other, as indicated in the figure. (To print an enlarged copy of the graph, go to the website *www.mathgraphs.com*.) Suppose the listener is free to move about the room to find those positions that receive equal amounts of sound from both speakers. Such a location satisfies two conditions: (1) the sound intensity at the listener's position is directly proportional to the sound level of a source, and (2) the sound intensity is inversely proportional to the square of the distance from the source.

(a) Find the points on the x-axis that receive equal amounts of sound from both speakers.

(b) Find and graph the equation of all locations (x, y) where one could stand and receive equal amounts of sound from both speakers.

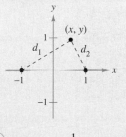

Figure for 11   **Figure for 12**

12. Suppose the speakers in Exercise 11 are 4 meters apart and the sound intensity of one speaker is k times that of the other, as indicated in the figure. To print an enlarged copy of the graph, go to the website *www.mathgraphs.com*.

(a) Find the equation of all locations (x, y) where one could stand and receive equal amounts of sound from both speakers.

(b) Graph the equation for the case $k = 3$.

(c) Describe the set of locations of equal sound as k becomes very large.

13. Let d_1 and d_2 be the distances from the point (x, y) to the points $(-1, 0)$ and $(1, 0)$, respectively, as indicated in the figure. Show that the equation of the graph of all points (x, y) satisfying $d_1 d_2 = 1$ is $(x^2 + y^2)^2 = 2(x^2 - y^2)$. This curve is called a **lemniscate.** Graph the lemniscate and identify three points on the graph.

14. Let $f(x) = \dfrac{1}{1 - x}$.

(a) What are the domain and range of f?

(b) Find the composition $f(f(x))$. What is the domain of this function?

(c) Find $f(f(f(x)))$. What is the domain of this function?

(d) Graph $f(f(f(x)))$. Is the graph a line? Why or why not?

Swimming Speed: Taking It to the Limit

A look at records set in various sports over the past century shows that humans continue to run faster, jump higher, and throw farther than ever before. What is allowing this to occur?

One factor is training. Physiologists are working to identify which systems in the human body limit performance, and to create training techniques that develop those systems. Similarly, sports psychologists work with individual and team members to help them develop the mental "flow" that will allow them to deliver peak performances. Moreover, trainers have developed devices to monitor athletes' bodies and provide them with more feedback on their performance than was available even 20 years ago.

Equipment has also improved vastly over the years. In some sports, the advancement is obvious. Bicycles are lighter and more aerodynamic than ever before. Improved track surfaces have boosted runners' speeds and aluminum poles have drastically increased vault heights.

Even sports such as swimming, with no obvious equipment, have benefited from technology. Shaving body hair cut a full second from male swimmers' times in the 100-meter freestyle, and new styles of swimsuits are expected to reduce drag and improve time even more. The two scatter plots below show the successive world records (in seconds) for two men's swimming events.

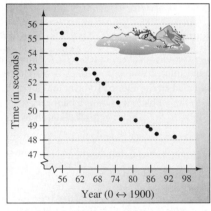

Men's 100-meter freestyle

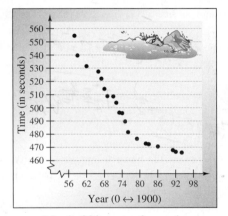

Men's 800-meter freestyle

QUESTIONS

1. From the scatter plots shown above, can you determine which year body shaving was started? Explain your reasoning.

2. In which other years do you think there may have been technological advances in swimming? Explain your reasoning.

3. What does the lower limit appear to be for a man to swim 100 meters? To swim 800 meters? How did you determine this?

4. Copy the two scatter plots and draw a curve that seems to fit the data best. What type of equation do you think would produce the curve you have drawn?

5. Read the excerpt from *Newsweek* on the next page. What do the authors mean by the phrase "approach a limit asymptotically"?

The concepts presented here will be explored further in this chapter. For an extension of this application, see Lab 2 in the lab series that accompanies this text at college.hmco.com.

Limits and Their Properties

Nik Wilson/Allsport

By the age of 17, Australian swimmer Ian Thorpe had set ten world records. At the 2000 Summer Olympics in Sydney, Australia, he broke his own world record in the 400 meter freestyle.

How High? How Fast?

Excerpted from Sharon Begley and Adam Rogers, "How High? How Fast?" from *Newsweek*, July 22, 1996.

Look more closely at the march of winning times and record distances, of gold-medal weights and precedent-setting heights. The law of diminishing returns has set in. The world-record time in the women's 400-meter freestyle, for example, dropped more than two minutes —a full 33 percent—from 1921 (6:16.6) to 1976 (4:11.69). In the 20 years since, it has fallen just eight seconds, to Janet Evans's 4:03.85 at the 1988 Seoul Olympics. If you were to plot world records on graph paper, you would get curves that seem to approach a limit asymptotically, coming tantalizingly closer but never quite reaching it. It is as if the curves were little south-pole magnets and the limit an imposing bar of north polarity. But what is the limit?

Excerpted from John MacDonald, "Carlile calls for hold on use of bodysuits" from Sports.com, September 4, 2000

Stager (Joel Stager, Indiana University's Councilman Centre for the Science of Swimming) did an analysis of times and the recent U.S. Swimming Trials, where 90 percent of the 1309 competitors wore Speedo suits.

He found there was only a 0.34 percent improvement compared with predictions made based on performances from the past 25 years.

This compared with manufacturers' claims of between 3 and 7 percent.

Al Bello/Allsport

The recent development of a swimming bodysuit proves to be a controversial issue.

Section 1.1	A Preview of Calculus

- Understand what calculus is and how it compares to precalculus.
- Understand that the tangent line problem is basic to calculus.
- Understand that the area problem is also basic to calculus.

What Is Calculus?

Calculus is the mathematics of change—velocities and accelerations. Calculus is also the mathematics of tangent lines, slopes, areas, volumes, arc lengths, centroids, curvatures, and a variety of other concepts that have enabled scientists, engineers, and economists to model real-life situations.

Although precalculus mathematics also deals with velocities, accelerations, tangent lines, slopes, and so on, there is a fundamental difference between precalculus mathematics and calculus. Precalculus mathematics is more static, whereas calculus is more dynamic. Here are some examples.

- An object traveling at a constant velocity can be analyzed with precalculus mathematics. To analyze the velocity of an accelerating object, you need calculus.
- The slope of a line can be analyzed with precalculus mathematics. To analyze the slope of a curve, you need calculus.
- A tangent line to a circle can be analyzed with precalculus mathematics. To analyze a tangent line of a general graph, you need calculus.
- The area of a rectangle can be analyzed with precalculus mathematics. To analyze the area under a general curve, you need calculus.

Each of these situations involves the same general strategy—the reformulation of precalculus mathematics through the use of a limit process. So, one way to answer the question "What is calculus?" is to say that calculus is a "limit machine" that involves three stages. The first stage is precalculus mathematics, such as the slope of a line or the area of a rectangle. The second stage is the limit process, and the third stage is a new calculus formulation, such as a derivative or integral.

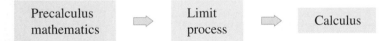

Some students try to learn calculus as if it were simply a collection of new formulas. This is unfortunate. If you reduce calculus to the memorization of differentiation and integration formulas, you will miss a great deal of understanding, self-confidence, and satisfaction.

On the following two pages we have listed some familiar precalculus concepts coupled with their calculus counterparts. Throughout the text, your goal should be to learn how precalculus formulas and techniques are used as building blocks to produce the more general calculus formulas and techniques. Don't worry if you are unfamiliar with some of the "old formulas" listed on the following two pages—we will be reviewing all of them.

As you proceed through this text, we suggest that you come back to this discussion repeatedly. Try to keep track of where you are relative to the three stages involved in the study of calculus. For example, the first three chapters break down as follows.

Chapter P: Preparation for Calculus	Precalculus
Chapter 1: Limits and Their Properties	Limit process
Chapter 2: Differentiation	Calculus

STUDY TIP As you progress through this course, remember that learning calculus is just one of your goals. Your most important goal is to learn how to use calculus to model and solve real-life problems. Here are a few problem-solving strategies that may help you.

- Be sure you understand the question. What is given? What are you asked to find?
- Outline a plan. There are many approaches you could use: look for a pattern, solve a simpler problem, work backwards, draw a diagram, use technology, or any of many other approaches.
- Complete your plan. Be sure to answer the question. Verbalize your answer. For example, rather than writing the answer as $x = 4.6$, it would be better to write the answer as "The area of the region is 4.6 meters."
- Look back at your work. Does your answer make sense? Is there a way you can check the reasonableness of your answer?

GRACE CHISHOLM YOUNG (1868–1944)

Grace Chisholm Young received her degree in mathematics from Girton College in Cambridge, England. Her early work was published under the name of William Young, her husband. Between 1914 and 1916, Grace Young published work on the foundations of calculus that won her the Gamble Prize from Girton College.

Without Calculus	**With Differential Calculus**
Value of $f(x)$ when $x = c$	Limit of $f(x)$ as x approaches c
Slope of a line	Slope of a curve
Secant line to a curve	Tangent line to a curve
Average rate of change between $t = a$ and $t = b$	Instantaneous rate of change at $t = c$
Curvature of a circle	Curvature of a curve
Height of a curve when $x = c$	Maximum height of a curve on an interval
Tangent plane to a sphere	Tangent plane to a surface
Direction of motion along a straight line	Direction of motion along a curved line

Without Calculus		With Integral Calculus	
Area of a rectangle		Area under a curve	
Work done by a constant force		Work done by a variable force	
Center of a rectangle		Centroid of a region	
Length of a line segment		Length of an arc	
Surface area of a cylinder		Surface area of a solid of revolution	
Mass of a solid of constant density		Mass of a solid of variable density	
Volume of a rectangular solid		Volume of a region under a surface	
Sum of a finite number of terms	$a_1 + a_2 + \cdots + a_n = S$	Sum of an infinite number of terms	$a_1 + a_2 + a_3 + \cdots = S$

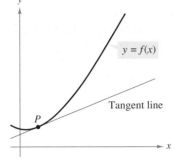

The tangent line to the graph of f at P
Figure 1.1

The Tangent Line Problem

The notion of a limit is fundamental to the study of calculus. The following brief descriptions of two classic problems in calculus—*the tangent line problem* and *the area problem*—should give you some idea of the way limits are used in calculus.

In the tangent line problem, you are given a function f and a point P on its graph and are asked to find an equation of the tangent line to the graph at point P, as shown in Figure 1.1.

Except for cases involving a vertical tangent line, the problem of finding the **tangent line** at a point P is equivalent to finding the *slope* of the tangent line at P. You can approximate this slope by using a line through the point of tangency and a second point on the curve, as shown in Figure 1.2(a). Such a line is called a **secant line.** If $P(c, f(c))$ is the point of tangency and

$$Q(c + \Delta x, f(c + \Delta x))$$

is a second point on the graph of f, the slope of the secant line through these two points is given by

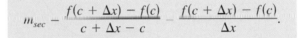

$$m_{sec} = \frac{f(c + \Delta x) - f(c)}{c + \Delta x - c} = \frac{f(c + \Delta x) - f(c)}{\Delta x}.$$

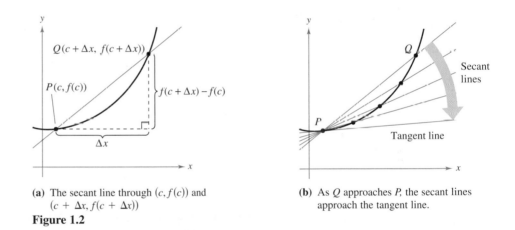

(a) The secant line through $(c, f(c))$ and $(c + \Delta x, f(c + \Delta x))$

(b) As Q approaches P, the secant lines approach the tangent line.

Figure 1.2

As point Q approaches point P, the slope of the secant line approaches the slope of the tangent line, as shown in Figure 1.2(b). When such a "limiting position" exists, the slope of the tangent line is said to be the **limit** of the slope of the secant line. (Much more will be said about this important problem in Chapter 2.)

EXPLORATION

The following points lie on the graph of $f(x) = x^2$.

$Q_1(1.5, f(1.5))$, $Q_2(1.1, f(1.1))$, $Q_3(1.01, f(1.01))$,
$Q_4(1.001, f(1.001))$, $Q_5(1.0001, f(1.0001))$

Each successive point gets closer to the point $P(1, 1)$. Find the slope of the secant line through Q_1 and P, Q_2 and P, and so on. Graph these secant lines on a graphing utility. Then use your results to estimate the slope of the tangent line to the graph of f at the point P.

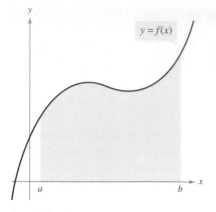

Area under a curve
Figure 1.3

The Area Problem

In the tangent line problem, you saw how the limit process can be applied to the slope of a line to find the slope of a general curve. A second classic problem in calculus is finding the area of a plane region that is bounded by the graphs of functions. This problem can also be solved with a limit process. In this case, the limit process is applied to the area of a rectangle to find the area of a general region.

As a simple example, consider the region bounded by the graph of the function $y = f(x)$, the x-axis, and the vertical lines $x = a$ and $x = b$, as shown in Figure 1.3. You can approximate the area of the region with several rectangular regions, as shown in Figure 1.4. As you increase the number of rectangles, the approximation tends to become better and better because the amount of area missed by the rectangles decreases. Your goal is to determine the limit of the sum of the areas of the rectangles as the number of rectangles increases without bound.

HISTORICAL NOTE

In one of the most astounding events ever to occur in mathematics, it was discovered that the tangent line problem and the area problem are closely related. This discovery led to the birth of calculus. You will learn about the relationship between these two problems when you study the Fundamental Theorem of Calculus in Chapter 4.

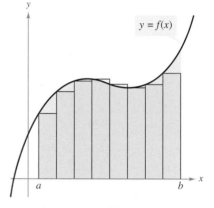

Approximation using four rectangles   Approximation using eight rectangles
Figure 1.4

EXPLORATION

Consider the region bounded by the graphs of $f(x) = x^2$, $y = 0$, and $x = 1$, as shown in part (a) of the figure. The area of the region can be approximated by two sets of rectangles—one set inscribed within the region and the other set circumscribed over the region, as shown in parts (b) and (c). Find the sum of the areas of each set of rectangles. Then use your results to approximate the area of the region.

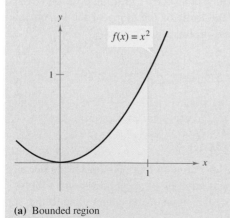

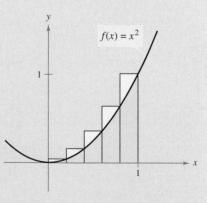

(a) Bounded region   **(b)** Inscribed rectangles   **(c)** Circumscribed rectangles

EXERCISES FOR SECTION 1.1

In Exercises 1–4, decide whether the problem can be solved using precalculus, or whether calculus is required. If the problem can be solved using precalculus, solve it. If the problem seems to require calculus, explain your reasoning and use a graphical or numerical approach to estimate the solution.

1. Find the distance traveled in 15 seconds by an object traveling at a constant velocity of 20 feet per second.

2. Find the distance traveled in 15 seconds by an object moving with a velocity of $v(t) = 20 + 7 \cos t$ feet per second.

3. A bicyclist is riding on a path modeled by the function $f(x) = 0.04(8x - x^2)$, where x and $f(x)$ are measured in miles. Find the rate of change of elevation when $x = 2$.

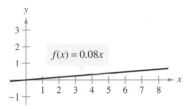

4. A bicyclist is riding on a path modeled by the function $f(x) = 0.08x$, where x and $f(x)$ are measured in miles. Find the rate of change of elevation when $x = 2$.

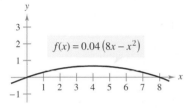

In Exercises 5 and 6, find the area of the shaded region.

5.
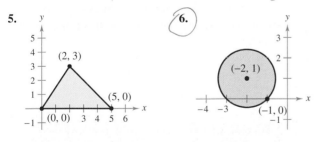

6.

In Exercises 7 and 8, find the volume of the solid shown.

7.

8.
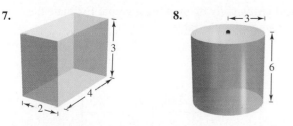

Getting at the Concept

9. (a) Use the *list* feature of a graphing utility to graph the following.

$$y_1 = 4x - x^2$$

$$y_2 = \left[\frac{y_1(1 + \{2, 1.5, 1, 0.5\}) - y_1(1)}{\{2, 1.5, 1, 0.5\}} \right](x - 1) + y_1(1)$$

(*Note:* If you cannot use lists on your graphing utility, graph y_2 four times using 2, 1.5, 1, and 0.5.)

(b) Give a written description of the graphs of y_2 relative to the graph of y_1.

(c) Use the results in part (a) to estimate the slope of the tangent line to the graph of y_1 at $(1, 3)$. If you want to improve your approximation of the slope, how could you change the list in the formula for y_2?

10. (a) Use the rectangles in each graph to approximate the area of the region bounded by $y = 5/x$, $y = 0$, $x = 1$, and $x = 5$.

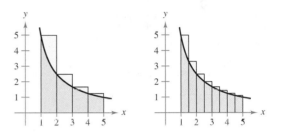

(b) Describe how you could continue this process to obtain a more accurate approximation of the area.

11. Consider the length of the graph of $f(x) = 5/x$ from $(1, 5)$ to $(5, 1)$.

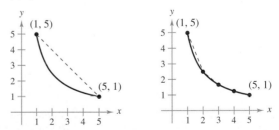

(a) Approximate the length of the curve by finding the distance between its two endpoints, as shown in the first figure.

(b) Approximate the length of the curve by finding the lengths of four line segments, as shown in the second figure.

(c) Describe how you could continue this process to obtain a more accurate approximation of the length of the curve.

The symbol ⚙ *indicates an exercise in which you are instructed to use graphing technology or a symbolic computer algebra system. The solutions of other exercises may also be facilitated by use of appropriate technology.*

- Estimate a limit using a numerical or graphical approach.
- Learn different ways that a limit can fail to exist.
- Study and use a formal definition of a limit.

An Introduction to Limits

Suppose you are asked to sketch the graph of the function f given by

$$f(x) = \frac{x^3 - 1}{x - 1}, \qquad x \neq 1.$$

For all values other than $x = 1$, you can use standard curve-sketching techniques. However, at $x = 1$, it is not clear what to expect. To get an idea of the behavior of the graph of f near $x = 1$, you can use two sets of x-values—one set that approaches 1 from the left and one that approaches 1 from the right, as shown in the table.

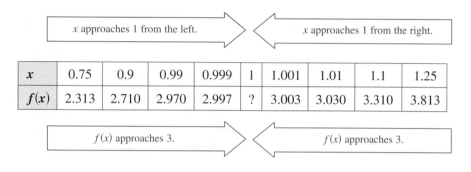

	x approaches 1 from the left.					x approaches 1 from the right.			
x	0.75	0.9	0.99	0.999	1	1.001	1.01	1.1	1.25
$f(x)$	2.313	2.710	2.970	2.997	?	3.003	3.030	3.310	3.813

	$f(x)$ approaches 3.	$f(x)$ approaches 3.

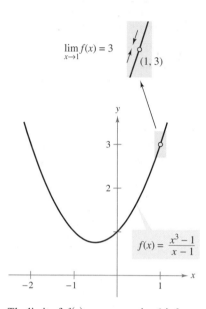

$\lim\limits_{x \to 1} f(x) = 3$

$(1, 3)$

$f(x) = \dfrac{x^3 - 1}{x - 1}$

The limit of $f(x)$ as x approaches 1 is 3.
Figure 1.5

The graph of f is a parabola that has a gap at the point $(1, 3)$, as shown in Figure 1.5. Although x cannot equal 1, you can move arbitrarily close to 1, and as a result $f(x)$ moves arbitrarily close to 3. Using limit notation, you can write

$$\lim_{x \to 1} f(x) = 3. \qquad \text{This is read as "the limit of } f(x) \text{ as } x \text{ approaches 1 is 3."}$$

This discussion leads to an informal description of a limit. If $f(x)$ becomes arbitrarily close to a single number L as x approaches c from either side, the **limit** of $f(x)$, as x approaches c, is L. This limit is written as

$$\lim_{x \to c} f(x) = L.$$

EXPLORATION

The discussion above gives an example of how you can estimate a limit *numerically* by constructing a table and *graphically* by drawing a graph. Estimate the following limit numerically by completing the table.

$$\lim_{x \to 2} \frac{x^2 - 3x + 2}{x - 2}$$

x	1.75	1.9	1.99	1.999	2	2.001	2.01	2.1	2.25
$f(x)$	?	?	?	?	?	?	?	?	?

Then use a graphing utility to estimate the limit graphically.

Example 1   **Estimating a Limit Numerically**

Evaluate the function $f(x) = x/(\sqrt{x+1} - 1)$ at several points near $x = 0$ and use the result to estimate the limit

$$\lim_{x \to 0} \frac{x}{\sqrt{x+1} - 1}.$$

Solution   The table lists the values of $f(x)$ for several x-values near 0.

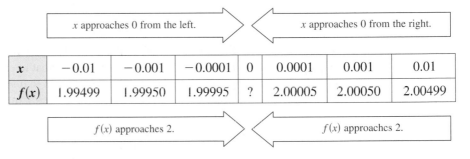

	x approaches 0 from the left.				*x* approaches 0 from the right.		
x	-0.01	-0.001	-0.0001	0	0.0001	0.001	0.01
f(x)	1.99499	1.99950	1.99995	?	2.00005	2.00050	2.00499
	f(x) approaches 2.				*f(x)* approaches 2.		

From the results shown in the table, you can estimate the limit to be 2. This limit is reinforced by the graph of f (see Figure 1.6).

In Example 1, note that the function is undefined at $x = 0$ and yet $f(x)$ appears to be approaching a limit as x approaches 0. This often happens, and it is important to realize that *the existence or nonexistence of $f(x)$ at $x = c$ has no bearing on the existence of the limit of $f(x)$ as x approaches c.*

Example 2   **Finding a Limit**

Find the limit of $f(x)$ as x approaches 2 where f is defined as

$$f(x) = \begin{cases} 1, & x \neq 2 \\ 0, & x = 2. \end{cases}$$

Solution   Because $f(x) = 1$ for all x other than $x = 2$, you can conclude that the limit is 1, as shown in Figure 1.7. So, you can write

$$\lim_{x \to 2} f(x) = 1.$$

The fact that $f(2) = 0$ has no bearing on the existence or value of the limit as x approaches 2. For instance, if the function were defined as

$$f(x) = \begin{cases} 1, & x \neq 2 \\ 2, & x = 2 \end{cases}$$

the limit would be the same.

So far in this section, you have been estimating limits numerically and graphically. Each of these approaches produces an estimate of the limit. In Section 1.3, you will study analytic techniques for evaluating limits. Throughout the course, try to develop a habit of using this three-pronged approach to problem solving.

1. Numerical approach   Construct a table of values.

2. Graphical approach   Draw a graph by hand or using technology.

3. Analytic approach   Use algebra or calculus.

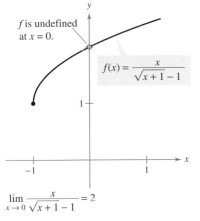

f is undefined at $x = 0$.

$$f(x) = \frac{x}{\sqrt{x+1} - 1}$$

$$\lim_{x \to 0} \frac{x}{\sqrt{x+1} - 1} = 2$$

The limit of $f(x)$ as x approaches 0 is 2.
Figure 1.6

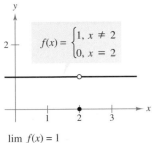

$$f(x) = \begin{cases} 1, x \neq 2 \\ 0, x = 2 \end{cases}$$

$$\lim_{x \to 2} f(x) = 1$$

The limit of $f(x)$ as x approaches 2 is 1.
Figure 1.7

Limits That Fail to Exist

In the next three examples you will examine some limits that fail to exist.

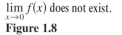

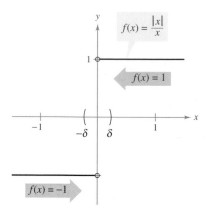

$\lim\limits_{x \to 0} f(x)$ does not exist.

Figure 1.8

Example 3   **Behavior That Differs from the Right and Left**

Show that the limit does not exist.

$$\lim_{x \to 0} \frac{|x|}{x}$$

Solution   Consider the graph of the function $f(x) = |x|/x$. From Figure 1.8, you can see that for positive x-values

$$\frac{|x|}{x} = 1, \qquad x > 0$$

and for negative x-values

$$\frac{|x|}{x} = -1, \qquad x < 0.$$

This means that no matter how close x gets to 0, there will be both positive and negative x-values that yield $f(x) = 1$ and $f(x) = -1$. Specifically, if δ (the lowercase Greek letter *delta*) is a positive number, then for x-values satisfying the inequality $0 < |x| < \delta$, you can classify the values of $|x|/x$ as follows.

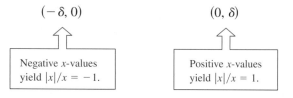

This implies that the limit does not exist.

Example 4   **Unbounded Behavior**

Discuss the existence of the limit

$$\lim_{x \to 0} \frac{1}{x^2}.$$

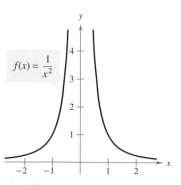

$\lim\limits_{x \to 0} f(x)$ does not exist.

Figure 1.9

Solution   Let $f(x) = 1/x^2$. In Figure 1.9, you can see that as x approaches 0 from either the right or the left, $f(x)$ increases without bound. This means that by choosing x close enough to 0, you can force $f(x)$ to be as large as you want. For instance, $f(x)$ will be larger than 100 if you choose x that is within $\frac{1}{10}$ of 0. That is,

$$0 < |x| < \frac{1}{10} \quad \Longrightarrow \quad f(x) = \frac{1}{x^2} > 100.$$

Similarly, you can force $f(x)$ to be larger than 1,000,000, as follows.

$$0 < |x| < \frac{1}{1000} \quad \Longrightarrow \quad f(x) = \frac{1}{x^2} > 1,000,000$$

Because $f(x)$ is not approaching a real number L as x approaches 0, you can conclude that the limit does not exist.

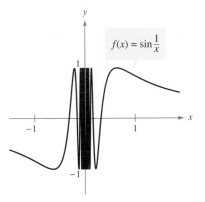

$$f(x) = \sin\frac{1}{x}$$

$\lim\limits_{x\to 0} f(x)$ does not exist.
Figure 1.10

The Granger Collection

PETER GUSTAV DIRICHLET (1805–1859)

In the early development of calculus, the definition of a function was much more restricted than it is today, and "functions" such as the Dirichlet function would not have been considered. The modern definition of a function was given by the German mathematician Peter Gustav Dirichlet.

Example 5   Oscillating Behavior

Discuss the existence of the limit $\lim\limits_{x\to 0} \sin \dfrac{1}{x}$.

Solution   Let $f(x) = \sin(1/x)$. In Figure 1.10, you can see that as x approaches 0, $f(x)$ oscillates between -1 and 1. Therefore, the limit does not exist because no matter how small you choose δ, it is possible to choose x_1 and x_2 within δ units of 0 such that $\sin(1/x_1) = 1$ and $\sin(1/x_2) = -1$, as indicated in the table.

x	$\dfrac{2}{\pi}$	$\dfrac{2}{3\pi}$	$\dfrac{2}{5\pi}$	$\dfrac{2}{7\pi}$	$\dfrac{2}{9\pi}$	$\dfrac{2}{11\pi}$	$x \to 0$
$\sin\dfrac{1}{x}$	1	-1	1	-1	1	-1	Limit does not exist.

Common Types of Behavior Associated with the Nonexistence of a Limit

1. $f(x)$ approaches a different number from the right side of c than it approaches from the left side.
2. $f(x)$ increases or decreases without bound as x approaches c.
3. $f(x)$ oscillates between two fixed values as x approaches c.

There are many other interesting functions that have unusual limit behavior. An often cited one is the *Dirichlet function*

$$f(x) = \begin{cases} 0, & \text{if } x \text{ is rational.} \\ 1, & \text{if } x \text{ is irrational.} \end{cases}$$

This function has *no limit* at any real number c.

TECHNOLOGY PITFALL   When you use a graphing utility to investigate the behavior of a function near the x-value at which you are trying to evaluate a limit, remember that you can't always trust the pictures that graphing utilities draw. For instance, if you use a graphing utility to sketch the graph of the function in Example 5 over an interval containing 0, you will most likely obtain an incorrect graph such as that shown in Figure 1.11. The reason that a graphing utility can't show the correct graph is that the graph has infinitely many oscillations over any interval that contains 0.

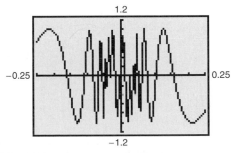

Incorrect graph of $f(x) = \sin(1/x)$.
Figure 1.11

---

indicates that in the Interactive 3.0 CD-ROM and Internet 3.0 versions of this text (available at college.hmco.com) *you will find an Open Exploration, which further explores this example using the computer algebra systems Maple, Mathcad, Mathematica, and Derive.*

A Formal Definition of a Limit

Let's take another look at the informal description of a limit. If $f(x)$ becomes arbitrarily close to a single number L as x approaches c from either side, we say that the limit of $f(x)$ as x approaches c is L, written as

$$\lim_{x \to c} f(x) = L.$$

At first glance, this description looks fairly technical. Even so, we call it informal because we have yet to give exact meanings to the two phrases

"$f(x)$ becomes arbitrarily close to L"

and

"x approaches c."

The first person to assign mathematically rigorous meanings to these two phrases was Augustin-Louis Cauchy. His ε-δ **definition of a limit** is the standard used today.

In Figure 1.12, let ε (the lowercase Greek letter *epsilon*) represent a (small) positive number. Then the phrase "$f(x)$ becomes arbitrarily close to L" means that $f(x)$ lies in the interval $(L - \varepsilon, L + \varepsilon)$. Using absolute value, you can write this as

$$|f(x) - L| < \varepsilon.$$

Similarly, the phrase "x approaches c" means that there exists a positive number δ such that x lies in either the interval $(c - \delta, c)$ or the interval $(c, c + \delta)$. This fact can be concisely expressed by the double inequality

$$0 < |x - c| < \delta.$$

The first inequality

$$0 < |x - c| \qquad \text{The difference between } x \text{ and } c \text{ is more than 0.}$$

expresses the fact that $x \neq c$. The second inequality

$$|x - c| < \delta \qquad x \text{ is within } \delta \text{ units of } c.$$

says that x is within a distance δ of c.

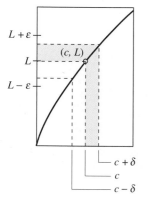

The ε-δ definition of the limit of $f(x)$ as x approaches c

Figure 1.12

Definition of Limit

Let f be a function defined on an open interval containing c (except possibly at c) and let L be a real number. The statement

$$\lim_{x \to c} f(x) = L$$

means that for each $\varepsilon > 0$ there exists a $\delta > 0$ such that if

$$0 < |x - c| < \delta, \quad \text{then} \quad |f(x) - L| < \varepsilon.$$

NOTE   Throughout this text, when we write

$$\lim_{x \to c} f(x) = L$$

we imply two statements—the limit **exists** *and* the limit is L.

Some functions do not have limits as $x \to c$, but those that do cannot have two different limits as $x \to c$. That is, *if the limit of a function exists, it is unique* (see Exercise 55).

FOR FURTHER INFORMATION For more on the introduction of rigor to calculus, see "Who Gave You the Epsilon? Cauchy and the Origins of Rigorous Calculus" by Judith V. Grabiner in *The American Mathematical Monthly*. To view this article, go to the website *www.matharticles.com*.

The next three examples should help you develop a better understanding of the ε-δ definition of a limit.

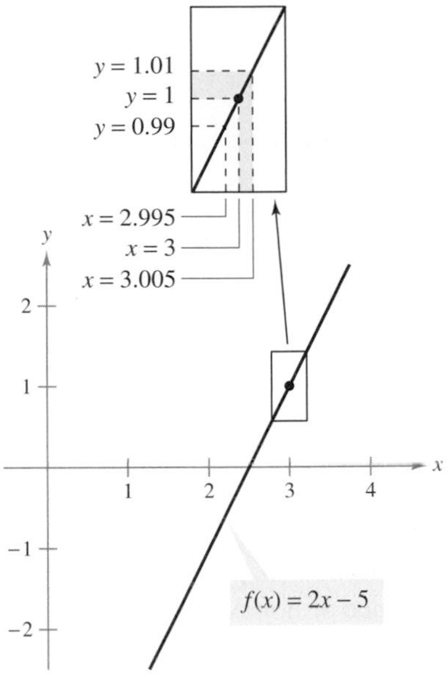

Example 6   Finding a δ for a Given ε

Given the limit

$$\lim_{x \to 3} (2x - 5) = 1$$

find δ such that $|(2x - 5) - 1| < 0.01$ whenever $0 < |x - 3| < \delta$.

Solution   In this problem, you are working with a given value of ε—namely, $\varepsilon = 0.01$. To find an appropriate δ, notice that

$$|(2x - 5) - 1| = |2x - 6| = 2|x - 3|.$$

Because the inequality $|(2x - 5) - 1| < 0.01$ is equivalent to $2|x - 3| < 0.01$, you can choose $\delta = \frac{1}{2}(0.01) = 0.005$. This choice works because

$$0 < |x - 3| < 0.005$$

implies that

$$|(2x - 5) - 1| = 2|x - 3| < 2(0.005) = 0.01$$

as shown in Figure 1.13.

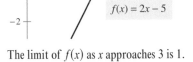

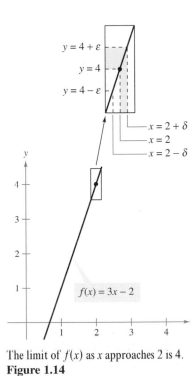

The limit of $f(x)$ as x approaches 3 is 1.
Figure 1.13

NOTE   In Example 6, note that 0.005 is the *largest* value of δ that will guarantee $|(2x - 5) - 1| < 0.01$ whenever $0 < |x - 3| < \delta$. Any *smaller* positive value of δ would, of course, also work.

In Example 6, you found a δ-value for a given ε. This does not prove the existence of the limit. To do that, you must prove that you can find a δ for any ε, as demonstrated in the next example.

Example 7   Using the ε-δ Definition of a Limit

Use the ε-δ definition of a limit to prove that

$$\lim_{x \to 2} (3x - 2) = 4.$$

Solution   You must show that for each $\varepsilon > 0$, there exists a $\delta > 0$ such that $|(3x - 2) - 4| < \varepsilon$ whenever $0 < |x - 2| < \delta$. Because your choice of δ depends on ε, you need to establish a connection between the absolute values $|(3x - 2) - 4|$ and $|x - 2|$.

$$|(3x - 2) - 4| = |3x - 6| = 3|x - 2|$$

So, for a given $\varepsilon > 0$ you can choose $\delta = \varepsilon/3$. This choice works because

$$0 < |x - 2| < \delta = \frac{\varepsilon}{3}$$

implies that

$$|(3x - 2) - 4| = 3|x - 2| < 3\left(\frac{\varepsilon}{3}\right) = \varepsilon$$

The limit of $f(x)$ as x approaches 2 is 4.
Figure 1.14

as shown in Figure 1.14.

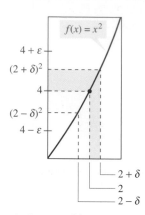

$f(x) = x^2$

The limit of $f(x)$ as x approaches 2 is 4.
Figure 1.15

Example 8   **Using the ε-δ Definition of a Limit**

Use the ε-δ definition of a limit to prove that

$$\lim_{x \to 2} x^2 = 4.$$

Solution   You must show that for each $\varepsilon > 0$, there exists a $\delta > 0$ such that

$$|x^2 - 4| < \varepsilon \text{ when } 0 < |x - 2| < \delta.$$

To find an appropriate δ, begin by writing $|x^2 - 4| = |x - 2||x + 2|$. For all x in the interval $(1, 3)$, you know that $|x + 2| < 5$. So, letting δ be the minimum of $\varepsilon/5$ and 1, it follows that, whenever $0 < |x - 2| < \delta$, you have

$$|x^2 - 4| = |x - 2||x + 2| < \left(\frac{\varepsilon}{5}\right)(5) = \varepsilon$$

as shown in Figure 1.15.

Throughout this chapter you will use the ε-δ definition of a limit primarily to prove theorems about limits and to establish the existence or nonexistence of particular types of limits. For *finding* limits, you will learn techniques that are easier to use than the ε-δ definition of a limit.

EXERCISES FOR SECTION 1.2

In Exercises 1–8, complete the table and use the result to estimate the limit. Use a graphing utility to graph the function to confirm your result.

1. $\lim\limits_{x \to 2} \dfrac{x - 2}{x^2 - x - 2}$

x	1.9	1.99	1.999	2.001	2.01	2.1
$f(x)$						

2. $\lim\limits_{x \to 2} \dfrac{x - 2}{x^2 - 4}$

x	1.9	1.99	1.999	2.001	2.01	2.1
$f(x)$						

3. $\lim\limits_{x \to 0} \dfrac{\sqrt{x + 3} - \sqrt{3}}{x}$

x	-0.1	-0.01	-0.001	0.001	0.01	0.1
$f(x)$						

4. $\lim\limits_{x \to -3} \dfrac{\sqrt{1 - x} - 2}{x + 3}$

x	-3.1	-3.01	-3.001	-2.999	-2.99	-2.9
$f(x)$						

5. $\lim\limits_{x \to 3} \dfrac{[1/(x + 1)] - (1/4)}{x - 3}$

x	2.9	2.99	2.999	3.001	3.01	3.1
$f(x)$						

6. $\lim\limits_{x \to 4} \dfrac{[x/(x + 1)] - (4/5)}{x - 4}$

x	3.9	3.99	3.999	4.001	4.01	4.1
$f(x)$						

7. $\lim\limits_{x \to 0} \dfrac{\sin x}{x}$

x	-0.1	-0.01	-0.001	0.001	0.01	0.1
$f(x)$						

8. $\lim\limits_{x \to 0} \dfrac{\cos x - 1}{x}$

x	-0.1	-0.01	-0.001	0.001	0.01	0.1
$f(x)$						

In Exercises 9–18, use the graph to find the limit (if it exists). If the limit does not exist, explain why.

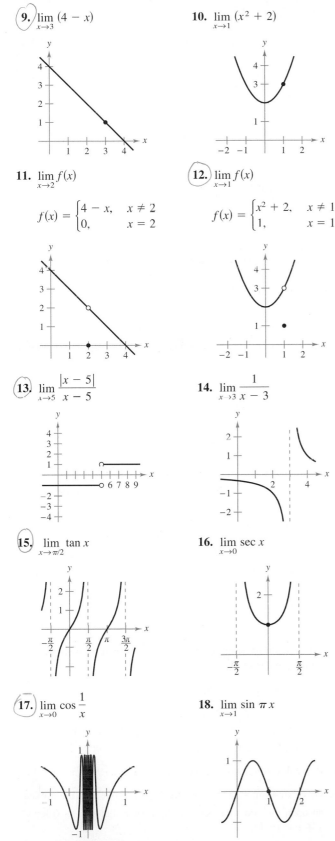

9. $\lim\limits_{x \to 3} (4 - x)$

10. $\lim\limits_{x \to 1} (x^2 + 2)$

11. $\lim\limits_{x \to 2} f(x)$

$f(x) = \begin{cases} 4 - x, & x \neq 2 \\ 0, & x = 2 \end{cases}$

12. $\lim\limits_{x \to 1} f(x)$

$f(x) = \begin{cases} x^2 + 2, & x \neq 1 \\ 1, & x = 1 \end{cases}$

13. $\lim\limits_{x \to 5} \dfrac{|x - 5|}{x - 5}$

14. $\lim\limits_{x \to 3} \dfrac{1}{x - 3}$

15. $\lim\limits_{x \to \pi/2} \tan x$

16. $\lim\limits_{x \to 0} \sec x$

17. $\lim\limits_{x \to 0} \cos \dfrac{1}{x}$

18. $\lim\limits_{x \to 1} \sin \pi x$

19. *Modeling Data*   The cost of a telephone call between two cities is $0.75 for the first minute and $0.50 for each additional minute. A formula for the cost is given by

$$C(t) = 0.75 - 0.50 \, [\![-(t - 1)]\!]$$

where t is the time in minutes.

(*Note:* $[\![x]\!]$ = greatest integer n such that $n \leq x$. For example, $[\![3.2]\!] = 3$ and $[\![-1.6]\!] = -2$.)

(a) Use a graphing utility to graph the cost function for $0 < t \leq 5$.

(b) Use the graph to complete the table and observe the behavior of the function as t approaches 3.5. Use the graph and the table to find

$$\lim_{t \to 3.5} C(t).$$

t	3	3.3	3.4	3.5	3.6	3.7	4
C				?			

(c) Use the graph to complete the table and observe the behavior of the function as t approaches 3.

t	2	2.5	2.9	3	3.1	3.5	4
C				?			

Does the limit of $C(t)$ as t approaches 3 exist? Explain.

20. Repeat Exercise 19 if $C(t) = 0.35 - 0.12 [\![-(t - 1)]\!]$.

21. The graph of $f(x) = 2 - 1/x$ is shown in the figure. Find δ such that if $0 < |x - 1| < \delta$ then $|f(x) - 1| < 0.1$.

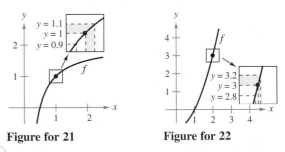

Figure for 21   **Figure for 22**

22. The graph of $f(x) = x^2 - 1$ is shown in the figure. Find δ such that if $0 < |x - 2| < \delta$ then $|f(x) - 3| < 0.2$.

In Exercises 23–26, find the limit L. Then find $\delta > 0$ such that $|f(x) - L| < 0.01$ whenever $0 < |x - c| < \delta$.

23. $\lim\limits_{x \to 2} (3x + 2)$

24. $\lim\limits_{x \to 4} \left(4 - \dfrac{x}{2} \right)$

25. $\lim\limits_{x \to 2} (x^2 - 3)$

26. $\lim\limits_{x \to 5} (x^2 + 4)$

In Exercises 27–38, find the limit L. Then use the ε-δ definition to prove that the limit is L.

27. $\lim\limits_{x \to 2} (x + 3)$

28. $\lim\limits_{x \to -3} (2x + 5)$

29. $\lim\limits_{x \to -4} \left(\frac{1}{2}x - 1\right)$

30. $\lim\limits_{x \to 1} \left(\frac{2}{3}x + 9\right)$

31. $\lim\limits_{x \to 6} 3$

32. $\lim\limits_{x \to 2} (-1)$

33. $\lim\limits_{x \to 0} \sqrt[3]{x}$

34. $\lim\limits_{x \to 4} \sqrt{x}$

35. $\lim\limits_{x \to -2} |x - 2|$

36. $\lim\limits_{x \to 3} |x - 3|$

37. $\lim\limits_{x \to 1} (x^2 + 1)$

38. $\lim\limits_{x \to -3} (x^2 + 3x)$

Writing **In Exercises 39–42, use a graphing utility to graph the function and estimate the limit (if it exists). What is the domain of the function? Can you detect a possible error in determining the domain of a function solely by analyzing the graph generated by a graphing utility? Write a short paragraph about the importance of examining a function analytically as well as graphically.**

39. $f(x) = \dfrac{\sqrt{x + 5} - 3}{x - 4}$

$\lim\limits_{x \to 4} f(x)$

40. $f(x) = \dfrac{x - 3}{x^2 - 4x + 3}$

$\lim\limits_{x \to 3} f(x)$

41. $f(x) = \dfrac{x - 9}{\sqrt{x} - 3}$

$\lim\limits_{x \to 9} f(x)$

42. $f(x) = \dfrac{x - 3}{x^2 - 9}$

$\lim\limits_{x \to 3} f(x)$

Getting at the Concept

43. Write a brief description of the meaning of the notation

$$\lim\limits_{x \to 8} f(x) = 25.$$

44. (a) If $f(2) = 4$, can you conclude anything about the limit of $f(x)$ as x approaches 2? Explain your reasoning.

 (b) If the limit of $f(x)$ as x approaches 2 is 4, can you conclude anything about $f(2)$? Explain your reasoning.

45. Identify three types of behavior associated with the nonexistence of a limit. Illustrate each type with a graph of a function.

46. Determine the limit of the function describing the atmospheric pressure on a plane as it descends from 32,000 feet to land at Honolulu, located at sea level. (The atmospheric pressure at sea level is 14.7 lb/in.^{2}.)

47. Consider the function $f(x) = (1 + x)^{1/x}$. Estimate the limit

$$\lim\limits_{x \to 0} (1 + x)^{1/x}$$

by evaluating f at x-values near 0. Sketch the graph of f.

48. *Graphical Analysis* The statement

$$\lim\limits_{x \to 2} \frac{x^2 - 4}{x - 2} = 4$$

means that for each $\varepsilon > 0$ there corresponds a $\delta > 0$ such that if $0 < |x - 2| < \delta$, then

$$\left| \frac{x^2 - 4}{x - 2} - 4 \right| < \varepsilon.$$

If $\varepsilon = 0.001$, then

$$\left| \frac{x^2 - 4}{x - 2} - 4 \right| < 0.001.$$

Use a graphing utility to graph each side of this inequality. Use the *zoom* feature to find an interval $(2 - \delta, 2 + \delta)$ such that the graph of the left side is below the graph of the right side of the inequality.

True or False? **In Exercises 49–52, determine whether the statement is true or false. If it is false, explain why or give an example that shows it is false.**

49. If f is undefined at $x = c$, then the limit of $f(x)$ as x approaches c does not exist.

50. If the limit of $f(x)$ as x approaches c is 0, then there must exist a number k such that $f(k) < 0.001$.

51. If $f(c) = L$, then $\lim\limits_{x \to c} f(x) = L$.

52. If $\lim\limits_{x \to c} f(x) = L$, then $f(c) = L$.

53. *Programming* Use the programming capabilities of a graphing utility to write a program for approximating $\lim\limits_{x \to c} f(x)$.

Assume the program will be applied only to functions whose limits exist as x approaches c. Let $y_1 = f(x)$ and generate two lists whose entries form the ordered pairs

$$(c \pm [0.1]^n, \, f(c \pm [0.1]^n))$$

for $n = 0, 1, 2, 3,$ and 4.

54. Use the program you created in Exercise 53 to approximate the limit

$$\lim\limits_{x \to 4} \frac{x^2 - x - 12}{x - 4}.$$

55. Prove that if the limit of $f(x)$ as $x \to c$ exists, then the limit must be unique. [*Hint:* Let

$$\lim\limits_{x \to c} f(x) = L_1 \quad \text{and} \quad \lim\limits_{x \to c} f(x) = L_2$$

and prove that $L_1 = L_2$.]

56. Consider the line $f(x) = mx + b$, where $m \neq 0$. Use the ε-δ definition of a limit to prove that $\lim\limits_{x \to c} f(x) = mc + b$.

57. Prove that $\lim\limits_{x \to c} f(x) = L$ is equivalent to $\lim\limits_{x \to c} [f(x) - L] = 0$.

58. Given that $\lim\limits_{x \to c} g(x) = L$, where $L > 0$, prove that there exists an open interval (a, b) containing c such that $g(x) > 0$ for all $x \neq c$ in (a, b).

- Evaluate a limit using properties of limits.
- Develop and use a strategy for finding limits.
- Evaluate a limit using dividing out and rationalizing techniques.
- Evaluate a limit using the Squeeze Theorem.

Properties of Limits

In Section 1.2, you learned that the limit of $f(x)$ as x approaches c does not depend on the value of f at $x = c$. It may happen, however, that the limit is precisely $f(c)$. In such cases, the limit can be evaluated by **direct substitution.** That is,

$$\lim_{x \to c} f(x) = f(c). \qquad \text{Substitute } c \text{ for } x.$$

Such *well-behaved* functions are **continuous at c.** You will examine this concept more closely in Section 1.4.

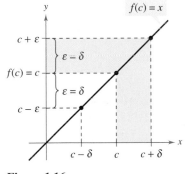

Figure 1.16

NOTE   When you encounter new notations or symbols in mathematics, be sure you know how the notations are read. For instance, the limit in Example 1c is read as "the limit of x^2 as x approaches 2 is 4."

THEOREM 1.1   Some Basic Limits

Let b and c be real numbers and let n be a positive integer.

1. $\lim\limits_{x \to c} b = b$ **2.** $\lim\limits_{x \to c} x = c$ **3.** $\lim\limits_{x \to c} x^n = c^n$

Proof   To prove Property 2 of Theorem 1.1, you need to show that for each $\varepsilon > 0$ there exists a $\delta > 0$ such that $|x - c| < \varepsilon$ whenever $0 < |x - c| < \delta$. Because the second inequality is a stricter version of the first, you can simply choose $\delta = \varepsilon$, as shown in Figure 1.16. This completes the proof. (Proofs of the other properties of limits in this section are listed in Appendix B or are discussed in the exercises.)

Example 1   Evaluating Basic Limits

a. $\lim\limits_{x \to 2} 3 = 3$ **b.** $\lim\limits_{x \to -4} x = -4$ **c.** $\lim\limits_{x \to 2} x^2 = 2^2 = 4$

THEOREM 1.2   Properties of Limits

Let b and c be real numbers, let n be a positive integer, and let f and g be functions with the following limits.

$$\lim_{x \to c} f(x) = L \qquad \text{and} \qquad \lim_{x \to c} g(x) = K$$

1. Scalar multiple: $\lim\limits_{x \to c} [b f(x)] = bL$

2. Sum or difference: $\lim\limits_{x \to c} [f(x) \pm g(x)] = L \pm K$

3. Product: $\lim\limits_{x \to c} [f(x)g(x)] = LK$

4. Quotient: $\lim\limits_{x \to c} \dfrac{f(x)}{g(x)} = \dfrac{L}{K}, \qquad$ provided $K \neq 0$

5. Power: $\lim\limits_{x \to c} [f(x)]^n = L^n$

Example 2 **The Limit of a Polynomial**

$$\lim_{x \to 2} (4x^2 + 3) = \lim_{x \to 2} 4x^2 + \lim_{x \to 2} 3 \qquad \text{Property 2}$$

$$= 4\left(\lim_{x \to 2} x^2\right) + \lim_{x \to 2} 3 \qquad \text{Property 1}$$

$$= 4(2^2) + 3 \qquad \text{Example 1}$$

$$= 19 \qquad \text{Simplify.}$$

In Example 2, note that the limit (as $x \to 2$) of the *polynomial function* $p(x) = 4x^2 + 3$ is simply the value of p at $x = 2$.

$$\lim_{x \to 2} p(x) = p(2) = 4(2^2) + 3 = 19$$

This *direct substitution* property is valid for all polynomial and rational functions with nonzero denominators.

THE SQUARE ROOT SYMBOL

The first use of a symbol to denote the square root can be traced to the sixteenth century. Mathematicians first used the symbol $\sqrt{\ }$, which had only two strokes. This symbol was chosen because it resembled a lowercase r, to stand for the Latin word *radix*, meaning root.

THEOREM 1.3   Limits of Polynomial and Rational Functions

If p is a polynomial function and c is a real number, then

$$\lim_{x \to c} p(x) = p(c).$$

If r is a rational function given by $r(x) = p(x)/q(x)$ and c is a real number such that $q(c) \neq 0$, then

$$\lim_{x \to c} r(x) = r(c) = \frac{p(c)}{q(c)}.$$

Example 3 **The Limit of a Rational Function**

Find the limit: $\displaystyle\lim_{x \to 1} \frac{x^2 + x + 2}{x + 1}$.

Solution   Because the denominator is not 0 when $x = 1$, you can apply Theorem 1.3 to obtain

$$\lim_{x \to 1} \frac{x^2 + x + 2}{x + 1} = \frac{1^2 + 1 + 2}{1 + 1} = \frac{4}{2} = 2.$$

Polynomial functions and rational functions are two of the three basic types of algebraic functions. The following theorem deals with the limit of the third type of algebraic function—one that involves a radical. See Appendix B for a proof of this theorem.

THEOREM 1.4   The Limit of a Function Involving a Radical

Let n be a positive integer. The following limit is valid for all c if n is odd, and is valid for $c > 0$ if n is even.

$$\lim_{x \to c} \sqrt[n]{x} = \sqrt[n]{c}$$

The following theorem greatly expands your ability to evaluate limits because it shows how to analyze the limit of a composite function. See Appendix B for a proof of this theorem.

THEOREM 1.5   The Limit of a Composite Function

If f and g are functions such that $\lim_{x \to c} g(x) = L$ and $\lim_{x \to L} f(x) = f(L)$, then

$$\lim_{x \to c} f(g(x)) = f\left(\lim_{x \to c} g(x)\right) = f(L).$$

 Example 4   **The Limit of a Composite Function**

a. Because

$$\lim_{x \to 0} (x^2 + 4) = 0^2 + 4 = 4 \qquad \text{and} \qquad \lim_{x \to 4} \sqrt{x} = 2$$

it follows that

$$\lim_{x \to 0} \sqrt{x^2 + 4} = \sqrt{4} = 2.$$

b. Because

$$\lim_{x \to 3} (2x^2 - 10) = 2(3^2) - 10 = 8 \quad \text{and} \quad \lim_{x \to 8} \sqrt[3]{x} = 2$$

it follows that

$$\lim_{x \to 3} \sqrt[3]{2x^2 - 10} = \sqrt[3]{8} = 2.$$

You have seen that the limits of many algebraic functions can be evaluated by direct substitution. Each of the six basic trigonometric functions also possesses this desirable quality, as shown in the next theorem (presented without proof).

THEOREM 1.6   Limits of Trigonometric Functions

Let c be a real number in the domain of the given trigonometric function.

1. $\lim_{x \to c} \sin x = \sin c$ **2.** $\lim_{x \to c} \cos x = \cos c$

3. $\lim_{x \to c} \tan x = \tan c$ **4.** $\lim_{x \to c} \cot x = \cot c$

5. $\lim_{x \to c} \sec x = \sec c$ **6.** $\lim_{x \to c} \csc x = \csc c$

Example 5   **Limits of Trigonometric Functions**

a. $\lim_{x \to 0} \tan x = \tan(0) = 0$

b. $\lim_{x \to \pi} (x \cos x) = \left(\lim_{x \to \pi} x\right)\left(\lim_{x \to \pi} \cos x\right) = \pi \cos(\pi) = -\pi$

c. $\lim_{x \to 0} \sin^2 x = \lim_{x \to 0} (\sin x)^2 = 0^2 = 0$

A Strategy for Finding Limits

On the previous three pages, you studied several types of functions whose limits can be evaluated by direct substitution. This knowledge, together with the following theorem, can be used to develop a strategy for finding limits. A proof of this theorem is given in Appendix B.

> **THEOREM 1.7** **Functions That Agree at All But One Point**
>
> Let c be a real number and let $f(x) = g(x)$ for all $x \neq c$ in an open interval containing c. If the limit of $g(x)$ as x approaches c exists, then the limit of $f(x)$ also exists and
>
> $$\lim_{x \to c} f(x) = \lim_{x \to c} g(x).$$

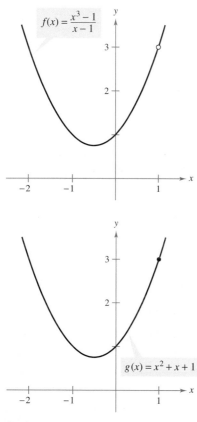

$f(x) = \dfrac{x^3 - 1}{x - 1}$

$g(x) = x^2 + x + 1$

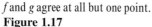

f and *g* agree at all but one point.
Figure 1.17

Example 6 **Finding the Limit of a Function**

Find the limit: $\lim\limits_{x \to 1} \dfrac{x^3 - 1}{x - 1}$.

Solution Let $f(x) = (x^3 - 1)/(x - 1)$. By factoring and dividing out like factors, you can rewrite f as

$$f(x) = \frac{(x - 1)(x^2 + x + 1)}{(x - 1)} = x^2 + x + 1 = g(x), \qquad x \neq 1.$$

So, for all x-values other than $x = 1$, the functions f and g agree, as shown in Figure 1.17. Because $\lim\limits_{x \to 1} g(x)$ exists, you can apply Theorem 1.7 to conclude that f and g have the same limit at $x = 1$.

$$\lim_{x \to 1} \frac{x^3 - 1}{x - 1} = \lim_{x \to 1} \frac{(x - 1)(x^2 + x + 1)}{x - 1} \qquad \text{Factor.}$$

$$= \lim_{x \to 1} \frac{(x - 1)(x^2 + x + 1)}{x - 1} \qquad \text{Divide out like factors.}$$

$$= \lim_{x \to 1} (x^2 + x + 1) \qquad \text{Apply Theorem 1.7.}$$

$$= 1^2 + 1 + 1 \qquad \text{Use direct substitution.}$$

$$= 3 \qquad \text{Simplify.}$$

STUDY TIP When applying this strategy for finding a limit, remember that some functions do not have a limit (as x approaches c). For instance, the following limit does not exist.

$$\lim_{x \to 1} \frac{x^3 + 1}{x - 1}$$

A Strategy for Finding Limits

1. Learn to recognize which limits can be evaluated by direct substitution. (These limits are listed in Theorems 1.1 through 1.6.)

2. If the limit of $f(x)$ as x approaches c *cannot* be evaluated by direct substitution, try to find a function g that agrees with f for all x other than $x = c$. [Choose g such that the limit of $g(x)$ *can* be evaluated by direct substitution.]

3. Apply Theorem 1.7 to conclude *analytically* that

   $$\lim_{x \to c} f(x) = \lim_{x \to c} g(x) = g(c).$$

4. Use a *graph* or *table* to reinforce your conclusion.

Dividing Out and Rationalizing Techniques

Two techniques for finding limits analytically are shown in Examples 7 and 8. The first technique involves dividing out common factors, and the second technique involves rationalizing the numerator of a fractional expression.

Example 7   Dividing Out Technique

Find the limit: $\lim\limits_{x \to -3} \dfrac{x^2 + x - 6}{x + 3}$.

Solution   Although you are taking the limit of a rational function, you *cannot* apply Theorem 1.3 because the limit of the denominator is 0.

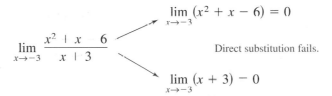

Because the limit of the numerator is also 0, the numerator and denominator have a *common factor* of $(x + 3)$. Thus, for all $x \neq -3$, you can divide out this factor to obtain

$$f(x) = \frac{x^2 + x - 6}{x + 3} = \frac{(x + 3)(x - 2)}{x + 3} = x - 2 = g(x), \qquad x \neq -3.$$

Using Theorem 1.7, it follows that

$$\lim_{x \to -3} \frac{x^2 + x - 6}{x + 3} = \lim_{x \to -3} (x - 2) \qquad \text{Apply Theorem 1.7.}$$

$$= -5. \qquad \text{Use direct substitution.}$$

This result is shown graphically in Figure 1.18. Note that the graph of the function f coincides with the graph of the function $g(x) = x - 2$, except that the graph of f has a gap at the point $(-3, -5)$.

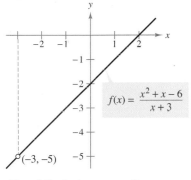

f is undefined when $x = -3$.
Figure 1.18

NOTE   In the solution of Example 7, be sure you see the usefulness of the Factor Theorem of Algebra. This theorem states that if c is a zero of a polynomial function, $(x - c)$ is a factor of the polynomial. Thus, if you apply direct substitution to a rational function and obtain

$$r(c) = \frac{p(c)}{q(c)} = \frac{0}{0},$$

you can conclude that $(x - c)$ must be a common factor to both $p(x)$ and $q(x)$.

In Example 7, direct substitution produced the meaningless fractional form $0/0$. An expression such as $0/0$ is called an **indeterminate form** because you cannot (from the form alone) determine the limit. When you try to evaluate a limit and encounter this form, remember that you must rewrite the fraction so that the new denominator does not have 0 as its limit. One way to do this is to *divide out like factors*, as shown in Example 7. A second way is to *rationalize the numerator*, as shown in Example 8.

TECHNOLOGY PITFALL   Because the graphs of

$$f(x) = \frac{x^2 + x - 6}{x + 3} \qquad \text{and} \qquad g(x) = x - 2$$

differ only at the point $(-3, -5)$, a standard graphing utility setting may not distinguish clearly between these graphs. However, because of the pixel configuration and rounding error of a graphing utility, it may be possible to find screen settings that distinguish between the graphs. Specifically, by repeatedly zooming in near the point $(-3, -5)$ on the graph of f, your graphing utility may show glitches or irregularities that do not exist on the actual graph. (See Figure 1.19.) By changing the screen settings on your graphing utility you may obtain the correct graph of f.

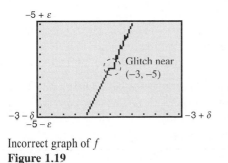

Incorrect graph of f
Figure 1.19

Example 8   **Rationalizing Technique**

Find the limit: $\displaystyle\lim_{x\to 0}\frac{\sqrt{x+1}-1}{x}$.

Solution   By direct substitution, you obtain the indeterminate form $0/0$.

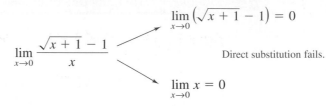

$$\lim_{x\to 0}\left(\sqrt{x+1}-1\right)=0$$

$$\lim_{x\to 0}\frac{\sqrt{x+1}-1}{x}$$

Direct substitution fails.

$$\lim_{x\to 0}x=0$$

In this case, you can rewrite the fraction by rationalizing the numerator.

$$\frac{\sqrt{x+1}-1}{x}=\left(\frac{\sqrt{x+1}-1}{x}\right)\left(\frac{\sqrt{x+1}+1}{\sqrt{x+1}+1}\right)$$

$$=\frac{(x+1)-1}{x\left(\sqrt{x+1}+1\right)}$$

$$=\frac{\not{x}}{\not{x}\left(\sqrt{x+1}+1\right)}$$

$$=\frac{1}{\sqrt{x+1}+1},\qquad x\neq 0$$

Now, using Theorem 1.7, you can evaluate the limit as follows.

$$\lim_{x\to 0}\frac{\sqrt{x+1}-1}{x}=\lim_{x\to 0}\frac{1}{\sqrt{x+1}+1}$$

$$=\frac{1}{1+1}$$

$$=\frac{1}{2}$$

A table or a graph can reinforce your conclusion that the limit is $\frac{1}{2}$. (See Figure 1.20.)

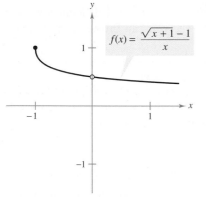

$f(x)=\dfrac{\sqrt{x+1}-1}{x}$

The limit of $f(x)$ as x approaches 0 is $\frac{1}{2}$.
Figure 1.20

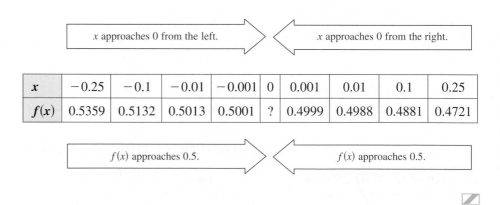

x approaches 0 from the left.					x approaches 0 from the right.			
x	-0.25	-0.1	-0.01	-0.001	0	0.001	0.01	0.1	0.25
$f(x)$	0.5359	0.5132	0.5013	0.5001	?	0.4999	0.4988	0.4881	0.4721

$f(x)$ approaches 0.5. $f(x)$ approaches 0.5.

NOTE   The rationalizing technique for evaluating limits is based on multiplication by a convenient form of 1. In Example 8, the convenient form is

$$1=\frac{\sqrt{x+1}+1}{\sqrt{x+1}+1}.$$

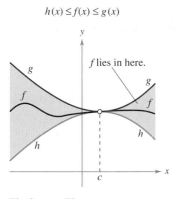

$h(x) \le f(x) \le g(x)$

f lies in here.

The Squeeze Theorem
Figure 1.21

The Squeeze Theorem

The next theorem concerns the limit of a function that is squeezed between two other functions, each of which has the same limit at a given *x*-value, as shown in Figure 1.21. (The proof of this theorem is given in Appendix B.)

> **THEOREM 1.8   The Squeeze Theorem**
>
> If $h(x) \le f(x) \le g(x)$ for all *x* in an open interval containing *c*, except possibly at *c* itself, and if
>
> $$\lim_{x \to c} h(x) = L = \lim_{x \to c} g(x)$$
>
> then $\lim_{x \to c} f(x)$ exists and is equal to *L*.

You can see the usefulness of the Squeeze Theorem in the proof of Theorem 1.9.

> **THEOREM 1.9   Two Special Trigonometric Limits**
>
> **1.** $\lim_{x \to 0} \dfrac{\sin x}{x} = 1$ **2.** $\lim_{x \to 0} \dfrac{1 - \cos x}{x} = 0$

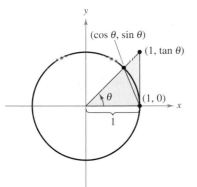

A circular sector is used to prove Theorem 1.9.
Figure 1.22

Proof   To avoid the confusion of two different uses of *x*, the proof is presented using the variable θ, where θ is an acute positive angle *measured in radians*. Figure 1.22 shows a circular sector that is squeezed between two triangles.

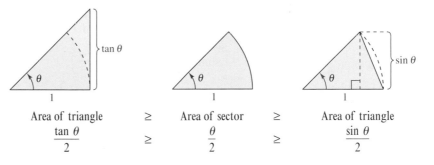

Area of triangle	$\ge$	Area of sector	$\ge$	Area of triangle
$\dfrac{\tan \theta}{2}$	$\ge$	$\dfrac{\theta}{2}$	$\ge$	$\dfrac{\sin \theta}{2}$

Multiplying each expression by $2/\sin \theta$ produces

$$\frac{1}{\cos \theta} \ge \frac{\theta}{\sin \theta} \ge 1$$

and taking reciprocals and reversing the inequalities yields

$$\cos \theta \le \frac{\sin \theta}{\theta} \le 1.$$

Because $\cos \theta = \cos(-\theta)$ and $(\sin \theta)/\theta = [\sin(-\theta)]/(-\theta)$, we can conclude that this inequality is valid for *all* nonzero θ in the open interval $(-\pi/2, \pi/2)$. Finally, because $\lim_{\theta \to 0} \cos \theta = 1$ and $\lim_{\theta \to 0} 1 = 1$, you can apply the Squeeze Theorem to conclude that $\lim_{\theta \to 0} (\sin \theta)/\theta = 1$. The proof of the second limit is left as an exercise (see Exercise 120).

FOR FURTHER INFORMATION For more information on the function $f(x) = (\sin x)/x$, see the article "The Function $(\sin x)/x$" by William B. Gearhart and Harris S. Shultz in *The College Mathematics Journal*. To view this article, go to the website *www.matharticles.com*.

Example 9   **A Limit Involving a Trigonometric Function**

Find the limit: $\displaystyle\lim_{x \to 0} \frac{\tan x}{x}$.

Solution   Direct substitution yields the indeterminate form 0/0. To solve this problem, you can write $\tan x$ as $(\sin x)/(\cos x)$ and obtain

$$\lim_{x \to 0} \frac{\tan x}{x} = \lim_{x \to 0} \left(\frac{\sin x}{x}\right)\left(\frac{1}{\cos x}\right).$$

Now, because

$$\lim_{x \to 0} \frac{\sin x}{x} = 1 \qquad \text{and} \qquad \lim_{x \to 0} \frac{1}{\cos x} = 1$$

you can obtain

$$\lim_{x \to 0} \frac{\tan x}{x} = \left(\lim_{x \to 0} \frac{\sin x}{x}\right)\left(\lim_{x \to 0} \frac{1}{\cos x}\right)$$
$$= (1)(1)$$
$$= 1.$$

(See Figure 1.23.)

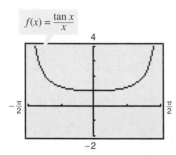

$f(x) = \dfrac{\tan x}{x}$

The limit of $f(x)$ as x approaches 0 is 1.
Figure 1.23

Example 10   **A Limit Involving a Trigonometric Function**

Find the limit: $\displaystyle\lim_{x \to 0} \frac{\sin 4x}{x}$.

Solution   Direct substitution yields the indeterminate form 0/0. To solve this problem, you can rewrite the limit as

$$\lim_{x \to 0} \frac{\sin 4x}{x} = 4\left(\lim_{x \to 0} \frac{\sin 4x}{4x}\right).$$

Now, by letting $y = 4x$ and observing that $x \to 0$ if and only if $y \to 0$, you can write

$$\lim_{x \to 0} \frac{\sin 4x}{x} = 4\left(\lim_{x \to 0} \frac{\sin 4x}{4x}\right)$$
$$= 4\left(\lim_{y \to 0} \frac{\sin y}{y}\right)$$
$$= 4(1)$$
$$= 4.$$

(See Figure 1.24.)

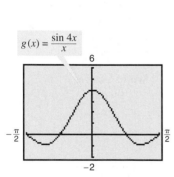

$g(x) = \dfrac{\sin 4x}{x}$

The limit of $g(x)$ as x approaches 0 is 4.
Figure 1.24

TECHNOLOGY   Try using a graphing utility to confirm the limits in the examples and exercise set. For instance, Figures 1.23 and 1.24 show the graphs of

$$f(x) = \frac{\tan x}{x} \qquad \text{and} \qquad g(x) = \frac{\sin 4x}{x}.$$

Note that the first graph appears to contain the point (0, 1) and the second graph appears to contain the point (0, 4), which lends support to the conclusions obtained in Examples 9 and 10.

EXERCISES FOR SECTION 1.3

In Exercises 1–4, use a graphing utility to graph the function and visually estimate the limits.

1. $h(x) = x^2 - 5x$

  (a) $\lim\limits_{x \to 5} h(x)$

  (b) $\lim\limits_{x \to -1} h(x)$

2. $g(x) = \dfrac{12(\sqrt{x} - 3)}{x - 9}$

  (a) $\lim\limits_{x \to 4} g(x)$

  (b) $\lim\limits_{x \to 0} g(x)$

3. $f(x) = x \cos x$

  (a) $\lim\limits_{x \to 0} f(x)$

  (b) $\lim\limits_{x \to \pi/3} f(x)$

4. $f(t) = t|t - 4|$

  (a) $\lim\limits_{t \to 4} f(t)$

  (b) $\lim\limits_{t \to -1} f(t)$

In Exercises 5–22, find the limit.

5. $\lim\limits_{x \to 2} x^4$

6. $\lim\limits_{x \to -2} x^3$

7. $\lim\limits_{x \to 0} (2x - 1)$

8. $\lim\limits_{x \to 3} (3x + 2)$

9. $\lim\limits_{x \to -3} (x^2 + 3x)$

10. $\lim\limits_{x \to 1} (-x^2 + 1)$

11. $\lim\limits_{x \to -3} (2x^2 + 4x + 1)$

12. $\lim\limits_{x \to 1} (3x^3 - 2x^2 + 4)$

13. $\lim\limits_{x \to 2} \dfrac{1}{x}$

14. $\lim\limits_{x \to -3} \dfrac{2}{x + 2}$

15. $\lim\limits_{x \to 1} \dfrac{x - 3}{x^2 + 4}$

16. $\lim\limits_{x \to 3} \dfrac{2x - 3}{x + 5}$

17. $\lim\limits_{x \to 7} \dfrac{5x}{\sqrt{x + 2}}$

18. $\lim\limits_{x \to 3} \dfrac{\sqrt{x + 1}}{x - 4}$

19. $\lim\limits_{x \to 3} \sqrt{x + 1}$

20. $\lim\limits_{x \to 4} \sqrt[3]{x + 4}$

21. $\lim\limits_{x \to -4} (x + 3)^2$

22. $\lim\limits_{x \to 0} (2x - 1)^3$

In Exercises 23–26, find the limits.

23. $f(x) = 5 - x$, $g(x) = x^3$

  (a) $\lim\limits_{x \to 1} f(x)$ (b) $\lim\limits_{x \to 4} g(x)$ (c) $\lim\limits_{x \to 1} g(f(x))$

24. $f(x) = x + 7$, $g(x) = x^2$

  (a) $\lim\limits_{x \to -3} f(x)$ (b) $\lim\limits_{x \to 4} g(x)$ (c) $\lim\limits_{x \to -3} g(f(x))$

25. $f(x) = 4 - x^2$, $g(x) = \sqrt{x + 1}$

  (a) $\lim\limits_{x \to 1} f(x)$ (b) $\lim\limits_{x \to 3} g(x)$ (c) $\lim\limits_{x \to 1} g(f(x))$

26. $f(x) = 2x^2 - 3x + 1$, $g(x) = \sqrt[3]{x + 6}$

  (a) $\lim\limits_{x \to 4} f(x)$ (b) $\lim\limits_{x \to 21} g(x)$ (c) $\lim\limits_{x \to 4} g(f(x))$

In Exercises 27–36, find the limit of the trigonometric function.

27. $\lim\limits_{x \to \pi/2} \sin x$

28. $\lim\limits_{x \to \pi} \tan x$

29. $\lim\limits_{x \to 2} \cos \dfrac{\pi x}{3}$

30. $\lim\limits_{x \to 1} \sin \dfrac{\pi x}{2}$

31. $\lim\limits_{x \to 0} \sec 2x$

32. $\lim\limits_{x \to \pi} \cos 3x$

33. $\lim\limits_{x \to 5\pi/6} \sin x$

34. $\lim\limits_{x \to 5\pi/3} \cos x$

35. $\lim\limits_{x \to 3} \tan\left(\dfrac{\pi x}{4}\right)$

36. $\lim\limits_{x \to 7} \sec\left(\dfrac{\pi x}{6}\right)$

In Exercises 37–40, use the information to evaluate the limits.

37. $\lim\limits_{x \to c} f(x) = 2$

  $\lim\limits_{x \to c} g(x) = 3$

  (a) $\lim\limits_{x \to c} [5g(x)]$

  (b) $\lim\limits_{x \to c} [f(x) + g(x)]$

  (c) $\lim\limits_{x \to c} [f(x)g(x)]$

  (d) $\lim\limits_{x \to c} \dfrac{f(x)}{g(x)}$

38. $\lim\limits_{x \to c} f(x) = \dfrac{3}{2}$

  $\lim\limits_{x \to c} g(x) = \dfrac{1}{2}$

  (a) $\lim\limits_{x \to c} [4f(x)]$

  (b) $\lim\limits_{x \to c} [f(x) + g(x)]$

  (c) $\lim\limits_{x \to c} [f(x)g(x)]$

  (d) $\lim\limits_{x \to c} \dfrac{f(x)}{g(x)}$

39. $\lim\limits_{x \to c} f(x) = 4$

  (a) $\lim\limits_{x \to c} [f(x)]^3$

  (b) $\lim\limits_{x \to c} \sqrt{f(x)}$

  (c) $\lim\limits_{x \to c} [3f(x)]$

  (d) $\lim\limits_{x \to c} [f(x)]^{3/2}$

40. $\lim\limits_{x \to c} f(x) = 27$

  (a) $\lim\limits_{x \to c} \sqrt[3]{f(x)}$

  (b) $\lim\limits_{x \to c} \dfrac{f(x)}{18}$

  (c) $\lim\limits_{x \to c} [f(x)]^2$

  (d) $\lim\limits_{x \to c} [f(x)]^{2/3}$

In Exercises 41–44, use the graph to determine the limit visually (if it exists). Write a simpler function that agrees with the given function at all but one point.

41. $g(x) = \dfrac{-2x^2 + x}{x}$

42. $h(x) = \dfrac{x^2 - 3x}{x}$

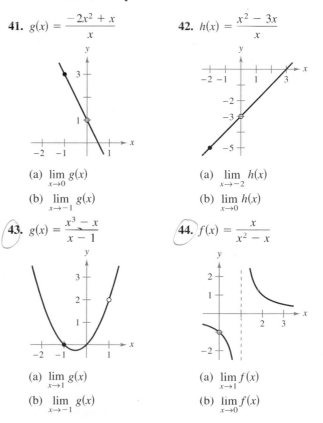

  (a) $\lim\limits_{x \to 0} g(x)$

  (b) $\lim\limits_{x \to -1} g(x)$

  (a) $\lim\limits_{x \to -2} h(x)$

  (b) $\lim\limits_{x \to 0} h(x)$

43. $g(x) = \dfrac{x^3 - x}{x - 1}$

44. $f(x) = \dfrac{x}{x^2 - x}$

  (a) $\lim\limits_{x \to 1} g(x)$

  (b) $\lim\limits_{x \to -1} g(x)$

  (a) $\lim\limits_{x \to 1} f(x)$

  (b) $\lim\limits_{x \to 0} f(x)$

In Exercises 45–48, find the limit of the function (if it exists). Write a simpler function that agrees with the given function at all but one point. Use a graphing utility to confirm your result.

45. $\lim\limits_{x \to -1} \dfrac{x^2 - 1}{x + 1}$

46. $\lim\limits_{x \to -1} \dfrac{2x^2 - x - 3}{x + 1}$

47. $\lim\limits_{x \to 2} \dfrac{x^3 - 8}{x - 2}$

48. $\lim\limits_{x \to -1} \dfrac{x^3 + 1}{x + 1}$

In Exercises 49–62, find the limit (if it exists).

49. $\lim\limits_{x \to 5} \dfrac{x - 5}{x^2 - 25}$

50. $\lim\limits_{x \to 2} \dfrac{2 - x}{x^2 - 4}$

51. $\lim\limits_{x \to -3} \dfrac{x^2 + x - 6}{x^2 - 9}$

52. $\lim\limits_{x \to 4} \dfrac{x^2 - 5x + 4}{x^2 - 2x - 8}$

53. $\lim\limits_{x \to 0} \dfrac{\sqrt{x + 5} - \sqrt{5}}{x}$

54. $\lim\limits_{x \to 0} \dfrac{\sqrt{2 + x} - \sqrt{2}}{x}$

55. $\lim\limits_{x \to 4} \dfrac{\sqrt{x + 5} - 3}{x - 4}$

56. $\lim\limits_{x \to 3} \dfrac{\sqrt{x + 1} - 2}{x - 3}$

57. $\lim\limits_{x \to 0} \dfrac{[1/(3 + x)] - (1/3)}{x}$

58. $\lim\limits_{x \to 0} \dfrac{[1/(x + 4)] - (1/4)}{x}$

59. $\lim\limits_{\Delta x \to 0} \dfrac{2(x + \Delta x) - 2x}{\Delta x}$

60. $\lim\limits_{\Delta x \to 0} \dfrac{(x + \Delta x)^2 - x^2}{\Delta x}$

61. $\lim\limits_{\Delta x \to 0} \dfrac{(x + \Delta x)^2 - 2(x + \Delta x) + 1 - (x^2 - 2x + 1)}{\Delta x}$

62. $\lim\limits_{\Delta x \to 0} \dfrac{(x + \Delta x)^3 - x^3}{\Delta x}$

Graphical, Numerical, and Analytic Analysis In Exercises 63–66, use a graphing utility to graph the function and estimate the limit. Use a table to reinforce your conclusion. Then find the limit by analytic methods.

63. $\lim\limits_{x \to 0} \dfrac{\sqrt{x + 2} - \sqrt{2}}{x}$

64. $\lim\limits_{x \to 16} \dfrac{4 - \sqrt{x}}{x - 16}$

65. $\lim\limits_{x \to 0} \dfrac{[1/(2 + x)] - (1/2)}{x}$

66. $\lim\limits_{x \to 2} \dfrac{x^5 - 32}{x - 2}$

In Exercises 67–78, determine the limit of the trigonometric function (if it exists).

67. $\lim\limits_{x \to 0} \dfrac{\sin x}{5x}$

68. $\lim\limits_{x \to 0} \dfrac{3(1 - \cos x)}{x}$

69. $\lim\limits_{x \to 0} \dfrac{\sin x(1 - \cos x)}{2x^2}$

70. $\lim\limits_{\theta \to 0} \dfrac{\cos \theta \tan \theta}{\theta}$

71. $\lim\limits_{x \to 0} \dfrac{\sin^2 x}{x}$

72. $\lim\limits_{x \to 0} \dfrac{\tan^2 x}{x}$

73. $\lim\limits_{h \to 0} \dfrac{(1 - \cos h)^2}{h}$

74. $\lim\limits_{\phi \to \pi} \phi \sec \phi$

75. $\lim\limits_{x \to \pi/2} \dfrac{\cos x}{\cot x}$

76. $\lim\limits_{x \to \pi/4} \dfrac{1 - \tan x}{\sin x - \cos x}$

77. $\lim\limits_{t \to 0} \dfrac{\sin 3t}{2t}$

78. $\lim\limits_{x \to 0} \dfrac{\sin 2x}{\sin 3x}$ $\left[\textit{Hint: Find } \lim\limits_{x \to 0} \left(\dfrac{2 \sin 2x}{2x} \right) \left(\dfrac{3x}{3 \sin 3x} \right). \right]$

Graphical, Numerical, and Analytic Analysis In Exercises 79–82, use a graphing utility to graph the function and estimate the limit. Use a table to reinforce your conclusion. Then find the limit by analytic methods.

79. $\lim\limits_{t \to 0} \dfrac{\sin 3t}{t}$

80. $\lim\limits_{h \to 0} (1 + \cos 2h)$

81. $\lim\limits_{x \to 0} \dfrac{\sin x^2}{x}$

82. $\lim\limits_{x \to 0} \dfrac{\sin x}{\sqrt[3]{x}}$

In Exercises 83–86, find $\lim\limits_{\Delta x \to 0} \dfrac{f(x + \Delta x) - f(x)}{\Delta x}$.

83. $f(x) = 2x + 3$

84. $f(x) = \sqrt{x}$

85. $f(x) = \dfrac{4}{x}$

86. $f(x) = x^2 - 4x$

In Exercises 87 and 88, use the Squeeze Theorem to find $\lim\limits_{x \to c} f(x)$.

87. $c = 0$

$4 - x^2 \leq f(x) \leq 4 + x^2$

88. $c = a$

$b - |x - a| \leq f(x) \leq b + |x - a|$

In Exercises 89–94, use a graphing utility to graph the given function and the equations $y = |x|$ and $y = -|x|$ in the same viewing window. Using the graphs to visually observe the Squeeze Theorem, find $\lim\limits_{x \to 0} f(x)$.

89. $f(x) = x \cos x$

90. $f(x) = |x \sin x|$

91. $f(x) = |x| \sin x$

92. $f(x) = |x| \cos x$

93. $f(x) = x \sin \dfrac{1}{x}$

94. $h(x) = x \cos \dfrac{1}{x}$

Getting at the Concept

95. In the context of finding limits, discuss what is meant by two functions that agree at all but one point.

96. Give an example of two functions that agree at all but one point.

97. What is meant by an indeterminate form?

98. In your own words, explain the Squeeze Theorem.

99. *Writing* Use a graphing utility to graph

$$f(x) = x, \quad g(x) = \sin x, \quad \text{and} \quad h(x) = \dfrac{\sin x}{x}$$

in the same viewing window. Compare the magnitudes of $f(x)$ and $g(x)$ when x is "close to" 0. Use the comparison to write a short paragraph explaining why

$$\lim\limits_{x \to 0} h(x) = 1.$$

100. *Writing*   Use a graphing utility to graph

$$f(x) = x, \quad g(x) = \sin^2 x, \quad \text{and } h(x) = \frac{\sin^2 x}{x}$$

in the same viewing window. Compare the magnitudes of $f(x)$ and $g(x)$ when x is "close to" 0. Use the comparison to write a short paragraph explaining why

$$\lim_{x \to 0} h(x) = 0.$$

Free-Falling Object   In Exercises 101 and 102, use the position function $s(t) = -16t^2 + 1000$, which gives the height (in feet) of an object that has fallen for t seconds from a height of 1000 feet. The velocity at time $t = a$ seconds is given by

$$\lim_{t \to a} \frac{s(a) - s(t)}{a - t}.$$

101. If a construction worker drops a wrench from a height of 1000 feet, how fast will the wrench be falling after 5 seconds?

102. If a construction worker drops a wrench from a height of 1000 feet, when will the wrench hit the ground? At what velocity will the wrench impact the ground?

Free-Falling Object   In Exercises 103 and 104, use the position function $s(t) = -4.9t^2 + 150$, which gives the height (in meters) of an object that has fallen from a height of 150 meters. The velocity at time $t = a$ seconds is given by

$$\lim_{t \to a} \frac{s(a) - s(t)}{a - t}.$$

103. Find the velocity of the object when $t = 3$.

104. At what velocity will the object impact the ground?

105. Find two functions f and g such that $\lim_{x \to 0} f(x)$ and $\lim_{x \to 0} g(x)$ do not exist, but $\lim_{x \to 0} [f(x) + g(x)]$ does exist.

106. Prove that if $\lim_{x \to c} f(x)$ exists and $\lim_{x \to c} [f(x) + g(x)]$ does not exist, then $\lim_{x \to c} g(x)$ does not exist.

107. Prove Property 1 of Theorem 1.1.

108. Prove Property 3 of Theorem 1.1. (You may use Property 3 of Theorem 1.2.)

109. Prove Property 1 of Theorem 1.2.

110. Prove that if $\lim_{x \to c} f(x) = 0$, then $\lim_{x \to c} |f(x)| = 0$.

111. Prove that if $\lim_{x \to c} f(x) = 0$ and $|g(x)| \le M$ for a fixed number M and all $x \ne c$, then $\lim_{x \to c} f(x)g(x) = 0$.

112. (a) Prove that if $\lim_{x \to c} |f(x)| = 0$, then $\lim_{x \to c} f(x) = 0$.
    (*Note:* This is the converse of Exercise 110.)

   (b) Prove that if $\lim_{x \to c} f(x) = L$, then $\lim_{x \to c} |f(x)| = |L|$.
    [*Hint:* Use the inequality $\|f(x)\| - |L\|\| \le |f(x) - L|.$]

True or False?   In Exercises 113–118, determine whether the statement is true or false. If it is false, explain why or give an example that shows it is false.

113. $\lim_{x \to 0} \dfrac{|x|}{x} = 1$

114. $\lim_{x \to 0} x^3 = 0$

115. If $f(x) = g(x)$ for all real numbers other than $x = 0$, and

$$\lim_{x \to 0} f(x) = L$$

then

$$\lim_{x \to 0} g(x) = L.$$

116. If $\lim_{x \to c} f(x) = L$, then $f(c) = L$.

117. $\lim_{x \to 2} f(x) = 3$, where $f(x) = \begin{cases} 3, & x \le 2 \\ 0, & x > 2 \end{cases}$

118. If $f(x) < g(x)$ for all $x \ne a$, then

$$\lim_{x \to a} f(x) < \lim_{x \to a} g(x).$$

119. *Think About It*   Find a function f to show that the converse of Exercise 112(b) is not true. [*Hint:* Find a function f such that $\lim_{x \to c} |f(x)| = |L|$ but $\lim_{x \to c} f(x)$ does not exist.]

120. Prove the second part of Theorem 1.9 by proving that

$$\lim_{x \to 0} \frac{1 - \cos x}{x} = 0.$$

121. Let $f(x) = \begin{cases} 0, & \text{if } x \text{ is rational} \\ 1, & \text{if } x \text{ is irrational} \end{cases}$

and

$$g(x) = \begin{cases} 0, & \text{if } x \text{ is rational} \\ x, & \text{if } x \text{ is irrational.} \end{cases}$$

Find (if possible) $\lim_{x \to 0} f(x)$ and $\lim_{x \to 0} g(x)$.

122. *Graphical Reasoning*   Consider $f(x) = \dfrac{\sec x - 1}{x^2}$.

   (a) Find the domain of f.

   (b) Use a graphing utility to graph f. Is the domain of f obvious from the graph? If not, explain.

   (c) Use the graph of f to approximate $\lim_{x \to 0} f(x)$.

   (d) Confirm the answer in part (c) analytically.

123. *Approximation*

   (a) Find $\lim_{x \to 0} \dfrac{1 - \cos x}{x^2}$.

   (b) Use the result in part (a) to derive the approximation $\cos x \approx 1 - \frac{1}{2}x^2$ for x near 0.

   (c) Use the result in part (b) to approximate $\cos(0.1)$.

   (d) Use a calculator to approximate $\cos(0.1)$ to four decimal places. Compare the result with part (c).

124. *Think About It*   When using a graphing utility to generate a table to approximate $\lim_{x \to 0} [(\sin x)/x]$, a student concluded that the limit was 0.01745 rather than 1. Determine the probable cause of the error.

| Section 1.4 | **Continuity and One-Sided Limits** |

- Determine continuity at a point and continuity on an open interval.
- Determine one-sided limits and continuity on a closed interval.
- Use properties of continuity.
- Understand and use the Intermediate Value Theorem.

Continuity at a Point and on an Open Interval

In mathematics, the term *continuous* has much the same meaning as it has in everyday usage. To say that a function f is continuous at $x = c$ means that there is no interruption in the graph of f at c. That is, its graph is unbroken at c and there are no holes, jumps, or gaps. Figure 1.25 identifies three values of x at which the graph of f is *not* continuous. At all other points in the interval (a, b), the graph of f is uninterrupted and **continuous.**

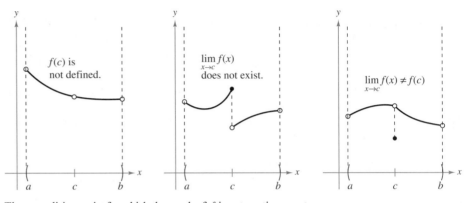

Three conditions exist for which the graph of f is not continuous at $x = c$.
Figure 1.25

In Figure 1.25, it appears that continuity at $x = c$ can be destroyed by any one of the following conditions.

1. The function is not defined at $x = c$.
2. The limit of $f(x)$ does not exist at $x = c$.
3. The limit of $f(x)$ exists at $x = c$, but it is not equal to $f(c)$.

If *none* of the above three conditions is true, the function f is called **continuous at c,** as indicated in the following important definition.

EXPLORATION

Informally, you might say that a function is *continuous* on an open interval if its graph can be drawn with a pencil without lifting the pencil from the paper. Use a graphing calculator to graph each of the following functions on the indicated interval. From the graphs, which functions would you say are continuous on the interval? Do you think you can trust the results you obtained graphically? Explain your reasoning.

Function	Interval
a. $y = x^2 + 1$	$(-3, 3)$
b. $y = \dfrac{1}{x - 2}$	$(-3, 3)$
c. $y = \dfrac{\sin x}{x}$	$(-\pi, \pi)$
d. $y = \dfrac{x^2 - 4}{x + 2}$	$(-3, 3)$
e. $y = \begin{cases} 2x - 4, & x \le 0 \\ x + 1, & x > 0 \end{cases}$	$(-3, 3)$

FOR FURTHER INFORMATION For more information on the concept of continuity, see the article "Leibniz and the Spell of the Continuous" by Hardy Grant in *The College Mathematics Journal.* To view this article, go to the website *www.matharticles.com.*

Definition of Continuity

Continuity at a Point: A function f is **continuous at c** if the following three conditions are met.

1. $f(c)$ is defined.
2. $\lim\limits_{x \to c} f(x)$ exists.
3. $\lim\limits_{x \to c} f(x) = f(c)$.

Continuity on an Open Interval: A function is **continuous on an open interval (a, b)** if it is continuous at each point in the interval. A function that is continuous on the entire real line $(-\infty, \infty)$ is **everywhere continuous.**

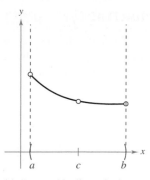

(a) Removable discontinuity

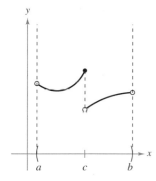

(b) Nonremovable discontinuity

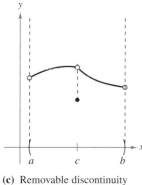

(c) Removable discontinuity
Figure 1.26

Consider an open interval I that contains a real number c. If a function f is defined on I (except possibly at c), and f is not continuous at c, then f is said to have a **discontinuity** at c. Discontinuities fall into two categories: **removable** and **nonremovable**. A discontinuity at c is called removable if f can be made continuous by appropriately defining (or redefining) $f(c)$. For instance, the functions shown in Figure 1.26(a) and (c) have removable discontinuities at c and the function shown in Figure 1.26(b) has a nonremovable discontinuity at c.

Example 1   **Continuity of a Function**

Discuss the continuity of each function.

a. $f(x) = \dfrac{1}{x}$ **b.** $g(x) = \dfrac{x^2 - 1}{x - 1}$ **c.** $h(x) = \begin{cases} x + 1, & x \le 0 \\ x^2 + 1, & x > 0 \end{cases}$ **d.** $y = \sin x$

Solution

a. The domain of f is all nonzero real numbers. From Theorem 1.3, you can conclude that f is continuous at every x-value in its domain. At $x = 0$, f has a nonremovable discontinuity, as shown in Figure 1.27(a). In other words, there is no way to define $f(0)$ so as to make the function continuous at $x = 0$.

b. The domain of g is all real numbers except $x = 1$. From Theorem 1.3, you can conclude that g is continuous at every x-value in its domain. At $x = 1$, the function has a removable discontinuity, as shown in Figure 1.27(b). If $g(1)$ is defined as 2, the "newly defined" function is continuous for all real numbers.

c. The domain of h is all real numbers. The function h is continuous on $(-\infty, 0)$ and $(0, \infty)$, and, because $\lim\limits_{x \to 0} h(x) = 1$, h is continuous on the entire real line, as shown in Figure 1.27(c).

d. The domain of y is all real numbers. From Theorem 1.6, you can conclude that the function is continuous on its entire domain, $(-\infty, \infty)$, as shown in Figure 1.27(d).

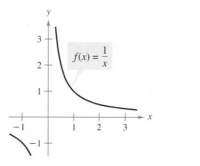

(a) Nonremovable discontinuity at $x = 0$

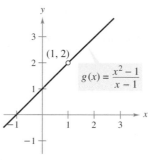

(b) Removable discontinuity at $x = 1$

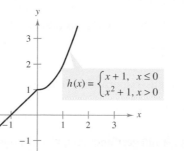

(c) Continuous on entire real line
Figure 1.27

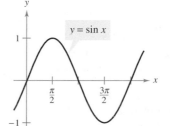

(d) Continuous on entire real line

STUDY TIP   Some people may refer to the function in Example 1a as "discontinuous." We have found that this terminology can be confusing. Rather than saying the function is discontinuous, we prefer to say that it has a discontinuity at $x = 0$.

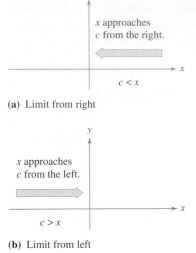

(a) Limit from right

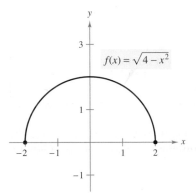

(b) Limit from left
Figure 1.28

One-Sided Limits and Continuity on a Closed Interval

To understand continuity on a closed interval, you first need to look at a different type of limit called a **one-sided limit.** For example, the **limit from the right** means that x approaches c from values greater than c [see Figure 1.28(a)]. This limit is denoted as

$$\lim_{x \to c^+} f(x) = L. \qquad \text{Limit from the right}$$

Similarly, the **limit from the left** means that x approaches c from values less than c [see Figure 1.28(b)]. This limit is denoted as

$$\lim_{x \to c^-} f(x) = L. \qquad \text{Limit from the left}$$

One-sided limits are useful in taking limits of functions involving radicals. For instance, if n is an even integer,

$$\lim_{x \to 0^+} \sqrt[n]{x} = 0.$$

Example 2   A One-Sided Limit

Find the limit of $f(x) = \sqrt{4 - x^2}$ as x approaches -2 from the right.

Solution   As indicated in Figure 1.29, the limit as x approaches -2 from the right is

$$\lim_{x \to -2^+} \sqrt{4 - x^2} = 0.$$

One-sided limits can be used to investigate the behavior of **step functions.** One common type of step function is the **greatest integer function** $[\![x]\!]$, defined by

$$[\![x]\!] = \text{greatest integer } n \text{ such that } n \leq x. \qquad \text{Greatest integer function}$$

For instance, $[\![2.5]\!] = 2$ and $[\![-2.5]\!] = -3$.

Example 3   The Greatest Integer Function

Find the limit of the greatest integer function $f(x) = [\![x]\!]$ as x approaches 0 from the left and from the right.

Solution   As shown in Figure 1.30, the limit as x approaches 0 *from the left* is given by

$$\lim_{x \to 0^-} [\![x]\!] = -1$$

and the limit as x approaches 0 *from the right* is given by

$$\lim_{x \to 0^+} [\![x]\!] = 0.$$

The greatest integer function has a discontinuity at 0 because the left and right limits at zero are different. By similar reasoning, you can see that the greatest integer function has a discontinuity at any integer n.

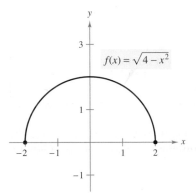

The limit of $f(x)$ as x approaches -2 from the right is 0.
Figure 1.29

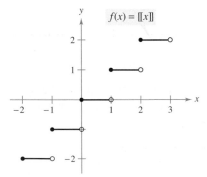

Greatest integer function
Figure 1.30

When the limit from the left is not equal to the limit from the right, the (two-sided) limit *does not exist*. The next theorem makes this more explicit. The proof of this theorem follows directly from the definition of a one-sided limit.

THEOREM 1.10   The Existence of a Limit

Let *f* be a function and let *c* and *L* be real numbers. The limit of $f(x)$ as *x* approaches *c* is *L* if and only if

$$\lim_{x \to c^-} f(x) = L \quad \text{and} \quad \lim_{x \to c^+} f(x) = L.$$

The concept of a one-sided limit allows you to extend the definition of continuity to closed intervals. Basically, a function is continuous on a closed interval if it is continuous in the interior of the interval and possesses one-sided continuity at the endpoints. We state this formally as follows.

Definition of Continuity on a Closed Interval

A function *f* is **continuous on the closed interval** $[a, b]$ if it is continuous on the open interval (a, b) and

$$\lim_{x \to a^+} f(x) = f(a) \quad \text{and} \quad \lim_{x \to b^-} f(x) - f(b).$$

The function *f* is **continuous from the right** at *a* and **continuous from the left** at *b* (see Figure 1.31).

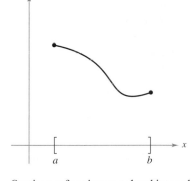

Continuous function on a closed interval
Figure 1.31

Similar definitions can be made to cover continuity on intervals of the form $(a, b]$ and $[a, b)$ that are neither open nor closed, or on infinite intervals. For example, the function

$$f(x) = \sqrt{x}$$

is continuous on the infinite interval $[0, \infty)$, and the function

$$g(x) = \sqrt{2 - x}$$

is continuous on the infinite interval $(-\infty, 2]$.

Example 4   **Continuity on a Closed Interval**

Discuss the continuity of $f(x) = \sqrt{1 - x^2}$.

Solution   The domain of *f* is the closed interval $[-1, 1]$. At all points in the open interval $(-1, 1)$, the continuity of *f* follows from Theorems 1.4 and 1.5. Moreover, because

$$\lim_{x \to -1^+} \sqrt{1 - x^2} = 0 = f(-1) \qquad \text{Continuous from the right}$$

and

$$\lim_{x \to 1^-} \sqrt{1 - x^2} = 0 = f(1) \qquad \text{Continuous from the left}$$

you can conclude that *f* is continuous on the closed interval $[-1, 1]$, as shown in Figure 1.32.

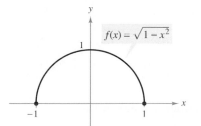

f is continuous on $[-1, 1]$.
Figure 1.32

The next example shows how a one-sided limit can be used to determine the value of absolute zero on the Kelvin scale.

Example 5   Charles's Law and Absolute Zero

On the Kelvin scale, *absolute zero* is the temperature 0 K. Although temperatures of approximately 0.0001 K have been produced in laboratories, absolute zero has never been attained. In fact, evidence suggests that absolute zero *cannot* be attained. How did scientists determine that 0 K is the "lower limit" of the temperature of matter? What is absolute zero on the Celsius scale?

Solution   The determination of absolute zero stems from the work of the French physicist Jacques Charles (1746–1823). Charles discovered that the volume of gas at a constant pressure increases linearly with the temperature of the gas. The table illustrates this relationship between volume and temperature. In the table, one mole of hydrogen is held at a constant pressure of one atmosphere. The volume V is measured in liters and the temperature T is measured in degrees Celsius.

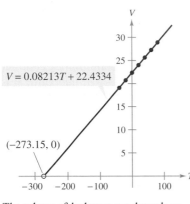

$V = 0.08213T + 22.4334$

$(-273.15, 0)$

The volume of hydrogen gas depends on its temperature.
Figure 1.33

T	-40	-20	0	20	40	60	80
V	19.1482	20.7908	22.4334	24.0760	25.7186	27.3612	29.0038

The points represented by the table are shown in Figure 1.33. Moreover, by using the points in the table, you can determine that T and V are related by the linear equation

$$V = 0.08213T + 22.4334 \qquad \text{or} \qquad T = \frac{V - 22.4334}{0.08213}.$$

By reasoning that the volume of the gas can approach 0 (but never equal or go below 0) you can determine that the "least possible temperature" is given by

$$\lim_{V \to 0^+} T = \lim_{V \to 0^+} \frac{V - 22.4334}{0.08213}$$

$$= \frac{0 - 22.4334}{0.08213} \qquad \text{Use direct substitution.}$$

$$\approx -273.15.$$

So, absolute zero on the Kelvin scale (0 K) is approximately $-273.15°$ on the Celsius scale.

The following table shows the temperatures in Example 5, converted to the Fahrenheit scale. Try repeating the solution shown in Example 5 using these temperatures and volumes. Use the result to find the value of absolute zero on the Fahrenheit scale.

T	-40	-4	32	68	104	140	176
V	19.1482	20.7908	22.4334	24.0760	25.7186	27.3612	29.0038

NOTE   Charles's Law for gases (assuming constant pressure) can be stated as

$$V = RT \qquad \qquad \text{Charles's Law}$$

where V is volume, R is constant, and T is temperature. In the statement of this law, what property must the temperature scale have?

In 1995, physicists Carl Wieman and Eric Cornell of the University of Colorado at Boulder used lasers and evaporation to produce a supercold gas in which atoms overlap. This gas is called a Bose-Einstein condensate. "We get to within a billionth of a degree of absolute zero," reported Wieman.   *(Source:* Time magazine, April 10, 2000)*

Properties of Continuity

In Section 1.3, you studied several properties of limits. Each of those properties yields a corresponding property pertaining to the continuity of a function. For instance, Theorem 1.11 follows directly from Theorem 1.2.

THEOREM 1.11   Properties of Continuity

If b is a real number and f and g are continuous at $x = c$, then the following functions are also continuous at c.

1. Scalar multiple: bf 2. Sum and difference: $f \pm g$

3. Product: fg 4. Quotient: $\dfrac{f}{g}$,   if $g(c) \neq 0$

AUGUSTIN-LOUIS CAUCHY (1789–1857)

The concept of a continuous function was first introduced by Augustin-Louis Cauchy in 1821. The definition given in his text *Cours d'Analyse* stated that indefinite small changes in y were the result of indefinite small changes in x. "...$f(x)$ will be called a *continuous* function if ... the numerical values of the difference $f(x + \alpha) - f(x)$ decrease indefinitely with those of α ...."

The following types of functions are continuous at every point in their domains.

1. Polynomial functions: $p(x) = a_n x^n + a_{n-1} x^{n-1} + \cdots + a_1 x + a_0$

2. Rational functions: $r(x) = \dfrac{p(x)}{q(x)}, \quad q(x) \neq 0$

3. Radical functions: $f(x) = \sqrt[n]{x}$

4. Trigonometric functions: $\sin x, \ \cos x, \ \tan x, \ \cot x, \ \sec x, \ \csc x$

By combining Theorem 1.11 with this summary, you can conclude that a wide variety of elementary functions are continuous at every point in their domains.

Example 6   **Applying Properties of Continuity**

By Theorem 1.11, it follows that each of the following functions is continuous at every point in its domain.

$$f(x) = x + \sin x, \quad f(x) = 3 \tan x, \quad f(x) = \frac{x^2 + 1}{\cos x}$$

The next theorem, which is a consequence of Theorem 1.5, allows you to determine the continuity of *composite* functions such as

$$f(x) = \sin 3x, \quad f(x) = \sqrt{x^2 + 1}, \quad f(x) = \tan \frac{1}{x}.$$

THEOREM 1.12   Continuity of a Composite Function

If g is continuous at c and f is continuous at $g(c)$, then the composite function given by $(f \circ g)(x) = f(g(x))$ is continuous at c.

One consequence of Theorem 1.12 is that if f and g satisfy the given conditions, you can determine the limit of $f(g(x))$ as x approaches c to be

$$\lim_{x \to c} f(g(x)) = f(g(c)).$$

Example 7 **Testing for Continuity**

Describe the interval(s) on which each function is continuous.

a. $f(x) = \tan x$ **b.** $g(x) = \begin{cases} \sin\dfrac{1}{x}, & x \neq 0 \\ 0, & x = 0 \end{cases}$ **c.** $h(x) = \begin{cases} x \sin\dfrac{1}{x}, & x \neq 0 \\ 0, & x = 0 \end{cases}$

Solution

a. The tangent function $f(x) = \tan x$ is undefined at

$$x = \frac{\pi}{2} + n\pi, \qquad n \text{ is an integer.}$$

At all other points it is continuous. So, $f(x) = \tan x$ is continuous on the open intervals

$$\ldots, \left(-\frac{3\pi}{2}, -\frac{\pi}{2}\right), \left(-\frac{\pi}{2}, \frac{\pi}{2}\right), \left(\frac{\pi}{2}, \frac{3\pi}{2}\right), \ldots$$

as shown in Figure 1.34(a).

b. Because $y = 1/x$ is continuous except at $x = 0$ and the sine function is continuous for all real values of x, it follows that $y = \sin(1/x)$ is continuous at all real values except $x = 0$. At $x = 0$, the limit of $g(x)$ does not exist (see Example 5, Section 1.2). So, g is continuous on the intervals $(-\infty, 0)$ and $(0, \infty)$, as indicated in Figure 1.34(b).

c. This function is similar to that in part (b) except that the oscillations are damped by the factor x. Using the Squeeze Theorem, you obtain

$$-|x| \leq x \sin\frac{1}{x} \leq |x|, \qquad x \neq 0$$

and you can conclude that

$$\lim_{x \to 0} h(x) = 0.$$

So, h is continuous on the entire real line, as indicated in Figure 1.34(c).

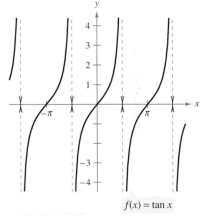

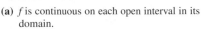

(a) f is continuous on each open interval in its domain.

Figure 1.34

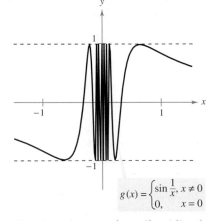

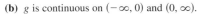

(b) g is continuous on $(-\infty, 0)$ and $(0, \infty)$.

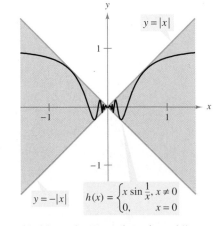

(c) h is continuous on the entire real line.

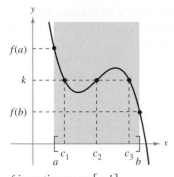

f is continuous on $[a, b]$.
[There exist three c's such that $f(c) = k$.]
Figure 1.35

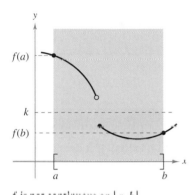

f is not continuous on $[a, b]$.
[There are no c's such that $f(c) = k$.]
Figure 1.36

The Intermediate Value Theorem

We conclude this section with an important theorem concerning the behavior of functions that are continuous on a closed interval.

THEOREM 1.13   Intermediate Value Theorem

If f is continuous on the closed interval $[a, b]$ and k is any number between $f(a)$ and $f(b)$, then there is at least one number c in $[a, b]$ such that $f(c) = k$.

NOTE   The Intermediate Value Theorem tells you that at least one c exists, but it does not give a method for finding c. Such theorems are called **existence theorems.**

By referring to a text on advanced calculus, you will find that a proof of this theorem is based on a property of real numbers called *completeness*. The Intermediate Value Theorem states that for a continuous function f, if x takes on all values between a and b, $f(x)$ must take on all values between $f(a)$ and $f(b)$.

As a simple example of this theorem, consider a person's height. Suppose that a girl is 5 feet tall on her thirteenth birthday and 5 feet 7 inches tall on her fourteenth birthday. Then, for any height h between 5 feet and 5 feet 7 inches, there must have been a time t when her height was exactly h. This seems reasonable because human growth is continuous and a person's height does not abruptly change from one value to another.

The Intermediate Value Theorem guarantees the existence of *at least one* number c in the closed interval $[a, b]$. There may, of course, be more than one number c such that $f(c) = k$, as shown in Figure 1.35. A function that is not continuous does not necessarily possess the intermediate value property. For example, the graph of the function shown in Figure 1.36 jumps over the horizontal line given by $y = k$, and for this function there is no value of c in $[a, b]$ such that $f(c) = k$.

The Intermediate Value Theorem often can be used to locate the zeros of a function that is continuous on a closed interval. Specifically, if f is continuous on $[a, b]$ and $f(a)$ and $f(b)$ differ in sign, the Intermediate Value Theorem guarantees the existence of at least one zero of f in the closed interval $[a, b]$.

Example 8   **An Application of the Intermediate Value Theorem**

Use the Intermediate Value Theorem to show that the polynomial function

$$f(x) = x^3 + 2x - 1$$

has a zero in the interval $[0, 1]$.

Solution   Note that f is continuous on the closed interval $[0, 1]$. Because

$$f(0) = 0^3 + 2(0) - 1 = -1$$

and

$$f(1) = 1^3 + 2(1) - 1 = 2$$

it follows that $f(0) < 0$ and $f(1) > 0$. You can therefore apply the Intermediate Value Theorem to conclude that there must be some c in $[0, 1]$ such that

$$f(c) = 0 \qquad \text{\small f has a zero in the closed interval $[0, 1]$.}$$

as shown in Figure 1.37.

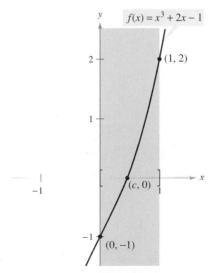

$f(x) = x^3 + 2x - 1$

f is continuous on $[0, 1]$ with $f(0) < 0$ and $f(1) > 0$.
Figure 1.37

The **bisection method** for approximating the real zeros of a continuous function is similar to the method used in Example 8. If you know that a zero exists in the closed interval $[a, b]$, the zero must lie in the interval $[a, (a + b)/2]$ or $[(a + b)/2, b]$. From the sign of $f([a + b]/2)$, you can determine which interval contains the zero. By repeatedly bisecting the interval, you can "close in" on the zero of the function.

TECHNOLOGY   You can also use the *zoom* feature of a graphing utility to approximate the real zeros of a continuous function. By repeatedly zooming in on the point where the graph crosses the *x*-axis, and adjusting the *x*-axis scale, you can approximate the zero of the function to any desired accuracy. The zero of $x^3 + 2x - 1$ is approximately 0.453, as shown in Figure 1.38.

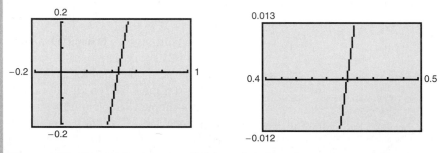

Figure 1.38   Zooming in on the zero of $f(x) = x^3 + 2x - 1$

EXERCISES FOR SECTION 1.4

In Exercises 1–6, use the graph to determine the limit, and discuss the continuity of the function.

(a) $\lim\limits_{x \to c^+} f(x)$ **(b)** $\lim\limits_{x \to c^-} f(x)$ **(c)** $\lim\limits_{x \to c} f(x)$

1.
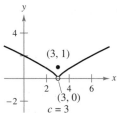

2.
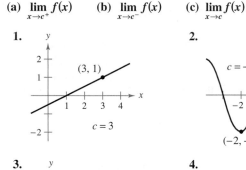

3.
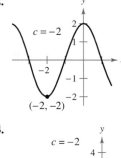

4.
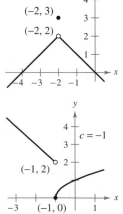

5.
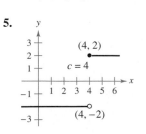

6.

In Exercises 7–24, find the limit (if it exists). If it does not exist, explain why.

7. $\lim\limits_{x \to 5^+} \dfrac{x - 5}{x^2 - 25}$

8. $\lim\limits_{x \to 2^+} \dfrac{2 - x}{x^2 - 4}$

9. $\lim\limits_{x \to -3^-} \dfrac{x}{\sqrt{x^2 - 9}}$

10. $\lim\limits_{x \to 4^-} \dfrac{\sqrt{x} - 2}{x - 4}$

11. $\lim\limits_{x \to 0^-} \dfrac{|x|}{x}$

12. $\lim\limits_{x \to 2^+} \dfrac{|x - 2|}{x - 2}$

13. $\lim\limits_{\Delta x \to 0^-} \dfrac{\dfrac{1}{x + \Delta x} - \dfrac{1}{x}}{\Delta x}$

14. $\lim\limits_{\Delta x \to 0^+} \dfrac{(x + \Delta x)^2 + x + \Delta x - (x^2 + x)}{\Delta x}$

15. $\lim\limits_{x \to 3^-} f(x)$, where $f(x) = \begin{cases} \dfrac{x + 2}{2}, & x \leq 3 \\ \dfrac{12 - 2x}{3}, & x > 3 \end{cases}$

16. $\lim\limits_{x \to 2} f(x)$, where $f(x) = \begin{cases} x^2 - 4x + 6, & x < 2 \\ -x^2 + 4x - 2, & x \geq 2 \end{cases}$

17. $\lim\limits_{x \to 1} f(x)$, where $f(x) = \begin{cases} x^3 + 1, & x < 1 \\ x + 1, & x \geq 1 \end{cases}$

18. $\lim\limits_{x \to 1^+} f(x)$, where $f(x) = \begin{cases} x, & x \leq 1 \\ 1 - x, & x > 1 \end{cases}$

19. $\lim\limits_{x \to \pi} \cot x$

20. $\lim\limits_{x \to \pi/2} \sec x$

21. $\lim\limits_{x \to 4^-} (3[\![x]\!] - 5)$

22. $\lim\limits_{x \to 2^+} (2x - [\![x]\!])$

23. $\lim\limits_{x \to 3} (2 - [\![-x]\!])$

24. $\lim\limits_{x \to 1} \left(1 - \left[\!\left[-\dfrac{x}{2}\right]\!\right]\right)$

In Exercises 25–28, discuss the continuity of each function.

25. $f(x) = \dfrac{1}{x^2 - 4}$

26. $f(x) = \dfrac{x^2 - 1}{x + 1}$

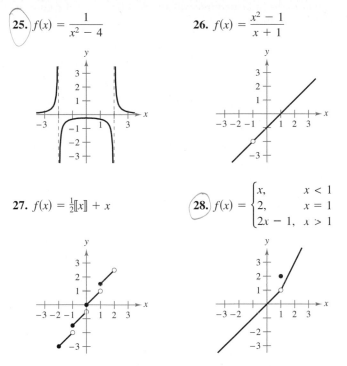

27. $f(x) = \frac{1}{2}[\![x]\!] + x$

28. $f(x) = \begin{cases} x, & x < 1 \\ 2, & x = 1 \\ 2x - 1, & x > 1 \end{cases}$

In Exercises 29–32, discuss the continuity of the function on the closed interval.

29. $g(x) = \sqrt{25 - x^2}, \quad [-5, 5]$

30. $f(t) = 3 - \sqrt{9 - t^2}, \quad [-3, 3]$

31. $f(x) = \begin{cases} 3 - x, & x \le 0 \\ 3 + \frac{1}{2}x, & x > 0 \end{cases} \quad [-1, 4]$

32. $g(x) = \dfrac{1}{x^2 - 4}, \quad [-1, 2]$

In Exercises 33–54, find the x-values (if any) at which f is not continuous. Which of the discontinuities are removable?

33. $f(x) = x^2 - 2x + 1$

34. $f(x) = \dfrac{1}{x^2 + 1}$

35. $f(x) = 3x - \cos x$

36. $f(x) = \cos \dfrac{\pi x}{2}$

37. $f(x) = \dfrac{x}{x^2 - x}$

38. $f(x) = \dfrac{x}{x^2 - 1}$

39. $f(x) = \dfrac{x}{x^2 + 1}$

40. $f(x) = \dfrac{x - 3}{x^2 - 9}$

41. $f(x) = \dfrac{x + 2}{x^2 - 3x - 10}$

42. $f(x) = \dfrac{x - 1}{x^2 + x - 2}$

43. $f(x) = \dfrac{|x + 2|}{x + 2}$

44. $f(x) = \dfrac{|x - 3|}{x - 3}$

45. $f(x) = \begin{cases} x, & x \le 1 \\ x^2, & x > 1 \end{cases}$

46. $f(x) = \begin{cases} -2x + 3, & x < 1 \\ x^2, & x \ge 1 \end{cases}$

47. $f(x) = \begin{cases} \frac{1}{2}x + 1, & x \le 2 \\ 3 - x, & x > 2 \end{cases}$

48. $f(x) = \begin{cases} -2x, & x \le 2 \\ x^2 - 4x + 1, & x > 2 \end{cases}$

49. $f(x) = \begin{cases} \tan \dfrac{\pi x}{4}, & |x| < 1 \\ x, & |x| \ge 1 \end{cases}$

50. $f(x) = \begin{cases} \csc \dfrac{\pi x}{6}, & |x - 3| \le 2 \\ 2, & |x - 3| > 2 \end{cases}$

51. $f(x) = \csc 2x$

52. $f(x) = \tan \dfrac{\pi x}{2}$

53. $f(x) = [\![x - 1]\!]$

54. $f(x) = 3 - [\![x]\!]$

In Exercises 55 and 56, use a graphing utility to graph the function. From the graph, estimate

$$\lim_{x \to 0^+} f(x) \qquad \text{and} \qquad \lim_{x \to 0^-} f(x).$$

Is the function continuous on the entire real line? Explain.

55. $f(x) = \dfrac{|x^2 - 4|x}{x + 2}$

56. $f(x) = \dfrac{|x^2 + 4x|(x + 2)}{x + 4}$

In Exercises 57–60, find the constants a and b such that the function is continuous on the entire real line.

57. $f(x) = \begin{cases} x^3, & x \le 2 \\ ax^2, & x > 2 \end{cases}$

58. $g(x) = \begin{cases} \dfrac{4 \sin x}{x}, & x < 0 \\ a - 2x, & x \ge 0 \end{cases}$

59. $f(x) = \begin{cases} 2, & x \le -1 \\ ax + b, & -1 < x < 3 \\ -2, & x \ge 3 \end{cases}$

60. $g(x) = \begin{cases} \dfrac{x^2 - a^2}{x - a}, & x \ne a \\ 8, & x = a \end{cases}$

In Exercises 61–64, discuss the continuity of the composite function $h(x) = f(g(x))$.

61. $f(x) = x^2$
$g(x) = x - 1$

62. $f(x) = \dfrac{1}{\sqrt{x}}$
$g(x) = x - 1$

63. $f(x) = \dfrac{1}{x - 6}$
$g(x) = x^2 + 5$

64. $f(x) = \sin x$
$g(x) = x^2$

In Exercises 65–68, use a graphing utility to graph the function. Use the graph to determine any x-values at which the function is not continuous.

65. $f(x) = [\![x]\!] - x$

66. $h(x) = \dfrac{1}{x^2 - x - 2}$

67. $g(x) = \begin{cases} 2x - 4, & x \le 3 \\ x^2 - 2x, & x > 3 \end{cases}$

68. $f(x) = \begin{cases} \dfrac{\cos x - 1}{x}, & x < 0 \\ 5x, & x \ge 0 \end{cases}$

In Exercises 69–72, describe the interval(s) on which the function is continuous.

69. $f(x) = \dfrac{x}{x^2 + 1}$

70. $f(x) = x\sqrt{x + 3}$

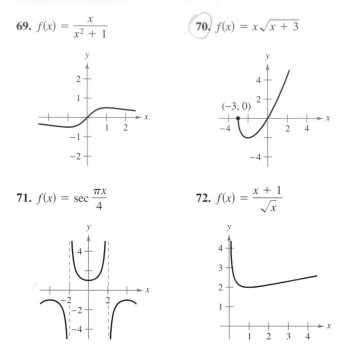

71. $f(x) = \sec \dfrac{\pi x}{4}$

72. $f(x) = \dfrac{x + 1}{\sqrt{x}}$

Writing In Exercises 73 and 74, use a graphing utility to graph the function on the interval $[-4, 4]$. Does the graph of the function appear continuous on this interval? Is the function continuous on $[-4, 4]$? Write a short paragraph about the importance of examining a function analytically as well as graphically.

73. $f(x) = \dfrac{\sin x}{x}$

74. $f(x) = \dfrac{x^3 - 8}{x - 2}$

Writing In Exercises 75–78, explain why the function has a zero in the specified interval.

75. $f(x) = \frac{1}{16}x^4 - x^3 + 3$, $[1, 2]$

76. $f(x) = x^3 + 3x - 2$, $[0, 1]$

77. $f(x) = x^2 - 2 - \cos x$, $[0, \pi]$

78. $f(x) = -\dfrac{4}{x} + \tan\left(\dfrac{\pi x}{8}\right)$, $[1, 3]$

In Exercises 79–82, use the Intermediate Value Theorem and a graphing utility to approximate the zero of the function in the interval $[0, 1]$. Repeatedly "zoom in" on the graph of the function to approximate the zero accurate to two decimal places. Use the root-finding capabilities of the graphing utility to approximate the zero accurate to four decimal places.

79. $f(x) = x^3 + x - 1$

80. $f(x) = x^3 + 3x - 2$

81. $g(t) = 2 \cos t - 3t$

82. $h(\theta) = 1 + \theta - 3 \tan \theta$

In Exercises 83–86, verify that the Intermediate Value Theorem applies to the indicated interval and find the value of c guaranteed by the theorem.

83. $f(x) = x^2 + x - 1$, $[0, 5]$, $f(c) = 11$

84. $f(x) = x^2 - 6x + 8$, $[0, 3]$, $f(c) = 0$

85. $f(x) = x^3 - x^2 + x - 2$, $[0, 3]$, $f(c) = 4$

86. $f(x) = \dfrac{x^2 + x}{x - 1}$, $\left[\dfrac{5}{2}, 4\right]$, $f(c) = 6$

Getting at the Concept

87. State how continuity is destroyed at $x = c$ for each of the following.

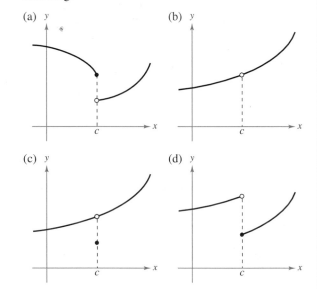

(a)   (b)   (c)   (d)

88. Describe the difference between a discontinuity that is removable and one that is nonremovable. In your explanation, give examples of the following.

   (a) A function with a nonremovable discontinuity at $x = 2$.

   (b) A function with a removable discontinuity at $x = -2$.

   (c) A function that has both of the characteristics described in parts (a) and (b).

89. Sketch the graph of any function f such that

   $$\lim_{x \to 3^+} f(x) = 1 \quad \text{and} \quad \lim_{x \to 3^-} f(x) = 0.$$

   Is the function continuous at $x = 3$? Explain.

90. If the functions f and g are continuous for all real x, is $f + g$ always continuous for all real x? Is f/g always continuous for all real x? If either is not continuous, give an example to verify your conclusion.

91. **Think About It** Describe how the functions $f(x) = 3 + [\![x]\!]$ and $g(x) = 3 - [\![-x]\!]$ differ.

92. Telephone Charges A dial-direct long distance call between two cities costs $1.04 for the first 2 minutes and $0.36 for each additional minute or fraction thereof. Use the greatest integer function to write the cost C of a call in terms of the time t (in minutes). Sketch a graph of this function and discuss its continuity.

93. Inventory Management The number of units in inventory in a small company is given by

$$N(t) = 25\left(2\left[\!\left[\frac{t+2}{2}\right]\!\right] - t\right)$$

where t is the time in months. Sketch the graph of this function and discuss its continuity. How often must this company replenish its inventory?

94. Déjà Vu At 8:00 A.M. on Saturday a man begins running up the side of a mountain to his weekend campsite (see figure). On Sunday morning at 8:00 A.M. he runs back down the mountain. It takes him 20 minutes to run up, but only 10 minutes to run down. At some point on the way down, he realizes that he passed the same place at exactly the same time on Saturday. Prove that he is correct. [*Hint:* Let $s(t)$ and $r(t)$ be the position functions for the runs up and down, and apply the Intermediate Value Theorem to the function $f(t) = s(t) - r(t)$.]

Saturday 8:00 A.M.   Sunday 8:00 A.M.
Not drawn to scale

95. Volume Use the Intermediate Value Theorem to show that for all spheres with radii in the interval $[1, 5]$, there is one with a volume of 275 cubic centimeters.

96. Prove that if f is continuous and has no zeros on $[a, b]$, then either

$f(x) > 0$ for all x in $[a, b]$ or $f(x) < 0$ for all x in $[a, b]$.

97. Show that the Dirichlet function

$$f(x) = \begin{cases} 0, & \text{if } x \text{ is rational} \\ 1, & \text{if } x \text{ is irrational} \end{cases}$$

is not continuous at any real number.

98. Show that the function

$$f(x) = \begin{cases} 0, & \text{if } x \text{ is rational} \\ kx, & \text{if } x \text{ is irrational} \end{cases}$$

is continuous only at $x = 0$. (Assume that k is any nonzero real number.)

99. The **signum function** is defined by

$$\text{sgn}(x) = \begin{cases} -1, & x < 0 \\ 0, & x = 0 \\ 1, & x > 0. \end{cases}$$

Sketch a graph of $\text{sgn}(x)$ and find the following (if possible).

(a) $\lim\limits_{x \to 0^-} \text{sgn}(x)$ (b) $\lim\limits_{x \to 0^+} \text{sgn}(x)$ (c) $\lim\limits_{x \to 0} \text{sgn}(x)$

True or False? In Exercises 100–103, determine whether the statement is true or false. If it is false, explain why or give an example that shows it is false.

100. If $\lim\limits_{x \to c} f(x) = L$ and $f(c) = L$, then f is continuous at c.

101. If $f(x) = g(x)$ for $x \neq c$ and $f(c) \neq g(c)$, then either f or g is not continuous at c.

102. A rational function can have infinitely many x-values at which it is not continuous.

103. The function $f(x) = |x - 1|/(x - 1)$ is continuous on $(-\infty, \infty)$.

104. Modeling Data After an object falls for t seconds, the speed S (in feet per second) of the object is recorded in the table.

t	0	5	10	15	20	25	30
S	0	48.2	53.5	55.2	55.9	56.2	56.3

(a) Create a line graph of the data.

(b) Does there appear to be a limiting speed of the object? If there is a limiting speed, identify a possible cause.

105. Creating Models A swimmer crosses a pool of width b by swimming in a straight line from $(0, 0)$ to $(2b, b)$. (See figure.)

(a) Let f be a function defined as the y-coordinate of the point on the long side of the pool that is nearest the swimmer at any given time during the swimmer's path across the pool. Determine the function f and sketch its graph. Is it continuous? Explain.

(b) Let g be the minimum distance between the swimmer and the long sides of the pool. Determine the function g and sketch its graph. Is it continuous? Explain.

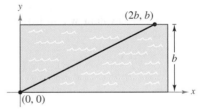

106. Prove that for any real number y there exists x in $(-\pi/2, \pi/2)$ such that $\tan x = y$.

107. Let $f(x) = \left(\sqrt{x + c^2} - c\right)/x, \ c > 0$. What is the domain of f? How can you define f at $x = 0$ in order for f to be continuous there?

108. Prove that if $\lim\limits_{\Delta x \to 0} f(c + \Delta x) = f(c)$, then f is continuous at c.

109. Discuss the continuity of the function $h(x) = x[\![x]\!]$.

110. Let $f_1(x)$ and $f_2(x)$ be continuous on the closed interval $[a, b]$. If $f_1(a) < f_2(a)$ and $f_1(b) > f_2(b)$, prove that there exists c between a and b such that $f_1(c) = f_2(c)$.

Section 1.5	**Infinite Limits**

- Determine infinite limits from the left and from the right.
- Find and sketch the vertical asymptotes of the graph of a function.

Infinite Limits

Let f be the function given by

$$f(x) = \frac{3}{x-2}.$$

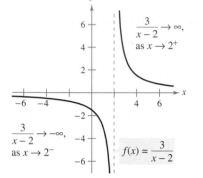

$f(x)$ increases and decreases without bound as x approaches 2.
Figure 1.39

From Figure 1.39 and the table, you can see that $f(x)$ *decreases without bound* as x approaches 2 from the left, and $f(x)$ *increases without bound* as x approaches 2 from the right. This behavior is denoted as

$$\lim_{x \to 2^-} \frac{3}{x-2} = -\infty \qquad f(x) \text{ decreases without bound as } x \text{ approaches 2 from the left.}$$

and

$$\lim_{x \to 2^+} \frac{3}{x-2} = \infty \qquad f(x) \text{ increases without bound as } x \text{ approaches 2 from the right.}$$

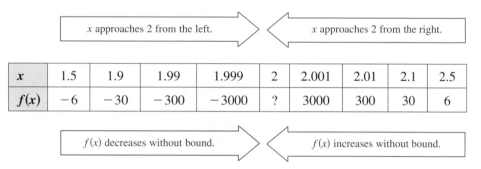

x	1.5	1.9	1.99	1.999	2	2.001	2.01	2.1	2.5
$f(x)$	-6	-30	-300	-3000	?	3000	300	30	6

A limit in which $f(x)$ increases or decreases without bound as x approaches c is called an **infinite limit.**

Definition of Infinite Limits

Let f be a function that is defined at every real number in some open interval containing c (except possibly at c itself). The statement

$$\lim_{x \to c} f(x) = \infty$$

means that for each $M > 0$ there exists a $\delta > 0$ such that $f(x) > M$ whenever $0 < |x - c| < \delta$ (see Figure 1.40). Similarly, the statement

$$\lim_{x \to c} f(x) = -\infty$$

means that for each $N < 0$ there exists a $\delta > 0$ such that $f(x) < N$ whenever $0 < |x - c| < \delta$. To define the **infinite limit from the left,** replace $0 < |x - c| < \delta$ by $c - \delta < x < c$. To define the **infinite limit from the right,** replace $0 < |x - c| < \delta$ by $c < x < c + \delta$.

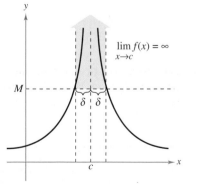

Infinite limits
Figure 1.40

Be sure you see that the equal sign in the statement $\lim f(x) = \infty$ does not mean that the limit exists! On the contrary, it tells you how the limit *fails to exist* by denoting the unbounded behavior of $f(x)$ as x approaches c.

EXPLORATION

Use a graphing utility to graph each function. For each function, analytically find the single real number c that is not in the domain. Then graphically find the limit of $f(x)$ as x approaches c from the left and from the right.

a. $f(x) = \dfrac{3}{x - 4}$ **b.** $f(x) = \dfrac{1}{2 - x}$

c. $f(x) = \dfrac{2}{(x - 3)^2}$ **d.** $f(x) = \dfrac{-3}{(x + 2)^2}$

Example 1   **Determining Infinite Limits from a Graph**

Use Figure 1.41 to determine the limit of each function as x approaches 1 from the left and from the right.

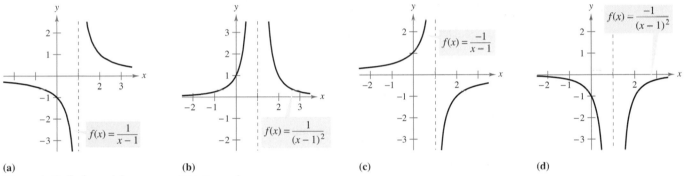

(a)   (b)   (c)   (d)

Figure 1.41 Each graph has an asymptote at $x = 1$.

Solution

a. $\displaystyle\lim_{x \to 1^-} \frac{1}{x - 1} = -\infty$ and $\displaystyle\lim_{x \to 1^+} \frac{1}{x - 1} = \infty$

b. $\displaystyle\lim_{x \to 1} \frac{1}{(x - 1)^2} = \infty$ Limit from each side is ∞.

c. $\displaystyle\lim_{x \to 1^-} \frac{-1}{x - 1} = \infty$ and $\displaystyle\lim_{x \to 1^+} \frac{-1}{x - 1} = -\infty$

d. $\displaystyle\lim_{x \to 1} \frac{-1}{(x - 1)^2} = -\infty$ Limit from each side is $-\infty$.

Vertical Asymptotes

If it were possible to extend the graphs in Figure 1.41 toward positive and negative infinity, you would see that each graph becomes arbitrarily close to the vertical line $x = 1$. This line is a **vertical asymptote** of the graph of f. (You will study other types of asymptotes in Sections 3.5 and 3.6.)

NOTE   If a function f has a vertical asymptote at $x = c$, then f is *not* continuous at c.

Definition of a Vertical Asymptote

If $f(x)$ approaches infinity (or negative infinity) as x approaches c from the right or the left, then the line $x = c$ is a **vertical asymptote** of the graph of f.

In Example 1, note that each of the functions is a *quotient* and that the vertical asymptote occurs at a number where the denominator is 0 (and the numerator is not 0). The next theorem generalizes this observation. (A proof of this theorem is given in Appendix B.)

THEOREM 1.14   Vertical Asymptotes

Let f and g be continuous on an open interval containing c. If $f(c) \neq 0$, $g(c) = 0$, and there exists an open interval containing c such that $g(x) \neq 0$ for all $x \neq c$ in the interval, then the graph of the function given by

$$h(x) = \frac{f(x)}{g(x)}$$

has a vertical asymptote at $x = c$.

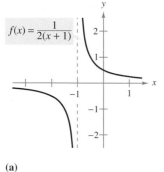

$f(x) = \dfrac{1}{2(x+1)}$

(a)

Example 2   **Finding Vertical Asymptotes**

Determine all vertical asymptotes of the graph of each function.

a. $f(x) = \dfrac{1}{2(x + 1)}$ **b.** $f(x) = \dfrac{x^2 + 1}{x^2 - 1}$ **c.** $f(x) = \cot x$

Solution

a. When $x = -1$, the denominator of

$$f(x) = \frac{1}{2(x + 1)}$$

is 0 and the numerator is not 0. Hence, by Theorem 1.14, you can conclude that $x = -1$ is a vertical asymptote, as shown in Figure 1.42(a).

b. By factoring the denominator as

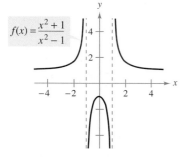

$f(x) = \dfrac{x^2+1}{x^2-1}$

(b)

$$f(x) = \frac{x^2 + 1}{x^2 - 1} = \frac{x^2 + 1}{(x - 1)(x + 1)}$$

you can see that the denominator is 0 at $x = -1$ and $x = 1$. Moreover, because the numerator is not 0 at these two points, you can apply Theorem 1.14 to conclude that the graph of f has two vertical asymptotes, as shown in Figure 1.42(b).

c. By writing the cotangent function in the form

$$f(x) = \cot x = \frac{\cos x}{\sin x}$$

you can apply Theorem 1.14 to conclude that vertical asymptotes occur at all values of x such that $\sin x = 0$ and $\cos x \neq 0$, as shown in Figure 1.42(c). Hence, the graph of this function has infinitely many vertical asymptotes. These asymptotes occur when $x = n\pi$, where n is an integer.

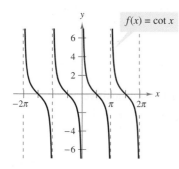

$f(x) = \cot x$

(c)

Functions with vertical asymptotes

Figure 1.42

Theorem 1.14 requires that the value of the numerator at $x = c$ be nonzero. If both the numerator and the denominator are 0 at $x = c$, you obtain the *indeterminate form* $0/0$, and you cannot determine the limit behavior at $x = c$ without further investigation, as illustrated in Example 3.

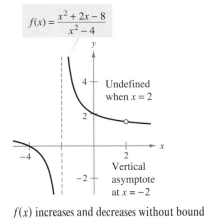

$$f(x) = \frac{x^2 + 2x - 8}{x^2 - 4}$$

$f(x)$ increases and decreases without bound as x approaches -2.
Figure 1.43

Example 3   A Rational Function with Common Factors

Determine all vertical asymptotes of the graph of

$$f(x) = \frac{x^2 + 2x - 8}{x^2 - 4}.$$

Solution   Begin by simplifying the expression, as follows.

$$f(x) = \frac{x^2 + 2x - 8}{x^2 - 4}$$

$$= \frac{(x + 4)(x - 2)}{(x + 2)(x - 2)}$$

$$= \frac{x + 4}{x + 2}, \quad x \neq 2$$

At all x-values other than $x = 2$, the graph of f coincides with the graph of $g(x) = (x + 4)/(x + 2)$. So, you can apply Theorem 1.14 to g to conclude that there is a vertical asymptote at $x = -2$, as shown in Figure 1.43. From the graph, you can see that

$$\lim_{x \to -2^-} \frac{x^2 + 2x - 8}{x^2 - 4} = -\infty \quad \text{and} \quad \lim_{x \to -2^+} \frac{x^2 + 2x - 8}{x^2 - 4} = \infty.$$

Note that $x = 2$ is *not* a vertical asymptote.

Example 4   Determining Infinite Limits

Find each limit.

$$\lim_{x \to 1^-} \frac{x^2 - 3x}{x - 1} \quad \text{and} \quad \lim_{x \to 1^+} \frac{x^2 - 3x}{x - 1}$$

Solution   Because the denominator is 0 when $x = 1$ (and the numerator is not zero), you know that the graph of

$$f(x) = \frac{x^2 - 3x}{x - 1}$$

has a vertical asymptote at $x = 1$. This means that each of the given limits is either ∞ or $-\infty$. A graphing utility can help determine the result. From the graph of f shown in Figure 1.44, you can see that the graph approaches ∞ from the left of $x = 1$ and approaches $-\infty$ from the right of $x = 1$. So, you can conclude that

$$\lim_{x \to 1^-} \frac{x^2 - 3x}{x - 1} = \infty \qquad \text{The limit from the left is infinity.}$$

and

$$\lim_{x \to 1^+} \frac{x^2 - 3x}{x - 1} = -\infty. \qquad \text{The limit from the right is negative infinity.}$$

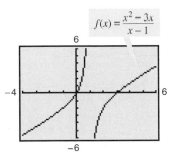

$$f(x) = \frac{x^2 - 3x}{x - 1}$$

f has a vertical asymptote at $x = 1$.
Figure 1.44

TECHNOLOGY PITFALL   When using a graphing calculator or graphing software, be careful to correctly interpret the graph of a function with a vertical asymptote—graphing utilities often have difficulty drawing this type of graph.

THEOREM 1.15   Properties of Infinite Limits

Let c and L be real numbers and let f and g be functions such that

$$\lim_{x \to c} f(x) = \infty \quad \text{and} \quad \lim_{x \to c} g(x) = L.$$

1. Sum or difference: $\lim\limits_{x \to c} [f(x) \pm g(x)] = \infty$

2. Product: $\lim\limits_{x \to c} [f(x)g(x)] = \infty, \quad L > 0$

$\lim\limits_{x \to c} [f(x)g(x)] = -\infty, \quad L < 0$

3. Quotient: $\lim\limits_{x \to c} \dfrac{g(x)}{f(x)} = 0$

Similar properties hold for one-sided limits and for functions for which the limit of $f(x)$ as x approaches c is $-\infty$.

Proof   To show that the limit of $f(x) + g(x)$ is infinite, choose $M > 0$. You then need to find $\delta > 0$ such that

$$[f(x) + g(x)] > M$$

whenever $0 < |x - c| < \delta$. For simplicity's sake, you can assume L is positive and let $M_1 = M + 1$. Because the limit of $f(x)$ is infinite, there exists δ_1 such that $f(x) > M_1$ whenever $0 < |x - c| < \delta_1$. Also, because the limit of $g(x)$ is L, there exists δ_2 such that $|g(x) - L| < 1$ whenever $0 < |x - c| < \delta_2$. By letting δ be the smaller of δ_1 and δ_2, you can conclude that $0 < |x - c| < \delta$ implies $f(x) > M + 1$ and $|g(x) - L| < 1$. The second of these two inequalities implies that $g(x) > L - 1$, and, adding this to the first inequality, you can write

$$f(x) + g(x) > (M + 1) + (L - 1) = M + L > M.$$

So, you can conclude that

$$\lim_{x \to c} [f(x) + g(x)] = \infty.$$

We leave the proofs of the remaining properties as exercises (see Exercise 73).

Example 5   **Determining Limits**

a. Because $\lim\limits_{x \to 0} 1 = 1$ and $\lim\limits_{x \to 0} \dfrac{1}{x^2} = \infty$, you can write

$$\lim_{x \to 0} \left(1 + \frac{1}{x^2}\right) = \infty. \qquad \text{Property 1, Theorem 1.15}$$

b. Because $\lim\limits_{x \to 1^-} (x^2 + 1) = 2$ and $\lim\limits_{x \to 1^-} (\cot \pi x) = -\infty$, you can write

$$\lim_{x \to 1^-} \frac{x^2 + 1}{\cot \pi x} = 0. \qquad \text{Property 3, Theorem 1.15}$$

c. Because $\lim\limits_{x \to 0^+} 3 = 3$ and $\lim\limits_{x \to 0^+} \cot x = \infty$, you can write

$$\lim_{x \to 0^+} 3 \cot x = \infty. \qquad \text{Property 2, Theorem 1.15}$$

EXERCISES FOR SECTION 1.5

In Exercises 1–4, determine whether $f(x)$ approaches ∞ or $-\infty$ as x approaches -2 from the left and from the right.

1. $f(x) = 2\left|\dfrac{x}{x^2 - 4}\right|$

2. $f(x) = \dfrac{1}{x + 2}$

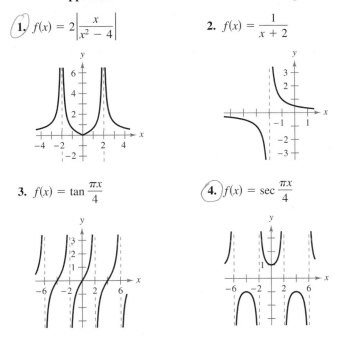

3. $f(x) = \tan \dfrac{\pi x}{4}$

4. $f(x) = \sec \dfrac{\pi x}{4}$

Numerical and Graphical Analysis   **In Exercises 5–8, determine whether $f(x)$ approaches ∞ or $-\infty$ as x approaches -3 from the left and from the right by completing the table. Use a graphing utility to graph the function and confirm your answer.**

x	-3.5	-3.1	3.01	-3.001
$f(x)$				

x	-2.999	-2.99	-2.9	-2.5
$f(x)$				

5. $f(x) = \dfrac{1}{x^2 - 9}$

6. $f(x) = \dfrac{x}{x^2 - 9}$

7. $f(x) = \dfrac{x^2}{x^2 - 9}$

8. $f(x) = \sec \dfrac{\pi x}{6}$

In Exercises 9–28, find the vertical asymptotes (if any) of the function.

9. $f(x) = \dfrac{1}{x^2}$

10. $f(x) = \dfrac{4}{(x - 2)^3}$

11. $h(x) = \dfrac{x^2 - 2}{x^2 - x - 2}$

12. $g(x) = \dfrac{2 + x}{x^2(1 - x)}$

13. $f(x) = \dfrac{x^2}{x^2 - 4}$

14. $f(x) = \dfrac{-4x}{x^2 + 4}$

15. $g(t) = \dfrac{t - 1}{t^2 + 1}$

16. $h(s) = \dfrac{2s - 3}{s^2 - 25}$

17. $f(x) = \tan 2x$

18. $f(x) = \sec \pi x$

19. $T(t) = 1 - \dfrac{4}{t^2}$

20. $g(x) = \dfrac{\frac{1}{2}x^3 - x^2 - 4x}{3x^2 - 6x - 24}$

21. $f(x) = \dfrac{x}{x^2 + x - 2}$

22. $f(x) = \dfrac{4x^2 + 4x - 24}{x^4 - 2x^3 - 9x^2 + 18x}$

23. $g(x) = \dfrac{x^3 + 1}{x + 1}$

24. $h(x) = \dfrac{x^2 - 4}{x^3 + 2x^2 + x + 2}$ -2

25. $f(x) = \dfrac{x^2 - 2x - 15}{x^3 - 5x^2 + x - 5}$

26. $h(t) = \dfrac{t^2 - 2t}{t^4 - 16}$

27. $s(t) = \dfrac{t}{\sin t}$

28. $g(\theta) = \dfrac{\tan \theta}{\theta}$

In Exercises 29–32, determine whether the function has a vertical asymptote or a removable discontinuity at $x = -1$. Graph the function using a graphing utility to confirm your answer.

29. $f(x) = \dfrac{x^2 - 1}{x + 1}$

30. $f(x) = \dfrac{x^2 - 6x - 7}{x + 1}$

31. $f(x) = \dfrac{x^2 + 1}{x + 1}$

32. $f(x) = \dfrac{\sin(x + 1)}{x + 1}$

In Exercises 33–48, find the limit.

33. $\displaystyle\lim_{x \to 2^+} \dfrac{x - 3}{x - 2}$

34. $\displaystyle\lim_{x \to 1^+} \dfrac{2 + x}{1 - x}$

35. $\displaystyle\lim_{x \to 3^+} \dfrac{x^2}{x^2 - 9}$

36. $\displaystyle\lim_{x \to 4^-} \dfrac{x^2}{x^2 + 16}$

37. $\displaystyle\lim_{x \to -3^-} \dfrac{x^2 + 2x - 3}{x^2 + x - 6}$

38. $\displaystyle\lim_{x \to (-1/2)^+} \dfrac{6x^2 + x - 1}{4x^2 - 4x - 3}$

39. $\displaystyle\lim_{x \to 1} \dfrac{x^2 - x}{(x^2 + 1)(x - 1)}$

40. $\displaystyle\lim_{x \to 3} \dfrac{x - 2}{x^2}$

41. $\displaystyle\lim_{x \to 0^-} \left(1 + \dfrac{1}{x}\right)$

42. $\displaystyle\lim_{x \to 0^-} \left(x^2 - \dfrac{1}{x}\right)$

43. $\displaystyle\lim_{x \to 0^+} \dfrac{2}{\sin x}$

44. $\displaystyle\lim_{x \to (\pi/2)^+} \dfrac{-2}{\cos x}$

45. $\displaystyle\lim_{x \to \pi} \dfrac{\sqrt{x}}{\csc x}$

46. $\displaystyle\lim_{x \to 0} \dfrac{x + 2}{\cot x}$

47. $\displaystyle\lim_{x \to 1/2} x \sec \pi x$

48. $\displaystyle\lim_{x \to 1/2} x^2 \tan \pi x$

In Exercises 49–52, use a graphing utility to graph the function and determine the one-sided limit.

49. $f(x) = \dfrac{x^2 + x + 1}{x^3 - 1}$

$\displaystyle\lim_{x \to 1^+} f(x)$

50. $f(x) = \dfrac{x^3 - 1}{x^2 + x + 1}$

$\displaystyle\lim_{x \to 1^-} f(x)$

51. $f(x) = \dfrac{1}{x^2 - 25}$

$\displaystyle\lim_{x \to 5^-} f(x)$

52. $f(x) = \sec \dfrac{\pi x}{6}$

$\displaystyle\lim_{x \to 3^+} f(x)$

Getting at the Concept

53. In your own words, describe the meaning of an infinite limit. Is ∞ a real number?

54. In your own words, describe what is meant by an asymptote of a graph.

55. Write a rational function with vertical asymptotes at $x = 6$ and $x = -2$, and with a zero at $x = 3$.

56. Does every rational function have a vertical asymptote? Explain.

57. Use the graph of the function f (see figure) to sketch the graph of $g(x) = 1/f(x)$ on the interval $[-2, 3]$. To print an enlarged copy of the graph, go to the website *www.mathgraphs.com*.

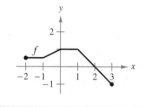

58. *Boyle's Law* For a quantity of gas at a constant temperature, the pressure P is inversely proportional to the volume V. Find the limit of P as $V \to 0^+$.

59. A given sum S is inversely proportional to $1 - r$, where $0 < |r| < 1$. Find the limit of S as $r \to 1^-$.

60. *Rate of Change* A patrol car is parked 50 feet from a long warehouse (see figure). The revolving light on top of the car turns at a rate of $\frac{1}{2}$ revolution per second. The rate at which the light beam moves along the wall is

$$r = 50\pi \sec^2 \theta \text{ ft/sec.}$$

(a) Find the rate r when θ is $\pi/6$.

(b) Find the rate r when θ is $\pi/3$.

(c) Find the limit of r as $\theta \to (\pi/2)^-$.

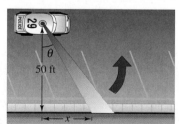

61. *Illegal Drugs* The cost in millions of dollars for a governmental agency to seize $x\%$ of an illegal drug is

$$C = \frac{528x}{100 - x}, \quad 0 \le x < 100.$$

(a) Find the cost of seizing 25% of the drug.

(b) Find the cost of seizing 50% of the drug.

(c) Find the cost of seizing 75% of the drug.

(d) Find the limit of C as $x \to 100^-$ and interpret its meaning.

62. *Relativity* According to the theory of relativity, the mass m of a particle depends on its velocity v. That is,

$$m = \frac{m_0}{\sqrt{1 - (v^2/c^2)}}$$

where m_0 is the mass when the particle is at rest and c is the speed of light. Find the limit of the mass as v approaches c^-.

63. *Rate of Change* A 25-foot ladder is leaning against a house (see figure). If the base of the ladder is pulled away from the house at a rate of 2 feet per second, the top will move down the wall at a rate of

$$r = \frac{2x}{\sqrt{625 - x^2}} \text{ ft/sec}$$

where x is the distance between the base of the ladder and the house.

(a) Find the rate r when x is 7 feet.

(b) Find the rate r when x is 15 feet.

(c) Find the limit of r as $x \to 25^-$.

64. *Average Speed* On a trip of d miles to another city, a truck driver's average speed was x miles per hour. On the return trip the average speed was y miles per hour. The average speed for the round trip was 50 miles per hour.

(a) Verify that $y = \frac{25x}{x - 25}$. What is the domain?

(b) Complete the table.

x	30	40	50	60
y				

Are the values of y different than you expected? Explain.

(c) Find the limit of y as $x \to 25^+$ and interpret its meaning.

65. *Numerical and Graphical Analysis* Use a graphing utility to complete the table for each function and graph each function to estimate the limit. What is the value of the limit when the power on x in the denominator is greater than 3?

x	1	0.5	0.2	0.1	0.01	0.001	0.0001
$f(x)$							

(a) $\lim\limits_{x \to 0^+} \dfrac{x - \sin x}{x}$

(b) $\lim\limits_{x \to 0^+} \dfrac{x - \sin x}{x^2}$

(c) $\lim\limits_{x \to 0^+} \dfrac{x - \sin x}{x^3}$

(d) $\lim\limits_{x \to 0^+} \dfrac{x - \sin x}{x^4}$

66. *Numerical and Graphical Analysis* Consider the shaded region outside the sector of a circle of radius 10 meters and inside a right triangle (see figure).

(a) Write the area $A = f(\theta)$ of the region as a function of θ. Determine the domain of the function.

(b) Use a graphing utility to complete the table.

θ	0.3	0.6	0.9	1.2	1.5
$f(\theta)$					

(c) Use a graphing utility to graph the function over the appropriate domain.

(d) Find the limit of A as $\theta \to (\pi/2)^{-}$.

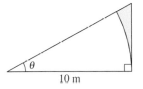

10 m

67. *Numerical and Graphical Reasoning* A crossed belt connects a 20-centimeter pulley (10-cm radius) on an electric motor with a 40-centimeter pulley (20-cm radius) on a saw arbor (see figure). The electric motor runs at 1700 revolutions per minute.

(a) Determine the number of revolutions per minute of the saw.

(b) How does crossing the belt affect the saw in relation to the motor?

(c) Let L be the total length of the belt. Write L as a function of ϕ, where ϕ is measured in radians. What is the domain of the function? (*Hint:* Add the lengths of the straight sections of the belt and the length of the belt around each pulley.)

(d) Use a graphing utility to complete the table.

ϕ	0.3	0.6	0.9	1.2	1.5
L					

(e) Use a graphing utility to graph the function over the appropriate domain.

(f) Find $\displaystyle\lim_{\phi \to (\pi/2)^{-}} L$.

Use a geometric argument as the basis of a second method of finding this limit.

(g) Find $\displaystyle\lim_{\phi \to 0^{+}} L$.

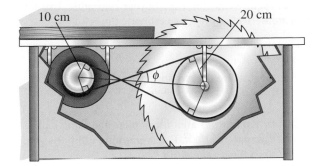

10 cm   20 cm

ϕ

True or False? **In Exercises 68–71, determine whether the statement is true or false. If it is false, explain why or give an example that shows it is false.**

68. If $p(x)$ is a polynomial, then the function given by

$$f(x) = \frac{p(x)}{x - 1}$$

has a vertical asymptote at $x = 1$.

69. A rational function has at least one vertical asymptote.

70. Polynomial functions have no vertical asymptotes.

71. If f has a vertical asymptote at $x = 0$, then f is undefined at $x = 0$.

72. Find functions f and g such that

$$\lim_{x \to c} f(x) = \infty \quad \text{and} \quad \lim_{x \to c} g(x) = \infty$$

but $\displaystyle\lim_{x \to c} [f(x) - g(x)] \neq 0$.

73. Prove the remaining properties of Theorem 1.15.

74. Prove that if $\displaystyle\lim_{x \to c} f(x) = \infty$ then $\displaystyle\lim_{x \to c} \frac{1}{f(x)} = 0$.

75. Prove that if $\displaystyle\lim_{x \to c} \frac{1}{f(x)} = 0$ then $\displaystyle\lim_{x \to c} f(x)$ does not exist.

SECTION PROJECT **GRAPHS AND LIMITS OF TRIGONOMETRIC FUNCTIONS**

Recall from Theorem 1.9 that the limit of $f(x) = (\sin x)/x$ as x approaches 0 is 1.

(a) Use a graphing utility to graph the function f on the interval $-\pi \leq 0 \leq \pi$. Explain how this graph helps confirm that

$$\lim_{x \to 0} \frac{\sin x}{x} = 1.$$

(b) Explain how you could use a table of values to confirm the value of this limit numerically.

(c) Graph $g(x) = \sin x$ by hand. Sketch a tangent line at the point $(0, 0)$ and visually estimate the slope of this tangent line.

(d) Let $(x, \sin x)$ be a point on the graph of g near $(0, 0)$, and write a formula for the slope of the secant line joining $(x, \sin x)$ and $(0, 0)$. Evaluate this formula for $x = 0.1$ and $x = 0.01$. Then find the exact slope of the tangent line to g at the point $(0, 0)$.

(e) Sketch the graph of the cosine function $h(x) = \cos x$. What is the slope of the tangent line at the point $(0, 1)$? Use limits to find this slope analytically.

(f) Find the slope of the tangent line to $k(x) = \tan x$ at $(0, 0)$.

REVIEW EXERCISES FOR CHAPTER 1

1.1   **In Exercises 1 and 2, determine whether the problem can be solved using precalculus or if calculus is required. If the problem can be solved using precalculus, solve it. If the problem seems to require calculus, explain your reasoning. Use a graphical or numerical approach to estimate the solution.**

1. Find the distance between the points $(1, 1)$ and $(3, 9)$ along the curve $y = x^2$.

2. Find the distance between the points $(1, 1)$ and $(3, 9)$ along the line $y = 4x - 3$.

1.2   **In Exercises 3 and 4, complete the table and use the result to estimate the limit. Use a graphing utility to graph the function to confirm your result.**

x	-0.1	-0.01	-0.001	0.001	0.01	0.1
$f(x)$						

3. $\lim\limits_{x\to 0} \dfrac{[4/(x+2)] - 2}{x}$

4. $\lim\limits_{x\to 0} \dfrac{4\left(\sqrt{x+2} - \sqrt{2}\right)}{x}$

In Exercises 5 and 6, use the graph to determine each limit.

5. $h(x) = \dfrac{x^2 - 2x}{x}$

6. $g(x) = \dfrac{3x}{x - 2}$

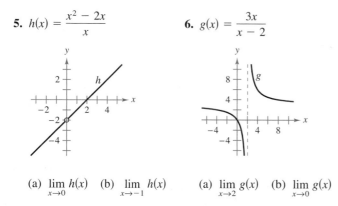

(a) $\lim\limits_{x\to 0} h(x)$ (b) $\lim\limits_{x\to -1} h(x)$ (a) $\lim\limits_{x\to 2} g(x)$ (b) $\lim\limits_{x\to 0} g(x)$

In Exercises 7–10, find the limit L. Then use the ε-δ definition to prove that the limit is L.

7. $\lim\limits_{x\to 1} (3 - x)$ 8. $\lim\limits_{x\to 9} \sqrt{x}$

9. $\lim\limits_{x\to 2} (x^2 - 3)$ 10. $\lim\limits_{x\to 5} 9$

1.3   **In Exercises 11–24, find the limit (if it exists).**

11. $\lim\limits_{t\to 4} \sqrt{t + 2}$ 12. $\lim\limits_{y\to 4} 3|y - 1|$

13. $\lim\limits_{t\to -2} \dfrac{t + 2}{t^2 - 4}$ 14. $\lim\limits_{t\to 3} \dfrac{t^2 - 9}{t - 3}$

15. $\lim\limits_{x\to 4} \dfrac{\sqrt{x} - 2}{x - 4}$ 16. $\lim\limits_{x\to 0} \dfrac{\sqrt{4 + x} - 2}{x}$

17. $\lim\limits_{x\to 0} \dfrac{[1/(x+1)] - 1}{x}$ 18. $\lim\limits_{s\to 0} \dfrac{(1/\sqrt{1 + s}) - 1}{s}$

19. $\lim\limits_{x\to -5} \dfrac{x^3 + 125}{x + 5}$ 20. $\lim\limits_{x\to -2} \dfrac{x^2 - 4}{x^3 + 8}$

21. $\lim\limits_{x\to 0} \dfrac{1 - \cos x}{\sin x}$ 22. $\lim\limits_{x\to \pi/4} \dfrac{4x}{\tan x}$

23. $\lim\limits_{\Delta x\to 0} \dfrac{\sin[(\pi/6) + \Delta x] - (1/2)}{\Delta x}$

[*Hint:* $\sin(\theta + \phi) = \sin\theta\cos\phi + \cos\theta\sin\phi$]

24. $\lim\limits_{\Delta x\to 0} \dfrac{\cos(\pi + \Delta x) + 1}{\Delta x}$

[*Hint:* $\cos(\theta + \phi) = \cos\theta\cos\phi - \sin\theta\sin\phi$]

In Exercises 25 and 26, evaluate the limit given $\lim\limits_{x\to c} f(x) = -\frac{3}{4}$ and $\lim\limits_{x\to c} g(x) = \frac{2}{3}$.

25. $\lim\limits_{x\to c} [f(x)g(x)]$ 26. $\lim\limits_{x\to c} [f(x) + 2g(x)]$

Numerical, Graphical, and Analytic Analysis   **In Exercises 27 and 28, consider**

$$\lim\limits_{x\to 1^+} f(x).$$

(a) **Complete the table to estimate the limit.**

(b) **Use a graphing utility to graph the function and use the graph to estimate the limit.**

(c) **Rationalize the numerator to find the exact value of the limit analytically.**

x	1.1	1.01	1.001	1.0001
$f(x)$				

27. $f(x) = \dfrac{\sqrt{2x + 1} - \sqrt{3}}{x - 1}$

28. $f(x) = \dfrac{1 - \sqrt[3]{x}}{x - 1}$

[*Hint:* $a^3 - b^3 = (a - b)(a^2 + ab + b^2)$]

Free-Falling Object   **In Exercises 29 and 30, use the position function**

$$s(t) = -4.9t^2 + 200$$

which gives the height (in meters) of an object that has fallen from a height of 200 meters. The velocity at time $t = a$ seconds is given by

$$\lim\limits_{t\to a} \dfrac{s(a) - s(t)}{a - t}.$$

29. Find the velocity of the object when $t = 4$.

30. At what velocity will the object impact the ground?

1.4   In Exercises 31–36, find the limit (if it exists). If the limit does not exist, explain why.

31. $\lim\limits_{x \to 3^-} \dfrac{|x - 3|}{x - 3}$

32. $\lim\limits_{x \to 4} [\![x - 1]\!]$

33. $\lim\limits_{x \to 2} f(x)$, where $f(x) = \begin{cases} (x - 2)^2, & x \le 2 \\ 2 - x, & x > 2 \end{cases}$

34. $\lim\limits_{x \to 1^+} g(x)$, where $g(x) = \begin{cases} \sqrt{1 - x}, & x \le 1 \\ x + 1, & x > 1 \end{cases}$

35. $\lim\limits_{t \to 1} h(t)$, where $h(t) = \begin{cases} t^3 + 1, & t < 1 \\ \frac{1}{2}(t + 1), & t \ge 1 \end{cases}$

36. $\lim\limits_{s \to -2} f(s)$, where $f(s) = \begin{cases} -s^2 - 4s - 2, & s \le -2 \\ s^2 + 4s + 6, & s > -2 \end{cases}$

In Exercises 37–46, determine the intervals on which the function is continuous.

37. $f(x) = [\![x + 3]\!]$

38. $f(x) = \dfrac{3x^2 - x - 2}{x - 1}$

39. $f(x) = \begin{cases} \dfrac{3x^2 - x - 2}{x - 1}, & x \ne 1 \\ 0, & x = 1 \end{cases}$

40. $f(x) = \begin{cases} 5 - x, & x \le 2 \\ 2x - 3, & x > 2 \end{cases}$

41. $f(x) = \dfrac{1}{(x - 2)^2}$

42. $f(x) = \sqrt{\dfrac{x + 1}{x}}$

43. $f(x) = \dfrac{3}{x + 1}$

44. $f(x) = \dfrac{x + 1}{2x + 2}$

45. $f(x) = \csc \dfrac{\pi x}{2}$

46. $f(x) = \tan 2x$

47. Determine the value of c such that the function is continuous on the entire real line.

$$f(x) = \begin{cases} x + 3, & x \le 2 \\ cx + 6, & x > 2 \end{cases}$$

48. Determine the values of b and c such that the function is continuous on the entire real line.

$$f(x) = \begin{cases} x + 1, & 1 < x < 3 \\ x^2 + bx + c, & |x - 2| \ge 1 \end{cases}$$

49. Use the Intermediate Value Theorem to show that $f(x) = 2x^3 - 3$ has a zero in the interval $[1, 2]$.

50. *Cost of Overnight Delivery*   The cost of sending an overnight package from New York to Atlanta is $9.80 for the first pound and $2.50 for each additional pound. Use the greatest integer function to create a model for the cost C of overnight delivery of a package weighing x pounds. Use a graphing utility to graph the function and discuss its continuity.

51. Let $f(x) = \dfrac{x^2 - 4}{|x - 2|}$. Find each limit (if possible).

(a) $\lim\limits_{x \to 2^-} f(x)$

(b) $\lim\limits_{x \to 2^+} f(x)$

(c) $\lim\limits_{x \to 2} f(x)$

52. Let $f(x) = \sqrt{x(x - 1)}$.

(a) Find the domain of f.

(b) Find $\lim\limits_{x \to 0^-} f(x)$.

(c) Find $\lim\limits_{x \to 1^+} f(x)$.

1.5   In Exercises 53–56, find the vertical asymptotes (if any) of the function.

53. $g(x) = 1 + \dfrac{2}{x}$

54. $h(x) = \dfrac{4x}{4 - x^2}$

55. $f(x) = \dfrac{8}{(x - 10)^2}$

56. $f(x) = \csc \pi x$

In Exercises 57–68, find the one-sided limit.

57. $\lim\limits_{x \to -2} \dfrac{2x^2 + x + 1}{x + 2}$

58. $\lim\limits_{x \to (1/2)^+} \dfrac{x}{2x - 1}$

59. $\lim\limits_{x \to -1^+} \dfrac{x + 1}{x^3 + 1}$

60. $\lim\limits_{x \to -1} \dfrac{x + 1}{x^4 - 1}$

61. $\lim\limits_{x \to 1} \dfrac{x^2 + 2x + 1}{x - 1}$

62. $\lim\limits_{x \to -1^+} \dfrac{x^2 - 2x + 1}{x + 1}$

63. $\lim\limits_{x \to 0^+} \left(x - \dfrac{1}{x^3}\right)$

64. $\lim\limits_{x \to 2^-} \dfrac{1}{\sqrt[3]{x^2 - 4}}$

65. $\lim\limits_{x \to 0^+} \dfrac{\sin 4x}{5x}$

66. $\lim\limits_{x \to 0^+} \dfrac{\sec x}{x}$

67. $\lim\limits_{x \to 0^+} \dfrac{\csc 2x}{x}$

68. $\lim\limits_{x \to 0^-} \dfrac{\cos^2 x}{x}$

69. *Cost of Clean Air*   A utility company burns coal to generate electricity. The cost C in dollars of removing $p\%$ of the air pollutants in the stack emissions is

$$C = \dfrac{80{,}000p}{100 - p}, \qquad 0 \le p < 100.$$

Find the cost of removing (a) 15%, (b) 50%, and (c) 90% of the pollutants. (d) Find the limit of C as $p \to 100^-$.

70. The function f is defined as follows.

$$f(x) = \dfrac{\tan 2x}{x}, \qquad x \ne 0$$

(a) Find $\lim\limits_{x \to 0} \dfrac{\tan 2x}{x}$ (if it exists).

(b) Can the function f be defined at $x = 0$ such that it is continuous at $x = 0$?

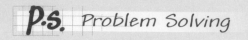

1. Let $P(x, y)$ be a point on the parabola $y = x^2$ in the first quadrant. Consider the triangle $\triangle PAO$ formed by P, $A(0, 1)$, and the origin $O(0, 0)$, and the triangle $\triangle PBO$ formed by P, $B(1, 0)$, and the origin.

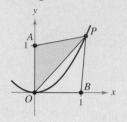

(a) Express the perimeter of each triangle in terms of x.

(b) Let $r(x)$ be the ratio of the perimeters of the two triangles,

$$r(x) = \frac{\text{Perimeter } \triangle PAO}{\text{Perimeter } \triangle PBO}.$$

Complete the table.

x	4	2	1	0.1	0.01
Perimeter $\triangle PAO$					
Perimeter $\triangle PBO$					
$r(x)$					

(c) Calculate $\lim\limits_{x \to 0^+} r(x)$.

2. Let $P(x, y)$ be a point on the parabola $y = x^2$ in the first quadrant. Consider the triangle $\triangle PAO$ formed by P, $A(0, 1)$, and the origin $O(0, 0)$, and the triangle $\triangle PBO$ formed by P, $B(1, 0)$, and the origin.

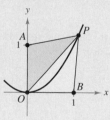

(a) Express the area of each triangle in terms of x.

(b) Let $a(x)$ be the ratio of the areas of the two triangles,

$$a(x) = \frac{\text{Area } \triangle PBO}{\text{Area } \triangle PAO}.$$

Complete the table.

x	4	2	1	0.1	0.01
Area $\triangle PAO$					
Area $\triangle PBO$					
$a(x)$					

(c) Calculate $\lim\limits_{x \to 0^+} a(x)$.

3. (a) Find the area of a regular hexagon inscribed in a circle of radius 1. How close is this area to that of the circle?

(b) Find the area A_n of an n-sided regular polygon inscribed in a circle of radius 1. Express your answer as a function of n.

(c) Complete the table.

n	6	12	24	48	96
A_n					

(d) What number does A_n approach as n gets larger and larger?

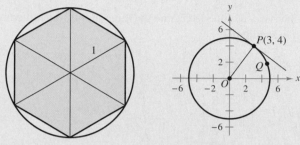

Figure for 3   **Figure for 4**

4. Let $P(3, 4)$ be a point on the circle $x^2 + y^2 = 25$.

(a) What is the slope of the line joining P and $O(0, 0)$?

(b) Find an equation of the tangent line to the circle at P.

(c) Let $Q(x, y)$ be another point on the circle in the first quadrant. Find the slope m_x of the line joining P and Q in terms of x.

(d) Calculate $\lim\limits_{x \to 3} m_x$.
How does this number relate to your answer in part (b)?

5. Let $P(5, -12)$ be a point on the circle $x^2 + y^2 = 169$.

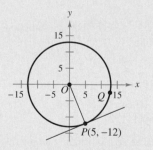

(a) What is the slope of the line joining P and $O(0, 0)$?

(b) Find an equation of the tangent line to the circle at P.

(c) Let $Q(x, y)$ be another point on the circle in the fourth quadrant. Find the slope m_x of the line joining P and Q in terms of x.

(d) Calculate $\lim\limits_{x \to 5} m_x$.
How does this number relate to your answer in part (b)?

6. Find the values for the constants a and b such that

$$\lim_{x \to 0} \frac{\sqrt{a + bx} - \sqrt{3}}{x} = \sqrt{3}.$$

7. Consider the function $f(x) = \dfrac{\sqrt{3 + x^{1/3}} - 2}{x - 1}$.

(a) Find the domain of f.

(b) Use a graphing utility to graph the function.

(c) Calculate $\lim\limits_{x \to -27^+} f(x)$.

(d) Calculate $\lim\limits_{x \to 1} f(x)$.

8. Determine all values of the constant a such that the following function is continuous for all real numbers.

$$f(x) = \begin{cases} \dfrac{ax}{\tan x}, & x \geq 0 \\ a^2 - 2, & x < 0 \end{cases}$$

9. Consider the graphs of the four functions g_1, g_2, g_3, and g_4.

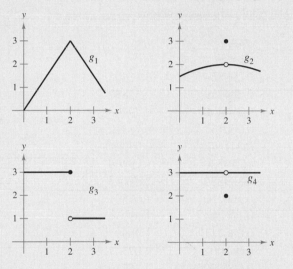

For the given condition of the function f, which of the graphs could be the graph of f?

(a) $\lim\limits_{x \to 2} f(x) = 3$ (b) f is continuous at 2.

(c) $\lim\limits_{x \to 2^-} f(x) = 3$

10. Sketch the graph of the function $f(x) = \left[\!\left[\dfrac{1}{x}\right]\!\right]$.

(a) Evaluate $f\left(\frac{1}{4}\right)$, $f(3)$, and $f(1)$.

(b) Evaluate the limits $\lim\limits_{x \to 1^-} f(x)$, $\lim\limits_{x \to 1^+} f(x)$, $\lim\limits_{x \to 0^-} f(x)$, and $\lim\limits_{x \to 0^+} f(x)$.

(c) Discuss the continuity of the function.

11. Sketch the graph of the function $f(x) = [\![x]\!] + [\![-x]\!]$.

(a) Evaluate $f(1)$, $f(0)$, $f\left(\frac{1}{2}\right)$, and $f(-2.7)$.

(b) Evaluate the limits $\lim\limits_{x \to 1^-} f(x)$, $\lim\limits_{x \to 1^+} f(x)$, and $\lim\limits_{x \to \frac{1}{2}} f(x)$.

(c) Discuss the continuity of the function.

12. To escape earth's gravitational field, a rocket must be launched with an initial velocity called the **escape velocity.** A rocket launched from the surface of earth has velocity v (in miles per second) given by

$$v = \sqrt{\frac{2GM}{r} + v_0{}^2 - \frac{2GM}{R}} \approx \sqrt{\frac{192{,}000}{r} + v_0{}^2 - 48}$$

where v_0 is the initial velocity, r is the distance from the rocket to the center of earth, G is the gravitational constant, M is the mass of earth, and R is the radius of earth (approximately 4000 miles).

(a) Find the value of v_0 for which you obtain an infinite limit for r as v tends to zero. This value of v_0 is the escape velocity for earth.

(b) A rocket launched from the surface of the moon has velocity v (in miles per second) given by

$$v = \sqrt{\frac{1920}{r} + v_0{}^2 - 2.17}.$$

Find the escape velocity for the moon.

(c) A rocket launched from the surface of a certain planet has velocity v (in miles per second) given by

$$v = \sqrt{\frac{10{,}600}{r} + v_0{}^2 - 6.99}.$$

Find the escape velocity for this planet. Is the mass of this planet larger or smaller than that of earth? (Assume that the mean density of this planet is the same as that of earth.)

13. For positive numbers $a < b$, the **pulse function** is defined as

$$P_{a,b}(x) = H(x - a) - H(x - b) = \begin{cases} 0, & x < a \\ 1, & a \leq x < b \\ 0, & x \geq b \end{cases}$$

where $H(x) = \begin{cases} 1, & x \geq 0 \\ 0, & x < 0 \end{cases}$ is the Heaviside function.

(a) Sketch the graph of the pulse function.

(b) Find the following limits:

(i) $\lim\limits_{x \to a^+} P_{a,b}(x)$ (ii) $\lim\limits_{x \to a^-} P_{a,b}(x)$

(iii) $\lim\limits_{x \to b^+} P_{a,b}(x)$ (iv) $\lim\limits_{x \to b^-} P_{a,b}(x)$

(c) Discuss the continuity of the pulse function.

(d) Why is $U(x) = \dfrac{1}{b - a} P_{a,b}(x)$ called the **unit** pulse function?

14. Let a be a nonzero constant. Prove that if

$$\lim\limits_{x \to 0} f(x) = L$$

then

$$\lim\limits_{x \to 0} f(ax) = L.$$

Show by means of an example that a must be nonzero.

Gravity: Finding It Experimentally

The study of dynamics dates back to the sixteenth century. As the Dark Ages gave way to the Renaissance, Galileo Galilei (1564–1642) was one of the first to take steps toward understanding the motion of objects under the influence of gravity.

Up until Galileo's time, it was recognized that a falling object moved faster and faster as it fell, but what mathematical law governed this accelerating motion was unknown. Free-falling objects move too fast to have been measured with any of the equipment available at that time. Galileo solved this problem with a rather ingenious setup. He reasoned that gravity could be "diluted" by rolling a ball down an inclined plane. He used a water clock, which kept track of time by measuring the amount of water that poured through a small opening at the bottom.

We now have relatively inexpensive instruments, such as the *Texas Instruments Calculator-Based Laboratory (CBL) System*, that allow accurate position data to be gathered on a free-falling object. A CBL System was used to track the positions of a falling ball at time intervals of 0.02 second. The results are shown below.

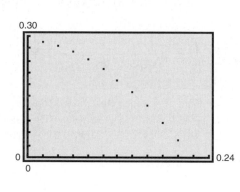

Time (sec)	Height (meters)	Velocity (meters/sec)
0.00	0.290864	−0.16405
0.02	0.284279	−0.32857
0.04	0.274400	−0.49403
0.06	0.260131	−0.71322
0.08	0.241472	−0.93309
0.10	0.219520	−1.09409
0.12	0.189885	−1.47655
0.14	0.160250	−1.47891
0.16	0.126224	−1.69994
0.18	0.086711	−1.96997
0.20	0.045002	−2.07747
0.22	0.000000	−2.25010

QUESTIONS

1. Use a graphing utility to sketch a scatter plot of the positions of the falling ball. What type of model seems to be the best fit? Use the regression features of the graphing utility to find the best-fitting model.

2. Repeat the procedure in Question 1 for the velocities of the falling ball. Describe any relationships between the two models.

3. In theory, the position of a free-falling object in a vacuum is given by $s = \frac{1}{2}gt^2 + v_0 t + s_0$, where g is the acceleration due to gravity (meters per second per second), t is the time (seconds), v_0 is the initial velocity (meters per second), and s_0 is the initial height (meters). From this experiment, estimate the value of g. Do you think your estimate is too great or too small? Explain your reasoning.

The concepts presented here will be explored further in this chapter. For an extension of this application, see Lab 3 in the lab series that accompanies this text at college.hmco.com.

Differentiation 2

Aerial cinematographers must have a thorough understanding of gravity's effect on a falling object in order to control the camera mounted on their helmets.

Bruno Brokkens/Allsport

Courtesy of Joe Jennings

The work of Joe Jennings, a renowned aerial cameraman, can be seen in many films, television shows, and commercials.

Excerpted from "Into the Stratosphere: Skysurfing Over Mission Bay" from wildca.com

Who would dare jump out of a plane with a bulky, 75-pound IMAX camera strapped to their chest? The answer turned out to be Joe Jennings, who is not only a skysurfer but also an innovative aerial cinematographer in his own right. Jennings designed a special harness to hold the camera, as well as a massive wing-suit—with fabric spanning from his knees to his wrists—to slow his rate of descent.

The pitfalls were enormous. Explains Krenzien: (Mark Krenzien, writer/producer) "One of the major problems is how do you balance the fall-rate of a photographer with the fall-rate of the surfer. Obviously, they have to be at fairly close levels to one another in the sky. In this case, Joe's winged suit and extraordinary skill made the difference."

The Derivative and the Tangent Line Problem

- Find the slope of the tangent line to a curve at a point.
- Use the limit definition to find the derivative of a function.
- Understand the relationship between differentiability and continuity.

The Tangent Line Problem

Calculus grew out of four major problems that European mathematicians were working on during the seventeenth century.

1. The tangent line problem (Section 1.1 and this section)
2. The velocity and acceleration problem (Sections 2.2 and 2.3)
3. The minimum and maximum problem (Section 3.1)
4. The area problem (Sections 1.1 and 4.2)

Each problem involves the notion of a limit, and we could introduce calculus with any of the four problems.

We gave a brief introduction to the tangent line problem in Section 1.1. Although partial solutions to this problem were given by Pierre de Fermat (1601–1665), René Descartes (1596–1650), Christian Huygens (1629–1695), and Isaac Barrow (1630–1677), credit for the first general solution is usually given to Isaac Newton (1642–1727) and Gottfried Leibniz (1646–1716). Newton's work on this problem stemmed from his interest in optics and light refraction.

What does it mean to say that a line is tangent to a curve at a point? For a circle, the tangent line at a point P is the line that is perpendicular to the radial line at point P, as shown in Figure 2.1.

For a general curve, however, the problem is more difficult. For example, how would you define the tangent lines shown in Figure 2.2? You might say that a line is tangent to a curve at a point P if it touches, but does not cross, the curve at point P. This definition would work for the first curve shown in Figure 2.2, but not for the second. *Or* you might say that a line is tangent to a curve if the line touches or intersects the curve at exactly one point. This definition would work for a circle but not for more general curves, as the third curve in Figure 2.2 shows.

ISAAC NEWTON (1642–1727)

In addition to his work in calculus, Newton made revolutionary contributions to physics, including the Universal Law of Gravitation and his three laws of motion.

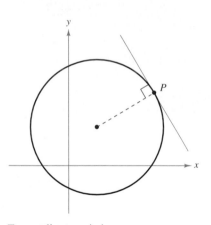

Tangent line to a circle
Figure 2.1

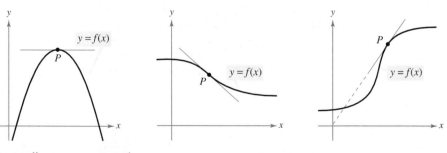

Tangent line to a curve at a point
Figure 2.2

FOR FURTHER INFORMATION For more information on the crediting of mathematical discoveries to the first "discoverer," see the article "Mathematical Firsts—Who Done It?" by Richard H. Williams and Roy D. Mazzagatti in *Mathematics Teacher*. To view this article, go to the website *www.matharticles.com*.

EXPLORATION

Identifying a Tangent Line Use a graphing utility to sketch the graph of $f(x) = 2x^3 - 4x^2 + 3x - 5$. On the same screen, sketch the graphs of $y = x - 5$, $y = 2x - 5$, and $y = 3x - 5$. Which of these lines, if any, appears to be tangent to the graph of f at the point $(0, -5)$? Explain your reasoning.

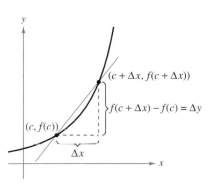

The secant line through $(c, f(c))$ and $(c + \Delta x, f(c + \Delta x))$

Figure 2.3

Essentially, the problem of finding the tangent line at a point P boils down to the problem of finding the *slope* of the tangent line at point P. You can approximate this slope using a **secant line*** through the point of tangency and a second point on the curve, as shown in Figure 2.3. If $(c, f(c))$ is the point of tangency and $(c + \Delta x, f(c + \Delta x))$ is a second point on the graph of f, the slope of the secant line through the two points is given by substitution into the slope formula

$$m = \frac{y_2 - y_1}{x_2 - x_1}$$

$$m_{\text{sec}} = \frac{f(c + \Delta x) - f(c)}{(c + \Delta x) - c} \qquad \text{Change in } y \\ \text{Change in } x$$

$$m_{\text{sec}} = \frac{f(c + \Delta x) - f(c)}{\Delta x}. \qquad \text{Slope of secant line}$$

The right-hand side of this equation is a **difference quotient.** The denominator Δx is the **change in x,** and the numerator $\Delta y = f(c + \Delta x) - f(c)$ is the **change in y.**

The beauty of this procedure is that you can obtain more and more accurate approximations to the slope of the tangent line by choosing points closer and closer to the point of tangency, as shown in Figure 2.4.

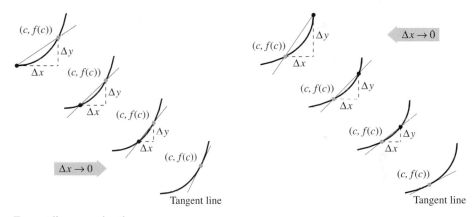

Tangent line approximations

Figure 2.4

Definition of Tangent Line with Slope m

If f is defined on an open interval containing c, and if the limit

$$\lim_{\Delta x \to 0} \frac{\Delta y}{\Delta x} = \lim_{\Delta x \to 0} \frac{f(c + \Delta x) - f(c)}{\Delta x} = m$$

exists, then the line passing through $(c, f(c))$ with slope m is the tangent line to the graph of f at the point $(c, f(c))$.

The slope of the tangent line to the graph of f at the point $(c, f(c))$ is also called the **slope of the graph of f at $x = c$.**

---

* *This use of the word* secant *comes from the Latin* secare, *meaning to cut, and is not a reference to the trigonometric function of the same name.*

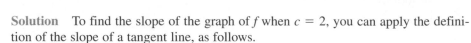

Example 1 The Slope of the Graph of a Linear Function

Find the slope of the graph of

$$f(x) = 2x - 3$$

at the point $(2, 1)$.

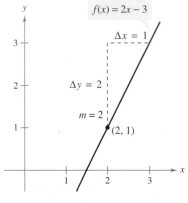

The slope of f at $(2, 1)$ is $m = 2$.
Figure 2.5

Solution To find the slope of the graph of f when $c = 2$, you can apply the definition of the slope of a tangent line, as follows.

$$\lim_{\Delta x \to 0} \frac{f(2 + \Delta x) - f(2)}{\Delta x} = \lim_{\Delta x \to 0} \frac{[2(2 + \Delta x) - 3] - [2(2) - 3]}{\Delta x}$$

$$= \lim_{\Delta x \to 0} \frac{4 + 2\Delta x - 3 - 4 + 3}{\Delta x}$$

$$= \lim_{\Delta x \to 0} \frac{2\Delta x}{\Delta x}$$

$$= \lim_{\Delta x \to 0} 2$$

$$= 2$$

The slope of f at $(c, f(c)) = (2, 1)$ is $m = 2$, as shown in Figure 2.5.

NOTE In Example 1, the limit definition of the slope of f agrees with the definition of the slope of a line as discussed in Section P.2.

The graph of a linear function has the same slope at any point. This is not true of nonlinear functions, as can be seen in the following example.

Example 2 Tangent Lines to the Graph of a Nonlinear Function

Find the slopes of the tangent lines to the graph of

$$f(x) = x^2 + 1$$

at the points $(0, 1)$ and $(-1, 2)$, as shown in Figure 2.6.

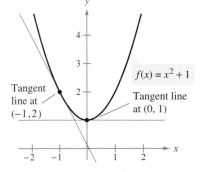

The slope of f at any point $(c, f(c))$ is $m = 2c$.
Figure 2.6

Solution Let $(c, f(c))$ represent an arbitrary point on the graph of f. Then the slope of the tangent line at $(c, f(c))$ is given by

$$\lim_{\Delta x \to 0} \frac{f(c + \Delta x) - f(c)}{\Delta x} = \lim_{\Delta x \to 0} \frac{[(c + \Delta x)^2 + 1] - (c^2 + 1)}{\Delta x}$$

$$= \lim_{\Delta x \to 0} \frac{c^2 + 2c(\Delta x) + (\Delta x)^2 + 1 - c^2 - 1}{\Delta x}$$

$$= \lim_{\Delta x \to 0} \frac{2c(\Delta x) + (\Delta x)^2}{\Delta x}$$

$$= \lim_{\Delta x \to 0} (2c + \Delta x)$$

$$= 2c.$$

So, the slope at *any* point $(c, f(c))$ on the graph of f is $m = 2c$. At the point $(0, 1)$, the slope is $m = 2(0) = 0$, and at $(-1, 2)$, the slope is $m = 2(-1) = -2$.

NOTE In Example 2, note that c is held constant in the limit process (as $\Delta x \to 0$).

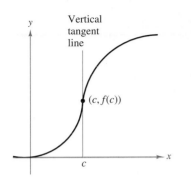

Vertical tangent line

$(c, f(c))$

The graph of f has a vertical tangent line at $(c, f(c))$.

Figure 2.7

The definition of a tangent line to a curve does not cover the possibility of a vertical tangent line. For vertical tangent lines, you can use the following definition. If f is continuous at c and

$$\lim_{\Delta x \to 0} \frac{f(c + \Delta x) - f(c)}{\Delta x} = \infty \quad \text{or} \quad \lim_{\Delta x \to 0} \frac{f(c + \Delta x) - f(c)}{\Delta x} = -\infty$$

the vertical line $x = c$ passing through $(c, f(c))$ is a **vertical tangent line** to the graph of f. For example, the function shown in Figure 2.7 has a vertical tangent line at $(c, f(c))$. If the domain of f is the closed interval $[a, b]$, you can extend the definition of a vertical tangent line to include the endpoints by considering continuity and limits from the right (for $x = a$) and from the left (for $x = b$).

The Derivative of a Function

You have now arrived at a crucial point in the study of calculus. The limit used to define the slope of a tangent line is also used to define one of the two fundamental operations of calculus—**differentiation.**

Definition of the Derivative of a Function

The **derivative** of f at x is given by

$$f'(x) = \lim_{\Delta x \to 0} \frac{f(x + \Delta x) - f(x)}{\Delta x}$$

provided the limit exists. For all x for which this limit exists, f' is a function of x.

Be sure you see that the derivative of a function of x is also a function of x. This "new" function gives the slope of the tangent line to the graph of f at the point $(x, f(x))$, provided that the graph has a tangent line at this point.

The process of finding the derivative of a function is called **differentiation.** A function is **differentiable** at x if its derivative exists at x and **differentiable on an open interval (a, b)** if it is differentiable at every point in the interval.

In addition to $f'(x)$, which is read as "f prime of x," other notations are used to denote the derivative of $y = f(x)$. The most common are

$$f'(x), \quad \frac{dy}{dx}, \quad y', \quad \frac{d}{dx}[f(x)], \quad D_x[y]. \qquad \text{Notation for derivatives}$$

The notation dy/dx is read as "the derivative of y *with respect to x.*" Using limit notation, you can write

$$\frac{dy}{dx} = \lim_{\Delta x \to 0} \frac{\Delta y}{\Delta x}$$

$$= \lim_{\Delta x \to 0} \frac{f(x + \Delta x) - f(x)}{\Delta x}$$

$$= f'(x).$$

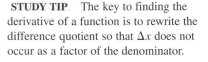 *Example 3* **Finding the Derivative by the Limit Process**

Find the derivative of $f(x) = x^3 + 2x$.

Solution

$$f'(x) = \lim_{\Delta x \to 0} \frac{f(x + \Delta x) - f(x)}{\Delta x} \qquad \text{Definition of derivative}$$

$$= \lim_{\Delta x \to 0} \frac{(x + \Delta x)^3 + 2(x + \Delta x) - (x^3 + 2x)}{\Delta x}$$

$$= \lim_{\Delta x \to 0} \frac{x^3 + 3x^2\Delta x + 3x(\Delta x)^2 + (\Delta x)^3 + 2x + 2\Delta x - x^3 - 2x}{\Delta x}$$

$$= \lim_{\Delta x \to 0} \frac{3x^2\Delta x + 3x(\Delta x)^2 + (\Delta x)^3 + 2\Delta x}{\Delta x}$$

$$= \lim_{\Delta x \to 0} \frac{\Delta x[3x^2 + 3x\Delta x + (\Delta x)^2 + 2]}{\Delta x}$$

$$= \lim_{\Delta x \to 0} [3x^2 + 3x\Delta x + (\Delta x)^2 + 2]$$

$$= 3x^2 + 2$$

STUDY TIP The key to finding the derivative of a function is to rewrite the difference quotient so that Δx does not occur as a factor of the denominator.

Remember that the derivative of a function f is itself a function, which can be used to find the slope of the tangent line at the point $(x, f(x))$ on the graph of f.

Example 4 **Using the Derivative to Find the Slope at a Point**

Find $f'(x)$ for $f(x) = \sqrt{x}$. Then find the slope of the graph of f at the points $(1, 1)$ and $(4, 2)$. Discuss the behavior of f at $(0, 0)$.

Solution Use the procedure for rationalizing numerators, as discussed in Section 1.3.

$$f'(x) = \lim_{\Delta x \to 0} \frac{f(x + \Delta x) - f(x)}{\Delta x} \qquad \text{Definition of derivative}$$

$$= \lim_{\Delta x \to 0} \frac{\sqrt{x + \Delta x} - \sqrt{x}}{\Delta x}$$

$$= \lim_{\Delta x \to 0} \left(\frac{\sqrt{x + \Delta x} - \sqrt{x}}{\Delta x}\right)\left(\frac{\sqrt{x + \Delta x} + \sqrt{x}}{\sqrt{x + \Delta x} + \sqrt{x}}\right)$$

$$= \lim_{\Delta x \to 0} \frac{(x + \Delta x) - x}{\Delta x(\sqrt{x + \Delta x} + \sqrt{x})}$$

$$= \lim_{\Delta x \to 0} \frac{\Delta x}{\Delta x(\sqrt{x + \Delta x} + \sqrt{x})}$$

$$= \lim_{\Delta x \to 0} \frac{1}{\sqrt{x + \Delta x} + \sqrt{x}}$$

$$= \frac{1}{2\sqrt{x}}$$

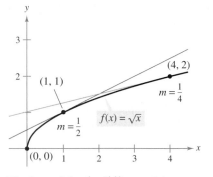

The slope of f at $(x, f(x))$, $x > 0$, is $m = 1/(2\sqrt{x})$.

Figure 2.8

At the point $(1, 1)$, the slope is $f'(1) = \frac{1}{2}$. At the point $(4, 2)$, the slope is $f'(4) = \frac{1}{4}$. (See Figure 2.8.) At the point $(0, 0)$ the slope is undefined. Moreover, because the limit of $f'(x)$ as $x \to 0$ from the right is infinite, the graph of f has a vertical tangent line at $(0, 0)$.

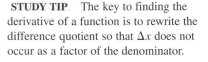 *indicates that in the* Interactive 3.0 *CD-ROM and* Internet 3.0 *versions of this text (available at* college.hmco.com*) you will find an Open Exploration, which further explores this example using the computer algebra systems* Maple, Mathcad, Mathematica, *and* Derive.

In many applications, it is convenient to use a variable other than x as the independent variable, as shown in Example 5.

Example 5   Finding the Derivative of a Function

Find the derivative with respect to t for the function $y = 2/t$.

Solution   Considering $y = f(t)$, you obtain the following.

$$\frac{dy}{dt} = \lim_{\Delta t \to 0} \frac{f(t + \Delta t) - f(t)}{\Delta t} \qquad \text{Definition of derivative}$$

$$= \lim_{\Delta t \to 0} \frac{\dfrac{2}{t + \Delta t} - \dfrac{2}{t}}{\Delta t} \qquad f(t + \Delta t) = 2/(t + \Delta t) \text{ and } f(t) = 2/t$$

$$= \lim_{\Delta t \to 0} \frac{\dfrac{2t - 2(t + \Delta t)}{t(t + \Delta t)}}{\Delta t} \qquad \text{Combine fractions in numerator.}$$

$$= \lim_{\Delta t \to 0} \frac{-2\Delta t}{\Delta t(t)(t + \Delta t)} \qquad \text{Divide out common factor of } \Delta t.$$

$$= \lim_{\Delta t \to 0} \frac{-2}{t(t + \Delta t)} \qquad \text{Simplify.}$$

$$= -\frac{2}{t^2} \qquad \text{Evaluate limit as } \Delta t \to 0.$$

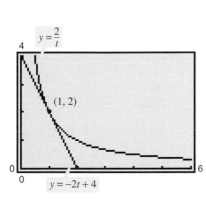

$y = \dfrac{2}{t}$

$(1, 2)$

$y = -2t + 4$

At the point $(1, 2)$ the line $y = -2t + 4$ is tangent to the graph of $y = 2/t$.

Figure 2.9

TECHNOLOGY   A graphing utility can be used to reinforce the result given in Example 5. For instance, using the formula $dy/dt = -2/t^2$, you know that the slope of the graph of $y = 2/t$ at the point $(1, 2)$ is $m = -2$. This implies that an equation of the tangent line to the graph at $(1, 2)$ is $y - 2 = -2(t - 1)$ or $y = -2t + 4$, as shown in Figure 2.9.

Differentiability and Continuity

The following alternative limit form of the derivative is useful in investigating the relationship between differentiability and continuity. The derivative of f at c is

$$f'(c) = \lim_{x \to c} \frac{f(x) - f(c)}{x - c} \qquad \text{Alternative form of derivative}$$

provided this limit exists (see Figure 2.10). (A proof of the equivalence of this form is given in Appendix B.) Note that the existence of the limit in this alternative form requires that the one-sided limits

$$\lim_{x \to c^-} \frac{f(x) - f(c)}{x - c} \quad \text{and} \quad \lim_{x \to c^+} \frac{f(x) - f(c)}{x - c}$$

exist and are equal. These one-sided limits are called the **derivatives from the left and from the right,** respectively. We say that f is **differentiable on the closed interval $[a, b]$** if it is differentiable on (a, b) and if the derivative from the right at a and the derivative from the left at b both exist.

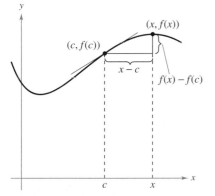

y

$(x, f(x))$

$(c, f(c))$

$x - c$

$f(x) - f(c)$

c x

As x approaches c, the secant line approaches the tangent line.

Figure 2.10

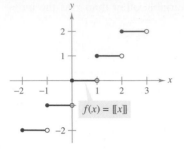

The greatest integer function is not differentiable at $x = 0$, because it is not continuous at $x = 0$.

Figure 2.11

If a function is not continuous at $x = c$, it is also not differentiable at $x = c$. For instance, the greatest integer function

$$f(x) = [\![x]\!]$$

is not continuous at $x = 0$—hence, it is not differentiable at $x = 0$ (see Figure 2.11). You can verify this by observing that

$$\lim_{x \to 0^-} \frac{f(x) - f(0)}{x - 0} = \lim_{x \to 0^-} \frac{[\![x]\!] - 0}{x} = \infty \qquad \text{Derivative from the left}$$

and

$$\lim_{x \to 0^+} \frac{f(x) - f(0)}{x - 0} = \lim_{x \to 0^+} \frac{[\![x]\!] - 0}{x} = 0. \qquad \text{Derivative from the right}$$

Although it is true that differentiability implies continuity (as we will show in Theorem 2.1), the converse is not true. That is, it is possible for a function to be continuous at $x = c$ and *not* differentiable at $x = c$. Examples 6 and 7 illustrate this possibility.

Example 6   **A Graph with a Sharp Turn**

The function

$$f(x) = |x - 2|$$

shown in Figure 2.12 is continuous at $x = 2$. However, the one-sided limits

$$\lim_{x \to 2^-} \frac{f(x) - f(2)}{x - 2} = \lim_{x \to 2^-} \frac{|x - 2| - 0}{x - 2} = -1 \qquad \text{Derivative from the left}$$

and

$$\lim_{x \to 2^+} \frac{f(x) - f(2)}{x - 2} = \lim_{x \to 2^+} \frac{|x - 2| - 0}{x - 2} = 1 \qquad \text{Derivative from the right}$$

are not equal. So, f is not differentiable at $x = 2$ and the graph of f does not have a tangent line at the point $(2, 0)$.

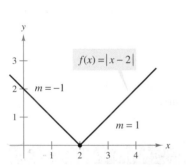

f is not differentiable at $x = 2$, because the derivatives from the left and from the right are not equal.

Figure 2.12

Example 7   **A Graph with a Vertical Tangent Line**

The function

$$f(x) = x^{1/3}$$

is continuous at $x = 0$, as shown in Figure 2.13. However, because the limit

$$\lim_{x \to 0} \frac{f(x) - f(0)}{x - 0} = \lim_{x \to 0} \frac{x^{1/3} - 0}{x}$$

$$= \lim_{x \to 0} \frac{1}{x^{2/3}}$$

$$= \infty$$

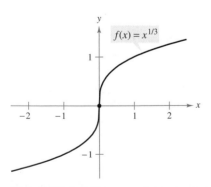

f is not differentiable at $x = 0$, because f has a vertical tangent at $x = 0$.

Figure 2.13

is infinite, you can conclude that the tangent line is vertical at $x = 0$. So, f is not differentiable at $x = 0$.

From Examples 6 and 7, you can see that a function is not differentiable at a point at which its graph has a sharp turn *or* a vertical tangent.

TECHNOLOGY Some graphing utilities, such as *Derive, Maple, Mathcad, Mathematica,* and the *TI-89,* perform symbolic differentiation. Others perform *numerical differentiation* by finding values of derivatives using the formula

$$f'(x) \approx \frac{f(x + \Delta x) - f(x - \Delta x)}{2\Delta x}$$

where Δx is a small number such as 0.001. Can you see any problems with this definition? For instance, using this definition, what is the value of the derivative of $f(x) = |x|$ when $x = 0$?

THEOREM 2.1   Differentiability Implies Continuity

If f is differentiable at $x = c$, then f is continuous at $x = c$.

Proof You can prove that f is continuous at $x = c$ by showing that $f(x)$ approaches $f(c)$ as $x \to c$. To do this, use the differentiability of f at $x = c$ and consider the following limit.

$$\lim_{x \to c} [f(x) - f(c)] = \lim_{x \to c} \left[(x - c)\left(\frac{f(x) - f(c)}{x - c}\right)\right]$$

$$= \left[\lim_{x \to c} (x - c)\right]\left[\lim_{x \to c} \frac{f(x) - f(c)}{x - c}\right]$$

$$= (0)[f'(c)]$$

$$= 0$$

Because the difference $f(x) - f(c)$ approaches zero as $x \to c$, you can conclude that $\lim_{x \to c} f(x) = f(c)$. So, f is continuous at $x = c$.

You can summarize the relationship between continuity and differentiability as follows.

1. If a function is differentiable at $x = c$, then it is continuous at $x = c$. So, differentiability implies continuity.

2. It is possible for a function to be continuous at $x = c$ and not be differentiable at $x = c$. So, continuity does not imply differentiability.

EXERCISES FOR SECTION 2.1

In Exercises 1 and 2, estimate the slope of the graph at the point (x, y).

1. (a)   (b)

2. (a)   (b)

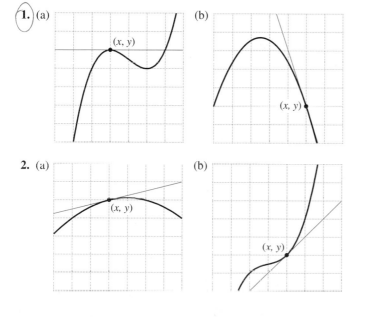

In Exercises 3 and 4, use the graph shown in the figure. To print an enlarged copy of the graph, go to the website *www.mathgraphs.com.*

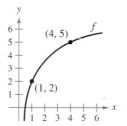

3. Identify or sketch each of the quantities on the figure.

(a) $f(1)$ and $f(4)$ (b) $f(4) - f(1)$

(c) $y = \dfrac{f(4) - f(1)}{4 - 1}(x - 1) + f(1)$

4. Insert the proper inequality symbol (< or >) between the given quantities.

(a) $\dfrac{f(4) - f(1)}{4 - 1}$ $\dfrac{f(4) - f(3)}{4 - 3}$

(b) $\dfrac{f(4) - f(1)}{4 - 1}$ $f'(1)$

In Exercises 5–10, find the slope of the tangent line to the graph of the function at the specified point.

5. $f(x) = 3 - 2x$, $(-1, 5)$

6. $g(x) = \frac{3}{2}x + 1$, $(-2, -2)$

7. $g(x) = x^2 - 4$, $(1, -3)$

8. $g(x) = 5 - x^2$, $(2, 1)$

9. $f(t) = 3t - t^2$, $(0, 0)$

10. $h(t) = t^2 + 3$, $(-2, 7)$

In Exercises 11–24, find the derivative by the limit process.

11. $f(x) = 3$

12. $g(x) = -5$

13. $f(x) = -5x$

14. $f(x) = 3x + 2$

15. $h(s) = 3 + \frac{2}{3}s$

16. $f(x) = 9 - \frac{1}{2}x$

17. $f(x) = 2x^2 + x - 1$

18. $f(x) = 1 - x^2$

19. $f(x) = x^3 - 12x$

20. $f(x) = x^3 + x^2$

21. $f(x) = \dfrac{1}{x - 1}$

22. $f(x) = \dfrac{1}{x^2}$

23. $f(x) = \sqrt{x + 1}$

24. $f(x) = \dfrac{4}{\sqrt{x}}$

In Exercises 25–32, (a) find an equation of the tangent line to the graph of f at the indicated point, (b) use a graphing utility to graph the function and its tangent line at the point, and (c) use the *derivative* feature of a graphing utility to confirm your results.

25. $f(x) = x^2 + 1$, $(2, 5)$

26. $f(x) = x^2 + 2x + 1$, $(-3, 4)$

27. $f(x) = x^3$, $(2, 8)$

28. $f(x) = x^3 + 1$, $(1, 2)$

29. $f(x) = \sqrt{x}$, $(1, 1)$

30. $f(x) = \sqrt{x - 1}$, $(5, 2)$

31. $f(x) = x + \dfrac{4}{x}$, $(4, 5)$

32. $f(x) = \dfrac{1}{x + 1}$, $(0, 1)$

In Exercises 33–36, find an equation of the line that is tangent to the graph of f *and* parallel to the given line.

Function	Line
33. $f(x) = x^3$	$3x - y + 1 = 0$
34. $f(x) = x^3 + 2$	$3x - y - 4 = 0$
35. $f(x) = \dfrac{1}{\sqrt{x}}$	$x + 2y - 6 = 0$
36. $f(x) = \dfrac{1}{\sqrt{x - 1}}$	$x + 2y + 7 = 0$

37. The tangent line to the graph of $y = g(x)$ at the point $(5, 2)$ passes through the point $(9, 0)$. Find $g(5)$ and $g'(5)$.

38. The tangent line to the graph of $y = h(x)$ at the point $(-1, 4)$ passes through the point $(3, 6)$. Find $h(-1)$ and $h'(-1)$.

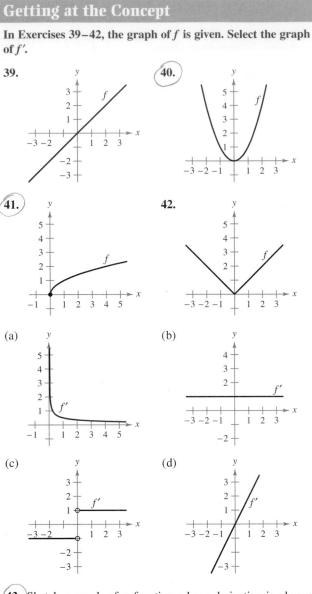

Getting at the Concept

In Exercises 39–42, the graph of f is given. Select the graph of f'.

39.

40.

41.

42.

(a)

(b)

(c)

(d)

43. Sketch a graph of a function whose derivative is always negative.

44. Sketch a graph of a function whose derivative is always positive.

45. Assume that $f'(c) = 3$. Find $f'(-c)$ if (a) f is an odd function and if (b) f is an even function.

46. Determine whether the limit yields the derivative of a differentiable function f. Explain.

(a) $\lim\limits_{\Delta x \to 0} \dfrac{f(x + 2\Delta x) - f(x)}{2\Delta x}$

(b) $\lim\limits_{\Delta x \to 0} \dfrac{f(x + 2) - f(x)}{\Delta x}$

(c) $\lim\limits_{\Delta x \to 0} \dfrac{f(x + \Delta x) - f(x - \Delta x)}{2\Delta x}$

(d) $\lim\limits_{\Delta x \to 0} \dfrac{f(x + \Delta x) - f(x)}{\Delta x}$

In Exercises 47 and 48, find equations of the two tangent lines to the graph of f that pass through the indicated point.

47. $f(x) = 4x - x^2$

48. $f(x) = x^2$

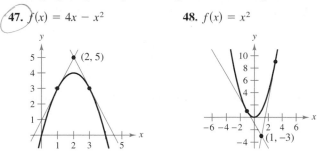

49. Graphical Reasoning   The figure shows the graph of g'.

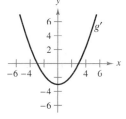

(a) $g'(0) = $ [ ]

(b) $g'(3) = $ [ ]

(c) What can you conclude about the graph of g knowing that $g'(1) = -\frac{8}{3}$?

(d) What can you conclude about the graph of g knowing that $g'(-4) = \frac{7}{3}$?

(e) Is $g(6) - g(4)$ positive or negative? Explain.

(f) Is it possible to find $g(2)$ from the graph? Explain.

50. Graphical Reasoning   Use a graphing utility to graph each function and its tangent lines when $x = -1$, $x = 0$, and $x = 1$. Based on the results, determine whether the slope of a tangent line to the graph of a function is always distinct for different values of x.

(a) $f(x) = x^2$ (b) $g(x) = x^3$

Graphical, Numerical, and Analytic Analysis   **In Exercises 51 and 52, use a graphing utility to graph f on the interval $[-2, 2]$. Complete the table by graphically estimating the slopes of the graph at the indicated points. Then evaluate the slopes analytically and compare your results with those obtained graphically.**

x	-2	-1.5	-1	-0.5	0	0.5	1	1.5	2
$f(x)$									
$f'(x)$									

51. $f(x) = \frac{1}{4}x^3$

52. $f(x) = \frac{1}{2}x^2$

Graphical Reasoning   **In Exercises 53 and 54, use a graphing utility to graph the functions f and g in the same viewing window where**

$$g(x) = \frac{f(x + 0.01) - f(x)}{0.01}.$$

Label the graphs and describe the relationship between them.

53. $f(x) = 2x - x^2$

54. $f(x) = 3\sqrt{x}$

In Exercises 55 and 56, evaluate $f(2)$ and $f(2.1)$ and use the results to approximate $f'(2)$.

55. $f(x) = x(4 - x)$

56. $f(x) = \frac{1}{4}x^3$

Graphical Reasoning   **In Exercises 57 and 58, use a graphing utility to graph the function and its derivative in the same viewing window. Label the graphs and describe the relationship between them.**

57. $f(x) = \dfrac{1}{\sqrt{x}}$

58. $f(x) = \dfrac{x^3}{4} - 3x$

Writing   **In Exercises 59 and 60, consider the functions f and $S_{\Delta x}$ where**

$$S_{\Delta x}(x) = \frac{f(2 + \Delta x) - f(2)}{\Delta x}(x - 2) + f(2).$$

(a) Use a graphing utility to graph f and $S_{\Delta x}$ in the same viewing window for $\Delta x = 1$, 0.5, and 0.1.

(b) Give a written description of the graphs of S for the different values of Δx in part (a).

59. $f(x) = 4 - (x - 3)^2$

60. $f(x) = x + \dfrac{1}{x}$

In Exercises 61–70, use the alternative form of the derivative to find the derivative at $x = c$ (if it exists).

61. $f(x) = x^2 - 1$, $c = 2$

62. $g(x) = x(x - 1)$, $c = 1$

63. $f(x) = x^3 + 2x^2 + 1$, $c = -2$

64. $f(x) = x^3 + 2x$, $c = 1$

65. $g(x) = \sqrt{|x|}$, $c = 0$

66. $f(x) = 1/x$, $c = 3$

67. $f(x) = (x - 6)^{2/3}$, $c = 6$

68. $g(x) = (x + 3)^{1/3}$, $c = -3$

69. $h(x) = |x + 5|$, $c = -5$

70. $f(x) = |x - 4|$, $c = 4$

In Exercises 71–80, describe the x-values at which f is differentiable.

71. $f(x) = |x + 3|$

72. $f(x) = |x^2 - 9|$

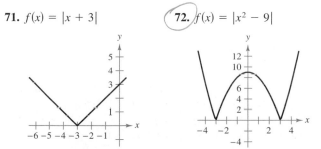

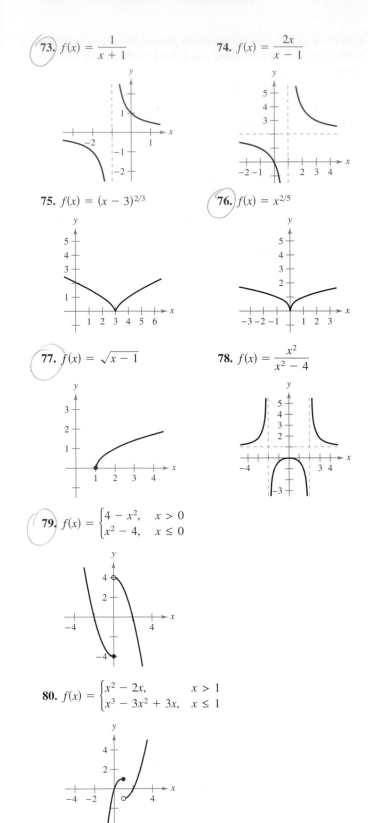

73. $f(x) = \dfrac{1}{x+1}$

74. $f(x) = \dfrac{2x}{x-1}$

75. $f(x) = (x-3)^{2/3}$

76. $f(x) = x^{2/5}$

77. $f(x) = \sqrt{x-1}$

78. $f(x) = \dfrac{x^2}{x^2-4}$

79. $f(x) = \begin{cases} 4-x^2, & x > 0 \\ x^2-4, & x \le 0 \end{cases}$

80. $f(x) = \begin{cases} x^2 - 2x, & x > 1 \\ x^3 - 3x^2 + 3x, & x \le 1 \end{cases}$

In Exercises 81–84, find the derivatives from the left and from the right at $x = 1$ (if they exist). Is the function differentiable at $x = 1$?

81. $f(x) = |x-1|$

82. $f(x) = \sqrt{1-x^2}$

83. $f(x) = \begin{cases} (x-1)^3, & x \le 1 \\ (x-1)^2, & x > 1 \end{cases}$

84. $f(x) = \begin{cases} x, & x \le 1 \\ x^2, & x > 1 \end{cases}$

In Exercises 85 and 86, determine whether the function is differentiable at $x = 2$.

85. $f(x) = \begin{cases} x^2 + 1, & x \le 2 \\ 4x - 3, & x > 2 \end{cases}$

86. $f(x) = \begin{cases} \frac{1}{2}x + 1, & x < 2 \\ \sqrt{2x}, & x \ge 2 \end{cases}$

87. *Graphical Reasoning* A line with slope m passes through the point $(0, 4)$ and has the equation $y = mx + 4$.

(a) Write the distance d between the line and the point $(3, 1)$ as a function of m.

(b) Use a graphing utility to graph the function d in part (a). Based on the graph, is the function differentiable at every value of m? If not, where is it not differentiable?

88. *Conjecture* Consider the functions $f(x) = x^2$ and $g(x) = x^3$.

(a) Graph f and f' on the same set of axes.

(b) Graph g and g' on the same set of axes.

(c) Identify any pattern between the functions f and g and their respective derivatives. Use the pattern to make a conjecture about $h'(x)$ if $h(x) = x^n$, where n is an integer and $n \ge 2$.

(d) Find $f'(x)$ if $f(x) = x^4$. Compare the result with the conjecture in part (c). Is this a proof of your conjecture? Explain.

True or False? **In Exercises 89–92, determine whether the statement is true or false. If it is false, explain why or give an example that shows it is false.**

89. The slope of the tangent line to the differentiable function f at the point $(2, f(2))$ is

$$\frac{f(x + \Delta x) - f(x)}{\Delta x}.$$

90. If a function is continuous at a point, then it is differentiable at that point.

91. If a function has derivatives from both the right and the left at a point, then it is differentiable at that point.

92. If a function is differentiable at a point, then it is continuous at that point.

93. Let $f(x) = \begin{cases} x \sin \frac{1}{x}, & x \ne 0 \\ 0, & x = 0 \end{cases}$ and $g(x) = \begin{cases} x^2 \sin \frac{1}{x}, & x \ne 0 \\ 0, & x = 0 \end{cases}$.

Show that f is continuous, but not differentiable, at $x = 0$. Show that g is differentiable at 0, and find $g'(0)$.

94. *Writing* Use a graphing utility to graph the two functions $f(x) = x^2 + 1$ and $g(x) = |x| + 1$ in the same viewing window. Use the *zoom* and *trace* features to analyze the graphs near the point $(0, 1)$. What do you observe? Which function is differentiable at this point? Write a short paragraph describing the geometric significance of differentiability at a point.

Section 2.2   **Basic Differentiation Rules and Rates of Change**

- Find the derivative of a function using the Constant Rule.
- Find the derivative of a function using the Power Rule.
- Find the derivative of a function using the Constant Multiple Rule.
- Find the derivative of a function using the Sum and Difference Rules.
- Find the derivative of the sine function and of the cosine function.
- Use derivatives to find rates of change.

The Constant Rule

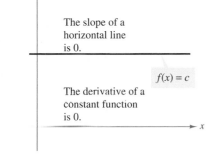

The slope of a horizontal line is 0.

$f(x) = c$

The derivative of a constant function is 0.

The Constant Rule
Figure 2.14

In Section 2.1 you used the limit definition to find derivatives. In this and the next two sections you will be introduced to several "differentiation rules" that allow you to find derivatives without the *direct* use of the limit definition.

THEOREM 2.2   The Constant Rule

The derivative of a constant function is 0. That is, if c is a real number, then

$$\frac{d}{dx}[c] = 0.$$

Proof   Let $f(x) = c$. Then, by the limit definition of the derivative,

$$\frac{d}{dx}[c] = f'(x)$$

$$= \lim_{\Delta x \to 0} \frac{f(x + \Delta x) - f(x)}{\Delta x}$$

$$= \lim_{\Delta x \to 0} \frac{c - c}{\Delta x}$$

$$= 0.$$

NOTE   In Figure 2.14, note that the Constant Rule is equivalent to saying that the slope of a horizontal line is 0. This demonstrates the relationship between slope and derivative.

Example 1   **Using the Constant Rule**

Function	Derivative
a. $y = 7$	$\dfrac{dy}{dx} = 0$
b. $f(x) = 0$	$f'(x) = 0$
c. $s(t) = -3$	$s'(t) = 0$
d. $y = k\pi^2$, k is constant	$y' = 0$

EXPLORATION

Writing a Conjecture   Use the definition of the derivative given in Section 2.1 to find the derivative of each of the following. What patterns do you see? Use your results to write a conjecture about the derivative of $f(x) = x^n$.

a. $f(x) = x^1$ **b.** $f(x) = x^2$ **c.** $f(x) = x^3$

d. $f(x) = x^4$ **e.** $f(x) = x^{1/2}$ **f.** $f(x) = x^{-1}$

The Power Rule

Before proving the next rule, we review the procedure for expanding a binomial.

$$(x + \Delta x)^2 = x^2 + 2x\Delta x + (\Delta x)^2$$
$$(x + \Delta x)^3 = x^3 + 3x^2\Delta x + 3x(\Delta x)^2 + (\Delta x)^3$$

The general binomial expansion for a positive integer n is

$$(x + \Delta x)^n = x^n + nx^{n-1}(\Delta x) + \underbrace{\frac{n(n-1)x^{n-2}}{2}(\Delta x)^2 + \cdots + (\Delta x)^n}.$$

$(\Delta x)^2$ is a factor of these terms.

This binomial expansion is used in proving a special case of the Power Rule.

THEOREM 2.3   The Power Rule

If n is a rational number, then the function $f(x) = x^n$ is differentiable and

$$\frac{d}{dx}[x^n] = nx^{n-1}.$$

For f to be differentiable at $x = 0$, n must be a number such that x^{n-1} is defined on an interval containing 0.

Proof   If n is a positive integer greater than 1, then the binomial expansion produces the following.

$$\frac{d}{dx}[x^n] = \lim_{\Delta x \to 0} \frac{(x + \Delta x)^n - x^n}{\Delta x}$$

$$= \lim_{\Delta x \to 0} \frac{x^n + nx^{n-1}(\Delta x) + \frac{n(n-1)x^{n-2}}{2}(\Delta x)^2 + \cdots + (\Delta x)^n - x^n}{\Delta x}$$

$$= \lim_{\Delta x \to 0} \left[nx^{n-1} + \frac{n(n-1)x^{n-2}}{2}(\Delta x) + \cdots + (\Delta x)^{n-1} \right]$$

$$= nx^{n-1} + 0 + \cdots + 0$$

$$= nx^{n-1}$$

This proves the case for which n is a positive integer greater than 1. We leave it to you to prove the case for $n = 1$. Example 7 in Section 2.3 proves the case for which n is a negative integer. In Exercise 63 in Section 2.5 you are asked to prove the case for which n is rational. (In Section 5.5, the Power Rule will be extended to cover irrational values of n.)

When using the Power Rule, the case for which $n = 1$ is best thought of as a separate differentiation rule. That is,

$$\frac{d}{dx}[x] = 1.$$ Power Rule when $n = 1$

This rule is consistent with the fact that the slope of the line $y = x$ is 1, as shown in Figure 2.15.

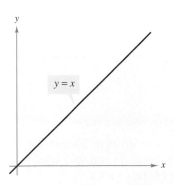

$y = x$

The slope of the line $y = x$ is 1.
Figure 2.15

Example 2   **Using the Power Rule**

Function	*Derivative*
a. $f(x) = x^3$	$f'(x) = 3x^2$
b. $g(x) = \sqrt[3]{x}$	$g'(x) = \dfrac{d}{dx}[x^{1/3}] = \dfrac{1}{3}x^{-2/3} = \dfrac{1}{3x^{2/3}}$
c. $y = \dfrac{1}{x^2}$	$\dfrac{dy}{dx} = \dfrac{d}{dx}[x^{-2}] = (-2)x^{-3} = -\dfrac{2}{x^3}$

In Example 2c, note that *before* differentiating, $1/x^2$ was rewritten as x^{-2}. Rewriting is the first step in *many* differentiation problems.

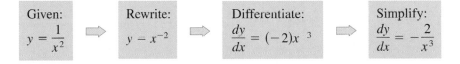

Given: $y = \dfrac{1}{x^2}$ ⟹ Rewrite: $y = x^{-2}$ ⟹ Differentiate: $\dfrac{dy}{dx} = (-2)x^{-3}$ ⟹ Simplify: $\dfrac{dy}{dx} = -\dfrac{2}{x^3}$

Example 3   **Finding the Slope of a Graph**

Find the slope of the graph of $f(x) = x^4$ when

a. $x = -1$ **b.** $x = 0$ **c.** $x = 1$.

Solution   The derivative of f is $f'(x) = 4x^3$.

a. When $x = -1$, the slope is $f'(-1) = 4(-1)^3 = -4$.   Slope is negative.
b. When $x = 0$, the slope is $f'(0) = 4(0)^3 = 0$.   Slope is zero.
c. When $x = 1$, the slope is $f'(1) = 4(1)^3 = 4$.   Slope is positive.

In Figure 2.16, note that the slope of the graph is negative at the point $(-1, 1)$, the slope is zero at the point $(0, 0)$, and the slope is positive at the point $(1, 1)$.

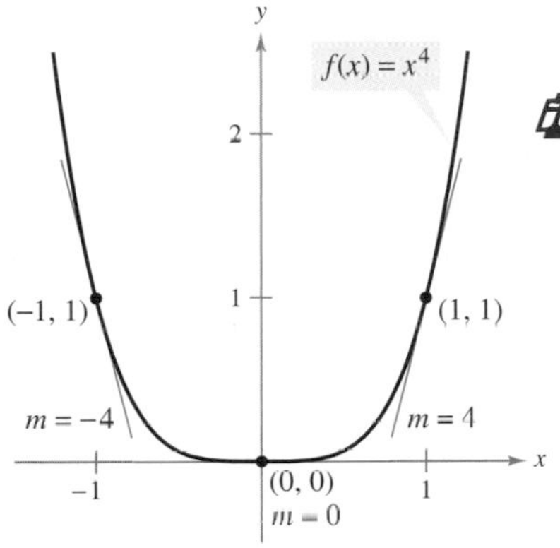

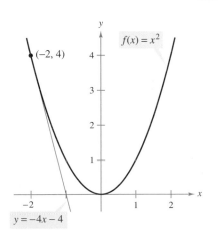

The slope of a graph at a point is the value of the derivative at that point.
Figure 2.16

Example 4   **Finding an Equation of a Tangent Line**

Find an equation of the tangent line to the graph of $f(x) = x^2$ when $x = -2$.

Solution   To find the *point* on the graph of f, evaluate the original function at $x = -2$.

$$(-2, f(-2)) = (-2, 4)   \text{Point on graph}$$

To find the *slope* of the graph when $x = -2$, evaluate the derivative, $f'(x) = 2x$, at $x = -2$.

$$m = f'(-2) = -4   \text{Slope of graph at } (-2, 4)$$

Now, using the point-slope form of the equation of a line, you can write

$$y - y_1 = m(x - x_1)   \text{Point-slope form}$$
$$y - 4 = -4[x - (-2)]   \text{Substitute for } y_1, m, \text{ and } x_1.$$
$$y = -4x - 4.   \text{Simplify.}$$

(See Figure 2.17.)

The line $y = -4x - 4$ is tangent to the graph of $f(x) = x^2$ at the point $(-2, 4)$.
Figure 2.17

The Constant Multiple Rule

THEOREM 2.4   The Constant Multiple Rule

If f is a differentiable function and c is a real number, then cf is also differentiable and

$$\frac{d}{dx}[cf(x)] = cf'(x).$$

Proof

$$\frac{d}{dx}[cf(x)] = \lim_{\Delta x \to 0} \frac{cf(x + \Delta x) - cf(x)}{\Delta x} \qquad \text{Definition of derivative}$$

$$= \lim_{\Delta x \to 0} c\left[\frac{f(x + \Delta x) - f(x)}{\Delta x}\right]$$

$$= c\left[\lim_{\Delta x \to 0} \frac{f(x + \Delta x) - f(x)}{\Delta x}\right]$$

$$= cf'(x) \qquad \blacksquare$$

Informally, the Constant Multiple Rule states that constants can be factored out of the differentiation process, even if the constants appear in the denominator.

$$\frac{d}{dx}[cf(x)] = c\frac{d}{dx}[f(x)] = cf'(x)$$

$$\frac{d}{dx}\left[\frac{f(x)}{c}\right] = \frac{d}{dx}\left[\left(\frac{1}{c}\right)f(x)\right]$$

$$= \left(\frac{1}{c}\right)\frac{d}{dx}[f(x)] = \left(\frac{1}{c}\right)f'(x)$$

Example 5   **Using the Constant Multiple Rule**

Function	*Derivative*
a. $y = \dfrac{2}{x}$	$\dfrac{dy}{dx} = \dfrac{d}{dx}[2x^{-1}] = 2\dfrac{d}{dx}[x^{-1}] = 2(-1)x^{-2} = -\dfrac{2}{x^2}$
b. $f(t) = \dfrac{4t^2}{5}$	$f'(t) = \dfrac{d}{dt}\left[\dfrac{4}{5}t^2\right] = \dfrac{4}{5}\dfrac{d}{dt}[t^2] = \dfrac{4}{5}(2t) = \dfrac{8}{5}t$
c. $y = 2\sqrt{x}$	$\dfrac{dy}{dx} = \dfrac{d}{dx}[2x^{1/2}] = 2\left(\dfrac{1}{2}x^{-1/2}\right) = x^{-1/2} = \dfrac{1}{\sqrt{x}}$
d. $y = \dfrac{1}{2\sqrt[3]{x^2}}$	$\dfrac{dy}{dx} = \dfrac{d}{dx}\left[\dfrac{1}{2}x^{-2/3}\right] = \dfrac{1}{2}\left(-\dfrac{2}{3}\right)x^{-5/3} = -\dfrac{1}{3x^{5/3}}$
e. $y = -\dfrac{3x}{2}$	$y' = \dfrac{d}{dx}\left[-\dfrac{3}{2}x\right] = -\dfrac{3}{2}(1) = -\dfrac{3}{2}$

NOTE   The Constant Multiple Rule and the Power Rule can be combined into one rule. The combination rule is $D_x[cx^n] = cnx^{n-1}$.

Example 6   **Using Parentheses When Differentiating**

Original Function	*Rewrite*	*Differentiate*	*Simplify*
a. $y = \dfrac{5}{2x^3}$	$y = \dfrac{5}{2}(x^{-3})$	$y' = \dfrac{5}{2}(-3x^{-4})$	$y' = -\dfrac{15}{2x^4}$
b. $y = \dfrac{5}{(2x)^3}$	$y = \dfrac{5}{8}(x^{-3})$	$y' = \dfrac{5}{8}(-3x^{-4})$	$y' = -\dfrac{15}{8x^4}$
c. $y = \dfrac{7}{3x^{-2}}$	$y = \dfrac{7}{3}(x^2)$	$y' = \dfrac{7}{3}(2x)$	$y' = \dfrac{14x}{3}$
d. $y = \dfrac{7}{(3x)^{-2}}$	$y = 63(x^2)$	$y' = 63(2x)$	$y' = 126x$

The Sum and Difference Rules

> **THEOREM 2.5   The Sum and Difference Rules**
>
> The sum (or difference) of two differentiable functions is differentiable and is the sum (or difference) of their derivatives.
>
> $$\frac{d}{dx}[f(x) + g(x)] = f'(x) + g'(x) \qquad \text{Sum Rule}$$
>
> $$\frac{d}{dx}[f(x) - g(x)] = f'(x) - g'(x) \qquad \text{Difference Rule}$$

Proof A proof of the Sum Rule follows from Theorem 1.2. (The Difference Rule can be proved in a similar way.)

$$\frac{d}{dx}[f(x) + g(x)] = \lim_{\Delta x \to 0} \frac{[f(x + \Delta x) + g(x + \Delta x)] - [f(x) + g(x)]}{\Delta x}$$

$$= \lim_{\Delta x \to 0} \frac{f(x + \Delta x) + g(x + \Delta x) - f(x) - g(x)}{\Delta x}$$

$$= \lim_{\Delta x \to 0} \left[\frac{f(x + \Delta x) - f(x)}{\Delta x} + \frac{g(x + \Delta x) - g(x)}{\Delta x} \right]$$

$$= \lim_{\Delta x \to 0} \frac{f(x + \Delta x) - f(x)}{\Delta x} + \lim_{\Delta x \to 0} \frac{g(x + \Delta x) - g(x)}{\Delta x}$$

$$= f'(x) + g'(x)$$

The Sum and Difference Rules can be extended to any finite number of functions. For instance, if $F(x) = f(x) + g(x) - h(x)$, then $F'(x) = f'(x) + g'(x) - h'(x)$.

Example 7   **Using the Sum and Difference Rules**

Function	*Derivative*
a. $f(x) = x^3 - 4x + 5$	$f'(x) = 3x^2 - 4$
b. $g(x) = -\dfrac{x^4}{2} + 3x^3 - 2x$	$g'(x) = -2x^3 + 9x^2 - 2$

FOR FURTHER INFORMATION For the outline of a geometric proof of the derivatives of the sine and cosine functions, see the article "The Spider's Spacewalk Derivation of sin′ and cos′" by Tim Hesterberg in *The College Mathematics Journal*. To view this article, go to the website *www.matharticles.com*.

Derivatives of Sine and Cosine Functions

In Section 1.3, you studied the following limits.

$$\lim_{\Delta x \to 0} \frac{\sin \Delta x}{\Delta x} = 1 \quad \text{and} \quad \lim_{\Delta x \to 0} \frac{1 - \cos \Delta x}{\Delta x} = 0$$

These two limits can be used to prove differentiation rules for the sine and cosine functions. (The derivatives of the other four trigonometric functions are discussed in Section 2.3.)

THEOREM 2.6   Derivatives of Sine and Cosine Functions

$$\frac{d}{dx}[\sin x] = \cos x \qquad \qquad \frac{d}{dx}[\cos x] = -\sin x$$

Proof

$$\frac{d}{dx}[\sin x] = \lim_{\Delta x \to 0} \frac{\sin(x + \Delta x) - \sin x}{\Delta x} \qquad \text{Definition of derivative}$$

$$= \lim_{\Delta x \to 0} \frac{\sin x \cos \Delta x + \cos x \sin \Delta x - \sin x}{\Delta x}$$

$$= \lim_{\Delta x \to 0} \frac{\cos x \sin \Delta x - (\sin x)(1 - \cos \Delta x)}{\Delta x}$$

$$= \lim_{\Delta x \to 0} \left[(\cos x)\left(\frac{\sin \Delta x}{\Delta x}\right) - (\sin x)\left(\frac{1 - \cos \Delta x}{\Delta x}\right) \right]$$

$$= \cos x \left(\lim_{\Delta x \to 0} \frac{\sin \Delta x}{\Delta x} \right) - \sin x \left(\lim_{\Delta x \to 0} \frac{1 - \cos \Delta x}{\Delta x} \right)$$

$$= (\cos x)(1) - (\sin x)(0)$$

$$= \cos x$$

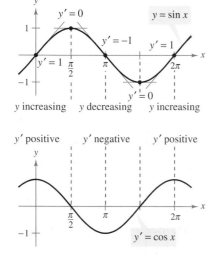

The derivative of the sine function is the cosine function.
Figure 2.18

This differentiation rule is shown graphically in Figure 2.18. Note that for each x, the *slope* of the sine curve is equal to the value of the cosine. The proof of the second rule is left as an exercise (see Exercise 113).

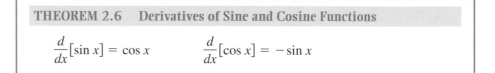

Example 8   Derivatives Involving Sines and Cosines

Function	*Derivative*
a. $y = 2 \sin x$	$y' = 2 \cos x$
b. $y = \dfrac{\sin x}{2} = \dfrac{1}{2} \sin x$	$y' = \dfrac{1}{2} \cos x = \dfrac{\cos x}{2}$
c. $y = x + \cos x$	$y' = 1 - \sin x$

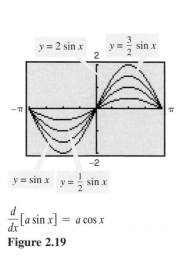

$$\frac{d}{dx}[a \sin x] = a \cos x$$
Figure 2.19

> **TECHNOLOGY**   A graphing utility can provide insight into the interpretation of a derivative. For instance, Figure 2.19 shows the graphs of
>
> $$y = a \sin x$$
>
> for $a = \frac{1}{2}$, 1, $\frac{3}{2}$, and 2. Estimate the slope of each graph at the point $(0, 0)$. Then verify your estimates analytically by evaluating the derivative of each function when $x = 0$.

Rates of Change

You have seen how the derivative is used to determine slope. The derivative can also be used to determine the rate of change of one variable with respect to another. Applications involving rates of change occur in a wide variety of fields. A few examples are population growth rates, production rates, water flow rates, velocity, and acceleration.

A common use of rate of change is to describe the motion of an object moving in a straight line. In such problems, it is customary to use either a horizontal or a vertical line with a designated origin to represent the line of motion. On such lines, movement to the right (or upward) is considered to be in the positive direction, and movement to the left (or downward) is considered to be in the negative direction.

The function s that gives the position (relative to the origin) of an object as a function of time t is called a **position function.** If, over a period of time Δt, the object changes its position by the amount $\Delta s = s(t + \Delta t) - s(t)$, then, by the familiar formula

$$\text{Rate} = \frac{\text{distance}}{\text{time}}$$

the **average velocity** is

$$\frac{\text{Change in distance}}{\text{Change in time}} = \frac{\Delta s}{\Delta t}. \qquad \text{Average velocity}$$

Example 9   **Finding Average Velocity of a Falling Object**

If a billiard ball is dropped from a height of 100 feet, its height s at time t is given by the position function

$$s = -16t^2 + 100 \qquad \text{Position function}$$

where s is measured in feet and t is measured in seconds. Find the average velocity over each of the following time intervals.

a. $[1, 2]$ **b.** $[1, 1.5]$ **c.** $[1, 1.1]$

Solution

a. For the interval $[1, 2]$ the object falls from a height of $s(1) = -16(1)^2 + 100 = 84$ feet to a height of $s(2) = -16(2)^2 + 100 = 36$ feet. The average velocity is

$$\frac{\Delta s}{\Delta t} = \frac{36 - 84}{2 - 1} = \frac{-48}{1} = -48 \text{ feet per second.}$$

b. For the interval $[1, 1.5]$, the object falls from a height of 84 feet to a height of 64 feet. The average velocity is

$$\frac{\Delta s}{\Delta t} = \frac{64 - 84}{1.5 - 1} = \frac{-20}{0.5} = -40 \text{ feet per second.}$$

c. For the interval $[1, 1.1]$, the object falls from a height of 84 feet to a height of 80.64 feet. The average velocity is

$$\frac{\Delta s}{\Delta t} = \frac{80.64 - 84}{1.1 - 1} = \frac{-3.36}{0.1} = -33.6 \text{ feet per second.}$$

Note that the average velocities are *negative*, indicating that the object is moving downward.

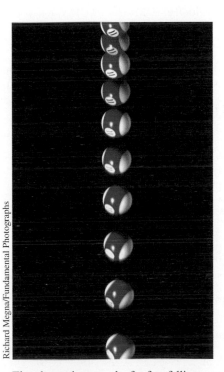

Time-lapse photograph of a free-falling billiard ball

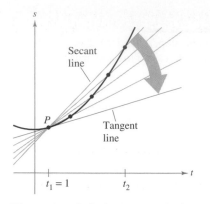

The average velocity between t_1 and t_2 is the slope of the secant line, and the instantaneous velocity at t_1 is the slope of the tangent line.
Figure 2.20

Suppose that in Example 9 you wanted to find the *instantaneous* velocity (or simply the velocity) of the object when $t = 1$. Just as you can approximate the slope of the tangent line by calculating the slope of the secant line, you can approximate the velocity at $t = 1$ by calculating the average velocity over a small interval $[1, 1 + \Delta t]$ (see Figure 2.20). By taking the limit as Δt approaches zero, you obtain the velocity when $t = 1$. Try doing this—you will find that the velocity when $t = 1$ is -32 feet per second.

In general, if $s = s(t)$ is the position function for an object moving along a straight line, the **velocity** of the object at time t is

$$v(t) = \lim_{\Delta t \to 0} \frac{s(t + \Delta t) - s(t)}{\Delta t} = s'(t).$$ Velocity function

In other words, the velocity function is the derivative of the position function. Velocity can be negative, zero, or positive. The **speed** of an object is the absolute value of its velocity. Speed cannot be negative.

The position of a free-falling object (neglecting air resistance) under the influence of gravity can be represented by the equation

$$s(t) = \frac{1}{2}gt^2 + v_0 t + s_0$$ Position function

where s_0 is the initial height of the object, v_0 is the initial velocity of the object, and g is the acceleration due to gravity. On earth, the value of g is approximately -32 feet per second per second or -9.8 meters per second per second.

Example 10   **Using the Derivative to Find Velocity**

At time $t = 0$, a diver jumps from a platform diving board that is 32 feet above the water (see Figure 2.21). The position of the diver is given by

$$s(t) = -16t^2 + 16t + 32$$ Position function

where s is measured in feet and t is measured in seconds.

a. When does the diver hit the water?

b. What is the diver's velocity at impact?

Solution

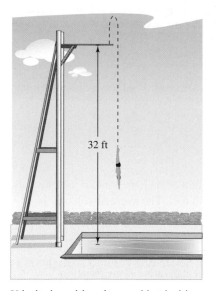

a. To find the time t when the diver hits the water, let $s = 0$ and solve for t.

$-16t^2 + 16t + 32 = 0$	Set position function equal to 0.
$-16(t + 1)(t - 2) = 0$	Factor.
$t = -1$ or 2	Solve for t.

Velocity is positive when an object is rising, and is negative when an object is falling.
Figure 2.21

Because $t \geq 0$, choose the positive value to conclude that the diver hits the water at $t = 2$ seconds.

b. The velocity at time t is given by the derivative $s'(t) = -32t + 16$. Therefore, the velocity at time $t = 2$ is

$$s'(2) = -32(2) + 16 = -48 \text{ feet per second.}$$

NOTE   In Figure 2.21, note that the diver moves upward for the first half-second because the velocity is positive for $0 < t < \frac{1}{2}$. When the velocity is 0, the diver has reached the maximum height of the dive.

EXERCISES FOR SECTION 2.2

In Exercises 1 and 2, use the graph to estimate the slope of the tangent line to $y = x^n$ at the point $(1, 1)$. Verify your answer analytically. To print an enlarged copy of the graph, go to the website *www.mathgraphs.com*.

1. (a) $y = x^{1/2}$ (b) $y = x^{3/2}$

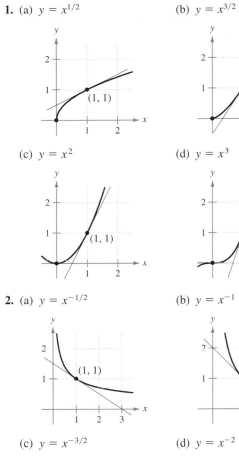

(c) $y = x^2$ (d) $y = x^3$

2. (a) $y = x^{-1/2}$ (b) $y = x^{-1}$

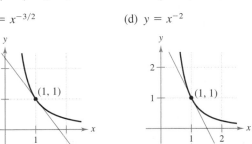

(c) $y = x^{-3/2}$ (d) $y = x^{-2}$

In Exercises 3–24, find the derivative of the function.

3. $y = 8$ **4.** $f(x) = -2$

5. $y = x^6$ **6.** $y = x^8$

7. $y = \dfrac{1}{x^7}$ **8.** $y = \dfrac{1}{x^8}$

9. $f(x) = \sqrt[5]{x}$ **10.** $g(x) = \sqrt[4]{x}$

11. $f(x) = x + 1$ **12.** $g(x) = 3x - 1$

13. $f(t) = -2t^2 + 3t - 6$ **14.** $y = t^2 + 2t - 3$

15. $g(x) = x^2 + 4x^3$ **16.** $y = 8 - x^3$

17. $s(t) = t^3 - 2t + 4$ **18.** $f(x) = 2x^3 - x^2 + 3x$

19. $y = \dfrac{\pi}{2} \sin \theta - \cos \theta$ **20.** $g(t) = \pi \cos t$

21. $y = x^2 - \tfrac{1}{2} \cos x$ **22.** $y = 5 + \sin x$

23. $y = \dfrac{1}{x} - 3 \sin x$ **24.** $y = \dfrac{5}{(2x)^3} + 2 \cos x$

In Exercises 25–30, complete the table, using Example 6 as a model.

Original Function	Rewrite	Differentiate	Simplify
25. $y = \dfrac{5}{2x^2}$			
26. $y = \dfrac{2}{3x^2}$			
27. $y = \dfrac{3}{(2x)^3}$			
28. $y = \dfrac{\pi}{(3x)^2}$			
29. $y = \dfrac{\sqrt{x}}{x}$			
30. $y = \dfrac{4}{x^{-3}}$			

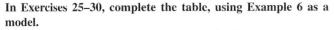 In Exercises 31–38, find the slope of the graph of the function at the indicated point. Use the *derivative* feature of a graphing utility to confirm your results.

Function	Point
31. $f(x) = \dfrac{3}{x^2}$	$(1, 3)$
32. $f(t) = 3 - \dfrac{3}{5t}$	$\left(\tfrac{3}{5}, 2\right)$
33. $f(x) = -\tfrac{1}{2} + \tfrac{7}{5}x^3$	$\left(0, -\tfrac{1}{2}\right)$
34. $y = 3x^3 - 6$	$(2, 18)$
35. $y = (2x + 1)^2$	$(0, 1)$
36. $f(x) = 3(5 - x)^2$	$(5, 0)$
37. $f(\theta) = 4 \sin \theta - \theta$	$(0, 0)$
38. $g(t) = 2 + 3 \cos t$	$(\pi, -1)$

In Exercises 39–52, find the derivative of the function.

39. $f(x) = x^2 + 5 - 3x^{-2}$ **40.** $f(x) = x^2 - 3x - 3x^{-2}$

41. $g(t) = t^2 - \dfrac{4}{t^3}$ **42.** $f(x) = x + \dfrac{1}{x^2}$

43. $f(x) = \dfrac{x^3 - 3x^2 + 4}{x^2}$ **44.** $h(x) = \dfrac{2x^2 - 3x + 1}{x}$

45. $y = x(x^2 + 1)$ **46.** $y = 3x(6x - 5x^2)$

47. $f(x) = \sqrt{x} - 6\sqrt[3]{x}$ **48.** $f(x) = \sqrt[3]{x} + \sqrt[5]{x}$

49. $h(s) = s^{4/5} - s^{2/3}$ **50.** $f(t) = t^{2/3} - t^{1/3} + 4$

51. $f(x) = 6\sqrt{x} + 5 \cos x$ **52.** $f(x) = \dfrac{2}{\sqrt[3]{x}} + 3 \cos x$

In Exercises 53–56, (a) find an equation of the tangent line to the graph of f at the indicated point, (b) use a graphing utility to graph the function and its tangent line at the point, and (c) use the *derivative* feature of a graphing utility to confirm your results.

Function	Point
53. $y = x^4 - 3x^2 + 2$	$(1, 0)$
54. $y = x^3 + x$	$(-1, -2)$
55. $f(x) = \dfrac{2}{\sqrt[4]{x^3}}$	$(1, 2)$
56. $y = (x^2 + 2x)(x + 1)$	$(1, 6)$

In Exercises 57–62, determine the point(s) (if any) at which the graph of the function has a horizontal tangent line.

57. $y = x^4 - 8x^2 + 2$ **58.** $y = x^3 + x$

59. $y = \dfrac{1}{x^2}$ **60.** $y = x^2 + 1$

61. $y = x + \sin x, \quad 0 \le x < 2\pi$

62. $y = \sqrt{3}x + 2\cos x, \quad 0 \le x < 2\pi$

In Exercises 63–66, find k such that the line is tangent to the graph of the function.

Function	Line
63. $f(x) = x^2 - kx$	$y = 4x - 9$
64. $f(x) = k - x^2$	$y = -4x + 7$
65. $f(x) = \dfrac{k}{x}$	$y = -\dfrac{3}{4}x + 3$
66. $f(x) = k\sqrt{x}$	$y = x + 4$

Getting at the Concept

67. Use the graph of f to answer each question. To print an enlarged copy of the graph, go to the website *www.mathgraphs.com*.

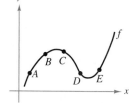

(a) Between which two consecutive points is the average rate of change of the function greatest?

(b) Is the average rate of change of the function between A and B greater than or less than the instantaneous rate of change at B?

(c) Sketch a tangent line to the graph between C and D such that the slope of the tangent line is the same as the average rate of change of the function between C and D.

68. Sketch the graph of a function f such that $f' > 0$ for all x and the rate of change of the function is decreasing.

Getting at the Concept *(continued)*

In Exercises 69 and 70, the relationship between f and g is given. Give the relationship between f' and g'.

69. $g(x) = f(x) + 6$ **70.** $g(x) = -5f(x)$

In Exercises 71 and 72, the graphs of a function f and its derivative f' are shown on the same set of coordinate axes. Label the graphs as f or f' and write a short paragraph stating the criteria used in making the selection. To print an enlarged copy of the graph, go to the website *www.mathgraphs.com*.

71.   **72.**

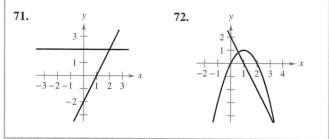

73. Sketch the graphs of $y = x^2$ and $y = -x^2 + 6x - 5$, and sketch the two lines that are tangent to both graphs. Find equations of these lines.

74. Show that the graphs of the two equations $y = x$ and $y = 1/x$ have tangent lines that are perpendicular to each other at their point of intersection.

In Exercises 75 and 76, find an equation of the tangent line to the graph of the function f through the point (x_0, y_0) not on the graph. To find the point of tangency (x, y) on the graph of f, solve the equation

$$f'(x) = \frac{y_0 - y}{x_0 - x}.$$

75. $f(x) = \sqrt{x}$ **76.** $f(x) = \dfrac{2}{x}$

$(x_0, y_0) = (-4, 0)$ $(x_0, y_0) = (5, 0)$

77. *Linear Approximation* Use a graphing utility (in square mode) to zoom in on the graph of $f(x) = 4 - \frac{1}{2}x^2$ to approximate $f'(1)$. Use the derivative to find $f'(1)$.

78. *Linear Approximation* Use a graphing utility (in square mode) to zoom in on the graph of $f(x) = 4\sqrt{x} + 1$ to approximate $f'(4)$. Use the derivative to find $f'(4)$.

79. *Linear Approximation* Consider the function $f(x) = x^{3/2}$ with the solution point $(4, 8)$.

(a) Use a graphing utility to obtain the graph of f. Use the *zoom* feature to obtain successive magnifications of the graph in the neighborhood of the point $(4, 8)$. After zooming in a few times, the graph should appear nearly linear. Use the *trace* feature to determine the coordinates of a point "near" $(4, 8)$. Find an equation of the secant line $S(x)$ through the two points.

(b) Find the equation of the line

$$T(x) = f'(4)(x - 4) + f(4)$$

tangent to the graph of f passing through the given point. Why are the linear functions S and T nearly the same?

(c) Use a graphing utility to graph f and T on the same set of coordinate axes. Note that T is a "good" approximation of f when x is "close to" 4. What happens to the accuracy of the approximation as you move farther away from the point of tangency?

(d) Demonstrate the conclusion in part (c) by completing the table.

Δx	-3	-2	-1	-0.5	-0.1	0
$f(4 + \Delta x)$						
$T(4 + \Delta x)$						

Δx	0.1	0.5	1	2	3
$f(4 + \Delta x)$					
$T(4 + \Delta x)$					

80. Linear Approximation Repeat Exercise 79 for the function $f(x) = x^3$ where $T(x)$ is the line tangent to the graph at the point $(1, 1)$. Explain why the accuracy of the linear approximation decreases more rapidly than in Exercise 79.

True or False? In Exercises 81–86, determine whether the statement is true or false. If it is false, explain why or give an example that shows it is false.

81. If $f'(x) = g'(x)$, then $f(x) = g(x)$.

82. If $f(x) = g(x) + c$, then $f'(x) = g'(x)$.

83. If $y = \pi^2$, then $dy/dx = 2\pi$.

84. If $y = x/\pi$, then $dy/dx = 1/\pi$.

85. If $g(x) = 3f(x)$, then $g'(x) = 3f'(x)$.

86. If $f(x) = 1/x^n$, then $f'(x) = 1/(nx^{n-1})$.

In Exercises 87–90, find the average rate of change of the function over the indicated interval. Compare this average rate of change with the instantaneous rates of change at the endpoints of the interval.

Function	Interval
87. $f(t) = 2t + 7$	$[1, 2]$
88. $f(t) = t^2 - 3$	$[2, 2.1]$
89. $f(x) = \dfrac{-1}{x}$	$[1, 2]$
90. $f(x) = \sin x$	$\left[0, \dfrac{\pi}{6}\right]$

Vertical Motion In Exercises 91 and 92, use the position function $s(t) = -16t^2 + v_0 t + s_0$ for free-falling objects.

91. A silver dollar is dropped from the top of a building that is 1362 feet tall.

(a) Determine the position and velocity functions for the coin.

(b) Determine the average velocity on the interval $[1, 2]$.

(c) Find the instantaneous velocities when $t = 1$ and $t = 2$.

(d) Find the time required for the coin to reach ground level.

(e) Find the velocity of the coin at impact.

92. A ball is thrown straight down from the top of a 220-foot building with an initial velocity of -22 feet per second. What is its velocity after 3 seconds? What is its velocity after falling 108 feet?

Vertical Motion In Exercises 93 and 94, use the position function $s(t) = -4.9t^2 + v_0 t + s_0$ for free-falling objects.

93. A projectile is shot upward from the surface of earth with an initial velocity of 120 meters per second. What is its velocity after 5 seconds? After 10 seconds?

94. To estimate the height of a building, a stone is dropped from the top of the building into a pool of water at ground level. How high is the building if the splash is seen 6.8 seconds after the stone is dropped?

Think About It In Exercises 95 and 96, the graph of a position function is shown. It represents the distance in miles that a person drives during a 10-minute trip to work. Make a sketch of the corresponding velocity function.

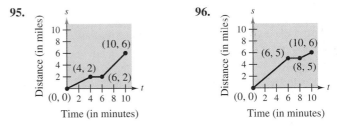

Think About It In Exercises 97 and 98, the graph of a velocity function is shown. It represents the velocity in miles per hour during a 10-minute drive to work. Make a sketch of the corresponding position function.

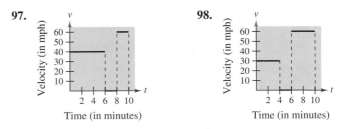

99. Modeling Data The stopping distance of an automobile traveling at a speed v (kilometers per hour) is the distance R (meters) the car travels during the reaction time of the driver plus the distance B (meters) the car travels after the brakes are applied (see figure). The table shows the results of an experiment.

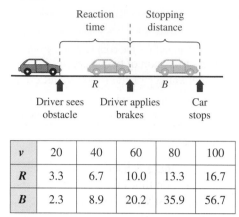

v	20	40	60	80	100
R	3.3	6.7	10.0	13.3	16.7
B	2.3	8.9	20.2	35.9	56.7

(a) Use the regression capabilities of a graphing utility to find a linear model for reaction time.

(b) Use the regression capabilities of a graphing utility to find a quadratic model for braking time.

(c) Determine the polynomial giving the total stopping distance T.

(d) Use a graphing utility to graph the functions R, B, and T in the same viewing window.

(e) Find the derivative of T and the rate of change of the total stopping distance for $v = 40$, $v = 80$, and $v = 100$.

(f) Use the results of this exercise to draw conclusions about the total stopping distance as speed increases.

100. Velocity Verify that the average velocity over the time interval $[t_0 - \Delta t, t_0 + \Delta t]$ is the same as the instantaneous velocity at $t = t_0$ for the position function

$$s(t) = -\tfrac{1}{2}at^2 + c.$$

101. Area The area of a square with sides of length s is given by $A = s^2$. Find the rate of change of the area with respect to s when $s = 4$ meters.

102. Volume The volume of a cube with sides of length s is given by $V = s^3$. Find the rate of change of the volume with respect to s when $s = 4$ centimeters.

103. Inventory Management The annual inventory cost C for a certain manufacturer is

$$C = \frac{1,008,000}{Q} + 6.3Q$$

where Q is the order size when the inventory is replenished. Find the change in annual cost when Q is increased from 350 to 351, and compare this with the instantaneous rate of change when $Q = 350$.

104. Fuel Cost A car is driven 15,000 miles a year and gets x miles per gallon. Assume that the average fuel cost is $1.25 per gallon. Find the annual cost of fuel C as a function of x and use this function to complete the table.

x	10	15	20	25	30	35	40
C							
$\dfrac{dC}{dx}$							

Who would benefit more from a 1-mile-per-gallon increase in fuel efficiency—the driver of a car that gets 15 miles per gallon or the driver of a car that gets 35 miles per gallon? Explain.

105. Writing The number of gallons N of regular unleaded gasoline sold by a gasoline station at a price of p dollars per gallon is given by $N = f(p)$.

(a) Describe the meaning of $f'(1.479)$.

(b) Is $f'(1.479)$ usually positive or negative? Explain.

106. Newton's Law of Cooling This law states that the rate of change of the temperature of an object is proportional to the difference between the object's temperature T and the temperature T_a of the surrounding medium. Write an equation for this law.

107. Find an equation of the parabola $y = ax^2 + bx + c$ that passes through $(0, 1)$ and is tangent to the line $y = x - 1$ at $(1, 0)$.

108. Let (a, b) be an arbitrary point on the graph of $y = 1/x$, $x > 0$. Prove that the area of the triangle formed by the tangent line through (a, b) and the coordinate axes is 2.

109. Find the tangent line(s) to the curve $y = x^3 - 9x$ through the point $(1, -9)$.

110. Find the equation(s) of the tangent line(s) to the parabola $y = x^2$ through the given point.

(a) $(0, a)$ (b) $(a, 0)$

Are there any restrictions on the constant a?

111. Find a and b such that

$$f(x) = \begin{cases} ax^3, & x \le 2 \\ x^2 + b, & x > 2 \end{cases}$$

is differentiable everywhere.

112. Where are the functions $f_1(x) = |\sin x|$ and $f_2(x) = \sin |x|$ differentiable?

113. Prove that $\dfrac{d}{dx}[\cos x] = -\sin x$.

FOR FURTHER INFORMATION For a geometric interpretation of the derivatives of trigonometric functions, see the article "Sines and Cosines of the Times" by Victor J. Katz in *Math Horizons*. To view this article, go to the website *www.matharticles.com*.

- Find the derivative of a function using the Product Rule.
- Find the derivative of a function using the Quotient Rule.
- Find the derivative of a trigonometric function.
- Find a higher-order derivative of a function.

The Product Rule

In Section 2.2 you learned that the derivative of the sum of two functions is simply the sum of their derivatives. The rules for the derivatives of the product and quotient of two functions are not as simple.

THEOREM 2.7   The Product Rule

The product of two differentiable functions f and g is itself differentiable. Moreover, the derivative of fg is the first function times the derivative of the second, plus the second function times the derivative of the first.

$$\frac{d}{dx}[f(x)g(x)] = f(x)g'(x) + g(x)f'(x)$$

Proof   Some mathematical proofs, such as the proof of the Sum Rule, are straightforward. Others involve clever steps that may appear unmotivated to a reader. This proof involves such a step—subtracting and adding the same quantity—which is shown in color.

$$\frac{d}{dx}[f(x)g(x)] = \lim_{\Delta x \to 0} \frac{f(x + \Delta x)g(x + \Delta x) - f(x)g(x)}{\Delta x}$$

$$= \lim_{\Delta x \to 0} \frac{f(x + \Delta x)g(x + \Delta x) - f(x + \Delta x)g(x) + f(x + \Delta x)g(x) - f(x)g(x)}{\Delta x}$$

$$= \lim_{\Delta x \to 0} \left[f(x + \Delta x)\frac{g(x + \Delta x) - g(x)}{\Delta x} + g(x)\frac{f(x + \Delta x) - f(x)}{\Delta x} \right]$$

$$= \lim_{\Delta x \to 0} \left[f(x + \Delta x)\frac{g(x + \Delta x) - g(x)}{\Delta x} \right] + \lim_{\Delta x \to 0} \left[g(x)\frac{f(x + \Delta x) - f(x)}{\Delta x} \right]$$

$$= \lim_{\Delta x \to 0} f(x + \Delta x) \cdot \lim_{\Delta x \to 0} \frac{g(x + \Delta x) - g(x)}{\Delta x} + \lim_{\Delta x \to 0} g(x) \cdot \lim_{\Delta x \to 0} \frac{f(x + \Delta x) - f(x)}{\Delta x}$$

$$= f(x)g'(x) + g(x)f'(x)$$

THE PRODUCT RULE

When Leibniz originally wrote a formula for the Product Rule, he was motivated by the expression

$$(x + dx)(y + dy) - xy$$

from which he subtracted $dx\, dy$ (as being negligible) and obtained the differential form $x\, dy + y\, dx$. This derivation resulted in the traditional form of the Product Rule. (*Source: The History of Mathematics by David M. Burton*)

A version of the Product Rule that some people prefer is

$$\frac{d}{dx}[f(x)g(x)] = f'(x)g(x) + f(x)g'(x).$$

The advantage of this form is that it generalizes easily to products involving three or more factors.

The Product Rule can be extended to cover products involving more than two factors. For example, if f, g, and h are differentiable functions of x, then

$$\frac{d}{dx}[f(x)g(x)h(x)] = f'(x)g(x)h(x) + f(x)g'(x)h(x) + f(x)g(x)h'(x).$$

For instance, the derivative of $y = x^2 \sin x \cos x$ is

$$\frac{dy}{dx} = 2x \sin x \cos x + x^2 \cos x \cos x + x^2 \sin x(-\sin x)$$

$$= 2x \sin x \cos x + x^2(\cos^2 x - \sin^2 x).$$

The derivative of a product of two functions is not (in general) given by the product of the derivatives of the two functions. To see this, try comparing the product of the derivatives of $f(x) = 3x - 2x^2$ and $g(x) = 5 + 4x$ with the derivative in Example 1.

Example 1   Using the Product Rule

Find the derivative of $h(x) = (3x - 2x^2)(5 + 4x)$.

Solution

$$h'(x) = \overbrace{(3x - 2x^2)}^{\text{First}} \overbrace{\frac{d}{dx}[5 + 4x]}^{\substack{\text{Derivative} \\ \text{of second}}} + \overbrace{(5 + 4x)}^{\text{Second}} \overbrace{\frac{d}{dx}[3x - 2x^2]}^{\substack{\text{Derivative} \\ \text{of first}}} \qquad \text{Apply Product Rule.}$$

$$= (3x - 2x^2)(4) + (5 + 4x)(3 - 4x)$$
$$= (12x - 8x^2) + (15 - 8x - 16x^2)$$
$$= -24x^2 + 4x + 15$$

In Example 1, you have the option of finding the derivative with or without the Product Rule. To find the derivative without the Product Rule, you can write

$$D_x[(3x - 2x^2)(5 + 4x)] = D_x[-8x^3 + 2x^2 + 15x]$$
$$= -24x^2 + 4x + 15.$$

In the next example, you must use the Product Rule.

Example 2   Using the Product Rule

Find the derivative of $y = x \sin x$.

Solution

$$\frac{d}{dx}[x \sin x] = x\frac{d}{dx}[\sin x] + \sin x \frac{d}{dx}[x] \qquad \text{Apply Product Rule.}$$

$$= x \cos x + (\sin x)(1)$$
$$= x \cos x + \sin x$$

Example 3   Using the Product Rule

Find the derivative of $y = 2x \cos x - 2 \sin x$.

Solution

$$\frac{dy}{dx} = \overbrace{(2x)\left(\frac{d}{dx}[\cos x]\right) + (\cos x)\left(\frac{d}{dx}[2x]\right)}^{\text{Product Rule}} - \overbrace{2\frac{d}{dx}[\sin x]}^{\text{Constant Multiple Rule}}$$

$$= (2x)(-\sin x) + (\cos x)(2) - 2(\cos x)$$
$$= -2x \sin x$$

NOTE   In Example 3, notice that you use the Product Rule when both factors of the product are variable, and you use the Constant Multiple Rule when one of the factors is a constant.

The Quotient Rule

THEOREM 2.8   The Quotient Rule

The quotient f/g of two differentiable functions f and g is itself differentiable at all values of x for which $g(x) \neq 0$. Moreover, the derivative of f/g is given by the denominator times the derivative of the numerator minus the numerator times the derivative of the denominator, all divided by the square of the denominator.

$$\frac{d}{dx}\left[\frac{f(x)}{g(x)}\right] = \frac{g(x)f'(x) - f(x)g'(x)}{[g(x)]^2}, \qquad g(x) \neq 0$$

Proof   As with the proof of Theorem 2.7, the key to this proof is subtracting and adding the same quantity.

$$\frac{d}{dx}\left[\frac{f(x)}{g(x)}\right] = \lim_{\Delta x \to 0} \frac{\dfrac{f(x + \Delta x)}{g(x + \Delta x)} - \dfrac{f(x)}{g(x)}}{\Delta x} \qquad \text{Definition of derivative}$$

$$= \lim_{\Delta x \to 0} \frac{g(x)f(x + \Delta x) - f(x)g(x + \Delta x)}{\Delta x g(x)g(x + \Delta x)}$$

$$= \lim_{\Delta x \to 0} \frac{g(x)f(x + \Delta x) - f(x)g(x) + f(x)g(x) - f(x)g(x + \Delta x)}{\Delta x g(x)g(x + \Delta x)}$$

$$= \frac{\displaystyle\lim_{\Delta x \to 0} \frac{g(x)[f(x + \Delta x) - f(x)]}{\Delta x} - \lim_{\Delta x \to 0} \frac{f(x)[g(x + \Delta x) - g(x)]}{\Delta x}}{\displaystyle\lim_{\Delta x \to 0} [g(x)g(x + \Delta x)]}$$

$$= \frac{\displaystyle g(x)\left[\lim_{\Delta x \to 0} \frac{f(x + \Delta x) - f(x)}{\Delta x}\right] - f(x)\left[\lim_{\Delta x \to 0} \frac{g(x + \Delta x) - g(x)}{\Delta x}\right]}{\displaystyle\lim_{\Delta x \to 0} [g(x)g(x + \Delta x)]}$$

$$= \frac{g(x)f'(x) - f(x)g'(x)}{[g(x)]^2}$$

TECHNOLOGY   Graphing utilities can be used to compare the graph of a function with the graph of its derivative. For instance, in Figure 2.22, the graph of the function in Example 4 appears to have two points that have horizontal tangent lines. What are the values of y' at these two points?

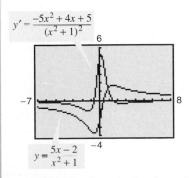

$$y' = \frac{-5x^2 + 4x + 5}{(x^2 + 1)^2}$$

$$y = \frac{5x - 2}{x^2 + 1}$$

Graphical comparison of a function and its derivative
Figure 2.22

Example 4   **Using the Quotient Rule**

Find the derivative of $y = \dfrac{5x - 2}{x^2 + 1}$.

Solution

$$\frac{d}{dx}\left[\frac{5x - 2}{x^2 + 1}\right] = \frac{(x^2 + 1)\dfrac{d}{dx}[5x - 2] - (5x - 2)\dfrac{d}{dx}[x^2 + 1]}{(x^2 + 1)^2} \qquad \text{Apply Quotient Rule.}$$

$$= \frac{(x^2 + 1)(5) - (5x - 2)(2x)}{(x^2 + 1)^2}$$

$$= \frac{(5x^2 + 5) - (10x^2 - 4x)}{(x^2 + 1)^2}$$

$$= \frac{-5x^2 + 4x + 5}{(x^2 + 1)^2}$$

Note the use of parentheses in Example 4. A liberal use of parentheses is recommended for *all* types of differentiation problems. For instance, with the Quotient Rule, it is a good idea to enclose all factors and derivatives in parentheses, and to pay special attention to the subtraction required in the numerator.

When we introduced differentiation rules in the preceding section, we emphasized the need for rewriting *before* differentiating. The next example illustrates this point with the Quotient Rule.

Example 5   Rewriting Before Differentiating

Find the derivative of $y = \dfrac{3 - (1/x)}{x + 5}$.

Solution   Begin by rewriting the function.

$$y = \frac{3 - (1/x)}{x + 5} \qquad \text{Write original function.}$$

$$= \frac{x\left(3 - \dfrac{1}{x}\right)}{x(x + 5)} \qquad \text{Multiply numerator and denominator by } x.$$

$$= \frac{3x - 1}{x^2 + 5x} \qquad \text{Rewrite.}$$

$$\frac{dy}{dx} = \frac{(x^2 + 5x)(3) - (3x - 1)(2x + 5)}{(x^2 + 5x)^2} \qquad \text{Quotient Rule}$$

$$= \frac{(3x^2 + 15x) - (6x^2 + 13x - 5)}{(x^2 + 5x)^2}$$

$$= \frac{-3x^2 + 2x + 5}{(x^2 + 5x)^2} \qquad \text{Simplify.}$$

Not every quotient needs to be differentiated by the Quotient Rule. For example, each quotient in the next example can be considered as the product of a constant times a function of x. In such cases it is more convenient to use the Constant Multiple Rule.

Example 6   Using the Constant Multiple Rule

Original Function	Rewrite	Differentiate	Simplify
a. $y = \dfrac{x^2 + 3x}{6}$	$y = \dfrac{1}{6}(x^2 + 3x)$	$y' = \dfrac{1}{6}(2x + 3)$	$y' = \dfrac{2x + 3}{6}$
b. $y = \dfrac{5x^4}{8}$	$y = \dfrac{5}{8}x^4$	$y' = \dfrac{5}{8}(4x^3)$	$y' = \dfrac{5}{2}x^3$
c. $y = \dfrac{-3(3x - 2x^2)}{7x}$	$y = -\dfrac{3}{7}(3 - 2x)$	$y' = -\dfrac{3}{7}(-2)$	$y' = \dfrac{6}{7}$
d. $y = \dfrac{9}{5x^2}$	$y = \dfrac{9}{5}(x^{-2})$	$y' = \dfrac{9}{5}(-2x^{-3})$	$y' = -\dfrac{18}{5x^3}$

NOTE   To see the benefit of using the Constant Multiple Rule for some quotients, try using the Quotient Rule to differentiate the functions in Example 6—you should obtain the same results, but with more work.

In Section 2.2, we proved the Power Rule only for the case where the exponent n is a positive integer greater than 1. The next example extends the proof to include negative integer exponents.

Example 7   **Proof of the Power Rule (Negative Integer Exponents)**

If n is a negative integer, there exists a positive integer k such that $n = -k$. So, by the Quotient Rule, you can write

$$\frac{d}{dx}[x^n] = \frac{d}{dx}\left[\frac{1}{x^k}\right]$$

$$= \frac{x^k(0) - (1)(kx^{k-1})}{(x^k)^2} \qquad \text{Quotient Rule}$$

$$= \frac{0 - kx^{k-1}}{x^{2k}}$$

$$= -kx^{-k-1}$$

$$= nx^{n-1}. \qquad n = -k$$

So, the Power Rule

$$D_x[x^n] = nx^{n-1} \qquad \text{Power Rule}$$

is valid for any integer. In Exercise 63 in Section 2.5, you are asked to prove the case for which n is any rational number.

Derivatives of Trigonometric Functions

Knowing the derivatives of the sine and cosine functions, you can use the Quotient Rule to find the derivatives of the four remaining trigonometric functions.

THEOREM 2.9   Derivatives of Trigonometric Functions

$$\frac{d}{dx}[\tan x] = \sec^2 x \qquad\qquad \frac{d}{dx}[\cot x] = -\csc^2 x$$

$$\frac{d}{dx}[\sec x] = \sec x \tan x \qquad\qquad \frac{d}{dx}[\csc x] = -\csc x \cot x$$

Proof   Considering $\tan x = (\sin x)/(\cos x)$ and applying the Quotient Rule, you obtain

$$\frac{d}{dx}[\tan x] = \frac{(\cos x)(\cos x) - (\sin x)(-\sin x)}{\cos^2 x} \qquad \text{Apply Quotient Rule.}$$

$$= \frac{\cos^2 x + \sin^2 x}{\cos^2 x}$$

$$= \frac{1}{\cos^2 x}$$

$$= \sec^2 x.$$

The proofs of the other three parts of the theorem are left as an exercise (see Exercise 81).

Example 8   Differentiating Trigonometric Functions

NOTE Because of trigonometric identities, the derivative of a trigonometric function can take many forms. This presents a challenge when you are trying to match your answers to those given in the back of the text.

Function	Derivative
a. $y = x - \tan x$	$\dfrac{dy}{dx} = 1 - \sec^2 x$
b. $y = x \sec x$	$y' = x(\sec x \tan x) + (\sec x)(1)$
	$\qquad = (\sec x)(1 + x \tan x)$

Example 9   Different Forms of a Derivative

Differentiate both forms of $y = \dfrac{1 - \cos x}{\sin x} = \csc x - \cot x$.

Solution

First form: $\quad y = \dfrac{1 - \cos x}{\sin x}$

$$y' = \frac{(\sin x)(\sin x) - (1 - \cos x)(\cos x)}{\sin^2 x}$$

$$= \frac{\sin^2 x + \cos^2 x - \cos x}{\sin^2 x}$$

$$= \frac{1 - \cos x}{\sin^2 x}$$

Second form: $\quad y = \csc x - \cot x$

$$y' = -\csc x \cot x + \csc^2 x$$

To show that the two derivatives are equal, you can write

$$\frac{1 - \cos x}{\sin^2 x} = \frac{1}{\sin^2 x} - \left(\frac{1}{\sin x}\right)\left(\frac{\cos x}{\sin x}\right) = \csc^2 x - \csc x \cot x.$$

The following summary shows that much of the work in obtaining a simplified form of a derivative occurs *after* differentiating. Note that two characteristics of a simplified form are the absence of negative exponents and the combining of like terms.

	$f'(x)$ After Differentiating	**$f'(x)$ After Simplifying**
Example 1	$(3x - 2x^2)(4) + (5 + 4x)(3 - 4x)$	$-24x^2 + 4x + 15$
Example 3	$(2x)(-\sin x) + (\cos x)(2) - 2(\cos x)$	$-2x \sin x$
Example 4	$\dfrac{(x^2 + 1)(5) - (5x - 2)(2x)}{(x^2 + 1)^2}$	$\dfrac{-5x^2 + 4x + 5}{(x^2 + 1)^2}$
Example 5	$\dfrac{(x^2 + 5x)(3) - (3x - 1)(2x + 5)}{(x^2 + 5x)^2}$	$\dfrac{-3x^2 + 2x + 5}{(x^2 + 5x)^2}$
Example 9	$\dfrac{(\sin x)(\sin x) - (1 - \cos x)(\cos x)}{\sin^2 x}$	$\dfrac{1 - \cos x}{\sin^2 x}$

Higher-Order Derivatives

Just as you can obtain a velocity function by differentiating a position function, you can obtain an **acceleration** function by differentiating a velocity function. Another way of looking at this is that you can obtain an acceleration function by differentiating a position function *twice*.

$$s(t) \qquad \text{Position function}$$
$$v(t) = s'(t) \qquad \text{Velocity function}$$
$$a(t) = v'(t) = s''(t) \qquad \text{Acceleration function}$$

The function given by $a(t)$ is the **second derivative** of $s(t)$ and is denoted by $s''(t)$.

The second derivative is an example of a **higher-order derivative.** You can define derivatives of any positive integer order. For instance, the **third derivative** is the derivative of the second derivative. Higher-order derivatives are denoted as follows.

First derivative: y', $f'(x)$, $\dfrac{dy}{dx}$, $\dfrac{d}{dx}[f(x)]$, $D_x[y]$

Second derivative: y'', $f''(x)$, $\dfrac{d^2y}{dx^2}$, $\dfrac{d^2}{dx^2}[f(x)]$, $D_x^2[y]$

Third derivative: y''', $f'''(x)$, $\dfrac{d^3y}{dx^3}$, $\dfrac{d^3}{dx^3}[f(x)]$, $D_x^3[y]$

Fourth derivative: $y^{(4)}$, $f^{(4)}(x)$, $\dfrac{d^4y}{dx^4}$, $\dfrac{d^4}{dx^4}[f(x)]$, $D_x^4[y]$

$$\vdots$$

nth derivative: $y^{(n)}$, $f^{(n)}(x)$, $\dfrac{d^ny}{dx^n}$, $\dfrac{d^n}{dx^n}[f(x)]$, $D_x^n[y]$

Example 10   **Finding the Acceleration Due to Gravity**

Because the moon has no atmosphere, a falling object on the moon encounters no air resistance. In 1971, astronaut David Scott demonstrated that a feather and a hammer fall at the same rate on the moon. The position function for each of these falling objects is given by

$$s(t) = -0.81t^2 + 2$$

where $s(t)$ is the height in meters and t is the time in seconds. What is the ratio of earth's gravitational force to the moon's?

Solution   To find the acceleration, differentiate the position function twice.

$$s(t) = -0.81t^2 + 2 \qquad \text{Position function}$$
$$s'(t) = -1.62t \qquad \text{Velocity function}$$
$$s''(t) = -1.62 \qquad \text{Acceleration function}$$

So, the acceleration due to gravity on the moon is -1.62 meters per second per second. Because the acceleration due to gravity on earth is -9.8 meters per second per second, the ratio of earth's gravitational force to the moon's is

$$\frac{\text{Earth's gravitational force}}{\text{Moon's gravitational force}} = \frac{-9.8}{1.62} \approx 6.05.$$

NASA

THE MOON

The moon's mass is 7.354×10^{22} kilograms, and earth's mass is 5.979×10^{24} kilograms. The moon's radius is 1738 kilometers, and earth's radius is 6371 kilometers. Because the gravitational force on the surface of a planet is directly proportional to its mass and inversely proportional to the square of its radius, the ratio of the gravitational force on earth to the gravitational force on the moon is

$$\frac{(5.979 \times 10^{24})/6371^2}{(7.354 \times 10^{22})/1738^2} \approx 6.05.$$

Lab Series   LAB 3

EXERCISES FOR SECTION 2.3

In Exercises 1–6, use the Product Rule to differentiate the function.

1. $g(x) = (x^2 + 1)(x^2 - 2x)$

2. $f(x) = (6x + 5)(x^3 - 2)$

3. $h(t) = \sqrt[3]{t}(t^2 + 4)$

4. $g(s) = \sqrt{s}(4 - s^2)$

5. $f(x) = x^3 \cos x$

6. $g(x) = \sqrt{x} \sin x$

In Exercises 7–12, use the Quotient Rule to differentiate the function.

7. $f(x) = \dfrac{x}{x^2 + 1}$

8. $g(t) = \dfrac{t^2 + 2}{2t - 7}$

9. $h(x) = \dfrac{\sqrt[3]{x}}{x^3 + 1}$

10. $h(s) = \dfrac{s}{\sqrt{s} - 1}$

11. $g(x) = \dfrac{\sin x}{x^2}$

12. $f(t) = \dfrac{\cos t}{t^3}$

In Exercises 13–18, find $f'(x)$ and $f'(c)$.

Function	Value of c
13. $f(x) = (x^3 - 3x)(2x^2 + 3x + 5)$	$c = 0$
14. $f(x) = (x^2 - 2x + 1)(x^3 - 1)$	$c = 1$
15. $f(x) = \dfrac{x^2 - 4}{x - 3}$	$c = 1$
16. $f(x) = \dfrac{x + 1}{x - 1}$	$c = 2$
17. $f(x) = x \cos x$	$c = \dfrac{\pi}{4}$
18. $f(x) = \dfrac{\sin x}{x}$	$c = \dfrac{\pi}{6}$

In Exercises 19–24, complete the table without using the Quotient Rule (see Example 6).

Function	Rewrite	Differentiate	Simplify
19. $y = \dfrac{x^2 + 2x}{3}$			
20. $y = \dfrac{5x^2 - 3}{4}$			
21. $y = \dfrac{7}{3x^3}$			
22. $y = \dfrac{4}{5x^2}$			
23. $y = \dfrac{4x^{3/2}}{x}$			
24. $y = \dfrac{3x^2 - 5}{7}$			

In Exercises 25–38, find the derivative of the algebraic function.

25. $f(x) = \dfrac{3 - 2x - x^2}{x^2 - 1}$

26. $f(x) = \dfrac{x^3 + 3x + 2}{x^2 - 1}$

27. $f(x) = x\left(1 - \dfrac{4}{x + 3}\right)$

28. $f(x) = x^4\left(1 - \dfrac{2}{x + 1}\right)$

29. $f(x) = \dfrac{2x + 5}{\sqrt{x}}$

30. $f(x) = \sqrt[3]{x}(\sqrt{x} + 3)$

31. $h(s) = (s^3 - 2)^2$

32. $h(x) = (x^2 - 1)^2$

33. $f(x) = \dfrac{2 - \dfrac{1}{x}}{x - 3}$

34. $g(x) = x^2\left(\dfrac{2}{x} - \dfrac{1}{x + 1}\right)$

35. $f(x) = (3x^3 + 4x)(x - 5)(x + 1)$

36. $f(x) = (x^2 - x)(x^2 + 1)(x^2 + x + 1)$

37. $f(x) = \dfrac{x^2 + c^2}{x^2 - c^2}$, c is a constant

38. $f(x) = \dfrac{c^2 - x^2}{c^2 + x^2}$, c is a constant

In Exercises 39–54, find the derivative of the trigonometric function.

39. $f(t) = t^2 \sin t$

40. $f(\theta) = (\theta + 1) \cos \theta$

41. $f(t) = \dfrac{\cos t}{t}$

42. $f(x) = \dfrac{\sin x}{x}$

43. $f(x) = -x + \tan x$

44. $y = x + \cot x$

45. $g(t) = \sqrt[4]{t} + 8 \sec t$

46. $h(s) = \dfrac{1}{s} - 10 \csc s$

47. $y = \dfrac{3(1 - \sin x)}{2 \cos x}$

48. $y = \dfrac{\sec x}{x}$

49. $y = -\csc x - \sin x$

50. $y = x \sin x + \cos x$

51. $f(x) = x^2 \tan x$

52. $f(x) = \sin x \cos x$

53. $y = 2x \sin x + x^2 \cos x$

54. $h(\theta) = 5\theta \sec \theta + \theta \tan \theta$

In Exercises 55–58, use a computer algebra system to differentiate the function.

55. $g(x) = \left(\dfrac{x + 1}{x + 2}\right)(2x - 5)$

56. $f(x) = \left(\dfrac{x^2 - x - 3}{x^2 + 1}\right)(x^2 + x + 1)$

57. $g(\theta) = \dfrac{\theta}{1 - \sin \theta}$

58. $f(\theta) = \dfrac{\sin \theta}{1 - \cos \theta}$

In Exercises 59–62, evaluate the derivative of the function at the indicated point. Use a graphing utility to verify your result.

Function	Point
59. $y = \dfrac{1 + \csc x}{1 - \csc x}$	$\left(\dfrac{\pi}{6}, -3\right)$
60. $f(x) = \tan x \cot x$	$(1, 1)$
61. $h(t) = \dfrac{\sec t}{t}$	$\left(\pi, -\dfrac{1}{\pi}\right)$
62. $f(x) = \sin x(\sin x + \cos x)$	$\left(\dfrac{\pi}{4}, 1\right)$

In Exercises 63–68, (a) find an equation of the tangent line to the graph of f at the indicated point, (b) use a graphing utility to graph the function and its tangent line at the point, and (c) use the *derivative* feature of a graphing utility to confirm your results.

Function	Point
63. $f(x) = (x^3 - 3x + 1)(x + 2)$	$(1, -3)$
64. $f(x) = (x - 1)(x^2 - 2)$	$(0, 2)$
65. $f(x) = \dfrac{x}{x - 1}$	$(2, 2)$
66. $f(x) = \dfrac{(x - 1)}{(x + 1)}$	$\left(2, \dfrac{1}{3}\right)$
67. $f(x) = \tan x$	$\left(\dfrac{\pi}{4}, 1\right)$
68. $f(x) = \sec x$	$\left(\dfrac{\pi}{3}, 2\right)$

In Exercises 69 and 70, determine the point(s) at which the graph of the function has a horizontal tangent.

69. $f(x) = \dfrac{x^2}{x - 1}$

70. $f(x) = \dfrac{x^2}{x^2 + 1}$

In Exercises 71 and 72, verify that $f'(x) = g'(x)$, and explain the relationship between f and g.

71. $f(x) = \dfrac{3x}{x + 2}, \quad g(x) = \dfrac{5x + 4}{x + 2} = \dfrac{3x}{x+2} + \dfrac{2x+4}{x+2}$

72. $f(x) = \dfrac{\sin x - 3x}{x}, \quad g(x) = \dfrac{\sin x + 2x}{x}$

In Exercises 73 and 74, find the derivative of the function f for $n = 1, 2, 3,$ and 4. Use the result to write a general rule for $f'(x)$ in terms of n.

73. $f(x) = x^n \sin x$

74. $f(x) = \dfrac{\cos x}{x^n}$

75. Area The length of a rectangle is given by $2t + 1$ and its height is $\sqrt{t}$, where t is time in seconds and the dimensions are in centimeters. Find the rate of change of the area with respect to time.

76. Volume The radius of a right circular cylinder is given by $\sqrt{t + 2}$ and its height is $\frac{1}{2}\sqrt{t}$, where t is time in seconds and the dimensions are in inches. Find the rate of change of the volume with respect to time.

77. Inventory Replenishment The ordering and transportation cost C for the components used in manufacturing a certain product is

$$C = 100\left(\dfrac{200}{x^2} + \dfrac{x}{x + 30}\right), \quad x \geq 1$$

where C is measured in thousands of dollars and x is the order size in hundreds. Find the rate of change of C with respect to x when (a) $x = 10$, (b) $x = 15$, and (c) $x = 20$. What do these rates of change imply about increasing order size?

78. Boyle's Law This law states that if the temperature of a gas remains constant, its pressure is inversely proportional to its volume. Use the derivative to show that the rate of change of the pressure is inversely proportional to the square of the volume.

79. Population Growth A population of 500 bacteria is introduced into a culture and grows in number according to the equation

$$P(t) = 500\left(1 + \dfrac{4t}{50 + t^2}\right)$$

where t is measured in hours. Find the rate at which the population is growing when $t = 2$.

80. Rate of Change Determine whether there exist any values of x in the interval $[0, 2\pi)$ such that the rate of change of $f(x) = \sec x$ and the rate of change of $g(x) = \csc x$ are equal.

81. Prove the following differentiation rules.

(a) $\dfrac{d}{dx}[\sec x] = \sec x \tan x$

(b) $\dfrac{d}{dx}[\csc x] = -\csc x \cot x$

(c) $\dfrac{d}{dx}[\cot x] = -\csc^2 x$

82. Modeling Data The table shows the number of motor homes n (in thousands) in the United States and the retail value v (in millions of dollars) of these motor homes for the years 1992 through 1997. The year is represented by t, with $t = 2$ corresponding to 1992. (*Source: Recreation Vehicle Industry Association*)

Year	1992	1993	1994	1995	1996	1997
n	226.3	243.8	306.7	281.0	274.6	239.3
v	\$6963	\$7544	\$9897	\$9768	\$9788	\$9139

(a) Use a graphing utility to find quadratic models for the number of motor homes $n(t)$ and the total retail value $v(t)$ of the motor homes.

(b) Find $A = v(t)/n(t)$. What does this function represent?

(c) Find $A'(t)$. Interpret the derivative in the context of these data.

In Exercises 83–88, find the second derivative of the function.

83. $f(x) = 4x^{3/2}$

84. $f(x) = x + 32x^{-2}$

85. $f(x) = \dfrac{x}{x - 1}$

86. $f(x) = \dfrac{x^2 + 2x - 1}{x}$

87. $f(x) = 3 \sin x$

88. $f(x) = \sec x$

In Exercises 89–92, find the higher-order derivative.

Given	Find
89. $f'(x) = x^2$	$f''(x)$
90. $f''(x) = 2 - \dfrac{2}{x}$	$f'''(x)$
91. $f'''(x) = 2\sqrt{x}$	$f^{(4)}(x)$
92. $f^{(4)}(x) = 2x + 1$	$f^{(6)}(x)$

Getting at the Concept

93. Sketch the graph of a differentiable function f such that $f(2) = 0$, $f' < 0$ for $-\infty < x < 2$, and $f' > 0$ for $2 < x < \infty$.

94. Sketch the graph of a differentiable function f such that $f > 0$ and $f' < 0$ for all real numbers x.

In Exercises 95–98, find $f'(2)$ given the following.

$$g(2) = 3 \quad \text{and} \quad g'(2) = -2$$

$$h(2) = -1 \quad \text{and} \quad h'(2) = 4$$

95. $f(x) = 2g(x) + h(x)$ 96. $f(x) = 4 - h(x)$

97. $f(x) = \dfrac{g(x)}{h(x)}$ 98. $f(x) = g(x)h(x)$

In Exercises 99 and 100, the graphs of f, f', and f'' are shown on the same set of coordinate axes. Which is which? To print an enlarged copy of the graph, go to the website www.mathgraphs.com.

99.   100.

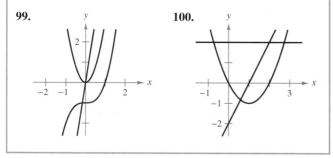

101. *Acceleration*   The velocity of an object in meters per second is

$$v(t) = 36 - t^2, \quad 0 \le t \le 6.$$

Find the velocity and acceleration of the object when $t = 3$. What can be said about the speed of the object when the velocity and acceleration have opposite signs?

102. *Stopping Distance*   A car is traveling at a rate of 66 feet per second (45 miles per hour) when the brakes are applied. The position function for the car is

$$s(t) = -8.25t^2 + 66t$$

where s is measured in feet and t is measured in seconds. Use this function to complete the table, and find the average velocity during each time interval.

t	0	1	2	3	4
$s(t)$					
$v(t)$					
$a(t)$					

103. *Acceleration*   An automobile's velocity starting from rest is

$$v(t) = \frac{100t}{2t + 15}$$

where v is measured in feet per second. Find the acceleration at each of the following times.

(a) 5 seconds   (b) 10 seconds   (c) 20 seconds

104. *Finding a Pattern*   Develop a general rule for $f^{(n)}(x)$ if

(a) $f(x) = x^n$ and   (b) $f(x) = \dfrac{1}{x}$.

105. *Finding a Pattern*   Consider the function $f(x) = g(x)h(x)$.

(a) Use the product rule to generate rules for finding $f''(x)$, $f'''(x)$, and $f^{(4)}(x)$.

(b) Use the results in part (a) to write a general rule for $f^{(n)}(x)$.

106. *Finding a Pattern*   Develop a general rule for $[xf(x)]^{(n)}$ where f is a differentiable function of x.

Linear and Quadratic Approximations   The linear and quadratic approximations of a function f at $x = a$ are

$$P_1(x) = f'(a)(x - a) + f(a) \text{ and}$$

$$P_2(x) = \tfrac{1}{2}f''(a)(x - a)^2 + f'(a)(x - a) + f(a).$$

In Exercises 107 and 108, (a) find the specified linear and quadratic approximations of f, (b) use a graphing utility to graph f and the approximations, (c) determine whether P_1 or P_2 is the better approximation, and (d) state how the accuracy changes as you move farther from $x = a$.

107. $f(x) = \cos x$ 108. $f(x) = \sin x$

$$a = \frac{\pi}{3} \qquad\qquad a = \frac{\pi}{2}$$

True or False?   **In Exercises 109–114, determine whether the statement is true or false. If it is false, explain why or give an example that shows it is false.**

109. If $y = f(x)g(x)$, then $dy/dx = f'(x)g'(x)$.

110. If $y = (x + 1)(x + 2)(x + 3)(x + 4)$, then $d^5y/dx^5 = 0$.

111. If $f'(c)$ and $g'(c)$ are zero and $h(x) = f(x)g(x)$, then $h'(c) = 0$.

112. If $f(x)$ is an nth-degree polynomial, then $f^{(n+1)}(x) = 0$.

113. The second derivative represents the rate of change of the first derivative.

114. If the velocity of an object is constant, then its acceleration is zero.

115. Find the derivative of $f(x) = x|x|$. Does $f''(0)$ exist?

116. *Think About It*   Let f and g be functions whose first and second derivatives exist on an interval I. Which of the following formulas is (are) true?

(a) $fg'' - f''g = (fg' - f'g)'$

(b) $fg'' + f''g = (fg)''$

- Find the derivative of a composite function using the Chain Rule.
- Find the derivative of a function using the General Power Rule.
- Simplify the derivative of a function using algebra.
- Find the derivative of a trigonometric function using the Chain Rule.

The Chain Rule

We have yet to discuss one of the most powerful differentiation rules—the **Chain Rule.** This rule deals with composite functions and adds a surprising versatility to the rules discussed in the two previous sections. For example, compare the following functions. Those on the left can be differentiated without the Chain Rule, and those on the right are best done with the Chain Rule.

Without the Chain Rule	*With the Chain Rule*
$y = x^2 + 1$	$y = \sqrt{x^2 + 1}$
$y = \sin x$	$y = \sin 6x$
$y = 3x + 2$	$y = (3x + 2)^5$
$y = x + \tan x$	$y = x + \tan x^2$

Basically, the Chain Rule states that if y changes dy/du times as fast as u, and u changes du/dx times as fast as x, then y changes $(dy/du)(du/dx)$ times as fast as x.

Example 1   **The Derivative of a Composite Function**

A set of gears is constructed, as shown in Figure 2.23, such that the second and third gears are on the same axle. As the first axle revolves, it drives the second axle, which in turn drives the third axle. Let y, u, and x represent the numbers of revolutions per minute of the first, second, and third axles. Find dy/du, du/dx, and dy/dx, and show that

$$\frac{dy}{dx} = \frac{dy}{du} \cdot \frac{du}{dx}.$$

Solution   Because the circumference of the second gear is three times that of the first, the first axle must make three revolutions to turn the second axle once. Similarly, the second axle must make two revolutions to turn the third axle once, and you can write

$$\frac{dy}{du} = 3 \quad \text{and} \quad \frac{du}{dx} = 2.$$

Combining these two results, you know that the first axle must make six revolutions to turn the third axle once. So, you can write

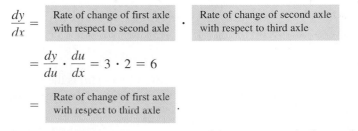

$$= \frac{dy}{du} \cdot \frac{du}{dx} = 3 \cdot 2 = 6$$

In other words, the rate of change of y with respect to x is the product of the rate of change of y with respect to u and the rate of change of u with respect to x.

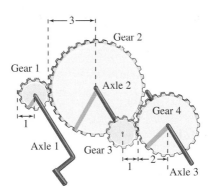

Axle 1: y revolutions per minute
Axle 2: u revolutions per minute
Axle 3: x revolutions per minute
Figure 2.23

Example 1 illustrates a simple case of the Chain Rule. The general rule is stated below.

THEOREM 2.10   The Chain Rule

If $y = f(u)$ is a differentiable function of u and $u = g(x)$ is a differentiable function of x, then $y = f(g(x))$ is a differentiable function of x and

$$\frac{dy}{dx} = \frac{dy}{du} \cdot \frac{du}{dx}$$

or, equivalently,

$$\frac{d}{dx}[f(g(x))] = f'(g(x))g'(x).$$

Proof Let $h(x) = f(g(x))$. Then, using the alternative form of the derivative, you need to show that, for $x = c$,

$$h'(c) = f'(g(c))g'(c).$$

An important consideration in this proof is the behavior of g as x approaches c. A problem occurs if there are values of x, other than c, such that $g(x) = g(c)$. In Appendix B we show how to use the differentiability of f and g to overcome this problem. For now, assume that $g(x) \neq g(c)$ for values of x other than c. In the proofs of the Product Rule and the Quotient Rule, we added and subtracted the same quantity to obtain the desired form. This proof uses a similar technique—multiplying and dividing by the same (nonzero) quantity. Note that because g is differentiable, it is also continuous, and it follows that $g(x) \to g(c)$ as $x \to c$.

$$h'(c) = \lim_{x \to c} \frac{f(g(x)) - f(g(c))}{x - c}$$

$$= \lim_{x \to c} \left[\frac{f(g(x)) - f(g(c))}{g(x) - g(c)} \cdot \frac{g(x) - g(c)}{x - c} \right], \quad g(x) \neq g(c)$$

$$= \left[\lim_{x \to c} \frac{f(g(x)) - f(g(c))}{g(x) - g(c)} \right]\left[\lim_{x \to c} \frac{g(x) - g(c)}{x - c} \right]$$

$$= f'(g(c))g'(c)$$

When applying the Chain Rule, it is helpful to think of the composite function $f \circ g$ as having two parts—an inner part and an outer part.

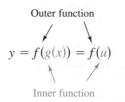

Outer function

$$y = f(g(x)) = f(u)$$

Inner function

The derivative of $y = f(u)$ is the derivative of the outer function (at the inner function u) *times* the derivative of the inner function.

$$y' = f'(u) \cdot u'$$

Example 2   **Decomposition of a Composite Function**

$y = f(g(x))$	$u = g(x)$	$y = f(u)$
a. $y = \dfrac{1}{x + 1}$	$u = x + 1$	$y = \dfrac{1}{u}$
b. $y = \sin 2x$	$u = 2x$	$y = \sin u$
c. $y = \sqrt{3x^2 - x + 1}$	$u = 3x^2 - x + 1$	$y = \sqrt{u}$
d. $y = \tan^2 x$	$u = \tan x$	$y = u^2$

Example 3   **Using the Chain Rule**

Find dy/dx for $y = (x^2 + 1)^3$.

STUDY TIP   You could also solve the problem in Example 3 without using the Chain Rule by observing that

$$y = x^6 + 3x^4 + 3x^2 + 1$$

and

$$y' = 6x^5 + 12x^3 + 6x.$$

Verify that this is the same as the derivative in Example 3. Which method would you use to find

$$\frac{d}{dx}(x^2 + 1)^{50}?$$

Solution   For this function, you can consider the inside function to be $u = x^2 + 1$. By the Chain Rule, you obtain

$$\frac{dy}{dx} = \underbrace{3(x^2 + 1)^2}_{\frac{dy}{du}}\underbrace{(2x)}_{\frac{du}{dx}} = 6x(x^2 + 1)^2.$$

The General Power Rule

The function in Example 3 is an example of one of the most common types of composite functions, $y = [u(x)]^n$. The rule for differentiating such functions is called the **General Power Rule,** and it is a special case of the Chain Rule.

THEOREM 2.11   The General Power Rule

If $y = [u(x)]^n$, where u is a differentiable function of x and n is a rational number, then

$$\frac{dy}{dx} = n[u(x)]^{n-1}\frac{du}{dx}$$

or, equivalently,

$$\frac{d}{dx}[u^n] = nu^{n-1}u'.$$

Proof   Because $y = u^n$, you apply the Chain Rule to obtain

$$\frac{dy}{dx} = \left(\frac{dy}{du}\right)\left(\frac{du}{dx}\right)$$

$$= \frac{d}{du}[u^n]\frac{du}{dx}.$$

By the (Simple) Power Rule in Section 2.2, you have $D_u[u^n] = nu^{n-1}$, and it follows that $dy/dx = n[u(x)]^{n-1}(du/dx)$.

Example 4   **Applying the General Power Rule**

Find the derivative of $f(x) = (3x - 2x^2)^3$.

Solution   Let $u = 3x - 2x^2$. Then

$$f(x) = (3x - 2x^2)^3 = u^3$$

and, by the General Power Rule, the derivative is

$$f'(x) = \overset{n}{3}\,\overbrace{(3x - 2x^2)^2}^{u^{n-1}}\,\overbrace{\frac{d}{dx}[3x - 2x^2]}^{u'} \qquad \text{Apply General Power Rule.}$$

$$= 3(3x - 2x^2)^2(3 - 4x). \qquad \text{Differentiate } 3x - 2x^2.$$

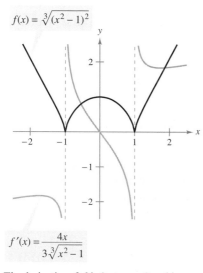

$f(x) = \sqrt[3]{(x^2 - 1)^2}$

$f'(x) = \dfrac{4x}{3\sqrt[3]{x^2 - 1}}$

The derivative of f is 0 at $x = 0$ and is undefined at $x = \pm 1$.
Figure 2.24

Example 5   **Differentiating Functions Involving Radicals**

Find all points on the graph of $f(x) = \sqrt[3]{(x^2 - 1)^2}$ for which $f'(x) = 0$ and those for which $f'(x)$ does not exist.

Solution   Begin by rewriting the function as

$$f(x) = (x^2 - 1)^{2/3}.$$

Then, applying the General Power Rule (with $u = x^2 - 1$) produces

$$f'(x) = \overset{n}{\tfrac{2}{3}}\,\overbrace{(x^2 - 1)^{-1/3}}^{u^{n-1}}\,\overbrace{(2x)}^{u'} \qquad \text{Apply General Power Rule.}$$

$$= \frac{4x}{3\sqrt[3]{x^2 - 1}}. \qquad \text{Write in radical form.}$$

So, $f'(x) = 0$ when $x = 0$ and $f'(x)$ does not exist when $x = \pm 1$, as indicated in Figure 2.24.

Example 6   **Differentiating Quotients with Constant Numerators**

Differentiate $g(t) = \dfrac{-7}{(2t - 3)^2}$.

Solution   Begin by rewriting the function as

$$g(t) = -7(2t - 3)^{-2}.$$

Then, applying the General Power Rule produces

$$g'(t) = \underbrace{(-7)}_{\substack{\text{Constant} \\ \text{Multiple Rule}}}\overset{n}{(-2)}\overbrace{(2t - 3)^{-3}}^{u^{n-1}}\overbrace{(2)}^{u'} \qquad \text{Apply General Power Rule.}$$

$$= 28(2t - 3)^{-3} \qquad \text{Simplify.}$$

$$= \frac{28}{(2t - 3)^3}. \qquad \text{Write with positive exponent.}$$

NOTE   Try differentiating the function in Example 6 using the Quotient Rule. You should obtain the same result, but using the Quotient Rule is less efficient than using the General Power Rule.

Simplifying Derivatives

The next three examples illustrate some techniques for simplifying the "raw derivatives" of functions involving products, quotients, and composites.

Example 7   Simplifying by Factoring Out the Least Powers

$$f(x) = x^2\sqrt{1 - x^2}$$ Original function

$$= x^2(1 - x^2)^{1/2}$$ Rewrite.

$$f'(x) = x^2\frac{d}{dx}\left[(1 - x^2)^{1/2}\right] + (1 - x^2)^{1/2}\frac{d}{dx}\left[x^2\right]$$ Product Rule

$$= x^2\left[\frac{1}{2}(1 - x^2)^{-1/2}(-2x)\right] + (1 - x^2)^{1/2}(2x)$$ General Power Rule

$$= -x^3(1 - x^2)^{-1/2} + 2x(1 - x^2)^{1/2}$$ Simplify.

$$= x(1 - x^2)^{-1/2}\left[-x^2(1) + 2(1 - x^2)\right]$$ Factor.

$$= \frac{x(2 - 3x^2)}{\sqrt{1 - x^2}}$$ Simplify.

Example 8   Simplifying the Derivative of a Quotient

TECHNOLOGY   Symbolic differentiation utilities are capable of differentiating very complicated functions. Often, however, the result is given in unsimplified form. If you have access to such a utility, use it to find the derivatives of the functions given in Examples 7, 8, and 9. Then compare the results with those given on this page.

$$f(x) = \frac{x}{\sqrt[3]{x^2 + 4}}$$ Original function

$$= \frac{x}{(x^2 + 4)^{1/3}}$$ Rewrite.

$$f'(x) = \frac{(x^2 + 4)^{1/3}(1) - x(1/3)(x^2 + 4)^{-2/3}(2x)}{(x^2 + 4)^{2/3}}$$ Quotient Rule

$$= \frac{1}{3}(x^2 + 4)^{-2/3}\left[\frac{3(x^2 + 4) - (2x^2)(1)}{(x^2 + 4)^{2/3}}\right]$$ Factor.

$$= \frac{x^2 + 12}{3(x^2 + 4)^{4/3}}$$ Simplify.

Example 9   Simplifying the Derivative of a Power

$$y = \left(\frac{3x - 1}{x^2 + 3}\right)^2$$ Original function

$$y' = 2\overbrace{\left(\frac{3x - 1}{x^2 + 3}\right)}^{u^{n-1}}\underbrace{\frac{d}{dx}\left[\frac{3x - 1}{x^2 + 3}\right]}_{u'}$$ General Power Rule

$$= \left[\frac{2(3x - 1)}{x^2 + 3}\right]\left[\frac{(x^2 + 3)(3) - (3x - 1)(2x)}{(x^2 + 3)^2}\right]$$ Quotient Rule

$$= \frac{2(3x - 1)(3x^2 + 9 - 6x^2 + 2x)}{(x^2 + 3)^3}$$ Multiply.

$$= \frac{2(3x - 1)(-3x^2 + 2x + 9)}{(x^2 + 3)^3}$$ Simplify.

Trigonometric Functions and the Chain Rule

The "Chain Rule versions" of the derivatives of the six trigonometric functions are as follows.

$$\frac{d}{dx}[\sin u] = (\cos u)\,u' \qquad \frac{d}{dx}[\cos u] = -(\sin u)\,u'$$

$$\frac{d}{dx}[\tan u] = (\sec^2 u)\,u' \qquad \frac{d}{dx}[\cot u] = -(\csc^2 u)\,u'$$

$$\frac{d}{dx}[\sec u] = (\sec u \tan u)\,u' \qquad \frac{d}{dx}[\csc u] = -(\csc u \cot u)\,u'$$

Example 10   Applying the Chain Rule to Trigonometric Functions

$$\overset{u}{\overbrace{}}$$

a. $y = \sin 2x$ $y' = \overset{\cos u}{\overbrace{\cos 2x}}\ \overset{u'}{\overbrace{\frac{d}{dx}[2x]}} = (\cos 2x)(2) = 2\cos 2x$

b. $y = \cos(x - 1)$ $y' = -\sin(x - 1)$

c. $y = \tan 3x$ $y' = 3\sec^2 3x$ ▢

Be sure that you understand the mathematical conventions regarding parentheses and trigonometric functions. For instance, in Example 10a, $\sin 2x$ is written to mean $\sin(2x)$.

Example 11   Parentheses and Trigonometric Functions

a. $y = \cos 3x^2 = \cos(3x^2)$ $y' = (-\sin 3x^2)(6x) = -6x\sin 3x^2$

b. $y = (\cos 3)x^2$ $y' = (\cos 3)(2x) = 2x\cos 3$

c. $y = \cos(3x)^2 = \cos(9x^2)$ $y' = (-\sin 9x^2)(18x) = -18x\sin 9x^2$

d. $y = \cos^2 x = (\cos x)^2$ $y' = 2(\cos x)(-\sin x) = -2\cos x \sin x$

e. $y = \sqrt{\cos x} = (\cos x)^{1/2}$ $y' = \frac{1}{2}(\cos x)^{-1/2}(-\sin x) = -\dfrac{\sin x}{2\sqrt{\cos x}}$ ▢

To find the derivative of a function of the form $k(x) = f(g(h(x)))$, you need to apply the Chain Rule twice, as shown in Example 12.

Example 12   Repeated Application of the Chain Rule

$$\begin{aligned}
f(t) &= \sin^3 4t & &\text{Original function} \\
&= (\sin 4t)^3 & &\text{Rewrite.} \\
f'(t) &= 3(\sin 4t)^2 \frac{d}{dt}[\sin 4t] & &\text{Apply Chain Rule once.} \\
&= 3(\sin 4t)^2(\cos 4t)\frac{d}{dt}[4t] & &\text{Apply Chain Rule a second time.} \\
&= 3(\sin 4t)^2(\cos 4t)(4) \\
&= 12\sin^2 4t \cos 4t & &\text{Simplify.}
\end{aligned}$$

    ▢

We conclude this section with a summary of the differentiation rules studied so far. To become skilled at differentiation, you should memorize each rule.

Summary of Differentiation Rules

General Differentiation Rules   Let f, g, and u be differentiable functions of x.

Constant Multiple Rule:

$$\frac{d}{dx}[cf] = cf'$$

Sum or Difference Rule:

$$\frac{d}{dx}[f \pm g] = f' \pm g'$$

Product Rule:

$$\frac{d}{dx}[fg] = fg' + gf'$$

Quotient Rule:

$$\frac{d}{dx}\left[\frac{f}{g}\right] = \frac{gf' - fg'}{g^2}$$

Derivatives of Algebraic Functions

Constant Rule:

$$\frac{d}{dx}[c] = 0$$

(Simple) Power Rule:

$$\frac{d}{dx}[x^n] = nx^{n-1}, \quad \frac{d}{dx}[x] = 1$$

Derivatives of Trigonometric Functions

$$\frac{d}{dx}[\sin x] = \cos x$$

$$\frac{d}{dx}[\cos x] = -\sin x$$

$$\frac{d}{dx}[\tan x] = \sec^2 x$$

$$\frac{d}{dx}[\cot x] = -\csc^2 x$$

$$\frac{d}{dx}[\sec x] = \sec x \tan x$$

$$\frac{d}{dx}[\csc x] = -\csc x \cot x$$

Chain Rule

Chain Rule:

$$\frac{d}{dx}[f(u)] = f'(u)\, u'$$

General Power Rule:

$$\frac{d}{dx}[u^n] = nu^{n-1}\, u'$$

STUDY TIP   As an aid to memorization, note that the cofunctions (cosine, cotangent, and cosecant) require a negative sign as part of their derivatives.

EXERCISES FOR SECTION 2.4

In Exercises 1–6, complete the table using Example 2 as a model.

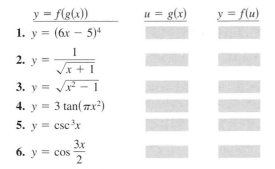

$y = f(g(x))$	$u = g(x)$	$y = f(u)$
1. $y = (6x - 5)^4$		
2. $y = \dfrac{1}{\sqrt{x+1}}$		
3. $y = \sqrt{x^2 - 1}$		
4. $y = 3\tan(\pi x^2)$		
5. $y = \csc^3 x$		
6. $y = \cos\dfrac{3x}{2}$		

In Exercises 7–34, find the derivative of the function.

7. $y = (2x - 7)^3$

8. $y = (2x^3 + 1)^2$

9. $g(x) = 3(4 - 9x)^4$

10. $y = 3(4 - x^2)^5$

11. $f(x) = (9 - x^2)^{2/3}$

12. $f(t) = (9t + 2)^{2/3}$

13. $f(t) = \sqrt{1 - t}$

14. $g(x) = \sqrt{5 - 3x}$

15. $y = \sqrt[3]{9x^2 + 4}$

16. $g(x) = \sqrt{x^2 - 2x + 1}$

17. $y = 2\sqrt[4]{4 - x^2}$

18. $f(x) = -3\sqrt[4]{2 - 9x}$

19. $y = \dfrac{1}{x - 2}$

20. $s(t) = \dfrac{1}{t^2 + 3t - 1}$

21. $f(t) = \left(\dfrac{1}{t - 3}\right)^2$

22. $y = -\dfrac{5}{(t + 3)^3}$

23. $y = \dfrac{1}{\sqrt{x + 2}}$

24. $g(t) = \sqrt{\dfrac{1}{t^2 - 2}}$

25. $f(x) = x^2(x - 2)^4$

26. $f(x) = x(3x - 9)^3$

27. $y = x\sqrt{1 - x^2}$

28. $y = \frac{1}{2}x^2\sqrt{16 - x^2}$

29. $y = \dfrac{x}{\sqrt{x^2 + 1}}$

30. $y = \dfrac{x}{\sqrt{x^4 + 4}}$

31. $g(x) = \left(\dfrac{x + 5}{x^2 + 2}\right)^2$

32. $h(t) = \left(\dfrac{t^2}{t^3 + 2}\right)^2$

33. $f(v) = \left(\dfrac{1 - 2v}{1 + v}\right)^3$

34. $g(x) = \left(\dfrac{3x^2 - 2}{2x + 3}\right)^3$

In Exercises 35–44, use a computer algebra system to find the derivative of the function. Then use the utility to graph the function and its derivative on the same set of coordinate axes. Describe the behavior of the function that corresponds to any zeros of the graph of the derivative.

35. $y = \dfrac{\sqrt{x} + 1}{x^2 + 1}$

36. $y = \sqrt{\dfrac{2x}{x + 1}}$

37. $g(t) = \dfrac{3t^2}{\sqrt{t^2 + 2t - 1}}$

38. $f(x) = \sqrt{x}(2 - x)^2$

39. $y = \sqrt{\dfrac{x + 1}{x}}$

40. $y = (t^2 - 9)\sqrt{t + 2}$

41. $s(t) = \dfrac{-2(2 - t)\sqrt{1 + t}}{3}$

42. $g(x) = \sqrt{x - 1} + \sqrt{x + 1}$

43. $y = \dfrac{\cos \pi x + 1}{x}$

44. $y = x^2 \tan \dfrac{1}{x}$

In Exercises 45 and 46, find the slope of the tangent line to the sine function at the origin. Compare this value with the number of complete cycles in the interval $[0, 2\pi]$. What can you conclude about the slope of the sine function $\sin ax$ at the origin?

45. (a)   (b)

46. (a)   (b)

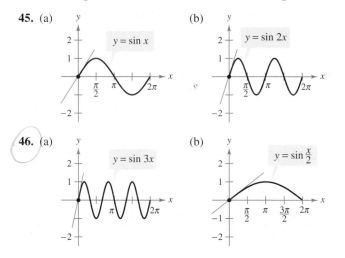

In Exercises 47–66, find the derivative of the function.

47. $y = \cos 3x$

48. $y = \sin \pi x$

49. $g(x) = 3 \tan 4x$

50. $h(x) = \sec x^2$

51. $y = \sin(\pi x)^2$

52. $y = \cos(1 - 2x)^2$

53. $h(x) = \sin 2x \cos 2x$

54. $g(\theta) = \sec(\tfrac{1}{2}\theta) \tan(\tfrac{1}{2}\theta)$

55. $f(x) = \dfrac{\cot x}{\sin x}$

56. $g(v) = \dfrac{\cos v}{\csc v}$

• 57. $y = 4 \sec^2 x$

58. $y = 2 \tan^3 x$

59. $f(\theta) = \tfrac{1}{4} \sin^2 2\theta$

60. $g(t) = 5 \cos^2 \pi t$

61. $f(t) = 3 \sec^2(\pi t - 1)$

62. $h(t) = 2 \cot^2(\pi t + 2)$

63. $y = \sqrt{x} + \tfrac{1}{4} \sin(2x)^2$

64. $y = 3x - 5 \cos(\pi x)^2$

65. $y = \sin(\cos x)$

66. $y = \sin \sqrt[3]{x} + \sqrt[3]{\sin x}$

In Exercises 67–74, evaluate the derivative of the function at the indicated point. Use a graphing utility to verify your result.

Function	Point
67. $s(t) = \sqrt{t^2 + 2t + 8}$	$(2, 4)$
68. $y = \sqrt[5]{3x^3 + 4x}$	$(2, 2)$
69. $f(x) = \dfrac{3}{x^3 - 4}$	$\left(-1, -\dfrac{3}{5}\right)$
70. $f(x) = \dfrac{1}{(x^2 - 3x)^2}$	$\left(4, \dfrac{1}{16}\right)$
71. $f(t) = \dfrac{3t + 2}{t - 1}$	$(0, -2)$
72. $f(x) = \dfrac{x + 1}{2x - 3}$	$(2, 3)$
73. $y = 37 - \sec^3(2x)$	$(0, 36)$
74. $y = \dfrac{1}{x} + \sqrt{\cos x}$	$\left(\dfrac{\pi}{2}, \dfrac{2}{\pi}\right)$

In Exercises 75–78, (a) find an equation of the tangent line to the graph of f at the indicated point, (b) use a graphing utility to graph the function and its tangent line at the point, and (c) use the *derivative* feature of a graphing utility to confirm your results.

Function	Point
75. $f(x) = \sqrt{3x^2 - 2}$	$(3, 5)$
76. $f(x) = \tfrac{1}{3}x\sqrt{x^2 + 5}$	$(2, 2)$
77. $f(x) = \sin 2x$	$(\pi, 0)$
78. $f(x) = \tan^2 x$	$\left(\dfrac{\pi}{4}, 1\right)$

In Exercises 79–82, find the second derivative of the function.

79. $f(x) = 2(x^2 - 1)^3$

80. $f(x) = \dfrac{1}{x - 2}$

81. $f(x) = \sin x^2$

82. $f(x) = \sec^2 \pi x$

Getting at the Concept

In Exercises 83–86, the graphs of a function f and its derivative f' are shown. Label the graphs as f or f' and write a short paragraph stating the criteria used in making the selection. To print an enlarged copy of the graph, go to the website *www.mathgraphs.com*.

83.   84.

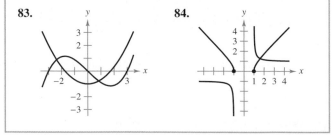

Getting at the Concept   *(continued)*

85.

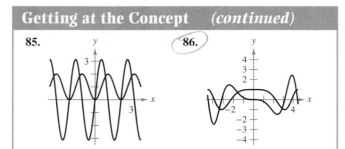

86.

In Exercises 87 and 88, the relationship between f and g is given. State the relationship between f' and g'.

87. $g(x) = f(3x)$

88. $g(x) = f(x^2)$

89. Given that $g(5) = 3$, $g'(5) = 6$, $h(5) = 3$, and $h'(5) = -2$, find $f'(5)$ (if possible) for each of the following. If it is not possible, state what additional information is required.

(a) $f(x) = g(x)h(x)$ (b) $f(x) = g(h(x))$

(c) $f(x) = \dfrac{g(x)}{h(x)}$ (d) $f(x) = [g(x)]^3$

90. (a) Find the derivative of the function $g(x) = \sin^2 x + \cos^2 x$ in two ways.

(b) For $f(x) = \sec^2 x$ and $g(x) = \tan^2 x$, show that $f'(x) = g'(x)$.

91. Doppler Effect   The frequency F of a fire truck siren heard by a stationary observer is

$$F = \frac{132{,}400}{331 \pm v}$$

where $\pm v$ represents the velocity of the accelerating fire truck in meters per second (see figure). Find the rate of change of F with respect to v when

(a) the fire truck is approaching at a velocity of 30 meters per second (use $-v$).

(b) the fire truck is moving away at a velocity of 30 meters per second (use $+v$).

$$F = \frac{132{,}400}{331 + v} \qquad\qquad F = \frac{132{,}400}{331 - v}$$

92. Harmonic Motion   The displacement from equilibrium of an object in harmonic motion on the end of a spring is

$$y = \tfrac{1}{3} \cos 12t - \tfrac{1}{4} \sin 12t$$

where y is measured in feet and t is the time in seconds. Determine the position and velocity of the object when $t = \pi/8$.

93. Pendulum   A 15-centimeter pendulum moves according to the equation

$$\theta = 0.2 \cos 8t$$

where θ is the angular displacement from the vertical in radians and t is the time in seconds. Determine the maximum angular displacement and the rate of change of θ when $t = 3$ seconds.

94. Wave Motion   A buoy oscillates in simple harmonic motion

$$y = A \cos \omega t$$

as waves move past it. The buoy moves a total of 3.5 feet (vertically) from its low point to its high point. It returns to its high point every 10 seconds.

(a) Write an equation describing the motion of the buoy if it is at its high point at $t = 0$.

(b) Determine the velocity of the buoy as a function of t.

95. Circulatory System   The speed S of blood that is r centimeters from the center of an artery is

$$S = C(R^2 - r^2)$$

where C is a constant, R is the radius of the artery, and S is measured in centimeters per second. Suppose a drug is administered and the artery begins to dilate at a rate of dR/dt. At a constant distance r, find the rate at which S changes with respect to t for $C = 1.76 \times 10^5$, $R = 1.2 \times 10^{-2}$, and $dR/dt = 10^{-5}$.

96. Modeling Data   The normal daily maximum temperature T (in degrees Fahrenheit) for Denver, Colorado, is shown in the table. *(Source: National Oceanic and Atmospheric Administration)*

Month	Jan	Feb	Mar	Apr	May	Jun
Temperature	43.2	46.6	52.2	61.8	70.8	81.4

Month	Jul	Aug	Sep	Oct	Nov	Dec
Temperature	88.2	85.8	76.9	66.3	52.5	44.5

(a) Use a graphing utility to plot the data and find a model for the data of the form

$$T(t) = a + b \sin(\pi t/6 - c)$$

where T is the temperature and t is the time in months, with $t = 1$ corresponding to January.

(b) Use a graphing utility to graph the model. How well does the model fit the data?

(c) Find T' and use a graphing utility to graph the derivative.

(d) Based on the graph of the derivative, during what times does the temperature change most rapidly? Most slowly? Do your answers agree with your observations of the temperature changes? Explain.

97. Modeling Data   The cost of producing x units of a product is $C = 60x + 1350$. For one week management determined the number of units produced at the end of t hours during an 8-hour shift. The average values of x for the week are shown in the table.

t	0	1	2	3	4	5	6	7	8
x	0	16	60	130	205	271	336	384	392

(a) Use a graphing utility to fit a cubic model to the data.

(b) Use the Chain Rule to find dC/dt.

(c) Explain why the cost function is not increasing at a constant rate during the 8-hour shift.

98. Think About It   The table shows some values of the derivative of an unknown function f. Complete the table by finding (if possible) the derivative of each transformation of f.

(a) $g(x) = f(x) - 2$ (b) $h(x) = 2f(x)$

(c) $r(x) = f(-3x)$ (d) $s(x) = f(x + 2)$

x	-2	-1	0	1	2	3
$f'(x)$	4	$\frac{2}{3}$	$-\frac{1}{3}$	-1	-2	-4
$g'(x)$						
$h'(x)$						
$r'(x)$						
$s'(x)$						

99. Finding a Pattern   Consider the function $f(x) = \sin \beta x$, where β is a constant.

(a) Find the first-, second-, third-, and fourth-order derivatives of the function.

(b) Verify that the function and its second derivative satisfy the equation $f''(x) + \beta^2 f(x) = 0$.

(c) Use the results in part (a) to write general rules for the even- and odd-order derivatives

$$f^{(2k)}(x) \text{ and } f^{(2k-1)}(x).$$

[Hint: $(-1)^k$ is positive if k is even and negative if k is odd.]

100. Conjecture   Let f be a differentiable function of period p.

(a) Is the function f' periodic? Verify your answer.

(b) Consider the function $g(x) = f(2x)$. Is the function $g'(x)$ periodic? Verify your answer.

101. Think About It   Let $r(x) = f(g(x))$ and $s(x) = g(f(x))$ where f and g are shown in the figure. Find (a) $r'(1)$ and (b) $s'(4)$.

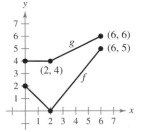

102. Show that the derivative of an odd function is even. That is, if $f(-x) = -f(x)$, then $f'(-x) = f'(x)$.

103. The geometric mean of x and $x + n$ is $g = \sqrt{x(x + n)}$, and the arithmetic mean is $a = [x + (x + n)]/2$. Show that

$$\frac{dg}{dx} = \frac{a}{g}.$$

104. Let u be a differentiable function of x. Use the fact that $|u| = \sqrt{u^2}$ to prove that

$$\frac{d}{dx}[|u|] = u'\frac{u}{|u|}, \quad u \neq 0.$$

In Exercises 105–108, use the result of Exercise 104 to find the derivative of the function.

105. $g(x) = |2x - 3|$

106. $f(x) = |x^2 - 4|$

107. $h(x) = |x| \cos x$

108. $f(x) = |\sin x|$

Linear and Quadratic Approximations   The linear and quadratic approximations of a function f at $x = a$ are

$$P_1(x) = f'(a)(x - a) + f(a) \text{ and}$$
$$P_2(x) = \tfrac{1}{2}f''(a)(x - a)^2 + f'(a)(x - a) + f(a).$$

In Exercises 109 and 110, (a) find the specified linear and quadratic approximations of f, (b) use a graphing utility to graph f and the approximations, (c) determine whether P_1 or P_2 is the better approximation, and (d) state how the accuracy changes as you move farther from $x = a$.

109. $f(x) = \tan \dfrac{\pi x}{4}$

$a = 1$

110. $f(x) = \sec 2x$

$a = \dfrac{\pi}{6}$

True or False?   **In Exercises 111–114, determine whether the statement is true or false. If it is false, explain why or give an example that shows it is false.**

111. If $y = (1 - x)^{1/2}$, then $y' = \frac{1}{2}(1 - x)^{-1/2}$.

112. If $f(x) = \sin^2(2x)$, then $f'(x) = 2(\sin 2x)(\cos 2x)$.

113. If y is a differentiable function of u, u is a differentiable function of v, and v is a differentiable function of x, then

$$\frac{dy}{dx} = \frac{dy}{du}\frac{du}{dv}\frac{dv}{dx}.$$

114. You would first apply the General Power Rule to find the derivative of $y = x \sin^3 x$.

Section 2.5	Implicit Differentiation

- Distinguish between functions written in implicit form and explicit form.
- Use implicit differentiation to find the derivative of a function.

EXPLORATION

Graphing an Implicit Equation How could you use a graphing utility to sketch the graph of the equation

$$x^2 - 2y^3 + 4y = 2?$$

Here are two possible approaches.

a. Solve the equation for x. Switch the roles of x and y and graph the two resulting equations. The combined graphs will show a 90° rotation of the graph of the original equation.

b. Set the graphing utility to *parametric mode* and graph the equations

$$x = -\sqrt{2t^3 - 4t + 2}$$
$$y = t$$

and

$$x = \sqrt{2t^3 - 4t + 2}$$
$$y = t.$$

From either of these two approaches, can you decide whether the graph has a tangent line at the point $(0, 1)$? Explain your reasoning.

Implicit and Explicit Functions

Up to this point in the text, most functions have been expressed in **explicit form.** For example, in the equation

$$y = 3x^2 - 5 \qquad\qquad \text{Explicit form}$$

the variable y is explicitly written as a function of x. Some functions, however, are only implied by an equation. For instance, the function $y = 1/x$ is defined **implicitly** by the equation $xy = 1$. Suppose you were asked to find dy/dx for this equation. For this equation, you could begin by writing y explicitly as a function of x and then differentiating.

Implicit Form	*Explicit Form*	*Derivative*
$xy = 1$	$y = \dfrac{1}{x} = x^{-1}$	$\dfrac{dy}{dx} = -x^{-2} = -\dfrac{1}{x^2}$

This strategy works well whenever you can solve for the function explicitly. You cannot, however, use this procedure when you are unable to solve for y as a function of x. For instance, how would you find dy/dx for the equation

$$x^2 - 2y^3 + 4y = 2$$

where it is very difficult to express y as a function of x explicitly? To do this, you can use **implicit differentiation.**

To understand how to find dy/dx implicitly, you must realize that the differentiation is taking place *with respect to x.* This means that when you differentiate terms involving x alone, you can differentiate as usual. However, when you differentiate terms involving y, you must apply the Chain Rule, because you are assuming that y is defined implicitly as a differentiable function of x.

Example 1   **Differentiating with Respect to x**

a. $\dfrac{d}{dx}[x^3] = 3x^2$ 　　　　　Variables agree: use Simple Power Rule.

Variables agree

b. $\dfrac{d}{dx}\overbrace{[y^3]}^{u^n} = \overbrace{3y^2}^{nu^{n-1}u'}\,\dfrac{dy}{dx}$ 　　　　　Variables disagree: use Chain Rule.

Variables disagree

c. $\dfrac{d}{dx}[x + 3y] = 1 + 3\dfrac{dy}{dx}$ 　　　　　Chain Rule: $\dfrac{d}{dx}[3y] = 3y'$

d. $\dfrac{d}{dx}[xy^2] = x\dfrac{d}{dx}[y^2] + y^2\dfrac{d}{dx}[x]$ 　　　　　Product Rule

$\qquad\qquad = x\left(2y\dfrac{dy}{dx}\right) + y^2(1)$ 　　　　　Chain Rule

$\qquad\qquad = 2xy\dfrac{dy}{dx} + y^2$ 　　　　　Simplify.

Implicit Differentiation

Guidelines for Implicit Differentiation

1. Differentiate both sides of the equation *with respect to x.*

2. Collect all terms involving dy/dx on the left side of the equation and move all other terms to the right side of the equation.

3. Factor dy/dx out of the left side of the equation.

4. Solve for dy/dx by dividing both sides of the equation by the left-hand factor that does not contain dy/dx.

Example 2 Implicit Differentiation

Find dy/dx given that $y^3 + y^2 - 5y - x^2 = -4$.

NOTE In Example 2, note that implicit differentiation can produce an expression for dy/dx that contains both x and y.

Solution

1. Differentiate both sides of the equation with respect to x.

$$\frac{d}{dx}[y^3 + y^2 - 5y - x^2] = \frac{d}{dx}[-4]$$

$$\frac{d}{dx}[y^3] + \frac{d}{dx}[y^2] - \frac{d}{dx}[5y] - \frac{d}{dx}[x^2] = \frac{d}{dx}[-4]$$

$$3y^2\frac{dy}{dx} + 2y\frac{dy}{dx} - 5\frac{dy}{dx} - 2x = 0$$

2. Collect the dy/dx terms on the left side of the equation.

$$3y^2\frac{dy}{dx} + 2y\frac{dy}{dx} - 5\frac{dy}{dx} = 2x$$

3. Factor dy/dx out of the left side of the equation.

$$\frac{dy}{dx}(3y^2 + 2y - 5) = 2x$$

4. Solve for dy/dx by dividing by $(3y^2 + 2y - 5)$.

$$\frac{dy}{dx} = \frac{2x}{3y^2 + 2y - 5}$$

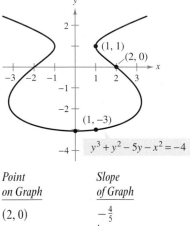

Point on Graph	Slope of Graph
$(2, 0)$	$-\frac{4}{5}$
$(1, -3)$	$\frac{1}{8}$
$x = 0$	0
$(1, 1)$	Undefined

The implicit equation

$$y^3 + y^2 - 5y - x^2 = -4$$

has the derivative

$$\frac{dy}{dx} = \frac{2x}{3y^2 + 2y - 5}.$$

Figure 2.25

   To see how you can use an *implicit derivative,* consider the graph shown in Figure 2.25. From the graph, you can see that y is not a function of x. Even so, the derivative found in Example 2 gives a formula for the slope of the tangent line at a point on this graph. The slopes at several points on the graph are shown below the graph.

TECHNOLOGY With most graphing utilities, it is easy to sketch the graph of an equation that explicitly represents y as a function of x. Sketching graphs of other equations, however, can require some ingenuity. For instance, to sketch the graph of the equation given in Example 2, try using a graphing utility, set in parametric mode, to sketch the graphs given by $x = \sqrt{t^3 + t^2 - 5t + 4}$, $y = t$, and $x = -\sqrt{t^3 + t^2 - 5t + 4}$, $y = t$, for $-5 \le t \le 5$. How does the result compare with the graph shown in Figure 2.25?

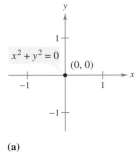

$x^2 + y^2 = 0$

(0, 0)

(a)

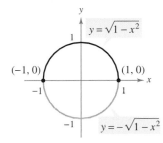

$y = \sqrt{1 - x^2}$

(−1, 0)   (1, 0)

$y = -\sqrt{1 - x^2}$

(b)

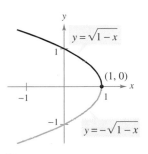

$y = \sqrt{1 - x}$

(1, 0)

$y = -\sqrt{1 - x}$

(c)

Some graph segments can be represented by differentiable functions.

Figure 2.26

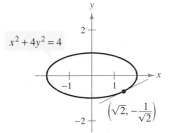

$x^2 + 4y^2 = 4$

$\left(\sqrt{2}, -\dfrac{1}{\sqrt{2}}\right)$

Slope of tangent line is $\frac{1}{2}$.

Figure 2.27

It is meaningless to solve for dy/dx in an equation that has no solution points. (For example, $x^2 + y^2 = -4$ has no solution points.) If, however, a segment of a graph can be represented by a differentiable function, dy/dx will have meaning as the slope at each point on the segment. Recall that a function is not differentiable at (1) points with vertical tangents and (2) points at which the function is not continuous.

Example 3   Representing a Graph by Differentiable Functions

If possible, represent y as a differentiable function of x (see Figure 2.26).

a. $x^2 + y^2 = 0$ **b.** $x^2 + y^2 = 1$ **c.** $x + y^2 = 1$

Solution

a. The graph of this equation is a single point. Therefore, it does not define y as a differentiable function of x.

b. The graph of this equation is the unit circle, centered at $(0, 0)$. The upper semicircle is given by the differentiable function

$$y = \sqrt{1 - x^2}, \quad -1 < x < 1$$

and the lower semicircle is given by the differentiable function

$$y = -\sqrt{1 - x^2}, \quad -1 < x < 1.$$

At the points $(-1, 0)$ and $(1, 0)$, the slope of the graph is undefined.

c. The upper half of this parabola is given by the differentiable function

$$y = \sqrt{1 - x}, \quad x < 1$$

and the lower half of this parabola is given by the differentiable function

$$y = -\sqrt{1 - x}, \quad x < 1.$$

At the point $(1, 0)$, the slope of the graph is undefined.

Example 4   Finding the Slope of a Graph Implicitly

Determine the slope of the tangent line to the graph of

$$x^2 + 4y^2 = 4$$

at the point $\left(\sqrt{2}, -1/\sqrt{2}\right)$. (See Figure 2.27.)

Solution

$$x^2 + 4y^2 = 4 \qquad \text{Write original equation.}$$

$$2x + 8y\frac{dy}{dx} = 0 \qquad \text{Differentiate with respect to } x.$$

$$\frac{dy}{dx} = \frac{-2x}{8y} = \frac{-x}{4y}. \qquad \text{Solve for } \frac{dy}{dx}.$$

So, at $\left(\sqrt{2}, -1/\sqrt{2}\right)$, the slope is

$$\frac{dy}{dx} = \frac{-\sqrt{2}}{-4/\sqrt{2}} = \frac{1}{2}. \qquad \text{Evaluate } \frac{dy}{dx} \text{ when } x = \sqrt{2} \text{ and } y = -\frac{1}{\sqrt{2}}.$$

NOTE   To see the benefit of implicit differentiation, try doing Example 4 using the explicit function $y = -\frac{1}{2}\sqrt{4 - x^2}$.

Example 5   **Finding the Slope of a Graph Implicitly**

Determine the slope of the graph of $3(x^2 + y^2)^2 = 100xy$ at the point $(3, 1)$.

Solution

$$\frac{d}{dx}[3(x^2 + y^2)^2] = \frac{d}{dx}[100xy]$$

$$3(2)(x^2 + y^2)\left(2x + 2y\frac{dy}{dx}\right) = 100\left[x\frac{dy}{dx} + y(1)\right]$$

$$12y(x^2 + y^2)\frac{dy}{dx} - 100x\frac{dy}{dx} = 100y - 12x(x^2 + y^2)$$

$$[12y(x^2 + y^2) - 100x]\frac{dy}{dx} = 100y - 12x(x^2 + y^2)$$

$$\frac{dy}{dx} = \frac{100y - 12x(x^2 + y^2)}{-100x + 12y(x^2 + y^2)}$$

$$= \frac{25y - 3x(x^2 + y^2)}{-25x + 3y(x^2 + y^2)}$$

At the point $(3, 1)$, the slope of the graph is

$$\frac{dy}{dx} = \frac{25(1) - 3(3)(3^2 + 1^2)}{-25(3) + 3(1)(3^2 + 1^2)} = \frac{25 - 90}{-75 + 30} = \frac{-65}{-45} = \frac{13}{9}$$

as shown in Figure 2.28. This graph is called a **lemniscate.**

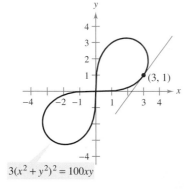

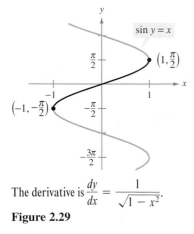

$3(x^2 + y^2)^2 = 100xy$

Lemniscate
Figure 2.28

Example 6   **Determining a Differentiable Function**

Find dy/dx implicitly for the equation $\sin y = x$. Then find the largest interval of the form $-a < y < a$ such that y is a differentiable function of x (see Figure 2.29).

Solution

$$\frac{d}{dx}[\sin y] = \frac{d}{dx}[x]$$

$$\cos y\frac{dy}{dx} = 1$$

$$\frac{dy}{dx} = \frac{1}{\cos y}$$

The largest interval about the origin for which y is a differentiable function of x is $-\pi/2 < y < \pi/2$. To see this, note that $\cos y$ is positive for all y in this interval and is 0 at the endpoints. If you restrict y to the interval $-\pi/2 < y < \pi/2$, you should be able to write dy/dx explicitly as a function of x. To do this, you can use

$$\cos y = \sqrt{1 - \sin^2 y}$$

$$= \sqrt{1 - x^2}, \quad -\frac{\pi}{2} < y < \frac{\pi}{2}$$

and conclude that

$$\frac{dy}{dx} = \frac{1}{\sqrt{1 - x^2}}.$$

The derivative is $\dfrac{dy}{dx} = \dfrac{1}{\sqrt{1 - x^2}}$.

Figure 2.29

With implicit differentiation, the form of the derivative often can be simplified (as in Example 6) by an appropriate use of the *original* equation. A similar technique can be used to find and simplify higher-order derivatives obtained implicitly.

ISAAC BARROW (1630–1677)

The graph in Example 8 is called the **kappa curve** because it resembles the Greek letter kappa, κ. The general solution for the tangent line to this curve was discovered by the English mathematician Isaac Barrow. Newton was Barrow's student and they corresponded frequently regarding their work in the early development of calculus.

Example 7   Finding the Second Derivative Implicitly

Given $x^2 + y^2 = 25$, find $\dfrac{d^2y}{dx^2}$.

Solution   Differentiating each term with respect to x produces

$$2x + 2y\frac{dy}{dx} = 0$$

$$2y\frac{dy}{dx} = -2x$$

$$\frac{dy}{dx} = \frac{-2x}{2y} = -\frac{x}{y}.$$

Differentiating a second time with respect to x yields

$$\frac{d^2y}{dx^2} = -\frac{(y)(1) - (x)(dy/dx)}{y^2} \qquad \text{Quotient Rule}$$

$$= -\frac{y - (x)(-x/y)}{y^2} \qquad \text{Substitute } -x/y \text{ for } \frac{dy}{dx}.$$

$$= -\frac{y^2 + x^2}{y^3} \qquad \text{Simplify.}$$

$$= -\frac{25}{y^3}. \qquad \text{Substitute 25 for } x^2 + y^2.$$

Example 8   Finding a Tangent Line to a Graph

Find the tangent line to the graph given by $x^2(x^2 + y^2) = y^2$ at the point $\left(\sqrt{2}/2, \sqrt{2}/2\right)$, as shown in Figure 2.30.

Solution   By rewriting and differentiating implicitly, you obtain

$$x^4 + x^2y^2 - y^2 = 0$$

$$4x^3 + x^2\left(2y\frac{dy}{dx}\right) + 2xy^2 - 2y\frac{dy}{dx} = 0$$

$$2y(x^2 - 1)\frac{dy}{dx} = -2x(2x^2 + y^2)$$

$$\frac{dy}{dx} = \frac{x(2x^2 + y^2)}{y(1 - x^2)}.$$

At the point $\left(\sqrt{2}/2, \sqrt{2}/2\right)$, the slope is

$$\frac{dy}{dx} = \frac{\left(\sqrt{2}/2\right)[2(1/2) + (1/2)]}{\left(\sqrt{2}/2\right)[1 - (1/2)]} = \frac{3/2}{1/2} = 3$$

and the equation of the tangent line at this point is

$$y - \frac{\sqrt{2}}{2} = 3\left(x - \frac{\sqrt{2}}{2}\right)$$

$$y = 3x - \sqrt{2}.$$

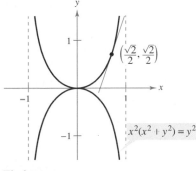

The kappa curve
Figure 2.30

EXERCISES FOR SECTION 2.5

In Exercises 1–16, find dy/dx by implicit differentiation.

1. $x^2 + y^2 = 36$
2. $x^2 - y^2 = 16$
3. $x^{1/2} + y^{1/2} = 9$
4. $x^3 + y^3 = 8$
5. $x^3 - xy + y^2 = 4$
6. $x^2y + y^2x = -2$
7. $x^3y^3 - y = x$
8. $\sqrt{xy} = x - 2y$
9. $x^3 - 3x^2y + 2xy^2 = 12$
10. $2 \sin x \cos y = 1$
11. $\sin x + 2 \cos 2y = 1$
12. $(\sin \pi x + \cos \pi y)^2 = 2$
13. $\sin x = x(1 + \tan y)$
14. $\cot y = x - y$
15. $y = \sin(xy)$
16. $x = \sec \dfrac{1}{y}$

In Exercises 17–20, (a) find two explicit functions by solving the equation for y in terms of x, (b) sketch the graph of the equation and label the parts given by the corresponding explicit functions, (c) differentiate the explicit functions, and (d) find dy/dx implicitly and show that the result is equivalent to that of part (c).

17. $x^2 + y^2 = 16$
18. $x^2 + y^2 - 4x + 6y + 9 = 0$
19. $9x^2 + 16y^2 = 144$
20. $9y^2 - x^2 = 9$

In Exercises 21–28, find dy/dx by implicit differentiation and evaluate the derivative at the indicated point.

	Equation	*Point*
21.	$xy = 4$	$(-4, -1)$
22.	$x^2 - y^3 = 0$	$(1, 1)$
23.	$y^2 = \dfrac{x^2 - 4}{x^2 + 4}$	$(2, 0)$
24.	$(x + y)^3 = x^3 + y^3$	$(-1, 1)$
25.	$x^{2/3} + y^{2/3} = 5$	$(8, 1)$
26.	$x^3 + y^3 = 4xy + 1$	$(2, 1)$
27.	$\tan(x + y) = x$	$(0, 0)$
28.	$x \cos y = 1$	$\left(2, \dfrac{\pi}{3}\right)$

In Exercises 29–32, find the slope of the tangent line to the graph at the indicated point.

29. Witch of Agnesi:

$(x^2 + 4)y = 8$

Point: $(2, 1)$

30. Cissoid:

$(4 - x)y^2 = x^3$

Point: $(2, 2)$

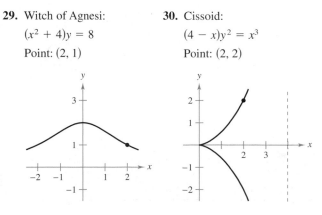

31. Bifolium:

$(x^2 + y^2)^2 = 4x^2y$

Point: $(1, 1)$

32. Folium of Descartes:

$x^3 + y^3 - 6xy = 0$

Point: $\left(\frac{4}{3}, \frac{8}{3}\right)$

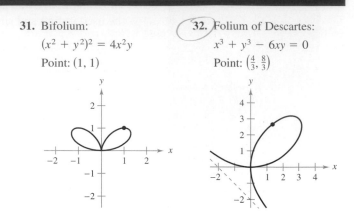

In Exercises 33 and 34, find dy/dx implicitly and find the largest interval of the form $-a < y < a$ such that y is a differentiable function of x. Then express dy/dx as a function of x.

33. $\tan y = x$
34. $\cos y = x$

In Exercises 35–40, find d^2y/dx^2 in terms of x and y.

35. $x^2 + y^2 = 36$
36. $x^2y^2 - 2x = 3$
37. $x^2 - y^2 = 16$
38. $1 - xy = x - y$
39. $y^2 = x^3$
40. $y^2 = 4x$

In Exercises 41 and 42, use a graphing utility to graph the equation. Find an equation of the tangent line to the graph at the indicated point and sketch its graph.

41. $\sqrt{x} + \sqrt{y} = 4$, $(9, 1)$
42. $y^2 = \dfrac{x - 1}{x^2 + 1}$, $\left(2, \dfrac{\sqrt{5}}{5}\right)$

In Exercises 43 and 44, find equations for the tangent line and normal line to the circle at the indicated points. (The *normal line* at a point is perpendicular to the tangent line at the point.) Use a graphing utility to graph the equation, tangent line, and normal line.

43. $x^2 + y^2 = 25$
$(4, 3), (-3, 4)$
44. $x^2 + y^2 = 9$
$(0, 3), \left(2, \sqrt{5}\right)$

45. Show that the normal line at any point on the circle $x^2 + y^2 = r^2$ passes through the origin.

46. Two circles of radius 4 are tangent to the graph of $y^2 = 4x$ at the point $(1, 2)$. Find equations of these two circles.

In Exercises 47 and 48, find the points at which the graph of the equation has a vertical or horizontal tangent line.

47. $25x^2 + 16y^2 + 200x - 160y + 400 = 0$
48. $4x^2 + y^2 - 8x + 4y + 4 = 0$

Orthogonal Trajectories   In Exercises 49–52, use a graphing utility to sketch the intersecting graphs of the equations and show that they are orthogonal. [Two graphs are *orthogonal* if at their point(s) of intersection their tangent lines are perpendicular to each other.]

49. $2x^2 + y^2 = 6$
$y^2 = 4x$

50. $y^2 = x^3$
$2x^2 + 3y^2 = 5$

51. $x + y = 0$
$x = \sin y$

52. $x^3 = 3(y - 1)$
$x(3y - 29) = 3$

Orthogonal Trajectories   In Exercises 53 and 54, verify that the two families of curves are orthogonal where C and K are real numbers. Use a graphing utility to graph the two families for two values of C and two values of K.

53. $xy = C$
$x^2 - y^2 = K$

54. $x^2 + y^2 = C^2$
$y = Kx$

In Exercises 55–58, differentiate (a) with respect to x (y is a function of x) and (b) with respect to t (x and y are functions of t).

55. $2y^2 - 3x^4 = 0$

56. $x^2 - 3xy^2 + y^3 = 10$

57. $\cos \pi y - 3 \sin \pi x = 1$

58. $4 \sin x \cos y = 1$

Getting at the Concept

59. Describe the difference between the explicit form of a function and an implicit equation. Give an example of each.

60. In your own words, state the guidelines for implicit differentiation.

61. Consider the equation $x^4 = 4(4x^2 - y^2)$.

(a) Use a graphing utility to graph the equation.

(b) Find and graph the four tangent lines to the curve for $y = 3$.

(c) Find the exact coordinates of the point of intersection of the two tangent lines in the first quadrant.

62. *Orthogonal Trajectories*   The figure below gives the topographic map carried by a group of hikers. The hikers are in a wooded area on top of the hill shown on the map and they decide to follow a path of steepest descent (orthogonal trajectories to the contours on the map). Draw their routes if they start from point A and if they start from point B. If their goal is to reach the road along the top of the map, which starting point should they use? To print an enlarged copy of the graph, go to the website *www.mathgraphs.com*.

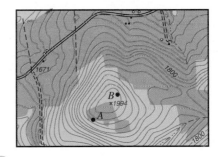

63. Prove (Theorem 2.3) that

$$\frac{d}{dx}[x^n] = nx^{n-1}$$

for the case in which n is a rational number. (*Hint:* Write $y = x^{p/q}$ in the form $y^q - x^p$ and differentiate implicitly. Assume that p and q are integers, where $q > 0$.)

64. Let L be any tangent line to the curve $\sqrt{x} + \sqrt{y} = \sqrt{c}$. Show that the sum of the x- and y-intercepts of L is c.

SECTION PROJECT   OPTICAL ILLUSIONS

In each graph below, an optical illusion is created by having lines intersect a family of curves. In each case, the lines appear to be curved. Find the value of dy/dx for the indicated values of x and y.

(a) Circles: $x^2 + y^2 = C^2$
$x = 3, y = 4, C = 5$

(b) Hyperbolas: $xy = C$
$x = 1, y = 4, C = 4$

(c) Lines: $ax = by$
$x = \sqrt{3}, y = 3,$
$a = \sqrt{3}, b = 1$

(d) Cosine curves: $y = C \cos x$
$x = \frac{\pi}{3}, y = \frac{1}{3}, C = \frac{2}{3}$

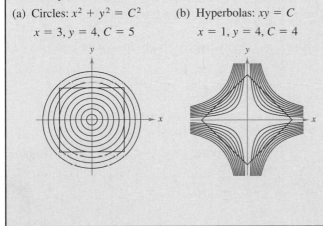

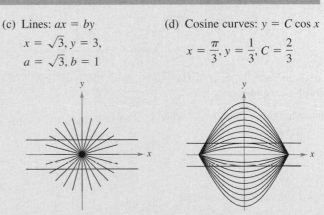

FOR FURTHER INFORMATION For more information on the mathematics of optical illusions, see the article "Descriptive Models for Perception of Optical Illusions" by David A. Smith in *The UMAP Journal*. To view this article, go to the website *www.matharticles.com*.

Section 2.6   Related Rates

- Find a related rate.
- Use related rates to solve real-life problems.

Finding Related Rates

You have seen how the Chain Rule can be used to find dy/dx implicitly. Another important use of the Chain Rule is to find the rates of change of two or more related variables that are changing with respect to *time*.

For example, when water is drained out of a conical tank (see Figure 2.31), the volume V, the radius r, and the height h of the water level are all functions of time t. Knowing that these variables are related by the equation

$$V = \frac{\pi}{3} r^2 h \qquad\qquad \text{Original equation}$$

you can differentiate implicitly with respect to t to obtain the **related-rate** equation

$$\frac{d}{dt}(V) = \frac{d}{dt}\left(\frac{\pi}{3} r^2 h\right)$$

$$\frac{dV}{dt} = \frac{\pi}{3}\left[r^2 \frac{dh}{dt} + h\left(2r \frac{dr}{dt}\right)\right] \qquad \text{Differentiate with respect to } t.$$

$$= \frac{\pi}{3}\left(r^2 \frac{dh}{dt} + 2rh \frac{dr}{dt}\right).$$

From this equation you can see that the rate of change of V is related to the rates of change of both h and r.

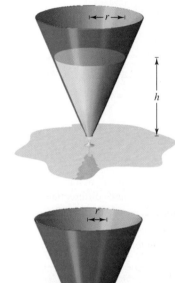

Volume is related to radius and height.
Figure 2.31

EXPLORATION

Finding a Related Rate   In the conical tank shown in Figure 2.31, suppose that the height is changing at a rate of -0.2 foot per minute and the radius is changing at a rate of -0.1 foot per minute. What is the rate of change in the volume when the radius is $r = 1$ foot and the height is $h = 2$ feet? Does the rate of change in the volume depend on the values of r and h? Explain.

Example 1   **Two Rates That Are Related**

Suppose x and y are both differentiable functions of t and are related by the equation $y = x^2 + 3$. Find dy/dt when $x = 1$, given that $dx/dt = 2$ when $x = 1$.

Solution   Using the Chain Rule, you can differentiate both sides of the equation *with respect to t.*

$$y = x^2 + 3 \qquad\qquad \text{Write original equation.}$$

$$\frac{d}{dt}[y] = \frac{d}{dt}[x^2 + 3] \qquad\qquad \text{Differentiate with respect to } t.$$

$$\frac{dy}{dt} = 2x \frac{dx}{dt} \qquad\qquad \text{Chain Rule}$$

When $x = 1$ and $dx/dt = 2$, you have

$$\frac{dy}{dt} = 2(1)(2) = 4.$$

FOR FURTHER INFORMATION To learn more about the history of related-rate problems, see the article "The Lengthening Shadow: The Story of Related Rates" by Bill Austin, Don Barry, and David Berman in *Mathematics Magazine*. To view this article, go to the website *www.matharticles.com*.

Problem Solving with Related Rates

In Example 1, you were *given* an equation that related the variables x and y and were asked to find the rate of change of y when $x = 1$.

Equation: $y = x^2 + 3$

Given rate: $\dfrac{dx}{dt} = 2$ when $x = 1$

Find: $\dfrac{dy}{dt}$ when $x = 1$

In each of the remaining examples in this section, you must *create* a mathematical model from a verbal description.

Example 2   Ripples in a Pond

A pebble is dropped into a calm pond, causing ripples in the form of concentric circles, as shown in Figure 2.32. The radius r of the outer ripple is increasing at a constant rate of 1 foot per second. When the radius is 4 feet, at what rate is the total area A of the disturbed water changing?

Solution   The variables r and A are related by $A = \pi r^2$. The rate of change of the radius r is $dr/dt = 1$.

Equation: $A = \pi r^2$

Given rate: $\dfrac{dr}{dt} = 1$

Find: $\dfrac{dA}{dt}$ when $r = 4$

With this information, you can proceed as in Example 1.

$$\frac{d}{dt}[A] = \frac{d}{dt}[\pi r^2] \qquad \text{Differentiate with respect to } t.$$

$$\frac{dA}{dt} = 2\pi r \frac{dr}{dt} \qquad \text{Chain Rule}$$

$$\frac{dA}{dt} = 2\pi(4)(1) = 8\pi \qquad \text{Substitute 4 for } r \text{ and 1 for } dr/dt.$$

When the radius is 4 feet, the area is changing at a rate of 8π square feet per second.

Total area increases as the outer radius increases.
Figure 2.32

Guidelines For Solving Related-Rate Problems

1. Identify all *given* quantities and quantities *to be determined*. Make a sketch and label the quantities.

2. Write an equation involving the variables whose rates of change either are given or are to be determined.

3. Using the Chain Rule, implicitly differentiate both sides of the equation *with respect to time t.*

4. *After* completing Step 3, substitute into the resulting equation all known values for the variables and their rates of change. Then solve for the required rate of change.

NOTE   When using these guidelines, be sure you perform Step 3 *before* Step 4. Substituting the known values of the variables before differentiating will produce an inappropriate derivative.

The following table lists examples of mathematical models involving rates of change. For instance, the rate of change in the first example is the velocity of a car.

Verbal Statement	Mathematical Model
The velocity of a car after traveling for 1 hour is 50 miles per hour.	x = distance traveled $\dfrac{dx}{dt} = 50$ when $t = 1$
Water is being pumped into a swimming pool at a rate of 10 cubic meters per hour.	V = volume of water in pool $\dfrac{dV}{dt} = 10\text{m}^3/\text{hr}$
A gear is revolving at a rate of 25 revolutions per minute (1 revolution = 2π rad).	θ = angle of revolution $\dfrac{d\theta}{dt} = 25(2\pi)$ rad/min

Example 3   An Inflating Balloon

Air is being pumped into a spherical balloon (see Figure 2.33) at a rate of 4.5 cubic feet per minute. Find the rate of change of the radius when the radius is 2 feet.

Solution   Let V be the volume of the balloon and let r be its radius. Because the volume is increasing at a rate of 4.5 cubic feet per minute, you know that at time t the rate of change of the volume is $dV/dt = \frac{9}{2}$. So, the problem can be stated as follows.

Given rate: $\dfrac{dV}{dt} = \dfrac{9}{2}$ (constant rate)

Find: $\dfrac{dr}{dt}$ when $r = 2$

To find the rate of change of the radius, you must find an equation that relates the radius r to the volume V.

Equation: $V = \dfrac{4}{3}\pi r^3$ Volume of a sphere

Implicit differentiation with respect to t produces

$\dfrac{dV}{dt} = 4\pi r^2 \dfrac{dr}{dt}$ Differentiate with respect to t.

$\dfrac{dr}{dt} = \dfrac{1}{4\pi r^2}\left(\dfrac{dV}{dt}\right).$ Solve for dr/dt.

Finally, when $r = 2$, the rate of change of the radius is

$\dfrac{dr}{dt} = \dfrac{1}{16\pi}\left(\dfrac{9}{2}\right) \approx 0.09$ foot per minute.

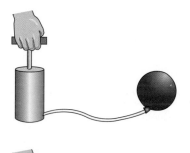

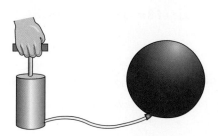

Inflating a balloon
Figure 2.33

In Example 3, note that the volume is increasing at a *constant* rate but the radius is increasing at a *variable* rate. Just because two rates are related does not mean that they are proportional. In this particular case, the radius is growing more and more slowly as t increases. Do you see why?

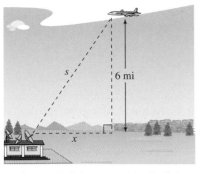

An airplane is flying at an altitude of 6 miles, s miles from the station.
Figure 2.34

📁 *Example 4*   **The Speed of an Airplane Tracked by Radar**

An airplane is flying on a flight path that will take it directly over a radar tracking station, as shown in Figure 2.34. If s is decreasing at a rate of 400 miles per hour when $s = 10$ miles, what is the speed of the plane?

Solution   Let x be the horizontal distance from the station, as shown in Figure 2.34. Notice that when $s = 10$, $x = \sqrt{10^2 - 36} = 8$.

Given rate: $ds/dt = -400$ when $s = 10$
Find: dx/dt when $s = 10$ and $x = 8$

You can find the velocity of the plane as follows.

Equation: $x^2 + 6^2 = s^2$ Pythagorean Theorem

$$2x\frac{dx}{dt} = 2s\frac{ds}{dt}$$ Differentiate with respect to t.

$$\frac{dx}{dt} = \frac{s}{x}\left(\frac{ds}{dt}\right)$$ Solve for dx/dt.

$$\frac{dx}{dt} = \frac{10}{8}(-400)$$ Substitute for s, x, and ds/dt.

$$= -500 \text{ miles per hour}$$ Simplify.

Because the velocity is -500 miles per hour, the *speed* is 500 miles per hour.

Example 5   **A Changing Angle of Elevation**

Find the rate of change in the angle of elevation of the camera shown in Figure 2.35 at 10 seconds after lift-off.

Solution   Let θ be the angle of elevation, as shown in Figure 2.35. When $t = 10$, the height s of the rocket is $s = 50t^2 = 50(10)^2 = 5000$ feet.

Given rate: $ds/dt = 100t = $ velocity of rocket
Find: $d\theta/dt$ when $t = 10$ and $s = 5000$

Using Figure 2.35, you can relate s and θ by the equation $\tan \theta = s/2000$.

Equation: $\tan \theta = \dfrac{s}{2000}$ See Figure 2.35.

$$(\sec^2\theta)\frac{d\theta}{dt} = \frac{1}{2000}\left(\frac{ds}{dt}\right)$$ Differentiate with respect to t.

$$\frac{d\theta}{dt} = \cos^2\theta\frac{100t}{2000}$$ Substitute $100t$ for ds/dt.

$$= \left(\frac{2000}{\sqrt{s^2 + 2000^2}}\right)^2 \frac{100t}{2000}$$ $\cos\theta = 2000/\sqrt{s^2 + 2000^2}$

When $t = 10$ and $s = 5000$, you have

$$\frac{d\theta}{dt} = \frac{2000(100)(10)}{5000^2 + 2000^2} = \frac{2}{29} \text{ radian per second.}$$

So, when $t = 10$, θ is changing at a rate of $\frac{2}{29}$ radian per second.   ◪

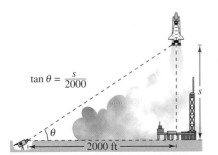

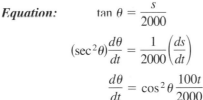

A television camera at ground level is filming the lift-off of a space shuttle that is rising vertically according to the position equation $s = 50t^2$, where s is measured in feet and t is measured in seconds. The camera is 2000 feet from the launch pad.
Figure 2.35

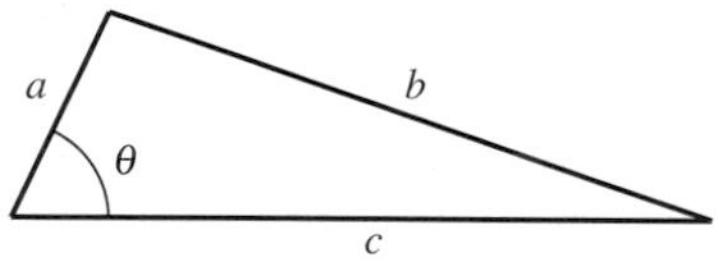

Example 6 **The Velocity of a Piston**

In the engine shown in Figure 2.36, a 7-inch connecting rod is fastened to a crank of radius 3 inches. The crankshaft rotates counterclockwise at a constant rate of 200 revolutions per minute. Find the velocity of the piston when $\theta = \pi/3$.

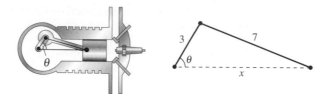

The velocity of a piston is related to the angle of the crankshaft.
Figure 2.36

Solution   Label the distances as shown in Figure 2.36. Because a complete revolution corresponds to 2π radians, it follows that $d\theta/dt = 200(2\pi) = 400\pi$ radians per minute.

Given rate: $\dfrac{d\theta}{dt} = 400\pi$ (constant rate)

Find: $\dfrac{dx}{dt}$ when $\theta = \dfrac{\pi}{3}$

You can use the Law of Cosines (Figure 2.37) to find an equation that relates x and θ.

Equation:
$$7^2 = 3^2 + x^2 - 2(3)(x)\cos\theta$$
$$0 = 2x\frac{dx}{dt} - 6\left(-x\sin\theta\frac{d\theta}{dt} + \cos\theta\frac{dx}{dt}\right)$$
$$(6\cos\theta - 2x)\frac{dx}{dt} = 6x\sin\theta\frac{d\theta}{dt}$$
$$\frac{dx}{dt} = \frac{6x\sin\theta}{6\cos\theta - 2x}\left(\frac{d\theta}{dt}\right)$$

When $\theta = \pi/3$, you can solve for x as follows.

$$7^2 = 3^2 + x^2 - 2(3)(x)\cos\frac{\pi}{3}$$
$$49 = 9 + x^2 - 6x\left(\frac{1}{2}\right)$$
$$0 = x^2 - 3x - 40$$
$$0 = (x - 8)(x + 5)$$
$$x = 8 \qquad\qquad \text{Choose positive solution.}$$

So, when $x = 8$ and $\theta = \pi/3$, the velocity of the piston is

$$\frac{dx}{dt} = \frac{6(8)\left(\sqrt{3}/2\right)}{6(1/2) - 16}(400\pi)$$
$$= \frac{9600\pi\sqrt{3}}{-13}$$
$$\approx -4018 \text{ inches per minute.}$$

NOTE   Note that the velocity in Example 6 is negative because x represents a distance that is decreasing.

Law of Cosines:
$b^2 = a^2 + c^2 - 2ac\cos\theta$
Figure 2.37

EXERCISES FOR SECTION 2.6

In Exercises 1–4, assume that x and y are both differentiable functions of t and find the required values of dy/dt and dx/dt.

Equation	Find	Given
1. $y = \sqrt{x}$	(a) $\dfrac{dy}{dt}$ when $x = 4$	$\dfrac{dx}{dt} = 3$
	(b) $\dfrac{dx}{dt}$ when $x = 25$	$\dfrac{dy}{dt} = 2$
2. $y = 2(x^2 - 3x)$	(a) $\dfrac{dy}{dt}$ when $x = 3$	$\dfrac{dx}{dt} = 2$
	(b) $\dfrac{dx}{dt}$ when $x = 1$	$\dfrac{dy}{dt} = 5$
3. $xy = 4$	(a) $\dfrac{dy}{dt}$ when $x = 8$	$\dfrac{dx}{dt} = 10$
	(b) $\dfrac{dx}{dt}$ when $x = 1$	$\dfrac{dy}{dt} = -6$
4. $x^2 + y^2 = 25$	(a) $\dfrac{dy}{dt}$ when $x = 3, y = 4$	$\dfrac{dx}{dt} = 8$
	(b) $\dfrac{dx}{dt}$ when $x = 4, y = 3$	$\dfrac{dy}{dt} = -2$

In Exercises 5–8, a point is moving along the graph of the function such that dx/dt is 2 centimeters per second. Find dy/dt for the specified values of x.

Function	Values of x		
5. $y = x^2 + 1$	(a) $x = -1$	(b) $x = 0$	(c) $x = 1$
6. $y = \dfrac{1}{1 + x^2}$	(a) $x = -2$	(b) $x = 0$	(c) $x = 2$
7. $y = \tan x$	(a) $x = -\dfrac{\pi}{3}$	(b) $x = -\dfrac{\pi}{4}$	(c) $x = 0$
8. $y = \sin x$	(a) $x = \dfrac{\pi}{6}$	(b) $x = \dfrac{\pi}{4}$	(c) $x = \dfrac{\pi}{3}$

Getting at the Concept

In Exercises 9 and 10, using the graph of f, (a) determine whether dy/dt is positive or negative given that dx/dt is negative, and (b) determine whether dx/dt is positive or negative given that dy/dt is positive.

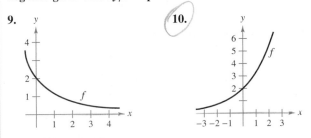

9.   **10.**

11. Consider the linear function $y = ax + b$. If x changes at a constant rate, does y change at a constant rate? If so, does it change at the same rate as x? Explain.

12. In your own words, state the guidelines for solving related rate problems.

13. Find the rate of change of the distance between the origin and a moving point on the graph of $y = x^2 + 1$ if $dx/dt = 2$ centimeters per second.

14. Find the rate of change of the distance between the origin and a moving point on the graph of $y = \sin x$ if $dx/dt = 2$ centimeters per second.

15. *Area* The radius r of a circle is increasing at a rate of 3 centimeters per minute. Find the rate of change of the area when (a) $r = 6$ centimeters and (b) $r = 24$ centimeters.

16. *Area* Let A be the area of a circle of radius r that is changing with respect to time. If dr/dt is constant, is dA/dt constant? Explain.

17. *Area* The included angle of the two sides of constant equal length s of an isosceles triangle is θ.

(a) Show that the area of the triangle is given by $A = \frac{1}{2}s^2 \sin \theta$.

(b) If θ is increasing at the rate of $\frac{1}{2}$ radian per minute, find the rate of change of the area when $\theta = \pi/6$ and $\theta = \pi/3$.

(c) Explain why the rate of change of the area of the triangle is not constant even though $d\theta/dt$ is constant.

18. *Volume* The radius r of a sphere is increasing at a rate of 2 inches per minute.

(a) Find the rate of change of the volume when $r = 6$ inches and $r = 24$ inches.

(b) Explain why the rate of change of the volume of the sphere is not constant even though dr/dt is constant.

19. *Volume* A spherical balloon is inflated with gas at the rate of 800 cubic centimeters per minute. How fast is the radius of the balloon increasing at the instant the radius is (a) 30 centimeters and (b) 60 centimeters?

20. *Volume* All edges of a cube are expanding at a rate of 3 centimeters per second. How fast is the volume changing when each edge is (a) 1 centimeter and (b) 10 centimeters?

21. *Surface Area* The conditions are the same as in Exercise 20. Determine how fast the *surface area* is changing when each edge is (a) 1 centimeter and (b) 10 centimeters.

22. *Volume* The formula for the volume of a cone is $V = \frac{1}{3}\pi r^2 h$. Find the rate of change of the volume if dr/dt is 2 inches per minute and $h = 3r$ when (a) $r = 6$ inches and (b) $r = 24$ inches.

23. *Volume* At a sand and gravel plant, sand is falling off a conveyor and onto a conical pile at a rate of 10 cubic feet per minute. The diameter of the base of the cone is approximately three times the altitude. At what rate is the height of the pile changing when the pile is 15 feet high?

24. *Depth* A conical tank (with vertex down) is 10 feet across the top and 12 feet deep. If water is flowing into the tank at a rate of 10 cubic feet per minute, find the rate of change of the depth of the water when the water is 8 feet deep.

25. Depth   A swimming pool is 12 meters long, 6 meters wide, 1 meter deep at the shallow end, and 3 meters deep at the deep end (see figure). Water is being pumped into the pool at $\frac{1}{4}$ cubic meter per minute, and there is 1 meter of water at the deep end.

(a) What percent of the pool is filled?

(b) At what rate is the water level rising?

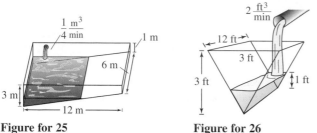

Figure for 25   **Figure for 26**

26. Depth   A trough is 12 feet long and 3 feet across the top (see figure). Its ends are isosceles triangles with altitudes of 3 feet.

(a) If water is being pumped into the trough at 2 cubic feet per minute, how fast is the water level rising when it is 1 foot deep?

(b) If the water is rising at a rate of $\frac{3}{8}$ inch per minute when $h = 2$, determine the rate at which water is being pumped into the trough.

27. Moving Ladder   A ladder 25 feet long is leaning against the wall of a house (see figure). The base of the ladder is pulled away from the wall at a rate of 2 feet per second.

(a) How fast is the top moving down the wall when the base of the ladder is 7 feet, 15 feet, and 24 feet from the wall?

(b) Consider the triangle formed by the side of the house, the ladder, and the ground. Find the rate at which the area of the triangle is changing when the base of the ladder is 7 feet from the wall.

(c) Find the rate at which the angle between the ladder and the wall of the house is changing when the base of the ladder is 7 feet from the wall.

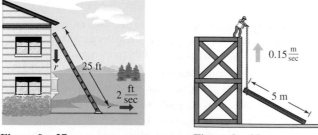

Figure for 27   **Figure for 28**

FOR FURTHER INFORMATION   For more information on the mathematics of moving ladders, see the article "The Falling Ladder Paradox" by Paul Scholten and Andrew Simoson in *The College Mathematics Journal*. To view this article, go to the website *www.matharticles.com*.

28. Construction   A construction worker pulls a 5-meter plank up the side of a building under construction by means of a rope tied to one end of the plank (see figure). Assume the opposite end of the plank follows a path perpendicular to the wall of the

building and the worker pulls the rope at a rate of 0.15 meter per second. How fast is the end of the plank sliding along the ground when it is 2.5 meters from the wall of the building?

29. Construction   A winch at the top of a 12-meter building pulls a pipe of the same length to a vertical position, as shown in the figure. The winch pulls in rope at a rate of -0.2 meter per second. Find the rate of vertical change and the rate of horizontal change at the end of the pipe when $y = 6$.

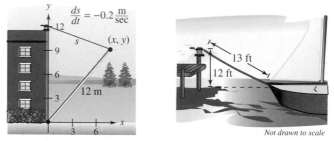

Figure for 29   **Figure for 30**

30. Boating   A boat is pulled into a dock by means of a winch 12 feet above the deck of the boat (see figure).

(a) The winch pulls in rope at a rate of 4 feet per second. Determine the speed of the boat when there is 13 feet of rope out. What happens to the speed of the boat as it gets closer to the dock?

(b) Suppose the boat is moving at a constant rate of 4 feet per second. Determine the speed at which the winch pulls in rope when there is a total of 13 feet of rope out. What happens to the speed at which the winch pulls in rope as the boat gets closer to the dock?

31. Air Traffic Control   An air traffic controller spots two planes at the same altitude converging on a point as they fly at right angles to each other (see figure). One plane is 150 miles from the point moving at 450 miles per hour. The other plane is 200 miles from the point moving at 600 miles per hour.

(a) At what rate is the distance between the planes decreasing?

(b) How much time does the air traffic controller have to get one of the planes on a different flight path?

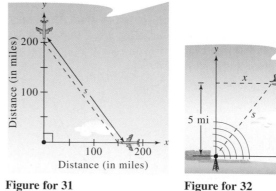

Figure for 31   **Figure for 32**

32. Air Traffic Control   An airplane is flying at an altitude of 5 miles and passes directly over a radar antenna (see figure). When the plane is 10 miles away ($s = 10$), the radar detects that the distance s is changing at a rate of 240 miles per hour. What is the speed of the plane?

33. Baseball   A baseball diamond has the shape of a square with sides 90 feet long (see figure). A player running from second base to third base at a speed of 28 feet per second is 30 feet from third base. At what rate is the player's distance s from home plate changing?

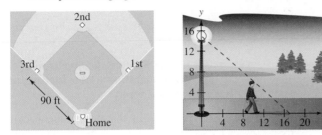

Figure for 33 and 34   **Figure for 35**

34. Baseball   For the baseball diamond in Exercise 33, suppose the player is running from first to second at a speed of 28 feet per second. Find the rate at which the distance from home plate is changing when the player is 30 feet from second base.

35. Shadow Length   A man 6 feet tall walks at a rate of 5 feet per second away from a light that is 15 feet above the ground (see figure). When he is 10 feet from the base of the light,

   (a) at what rate is the tip of his shadow moving?

   (b) at what rate is the length of his shadow changing?

36. Shadow Length   Repeat Exercise 35 for a man 6 feet tall walking at a rate of 5 feet per second *toward* a light that is 20 feet above the ground (see figure).

   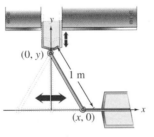

Figure for 36   **Figure for 37**

37. Machine Design   The endpoints of a movable rod of length 1 meter have coordinates $(x, 0)$ and $(0, y)$ (see figure). The position of the end on the x-axis is

$$x(t) = \frac{1}{2} \sin \frac{\pi t}{6}$$

where t is the time in seconds.

   (a) Find the time of one complete cycle of the rod.

   (b) What is the lowest point reached by the end of the rod on the y-axis?

   (c) Find the speed of the y-axis endpoint when the x-axis endpoint is $\left(\frac{1}{4}, 0\right)$.

38. Machine Design   Repeat Exercise 37 for a position function of $x(t) = \frac{3}{5} \sin \pi t$. Use the point $\left(\frac{3}{10}, 0\right)$ for part (c).

39. Evaporation   As a spherical raindrop falls, it reaches a layer of dry air and begins to evaporate at a rate that is proportional to its surface area $(S = 4\pi r^2)$. Show that the radius of the raindrop decreases at a constant rate.

40. Electricity   The combined electrical resistance R of R_1 and R_2, connected in parallel, is given by

$$\frac{1}{R} = \frac{1}{R_1} + \frac{1}{R_2}$$

where R, R_1, and R_2 are measured in ohms. R_1 and R_2 are increasing at rates of 1 and 1.5 ohms per second, respectively. At what rate is R changing when $R_1 = 50$ ohms and $R_2 = 75$ ohms?

41. Adiabatic Expansion   When a certain polyatomic gas undergoes adiabatic expansion, its pressure p and volume V satisfy the equation

$$pV^{1.3} = k$$

where k is a constant. Find the relationship between the related rates dp/dt and dV/dt.

42. Roadway Design   Cars on a certain roadway travel on a circular arc of radius r. In order not to rely on friction alone to overcome the centrifugal force, the road is banked at an angle of magnitude θ from the horizontal (see figure). The banking angle must satisfy the equation

$$rg \tan \theta = v^2$$

where v is the velocity of the cars and $g = 32$ feet per second per second is the acceleration due to gravity. Find the relationship between the related rates dv/dt and $d\theta/dt$.

43. Angle of Elevation   A balloon rises at a rate of 3 meters per second from a point on the ground 30 meters from an observer. Find the rate of change of the angle of elevation of the balloon from the observer when the balloon is 30 meters above the ground.

44. Angle of Elevation   A fish is reeled in at a rate of 1 foot per second from a point 10 feet above the water (see figure). At what rate is the angle between the line and the water changing when there is a total of 25 feet of line out?

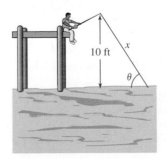

45. *Angle of Elevation*   An airplane flies at an altitude of 5 miles toward a point directly over an observer (see figure). The speed of the plane is 600 miles per hour. Find the rate at which the angle of elevation θ is changing when the angle is (a) $\theta = 30°$, (b) $\theta = 60°$, and (c) $\theta = 75°$.

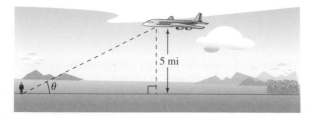

46. *Linear vs. Angular Speed*   A patrol car is parked 50 feet from a long warehouse (see figure). The revolving light on top of the car turns at a rate of 30 revolutions per minute. How fast is the light beam moving along the wall when the beam makes angles of (a) $\theta = 30°$, (b) $\theta = 60°$, and (c) $\theta = 70°$ with the line perpendicular from the light to the wall?

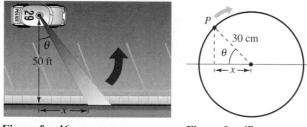

Figure for 46   **Figure for 47**

47. *Linear vs. Angular Speed*   A wheel of radius 30 centimeters revolves at a rate of 10 revolutions per second. A dot is painted at a point P on the rim of the wheel (see figure).

(a) Find dx/dt as a function of θ.

(b) Use a graphing utility to graph the function in part (a).

(c) When is the absolute value of the rate of change of x greatest? When is it least?

(d) Find dx/dt when $\theta = 30°$ and $\theta = 60°$.

48. *Flight Control*   An airplane is flying in still air with an airspeed of 240 miles per hour. If it is climbing at an angle of 22°, find the rate at which it is gaining altitude.

49. *Security Camera*   A security camera is centered 50 feet above a 100-foot hallway (see figure). It is easiest to design the camera with a constant angular rate of rotation, but this results in a variable rate at which the images of the surveillance area are recorded. Therefore, it is desirable to design a system with a variable rate of rotation and a constant rate of movement of the scanning beam along the hallway. Find a model for the variable rate of rotation if $|dx/dt| = 2$ feet per second.

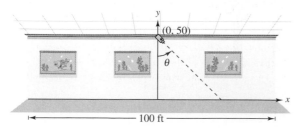

50. *Think About It*   Describe the relationship between the rate of change of y and the rate of change of x in each of the following. Assume all variables and derivatives are positive.

(a) $\dfrac{dy}{dt} = 3\dfrac{dx}{dt}$

(b) $\dfrac{dy}{dt} = x(L - x)\dfrac{dx}{dt}, \quad 0 \le x \le L$

Acceleration   In Exercises 51 and 52, find the acceleration of the specified object. (*Hint:* Recall that if a variable is changing at a constant rate, its acceleration is zero.)

51. Find the acceleration of the top of the ladder described in Exercise 27 when the base of the ladder is 7 feet from the wall.

52. Find the acceleration of the boat in Exercise 30(a) when there is a total of 13 feet of rope out.

53. *Modeling Data*   The table shows the numbers (in millions) of single women s and married women m in the civilian work force in the United States for the years 1990 through 1998. (*Source: U.S. Bureau of Labor Statistics*)

Year	1990	1991	1992	1993	1994	1995	1996	1997	1998
s	14.6	14.7	14.9	15.0	15.3	15.5	15.8	16.5	17.1
m	30.9	31.1	31.7	32.0	32.9	33.4	33.6	33.8	33.9

(a) Use the regression capabilities of a graphing utility to find a model of the form $m(s) = as^2 + bs + c$ for the data, where t is the time in years, with $t = 0$ corresponding to 1990.

(b) Find $\dfrac{dm}{dt}$.

(c) Use the model to estimate dm/dt for $t = 5$ if it is predicted that the number of single women in the work force will increase at the rate of 1.2 million per year.

54. A ball is dropped from a height of 20 meters, 12 meters away from the top of a 20-meter lamppost (see figure). The ball's shadow, caused by the light at the top of the lamppost, is moving along the level ground. How fast is the shadow moving 1 second after the ball is released? (*Submitted by Dennis Gittinger, St. Philips College, San Antonio, TX*)

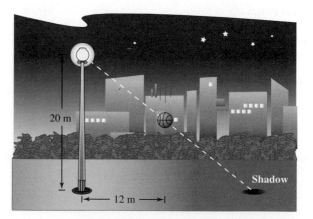

REVIEW EXERCISES FOR CHAPTER 2

2.1 In Exercises 1–4, find the derivative of the function by using the definition of the derivative.

1. $f(x) = x^2 - 2x + 3$

2. $f(x) = \dfrac{x + 1}{x - 1}$

3. $f(x) = \sqrt{x} + 1$

4. $f(x) = \dfrac{2}{x}$

In Exercises 5 and 6, describe the x-values at which f is differentiable.

5. $f(x) = (x + 1)^{2/3}$

6. $f(x) = \dfrac{4x}{x + 3}$

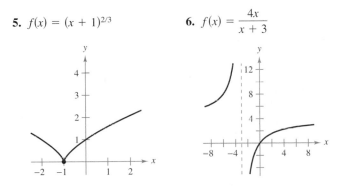

7. Sketch the graph of $f(x) = 4 - |x - 2|$.

    (a) Is f continuous at $x = 2$?

    (b) Is f differentiable at $x = 2$? Explain.

8. Sketch the graph of $f(x) = \begin{cases} x^2 + 4x + 2, & x < -2 \\ 1 - 4x - x^2, & x \geq -2. \end{cases}$

    (a) Is f continuous at $x = -2$?

    (b) Is f differentiable at $x = -2$? Explain.

In Exercises 9 and 10, find the slope of the tangent line to the graph of the function at the specified point.

9. $g(x) = \dfrac{2}{3}x^2 - \dfrac{x}{6}, \quad \left(-1, \dfrac{5}{6}\right)$

10. $h(x) = \dfrac{3x}{8} - 2x^2, \quad \left(-2, -\dfrac{35}{4}\right)$

In Exercises 11 and 12, (a) find an equation of the tangent line to the graph of f at the indicated point, (b) use a graphing utility to graph the function and its tangent line at the point, and (c) use the *derivative* feature of the graphing utility to confirm your results.

11. $f(x) = x^3 - 1, \quad (-1, -2)$

12. $f(x) = \dfrac{2}{x + 1}, \quad (0, 2)$

In Exercises 13 and 14, use the alternative form of the derivative to find the derivative at $x = c$ (if it exists).

13. $g(x) = x^2(x - 1), \quad c = 2$

14. $f(x) = \dfrac{1}{x + 1}, \quad c = 2$

Writing In Exercises 15 and 16, the figure shows the graphs of a function and its derivative. Label the graphs as f or f' and write a short paragraph stating the criteria used in making the selection. To print an enlarged copy of the graph, go to the website *www.mathgraphs.com*.

15.

16.

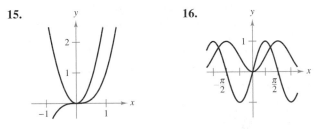

2.2 In Exercises 17–32, find the derivative of the function.

17. $y = 25$

18. $y = -12$

19. $f(x) = x^8$

20. $g(x) = x^{12}$

21. $h(t) = 3t^4$

22. $f(t) = -8t^5$

23. $f(x) = x^3 - 3x^2$

24. $g(s) = 4s^4 - 5s^2$

25. $h(x) = 6\sqrt{x} + 3\sqrt[3]{x}$

26. $f(x) = x^{1/2} - x^{-1/2}$

27. $g(t) = \dfrac{2}{3t^2}$

28. $h(x) = \dfrac{2}{(3x)^2}$

29. $f(\theta) = 2\theta - 3\sin\theta$

30. $g(\alpha) = 4\cos\alpha + 6$

31. $f(\theta) = 3\cos\theta - \dfrac{\sin\theta}{4}$

32. $g(\alpha) = \dfrac{5\sin\alpha}{3} - 2\alpha$

33. *Vibrating String* When a guitar string is plucked, it vibrates with a frequency of $F = 200\sqrt{T}$, where F is measured in vibrations per second and the tension T is measured in pounds. Find the rate of change of F when (a) $T = 4$ and (b) $T = 9$.

34. *Vertical Motion* A ball is dropped from a height of 100 feet. One second later, another ball is dropped from a height of 75 feet. Which ball hits the ground first?

35. *Vertical Motion* To estimate the height of a building, a weight is dropped from the top of the building into a pool at ground level. How high is the building if the splash is seen 9.2 seconds after the weight is dropped?

36. *Vertical Motion* A bomb is dropped from an airplane at an altitude of 14,400 feet. How long will it take for the bomb to reach the ground? (Because of the motion of the plane, the fall will not be vertical, but the time will be the same as that for a vertical fall.) The plane is moving at 600 miles per hour. How far will the bomb move horizontally after it is released from the plane?

37. Projectile Motion A ball thrown follows a path described by $y = x - 0.02x^2$.

(a) Sketch a graph of the path.

(b) Find the total horizontal distance the ball was thrown.

(c) At what x-value does the ball reach its maximum height? (Use the symmetry of the path.)

(d) Find an equation that gives the instantaneous rate of change of the height of the ball with respect to the horizontal change. Evaluate the equation at $x = 0, 10, 25, 30$, and 50.

(e) What is the instantaneous rate of change of the height when the ball reaches its maximum height?

38. Projectile Motion The path of a projectile thrown at an angle of $45°$ with level ground is

$$y = x - \frac{32}{v_0^2}(x^2)$$

where the initial velocity is v_0 feet per second.

(a) Find the x-coordinate of the point where the projectile strikes the ground. Use the symmetry of the path of the projectile to locate the x-coordinate of the point where the projectile reaches its maximum height.

(b) What is the instantaneous rate of change of the height when the projectile is at its maximum height?

(c) Show that doubling the initial velocity of the projectile multiplies both the maximum height and the range by a factor of 4.

(d) Find the maximum height and range of a projectile thrown with an initial velocity of 70 feet per second. Use a graphing utility to sketch the path of the projectile.

39. Horizontal Motion The position function of a particle moving along the x-axis is

$$x(t) = t^2 - 3t + 2$$

for $-\infty < t < \infty$.

(a) Find the velocity of the particle.

(b) Find the open t-interval(s) in which the particle is moving to the left.

(c) Find the position of the particle when the velocity is 0.

(d) Find the speed of the particle when the position is 0.

40. Modeling Data The speed of a car in miles per hour and the stopping distance in feet are recorded in the table.

Speed (x)	20	30	40	50	60
Stopping Distance (y)	25	55	105	188	300

(a) Use the regression capabilities of a graphing utility to find a quadratic model for the data.

(b) Use a graphing utility to plot the data and graph the model.

(c) Use a graphing utility to graph dy/dx.

(d) Use the model to approximate the stopping distance at a speed of 65 miles per hour.

(e) Use the graphs in parts (b) and (c) to explain the change in stopping distance as the speed increases.

2.3 **In Exercises 41–57, find the derivative of the function.**

41. $f(x) = (3x^2 + 7)(x^2 - 2x + 3)$

42. $g(x) = (x^3 - 3x)(x + 2)$

43. $h(x) = \sqrt{x} \sin x$

44. $f(t) = t^3 \cos t$

45. $f(x) = \dfrac{2x^3 - 1}{x^2}$

46. $f(x) = \dfrac{x + 1}{x - 1}$

47. $f(x) = \dfrac{x^2 + x - 1}{x^2 - 1}$

48. $f(x) = \dfrac{6x - 5}{x^2 + 1}$

49. $f(x) = \dfrac{1}{4 - 3x^2}$

50. $f(x) = \dfrac{9}{3x^2 - 2x}$

51. $y = \dfrac{x^2}{\cos x}$

52. $y = \dfrac{\sin x}{x^2}$

53. $y = 3x^2 \sec x$

54. $y = 2x - x^2 \tan x$

55. $y = -x \tan x$

56. $y = \dfrac{1 + \sin x}{1 - \sin x}$

57. $y = x \cos x - \sin x$

58. Acceleration The velocity of an object in meters per second is $v(t) = 36 - t^2$, $0 \le t \le 6$. Find the velocity and acceleration of the object when $t = 4$.

In Exercises 59–62, find the second derivative of the function.

59. $g(t) = t^3 - 3t + 2$

60. $f(x) = 12\sqrt[4]{x}$

61. $f(\theta) = 3 \tan \theta$

62. $h(t) = 4 \sin t - 5 \cos t$

In Exercises 63 and 64, show that the function satisfies the equation.

Function	Equation
63. $y = 2 \sin x + 3 \cos x$	$y'' + y = 0$
64. $y = \dfrac{10 - \cos x}{x}$	$xy' + y = \sin x$

2.4 **In Exercises 65–80, find the derivative of the function.**

65. $f(x) = \sqrt{1 - x^3}$

66. $f(x) = \sqrt[3]{x^2 - 1}$

67. $h(x) = \left(\dfrac{x - 3}{x^2 + 1}\right)^2$

68. $f(x) = \left(x^2 + \dfrac{1}{x}\right)^5$

69. $f(s) = (s^2 - 1)^{5/2}(s^3 + 5)$

70. $h(\theta) = \dfrac{\theta}{(1 - \theta)^3}$

71. $y = 3 \cos(3x + 1)$

72. $y = 1 - \cos 2x + 2 \cos^2 x$

73. $y = \frac{1}{2} \csc 2x$

74. $y = \csc 3x + \cot 3x$

75. $y = \dfrac{x}{2} - \dfrac{\sin 2x}{4}$

76. $y = \dfrac{\sec^7 x}{7} - \dfrac{\sec^5 x}{5}$

77. $y = \frac{2}{3} \sin^{3/2} x - \frac{2}{7} \sin^{7/2} x$

78. $f(x) = \dfrac{3x}{\sqrt{x^2 + 1}}$

79. $y = \dfrac{\sin \pi x}{x + 2}$

80. $y = \dfrac{\cos(x - 1)}{x - 1}$

In Exercises 81–88, use a computer algebra system to find the derivative of the function. Use the utility to graph the function and its derivative on the same set of coordinate axes. Describe the behavior of the function that corresponds to any zeros of the graph of the derivative.

81. $f(t) = t^2(t - 1)^5$

82. $f(x) = [(x - 2)(x + 4)]^2$

83. $g(x) = \dfrac{2x}{\sqrt{x + 1}}$

84. $g(x) = x\sqrt{x^2 + 1}$

85. $f(t) = \sqrt{t + 1} \sqrt[3]{t + 1}$

86. $y = \sqrt{3x}(x + 2)^3$

87. $y = \tan\sqrt{1 - x}$

88. $y = 2\csc^3(\sqrt{x})$

In Exercises 89–92, find the second derivative of the function.

89. $y = 2x^2 + \sin 2x$

90. $y = \dfrac{1}{x} + \tan x$

91. $f(x) = \cot x$

92. $y = \sin^2 x$

In Exercises 93–96, use a computer algebra system to find the second derivative of the function.

93. $f(t) = \dfrac{t}{(1 - t)^2}$

94. $g(x) = \dfrac{6x - 5}{x^2 + 1}$

95. $g(\theta) = \tan 3\theta - \sin(\theta - 1)$

96. $h(x) = x\sqrt{x^2 - 1}$

97. *Refrigeration* The temperature T of food put in a freezer is

$$T = \frac{700}{t^2 + 4t + 10}$$

where t is the time in hours. Find the rate of change of T with respect to t at each of the following times.

(a) $t = 1$ (b) $t = 3$ (c) $t = 5$ (d) $t = 10$

98. *Fluid Flow* The emergent velocity v of a liquid flowing from a hole in the bottom of a tank is given by $v = \sqrt{2gh}$, where g is the acceleration due to gravity (32 feet per second per second) and h is the depth of the liquid in the tank. Find the rate of change of v with respect to h when (a) $h = 9$ and (b) $h = 4$. (Note that $g = +32$ feet per second per second. The sign of g depends on how a problem is modeled. In this case, letting g be negative would produce an imaginary value for v.)

2.5 In Exercises 99–104, use implicit differentiation to find dy/dx.

99. $x^2 + 3xy + y^3 = 10$

100. $x^2 + 9y^2 - 4x + 3y = 0$

101. $y\sqrt{x} - x\sqrt{y} = 16$

102. $y^2 = (x - y)(x^2 + y)$

103. $x \sin y = y \cos x$

104. $\cos(x + y) = x$

In Exercises 105 and 106, find the equations of the tangent line and the normal line to the graph of the equation at the indicated point. Use a graphing utility to graph the equation, the tangent line, and the normal line.

105. $x^2 + y^2 = 20$, $(2, 4)$

106. $x^2 - y^2 = 16$, $(5, 3)$

2.6

107. A point moves along the curve $y = \sqrt{x}$ in such a way that the y-value is increasing at a rate of 2 units per second. At what rate is x changing for each of the following values?

(a) $x = \frac{1}{2}$ (b) $x = 1$ (c) $x = 4$

108. *Surface Area* The edges of a cube are expanding at a rate of 5 centimeters per second. How fast is the surface area changing when each edge is 4.5 centimeters?

109. *Changing Depth* The cross section of a 5-meter trough is an isosceles trapezoid with a 2-meter lower base, a 3-meter upper base, and an altitude of 2 meters. Water is running into the trough at a rate of 1 cubic meter per minute. How fast is the water level rising when the water is 1 meter deep?

110. *Linear and Angular Velocity* A rotating beacon is located 1 kilometer off a straight shoreline (see figure). If the beacon rotates at a rate of 3 revolutions per minute, how fast (in kilometers per hour) does the beam of light appear to be moving to a viewer who is $\frac{1}{2}$ kilometer down the shoreline?

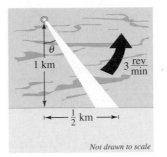

Not drawn to scale

111. *Moving Shadow* A sandbag is dropped from a balloon at a height of 60 meters when the angle of elevation to the sun is 30° (see figure). Find the rate at which the shadow of the sandbag is traveling along the ground when the sandbag is at a height of 35 meters. [*Hint:* The position of the sandbag is given by $s(t) = 60 - 4.9t^2$.]

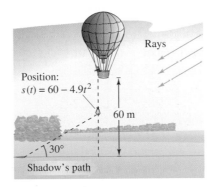

P.S. Problem Solving

1. Consider the graph of the parabola $y = x^2$.

   (a) Find the radius r of the largest possible circle centered on the y-axis that is tangent to the parabola at the origin, as indicated in the figure. This circle is called the **circle of curvature** (see Section 11.5). Use a graphing utility to graph the circle and parabola in the same viewing window.

   (b) Find the center $(0, b)$ of the circle of radius 1 centered on the y-axis that is tangent to the parabola at two points, as indicated in the figure. Use a graphing utility to graph the circle and parabola in the same viewing window.

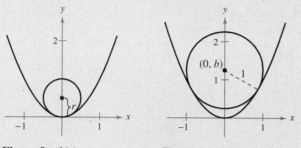

   Figure for 1(a)   **Figure for 1(b)**

2. Graph the two parabolas $y = x^2$ and $y = -x^2 + 2x - 5$ in the same coordinate plane. Find equations of the two lines simultaneously tangent to both parabolas.

3. (a) Find the polynomial $P_1(x) = a_0 + a_1 x$ whose value and slope agree with the value and slope of $f(x) = \cos x$ at the point $x = 0$.

   (b) Find the polynomial $P_2(x) = a_0 + a_1 x + a_2 x^2$ whose value and first two derivatives agree with the value and first two derivatives of $f(x) = \cos x$ at the point $x = 0$. This polynomial is called the second-degree **Taylor polynomial** of $f(x) = \cos x$ at $x = 0$.

   (c) Complete the table comparing the values of f and P_2. What do you observe?

x	-1.0	-0.1	-0.001	0	0.001	0.1	1.0
$\cos x$							
$P_2(x)$							

   (d) Find the third-degree Taylor polynomial of $f(x) = \sin x$ at $x = 0$.

4. (a) Find an equation of the tangent line to the parabola $y = x^2$ at the point $(2, 4)$.

   (b) Find an equation of the normal line to $y = x^2$ at the point $(2, 4)$. (The normal line is perpendicular to the tangent line.) Where does this line intersect the parabola a second time?

   (c) Find equations of the tangent line and normal line to $y = x^2$ at the point $(0, 0)$.

   (d) Prove that for any point $(a, b) \neq (0, 0)$ on the parabola $y = x^2$, the normal line intersects the graph a second time.

5. Find a third-degree polynomial $p(x)$ that is tangent to the line $y = 14x - 13$ at the point $(1, 1)$, and tangent to the line $y = -2x - 5$ at the point $(-1, -3)$.

6. Find a function of the form $f(x) = a + b \cos cx$ that is tangent to the line $y = 1$ at the point $(0, 1)$, and tangent to the line

$$y = x + \frac{3}{2} - \frac{\pi}{4}$$

   at the point $\left(\frac{\pi}{4}, \frac{3}{2}\right)$.

7. The graph of the **eight curve,**

$$x^4 = a^2(x^2 - y^2), \quad a \neq 0,$$

   is shown below.

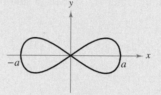

   (a) Explain how you could use a graphing utility to obtain the graph of this curve.

   (b) Use a graphing utility to graph the curve for various values of the constant a. Describe how a affects the shape of the curve.

   (c) Determine the points on the curve where the tangent line is horizontal.

8. The graph of the **pear-shaped quartic,**

$$b^2 y^2 = x^3(a - x), \quad a, b > 0,$$

   is shown below.

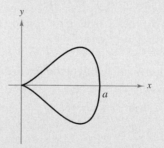

   (a) Explain how you could use a graphing utility to obtain the graph of this curve.

   (b) Use a graphing utility to graph the curve for various values of the constants a and b. Describe how a and b affect the shape of the curve.

   (c) Determine the points on the curve where the tangent line is horizontal.

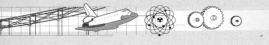

9. A man 6 feet tall walks at a rate of 5 feet per second toward a street light that is 30 feet high. The man's 3-foot-tall child follows at the same speed, but 10 feet behind the man. At times, the shadow behind the child is caused by the man, and at other times, by the child.

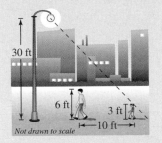

30 ft

6 ft   3 ft

←—10 ft—→

Not drawn to scale

(a) Suppose the man is 90 feet from the street light. Show that the man's shadow extends beyond the child's shadow.

(b) Suppose the man is 60 feet from the street light. Show that the child's shadow extends beyond the man's shadow.

(c) Determine the distance d from the man to the street light at which the tips of the two shadows are exactly the same distance from the street light.

(d) Determine how fast the tip of the shadow is moving as a function of x, the distance between the man and the street light. Discuss the continuity of this shadow speed function.

10. A particle is moving along the graph of $y = \sqrt[3]{x}$. When $x = 8$, the y-component of its position is increasing at the rate of 1 centimeter per second.

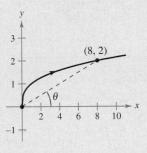

(8, 2)

θ

(a) How fast is the x-component changing at this moment?

(b) How fast is the distance from the origin changing at this moment?

(c) How fast is the angle of inclination θ changing at this moment?

11. Let L be a differentiable function for all x. Prove that if $L(a + b) = L(a) + L(b)$ for all a and b, then $L'(x) = L'(0)$ for all x. What does the graph of L look like?

12. Let E be a function satisfying $E(0) = E'(0) = 1$. Prove that if $E(a + b) = E(a)E(b)$ for all a and b, then E is differentiable and $E'(x) = E(x)$ for all x. Find an example of a function satisfying $E(a + b) = E(a)E(b)$.

13. The fundamental limit

$$\lim_{x \to 0} \frac{\sin x}{x} = 1$$

assumes that x is measured in radians. What happens if we assume that x is measured in degrees instead of radians?

(a) Set your calculator to degree mode and complete the table.

z (in degrees)	0.1	0.01	0.0001
$\dfrac{\sin z}{z}$			

(b) Use the table to estimate

$$\lim_{z \to 0} \frac{\sin z}{z}$$

for z in degrees. What is the exact value of this limit? (*Hint:* $180° = \pi$ radians)

(c) Use the limit definition of the derivative to find

$$\frac{d}{dz} \sin z$$

for z in degrees.

(d) Define the new functions $S(z) = \sin(cz)$ and $C(z) = \cos(cz)$, where $c = \pi/180$. Find $S(90)$ and $C(180)$. Use the Chain Rule to calculate

$$\frac{d}{dz} S(z).$$

(e) Explain why calculus is made easier by using radians instead of degrees.

14. An astronaut standing on the moon throws a rock into the air. The height of the rock is

$$s = -\frac{27}{10}t^2 + 27t + 6$$

where s is measured in feet and t is measured in seconds.

(a) Find expressions for the velocity and acceleration of the rock.

(b) Find the time when the rock is at its highest point by finding the time when the velocity is zero. What is its height at this time?

(c) How does the acceleration of the rock compare with the acceleration due to gravity on earth?

15. If a is the acceleration of an object, the *jerk* j is defined by $j = a'(t)$.

(a) Use this definition to give a physical interpretation of j.

(b) Find j for the slowing vehicle in Exercise 102 in Section 2.3 and interpret the result.

Packaging: The Optimal Form

Many people are involved in deciding how to package the products you see in grocery stores. Packaging engineers select materials and package shapes to adequately protect the product through shipping at a reasonable cost.

A container's shape, as well as its material, is important in determining its strength. From an engineering perspective, the sphere is the strongest form, followed by the circular cylinder. The rectangular box comes in a poor third. From a cost perspective, it is preferable to use the smallest amount of material possible.

The table gives the approximate measurements in inches of several common items packed in cylindrical containers.

Product	Radius (in.)	Height (in.)	Volume (in.³)
Coffee creamer	1.50	6.85	48.42
Cleanser	1.45	7.50	49.54
Coffee	1.95	5.20	62.12
Pineapple juice	2.10	6.70	92.82
Frosting	1.63	3.60	30.05
Soup	1.30	3.80	20.18
Tomato puree	1.95	4.40	52.56
Baking powder	1.25	3.65	17.92

An infinite number of dimensions can be used to construct a right circular container of a given volume. The graph at the right shows the relationship between the radius and surface area for containers that have a volume of 48.4 cubic inches.

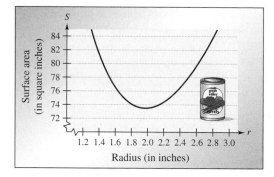

QUESTIONS

1. Create a table of values for the dimensions of a cylinder with a volume of 49.54 cubic inches. Does it appear that the cleanser container minimizes surface area?

2. Suppose you are designing a coffee creamer container that has a volume of 48.42 cubic inches. Use the equations for the surface area of a cylinder and the volume of a cylinder to develop an equation relating the radius r and surface area S.

   $S = 2\pi r^2 + 2\pi rh$ Surface area of a right circular cylinder

   $V = \pi r^2 h$ Volume of a right circular cylinder

3. Repeat Question 2 for each of the other containers in the table. Use a graphing utility to plot each equation. Determine whether the radius of each container is larger than, smaller than, or equal to the "optimal" radius.

4. Suppose, in order to fit more writing on the cylinder, you want to maximize the surface area of a cylinder that holds 49.5 cubic inches. Can you do this? Explain.

The concepts presented here will be explored further in this chapter. For an extension of this application, see Lab 5 in the lab series that accompanies this text at college.hmco.com.

Applications of Differentiation 3

Sally and Derk Kuyper

In addition to strength, engineers must consider not only how a package will fit into a shipping container but also how it will be displayed on a store shelf. The sphere may be the strongest form, but it would surely be impractical to use for product packaging

By the time packaging engineers begin work on a container, design specialists have already done their work. Designers use color, shape, and words to create an image that they think will appeal to their targeted market. Many designers believe that the package is at least as important as the product inside.

Successful designer Primo Angeli feels so strongly about the importance of packaging that he has designed entire lines of packaged product ideas in realistic packages so that consumer response to these ideas can be measured before massive investments are made in product development.

Sally anc Derk Kuyper

Primc Angeli

159

Section 3.1   Extrema on an Interval

- Understand the definition of extrema of a function on an interval.
- Understand the definition of relative extrema of a function on an open interval.
- Find extrema on a closed interval.

Extrema of a Function

In calculus, much effort is devoted to determining the behavior of a function f on an interval I. Does f have a maximum value on I? Does it have a minimum value? Where is the function increasing? Where is it decreasing? In this chapter you will learn how derivatives can be used to answer these questions. You will also see why these questions are important in real-life applications.

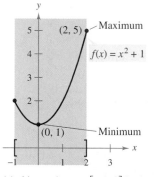

(a) f is continuous, $[-1, 2]$ is closed.

> ### Definition of Extrema
>
> Let f be defined on an interval I containing c.
>
> 1. $f(c)$ is the **minimum of f on I** if $f(c) \le f(x)$ for all x in I.
> 2. $f(c)$ is the **maximum of f on I** if $f(c) \ge f(x)$ for all x in I.
>
> The minimum and maximum of a function on an interval are the **extreme values,** or **extrema,** of the function on the interval. The minimum and maximum of a function on an interval are also called the **absolute minimum** and **absolute maximum** on the interval.

A function need not have a minimum or a maximum on an interval. For instance, in Figure 3.1(a) and (b), you can see that the function $f(x) = x^2 + 1$ has both a minimum and a maximum on the closed interval $[-1, 2]$, but does not have a maximum on the open interval $(-1, 2)$. Moreover, in Figure 3.1(c), you can see that continuity (or the lack of it) can affect the existence of an extremum on the interval. This suggests the following theorem. (Although the Extreme Value Theorem is intuitively plausible, a proof of this theorem is not within the scope of this text.)

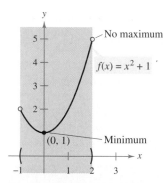

(b) f is continuous, $(-1, 2)$ is open.

> ### THEOREM 3.1   The Extreme Value Theorem
>
> If f is continuous on a closed interval $[a, b]$, then f has both a minimum and a maximum on the interval.

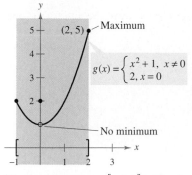

(c) g is not continuous, $[-1, 2]$ is closed.

Extrema can occur at interior points or endpoints of an interval. Extrema that occur at the endpoints are called **endpoint extrema.**
Figure 3.1

EXPLORATION

Finding Minimum and Maximum Values   The Extreme Value Theorem (like the Intermediate Value Theorem) is an *existence theorem* because it tells of the existence of minimum and maximum values but does not show how to find these values. Use the extreme-value capability of a graphing utility to find the minimum and maximum values of each of the following. In each case, do you think the x-values are exact or approximate? Explain your reasoning.

a. $f(x) = x^2 - 4x + 5$ on the closed interval $[-1, 3]$
b. $f(x) = x^3 - 2x^2 - 3x - 2$ on the closed interval $[-1, 3]$

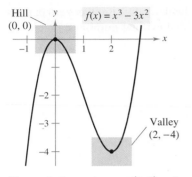

f has a relative maximum at $(0, 0)$ and a relative minimum at $(2, -4)$.
Figure 3.2

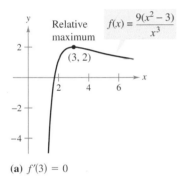

(a) $f'(3) = 0$

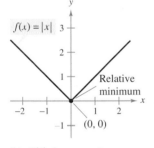

(b) $f'(0)$ does not exist.

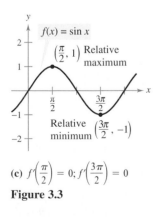

(c) $f'\left(\dfrac{\pi}{2}\right) = 0;\ f'\left(\dfrac{3\pi}{2}\right) = 0$

Figure 3.3

Relative Extrema and Critical Numbers

In Figure 3.2, the graph of $f(x) = x^3 - 3x^2$ has a **relative maximum** at the point $(0, 0)$ and a **relative minimum** at the point $(2, -4)$. Informally, you can think of a relative maximum as occurring on a "hill" on the graph, and a relative minimum as occurring in a "valley" on the graph. Such a hill and valley can occur in two ways. If the hill (or valley) is smooth and rounded, the graph has a horizontal tangent line at the high point (or low point). If the hill (or valley) is sharp and peaked, the graph represents a function that is not differentiable at the high point (or low point).

Definition of Relative Extrema

1. If there is an open interval containing c on which $f(c)$ is a maximum, then $f(c)$ is called a **relative maximum** of f.

2. If there is an open interval containing c on which $f(c)$ is a minimum, then $f(c)$ is called a **relative minimum** of f.

The plural of relative maximum is relative maxima, and the plural of relative minimum is relative minima.

Example 1 examines the derivatives of functions at *given* relative extrema. (Much more is said about *finding* the relative extrema of a function in Section 3.3.)

Example 1   **The Value of the Derivative at Relative Extrema**

Find the value of the derivative at each of the relative extrema shown in Figure 3.3.

Solution

a. The derivative of $f(x) = \dfrac{9(x^2 - 3)}{x^3}$ is

$$f'(x) = \frac{x^3(18x) - (9)(x^2 - 3)(3x^2)}{(x^3)^2} \qquad \text{Differentiate using Quotient Rule.}$$

$$= \frac{9(9 - x^2)}{x^4}. \qquad \text{Simplify.}$$

At the point $(3, 2)$, the value of the derivative is $f'(3) = 0$ (see Figure 3.3a).

b. At $x = 0$, the derivative of $f(x) = |x|$ *does not exist* because the following one-sided limits differ (see Figure 3.3b).

$$\lim_{x \to 0^-} \frac{f(x) - f(0)}{x - 0} = \lim_{x \to 0^-} \frac{|x|}{x} = -1 \qquad \text{Limit from the left}$$

$$\lim_{x \to 0^+} \frac{f(x) - f(0)}{x - 0} = \lim_{x \to 0^+} \frac{|x|}{x} = 1 \qquad \text{Limit from the right}$$

c. The derivative of $f(x) = \sin x$ is

$$f'(x) = \cos x.$$

At the point $(\pi/2, 1)$, the value of the derivative is $f'(\pi/2) = \cos(\pi/2) = 0$. At the point $(3\pi/2, -1)$, the value of the derivative is $f'(3\pi/2) = \cos(3\pi/2) = 0$ (see Figure 3.3c).

Note in Example 1 that at the relative extrema, the derivative is either zero or does not exist. The x-values at these special points are called **critical numbers.** Figure 3.4 illustrates the two types of critical numbers.

Definition of a Critical Number

Let f be defined at c. If $f'(c) = 0$ or if f is not differentiable at c, then c is a **critical number** of f.

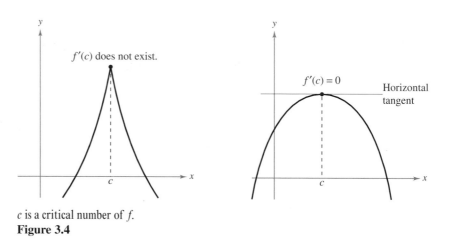

$f'(c)$ does not exist.

$f'(c) = 0$

Horizontal tangent

c is a critical number of f.
Figure 3.4

THEOREM 3.2   Relative Extrema Occur Only at Critical Numbers

If f has a relative minimum or relative maximum at $x = c$, then c is a critical number of f.

PIERRE DE FERMAT (1601–1665)

For Fermat, who was trained as a lawyer, mathematics was more of a hobby than a profession. Nevertheless, Fermat made many contributions to analytic geometry, number theory, calculus, and probability. In letters to friends, he wrote of many of the fundamental ideas of calculus, long before Newton or Leibniz. For instance, the theorem at the right is sometimes attributed to Fermat.

Proof

Case 1: If f is *not* differentiable at $x = c$, then, by definition, c is a critical number of f and the theorem is valid.

Case 2: If f is differentiable at $x = c$, then $f'(c)$ must be positive, negative, or 0. Suppose $f'(c)$ is positive. Then

$$f'(c) = \lim_{x \to c} \frac{f(x) - f(c)}{x - c} > 0$$

which implies that there exists an interval (a, b) containing c such that

$$\frac{f(x) - f(c)}{x - c} > 0, \text{ for all } x \neq c \text{ in } (a, b). \qquad \text{(See Exercise 58, Section 1.2)}$$

Because this quotient is positive, the signs of the denominator and numerator must agree. This produces the following inequalities for x-values in the interval (a, b).

Left of c: $x < c$ and $f(x) < f(c)$ $\implies$ $f(c)$ is not a relative minimum

Right of c: $x > c$ and $f(x) > f(c)$ $\implies$ $f(c)$ is not a relative maximum

So, the assumption that $f'(c) > 0$ contradicts the hypothesis that $f(c)$ is a relative extremum. Assuming that $f'(c) < 0$ produces a similar contradiction, you are left with only one possibility—namely, $f'(c) = 0$. So, by definition, c is a critical number of f and the theorem is valid.

Finding Extrema on a Closed Interval

Theorem 3.2 states that the relative extrema of a function can occur *only* at the critical numbers of the function. Knowing this, you can use the following guidelines to find extrema on a closed interval.

Guidelines for Finding Extrema on a Closed Interval

To find the extrema of a continuous function f on a closed interval $[a, b]$, use the following steps.

1. Find the critical numbers of f in (a, b).
2. Evaluate f at each critical number in (a, b).
3. Evaluate f at each endpoint of $[a, b]$.
4. The least of these values is the minimum. The greatest is the maximum.

The next three examples show how to apply these guidelines. Be sure you see that finding the critical numbers of the function is only part of the procedure. Evaluating the function at the critical numbers *and* the endpoints is the other part.

Example 2   **Finding Extrema on a Closed Interval**

Find the extrema of $f(x) = 3x^4 - 4x^3$ on the interval $[-1, 2]$.

Solution   Begin by differentiating the function.

$$f(x) = 3x^4 - 4x^3 \qquad \text{Write original function.}$$
$$f'(x) = 12x^3 - 12x^2 \qquad \text{Differentiate.}$$

To find the critical numbers of f, you must find all x-values for which $f'(x) = 0$ and all x-values for which $f'(x)$ does not exist.

$$f'(x) = 12x^3 - 12x^2 = 0 \qquad \text{Set } f'(x) \text{ equal to 0.}$$
$$12x^2(x - 1) = 0 \qquad \text{Factor.}$$
$$x = 0, 1 \qquad \text{Critical numbers}$$

Because f' is defined for all x, you can conclude that these are the only critical numbers of f. By evaluating f at these two critical numbers and at the endpoints of $[-1, 2]$, you can determine that the maximum is $f(2) = 16$ and the minimum is $f(1) = -1$, as indicated in the table. The graph of f is shown in Figure 3.5.

Left Endpoint	Critical Number	Critical Number	Right Endpoint
$f(-1) = 7$	$f(0) = 0$	$f(1) = -1$	$f(2) = 16$
		Minimum	Maximum

In Figure 3.5, note that the critical number $x = 0$ does not yield a relative minimum or a relative maximum. This tells you that the converse of Theorem 3.2 is not true. In other words, *the critical numbers of a function need not produce relative extrema.*

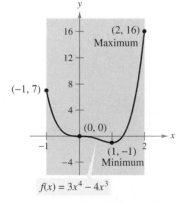

$f(x) = 3x^4 - 4x^3$

On the closed interval $[-1, 2]$, f has a minimum at $(1, -1)$ and a maximum at $(2, 16)$.

Figure 3.5

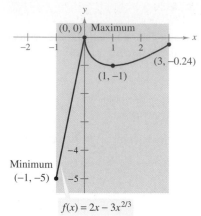

On the closed interval $[-1, 3]$, f has a minimum at $(-1, -5)$ and a maximum at $(0, 0)$.

Figure 3.6

Example 3   **Finding Extrema on a Closed Interval**

Find the extrema of $f(x) = 2x - 3x^{2/3}$ on the interval $[-1, 3]$.

Solution   Differentiating produces the following.

$$f(x) = 2x - 3x^{2/3}$$ Write original function.

$$f'(x) = 2 - \frac{2}{x^{1/3}} = 2\left(\frac{x^{1/3} - 1}{x^{1/3}}\right)$$ Differentiate.

From this derivative, you can see that the function has two critical numbers in the interval $[-1, 3]$. The number 1 is a critical number because $f'(1) = 0$, and the number 0 is a critical number because $f'(0)$ does not exist. By evaluating f at these two numbers and at the endpoints of the interval, you can conclude that the minimum is $f(-1) = -5$ and the maximum is $f(0) = 0$, as indicated in the table. The graph of f is shown in Figure 3.6.

Left Endpoint	Critical Number	Critical Number	Right Endpoint
$f(-1) = -5$	$f(0) = 0$	$f(1) = -1$	$f(3) = 6 - 3\sqrt[3]{9} \approx -0.24$
Minimum	Maximum		

Example 4   Finding Extrema on a Closed Interval

Find the extrema of $f(x) = 2 \sin x - \cos 2x$ on the interval $[0, 2\pi]$.

Solution   This function is differentiable for all real x, so you can find all critical numbers by differentiating the function and setting $f'(x)$ equal to zero, as follows.

$$f(x) = 2 \sin x - \cos 2x$$ Write original function.

$$f'(x) = 2 \cos x + 2 \sin 2x = 0$$ Set $f'(x)$ equal to 0.

$$2 \cos x + 4 \cos x \sin x = 0$$ $\sin 2x = 2 \cos x \sin x$

$$2(\cos x)(1 + 2 \sin x) = 0$$ Factor.

In the interval $[0, 2\pi]$, the factor $\cos x$ is zero when $x = \pi/2$ and when $x = 3\pi/2$. The factor $(1 + 2 \sin x)$ is zero when $x = 7\pi/6$ and when $x = 11\pi/6$. By evaluating f at these four critical numbers and at the endpoints of the interval, you can conclude that the maximum is $f(\pi/2) = 3$ and the minimum occurs at *two* points, $f(7\pi/6) = -3/2$ and $f(11\pi/6) = -3/2$, as indicated in the table. The graph is shown in Figure 3.7.

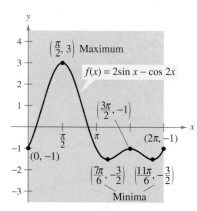

On the closed interval $[0, 2\pi]$, f has two minima at $(7\pi/6, -3/2)$ and $(11\pi/6, -3/2)$ and a maximum at $(\pi/2, 3)$.

Figure 3.7

Left Endpoint	Critical Number	Critical Number	Critical Number	Critical Number	Right Endpoint
$f(0) = -1$	$f\left(\dfrac{\pi}{2}\right) = 3$	$f\left(\dfrac{7\pi}{6}\right) = -\dfrac{3}{2}$	$f\left(\dfrac{3\pi}{2}\right) = -1$	$f\left(\dfrac{11\pi}{6}\right) = -\dfrac{3}{2}$	$f(2\pi) = -1$
	Maximum	Minimum		Minimum	

indicates that in the Interactive 3.0 *CD-ROM and* Internet 3.0 *versions of this text (available at* college.hmco.com*) you will find an Open Exploration, which further explores this example using the computer algebra systems* Maple, Mathcad, Mathematica, *and* Derive.

EXERCISES FOR SECTION 3.1

In Exercises 1–6, find the value of the derivative (if it exists) at each indicated extremum.

1. $f(x) = \dfrac{x^2}{x^2 + 4}$

2. $f(x) = \cos \dfrac{\pi x}{2}$

3. $f(x) = x + \dfrac{27}{2x^2}$

4. $f(x) = -3x\sqrt{x + 1}$

5. $f(x) = (x + 2)^{2/3}$

6. $f(x) = 4 - |x|$

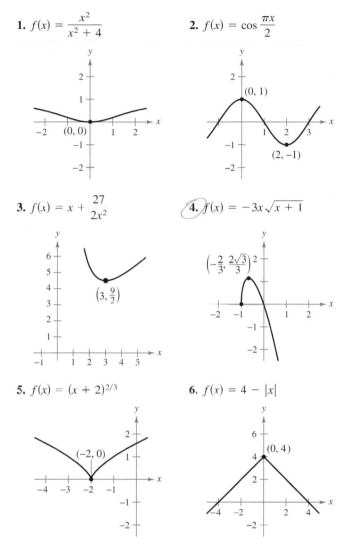

In Exercises 7–10, approximate the critical numbers of the function shown in the graph. Determine whether the function has a relative maximum, relative minimum, absolute maximum, absolute minimum, or none of these at each critical number on the interval shown.

7.

8.

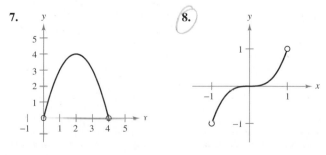

9.

10.

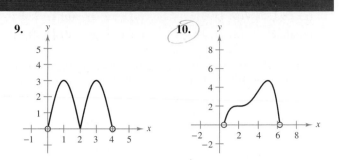

In Exercises 11–16, find any critical numbers of the function.

11. $f(x) = x^2(x - 3)$

12. $g(x) = x^2(x^2 - 4)$

13. $g(t) = t\sqrt{4 - t},\ t < 3$

14. $f(x) = \dfrac{4x}{x^2 + 1}$

15. $h(x) = \sin^2 x + \cos x$
$0 < x < 2\pi$

16. $f(\theta) = 2 \sec \theta + \tan \theta$
$0 < \theta < 2\pi$

In Exercises 17–32, locate the absolute extrema of the function on the closed interval.

17. $f(x) = 2(3 - x),\ [-1, 2]$

18. $f(x) = \dfrac{2x + 5}{3},\ [0, 5]$

19. $f(x) = -x^2 + 3x,\ [0, 3]$

20. $f(x) = x^2 + 2x - 4,\ [-1, 1]$

21. $f(x) = x^3 - \dfrac{3}{2}x^2,\ [-1, 2]$

22. $f(x) = x^3 - 12x,\ [0, 4]$

23. $y = 3x^{2/3} - 2x,\ [-1, 1]$

24. $g(x) = \sqrt[3]{x},\ [-1, 1]$

25. $g(t) = \dfrac{t^2}{t^2 + 3},\ [-1, 1]$

26. $y = 3 - |t - 3|,\ [-1, 5]$

27. $h(s) = \dfrac{1}{s - 2},\ [0, 1]$

28. $h(t) = \dfrac{t}{t - 2},\ [3, 5]$

29. $f(x) = \cos \pi x,\ \left[0, \dfrac{1}{6}\right]$

30. $g(x) = \sec x,\ \left[-\dfrac{\pi}{6}, \dfrac{\pi}{3}\right]$

31. $y = \dfrac{4}{x} + \tan\left(\dfrac{\pi x}{8}\right),\ [1, 2]$

32. $y = x^2 - 2 - \cos x,\ [-1, 3]$

In Exercises 33–36, locate the absolute extrema of the function (if any exist) over the indicated intervals.

33. $f(x) = 2x - 3$
(a) $[0, 2]$ (b) $[0, 2)$
(c) $(0, 2]$ (d) $(0, 2)$

34. $f(x) = 5 - x$
(a) $[1, 4]$ (b) $[1, 4)$
(c) $(1, 4]$ (d) $(1, 4)$

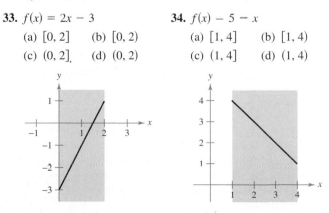

35. $f(x) = x^2 - 2x$

   (a) $[-1, 2]$ (b) $(1, 3]$

   (c) $(0, 2)$ (d) $[1, 4)$

36. $f(x) = \sqrt{4 - x^2}$

   (a) $[-2, 2]$ (b) $[-2, 0)$

   (c) $(-2, 2)$ (d) $[1, 2)$

In Exercises 37–40, use a graphing utility to graph the function. Locate the absolute extrema of the function on the closed interval.

Function	Interval
37. $f(x) = \begin{cases} 2x + 2, & 0 \le x \le 1 \\ 4x^2, & 1 < x \le 3 \end{cases}$	$[0, 3]$
38. $f(x) = \begin{cases} 2 - x^2, & 1 \le x < 3 \\ 2 - 3x, & 3 \le x \le 5 \end{cases}$	$[1, 5]$
39. $f(x) = \dfrac{3}{x - 1}$	$(1, 4]$
40. $f(x) = \dfrac{2}{2 - x}$	$[0, 2)$

In Exercises 41 and 42, (a) use a computer algebra system to graph the function and approximate any absolute extrema on the indicated interval. (b) Use the utility to find any critical numbers, and use them to find any absolute extrema not located at the endpoints. Compare the results with those in part (a).

Function	Interval
41. $f(x) = 3.2x^5 + 5x^3 - 3.5x$	$[0, 1]$
42. $f(x) = \dfrac{4}{3}x\sqrt{3 - x}$	$[0, 3]$

In Exercises 43 and 44, use a computer algebra system to find the maximum value of $|f''(x)|$ on the closed interval. (This value is used in the error estimate for the Trapezoidal Rule, as discussed in Section 4.6.)

Function	Interval
43. $f(x) = \sqrt{1 + x^3}$	$[0, 2]$
44. $f(x) = \dfrac{1}{x^2 + 1}$	$\left[\dfrac{1}{2}, 3\right]$

In Exercises 45 and 46, use a computer algebra system to find the maximum value of $|f^4(x)|$ on the closed interval. (This value is used in the error estimate for Simpson's Rule, as discussed in Section 4.6.)

Function	Interval
45. $f(x) = (x + 1)^{2/3}$	$[0, 2]$
46. $f(x) = \dfrac{1}{x^2 + 1}$	$[-1, 1]$

47. Explain why the function $f(x) = \tan x$ has a maximum on $[0, \pi/4]$ but not on $[0, \pi]$.

48. *Writing* Write a short paragraph explaining why a continuous function on an open interval may not have a maximum or minimum. Illustrate your explanation with a sketch of the graph of a function.

Getting at the Concept

In Exercises 49 and 50, graph a function on the interval $[-2, 5]$ having the given characteristics.

49. Absolute maximum at $x = -2$

   Absolute minimum at $x = 1$

   Relative maximum at $x = 3$

50. Relative minimum at $x = -1$

   Critical number at $x = 0$, but no extrema

   Absolute maximum at $x = 2$

   Absolute minimum at $x = 5$

In Exercises 51–54, determine from the graph whether f has a minimum in the open interval (a, b).

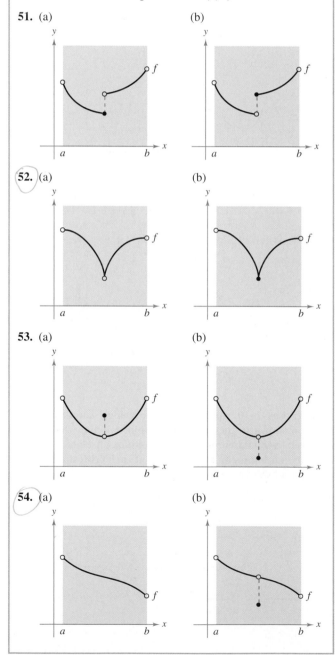

51. (a)   (b)

52. (a)   (b)

53. (a)   (b)

54. (a)   (b)

55. Power   The formula for the power output P of a battery is $P = VI - RI^2$ where V is the electromotive force in volts, R is the resistance, and I is the current. Find the current (measured in amperes) that corresponds to a maximum value of P in a battery for which $V = 12$ volts and $R = 0.5$ ohm. Assume that a 15-ampere fuse bounds the output in the interval $0 \le I \le 15$. Could the power output be increased by replacing the 15-ampere fuse with a 20-ampere fuse? Explain.

56. Lawn Sprinkler   A lawn sprinkler is constructed in such a way that $d\theta/dt$ is constant, where θ ranges between $45°$ and $135°$ (see figure). The distance the water travels horizontally is

$$x = \frac{v^2 \sin 2\theta}{32}, \qquad 45° \le \theta \le 135°$$

where v is the speed of the water. Find dx/dt and explain why this lawn sprinkler does not water evenly. What part of the lawn receives the most water?

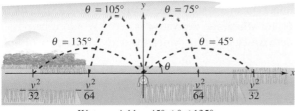

Water sprinkler: $45° \le \theta \le 135°$

FOR FURTHER INFORMATION   For more information on the "calculus of lawn sprinklers," see the article "Design of an Oscillating Sprinkler" by Bart Braden in *Mathematics Magazine*. To view this article, go to the website *www.mathurticles.com*.

57. Honeycomb   The surface area of a cell in a honeycomb is

$$S = 6hs + \frac{3s^2}{2}\left(\frac{\sqrt{3} - \cos\theta}{\sin\theta}\right)$$

where h and s are positive constants and θ is the angle at which the upper faces meet the altitude of the cell. Find the angle θ $(\pi/6 \le \theta \le \pi/2)$ that minimizes the surface area S.

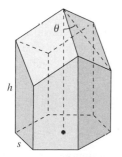

FOR FURTHER INFORMATION   For more information on the geometric structure of a honeycomb cell, see the article "The Design of Honeycombs" by Anthony L. Peressini in UMAP Module 502, published by COMAP, Inc., Suite 210, 57 Bedford Street, Lexington, MA. To view this article, go to the website *www.mathurticles.com*.

58. Inventory Cost   A retailer has determined that the cost C of ordering and storing x units of a certain product is

$$C = 2x + \frac{300{,}000}{x}, \qquad 1 \le x \le 300.$$

The delivery truck can bring at most 300 units per order. Find the order size that will minimize cost. Could the cost be decreased if the truck were replaced with one that could bring at most 400 units? Explain.

59. Highway Design   In order to build a highway it is necessary to fill a section of a valley where the grades (slopes) of the sides are 9% and 6% (see figure). The top of the filled region will have the shape of a parabolic arc that is tangent to the two slopes at the points A and B. The horizontal distance between the points A and B is 1000 feet.

(a) Find a quadratic function $y = ax^2 + bx + c$, $-500 \le x \le 500$, that describes the top of the filled region.

(b) Complete the table giving the depths d of the fill at the specified values of x.

x	-500	-400	-300	-200	-100
d					

x	0	100	200	300	400	500
d						

(c) What will be the lowest point on the completed highway? Will it be directly over the point where the two hillsides come together?

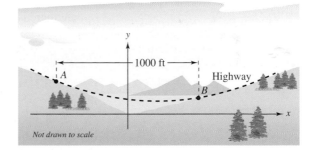

Not drawn to scale

60. Find all critical numbers of the greatest integer function $f(x) = [\![x]\!]$.

True or False?   **In Exercises 61–64, determine whether the statement is true or false. If it is false, explain why or give an example that shows it is false.**

61. The maximum of a function that is continuous on a closed interval can occur at two different values in the interval.

62. If a function is continuous on a closed interval, then it must have a minimum on the interval.

63. If $x = c$ is a critical number of the function f, then it is also a critical number of the function $g(x) = f(x) + k$, where k is a constant.

64. If $x = c$ is a critical number of the function f, then it is also a critical number of the function $g(x) = f(x - k)$, where k is a constant.

- Understand and use Rolle's Theorem.
- Understand and use the Mean Value Theorem.

Rolle's Theorem

The Extreme Value Theorem (Section 3.1) states that a continuous function on a closed interval $[a, b]$ must have both a minimum and a maximum on the interval. Both of these values, however, can occur at the endpoints. **Rolle's Theorem,** named after the French mathematician Michel Rolle (1652–1719), gives conditions that guarantee the existence of an extreme value in the *interior* of a closed interval.

EXPLORATION

Extreme Values in a Closed Interval   Sketch a rectangular coordinate plane on a piece of paper. Label the points $(1, 3)$ and $(5, 3)$. Using a pencil or pen, draw the graph of a differentiable function f that starts at $(1, 3)$ and ends at $(5, 3)$. Is there at least one point on the graph for which the derivative is zero? Would it be possible to draw the graph so that there *isn't* a point for which the derivative is zero? Explain your reasoning.

THEOREM 3.3   Rolle's Theorem

Let f be continuous on the closed interval $[a, b]$ and differentiable on the open interval (a, b). If

$$f(a) = f(b)$$

then there is at least one number c in (a, b) such that $f'(c) = 0$.

Proof   Let $f(a) = d = f(b)$.

Case 1: If $f(x) = d$ for all x in $[a, b]$, f is constant on the interval and, by Theorem 2.2, $f'(x) = 0$ for all x in (a, b).

Case 2: Suppose $f(x) > d$ for some x in (a, b). By the Extreme Value Theorem, you know that f has a maximum at some c in the interval. Moreover, because $f(c) > d$, this maximum does not occur at either endpoint. So, f has a maximum in the *open* interval (a, b). This implies that $f(c)$ is a *relative* maximum and, by Theorem 3.2, c is a critical number of f. Finally, because f is differentiable at c, you can conclude that $f'(c) = 0$.

Case 3: If $f(x) < d$ for some x in (a, b), you can use an argument similar to that in Case 2, but involving the minimum instead of the maximum.   ▨

From Rolle's Theorem, you can see that if a function f is continuous on $[a, b]$ and differentiable on (a, b), and if $f(a) = f(b)$, there must be at least one x-value between a and b at which the graph of f has a horizontal tangent, as shown in Figure 3.8(a). If the differentiability requirement is dropped from Rolle's Theorem, f will still have a critical number in (a, b), but it may not yield a horizontal tangent. Such a case is shown in Figure 3.8(b).

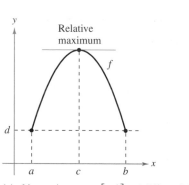

(a) f is continuous on $[a, b]$ and differentiable on (a, b).

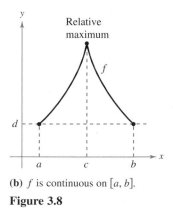

(b) f is continuous on $[a, b]$.

Figure 3.8

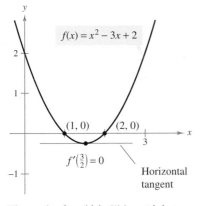

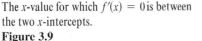

The x-value for which $f'(x) = 0$ is between the two x-intercepts.
Figure 3.9

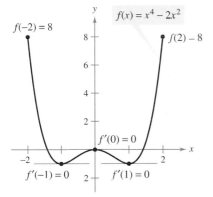

$f'(x) = 0$ for more than one x-value in the interval $(-2, 2)$.
Figure 3.10

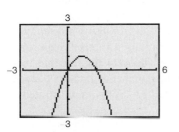

Figure 3.11

Example 1   **Illustrating Rolle's Theorem**

Find the two x-intercepts of

$$f(x) = x^2 - 3x + 2$$

and show that $f'(x) = 0$ at some point between the two intercepts.

Solution   Note that f is differentiable on the entire real line. Setting $f(x)$ equal to 0 produces

$$x^2 - 3x + 2 = 0 \qquad \text{Set } f(x) \text{ equal to 0.}$$
$$(x - 1)(x - 2) = 0. \qquad \text{Factor.}$$

So, $f(1) = f(2) = 0$, and from Rolle's Theorem you know that there *exists* at least one c in the interval $(1, 2)$ such that $f'(c) = 0$. To *find* such a c, you can solve the equation

$$f'(x) = 2x - 3 = 0 \qquad \text{Set } f'(x) \text{ equal to 0.}$$

and determine that $f'(x) = 0$ when $x = \frac{3}{2}$. Note that the x-value lies in the open interval $(1, 2)$, as shown in Figure 3.9.

Rolle's Theorem states that if f satisfies the conditions of the theorem, there must be *at least* one point between a and b at which the derivative is 0. There may of course be more than one such point, as illustrated in the next example.

Example 2   **Illustrating Rolle's Theorem**

Let $f(x) = x^4 - 2x^2$. Find all values of c in the interval $(-2, 2)$ such that $f'(c) = 0$.

Solution   To begin, note that the function satisfies the conditions of Rolle's Theorem. That is, f is continuous on the interval $[-2, 2]$ and differentiable on the interval $(-2, 2)$. Moreover, because $f(-2) = 8 = f(2)$, you can conclude that there exists at least one c in $(-2, 2)$ such that $f'(c) = 0$. Setting the derivative equal to 0 produces

$$f'(x) = 4x^3 - 4x = 0 \qquad \text{Set } f'(x) \text{ equal to 0.}$$
$$4x(x^2 - 1) = 0 \qquad \text{Factor.}$$
$$x = 0, 1, -1. \qquad x\text{-values for which } f'(x) = 0$$

So, in the interval $(-2, 2)$, the derivative is zero at three different values of x, as shown in Figure 3.10.

TECHNOLOGY PITFALL   A graphing utility can be used to indicate whether the points on the graphs in Examples 1 and 2 are relative minima or relative maxima of the functions. When using a graphing utility, however, you should keep in mind that it can give misleading pictures of graphs. For example, try using a graphing utility to graph

$$f(x) = 1 - (x - 1)^2 - \frac{1}{1000(x - 1)^{1/7} + 1}.$$

With most viewing windows, it appears that the function has a maximum of 1 when $x = 1$ (see Figure 3.11). By evaluating the function at $x = 1$, however, you can see that $f(1) = 0$. To determine the behavior of this function near $x = 1$, you need to examine the graph analytically to get the complete picture.

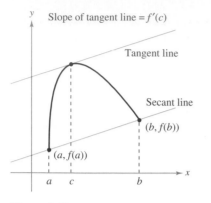

Figure 3.12

JOSEPH-LOUIS LAGRANGE (1736–1813)

The Mean Value Theorem was first proved by the famous mathematician Joseph-Louis Lagrange. Born in Italy, Lagrange held a position in the court of Frederick the Great in Berlin for 20 years. Afterward, he moved to France, where he met emperor Napoleon Bonaparte, who is quoted as saying, "Lagrange is the lofty pyramid of the mathematical sciences."

The Mean Value Theorem

Rolle's Theorem can be used to prove another theorem—the **Mean Value Theorem.**

> **THEOREM 3.4   The Mean Value Theorem**
>
> If f is continuous on the closed interval $[a, b]$ and differentiable on the open interval (a, b), then there exists a number c in (a, b) such that
>
> $$f'(c) = \frac{f(b) - f(a)}{b - a}.$$

Proof   Refer to Figure 3.12. The equation of the secant line containing the points $(a, f(a))$ and $(b, f(b))$ is

$$y = \left[\frac{f(b) - f(a)}{b - a} \right](x - a) + f(a).$$

Let $g(x)$ be the difference between $f(x)$ and y. Then

$$g(x) = f(x) - y$$
$$= f(x) - \left[\frac{f(b) - f(a)}{b - a} \right](x - a) - f(a).$$

By evaluating g at a and b, you can see that $g(a) = 0 = g(b)$. Furthermore, because f is differentiable, g is also differentiable, and you can apply Rolle's Theorem to the function g. So, there exists a number c in (a, b) such that $g'(c) = 0$, which implies that

$$0 = g'(c)$$
$$= f'(c) - \frac{f(b) - f(a)}{b - a}.$$

Therefore, there exists a number c in (a, b) such that

$$f'(c) = \frac{f(b) - f(a)}{b - a}.$$

NOTE   The "mean" in the Mean Value Theorem refers to the mean (or average) rate of change of f in the interval $[a, b]$.

Although the Mean Value Theorem can be used directly in problem solving, it is used more often to prove other theorems. In fact, some people consider this to be the most important theorem in calculus—it is closely related to the Fundamental Theorem of Calculus discussed in Chapter 4. For now, you can get an idea of the versatility of this theorem by looking at the results stated in Exercises 57–62 in this section.

The Mean Value Theorem has implications for both basic interpretations of the derivative. Geometrically, the theorem guarantees the existence of a tangent line that is parallel to the secant line through the points $(a, f(a))$ and $(b, f(b))$, as shown in Figure 3.12. Example 3 illustrates this geometric interpretation of the Mean Value Theorem. In terms of rates of change, the Mean Value Theorem implies that there must be a point in the open interval (a, b) at which the instantaneous rate of change is equal to the average rate of change over the interval $[a, b]$. This is illustrated in Example 4.

□ *Example 3*   Finding a Tangent Line

Given $f(x) = 5 - (4/x)$, find all values of c in the open interval $(1, 4)$ such that

$$f'(c) = \frac{f(4) - f(1)}{4 - 1}.$$

Solution   The slope of the secant line through $(1, f(1))$ and $(4, f(4))$ is

$$\frac{f(4) - f(1)}{4 - 1} = \frac{4 - 1}{4 - 1} = 1.$$

Because f satisfies the conditions of the Mean Value Theorem, there exists at least one number c in $(1, 4)$ such that $f'(c) = 1$. Solving the equation $f'(x) = 1$ yields

$$f'(x) = \frac{4}{x^2} = 1$$

which implies that $x = \pm 2$. So, in the interval $(1, 4)$, you can conclude that $c = 2$, as shown in Figure 3.13.

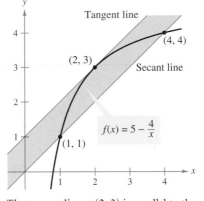

The tangent line at $(2, 3)$ is parallel to the secant line through $(1, 1)$ and $(4, 4)$.
Figure 3.13

□ *Example 4*   Finding an Instantaneous Rate of Change

Two stationary patrol cars equipped with radar are 5 miles apart on a highway, as shown in Figure 3.14. As a truck passes the first patrol car, its speed is clocked at 55 miles per hour. Four minutes later, when the truck passes the second patrol car, its speed is clocked at 50 miles per hour. Prove that the truck must have exceeded the speed limit (of 55 miles per hour) at some time during the four minutes.

Solution   Let $t = 0$ be the time (in hours) when the truck passes the first patrol car. The time when the truck passes the second patrol car is

$$t = \frac{4}{60} = \frac{1}{15} \text{ hour.}$$

By letting $s(t)$ represent the distance (in miles) traveled by the truck, you have $s(0) = 0$ and $s\left(\frac{1}{15}\right) = 5$. So, the average velocity of the truck over the 5-mile stretch of highway is

$$\text{Average velocity} = \frac{s(1/15) - s(0)}{(1/15) - 0}$$

$$= \frac{5}{1/15} = 75 \text{ mph.}$$

Assuming that the position function is differentiable, you can apply the Mean Value Theorem to conclude that the truck must have been traveling at a rate of 75 miles per hour sometime during the four minutes.   ◢

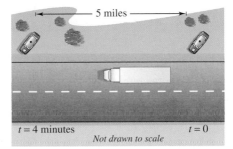

At some time t, the instantaneous velocity is equal to the average velocity over four minutes.
Figure 3.14

A useful alternative form of the Mean Value Theorem is as follows: If f is continuous on $[a, b]$ and differentiable on (a, b), then there exists a number c in (a, b) such that

$$f(b) = f(a) + (b - a)f'(c). \qquad \text{Alternative form of Mean Value Theorem}$$

NOTE   When working the exercises for this section, keep in mind that polynomial functions, rational functions, and trigonometric functions are differentiable at all points in their domains.

EXERCISES FOR SECTION 3.2

In Exercises 1 and 2, explain why Rolle's Theorem does not apply to the function even though there exist *a* and *b* such that $f(a) = f(b)$.

1. $f(x) = 1 - |x - 1|$

2. $f(x) = \cot \dfrac{x}{2}$

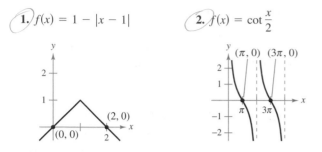

In Exercises 3–6, find the two *x*-intercepts of the function *f* and show that $f'(x) = 0$ at some point between the two intercepts.

3. $f(x) = x^2 - x - 2$

4. $f(x) = x(x - 3)$

5. $f(x) = x\sqrt{x + 4}$

6. $f(x) = -3x\sqrt{x + 1}$

In Exercises 7–20, determine whether Rolle's Theorem can be applied to *f* on the closed interval $[a, b]$. If Rolle's Theorem can be applied, find all values of *c* in the open interval (a, b) such that $f'(c) = 0$.

7. $f(x) = x^2 - 2x, [0, 2]$

8. $f(x) = x^2 - 5x + 4, [1, 4]$

9. $f(x) = (x - 1)(x - 2)(x - 3), [1, 3]$

10. $f(x) = (x - 3)(x + 1)^2, [-1, 3]$

11. $f(x) = x^{2/3} - 1, [-8, 8]$

12. $f(x) = 3 - |x - 3|, [0, 6]$

13. $f(x) = \dfrac{x^2 - 2x - 3}{x + 2}, [-1, 3]$

14. $f(x) = \dfrac{x^2 - 1}{x}, [-1, 1]$

15. $f(x) = \sin x, [0, 2\pi]$

16. $f(x) = \cos x, [0, 2\pi]$

17. $f(x) = \dfrac{6x}{\pi} - 4\sin^2 x, \left[0, \dfrac{\pi}{6}\right]$

18. $f(x) = \cos 2x, \left[-\dfrac{\pi}{12}, \dfrac{\pi}{6}\right]$

19. $f(x) = \tan x, [0, \pi]$

20. $f(x) = \sec x, \left[-\dfrac{\pi}{4}, \dfrac{\pi}{4}\right]$

In Exercises 21–24, use a graphing utility to graph the function on the closed interval $[a, b]$. Determine whether Rolle's Theorem can be applied to *f* on the interval and, if so, find all values of *c* in the open interval (a, b) such that $f'(c) = 0$.

21. $f(x) = |x| - 1, [-1, 1]$

22. $f(x) = x - x^{1/3}, [0, 1]$

23. $f(x) = 4x - \tan \pi x, \left[-\dfrac{1}{4}, \dfrac{1}{4}\right]$

24. $f(x) = \dfrac{x}{2} - \sin \dfrac{\pi x}{6}, [-1, 0]$

25. *Vertical Motion*   The height of a ball *t* seconds after it is thrown upward from a height of 32 feet and with an initial velocity of 48 feet per second is $f(t) = -16t^2 + 48t + 32$.

(a) Verify that $f(1) = f(2)$.

(b) According to Rolle's Theorem, what must be the velocity at some time in the interval $(1, 2)$? Find that time.

26. *Reorder Costs*   The ordering and transportation cost *C* of components used in a manufacturing process is approximated by

$$C(x) = 10\left(\dfrac{1}{x} + \dfrac{x}{x + 3}\right)$$

where *C* is measured in thousands of dollars and *x* is the order size in hundreds.

(a) Verify that $C(3) = C(6)$.

(b) According to Rolle's Theorem, the rate of change of cost must be 0 for some order size in the interval $(3, 6)$. Find that order size.

In Exercises 27 and 28, copy the graph and sketch the secant line to the graph through the points $(a, f(a))$ and $(b, f(b))$. Then sketch any tangent lines to the graph for each value of *c* guaranteed by the Mean Value Theorem. To print an enlarged copy of the graph, go to the website *www.mathgraphs.com*.

27.

28.

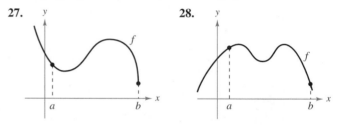

In Exercises 29 and 30, explain why the Mean Value Theorem does not apply to the function on the interval $[0, 6]$.

29. $f(x) = \dfrac{1}{x - 3}$

30. $f(x) = |x - 3|$

In Exercises 31–38, determine whether the Mean Value Theorem can be applied to *f* on the closed interval $[a, b]$. If the Mean Value Theorem can be applied, find all values of *c* in the open interval (a, b) such that

$$f'(c) = \dfrac{f(b) - f(a)}{b - a}.$$

31. $f(x) = x^2, [-2, 1]$

32. $f(x) = x(x^2 - x - 2), [-1, 1]$

33. $f(x) = x^{2/3}, [0, 1]$

34. $f(x) = \dfrac{x + 1}{x}, \left[\dfrac{1}{2}, 2\right]$

35. $f(x) = \sqrt{2 - x}, [-7, 2]$

36. $f(x) = x^3, [0, 1]$

37. $f(x) = \sin x, [0, \pi]$

38. $f(x) = 2\sin x + \sin 2x, [0, \pi]$

In Exercises 39–42, use a graphing utility to (a) graph the function *f* on the indicated interval, (b) find and graph the secant line through points on the graph of *f* at the endpoints of the indicated interval, and (c) find and graph any tangent lines to the graph of *f* that are parallel to the secant line.

39. $f(x) = \dfrac{x}{x + 1}, \left[-\dfrac{1}{2}, 2\right]$

40. $f(x) = x - 2\sin x, [-\pi, \pi]$

41. $f(x) = \sqrt{x}, [1, 9]$

42. $f(x) = -x^4 + 4x^3 + 8x^2 + 5, [0, 5]$

43. Vertical Motion   The height of an object t seconds after it is dropped from a height of 500 meters is $s(t) = -4.9t^2 + 500$.

(a) Find the average velocity of the object during the first 3 seconds.

(b) Use the Mean Value Theorem to verify that at some time during the first 3 seconds of fall the instantaneous velocity equals the average velocity. Find that time.

44. Sales   A company introduces a new product for which the number of units sold S is

$$S(t) = 200\left(5 - \frac{9}{2 + t}\right)$$

where t is the time in months.

(a) Find the average value of $S(t)$ during the first year.

(b) During what month does $S'(t)$ equal the average value during the first year?

Getting at the Concept

45. Let f be continuous on $[a, b]$ and differentiable on (a, b). If there exists c in (a, b) such that $f'(c) = 0$, does it follow that $f(a) = f(b)$? Explain.

46. Let f be continuous on the closed interval $[a, b]$ and differentiable on the open interval (a, b). Also, suppose that $f(a) = f(b)$ and that c is a real number in the interval such that $f'(c) = 0$. Find an interval for the function g over which Rolle's Theorem can be applied, and find the corresponding critical number of g (k is a constant).

(a) $g(x) = f(x) + k$ (b) $g(x) = f(x - k)$

(c) $g(x) = f(kx)$

47. A plane begins its takeoff at 2:00 P.M. on a 2500-mile flight. The plane arrives at its destination at 7:30 P.M. Explain why there were at least two times during the flight when the speed of the plane was 400 miles per hour.

48. When an object is removed from a furnace and placed in an environment with a constant temperature of 90°F, its core temperature is 1500°F. Five hours later the core temperature is 390°F. Explain why there must exist a time in the interval when the temperature is decreasing at a rate of 222°F per hour.

49. Graphical Reasoning   The figure gives two parts of the graph of a continuous differentiable function f on $[-10, 4]$. The derivative f' is also continuous. To print an enlarged copy of the graph, go to the website *www.mathgraphs.com*.

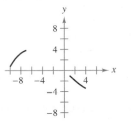

(a) Explain why f must have at least one zero in $[-10, 4]$.

(b) Explain why f' must also have at least one zero in the interval $[-10, 4]$. What are these zeros called?

(c) Make a possible sketch of the function with one zero of f' on the interval $[-10, 4]$.

(d) Make a possible sketch of the function with two zeros of f' on the interval $[-10, 4]$.

(e) Were the conditions of continuity of f and f' necessary to do parts (a) through (d)? Explain.

50. Consider the function $f(x) = 3 \cos^2\left(\frac{\pi x}{2}\right)$.

(a) Use a graphing utility to graph f and f'.

(b) Is f a continuous function? Is f' a continuous function?

(c) Does Rolle's Theorem apply on the interval $[-1, 1]$? Does it apply on the interval $[1, 2]$? Explain.

(d) Evaluate, if possible, $\lim_{x \to 3^-} f'(x)$ and $\lim_{x \to 3^+} f'(x)$.

Think About It   In Exercises 51 and 52, sketch the graph of an arbitrary function f that satisfies the given condition but does not satisfy the conditions of the Mean Value Theorem on the interval $[-5, 5]$.

51. f is continuous on $[-5, 5]$.

52. f is not continuous on $[-5, 5]$.

True or False?   In Exercises 53–56, determine whether the statement is true or false. If it is false, explain why or give an example that shows it is false.

53. The Mean Value Theorem can be applied to $f(x) = 1/x$ on the interval $[-1, 1]$.

54. If the graph of a function has three x-intercepts, then it must have at least two points at which its tangent line is horizontal.

55. If the graph of a polynomial function has three x-intercepts, then it must have at least two points at which its tangent line is horizontal.

56. If $f'(x) = 0$ for all x in the domain of f, then f is a constant function.

57. Prove that if $a > 0$ and n is any positive integer, then the polynomial function $p(x) = x^{2n+1} + ax + b$ cannot have two real roots.

58. Prove that if $f'(x) = 0$ for all x in an interval (a, b), then f is constant on (a, b).

59. Let $p(x) = Ax^2 + Bx + C$. Prove that for any interval $[a, b]$, the value c guaranteed by the Mean Value Theorem is the midpoint of the interval.

60. Prove that if f is differentiable on $(-\infty, \infty)$ and $f'(x) < 1$ for all real numbers, then f has at most one fixed point. A fixed point of a function f is a real number c such that $f(c) = c$.

61. Use the result of Exercise 60 to show that $f(x) = \frac{1}{2} \cos x$ has at most one fixed point.

62. Prove that $|\cos x - \cos y| \le |x - y|$ for all x and y.

- Determine intervals on which a function is increasing or decreasing.
- Apply the First Derivative Test to find relative extrema of a function.

Increasing and Decreasing Functions

In this section you will learn how derivatives can be used to *classify* relative extrema as either relative minima or relative maxima. We begin by defining increasing and decreasing functions.

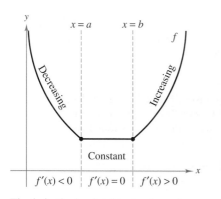

The derivative is related to the slope of a function.

Figure 3.15

Definitions of Increasing and Decreasing Functions

A function f is **increasing** on an interval if for any two numbers x_1 and x_2 in the interval, $x_1 < x_2$ implies $f(x_1) < f(x_2)$.

A function f is **decreasing** on an interval if for any two numbers x_1 and x_2 in the interval, $x_1 < x_2$ implies $f(x_1) > f(x_2)$.

A function is increasing if, *as x moves to the right*, its graph moves up, and is decreasing if its graph moves down. For example, the function in Figure 3.15 is decreasing on the interval $(-\infty, a)$, is constant on the interval (a, b), and is increasing on the interval (b, ∞). As shown in Theorem 3.5 below, a positive derivative implies that the function is increasing; a negative derivative implies that the function is decreasing; and a zero derivative on an entire interval implies that the function is constant on that interval.

THEOREM 3.5   Test for Increasing and Decreasing Functions

Let f be a function that is continuous on the closed interval $[a, b]$ and differentiable on the open interval (a, b).

1. If $f'(x) > 0$ for all x in (a, b), then f is increasing on $[a, b]$.
2. If $f'(x) < 0$ for all x in (a, b), then f is decreasing on $[a, b]$.
3. If $f'(x) = 0$ for all x in (a, b), then f is constant on $[a, b]$.

Proof   To prove the first case, assume that $f'(x) > 0$ for all x in the interval (a, b) and let $x_1 < x_2$ be any two points in the interval. By the Mean Value Theorem, you know that there exists a number c such that $x_1 < c < x_2$, and

$$f'(c) = \frac{f(x_2) - f(x_1)}{x_2 - x_1}.$$

Because $f'(c) > 0$ and $x_2 - x_1 > 0$, you know that

$$f(x_2) - f(x_1) > 0$$

which implies that $f(x_1) < f(x_2)$. So, f is increasing on the interval. The second case has a similar proof (see Exercise 77), and the third case was given as Exercise 58 in Section 3.2.

NOTE   The conclusions in the first two cases of Theorem 3.5 are valid even if $f'(x) = 0$ at a finite number of x-values in (a, b).

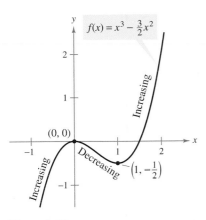

$f(x) = x^3 - \frac{3}{2}x^2$

Figure 3.16

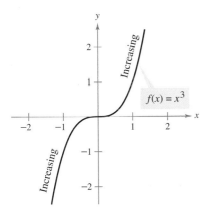

(a) Strictly monotonic function

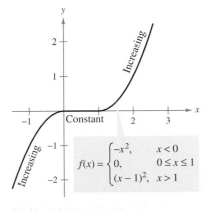

$f(x) = \begin{cases} -x^2, & x < 0 \\ 0, & 0 \le x \le 1 \\ (x-1)^2, & x > 1 \end{cases}$

(b) Not strictly monotonic
Figure 3.17

Example 1   Intervals on Which f is Increasing or Decreasing

Find the open intervals on which $f(x) = x^3 - \frac{3}{2}x^2$ is increasing or decreasing.

Solution   Note that f is continuous on the entire real line. To determine the critical numbers of f, set $f'(x)$ equal to zero.

$$f(x) = x^3 - \frac{3}{2}x^2 \qquad \text{Write original function.}$$
$$f'(x) = 3x^2 - 3x = 0 \qquad \text{Differentiate and set } f'(x) \text{ equal to 0.}$$
$$3(x)(x-1) = 0 \qquad \text{Factor.}$$
$$x = 0, 1 \qquad \text{Critical numbers}$$

Because there are no points for which f' does not exist, you can conclude that $x = 0$ and $x = 1$ are the only critical numbers. The table summarizes the testing of the three intervals determined by these two critical numbers.

Interval	$-\infty < x < 0$	$0 < x < 1$	$1 < x < \infty$
Test Value	$x = -1$	$x = \frac{1}{2}$	$x = 2$
Sign of $f'(x)$	$f'(-1) = 6 > 0$	$f'\left(\frac{1}{2}\right) = -\frac{3}{4} < 0$	$f'(2) = 6 > 0$
Conclusion	Increasing	Decreasing	Increasing

So, f is increasing on the intervals $(-\infty, 0)$ and $(1, \infty)$ and decreasing on the interval $(0, 1)$, as shown in Figure 3.16.

Example 1 gives you one example of how to find intervals on which a function is increasing or decreasing. The guidelines below summarize the steps followed in the example.

Guidelines for Finding Intervals on Which a Function Is Increasing or Decreasing

Let f be continuous on the interval (a, b). To find the open intervals on which f is increasing or decreasing, use the following steps.

1. Locate the critical numbers of f in (a, b), and use these numbers to determine test intervals.
2. Determine the sign of $f'(x)$ at one test value in each of the intervals.
3. Use Theorem 3.5 to determine whether f is increasing or decreasing on each interval.

These guidelines are also valid if the interval (a, b) is replaced by an interval of the form $(-\infty, b)$, (a, ∞), or $(-\infty, \infty)$.

A function is **strictly monotonic** on an interval if it is either increasing on the entire interval or decreasing on the entire interval. For instance, the function $f(x) = x^3$ is strictly monotonic on the entire real line because it is increasing on the entire real line, as shown in Figure 3.17(a). The function shown in Figure 3.17(b) is not strictly monotonic on the entire real line because it is constant on the interval $[0, 1]$.

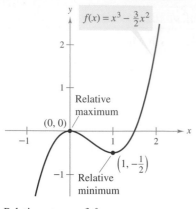

Relative extrema of f
Figure 3.18

The First Derivative Test

After you have determined the intervals on which a function is increasing or decreasing, it is not difficult to locate the relative extrema of the function. For instance, in Figure 3.18 (from Example 1), the function

$$f(x) = x^3 - \frac{3}{2}x^2$$

has a relative maximum at the point $(0, 0)$ because f is increasing immediately to the left of $x = 0$ and decreasing immediately to the right of $x = 0$. Similarly, f has a relative minimum at the point $\left(1, -\frac{1}{2}\right)$ because f is decreasing immediately to the left of $x = 1$ and increasing immediately to the right of $x = 1$. The following theorem, called the First Derivative Test, makes this more explicit.

THEOREM 3.6   The First Derivative Test

Let c be a critical number of a function f that is continuous on an open interval I containing c. If f is differentiable on the interval, except possibly at c, then $f(c)$ can be classified as follows.

1. If $f'(x)$ changes from negative to positive at c, then $f(c)$ is a *relative minimum* of f.

2. If $f'(x)$ changes from positive to negative at c, then $f(c)$ is a *relative maximum* of f.

3. If $f'(x)$ does not change sign at c, then $f(c)$ is neither a relative minimum nor a relative maximum.

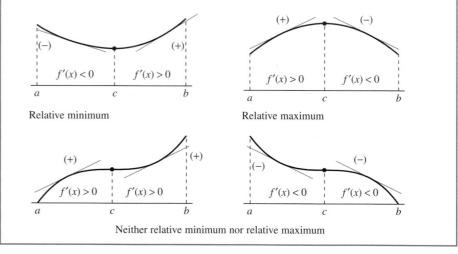

Proof   Assume that $f'(x)$ changes from negative to positive at c. Then there exist a and b in I such that

$$f'(x) < 0 \text{ for all } x \text{ in } (a, c)$$

and

$$f'(x) > 0 \text{ for all } x \text{ in } (c, b).$$

By Theorem 3.5, f is decreasing on (a, c) and increasing on (c, b). So, $f(c)$ is a minimum of f on the open interval (a, b) and, consequently, a relative minimum of f. This proves the first case of the theorem. The second case can be proved in a similar way (see Exercise 78).

Example 2   Applying the First Derivative Test

Find the relative extrema of the function $f(x) = \frac{1}{2}x - \sin x$ in the interval $(0, 2\pi)$.

Solution   Note that f is continuous on the interval $(0, 2\pi)$. To determine the critical numbers of f in this interval, set $f'(x)$ equal to 0.

$$f'(x) = \frac{1}{2} - \cos x = 0 \qquad \text{Set } f'(x) \text{ equal to 0.}$$

$$\cos x = \frac{1}{2}$$

$$x = \frac{\pi}{3}, \frac{5\pi}{3} \qquad \text{Critical numbers}$$

Because there are no points for which f' does not exist, you can conclude that $x = \pi/3$ and $x = 5\pi/3$ are the only critical numbers. The table summarizes the testing of the three intervals determined by these two critical numbers.

Interval	$0 < x < \dfrac{\pi}{3}$	$\dfrac{\pi}{3} < x < \dfrac{5\pi}{3}$	$\dfrac{5\pi}{3} < x < 2\pi$
Test Value	$x = \dfrac{\pi}{4}$	$x = \pi$	$x = \dfrac{7\pi}{4}$
Sign of $f'(x)$	$f'\left(\dfrac{\pi}{4}\right) < 0$	$f'(\pi) > 0$	$f'\left(\dfrac{7\pi}{4}\right) < 0$
Conclusion	Decreasing	Increasing	Decreasing

By applying the First Derivative Test, you can conclude that f has a relative minimum at

$$x = \frac{\pi}{3} \qquad \text{x-value of relative minimum}$$

and a relative maximum at

$$x = \frac{5\pi}{3} \qquad \text{x-value of relative maximum}$$

as shown in Figure 3.19.

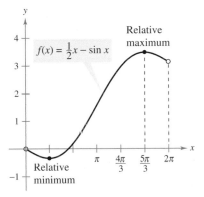

$f(x) = \frac{1}{2}x - \sin x$

A relative minimum occurs where f changes from decreasing to increasing, and a relative maximum occurs where f changes from increasing to decreasing.

Figure 3.19

EXPLORATION

Comparing Graphical and Analytical Approaches   In Section 3.2, we pointed out that, *by itself*, a graphing utility can give misleading information about the relative extrema of a graph. *Used in conjunction with an analytical approach*, however, a graphing utility can provide a good way to reinforce your conclusions. Try using a graphing utility to graph the function in Example 2. Then use the *zoom* and *trace* features to estimate the relative extrema. How close are your graphical approximations?

Note that in Examples 1 and 2 the given functions are differentiable on the entire real line. For such functions, the only critical numbers are those for which $f'(x) = 0$. Example 3 concerns a function that has two types of critical numbers—those for which $f'(x) = 0$ and those for which f is not differentiable.

Example 3   **Applying the First Derivative Test**

Find the relative extrema of

$$f(x) = (x^2 - 4)^{2/3}.$$

Solution   Begin by noting that f is continuous on the entire real line. The derivative of f

$$f'(x) = \frac{2}{3}(x^2 - 4)^{-1/3}(2x) \qquad \text{General Power Rule}$$

$$= \frac{4x}{3(x^2 - 4)^{1/3}} \qquad \text{Simplify.}$$

is 0 when $x = 0$ and does not exist when $x = \pm 2$. So, the critical numbers are $x = -2$, $x = 0$, and $x = 2$. The table summarizes the testing of the four intervals determined by these three critical numbers.

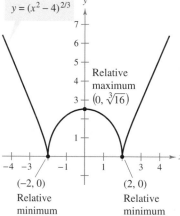

$y = (x^2 - 4)^{2/3}$

Relative
maximum
$(0, \sqrt[3]{16})$

$(-2, 0)$
Relative
minimum

$(2, 0)$
Relative
minimum

You can apply the First Derivative Test to find relative extrema.
Figure 3.20

Interval	$-\infty < x < -2$	$-2 < x < 0$	$0 < x < 2$	$2 < x < \infty$
Test Value	$x = -3$	$x = -1$	$x = 1$	$x = 3$
Sign of $f'(x)$	$f'(-3) < 0$	$f'(-1) > 0$	$f'(1) < 0$	$f'(3) > 0$
Conclusion	Decreasing	Increasing	Decreasing	Increasing

By applying the First Derivative Test, you can conclude that f has a relative minimum at the point $(-2, 0)$, a relative maximum at the point $\left(0, \sqrt[3]{16}\right)$, and another relative minimum at the point $(2, 0)$, as shown in Figure 3.20.

TECHNOLOGY PITFALL   When using a graphing utility to graph a function involving radicals or rational exponents, be sure you understand the way the utility evaluates radical expressions. For instance, even though

$$f(x) = (x^2 - 4)^{2/3}$$

and

$$g(x) = [(x^2 - 4)^2]^{1/3}$$

are the same algebraically, many graphing utilities distinguish between these two functions. Which of the graphs shown in Figure 3.21 is incorrect? Why did the graphing utility produce an incorrect graph?

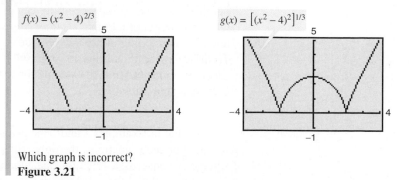

Which graph is incorrect?
Figure 3.21

When using the First Derivative Test, be sure to consider the domain of the function. For instance, in the next example, the function

$$f(x) = \frac{x^4 + 1}{x^2}$$

is not defined when $x = 0$. This x-value must be used with the critical numbers to determine the test intervals.

Example 4   Applying the First Derivative Test

Find the relative extrema of $f(x) = \dfrac{x^4 + 1}{x^2}$.

Solution

$$f(x) = x^2 + x^{-2} \qquad\qquad \text{Rewrite original function.}$$
$$f'(x) = 2x - 2x^{-3} \qquad\qquad \text{Differentiate.}$$
$$= 2x - \frac{2}{x^3}$$
$$= \frac{2(x^4 - 1)}{x^3} \qquad\qquad \text{Simplify.}$$
$$= \frac{2(x^2 + 1)(x - 1)(x + 1)}{x^3} \qquad\qquad \text{Factor.}$$

So, $f'(x)$ is zero at $x = \pm 1$. Moreover, because $x = 0$ is not in the domain of f, you should use this x-value along with the critical numbers to determine the test intervals.

$$x = \pm 1 \qquad\qquad \text{Critical numbers, } f'(\pm 1) = 0$$
$$x = 0 \qquad\qquad 0 \text{ is not in the domain of } f.$$

The table summarizes the testing of the four intervals determined by these three x-values.

Interval	$-\infty < x < -1$	$-1 < x < 0$	$0 < x < 1$	$1 < x < \infty$
Test Value	$x = -2$	$x = -\frac{1}{2}$	$x = \frac{1}{2}$	$x = 2$
Sign of $f'(x)$	$f'(-2) < 0$	$f'(-\frac{1}{2}) > 0$	$f'(\frac{1}{2}) < 0$	$f'(2) > 0$
Conclusion	Decreasing	Increasing	Decreasing	Increasing

By applying the First Derivative Test, you can conclude that f has one relative minimum at the point $(-1, 2)$ and another at the point $(1, 2)$, as shown in Figure 3.22.

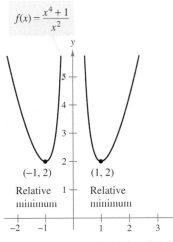

$f(x) = \dfrac{x^4 + 1}{x^2}$

$(-1, 2)$ $(1, 2)$

Relative   Relative
minimum   minimum

x-values that are not in the domain of f, as well as critical numbers, determine test intervals for f'.

Figure 3.22

TECHNOLOGY   The most difficult step in applying the First Derivative Test is finding the values for which the derivative is equal to 0. For instance, the values of x for which the derivative of

$$f(x) = \frac{x^4 + 1}{x^2 + 1}$$

is equal to zero are 0 and $\pm\sqrt{\sqrt{2} - 1}$. If you have access to technology that can perform symbolic differentiation and solve equations, try using it to apply the First Derivative Test to this function.

If a projectile is propelled from ground level and air resistance is neglected, the object will travel farthest with an initial angle of 45°. If, however, the projectile is propelled from a point above ground level, the angle that yields a maximum horizontal distance is not 45° (see Example 5).

Example 5   **The Path of a Projectile**

Neglecting air resistance, the path of a projectile that is propelled at an angle θ is

$$y = \frac{g \sec^2 \theta}{2v_0^2}x^2 + (\tan \theta)x + h, \qquad 0 \le \theta \le \frac{\pi}{2}$$

where y is the height, x is the horizontal distance, g is the acceleration due to gravity, v_0 is the initial velocity, and h is the initial height. (This equation is derived in Section 11.3.) Let $g = -32$ feet per second per second, $v_0 = 24$ feet per second, and $h = 9$ feet. What value of θ will produce a maximum horizontal distance?

Solution   To find the distance the projectile travels, let $y = 0$, and use the quadratic formula to solve for x.

$$\frac{g \sec^2 \theta}{2v_0^2}x^2 + (\tan \theta)x + h = 0$$

$$\frac{-32 \sec^2 \theta}{2(24^2)}x^2 + (\tan \theta)x + 9 = 0$$

$$-\frac{\sec^2 \theta}{36}x^2 + (\tan \theta)x + 9 = 0$$

$$x = \frac{-\tan \theta \pm \sqrt{\tan^2 \theta + \sec^2 \theta}}{-\sec^2 \theta/18}$$

$$x = 18 \cos \theta \left(\sin \theta + \sqrt{\sin^2 \theta + 1}\right), \qquad x \ge 0$$

At this point, you need to find the value of θ that produces a maximum value of x. Applying the First Derivative Test by hand would be very tedious. Using technology to solve the equation $dx/d\theta = 0$, however, eliminates most of the messy computations. The result is that the maximum value of x occurs when

$$\theta \approx 0.61548 \text{ radians, or } 35.3°.$$

This conclusion is reinforced by sketching the path of the projectile for different values of θ, as shown in Figure 3.23. Of the three paths shown, note that the distance traveled is greatest for $\theta = 35°$.

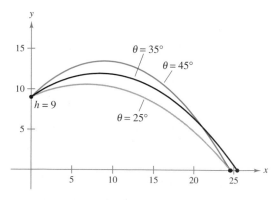

The path of a projectile with initial angle θ
Figure 3.23

NOTE   A computer simulation of this example is given in the *Interactive* CD-ROM version of this text (available at *college.hmco.com*). Using that simulation, you can experimentally discover that the maximum value of x occurs when $\theta \approx 35.3°$.

Brian Bailey/Tony Stone Images

EXERCISES FOR SECTION 3.3

In Exercises 1–10, identify the open intervals on which the function is increasing or decreasing.

1. $f(x) = x^2 - 6x + 8$

2. $y = -(x + 1)^2$

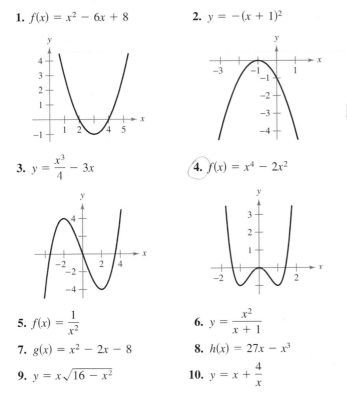

3. $y = \dfrac{x^3}{4} - 3x$

4. $f(x) = x^4 - 2x^2$

5. $f(x) = \dfrac{1}{x^2}$

6. $y = \dfrac{x^2}{x + 1}$

7. $g(x) = x^2 - 2x - 8$

8. $h(x) = 27x - x^3$

9. $y = x\sqrt{16 - x^2}$

10. $y = x + \dfrac{4}{x}$

In Exercises 11–32, find the critical numbers of f (if any). Find the open intervals on which the function is increasing or decreasing and locate all relative extrema. Use a graphing utility to confirm your results.

11. $f(x) = x^2 - 6x$

12. $f(x) = x^2 + 8x + 10$

13. $f(x) = -2x^2 + 4x + 3$

14. $f(x) = -(x^2 + 8x + 12)$

15. $f(x) = 2x^3 + 3x^2 - 12x$

16. $f(x) = x^3 - 6x^2 + 15$

17. $f(x) = x^2(3 - x)$

18. $f(x) = (x + 2)^2(x - 1)$

19. $f(x) = \dfrac{x^5 - 5x}{5}$

20. $f(x) = x^4 - 32x + 4$

21. $f(x) = x^{1/3} + 1$

22. $f(x) = x^{2/3} - 4$

23. $f(x) = (x - 1)^{2/3}$

24. $f(x) = (x - 1)^{1/3}$

25. $f(x) = 5 - |x - 5|$

26. $f(x) = |x + 3| - 1$

27. $f(x) = x + \dfrac{1}{x}$

28. $f(x) = \dfrac{x}{x + 1}$

29. $f(x) = \dfrac{x^2}{x^2 - 9}$

30. $f(x) = \dfrac{x + 3}{x^2}$

31. $f(x) = \dfrac{x^2 - 2x + 1}{x + 1}$

32. $f(x) = \dfrac{x^2 - 3x - 4}{x - 2}$

In Exercises 33–36, consider the function on the interval $(0, 2\pi)$. Find the open intervals on which the function is increasing or decreasing and locate all relative extrema. Use a graphing utility to confirm your results.

33. $f(x) = \dfrac{x}{2} + \cos x$

34. $f(x) = \sin x \cos x$

35. $f(x) = \sin^2 x + \sin x$

36. $f(x) = \dfrac{\sin x}{1 + \cos^2 x}$

In Exercises 37–40, (a) use a computer algebra system to differentiate the function, (b) sketch the graphs of f and f' on the same set of coordinate axes over the indicated interval, (c) find the critical numbers of f in the open interval, and (d) find the interval(s) on which f' is positive and the interval(s) on which it is negative. Compare the behavior of f and the sign of f'.

37. $f(x) = 2x\sqrt{9 - x^2}, [-3, 3]$

38. $f(x) = 10(5 - \sqrt{x^2 - 3x + 16}), [0, 5]$

39. $f(t) = t^2 \sin t, [0, 2\pi]$

40. $f(x) = \dfrac{x}{2} + \cos\dfrac{x}{2}, [0, 4\pi]$

In Exercises 41 and 42, use symmetry, extrema, and zeros to sketch the graph of f. How do the functions f and g differ? Explain.

41. $f(x) = \dfrac{x^5 - 4x^3 + 3x}{x^2 - 1}, \quad g(x) = x(x^2 - 3)$

42. $f(t) = \cos^2 t - \sin^2 t, \quad g(t) = 1 - 2\sin^2 t, \quad (-2, 2)$

Think About It In Exercises 43–48, the graph of f is shown in the figure. Sketch a graph of the derivative of f. To print an enlarged copy of the graph, go to the website *www.mathgraphs.com.*

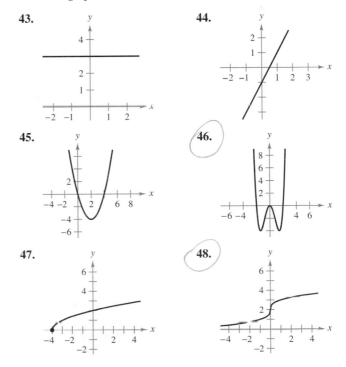

43.

44.

45.

46.

47.

48.

Getting at the Concept

In Exercises 49–54, assume that f is differentiable for all x. The sign of f' is as follows.

$f'(x) > 0$ on $(-\infty, -4)$

$f'(x) < 0$ on $(-4, 6)$

$f'(x) > 0$ on $(6, \infty)$

Supply the appropriate inequality for the indicated value of c.

Function	Sign of $g'(c)$
49. $g(x) = f(x) + 5$	$g'(0)$ ▢ 0
50. $g(x) = 3f(x) - 3$	$g'(-5)$ ▢ 0
51. $g(x) = -f(x)$	$g'(-6)$ ▢ 0
52. $g(x) = -f(x)$	$g'(0)$ ▢ 0
53. $g(x) = f(x - 10)$	$g'(0)$ ▢ 0
54. $g(x) = f(x - 10)$	$g'(8)$ ▢ 0

55. Sketch the graph of an arbitrary function f such that

$$f'(x) \begin{cases} > 0, & x < 4 \\ \text{undefined}, & x = 4 \\ < 0, & x > 4. \end{cases}$$

56. A differentiable function f has one critical number at $x = 5$. Identify the relative extrema of f at the critical number if $f'(4) = -2.5$ and $f'(6) = 3$.

57. Think About It   The function f is differentiable on the interval $[-1, 1]$. The table shows the values of f' for selected values of x. Sketch the graph of f, approximate the critical numbers, and identify the relative extrema.

x	-1	-0.75	-0.50	-0.25
$f'(x)$	-10	-3.2	-0.5	0.8

x	0	0.25	0.50	0.75	1
$f'(x)$	5.6	3.6	-0.2	-6.7	-20.1

58. Rolling a Ball Bearing   A ball bearing is placed on an inclined plane and begins to roll. The angle of elevation of the plane is θ. The distance (in meters) the ball bearing rolls in t seconds is

$$s(t) = 4.9(\sin\theta)t^2.$$

(a) Determine the speed of the ball bearing after t seconds.

(b) What value of θ will produce the maximum speed at a particular time?

59. Numerical, Graphical, and Analytic Analysis   Consider the functions $f(x) = x$ and $g(x) = \sin x$ on the interval $(0, \pi)$.

(a) Complete the table and make a conjecture about which is the greater function on the interval $(0, \pi)$.

x	0.5	1	1.5	2	2.5	3
$f(x)$						
$g(x)$						

(b) Use a graphing utility to graph the functions and use the graphs to make a conjecture about which is the greater function on the interval $(0, \pi)$.

(c) Prove that $f(x) > g(x)$ on the interval $(0, \pi)$. [*Hint:* Show that $h'(x) > 0$ where $h = f - g$.]

60. Numerical, Graphical, and Analytic Analysis   The concentration C of a chemical in the bloodstream t hours after injection into muscle tissue is

$$C(t) = \frac{3t}{27 + t^3}, \quad t \geq 0.$$

(a) Complete the table and use the table to approximate the time when the concentration is greatest.

t	0	0.5	1	1.5	2	2.5	3
$C(t)$							

(b) Use a graphing utility to graph the concentration function and use the graph to approximate the time when the concentration is greatest.

(c) Use calculus to determine analytically the time when the concentration is greatest.

61. Trachea Contraction   Coughing forces the trachea (windpipe) to contract, which affects the velocity v of the air passing through the trachea. Suppose the velocity of the air during coughing is

$$v = k(R - r)r^2, \quad 0 \leq r < R$$

where k is constant, R is the normal radius of the trachea, and r is the radius during coughing. What radius will produce the maximum air velocity?

62. Profit   The profit P (in dollars) made by a fast-food restaurant selling x hamburgers is

$$P = 2.44x - \frac{x^2}{20{,}000} - 5000, \quad 0 \leq x \leq 35{,}000.$$

Find the open intervals on which P is increasing or decreasing.

63. Power   The electric power P in watts in a direct-current circuit with two resistors R_1 and R_2 connected in series is

$$P = \frac{vR_1R_2}{(R_1 + R_2)^2}$$

where v is the voltage. If v and R_1 are held constant, what resistance R_2 produces maximum power?

64. *Electrical Resistance*   The resistance R of a certain type of resistor is

$$R = \sqrt{0.001T^4 - 4T + 100}$$

where R is measured in ohms and the temperature T is measured in degrees Celsius.

(a) Use a computer algebra system to find dR/dT and the critical number of the function. Determine the minimum resistance for this type of resistor.

(b) Use a graphing utility to graph the function R and use the graph to approximate the minimum resistance for this type of resistor.

65. *Modeling Data*   The number of bankruptcies (in thousands) for the years 1981 through 1998 are as follows.

1981: 360.3; 1982: 367.9; 1983: 374.7; 1984: 344.3;

1985: 364.5; 1986: 477.9; 1987: 561.3; 1988: 594.6;

1989: 643.0; 1990: 725.5; 1991: 880.4; 1992: 972.5;

1993: 918.7; 1994: 845.3; 1995: 858.1; 1996: 1042.1;

1997: 1317.0; 1998: 1411.4

(Source: Administrative Office of the U.S. Courts)

(a) Use the regression capabilities of a graphing utility to find a model of the form

$$B = at^4 + bt^3 + ct^2 + dt + e$$

for the data. (Let $t = 1$ represent 1981.)

(b) Use a graphing utility to plot the data and graph the model.

(c) Analytically find the minimum of the model and compare the result with the actual data.

66. Use a graphing utility to graph $f(x) = 2 \sin 3x + 4 \cos 3x$. Find the maximum value of f. How could you use calculus to estimate the maximum?

Creating Polynomial Functions   In Exercises 67–70, find a polynomial function

$$f(x) = a_n x^n + a_{n-1} x^{n-1} + \cdots + a_2 x^2 + a_1 x + a_0$$

that has only the specified extrema. (a) Determine the minimum degree of the function and give the criteria you used in determining the degree. (b) Using the fact that the coordinates of the extrema are solution points of the function, and that the x-coordinates are critical numbers, determine a system of linear equations whose solution yields the coefficients of the required function. (c) Use a graphing utility to solve the system of equations and determine the function. (d) Use a graphing utility to confirm your result graphically.

67. Relative minimum: $(0, 0)$; Relative maximum: $(2, 2)$

68. Relative minimum: $(0, 0)$; Relative maximum: $(4, 1000)$

69. Relative minima: $(0, 0)$, $(4, 0)$

Relative maximum: $(2, 4)$

70. Relative minimum: $(1, 2)$

Relative maxima: $(-1, 4)$, $(3, 4)$

True or False?   **In Exercises 71–76, determine whether the statement is true or false. If it is false, explain why or give an example that shows it is false.**

71. The sum of two increasing functions is increasing.

72. The product of two increasing functions is increasing.

73. Every nth-degree polynomial has $(n - 1)$ critical numbers.

74. An nth-degree polynomial has at most $(n - 1)$ critical numbers.

75. There is a relative maximum or minimum at each critical number.

76. The relative maxima of the function f are $f(1) = 4$ and $f(3) = 10$. Therefore, f has at least one minimum for some x in the interval $(1, 3)$.

77. Prove the second case of Theorem 3.5.

78. Prove the second case of Theorem 3.6.

79. Let $x > 0$ and $n > 1$ be real numbers. Prove that $(1 + x)^n > 1 + nx$.

SECTION PROJECT   RAINBOWS

Rainbows are formed when light strikes raindrops and is reflected and refracted, as shown in the figure. (This figure shows a cross section of a spherical raindrop.) The Law of Refraction states that $(\sin \alpha)/(\sin \beta) = k$, where $k \approx 1.33$ (for water). The angle of deflection is given by $D = \pi + 2\alpha - 4\beta$.

(a) Sketch the graph of D for $0 \leq \alpha \leq \pi/2$. Use a graphing utility with

$$D = \pi + 2\alpha - 4 \sin^{-1}\left(\frac{1}{k} \sin \alpha\right).$$

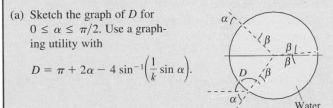

(b) Prove that the minimum angle of deflection occurs when

$$\cos \alpha = \sqrt{\frac{k^2 - 1}{3}}.$$

For water, what is the minimum angle of deflection, D_{min}? (The angle $\pi - D_{min}$ is called the *rainbow angle*.) What value of α produces this minimum angle? (A ray of sunlight that strikes a raindrop at this angle, α, is called a *rainbow ray*.)

FOR FURTHER INFORMATION   For more information about the mathematics of rainbows, see the article "Somewhere Within the Rainbow" by Steven Janke in *The UMAP Journal*. To view this article, go to the website *www.matharticles.com*.

- Determine intervals on which a function is concave upward or concave downward.
- Find any points of inflection of the graph of a function.
- Apply the Second Derivative Test to find relative extrema of a function.

Concavity

You have already seen that locating the intervals in which a function f increases or decreases helps to describe its graph. In this section, you will see how locating the intervals in which f' increases or decreases can be used to determine where the graph of f is *curving upward* or *curving downward*.

Definition of Concavity

Let f be differentiable on an open interval I. The graph of f is **concave upward** on I if f' is increasing on the interval and **concave downward** on I if f' is decreasing on the interval.

The following graphical interpretation of concavity is useful. (See Appendix B for a proof of these results.)

1. Let f be differentiable on an open interval I. If the graph of f is concave *upward* on I, then the graph of f lies *above* all of its tangent lines on I. (See Figure 3.24a.)

2. Let f be differentiable on an open interval I. If the graph of f is concave *downward* on I, then the graph of f lies *below* all of its tangent lines on I. (See Figure 3.24b.)

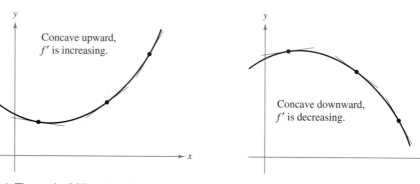

(a) The graph of f lies above its tangent lines.

(b) The graph of f lies below its tangent lines.

Figure 3.24

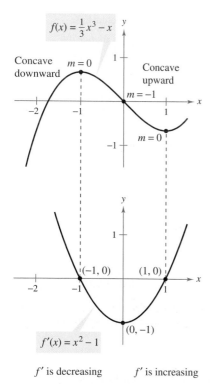

The concavity of f is related to the slope of the derivative.

Figure 3.25

To find the open intervals on which the graph of a function f is concave upward or downward, you need to find the intervals on which f' is increasing or decreasing. For instance, the graph of

$$f(x) = \tfrac{1}{3}x^3 - x$$

is concave downward on the open interval $(-\infty, 0)$ because $f'(x) = x^2 - 1$ is decreasing there. (See Figure 3.25.) Similarly, the graph of f is concave upward on the interval $(0, \infty)$ because f' is increasing on $(0, \infty)$.

The following theorem shows how to use the *second* derivative of a function f to determine intervals on which the graph of f is concave upward or downward. A proof of this theorem follows directly from Theorem 3.5 and the definition of concavity.

THEOREM 3.7   Test for Concavity

Let f be a function whose second derivative exists on an open interval I.

1. If $f''(x) > 0$ for all x in I, then the graph of f is concave upward in I.
2. If $f''(x) < 0$ for all x in I, then the graph of f is concave downward in I.

NOTE   A third case of Theorem 3.7 could be that if $f''(x) = 0$ for all x in I, then f is linear. Note, however, that concavity is not defined for a line. In other words, a straight line is neither concave upward nor concave downward.

To apply Theorem 3.7, locate the x-values at which $f''(x) = 0$ or f'' does not exist. Second, use these x-values to determine test intervals. Finally, test the sign of $f''(x)$ in each of the test intervals.

Example 1   **Determining Concavity**

Determine the open intervals on which the graph of

$$f(x) = \frac{6}{x^2 + 3}$$

is concave upward or downward.

Solution   Begin by observing that f is continuous on the entire real line. Next, find the second derivative of f.

$$f(x) = 6(x^2 + 3)^{-1} \qquad \text{Rewrite original function.}$$

$$f'(x) = (-6)(x^2 + 3)^{-2}(2x) \qquad \text{Differentiate.}$$

$$= \frac{-12x}{(x^2 + 3)^2} \qquad \text{First derivative}$$

$$f''(x) = \frac{(x^2 + 3)^2(-12) - (-12x)(2)(x^2 + 3)(2x)}{(x^2 + 3)^4} \qquad \text{Differentiate.}$$

$$= \frac{36(x^2 - 1)}{(x^2 + 3)^3} \qquad \text{Second derivative}$$

Because $f''(x) = 0$ when $x = \pm 1$ and f'' is defined on the entire real line, you should test f'' in the intervals $(-\infty, -1)$, $(-1, 1)$, and $(1, \infty)$. The results are shown in the table and in Figure 3.26.

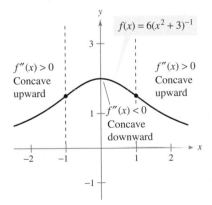

$f(x) = 6(x^2 + 3)^{-1}$

$f''(x) > 0$ Concave upward

$f''(x) > 0$ Concave upward

$f''(x) < 0$ Concave downward

From the sign of f'' you can determine the concavity of the graph of f.
Figure 3.26

Interval	$-\infty < x < -1$	$-1 < x < 1$	$1 < x < \infty$
Test value	$x = -2$	$x = 0$	$x = 2$
Sign of $f''(x)$	$f''(-2) > 0$	$f''(0) < 0$	$f''(2) > 0$
Conclusion	Concave upward	Concave downward	Concave upward

The function given in Example 1 is continuous on the entire real line. If there are *x*-values at which the function is not continuous, these values should be used along with the points at which $f''(x) = 0$ or points at which f is not differentiable to form the test intervals.

Example 2   Determining Concavity

Determine the open intervals in which the graph of $f(x) = \dfrac{x^2 + 1}{x^2 - 4}$ is concave upward or downward.

Solution   Differentiating twice produces the following.

$$f(x) = \frac{x^2 + 1}{x^2 - 4} \qquad\qquad \text{Write original equation.}$$

$$f'(x) = \frac{(x^2 - 4)(2x) - (x^2 + 1)(2x)}{(x^2 - 4)^2} \qquad\qquad \text{Differentiate.}$$

$$= \frac{-10x}{(x^2 - 4)^2} \qquad\qquad \text{First derivative}$$

$$f''(x) = \frac{(x^2 - 4)^2(-10) - (-10x)(2)(x^2 - 4)(2x)}{(x^2 - 4)^4} \qquad\qquad \text{Differentiate.}$$

$$= \frac{10(3x^2 + 4)}{(x^2 - 4)^3} \qquad\qquad \text{Second derivative}$$

There are no points at which $f''(x) = 0$, but at $x = \pm 2$ the function f is not continuous, so you test for concavity in the intervals $(-\infty, -2)$, $(-2, 2)$, and $(2, \infty)$, as shown in the table. The graph of f is shown in Figure 3.27.

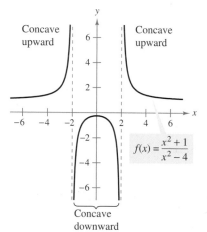

Concave upward

Concave upward

$f(x) = \dfrac{x^2 + 1}{x^2 - 4}$

Concave downward

Figure 3.27

Interval	$-\infty < x < -2$	$-2 < x < 2$	$2 < x < \infty$
Test value	$x = -3$	$x = 0$	$x = 3$
Sign of $f''(x)$	$f''(-3) > 0$	$f''(0) < 0$	$f''(3) > 0$
Conclusion	Concave upward	Concave downward	Concave upward

Points of Inflection

The graph in Figure 3.26 has two points at which the concavity changes. If the tangent line to the graph exists at such a point, that point is a **point of inflection.** Three types of points of inflection are shown in Figure 3.28. Note that a graph crosses its tangent line at a point of inflection.

NOTE: The definition of *point of inflection* given in this book requires that the tangent line exists at the point of inflection. Some books do not require this. For instance, we do not consider the function

$$f(x) = \begin{cases} x^3, & x < 0 \\ x^2 + 2x, & x \geq 0 \end{cases}$$

to have a point of inflection at the origin, even though the concavity of the graph changes from concave downward to concave upward.

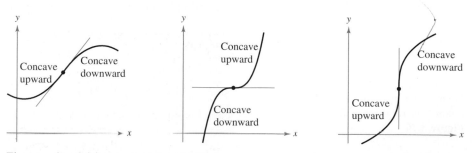

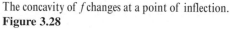

The concavity of *f* changes at a point of inflection.
Figure 3.28

To locate *possible* points of inflection, you need only determine the values of x for which $f''(x) = 0$ or for which f is not differentiable. This is similar to the procedure for locating relative extrema of f.

THEOREM 3.8   Points of Inflection

If $(c, f(c))$ is a point of inflection of the graph of f, then either $f''(c) = 0$ or f is not differentiable at $x = c$.

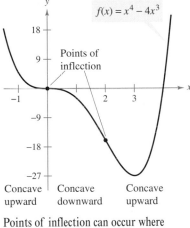

Points of inflection can occur where $f''(x) = 0$ or f'' does not exist.
Figure 3.29

Example 3   Finding Points of Inflection

Determine the points of inflection and discuss the concavity of the graph of

$$f(x) = x^4 - 4x^3.$$

Solution   Differentiating twice produces the following.

$f(x) = x^4 - 4x^3$ Write original function.

$f'(x) = 4x^3 - 12x^2$ Find first derivative.

$f''(x) = 12x^2 - 24x = 12x(x - 2)$ Find second derivative.

Setting $f''(x) = 0$, you can determine that the possible points of inflection occur at $x = 0$ and $x = 2$. By testing the intervals determined by these x-values, you can conclude that they both yield points of inflection. A summary of this testing is shown in the table, and the graph of f is shown in Figure 3.29.

Interval	$-\infty < x < 0$	$0 < x < 2$	$2 < x < \infty$
Test value	$x = -1$	$x = 1$	$x = 3$
Sign of $f''(x)$	$f''(-1) > 0$	$f''(1) < 0$	$f''(3) > 0$
Conclusion	Concave upward	Concave downward	Concave upward

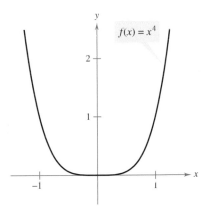

$f''(0) = 0$, but $(0, 0)$ is not a point of inflection.
Figure 3.30

The converse of Theorem 3.8 is not generally true. That is, it is possible for the second derivative to be 0 at a point that is *not* a point of inflection. For instance, the graph of $f(x) = x^4$ is shown in Figure 3.30. The second derivative is 0 when $x = 0$, but the point $(0, 0)$ is not a point of inflection because the graph of f is concave upward in both intervals $-\infty < x < 0$ and $0 < x < \infty$.

EXPLORATION

Consider a general cubic function of the form

$$f(x) = ax^3 + bx^2 + cx + d.$$

You know that the value of d has a bearing on the location of the graph but has no bearing on the value of the first derivative at given values of x. Graphically, this is true because changes in the value of d shift the graph up or down but do not change its basic shape. Use a graphing utility to graph several cubics with different values of c. Then give a graphical explanation of why changes in c do not affect the values of the second derivative.

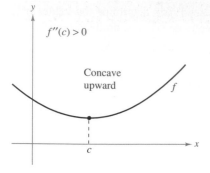

If $f'(c) = 0$ and $f''(c) > 0$, $f(c)$ is a relative minimum.

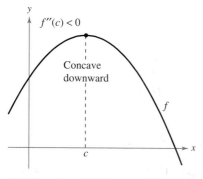

If $f'(c) = 0$ and $f''(c) < 0$, $f(c)$ is a relative maximum.
Figure 3.31

The Second Derivative Test

In addition to testing for concavity, the second derivative can be used to perform a simple test for relative maxima and minima. The test is based on the fact that if the graph of a function f is concave upward on an open interval containing c, and $f'(c) = 0, f(c)$ must be a relative minimum of f. Similarly, if the graph of a function f is concave downward on an open interval containing c, and $f'(c) = 0, f(c)$ must be a relative maximum of f (see Figure 3.31).

THEOREM 3.9   Second Derivative Test

Let f be a function such that $f'(c) = 0$ and the second derivative of f exists on an open interval containing c.

1. If $f''(c) > 0$, then $f(c)$ is a relative minimum.
2. If $f''(c) < 0$, then $f(c)$ is a relative maximum.

If $f''(c) = 0$, the test fails. In such cases, you can use the First Derivative Test.

Proof   If $f'(c) = 0$ and $f''(c) > 0$, there exists an open interval I containing c for which

$$\frac{f'(x) - f'(c)}{x - c} = \frac{f'(x)}{x - c} > 0$$

for all $x \neq c$ in I. If $x < c$, then $x - c < 0$ and $f'(x) < 0$. Also, if $x > c$, then $x - c > 0$ and $f'(x) > 0$. So, $f'(x)$ changes from negative to positive at c, and the First Derivative Test implies that $f(c)$ is a relative minimum. A proof of the second case is left to you.

Example 4   **Using the Second Derivative Test**

Find the relative extrema for $f(x) = -3x^5 + 5x^3$.

Solution   Begin by finding the critical numbers of f.

$$f'(x) = -15x^4 + 15x^2 = 15x^2(1 - x^2) = 0 \qquad \text{Set } f'(x) \text{ equal to 0.}$$
$$x = -1, 0, 1 \qquad \text{Critical numbers}$$

Using

$$f''(x) = -60x^3 + 30x = 30(-2x^3 + x),$$

you can apply the Second Derivative Test as follows.

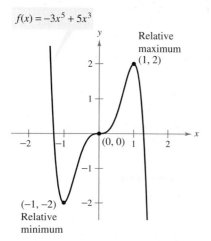

$f(x) = -3x^5 + 5x^3$

$(0, 0)$ is neither a relative minimum nor a relative maximum.
Figure 3.32

Point	$(-1, -2)$	$(1, 2)$	$(0, 0)$
Sign of $f''(x)$	$f''(-1) > 0$	$f''(1) < 0$	$f''(0) = 0$
Conclusion	Relative minimum	Relative maximum	Test fails

Because the Second Derivative Test fails at $(0, 0)$, you can use the First Derivative Test and observe that f increases to the left and right of $x = 0$. So, $(0, 0)$ is neither a relative minimum nor a relative maximum (even though the graph has a horizontal tangent line at this point). The graph of f is shown in Figure 3.32.

EXERCISES FOR SECTION 3.4

In Exercises 1–10, determine the open intervals on which the graph is concave upward or concave downward.

1. $y = x^2 - x - 2$

2. $y = -x^3 + 3x^2 - 2$

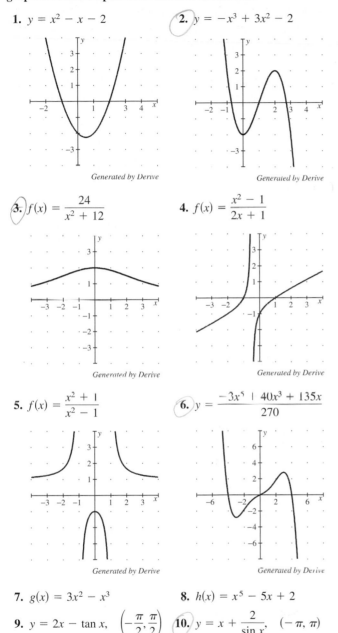

Generated by Derive

Generated by Derive

3. $f(x) = \dfrac{24}{x^2 + 12}$

4. $f(x) = \dfrac{x^2 - 1}{2x + 1}$

Generated by Derive

Generated by Derive

5. $f(x) = \dfrac{x^2 + 1}{x^2 - 1}$

6. $y = \dfrac{-3x^5 + 40x^3 + 135x}{270}$

Generated by Derive

Generated by Derive

7. $g(x) = 3x^2 - x^3$

8. $h(x) = x^5 - 5x + 2$

9. $y = 2x - \tan x, \quad \left(-\dfrac{\pi}{2}, \dfrac{\pi}{2}\right)$

10. $y = x + \dfrac{2}{\sin x}, \quad (-\pi, \pi)$

In Exercises 11–26, find the points of inflection and discuss the concavity of the graph of the function.

11. $f(x) = x^3 - 6x^2 + 12x$

12. $f(x) = 2x^3 - 3x^2 - 12x + 5$

13. $f(x) = \frac{1}{4}x^4 - 2x^2$

14. $f(x) = 2x^4 - 8x + 3$

15. $f(x) = x(x - 4)^3$

16. $f(x) = x^3(x - 4)$

17. $f(x) = x\sqrt{x + 3}$

18. $f(x) = x\sqrt{x + 1}$

19. $f(x) = \dfrac{x}{x^2 + 1}$

20. $f(x) = \dfrac{x + 1}{\sqrt{x}}$

21. $f(x) = \sin\dfrac{x}{2}, \quad [0, 4\pi]$

22. $f(x) = 2\csc\dfrac{3x}{2}, \quad (0, 2\pi)$

23. $f(x) = \sec\left(x - \dfrac{\pi}{2}\right), \quad (0, 4\pi)$

24. $f(x) = \sin x + \cos x, \quad [0, 2\pi]$

25. $f(x) = 2\sin x + \sin 2x, \quad [0, 2\pi]$

26. $f(x) = x + 2\cos x, \quad [0, 2\pi]$

In Exercises 27–40, find all relative extrema. Use the Second Derivative Test where applicable.

27. $f(x) = x^4 - 4x^3 + 2$

28. $f(x) = x^2 + 3x - 8$

29. $f(x) = (x - 5)^2$

30. $f(x) = -(x - 5)^2$

31. $f(x) = x^3 - 3x^2 + 3$

32. $f(x) = x^3 - 9x^2 + 27x$

33. $g(x) = x^2(6 - x)^3$

34. $g(x) = -\frac{1}{8}(x + 2)^2(x - 4)^2$

35. $f(x) = x^{2/3} - 3$

36. $f(x) = \sqrt{x^2 + 1}$

37. $f(x) = x + \dfrac{4}{x}$

38. $f(x) = \dfrac{x}{x - 1}$

39. $f(x) = \cos x - x, \quad [0, 4\pi]$

40. $f(x) = 2\sin x + \cos 2x, \quad [0, 2\pi]$

In Exercises 41–44, use a computer algebra system to analyze the function over the indicated interval. (a) Find the first and second derivatives of the function. (b) Find any relative extrema and points of inflection. (c) Graph f, f', and f'' on the same set of coordinate axes and state the relationship between the behavior of f and the signs of f' and f''.

41. $f(x) = 0.2x^2(x - 3)^3, \quad [-1, 4]$

42. $f(x) = x^2\sqrt{6 - x^2}, \quad \left[-\sqrt{6}, \sqrt{6}\right]$

43. $f(x) = \sin x - \frac{1}{3}\sin 3x + \frac{1}{5}\sin 5x, \quad [0, \pi]$

44. $f(x) = \sqrt{2x}\sin x, \quad [0, 2\pi]$

Getting at the Concept

45. Consider a function f such that f' is increasing. Sketch graphs of f for (a) $f' < 0$ and (b) $f' > 0$.

46. Consider a function f such that f' is decreasing. Sketch graphs of f for (a) $f' < 0$ and (b) $f' > 0$.

47. Sketch the graph of a function f that does *not* have a point of inflection at $(c, f(c))$ even though $f''(c) = 0$.

48. S represents weekly sales of a product. What can be said of S' and S'' for each of the following?

(a) The rate of change of sales is increasing.

(b) Sales are increasing at a slower rate.

(c) The rate of change of sales is constant.

(d) Sales are steady.

(e) Sales are declining, but at a slower rate.

(f) Sales have bottomed out and have started to rise.

In Exercises 49–52, graph f, f', and f'' on the same set of coordinate axes. To print an enlarged copy of the graph, go to the website www.mathgraphs.com.

49. **50.**

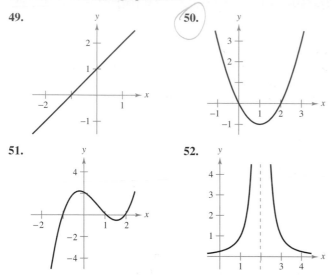

51. **52.**

Think About It **In Exercises 53–56, sketch the graph of a function f having the indicated characteristics.**

53. $f(2) = f(4) = 0$

$f(3)$ is defined.

$f'(x) < 0$ if $x < 3$

$f'(3)$ does not exist.

$f'(x) > 0$ if $x > 3$

$f''(x) < 0$, $x \neq 3$

54. $f(0) = f(2) = 0$

$f'(x) > 0$ if $x < 1$

$f'(1) = 0$

$f'(x) < 0$ if $x > 1$

$f''(x) < 0$

55. $f(2) = f(4) = 0$

$f'(x) > 0$ if $x < 3$

$f'(3)$ does not exist.

$f'(x) < 0$ if $x > 3$

$f''(x) > 0$, $x \neq 3$

56. $f(0) = f(2) = 0$

$f'(x) < 0$ if $x < 1$

$f'(1) = 0$

$f'(x) > 0$ if $x > 1$

$f''(x) > 0$

57. *Think About It* The figure shows the graph of f''. Sketch a graph of f. (The answer is not unique.)

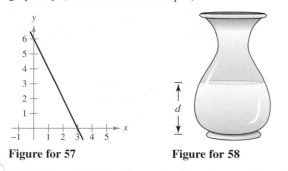

Figure for 57   Figure for 58

58. *Think About It* Water is running into the vase shown in the figure at a constant rate.

(a) Graph the depth d of water in the vase as a function of time.

(b) Does the function have any extrema? Explain.

(c) Interpret the inflection points of the graph of d.

59. *Conjecture* Consider the function $f(x) = (x - 2)^n$.

(a) Use a graphing utility to graph f for $n = 1, 2, 3$, and 4. Use the graphs to make a conjecture about the relationship between n and any inflection points of the graph of f.

(b) Verify your conjecture in part (a).

60. (a) Graph $f(x) = \sqrt[3]{x}$ and identify the inflection point.

(b) Does $f''(x)$ exist at the inflection point? Explain.

In Exercises 61 and 62, find a, b, c, and d such that the cubic $f(x) = ax^3 + bx^2 + cx + d$ satisfies the indicated conditions.

61. Relative maximum: $(3, 3)$

Relative minimum: $(5, 1)$

Inflection point: $(4, 2)$

62. Relative maximum: $(2, 4)$

Relative minimum: $(4, 2)$

Inflection point: $(3, 3)$

63. *Aircraft Glide Path* A small aircraft starts its descent from an altitude of 1 mile, 4 miles west of the runway (see figure).

(a) Find the cubic $f(x) = ax^3 + bx^2 + cx + d$ on the interval $[-4, 0]$ that describes a smooth glide path for the landing.

(b) The function in part (a) models the glide path of the plane. When would the plane be descending at the most rapid rate?

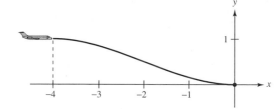

FOR FURTHER INFORMATION For more information on this type of modeling, see the article "How Not to Land at Lake Tahoe!" by Richard Barshinger in *The American Mathematical Monthly*. To view this article, go to the website www.matharticles.com.

64. *Highway Design* A section of highway connecting two hillsides with grades of 6% and 4% is to be built between two points that are separated by a horizontal distance of 2000 feet (see figure). At the point where the two hillsides come together, there is a 50-foot difference in elevation.

(a) Design a section of highway connecting the hillsides modeled by the function $f(x) = ax^3 + bx^2 + cx + d$ $(-1000 \leq x \leq 1000)$. At the points A and B, the slope of the model must match the grade of the hillside.

(b) Use a graphing utility to graph the model.

(c) Use a graphing utility to graph the derivative of the model.

(d) Determine the grade at the steepest part of the transitional section of the highway.

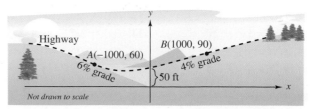

65. Beam Deflection   The deflection D of a particular beam of length L is

$$D = 2x^4 - 5Lx^3 + 3L^2x^2$$

where x is the distance from one end of the beam. Find the value of x that yields the maximum deflection.

66. Specific Gravity   A model for the specific gravity of water S is

$$S = \frac{5.755}{10^8}T^3 - \frac{8.521}{10^6}T^2 + \frac{6.540}{10^5}T + 0.99987, \ 0 < T < 25$$

where T is the water temperature in degrees Celsius.

(a) Use a computer algebra system to find the coordinates of the maximum value of the function.

(b) Sketch a graph of the function over the specified domain. (Use a setting in which $0.996 \le S \le 1.001$.)

(c) Estimate the specific gravity of water when $T = 20°$.

67. Average Cost   A manufacturer has determined that the total cost C of operating a factory is

$$C = 0.5x^2 + 15x + 5000$$

where x is the number of units produced. At what level of production will the average cost per unit be minimized? (The average cost per unit is C/x.)

68. Inventory Cost   The total cost C for ordering and storing x units is

$$C = 2x + \frac{300,000}{x}.$$

What order size will produce a minimum cost?

69. Sales Growth   The annual sales S of a new product is given by

$$S = \frac{5000t^2}{8 + t^2}$$

where t is time in years. Find the time when sales are increasing at the greatest rate.

70. Modeling Data   The average typing speed S of a typing student after t weeks of lessons is shown in the table.

t	5	10	15	20	25	30
S	38	56	79	90	93	94

A model for the data is $S = \dfrac{100t^2}{65 + t^2}$, $t > 0$.

(a) Use a graphing utility to plot the data and graph the model.

(b) Use the second derivative to determine the concavity of S. Compare the result with the graph in part (a).

(c) What is the sign of the first derivative for $t > 0$? Combining this information with the concavity of the model, what inferences can be made about the typing speed as t increases?

Linear and Quadratic Approximations   In Exercises 71–74, use a graphing utility to graph the function. Then graph the linear and quadratic approximations

$$P_1(x) = f(a) + f'(a)(x - a)$$

and

$$P_2(x) = f(a) + f'(a)(x - a) + \tfrac{1}{2}f''(a)(x - a)^2$$

in the same viewing window. Compare the values of f, P_1, and P_2 and their first derivatives at $x = a$. How do the approximations change as you move farther away from $x = a$?

Function	Value of a
71. $f(x) = 2(\sin x + \cos x)$	$a = \dfrac{\pi}{4}$
72. $f(x) = 2(\sin x + \cos x)$	$a = 0$
73. $f(x) = \sqrt{1 - x}$	$a = 0$
74. $f(x) = \dfrac{\sqrt{x}}{x - 1}$	$a = 2$

75. Use a graphing utility to graph $y = x \sin(1/x)$. Show that the graph is concave downward to the right of $x = 1/\pi$.

76. Show that the point of inflection of $f(x) = x(x - 6)^2$ lies midway between the relative extrema of f.

77. Prove that every cubic function with three distinct real zeros has a point of inflection whose x-coordinate is the average of the three zeros.

78. Show that the cubic polynomial $p(x) = ax^3 + bx^2 + cx + d$ has exactly one point of inflection (x_0, y_0), where

$$x_0 = \frac{-b}{3a} \quad \text{and} \quad y_0 - \frac{2b^3}{27a^2} - \frac{bc}{3a} + d.$$

Use this formula to find the point of inflection of $p(x) = x^3 - 3x^2 + 2$.

True or False?   In Exercises 79–84, determine whether the statement is true or false. If it is false, explain why or give an example that shows it is false.

79. The graph of every cubic polynomial has precisely one point of inflection.

80. The graph of $f(x) = 1/x$ is concave downward for $x < 0$ and concave upward for $x > 0$, and thus it has a point of inflection at $x = 0$.

81. The maximum value of $y = 3\sin x + 2\cos x$ is 5.

82. The maximum slope of the graph of $y = \sin(bx)$ is b.

83. If $f'(c) > 0$, then f is concave upward at $x = c$.

84. If $f''(2) = 0$, then the graph of f must have a point of inflection at $x = 2$.

- Determine (finite) limits at infinity.
- Determine the horizontal asymptotes, if any, of the graph of a function.
- Determine infinite limits at infinity.

Limits at Infinity

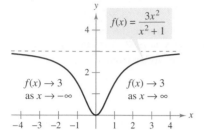

The limit of $f(x)$ as x approaches $-\infty$ or ∞ is 3.
Figure 3.33

This section discusses the "end behavior" of a function on an *infinite* interval. Consider the graph of

$$f(x) = \frac{3x^2}{x^2 + 1}$$

as shown in Figure 3.33. Graphically, you can see that the values of $f(x)$ appear to approach 3 as x increases without bound or decreases without bound. You can come to the same conclusions numerically, as shown in the table.

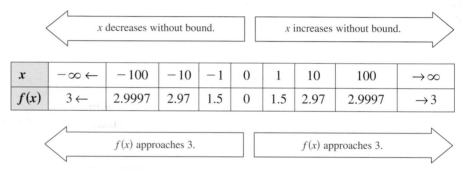

x	$-\infty \leftarrow$	-100	-10	-1	0	1	10	100	$\rightarrow \infty$
$f(x)$	$3 \leftarrow$	2.9997	2.97	1.5	0	1.5	2.97	2.9997	$\rightarrow 3$

The table suggests that the value of $f(x)$ approaches 3 as x increases without bound $(x \rightarrow \infty)$. Similarly, $f(x)$ approaches 3 as x decreases without bound $(x \rightarrow -\infty)$. These **limits at infinity** are denoted by

$$\lim_{x \to -\infty} f(x) = 3 \qquad \text{Limit at negative infinity}$$

and

$$\lim_{x \to \infty} f(x) = 3. \qquad \text{Limit at positive infinity}$$

NOTE By writing $\lim_{x \to -\infty} f(x) = L$ or $\lim_{x \to \infty} f(x) = L$, we mean that the limit exists *and* the limit is equal to L.

To say that a statement is true as x increases *without bound* means that for some (large) real number M, the statement is true for *all* x in the interval $\{x: x > M\}$. The following definition uses this concept.

Definition of Limits at Infinity

Let L be a real number.

1. The statement $\lim_{x \to \infty} f(x) = L$ means that for each $\varepsilon > 0$ there exists an $M > 0$ such that $|f(x) - L| < \varepsilon$ whenever $x > M$.
2. The statement $\lim_{x \to -\infty} f(x) = L$ means that for each $\varepsilon > 0$ there exists an $N < 0$ such that $|f(x) - L| < \varepsilon$ whenever $x < N$.

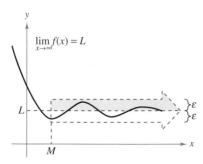

$f(x)$ is within ε units of L as $x \rightarrow \infty$.
Figure 3.34

The definition of a limit at infinity is illustrated in Figure 3.34. In this figure, note that for a given positive number ε there exists a positive number M such that, for $x > M$, the graph of f will lie between the horizontal lines given by $y = L + \varepsilon$ and $y = L - \varepsilon$.

Use a graphing utility to sketch the graph of

$$f(x) = \frac{2x^2 + 4x - 6}{3x^2 + 2x - 16}.$$

Describe all the important features of the graph. Can you find a single viewing window that shows all of these features clearly? Explain your reasoning.

What are the horizontal asymptotes of the graph? How far to the right do you have to move on the graph so that the graph is within 0.001 unit of its horizontal asymptote? Explain your reasoning.

Horizontal Asymptotes

In Figure 3.34, the graph of f approaches the line $y = L$ as x increases without bound. We call the line $y = L$ a **horizontal asymptote** of the graph of f.

Definition of a Horizontal Asymptote

The line $y = L$ is a **horizontal asymptote** of the graph of f if

$$\lim_{x \to -\infty} f(x) = L \quad \text{or} \quad \lim_{x \to \infty} f(x) = L.$$

Note that from this definition, it follows that the graph of a *function* of x can have at most two horizontal asymptotes—one to the right and one to the left.

Limits at infinity have many of the same properties of limits discussed in Section 1.3. For example, if $\lim_{x \to \infty} f(x)$ and $\lim_{x \to \infty} g(x)$ both exist, then

$$\lim_{x \to \infty} [f(x) + g(x)] = \lim_{x \to \infty} f(x) + \lim_{x \to \infty} g(x)$$

and

$$\lim_{x \to \infty} [f(x)g(x)] = \left[\lim_{x \to \infty} f(x)\right]\left[\lim_{x \to \infty} g(x)\right].$$

Similar properties hold for limits at $-\infty$.

When evaluating limits at infinity, the following theorem is helpful. (A proof of this theorem is given in Appendix B.)

THEOREM 3.10   Limits at Infinity

If r is a positive rational number and c is any real number, then

$$\lim_{x \to \infty} \frac{c}{x^r} = 0.$$

Furthermore, if x^r is defined when $x < 0$, then

$$\lim_{x \to -\infty} \frac{c}{x^r} = 0.$$

Example 1   **Evaluating a Limit at Infinity**

Find the limit: $\lim_{x \to \infty} \left(5 - \frac{2}{x^2}\right)$.

Solution   Using Theorem 3.10, you can write the following.

$$\lim_{x \to \infty} \left(5 - \frac{2}{x^2}\right) = \lim_{x \to \infty} 5 - \lim_{x \to \infty} \frac{2}{x^2} \qquad \text{Property of limits}$$

$$= 5 - 0$$

$$= 5$$

Example 2   **Evaluating a Limit at Infinity**

Find the limit: $\displaystyle\lim_{x\to\infty}\frac{2x-1}{x+1}$.

Solution   Note that both the numerator and the denominator approach infinity as x approaches infinity.

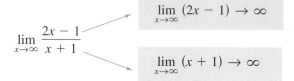

This results in $\dfrac{\infty}{\infty}$, an **indeterminate form.** To resolve this problem, you can divide both the numerator and the denominator by x. After dividing, the limit may be evaluated as follows.

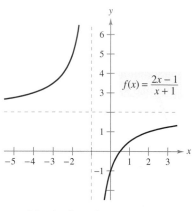

$y = 2$ is a horizontal asymptote.
Figure 3.35

NOTE   When you encounter an indeterminate form such as the one in Example 2, we suggest dividing the numerator and denominator by the highest power of x in the *denominator*.

$$\lim_{x\to\infty}\frac{2x-1}{x+1}=\lim_{x\to\infty}\frac{\dfrac{2x-1}{x}}{\dfrac{x+1}{x}}\qquad\text{Divide numerator and denominator by } x.$$

$$=\lim_{x\to\infty}\frac{2-\dfrac{1}{x}}{1+\dfrac{1}{x}}$$

$$=\frac{\displaystyle\lim_{x\to\infty}2-\lim_{x\to\infty}\dfrac{1}{x}}{\displaystyle\lim_{x\to\infty}1+\lim_{x\to\infty}\dfrac{1}{x}}\qquad\text{Take limits of numerator and denominator.}$$

$$=\frac{2-0}{1+0}\qquad\text{Apply Theorem 3.10.}$$

$$=2$$

So, the line $y = 2$ is a horizontal asymptote to the right. By taking the limit as $x\to-\infty$, you can see that $y = 2$ is also a horizontal asymptote to the left. The graph of the function is shown in Figure 3.35.

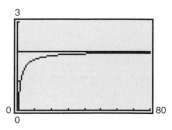

As x increases, the graph of f moves closer and closer to the line $y = 2$.
Figure 3.36

TECHNOLOGY   You can test the reasonableness of the limit found in Example 2 by evaluating $f(x)$ for a few large positive values of x. For instance,

$$f(100)\approx 1.9703,\quad f(1000)\approx 1.9970,\quad\text{and}\quad f(10,000)\approx 1.9997.$$

Another way to test the reasonableness of the limit is to use a graphing utility. For instance, in Figure 3.36, the graph of

$$f(x)=\frac{2x-1}{x+1}$$

is shown with the horizontal line $y = 2$. Note that as x increases, the graph of f moves closer and closer to its horizontal asymptote.

Example 3   **A Comparison of Three Rational Functions**

Find each of the limits.

a. $\lim\limits_{x \to \infty} \dfrac{2x + 5}{3x^2 + 1}$ **b.** $\lim\limits_{x \to \infty} \dfrac{2x^2 + 5}{3x^2 + 1}$ **c.** $\lim\limits_{x \to \infty} \dfrac{2x^3 + 5}{3x^2 + 1}$

Solution   In each case, attempting to evaluate the limit produces the indeterminate form ∞/∞.

a. Divide both the numerator and the denominator by x^2.

$$\lim_{x \to \infty} \frac{2x + 5}{3x^2 + 1} = \lim_{x \to \infty} \frac{(2/x) + (5/x^2)}{3 + (1/x^2)} = \frac{0 + 0}{3 + 0} = \frac{0}{3} = 0$$

b. Divide both the numerator and the denominator by x^2.

$$\lim_{x \to \infty} \frac{2x^2 + 5}{3x^2 + 1} = \lim_{x \to \infty} \frac{2 + (5/x^2)}{3 + (1/x^2)} = \frac{2 + 0}{3 + 0} = \frac{2}{3}$$

c. Divide both the numerator and the denominator by x^2.

$$\lim_{x \to \infty} \frac{2x^3 + 5}{3x^2 + 1} = \lim_{x \to \infty} \frac{2x + (5/x^2)}{3 + (1/x^2)} = \frac{\infty}{3}$$

You can conclude that the limit *does not exist* because the numerator increases without bound while the denominator approaches 3.

MARIA AGNESI (1718–1799)

Agnesi was one of a handful of women to receive credit for significant contributions to mathematics before the twentieth century. In her early twenties, she wrote the first text that included both differential and integral calculus. By age 30, she was an honorary member of the faculty at the University of Bologna.

Guidelines for Finding Limits at Infinity of Rational Functions

1. If the degree of the numerator is *less than* the degree of the denominator, then the limit of the rational function is 0.

2. If the degree of the numerator is *equal to* the degree of the denominator, then the limit of the rational function is the ratio of the leading coefficients.

3. If the degree of the numerator is *greater than* the degree of the denominator, then the limit of the rational function does not exist.

Use these guidelines to check the results in Example 3. These limits seem reasonable when you consider that for large values of x, the highest-power term of the rational function is the most "influential" in determining the limit. For instance, the limit as x approaches infinity of the function

$$f(x) = \frac{1}{x^2 + 1}$$

is 0 because the denominator overpowers the numerator as x increases or decreases without bound, as shown in Figure 3.37.

The function shown in Figure 3.37 is a special case of a type of curve studied by the Italian mathematician Maria Gaetana Agnesi. The general form of this function is

$$f(x) = \frac{8a^3}{x^2 + 4a^2} \qquad \text{Witch of Agnesi}$$

and, through a mistranslation of the Italian word *vertéré*, the curve has come to be known as the Witch of Agnesi. Agnesi's work with this curve first appeared in a comprehensive text on calculus that was published in 1748.

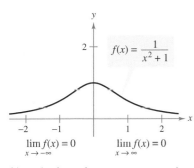

f has a horizontal asymptote at $y = 0$.
Figure 3.37

FOR FURTHER INFORMATION For more information on the contributions of women to mathematics, see the article "Why Women Succeed in Mathematics" by Mona Fabricant, Sylvia Svitak, and Patricia Clark Kenschaft in *Mathematics Teacher*. To view this article, go to the website *www.matharticles.com*.

In Figure 3.37, you can see that the function $f(x) = 1/(x^2 + 1)$ approaches the same horizontal asymptote to the right and to the left. This is always true of rational functions. Functions that are not rational, however, may approach different horizontal asymptotes to the right and to the left. This is demonstrated in Example 4.

Example 4   A Function with Two Horizontal Asymptotes

Determine each of the limits.

a. $\displaystyle \lim_{x \to \infty} \frac{3x - 2}{\sqrt{2x^2 + 1}}$ **b.** $\displaystyle \lim_{x \to -\infty} \frac{3x - 2}{\sqrt{2x^2 + 1}}$

Solution

a. For $x > 0$, you can write $x = \sqrt{x^2}$. So, dividing both the numerator and the denominator by x produces

$$\frac{3x - 2}{\sqrt{2x^2 + 1}} = \frac{\dfrac{3x - 2}{x}}{\dfrac{\sqrt{2x^2 + 1}}{\sqrt{x^2}}} = \frac{3 - \dfrac{2}{x}}{\sqrt{\dfrac{2x^2 + 1}{x^2}}} = \frac{3 - \dfrac{2}{x}}{\sqrt{2 + \dfrac{1}{x^2}}}$$

and you can take the limit as follows.

$$\lim_{x \to \infty} \frac{3x - 2}{\sqrt{2x^2 + 1}} = \lim_{x \to \infty} \frac{3 - \dfrac{2}{x}}{\sqrt{2 + \dfrac{1}{x^2}}} = \frac{3 - 0}{\sqrt{2 + 0}} = \frac{3}{\sqrt{2}}$$

b. For $x < 0$, you can write $x = -\sqrt{x^2}$. So, dividing both the numerator and the denominator by x produces

$$\frac{3x - 2}{\sqrt{2x^2 + 1}} = \frac{\dfrac{3x - 2}{x}}{\dfrac{\sqrt{2x^2 + 1}}{-\sqrt{x^2}}} = \frac{3 - \dfrac{2}{x}}{-\sqrt{\dfrac{2x^2 + 1}{x^2}}} = \frac{3 - \dfrac{2}{x}}{-\sqrt{2 + \dfrac{1}{x^2}}}$$

and you can take the limit as follows.

$$\lim_{x \to -\infty} \frac{3x - 2}{\sqrt{2x^2 + 1}} = \lim_{x \to -\infty} \frac{3 - \dfrac{2}{x}}{-\sqrt{2 + \dfrac{1}{x^2}}} = \frac{3 - 0}{-\sqrt{2 + 0}} = -\frac{3}{\sqrt{2}}$$

The graph of $f(x) = (3x - 2)/\sqrt{2x^2 + 1}$ is shown in Figure 3.38.

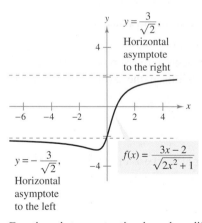

$y = \dfrac{3}{\sqrt{2}}$,
Horizontal asymptote to the right

$y = -\dfrac{3}{\sqrt{2}}$,
Horizontal asymptote to the left

$f(x) = \dfrac{3x - 2}{\sqrt{2x^2 + 1}}$

Functions that are not rational may have different right and left horizontal asymptotes.
Figure 3.38

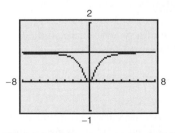

The horizontal asymptote appears to be the line $y = 1$ but is actually the line $y = 2$.
Figure 3.39

TECHNOLOGY PITFALL   If you use a graphing utility to help estimate a limit, be sure that you also confirm the estimate analytically—the pictures shown by a graphing utility can be misleading. For instance, Figure 3.39 shows one view of the graph of

$$y = \frac{2x^3 + 1000x^2 + x}{x^3 + 1000x^2 + x + 1000}.$$

From this view, one could be convinced that the graph has $y = 1$ as a horizontal asymptote. An analytical approach shows that the horizontal asymptote is actually $y = 2$. Confirm this by enlarging the viewing window on the graphing utility.

In Section 1.3 (Example 9), you saw how the Squeeze Theorem can be used to evaluate limits involving trigonometric functions. This theorem is also valid for limits at infinity.

Example 5   **Limits Involving Trigonometric Functions**

Determine each of the limits.

a. $\lim\limits_{x \to \infty} \sin x$ **b.** $\lim\limits_{x \to \infty} \dfrac{\sin x}{x}$

Solution

a. As x approaches infinity, the sine function oscillates between 1 and -1. So, this limit does not exist.

b. Because $-1 \le \sin x \le 1$, it follows that for $x > 0$,

$$-\frac{1}{x} \le \frac{\sin x}{x} \le \frac{1}{x}$$

where $\lim\limits_{x \to \infty} (-1/x) = 0$ and $\lim\limits_{x \to \infty} (1/x) = 0$. Therefore, by the Squeeze Theorem, you can obtain

$$\lim\limits_{x \to \infty} \frac{\sin x}{x} = 0$$

as indicated in Figure 3.40.

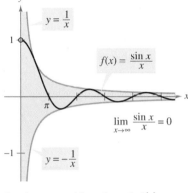

$y = \dfrac{1}{x}$

$f(x) = \dfrac{\sin x}{x}$

$\lim\limits_{x \to \infty} \dfrac{\sin x}{x} = 0$

$y = -\dfrac{1}{x}$

As x increases without bound, $f(x)$ approaches 0.
Figure 3.40

Example 6   **Oxygen Level in a Pond**

Suppose that $f(t)$ measures the level of oxygen in a pond, where $f(t) = 1$ is the normal (unpolluted) level and the time t is measured in weeks. When $t = 0$, organic waste is dumped into the pond, and as the waste material oxidizes, the level of oxygen in the pond is

$$f(t) = \frac{t^2 - t + 1}{t^2 + 1}.$$

What percent of the normal level of oxygen exists in the pond after 1 week? After 2 weeks? After 10 weeks? What is the limit as t approaches infinity?

Solution   When $t = 1, 2,$ and 10, the levels of oxygen are as follows.

$$f(1) = \frac{1^2 - 1 + 1}{1^2 + 1} = \frac{1}{2} = 50\% \qquad \text{1 week}$$

$$f(2) = \frac{2^2 - 2 + 1}{2^2 + 1} = \frac{3}{5} = 60\% \qquad \text{2 weeks}$$

$$f(10) = \frac{10^2 - 10 + 1}{10^2 + 1} = \frac{91}{101} \approx 90.1\% \qquad \text{10 weeks}$$

To take the limit as t approaches infinity, divide the numerator and the denominator by t^2 to obtain

$$\lim\limits_{x \to \infty} \frac{t^2 - t + 1}{t^2 + 1} = \lim\limits_{x \to \infty} \frac{1 - (1/t) + (1/t^2)}{1 + (1/t^2)} = \frac{1 - 0 + 0}{1 + 0} = 1 = 100\%.$$

(See Figure 3.41.)

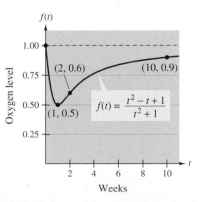

(2, 0.6)   (10, 0.9)

$f(t) = \dfrac{t^2 - t + 1}{t^2 + 1}$

(1, 0.5)

Weeks

The level of oxygen in a pond approaches the normal level of 1 as t approaches ∞.
Figure 3.41

Infinite Limits at Infinity

Many functions do not approach a finite limit as x increases (or decreases) without bound. For instance, no polynomial function has a finite limit at infinity. To describe the behavior of polynomial and other functions at infinity, we use the following definition.

NOTE Determining whether a function has an infinite limit at infinity is useful in analyzing the "end behavior" of its graph. You will see examples of this in Section 3.6 on curve sketching.

Definition of Infinite Limits at Infinity

Let f be a function defined on the interval (a, ∞).

1. The statement $\lim\limits_{x \to \infty} f(x) = \infty$ means that for each positive number M, there is a corresponding number $N > 0$ such that $f(x) > M$ whenever $x > N$.
2. The statement $\lim\limits_{x \to \infty} f(x) = -\infty$ means that for each negative number M, there is a corresponding number $N > 0$ such that $f(x) < M$ whenever $x > N$.

Similar statements can be made about the notations $\lim\limits_{x \to -\infty} f(x) = \infty$ and $\lim\limits_{x \to -\infty} f(x) = -\infty$.

Example 7 **Finding Infinite Limits at Infinity**

Find each limit.

a. $\lim\limits_{x \to \infty} x^3$ **b.** $\lim\limits_{x \to -\infty} x^3$

Solution

a. As x increases without bound, x^3 also increases without bound. So, you can write
$$\lim_{x \to \infty} x^3 = \infty.$$

b. As x decreases without bound, x^3 also decreases without bound. So, you can write
$$\lim_{x \to -\infty} x^3 = -\infty.$$

The graph of $f(x) = x^3$ in Figure 3.42 illustrates these two results. These results agree with the Leading Coefficient Test for polynomial functions as described in Section P.3.

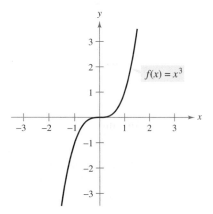

$f(x) = x^3$

Figure 3.42

Example 8 **Finding Infinite Limits at Infinity**

Find each limit.

a. $\lim\limits_{x \to \infty} \dfrac{2x^2 - 4x}{x + 1}$ **b.** $\lim\limits_{x \to -\infty} \dfrac{2x^2 - 4x}{x + 1}$

Solution One way to evaluate these limits is to use long division to rewrite the improper rational function as the sum of a polynomial and a rational function.

a. $\lim\limits_{x \to \infty} \dfrac{2x^2 - 4x}{x + 1} = \lim\limits_{x \to \infty} \left(2x - 6 + \dfrac{6}{x + 1}\right) = \infty$

b. $\lim\limits_{x \to -\infty} \dfrac{2x^2 - 4x}{x + 1} = \lim\limits_{x \to -\infty} \left(2x - 6 + \dfrac{6}{x + 1}\right) = -\infty$

The above statements can be interpreted as saying that as x approaches $\pm\infty$, the function $f(x) = (2x^2 - 4x)/(x + 1)$ behaves like the function $g(x) = 2x - 6$. In Section 3.6, you will see that this is graphically described by saying that the line $y = 2x - 6$ is a slant asymptote of the graph of f, as shown in Figure 3.43.

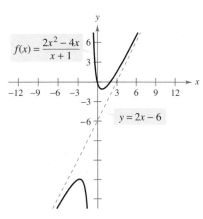

$f(x) = \dfrac{2x^2 - 4x}{x + 1}$

$y = 2x - 6$

Figure 3.43

EXERCISES FOR SECTION 3.5

In Exercises 1–6, match the function with one of the graphs [(a), (b), (c), (d), (e), or (f)] using horizontal asymptotes as an aid.

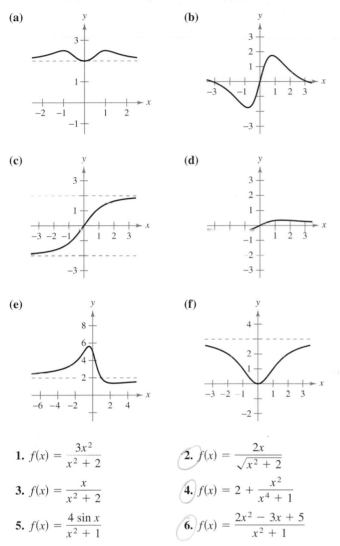

(a)

(b)

(c)

(d)

(e)

(f)

1. $f(x) = \dfrac{3x^2}{x^2 + 2}$

2. $f(x) = \dfrac{2x}{\sqrt{x^2 + 2}}$

3. $f(x) = \dfrac{x}{x^2 + 2}$

4. $f(x) = 2 + \dfrac{x^2}{x^4 + 1}$

5. $f(x) = \dfrac{4 \sin x}{x^2 + 1}$

6. $f(x) = \dfrac{2x^2 - 3x + 5}{x^2 + 1}$

Numerical and Graphical Analysis In Exercises 7–12, use a graphing utility to complete the table and estimate the limit as *x* approaches infinity. Then use a graphing utility to graph the function and estimate the limit graphically.

x	10^0	10^1	10^2	10^3	10^4	10^5	10^6
$f(x)$							

7. $f(x) = \dfrac{4x + 3}{2x - 1}$

8. $f(x) = \dfrac{2x^2}{x + 1}$

9. $f(x) = \dfrac{-6x}{\sqrt{4x^2 + 5}}$

10. $f(x) = \dfrac{8x}{\sqrt{x^2 - 3}}$

11. $f(x) = 5 - \dfrac{1}{x^2 + 1}$

12. $f(x) = 4 + \dfrac{3}{x^2 + 2}$

In Exercises 13 and 14, find $\lim\limits_{x \to \infty} h(x)$, if possible.

13. $f(x) = 5x^3 - 3x^2 + 10$

    (a) $h(x) = \dfrac{f(x)}{x^2}$ (b) $h(x) = \dfrac{f(x)}{x^3}$

    (c) $h(x) = \dfrac{f(x)}{x^4}$

14. $f(x) = 5x^2 - 3x + 7$

    (a) $h(x) = \dfrac{f(x)}{x}$ (b) $h(x) = \dfrac{f(x)}{x^2}$

    (c) $h(x) = \dfrac{f(x)}{x^3}$

In Exercises 15–18, find each of the limits, if possible.

15. (a) $\lim\limits_{x \to \infty} \dfrac{x^2 + 2}{x^3 - 1}$

    (b) $\lim\limits_{x \to \infty} \dfrac{x^2 + 2}{x^2 - 1}$

    (c) $\lim\limits_{x \to \infty} \dfrac{x^2 + 2}{x - 1}$

16. (a) $\lim\limits_{x \to \infty} \dfrac{3 - 2x}{3x^3 - 1}$

    (b) $\lim\limits_{x \to \infty} \dfrac{3 - 2x}{3x - 1}$

    (c) $\lim\limits_{x \to \infty} \dfrac{3 - 2x^2}{3x - 1}$

17. (a) $\lim\limits_{x \to \infty} \dfrac{5 - 2x^{3/2}}{3x^2 - 4}$

    (b) $\lim\limits_{x \to \infty} \dfrac{5 - 2x^{3/2}}{3x^{3/2} - 4}$

    (c) $\lim\limits_{x \to \infty} \dfrac{5 - 2x^{3/2}}{3x - 4}$

18. (a) $\lim\limits_{x \to \infty} \dfrac{5x^{3/2}}{4x^2 + 1}$

    (b) $\lim\limits_{x \to \infty} \dfrac{5x^{3/2}}{4x^{3/2} + 1}$

    (c) $\lim\limits_{x \to \infty} \dfrac{5x^{3/2}}{4\sqrt{x} + 1}$

In Exercises 19–32, find the limit.

19. $\lim\limits_{x \to \infty} \dfrac{2x - 1}{3x + 2}$

20. $\lim\limits_{x \to \infty} \dfrac{3x^3 + 2}{9x^3 - 2x^2 + 7}$

21. $\lim\limits_{x \to \infty} \dfrac{x}{x^2 - 1}$

22. $\lim\limits_{x \to \infty} \left(4 + \dfrac{3}{x}\right)$

23. $\lim\limits_{x \to -\infty} \dfrac{5x^2}{x + 3}$

24. $\lim\limits_{x \to -\infty} \left(\dfrac{1}{2}x - \dfrac{4}{x^2}\right)$

25. $\lim\limits_{x \to -\infty} \dfrac{x}{\sqrt{x^2 - x}}$

26. $\lim\limits_{x \to -\infty} \dfrac{x}{\sqrt{x^2 + 1}}$

27. $\lim\limits_{x \to -\infty} \dfrac{2x + 1}{\sqrt{x^2 - x}}$

28. $\lim\limits_{x \to -\infty} \dfrac{-3x + 1}{\sqrt{x^2 + x}}$

29. $\lim\limits_{x \to \infty} \dfrac{\sin 2x}{x}$

30. $\lim\limits_{x \to \infty} \dfrac{x - \cos x}{x}$

31. $\lim\limits_{x \to \infty} \dfrac{1}{2x + \sin x}$

32. $\lim\limits_{x \to \infty} \cos \dfrac{1}{x}$

In Exercises 33 and 34, use a graphing utility to graph the function and verify that it has two horizontal asymptotes.

33. $f(x) = \dfrac{|x|}{x + 1}$

34. $f(x) = \dfrac{3x}{\sqrt{x^2 + 2}}$

In Exercises 35 and 36, find the limit. (*Hint:* Let $x = 1/t$ and find the limit as $t \to 0^+$.)

35. $\displaystyle \lim_{x \to \infty} x \sin \frac{1}{x}$

36. $\displaystyle \lim_{x \to \infty} x \tan \frac{1}{x}$

In Exercises 37–40, find the limit. (*Hint:* Treat the expression as a fraction whose denominator is 1, and rationalize the numerator.) Use a graphing utility to verify your result.

37. $\displaystyle \lim_{x \to -\infty} \left(x + \sqrt{x^2 + 3} \right)$

38. $\displaystyle \lim_{x \to \infty} \left(2x - \sqrt{4x^2 + 1} \right)$

39. $\displaystyle \lim_{x \to \infty} \left(x - \sqrt{x^2 + x} \right)$

40. $\displaystyle \lim_{x \to -\infty} \left(3x + \sqrt{9x^2 - x} \right)$

Numerical, Graphical, and Analytic Analysis   In Exercises 41–44, use a graphing utility to complete the table and estimate the limit as x approaches infinity. Then use a graphing utility to graph the function and estimate the limit graphically. Finally, find the limit analytically and compare your results with the estimates.

x	10^0	10^1	10^2	10^3	10^4	10^5	10^6
$f(x)$							

41. $f(x) = x - \sqrt{x(x-1)}$

42. $f(x) = x^2 - x\sqrt{x(x-1)}$

43. $f(x) = x \sin \dfrac{1}{2x}$

44. $f(x) = \dfrac{x+1}{x\sqrt{x}}$

Getting at the Concept

45. The graph of a function f is shown below. To print an enlarged copy of the graph, go to the website *www.mathgraphs.com*.

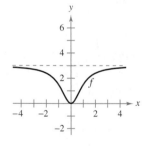

(a) Sketch f'.

(b) Use the graphs to estimate $\displaystyle \lim_{x \to \infty} f(x)$ and $\displaystyle \lim_{x \to \infty} f'(x)$.

(c) Explain the answers you gave in part (b).

46. Sketch a graph of a differentiable function f that satisfies the following conditions and has $x = 2$ as its only critical number.

$f'(x) < 0$ for $x < 2$

$f'(x) > 0$ for $x > 2$

$\displaystyle \lim_{x \to -\infty} f(x) = \lim_{x \to \infty} f(x) = 6$

47. Is it possible to sketch a graph of a function that satisfies the conditions of Exercise 46 and has *no* points of inflection? Explain.

Getting at the Concept (continued)

48. If f is a continuous function such that $\displaystyle \lim_{x \to \infty} f(x) = 5$, find, if possible, $\displaystyle \lim_{x \to -\infty} f(x)$ for each specified condition.

(a) The graph of f is symmetric to the y-axis.

(b) The graph of f is symmetric to the origin.

In Exercises 49–66, sketch the graph of the equation. Look for extrema, intercepts, symmetry, and asymptotes as necessary. Use a graphing utility to verify your result.

49. $y = \dfrac{2+x}{1-x}$

50. $y = \dfrac{x-3}{x-2}$

51. $y = \dfrac{x}{x^2 - 4}$

52. $y = \dfrac{2x}{9 - x^2}$

53. $y = \dfrac{x^2}{x^2 + 9}$

54. $y = \dfrac{x^2}{x^2 - 9}$

55. $y = \dfrac{2x^2}{x^2 - 4}$

56. $y = \dfrac{2x^2}{x^2 + 4}$

57. $xy^2 = 4$

58. $x^2 y = 4$

59. $y = \dfrac{2x}{1-x}$

60. $y = \dfrac{2x}{1 - x^2}$

61. $y = 2 - \dfrac{3}{x^2}$

62. $y = 1 + \dfrac{1}{x}$

63. $y = 3 + \dfrac{2}{x}$

64. $y = 4\left(1 - \dfrac{1}{x^2} \right)$

65. $y = \dfrac{x^3}{\sqrt{x^2 - 4}}$

66. $y = \dfrac{x}{\sqrt{x^2 - 4}}$

In Exercises 67–76, use a computer algebra system to analyze the graph of the function. Label any extrema and/or asymptotes that exist.

67. $f(x) = 5 - \dfrac{1}{x^2}$

68. $f(x) = \dfrac{x^2}{x^2 - 1}$

69. $f(x) = \dfrac{x}{x^2 - 4}$

70. $f(x) = \dfrac{1}{x^2 - x - 2}$

71. $f(x) = \dfrac{x-2}{x^2 - 4x + 3}$

72. $f(x) = \dfrac{x+1}{x^2 + x + 1}$

73. $f(x) = \dfrac{3x}{\sqrt{4x^2 + 1}}$

74. $g(x) = \dfrac{2x}{\sqrt{3x^2 + 1}}$

75. $g(x) = \sin \dfrac{x}{x-2}, \quad 3 < x < \infty$

76. $f(x) = \dfrac{2 \sin 2x}{x}$

In Exercises 77 and 78, (a) use a graphing utility to graph f and g in the same viewing window, (b) verify algebraically that f and g represent the same function, and (c) zoom out sufficiently far so that the graph appears as a line. What equation does this line appear to have? (Note that the points at which the function is not continuous are not readily seen when you zoom out.)

77. $f(x) = \dfrac{x^3 - 3x^2 + 2}{x(x - 3)}$

$g(x) = x + \dfrac{2}{x(x - 3)}$

78. $f(x) = -\dfrac{x^3 - 2x^2 + 2}{2x^2}$

$g(x) = -\dfrac{1}{2}x + 1 - \dfrac{1}{x^2}$

79. *Average Cost*   A business has a cost of $C = 0.5x + 500$ for producing x units. The average cost per unit is

$$\overline{C} = \frac{C}{x}.$$

Find the limit of $\overline{C}$ as x approaches infinity.

80. *Engine Efficiency*   The efficiency of an internal combustion engine is

$$\text{Efficiency (\%)} = 100\left[1 - \frac{1}{(v_1/v_2)^c}\right]$$

where v_1/v_2 is the ratio of the uncompressed gas to the compressed gas and c is a positive constant dependent on the engine design. Find the limit of the efficiency as the compression ratio approaches infinity.

81. A line with slope m passes through the point $(0, 4)$.

(a) Write the distance d between the line and the point $(3, 1)$ as a function of m.

(b) Use a graphing utility to graph the equation in part (a).

(c) Find $\displaystyle\lim_{m\to\infty} d(m)$ and $\displaystyle\lim_{m\to-\infty} d(m)$. Interpret the results geometrically.

82. *Modeling Data*   The table shows the world record times for running one mile, where t represents the year with $t = 0$ corresponding to 1900, and y is the time in minutes and seconds.

t	23	33	45	54
y	4:10.4	4:07.6	4:01.3	3:59.4

t	58	66	79	85	99
y	3:54.5	3:51.3	3:48.9	3:46.3	3:43.1

A model for the data is

$$y = \frac{3.351t^2 + 42.461t - 543.730}{t^2}$$

where the seconds have been changed to a decimal part of a minute.

(a) Use a graphing utility to plot the data and graph the model.

(b) Does there appear to be a limiting time for running one mile? Explain.

83. *Modeling Data*   A heat probe is attached to the heat exchanger of a heating system. The temperature T (degrees Celsius) is recorded t seconds after the furnace is started. The results for the first 2 minutes are recorded in the table.

t	0	15	30	45	60
T	25.2°	36.9°	45.5°	51.4°	56.0°

t	75	90	105	120
T	59.6°	62.0°	64.0°	65.2°

(a) Use the regression capabilities of a graphing utility to find a model of the form $T_1 = at^2 + bt + c$ for the data.

(b) Use a graphing utility to graph T_1.

(c) A rational model for the data is

$$T_2 = \frac{1451 + 86t}{58 + t}.$$

Use a graphing utility to graph the model.

(d) Find $T_1(0)$ and $T_2(0)$.

(e) Find $\displaystyle\lim_{t\to\infty} T_2$.

(f) Interpret the result in part (e) in the context of the problem. Is it possible to do this type of analysis using T_1? Explain.

84. *Modeling Data*   The average typing speed S of a typing student after t weeks of lessons is shown in the table.

t	5	10	15	20	25	30
S	28	56	79	90	93	94

A model for the data is $S = \dfrac{100t^2}{65 + t^2}$, $t > 0$.

(a) Use a graphing utility to plot the data and graph the model.

(b) Does there appear to be a limiting typing speed? Explain.

85. In your own words, state the guidelines for finding the limit of a rational function. Give examples.

86. Prove that if $p(x) = a_n x^n + \cdots + a_1 x + a_0$ and $q(x) = b_m x^m + \cdots + b_1 x + b_0$ $(a_n \neq 0, b_m \neq 0)$, then

$$\lim_{x\to\infty} \frac{p(x)}{q(x)} = \begin{cases} 0, & n < m \\ \dfrac{a_n}{b_m}, & n = m \\ \pm\infty, & n > m. \end{cases}$$

True or False?   In Exercises 87 and 88, determine whether the statement is true or false. If it is false, explain why or give an example that shows it is false.

87. If $f'(x) > 0$ for all real numbers x, then f increases without bound.

88. If $f''(x) < 0$ for all real numbers x, then f decreases without bound.

• Analyze and sketch the graph of a function.

Analyzing the Graph of a Function

It would be difficult to overstate the importance of using graphs in mathematics. Descartes's introduction of analytic geometry contributed significantly to the rapid advances in calculus that began during the mid-seventeenth century. In the words of Lagrange, "As long as algebra and geometry traveled separate paths their advance was slow and their applications limited. But when these two sciences joined company, they drew from each other fresh vitality and thenceforth marched on at a rapid pace toward perfection."

So far, you have studied several concepts that are useful in analyzing the graph of a function.

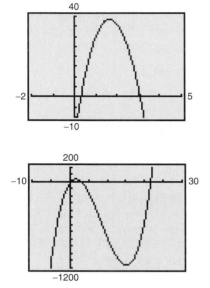

Different viewing windows for the same graph
Figure 3.44

• x-intercepts and y-intercepts	(Section P.1)
• Symmetry	(Section P.1)
• Domain and range	(Section P.3)
• Continuity	(Section 1.4)
• Vertical asymptotes	(Section 1.5)
• Differentiability	(Section 2.1)
• Relative extrema	(Section 3.1)
• Concavity	(Section 3.4)
• Points of inflection	(Section 3.4)
• Horizontal asymptotes	(Section 3.5)
• Infinite limits at infinity	(Section 3.5)

When you are sketching the graph of a function, either by hand or with a graphing utility, remember that normally you cannot show the *entire* graph. The decision as to which part of the graph you choose to show is often crucial. For instance, which of the viewing windows in Figure 3.44 better represents the graph of

$$f(x) = x^3 - 25x^2 + 74x - 20?$$

By seeing both views, it is clear that the second viewing window gives a more complete representation of the graph. But would a third viewing window reveal other interesting portions of the graph? To answer this, you need to use calculus to interpret the first and second derivatives. Here are some guidelines for determining a good viewing window for the graph of a function.

Guidelines for Analyzing the Graph of a Function

1. Determine the domain and range of the function.
2. Determine the intercepts, asymptotes, and symmetry of the graph.
3. Locate the x-values for which $f'(x)$ and $f''(x)$ are either zero or do not exist. Use the results to determine relative extrema and points of inflection.

NOTE   In these guidelines, note the importance of *algebra* (as well as calculus) for solving the equations $f(x) = 0$, $f'(x) = 0$, and $f''(x) = 0$.

Example 1   Analyzing and Sketching the Graph of a Rational Function

Analyze and sketch the graph of $f(x) = \dfrac{2(x^2 - 9)}{x^2 - 4}$.

Solution

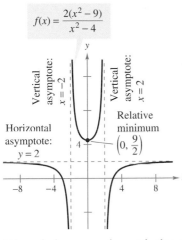

$f(x) = \dfrac{2(x^2 - 9)}{x^2 - 4}$

Vertical asymptote: $x = -2$

Vertical asymptote: $x = 2$

Horizontal asymptote: $y = 2$

Relative minimum $\left(0, \frac{9}{2}\right)$

Using calculus, you can be certain that you have determined all characteristics of the graph of f.

Figure 3.45

First derivative:	$f'(x) = \dfrac{20x}{(x^2 - 4)^2}$
Second derivative:	$f''(x) = \dfrac{-20(3x^2 + 4)}{(x^2 - 4)^3}$
x-intercepts:	$(-3, 0), (3, 0)$
y-intercept:	$\left(0, \frac{9}{2}\right)$
Vertical asymptotes:	$x = -2, x = 2$
Horizontal asymptote:	$y = 2$
Critical number:	$x = 0$
Possible points of inflection:	None
Domain:	All real numbers except $x = \pm 2$
Symmetry:	With respect to y-axis
Test intervals:	$(-\infty, -2), (-2, 0), (0, 2), (2, \infty)$

The table shows how the test intervals are used to determine several characteristics of the graph. The graph of f is shown in Figure 3.45.

	$f(x)$	$f'(x)$	$f''(x)$	**Characteristic of Graph**
$\infty < x < -2$		$-$	$-$	Decreasing, concave downward
$x = -2$	Undef.	Undef.	Undef.	Vertical asymptote
$-2 < x < 0$		$-$	$+$	Decreasing, concave upward
$x = 0$	$\frac{9}{2}$	0	$+$	Relative minimum
$0 < x < 2$		$+$	$+$	Increasing, concave upward
$x = 2$	Undef.	Undef.	Undef.	Vertical asymptote
$2 < x < \infty$		$+$	$-$	Increasing, concave downward

FOR FURTHER INFORMATION For more information on the use of technology to graph rational functions, see the article "Graphs of Rational Functions for Computer Assisted Calculus" by Stan Byrd and Terry Walters in *The College Mathematics Journal*. To view this article, go to the website *www.matharticles.com*.

Be sure you understand all of the implications of creating a table such as that shown in Example 1. Because of the use of calculus, you can *be sure* that the graph has no relative extrema or points of inflection other than those indicated in Figure 3.45.

TECHNOLOGY PITFALL   Without using the type of analysis outlined in Example 1, it is easy to obtain an incomplete view of a graph's basic characteristics. For instance, Figure 3.46 shows a view of the graph of

$$g(x) = \frac{2(x^2 - 9)(x - 20)}{(x^2 - 4)(x - 21)}.$$

From this view, it appears that the graph of g is about the same as the graph of f shown in Figure 3.45. The graphs of these two functions, however, differ significantly. Try enlarging the viewing window to see the differences.

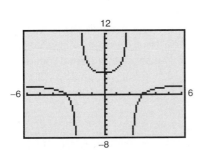

By not using calculus you may overlook important characteristics of the graph of g.

Figure 3.46

Example 2   Analyzing and Sketching the Graph of a Rational Function

Analyze and sketch the graph of $f(x) = \dfrac{x^2 - 2x + 4}{x - 2}$.

Solution

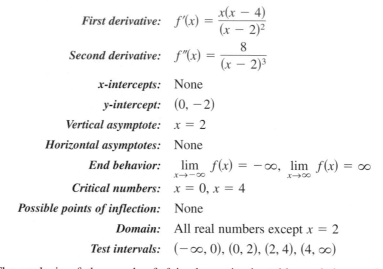

First derivative:	$f'(x) = \dfrac{x(x - 4)}{(x - 2)^2}$
Second derivative:	$f''(x) = \dfrac{8}{(x - 2)^3}$
x-intercepts:	None
y-intercept:	$(0, -2)$
Vertical asymptote:	$x = 2$
Horizontal asymptotes:	None
End behavior:	$\displaystyle\lim_{x \to -\infty} f(x) = -\infty, \ \lim_{x \to \infty} f(x) = \infty$
Critical numbers:	$x = 0, \ x = 4$
Possible points of inflection:	None
Domain:	All real numbers except $x = 2$
Test intervals:	$(-\infty, 0), \ (0, 2), \ (2, 4), \ (4, \infty)$

The analysis of the graph of f is shown in the table, and the graph is shown in Figure 3.47.

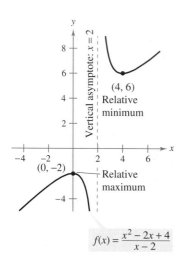

$f(x) = \dfrac{x^2 - 2x + 4}{x - 2}$

Figure 3.47

	$f(x)$	$f'(x)$	$f''(x)$	**Characteristic of Graph**
$-\infty < x < 0$		$+$	$-$	Increasing, concave downward
$x = 0$	-2	0	$-$	Relative maximum
$0 < x < 2$		$-$	$-$	Decreasing, concave downward
$x = 2$	Undef.	Undef.	Undef.	Vertical asymptote
$2 < x < 4$		$-$	$+$	Decreasing, concave upward
$x = 4$	6	0	$+$	Relative minimum
$4 < x < \infty$		$+$	$+$	Increasing, concave upward

Although the graph of the function in Example 2 has no horizontal asymptote, it does have a slant asymptote. The graph of a rational function (having no common factors) has a **slant asymptote** if the degree of the numerator exceeds the degree of the denominator by 1. To find the slant asymptote, use long division to rewrite the rational function as the sum of a first-degree polynomial and another rational function.

$$f(x) = \frac{x^2 - 2x + 4}{x - 2} \qquad \text{Rewrite using long division.}$$

$$= x + \frac{4}{x - 2} \qquad y = x \text{ is a slant asymptote.}$$

In Figure 3.48, note that the graph of f approaches the slant asymptote $y = x$ as x approaches $-\infty$ or ∞.

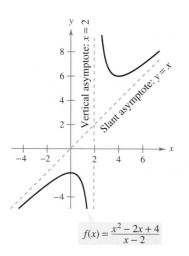

$f(x) = \dfrac{x^2 - 2x + 4}{x - 2}$

A slant asymptote
Figure 3.48

Example 3   **Analyzing and Sketching the Graph of a Radical Function**

Analyze and sketch the graph of $f(x) = \dfrac{x}{\sqrt{x^2 + 2}}$.

Solution

$$f'(x) = \frac{2}{(x^2 + 2)^{3/2}} \qquad f''(x) = -\frac{6x}{(x^2 + 2)^{5/2}}$$

The graph has only one intercept, $(0, 0)$. It has no vertical asymptotes, but it has two horizontal asymptotes: $y = 1$ (to the right) and $y = -1$ (to the left). The function has no critical numbers and one possible point of inflection (at $x = 0$). The domain of the function is all real numbers, and the graph is symmetric with respect to the origin. The analysis of the graph of f is shown in the table, and the graph is shown in Figure 3.49.

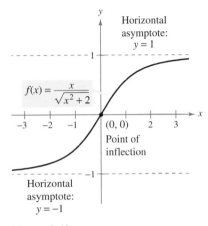

$f(x) = \dfrac{x}{\sqrt{x^2 + 2}}$

Figure 3.49

	$f(x)$	$f'(x)$	$f''(x)$	**Characteristic of Graph**
$-\infty < x < 0$		$+$	$+$	Increasing, concave upward
$x = 0$	0	$\dfrac{1}{\sqrt{2}}$	0	Point of inflection
$0 < x < \infty$		$+$	$-$	Increasing, concave downward

Example 4   **Analyzing and Sketching the Graph of a Radical Function**

Analyze and sketch the graph of $f(x) = 2x^{5/3} - 5x^{4/3}$.

Solution

$$f'(x) = \frac{10}{3}x^{1/3}(x^{1/3} - 2) \qquad f''(x) = \frac{20(x^{1/3} - 1)}{9x^{2/3}}$$

The function has two intercepts: $(0, 0)$ and $\left(\frac{125}{8}, 0\right)$. There are no horizontal or vertical asymptotes. The function has two critical numbers ($x = 0$ and $x = 8$) and two possible points of inflection ($x = 0$ and $x = 1$). The domain is all real numbers. The analysis of the graph of f is shown in the table, and the graph is shown in Figure 3.50.

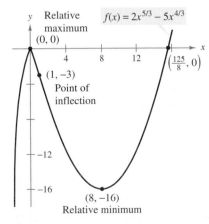

$f(x) = 2x^{5/3} - 5x^{4/3}$

Figure 3.50

	$f(x)$	$f'(x)$	$f''(x)$	**Characteristic of Graph**
$-\infty < x < 0$		$+$	$-$	Increasing, concave downward
$x = 0$	0	0	Undef.	Relative maximum
$0 < x < 1$		$-$	$-$	Decreasing, concave downward
$x = 1$	-3	$-$	0	Point of inflection
$1 < x < 8$		$-$	$+$	Decreasing, concave upward
$x = 8$	-16	0	$+$	Relative minimum
$8 < x < \infty$		$+$	$+$	Increasing, concave upward

Example 5 **Analyzing and Sketching the Graph of a Polynomial Function**

Analyze and sketch the graph of $f(x) = x^4 - 12x^3 + 48x^2 - 64x$.

Solution Begin by factoring to obtain

$$f(x) = x^4 - 12x^3 + 48x^2 - 64x$$
$$= x(x - 4)^3.$$

Then, using the factored form of $f(x)$, you can perform the following analysis.

First derivative:	$f'(x) = 4(x - 1)(x - 4)^2$
Second derivative:	$f''(x) = 12(x - 4)(x - 2)$
x-intercepts:	$(0, 0), (4, 0)$
y-intercept:	$(0, 0)$
Vertical asymptotes:	None
Horizontal asymptotes:	None
End behavior:	$\lim\limits_{x \to -\infty} f(x) = \infty, \ \lim\limits_{x \to \infty} f(x) = \infty$
Critical numbers:	$x = 1, x = 4$
Possible points of inflection:	$x = 2, x = 4$
Domain:	All real numbers
Test intervals:	$(-\infty, 1), (1, 2), (2, 4), (4, \infty)$

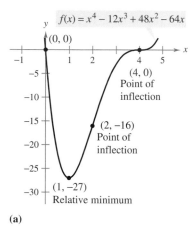

$f(x) = x^4 - 12x^3 + 48x^2 - 64x$

(0, 0)

(4, 0)
Point of
inflection

(2, -16)
Point of
inflection

(1, -27)
Relative minimum

(a)

The analysis of the graph of f is shown in the table, and the graph is shown in Figure 3.51(a). Using a computer algebra system such as Derive (see Figure 3.51b) can help you verify your analysis.

	$f(x)$	$f'(x)$	$f''(x)$	**Characteristic of Graph**
$-\infty < x < 1$		$-$	$+$	Decreasing, concave upward
$x = 1$	-27	0	$+$	Relative minimum
$1 < x < 2$		$+$	$+$	Increasing, concave upward
$x = 2$	-16	$+$	0	Point of inflection
$2 < x < 4$		$+$	$-$	Increasing, concave downward
$x = 4$	0	0	0	Point of inflection
$4 < x < \infty$		$+$	$+$	Increasing, concave upward

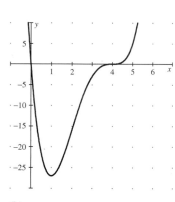

(b)
A polynomial function of even degree must have at least one relative extremum.
Figure 3.51

The fourth-degree polynomial function in Example 5 has one relative minimum and no relative maxima. In general, a polynomial function of degree n can have *at most* $n - 1$ relative extrema, and *at most* $n - 2$ points of inflection. Moreover, polynomial functions of even degree must have *at least* one relative extremum.

Remember from the leading coefficient test described in Section P.3 that the "end behavior" of the graph of a polynomial function is determined by its leading coefficient and its degree. For instance, because the polynomial in Example 5 has a positive leading coefficient, the graph rises to the right. Moreover, because the degree is even, the graph also rises to the left.

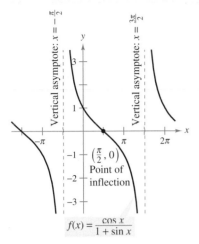

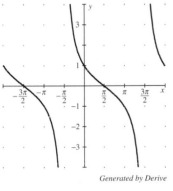

$$f(x) = \frac{\cos x}{1 + \sin x}$$

(a)

Generated by Derive

(b)

Figure 3.52

Example 6   **Analyzing and Sketching the Graph of a Trigonometric Function**

Analyze and sketch the graph of $f(x) = \dfrac{\cos x}{1 + \sin x}$.

Solution   Because the function has a period of 2π, you can restrict the analysis of the graph to the interval $(-\pi/2, 3\pi/2)$.

First derivative:	$f'(x) = -\dfrac{1}{1 + \sin x}$
Second derivative:	$f''(x) = \dfrac{\cos x}{(1 + \sin x)^2}$
Period:	2π
x-intercept:	$\left(\dfrac{\pi}{2}, 0\right)$
y-intercept:	$(0, 1)$
Vertical asymptotes:	$x = -\dfrac{\pi}{2}, x = \dfrac{3\pi}{2}$
Horizontal asymptotes:	None
Critical numbers:	None
Possible points of inflection:	$x = \dfrac{\pi}{2}$
Domain:	All real numbers except $x = \dfrac{3 + 4n}{2}\pi$
Test intervals:	$\left(-\dfrac{\pi}{2}, \dfrac{\pi}{2}\right), \left(\dfrac{\pi}{2}, \dfrac{3\pi}{2}\right)$

The analysis of the graph of f on the interval $(-\pi/2, 3\pi/2)$ is shown in the table, and the graph is shown in Figure 3.52(a). Compare this with the graph generated by the computer algebra system Derive in Figure 3.52(b).

	$f(x)$	$f'(x)$	$f''(x)$	**Characteristic of Graph**
$x = -\dfrac{\pi}{2}$	Undef.	Undef.	Undef.	Vertical asymptote
$-\dfrac{\pi}{2} < x < \dfrac{\pi}{2}$		$-$	$+$	Decreasing, concave upward
$x = \dfrac{\pi}{2}$	0	$-\dfrac{1}{2}$	0	Point of inflection
$\dfrac{\pi}{2} < x < \dfrac{3\pi}{2}$		$-$	$-$	Decreasing, concave downward
$x = \dfrac{3\pi}{2}$	Undef.	Undef.	Undef.	Vertical asymptote

NOTE   The work involved in sketching the graph of a trigonometric function can be lessened sometimes by using trigonometric identities. For instance, the function in Example 6 can be rewritten as

$$f(x) = \frac{\cos x}{1 + \sin x} = \cot\left(\frac{x}{2} + \frac{\pi}{4}\right).$$

In this form, you can recognize the familiar cotangent graph shown in Figure 3.52.

EXERCISES FOR SECTION 3.6

In Exercises 1–4, match the graph of f in the left column with that of its derivative in the right column.

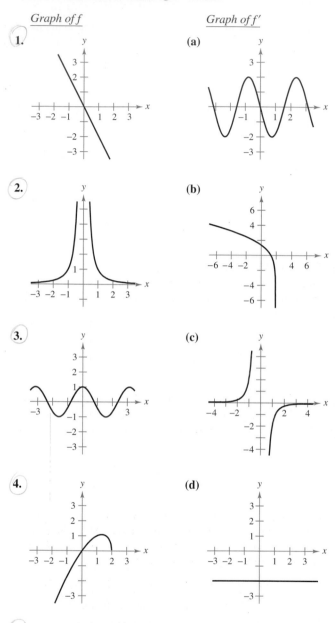

Graph of f   *Graph of f'*

1.   (a)

2.   (b)

3.   (c)

4.   (d)

5. *Graphical Reasoning*   The graph of f is given in the figure.

  (a) For which values of x is $f'(x)$ zero? Positive? Negative?

  (b) For which values of x is $f''(x)$ zero? Positive? Negative?

  (c) On what interval is f' an increasing function?

  (d) For which value of x is $f'(x)$ minimum? For this value of x, how does the rate of change of f compare with the rate of change of f for other values of x? Explain.

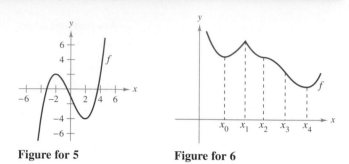

Figure for 5   **Figure for 6**

6. *Graphical Reasoning*   Identify the real numbers x_0, x_1, x_2, x_3, and x_4 in the figure such that each of the following is true.

  (a) $f'(x) = 0$ (b) $f''(x) = 0$

  (c) $f'(x)$ does not exist.   (d) f has a relative maximum.

  (e) f has a point of inflection.

In Exercises 7–38, analyze and sketch a graph of the function. Label any intercepts, relative extrema, points of inflection, and asymptotes. Use a graphing utility to verify your results.

7. $y = \dfrac{x^2}{x^2 + 3}$

8. $y = \dfrac{x}{x^2 + 1}$

9. $y = \dfrac{1}{x - 2} - 3$

10. $y = \dfrac{x^2 + 1}{x^2 - 9}$

11. $y = \dfrac{2x}{x^2 - 1}$

12. $f(x) = \dfrac{x + 2}{x}$

13. $g(x) = x + \dfrac{4}{x^2 + 1}$

14. $f(x) = x + \dfrac{32}{x^2}$

15. $f(x) = \dfrac{x^2 + 1}{x}$

16. $f(x) = \dfrac{x^3}{x^2 - 4}$

17. $y = \dfrac{x^2 - 6x + 12}{x - 4}$

18. $y = \dfrac{2x^2 - 5x + 5}{x - 2}$

19. $y = x\sqrt{4 - x}$

20. $g(x) = x\sqrt{9 - x}$

21. $h(x) = x\sqrt{9 - x^2}$

22. $y = x\sqrt{16 - x^2}$

23. $y = 3x^{2/3} - 2x$

24. $y = 3(x - 1)^{2/3} - (x - 1)^2$

25. $y = x^3 - 3x^2 + 3$

26. $y = -\frac{1}{3}(x^3 - 3x + 2)$

27. $y = 2 - x - x^3$

28. $f(x) = \frac{1}{3}(x - 1)^3 + 2$

29. $f(x) = 3x^3 - 9x + 1$

30. $f(x) = (x + 1)(x - 2)(x - 5)$

31. $y = 3x^4 + 4x^3$

32. $y = 3x^4 - 6x^2 + \frac{5}{3}$

33. $f(x) = x^4 - 4x^3 + 16x$

34. $f(x) = x^4 - 8x^3 + 18x^2 - 16x + 5$

35. $y = x^5 - 5x$

36. $y = (x - 1)^5$

37. $y = |2x - 3|$

38. $y = |x^2 - 6x + 5|$

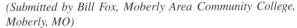

In Exercises 39–46, sketch a graph of the function over the indicated interval. Use a graphing utility to verify your graph.

Function	Interval
39. $y = \sin x - \frac{1}{18} \sin 3x$	$0 \le x \le 2\pi$
40. $y = \cos x - \frac{1}{2} \cos 2x$	$0 \le x \le 2\pi$
41. $y = 2x - \tan x$	$-\frac{\pi}{2} < x < \frac{\pi}{2}$
42. $y = 2(x - 2) + \cot x$	$0 < x < \pi$
43. $y = 2(\csc x + \sec x)$	$0 < x < \frac{\pi}{2}$
44. $y = \sec^2\left(\frac{\pi x}{8}\right) - 2\tan\left(\frac{\pi x}{8}\right) - 1$	$-3 < x < 3$
45. $g(x) = x \tan x$	$-\frac{3\pi}{2} < x < \frac{3\pi}{2}$
46. $g(x) = x \cot x$	$-2\pi < x < 2\pi$

In Exercises 47–50, use a computer algebra system to analyze and graph the function. Identify any relative extrema, points of inflection, and asymptotes.

47. $f(x) = \dfrac{20x}{x^2 + 1} - \dfrac{1}{x}$ **48.** $f(x) = 5\left(\dfrac{1}{x - 4} - \dfrac{1}{x + 2}\right)$

49. $f(x) = \dfrac{x}{\sqrt{x^2 + 7}}$ **50.** $f(x) = \dfrac{4x}{\sqrt{x^2 + 15}}$

Getting at the Concept

In Exercises 51 and 52, the graphs of f, f', and f'' are shown on the same set of coordinate axes. Which is which? To print an enlarged copy of the graph, go to the website *www.mathgraphs.com.*

51.   **52.**

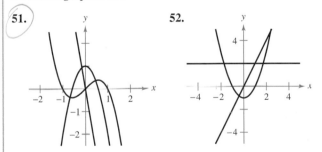

In Exercises 53–56, use the graph of f' to sketch a graph of f and the graph of f''. To print an enlarged copy of the graph, go to the website *www.mathgraphs.com.*

53.   **54.**

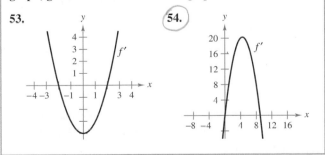

Getting at the Concept *(continued)*

55.   **56.**

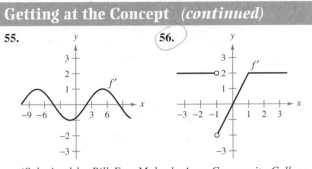

(Submitted by Bill Fox, Moberly Area Community College, Moberly, MO)

57. Suppose $f'(t) < 0$ for all t in the interval $(2, 8)$. Explain why $f(3) > f(5)$.

58. Suppose $f(0) = 3$ and $2 \le f'(x) \le 4$ for all x in the interval $[-5, 5]$. Determine the greatest and least possible values of $f(2)$.

In Exercises 59 and 60, use a graphing utility to graph the function. Use the graph to determine whether it is possible for the graph of a function to cross its horizontal asymptote. Do you think it is possible for the graph of a function to cross its vertical asymptote? Why or why not?

59. $f(x) = \dfrac{4(x - 1)^2}{x^2 - 4x + 5}$ **60.** $g(x) = \dfrac{3x^4 - 5x + 3}{x^4 + 1}$

Writing   In Exercises 61 and 62, use a graphing utility to graph the function. Explain why there is no vertical asymptote when a superficial examination of the function may indicate that there should be one.

61. $h(x) = \dfrac{6 - 2x}{3 - x}$ **62.** $g(x) = \dfrac{x^2 + x - 2}{x - 1}$

Writing   In Exercises 63 and 64, use a graphing utility to graph the function and determine the slant asymptote of the graph. Zoom out repeatedly and describe how the graph on the display appears to change. Why does this occur?

63. $f(x) = -\dfrac{x^2 - 3x - 1}{x - 2}$ **64.** $g(x) = \dfrac{2x^2 - 8x - 15}{x - 5}$

65. *Graphical Reasoning*   Consider the function

$$f(x) = \dfrac{\cos^2 \pi x}{\sqrt{x^2 + 1}}, \quad 0 < x < 4.$$

(a) Use a computer algebra system to graph the function and use the graph to visually approximate the critical numbers.

(b) Use a computer algebra system to find f' and approximate the critical numbers. Are the results the same as the visual approximation in part (a)? Explain.

66. Graphical Reasoning Consider the function

$$f(x) = \tan(\sin \pi x).$$

(a) Use a graphing utility to graph the function.

(b) Identify any symmetry of the graph.

(c) Is the function periodic? If so, what is the period?

(d) Identify any extrema on $(-1, 1)$.

(e) Use a graphing utility to determine the concavity of the graph on $(0, 1)$.

Think About It In Exercises 67–70, create a function whose graph has the indicated characteristics. (There is more than one correct answer.)

67. Vertical asymptote: $x = 5$

Horizontal asymptote: $y = 0$

68. Vertical asymptote: $x = -3$

Horizontal asymptote: None

69. Vertical asymptote: $x = 5$

Slant asymptote: $y = 3x + 2$

70. Vertical asymptote: $x = 0$

Slant asymptote: $y = -x$

71. Graphical Reasoning Consider the function

$$f(x) = \frac{ax}{(x - b)^2}.$$

(a) Determine the effect on the graph of f if $b \neq 0$ and a is varied. Consider cases where a is positive and a is negative.

(b) Determine the effect on the graph of f if $a \neq 0$ and b is varied.

72. Consider the function

$$f(x) = \tfrac{1}{2}(ax)^2 - (ax), \quad a \neq 0.$$

(a) Determine the changes (if any) in the intercepts, extrema, and concavity of the graph of f when a is varied.

(b) In the same viewing window, use a graphing utility to graph the function for four different values of a.

73. Investigation Consider the function

$$f(x) = \frac{3x^n}{x^4 + 1}$$

for nonnegative integer values of n.

(a) Discuss the relationship between the value of n and the symmetry of the graph.

(b) For which values of n will the x-axis be the horizontal asymptote?

(c) For which value of n will $y = 3$ be the horizontal asymptote?

(d) What is the asymptote of the graph when $n = 5$?

(e) Use a graphing utility to graph f for the indicated values of n in the table. Use the graph to determine the number of extrema M and the number of inflection points N of the graph.

n	0	1	2	3	4	5
M						
N						

Table for 73(e)

74. Investigation Let $P(x_0, y_0)$ be an arbitrary point on the graph of f such that $f'(x_0) \neq 0$, as indicated in the figure. Verify each of the following.

(a) The x-intercept of the tangent line is $\left(x_0 - \dfrac{f(x_0)}{f'(x_0)}, 0\right)$.

(b) The y-intercept of the tangent line is $(0, f(x_0) - x_0 f'(x_0))$.

(c) The x-intercept of the normal line is $(x_0 + f(x_0)f'(x_0), 0)$.

(d) The y-intercept of the normal line is $\left(0, y_0 + \dfrac{x_0}{f'(x_0)}\right)$.

(e) $|BC| = \left|\dfrac{f(x_0)}{f'(x_0)}\right|$

(f) $|PC| = \left|\dfrac{f(x_0)\sqrt{1 + [f'(x_0)]^2}}{f'(x_0)}\right|$

(g) $|AB| = |f(x_0)f'(x_0)|$

(h) $|AP| = |f(x_0)|\sqrt{1 + [f'(x_0)]^2}$

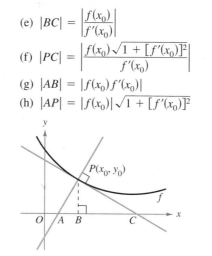

75. Modeling Data The data in the table show the number N of bacteria in a culture at time t, where t is measured in days.

t	1	2	3	4
N	25	200	804	1756

t	5	6	7	8
N	2296	2434	2467	2473

A model for this data is given by

$$N = \frac{24{,}670 - 35{,}153t + 13{,}250t^2}{100 - 39t + 7t^2}, \quad 1 \le t \le 8.$$

(a) Use a graphing utility to plot the data and graph the model.

(b) Use the model to estimate the number of bacteria when $t = 10$.

(c) Approximate the day when the number of bacteria was greatest.

(d) Use a computer algebra system to determine the time when the rate of increase in the number of bacteria was greatest.

(e) Find $\lim_{t \to \infty} N(t)$.

Section 3.7	**Optimization Problems**

• Solve applied minimum and maximum problems.

Applied Minimum and Maximum Problems

One of the most common applications of calculus involves the determination of minimum and maximum values. Consider how frequently you hear or read terms such as greatest profit, least cost, least time, greatest voltage, optimum size, least size, greatest strength, and greatest distance. Before outlining a general problem-solving strategy for such problems, let's look at an example.

Example 1   Finding Maximum Volume

A manufacturer wants to design an open box having a square base and a surface area of 108 square inches, as shown in Figure 3.53. What dimensions will produce a box with maximum volume?

Solution   Because the box has a square base, its volume is

$$V = x^2h. \qquad \text{Primary equation}$$

(This equation is called the **primary equation** because it gives a formula for the quantity to be optimized.) The surface area of the box is

$$S = (\text{area of base}) + (\text{area of four sides})$$
$$S = x^2 + 4xh = 108. \qquad \text{Secondary equation}$$

Because V is to be maximized, you want to express V as a function of just one variable. To do this, you can solve the equation $x^2 + 4xh - 108$ for h in terms of x to obtain $h = (108 - x^2)/(4x)$. Substituting into the primary equation produces

$$V = x^2h \qquad \text{Function of two variables}$$
$$= x^2\left(\frac{108 - x^2}{4x}\right) \qquad \text{Substitute for } h.$$
$$= 27x - \frac{x^3}{4}. \qquad \text{Function of one variable}$$

Before finding which x-value will yield a maximum value of V, you should determine the *feasible domain*. That is, what values of x make sense in this problem? You know that $V \geq 0$. You also know that x must be nonnegative and that the area of the base ($A = x^2$) is at most 108. So, the feasible domain is

$$0 \leq x \leq \sqrt{108}. \qquad \text{Feasible domain}$$

To maximize V, you can find the critical numbers of the volume function.

$$\frac{dV}{dx} = 27 - \frac{3x^2}{4} = 0 \qquad \text{Set derivative equal to 0.}$$
$$3x^2 = 108$$
$$x = \pm 6 \qquad \text{Critical numbers}$$

So, the critical numbers are $x = \pm 6$. You do not need to consider -6 because it is outside the domain. Evaluating V at the critical number $x = 6$ and at the endpoints of the domain produces $V(0) - 0$, $V(6) - 108$, and $V(\sqrt{108}) = 0$. Thus, V is maximum when $x = 6$ and the dimensions of the box are $6 \times 6 \times 3$ inches.

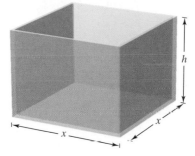

Open box with square base:
$S = x^2 + 4xh = 108$
Figure 3.53

TECHNOLOGY   You can verify your answer by using a graphing utility to graph the volume

$$V = 27x - \frac{x^3}{4}.$$

Use a viewing window in which $0 \leq x \leq \sqrt{108} \approx 10.4$ and $0 \leq y \leq 120$, and the *trace* feature to determine the maximum value of V.

In Example 1, you should realize that there are infinitely many open boxes having 108 square inches of surface area. To begin solving the problem, you might ask yourself which basic shape would seem to yield a maximum volume. Should the box be tall, squat, or nearly cubical?

You might even try calculating a few volumes, as shown in Figure 3.54, to see if you can get a better feeling for what the optimum dimensions should be. Remember that you are not ready to begin solving a problem until you have clearly identified what the problem is.

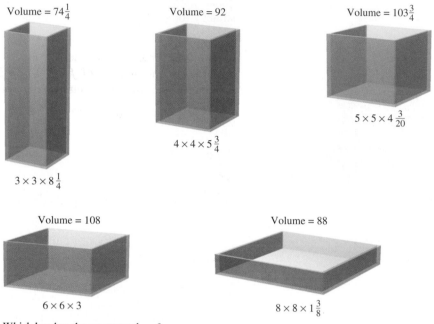

Volume = $74\frac{1}{4}$

$3 \times 3 \times 8\frac{1}{4}$

Volume = 92

$4 \times 4 \times 5\frac{3}{4}$

Volume = $103\frac{3}{4}$

$5 \times 5 \times 4\frac{3}{20}$

Volume = 108

$6 \times 6 \times 3$

Volume = 88

$8 \times 8 \times 1\frac{3}{8}$

Which box has the greatest volume?
Figure 3.54

Example 1 illustrates the following guidelines for solving applied minimum and maximum problems.

Guidelines for Solving Applied Minimum and Maximum Problems

1. Identify all *given* quantities and quantities *to be determined*. When feasible, make a sketch.

2. Write a **primary equation** for the quantity that is to be maximized (or minimized). (A review of several useful formulas from geometry is presented inside the front cover.)

3. Reduce the primary equation to one having a *single independent variable*. This may involve the use of **secondary equations** relating the independent variables of the primary equation.

4. Determine the feasible domain of the primary equation. That is, determine the values for which the stated problem makes sense.

5. Determine the desired maximum or minimum value by the calculus techniques discussed in Sections 3.1 through 3.4.

NOTE When performing Step 5, recall that to determine the maximum or minimum value of a continuous function f on a closed interval, you should compare the values of f at its critical numbers with the values of f at the endpoints of the interval.

⌐⌐   *Example 2   Finding Minimum Distance*

Which points on the graph of $y = 4 - x^2$ are closest to the point $(0, 2)$?

Solution   Figure 3.55 indicates that there are two points at a minimum distance from the point $(0, 2)$. The distance between the point $(0, 2)$ and a point (x, y) on the graph of $y = 4 - x^2$ is given by

$$d = \sqrt{(x - 0)^2 + (y - 2)^2}.$$ Primary equation

Using the secondary equation $y = 4 - x^2$, you can rewrite the primary equation as

$$d = \sqrt{x^2 + (4 - x^2 - 2)^2} = \sqrt{x^4 - 3x^2 + 4}.$$

Because d is smallest when the expression inside the radical is smallest, you need only find the critical numbers of $f(x) = x^4 - 3x^2 + 4$. Note that the domain of f is the entire real line. Moreover, setting $f'(x)$ equal to 0 yields

$$f'(x) = 4x^3 - 6x = 2x(2x^2 - 3) = 0$$

$$x = 0, \sqrt{\frac{3}{2}}, -\sqrt{\frac{3}{2}}.$$

The First Derivative Test verifies that $x = 0$ yields a relative maximum, whereas both $x = \sqrt{3/2}$ and $x = -\sqrt{3/2}$ yield a minimum distance. Hence, the closest points are $\left(\sqrt{3/2}, 5/2\right)$ and $\left(-\sqrt{3/2}, 5/2\right)$.

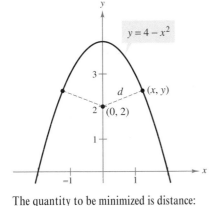

The quantity to be minimized is distance:
$d = \sqrt{(x - 0)^2 + (y - 2)^2}.$
Figure 3.55

Example 3   Finding Minimum Area

A rectangular page is to contain 24 square inches of print. The margins at the top and bottom of the page are to be $1\frac{1}{2}$ inches, and the margins on the left and right are to be 1 inch (see Figure 3.56). What should the dimensions of the page be so that the least amount of paper is used?

Solution   Let A be the area to be minimized.

$$A = (x + 3)(y + 2)$$ Primary equation

The printed area inside the margins is given by

$$24 = xy.$$ Secondary equation

Solving this equation for y produces $y = 24/x$. Substitution into the primary equation produces

$$A = (x + 3)\left(\frac{24}{x} + 2\right) = 30 + 2x + \frac{72}{x}.$$ Function of one variable

Because x must be positive, you are interested only in values of A for $x > 0$. To find the critical numbers, differentiate with respect to x.

$$\frac{dA}{dx} = 2 - \frac{72}{x^2} = 0 \implies x^2 = 36$$

So, the critical numbers are $x = \pm 6$. You do not have to consider -6 because it is outside the domain. The First Derivative Test confirms that A is a minimum when $x = 6$. Therefore, $y = \frac{24}{6} = 4$ and the dimensions of the page should be $x + 3 = 9$ inches by $y + 2 = 6$ inches.   ▨

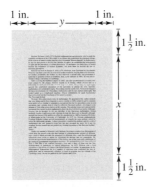

1 in.   1 in.
$1\frac{1}{2}$ in.

$1\frac{1}{2}$ in.

The quantity to be minimized is area:
$A = (x + 3)(y + 2).$
Figure 3.56

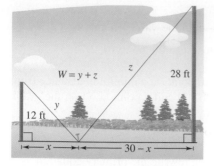

The quantity to be minimized is length. From the diagram, you can see that x varies between 0 and 30.

Figure 3.57

Example 4   **Finding Minimum Length**

Two posts, one 12 feet high and the other 28 feet high, stand 30 feet apart. They are to be stayed by two wires, attached to a single stake, running from ground level to the top of each post. Where should the stake be placed to use the least wire?

Solution   Let W be the wire length to be minimized. Using Figure 3.57, you can write

$W = y + z.$ Primary equation

In this problem, rather than solving for y in terms of z (or vice versa), you can solve for both y and z in terms of a third variable x, as shown in Figure 3.57. From the Pythagorean Theorem, you obtain

$$x^2 + 12^2 = y^2$$
$$(30 - x)^2 + 28^2 = z^2$$

which implies that

$$y = \sqrt{x^2 + 144}$$
$$z = \sqrt{x^2 - 60x + 1684}.$$

Thus, W is given by

$$W = y + z$$
$$= \sqrt{x^2 + 144} + \sqrt{x^2 - 60x + 1684}, \qquad 0 \le x \le 30.$$

Differentiating W with respect to x yields

$$\frac{dW}{dx} = \frac{x}{\sqrt{x^2 + 144}} + \frac{x - 30}{\sqrt{x^2 - 60x + 1684}}.$$

By letting $dW/dx = 0$, you obtain

$$\frac{x}{\sqrt{x^2 + 144}} + \frac{x - 30}{\sqrt{x^2 - 60x + 1684}} = 0$$
$$x\sqrt{x^2 - 60x + 1684} = (30 - x)\sqrt{x^2 + 144}$$
$$x^2(x^2 - 60x + 1684) = (30 - x)^2(x^2 + 144)$$
$$x^4 - 60x^3 + 1684x^2 = x^4 - 60x^3 + 1044x^2 - 8640x + 129{,}600$$
$$640x^2 + 8640x - 129{,}600 = 0$$
$$320(x - 9)(2x + 45) = 0$$
$$x = 9, \; -22.5.$$

Because $x = -22.5$ is not in the domain and

$$W(0) \approx 53.04, \qquad W(9) = 50, \qquad \text{and} \qquad W(30) \approx 60.31$$

you can conclude that the wire should be staked at 9 feet from the 12-foot pole.

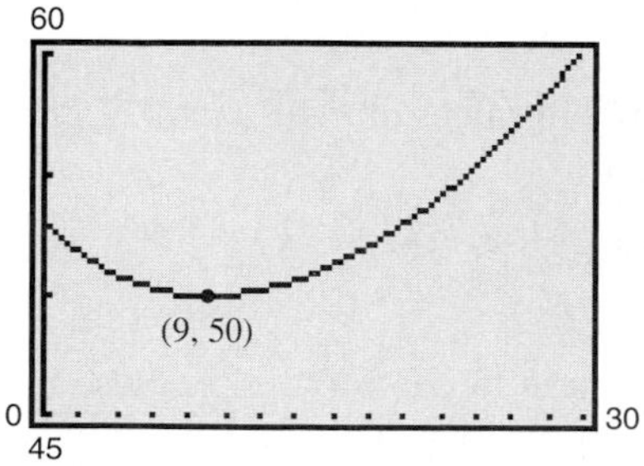

TECHNOLOGY   From Example 4, you can see that applied optimization problems can involve a lot of algebra. If you have access to a graphing utility, you can confirm that $x = 9$ yields a minimum value of W by sketching the graph of

$$W = \sqrt{x^2 + 144} + \sqrt{x^2 - 60x + 1684}$$

as shown in Figure 3.58.

You can confirm the minimum value of W with a graphing utility.

Figure 3.58

In each of the first four examples, the extreme value occurs at a critical number. Although this happens often, remember that an extreme value can also occur at an endpoint of an interval, as shown in Example 5.

Example 5   An Endpoint Maximum

Four feet of wire is to be used to form a square and a circle. How much of the wire should be used for the square and how much should be used for the circle to enclose the maximum total area?

Solution   The total area (see Figure 3.59) is given by

$$A = (\text{area of square}) + (\text{area of circle})$$
$$A = x^2 + \pi r^2. \qquad \text{Primary equation}$$

Because the total amount of wire is 4 feet, you obtain

$$4 = (\text{perimeter of square}) + (\text{circumference of circle})$$
$$4 = 4x + 2\pi r.$$

So, $r = 2(1 - x)/\pi$, and by substituting into the primary equation you have

$$A = x^2 + \pi \left[\frac{2(1 - x)}{\pi} \right]^2$$

$$= x^2 + \frac{4(1 - x)^2}{\pi}$$

$$= \frac{1}{\pi}[(\pi + 4)x^2 - 8x + 4].$$

The feasible domain is $0 \le x \le 1$ restricted by the square's perimeter. Because

$$\frac{dA}{dx} = \frac{2(\pi + 4)x - 8}{\pi}$$

the only critical number in $(0, 1)$ is $x = 4/(\pi + 4) \approx 0.56$. Therefore, using

$$A(0) \approx 1.273, \qquad A(0.56) \approx 0.56, \qquad \text{and} \qquad A(1) = 1$$

you can conclude that the maximum area occurs when $x = 0$. That is, *all* the wire is used for the circle.   ▨

Let's review the primary equations developed in the first five examples. As applications go, these five examples are fairly simple, and yet the resulting primary equations are quite complicated.

$$V = 27x - \frac{x^3}{4} \qquad\qquad W = \sqrt{x^2 + 144} + \sqrt{x^2 - 60x + 1684}$$

$$d = \sqrt{x^4 - 3x^2 + 4} \qquad A = \frac{1}{\pi}[(\pi + 4)x^2 - 8x + 4]$$

$$A = 30 + 2x + \frac{72}{x}$$

You must expect that real-life applications often involve equations that are *at least as complicated* as these five. Remember that one of the main goals of this course is to learn to use calculus to analyze equations that initially seem formidable.

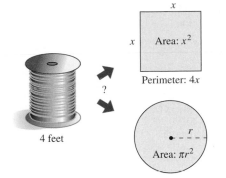

x

x   Area: x^2

Perimeter: $4x$

?

4 feet

r

Area: πr^2

Circumference: $2\pi r$

The quantity to be maximized is area:
$A = x^2 + \pi r^2.$
Figure 3.59

EXPLORATION

What would the answer be if Example 5 asked for the dimensions needed to enclose the *minimum* total area?

EXERCISES FOR SECTION 3.7

1. *Numerical, Graphical, and Analytic Analysis*   Find two positive numbers whose sum is 110 and whose product is a maximum.

(a) Analytically complete six rows of a table such as the one below. (The first two rows are shown.)

First Number x	Second Number	Product P
10	$110 - 10$	$10(110 - 10) = 1000$
20	$110 - 20$	$20(110 - 20) = 1800$

(b) Use a graphing utility to generate additional rows of the table. Use the table to estimate the solution. (*Hint:* Use the *table* feature of the graphing utility.)

(c) Write the product P as a function of x.

(d) Use a graphing utility to graph the function in part (c) and estimate the solution from the graph.

(e) Use calculus to find the critical number of the function in part (c). Then find the two numbers.

In Exercises 2–6, find two positive numbers that satisfy the given requirements.

2. The sum is S and the product is a maximum.

3. The product is 192 and the sum is a minimum.

4. The product is 192 and the sum of the first plus three times the second is a minimum.

5. The second number is the reciprocal of the first and the sum is a minimum.

6. The sum of the first and twice the second is 100 and the product is a maximum.

In Exercises 7 and 8, find the length and width of a rectangle that has the given perimeter and a maximum area.

7. Perimeter: 100 meters

8. Perimeter: P units

In Exercises 9 and 10, find the length and width of a rectangle that has the given area and a minimum perimeter.

9. Area: 64 square feet

10. Area: A square centimeters

In Exercises 11–14, find the point on the graph of the function that is closest to the given point.

Function	Point	Function	Point
11. $f(x) = \sqrt{x}$	$(4, 0)$	**12.** $f(x) = \sqrt{x - 8}$	$(2, 0)$
13. $f(x) = x^2$	$\left(2, \frac{1}{2}\right)$	**14.** $f(x) = (x + 1)^2$	$(5, 3)$

15. *Chemical Reaction*   In an autocatalytic chemical reaction, the product formed is a catalyst for the reaction. If Q_0 is the amount of the original substance and x is the amount of catalyst formed, the rate of chemical reaction is

$$\frac{dQ}{dx} = kx(Q_0 - x).$$

For what value of x will the rate of chemical reaction be greatest?

16. *Traffic Control*   On a given day, the flow rate F (cars per hour) on a congested roadway is

$$F = \frac{v}{22 + 0.02v^2}$$

where v is the speed of the traffic in miles per hour. What speed will maximize the flow rate on the road?

17. *Area*   A farmer plans to fence a rectangular pasture adjacent to a river. The pasture must contain 180,000 square meters in order to provide enough grass for the herd. What dimensions would require the least amount of fencing if no fencing is needed along the river?

18. *Area*   A rancher has 200 feet of fencing with which to enclose two adjacent rectangular corrals (see figure). What dimensions should be used so that the enclosed area will be a maximum?

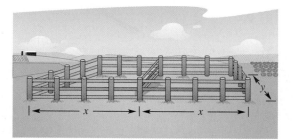

19. *Volume*

(a) Verify that each of the rectangular solids shown in the figure has a surface area of 150 square inches.

(b) Find the volume of each.

(c) Determine the dimensions of a rectangular solid (with a square base) of maximum volume if its surface area is 150 square inches.

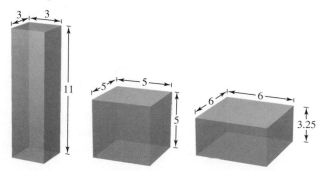

20. ***Numerical, Graphical, and Analytic Analysis*** An open box of maximum volume is to be made from a square piece of material, 24 inches on a side, by cutting equal squares from the corners and turning up the sides (see figure).

(a) Analytically complete six rows of a table such as the one below. (The first two rows are shown.) Use the table to guess the maximum volume.

Height	Length and Width	Volume
1	$24 - 2(1)$	$1[24 - 2(1)]^2 = 484$
2	$24 - 2(2)$	$2[24 - 2(2)]^2 = 800$

(b) Write the volume V as a function of x.

(c) Use calculus to find the critical number of the function in part (b) and find the maximum value.

(d) Use a graphing utility to graph the function in part (b) and verify the maximum volume from the graph.

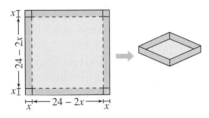

21. (a) Solve Exercise 20 given that the square piece of material is s meters on a side.

(b) If the dimensions of the square piece of material are doubled, how does the volume change?

22. ***Numerical, Graphical, and Analytic Analysis*** A physical fitness room consists of a rectangle with a semicircle on each end. A 200-meter running track runs around the outside of the room.

(a) Draw a figure to represent the problem. Let x and y represent the length and width of the rectangle.

(b) Analytically complete six rows of a table such as the one below. (The first two rows are shown.) Use the table to guess the maximum area of the rectangular region.

Length x	Width y	Area
10	$\frac{2}{\pi}(100 - 10)$	$(10)\frac{2}{\pi}(100 - 10) \approx 573$
20	$\frac{2}{\pi}(100 - 20)$	$(20)\frac{2}{\pi}(100 - 20) \approx 1019$

(c) Write the area A as a function of x.

(d) Use calculus to find the critical number of the function in part (c) and find the maximum value.

(e) Use a graphing utility to graph the function in part (c) and verify the maximum area from the graph.

23. ***Area*** A Norman window is constructed by adjoining a semicircle to the top of an ordinary rectangular window (see figure). Find the dimensions of a Norman window of maximum area if the total perimeter is 16 feet.

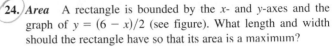

24. ***Area*** A rectangle is bounded by the x- and y-axes and the graph of $y = (6 - x)/2$ (see figure). What length and width should the rectangle have so that its area is a maximum?

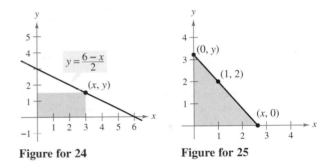

Figure for 24   **Figure for 25**

25. ***Length*** A right triangle is formed in the first quadrant by the x- and y-axes and a line through the point $(1, 2)$ (see figure).

(a) Write the length L of the hypotenuse as a function of x.

(b) Use a graphing utility to graphically approximate x such that the length of the hypotenuse is a minimum.

(c) Find the vertices of the triangle such that its area is a minimum.

26. ***Area*** Find the area of the largest isosceles triangle that can be inscribed in a circle of radius 4 (see figure).

(a) Solve by writing the area as a function of h.

(b) Solve by writing the area as a function of α.

(c) Identify the type of triangle of maximum area.

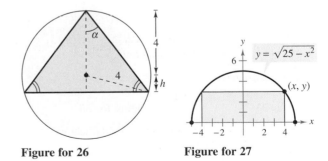

Figure for 26   **Figure for 27**

27. ***Area*** A rectangle is bounded by the x-axis and the semicircle $y = \sqrt{25 - x^2}$ (see figure). What length and width should the rectangle have so that its area is a maximum?

28. Area Find the dimensions of the largest rectangle that can be inscribed in a semicircle of radius r (see Exercise 27).

29. Area A rectangular page is to contain 30 square inches of print. The margins on each side are 1 inch. Find the dimensions of the page such that the least amount of paper is used.

30. Area A rectangular page is to contain 36 square inches of print. The margins on each side are to be $1\frac{1}{2}$ inches. Find the dimensions of the page such that the least amount of paper is used.

31. Numerical, Graphical, and Analytic Analysis A right circular cylinder is to be designed to hold 22 cubic inches of a soft drink (approximately 12 fluid ounces).

(a) Analytically complete six rows of a table such as the one below. (The first two rows are shown.)

Radius r	Height	Surface Area
0.2	$\dfrac{22}{\pi(0.2)^2}$	$2\pi(0.2)\left[0.2 + \dfrac{22}{\pi(0.2)^2}\right] \approx 220.3$
0.4	$\dfrac{22}{\pi(0.4)^2}$	$2\pi(0.4)\left[0.4 + \dfrac{22}{\pi(0.4)^2}\right] \approx 111.1$

(b) Use a graphing utility to generate additional rows of the table. Use the table to estimate the minimum surface area. (*Hint:* Use the *table* feature of the graphing utility.)

(c) Write the surface area S as a function of r.

(d) Use a graphing utility to graph the function in part (c) and estimate the minimum surface area from the graph.

(e) Use calculus to find the critical number of the function in part (c) and find dimensions that will yield the minimum surface area.

32. Surface Area Use calculus to find the required dimensions for the cylinder in Exercise 31 if its volume is V_0 cubic units.

33. Volume A rectangular package to be sent by a postal service can have a maximum combined length and girth (perimeter of a cross section) of 108 inches (see figure). Find the dimensions of the package of maximum volume that can be sent. (Assume the cross section is square.)

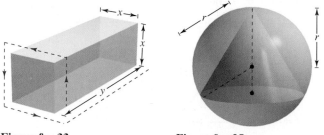

Figure for 33   **Figure for 35**

34. Volume Rework Exercise 33 for a cylindrical package. (The cross section is circular.)

35. Volume Find the volume of the largest right circular cone that can be inscribed in a sphere of radius r (see figure).

36. Volume Find the volume of the largest right circular cylinder that can be inscribed in a sphere of radius r.

37. The perimeter of a rectangle is 20 feet. Of all possible dimensions, the maximum area is 25 square feet when its length and width are both 5 feet. Are there dimensions that yield a minimum area? Explain.

38. A plastic shampoo bottle is a right circular cylinder. Because the surface area of the bottle does not change when it is squeezed, is it true that the volume remains the same? Explain.

39. Surface Area A solid is formed by adjoining two hemispheres to the ends of a right circular cylinder. The total volume of the solid is 12 cubic centimeters. Find the radius of the cylinder that produces the minimum surface area.

40. Cost An industrial tank of the shape described in Exercise 39 must have a volume of 3000 cubic feet. The hemispherical ends cost twice as much per square foot of surface area as the sides. Find the dimensions that will minimize cost.

41. Area The sum of the perimeters of an equilateral triangle and a square is 10. Find the dimensions of the triangle and the square that produce a minimum total area.

42. Area Twenty feet of wire is to be used to form two figures. In each of the following cases, how much should be used for each figure so that the total enclosed area is maximum?

(a) Equilateral triangle and square

(b) Square and regular pentagon

(c) Regular pentagon and regular hexagon

(d) Regular hexagon and circle

What can you conclude from this pattern? {*Hint:* The area of a regular polygon with n sides of length x is $A = (n/4)[\cot(\pi/n)]x^2$. }

43. Beam Strength A wooden beam has a rectangular cross section of height h and width w (see figure). The strength S of the beam is directly proportional to the width and the square of the height. What are the dimensions of the strongest beam that can be cut from a round log of diameter 24 inches? (*Hint:* $S = kh^2w$, where k is the proportionality constant.)

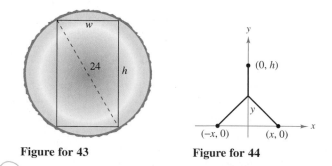

Figure for 43   **Figure for 44**

44. Minimum Length Two factories are located at the coordinates $(-x, 0)$ and $(x, 0)$ with their power supply located at the point $(0, h)$ (see figure). Find y such that the total amount of power line from the power supply to the factories is a minimum.

45. Projectile Range The range R of a projectile fired with an initial velocity v_0 at an angle θ with the horizontal is

$$R = \frac{v_0^2 \sin 2\theta}{g}$$

where g is the acceleration due to gravity. Find the angle θ such that the range is a maximum.

46. Conjecture Consider the functions $f(x) = \frac{1}{2}x^2$ and $g(x) = \frac{1}{16}x^4 - \frac{1}{2}x^2$ on the domain $[0, 4]$.

(a) Use a graphing utility to graph the functions on the specified domain.

(b) Write the vertical distance d between the functions as a function of x and use calculus to find the value of x for which d is maximum.

(c) Find the equations of the tangent lines to the graphs of f and g at the critical number found in part (b). Graph the tangent lines. What is the relationship between the lines?

(d) Make a conjecture about the relationship between tangent lines to the graphs of two functions at the value of x at which the vertical distance between the functions is greatest, and prove your conjecture.

47. Illumination A light source is located over the center of a circular table of diameter 4 feet (see figure). Find the height h of the light source such that the illumination I at the perimeter of the table is maximum if $I = k(\sin \alpha)/s^2$, where s is the slant height, α is the angle at which the light strikes the table, and k is a constant.

48. Illumination The illumination from a light source is directly proportional to the strength of the source and inversely proportional to the square of the distance from the source. Two light sources of intensities I_1 and I_2 are d units apart. What point on the line segment joining the two sources has the least illumination?

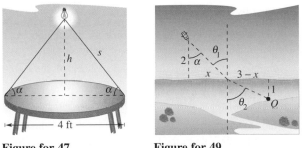

Figure for 47   Figure for 49

49. Minimum Time A man is in a boat 2 miles from the nearest point on the coast. He is to go to a point Q, located 3 miles down the coast and 1 mile inland (see figure). If he can row at 2 miles per hour and walk at 4 miles per hour, toward what point on the coast should he row in order to reach point Q in the least time?

50. Minimum Time Consider Exercise 49 if the point Q is on the shoreline rather than 1 mile inland.

(a) Write the travel time T as a function of α.

(b) Use the result of part (a) to find the minimum time to reach Q.

(c) Suppose the man can row at v_1 miles per hour and walk at v_2 miles per hour. Write the time T as a function of α. Show that the critical number of T depends only on v_1 and v_2 and not the distances. Explain how this result would be more beneficial to the man than the result of Exercise 49.

(d) Describe how to apply the result of part (c) to minimizing the cost of constructing a power transmission cable that costs c_1 dollars per mile under water and c_2 dollars per mile over land.

51. Minimum Time The conditions are the same as in Exercise 49 except that the man can row at v_1 miles per hour and walk at v_2 miles per hour. If θ_1 and θ_2 are the magnitudes of the angles, show that the man will reach point Q in the least time when

$$\frac{\sin \theta_1}{v_1} = \frac{\sin \theta_2}{v_2}.$$

52. Minimum Time When light waves, traveling in a transparent medium, strike the surface of a second transparent medium, they change directions. This change of direction is called *refraction* and is defined by **Snell's Law of Refraction,**

$$\frac{\sin \theta_1}{v_1} = \frac{\sin \theta_2}{v_2}$$

where θ_1 and θ_2 are the magnitudes of the angles shown in the figure and v_1 and v_2 are the velocities of light in the two media. Show that this problem is equivalent to that of Exercise 51, and that light waves traveling from P to Q follow the path of minimum time.

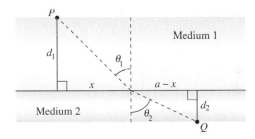

53. Sketch the graph of $f(x) = 2 - 2 \sin x$ on the interval $[0, \pi/2]$.

(a) Find the distance from the origin to the y-intercept and the distance from the origin to the x-intercept.

(b) Express the distance d from the origin to a point on the graph of f as a function of x. Use your graphing utility to graph d and find the minimum distance.

(c) Use calculus and the root finding capabilities of a graphing utility to find the value of x that minimizes the function d on the interval $[0, \pi/2]$. What is the minimum distance?

(Submitted by Tim Chapell, Penn Valley Community College, Kansas City, MO.)

54. Minimum Cost An offshore oil well is 2 kilometers off the coast. The refinery is 4 kilometers down the coast. If laying pipe in the ocean is twice as expensive as on land, what path should the pipe follow in order to minimize the cost?

55. Minimum Force A component is designed to slide a block of steel with weight W across a table and into a chute (see figure.) The motion of the block is resisted by a frictional force proportional to its apparent weight. (Let k be the constant of proportionality.) Find the minimum force F needed to slide the block, and find the corresponding value of θ. (*Hint:* $F \cos \theta$ is the force in the direction of motion, and $F \sin \theta$ is the amount of force tending to lift the block. Therefore, the apparent weight of the block is $W - F \sin \theta$.)

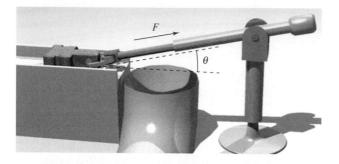

56. Volume A sector with central angle θ is cut from a circle of radius 12 inches (see figure), and the edges of the sector are brought together to form a cone. Find the magnitude of θ such that the volume of the cone is a maximum.

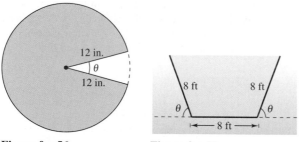

Figure for 56   Figure for 57

57. Numerical, Graphical, and Analytic Analysis The cross sections of an irrigation canal are isosceles trapezoids of which three sides are 8 feet long (see figure). Determine the angle of elevation θ of the sides such that the area of the cross section is a maximum by completing the following.

(a) Analytically complete six rows of a table such as the one below. (The first two rows are shown.)

Base 1	Base 2	Altitude	Area
8	$8 + 16 \cos 10°$	$8 \sin 10°$	≈ 22.1
8	$8 + 16 \cos 20°$	$8 \sin 20°$	≈ 42.5

(b) Use a graphing utility to generate additional rows of the table and estimate the maximum cross-sectional area. (*Hint:* Use the *table* feature of the graphing utility.)

(c) Write the cross-sectional area A as a function of θ.

(d) Use calculus to find the critical number of the function in part (c) and find the angle that will yield the maximum cross-sectional area.

(e) Use a graphing utility to graph the function in part (c) and verify the maximum cross-sectional area.

58. Maximum Profit Assume that the amount of money deposited in a bank is proportional to the square of the interest rate the bank pays on this money. Furthermore, the bank can reinvest this money at 12%. Find the interest rate the bank should pay to maximize profit. (Use the simple interest formula.)

59. Minimum Cost The ordering and transportation cost C of the components used in manufacturing a certain product is

$$C = 100\left(\frac{200}{x^2} + \frac{x}{x + 30}\right), \qquad x \geq 1$$

where C is measured in thousands of dollars and x is the order size in hundreds. Find the order size that minimizes the cost. (*Hint:* Use the *root* feature of a graphing utility.)

60. Diminishing Returns The profit P (in thousands of dollars) for a company spending an amount s (in thousands of dollars) on advertising is

$$P = -\tfrac{1}{10}s^3 + 6s^2 + 400.$$

(a) Find the amount of money the company should spend on advertising in order to yield a maximum profit.

(b) The *point of diminishing returns* is the point at which the rate of growth of the profit function begins to decline. Find the point of diminishing returns.

Minimum Distance In Exercises 61–63, consider a fuel distribution center located at the origin of the rectangular coordinate system (units in miles; see figures). The center supplies three factories with coordinates (4, 1), (5, 6), and (10, 3). A trunk line will run from the distribution center along the line $y = mx$, and feeder lines will run to the three factories. The objective is to find m such that the lengths of the feeder lines are minimized.

61. Minimize the sum of the squares of the lengths of vertical feeder lines given by

$$S_1 = (4m - 1)^2 + (5m - 6)^2 + (10m - 3)^2.$$

Find the equation for the trunk line by this method and then determine the sum of the lengths of the feeder lines.

62. Minimize the sum of the absolute values of the lengths of vertical feeder lines given by

$$S_2 = |4m - 1| + |5m - 6| + |10m - 3|.$$

Find the equation for the trunk line by this method and then determine the sum of the lengths of the feeder lines. (*Hint:* Use a graphing utility to graph the function S_2 and approximate the required critical number.)

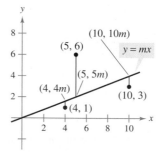

Figure for 61 and 62

63. Minimize the sum of the perpendicular distances (see Exercises 83–88 in Section P.2) from the trunk line to the factories given by

$$S_3 = \frac{|4m - 1|}{\sqrt{m^2 + 1}} + \frac{|5m - 6|}{\sqrt{m^2 + 1}} + \frac{|10m - 3|}{\sqrt{m^2 + 1}}.$$

Find the equation for the trunk line by this method and then determine the sum of the lengths of the feeder lines. (*Hint:* Use a graphing utility to graph the function S_3 and approximate the required critical number.)

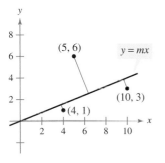

64. *Area*   Consider a symmetric cross inscribed in a circle of radius r (see figure).

(a) Write the area A of the cross as a function of x and find the value of x that maximizes the area.

(b) Write the area A of the cross as a function of θ and find the value of θ that maximizes the area.

(c) Show that the critical numbers of parts (a) and (b) yield the same maximum area. What is that area?

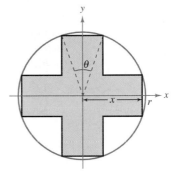

SECTION PROJECT   CONNECTICUT RIVER

Whenever the Connecticut River reaches a level of 105 feet above sea level, two Northampton, Massachusetts flood control station operators begin a round-the-clock river watch. Every two hours, they check the height of the river, using a scale marked off in tenths of a foot, and record the data in a log book. In the spring of 1996, the flood watch lasted from April 4, when the river reached 105 feet and was rising at 0.2 foot per hour, until April 25, when the level subsided again to 105 feet. Between those dates, their log shows that the river rose and fell several times, at one point coming close to the 115-foot mark. If the river had reached 115 feet, the city would have closed down Mount Tom Road (Route 5, south of Northampton).

The graph below shows the *rate of change* of the level of the river during one portion of the flood watch. Use the graph to answer the following questions.

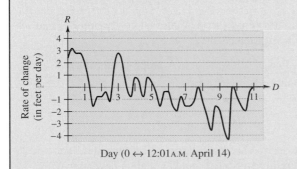

Day (0 ↔ 12:01 A.M. April 14)

(a) On what date was the river rising most rapidly? How do you know?

(b) On what date was the river falling most rapidly? How do you know?

(c) There were two dates in a row on which the river rose, then fell, then rose again during the course of the day. On which days did this occur, and how do you know?

(d) At one minute past midnight, April 14, the river level was 111.0 feet. Estimate its height 24 hours later and 48 hours later. Explain how you made your estimates.

(e) The river crested at 114.4 feet. On what date do you think this occurred?

(Submitted by Mary Murphy, Smith College, Northampton, MA)

UPI/Corbis-Bettmann

• Approximate a zero of a function using Newton's Method.

Newton's Method

In this section you will study a technique for approximating the real zeros of a function. The technique is called **Newton's Method,** and it uses tangent lines to approximate the graph of the function near its x-intercepts.

To see how Newton's Method works, consider a function f that is continuous on the interval $[a, b]$ and differentiable on the interval (a, b). If $f(a)$ and $f(b)$ differ in sign, then, by the Intermediate Value Theorem, f must have at least one zero in the interval (a, b). Suppose you estimate this zero to occur at

$$x = x_1 \qquad \text{First estimate}$$

as shown in Figure 3.60(a). Newton's Method is based on the assumption that the graph of f and the tangent line at $(x_1, f(x_1))$ both cross the x-axis at *about* the same point. Because you can easily calculate the x-intercept for this tangent line, you can use it as a second (and, usually, better) estimate for the zero of f. The tangent line passes through the point $(x_1, f(x_1))$ with a slope of $f'(x_1)$. In point-slope form, the equation of the tangent line is therefore

$$y - f(x_1) = f'(x_1)(x - x_1)$$
$$y = f'(x_1)(x - x_1) + f(x_1).$$

Letting $y = 0$ and solving for x produces

$$x = x_1 - \frac{f(x_1)}{f'(x_1)}.$$

So, from the initial estimate x_1 you obtain a new estimate

$$x_2 = x_1 - \frac{f(x_1)}{f'(x_1)}. \qquad \text{Second estimate (see Figure 3.60b)}$$

You can improve on x_2 and calculate yet a third estimate

$$x_3 = x_2 - \frac{f(x_2)}{f'(x_2)}. \qquad \text{Third estimate}$$

Repeated application of this process is called Newton's Method.

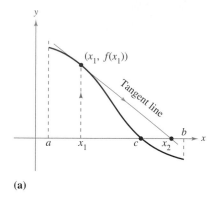

(a)

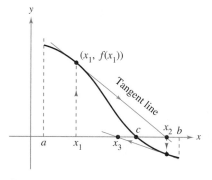

(b)

The x-intercept of the tangent line approximates the zero of f.

Figure 3.60

NEWTON'S METHOD

Isaac Newton first described the method for approximating the real zeros of a function in his text *Method of Fluxions.* Although the book was written in 1671, it was not published until 1736. Meanwhile, in 1690, Joseph Raphson (1648–1715) published a paper describing a method for approximating the real zeros of a function that was very similar to Newton's. For this reason, the method is often referred to as the Newton-Raphson method.

Newton's Method for Approximating the Zeros of a Function

Let $f(c) = 0$, where f is differentiable on an open interval containing c. Then, to approximate c, use the following steps.

1. Make an initial estimate x_1 that is "close to" c. (A graph is helpful.)
2. Determine a new approximation

$$x_{n+1} = x_n - \frac{f(x_n)}{f'(x_n)}.$$

3. If $|x_n - x_{n+1}|$ is within the desired accuracy, let x_{n+1} serve as the final approximation. Otherwise, return to Step 2 and calculate a new approximation.

Each successive application of this procedure is called an **iteration.**

NOTE   For many functions, just a few iterations of Newton's Method will produce approximations having very small errors, as shown in Example 1.

Example 1   Using Newton's Method

Calculate three iterations of Newton's Method to approximate a zero of $f(x) = x^2 - 2$. Use $x_1 = 1$ as the initial guess.

Solution   Because $f(x) = x^2 - 2$, you have $f'(x) = 2x$, and the iterative process is given by the formula

$$x_{n+1} = x_n - \frac{f(x_n)}{f'(x_n)} = x_n - \frac{x_n^2 - 2}{2x_n}.$$

The calculations for three iterations are shown in the table.

n	x_n	$f(x_n)$	$f'(x_n)$	$\dfrac{f(x_n)}{f'(x_n)}$	$x_n - \dfrac{f(x_n)}{f'(x_n)}$
1	1.000000	-1.000000	2.000000	-0.500000	1.500000
2	1.500000	0.250000	3.000000	0.083333	1.416667
3	1.416667	0.006945	2.833334	0.002451	1.414216
4	1.414216				

Of course, in this case you know that the two zeros of the function are $\pm\sqrt{2}$. To six decimal places, $\sqrt{2} = 1.414214$. So, after only three iterations of Newton's Method, you have obtained an approximation that is within 0.000002 of an actual root. The first iteration of this process is shown in Figure 3.61.

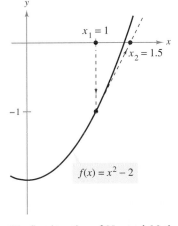

The first iteration of Newton's Method
Figure 3.61

Example 2   Using Newton's Method

Use Newton's Method to approximate the zeros of

$$f(x) = 2x^3 + x^2 - x + 1.$$

Continue the iterations until two successive approximations differ by less than 0.0001.

Solution   Begin by sketching a graph of f, as shown in Figure 3.62. From the graph, you can observe that the function has only one zero, which occurs near $x = -1.2$. Next, differentiate f and form the iterative formula

$$x_{n+1} = x_n - \frac{f(x_n)}{f'(x_n)} = x_n - \frac{2x_n^3 + x_n^2 - x_n + 1}{6x_n^2 + 2x_n - 1}.$$

The calculations are shown in the table.

n	x_n	$f(x_n)$	$f'(x_n)$	$\dfrac{f(x_n)}{f'(x_n)}$	$x_n - \dfrac{f(x_n)}{f'(x_n)}$
1	-1.20000	0.18400	5.24000	0.03511	-1.23511
2	-1.23511	-0.00771	5.68276	-0.00136	-1.23375
3	-1.23375	0.00001	5.66533	0.00000	-1.23375
4	-1.23375				

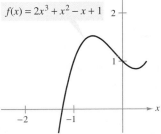

After three iterations of Newton's Method, the zero of f is approximated to the desired accuracy.
Figure 3.62

Because two successive approximations differ by less than the required 0.0001, you can estimate the zero of f to be -1.23375.

When, as in Examples 1 and 2, the approximations approach a limit, the sequence $x_1, x_2, x_3, \ldots, x_n, \ldots$ is said to **converge.** Moreover, if the limit is c, it can be shown that c must be a zero of f.

Newton's Method does not always yield a convergent sequence. One way this can happen is shown in Figure 3.63. Because Newton's Method involves division by $f'(x_n)$, it is clear that the method will fail if the derivative is zero for any x_n in the sequence. When you encounter this problem, you can usually overcome it by choosing a different value for x_1. Another way Newton's Method can fail is illustrated in the next example.

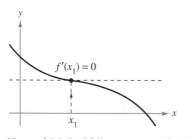

Newton's Method fails to converge if $f'(x_n) = 0$.
Figure 3.63

Example 3 **An Example in Which Newton's Method Fails**

Using $x_1 = 0.1$, show that Newton's Method fails to converge for $f(x) = x^{1/3}$.

Solution Because $f'(x) = \frac{1}{3}x^{-2/3}$, the iterative formula is

$$
\begin{aligned}
x_{n+1} &= x_n - \frac{f(x_n)}{f'(x_n)} \\
&= x_n - \frac{x_n^{1/3}}{\frac{1}{3}x_n^{-2/3}} \\
&= x_n - 3x_n \\
&= -2x_n.
\end{aligned}
$$

The calculations are shown in the table. This table and Figure 3.64 indicate that x_n continues to increase in magnitude as $n \to \infty$, and thus the limit of the sequence does not exist.

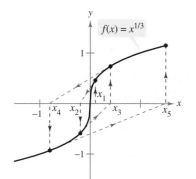

Newton's Method fails to converge for every x-value other than the actual zero of f.
Figure 3.64

n	x_n	$f(x_n)$	$f'(x_n)$	$\dfrac{f(x_n)}{f'(x_n)}$	$x_n - \dfrac{f(x_n)}{f'(x_n)}$
1	0.10000	0.46416	1.54720	0.30000	−0.20000
2	−0.20000	−0.58480	0.97467	−0.60000	0.40000
3	0.40000	0.73681	0.61401	1.20000	−0.80000
4	−0.80000	−0.92832	0.38680	−2.40000	1.60000

NOTE In Example 3, the initial estimate $x_1 = 0.1$ fails to produce a convergent sequence. Try showing that Newton's Method also fails for every other choice of x_1 (other than the actual zero).

It can be shown that a condition sufficient to produce convergence of Newton's Method to a zero of f is that

$$\left| \frac{f(x)\,f''(x)}{[f'(x)]^2} \right| < 1 \qquad \text{Condition for convergence}$$

on an open interval containing the zero. For instance, in Example 1 this test would yield $f(x) = x^2 - 2,\, f'(x) = 2x,\, f''(x) = 2$, and

$$\left| \frac{f(x)\,f''(x)}{[f'(x)]^2} \right| = \left| \frac{(x^2 - 2)(2)}{4x^2} \right| = \left| \frac{1}{2} - \frac{1}{x^2} \right|. \qquad \text{Example 1}$$

On the interval $(1, 3)$, this quantity is less than 1 and therefore the convergence of Newton's Method is guaranteed. On the other hand, in Example 3, you have $f(x) = x^{1/3},\, f'(x) = \frac{1}{3}x^{-2/3},\, f''(x) = -\frac{2}{9}x^{-5/3}$, and

$$\left| \frac{f(x)\,f''(x)}{[f'(x)]^2} \right| = \left| \frac{x^{1/3}(-2/9)(x^{-5/3})}{(1/9)(x^{-4/3})} \right| = 2 \qquad \text{Example 3}$$

which is not less than 1 for any value of x, so you cannot conclude that Newton's Method will converge.

Algebraic Solutions of Polynomial Equations

The zeros of some functions, such as

$$f(x) = x^3 - 2x^2 - x + 2$$

can be found by simple algebraic techniques, such as factoring. The zeros of other functions, such as

$$f(x) = x^3 - x + 1$$

cannot be found by *elementary* algebraic methods. This particular function has only one real zero, and by using more advanced algebraic techniques you can determine the zero to be

$$x = -\sqrt[3]{\frac{3 - \sqrt{23/3}}{6}} - \sqrt[3]{\frac{3 + \sqrt{23/3}}{6}}.$$

Because the *exact* solution is written in terms of square roots and cube roots, it is called a **solution by radicals.**

NOTE   Try approximating the real zero of $f(x) = x^3 - x + 1$ and compare your result with the exact solution shown above.

The determination of radical solutions of a polynomial equation is one of the fundamental problems of algebra. The earliest such result is the Quadratic Formula, which dates back at least to Babylonian times. The general formula for the zeros of a cubic function was developed much later. In the sixteenth century an Italian mathematician, Jerome Cardan, published a method for finding radical solutions to cubic and quartic equations. Then, for 300 years, the problem of finding a general quintic formula remained open. Finally, in the nineteenth century, the problem was answered independently by two young mathematicians. Niels Henrik Abel, a Norwegian mathematician, and Evariste Galois, a French mathematician, proved that it is not possible to solve a *general* fifth- (or higher-) degree polynomial equation by radicals. Of course, you can solve particular fifth-degree equations such as $x^5 - 1 = 0$, but Abel and Galois were able to show that no general *radical* solution exists.

The Granger Collection

NIELS HENRIK ABEL (1802–1829)

The Granger Collection

EVARISTE GALOIS (1811–1832)

Although the lives of both Abel and Galois were brief, their work in the fields of analysis and abstract algebra was far-reaching.

EXERCISES FOR SECTION 3.8

In Exercises 1–4, complete two iterations of Newton's Method for the function using the indicated initial guess.

1. $f(x) = x^2 - 3$, $x_1 = 1.7$ **2.** $f(x) = 2x^2 - 3$, $x_1 = 1$

3. $f(x) = \sin x$, $x_1 = 3$ **4.** $f(x) = \tan x$, $x_1 = 0.1$

In Exercises 5–14, approximate the zero(s) of the function. Use Newton's Method and continue the process until two successive approximations differ by less than 0.001. Then find the zero(s) using a graphing utility and compare the results.

5. $f(x) = x^3 + x - 1$ **6.** $f(x) = x^5 + x - 1$

7. $f(x) = 3\sqrt{x-1} - x$ **8.** $f(x) = x - 2\sqrt{x+1}$

9. $f(x) = x^3 + 3$ **10.** $f(x) = 1 - 2x^3$

11. $f(x) = x^3 - 3.9x^2 + 4.79x - 1.881$

12. $f(x) = \frac{1}{2}x^4 - 3x - 3$

13. $f(x) = x + \sin(x + 1)$ **14.** $f(x) = x^3 - \cos x$

In Exercises 15–18, apply Newton's Method to approximate the x-value of the indicated point(s) of intersection of the two graphs. Continue the process until two successive approximations differ by less than 0.001. [*Hint:* Let $h(x) = f(x) - g(x)$.]

15. $f(x) = 2x + 1$
$\quad g(x) = \sqrt{x + 4}$

16. $f(x) = 3 - x$
$\quad g(x) = 1/(x^2 + 1)$

17. $f(x) = x$
$\quad g(x) = \tan x$

18. $f(x) = x^2$
$\quad g(x) = \cos x$

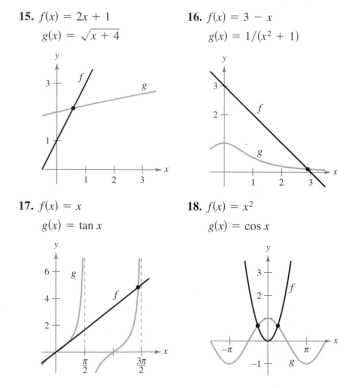

In Exercises 19 and 20, use Newton's Method to obtain a general rule for approximating the required radical.

19. $x = \sqrt{a}$ [*Hint:* Consider $f(x) = x^2 - a$.]

20. $x = \sqrt[n]{a}$ [*Hint:* Consider $f(x) = x^n - a$.]

In Exercises 21–24, use the results of Exercises 19 and 20 to approximate the indicated radical to three decimal places.

21. $\sqrt{7}$ **22.** $\sqrt{5}$

23. $\sqrt[4]{6}$ **24.** $\sqrt[3]{15}$

In Exercises 25 and 26, approximate π to three decimal places using Newton's Method and the given function.

25. $f(x) = 1 + \cos x$ **26.** $f(x) = \tan x$

In Exercises 27–30, apply Newton's Method using the indicated initial guess, and explain why the method fails.

27. $y = 2x^3 - 6x^2 + 6x - 1$, $x_1 = 1$

28. $y = 4x^3 - 12x^2 + 12x - 3$, $x_1 = \frac{3}{2}$

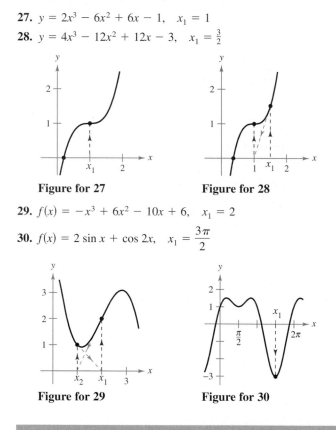

Figure for 27   Figure for 28

29. $f(x) = -x^3 + 6x^2 - 10x + 6$, $x_1 = 2$

30. $f(x) = 2\sin x + \cos 2x$, $x_1 = \dfrac{3\pi}{2}$

Figure for 29   Figure for 30

Getting at the Concept

31. In your own words and using a sketch, describe Newton's Method for approximating the zeros of a function.

32. Under what conditions will Newton's Method fail?

Fixed Point In Exercises 33 and 34, approximate the fixed point of the function to two decimal places. [A *fixed point* x_0 of a function f is a value of x such that $f(x_0) = x_0$.]

33. $f(x) = \cos x$ **34.** $f(x) = \cot x$, $0 < x < \pi$

35. Writing   Consider the function $f(x) = x^3 - 3x^2 + 3$.

(a) Use a graphing utility to obtain the graph of f.

(b) Use Newton's Method with $x_1 = 1$ as an initial guess.

(c) Repeat part (b) using $x_1 = \frac{1}{4}$ as an initial guess and observe that the result is different.

(d) To understand why the results in parts (b) and (c) are different, sketch the tangent lines to the graph of f at the points $(1, f(1))$ and $\left(\frac{1}{4}, f\left(\frac{1}{4}\right)\right)$. Find the x-intercept of each tangent line and compare the intercepts with the first iteration of Newton's Method using the respective initial guesses.

(e) Write a short paragraph summarizing how Newton's Method works. Use the results of this exercise to describe why it is important to select the initial guess carefully.

36. Writing   Repeat the steps in Exercise 35 for the function $f(x) = \sin x$ with initial guesses of $x_1 = 1.8$ and $x_1 = 3$.

37. Use Newton's Method to show that the equation

$$x_{n+1} = x_n(2 - ax_n)$$

can be used to approximate $1/a$ if x_1 is an initial guess of the reciprocal of a. Note that this method of approximating reciprocals uses only the operations of multiplication and subtraction. [*Hint:* Consider $f(x) = (1/x) - a$.]

38. Use the result of Exercise 37 to approximate the indicated reciprocal to three decimal places.

(a) $\frac{1}{3}$ (b) $\frac{1}{11}$

In Exercises 39 and 40, approximate the critical number of f on the interval $(0, \pi)$. Sketch the graph of f, labeling any extrema.

39. $f(x) = x \cos x$

40. $f(x) = x \sin x$

In Exercises 41–44, we review some typical problems from the previous sections of this chapter. In each case, use Newton's Method to approximate the solution.

41. Minimum Distance   Find the point on the graph of $f(x) = 4 - x^2$ that is closest to the point $(1, 0)$.

42. Minimum Distance   Find the point on the graph of $f(x) = x^2$ that is closest to the point $(4, -3)$.

43. Minimum Time   You are in a boat 2 miles from the nearest point on the coast (see figure). You are to go to a point Q, which is 3 miles down the coast and 1 mile inland. You can row at 3 miles per hour and walk at 4 miles per hour. Toward what point on the coast should you row in order to reach Q in the least time?

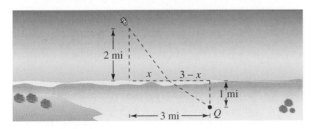

44. Medicine   The concentration C of a certain chemical in the bloodstream t hours after injection into muscle tissue is given by $C = (3t^2 + t)/(50 + t^3)$. When is the concentration greatest?

45. Advertising Costs   A company that produces portable cassette players estimates that the profit for selling a particular model is

$$P = -76x^3 + 4830x^2 - 320{,}000, \quad 0 \le x \le 60$$

where P is the profit in dollars and x is the advertising expense in 10,000s of dollars (see figure). According to this model, find the smaller of two advertising amounts that yield a profit P of $2,500,000.

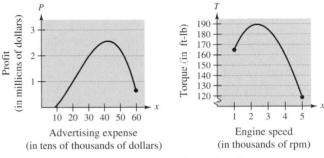

Figure for 45   Figure for 46

46. Engine Power   The torque produced by a compact automobile engine is approximated by the model

$$T = 0.808x^3 - 17.974x^2 + 71.248x + 110.843, \quad 1 \le x \le 5$$

where T is the torque in foot-pounds and x is the engine speed in thousands of revolutions per minute (see figure). Approximate the two engine speeds that yield a torque T of 170 foot-pounds.

True or False?   **In Exercises 47–50, determine whether the statement is true or false. If it is false, explain why or give an example that shows it is false.**

47. The zeros of $f(x) = p(x)/q(x)$ coincide with the zeros of $p(x)$.

48. If the coefficients of a polynomial function are all positive, then the polynomial has no positive zeros.

49. If $f(x)$ is a cubic polynomial such that $f'(x)$ is never zero, then any initial guess will force Newton's Method to converge to the zero of f.

50. The roots of $\sqrt{f(x)} = 0$ coincide with the roots of $f(x) = 0$.

In Exercises 51 and 52, write a computer program or use a spreadsheet to find the zeros of a function using Newton's Method. Approximate the zeros of the function accurate to three decimal places. The output should be a table with the following headings.

$$n, \quad x_n, \quad f(x_n), \quad f'(x_n), \quad \frac{f(x_n)}{f'(x_n)}, \quad x_n - \frac{f(x_n)}{f'(x_n)}$$

51. $f(x) = \frac{1}{4}x^3 - 3x^2 + \frac{3}{4}x - 2$

52. $f(x) = \sqrt{4 - x^2}\, \sin(x - 2)$

Differentials

- Understand the concept of a tangent line approximation.
- Compare the value of the differential, dy, with the actual change in y, Δy.
- Estimate a propagated error using a differential.
- Find the differential of a function using differentiation formulas.

EXPLORATION

Tangent Line Approximation
Use a graphing utility to sketch the graph of

$$f(x) = x^2.$$

In the same viewing window, sketch the graph of the tangent line to the graph of f at the point $(1, 1)$. Zoom in twice on the point of tangency. Does your graphing utility distinguish between the two graphs? Use the *trace* feature to compare the two graphs. As the x-values get closer to 1, what can you say about the y-values?

Linear Approximations

Newton's Method (Section 3.8) is an example of the use of a tangent line to a graph to approximate the graph. In this section, you will study other situations in which the graph of a function can be approximated by a straight line.

To begin, consider a function f that is differentiable at c. The equation for the tangent line at the point $(c, f(c))$ is given by

$$y - f(c) = f'(c)(x - c)$$
$$y = f(c) + f'(c)(x - c).$$

Because c is a constant, y is a linear function of x. Moreover, by restricting the values of x to be sufficiently close to c, the values of y can be used as approximations (to any desired accuracy) of the values of the function f. In other words, as $x \to c$, the limit of y is $f(c)$.

Example 1   Using a Tangent Line Approximation

Find the tangent line approximation of

$$f(x) = 1 + \sin x$$

at the point $(0, 1)$. Then use a table to compare the y-values of the linear function with those of $f(x)$ in an open interval containing $x = 0$.

Solution   The derivative of f is

$$f'(x) = \cos x. \qquad \text{First derivative}$$

So, the equation of the tangent line to the graph of f at the point $(0, 1)$ is

$$y - f(0) = f'(0)(x - 0)$$
$$y - 1 = (1)(x - 0)$$
$$y = 1 + x. \qquad \text{Tangent line approximation}$$

The table compares the values of y given by this linear approximation with the values of $f(x)$ near $x = 0$. Notice that the closer x is to 0, the better the approximation is. This conclusion is reinforced by the graph shown in Figure 3.65.

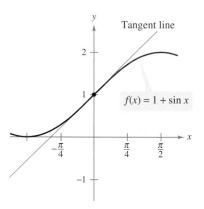

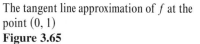

The tangent line approximation of f at the point $(0, 1)$
Figure 3.65

x	-0.5	-0.1	-0.01	0	0.01	0.1	0.5
$f(x) = 1 + \sin x$	0.521	0.9002	0.9900002	1	1.0099998	1.0998	1.479
$y = 1 + x$	0.5	0.9	0.99	1	1.01	1.1	1.5

NOTE   Be sure you see that this linear approximation of $f(x) = 1 + \sin x$ depends on the point of tangency. At a different point on the graph of f, you would obtain a different tangent line approximation.

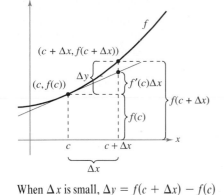

When Δx is small, $\Delta y = f(c + \Delta x) - f(c)$ is approximated by $f'(c)\Delta x$.

Figure 3.66

Differentials

When the tangent line to the graph of f at the point $(c, f(c))$

$$y = f(c) + f'(c)(x - c) \qquad \text{Tangent line at } (c, f(c))$$

is used as an approximation to the graph of f, the quantity $x - c$ is called the change in x, and is denoted by Δx, as shown in Figure 3.66. When Δx is small, the change in y (denoted by Δy) can be approximated as follows.

$$\Delta y = f(c + \Delta x) - f(c) \qquad \text{Actual change in } y$$
$$\approx f'(c)\Delta x \qquad \text{Approximate change in } y$$

For such an approximation, the quantity Δx is traditionally denoted by dx, and is called the **differential of x.** The expression $f'(x)\,dx$ is denoted by dy, and is called the **differential of y.**

Definition of Differentials

Let $y = f(x)$ represent a function that is differentiable in an open interval containing x. The **differential of x** (denoted by dx) is any nonzero real number. The **differential of y** (denoted by dy) is

$$dy = f'(x)\,dx.$$

In many types of applications, the differential of y can be used as an approximation of the change in y. That is,

$$\Delta y \approx dy \qquad \text{or} \qquad \Delta y \approx f'(x)dx.$$

Example 2   Comparing Δy and dy

Let $y = x^2$. Find dy when $x = 1$ and $dx = 0.01$. Compare this value with Δy for $x = 1$ and $\Delta x = 0.01$.

Solution   Because $y = f(x) = x^2$, you have $f'(x) = 2x$, and the differential dy is given by

$$dy = f'(x)\,dx = f'(1)(0.01) = 2(0.01) = 0.02. \qquad \text{Differential of } y$$

Now, using $\Delta x = 0.01$, the change in y is

$$\Delta y = f(x + \Delta x) - f(x) = f(1.01) - f(1) = (1.01)^2 - 1^2 = 0.0201.$$

Figure 3.67 shows the geometric comparison of dy and Δy. Try comparing other values of dy and Δy. You will see that the values become closer to each other as dx (or Δx) approaches 0.

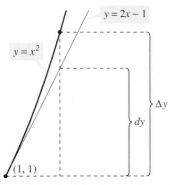

The change in y, Δy, is approximated by the differential of y, dy.

Figure 3.67

In Example 2, the tangent line to the graph of $f(x) = x^2$ at $x = 1$ is

$$y = 2x - 1 \qquad \text{or} \qquad g(x) = 2x - 1. \qquad \text{Tangent line to the graph of } f \text{ at } x = 1.$$

For x-values near 1, this line is close to the graph of f, as shown in Figure 3.67. For instance,

$$f(1.01) = 1.01^2 = 1.0201 \qquad \text{and} \qquad g(1.01) = 2(1.01) - 1 = 1.02.$$

We say that the line $y = 2x - 1$ is the **linear approximation** or **tangent line approximation** to the graph of $f(x) = x^2$ at $x = 1$.

Error Propagation

Physicists and engineers tend to make liberal use of the approximation of Δy by dy. One way this occurs in practice is in the estimation of errors propagated by physical measuring devices. For example, if you let x represent the measured value of a variable and let $x + \Delta x$ represent the exact value, then Δx is the *error in measurement*. Finally, if the measured value x is used to compute another value $f(x)$, the difference between $f(x + \Delta x)$ and $f(x)$ is the **propagated error.**

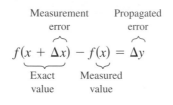

$$\underbrace{f(\overbrace{x + \Delta x}^{\text{Measurement error}}) - \underbrace{f(x)}_{\substack{\text{Measured} \\ \text{value}}}}_{\substack{\text{Exact} \\ \text{value}}} = \overbrace{\Delta y}^{\text{Propagated error}}$$

Example 3   **Estimation of Error**

The radius of a ball bearing is measured to be 0.7 inch, as shown in Figure 3.68. If the measurement is correct to within 0.01 inch, estimate the propagated error in the volume V of the ball bearing.

Solution   The formula for the volume of a sphere is $V = \frac{4}{3}\pi r^3$, where r is the radius of the sphere. So, you can write

$$r = 0.7 \qquad\qquad \text{Measured radius}$$

and

$$-0.01 \le \Delta r \le 0.01. \qquad \text{Possible error}$$

To approximate the propagated error in the volume, differentiate V to obtain $dV/dr = 4\pi r^2$ and write

$$\begin{aligned}
\Delta V &\approx dV && \text{Approximate } \Delta V \text{ by } dV.\\
&= 4\pi r^2\, dr\\
&= 4\pi(0.7)^2(\pm 0.01) && \text{Substitute for } r \text{ and } dr.\\
&\approx \pm 0.06158 \text{ in}^3.
\end{aligned}$$

So the volume has a propagated error of about 0.06 cubic inch.

Would you say that the propagated error in Example 3 is large or small? The answer is best given in *relative* terms by comparing dV with V. The ratio

$$\begin{aligned}
\frac{dV}{V} &= \frac{4\pi r^2\, dr}{\frac{4}{3}\pi r^3} && \text{Ratio of } dV \text{ to } V\\[2mm]
&= \frac{3\,dr}{r} && \text{Simplify.}\\[2mm]
&\approx \frac{3}{0.7}(\pm 0.01) && \text{Substitute for } dr \text{ and } r.\\[2mm]
&\approx \pm 0.0429
\end{aligned}$$

is called the **relative error.** The corresponding **percent error** is approximately 4.29%.

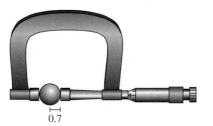

Ball bearing with measured radius that is correct to within 0.01 inch
Figure 3.68

Calculating Differentials

Each of the differentiation rules that you studied in Chapter 2 can be written in **differential form.** For example, suppose u and v are differentiable functions of x. By the definition of differentials, you have

$$du = u'\,dx \quad \text{and} \quad dv = v'\,dx.$$

Therefore, you can write the differential form of the Product Rule as follows.

$$d[uv] = \frac{d}{dx}[uv]\,dx \qquad \text{Differential of } uv$$

$$= [uv' + vu']\,dx \qquad \text{Product Rule}$$

$$= uv'\,dx + vu'\,dx$$

$$= u\,dv + v\,du$$

Differential Formulas

Let u and v be differentiable functions of x.

Constant multiple: $d[cu] = c\,du$

Sum or difference: $d[u \pm v] = du \pm dv$

Product: $d[uv] = u\,dv + v\,du$

Quotient: $d\left[\dfrac{u}{v}\right] = \dfrac{v\,du - u\,dv}{v^2}$

Example 4   **Finding Differentials**

Function	Derivative	Differential
a. $y = x^2$	$\dfrac{dy}{dx} = 2x$	$dy = 2x\,dx$
b. $y = 2\sin x$	$\dfrac{dy}{dx} = 2\cos x$	$dy = 2\cos x\,dx$
c. $y = x\cos x$	$\dfrac{dy}{dx} = -x\sin x + \cos x$	$dy = (-x\sin x + \cos x)\,dx$
d. $y = \dfrac{1}{x}$	$\dfrac{dy}{dx} = -\dfrac{1}{x^2}$	$dy = -\dfrac{dx}{x^2}$

The notation in Example 4 is called the **Leibniz notation** for derivatives and differentials, named after the German mathematician Gottfried Wilhelm Leibniz. The beauty of this notation is that it provides an easy way to remember several important calculus formulas by making it seem as though the formulas were derived from algebraic manipulations of differentials. For instance, in Leibniz notation, the *Chain Rule*

$$\frac{dy}{dx} = \frac{dy}{du}\frac{du}{dx}$$

would appear to be true because the du's cancel. Even though this reasoning is *incorrect*, the notation does help one remember the Chain Rule.

GOTTFRIED WILHELM LEIBNIZ (1646–1716)

Both Leibniz and Newton are credited with creating calculus. It was Leibniz, however, who tried to broaden calculus by developing rules and formal notation. He often spent days choosing an appropriate notation for a new concept.

Example 5   Finding the Differential of a Composite Function

$$y = f(x) = \sin 3x \qquad \text{Original function}$$
$$f'(x) = 3 \cos 3x \qquad \text{Apply Chain Rule.}$$
$$dy = f'(x)\, dx = 3 \cos 3x\, dx \qquad \text{Differential form}$$

Example 6   Finding the Differential of a Composite Function

$$y = f(x) = (x^2 + 1)^{1/2} \qquad \text{Original function}$$
$$f'(x) = \frac{1}{2}(x^2 + 1)^{-1/2}(2x) = \frac{x}{\sqrt{x^2 + 1}} \qquad \text{Apply Chain Rule.}$$
$$dy = f'(x)\, dx = \frac{x}{\sqrt{x^2 + 1}}\, dx \qquad \text{Differential form}$$

Differentials can be used to approximate function values. To do this for the function given by $y = f(x)$, you use the formula

$$f(x + \Delta x) \approx f(x) + dy = f(x) + f'(x)\, dx$$

which is derived from the approximation $\Delta y = f(x + \Delta x) - f(x) \approx dy$. The key to using this formula is to choose a value for x that makes the calculations easier, as shown in Example 7.

Example 7   Approximating Function Values

Use differentials to approximate $\sqrt{16.5}$.

Solution   Using $f(x) = \sqrt{x}$, you can write

$$f(x + \Delta x) \approx f(x) + f'(x)\, dx = \sqrt{x} + \frac{1}{2\sqrt{x}}\, dx.$$

Now, choosing $x = 16$ and $dx = 0.5$, you obtain the following approximation.

$$f(x + \Delta x) = \sqrt{16.5} \approx \sqrt{16} + \frac{1}{2\sqrt{16}}(0.5) = 4 + \left(\frac{1}{8}\right)\left(\frac{1}{2}\right) = 4.0625$$

The tangent line approximation to $f(x) = \sqrt{x}$ at $x = 16$ is the line $g(x) = \frac{1}{8}x + 2$. For x-values near 16, the graphs of f and g are close together, as shown in Figure 3.69. For instance,

$$f(16.5) = \sqrt{16.5} \approx 4.0620 \quad \text{and} \quad g(16.5) = \frac{1}{8}(16.5) + 2 = 4.0625.$$

In fact, if you use a graphing utility to zoom in near the point of tangency $(16, 4)$, you will see that the two graphs appear to coincide. Notice also that as you move farther away from the point of tangency, the linear approximation is less accurate.

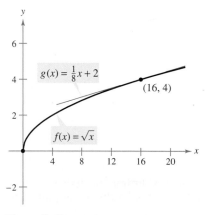

Figure 3.69

EXERCISES FOR SECTION 3.9

In Exercises 1–6, find the equation of the tangent line T to the graph of f at the indicated point. Use this linear approximation to complete the table.

x	1.9	1.99	2	2.01	2.1
$f(x)$					
$T(x)$					

	Function	Point
1.	$f(x) = x^2$	$(2, 4)$
2.	$f(x) = \dfrac{6}{x^2}$	$\left(2, \dfrac{3}{2}\right)$
3.	$f(x) = x^5$	$(2, 32)$
4.	$f(x) = \sqrt{x}$	$\left(2, \sqrt{2}\right)$
5.	$f(x) = \sin x$	$(2, \sin 2)$
6.	$f(x) = \csc x$	$(2, \csc 2)$

In Exercises 7–10, use the information to evaluate and compare Δy and dy.

7. $y = \frac{1}{2}x^3$ $x = 2$ $\Delta x = dx = 0.1$

8. $y = 1 - 2x^2$ $x = 0$ $\Delta x = dx = -0.1$

9. $y = x^4 + 1$ $x = -1$ $\Delta x = dx = 0.01$

10. $y = 2x + 1$ $x = 2$ $\Delta x = dx = 0.01$

In Exercises 11–20, find the differential dy of the given function.

11. $y = 3x^2 - 4$

12. $y = 3x^{2/3}$

13. $y = \dfrac{x + 1}{2x - 1}$

14. $y = \sqrt{9 - x^2}$

15. $y = x\sqrt{1 - x^2}$

16. $y = \sqrt{x} + \dfrac{1}{\sqrt{x}}$

17. $y = 2x - \cot^2 x$

18. $y = x \sin x$

19. $y = \dfrac{1}{3}\cos\left(\dfrac{6\pi x - 1}{2}\right)$

20. $y = \dfrac{\sec^2 x}{x^2 + 1}$

In Exercises 21–24, use differentials and the graph of f to approximate (a) $f(1.9)$ and (b) $f(2.04)$. To print an enlarged copy of the graph, go to the website *www.mathgraphs.com*.

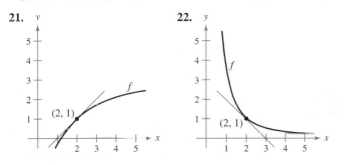

21.

22.

23.

24.

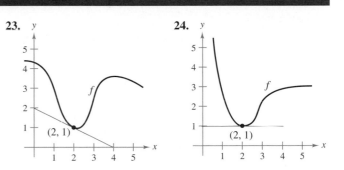

In Exercises 25–28, use differentials and the graph of g' to approximate (a) $g(2.93)$ and (b) $g(3.1)$ given that $g(3) = 8$.

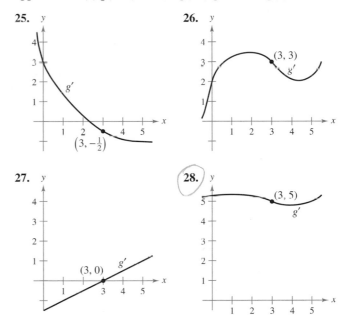

25.

26.

27.

28.

29. **Area** The measurement of the side of a square is found to be 12 inches, with a possible error of $\frac{1}{64}$ inch. Use differentials to approximate the possible propagated error in computing the area of the square.

30. **Area** The measurements of the base and altitude of a triangle are found to be 36 and 50 centimeters. The possible error in each measurement is 0.25 centimeter. Use differentials to approximate the possible propagated error in computing the area of the triangle.

31. **Area** The measurement of the radius of the end of a log is found to be 14 inches, with a possible error of $\frac{1}{4}$ inch. Use differentials to approximate the possible propagated error in computing the area of the end of the log.

32. **Volume and Surface Area** The measurement of the edge of a cube is found to be 12 inches, with a possible error of 0.03 inch. Use differentials to approximate the maximum possible propagated error in computing

   (a) the volume of the cube.

   (b) the surface area of the cube.

33. *Area* The measurement of a side of a square is found to be 15 centimeters. The possible error in measuring the side is 0.05 centimeter.

(a) Approximate the percent error in computing the area of the square.

(b) Estimate the maximum allowable percent error in measuring the side if the error in computing the area cannot exceed 2.5%.

34. *Circumference* The measurement of the circumference of a circle is found to be 56 centimeters. The possible error in measuring the circumference is 1.2 centimeters.

(a) Approximate the percent error in computing the area of the circle.

(b) Estimate the maximum allowable percent error in measuring the circumference if the error in computing the area cannot exceed 3%.

35. *Volume and Surface Area* The radius of a sphere is measured to be 6 inches, with a possible error of 0.02 inch. Use differentials to approximate the maximum possible error in calculating (a) the volume of the sphere, (b) the surface area of the sphere, and (c) the relative errors in parts (a) and (b).

36. *Profit* The profit P for a company is given by

$$P = (500x - x^2) - \left(\tfrac{1}{2}x^2 - 77x + 3000\right).$$

Approximate the change and percent change in profit as production changes from $x = 115$ to $x = 120$ units.

In Exercises 37 and 38, the thickness of the shell is 0.2 centimeter. Use differentials to approximate the volume of the shell.

37. A cylindrical shell with height 40 centimeters and radius 5 centimeters

38. A spherical shell of radius 100 centimeters

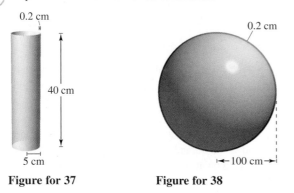

0.2 cm

0.2 cm

40 cm

5 cm

←100 cm→

Figure for 37   **Figure for 38**

39. *Pendulum* The period of a pendulum is given by

$$T = 2\pi\sqrt{\frac{L}{g}}$$

where L is the length of the pendulum in feet, g is the acceleration due to gravity, and T is the time in seconds. Suppose that the pendulum has been subjected to an increase in temperature such that the length has increased by $\tfrac{1}{2}\%$.

(a) Find the approximate percent change in the period.

(b) Using the result in part (a), find the approximate error in this pendulum clock in one day.

40. *Ohm's Law* A current of I amperes passes through a resistor of R ohms. **Ohm's Law** states that the voltage E applied to the resistor is

$$E = IR.$$

If the voltage is constant, show that the magnitude of the relative error in R caused by a change in I is equal in magnitude to the relative error in I.

41. *Triangle Measurements* The measurement of one side of a right triangle is found to be 9.5 inches, and the angle opposite that side is $26°45'$ with a possible error of $15'$.

(a) Approximate the percent error in computing the length of the hypotenuse.

(b) Estimate the maximum allowable percent error in measuring the angle if the error in computing the length of the hypotenuse cannot exceed 2%.

42. *Area* Approximate the percent error in computing the area of the triangle in Exercise 41.

43. *Projectile Motion* The range R of a projectile is

$$R = \frac{v_0^2}{32}(\sin 2\theta)$$

where v_0 is the initial velocity in feet per second and θ is the angle of elevation. If $v_0 = 2200$ feet per second and θ is changed from $10°$ to $11°$, use differentials to approximate the change in the range.

44. *Surveying* A surveyor standing 50 feet from the base of a large tree measures the angle of elevation to the top of the tree as $71.5°$. How accurately must the angle be measured if the percent error in estimating the height of the tree is to be less than 6%?

In Exercises 45–48, use differentials to approximate the value of the expression. Compare your answer with that of a calculator.

45. $\sqrt{99.4}$ **46.** $\sqrt[3]{26}$ **47.** $\sqrt[4]{624}$ **48.** $(2.99)^3$

Writing **In Exercises 49 and 50, give a short explanation of why the approximation is valid.**

49. $\sqrt{4.02} \approx 2 + \tfrac{1}{4}(0.02)$ **50.** $\tan 0.05 \approx 0 + 1(0.05)$

Getting at the Concept

51. Describe the change in accuracy of dy as an approximation for Δy when Δx is decreased.

52. When using differentials, what is meant by the terms propagated error, relative error, and percent error?

True or False? **In Exercises 53–56, determine whether the statement is true or false. If it is false, explain why or give an example that shows it is false.**

53. If $y = x + c$, then $dy = dx$.

54. If $y = ax + b$, then $\Delta y/\Delta x = dy/dx$.

55. If y is differentiable, then $\lim_{\Delta x \to 0} (\Delta y - dy) = 0$.

56. If $y = f(x)$, f is increasing and differentiable, and $\Delta x > 0$, then $\Delta y \geq dy$.

REVIEW EXERCISES FOR CHAPTER 3

3.1

1. Give the definition of a critical number, and graph a function f showing the different types of critical numbers.

2. Consider the odd function f that is continuous, differentiable, and has the functional values shown in the table.

x	-5	-4	-1	0	2	3	6
$f(x)$	1	3	2	0	-1	-4	0

(a) Determine $f(4)$.

(b) Determine $f(-3)$.

(c) Plot the points and make a possible sketch of the graph of f on the interval $[-6, 6]$. What is the smallest number of critical points in the interval? Explain.

(d) Does there exist at least one real number c in the interval $(-6, 6)$ where $f'(c) = -1$? Explain.

(e) Is it possible that $\lim\limits_{x \to 0} f(x)$ does not exist? Explain.

(f) Is it necessary that $f'(x)$ exists at $x = 2$? Explain.

In Exercises 3 and 4, find the absolute extrema of the function on the closed interval. Use a graphing utility to graph the function over the indicated interval to confirm your results.

• **3.** $g(x) = 2x + 5 \cos x$, $[0, 2\pi]$ **4.** $f(x) = \dfrac{x}{\sqrt{x^2 + 1}}$, $[0, 2]$

3.2

In Exercises 5 and 6, determine whether Rolle's Theorem can be applied to f on the closed interval $[a, b]$. If Rolle's Theorem can be applied, find all values of c in the open interval (a, b) such that $f'(c) = 0$.

5. $f(x) = (x - 2)(x + 3)^2$, $[-3, 2]$

6. $f(x) = |x - 2| - 2$, $[0, 4]$

7. Consider the function $f(x) = 3 - |x - 4|$.

(a) Graph the function and verify that $f(1) = f(7)$.

(b) Note that $f'(x)$ is not equal to zero for any x in $[1, 7]$. Explain why this does not contradict Rolle's Theorem.

✔ **8.** Can the Mean Value Theorem be applied to the function $f(x) = 1/x^2$ on the interval $[-2, 1]$? Explain.

In Exercises 9–12, find the point(s) guaranteed by the Mean Value Theorem for the closed interval $[a, b]$.

• **9.** $f(x) = x^{2/3}$, $[1, 8]$ **10.** $f(x) = \dfrac{1}{x}$, $[1, 4]$

• **11.** $f(x) = x - \cos x$, $\left[-\dfrac{\pi}{2}, \dfrac{\pi}{2}\right]$ **12.** $f(x) = \sqrt{x} - 2x$, $[0, 4]$

13. For the function $f(x) = Ax^2 + Bx + C$, determine the value of c guaranteed by the Mean Value Theorem on the interval $[x_1, x_2]$.

14. Demonstrate the result of Exercise 13 for $f(x) = 2x^2 - 3x + 1$ on the interval $[0, 4]$.

3.3

In Exercises 15–18, find the critical numbers (if any) and the open intervals on which the function is increasing or decreasing.

15. $f(x) = (x - 1)^2(x - 3)$

16. $g(x) = (x + 1)^3$

17. $h(x) = \sqrt{x}(x - 3)$, $x > 0$

18. $f(x) = \sin x + \cos x$, $[0, 2\pi]$

In Exercises 19 and 20, use the First Derivative Test to find any relative extrema of the function. Use a graphing utility to verify your results.

19. $h(t) = \dfrac{1}{4}t^4 - 8t$

20. $g(x) = \dfrac{3}{2}\sin\left(\dfrac{\pi x}{2} - 1\right)$, $[0, 4]$

21. *Harmonic Motion* The height of an object attached to a spring is given by the harmonic equation

$$y = \tfrac{1}{3}\cos 12t - \tfrac{1}{4}\sin 12t$$

where y is measured in inches and t is measured in seconds.

(a) Calculate the height and velocity of the object when $t = \pi/8$ second.

(b) Show that the maximum displacement of the object is $\dfrac{5}{12}$ inch.

(c) Find the period P of y. Also, find the frequency f (number of oscillations per second) if $f = 1/P$.

22. *Writing* The general equation giving the height of an oscillating object attached to a spring is

$$y = A \sin \sqrt{\dfrac{k}{m}}\, t + B \cos \sqrt{\dfrac{k}{m}}\, t$$

where k is the spring constant and m is the mass of the object.

(a) Show that the maximum displacement of the object is $\sqrt{A^2 + B^2}$.

(b) Show that the object oscillates with a frequency of

$$f = \dfrac{1}{2\pi}\sqrt{\dfrac{k}{m}}.$$

3.4

In Exercises 23 and 24, determine the points of inflection of the function.

23. $f(x) = x + \cos x$, $[0, 2\pi]$ **24.** $f(x) = (x + 2)^2(x - 4)$

In Exercises 25 and 26, use the Second Derivative Test to find all relative extrema.

25. $g(x) = 2x^2(1 - x^2)$ **26.** $h(t) = t - 4\sqrt{t + 1}$

Think About It In Exercises 27 and 28, sketch the graph of a function f having the indicated characteristics.

27. $f(0) = f(6) = 0$

$f'(3) = f'(5) = 0$

$f'(x) > 0$ if $x < 3$

$f'(x) > 0$ if $3 < x < 5$

$f'(x) < 0$ if $x > 5$

$f''(x) < 0$ if $x < 3$ and $x > 4$

$f''(x) > 0$, $3 < x < 4$

28. $f(0) = 4$, $f(6) = 0$

$f'(x) < 0$ if $x < 2$ and $x > 4$

$f'(2)$ does not exist.

$f'(4) = 0$

$f'(x) > 0$ if $2 < x < 4$

$f''(x) < 0$, $x \neq 2$

29. *Writing* A newspaper headline states that "The rate of growth of the national deficit is decreasing." What does this mean? What does it imply about the graph of the deficit as a function of time?

30. *Inventory Cost* The cost of inventory depends on the ordering and storage costs according to the inventory model

$$C = \left(\frac{Q}{x}\right)s + \left(\frac{x}{2}\right)r.$$

Determine the order size that will minimize the cost, assuming that sales occur at a constant rate, Q is the number of units sold per year, r is the cost of storing one unit for 1 year, s is the cost of placing an order, and x is the number of units per order.

31. *Modeling Data* Outlays for national defense D (in billions of dollars) for selected years from 1970 through 1999 are shown in the table, where t is time in years, with $t = 0$ corresponding to 1970. *(Source: U.S. Office of Management and Budget)*

t	0	5	10	15	20
D	90.4	103.1	155.1	279.0	328.3

t	25	26	27	28	29
D	309.9	302.7	309.8	310.3	320.2

(a) Use the regression capabilities of a graphing utility to fit a model of the form $D = at^4 + bt^3 + ct^2 + dt + e$ to the data.

(b) Use a graphing utility to plot the data and graph the model.

(c) For the years shown in the table, when does the model indicate that the outlay for national defense is at a maximum? When is it at a minimum?

(d) For the years shown in the table, when does the model indicate that the outlay for national defense is increasing at the greatest rate?

32. *Modeling Data* The manager of a store recorded the annual sales S (in thousands of dollars) of a product over a period of 7 years, as shown in the table, where t is the time in years, with $t = 1$ corresponding to 1991.

t	1	2	3	4	5	6	7
S	5.4	6.9	11.5	15.5	19.0	22.0	23.6

(a) Use the regression capabilities of a graphing utility to find a model of the form $S = at^3 + bt^2 + ct + d$ for the data.

(b) Use a graphing utility to plot the data and graph the model.

(c) Use calculus to find the time t when sales were increasing at the greatest rate.

(d) Do you think the model would be accurate for predicting future sales? Explain.

3.5 In Exercises 33–36, find the limit.

33. $\displaystyle\lim_{x \to \infty} \frac{2x^2}{3x^2 + 5}$

34. $\displaystyle\lim_{x \to \infty} \frac{2x}{3x^2 + 5}$

35. $\displaystyle\lim_{x \to \infty} \frac{5 \cos x}{x}$

36. $\displaystyle\lim_{x \to \infty} \frac{3x}{\sqrt{x^2 + 4}}$

In Exercises 37–40, find any vertical and horizontal asymptotes of the graph of the function. Use a graphing utility to verify your results.

37. $h(x) = \dfrac{2x + 3}{x - 4}$

38. $g(x) = \dfrac{5x^2}{x^2 + 2}$

39. $f(x) = \dfrac{3}{x} - 2$

40. $f(x) = \dfrac{3x}{\sqrt{x^2 + 2}}$

In Exercises 41–44, use a graphing utility to graph the function. Use the graph to approximate any relative extrema or asymptotes.

41. $f(x) = x^3 + \dfrac{243}{x}$

42. $f(x) = |x^3 - 3x^2 + 2x|$

43. $f(x) = \dfrac{x - 1}{1 + 3x^2}$

44. $g(x) = \dfrac{\pi^2}{3} - 4 \cos x + \cos 2x$

3.6 In Exercises 45–62, analyze and sketch the graph of the function.

45. $f(x) = 4x - x^2$

46. $f(x) = 4x^3 - x^4$

47. $f(x) = x\sqrt{16 - x^2}$

48. $f(x) = (x^2 - 4)^2$

49. $f(x) = (x - 1)^3(x - 3)^2$

50. $f(x) = (x - 3)(x + 2)^3$

51. $f(x) = x^{1/3}(x + 3)^{2/3}$

52. $f(x) = (x - 2)^{1/3}(x + 1)^{2/3}$

53. $f(x) = \dfrac{x + 1}{x - 1}$

54. $f(x) = \dfrac{2x}{1 + x^2}$

55. $f(x) = \dfrac{4}{1 + x^2}$

56. $f(x) = \dfrac{x^2}{1 + x^4}$

57. $f(x) = x^3 + x + \dfrac{4}{x}$

58. $f(x) = x^2 + \dfrac{1}{x}$

59. $f(x) = |x^2 - 9|$

60. $f(x) = |x - 1| + |x - 3|$

61. $f(x) = x + \cos x, \qquad 0 \le x \le 2\pi$

62. $f(x) = \dfrac{1}{\pi}(2 \sin \pi x - \sin 2\pi x), \qquad -1 \le x \le 1$

63. Find the maximum and minimum points on the graph of

$$x^2 + 4y^2 - 2x - 16y + 13 = 0$$

(a) without using calculus.

(b) using calculus.

64. Consider the function $f(x) = x^n$ for positive integer values of n.

(a) For what values of n does the function have a relative minimum at the origin?

(b) For what values of n does the function have a point of inflection at the origin?

3.7

65. *Minimum Distance* At noon, ship A is 100 kilometers due east of ship B. Ship A is sailing west at 12 kilometers per hour, and ship B is sailing south at 10 kilometers per hour. At what time will the ships be nearest to each other, and what will this distance be?

66. *Maximum Area* Find the dimensions of the rectangle of maximum area, with sides parallel to the coordinate axes, that can be inscribed in the ellipse given by

$$\dfrac{x^2}{144} + \dfrac{y^2}{16} = 1.$$

67. *Minimum Length* A right triangle in the first quadrant has the coordinate axes as sides, and the hypotenuse passes through the point $(1, 8)$. Find the vertices of the triangle such that the length of the hypotenuse is minimum.

68. *Minimum Length* The wall of a building is to be braced by a beam that must pass over a parallel fence 5 feet high and 4 feet from the building. Find the length of the shortest beam that can be used.

69. *Maximum Area* Three sides of a trapezoid have the same length s. Of all such possible trapezoids, show that the one of maximum area has a fourth side of length $2s$.

70. *Maximum Area* Show that the greatest area of any rectangle inscribed in a triangle is one half that of the triangle.

71. *Minimum Distance* Find the length of the longest pipe that can be carried level around a right-angle corner at the intersection of two corridors of widths 4 feet and 6 feet. (Do not use trigonometry.)

72. *Minimum Distance* Rework Exercise 71, given corridors of widths a meters and b meters.

73. *Minimum Distance* A hallway of width 6 feet meets a hallway of width 9 feet at right angles. Find the length of the longest pipe that can be carried level around this corner. [*Hint:* If L is the length of the pipe, show that

$$L = 6 \csc \theta + 9 \csc\left(\dfrac{\pi}{2} - \theta\right)$$

where θ is the angle between the pipe and the wall of the narrower hallway.]

74. *Minimum Distance* Rework Exercise 73, given that one hallway is of width a meters and the other is of width b meters. Show that the result is the same as in Exercise 72.

Minimum Cost **In Exercises 75 and 76, find the speed v, in miles per hour, that will minimize costs on a 110-mile delivery trip. The cost per hour for fuel is C dollars, and the driver is paid W dollars per hour. (Assume there are no costs other than wages and fuel.)**

75. Fuel cost: $C - \dfrac{v^2}{600}$

Driver: $W = \$5$

76. Fuel cost: $C - \dfrac{v^2}{500}$

Driver: $W = \$7.50$

3.8 **In Exercises 77 and 78, use Newton's Method to approximate any real zeros of the function accurate to three decimal places. Use the root-finding capabilities of a graphing utility to verify your results.**

77. $f(x) = x^3 - 3x - 1$

78. $f(x) = x^3 + 2x + 1$

In Exercises 79 and 80, use Newton's Method to approximate, to three decimal places, the x-value of the points of intersection of the equations. Use a graphing utility to verify your results.

79. $y = x^4$

$y = x + 3$

80. $y = \sin \pi x$

$y = 1 - x$

3.9 **In Exercises 81 and 82, find the differential dy.**

81. $y = x(1 - \cos x)$

82. $y = \sqrt{36 - x^2}$

83. *Surface Area and Volume* The diameter of a sphere is measured to be 18 centimeters, with a maximum possible error of 0.05 centimeter. Use differentials to approximate the possible propagated error and percent error in calculating the surface area and the volume of the sphere.

84. *Demand Function* A company finds that the demand for its commodity is $p = 75 - \frac{1}{4}x$. If x changes from 7 to 8, find and compare the values of Δp and dp.

P.S. Problem Solving

1. Prove Darboux's Theorem: Let f be differentiable on the closed interval $[a, b]$ such that $f'(a) = y_1$ and $f'(b) = y_2$. If d lies between y_1 and y_2, then there exists c in (a, b) such that $f'(c) = d$.

2. (a) Let $V = x^3$. Find dV and ΔV. Show that for small values of x, the difference $\Delta V - dV$ is very small in the sense that there exists ε such that $\Delta V - dV = \varepsilon \Delta x$, where $\varepsilon \to 0$ as $\Delta x \to 0$.

   (b) Generalize this result by showing that if $y = f(x)$ is a differentiable function, then $\Delta y - dy = \varepsilon \Delta x$, where $\varepsilon \to 0$ as $\Delta x \to 0$.

3. (a) Graph the fourth-degree polynomial $p(x) = ax^4 - 6x^2$ for $a = -3, -2, -1, 0, 1, 2,$ and 3. For what values of the constant a does p have a relative minimum or relative maximum?

   (b) Show that p has a relative maximum for all values of the constant a.

   (c) Determine analytically the values of a for which p has a relative minimum.

   (d) Let $(x, y) = (x, p(x))$ be a relative extremum of p. Show that (x, y) lies on the graph of $y = -3x^2$. Verify this result graphically by graphing $y = -3x^2$ together with the seven curves from part (a).

4. Let f and g be continuous functions on $[a, b]$ and differentiable on (a, b). Prove that if $f(a) = g(a)$ and $g'(x) > f'(x)$ for all x in (a, b), then $g(b) > f(b)$.

5. Graph the fourth-degree polynomial $p(x) = x^4 + ax^2 + 1$ for various values of the constant a.

   (a) Determine the values of a for which p has exactly one relative minimum.

   (b) Determine the values of a for which p has exactly one relative maximum.

   (c) Determine the values of a for which p has exactly two relative minima.

   (d) Show that the graph of p cannot have exactly two relative extrema.

6. (a) Let $f(x) = ax^2 + bx + c$, $a \neq 0$ be a quadratic polynomial. How many points of inflection does the graph of f have?

   (b) Let $f(x) = ax^3 + bx^2 + cx + d$, $a \neq 0$ be a cubic polynomial. How many points of inflection does the graph of f have?

   (c) Suppose the function $y = f(x)$ satisfies the equation $\dfrac{dy}{dx} = ky(L - y)$, where k and L are positive constants. Show that the graph of f has a point of inflection at the point where $y = \dfrac{L}{2}$. (This equation is called the **logistics differential equation.**)

7. Let $f(x) = \dfrac{c}{x} + x^2$. Determine all values of the constant c such that f has a relative minimum, but no relative maximum.

8. The amount of illumination of a surface is proportional to the intensity of the light source, inversely proportional to the square of the distance from the light source, and proportional to $\sin \theta$, where θ is the angle at which the light strikes the surface. A rectangular room measures 10 feet by 24 feet, with a 10-foot ceiling. Determine the height at which the light should be placed to allow the corners of the floor to receive as much light as possible.

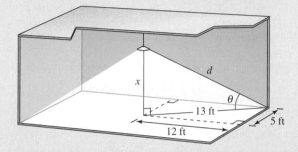

9. Prove the following **Extended Mean Value Theorem.** If f and f' are continuous on the closed interval $[a, b]$, and if f'' exists in the open interval (a, b), then there exists a number c in (a, b) such that

$$f(b) = f(a) + f'(a)(b - a) + \frac{1}{2}f''(c)(b - a)^2.$$

10. The line joining P and Q crosses the two parallel lines, as shown in the figure. The point R is d units from P. How far from Q should the point S be chosen so that the sum of the areas of the two shaded triangles is a minimum? So that the sum is a maximum?

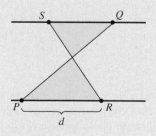

11. The efficiency E of a screw with square threads is

$$E = \frac{\tan \phi(1 - \mu \tan \phi)}{\mu + \tan \phi}$$

where μ is the coefficient of sliding friction and ϕ is the angle of inclination of the threads to a plane perpendicular to the axis of the screw. Find the angle ϕ that yields maximum efficiency when $\mu = 0.1$.

12. (a) Prove that $\lim\limits_{x \to \infty} x^2 = \infty$.

(b) Prove that $\lim\limits_{x \to \infty}\left(\dfrac{1}{x^2}\right) = 0$.

(c) Let L be a real number. Prove that if $\lim\limits_{x \to \infty} f(x) = L$, then

$$\lim_{y \to 0^+} f\left(\frac{1}{y}\right) = L.$$

13. In the engine shown in the figure, a connecting rod 18 centimeters long is fastened to a crank of radius 6 centimeters at point P. The crankshaft rotates counterclockwise at a constant rate of 200 revolutions per minute. The horizontal velocity (cm/min) of point P is

$$v = -2400\pi \sin\theta$$

where θ is the central angle of the crankshaft. What values of θ produce a maximum horizontal velocity?

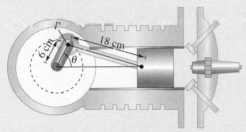

14. Consider a room in the shape of a cube, 4 meters on each side. A bug at point P wants to walk to point Q at the opposite corner, as indicated in the figure. Use calculus to determine the shortest path. Can you solve the problem without calculus?

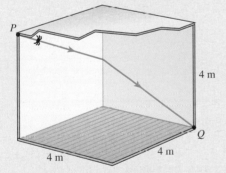

15. The figures show a rectangle, a circle, and a semicircle inscribed in a triangle bounded by the coordinate axes and the first quadrant portion of the line with intercepts $(3, 0)$ and $(0, 4)$. Find the dimensions of each inscribed figure such that its area is maximum. State whether calculus was helpful in finding the required dimensions. Explain your reasoning.

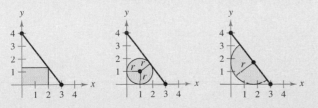

16. The police department must determine the speed limit on a bridge such that the flow rate of cars is maximum per unit time. The greater the speed limit, the farther apart the cars must be in order to keep a safe stopping distance. Experimental data on the stopping distance d (in meters) for various velocities v (in kilometers per hour) are shown in the table.

v	20	40	60	80	100
d	5.1	13.7	27.2	44.2	66.4

(a) Convert the speeds v in the table to the speeds s in meters per second. Use the regression capabilities of a graphing utility to find a model of the form $d(s) = as^2 + bs + c$ for the data.

(b) Consider two consecutive vehicles of average length 5.5 meters, traveling at a safe speed on the bridge. Let T be the difference between the times (in seconds) when the front bumpers of the vehicles pass a given point on the bridge. Verify that this difference in times is given by

$$T = \frac{d(s)}{s} + \frac{5.5}{s}.$$

(c) Use a graphing utility to graph the function T and estimate the speed s that minimizes the time between vehicles.

(d) Use calculus to determine the speed that minimizes T. What is the minimum value of T? Convert the required speed to kilometers per hour.

(e) Find the optimal distance between vehicles for the posted speed limit determined in part (d).

17. Find the point on the graph of $y = \dfrac{1}{1 + x^2}$ where the tangent line has the greatest slope, and the point where the tangent line has the least slope.

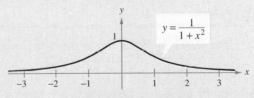

18. (a) Let x be a positive number. Use the *table* feature of a graphing utility to verify that $\sqrt{1 + x} < \frac{1}{2}x + 1$.

(b) Use the Mean Value Theorem to prove that $\sqrt{1 + x} < \frac{1}{2}x + 1$ for all positive real numbers x.

19. (a) Let x be a positive number. Use the *table* feature of a graphing utility to verify that $\sin x < x$.

(b) Use the Mean Value Theorem to prove that $\sin x < x$ for all positive real numbers x.

The Wankel Rotary Engine and Area

Named for Felix Wankel, who developed its basic principles in the 1950s, the Wankel rotary engine presents an alternative to the piston engine commonly used in automobiles. Many auto makers, including Mercedes-Benz, Citroën, and Ford, have experimented with rotary engines. By far the greatest number of Wankel-powered vehicles have been put on the road by Mazda, whose current rotary engine design is the RX-7.

The Wankel rotary engine has several advantages over the piston engine. A rotary engine is approximately half the size and weight of a piston engine of equivalent power. Compared with about 97 major moving parts in a V-8 engine, the typical two-rotor rotary engine has only three major moving parts. As a result, the Wankel engine has lower labor and material costs and less internal energy waste.

Although many different designs are possible for the rotary engine, the most common configuration is a two-lobed housing containing a three-sided rotor. The size of the rotor in comparison with the size of the housing cavity is critical in determining the compression ratio and thus the combustion efficiency.

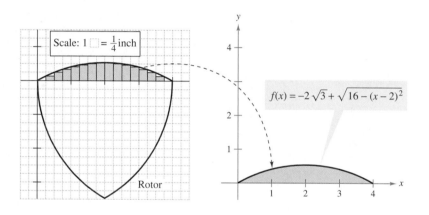

QUESTIONS

1. The region shown in the figure on the right above is bounded above by the graph of

   $$f(x) = -2\sqrt{3} + \sqrt{16 - (x - 2)^2}$$

   and below by the x-axis. Describe different ways in which you might approximate the area of the region. Then choose one of the ways and use it to obtain an approximation. What type of accuracy do you think your approximation has?

2. Now that you have found one approximation for the area of the region, describe a way that you can improve your approximation. Does your strategy allow you to obtain an approximation that is arbitrarily close to the actual area? Explain.

3. Use your approximation to estimate the area of the "bulged triangle" shown in the figure on the left above.

The concepts presented here will be explored further in this chapter. For an extension of this application, see Lab 6 in the lab series that accompanies this text at college.hmco.com.

Integration

In 2001, Mazda Motor Corporation unveiled the new RX-8 concept car at the North American International Auto Show in Detroit. The RX-8 is powered by an iteration of the Wankel rotary engine. Although the last mass-produced car equipped with a rotary engine sold worldwide was the RX-7, whose shipment to the United States ended in 1995, the Mazda Corporation intends to reestablish an interest in rotary-powered cars through new designs like the RX-8.

Reuters/Rebecca Cook/Archive Photos

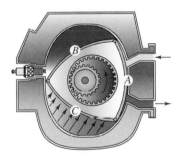

A sweeps out exhaust, *B* begins compression, and *C* is nearly finished with expansion.

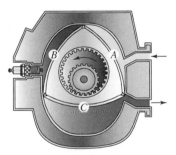

A moves back to allow intake while *B* continues compression. *C* begins to push out exhaust.

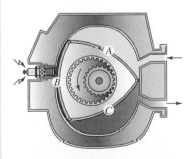

Ignition takes place in *B*, *A* continues intake, and *C* continues exhaust.

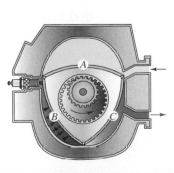

Intake is nearly complete in *A*, *B* expands following ignition, and *C* is nearly finished with exhaust.

Section 4.1   Antiderivatives and Indefinite Integration

- Write the general solution of a differential equation.
- Use indefinite integral notation for antiderivatives.
- Use basic integration rules to find antiderivatives.
- Find a particular solution of a differential equation.

Antiderivatives

Suppose you were asked to find a function F whose derivative is $f(x) = 3x^2$. From your knowledge of derivatives, you would probably say that

$$F(x) = x^3 \text{ because } \frac{d}{dx}[x^3] = 3x^2.$$

The function F is an *antiderivative* of f.

Definition of an Antiderivative

A function F is an **antiderivative** of f on an interval I if $F'(x) = f(x)$ for all x in I.

Note that F is called *an* antiderivative of f, rather than *the* antiderivative of f. To see why, observe that

$$F_1(x) = x^3, \quad F_2(x) = x^3 - 5, \quad \text{and} \quad F_3(x) = x^3 + 97$$

are all antiderivatives of $f(x) = 3x^2$. In fact, for any constant C, the function given by $F(x) = x^3 + C$ is an antiderivative of f.

THEOREM 4.1   Representation of Antiderivatives

If F is an antiderivative of f on an interval I, then G is an antiderivative of f on the interval I if and only if G is of the form

$$G(x) = F(x) + C, \text{ for all } x \text{ in } I$$

where C is a constant.

Proof   The proof of one direction is straightforward. That is, if $G(x) = F(x) + C$, $F'(x) = f(x)$, and C is a constant, then

$$G'(x) = \frac{d}{dx}[F(x) + C] = F'(x) + 0 = f(x).$$

To prove the other direction, you can define a function H such that

$$H(x) = G(x) - F(x).$$

If H is not constant on the interval I, there must exist a and b $(a < b)$ in the interval such that $H(a) \neq H(b)$. Moreover, because H is differentiable on (a, b), you can apply the Mean Value Theorem to conclude that there exists some c in (a, b) such that

$$H'(c) = \frac{H(b) - H(a)}{b - a}.$$

Because $H(b) \neq H(a)$, it follows that $H'(c) \neq 0$. However, because $G'(c) = F'(c)$, you know that $H'(c) = G'(c) - F'(c) = 0$, which contradicts the fact that $H'(c) \neq 0$. Consequently, you can conclude that $H(x)$ is a constant, C. Therefore, $G(x) - F(x) = C$ and it follows that $G(x) = F(x) + C$.

Using Theorem 4.1, you can represent the entire family of antiderivatives of a function by adding a constant to a *known* antiderivative. For example, knowing that $D_x[x^2] = 2x$, you can represent the family of *all* antiderivatives of $f(x) = 2x$ by

$G(x) = x^2 + C$ Family of all antiderivatives of $f(x) = 2x$

where C is a constant. The constant C is called the **constant of integration**. The family of functions represented by G is the **general antiderivative** of f, and $G(x) = x^2 + C$ is the **general solution** of the *differential equation*

$G'(x) = 2x.$ Differential equation

A **differential equation** in x and y is an equation that involves x, y, and derivatives of y. For instance, $y' = 3x$ and $y' = x^2 + 1$ are examples of differential equations.

Example 1   **Solving a Differential Equation**

Find the general solution of the differential equation $y' = 2$.

Solution   To begin, you need to find a function whose derivative is 2. One such function is

$y = 2x.$ 2x is *an* antiderivative of 2.

Now, you can use Theorem 4.1 to conclude that the general solution of the differential equation is

$y = 2x + C.$ General solution

The graphs of several functions of the form $y = 2x + C$ are shown in Figure 4.1.

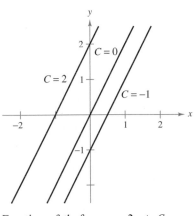

Functions of the form $y = 2x + C$
Figure 4.1

Notation for Antiderivatives

When solving a differential equation of the form

$$\frac{dy}{dx} = f(x)$$

it is convenient to write it in the equivalent differential form

$dy = f(x)\,dx.$

The operation of finding all solutions of this equation is called **antidifferentiation** (or **indefinite integration**) and is denoted by an integral sign $\int$. The general solution is denoted by

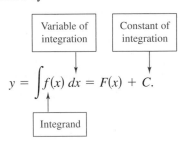

The expression $\int f(x)\,dx$ is read as the *antiderivative of f with respect to x*. So, the differential dx serves to identify x as the variable of integration. The term **indefinite integral** is a synonym for antiderivative.

NOTE   In this text, whenever we write $\int f(x)\,dx = F(x) + C$, we mean that F is an antiderivative of f *on an interval*.

Basic Integration Rules

The inverse nature of integration and differentiation can be verified by substituting $F'(x)$ for $f(x)$ in the indefinite integration definition to obtain

$$\int F'(x)\,dx = F(x) + C.$$ Integration is the "inverse" of differentiation.

Moreover, if $\int f(x)\,dx = F(x) + C$, then

$$\frac{d}{dx}\left[\int f(x)\,dx\right] = f(x).$$ Differentiation is the "inverse" of integration.

These two equations allow you to obtain integration formulas directly from differentiation formulas, as shown in the following summary.

Basic Integration Rules

Differentiation Formula	Integration Formula
$\dfrac{d}{dx}[C] = 0$	$\displaystyle\int 0\,dx = C$
$\dfrac{d}{dx}[kx] = k$	$\displaystyle\int k\,dx = kx + C$
$\dfrac{d}{dx}[kf(x)] = kf'(x)$	$\displaystyle\int kf(x)\,dx = k\int f(x)\,dx$
$\dfrac{d}{dx}[f(x) \pm g(x)] = f'(x) \pm g'(x)$	$\displaystyle\int [f(x) \pm g(x)]\,dx = \int f(x)\,dx \pm \int g(x)\,dx$
$\dfrac{d}{dx}[x^n] = nx^{n-1}$	$\displaystyle\int x^n\,dx = \frac{x^{n+1}}{n+1} + C, \quad n \neq -1$ Power Rule
$\dfrac{d}{dx}[\sin x] = \cos x$	$\displaystyle\int \cos x\,dx = \sin x + C$
$\dfrac{d}{dx}[\cos x] = -\sin x$	$\displaystyle\int \sin x\,dx = -\cos x + C$
$\dfrac{d}{dx}[\tan x] = \sec^2 x$	$\displaystyle\int \sec^2 x\,dx = \tan x + C$
$\dfrac{d}{dx}[\sec x] = \sec x \tan x$	$\displaystyle\int \sec x \tan x\,dx = \sec x + C$
$\dfrac{d}{dx}[\cot x] = -\csc^2 x$	$\displaystyle\int \csc^2 x\,dx = -\cot x + C$
$\dfrac{d}{dx}[\csc x] = -\csc x \cot x$	$\displaystyle\int \csc x \cot x\,dx = -\csc x + C$

NOTE Note that the Power Rule for integration has the restriction that $n \neq -1$. The evaluation of $\int 1/x\,dx$ must wait until the introduction of the natural logarithm function in Chapter 5.

Example 2   **Applying the Basic Integration Rules**

Describe the antiderivatives of $3x$.

Solution $\displaystyle \int 3x \, dx = 3 \int x \, dx$ Constant Multiple Rule

$\displaystyle = 3 \int x^1 \, dx$ Rewrite $(x = x^1)$.

$\displaystyle = 3\left(\frac{x^2}{2}\right) + C$ Power Rule $(n = 1)$

$\displaystyle = \frac{3}{2} x^2 + C$ Simplify.

When indefinite integrals are evaluated, a strict application of the basic integration rules tends to produce complicated constants of integration. For instance, in Example 2, we could have written

$$\int 3x \, dx = 3 \int x \, dx$$

$$= 3\left(\frac{x^2}{2} + C\right)$$

$$= \frac{3}{2} x^2 + 3C.$$

However, because C represents *any* constant, it is both cumbersome and unnecessary to write $3C$ as the constant of integration, and we choose the simpler form, $\frac{3}{2}x^2 + C$.

In Example 2, note that the general pattern of integration is similar to that of differentiation.

| Original integral | $\Rightarrow$ | Rewrite | $\Rightarrow$ | Integrate | $\Rightarrow$ | Simplify |

Example 3   **Rewriting Before Integrating**

TECHNOLOGY   Some software programs, such as *Derive, Maple, Mathcad, Mathematica,* and the *TI-89,* are capable of performing integration symbolically. If you have access to such a symbolic integration utility, try using it to evaluate the indefinite integrals in Example 3.

	Original Integral	Rewrite	Integrate	Simplify
a.	$\displaystyle \int \frac{1}{x^3} \, dx$	$\displaystyle \int x^{-3} \, dx$	$\displaystyle \frac{x^{-2}}{-2} + C$	$\displaystyle -\frac{1}{2x^2} + C$
b.	$\displaystyle \int \sqrt{x} \, dx$	$\displaystyle \int x^{1/2} \, dx$	$\displaystyle \frac{x^{3/2}}{3/2} + C$	$\displaystyle \frac{2}{3}x^{3/2} + C$
c.	$\displaystyle \int 2 \sin x \, dx$	$\displaystyle 2\int \sin x \, dx$	$2(-\cos x) + C$	$-2 \cos x + C$

Remember that you can check your answer to an antidifferentiation problem by differentiating. For instance, in Example 3b, you can check that $\frac{2}{3}x^{3/2} + C$ is the correct antiderivative by differentiating the answer to obtain

$$D_x\left[\frac{2}{3}x^{3/2} + C\right] = \left(\frac{2}{3}\right)\left(\frac{3}{2}\right)x^{1/2} = \sqrt{x}.$$ Use differentiation to check antiderivative.

---

indicates that in the Interactive 3.0 CD-ROM and Internet 3.0 versions of this text (available at college.hmco.com) *you will find an Open Exploration, which further explores this example using the computer algebra systems Maple, Mathcad, Mathematica, and Derive.*

The basic integration rules listed earlier in this section allow you to integrate *any* polynomial function, as demonstrated in Example 4.

Example 4   Integrating Polynomial Functions

a. $\displaystyle\int dx = \int 1\,dx$ Integrand is understood to be 1.

$\qquad\quad = x + C$ Integrate.

b. $\displaystyle\int (x + 2)\,dx = \int x\,dx + \int 2\,dx$

$\qquad\qquad\quad = \dfrac{x^2}{2} + C_1 + 2x + C_2$ Integrate.

$\qquad\qquad\quad = \dfrac{x^2}{2} + 2x + C$ $C = C_1 + C_2$

The second line in the solution is usually omitted.

c. $\displaystyle\int (3x^4 - 5x^2 + x)\,dx = 3\left(\dfrac{x^5}{5}\right) - 5\left(\dfrac{x^3}{3}\right) + \dfrac{x^2}{2} + C$ Integrate.

$\qquad\qquad\qquad\qquad\quad = \dfrac{3}{5}x^5 - \dfrac{5}{3}x^3 + \dfrac{1}{2}x^2 + C$ Simplify.

Example 5   Rewriting Before Integrating

$\displaystyle\int \dfrac{x + 1}{\sqrt{x}}\,dx = \int \left(\dfrac{x}{\sqrt{x}} + \dfrac{1}{\sqrt{x}}\right) dx$ Rewrite as two fractions.

$\qquad\qquad\quad = \int (x^{1/2} + x^{-1/2})\,dx$ Rewrite with fractional exponents.

$\qquad\qquad\quad = \dfrac{x^{3/2}}{3/2} + \dfrac{x^{1/2}}{1/2} + C$ Integrate.

$\qquad\qquad\quad = \dfrac{2}{3}x^{3/2} + 2x^{1/2} + C$ Simplify.

NOTE   When integrating quotients, do not integrate the numerator and denominator separately. This is no more valid in integration than it is in differentiation. For instance, in Example 5, be sure you understand that

$$\int \dfrac{x + 1}{\sqrt{x}}\,dx \ne \dfrac{\int (x + 1)\,dx}{\int \sqrt{x}\,dx}.$$

Example 6   Rewriting Before Integrating

$\displaystyle\int \dfrac{\sin x}{\cos^2 x}\,dx = \int \left(\dfrac{1}{\cos x}\right)\left(\dfrac{\sin x}{\cos x}\right) dx$ Rewrite as a product.

$\qquad\qquad\quad = \int \sec x \tan x\,dx$ Rewrite using trigonometric identities.

$\qquad\qquad\quad = \sec x + C$ Integrate.

Initial Conditions and Particular Solutions

You have already seen that the equation $y = \int f(x)\,dx$ has many solutions (each differing from the others by a constant). This means that the graphs of any two antiderivatives of f are vertical translations of each other. For example, Figure 4.2 shows the graphs of several antiderivatives of the form

$$y = \int (3x^2 - 1)\,dx = x^3 - x + C \qquad \text{General solution}$$

for various integer values of C. Each of these antiderivatives is a solution of the differential equation

$$\frac{dy}{dx} = 3x^2 - 1.$$

In many applications of integration, you are given enough information to determine a **particular solution**. To do this, you need only know the value of $y = F(x)$ for one value of x. (This information is called an **initial condition**.) For example, in Figure 4.2, only one curve passes through the point $(2, 4)$. To find this curve, you can use the following information.

$$F(x) = x^3 - x + C \qquad \text{General solution}$$
$$F(2) = 4 \qquad \text{Initial condition}$$

By using the initial condition in the general solution, you can determine that $F(2) = 8 - 2 + C = 4$, which implies that $C = -2$. So, you obtain

$$F(x) = x^3 - x - 2. \qquad \text{Particular solution}$$

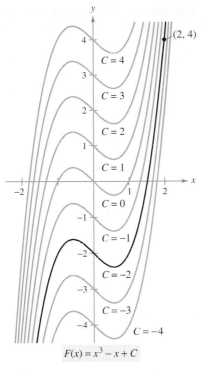

The particular solution that satisfies the initial condition $F(2) = 4$ is $F(x) = x^3 - x - 2$.

Figure 4.2

Example 7   **Finding a Particular Solution**

Find the general solution of

$$F'(x) = \frac{1}{x^2}, \quad x > 0$$

and find the particular solution that satisfies the initial condition $F(1) = 0$.

Solution   To find the general solution, integrate to obtain

$$F(x) = \int \frac{1}{x^2}\,dx \qquad F(x) = \int F'(x)\,dx$$

$$= \int x^{-2}\,dx \qquad \text{Rewrite as a power.}$$

$$= \frac{x^{-1}}{-1} + C \qquad \text{Integrate.}$$

$$= -\frac{1}{x} + C, \quad x > 0. \qquad \text{General solution}$$

Using the initial condition $F(1) = 0$, you can solve for C as follows.

$$F(1) = -\frac{1}{1} + C = 0 \quad \Longrightarrow \quad C = 1$$

So, the particular solution, as shown in Figure 4.3, is

$$F(x) = -\frac{1}{x} + 1, \quad x > 0. \qquad \text{Particular solution}$$

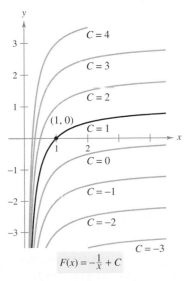

The particular solution that satisfies the initial condition $F(1) = 0$ is $F(x) = -(1/x) + 1, x > 0.$

Figure 4.3

So far in this section we have been using x as the variable of integration. In applications, it is often convenient to use a different variable. For instance, in the following example involving *time*, the variable of integration is t.

Example 8 Solving a Vertical Motion Problem

A ball is thrown upward with an initial velocity of 64 feet per second from an initial height of 80 feet. See Figure 4.4.

a. Find the position function giving the height s as a function of the time t.

b. When does the ball hit the ground?

Solution

a. Let $t = 0$ represent the initial time. The two given initial conditions can be written as follows.

$$s(0) = 80 \qquad \text{Initial height is 80 feet.}$$
$$s'(0) = 64 \qquad \text{Initial velocity is 64 feet per second.}$$

Using -32 feet per second per second as the acceleration due to gravity, you can write

$$s''(t) = -32$$
$$s'(t) = \int s''(t)\, dt = \int -32\, dt = -32t + C_1.$$

Using the initial velocity, you obtain $s'(0) = 64 = -32(0) + C_1$, which implies that $C_1 = 64$. Next, by integrating $s'(t)$, you obtain

$$s(t) = \int s'(t)\, dt = \int (-32t + 64)\, dt = -16t^2 + 64t + C_2.$$

Using the initial height, you obtain

$$s(0) = 80 = -16(0^2) + 64(0) + C_2$$

which implies that $C_2 = 80$. Therefore, the position function is

$$s(t) = -16t^2 + 64t + 80.$$

b. Using the position function found in part (a), you can find the time that the ball hits the ground by solving the equation $s(t) = 0$.

$$s(t) = -16t^2 + 64t + 80 = 0$$
$$-16(t + 1)(t - 5) = 0$$
$$t = -1, 5$$

Because t must be positive, you can conclude that the ball hits the ground 5 seconds after it was thrown.

Example 8 shows how to use calculus to analyze vertical motion problems in which the acceleration is determined by a gravitational force. You can use a similar strategy to analyze other linear motion problems (vertical or horizontal) in which the acceleration (or deceleration) is the result of some other force, as you can see in Exercises 77–88.

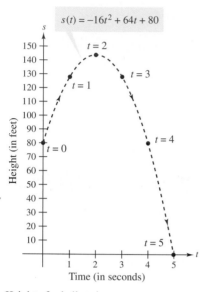

$s(t) = -16t^2 + 64t + 80$

Height of a ball at time t
Figure 4.4

NOTE In Example 8, note that the position function has the form

$$s(t) = \tfrac{1}{2}gt^2 + v_0 t + s_0,$$

where $g = -32$, v_0 is the initial velocity, and s_0 is the initial height, as presented in Section 2.2.

Before you begin the exercise set, be sure you realize that one of the most important steps in integration is *rewriting the integrand* in a form that fits the basic integration rules. To further illustrate this point, here are some additional examples.

Original Integral	*Rewrite*	*Integrate*	*Simplify*
$\int \dfrac{2}{\sqrt{x}}\,dx$	$2\int x^{-1/2}\,dx$	$2\left(\dfrac{x^{1/2}}{1/2}\right) + C$	$4x^{1/2} + C$
$\int (t^2 + 1)^2\,dt$	$\int (t^4 + 2t^2 + 1)\,dt$	$\dfrac{t^5}{5} + 2\left(\dfrac{t^3}{3}\right) + t + C$	$\dfrac{1}{5}t^5 + \dfrac{2}{3}t^3 + t + C$
$\int \dfrac{x^3 + 3}{x^2}\,dx$	$\int (x + 3x^{-2})\,dx$	$\dfrac{x^2}{2} + 3\left(\dfrac{x^{-1}}{-1}\right) + C$	$\dfrac{1}{2}x^2 - \dfrac{3}{x} + C$
$\int \sqrt[3]{x}(x - 4)\,dx$	$\int (x^{4/3} - 4x^{1/3})\,dx$	$\dfrac{x^{7/3}}{7/3} - 4\left(\dfrac{x^{4/3}}{4/3}\right) + C$	$\dfrac{3}{7}x^{4/3}(x - 7) + C$

EXERCISES FOR SECTION 4.1

In Exercises 1–4, verify the statement by showing that the derivative of the right side equals the integrand of the left side.

1. $\int \left(-\dfrac{9}{x^4}\right)dx = \dfrac{3}{x^3} + C$

2. $\int \left(4x^3 - \dfrac{1}{x^2}\right)dx = x^4 + \dfrac{1}{x} + C$

3. $\int (x - 2)(x + 2)\,dx = \tfrac{1}{3}x^3 - 4x + C$

4. $\int \dfrac{x^2 - 1}{x^{3/2}}\,dx = \dfrac{2(x^2 + 3)}{3\sqrt{x}} + C$

In Exercises 5–8, find the general solution of the differential equation and check the result by differentiation.

5. $\dfrac{dy}{dt} = 3t^2$

6. $\dfrac{dr}{d\theta} = \pi$

7. $\dfrac{dy}{dx} = x^{3/2}$

8. $\dfrac{dy}{dx} = 2x^{-3}$

In Exercises 9–14, complete the table using Example 3 and the examples at the top of this page as a model.

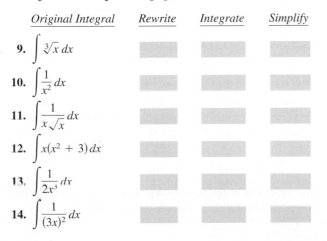

	Original Integral	*Rewrite*	*Integrate*	*Simplify*
9.	$\int \sqrt[3]{x}\,dx$			
10.	$\int \dfrac{1}{x^2}\,dx$			
11.	$\int \dfrac{1}{x\sqrt{x}}\,dx$			
12.	$\int x(x^2 + 3)\,dx$			
13.	$\int \dfrac{1}{2x^3}\,dx$			
14.	$\int \dfrac{1}{(3x)^2}\,dx$			

In Exercises 15–34, find the indefinite integral and check the result by differentiation.

15. $\int (x + 3)\,dx$

16. $\int (5 - x)\,dx$

17. $\int (2x - 3x^2)\,dx$

18. $\int (4x^3 + 6x^2 - 1)\,dx$

19. $\int (x^3 + 2)\,dx$

20. $\int (x^3 - 4x + 2)\,dx$

21. $\int (x^{3/2} + 2x + 1)\,dx$

22. $\int \left(\sqrt{x} + \dfrac{1}{2\sqrt{x}}\right)dx$

23. $\int \sqrt[3]{x^2}\,dx$

24. $\int \left(\sqrt[4]{x^3} + 1\right)dx$

25. $\int \dfrac{1}{x^3}\,dx$

26. $\int \dfrac{1}{x^4}\,dx$

27. $\int \dfrac{x^2 + x + 1}{\sqrt{x}}\,dx$

28. $\int \dfrac{x^2 + 2x - 3}{x^4}\,dx$

29. $\int (x + 1)(3x - 2)\,dx$

30. $\int (2t^2 - 1)^2\,dt$

31. $\int y^2\sqrt{y}\,dy$

32. $\int (1 + 3t)t^2\,dt$

33. $\int dx$

34. $\int 3\,dt$

In Exercises 35–42, find the indefinite integral and check the result by differentiation.

35. $\int (2\sin x + 3\cos x)\,dx$

36. $\int (t^2 - \sin t)\,dt$

37. $\int (1 - \csc t \cot t)\,dt$

38. $\int (\theta^2 + \sec^2 \theta)\,d\theta$

39. $\int (\sec^2 \theta - \sin \theta)\,d\theta$

40. $\int \sec y(\tan y - \sec y)\,dy$

41. $\int (\tan^2 y + 1)\,dy$

42. $\int \dfrac{\cos x}{1 - \cos^2 x}\,dx$

In Exercises 43 and 44, sketch the graphs of the function $g(x) = f(x) + C$ for $C = -2, C = 0$, and $C = 3$ on the same set of coordinate axes.

43. $f(x) = \cos x$ **44.** $f(x) = \sqrt{x}$

In Exercises 45–48, the graph of the derivative of a function is given. Sketch the graphs of *two* functions that have the given derivative. (There is more than one correct answer.) To print an enlarged copy of the graph, go to the website *www.mathgraphs.com.*

45.   **46.**

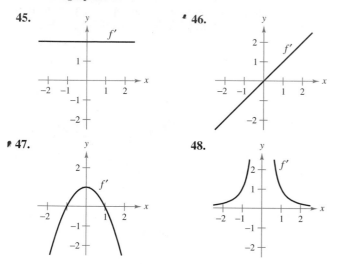

47.   **48.**

In Exercises 49–52, find the equation for y, given the derivative and the indicated point on the curve.

49. $\dfrac{dy}{dx} = 2x - 1$ **50.** $\dfrac{dy}{dx} = 2(x - 1)$

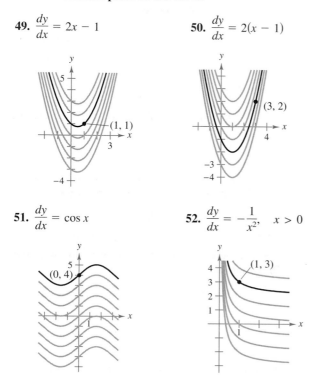

51. $\dfrac{dy}{dx} = \cos x$ **52.** $\dfrac{dy}{dx} = -\dfrac{1}{x^2}$, $\quad x > 0$

Slope Fields In Exercises 53 and 54, a differential equation, a point, and a slope field are given. A *slope field* (or *direction field*) consists of line segments with slopes given by the differential equation. These line segments give a visual perspective of the slopes of the solutions of the differential equation. (a) Sketch two approximate solutions of the differential equation on the slope field, one of which passes through the indicated point. (To print an enlarged copy of the graph, go to the website *www.mathgraphs.com.*) (b) Use integration to find the particular solution of the differential equation and use a graphing utility to graph the solution. Compare the result with the sketches in part (a).

53. $\dfrac{dy}{dx} = \dfrac{1}{2}x - 1$, $\quad (4, 2)$ **54.** $\dfrac{dy}{dx} = x^2 - 1$, $\quad (-1, 3)$

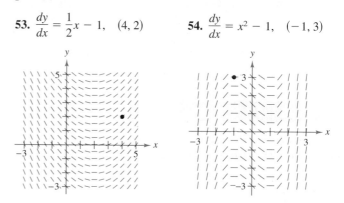

In Exercises 55–62, solve the differential equation.

55. $f'(x) = 4x, f(0) = 6$

56. $g'(x) = 6x^2, g(0) = -1$

57. $h'(t) = 8t^3 + 5, h(1) = -4$

58. $f'(s) = 6s - 8s^3, f(2) = 3$

59. $f''(x) = 2, f'(2) = 5, f(2) = 10$

60. $f''(x) = x^2, f'(0) = 6, f(0) = 3$

61. $f''(x) = x^{-3/2}, f'(4) = 2, f(0) = 0$

62. $f''(x) = \sin x, f'(0) = 1, f(0) = 6$

63. *Tree Growth* An evergreen nursery usually sells a certain shrub after 6 years of growth and shaping. The growth rate during those 6 years is approximated by

$$\frac{dh}{dt} = 1.5t + 5$$

where t is the time in years and h is the height in centimeters. The seedlings are 12 centimeters tall when planted ($t = 0$).

(a) Find the height after t years.

(b) How tall are the shrubs when they are sold?

64. *Population Growth* The rate of growth dP/dt of a population of bacteria is proportional to the square root of t, where P is the population size and t is the time in days ($0 \le t \le 10$). That is,

$$\frac{dP}{dt} = k\sqrt{t}.$$

The initial size of the population is 500. After 1 day the population has grown to 600. Estimate the population after 7 days.

Getting at the Concept

65. Use the graph of f' in the figure to answer the following, given that $f(0) = -4$.

(a) Approximate the slope of f at $x = 4$. Explain.

(b) Is it possible that $f(2) = -1$? Explain.

(c) Is $f(5) - f(4) > 0$? Explain.

(d) Approximate the value of x where f is maximum. Explain.

(e) Approximate any intervals in which the graph of f is concave upward and any intervals in which it is concave downward. Approximate the x-coordinates of any points of inflection.

(f) Approximate the x-coordinate of the minimum of $f''(x)$.

(g) Sketch an approximate graph of f. To print an enlarged copy of the graph, go to the website *www.mathgraphs.com*.

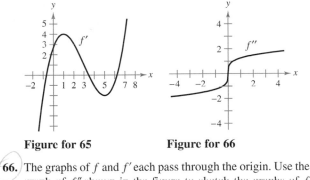

| Figure for 65 | Figure for 66 |

66. The graphs of f and f' each pass through the origin. Use the graph of f'' shown in the figure to sketch the graphs of f and f'. To print an enlarged copy of the graph, go to the website *www.mathgraphs.com*.

Vertical Motion   In Exercises 67–70, use $a(t) = -32$ feet per second per second as the acceleration due to gravity. (Neglect air resistance.)

67. A ball is thrown vertically upward from a height of 6 feet with an initial velocity of 60 feet per second. How high will the ball go?

68. Show that the height above the ground of an object thrown upward from a point s_0 feet above the ground with an initial velocity of v_0 feet per second is given by the function

$$f(t) = -16t^2 + v_0t + s_0.$$

69. With what initial velocity must an object be thrown upward (from ground level) to reach the top of the Washington Monument (approximately 550 feet)?

70. A balloon, rising vertically with a velocity of 16 feet per second, releases a sandbag at the instant it is 64 feet above the ground.

(a) How many seconds after its release will the bag strike the ground?

(b) At what velocity will it hit the ground?

Vertical Motion   In Exercises 71–74, use $a(t) = -9.8$ meters per second per second as the acceleration due to gravity. (Neglect air resistance.)

71. Show that the height above the ground of an object thrown upward from a point s_0 meters above the ground with an initial velocity of v_0 meters per second is given by the function

$$f(t) = -4.9t^2 + v_0t + s_0.$$

72. The Grand Canyon is 1600 meters deep at its deepest point. A rock is dropped from the rim above this point. Express the height of the rock as a function of the time t in seconds. How long will it take the rock to hit the canyon floor?

73. A baseball is thrown upward from a height of 2 meters with a velocity of 10 meters per second. Determine its maximum height.

74. With what initial velocity must an object be thrown upward (from a height of 2 meters) to reach a maximum height of 200 meters?

75. ***Lunar Gravity***   On the moon, the acceleration due to gravity is -1.6 meters per second per second. A stone is dropped from a cliff on the moon and hits the surface of the moon 20 seconds later. How far did it fall? What was its velocity at impact?

76. ***Escape Velocity***   The minimum velocity required for an object to escape earth's gravitational pull is obtained from the solution of the equation

$$\int v \, dv = -GM \int \frac{1}{y^2} \, dy$$

where v is the velocity of the object projected from earth, y is the distance from the center of earth, G is the gravitational constant, and M is the mass of earth. Show that v and y are related by the equation

$$v^2 = v_0^2 + 2GM\left(\frac{1}{y} - \frac{1}{R}\right)$$

where v_0 is the initial velocity of the object and R is the radius of earth.

Rectilinear Motion   In Exercises 77–80, consider a particle moving along the x-axis where $x(t)$ is the position of the particle at time t, $x'(t)$ is its velocity, and $x''(t)$ is its acceleration.

77. $x(t) = t^3 - 6t^2 + 9t - 2,$ $0 \le t \le 5$

(a) Find the velocity and acceleration of the particle.

(b) Find the open t-intervals on which the particle is moving to the right.

(c) Find the velocity of the particle when the acceleration is 0.

78. Repeat Exercise 77 for the position function

$$x(t) = (t - 1)(t - 3)^2,   0 \le t \le 5.$$

79. A particle moves along the x-axis at a velocity of $v(t) = 1/\sqrt{t}$, $t > 0$. At time $t = 1$, its position is $x = 4$. Find the acceleration and position functions for the particle.

80. A particle, initially at rest, moves along the x-axis such that its acceleration at time $t > 0$ is given by $a(t) = \cos t$. At the time $t = 0$, its position is $x = 3$.

(a) Find the velocity and position functions for the particle.

(b) Find the values of t for which the particle is at rest.

81. *Acceleration*   The maker of a certain automobile advertises that it takes 13 seconds to accelerate from 25 kilometers per hour to 80 kilometers per hour. Assuming constant acceleration, compute the following.

(a) The acceleration in meters per second per second

(b) The distance the car travels during the 13 seconds

82. *Deceleration*   A car traveling at 45 miles per hour is brought to a stop, at constant deceleration, 132 feet from where the brakes are applied.

(a) How far has the car moved when its speed has been reduced to 30 miles per hour?

(b) How far has the car moved when its speed has been reduced to 15 miles per hour?

(c) Draw the real number line from 0 to 132, and plot the points found in parts (a) and (b). What can you conclude?

83. *Acceleration*   At the instant the traffic light turns green, a car that has been waiting at an intersection starts with a constant acceleration of 6 feet per second per second. At the same instant, a truck traveling with a constant velocity of 30 feet per second passes the car.

(a) How far beyond its starting point will the car pass the truck?

(b) How fast will the car be traveling when it passes the truck?

84. *Think About It*   Two cars starting from rest accelerate to 65 miles per hour in 30 seconds. The velocity of each car is shown in the figure. Are the cars side by side at the end of the 30-second time interval? Explain.

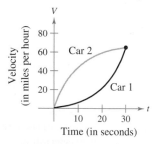

85. *Data Analysis*   The table shows the velocities (in miles per hour) of two cars on the entrance ramp of an interstate highway. The time t is in seconds.

t	0	5	10	15	20	25	30
v_1	0	2.5	7	16	29	45	65
v_2	0	21	38	51	60	64	65

(a) Rewrite the table converting miles per hour to feet per second.

(b) Use the regression capabilities of a graphing utility to fit quadratic models to the data in part (a).

(c) Approximate the distance traveled by each car during the 30 seconds. Explain the difference in distances.

86. *Data Analysis*   A vehicle slows to a stop from 45 miles per hour in 6 seconds. The table shows the velocities in feet per second.

t	0	1	2	3	4	5	6
v	66.0	61.1	48.9	33.0	17.1	4.8	0

(a) Use the regression capabilities of a graphing utility to fit a cubic model to the data.

(b) Approximate the distance traveled by the car during the 6 seconds.

87. *Acceleration*   Assume that a fully loaded plane starting from rest has a constant acceleration while moving down a runway. The plane requires 0.7 mile of runway and a speed of 160 miles per hour in order to lift off. What is the plane's acceleration?

88. *Airplane Separation*   Two airplanes are in a straight-line landing pattern and, according to FAA regulations, must keep at least a 3-mile separation. Airplane A is 10 miles from touchdown and is gradually slowing its speed from 150 miles per hour to a landing speed of 100 miles per hour. Airplane B is 17 miles from touchdown and is gradually slowing its speed from 250 miles per hour to a landing speed of 115 miles per hour.

(a) Assuming the deceleration of each airplane is constant, find the position functions s_1 and s_2 for Airplane A and Airplane B. Let $t = 0$ represent the times when the airplanes are 10 and 17 miles from the airport.

(b) Use a graphing utility to graph the position functions.

(c) Find a formula for the magnitude of the distance d between the two airplanes as a function of t. Use a graphing utility to graph d. Is $d < 3$ for some time prior to the landing of Airplane A? If so, find that time.

True or False?   **In Exercises 89–94, determine whether the statement is true or false. If it is false, explain why or give an example that shows it is false.**

89. Each antiderivative of an nth-degree polynomial function is an $(n + 1)$th-degree polynomial function.

90. If $p(x)$ is a polynomial function, then p has exactly one antiderivative whose graph contains the origin.

91. If $F(x)$ and $G(x)$ are antiderivatives of $f(x)$, then $F(x) = G(x) + C$.

92. If $f'(x) = g(x)$, then $\int g(x)\,dx = f(x) + C$.

93. $\int f(x)g(x)\,dx = \int f(x)\,dx \int g(x)\,dx$

94. The antiderivative of $f(x)$ is unique.

95. If $f'(x) = \begin{cases} 1, & 0 \le x < 2 \\ 3x, & 2 \le x \le 5 \end{cases}$, f is continuous, and $f(1) = 3$, find f. Is f differentiable at $x = 2$?

96. Let $s(x)$ and $c(x)$ be two functions satisfying $s'(x) = c(x)$ and $c'(x) = -s(x)$ for all x. If $s(0) = 0$ and $c(0) = 1$, prove that $[s(x)]^2 + [c(x)]^2 = 1$.

| **Section 4.2** | **Area** |

- Use sigma notation to write and evaluate a sum.
- Understand the concept of area.
- Approximate the area of a plane region.
- Find the area of a plane region using limits.

Sigma Notation

In the preceding section, you studied antidifferentiation. In this section, you will look further into a problem introduced in Section 1.1—that of finding the area of a region in the plane. At first glance, these two ideas may seem unrelated, but you will discover in Section 4.4 that they are closely related by an extremely important theorem called the Fundamental Theorem of Calculus.

We begin this section by introducing a concise notation for sums. This notation is called **sigma notation** because it uses the uppercase Greek letter sigma, written as Σ.

Sigma Notation

The sum of n terms $a_1, a_2, a_3, \ldots, a_n$ is written as

$$\sum_{i=1}^{n} a_i = a_1 + a_2 + a_3 + \cdots + a_n$$

where i is the **index of summation,** a_i is the **ith term** of the sum, and the **upper and lower bounds of summation** are n and 1.

NOTE   The upper and lower bounds must be constant with respect to the index of summation. However, the lower bound doesn't have to be 1. Any integer less than or equal to the upper bound is legitimate.

Example 1   **Examples of Sigma Notation**

---

a. $\displaystyle\sum_{i=1}^{6} i = 1 + 2 + 3 + 4 + 5 + 6$

b. $\displaystyle\sum_{i=0}^{5} (i + 1) = 1 + 2 + 3 + 4 + 5 + 6$

c. $\displaystyle\sum_{j=3}^{7} j^2 = 3^2 + 4^2 + 5^2 + 6^2 + 7^2$

d. $\displaystyle\sum_{k=1}^{n} \frac{1}{n}(k^2 + 1) = \frac{1}{n}(1^2 + 1) + \frac{1}{n}(2^2 + 1) + \cdots + \frac{1}{n}(n^2 + 1)$

e. $\displaystyle\sum_{i=1}^{n} f(x_i)\,\Delta x = f(x_1)\,\Delta x + f(x_2)\,\Delta x + \cdots + f(x_n)\,\Delta x$

From parts (a) and (b), notice that the same sum can be represented in different ways using sigma notation.   ◢

FOR FURTHER INFORMATION For a geometric interpretation of summation formulas, see the article, "Looking at $\displaystyle\sum_{k=1}^{n} k$ and $\displaystyle\sum_{k=1}^{n} k^2$ Geometrically" by Eric Hegblom in *Mathematics Teacher.* To view this article, go to the website *www.matharticles.com.*

Although any variable can be used as the index of summation $i, j,$ and k are often used. Notice in Example 1 that the index of summation does not appear in the terms of the expanded sum.

The following properties of summation can be derived using the associative and commutative properties of addition and the distributive property of addition over multiplication. (In the first property, k is a constant.)

1. $\displaystyle\sum_{i=1}^{n} k a_i = k \sum_{i=1}^{n} a_i$

2. $\displaystyle\sum_{i=1}^{n} (a_i \pm b_i) = \sum_{i=1}^{n} a_i \pm \sum_{i=1}^{n} b_i$

The next theorem lists some useful formulas for sums of powers. A proof of this theorem is given in Appendix B.

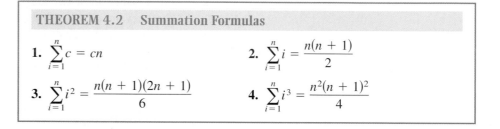

THEOREM 4.2   Summation Formulas

1. $\displaystyle\sum_{i=1}^{n} c = cn$ **2.** $\displaystyle\sum_{i=1}^{n} i = \frac{n(n+1)}{2}$

3. $\displaystyle\sum_{i=1}^{n} i^2 = \frac{n(n+1)(2n+1)}{6}$ **4.** $\displaystyle\sum_{i=1}^{n} i^3 = \frac{n^2(n+1)^2}{4}$

Example 2   **Evaluating a Sum**

Evaluate $\displaystyle\sum_{i=1}^{n} \frac{i+1}{n^2}$ for $n = 10, 100, 1000,$ and $10,000$.

Solution   Applying Theorem 4.2, you can write

$$\sum_{i=1}^{n} \frac{i+1}{n^2} = \frac{1}{n^2} \sum_{i=1}^{n} (i + 1) \qquad \text{Factor constant } 1/n^2 \text{ out of sum.}$$

$$= \frac{1}{n^2} \left(\sum_{i=1}^{n} i + \sum_{i=1}^{n} 1 \right) \qquad \text{Write as two sums.}$$

$$= \frac{1}{n^2} \left[\frac{n(n+1)}{2} + n \right] \qquad \text{Apply Theorem 4.2.}$$

$$= \frac{1}{n^2} \left[\frac{n^2 + 3n}{2} \right] \qquad \text{Simplify.}$$

$$= \frac{n+3}{2n}. \qquad \text{Simplify.}$$

Now you can evaluate the sum by substituting the appropriate values of n, as shown in the table at the left.

n	$\displaystyle\sum_{i=1}^{n} \frac{i+1}{n^2} = \frac{n+3}{2n}$
10	0.65000
100	0.51500
1,000	0.50150
10,000	0.50015

In the table, note that the sum appears to approach a limit as n increases. Although the discussion of limits at infinity in Section 3.5 applies to a variable x, where x can be any real number, many of the same results hold true for limits involving the variable n, where n is restricted to positive integer values. So, to find the limit of $(n + 3)/2n$ as n approaches infinity, you can write

$$\lim_{n \to \infty} \frac{n+3}{2n} = \frac{1}{2}.$$

Rectangle: $A = bh$
Figure 4.5

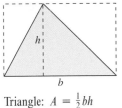

Triangle: $A = \frac{1}{2}bh$
Figure 4.6

ARCHIMEDES (287–212 B.C.)

Archimedes used the method of exhaustion to derive formulas for the areas of ellipses, parabolic segments, and sectors of a spiral. He is considered to have been the greatest applied mathematician of antiquity.

FOR FURTHER INFORMATION For an alternative development of the formula for the area of a circle, see the article "Proof Without Words: Area of a Disk is πR^2" by Russell Jay Hendel in *Mathematics Magazine*. To view this article, go to the website *www.matharticles.com*.

Area

In Euclidean geometry, the simplest type of plane region is a rectangle. Although people often say that the *formula* for the area of a rectangle is $A = bh$, as shown in Figure 4.5, it is actually more proper to say that this is the *definition* of the **area of a rectangle.**

From this definition, you can develop formulas for the areas of many other plane regions. For example, to determine the area of a triangle, you can form a rectangle whose area is twice that of the triangle, as shown in Figure 4.6. Once you know how to find the area of a triangle, you can determine the area of any polygon by subdividing the polygon into triangular regions, as shown in Figure 4.7.

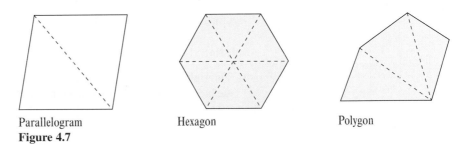

Parallelogram   Hexagon   Polygon
Figure 4.7

Finding the areas of regions other than polygons is more difficult. The ancient Greeks were able to determine formulas for the areas of some general regions (principally those bounded by conics) by the *exhaustion* method. The clearest description of this method was given by Archimedes. Essentially, the method is a limiting process in which the area is squeezed between two polygons—one inscribed in the region and one circumscribed about the region.

For instance, in Figure 4.8 the area of a circular region is approximated by an n-sided inscribed polygon and an n-sided circumscribed polygon. For each value of n the area of the inscribed polygon is less than the area of the circle, and the area of the circumscribed polygon is greater than the area of the circle. Moreover, as n increases, the areas of both polygons become better and better approximations of the area of the circle.

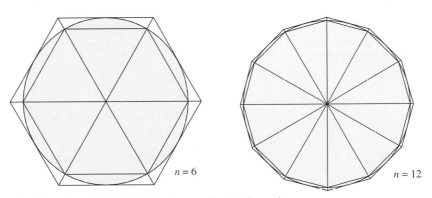

The exhaustion method for finding the area of a circular region
Figure 4.8

In the remaining examples in this section, we use a process that is similar to that used by Archimedes to determine the area of a plane region.

The Area of a Plane Region

Recall from Section 1.1 that the origins of calculus are connected to two classic problems: the tangent line problem and the area problem. We begin the investigation of the area problem with an example.

Example 3   Approximating the Area of a Plane Region

Use the five rectangles in Figure 4.9(a) and (b) to find *two* approximations of the area of the region lying between the graph of

$$f(x) = -x^2 + 5$$

and the *x*-axis between $x = 0$ and $x = 2$.

Solution

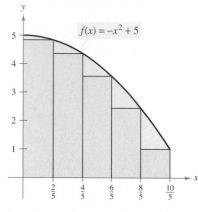

(a) The area of the parabolic region is greater than the area of the rectangles.

a. The right endpoints of the five intervals are $\frac{2}{5}i$, where $i = 1, 2, 3, 4, 5$. The width of each rectangle is $\frac{2}{5}$, and the height of each rectangle can be obtained by evaluating f at the right endpoint of each interval.

$$\left[0, \frac{2}{5}\right], \left[\frac{2}{5}, \frac{4}{5}\right], \left[\frac{4}{5}, \frac{6}{5}\right], \left[\frac{6}{5}, \frac{8}{5}\right], \left[\frac{8}{5}, \frac{10}{5}\right]$$

$$\uparrow \qquad \uparrow \qquad \uparrow \qquad \uparrow \qquad \uparrow$$

Evaluate f at the right endpoints of these intervals.

The sum of the areas of the five rectangles is

Height Width

$$\sum_{i=1}^{5} f\left(\frac{2i}{5}\right)\left(\frac{2}{5}\right) = \sum_{i=1}^{5}\left[-\left(\frac{2i}{5}\right)^2 + 5\right]\left(\frac{2}{5}\right) = \frac{162}{25} = 6.48.$$

Because each of the five rectangles lies inside the parabolic region, you can conclude that the area of the parabolic region is greater than 6.48.

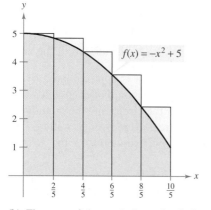

(b) The area of the parabolic region is less than the area of the rectangles.

Figure 4.9

b. The left endpoints of the five intervals are $\frac{2}{5}(i - 1)$, where $i = 1, 2, 3, 4, 5$. The width of each rectangle is $\frac{2}{5}$, and the height of each rectangle can be obtained by evaluating f at the left endpoint of each interval.

Height   Width

$$\sum_{i=1}^{5} f\left(\frac{2i - 2}{5}\right)\left(\frac{2}{5}\right) = \sum_{i=1}^{5}\left[-\left(\frac{2i - 2}{5}\right)^2 + 5\right]\left(\frac{2}{5}\right) = \frac{202}{25} = 8.08.$$

Because the parabolic region lies within the union of the five rectangular regions, you can conclude that the area of the parabolic region is less than 8.08.

By combining the results in parts (a) and (b), you can conclude that

$$6.48 < (\text{Area of region}) < 8.08.$$

NOTE   By increasing the number of rectangles used in Example 3, you can obtain closer and closer approximations of the area of the region. For instance, using 25 rectangles of width $\frac{2}{25}$ each, you can conclude that

$$7.17 < (\text{Area of region}) < 7.49.$$

Upper and Lower Sums

The procedure used in Example 3 can be generalized as follows. Consider a plane region bounded above by the graph of a nonnegative, continuous function $y = f(x)$, as shown in Figure 4.10. The region is bounded below by the x-axis, and the left and right boundaries of the region are the vertical lines $x = a$ and $x = b$.

To approximate the area of the region, begin by subdividing the interval $[a, b]$ into n subintervals, each of width $\Delta x = (b - a)/n$, as shown in Figure 4.11. The endpoints of the intervals are as follows.

$$\overbrace{a + 0(\Delta x)}^{a\,=\,x_0} < \overbrace{a + 1(\Delta x)}^{x_1} < \overbrace{a + 2(\Delta x)}^{x_2} < \cdots < \overbrace{a + n(\Delta x)}^{x_n\,=\,b}$$

Because f is continuous, the Extreme Value Theorem guarantees the existence of a minimum and a maximum value of $f(x)$ in *each* subinterval.

$f(m_i) =$ Minimum value of $f(x)$ in ith subinterval

$f(M_i) =$ Maximum value of $f(x)$ in ith subinterval

Next, define an **inscribed rectangle** lying *inside* the ith subregion and a **circumscribed rectangle** extending *outside* the ith subregion. The height of the ith inscribed rectangle is $f(m_i)$ and the height of the ith circumscribed rectangle is $f(M_i)$. For *each* i, the area of the inscribed rectangle is less than or equal to the area of the circumscribed rectangle.

$$\left(\begin{matrix}\text{Area of inscribed}\\ \text{rectangle}\end{matrix}\right) = f(m_i)\,\Delta x \le f(M_i)\,\Delta x = \left(\begin{matrix}\text{Area of circumscribed}\\ \text{rectangle}\end{matrix}\right)$$

The sum of the areas of the inscribed rectangles is called a **lower sum,** and the sum of the areas of the circumscribed rectangles is called an **upper sum.**

$$\text{Lower sum} = s(n) = \sum_{i=1}^{n} f(m_i)\,\Delta x \qquad \text{Area of inscribed rectangles}$$

$$\text{Upper sum} = S(n) = \sum_{i=1}^{n} f(M_i)\,\Delta x \qquad \text{Area of circumscribed rectangles}$$

From Figure 4.12, you can see that the lower sum $s(n)$ is less than or equal to the upper sum $S(n)$. Moreover, the actual area of the region lies between these two sums.

$$s(n) \le (\text{Area of region}) \le S(n)$$

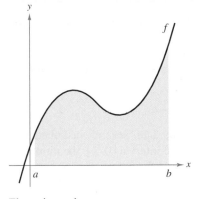

The region under a curve
Figure 4.10

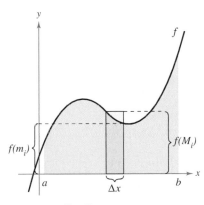

The interval $[a, b]$ is divided into n subintervals of width $\Delta x = \dfrac{b - a}{n}$.

Figure 4.11

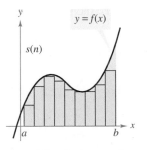

Area of inscribed rectangles is less than area of region.

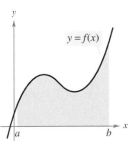

Area of region

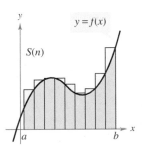

Area of circumscribed rectangles is greater than area of region.

Figure 4.12

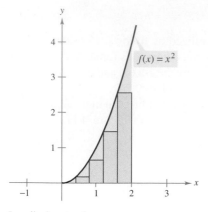

Inscribed rectangles

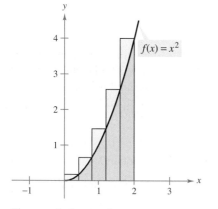

Circumscribed rectangles
Figure 4.13

Example 4   **Finding Upper and Lower Sums for a Region**

Find the upper and lower sums for the region bounded by the graph of $f(x) = x^2$ and the x-axis between $x = 0$ and $x = 2$.

Solution   To begin, partition the interval $[0, 2]$ into n subintervals, each of width

$$\Delta x = \frac{b - a}{n} = \frac{2 - 0}{n} = \frac{2}{n}.$$

Figure 4.13 shows the endpoints of the subintervals and several inscribed and circumscribed rectangles. Because f is increasing on the interval $[0, 2]$, the minimum value on each subinterval occurs at the left endpoint, and the maximum value occurs at the right endpoint.

Left Endpoints	*Right Endpoints*

$$m_i = 0 + (i - 1)\left(\frac{2}{n}\right) = \frac{2(i - 1)}{n} \qquad M_i = 0 + i\left(\frac{2}{n}\right) = \frac{2i}{n}$$

Using the left endpoints, the lower sum is

$$s(n) = \sum_{i=1}^{n} f(m_i)\, \Delta x = \sum_{i=1}^{n} f\left[\frac{2(i - 1)}{n}\right]\left(\frac{2}{n}\right)$$

$$= \sum_{i=1}^{n} \left[\frac{2(i - 1)}{n}\right]^2 \left(\frac{2}{n}\right)$$

$$= \sum_{i=1}^{n} \left(\frac{8}{n^3}\right)(i^2 - 2i + 1)$$

$$= \frac{8}{n^3}\left(\sum_{i=1}^{n} i^2 - 2\sum_{i=1}^{n} i + \sum_{i=1}^{n} 1\right)$$

$$= \frac{8}{n^3}\left\{\frac{n(n + 1)(2n + 1)}{6} - 2\left[\frac{n(n + 1)}{2}\right] + n\right\}$$

$$= \frac{4}{3n^3}(2n^3 - 3n^2 + n)$$

$$= \frac{8}{3} - \frac{4}{n} + \frac{4}{3n^2}. \qquad \text{Lower sum}$$

Using the right endpoints, the upper sum is

$$S(n) = \sum_{i=1}^{n} f(M_i)\, \Delta x = \sum_{i=1}^{n} f\left(\frac{2i}{n}\right)\left(\frac{2}{n}\right)$$

$$= \sum_{i=1}^{n} \left(\frac{2i}{n}\right)^2 \left(\frac{2}{n}\right)$$

$$= \sum_{i=1}^{n} \left(\frac{8}{n^3}\right)i^2$$

$$= \frac{8}{n^3}\left[\frac{n(n + 1)(2n + 1)}{6}\right]$$

$$= \frac{4}{3n^3}(2n^3 + 3n^2 + n)$$

$$= \frac{8}{3} + \frac{4}{n} + \frac{4}{3n^2}. \qquad \text{Upper sum}$$

EXPLORATION

For the region given in Example 4, evaluate the lower sum

$$s(n) = \frac{8}{3} - \frac{4}{n} + \frac{4}{3n^2}$$

and the upper sum

$$S(n) = \frac{8}{3} + \frac{4}{n} + \frac{4}{3n^2}$$

for $n = 10$, 100, and 1000. Use your results to determine the area of the region.

Example 4 illustrates some important things about lower and upper sums. First, notice that for any value of n, the lower sum is less than (or equal to) the upper sum.

$$s(n) = \frac{8}{3} - \frac{4}{n} + \frac{4}{3n^2} < \frac{8}{3} + \frac{4}{n} + \frac{4}{3n^2} = S(n)$$

Second, the difference between these two sums lessens as n increases. In fact, if you take the limit as $n \to \infty$, both the upper sum and the lower sum approach $\frac{8}{3}$.

$$\lim_{n \to \infty} s(n) = \lim_{n \to \infty} \left(\frac{8}{3} - \frac{4}{n} + \frac{4}{3n^2} \right) = \frac{8}{3} \qquad \text{Lower sum limit}$$

$$\lim_{n \to \infty} S(n) = \lim_{n \to \infty} \left(\frac{8}{3} + \frac{4}{n} + \frac{4}{3n^2} \right) = \frac{8}{3} \qquad \text{Upper sum limit}$$

The next theorem shows that the equivalence of the limits (as $n \to \infty$) of the upper and lower sums is not mere coincidence. It is true for all functions that are continuous and nonnegative on the closed interval $[a, b]$. The proof of this theorem is best left to a course in advanced calculus.

THEOREM 4.3   Limit of the Lower and Upper Sums

Let f be continuous and nonnegative on the interval $[a, b]$. The limits as $n \to \infty$ of both the lower and upper sums exist and are equal to each other. That is,

$$\lim_{n \to \infty} s(n) = \lim_{n \to \infty} \sum_{i=1}^{n} f(m_i) \, \Delta x$$

$$= \lim_{n \to \infty} \sum_{i=1}^{n} f(M_i) \, \Delta x$$

$$= \lim_{n \to \infty} S(n)$$

where $\Delta x = (b - a)/n$ and $f(m_i)$ and $f(M_i)$ are the minimum and maximum values of f on the subinterval.

Because the same limit is attained for both the minimum value $f(m_i)$ and the maximum value $f(M_i)$, it follows from the Squeeze Theorem (Theorem 1.8) that the choice of x in the ith subinterval does not affect the limit. This means that you are free to choose an *arbitrary* x-value in the ith subinterval, as in the following *definition of the area of a region in the plane.*

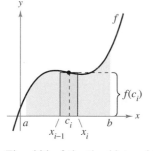

The width of the ith subinterval is $\Delta x = x_i - x_{i-1}$.
Figure 4.14

Definition of the Area of a Region in the Plane

Let f be continuous and nonnegative on the interval $[a, b]$. The area of the region bounded by the graph of f, the x-axis, and the vertical lines $x = a$ and $x = b$ is

$$\text{Area} = \lim_{n \to \infty} \sum_{i=1}^{n} f(c_i) \, \Delta x, \qquad x_{i-1} \le c_i \le x_i$$

where $\Delta x = (b - a)/n$ (see Figure 4.14).

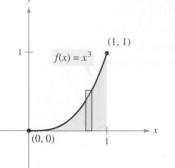

The area of the region bounded by the graph of f, the x-axis, $x = 0$, and $x = 1$ is $\frac{1}{4}$.
Figure 4.15

Example 5 Finding Area by the Limit Definition

Find the area of the region bounded by the graph $f(x) = x^3$, the x-axis, and the vertical lines $x = 0$ and $x = 1$, as shown in Figure 4.15.

Solution Begin by noting that f is continuous and nonnegative on the interval $[0, 1]$. Next, partition the interval $[0, 1]$ into n subintervals, each of width $\Delta x = 1/n$. According to the definition of area, you can choose any x-value in the ith subinterval. For this example, the right endpoints $c_i = i/n$ are convenient.

$$\text{Area} = \lim_{n \to \infty} \sum_{i=1}^{n} f(c_i)\,\Delta x = \lim_{n \to \infty} \sum_{i=1}^{n} \left(\frac{i}{n}\right)^3 \left(\frac{1}{n}\right) \qquad \text{Right endpoints: } c_i = \frac{i}{n}$$

$$= \lim_{n \to \infty} \frac{1}{n^4} \sum_{i=1}^{n} i^3$$

$$= \lim_{n \to \infty} \frac{1}{n^4} \left[\frac{n^2(n+1)^2}{4}\right]$$

$$= \lim_{n \to \infty} \left(\frac{1}{4} + \frac{1}{2n} + \frac{1}{4n^2}\right)$$

$$= \frac{1}{4}$$

The area of the region is $\frac{1}{4}$.

Example 6 Finding Area by the Limit Definition

Find the area of the region bounded by the graph of $f(x) = 4 - x^2$, the x-axis, and the vertical lines $x = 1$ and $x = 2$, as shown in Figure 4.16.

Solution The function f is continuous and nonnegative on the interval $[1, 2]$, and so you begin by partitioning the interval into n subintervals, each of width $\Delta x = 1/n$. Choosing the right endpoint,

$$c_i = a + i\Delta x = 1 + \frac{i}{n} \qquad \text{Right endpoints}$$

of each subinterval, you obtain the following.

$$\text{Area} = \lim_{n \to \infty} \sum_{i=1}^{n} f(c_i)\,\Delta x = \lim_{n \to \infty} \sum_{i=1}^{n} \left[4 - \left(1 + \frac{i}{n}\right)^2\right]\left(\frac{1}{n}\right)$$

$$= \lim_{n \to \infty} \sum_{i=1}^{n} \left(3 - \frac{2i}{n} - \frac{i^2}{n^2}\right)\left(\frac{1}{n}\right)$$

$$= \lim_{n \to \infty} \left(\frac{1}{n} \sum_{i=1}^{n} 3 - \frac{2}{n^2} \sum_{i=1}^{n} i - \frac{1}{n^3} \sum_{i=1}^{n} i^2\right)$$

$$= \lim_{n \to \infty} \left[3 - \left(1 + \frac{1}{n}\right) - \left(\frac{1}{3} + \frac{1}{2n} + \frac{1}{6n^2}\right)\right]$$

$$= 3 - 1 - \frac{1}{3}$$

$$= \frac{5}{3}$$

The area of the region is $\frac{5}{3}$.

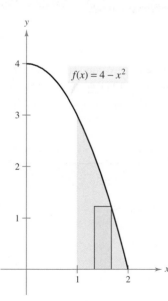

The area of the region bounded by the graph of f, the x-axis, $x = 1$, and $x = 2$ is $\frac{5}{3}$.
Figure 4.16

The last example in this section looks at a region that is bounded by the y-axis (rather than by the x-axis).

Example 7   **A Region Bounded by the y-Axis**

Find the area of the region bounded by the graph of $f(y) = y^2$ and the y-axis for $0 \le y \le 1$, as shown in Figure 4.17.

Solution   When f is a continuous, nonnegative function of y, you still can use the same basic procedure illustrated in Examples 5 and 6. Begin by partitioning the interval $[0, 1]$ into n subintervals, each of width $\Delta y = 1/n$. Then, using the upper endpoints $c_i = i/n$, you obtain the following.

$$\text{Area} = \lim_{n\to\infty} \sum_{i=1}^{n} f(c_i)\,\Delta y = \lim_{n\to\infty} \sum_{i=1}^{n} \left(\frac{i}{n}\right)^2\left(\frac{1}{n}\right) \qquad \text{Upper endpoints: } c_i = \frac{i}{n}$$

$$= \lim_{n\to\infty} \frac{1}{n^3}\sum_{i=1}^{n} i^2$$

$$= \lim_{n\to\infty} \frac{1}{n^3}\left[\frac{n(n+1)(2n+1)}{6}\right]$$

$$= \lim_{n\to\infty} \left(\frac{1}{3} + \frac{1}{2n} + \frac{1}{6n^2}\right)$$

$$= \frac{1}{3}$$

The area of the region is $\frac{1}{3}$.

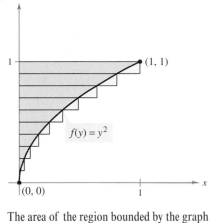

The area of the region bounded by the graph of f and the y-axis for $0 \le y \le 1$ is $\frac{1}{3}$.
Figure 4.17

EXERCISES FOR SECTION 4.2

In Exercises 1–6, find the sum. Use the summation capabilities of a graphing utility to verify your result.

1. $\displaystyle\sum_{i=1}^{5}(2i + 1)$

2. $\displaystyle\sum_{k=3}^{6}k(k - 2)$

3. $\displaystyle\sum_{k=0}^{4}\frac{1}{k^2 + 1}$

4. $\displaystyle\sum_{j=3}^{5}\frac{1}{j}$

5. $\displaystyle\sum_{k=1}^{4}c$

6. $\displaystyle\sum_{i=1}^{4}[(i - 1)^2 + (i + 1)^3]$

In Exercises 7–14, use sigma notation to write the sum.

7. $\dfrac{1}{3(1)} + \dfrac{1}{3(2)} + \dfrac{1}{3(3)} + \cdots + \dfrac{1}{3(9)}$

8. $\dfrac{5}{1 + 1} + \dfrac{5}{1 + 2} + \dfrac{5}{1 + 3} + \cdots + \dfrac{5}{1 + 15}$

9. $\left[5\left(\dfrac{1}{8}\right) + 3\right] + \left[5\left(\dfrac{2}{8}\right) + 3\right] + \cdots + \left[5\left(\dfrac{8}{8}\right) + 3\right]$

10. $\left[1 - \left(\dfrac{1}{4}\right)^2\right] + \left[1 - \left(\dfrac{2}{4}\right)^2\right] + \cdots + \left[1 - \left(\dfrac{4}{4}\right)^2\right]$

11. $\left[\left(\dfrac{2}{n}\right)^3 - \dfrac{2}{n}\right]\left(\dfrac{2}{n}\right) + \cdots + \left[\left(\dfrac{2n}{n}\right)^3 - \dfrac{2n}{n}\right]\left(\dfrac{2}{n}\right)$

12. $\left[1 - \left(\dfrac{2}{n} - 1\right)^2\right]\left(\dfrac{2}{n}\right) + \cdots + \left[1 - \left(\dfrac{2n}{n} - 1\right)^2\right]\left(\dfrac{2}{n}\right)$

13. $\left[2\left(1 + \dfrac{3}{n}\right)^2\right]\left(\dfrac{3}{n}\right) + \cdots + \left[2\left(1 + \dfrac{3n}{n}\right)^2\right]\left(\dfrac{3}{n}\right)$

14. $\left(\dfrac{1}{n}\right)\sqrt{1 - \left(\dfrac{0}{n}\right)^2} + \cdots + \left(\dfrac{1}{n}\right)\sqrt{1 - \left(\dfrac{n-1}{n}\right)^2}$

In Exercises 15–20, use the properties of summation and Theorem 4.2 to evaluate the sum. Use the summation capabilities of a graphing utility to verify your result.

15. $\displaystyle\sum_{i=1}^{20} 2i$

16. $\displaystyle\sum_{i=1}^{15}(2i - 3)$

17. $\displaystyle\sum_{i=1}^{20}(i - 1)^2$

18. $\displaystyle\sum_{i=1}^{10}(i^2 - 1)$

19. $\displaystyle\sum_{i=1}^{15} i(i - 1)^2$

20. $\displaystyle\sum_{i=1}^{10} i(i^2 + 1)$

In Exercises 21 and 22, use the summation capabilities of a graphing utility to evaluate the sum. Then use the properties of summation and Theorem 4.2 to verify the sum.

21. $\displaystyle\sum_{i=1}^{20}(i^2 + 3)$

22. $\displaystyle\sum_{i=1}^{15}(i^3 - 2i)$

In Exercises 23–26, bound the area of the shaded region by approximating the upper and lower sums. Use rectangles of width 1.

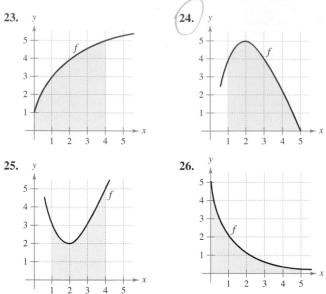

23.

24.

25.

26.

In Exercises 27–30, use upper and lower sums to approximate the area of the region using the indicated number of subintervals (of equal width).

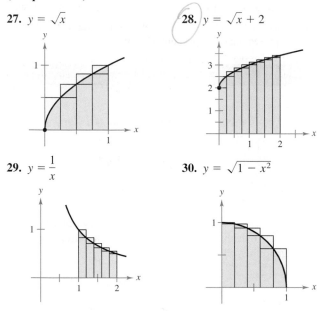

27. $y = \sqrt{x}$

28. $y = \sqrt{x} + 2$

29. $y = \dfrac{1}{x}$

30. $y = \sqrt{1 - x^2}$

In Exercises 31–34, find the limit of $s(n)$ as $n \to \infty$.

31. $s(n) = \dfrac{81}{n^4}\left[\dfrac{n^2(n+1)^2}{4}\right]$

32. $s(n) = \dfrac{64}{n^3}\left[\dfrac{n(n+1)(2n+1)}{6}\right]$

33. $s(n) = \dfrac{18}{n^2}\left[\dfrac{n(n+1)}{2}\right]$

34. $s(n) = \dfrac{1}{n^2}\left[\dfrac{n(n+1)}{2}\right]$

In Exercises 35–38, use the summation formulas to rewrite the expression without the summation notation. Use the result to find the sum for $n = 10, 100, 1000,$ and $10,000$.

35. $\displaystyle\sum_{i=1}^{n} \dfrac{2i+1}{n^2}$

36. $\displaystyle\sum_{j=1}^{n} \dfrac{4j+3}{n^2}$

37. $\displaystyle\sum_{k=1}^{n} \dfrac{6k(k-1)}{n^3}$

38. $\displaystyle\sum_{i=1}^{n} \dfrac{4i^2(i-1)}{n^4}$

In Exercises 39–44, find a formula for the sum of n terms. Use the formula to find the limit as $n \to \infty$.

39. $\displaystyle\lim_{n\to\infty} \sum_{i=1}^{n} \dfrac{16i}{n^2}$

40. $\displaystyle\lim_{n\to\infty} \sum_{i=1}^{n} \left(\dfrac{2i}{n}\right)\left(\dfrac{2}{n}\right)$

41. $\displaystyle\lim_{n\to\infty} \sum_{i=1}^{n} \dfrac{1}{n^3}(i-1)^2$

42. $\displaystyle\lim_{n\to\infty} \sum_{i=1}^{n} \left(1 + \dfrac{2i}{n}\right)^2\left(\dfrac{2}{n}\right)$

43. $\displaystyle\lim_{n\to\infty} \sum_{i=1}^{n} \left(1 + \dfrac{i}{n}\right)\left(\dfrac{2}{n}\right)$

44. $\displaystyle\lim_{n\to\infty} \sum_{i=1}^{n} \left(1 + \dfrac{2i}{n}\right)^3\left(\dfrac{2}{n}\right)$

45. *Numerical Reasoning* Consider a triangle of area 2 bounded by the graphs of $y = x$, $y = 0$, and $x = 2$.

(a) Sketch the region.

(b) Divide the interval $[0, 2]$ into n subintervals of equal width and show that the endpoints are
$$0 < 1\left(\dfrac{2}{n}\right) < \cdots < (n-1)\left(\dfrac{2}{n}\right) < n\left(\dfrac{2}{n}\right).$$

(c) Show that $s(n) = \displaystyle\sum_{i=1}^{n}\left[(i-1)\left(\dfrac{2}{n}\right)\right]\left(\dfrac{2}{n}\right).$

(d) Show that $S(n) = \displaystyle\sum_{i=1}^{n}\left[i\left(\dfrac{2}{n}\right)\right]\left(\dfrac{2}{n}\right).$

(e) Complete the table.

n	5	10	50	100
$s(n)$				
$S(n)$				

(f) Show that $\displaystyle\lim_{n\to\infty} s(n) = \lim_{n\to\infty} S(n) = 2.$

46. *Numerical Reasoning* Consider a trapezoid of area 4 bounded by the graphs of $y = x$, $y = 0$, $x = 1$, and $x = 3$.

(a) Sketch the region.

(b) Divide the interval $[1, 3]$ into n subintervals of equal width and show that the endpoints are
$$1 < 1 + 1\left(\dfrac{2}{n}\right) < \cdots < 1 + (n-1)\left(\dfrac{2}{n}\right) < 1 + n\left(\dfrac{2}{n}\right).$$

(c) Show that $s(n) = \displaystyle\sum_{i=1}^{n}\left[1 + (i-1)\left(\dfrac{2}{n}\right)\right]\left(\dfrac{2}{n}\right).$

(d) Show that $S(n) = \displaystyle\sum_{i=1}^{n}\left[1 + i\left(\dfrac{2}{n}\right)\right]\left(\dfrac{2}{n}\right).$

(e) Complete the table.

n	5	10	50	100
$s(n)$				
$S(n)$				

(f) Show that $\displaystyle\lim_{n\to\infty} s(n) = \lim_{n\to\infty} S(n) = 4.$

In Exercises 47–56, use the limit process to find the area of the region between the graph of the function and the x-axis over the indicated interval. Sketch the region.

	Function	Interval		Function	Interval
47.	$y = -2x + 3$	$[0, 1]$	**48.**	$y = 3x - 4$	$[2, 5]$
49.	$y = x^2 + 2$	$[0, 1]$	**50.**	$y = x^2 + 1$	$[0, 3]$
51.	$y = 16 - x^2$	$[1, 3]$	**52.**	$y = 1 - x^2$	$[-1, 1]$
53.	$y = 64 - x^3$	$[1, 4]$	**54.**	$y = 2x - x^3$	$[0, 1]$
55.	$y = x^2 - x^3$	$[-1, 1]$	**56.**	$y = x^2 - x^3$	$[-1, 0]$

In Exercises 57–62, use the limit process to find the area of the region between the graph of the function and the y-axis over the indicated y-interval. Sketch the region.

57. $f(y) = 3y, 0 \le y \le 2$ **58.** $g(y) = \frac{1}{2}y, 2 \le y \le 4$

59. $f(y) = y^2, 0 \le y \le 3$ **60.** $f(y) = 4y - y^2, 1 \le y \le 2$

61. $g(y) = 4y^2 - y^3, 1 \le y \le 3$

62. $h(y) = y^3 + 1, 1 \le y \le 2$

In Exercises 63–66 use the *Midpoint Rule*

$$\text{Area} \approx \sum_{i=1}^{n} f\left(\frac{x_i + x_{i-1}}{2}\right)\Delta x$$

with $n = 4$ to approximate the area of the region bounded by the graph of the function and the x-axis over the indicated interval.

	Function	Interval
63.	$f(x) = x^2 + 3$	$[0, 2]$
64.	$f(x) = x^2 + 4x$	$[0, 4]$
65.	$f(x) = \tan x$	$\left[0, \frac{\pi}{4}\right]$
66.	$f(x) = \sin x$	$\left[0, \frac{\pi}{2}\right]$

Write a program for a graphing utility to approximate areas by using the Midpoint Rule. Assume that the function is positive over the indicated interval and the subintervals are of equal width. In Exercises 67–70, use the program to approximate the area of the region between the graph of the function and the x-axis over the indicated interval, and complete the table.

n	4	8	12	16	20
Approximate area					

	Function	Interval		Function	Interval
67.	$f(x) = \sqrt{x}$	$[0, 4]$	**68.**	$f(x) = \dfrac{8}{x^2 + 1}$	$[2, 6]$
69.	$f(x) = \tan\left(\dfrac{\pi x}{8}\right)$	$[1, 3]$	**70.**	$f(x) = \cos \sqrt{x}$	$[0, 2]$

Getting at the Concept

71. In your own words and using appropriate figures, describe the methods of upper sums and lower sums in approximating the area of a region.

72. Give the definition of the area of a region in the plane.

73. *Graphical Reasoning*   Consider the region bounded by the graphs of

$$f(x) = \frac{8x}{x + 1},$$

$x = 0$, $x = 4$, and $y = 0$, as shown in the figure. To print an enlarged copy of the graph, go to the website *www.mathgraphs.com.*

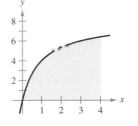

(a) Redraw the figure, and complete and shade the rectangles representing the lower sum when $n = 4$. Find this lower sum.

(b) Redraw the figure, and complete and shade the rectangles representing the upper sum when $n = 4$. Find this upper sum.

(c) Redraw the figure, and complete and shade the rectangles whose heights are determined by the functional values at the midpoint of each subinterval when $n = 4$. Find this sum using the Midpoint Rule.

(d) Verify the following formulas for approximating the area of the region using n subintervals of equal width.

Lower sum: $s(n) = \displaystyle\sum_{i=1}^{n} f\left[(i - 1)\frac{4}{n}\right]\left(\frac{4}{n}\right)$

Upper sum: $S(n) = \displaystyle\sum_{i=1}^{n} f\left[(i)\frac{4}{n}\right]\left(\frac{4}{n}\right)$

Midpoint Rule: $M(n) = \displaystyle\sum_{i=1}^{n} f\left[\left(i - \frac{1}{2}\right)\frac{4}{n}\right]\left(\frac{4}{n}\right)$

(e) Use a graphing utility and the formulas in part (d) to complete the table.

n	4	8	20	100	200
$s(n)$					
$S(n)$					
$M(n)$					

(f) Explain why $s(n)$ increases and $S(n)$ decreases for increasing values of n, as shown in the table in part (e).

74. Use a graphing utility to complete the table for approximations of the area of the region bounded by the graphs of $f(x) = \sqrt[3]{x}$, $x = 0$, $x = 8$, and $y = 0$.

n	10	20	50	100	200
$s(n)$					
$S(n)$					
$M(n)$					

Approximation In Exercises 75 and 76, determine which value best approximates the area of the region between the *x*-axis and the graph of the function over the indicated interval. (Make your selection on the basis of a sketch of the region and not by performing calculations.)

75. $f(x) = 4 - x^2$, $[0, 2]$

    (a) -2 (b) 6   (c) 10   (d) 3   (e) 8

76. $f(x) = \sin \dfrac{\pi x}{4}$, $[0, 4]$

    (a) 3   (b) 1   (c) -2 (d) 8   (e) 6

True or False? In Exercises 77 and 78, determine whether the statement is true or false. If it is false, explain why or give an example that shows it is false.

77. The sum of the first n positive integers is $n(n + 1)/2$.

78. If f is continuous and nonnegative on $[a, b]$, then the limits as $n \to \infty$ of its lower sum $s(n)$ and upper sum $S(n)$ both exist and are equal.

79. *Monte Carlo Method* The following computer program approximates the area of the region under the graph of a monotonic function and above the *x*-axis between $x = a$ and $x = b$. Run the program for $a = 0$ and $b = \pi/2$ for several values of N2. Explain why the Monte Carlo Method works. [*Adaptation of Monte Carlo Method program from James M. Sconyers, "Approximation of Area Under a Curve," MATHEMATICS TEACHER 77, no. 2 (February 1984). Copyright © 1984 by the National Council of Teachers of Mathematics. Reprinted with permission.*]

```
10   DEF FNF(X)=SIN(X)
20   A=0
30   B=1.570796
40   PRINT "Input Number of Random Points"
50   INPUT N2
60   N1=0
70   IF FNF(A)>FNF(B) THEN YMAX=FNF(A) ELSE
       YMAX=FNF(B)
80   FOR I=1 TO N2
90   X=A+(B-A)*RND(1)
100  Y=YMAX*RND(1)
110  IF Y>=FNF(X) THEN GOTO 130
120  N1=N1+1
130  NEXT I
140  AREA=(N1/N2)*(B-A)*YMAX
150  PRINT "Approximate Area:"; AREA
160  END
```

80. *Graphical Reasoning* Consider an *n*-sided regular polygon inscribed in a circle of radius *r*. Join the vertices of the polygon to the center of the circle, forming *n* congruent triangles (see figure).

    (a) Determine the central angle θ in terms of n.

    (b) Show that the area of each triangle is $\frac{1}{2}r^2 \sin \theta$.

    (c) Let A_n be the sum of the areas of the n triangles. Find $\lim\limits_{n\to\infty} A_n$.

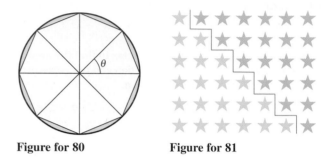

Figure for 80   **Figure for 81**

81. *Writing* Use the figure to write a short paragraph explaining why the formula

$$1 + 2 + \cdots + n = \tfrac{1}{2}n(n + 1)$$

is valid for all positive integers n.

82. Prove each of the formulas by mathematical induction. (You may need to review the method of proof by induction from a precalculus text.)

    (a) $\displaystyle\sum_{i=1}^{n} 2i = n(n + 1)$

    (b) $\displaystyle\sum_{i=1}^{n} i^3 = \dfrac{n^2(n + 1)^2}{4}$

83. *Modeling Data* The table lists the measurements of a lot bounded by a stream and two straight roads that meet at right angles, where *x* and *y* are measured in feet (see figure).

x	0	50	100	150	200	250	300
y	450	362	305	268	245	156	0

    (a) Use the regression capabilities of a graphing utility to find a model of the form

$$y = ax^3 + bx^2 + cx + d.$$

    (b) Use a graphing utility to plot the data and graph the model.

    (c) Use the model in part (a) to estimate the area of the lot.

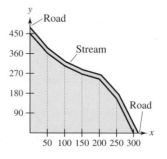

• Understand the definition of a Riemann sum.
• Evaluate a definite integral using limits.
• Evaluate a definite integral using properties of definite integrals.

Riemann Sums

In the definition of area given in Section 4.2, the partitions have subintervals of *equal width*. This was done only for computational convenience. We begin this section with an example that shows that it is not necessary to have subintervals of equal width.

Example 1   **A Partition with Subintervals of Unequal Widths**

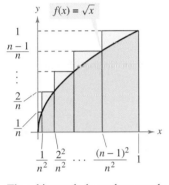

y $f(x) = \sqrt{x}$

The subintervals do not have equal widths.
Figure 4.18

Consider the region bounded by the graph of $f(x) = \sqrt{x}$ and the x-axis for $0 \le x \le 1$, as shown in Figure 4.18. Evaluate the limit

$$\lim_{n \to \infty} \sum_{i=1}^{n} f(c_i) \, \Delta x_i$$

where c_i is the right endpoint of the partition given by $x_i = i^2/n^2$ and Δx_i is the width of the ith interval.

Solution   The width of the ith interval is given by

$$\Delta x_i = \frac{i^2}{n^2} - \frac{(i-1)^2}{n^2}$$

$$= \frac{i^2 - i^2 + 2i - 1}{n^2}$$

$$= \frac{2i-1}{n^2}.$$

So, the limit is

$$\lim_{n \to \infty} \sum_{i=1}^{n} f(c_i) \, \Delta x_i = \lim_{n \to \infty} \sum_{i=1}^{n} \sqrt{\frac{i^2}{n^2}} \left(\frac{2i-1}{n^2} \right)$$

$$= \lim_{n \to \infty} \frac{1}{n^3} \sum_{i=1}^{n} (2i^2 - i)$$

$$= \lim_{n \to \infty} \frac{1}{n^3} \left[2 \left(\frac{n(n+1)(2n+1)}{6} \right) - \frac{n(n+1)}{2} \right]$$

$$= \lim_{n \to \infty} \frac{4n^3 + 3n^2 - n}{6n^3}$$

$$= \frac{2}{3}.$$

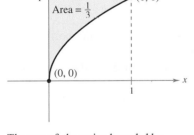

y

$x = y^2$

Area $= \frac{1}{3}$

$(1, 1)$

$(0, 0)$

The area of the region bounded by the graph of $x = y^2$ and the y-axis for $0 \le y \le 1$ is $\frac{1}{3}$.
Figure 4.19

From Example 7 in Section 4.2, you know that the region shown in Figure 4.19 has an area of $\frac{1}{3}$. Because the square bounded by $0 \le x \le 1$ and $0 \le y \le 1$ has an area of 1, you can conclude that the area of the region shown in Figure 4.18 has an area of $\frac{2}{3}$. This agrees with the limit found in Example 1, even though that example used a partition having subintervals of unequal widths. The reason this particular partition gave the proper area is that as n increases, the *width of the largest subinterval approaches zero*. This is a key feature of the development of definite integrals.

**GEORG FRIEDRICH BERNHARD RIEMANN
(1826–1866)**

German mathematician Riemann did his most famous work in the areas of non-Euclidean geometry, differential equations, and number theory. It was Riemann's results in physics and mathematics that formed the structure on which Einstein's theory of general relativity is based.

In the preceding section, the limit of a sum was used to define the area of a region in the plane. Finding area by this means is only one of *many* applications involving the limit of a sum. A similar approach can be used to determine quantities as diverse as arc length, average value, centroids, volumes, work, and surface areas. The following definition is named after Georg Friedrich Bernhard Riemann. Although the definite integral had been defined and used long before the time of Riemann, he generalized the concept to cover a broader category of functions.

In the following definition of a Riemann sum, note that the function f has no restrictions other than being defined on the interval $[a, b]$. (In the preceding section, the function f was assumed to be continuous and nonnegative because we were dealing with the area under a curve.)

Definition of a Riemann Sum

Let f be defined on the closed interval $[a, b]$, and let Δ be a partition of $[a, b]$ given by

$$a = x_0 < x_1 < x_2 < \cdots < x_{n-1} < x_n = b$$

where Δx_i is the width of the ith subinterval. If c_i is *any* point in the ith subinterval, then the sum

$$\sum_{i=1}^{n} f(c_i)\, \Delta x_i, \qquad x_{i-1} \le c_i \le x_i$$

is called a **Riemann sum** of f for the partition Δ.

NOTE The sums in Section 4.2 are examples of Riemann sums, but there are more general Riemann sums than those covered there.

The width of the largest subinterval of a partition Δ is the **norm** of the partition and is denoted by $\|\Delta\|$. If every subinterval is of equal width, the partition is **regular** and the norm is denoted by

$$\|\Delta\| = \Delta x = \frac{b-a}{n}. \qquad \text{Regular partition}$$

For a general partition, the norm is related to the number of subintervals of $[a, b]$ in the following way.

$$\frac{b-a}{\|\Delta\|} \le n \qquad \text{General partition}$$

So, the number of subintervals in a partition approaches infinity as the norm of the partition approaches 0. That is, $\|\Delta\| \to 0$ implies that $n \to \infty$.

The converse of this statement is not true. For example, let Δ_n be the partition of the interval $[0, 1]$ given by

$$0 < \frac{1}{2^n} < \frac{1}{2^{n-1}} < \cdots < \frac{1}{8} < \frac{1}{4} < \frac{1}{2} < 1.$$

As shown in Figure 4.20, for any positive value of n, the norm of the partition Δ_n is $\frac{1}{2}$. So, letting n approach infinity does not force $\|\Delta\|$ to approach 0. In a regular partition, however, the statements $\|\Delta\| \to 0$ and $n \to \infty$ are equivalent.

$\|\Delta\| = \frac{1}{2}$

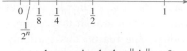

$n \to \infty$ does not imply that $\|\Delta\| \to 0$.
Figure 4.20

Definite Integrals

To define the definite integral, consider the following limit.

$$\lim_{\|\Delta\| \to 0} \sum_{i=1}^{n} f(c_i)\,\Delta x_i = L$$

To say that this limit exists means that for $\varepsilon > 0$ there exists a $\delta > 0$ such that for every partition with $\|\Delta\| < \delta$ it follows that

$$\left| L - \sum_{i=1}^{n} f(c_i)\,\Delta x_i \right| < \varepsilon.$$

(This must be true for any choice of c_i in the ith subinterval of Δ.)

FOR FURTHER INFORMATION For insight into the history of the definite integral, see the article "The Evolution of Integration" by A. Shenitzer and J. Steprāns in *The American Mathematical Monthly*. To view this article, go to the website *www.matharticles.com.*

Definition of a Definite Integral

If f is defined on the closed interval $[a, b]$ and the limit

$$\lim_{\|\Delta\| \to 0} \sum_{i=1}^{n} f(c_i)\,\Delta x_i$$

exists (as described above), then f is **integrable** on $[a, b]$ and the limit is denoted by

$$\lim_{\|\Delta\| \to 0} \sum_{i=1}^{n} f(c_i)\,\Delta x_i = \int_{a}^{b} f(x)\,dx.$$

The limit is called the **definite integral** of f from a to b. The number a is the **lower limit** of integration, and the number b is the **upper limit** of integration.

It is not a coincidence that the notation for definite integrals is similar to that used for indefinite integrals. You will see why in the next section when we discuss the Fundamental Theorem of Calculus. For now it is important to see that definite integrals and indefinite integrals are different identities. A definite integral is a *number*, whereas an indefinite integral is a *family of functions*.

A sufficient condition for a function f to be integrable on $[a, b]$ is that it is continuous on $[a, b]$. A proof of this theorem is beyond the scope of this text.

THEOREM 4.4   Continuity Implies Integrability

If a function f is continuous on the closed interval $[a, b]$, then f is integrable on $[a, b]$.

EXPLORATION

The Converse of Theorem 4.4   Is the converse of Theorem 4.4 true? That is, if a function is integrable, does it have to be continuous? Explain your reasoning and give examples.

Describe the relationships among continuity, differentiability, and integrability. Which is the strongest condition? Which is the weakest? Which conditions imply other conditions?

Example 2   **Evaluating a Definite Integral as a Limit**

Evaluate the definite integral $\int_{-2}^{1} 2x \, dx$.

Solution   The function $f(x) = 2x$ is integrable on the interval $[-2, 1]$ because it is continuous on $[-2, 1]$. Moreover, the definition of integrability implies that any partition whose norm approaches 0 can be used to determine the limit. For computational convenience, define Δ by subdividing $[-2, 1]$ into n subintervals of equal width

$$\Delta x_i = \Delta x = \frac{b - a}{n} = \frac{3}{n}.$$

Choosing c_i as the right endpoint of each subinterval produces

$$c_i = a + i(\Delta x) = -2 + \frac{3i}{n}.$$

So, the definite integral is given by

$$\int_{-2}^{1} 2x \, dx = \lim_{\|\Delta\| \to 0} \sum_{i=1}^{n} f(c_i) \, \Delta x_i$$

$$= \lim_{n \to \infty} \sum_{i=1}^{n} f(c_i) \, \Delta x$$

$$= \lim_{n \to \infty} \sum_{i=1}^{n} 2\left(-2 + \frac{3i}{n}\right)\left(\frac{3}{n}\right)$$

$$= \lim_{n \to \infty} \frac{6}{n} \sum_{i=1}^{n} \left(-2 + \frac{3i}{n}\right)$$

$$= \lim_{n \to \infty} \frac{6}{n} \left\{-2n + \frac{3}{n}\left[\frac{n(n + 1)}{2}\right]\right\}$$

$$= \lim_{n \to \infty} \left(-12 + 9 + \frac{9}{n}\right)$$

$$= -3.$$

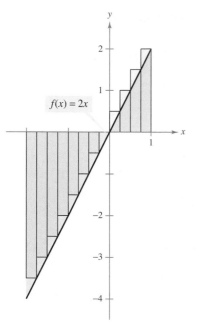

$f(x) = 2x$

Because the definite integral is negative, it does not represent the area of the region.
Figure 4.21

Because the definite integral in Example 2 is negative, it *does not* represent the area of the region shown in Figure 4.21. Definite integrals can be positive, negative, or zero. For a definite integral to be interpreted as an area (as defined in Section 4.2), the function f must be continuous and nonnegative on $[a, b]$, as stated in the following theorem. (The proof of this theorem is straightforward—you simply use the definition of area given in Section 4.2.)

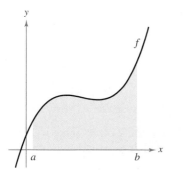

You can use a definite integral to find the area of the region bounded by the graph of f, the x-axis, $x = a$, and $x = b$.
Figure 4.22

THEOREM 4.5   The Definite Integral as the Area of a Region

If f is continuous and nonnegative on the closed interval $[a, b]$, then the area of the region bounded by the graph of f, the x-axis, and the vertical lines $x = a$ and $x = b$ is given by

$$\text{Area} = \int_{a}^{b} f(x) \, dx.$$

(See Figure 4.22.)

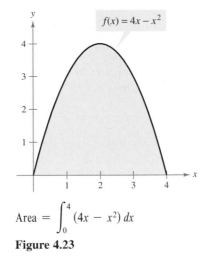

$$\text{Area} = \int_0^4 (4x - x^2)\, dx$$

Figure 4.23

As an example of Theorem 4.5, consider the region bounded by the graph of

$$f(x) = 4x - x^2$$

and the x-axis, as shown in Figure 4.23. Because f is continuous and nonnegative on the closed interval $[0, 4]$, the area of the region is

$$\text{Area} = \int_0^4 (4x - x^2)\, dx.$$

A straightforward technique for evaluating a definite integral such as this will be discussed in Section 4.4. For now, however, you can evaluate a definite integral in two ways—you can use the limit definition *or* you can check to see whether the definite integral represents the area of a common geometric region such as a rectangle, triangle, or semicircle.

Example 3   **Areas of Common Geometric Figures**

Sketch the region corresponding to each definite integral. Then evaluate each integral using a geometric formula.

a. $\displaystyle\int_1^3 4\, dx$ **b.** $\displaystyle\int_0^3 (x + 2)\, dx$ **c.** $\displaystyle\int_{-2}^2 \sqrt{4 - x^2}\, dx$

Solution   A sketch of each region is shown in Figure 4.24.

a. This region is a rectangle of height 4 and width 2.

$$\int_1^3 4\, dx = (\text{Area of rectangle}) = 4(2) = 8$$

b. This region is a trapezoid with an altitude of 3 and parallel bases of lengths 2 and 5. The formula for the area of a trapezoid is $\frac{1}{2}h(b_1 + b_2)$.

$$\int_0^3 (x + 2)\, dx = (\text{Area of trapezoid}) = \frac{1}{2}(3)(2 + 5) = \frac{21}{2}$$

c. This region is a semicircle of radius 2. The formula for the area of a semicircle is $\frac{1}{2}\pi r^2$.

$$\int_{-2}^2 \sqrt{4 - x^2}\, dx = (\text{Area of semicircle}) = \frac{1}{2}\pi(2^2) = 2\pi$$

NOTE   The variable of integration in a definite integral is sometimes called a *dummy variable* because it can be replaced by any other variable without changing the value of the integral. For instance, the definite integrals

$$\int_0^3 (x + 2)\, dx$$

and

$$\int_0^3 (t + 2)\, dt$$

have the same value.

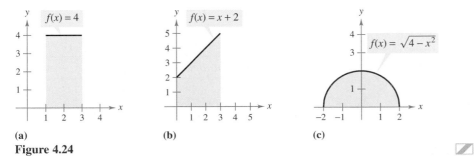

(a)   **(b)**   **(c)**

Figure 4.24

Properties of Definite Integrals

The definition of the definite integral of f on the interval $[a, b]$ specifies that $a < b$. Now, however, it is convenient to extend the definition to cover cases in which $a = b$ or $a > b$. Geometrically, the following two definitions seem reasonable. For instance, it makes sense to define the area of a region of zero width and finite height to be 0.

Definition of Two Special Definite Integrals

1. If f is defined at $x = a$, then $\displaystyle\int_a^a f(x)\, dx = 0$.

2. If f is integrable on $[a, b]$, then $\displaystyle\int_b^a f(x)\, dx = -\int_a^b f(x)\, dx$.

 Example 4   Evaluating Definite Integrals

a. Because the sine function is defined at $x = \pi$, and the upper and lower limits of integration are equal, you can write

$$\int_\pi^\pi \sin x\, dx = 0.$$

b. The integral $\int_3^0 (x + 2)\, dx$ is the same as that given in Example 3b except that the upper and lower limits are interchanged. Because the integral in Example 3b has a value of $\frac{21}{2}$, you can write

$$\int_3^0 (x + 2)\, dx = -\int_0^3 (x + 2)\, dx = -\frac{21}{2}.$$

In Figure 4.25, the larger region can be divided at $x = c$ into two subregions whose intersection is a line segment. Because the line segment has zero area, it follows that the area of the larger region is equal to the sum of the areas of the two smaller regions.

THEOREM 4.6   Additive Interval Property

If f is integrable on the three closed intervals determined by a, b, and c, then

$$\int_a^b f(x)\, dx = \int_a^c f(x)\, dx + \int_c^b f(x)\, dx.$$

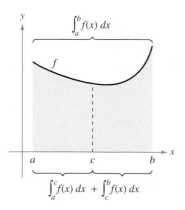

Figure 4.25

Example 5   Using the Additive Interval Property

$$\int_{-1}^1 |x|\, dx = \int_{-1}^0 -x\, dx + \int_0^1 x\, dx \qquad \text{Theorem 4.6}$$

$$= \frac{1}{2} + \frac{1}{2} \qquad\qquad \text{Area of a triangle}$$

$$= 1$$

Because the definite integral is defined as the limit of a sum, it inherits the properties of summation given at the top of page 254.

THEOREM 4.7   Properties of Definite Integrals

If f and g are integrable on $[a, b]$ and k is a constant, then the functions of kf and $f \pm g$ are integrable on $[a, b]$, and

1. $\displaystyle \int_a^b kf(x)\, dx = k \int_a^b f(x)\, dx$

2. $\displaystyle \int_a^b [f(x) \pm g(x)]\, dx = \int_a^b f(x)\, dx \pm \int_a^b g(x)\, dx.$

Note that Property 2 of Theorem 4.7 can be extended to cover any finite number of functions. For example,

$$\int_a^b [f(x) + g(x) + h(x)]\, dx = \int_a^b f(x)\, dx + \int_a^b g(x)\, dx + \int_a^b h(x)\, dx.$$

Example 6   **Evaluation of a Definite Integral**

Evaluate $\displaystyle \int_1^3 (-x^2 + 4x - 3)\, dx$ using each of the following values.

$$\int_1^3 x^2\, dx = \frac{26}{3}, \qquad \int_1^3 x\, dx - 4, \qquad \int_1^3 dx = 2$$

Solution

$$\int_1^3 (-x^2 + 4x - 3)\, dx = \int_1^3 (-x^2)\, dx + \int_1^3 4x\, dx + \int_1^3 (-3)\, dx$$

$$= -\int_1^3 x^2\, dx + 4\int_1^3 x\, dx - 3\int_1^3 dx$$

$$= -\left(\frac{26}{3}\right) + 4(4) - 3(2)$$

$$= \frac{4}{3}$$

If f and g are continuous on the closed interval $[a, b]$ and

$$0 \le f(x) \le g(x)$$

for $a \le x \le b$, the following properties are true. First, the area of the region bounded by the graph of f and the x-axis (between a and b) must be nonnegative. Second, this area must be less than or equal to the area of the region bounded by the graph of g and the x-axis (between a and b), as shown in Figure 4.26. These two results are generalized in Theorem 4.8. (A proof of this theorem is given in Appendix B.)

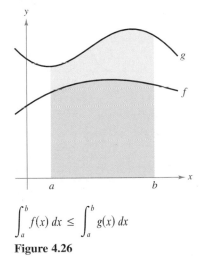

$$\int_a^b f(x)\, dx \le \int_a^b g(x)\, dx$$

Figure 4.26

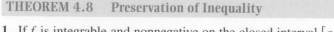

THEOREM 4.8   Preservation of Inequality

1. If f is integrable and nonnegative on the closed interval $[a, b]$, then

$$0 \le \int_a^b f(x)\,dx.$$

2. If f and g are integrable on the closed interval $[a, b]$ and $f(x) \le g(x)$ for every x in $[a, b]$, then

$$\int_a^b f(x)\,dx \int_a^b g(x)\,dx.$$

EXERCISES FOR SECTION 4.3

In Exercises 1 and 2, use Example 1 as a model to evaluate the limit

$$\lim_{n \to \infty} \sum_{i=1}^{n} f(c_i)\,\Delta x_i$$

over the region bounded by the graphs of the equations.

1. $f(x) = \sqrt{x}, \quad y = 0, \quad x = 0, \quad x = 3$

   (Hint: Let $c_i = 3i^2/n^2$.)

2. $f(x) = \sqrt[3]{x}, \quad y = 0, \quad x = 0, \quad x = 1$

   (Hint: Let $c_i = i^3/n^3$.)

In Exercises 3–8, evaluate the definite integral by the limit definition.

3. $\displaystyle\int_4^{10} 6\,dx$

4. $\displaystyle\int_{-2}^{3} x\,dx$

5. $\displaystyle\int_{-1}^{1} x^3\,dx$

6. $\displaystyle\int_1^3 3x^2\,dx$

7. $\displaystyle\int_1^2 (x^2 + 1)\,dx$

8. $\displaystyle\int_{-1}^{2} (3x^2 + 2)\,dx$

In Exercises 9–12, express the limit as a definite integral on the interval $[a, b]$, where c_i is any point in the ith subinterval.

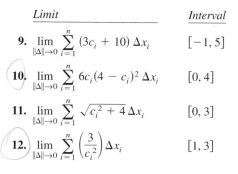

Limit	*Interval*
9. $\displaystyle\lim_{\|\Delta\|\to 0} \sum_{i=1}^{n} (3c_i + 10)\,\Delta x_i$	$[-1, 5]$
10. $\displaystyle\lim_{\|\Delta\|\to 0} \sum_{i=1}^{n} 6c_i(4 - c_i)^2\,\Delta x_i$	$[0, 4]$
11. $\displaystyle\lim_{\|\Delta\|\to 0} \sum_{i=1}^{n} \sqrt{c_i^2 + 4}\,\Delta x_i$	$[0, 3]$
12. $\displaystyle\lim_{\|\Delta\|\to 0} \sum_{i=1}^{n} \left(\frac{3}{c_i^2}\right)\Delta x_i$	$[1, 3]$

In Exercises 13–22, set up a definite integral that yields the area of the region. (Do not evaluate the integral.)

13. $f(x) = 3$

14. $f(x) = 4 - 2x$

15. $f(x) = 4 - |x|$

16. $f(x) = x^2$

17. $f(x) = 4 - x^2$

18. $f(x) = \dfrac{1}{x^2 + 1}$

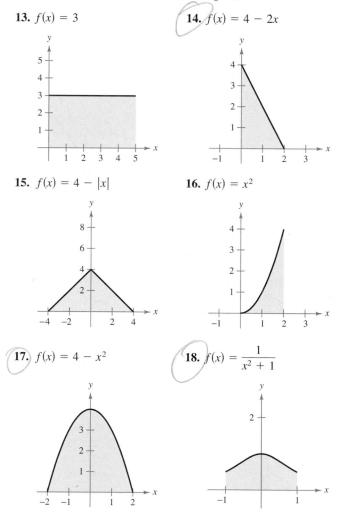

19. $f(x) = \sin x$

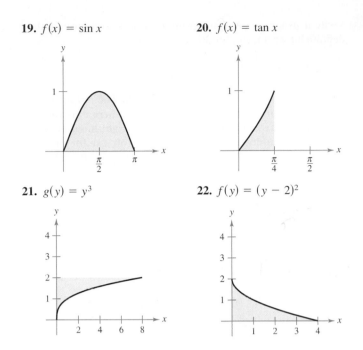

20. $f(x) = \tan x$

21. $g(y) = y^3$

22. $f(y) = (y - 2)^2$

In Exercises 23–32, sketch the region whose area is given by the definite integral. Then use a geometric formula to evaluate the integral ($a > 0, r > 0$).

23. $\displaystyle\int_0^3 4\,dx$

24. $\displaystyle\int_{-a}^a 4\,dx$

25. $\displaystyle\int_0^4 x\,dx$

26. $\displaystyle\int_0^4 \frac{x}{2}\,dx$

27. $\displaystyle\int_0^2 (2x + 5)\,dx$

28. $\displaystyle\int_0^8 (8 - x)\,dx$

29. $\displaystyle\int_{-1}^1 (1 - |x|)\,dx$

30. $\displaystyle\int_{-a}^a (a - |x|)\,dx$

31. $\displaystyle\int_{-3}^3 \sqrt{9 - x^2}\,dx$

32. $\displaystyle\int_{-r}^r \sqrt{r^2 - x^2}\,dx$

In Exercises 33–40, evaluate the integral using the following values.

$$\int_2^4 x^3\,dx = 60, \qquad \int_2^4 x\,dx = 6, \qquad \int_2^4 dx = 2$$

33. $\displaystyle\int_4^2 x\,dx$

34. $\displaystyle\int_2^2 x^3\,dx$

35. $\displaystyle\int_2^4 4x\,dx$

36. $\displaystyle\int_2^4 15\,dx$

37. $\displaystyle\int_2^4 (x - 8)\,dx$

38. $\displaystyle\int_2^4 (x^3 + 4)\,dx$

39. $\displaystyle\int_2^4 \left(\tfrac{1}{2}x^3 - 3x + 2\right)dx$

40. $\displaystyle\int_2^4 (6 + 2x - x^3)\,dx$

41. Given $\displaystyle\int_0^5 f(x)\,dx = 10$ and $\displaystyle\int_5^7 f(x)\,dx = 3$, find

(a) $\displaystyle\int_0^7 f(x)\,dx.$

(b) $\displaystyle\int_5^0 f(x)\,dx.$

(c) $\displaystyle\int_5^5 f(x)\,dx.$

(d) $\displaystyle\int_0^5 3f(x)\,dx.$

42. Given $\displaystyle\int_0^3 f(x)\,dx = 4$ and $\displaystyle\int_3^6 f(x)\,dx = -1$, find

(a) $\displaystyle\int_0^6 f(x)\,dx.$

(b) $\displaystyle\int_6^3 f(x)\,dx.$

(c) $\displaystyle\int_3^3 f(x)\,dx.$

(d) $\displaystyle\int_3^6 -5f(x)\,dx.$

43. Given $\displaystyle\int_2^6 f(x)\,dx = 10$ and $\displaystyle\int_2^6 g(x)\,dx = -2$, find

(a) $\displaystyle\int_2^6 [f(x) + g(x)]\,dx.$

(b) $\displaystyle\int_2^6 [g(x) - f(x)]\,dx.$

(c) $\displaystyle\int_2^6 2g(x)\,dx.$

(d) $\displaystyle\int_2^6 3f(x)\,dx.$

44. Given $\displaystyle\int_{-1}^1 f(x)\,dx = 0$ and $\displaystyle\int_0^1 f(x)\,dx = 5$, find

(a) $\displaystyle\int_{-1}^0 f(x)\,dx.$

(b) $\displaystyle\int_0^1 f(x)\,dx - \int_{-1}^0 f(x)\,dx.$

(c) $\displaystyle\int_{-1}^1 3f(x)\,dx.$

(d) $\displaystyle\int_0^1 3f(x)\,dx.$

45. *Think About It*   The graph of f consists of line segments and a semicircle, as shown in the figure. Evaluate each definite integral by using geometric formulas.

(a) $\displaystyle\int_0^2 f(x)\,dx$

(b) $\displaystyle\int_2^6 f(x)\,dx$

(c) $\displaystyle\int_{-4}^2 f(x)\,dx$

(d) $\displaystyle\int_{-4}^6 f(x)\,dx$

(e) $\displaystyle\int_{-4}^6 |f(x)|\,dx$

(f) $\displaystyle\int_{-4}^6 [f(x) + 2]\,dx$

$(4, 2)$

$(-4, -1)$

46. *Think About It*   Consider the function f that is continuous on the interval $[-5, 5]$ and for which

$$\int_0^5 f(x)\,dx = 4.$$

Evaluate each integral.

(a) $\displaystyle\int_0^5 [f(x) + 2]\,dx$

(b) $\displaystyle\int_{-2}^3 f(x + 2)\,dx$

(c) $\displaystyle\int_{-5}^5 f(x)\,dx$ (f is even.)

(d) $\displaystyle\int_{-5}^5 f(x)\,dx$ (f is odd.)

Getting at the Concept

In Exercises 47–50, use the figure to fill in the blank with the symbol $<$, $>$, or $=$.

47. The interval $[1, 5]$ is partitioned into n subintervals of equal width Δx, and x_i is the left endpoint of the ith subinterval.

$$\sum_{i=1}^{n} f(x_i) \, \Delta x \quad \boxed{} \quad \int_{1}^{5} f(x) \, dx$$

48. The interval $[1, 5]$ is partitioned into n subintervals of equal width Δx, and x_i is the right endpoint of the ith subinterval.

$$\sum_{i=1}^{n} f(x_i) \, \Delta x \quad \boxed{} \quad \int_{1}^{5} f(x) \, dx$$

49. The interval $[1, 5]$ is partitioned into n subintervals of equal width Δx, and x_i is the midpoint of the ith subinterval.

$$\sum_{i=1}^{n} f(x_i) \, \Delta x \quad \boxed{} \quad \int_{1}^{5} f(x) \, dx$$

50. Let T be the average of the results of Exercises 47 and 48.

$$T \quad \boxed{} \quad \int_{1}^{5} f(x) \, dx$$

51. Determine whether the function $f(x) = \dfrac{1}{x - 4}$ is integrable on the interval $[3, 5]$. Explain.

52. Give an example of a function that is integrable on the interval $[-1, 1]$, but not continuous on $[-1, 1]$.

In Exercises 53–56, determine which value best approximates the definite integral. Make your selection on the basis of a sketch.

53. $\displaystyle\int_{0}^{4} \sqrt{x} \, dx$

(a) 5   (b) -3 (c) 10   (d) 2   (e) 8

54. $\displaystyle\int_{0}^{1/2} 4 \cos \pi x \, dx$

(a) 4   (b) $\frac{4}{3}$ (c) 16   (d) 2π (e) -6

55. $\displaystyle\int_{0}^{1} 2 \sin \pi x \, dx$

(a) 6   (b) $\frac{1}{2}$ (c) 4   (d) $\frac{5}{4}$

56. $\displaystyle\int_{0}^{9} \left(1 + \sqrt{x}\right) dx$

(a) -3 (b) 9   (c) 27   (d) 3

Write a program for your graphing utility to approximate a definite integral using the Riemann sum

$$\sum_{i=1}^{n} f(c_i) \, \Delta x_i$$

where the subintervals are of equal width. The output should give three approximations of the integral where c_i is the left-hand endpoint $L(n)$, midpoint $M(n)$, and right-hand endpoint $R(n)$ of each subinterval. In Exercises 57–60, use the program to approximate the definite integral and complete the table.

n	4	8	12	16	20
$L(n)$					
$M(n)$					
$R(n)$					

57. $\displaystyle\int_{0}^{3} x\sqrt{3 - x} \, dx$

58. $\displaystyle\int_{0}^{3} \dfrac{5}{x^2 + 1} \, dx$

59. $\displaystyle\int_{0}^{\pi/2} \sin^2 x \, dx$

60. $\displaystyle\int_{0}^{3} x \sin x \, dx$

True or False? In Exercises 61–66, determine whether the statement is true or false. If it is false, explain why or give an example that shows it is false.

61. $\displaystyle\int_{a}^{b} [f(x) + g(x)] \, dx = \int_{a}^{b} f(x) \, dx + \int_{a}^{b} g(x) \, dx$

62. $\displaystyle\int_{a}^{b} f(x)g(x) \, dx = \left[\int_{a}^{b} f(x) \, dx\right]\left[\int_{a}^{b} g(x) \, dx\right]$

63. If the norm of a partition approaches zero, then the number of subintervals approaches infinity.

64. If f is increasing on $[a, b]$, then the minimum value of $f(x)$ on $[a, b]$ is $f(a)$.

65. The value of $\displaystyle\int_{a}^{b} f(x) \, dx$ must be positive.

66. If $\displaystyle\int_{a}^{b} f(x) \, dx > 0$, then f is nonnegative for all x in $[a, b]$.

67. Find the Riemann sum for $f(x) = x^2 + 3x$ over the interval $[0, 8]$, where $x_0 = 0$, $x_1 = 1$, $x_2 = 3$, $x_3 = 7$, and $x_4 = 8$, and where $c_1 = 1$, $c_2 = 2$, $c_3 = 5$, and $c_4 = 8$.

68. Find the Riemann sum for $f(x) = \sin x$ over the interval $[0, 2\pi]$, where $x_0 = 0$, $x_1 = \pi/4$, $x_2 = \pi/3$, $x_3 = \pi$, and $x_4 = 2\pi$, and where $c_1 = \pi/6$, $c_2 = \pi/3$, $c_3 = 2\pi/3$, and $c_4 = 3\pi/2$.

69. *Think About It* Determine whether the Dirichlet function

$$f(x) = \begin{cases} 1, & x \text{ is rational} \\ 0, & x \text{ is irrational} \end{cases}$$

is integrable on the interval $[0, 1]$. Explain.

70. Evaluate, if possible, the integral $\displaystyle\int_{0}^{2} [\![x]\!] \, dx$.

71. Determine $\displaystyle\lim_{n \to \infty} \frac{1}{n^3}[1^2 + 2^2 + 3^2 + \cdots + n^2]$ by using an appropriate Riemann sum.

Section 4.4	**The Fundamental Theorem of Calculus**

- Evaluate a definite integral using the Fundamental Theorem of Calculus.
- Understand and use the Mean Value Theorem for Integrals.
- Find the average value of a function over a closed interval.
- Understand and use the Second Fundamental Theorem of Calculus.

EXPLORATION

Integration and Antidifferentiation
Throughout this chapter, we have been using the integral sign to denote an antiderivative (a family of functions) and a definite integral (a number).

Antidifferentiation: $\displaystyle\int f(x)\, dx$

Definite integration: $\displaystyle\int_a^b f(x)\, dx$

The use of this same symbol for both operations makes it appear that they are related. In the early work with calculus, however, it was not known that the two operations were related. Do you think the symbol ∫ was first applied to antidifferentiation or to definite integration? Explain your reasoning. (*Hint:* The symbol was first used by Leibniz and was derived from the letter *S*.)

The Fundamental Theorem of Calculus

You have now been introduced to the two major branches of calculus: differential calculus (introduced with the tangent line problem) and integral calculus (introduced with the area problem). At this point, these two problems might seem unrelated—but there is a very close connection. The connection was discovered independently by Isaac Newton and Gottfried Leibniz and is stated in a theorem that is appropriately called the **Fundamental Theorem of Calculus.**

Informally, the theorem states that differentiation and (definite) integration are inverse operations, in the same sense that division and multiplication are inverse operations. To see how Newton and Leibniz might have anticipated this relationship, consider the approximations shown in Figure 4.27. When we defined the slope of the tangent line, we used the *quotient* $\Delta y/\Delta x$ (the slope of the secant line). Similarly, when we defined the area of a region under a curve, we used the *product* $\Delta y\Delta x$ (the area of a rectangle). So, at least in the primitive approximation stage, the operations of differentiation and definite integration appear to have an inverse relationship in the same sense that division and multiplication are inverse operations. The Fundamental Theorem of Calculus states that the limit processes (used to define the derivative and definite integral) preserve this inverse relationship.

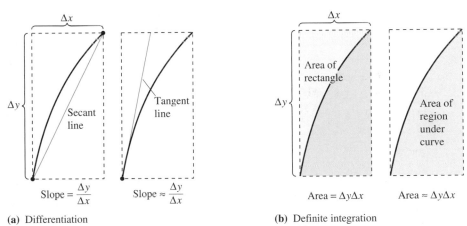

(a) Differentiation   (b) Definite integration

Differentiation and definite integration have an "inverse" relationship.

Figure 4.27

THEOREM 4.9   The Fundamental Theorem of Calculus

If a function f is continuous on the closed interval $[a, b]$ and F is an antiderivative of f on the interval $[a, b]$, then

$$\int_a^b f(x)\, dx = F(b) - F(a).$$

Proof The key to the proof is in writing the difference $F(b) - F(a)$ in a convenient form. Let Δ be the following partition of $[a, b]$.

$$a = x_0 < x_1 < x_2 < \cdots < x_{n-1} < x_n = b$$

By pairwise subtraction and addition of like terms, you can write

$$F(b) - F(a) = F(x_n) - F(x_{n-1}) + F(x_{n-1}) - \cdots - F(x_1) + F(x_1) - F(x_0)$$

$$= \sum_{i=1}^{n} [F(x_i) - F(x_{i-1})].$$

By the Mean Value Theorem, you know that there exists a number c_i in the ith subinterval such that

$$F'(c_i) = \frac{F(x_i) - F(x_{i-1})}{x_i - x_{i-1}}.$$

Because $F'(c_i) = f(c_i)$, you can let $\Delta x_i = x_i - x_{i-1}$ and obtain

$$F(b) - F(a) = \sum_{i=1}^{n} f(c_i) \Delta x_i.$$

This important equation tells you that by applying the Mean Value Theorem you can always find a collection of c_i's such that the *constant* $F(b) - F(a)$ is a Riemann sum of f on $[a, b]$. Taking the limit (as $\|\Delta\| \to 0$) produces

$$F(b) - F(a) = \int_a^b f(x)\, dx.$$

The following guidelines can help you understand the use of the Fundamental Theorem of Calculus.

Guidelines for Using the Fundamental Theorem of Calculus

1. *Provided you can find* an antiderivative of f, you now have a way to evaluate a definite integral without having to use the limit of a sum.
2. When applying the Fundamental Theorem of Calculus, the following notation is convenient.

$$\int_a^b f(x)\, dx = F(x) \Big]_a^b$$

$$= F(b) - F(a)$$

For instance, to evaluate $\int_1^3 x^3\, dx$, you can write

$$\int_1^3 x^3\, dx = \frac{x^4}{4} \Big]_1^3 = \frac{3^4}{4} - \frac{1^4}{4} = \frac{81}{4} - \frac{1}{4} = 20.$$

3. It is not necessary to include a constant of integration C in the antiderivative because

$$\int_a^b f(x)\, dx = \Big[F(x) + C \Big]_a^b$$

$$= [F(b) + C] - [F(a) + C]$$

$$= F(b) - F(a).$$

Example 1   Evaluating a Definite Integral

Evaluate each definite integral.

a. $\displaystyle\int_1^2 (x^2 - 3)\, dx$ **b.** $\displaystyle\int_1^4 3\sqrt{x}\, dx$ **c.** $\displaystyle\int_0^{\pi/4} \sec^2 x\, dx$

Solution

a. $\displaystyle\int_1^2 (x^2 - 3)\, dx = \left[\frac{x^3}{3} - 3x\right]_1^2 = \left(\frac{8}{3} - 6\right) - \left(\frac{1}{3} - 3\right) = -\frac{2}{3}$

b. $\displaystyle\int_1^4 3\sqrt{x}\, dx = 3\int_1^4 x^{1/2}\, dx = 3\left[\frac{x^{3/2}}{3/2}\right]_1^4 = 2(4)^{3/2} - 2(1)^{3/2} = 14$

c. $\displaystyle\int_0^{\pi/4} \sec^2 x\, dx = \tan x\Big]_0^{\pi/4} = 1 - 0 = 1$

Example 2   A Definite Integral Involving Absolute Value

Evaluate $\displaystyle\int_0^2 |2x - 1|\, dx.$

Solution   Using Figure 4.28 and the definition of absolute value, you can rewrite the integrand as follows.

$$|2x - 1| = \begin{cases} -(2x - 1), & x < \frac{1}{2} \\ 2x - 1, & x \geq \frac{1}{2} \end{cases}$$

From this, you can rewrite the integral in two parts.

$$\int_0^2 |2x - 1|\, dx = \int_0^{1/2} -(2x - 1)\, dx + \int_{1/2}^2 (2x - 1)\, dx$$

$$= \left[-x^2 + x\right]_0^{1/2} + \left[x^2 - x\right]_{1/2}^2$$

$$= \left(-\frac{1}{4} + \frac{1}{2}\right) - (0 + 0) + (4 - 2) - \left(\frac{1}{4} - \frac{1}{2}\right)$$

$$= \frac{5}{2}$$

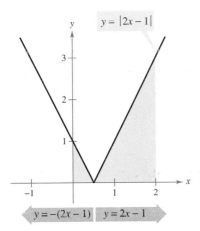

$y = |2x - 1|$

$y = -(2x - 1)$ $y = 2x - 1$

The definite integral of y on $[0, 2]$ is $\frac{5}{2}$.
Figure 4.28

Example 3   Using the Fundamental Theorem to Find Area

Find the area of the region bounded by the graph of $y = 2x^2 - 3x + 2$, the x-axis, and the vertical lines $x = 0$ and $x = 2$, as shown in Figure 4.29.

Solution   Note that $y > 0$ on the interval $[0, 2]$.

$$\text{Area} = \int_0^2 (2x^2 - 3x + 2)\, dx \qquad\text{Integrate between } x = 0 \text{ and } x = 2.$$

$$= \left[\frac{2x^3}{3} - \frac{3x^2}{2} + 2x\right]_0^2 \qquad\text{Find antiderivative.}$$

$$= \left(\frac{16}{3} - 6 + 4\right) - (0 - 0 + 0) \qquad\text{Apply Fundamental Theorem.}$$

$$= \frac{10}{3} \qquad\text{Simplify.}$$

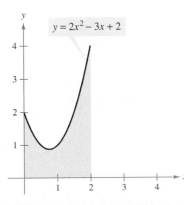

$y = 2x^2 - 3x + 2$

The area of the region bounded by the graph of y, the x-axis, $x = 0$, and $x = 2$ is $\frac{10}{3}$.
Figure 4.29

The Mean Value Theorem for Integrals

In Section 4.2, you saw that the area of a region under a curve is greater than the area of an inscribed rectangle and less than the area of a circumscribed rectangle. The Mean Value Theorem for Integrals states that somewhere "between" the inscribed and circumscribed rectangles there is a rectangle whose area is precisely equal to the area of the region under the curve, as shown in Figure 4.30.

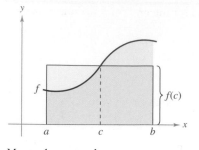

Mean value rectangle:

$$f(c)(b - a) = \int_a^b f(x)\, dx$$

Figure 4.30

THEOREM 4.10   Mean Value Theorem for Integrals

If f is continuous on the closed interval $[a, b]$, then there exists a number c in the closed interval $[a, b]$ such that

$$\int_a^b f(x)\, dx = f(c)(b - a).$$

Proof

Case 1: If f is constant on the interval $[a, b]$, the theorem is clearly valid because c can be any point in $[a, b]$.

Case 2: If f is not constant on $[a, b]$, then, by the Extreme Value Theorem, you can choose $f(m)$ and $f(M)$ to be the minimum and maximum values of f on $[a, b]$. Because $f(m) \le f(x) \le f(M)$ for all x in $[a, b]$, you can apply Theorem 4.8 to write the following.

$$\int_a^b f(m)\, dx \le \int_a^b f(x)\, dx \le \int_a^b f(M)\, dx \qquad \text{See Figure 4.31.}$$

$$f(m)(b - a) \le \int_a^b f(x)\, dx \le f(M)(b - a)$$

$$f(m) \le \frac{1}{b - a}\int_a^b f(x)\, dx \le f(M)$$

From the third inequality, you can apply the Intermediate Value Theorem to conclude that there exists some c in $[a, b]$ such that

$$f(c) = \frac{1}{b - a}\int_a^b f(x)\, dx \qquad \text{or} \qquad f(c)(b - a) = \int_a^b f(x)\, dx.$$

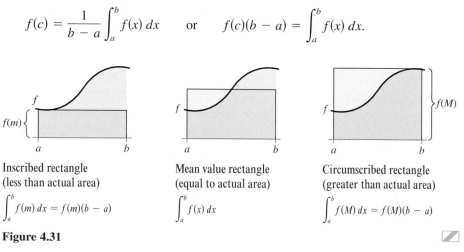

Inscribed rectangle
(less than actual area)

$$\int_a^b f(m)\, dx = f(m)(b - a)$$

Mean value rectangle
(equal to actual area)

$$\int_a^b f(x)\, dx$$

Circumscribed rectangle
(greater than actual area)

$$\int_a^b f(M)\, dx = f(M)(b - a)$$

Figure 4.31

NOTE   Notice that Theorem 4.10 does not specify how to determine c. It merely guarantees the existence of at least one number c in the interval.

Average Value of a Function

The value of $f(c)$ given in the Mean Value Theorem for Integrals is called the **average value** of f on the interval $[a, b]$.

Average value

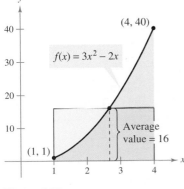

$$\text{Average value} = \frac{1}{b - a} \int_a^b f(x) \, dx$$

Figure 4.32

Definition of the Average Value of a Function on an Interval

If f is integrable on the closed interval $[a, b]$, then the **average value** of f on the interval is

$$\frac{1}{b - a} \int_a^b f(x) \, dx.$$

NOTE   Notice in Figure 4.32 that the area of the region under the graph of f is equal to the area of the rectangle whose height is the average value.

To see why the average value of f is defined in this way, suppose that you partition $[a, b]$ into n subintervals of equal width $\Delta x = (b - a)/n$. If c_i is any point in the ith subinterval, the arithmetic average (or mean) of the function values at the c_i's is given by

$$a_n = \frac{1}{n}\left[f(c_1) + f(c_2) + \cdots + f(c_n) \right]. \qquad \text{Average of } f(c_1), \ldots, f(c_n)$$

By multiplying and dividing by $(b - a)$, you can write the average as

$$a_n = \frac{1}{n}\sum_{i=1}^n f(c_i)\left(\frac{b - a}{b - a} \right) = \frac{1}{b - a}\sum_{i=1}^n f(c_i)\left(\frac{b - a}{n} \right)$$

$$= \frac{1}{b - a}\sum_{i=1}^n f(c_i)\, \Delta x.$$

Finally, taking the limit as $n \to \infty$ produces the average value of f on the interval $[a, b]$, as given in the definition above.

This development of the average value of a function on an interval is only one of many practical uses of definite integrals to represent summation processes. In Chapter 6, you will study other applications, such as volume, arc length, centers of mass, and work.

Example 4   **Finding the Average Value of a Function**

Find the average value of $f(x) = 3x^2 - 2x$ on the interval $[1, 4]$.

Solution   The average value is given by

$$\frac{1}{b - a}\int_a^b f(x)\, dx = \frac{1}{3}\int_1^4 (3x^2 - 2x)\, dx$$

$$= \frac{1}{3}\left[x^3 - x^2 \right]_1^4$$

$$= \frac{1}{3}[64 - 16 - (1 - 1)] = \frac{48}{3} = 16.$$

(See Figure 4.33.)

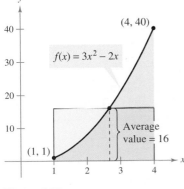

$f(x) = 3x^2 - 2x$

$(4, 40)$

Average value = 16

$(1, 1)$

Figure 4.33

The first person to fly at a speed greater than the speed of sound was Charles Yeager. On October 14, 1947, flying in an *X-1* rocket plane at an altitude of 12.8 kilometers, Yeager was clocked at 299.5 meters per second. If Yeager had been flying at an altitude under 10.375 kilometers, his speed of 299.5 meters per second would not have "broken the sound barrier." The photo above shows the *X-1* and its *B-29* mother plane.

Example 5 The Speed of Sound

At different altitudes in earth's atmosphere, sound travels at different speeds. The speed of sound $s(x)$ (in meters per second) can be modeled by

$$s(x) = \begin{cases} -4x + 341, & 0 \leq x < 11.5 \\ 295, & 11.5 \leq x < 22 \\ \frac{3}{4}x + 278.5, & 22 \leq x < 32 \\ \frac{3}{2}x + 254.5, & 32 \leq x < 50 \\ -\frac{3}{2}x + 404.5, & 50 \leq x \leq 80 \end{cases}$$

where x is the altitude in kilometers (see Figure 4.34). What is the average speed of sound over the interval $[0, 80]$?

Solution Begin by integrating $s(x)$ over the interval $[0, 80]$. To do this, you can break the integral into five parts.

$$\int_0^{11.5} s(x)\, dx = \int_0^{11.5} (-4x + 341)\, dx = \left[-2x^2 + 341x \right]_0^{11.5} = 3657$$

$$\int_{11.5}^{22} s(x)\, dx = \int_{11.5}^{22} (295)\, dx = \left[295x \right]_{11.5}^{22} = 3097.5$$

$$\int_{22}^{32} s(x)\, dx = \int_{22}^{32} \left(\tfrac{3}{4}x + 278.5\right) dx = \left[\tfrac{3}{8}x^2 + 278.5x \right]_{22}^{32} = 2987.5$$

$$\int_{32}^{50} s(x)\, dx = \int_{32}^{50} \left(\tfrac{3}{2}x + 254.5\right) dx = \left[\tfrac{3}{4}x^2 + 254.5x \right]_{32}^{50} = 5688$$

$$\int_{50}^{80} s(x)\, dx = \int_{50}^{80} \left(-\tfrac{3}{2}x + 404.5\right) dx = \left[-\tfrac{3}{4}x^2 + 404.5x \right]_{50}^{80} = 9210$$

By adding the values of the five integrals, you have

$$\int_0^{80} s(x)\, dx = 24{,}640.$$

Therefore, the average speed of sound from an altitude of 0 kilometers to an altitude of 80 kilometers is

$$\text{Average speed} = \frac{1}{80} \int_0^{80} s(x)\, dx = \frac{24{,}640}{80} = 308 \text{ meters per second.}$$

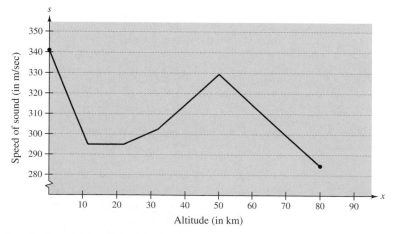

Speed of sound depends on altitude.
Figure 4.34

The Second Fundamental Theorem of Calculus

When we defined the definite integral of f on the interval $[a, b]$, we used the constant b as the upper limit of integration and x as the variable of integration. We now look at a slightly different situation in which the variable x is used as the upper limit of integration. To avoid the confusion of using x in two different ways, we temporarily switch to using t as the variable of integration. (Remember that the definite integral is *not* a function of its variable of integration.)

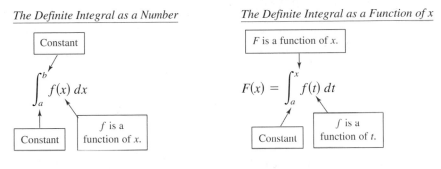

The Definite Integral as a Number

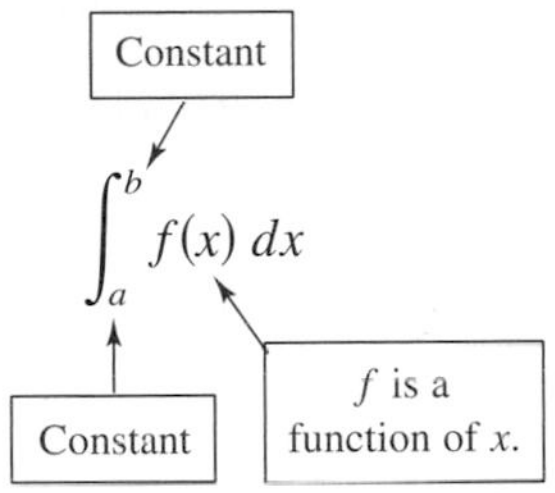
Constant
$\int_a^b f(x)\,dx$
Constant
f is a function of x.

The Definite Integral as a Function of x

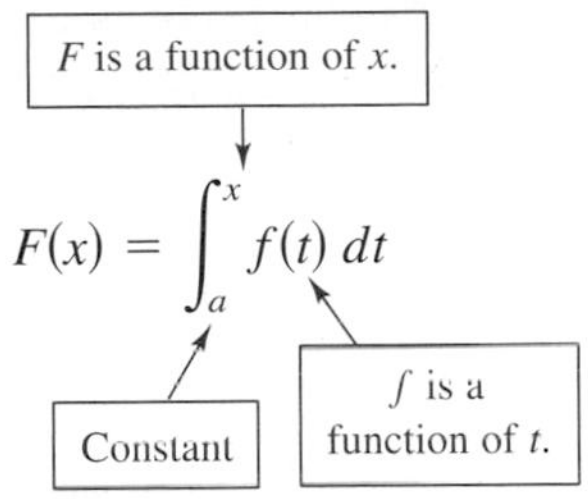
F is a function of x.
$F(x) = \int_a^x f(t)\,dt$
Constant
f is a function of t.

EXPLORATION

Use a graphing utility to graph the function

$$F(x) = \int_0^x \cos t\,dt$$

for $0 \le x \le \pi$. Do you recognize this graph? Explain.

Example 6 The Definite Integral as a Function

Evaluate the function

$$F(x) = \int_0^x \cos t\,dt$$

at $x = 0, \pi/6, \pi/4, \pi/3$, and $\pi/2$.

Solution You could evaluate five different definite integrals, one for each of the given upper limits. However, it is much simpler to fix x (as a constant) temporarily and apply the Fundamental Theorem once, to obtain

$$\int_0^x \cos t\,dt = \sin t\Big]_0^x = \sin x - \sin 0 = \sin x.$$

Now, using $F(x) = \sin x$, you can obtain the results shown in Figure 4.35.

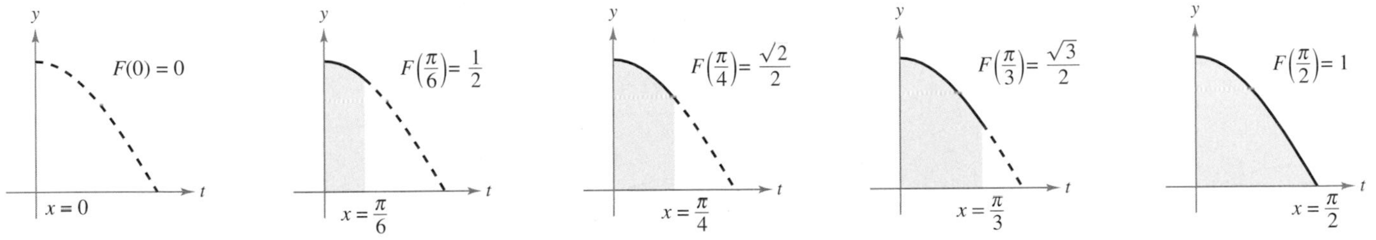
$F(0) = 0$
$x = 0$
$F\left(\frac{\pi}{6}\right) = \frac{1}{2}$
$x = \frac{\pi}{6}$
$F\left(\frac{\pi}{4}\right) = \frac{\sqrt{2}}{2}$
$x = \frac{\pi}{4}$
$F\left(\frac{\pi}{3}\right) = \frac{\sqrt{3}}{2}$
$x = \frac{\pi}{3}$
$F\left(\frac{\pi}{2}\right) = 1$
$x = \frac{\pi}{2}$

$F(x) = \int_0^x \cos t\,dt$ is the area under the curve $f(t) = \cos t$ from 0 to x.

Figure 4.35

You can think of the function $F(x)$ as *accumulating* the area under the curve $f(t) = \cos t$ from $t = 0$ to $t = x$. For $x = 0$, the area is 0 and $F(0) = 0$. For $x = \pi/2$, $F(\pi/2) = 1$ gives the accumulated area under the cosine curve on the entire interval $[0, \pi/2]$. This interpretation of an integral as an **accumulation function** is used often in applications of integration.

In Example 6, note that the derivative of F is the original integrand (with only the variable changed). That is,

$$\frac{d}{dx}[F(x)] = \frac{d}{dx}[\sin x] = \frac{d}{dx}\left[\int_0^x \cos t \, dt\right] = \cos x.$$

We generalize this result in the following theorem, called the **Second Fundamental Theorem of Calculus.**

THEOREM 4.11   The Second Fundamental Theorem of Calculus

If f is continuous on an open interval I containing a, then, for every x in the interval,

$$\frac{d}{dx}\left[\int_a^x f(t) \, dt\right] = f(x).$$

Proof   Begin by defining F as

$$F(x) = \int_a^x f(t) \, dt.$$

Then, by the definition of the derivative, you can write

$$F'(x) = \lim_{\Delta x \to 0} \frac{F(x + \Delta x) - F(x)}{\Delta x}$$

$$= \lim_{\Delta x \to 0} \frac{1}{\Delta x}\left[\int_a^{x+\Delta x} f(t) \, dt - \int_a^x f(t) \, dt\right]$$

$$= \lim_{\Delta x \to 0} \frac{1}{\Delta x}\left[\int_a^{x+\Delta x} f(t) \, dt + \int_x^a f(t) \, dt\right]$$

$$= \lim_{\Delta x \to 0} \frac{1}{\Delta x}\left[\int_x^{x+\Delta x} f(t) \, dt\right].$$

From the Mean Value Theorem for Integrals (assuming $\Delta x > 0$), you know there exists a number c in the interval $[x, x + \Delta x]$ such that the integral in the expression above is equal to $f(c) \, \Delta x$. Moreover, because $x \le c \le x + \Delta x$, it follows that $c \to x$ as $\Delta x \to 0$. So, you obtain

$$F'(x) = \lim_{\Delta x \to 0}\left[\frac{1}{\Delta x} f(c) \, \Delta x\right]$$

$$= \lim_{\Delta x \to 0} f(c)$$

$$= f(x).$$

A similar argument can be made for $\Delta x < 0$.

NOTE   Using the area model for definite integrals, you can view the approximation

$$f(x) \, \Delta x \approx \int_x^{x+\Delta x} f(t) \, dt$$

as saying that the area of the rectangle of height $f(x)$ and width Δx is approximately equal to the area of the region lying between the graph of f and the x-axis on the interval $[x, x + \Delta x]$, as shown in Figure 4.36.

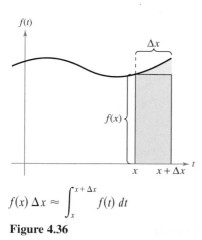

$$f(x) \, \Delta x \approx \int_x^{x+\Delta x} f(t) \, dt$$

Figure 4.36

Note that the Second Fundamental Theorem of Calculus tells you that if a function is continuous, you can be sure that it has an antiderivative. This antiderivative need not, however, be an elementary function. (Recall the discussion of elementary functions in Section P.3.)

Example 7   Using the Second Fundamental Theorem of Calculus

Evaluate $\dfrac{d}{dx}\left[\displaystyle\int_{0}^{x}\sqrt{t^2+1}\,dt\right]$.

Solution   Note that $f(t)=\sqrt{t^2+1}$ is continuous on the entire real line. So, using the Second Fundamental Theorem of Calculus, you can write

$$\frac{d}{dx}\left[\int_{0}^{x}\sqrt{t^2+1}\,dt\right]=\sqrt{x^2+1}.$$

The differentiation shown in Example 7 is a straightforward application of the Second Fundamental Theorem of Calculus. The next example shows how this theorem can be combined with the Chain Rule to find the derivative of a function.

Example 8   Using the Second Fundamental Theorem of Calculus

Find the derivative of $F(x)=\displaystyle\int_{\pi/2}^{x^3}\cos t\,dt$.

Solution   Using $u=x^3$, you can apply the Second Fundamental Theorem of Calculus with the Chain Rule as follows.

$$
\begin{aligned}
F'(x) &= \frac{dF}{du}\frac{du}{dx} && \text{Chain Rule}\\[2mm]
&= \frac{d}{du}[F(x)]\frac{du}{dx} && \text{Definition of }\frac{dF}{du}\\[2mm]
&= \frac{d}{du}\left[\int_{\pi/2}^{x^3}\cos t\,dt\right]\frac{du}{dx} && \text{Substitute }\int_{\pi/2}^{x^3}\cos t\,dt\text{ for }F(x).\\[2mm]
&= \frac{d}{du}\left[\int_{\pi/2}^{u}\cos t\,dt\right]\frac{du}{dx} && \text{Substitute }u\text{ for }x^3.\\[2mm]
&= (\cos u)(3x^2) && \text{Apply Second Fundamental Theorem of Calculus.}\\[2mm]
&= (\cos x^3)(3x^2) && \text{Rewrite as function of }x.
\end{aligned}
$$

Because the integrand in Example 8 is easily integrated, you can verify the derivative as follows.

$$F(x)=\int_{\pi/2}^{x^3}\cos t\,dt=\sin t\Big]_{\pi/2}^{x^3}=\sin x^3-\sin\frac{\pi}{2}=(\sin x^3)-1$$

In this form, you can apply the Power Rule to verify that the derivative is the same as that obtained in Example 8.

$$F'(x)=(\cos x^3)(3x^2)$$

EXERCISES FOR SECTION 4.4

Graphical Reasoning   In Exercises 1–4, use a graphing utility to graph the integrand. Use the graph to determine whether the definite integral is positive, negative, or zero.

1. $\displaystyle\int_0^\pi \frac{4}{x^2 + 1}\, dx$

2. $\displaystyle\int_0^\pi \cos x\, dx$

3. $\displaystyle\int_{-2}^2 x\sqrt{x^2 + 1}\, dx$

4. $\displaystyle\int_{-2}^2 x\sqrt{2 - x}\, dx$

In Exercises 5–26, evaluate the definite integral of the algebraic function. Use a graphing utility to verify your result.

5. $\displaystyle\int_0^1 2x\, dx$

6. $\displaystyle\int_2^7 3\, dv$

7. $\displaystyle\int_{-1}^0 (x - 2)\, dx$

8. $\displaystyle\int_2^5 (-3v + 4)\, dv$

9. $\displaystyle\int_{-1}^1 (t^2 - 2)\, dt$

10. $\displaystyle\int_1^3 (3x^2 + 5x - 4)\, dx$

11. $\displaystyle\int_0^1 (2t - 1)^2\, dt$

12. $\displaystyle\int_{-1}^1 (t^3 - 9t)\, dt$

13. $\displaystyle\int_1^2 \left(\frac{3}{x^2} - 1\right) dx$

14. $\displaystyle\int_{-2}^{-1} \left(u - \frac{1}{u^2}\right) du$

15. $\displaystyle\int_1^4 \frac{u - 2}{\sqrt{u}}\, du$

16. $\displaystyle\int_{-3}^3 v^{1/3}\, dv$

17. $\displaystyle\int_{-1}^1 (\sqrt[3]{t} - 2)\, dt$

18. $\displaystyle\int_1^8 \sqrt{\frac{2}{x}}\, dx$

19. $\displaystyle\int_0^1 \frac{x - \sqrt{x}}{3}\, dx$

20. $\displaystyle\int_0^2 (2 - t)\sqrt{t}\, dt$

21. $\displaystyle\int_{-1}^0 \left(t^{1/3} - t^{2/3}\right) dt$

22. $\displaystyle\int_{-8}^{-1} \frac{x - x^2}{2\sqrt[3]{x}}\, dx$

23. $\displaystyle\int_0^3 |2x - 3|\, dx$

24. $\displaystyle\int_1^4 (3 - |x - 3|)\, dx$

25. $\displaystyle\int_0^3 |x^2 - 4|\, dx$

26. $\displaystyle\int_0^4 |x^2 - 4x + 3|\, dx$

In Exercises 27–32, evaluate the definite integral of the trigonometric function. Use a graphing utility to verify your result.

27. $\displaystyle\int_0^\pi (1 + \sin x)\, dx$

28. $\displaystyle\int_0^{\pi/4} \frac{1 - \sin^2\theta}{\cos^2\theta}\, d\theta$

29. $\displaystyle\int_{-\pi/6}^{\pi/6} \sec^2 x\, dx$

30. $\displaystyle\int_{\pi/4}^{\pi/2} (2 - \csc^2 x)\, dx$

31. $\displaystyle\int_{-\pi/3}^{\pi/3} 4\sec\theta\tan\theta\, d\theta$

32. $\displaystyle\int_{-\pi/2}^{\pi/2} (2t + \cos t)\, dt$

33. *Depreciation*   A company purchases a new machine for which the rate of depreciation is $dV/dt = 10{,}000(t - 6)$, $0 \le t \le 5$, where V is the value of the machine after t years. Set up and evaluate the definite integral that yields the total loss of value of the machine over the first 3 years.

34. *Buffon's Needle Experiment*   A horizontal plane is ruled with parallel lines 2 inches apart. If a 2-inch needle is tossed randomly onto the plane, the probability that the needle will touch a line is

$$P = \frac{2}{\pi}\int_0^{\pi/2} \sin\theta\, d\theta$$

where θ is the acute angle between the needle and any one of the parallel lines. Find this probability.

In Exercises 35–40, determine the area of the indicated region.

35. $y = x - x^2$

36. $y = 1 - x^4$

37. $y = (3 - x)\sqrt{x}$

38. $y = \dfrac{1}{x^2}$

39. $y = \cos x$

40. $y = x + \sin x$

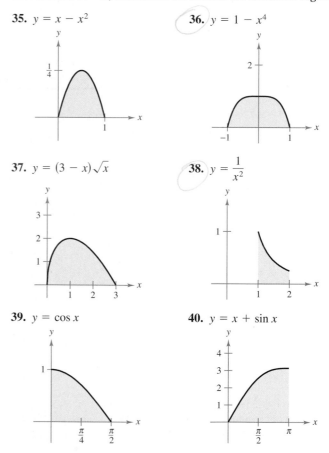

In Exercises 41–44, find the area of the region bounded by the graphs of the equations.

41. $y = 3x^2 + 1$, $x = 0$, $x = 2$, $y = 0$

42. $y = 1 + \sqrt[3]{x}$, $x = 0$, $x = 8$, $y = 0$

43. $y = x^3 + x$, $x = 2$, $y = 0$

44. $y = -x^2 + 3x$, $y = 0$

In Exercises 45–48, find the value(s) of c guaranteed by the Mean Value Theorem for Integrals for the function over the indicated interval.

Function	Interval
45. $f(x) = x - 2\sqrt{x}$	$[0, 2]$
46. $f(x) = \dfrac{9}{x^3}$	$[1, 3]$
47. $f(x) = 2\sec^2 x$	$[-\pi/4, \pi/4]$
48. $f(x) = \cos x$	$[-\pi/3, \pi/3]$

In Exercises 49–52, find the average value of the function over the interval and all values of x in the interval for which the function equals its average value.

Function	Interval
49. $f(x) = 4 - x^2$	$[-2, 2]$
50. $f(x) = \dfrac{4(x^2 + 1)}{x^2}$	$[1, 3]$
51. $f(x) = \sin x$	$[0, \pi]$
52. $f(x) = \cos x$	$[0, \pi/2]$

Getting at the Concept

53. State the Fundamental Theorem of Calculus.

54. The graph of f is given in the figure.

(a) Evaluate $\displaystyle\int_1^7 f(x)\, dx$.

(b) Determine the average value of f on the interval $[1, 7]$.

(c) Determine the answers to parts (a) and (b) if the graph is translated two units upward.

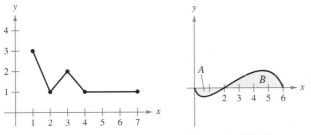

Figure for 54   **Figure for 55–60**

In Exercises 55–60, use the graph of f shown in the figure. The shaded region A has an area of 1.5, and $\int_0^6 f(x)\, dx = 3.5$. Use this information to fill in the blanks.

55. $\displaystyle\int_0^2 f(x)\, dx =$ ▩

56. $\displaystyle\int_2^6 f(x)\, dx =$ ▩

57. $\displaystyle\int_0^6 |f(x)|\, dx =$ ▩

58. $\displaystyle\int_0^2 -2f(x)\, dx =$ ▩

59. $\displaystyle\int_0^6 [2 + f(x)]\, dx =$ ▩

60. The average value of f over the interval $[0, 6]$ is ▩ .

61. *Force* The force F (in newtons) of a hydraulic cylinder in a press is proportional to the square of $\sec x$, where x is the distance (in meters) that the cylinder is extended in its cycle. The domain of F is $[0, \pi/3]$, and $F(0) = 500$.

(a) Find F as a function of x.

(b) Find the average force exerted by the press over the interval $[0, \pi/3]$.

62. *Blood Flow* The velocity v of the flow of blood at a distance r from the central axis of an artery of radius R is

$$v = k(R^2 - r^2)$$

where k is the constant of proportionality. Find the average rate of flow of blood along a radius of the artery. (Use 0 and R as the limits of integration.)

63. *Respiratory Cycle* The volume V in liters of air in the lungs during a 5-second respiratory cycle is approximated by the model

$$V = 0.1729t + 0.1522t^2 - 0.0374t^3$$

where t is the time in seconds. Approximate the average volume of air in the lungs during one cycle.

64. *Average Profit* A company introduces a new product, and the profit in thousands of dollars over the first 6 months is approximated by the model

$$P = 5(\sqrt{t} + 30), \qquad t = 1, 2, 3, 4, 5, 6.$$

(a) Use the model to complete the table and use the entries to calculate (arithmetically) the average profit over the first 6 months.

t	1	2	3	4	5	6
P						

(b) Find the average value of the profit function by integration and compare the result with that in part (a). (Integrate over the interval $[0.5, 6.5]$.)

(c) What, if any, is the advantage of using the approximation of the average given by the definite integral? (Note that the integral approximation utilizes all real values of t in the interval rather than just integers.)

65. *Average Sales* A company fit a model to the monthly sales data of a seasonal product. The model is

$$S(t) = \frac{t}{4} + 1.8 + 0.5\sin\left(\frac{\pi t}{6}\right), \qquad 0 \le t \le 24$$

where S is sales (in thousands) and t is time in months.

(a) Use a graphing utility to graph $f(t) = 0.5\sin(\pi t/6)$ for $0 \le t \le 24$. Use the graph to explain why the average value of $f(t)$ is 0 over the interval.

(b) Use a graphing utility to graph $S(t)$ and the line $g(t) = t/4 + 1.8$ in the same viewing window. Use the graph and the result of part (a) to explain why g is called the *trend line*.

66. *Modeling Data* In the manufacturing process of a product, there is a repetitive heating cycle of 4 minutes. During a review of the process, the flow R (cubic feet per minute) of natural gas was measured in 1-minute intervals and the results were recorded in the table.

t	0	1	2	3	4
R	0	62	76	38	0

(a) Use a graphing utility to find a model of the form $R = at^4 + bt^3 + ct^2 + dt + e$ for the data.

(b) Use a graphing utility to plot the data and graph the model.

(c) Use the Fundamental Theorem of Calculus to approximate the number of cubic feet of natural gas used in one heating cycle.

67. *Modeling Data* A radio-controlled experimental vehicle is tested on a straight track. It starts from rest, and its velocity v (meters per second) is recorded in the table every 10 seconds for 1 minute.

t	0	10	20	30	40	50	60
v	0	5	21	40	62	78	83

(a) Use a graphing utility to find a model of the form $v = at^3 + bt^2 + ct + d$ for the data.

(b) Use a graphing utility to plot the data and graph the model.

(c) Use the Fundamental Theorem of Calculus to approximate the distance traveled by the vehicle during the test.

68. *Modeling Data* A department store manager wants to estimate the number of customers that enter the store from noon until closing at 9 P.M. The table shows the number of customers N entering the store during a randomly selected minute each hour from $t - 1$ to t, with $t = 0$ corresponding to noon.

t	1	2	3	4	5	6	7	8	9
N	6	7	9	12	15	14	11	7	2

(a) Draw a histogram of the data.

(b) Estimate the total number of customers entering the store between noon and 9 P.M.

(c) Use the regression capabilities of a graphing utility to find a model of the form

$$N(t) = at^3 + bt^2 + ct + d$$

for the data.

(d) Use a graphing utility to plot the data and graph the model.

(e) Use a graphing utility to evaluate $\int_0^9 N(t)\,dt$, and use the result to estimate the number of customers entering the store between noon and 9 P.M. Compare this with your answer in part (b).

(f) Estimate the average number of customers entering the store per minute between 3 P.M. and 7 P.M.

In Exercises 69–74, find F as a function of x and evaluate it at $x = 2$, $x = 5$, and $x = 8$.

69. $F(x) = \int_0^x (t - 5)\,dt$

70. $F(x) = \int_2^x (t^3 + 2t - 2)\,dt$

71. $F(x) = \int_1^x \dfrac{10}{v^2}\,dv$

72. $F(x) = \int_2^x -\dfrac{2}{t^3}\,dt$

73. $F(x) = \int_1^x \cos \theta\,d\theta$

74. $F(x) = \int_0^x \sin \theta\,d\theta$

In Exercises 75–80, (a) integrate to find F as a function of x and (b) demonstrate the Second Fundamental Theorem of Calculus by differentiating the result in part (a).

75. $F(x) = \int_0^x (t + 2)\,dt$

76. $F(x) = \int_0^x t(t^2 + 1)\,dt$

77. $F(x) = \int_8^x \sqrt[3]{t}\,dt$

78. $F(x) = \int_4^x \sqrt{t}\,dt$

79. $F(x) = \int_{\pi/4}^x \sec^2 t\,dt$

80. $F(x) = \int_{\pi/3}^x \sec t \tan t\,dt$

In Exercises 81–86, use the Second Fundamental Theorem of Calculus to find $F'(x)$.

81. $F(x) = \int_{-2}^x (t^2 - 2t)\,dt$

82. $F(x) = \int_1^x \dfrac{t^2}{t^2 + 1}\,dt$

83. $F(x) = \int_{-1}^x \sqrt{t^4 + 1}\,dt$

84. $F(x) = \int_1^x \sqrt[4]{t}\,dt$

85. $F(x) = \int_0^x t \cos t\,dt$

86. $F(x) = \int_0^x \sec^3 t\,dt$

In Exercises 87–92, find $F'(x)$.

87. $F(x) = \int_x^{x+2} (4t + 1)\,dt$

88. $F(x) = \int_{-x}^x t^3\,dt$

89. $F(x) = \int_0^{\sin x} \sqrt{t}\,dt$

90. $F(x) = \int_2^{x^2} \dfrac{1}{t^3}\,dt$

91. $F(x) = \int_0^{x^3} \sin t^2\,dt$

92. $F(x) = \int_0^{x^2} \sin \theta^2\,d\theta$

93. *Graphical Analysis* Approximate the graph of g on the interval $0 \le x \le 4$ where $g(x) = \int_0^x f(t)\,dt$. Identify the x-coordinate of an extremum of g. To print an enlarged copy of the graph, go to the website *www.mathgraphs.com*.

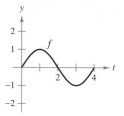

94. Use the function f in the figure and the function g defined by

$$g(x) = \int_0^x f(t)\,dt.$$

(a) Complete the table.

x	1	2	3	4	5	6	7	8	9	10
$g(x)$										

(b) Plot the points from the table in part (a) and graph g.

(c) Where does g have its minimum? Explain.

(d) Where does g have a maximum? Explain.

(e) On what interval does g increase at the greatest rate? Explain.

(f) Identify the zeros of g.

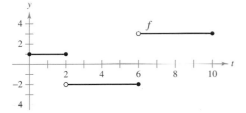

95. *Cost*   The total cost of purchasing and maintaining a piece of equipment for x years is

$$C(x) = 5000\left(25 + 3\int_0^x t^{1/4}\,dt\right).$$

(a) Perform the integration to write C as a function of x.

(b) Find $C(1)$, $C(5)$, and $C(10)$.

96. *Area*   The area A between the graph of the function $g(t) = 4 - 4/t^2$ and the t-axis over the interval $[1, x]$ is

$$A(x) = \int_1^x \left(4 - \frac{4}{t^2}\right)dt.$$

(a) Find the horizontal asymptote of the graph of g.

(b) Integrate to find A as a function of x. Does the graph of A have a horizontal asymptote? Explain.

True or False? In Exercises 97–99, determine whether the statement is true or false. If it is false, explain why or give an example that shows it is false.

97. If $F'(x) = G'(x)$ on the interval $[a, b]$, then $F(b) - F(a) = G(b) - G(a)$.

98. If f is continuous on $[a, b]$, then f is integrable on $[a, b]$.

99. $\displaystyle\int_{-1}^{1} x^{-2}\,dx = \left[-x^{-1}\right]_{-1}^{1} = (-1) - 1 = -2$

100. Prove: $\displaystyle\frac{d}{dx}\left[\int_{u(x)}^{v(x)} f(t)\,dt\right] = f(v(x))v'(x) - f(u(x))u'(x).$

101. Show that the function

$$f(x) = \int_0^{1/x} \frac{1}{t^2 + 1}\,dt + \int_0^x \frac{1}{t^2 + 1}\,dt$$

is constant for $x > 0$.

102. Let $G(x) = \displaystyle\int_0^x \left[s\int_0^s f(t)\,dt\right]ds$, where f is continuous for all real t. Find

(a) $G(0)$.   (b) $G'(0)$.

(c) $G''(x)$.   (d) $G''(0)$.

Rectilinear Motion   **In Exercises 103–105, consider a particle moving along the x-axis where $x(t)$ is the position of the particle at time t, $x'(t)$ is its velocity, and $\int_a^b |x'(t)|\,dt$ is the distance the particle travels in the interval of time.**

103. The position function is

$$x(t) = t^3 - 6t^2 + 9t - 2, \quad 0 \le t \le 5.$$

Find the total distance the particle travels in 5 units of time.

104. Repeat Exercise 103 for the position function given by $x(t) = (t - 1)(t - 3)^2$, $0 \le t \le 5$.

105. A particle moves along the x-axis with velocity $v(t) = 1/\sqrt{t}$, $t > 0$. At time $t = 1$, its position is $x = 4$. Find the total distance traveled by the particle on the interval $1 \le t \le 4$.

SECTION PROJECT **DEMONSTRATING THE FUNDAMENTAL THEOREM**

Use a graphing utility to graph the function $y_1 = \sin^2 t$ on the interval $0 \le t \le \pi$. Let $F(x)$ be the following function of x.

$$F(x) = \int_0^x \sin^2 t\,dt$$

(a) Complete the table and explain why the values of F are increasing.

x	0	$\pi/6$	$\pi/3$	$\pi/2$	$2\pi/3$	$5\pi/6$	π
$F(x)$							

(b) Use the integration capabilities of a graphing utility to graph F.

(c) Use the differentiation capabilities of a graphing utility to graph $F'(x)$. How is this graph related to the graph in part (b)?

(d) Verify that the derivative of $y = (1/2)t - (\sin 2t)/4$ is $\sin^2 t$. Graph y and write a short paragraph about how this graph is related to those in parts (b) and (c).

Integration by Substitution

- Use pattern recognition to evaluate an indefinite integral.
- Use a change of variables to evaluate an indefinite integral.
- Use the General Power Rule for Integration to evaluate an indefinite integral.
- Use a change of variables to evaluate a definite integral.
- Evaluate a definite integral involving an even or odd function.

Pattern Recognition

In this section you will study techniques for integrating composite functions. The discussion is split into two parts—*pattern recognition* and *change of variables*. Both techniques involve a **u-substitution.** With pattern recognition you perform the substitution mentally, and with change of variables you write the substitution steps.

   The role of substitution in integration is comparable to the role of the Chain Rule in differentiation. Recall that for differentiable functions given by $y = F(u)$ and $u = g(x)$, the Chain Rule states that

$$\frac{d}{dx}[F(g(x))] = F'(g(x))g'(x).$$

From the definition of an antiderivative, it follows that

$$\int F'(g(x))g'(x)\,dx = F(g(x)) + C$$

$$= F(u) + C.$$

These results are summarized in the following theorem.

NOTE   The statement of Theorem 4.12 doesn't tell how to distinguish between $f(g(x))$ and $g'(x)$ in the integrand. As you become more experienced at integration, your skill in doing this will increase. Of course, part of the key is familiarity with derivatives.

> **THEOREM 4.12   Antidifferentiation of a Composite Function**
>
> Let g be a function whose range is an interval I, and let f be a function that is continuous on I. If g is differentiable on its domain and F is an antiderivative of f on I, then
>
> $$\int f(g(x))g'(x)\,dx = F(g(x)) + C.$$
>
> If $u = g(x)$, then $du = g'(x)\,dx$ and
>
> $$\int f(u)\,du = F(u) + C.$$

 STUDY TIP   There are several techniques for applying substitution, each differing slightly from the others. However, you should remember that the goal is the same with every technique— *you are trying to find an antiderivative of the integrand.*

EXPLORATION

Recognizing Patterns   The integrand in each of the following integrals fits the pattern $f(g(x))g'(x)$. Identify the pattern and use the result to evaluate the integral.

a. $\displaystyle\int 2x(x^2 + 1)^4\,dx$ **b.** $\displaystyle\int 3x^2\sqrt{x^3 + 1}\,dx$ **c.** $\displaystyle\int \sec^2 x(\tan x + 3)\,dx$

The next three integrals are similar to the first three. Show how you can multiply and divide by a constant to evaluate these integrals.

d. $\displaystyle\int x(x^2 + 1)^4\,dx$ **e.** $\displaystyle\int x^2\sqrt{x^3 + 1}\,dx$ **f.** $\displaystyle\int 2\sec^2 x(\tan x + 3)\,dx$

Examples 1 and 2 show how to apply Theorem 4.12 *directly*, by recognizing the presence of $f(g(x))$ and $g'(x)$. Note that the composite function in the integrand has an *outside function f* and an *inside function g*. Moreover, the derivative $g'(x)$ is present as a factor of the integrand.

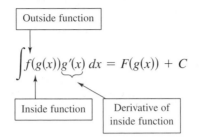

$$\int f(g(x))g'(x)\,dx = F(g(x)) + C$$

Example 1   Recognizing the $f(g(x))g'(x)$ Pattern

Find $\displaystyle\int (x^2 + 1)^2(2x)\,dx$.

Solution   Letting $g(x) = x^2 + 1$, you obtain

$$g'(x) = 2x$$

and

$$f(g(x)) = f(x^2 + 1) = (x^2 + 1)^2.$$

From this, you can recognize that the integrand follows the $f(g(x))g'(x)$ pattern. Using the Power Rule for integration and Theorem 4.12, you can write

$$\int \overbrace{(x^2 + 1)^2}^{f(g(x))}\,\overbrace{(2x)}^{g'(x)}\,dx = \frac{1}{3}(x^2 + 1)^3 + C.$$

Try using the Chain Rule to check that the derivative of $\frac{1}{3}(x^2 + 1)^3 + C$ is the integrand of the original integral.

Example 2   Recognizing the $f(g(x))g'(x)$ Pattern

Find $\displaystyle\int 5 \cos 5x\,dx$.

Solution   Letting $g(x) = 5x$, you obtain

$$g'(x) = 5$$

and

$$f(g(x)) = f(5x) = \cos 5x.$$

From this, you can recognize that the integrand follows the $f(g(x))g'(x)$ pattern. Using the Cosine Rule for integration and Theorem 4.12, you can write

$$\int \overbrace{(\cos 5x)}^{f(g(x))}\,\overbrace{(5)}^{g'(x)}\,dx = \sin 5x + C.$$

You can check this by differentiating $\sin 5x + C$ to obtain the original integrand.

The integrands in Examples 1 and 2 fit the $f(g(x))g'(x)$ pattern exactly—you only had to recognize the pattern. You can extend this technique considerably with the Constant Multiple Rule

$$\int kf(x)\, dx = k \int f(x)\, dx.$$

Many integrands contain the essential part (the variable part) of $g'(x)$ but are missing a constant multiple. In such cases, you can multiply and divide by the necessary constant multiple, as demonstrated in Example 3.

Example 3   **Multiplying and Dividing by a Constant**

Find $\int x(x^2 + 1)^2\, dx.$

Solution   This is similar to the integral given in Example 1, except that the integrand is missing a factor of 2. Recognizing that $2x$ is the derivative of $x^2 + 1$, you can let $g(x) = x^2 + 1$ and supply the $2x$ as follows.

$$\int x(x^2 + 1)^2\, dx = \int (x^2 + 1)^2 \left(\frac{1}{2}\right)(2x)\, dx \qquad \text{Multiply and divide by 2.}$$

$$= \frac{1}{2}\int \overbrace{(x^2 + 1)^2}^{f(g(x))} \overbrace{(2x)}^{g'(x)}\, dx \qquad \text{Constant Multiple Rule}$$

$$= \frac{1}{2}\left[\frac{(x^2 + 1)^3}{3}\right] + C \qquad \text{Integrate.}$$

$$= \frac{1}{6}(x^2 + 1)^3 + C$$

In practice, most people would not write as many steps as are shown in Example 3. For instance, you could evaluate the integral by simply writing

$$\int x(x^2 + 1)^2\, dx = \frac{1}{2}\int (x^2 + 1)^2\, 2x\, dx$$

$$= \frac{1}{2}\left[\frac{(x^2 + 1)^3}{3}\right] + C$$

$$= \frac{1}{6}(x^2 + 1)^3 + C.$$

NOTE   Be sure you see that the *Constant* Multiple Rule applies only to *constants*. You cannot multiply and divide by a variable and then move the variable outside the integral sign. For instance,

$$\int (x^2 + 1)^2\, dx \neq \frac{1}{2x}\int (x^2 + 1)^2\, (2x)\, dx.$$

After all, if it were legitimate to move variable quantities outside the integral sign, you could move the entire integrand out and simplify the whole process. But the result would be incorrect.

Change of Variables

With a formal **change of variables,** you completely rewrite the integral in terms of u and du (or any other convenient variable). Although this procedure can involve more written steps than the pattern recognition illustrated in Examples 1 to 3, it is useful for complicated integrands. The change of variable technique uses the Leibniz notation for the differential. That is, if $u = g(x)$, then $du = g'(x)\,dx$, and the integral in Theorem 4.12 takes the form

$$\int f(g(x))g'(x)\,dx = \int f(u)\,du = F(u) + C.$$

Example 4   **Change of Variables**

Find $\displaystyle\int \sqrt{2x - 1}\,dx$.

Solution   First, let u be the inner function, $u = 2x - 1$. Then calculate the differential du to be $du = 2\,dx$. Now, using $\sqrt{2x - 1} = \sqrt{u}$ and $dx = du/2$, substitute to obtain the following.

$$\int \sqrt{2x - 1}\,dx = \int \sqrt{u}\left(\frac{du}{2}\right) \qquad \text{Integral in terms of } u$$

$$= \frac{1}{2}\int u^{1/2}\,du$$

$$= \frac{1}{2}\left(\frac{u^{3/2}}{3/2}\right) + C \qquad \text{Antiderivative in terms of } u$$

$$= \frac{1}{3}u^{3/2} + C$$

$$= \frac{1}{3}(2x - 1)^{3/2} + C \qquad \text{Antiderivative in terms of } x$$

STUDY TIP   Because integration is usually more difficult than differentiation, you should always check your answer to an integration problem by differentiating. For instance, in Example 4 you should differentiate $\frac{1}{3}(2x - 1)^{3/2} + C$ to verify that you obtain the original integrand.

Example 5   **Change of Variables**

Find $\displaystyle\int x\sqrt{2x - 1}\,dx$.

Solution   As in the previous example, let $u = 2x - 1$ and obtain $dx = du/2$. Because the integrand contains a factor of x, you must also solve for x in terms of u, as follows.

$$u = 2x - 1 \quad \Longrightarrow \quad x = (u + 1)/2 \qquad \text{Solve for } x \text{ in terms of } u.$$

Now, using substitution, you obtain the following.

$$\int x\sqrt{2x - 1}\,dx = \int \left(\frac{u + 1}{2}\right)u^{1/2}\left(\frac{du}{2}\right)$$

$$= \frac{1}{4}\int (u^{3/2} + u^{1/2})\,du$$

$$= \frac{1}{4}\left(\frac{u^{5/2}}{5/2} + \frac{u^{3/2}}{3/2}\right) + C$$

$$= \frac{1}{10}(2x - 1)^{5/2} + \frac{1}{6}(2x - 1)^{3/2} + C$$

To complete the change of variables in Example 5, we solved for x in terms of u. Sometimes this is very difficult. Fortunately it is not always necessary, as shown in the next example.

Example 6 **Change of Variables**

Find $\displaystyle\int \sin^2 3x \cos 3x \, dx$.

Solution Because $\sin^2 3x = (\sin 3x)^2$, you can let $u = \sin 3x$. Then

$$du = (\cos 3x)(3) \, dx.$$

Now, because $\cos 3x \, dx$ is part of the given integral, you can write

$$\frac{du}{3} = \cos 3x \, dx.$$

Substituting u and $du/3$ in the given integral yields the following.

$$
\begin{aligned}
\int \sin^2 3x \cos 3x \, dx &= \int u^2 \frac{du}{3} \\
&= \frac{1}{3} \int u^2 \, du \\
&= \frac{1}{3}\left(\frac{u^3}{3}\right) + C \\
&= \frac{1}{9} \sin^3 3x + C
\end{aligned}
$$

You can check this by differentiating.

$$
\begin{aligned}
\frac{d}{dx}\left[\frac{1}{9}\sin^3 3x\right] &= \left(\frac{1}{9}\right)(3)(\sin 3x)^2(\cos 3x)(3) \\
&= \sin^2 3x \cos 3x
\end{aligned}
$$

Because differentiation produces the original integrand, you know that you have obtained the correct antiderivative.

STUDY TIP When making a change of variables, be sure that your answer is written using the same variables as in the original integrand. For instance, in Example 6, you should not leave your answer as

$$\tfrac{1}{9}u^3 + C$$

but rather, replace u by $\sin 3x$.

We summarize the steps used for integration by substitution in the following guidelines.

Guidelines for Making a Change of Variables

1. Choose a substitution $u = g(x)$. Usually, it is best to choose the *inner* part of a composite function, such as a quantity raised to a power.
2. Compute $du = g'(x) \, dx$.
3. Rewrite the integral in terms of the variable u.
4. Find the resulting integral in terms of u.
5. Replace u by $g(x)$ to obtain an antiderivative in terms of x.
6. Check your answer by differentiating.

The General Power Rule for Integration

One of the most common u-substitutions involves quantities in the integrand that are raised to a power. Because of the importance of this type of substitution, it is given a special name—the **General Power Rule** for integration. A proof of this rule follows directly from the (simple) Power Rule for integration, together with Theorem 4.12.

THEOREM 4.13   The General Power Rule for Integration

If g is a differentiable function of x, then

$$\int [g(x)]^n g'(x)\, dx = \frac{[g(x)]^{n+1}}{n+1} + C, \qquad n \neq -1.$$

Equivalently, if $u = g(x)$, then

$$\int u^n\, du = \frac{u^{n+1}}{n+1} + C, \qquad n \neq -1.$$

Example 7   **Substitution and the General Power Rule**

a. $\displaystyle \int 3(3x-1)^4\, dx = \int \overbrace{(3x-1)^4}^{u^4}\,\overbrace{(3)\, dx}^{du} = \overbrace{\frac{(3x-1)^5}{5}}^{u^5/5} + C$

b. $\displaystyle \int (2x+1)(x^2+x)\,dx = \int \overbrace{(x^2+x)^1}^{u^1}\,\overbrace{(2x+1)\, dx}^{du} = \overbrace{\frac{(x^2+x)^2}{2}}^{u^2/2} + C$

c. $\displaystyle \int 3x^2\sqrt{x^3-2}\, dx = \int \overbrace{(x^3-2)^{1/2}}^{u^{1/2}}\,\overbrace{(3x^2)\, dx}^{du} = \overbrace{\frac{(x^3-2)^{3/2}}{3/2}}^{u^{3/2}/(3/2)} + C = \frac{2}{3}(x^3-2)^{3/2} + C$

d. $\displaystyle \int \frac{-4x}{(1-2x^2)^2}\, dx = \int \overbrace{(1-2x^2)^{-2}}^{u^{-2}}\,\overbrace{(-4x)\, dx}^{du} - \overbrace{\frac{(1-2x^2)^{-1}}{-1}}^{u^{-1}/(-1)} + C$

e. $\displaystyle \int \cos^2 x \sin x\, dx = -\int \overbrace{(\cos x)^2}^{u^2}\,\overbrace{(-\sin x)\, dx}^{du} = -\overbrace{\frac{(\cos x)^3}{3}}^{u^3/3} + C$

Some integrals whose integrands involve a quantity raised to a power cannot be found by the General Power Rule. Consider the two integrals

$$\int x(x^2+1)^2\, dx \quad \text{and} \quad \int (x^2+1)^2\, dx.$$

The substitution $u = x^2 + 1$ works in the first integral but not in the second. (In the second, the substitution fails because the integrand lacks the factor x needed for du.) Fortunately, *for this particular integral,* you can expand the integrand as $(x^2+1)^2 = x^4 + 2x^2 + 1$ and use the (simple) Power Rule to integrate each term.

EXPLORATION

Suppose you were asked to find one of the following integrals. Which one would you choose? Explain your reasoning.

a. $\displaystyle \int \sqrt{x^3+1}\, dx$ or

$\displaystyle \int x^2\sqrt{x^3+1}\, dx$

b. $\displaystyle \int \tan(3x)\sec^2(3x)\, dx$ or

$\displaystyle \int \tan(3x)\, dx$

Change of Variables for Definite Integrals

When using u-substitution with a definite integral, it is often convenient to determine the limits of integration for the variable u rather than to convert the antiderivative back to the variable x and evaluate at the original limits. This change of variables is stated explicitly in the next theorem. The proof follows from Theorem 4.12 combined with the Fundamental Theorem of Calculus.

> **THEOREM 4.14   Change of Variables for Definite Integrals**
>
> If the function $u = g(x)$ has a continuous derivative on the closed interval $[a, b]$ and f is continuous on the range of g, then
>
> $$\int_a^b f(g(x))g'(x)\,dx = \int_{g(a)}^{g(b)} f(u)\,du.$$

Example 8   **Change of Variables**

Evaluate $\displaystyle\int_0^1 x(x^2 + 1)^3\,dx$.

Solution   To evaluate this integral, let $u = x^2 + 1$. Then, you obtain

$$u = x^2 + 1 \implies du = 2x\,dx.$$

Before substituting, determine the new upper and lower limits of integration.

Lower Limit	*Upper Limit*
When $x = 0$, $u = 0^2 + 1 = 1$.	When $x = 1$, $u = 1^2 + 1 = 2$.

Now, you can substitute to obtain

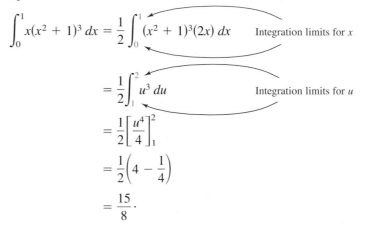

$$\int_0^1 x(x^2 + 1)^3\,dx = \frac{1}{2}\int_0^1 (x^2 + 1)^3(2x)\,dx \qquad \text{Integration limits for } x$$

$$= \frac{1}{2}\int_1^2 u^3\,du \qquad \text{Integration limits for } u$$

$$= \frac{1}{2}\left[\frac{u^4}{4}\right]_1^2$$

$$= \frac{1}{2}\left(4 - \frac{1}{4}\right)$$

$$= \frac{15}{8}.$$

Try converting the antiderivative $\frac{1}{2}(u^4/4)$ back to the variable x and evaluate the definite integral at the original limits of integration, as follows.

$$\frac{1}{2}\left[\frac{u^4}{4}\right]_1^2 = \frac{1}{2}\left[\frac{(x^2 + 1)^4}{4}\right]_0^1 = \frac{1}{2}\left(4 - \frac{1}{4}\right) = \frac{15}{8}$$

Notice that you obtain the same result.

Example 9   **Change of Variables**

Evaluate $A = \int_1^5 \dfrac{x}{\sqrt{2x-1}}\,dx$.

Solution   To evaluate this integral, let $u = \sqrt{2x-1}$. Then, you obtain

$$u^2 = 2x - 1$$
$$u^2 + 1 = 2x$$
$$\frac{u^2+1}{2} = x$$
$$u\,du = dx. \qquad \text{Differentiate both sides.}$$

Before substituting, determine the new upper and lower limits of integration.

Lower Limit	*Upper Limit*
When $x = 1$, $u = \sqrt{2-1} = 1$.	When $x = 5$, $u = \sqrt{10-1} = 3$.

Now, substitute to obtain

$$\int_1^5 \frac{x}{\sqrt{2x-1}}\,dx = \int_1^3 \frac{1}{u}\left(\frac{u^2+1}{2}\right)u\,du$$
$$= \frac{1}{2}\int_1^3 (u^2+1)\,du$$
$$= \frac{1}{2}\left[\frac{u^3}{3}+u\right]_1^3$$
$$= \frac{1}{2}\left(9+3-\frac{1}{3}-1\right)$$
$$= \frac{16}{3}.$$

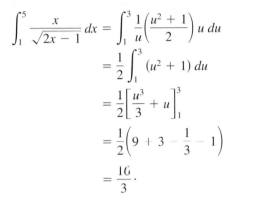

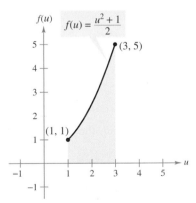

The region before substitution has an area of $\frac{16}{3}$.
Figure 4.37

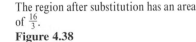

The region after substitution has an area of $\frac{16}{3}$.
Figure 4.38

Geometrically, you can interpret the equation

$$\int_1^5 \frac{x}{\sqrt{2x-1}}\,dx = \int_1^3 \frac{u^2+1}{2}\,du$$

to mean that the two *different* regions shown in Figures 4.37 and 4.38 have the *same* area.

When evaluating definite integrals by substitution, it is possible for the upper limit of integration of the u-variable form to be smaller than the lower limit. If this happens, don't rearrange the limits. Simply evaluate as usual. For example, after substituting $u = \sqrt{1-x}$ in the integral

$$\int_0^1 x^2(1-x)^{1/2}\,dx$$

you obtain $u = \sqrt{1-1} = 0$ when $x = 1$, and $u = \sqrt{1-0} = 1$ when $x = 0$. So, the correct u-variable form of this integral is

$$-2\int_1^0 (1-u^2)^2 u^2\,du.$$

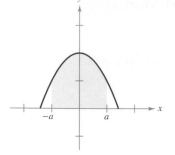

Even function

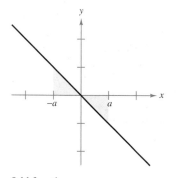

Odd function
Figure 4.39

Integration of Even and Odd Functions

Even with a change of variables, integration can be difficult. Occasionally, you can simplify the evaluation of a definite integral (over an interval that is symmetric about the y-axis or about the origin) by recognizing the integrand to be an even or odd function (see Figure 4.39).

> **THEOREM 4.15   Integration of Even and Odd Functions**
>
> Let f be integrable on the closed interval $[-a, a]$.
>
> **1.** If f is an *even* function, then $\displaystyle\int_{-a}^{a} f(x)\, dx = 2 \int_{0}^{a} f(x)\, dx$.
>
> **2.** If f is an *odd* function, then $\displaystyle\int_{-a}^{a} f(x)\, dx = 0$.

Proof   Because f is even, you know that $f(x) = f(-x)$. Using Theorem 4.12 with the substitution $u = -x$ produces

$$\int_{-a}^{0} f(x)\, dx = \int_{a}^{0} f(-u)(-du) = -\int_{a}^{0} f(u)\, du = \int_{0}^{a} f(u)\, du = \int_{0}^{a} f(x)\, dx.$$

Finally, using Theorem 4.6, you obtain

$$\int_{-a}^{a} f(x)\, dx = \int_{-a}^{0} f(x)\, dx + \int_{0}^{a} f(x)\, dx$$

$$= \int_{0}^{a} f(x)\, dx + \int_{0}^{a} f(x)\, dx = 2 \int_{0}^{a} f(x)\, dx.$$

This proves the first property. The proof of the second property is left to you (see Exercise 116).

Example 10   **Integration of an Odd Function**

Evaluate $\displaystyle\int_{-\pi/2}^{\pi/2} (\sin^3 x \cos x + \sin x \cos x)\, dx$.

Solution   Letting $f(x) = \sin^3 x \cos x + \sin x \cos x$ produces

$$f(-x) = \sin^3(-x) \cos(-x) + \sin(-x) \cos(-x)$$
$$= -\sin^3 x \cos x - \sin x \cos x = -f(x).$$

So, f is an odd function, and because f is symmetric about the origin over $[-\pi/2, \pi/2]$, you can apply Theorem 4.15 to conclude that

$$\int_{-\pi/2}^{\pi/2} (\sin^3 x \cos x + \sin x \cos x)\, dx = 0.$$

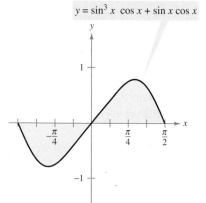

$y = \sin^3 x \, \cos x + \sin x \cos x$

Because f is an odd function,
$$\int_{-\pi/2}^{\pi/2} f(x)\, dx = 0.$$
Figure 4.40

NOTE   From Figure 4.40 you can see that the two regions on either side of the y-axis have the same area. However, because one lies below the x-axis and one lies above it, integration produces a cancellation effect. (We will say more about this in Section 6.1.)

EXERCISES ·FOR SECTION 4.5

In Exercises 1–6, complete the table by identifying *u* and *du* for the integral.

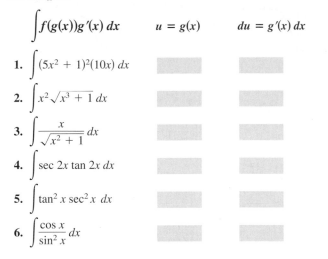

$\int f(g(x))g'(x)\,dx$	$u = g(x)$	$du = g'(x)\,dx$
1. $\int (5x^2 + 1)^2(10x)\,dx$		
2. $\int x^2\sqrt{x^3 + 1}\,dx$		
3. $\int \dfrac{x}{\sqrt{x^2 + 1}}\,dx$		
4. $\int \sec 2x \tan 2x\,dx$		
5. $\int \tan^2 x \sec^2 x\,dx$		
6. $\int \dfrac{\cos x}{\sin^2 x}\,dx$		

In Exercises 7–34, find the indefinite integral and check the result by differentiation.

7. $\int (1 + 2x)^4(2)\,dx$

8. $\int (x^2 - 9)^3(2x)\,dx$

9. $\int \sqrt{9 - x^2}\,(-2x)\,dx$

10. $\int \sqrt[3]{(1 - 2x^2)}(-4x)\,dx$

11. $\int x^3(x^4 + 3)^2\,dx$

12. $\int x^2(x^3 + 5)^4\,dx$

13. $\int x^2(x^3 - 1)^4\,dx$

14. $\int x(4x^2 + 3)^3\,dx$

15. $\int t\sqrt{t^2 + 2}\,dt$

16. $\int t^3\sqrt{t^4 + 5}\,dt$

17. $\int 5x\sqrt[3]{1 - x^2}\,dx$

18. $\int u^2\sqrt{u^3 + 2}\,du$

19. $\int \dfrac{x}{(1 - x^2)^3}\,dx$

20. $\int \dfrac{x^3}{(1 + x^4)^2}\,dx$

21. $\int \dfrac{x^2}{(1 + x^3)^2}\,dx$

22. $\int \dfrac{x^2}{(16 - x^3)^2}\,dx$

23. $\int \dfrac{x}{\sqrt{1 - x^2}}\,dx$

24. $\int \dfrac{x^3}{\sqrt{1 + x^4}}\,dx$

25. $\int \left(1 + \dfrac{1}{t}\right)^3\left(\dfrac{1}{t^2}\right)\,dt$

26. $\int \left[x^2 + \dfrac{1}{(3x)^2}\right]\,dx$

27. $\int \dfrac{1}{\sqrt{2x}}\,dx$

28. $\int \dfrac{1}{2\sqrt{x}}\,dx$

29. $\int \dfrac{x^2 + 3x + 7}{\sqrt{x}}\,dx$

30. $\int \dfrac{t + 2t^2}{\sqrt{t}}\,dt$

31. $\int t^2\left(t - \dfrac{2}{t}\right)\,dt$

32. $\int \left(\dfrac{t^3}{3} + \dfrac{1}{4t^2}\right)\,dt$

33. $\int (9 - y)\sqrt{y}\,dy$

34. $\int 2\pi y(8 - y^{3/2})\,dy$

In Exercises 35–38, solve the differential equation.

35. $\dfrac{dy}{dx} = 4x + \dfrac{4x}{\sqrt{16 - x^2}}$

36. $\dfrac{dy}{dx} = \dfrac{10x^2}{\sqrt{1 + x^3}}$

37. $\dfrac{dy}{dx} = \dfrac{x + 1}{(x^2 + 2x - 3)^2}$

38. $\dfrac{dy}{dx} = \dfrac{x - 4}{\sqrt{x^2 - 8x + 1}}$

Slope Fields   **In Exercises 39 and 40, a differential equation, a point, and a slope field are given. A *slope field* consists of line segments with slopes given by the differential equation. These line segments give a visual perspective of the directions of the solutions of the differential equation. (a) Sketch two approximate solutions of the differential equation on the slope field, one of which passes through the indicated point. (To print an enlarged copy of the graph, go to the website *www.mathgraphs.com*.) (b) Use integration to find the particular solution of the differential equation and use a graphing utility to graph the solution. Compare the result with the sketches in part (a).**

39. $\dfrac{dy}{dx} = x\sqrt{4 - x^2}$, $(2, 2)$

40. $\dfrac{dy}{dx} = x\cos x^2$, $(0, 1)$

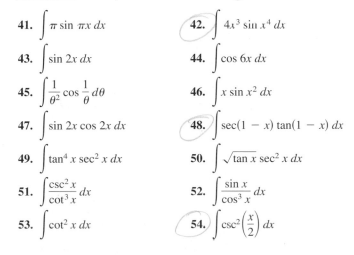

In Exercises 41–54, find the indefinite integral.

41. $\int \pi \sin \pi x\,dx$

42. $\int 4x^3 \sin x^4\,dx$

43. $\int \sin 2x\,dx$

44. $\int \cos 6x\,dx$

45. $\int \dfrac{1}{\theta^2}\cos \dfrac{1}{\theta}\,d\theta$

46. $\int x \sin x^2\,dx$

47. $\int \sin 2x \cos 2x\,dx$

48. $\int \sec(1 - x)\tan(1 - x)\,dx$

49. $\int \tan^4 x \sec^2 x\,dx$

50. $\int \sqrt{\tan x}\,\sec^2 x\,dx$

51. $\int \dfrac{\csc^2 x}{\cot^3 x}\,dx$

52. $\int \dfrac{\sin x}{\cos^3 x}\,dx$

53. $\int \cot^2 x\,dx$

54. $\int \csc^2\left(\dfrac{x}{2}\right)\,dx$

In Exercises 55 and 56, find an equation for the function f that has the indicated derivative and whose graph passes through the given point.

Derivative	Point
55. $f'(x) = \cos\dfrac{x}{2}$	$(0, 3)$
56. $f'(x) = \pi \sec \pi x \tan \pi x$	$\left(\dfrac{1}{3}, 1\right)$

In Exercises 57–64, find the indefinite integral by the method shown in Example 5.

57. $\displaystyle\int x\sqrt{x+2}\,dx$

58. $\displaystyle\int x\sqrt{2x+1}\,dx$

59. $\displaystyle\int x^2\sqrt{1-x}\,dx$

60. $\displaystyle\int (x+1)\sqrt{2-x}\,dx$

61. $\displaystyle\int \dfrac{x^2-1}{\sqrt{2x-1}}\,dx$

62. $\displaystyle\int \dfrac{2x+1}{\sqrt{x+4}}\,dx$

63. $\displaystyle\int \dfrac{-x}{(x+1)-\sqrt{x+1}}\,dx$

64. $\displaystyle\int t\sqrt[3]{t-4}\,dt$

In Exercises 65–76, evaluate the definite integral. Use a graphing utility to verify your result.

65. $\displaystyle\int_{-1}^{1} x(x^2+1)^3\,dx$

66. $\displaystyle\int_{-2}^{4} x^2(x^3+8)^2\,dx$

67. $\displaystyle\int_{1}^{2} 2x^2\sqrt{x^3+1}\,dx$

68. $\displaystyle\int_{0}^{1} x\sqrt{1-x^2}\,dx$

69. $\displaystyle\int_{0}^{4} \dfrac{1}{\sqrt{2x+1}}\,dx$

70. $\displaystyle\int_{0}^{2} \dfrac{x}{\sqrt{1+2x^2}}\,dx$

71. $\displaystyle\int_{1}^{9} \dfrac{1}{\sqrt{x}\left(1+\sqrt{x}\right)^2}\,dx$

72. $\displaystyle\int_{0}^{2} x\sqrt[3]{4+x^2}\,dx$

73. $\displaystyle\int_{1}^{2} (x-1)\sqrt{2-x}\,dx$

74. $\displaystyle\int_{1}^{5} \dfrac{x}{\sqrt{2x-1}}\,dx$

75. $\displaystyle\int_{0}^{\pi/2} \cos\left(\dfrac{2x}{3}\right)dx$

76. $\displaystyle\int_{\pi/3}^{\pi/2} (x+\cos x)\,dx$

In Exercises 77–82, find the area of the region. Use a graphing utility to verify your result.

77. $\displaystyle\int_{0}^{7} x\sqrt[3]{x+1}\,dx$

78. $\displaystyle\int_{-2}^{6} x^2\sqrt[3]{x+2}\,dx$

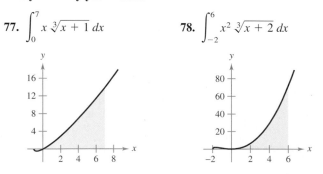

79. $y = 2\sin x + \sin 2x$

80. $y = \sin x + \cos 2x$

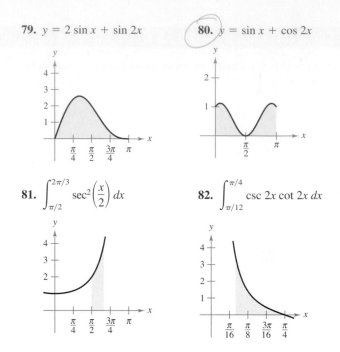

81. $\displaystyle\int_{\pi/2}^{2\pi/3} \sec^2\left(\dfrac{x}{2}\right)dx$

82. $\displaystyle\int_{\pi/12}^{\pi/4} \csc 2x \cot 2x\,dx$

In Exercises 83–88, use a graphing utility to evaluate the integral. Graph the region whose area is given by the definite integral.

83. $\displaystyle\int_{0}^{4} \dfrac{x}{\sqrt{2x+1}}\,dx$

84. $\displaystyle\int_{0}^{2} x^3\sqrt{x+2}\,dx$

85. $\displaystyle\int_{3}^{7} x\sqrt{x-3}\,dx$

86. $\displaystyle\int_{1}^{5} x^2\sqrt{x-1}\,dx$

87. $\displaystyle\int_{0}^{3} \left(\theta + \cos\dfrac{\theta}{6}\right)d\theta$

88. $\displaystyle\int_{0}^{\pi/2} \sin 2x\,dx$

Writing **In Exercises 89 and 90, find the indefinite integral in two ways. Explain any difference in the forms of the answers.**

89. $\displaystyle\int (2x-1)^2\,dx$

90. $\displaystyle\int \sin x \cos x\,dx$

In Exercises 91–94, evaluate the integral using the properties of even and odd functions as an aid.

91. $\displaystyle\int_{-2}^{2} x^2(x^2+1)\,dx$

92. $\displaystyle\int_{-\pi/2}^{\pi/2} \sin^2 x \cos x\,dx$

93. $\displaystyle\int_{-2}^{2} x(x^2+1)^3\,dx$

94. $\displaystyle\int_{-\pi/2}^{\pi/2} \sin x \cos x\,dx$

95. Use $\int_{0}^{2} x^2\,dx = \dfrac{8}{3}$ to evaluate the definite integrals without using the Fundamental Theorem of Calculus.

(a) $\displaystyle\int_{-2}^{0} x^2\,dx$

(b) $\displaystyle\int_{-2}^{2} x^2\,dx$

(c) $\displaystyle\int_{0}^{2} -x^2\,dx$

(d) $\displaystyle\int_{-2}^{0} 3x^2\,dx$

96. Use the symmetry of the graphs of the sine and cosine functions as an aid in evaluating each of the integrals.

(a) $\displaystyle\int_{-\pi/4}^{\pi/4} \sin x \, dx$

(b) $\displaystyle\int_{-\pi/4}^{\pi/4} \cos x \, dx$

(c) $\displaystyle\int_{-\pi/2}^{\pi/2} \cos x \, dx$

(d) $\displaystyle\int_{-\pi/2}^{\pi/2} \sin x \cos x \, dx$

In Exercises 97 and 98, write the integral as the sum of the integral of an odd function and the integral of an even function. Use this simplification to evaluate the integral.

97. $\displaystyle\int_{-4}^{4} (x^3 + 6x^2 - 2x - 3) \, dx$

98. $\displaystyle\int_{-\pi}^{\pi} (\sin 3x + \cos 3x) \, dx$

Getting at the Concept

99. In your own words, state the guidelines for making a change of variables when integrating.

100. Describe why $\int x(5 - x^2)^3 \, dx \neq \int u^3 \, du$ where $u = 5 - x^2$.

101. Without integrating, explain why $\int_{-2}^{2} x(x^2 + 1)^2 \, dx = 0$.

102. *Cash Flow*   The rate of disbursement dQ/dt of a 2 million dollar federal grant is proportional to the square of $100 - t$. Time t is measured in days $(0 \le t \le 100)$, and Q is the amount that remains to be disbursed. Find the amount that remains to be disbursed after 50 days. Assume that all the money will be disbursed in 100 days.

103. *Depreciation*   The rate of depreciation dV/dt of a machine is inversely proportional to the square of $t + 1$, where V is the value of the machine t years after it was purchased. If the initial value of the machine was $500,000, and its value decreased $100,000 in the first year, estimate its value after 4 years.

104. *Rainfall*   The normal monthly rainfall at the Seattle-Tacoma airport can be approximated by the model

$R = 3.121 + 2.399 \sin(0.524t + 1.377)$

where R is measured in inches and t is the time in months, with $t = 1$ corresponding to January.   *(Source: U.S. National Oceanic and Atmospheric Administration)*

(a) Determine the extrema of the function over a 1-year period.

(b) Use integration to approximate the normal annual rainfall. (*Hint:* Integrate over the interval $[0, 12]$.)

(c) Approximate the average monthly rainfall during the months of October, November, and December.

105. *Sales*   The sales of a seasonal product are given by the model

$S = 74.50 + 43.75 \sin \dfrac{\pi t}{6}$

where S is measured in thousands of units and t is the time in months, with $t = 1$ corresponding to January. Find the average sales for the following periods.

(a) The first quarter $(0 \le t \le 3)$

(b) The second quarter $(3 \le t \le 6)$

(c) The entire year $(0 \le t \le 12)$

106. *Water Supply*   A model for the flow rate of water at a pumping station on a given day is

$R(t) = 53 + 7 \sin\left(\dfrac{\pi t}{6} + 3.6\right) + 9 \cos\left(\dfrac{\pi t}{12} + 8.9\right)$

where $0 \le t \le 24$. R is the flow rate in thousands of gallons per hour, and t is the time in hours.

(a) Use a graphing utility to graph the rate function and approximate the maximum flow rate at the pumping station.

(b) Approximate the total volume of water pumped in 1 day.

107. *Electricity*   The oscillating current in an electrical circuit is

$I = 2 \sin(60\pi t) + \cos(120\pi t)$

where I is measured in amperes and t is measured in seconds. Find the average current for each time interval.

(a) $0 \le t \le \frac{1}{60}$

(b) $0 \le t \le \frac{1}{240}$

(c) $0 \le t \le \frac{1}{30}$

108. *Graphical Analysis*   Consider the functions f and g, where

$f(x) = 6 \sin x \cos^2 x \quad \text{and} \quad g(t) = \displaystyle\int_{0}^{t} f(x) \, dx.$

(a) Use a graphing utility to graph f and g in the same viewing window.

(b) Explain why g is nonnegative.

(c) Identify the points on the graph of g that correspond to the extrema of f.

(d) Does each of the zeros of f correspond to an extremum of g? Explain.

(e) Consider the function $h(t) = \displaystyle\int_{\pi/2}^{t} f(x) \, dx$. Use a graphing utility to graph h. What is the relationship between g and h? Verify your conjecture.

True or False?   **In Exercises 109–114, determine whether the statement is true or false. If it is false, explain why or give an example that shows it is false.**

109. $\displaystyle\int (2x + 1)^2 \, dx = \frac{1}{3}(2x + 1)^3 + C$

110. $\displaystyle\int x(x^2 + 1) \, dx = \frac{1}{2}x^2 \left(\frac{1}{3}x^3 + x\right) + C$

111. $\displaystyle\int_{-10}^{10} (ax^3 + bx^2 + cx + d) \, dx = 2 \int_{0}^{10} (bx^2 + d) \, dx$

112. $\displaystyle\int_{a}^{b} \sin x \, dx = \int_{a}^{b+2\pi} \sin x \, dx$

113. $\displaystyle 4\int \sin x \cos x \, dx = -\cos 2x + C$

114. $\displaystyle\int \sin^2 2x \cos 2x \, dx = \frac{1}{3} \sin^3 2x + C$

115. Show that if f is continuous on the entire real line, then

$\displaystyle\int_{a}^{b} f(x + h) \, dx = \int_{a+h}^{b+h} f(x) \, dx.$

116. Complete the proof of Theorem 4.15.

Section 4.6   Numerical Integration

- Approximate a definite integral using the Trapezoidal Rule.
- Approximate a definite integral using Simpson's Rule.
- Analyze the approximate error in the Trapezoidal Rule and in Simpson's Rule.

The Trapezoidal Rule

Some elementary functions simply do not have antiderivatives that are elementary functions. For example, there is no elementary function that has any of the following functions as its derivative.

$$\sqrt[3]{x}\sqrt{1-x}, \qquad \sqrt{x}\cos x, \qquad \frac{\cos x}{x}, \qquad \sqrt{1-x^3}, \qquad \sin x^2$$

If you need to evaluate a definite integral involving a function whose antiderivative cannot be found, the Fundamental Theorem of Calculus cannot be applied, and you must resort to an approximation technique. Two such techniques are described in this section.

One way to approximate a definite integral is to use n trapezoids, as shown in Figure 4.41. In the development of this method, assume that f is continuous and positive on the interval $[a, b]$. So, the definite integral

$$\int_a^b f(x)\,dx$$

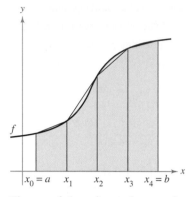

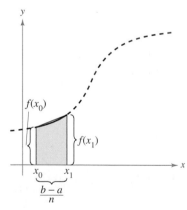

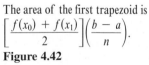

The area of the region can be approximated using four trapezoids.
Figure 4.41

represents the area of the region bounded by the graph of f and the x-axis, from $x = a$ to $x = b$. First, partition the interval $[a, b]$ into n subintervals, each of width $\Delta x = (b - a)/n$, such that

$$a = x_0 < x_1 < x_2 < \cdots < x_n = b.$$

Then form a trapezoid for each subinterval, as shown in Figure 4.42. The area of the ith trapezoid is

$$\text{Area of } i\text{th trapezoid} = \left[\frac{f(x_{i-1}) + f(x_i)}{2}\right]\left(\frac{b-a}{n}\right).$$

This implies that the sum of the areas of the n trapezoids is

$$\text{Area} = \left(\frac{b-a}{n}\right)\left[\frac{f(x_0) + f(x_1)}{2} + \cdots + \frac{f(x_{n-1}) + f(x_n)}{2}\right]$$

$$= \left(\frac{b-a}{2n}\right)[f(x_0) + f(x_1) + f(x_1) + f(x_2) + \cdots + f(x_{n-1}) + f(x_n)]$$

$$= \left(\frac{b-a}{2n}\right)[f(x_0) + 2f(x_1) + 2f(x_2) + \cdots + 2f(x_{n-1}) + f(x_n)].$$

The area of the first trapezoid is
$$\left[\frac{f(x_0) + f(x_1)}{2}\right]\left(\frac{b-a}{n}\right).$$
Figure 4.42

Letting $\Delta x = (b - a)/n$, you can take the limit as $n \to \infty$ to obtain

$$\lim_{n \to \infty} \left(\frac{b-a}{2n}\right)[f(x_0) + 2f(x_1) + \cdots + 2f(x_{n-1}) + f(x_n)]$$

$$= \lim_{n \to \infty}\left[\frac{[f(a) - f(b)]\,\Delta x}{2} + \sum_{i=1}^{n} f(x_i)\,\Delta x\right]$$

$$= \lim_{n \to \infty}\frac{[f(a) - f(b)](b - a)}{2n} + \lim_{n \to \infty}\sum_{i=1}^{n} f(x_i)\,\Delta x$$

$$= 0 + \int_a^b f(x)\,dx.$$

The result is summarized in the following theorem.

> **THEOREM 4.16   The Trapezoidal Rule**
>
> Let f be continuous on $[a, b]$. The Trapezoidal Rule for approximating $\int_a^b f(x)\,dx$ is given by
>
> $$\int_a^b f(x)\,dx \approx \frac{b-a}{2n}\left[f(x_0) + 2f(x_1) + 2f(x_2) + \cdots + 2f(x_{n-1}) + f(x_n)\right].$$
>
> Moreover, as $n \to \infty$, the right-hand side approaches $\int_a^b f(x)\,dx$.

NOTE   Observe that the coefficients in the Trapezoidal Rule have the following pattern.

$$1 \quad 2 \quad 2 \quad 2 \quad \cdots \quad 2 \quad 2 \quad 1$$

Example 1   Approximation with the Trapezoidal Rule

Use the Trapezoidal Rule to approximate

$$\int_0^\pi \sin x\,dx.$$

Compare the results for $n = 4$ and $n = 8$, as shown in Figure 4.43.

Solution   When $n = 4$, $\Delta x = \pi/4$, and you obtain

$$\int_0^\pi \sin x\,dx \approx \frac{\pi}{8}\left(\sin 0 + 2\sin\frac{\pi}{4} + 2\sin\frac{\pi}{2} + 2\sin\frac{3\pi}{4} + \sin\pi\right)$$

$$= \frac{\pi}{8}\left(0 + \sqrt{2} + 2 + \sqrt{2} + 0\right) = \frac{\pi(1 + \sqrt{2})}{4} \approx 1.896.$$

When $n = 8$, $\Delta x = \pi/8$, and you obtain

$$\int_0^\pi \sin x\,dx \approx \frac{\pi}{16}\left(\sin 0 + 2\sin\frac{\pi}{8} + 2\sin\frac{\pi}{4} + 2\sin\frac{3\pi}{8} + 2\sin\frac{\pi}{2}\right.$$

$$\left. + 2\sin\frac{5\pi}{8} + 2\sin\frac{3\pi}{4} + 2\sin\frac{7\pi}{8} + \sin\pi\right)$$

$$= \frac{\pi}{16}\left(2 + 2\sqrt{2} + 4\sin\frac{\pi}{8} + 4\sin\frac{3\pi}{8}\right) \approx 1.974.$$

For this particular integral, you could have found an antiderivative and determined that the exact area of the region is 2.

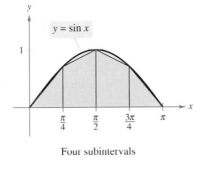

Four subintervals

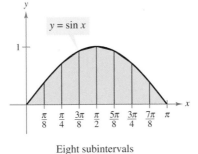

Eight subintervals

Trapezoidal approximations
Figure 4.43

> **TECHNOLOGY**   Most graphing utilities and computer algebra systems have built-in programs that can be used to approximate the value of a definite integral. Try using such a program to approximate the integral in Example 1. How close is your approximation?
>
> When you use such a program, you need to be aware of its limitations. Often, you are given no indication of the degree of accuracy of the approximation. Other times, you may be given an approximation that is completely wrong. For instance, try using a built-in numerical integration program to evaluate
>
> $$\int_{-1}^2 \frac{1}{x}\,dx.$$
>
> Your calculator should give an error message. Does yours?

It is interesting to compare the Trapezoidal Rule with the Midpoint Rule given in Section 4.2 (Exercises 63–66). For the Trapezoidal Rule, you average the function values at the endpoints of the subintervals, but for the Midpoint Rule you take the function values of the subinterval midpoints.

$$\int_a^b f(x)\, dx \approx \sum_{i=1}^{n} f\left(\frac{x_i + x_{i-1}}{2}\right) \Delta x \qquad \text{Midpoint Rule}$$

$$\int_a^b f(x)\, dx \approx \sum_{i=1}^{n} \left(\frac{f(x_i) + f(x_{i-1})}{2}\right) \Delta x \qquad \text{Trapezoidal Rule}$$

NOTE   There are two important points that should be made concerning the Trapezoidal Rule (or the Midpoint Rule). First, the approximation tends to become more accurate as n increases. For instance, in Example 1, if $n = 16$, the Trapezoidal Rule yields an approximation of 1.994. Second, although you could have used the Fundamental Theorem to evaluate the integral in Example 1, this theorem cannot be used to evaluate an integral as simple as $\int_0^\pi \sin x^2\, dx$ because $\sin x^2$ has no elementary antiderivative. Yet, the Trapezoidal Rule can be applied easily to this integral.

Simpson's Rule

One way to view the trapezoidal approximation of a definite integral is to say that on each subinterval you approximate f by a *first*-degree polynomial. In Simpson's Rule, named after the English mathematician Thomas Simpson (1710–1761), you take this procedure one step further and approximate f by *second*-degree polynomials.

Before presenting Simpson's Rule, we list a theorem for evaluating integrals of polynomials of degree 2 (or less).

THEOREM 4.17   Integral of $p(x) = Ax^2 + Bx + C$

If $p(x) = Ax^2 + Bx + C$, then

$$\int_a^b p(x)\, dx = \left(\frac{b-a}{6}\right)\left[p(a) + 4p\left(\frac{a+b}{2}\right) + p(b)\right].$$

Proof

$$\int_a^b p(x)\, dx = \int_a^b (Ax^2 + Bx + C)\, dx$$

$$= \left[\frac{Ax^3}{3} + \frac{Bx^2}{2} + Cx\right]_a^b$$

$$= \frac{A(b^3 - a^3)}{3} + \frac{B(b^2 - a^2)}{2} + C(b - a)$$

$$= \left(\frac{b-a}{6}\right)\left[2A(a^2 + ab + b^2) + 3B(b + a) + 6C\right]$$

By expansion and collection of terms, the expression inside the brackets becomes

$$\underbrace{(Aa^2 + Ba + C)}_{p(a)} + \underbrace{4\left[A\left(\frac{b+a}{2}\right)^2 + B\left(\frac{b+a}{2}\right) + C\right]}_{4p\left(\frac{a+b}{2}\right)} + \underbrace{(Ab^2 + Bb + C)}_{p(b)}$$

and you can write

$$\int_a^b p(x)\, dx = \left(\frac{b-a}{6}\right)\left[p(a) + 4p\left(\frac{a+b}{2}\right) + p(b)\right].$$

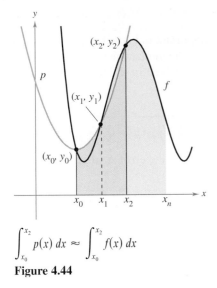

$$\int_{x_0}^{x_2} p(x)\,dx \approx \int_{x_0}^{x_2} f(x)\,dx$$

Figure 4.44

To develop Simpson's Rule for approximating a definite integral, you again partition the interval $[a, b]$ into n subintervals, each of width $\Delta x = (b - a)/n$. This time, however, n is required to be even, and the subintervals are grouped in pairs such that

$$a = \underbrace{x_0 < x_1 < x_2}_{[x_0, x_2]} < \underbrace{x_3 < x_4}_{[x_2, x_4]} < \cdots < \underbrace{x_{n-2} < x_{n-1} < x_n}_{[x_{n-2}, x_n]} = b.$$

On each (double) subinterval $[x_{i-2}, x_i]$, you can approximate f by a polynomial p of degree less than or equal to 2. (See Exercise 46.) For example, on the subinterval $[x_0, x_2]$, choose the polynomial of least degree passing through the points (x_0, y_0), (x_1, y_1), and (x_2, y_2), as shown in Figure 4.44. Now, using p as an approximation of f on this subinterval, you have, by Theorem 4.17,

$$\int_{x_0}^{x_2} f(x)\,dx \approx \int_{x_0}^{x_2} p(x)\,dx = \frac{x_2 - x_0}{6}\left[p(x_0) + 4p\left(\frac{x_0 + x_2}{2}\right) + p(x_2)\right]$$

$$= \frac{2[(b-a)/n]}{6}\left[p(x_0) + 4p(x_1) + p(x_2)\right]$$

$$= \frac{b-a}{3n}\left[f(x_0) + 4f(x_1) + f(x_2)\right].$$

Repeating this procedure on the entire interval $[a, b]$ produces the following theorem.

THEOREM 4.18 Simpson's Rule (n is even)

Let f be continuous on $[a, b]$. Simpson's Rule for approximating $\int_a^b f(x)\,dx$ is

$$\int_a^b f(x)\,dx \approx \frac{b-a}{3n}\lfloor f(x_0) + 4f(x_1) + 2f(x_2) + 4f(x_3) + \cdots$$

$$+ 4f(x_{n-1}) + f(x_n)\rfloor.$$

Moreover, as $n \to \infty$, the right-hand side approaches $\int_a^b f(x)\,dx$.

NOTE Observe that the coefficients in Simpson's Rule have the following pattern.

1 4 2 4 2 4 . . . 4 2 4 1

In Example 1, the Trapezoidal Rule was used to estimate $\int_0^\pi \sin x\,dx$. In the next example, Simpson's Rule is applied to the same integral.

 Example 2 **Approximation with Simpson's Rule**

NOTE In Example 1, the Trapezoidal Rule with $n = 8$ approximated $\int_0^\pi \sin x\,dx$ as 1.974. In Example 2, Simpson's Rule with $n = 8$ gave an approximation of 2.0003. The antiderivative would produce the true value of 2.

Use Simpson's Rule to approximate

$$\int_0^\pi \sin x\,dx.$$

Compare the results for $n = 4$ and $n = 8$.

Solution When $n = 4$, you have

$$\int_0^\pi \sin x\,dx \approx \frac{\pi}{12}\left(\sin 0 + 4\sin\frac{\pi}{4} + 2\sin\frac{\pi}{2} + 4\sin\frac{3\pi}{4} + \sin\pi\right)$$

$$\approx 2.005.$$

When $n = 8$, you have $\displaystyle\int_0^\pi \sin x\,dx \approx 2.0003.$

Error Analysis

If you must use an approximation technique, it is important to know how accurate you can expect the approximation to be. The following theorem, which we list without proof, gives the formulas for estimating the errors involved in the use of Simpson's Rule and the Trapezoidal Rule.

THEOREM 4.19   Errors in the Trapezoidal Rule and Simpson's Rule

If f has a continuous second derivative on $[a, b]$, then the error E in approximating $\int_a^b f(x)\, dx$ by the Trapezoidal Rule is

$$E \le \frac{(b-a)^3}{12n^2}\left[\max \left|f''(x)\right|\right], \quad a \le x \le b. \qquad \text{Trapezoidal Rule}$$

Moreover, if f has a continuous fourth derivative on $[a, b]$, then the error E in approximating $\int_a^b f(x)\, dx$ by Simpson's Rule is

$$E \le \frac{(b-a)^5}{180n^4}\left[\max \left|f^{(4)}(x)\right|\right], \quad a \le x \le b. \qquad \text{Simpson's Rule}$$

TECHNOLOGY If you have access to a computer algebra system, try using it to evaluate the definite integral in Example 3. You should obtain a value of

$$\int_0^1 \sqrt{1+x^2}\, dx = \tfrac{1}{2}\left[\sqrt{2} + \ln\left(1+\sqrt{2}\right)\right]$$

$$\approx 1.14779.$$

("ln" represents the natural logarithmic function, which you will study in Section 5.1.)

Theorem 4.19 states that the errors generated by the Trapezoidal Rule and Simpson's Rule have upper bounds dependent on the extreme values of $f''(x)$ and $f^{(4)}(x)$ in the interval $[a, b]$. Furthermore, these errors can be made arbitrarily small by *increasing n*, provided that f'' and $f^{(4)}$ are continuous and therefore bounded in $[a, b]$.

Example 3   **The Approximate Error in the Trapezoidal Rule**

Determine a value of n such that the Trapezoidal Rule will approximate the value of $\int_0^1 \sqrt{1+x^2}\, dx$ with an error that is less than 0.01.

Solution Begin by letting $f(x) = \sqrt{1+x^2}$ and finding the second derivative of f.

$$f'(x) = x(1+x^2)^{-1/2} \quad \text{and} \quad f''(x) = (1+x^2)^{-3/2}$$

The maximum value of $\left|f''(x)\right|$ on the interval $[0, 1]$ is $\left|f''(0)\right| = 1$. So, by Theorem 4.19, you can write

$$E \le \frac{(b-a)^3}{12n^2}\left|f''(0)\right| = \frac{1}{12n^2}(1) = \frac{1}{12n^2}.$$

To obtain an error E that is less than 0.01, you must choose n such that $1/(12n^2) \le 1/100$.

$$100 \le 12n^2 \quad \Longrightarrow \quad n \ge \sqrt{\tfrac{100}{12}} \approx 2.89$$

Therefore, you can choose $n = 3$ (because n must be greater than or equal to 2.89) and apply the Trapezoidal Rule, as shown in Figure 4.45, to obtain

$$\int_0^1 \sqrt{1+x^2}\, dx \approx \frac{1}{6}\sqrt{1+0^2} + 2\sqrt{1+\left(\tfrac{1}{3}\right)^2} + 2\sqrt{1+\left(\tfrac{2}{3}\right)^2} + \sqrt{1+1^2}$$

$$\approx 1.154.$$

So, with an error no larger than 0.01, you know that

$$1.144 \le \int_0^1 \sqrt{1+x^2}\, dx \le 1.164.$$

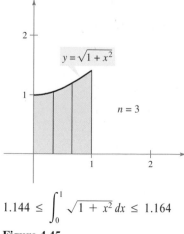

$$1.144 \le \int_0^1 \sqrt{1+x^2}\, dx \le 1.164$$

Figure 4.45

EXERCISES FOR SECTION 4.6

In Exercises 1–10, use the Trapezoidal Rule and Simpson's Rule to approximate the value of the definite integral for the indicated value of n. Round your answer to four decimal places and compare the results with the exact value of the definite integral.

1. $\int_0^2 x^2 \, dx, \quad n = 4$

2. $\int_0^1 \left(\frac{x^2}{2} + 1 \right) dx, \quad n = 4$

3. $\int_0^2 x^3 \, dx, \quad n = 4$

4. $\int_1^2 \frac{1}{x^2} \, dx, \quad n = 4$

5. $\int_0^2 x^3 \, dx, \quad n = 8$

6. $\int_0^8 \sqrt[3]{x} \, dx, \quad n = 8$

7. $\int_4^9 \sqrt{x} \, dx, \quad n = 8$

8. $\int_1^3 (4 - x^2) \, dx, \quad n = 4$

9. $\int_1^2 \frac{1}{(x + 1)^2} \, dx, \quad n = 4$

10. $\int_0^2 x\sqrt{x^2 + 1} \, dx, \quad n = 4$

In Exercises 11–20, approximate the definite integral using the Trapezoidal Rule and Simpson's Rule with $n = 4$. Compare these results with the approximation of the integral using a graphing utility.

9 - 24

11. $\int_0^2 \sqrt{1 + x^3} \, dx$

12. $\int_0^2 \frac{1}{\sqrt{1 + x^3}} \, dx$

13. $\int_0^1 \sqrt{x} \sqrt{1 - x} \, dx$

14. $\int_{\pi/2}^{\pi} \sqrt{x} \sin x \, dx$

15. $\int_0^{\sqrt{\pi/2}} \cos x^2 \, dx$

16. $\int_0^{\sqrt{\pi/4}} \tan x^2 \, dx$

17. $\int_1^{1.1} \sin x^2 \, dx$

18. $\int_0^{\pi/2} \sqrt{1 + \cos^2 x} \, dx$

19. $\int_0^{\pi/4} x \tan x \, dx$

20. $\int_0^{\pi} f(x) \, dx, \quad f(x) = \begin{cases} \dfrac{\sin x}{x}, & x > 0 \\ 1, & x = 0 \end{cases}$

Getting at the Concept

21. If the function f is concave upward on the interval $[a, b]$, will the Trapezoidal Rule yield a result greater than or less than $\int_a^b f(x) \, dx$? Explain.

22. The Trapezoidal Rule and Simpson's Rule yield approximations of a definite integral $\int_a^b f(x) \, dx$ based on polynomial approximations of f. What degree polynomial is used for each?

In Exercises 23 and 24, use the error formulas in Theorem 4.19 to estimate the error in approximating the integral, with $n = 4$, using (a) the Trapezoidal Rule and (b) Simpson's Rule.

23. $\int_0^2 x^3 \, dx$

24. $\int_0^1 \frac{1}{x + 1} \, dx$

In Exercises 25 and 26, use the error formulas in Theorem 4.19 to find n such that the error in the approximation of the definite integral is less than 0.00001 using (a) the Trapezoidal Rule and (b) Simpson's Rule.

25. $\int_1^3 \frac{1}{x} \, dx$

26. $\int_0^1 \frac{1}{1 + x} \, dx$

In Exercises 27–30, use a computer algebra system and the error formulas to find n such that the error in the approximation of the definite integral is less than 0.00001 using (a) the Trapezoidal Rule and (b) Simpson's Rule.

27. $\int_0^2 \sqrt{1 + x} \, dx$

28. $\int_0^2 (x + 1)^{2/3} \, dx$

29. $\int_0^1 \tan x^2 \, dx$

30. $\int_0^1 \sin x^2 \, dx$

31. Prove that Simpson's Rule is exact when approximating the integral of a cubic polynomial function, and demonstrate the result for

$$\int_0^1 x^3 \, dx, \quad n = 2.$$

32. Write a program for a graphing utility to approximate a definite integral using the Trapezoidal Rule and Simpson's Rule. Start with the program written in Section 4.3, Exercises 57–60, and note that the Trapezoidal Rule can be written as

$$T(n) = \frac{1}{2}[L(n) + R(n)]$$

and Simpson's Rule can be written as

$$S(n) = \frac{1}{3}[T(n/2) + 2M(n/2)].$$

[Recall that $L(n)$, $M(n)$, and $R(n)$ represent the Riemann sums using the left-hand endpoint, midpoint, and right-hand endpoint of subintervals of equal width.]

In Exercises 33–36, use the program in Exercise 32 to approximate the definite integral and complete the table.

n	$L(n)$	$M(n)$	$R(n)$	$T(n)$	$S(n)$
4					
8					
10					
12					
16					
20					

33. $\int_0^4 \sqrt{2 + 3x^2} \, dx$

34. $\int_0^1 \sqrt{1 - x^2} \, dx$

35. $\int_0^4 \sin \sqrt{x} \, dx$

36. $\int_1^2 \frac{\sin x}{x} \, dx$

37. *Area* Use Simpson's Rule with $n = 14$ to approximate the area of the region bounded by the graphs of $y = \sqrt{x} \cos x$, $y = 0$, $x = 0$, and $x = \pi/2$.

38. *Circumference* The **elliptic integral**

$$8\sqrt{3} \int_0^{\pi/2} \sqrt{1 - \tfrac{2}{3} \sin^2 \theta} \, d\theta$$

gives the circumference of an ellipse. Use Simpson's Rule with $n = 8$ to approximate the circumference.

39. *Work* To determine the size of the motor required to operate a press, a company must know the amount of work done when the press moves an object linearly 5 feet. The variable force to move the object is

$$F(x) = 100x\sqrt{125 - x^3}$$

where F is given in pounds and x gives the position of the unit in feet. Use Simpson's Rule with $n = 12$ to approximate the work W (in foot-pounds) done through one cycle if

$$W = \int_0^5 F(x) \, dx.$$

40. The table lists several measurements gathered in an experiment to approximate an unknown continuous function $y = f(x)$.

(a) Approximate the integral $\int_0^2 f(x) \, dx$ using the Trapezoidal Rule and Simpson's Rule.

x	0.00	0.25	0.50	0.75	1.00
y	4.32	4.36	4.58	5.79	6.14

x	1.25	1.50	1.75	2.00
y	7.25	7.64	8.08	8.14

(b) Use a graphing utility to find a model of the form $y = ax^3 + bx^2 + cx + d$ for the data. Integrate the resulting polynomial over $[0, 2]$ and compare the result with part (a).

Approximation of Pi In Exercises 41 and 42, use Simpson's Rule with $n = 6$ to approximate π using the given equation. (In Section 5.9, you will be able to evaluate the integral using inverse trigonometric functions.)

41. $\pi = \displaystyle\int_0^{1/2} \frac{6}{\sqrt{1 - x^2}} \, dx$

42. $\pi = \displaystyle\int_0^1 \frac{4}{1 + x^2} \, dx$

Area In Exercises 43 and 44, use the Trapezoidal Rule to estimate the number of square meters of land in a lot where x and y are measured in meters, as shown in the figures. The land is bounded by a stream and two straight roads that meet at right angles.

43.

x	y
0	125
100	125
200	120
300	112
400	90
500	90
600	95
700	88
800	75
900	35
1000	0

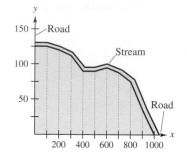

44.

x	y
0	75
10	81
20	84
30	76
40	67
50	68
60	69
70	72
80	68
90	56
100	42
110	23
120	0

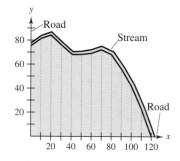

45. Use Simpson's Rule with $n = 10$ and a computer algebra system to approximate t in the integral equation

$$\int_0^t \sin \sqrt{x} \, dx = 2.$$

46. Prove that you can find a polynomial $p(x) = Ax^2 + Bx + C$ that passes through any three points (x_1, y_1), (x_2, y_2), and (x_3, y_3), where the x_i are distinct.

REVIEW EXERCISES FOR CHAPTER 4

4.1 In Exercises 1 and 2, use the graph of f' to sketch a graph of f. To print an enlarged copy of the graph, go to the website *www.mathgraphs.com*.

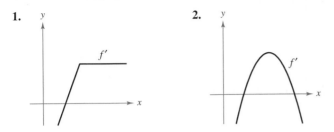

1.   2.

In Exercises 3–8, find the indefinite integral.

3. $\displaystyle\int (2x^2 + x - 1)\, dx$

4. $\displaystyle\int \frac{2}{\sqrt[3]{3x}}\, dx$

5. $\displaystyle\int \frac{x^3 + 1}{x^2}\, dx$

6. $\displaystyle\int \frac{x^3 - 2x^2 + 1}{x^2}\, dx$

7. $\displaystyle\int (4x - 3 \sin x)\, dx$

8. $\displaystyle\int (5 \cos x - 2 \sec^2 x)\, dx$

9. Find the particular solution of the differential equation $f'(x) = -2x$ whose graph passes through the point $(-1, 1)$.

10. Find the particular solution of the differential equation $f''(x) = 6(x - 1)$ whose graph passes through the point $(2, 1)$ and is tangent to the line $3x - y - 5 = 0$ at that point.

11. **Velocity and Acceleration** An airplane taking off from a runway travels 3600 feet before lifting off. If it starts from rest, moves with constant acceleration, and makes the run in 30 seconds, with what speed does it lift off?

12. **Velocity and Acceleration** The speed of a car traveling in a straight line is reduced from 45 to 30 miles per hour in a distance of 264 feet. Find the distance in which the car can be brought to rest from 30 miles per hour, assuming the same constant deceleration.

13. **Velocity and Acceleration** A ball is thrown vertically upward from ground level with an initial velocity of 96 feet per second.

    (a) How long will it take the ball to rise to its maximum height?

    (b) What is the maximum height?

    (c) When is the velocity of the ball one-half the initial velocity?

    (d) What is the height of the ball when its velocity is one-half the initial velocity?

14. **Velocity and Acceleration** Repeat Exercise 13 for an initial velocity of 40 meters per second.

4.2

15. Write in sigma notation (a) the sum of the first ten positive odd integers, (b) the sum of the cubes of the first n positive integers, and (c) $6 + 10 + 14 + 18 + \cdots + 42$.

16. Evaluate each sum for $x_1 = 2,\ x_2 = -1,\ x_3 = 5,\ x_4 = 3,$ and $x_5 = 7$.

    (a) $\displaystyle\frac{1}{5}\sum_{i=1}^{5} x_i$ (b) $\displaystyle\sum_{i=1}^{5} \frac{1}{x_i}$

    (c) $\displaystyle\sum_{i=1}^{5} (2x_i - x_i^2)$ (d) $\displaystyle\sum_{i=2}^{5} (x_i - x_{i-1})$

In Exercises 17 and 18, use upper and lower sums to approximate the area of the region using the indicated number of subintervals of equal width.

17. $y = \dfrac{10}{x^2 + 1}$ 18. $y = 9 - \tfrac{1}{4}x^2$

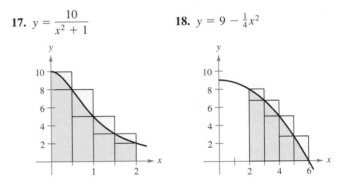

In Exercises 19–22, use the limit process to find the area of the region between the graph of the function and the x-axis over the indicated interval. Sketch the region.

Function	Interval
19. $y = 6 - x$	$[0, 4]$
20. $y = x^2 + 3$	$[0, 2]$
21. $y = 5 - x^2$	$[-2, 1]$
22. $y = \tfrac{1}{4}x^3$	$[2, 4]$

23. Use the limit process to find the area of the region bounded by $x = 5y - y^2,\ x = 0,\ y = 2,$ and $y = 5$.

24. Consider the region bounded by $y = mx,\ y = 0,\ x = 0,$ and $x = b$.

    (a) Find the upper and lower sums to approximate the area of the region when $\Delta x = b/4$.

    (b) Find the upper and lower sums to approximate the area of the region when $\Delta x = b/n$.

    (c) Find the area of the region by letting n approach infinity in both sums in part (b). Show that in each case you obtain the formula for the area of a triangle.

4.3 In Exercises 25 and 26, express the limit as a definite integral on the interval $[a, b]$, where c_i is any point in the ith subinterval.

Limit	Interval
25. $\displaystyle\lim_{\|\Delta\|\to 0}\sum_{i=1}^{n}(2c_i - 3)\Delta x_i$	$[4, 6]$
26. $\displaystyle\lim_{\|\Delta\|\to 0}\sum_{i=1}^{n}3c_i(9 - c_i^2)\Delta x_i$	$[1, 3]$

In Exercises 27 and 28, sketch the region whose area is given by the definite integral. Then use a geometric formula to evaluate the integral.

27-30

27. $\displaystyle\int_0^5 (5 - |x - 5|)\, dx$

28. $\displaystyle\int_{-4}^4 \sqrt{16 - x^2}\, dx$

In Exercises 29 and 30, use the given values to evaluate each definite integral.

29. If $\displaystyle\int_2^6 f(x)\, dx = 10$ and $\displaystyle\int_2^6 g(x)\, dx = 3$, find

(a) $\displaystyle\int_2^6 [f(x) + g(x)]\, dx.$

(b) $\displaystyle\int_2^6 [f(x) - g(x)]\, dx.$

(c) $\displaystyle\int_2^6 [2f(x) - 3g(x)]\, dx.$

(d) $\displaystyle\int_2^6 5f(x)\, dx.$

30. If $\displaystyle\int_0^3 f(x)\, dx = 4$ and $\displaystyle\int_3^6 f(x)\, dx = -1$, find

(a) $\displaystyle\int_0^6 f(x)\, dx.$

(b) $\displaystyle\int_6^3 f(x)\, dx.$

(c) $\displaystyle\int_4^4 f(x)\, dx.$

(d) $\displaystyle\int_3^6 -10 f(x)\, dx.$

4.4 In Exercises 31 and 32, select the correct value of the definite integral.

31. $\displaystyle\int_1^8 (\sqrt[3]{x} + 1)\, dx$

(a) $\frac{81}{4}$ (b) $\frac{331}{12}$

(c) $\frac{73}{4}$ (d) $\frac{355}{12}$

32. $\displaystyle\int_1^3 \frac{12}{x^3}\, dx$

(a) $\frac{320}{9}$ (b) $-\frac{16}{3}$

(c) $-\frac{5}{9}$ (d) $\frac{16}{3}$

In Exercises 33–40, use the Fundamental Theorem of Calculus to evaluate the definite integral.

33. $\displaystyle\int_0^4 (2 + x)\, dx$

34. $\displaystyle\int_{-1}^1 (t^2 + 2)\, dt$

35. $\displaystyle\int_{-1}^1 (4t^3 - 2t)\, dt$

36. $\displaystyle\int_{-2}^2 (x^4 + 2x^2 - 5)\, dx$

37. $\displaystyle\int_4^9 x\sqrt{x}\, dx$

38. $\displaystyle\int_1^2 \left(\frac{1}{x^2} - \frac{1}{x^3}\right) dx$

39. $\displaystyle\int_0^{3\pi/4} \sin\theta\, d\theta$

40. $\displaystyle\int_{-\pi/4}^{\pi/4} \sec^2 t\, dt$

33-40

In Exercises 41–46, sketch the graph of the region whose area is given by the integral, and find the area.

41. $\displaystyle\int_1^3 (2x - 1)\, dx$

42. $\displaystyle\int_0^2 (x + 4)\, dx$

43. $\displaystyle\int_3^4 (x^2 - 9)\, dx$

44. $\displaystyle\int_{-1}^2 (-x^2 + x + 2)\, dx$

45. $\displaystyle\int_0^1 (x - x^3)\, dx$

46. $\displaystyle\int_0^1 \sqrt{x}\,(1 - x)\, dx$

In Exercises 47 and 48, sketch the region bounded by the graphs of the equations, and determine its area.

47. $y = \dfrac{4}{\sqrt{x}}, \quad y = 0, \quad x = 1, \quad x = 9$

48. $y = \sec^2 x, \quad y = 0, \quad x = 0, \quad x = \dfrac{\pi}{3}$

In Exercises 49 and 50, find the average value of the function over the interval. Find the values of x at which the function assumes its average value, and graph the function.

Function	Interval
49. $f(x) = \dfrac{1}{\sqrt{x}}$	$[4, 9]$
50. $f(x) = x^3$	$[0, 2]$

49-54

In Exercises 51–54, use the Second Fundamental Theorem of Calculus to find $F'(x)$.

51. $F(x) = \displaystyle\int_0^x t^2\sqrt{1 + t^3}\, dt$

52. $F(x) = \displaystyle\int_1^x \frac{1}{t^2}\, dt$

53. $F(x) = \displaystyle\int_{-3}^x (t^2 + 3t + 2)\, dt$

54. $F(x) = \displaystyle\int_0^x \csc^2 t\, dt$

4.5 In Exercises 55–68, find the indefinite integral.

55. $\displaystyle\int (x^2 + 1)^3\, dx$

56. $\displaystyle\int \left(x + \frac{1}{x}\right)^2 dx$

57. $\displaystyle\int \frac{x^2}{\sqrt{x^3 + 3}}\, dx$

58. $\displaystyle\int x^2\sqrt{x^3 + 3}\, dx$

59. $\displaystyle\int x(1 - 3x^2)^4\, dx$

60. $\displaystyle\int \frac{x + 3}{(x^2 + 6x - 5)^2}\, dx$

61. $\displaystyle\int \sin^3 x \cos x\, dx$

62. $\displaystyle\int x \sin 3x^2\, dx$

63. $\displaystyle\int \frac{\sin\theta}{\sqrt{1 - \cos\theta}}\, d\theta$

64. $\displaystyle\int \frac{\cos x}{\sqrt{\sin x}}\, dx$

65. $\displaystyle\int \tan^n x \sec^2 x\, dx, \qquad n \neq -1$

66. $\displaystyle\int \sec 2x \tan 2x\, dx$

55-68

67. $\displaystyle\int (1 + \sec \pi x)^2 \sec \pi x\, \tan \pi x\, dx$

68. $\displaystyle\int \cot^4 \alpha \csc^2 \alpha\, d\alpha$

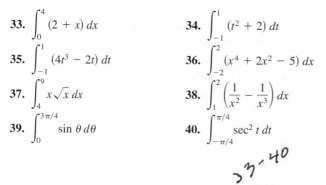

In Exercises 69–76, evaluate the definite integral. Use a graphing utility to verify your result.

69. $\displaystyle\int_{-1}^{2} x(x^2 - 4)\, dx$

70. $\displaystyle\int_{0}^{1} x^2(x^3 + 1)^3\, dx$

71. $\displaystyle\int_{0}^{3} \frac{1}{\sqrt{1 + x}}\, dx$

72. $\displaystyle\int_{3}^{6} \frac{x}{3\sqrt{x^2 - 8}}\, dx$

73. $2\pi\displaystyle\int_{0}^{1} (y + 1)\sqrt{1 - y}\, dy$

74. $2\pi\displaystyle\int_{-1}^{0} x^2\sqrt{x + 1}\, dx$

75. $\displaystyle\int_{0}^{\pi} \cos\frac{x}{2}\, dx$

76. $\displaystyle\int_{-\pi/4}^{\pi/4} \sin 2x\, dx$

Probability   **In Exercises 77 and 78, the function**

$$f(x) = kx^n\,(1 - x)^m, \qquad 0 \le x \le 1$$

where $n > 0$, $m > 0$, and k is a constant, can be used to represent various probability distributions. If k is chosen such that

$$\int_{0}^{1} f(x)\, dx = 1$$

the probability that x will fall between a and b ($0 \le a \le b \le 1$) is

$$P_{a,b} = \int_{a}^{b} f(x)\, dx.$$

77. The probability that a person will remember between $a\%$ and $b\%$ of material learned in a certain experiment is

$$P_{a,b} = \int_{a}^{b} \frac{15}{4} x\sqrt{1 - x}\, dx$$

where x represents the percent remembered. (See figure.)

(a) For a randomly chosen individual, what is the probability that he or she will recall between 50% and 75% of the material?

(b) What is the median percent recall? That is, for what value of b is it true that the probability from 0 to b is 0.5?

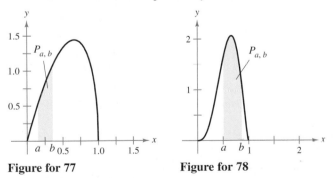

Figure for 77   **Figure for 78**

78. The probability that ore samples taken from a certain region contain between $a\%$ and $b\%$ iron is

$$P_{a,b} = \int_{a}^{b} \frac{1155}{32} x^3(1 - x)^{3/2}\, dx$$

where x represents the percent of iron. (See figure.) What is the probability that a sample will contain between

(a) 0% and 25% iron?

(b) 50% and 100% iron?

79. Suppose that gasoline is increasing in price according to the equation

$$p = 1.20 + 0.04t$$

where p is the dollar price per gallon and t is the time in years, with $t = 0$ representing 1990. If an automobile is driven 15,000 miles a year and gets M miles per gallon, the annual fuel cost is

$$C = \frac{15{,}000}{M} \int_{t}^{t+1} p\, dt.$$

Estimate the annual fuel cost for the year (a) 2000 and (b) 2005.

80. *Respiratory Cycle*   After exercising for a few minutes, a person has a respiratory cycle for which the rate of air intake is

$$v = 1.75 \sin\frac{\pi t}{2}.$$

Find the volume, in liters, of air inhaled during one cycle by integrating the function over the interval $[0, 2]$.

4.6   **In Exercises 81–84, use the Trapezoidal Rule and Simpson's Rule with $n = 4$, and use the integration capabilities of a graphing utility, to approximate the definite integral. Compare the results.**

81. $\displaystyle\int_{1}^{2} \frac{1}{1 + x^3}\, dx$

82. $\displaystyle\int_{0}^{1} \frac{x^{3/2}}{3 - x^2}\, dx$

83. $\displaystyle\int_{0}^{\pi/2} \sqrt{x}\cos x\, dx$

84. $\displaystyle\int_{0}^{\pi} \sqrt{1 + \sin^2 x}\, dx$

85. Let

$$I = \int_{0}^{4} f(x)\, dx$$

where f is shown in the figure. Let $L(n)$ and $R(n)$ represent the Riemann sums using the left-hand endpoint and right-hand endpoint of n subintervals of equal width. (Assume n is even.) Let $T(n)$ and $S(n)$ be the corresponding values of the Trapezoidal Rule and Simpson's Rule.

(a) For any n, list $L(n)$, $R(n)$, $T(n)$, and I in increasing order.

(b) Approximate $S(4)$.

81–85

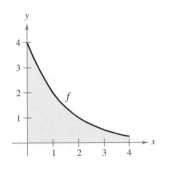

P.S. Problem Solving

1. Let $L(x) = \displaystyle\int_1^x \frac{1}{t}\,dt,\ x > 0$.

    (a) Find $L(1)$.

    (b) Find $L'(x)$ and $L'(1)$.

    (c) Use a graphing utility to approximate the value of x (to three decimal places) for which $L(x) = 1$.

    (d) Prove that $L(x_1 x_2) = L(x_1) + L(x_2)$ for all positive values of x_1 and x_2.

2. Let $F(x) = \displaystyle\int_2^x \sin t^2\,dt$.

    (a) Use a graphing utility to complete the table.

x	0	1.0	1.5	1.9	2.0
$F(x)$					
x	2.1	2.5	3.0	4.0	5.0
$F(x)$					

    (b) Let $G(x) = \dfrac{1}{x-2} F(x) = \dfrac{1}{x-2}\displaystyle\int_2^x \sin t^2\,dt$. Use a graphing utility to complete the table and estimate $\displaystyle\lim_{x\to 2} G(x)$.

x	1.9	1.95	1.99	2.01	2.1
$G(x)$					

    (c) Use the definition of the derivative to find the exact value of the limit $\displaystyle\lim_{x\to 2} G(x)$.

3. The **Fresnel function S** is defined by the integral

$$S(x) = \int_0^x \sin\!\left(\frac{\pi t^2}{2}\right) dt.$$

    (a) Graph the function $y = \sin\!\left(\dfrac{\pi x^2}{2}\right)$ on the interval $[0, 3]$.

    (b) Use the graph in part (a) to sketch the graph of S on the interval $[0, 3]$.

    (c) Locate all relative extrema of S on the interval $(0, 3)$.

    (d) Locate all points of inflection of S on the interval $(0, 3)$.

4. Galileo Galilei (1564–1642) stated the following proposition concerning falling objects:

    The time in which any space is traversed by a uniformly accelerating body is equal to the time in which that same space would be traversed by the same body moving at a uniform speed whose value is the mean of the highest speed of the accelerating body and the speed just before acceleration began.

    Use the techniques of this section to verify this proposition.

5. The graph of the function f consists of the three line segments joining the points $(0, 0)$, $(2, -2)$, $(6, 2)$, and $(8, 3)$. The function F is defined by the integral

$$F(x) = \int_0^x f(t)\,dt.$$

    (a) Sketch the graph of f.

    (b) Complete the table of values.

x	0	1	2	3	4	5	6	7	8
$F(x)$									

    (c) Find the extrema of F on the interval $[0, 8]$.

    (d) Determine all points on inflection of inflection of F on the interval $(0, 8)$.

6. A car is traveling in a straight line for one hour. Its velocity v in miles per hour at six-minute intervals is shown in the table.

t (hours)	0	0.1	0.2	0.3	0.4	0.5
v (mi/hr)	0	10	20	40	60	50

t (hours)	0.6	0.7	0.8	0.9	1.0
v (mi/hr)	40	35	40	50	65

    (a) Produce a reasonable graph of the velocity function v by graphing these points and connecting them with a smooth curve.

    (b) Find the open intervals over which the acceleration a is positive.

    (c) Find the average acceleration of the car (in miles per hour squared) over the interval $[0, 0.4]$.

    (d) What does the integral $\displaystyle\int_0^1 v(t)\,dt$ signify? Approximate this integral using the Trapezoidal Rule with five subintervals.

    (e) Approximate the acceleration at $t = 0.8$.

7. The **Two-Point Gaussian Quadrature Approximation** for f is

$$\int_{-1}^1 f(x)\,dx \approx f\!\left(-\frac{1}{\sqrt{3}}\right) + f\!\left(\frac{1}{\sqrt{3}}\right).$$

    (a) Use this formula to approximate

$$\int_{-1}^1 \cos x\,dx.$$

    Find the error of the approximation.

    (b) Use this formula to approximate

$$\int_{-1}^1 \frac{1}{1+x^2}\,dx.$$

    (c) Prove that the Two-Point Gaussian Quadrature Approximation is exact for all polynomials of degree 3 or less.

8. Prove $\displaystyle\int_0^x f(t)(x - t)\, dt = \int_0^x \left(\int_0^t f(v)\, dv \right) dt.$

9. Prove $\displaystyle\int_a^b f(x)f'(x)\, dx = \frac{1}{2}([f(b)]^2 - [f(a)]^2).$

10. Use an appropriate Riemann sum to evaluate the limit

$$\lim_{n\to\infty} \frac{\sqrt{1} + \sqrt{2} + \sqrt{3} + \cdots + \sqrt{n}}{n^{3/2}}.$$

11. Use an appropriate Riemann sum to evaluate the limit

$$\lim_{n\to\infty} \frac{1^5 + 2^5 + 3^5 + \cdots + n^5}{n^6}.$$

12. Archimedes showed that the area of a parabolic arch is equal to $\frac{2}{3}$ the product of the base and the height, as indicated in the figure.

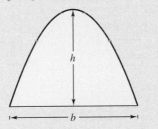

(a) Graph the parabolic arch bounded by $y = 9 - x^2$ and the x-axis. Use an appropriate integral to find the area A.

(b) Find the base and height of the arch and verify Archimedes' formula.

(c) Prove Archimedes' formula for a general parabola.

13. Suppose that f is integrable on $[a, b]$ and $0 < m \le f(x) \le M$ for all x in the interval $[a, b]$. Prove that

$$m(a - b) \le \int_a^b f(x)\, dx \le M(b - a).$$

Use this result to estimate $\displaystyle\int_0^1 \sqrt{1 + x^4}\, dx.$

14. Verify that

$$\sum_{i=1}^{n} i^2 = \frac{n(n + 1)(2n + 1)}{6}$$

by showing the following.

(a) $(1 + i)^3 - i^3 = 3i^2 + 3i + 1$

(b) $(n + 1)^3 = \displaystyle\sum_{i=1}^{n} (3i^2 + 3i + 1) + 1$

(c) $\displaystyle\sum_{i=1}^{n} i^2 = \frac{n(n + 1)(2n + 1)}{6}$

15. Prove that if f is a continuous function on a closed interval $[a, b]$, then

$$\left| \int_a^b f(x)\, dx \right| \le \int_a^b |f(x)|\, dx.$$

16. The temperature in degrees Fahrenheit is

$$T = 72 + 12 \sin\left[\frac{\pi(t - 8)}{12} \right]$$

where t is time in hours, with $t = 0$ representing midnight. Suppose the hourly cost of cooling a house is \$0.10 per degree.

(a) Find the cost C of cooling the house if its thermostat is set at 72°F by evaluating the integral

$$C = 0.1 \int_8^{20} \left[72 + 12 \sin \frac{\pi(t - 8)}{12} - 72 \right] dt. \text{ (See figure.)}$$

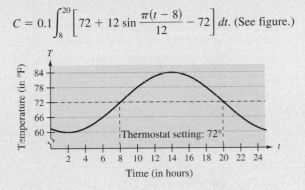

(b) Find the savings from resetting the thermostat to 78°F by evaluating the integral

$$C = 0.1 \int_{10}^{18} \left[72 + 12 \sin \frac{\pi(t - 8)}{12} - 78 \right] dt. \text{ (See figure.)}$$

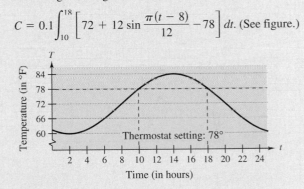

17. A manufacturer of fertilizer finds that national sales of fertilizer follow the seasonal pattern

$$F = 100{,}000 \left[1 + \sin \frac{2\pi(t - 60)}{365} \right]$$

where F is measured in pounds and t is time in days, with $t = 1$ representing January 1. The manufacturer wants to set up a schedule to produce a uniform amount of fertilizer each day. What should this amount be?

18. Let f be continuous on the interval $[0, b]$. Show that

$$\int_0^b \frac{f(x)}{f(x) + f(b - x)}\, dx = \frac{b}{2}.$$

Use this result to evaluate

$$\int_0^1 \frac{\sin x}{\sin(1 - x) + \sin x}\, dx.$$

Plastics and Cooling

What do Corvette fenders, panty hose, and garbage bags have in common? They are all made of plastic. The Greek word *plastikos*, meaning "able to be shaped," was modified to name the most versatile family of materials ever created. Since Bakelite was introduced in 1909, the plastics industry has steadily expanded to the point where today, plastics are used in nearly every aspect of our daily lives.

Several methods are used to shape plastic products, one of the most common being to pour hot, syrupy *plastic resin* into a mold or cast. The temperature of the molten resin is over 300°F. The mold is then cooled in a chiller system that is kept at 58°F before the part is ejected from the mold. To minimize the cost, it helps to eject the parts quickly, allowing the mold to be reused as soon as possible. Yet ejecting the part when it is too hot can cause warping or punctures. The rate at which objects cool is therefore of great interest.

To illustrate the rate of cooling, the *Texas Instruments Calculator-Based Laboratory (CBL) System* was used to measure the temperature of a cup of water over a 40-second period. The room temperature was measured at 69.55°F, and the water temperature at time $t = 0$ was measured at 165.58°F. The results are shown in the following scatter plot.

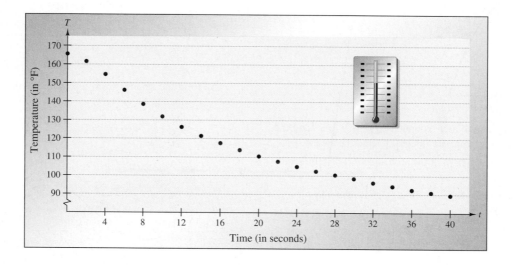

QUESTIONS

1. Describe the pattern of the temperature points over time. Does the *rate* at which the temperature changes seem to increase, decrease, or remain constant?

2. Imagine a curve running through the data points. How would you expect the curve to behave as the value of t increases? Would you expect the curve to intersect the line $T = 69.55$? Explain your reasoning.

3. Would the derivative of a function modeling the data points be increasing, decreasing, or constant? Explain your reasoning.

4. The data in the scatter plot can be modeled using a function of the form

$$T = a \cdot b^t + c.$$

Find values of a, b, and c that produce a reasonable model.

The concepts presented here will be explored further in this chapter. For an extension of this application, see Lab 7 in the lab series that accompanies this text at college.hmco.com.

Logarithmic, Exponential, and Other Transcendental Functions

5

Bob Krist/Tony Stone Worldwide

Plastic resin is produced in very small pieces that are easy to heat to a liquid state. The resin can then go through the injection molding process described on the facing page.

Ron Kimball

In 1953, Chevrolet introduced the Corvette, the first mass-produced automobile with a plastic body.

Bettmann/Corbis

Leo Hendrik Baekeland attempted to create a shellac by combining phenol and formaldehyde. The experiment "failed" in that it did not result in shellac, but it did form the first completely synthetic plastic resin. Bakelite is still used today in the automotive and electronics industries.

The Natural Logarithmic Function: Differentiation

- Develop and use properties of the natural logarithmic function.
- Understand the definition of the number e.
- Find derivatives of functions involving the natural logarithmic function.

JOHN NAPIER (1550–1617)

Logarithms were invented by the Scottish mathematician John Napier. Although he did not introduce the *natural* logarithmic function, it is sometimes called the *Napierian* logarithm.

The Natural Logarithmic Function

Recall that the General Power Rule

$$\int x^n \, dx = \frac{x^{n+1}}{n+1} + C, \qquad n \neq -1 \qquad \text{General Power Rule}$$

has an important disclaimer—it doesn't apply when $n = -1$. Consequently, we have not yet found an antiderivative for the function $f(x) = 1/x$. In this section, we will use the Second Fundamental Theorem of Calculus to *define* such a function. This antiderivative is a function that we have not encountered previously in the text. It is neither algebraic nor trigonometric, but falls into a new class of functions called *logarithmic functions*. This particular function is the **natural logarithmic function.**

Definition of the Natural Logarithmic Function

The **natural logarithmic function** is defined by

$$\ln x = \int_1^x \frac{1}{t} \, dt, \qquad x > 0.$$

The domain of the natural logarithmic function is the set of all positive real numbers.

From the definition, you can see that $\ln x$ is positive for $x > 1$ and negative for $0 < x < 1$, as shown in Figure 5.1. Moreover, $\ln(1) = 0$, because the upper and lower limits of integration are equal when $x = 1$.

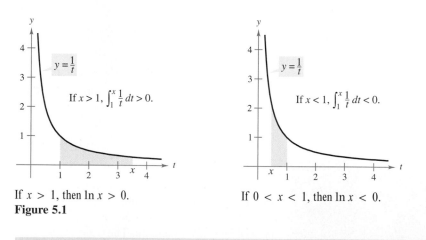

If $x > 1$, then $\ln x > 0$.   If $0 < x < 1$, then $\ln x < 0$.

Figure 5.1

EXPLORATION

Graphing the Natural Logarithmic Function Using *only* the definition of the natural logarithmic function, sketch a graph of the function. Explain your reasoning.

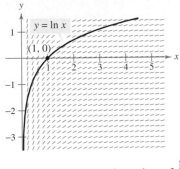

Each small line segment has a slope of $\dfrac{1}{x}$.

Figure 5.2

NOTE   Slope fields can be helpful in getting a visual perspective of the directions of the solutions of a differential equation.

To sketch the graph of $y = \ln x$, you can think of the natural logarithmic function as an *antiderivative* given by the differential equation

$$\frac{dy}{dx} = \frac{1}{x}.$$

Figure 5.2 is a computer-generated graph, called a *slope (or direction) field*, showing small line segments of slope $1/x$. The graph of $y = \ln x$ is the solution that passes through the point $(1, 0)$. For a complete discussion of slope fields, see Appendix A.

The following theorem lists some basic properties of the natural logarithmic function.

THEOREM 5.1   Properties of the Natural Logarithmic Function

The natural logarithmic function has the following properties.

1. The domain is $(0, \infty)$ and the range is $(-\infty, \infty)$.

2. The function is continuous, increasing, and one-to-one.

3. The graph is concave downward.

Proof   The domain of $f(x) = \ln x$ is $(0, \infty)$ by definition. Moreover, the function is continuous because it is differentiable. It is increasing because its derivative

$$f'(x) = \frac{1}{x} \qquad \text{First derivative}$$

is positive for $x > 0$, as shown in Figure 5.3. It is concave downward because

$$f''(x) = -\frac{1}{x^2} \qquad \text{Second derivative}$$

is negative for $x > 0$. We leave the proof that f is one-to-one as an exercise (see Exercise 104). The following limits imply that its range is the entire real line.

$$\lim_{x \to 0^+} \ln x = -\infty \qquad \text{and} \qquad \lim_{x \to \infty} \ln x = \infty$$

Verification of these two limits is given in Appendix B.

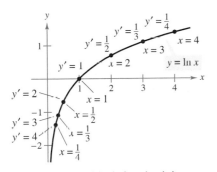

The natural logarithmic function is increasing, and its graph is concave downward.

Figure 5.3

Using the definition of the natural logarithmic function, you can prove several important properties involving operations with natural logarithms. If you are already familiar with logarithms, you will recognize that these properties are characteristic of all logarithms.

THEOREM 5.2   Logarithmic Properties

If a and b are positive numbers and n is rational, then the following properties are true.

1. $\ln(1) = 0$

2. $\ln(ab) = \ln a + \ln b$

3. $\ln(a^n) = n \ln a$

4. $\ln\!\left(\dfrac{a}{b}\right) = \ln a - \ln b$

LOGARITHMS

Napier coined the term *logarithm*, from the two Greek words *logos* (or ratio) and *arithmos* (or number), to describe the theory that he spent 20 years developing and that first appeared in the book *Mirifici Logarithmorum canonis descriptio* (A Description of the Marvelous Rule of Logarithms).

Proof We have already discussed the first property. The proof of the second property follows from the fact that two antiderivatives of the same function differ at most by a constant. From the Second Fundamental Theorem of Calculus and the definition of the natural logarithmic function, you know that

$$\frac{d}{dx}[\ln x] = \frac{d}{dx}\left[\int_1^x \frac{1}{t}\, dt\right] = \frac{1}{x}.$$

So, consider the two derivatives

$$\frac{d}{dx}[\ln(ax)] = \frac{a}{ax} = \frac{1}{x}$$

and

$$\frac{d}{dx}[\ln a + \ln x] = 0 + \frac{1}{x} = \frac{1}{x}.$$

Because $\ln(ax)$ and $(\ln a + \ln x)$ are both antiderivatives of $1/x$, they must differ at most by a constant.

$$\ln(ax) = \ln a + \ln x + C$$

By letting $x = 1$, you can see that $C = 0$. The third property can be proved similarly by comparing the derivatives of $\ln(x^n)$ and $n \ln x$. Finally, using the second and third properties, you can prove the fourth property.

$$\ln\left(\frac{a}{b}\right) = \ln[a(b^{-1})] = \ln a + \ln(b^{-1}) = \ln a - \ln b$$

Example 1 shows how logarithmic properties can be used to expand logarithmic expressions.

Example 1 **Expanding Logarithmic Expressions**

a. $\ln\dfrac{10}{9} = \ln 10 - \ln 9$ Property 4

b. $\ln\sqrt{3x + 2} = \ln(3x + 2)^{1/2}$ Rewrite with rational exponent.

     $= \dfrac{1}{2}\ln(3x + 2)$ Property 3

c. $\ln\dfrac{6x}{5} = \ln(6x) - \ln 5$ Property 4

     $= \ln 6 + \ln x - \ln 5$ Property 2

d. $\ln\dfrac{(x^2 + 3)^2}{x\sqrt[3]{x^2 + 1}} = \ln(x^2 + 3)^2 - \ln\left(x\sqrt[3]{x^2 + 1}\right)$

     $= 2\ln(x^2 + 3) - \left[\ln x + \ln(x^2 + 1)^{1/3}\right]$

     $= 2\ln(x^2 + 3) - \ln x - \ln(x^2 + 1)^{1/3}$

     $= 2\ln(x^2 + 3) - \ln x - \dfrac{1}{3}\ln(x^2 + 1)$

When using the properties of logarithms to rewrite logarithmic functions, you must check to see whether the domain of the rewritten function is the same as the domain of the original. For instance, the domain of $f(x) = \ln x^2$ is all real numbers except $x = 0$, and the domain of $g(x) = 2\ln x$ is all positive real numbers. (See Figure 5.4.)

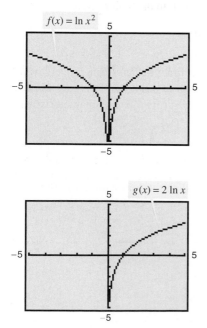

$f(x) = \ln x^2$

$g(x) = 2 \ln x$

Figure 5.4

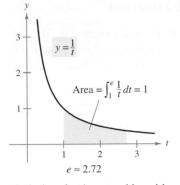

e is the base for the natural logarithm because $\ln e = 1$.
Figure 5.5

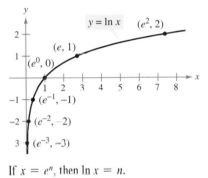

If $x = e^n$, then $\ln x = n$.
Figure 5.6

The Number e

It is likely that you have studied logarithms in an algebra course. There, without the benefit of calculus, logarithms would have been defined in terms of a **base** number. For example, common logarithms have a base of 10 because $\log_{10} 10 = 1$. (We will say more about this in Section 5.5.)

To define the **base for the natural logarithm,** we use the fact that the natural logarithmic function is continuous, is one-to-one, and has a range of $(-\infty, \infty)$. Hence, there must be a unique real number x such that $\ln x = 1$, as shown in Figure 5.5. This number is denoted by the letter e. It can be shown that e is irrational and has the following decimal approximation.

$$e \approx 2.71828182846$$

Definition of e

The letter e denotes the positive real number such that

$$\ln e = \int_1^e \frac{1}{t} \, dt = 1.$$

FOR FURTHER INFORMATION To learn more about the number e, see the article "Unexpected Occurrences of the Number e" by Harris S. Shultz and Bill Leonard in *Mathematics Magazine*. To view this article, go to the website *www.matharticles.com*.

Once you know that $\ln e = 1$, you can use logarithmic properties to evaluate the natural logarithms of several other numbers. For example, by using the property

$$\ln(e^n) = n \ln e$$
$$= n(1)$$
$$= n$$

you can evaluate $\ln(e^n)$ for various powers of n, as shown in the table and in Figure 5.6.

x	$\dfrac{1}{e^3} \approx 0.050$	$\dfrac{1}{e^2} \approx 0.135$	$\dfrac{1}{e} \approx 0.368$	$e^0 = 1$	$e \approx 2.718$	$e^2 \approx 7.389$
$\ln x$	-3	-2	-1	0	1	2

The logarithms given in the table above are convenient because the x-values are integer powers of e. Most logarithmic expressions are, however, best evaluated with a calculator.

Example 2   **Evaluating Natural Logarithmic Expressions**

a. $\ln 2 \approx 0.693$

b. $\ln 32 \approx 3.466$

c. $\ln 0.1 \approx -2.303$

The Derivative of the Natural Logarithmic Function

The derivative of the natural logarithmic function is given in Theorem 5.3. The first part of the theorem follows from the definition of the natural logarithmic function as an antiderivative. The second part of the theorem is simply the Chain Rule version of the first part.

THEOREM 5.3   Derivative of the Natural Logarithmic Function

Let u be a differentiable function of x.

1. $\dfrac{d}{dx}[\ln x] = \dfrac{1}{x}, \ x > 0$ **2.** $\dfrac{d}{dx}[\ln u] = \dfrac{1}{u}\dfrac{du}{dx} = \dfrac{u'}{u}, \ u > 0$

Example 3   Differentiation of Logarithmic Functions

a. $\dfrac{d}{dx}[\ln(2x)] = \dfrac{u'}{u} = \dfrac{2}{2x} = \dfrac{1}{x}$ $u = 2x$

b. $\dfrac{d}{dx}[\ln(x^2 + 1)] = \dfrac{u'}{u} = \dfrac{2x}{x^2 + 1}$ $u = x^2 + 1$

c. $\dfrac{d}{dx}[x \ln x] = x\left(\dfrac{d}{dx}[\ln x]\right) + (\ln x)\left(\dfrac{d}{dx}[x]\right)$ Product Rule

$\qquad = x\left(\dfrac{1}{x}\right) + (\ln x)(1)$

$\qquad = 1 + \ln x$

d. $\dfrac{d}{dx}[(\ln x)^3] = 3(\ln x)^2 \dfrac{d}{dx}[\ln x]$ Chain Rule

$\qquad = 3(\ln x)^2 \dfrac{1}{x}$

Napier used logarithmic properties to simplify *calculations* involving products, quotients, and powers. Of course, given the availability of calculators, there is now little need for this particular application of logarithms. However, there is great value in using logarithmic properties to simplify *differentiation* involving products, quotients, and powers.

Example 4   Logarithmic Properties as Aids to Differentiation

Differentiate $f(x) = \ln\sqrt{x + 1}$.

Solution Because

$$f(x) = \ln\sqrt{x + 1} = \ln(x + 1)^{1/2} = \frac{1}{2}\ln(x + 1) \qquad \text{Rewrite before differentiating.}$$

you can write the following.

$$f'(x) = \frac{1}{2}\left(\frac{1}{x + 1}\right) = \frac{1}{2(x + 1)} \qquad \text{Differentiate.}$$

---

indicates that in the Interactive 3.0 *CD-ROM and* Internet 3.0 *versions of this text (available at* college.hmco.com) *you will find an Open Exploration, which further explores this example using the computer algebra systems* Maple, Mathcad, Mathematica, *and* Derive.

Example 5 Logarithmic Properties as Aids to Differentiation

Differentiate $f(x) = \ln \dfrac{x(x^2 + 1)^2}{\sqrt{2x^3 - 1}}$.

Solution

$$f(x) = \ln \frac{x(x^2 + 1)^2}{\sqrt{2x^3 - 1}} \qquad \text{Write original function.}$$

$$= \ln x + 2\ln(x^2 + 1) - \frac{1}{2}\ln(2x^3 - 1) \qquad \text{Rewrite before differentiating.}$$

$$f'(x) = \frac{1}{x} + 2\left(\frac{2x}{x^2 + 1}\right) - \frac{1}{2}\left(\frac{6x^2}{2x^3 - 1}\right) \qquad \text{Differentiate.}$$

$$= \frac{1}{x} + \frac{4x}{x^2 + 1} - \frac{3x^2}{2x^3 - 1} \qquad \text{Simplify.}$$

NOTE In Examples 4 and 5, be sure that you see the benefit of applying logarithmic properties *before* differentiating. Consider, for instance, the difficulty of direct differentiation of the function given in Example 5.

On occasion, it is convenient to use logarithms as aids in differentiating *nonlogarithmic* functions. This procedure is called **logarithmic differentiation.**

Example 6 Logarithmic Differentiation

Find the derivative of $y = \dfrac{(x - 2)^2}{\sqrt{x^2 + 1}}, \ x \neq 2$.

Solution Note that $y > 0$ for all $x \neq 2$ and hence that $\ln y$ is defined. Begin by taking the natural logarithms of both sides of the equation. Then apply logarithmic properties and differentiate implicitly. Finally, solve for y'.

$$y = \frac{(x - 2)^2}{\sqrt{x^2 + 1}}, \quad x \neq 2 \qquad \text{Write original equation.}$$

$$\ln y = \ln \frac{(x - 2)^2}{\sqrt{x^2 + 1}} \qquad \text{Take natural log of both sides.}$$

$$\ln y = 2\ln(x - 2) - \frac{1}{2}\ln(x^2 + 1) \qquad \text{Logarithmic properties}$$

$$\frac{y'}{y} = 2\left(\frac{1}{x - 2}\right) - \frac{1}{2}\left(\frac{2x}{x^2 + 1}\right) \qquad \text{Differentiate.}$$

$$= \frac{2}{x - 2} - \frac{x}{x^2 + 1} \qquad \text{Simplify.}$$

$$y' = y\left(\frac{2}{x - 2} - \frac{x}{x^2 + 1}\right) \qquad \text{Solve for } y'.$$

$$= \frac{(x - 2)^2}{\sqrt{x^2 + 1}}\left[\frac{x^2 + 2x + 2}{(x - 2)(x^2 + 1)}\right] \qquad \text{Substitute for } y.$$

$$= \frac{(x - 2)(x^2 + 2x + 2)}{(x^2 + 1)^{3/2}} \qquad \text{Simplify.}$$

Because the natural logarithm is undefined for negative numbers, you will often encounter expressions of the form $\ln|u|$. The following theorem states that you can differentiate functions of the form $y = \ln|u|$ as if the absolute value sign were not present.

THEOREM 5.4   Derivative Involving Absolute Value

If u is a differentiable function of x such that $u \neq 0$, then

$$\frac{d}{dx}[\ln|u|] = \frac{u'}{u}.$$

Proof   If $u > 0$, then $|u| = u$, and the result follows from Theorem 5.3. If $u < 0$, then $|u| = -u$, and you have

$$\frac{d}{dx}[\ln|u|] = \frac{d}{dx}[\ln(-u)]$$

$$= \frac{-u'}{-u}$$

$$= \frac{u'}{u}.$$

Example 7   **Derivative Involving Absolute Value**

Find the derivative of

$$f(x) = \ln|\cos x|.$$

Solution   Using Theorem 5.4, let $u = \cos x$ and write

$$\frac{d}{dx}[\ln|\cos x|] = \frac{u'}{u} \qquad\qquad \frac{d}{dx}[\ln|u|] = \frac{u'}{u}$$

$$= \frac{-\sin x}{\cos x} \qquad u = \cos x$$

$$= -\tan x. \qquad \text{Simplify.}$$

Example 8   **Finding Relative Extrema**

Locate the relative extrema of

$$y = \ln(x^2 + 2x + 3).$$

Solution   Differentiating y, you obtain

$$\frac{dy}{dx} = \frac{2x + 2}{x^2 + 2x + 3}.$$

Because $dy/dx = 0$ when $x = -1$, you can apply the First Derivative Test and conclude that the point $(-1, \ln 2)$ is a relative minimum. Because there are no other critical points, it follows that this is the only relative extremum (see Figure 5.7).

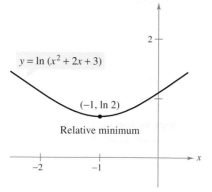

$y = \ln(x^2 + 2x + 3)$

$(-1, \ln 2)$

Relative minimum

The derivative of y changes from negative to positive at $x = -1$.
Figure 5.7

EXERCISES FOR SECTION 5.1

1. Complete the table below. Use a graphing utility and Simpson's Rule with $n = 10$ to approximate the integral

$$\int_1^x \frac{1}{t}\, dt.$$

x	0.5	1.5	2	2.5	3	3.5	4
$\int_1^x (1/t)\, dt$							

2. (a) Plot the points generated in Exercise 1 and connect them with a smooth curve. Compare the result with the graph of $y = \ln x$.

 (b) Use a graphing utility to graph $y = \int_1^x (1/t)\, dt$ for $0.2 \le x \le 4$. Compare the result with the graph of $y = \ln x$.

In Exercises 3–6, use a graphing utility to evaluate the logarithm by (a) using the natural logarithm key, and (b) using the integration capabilities to evaluate the integral

$$\int_1^x \frac{1}{t}\, dt.$$

3. $\ln 45$ **4.** $\ln 8.3$

5. $\ln 0.8$ **6.** $\ln 0.6$

In Exercises 7–10, match the function with its graph. [The graphs are labeled (a), (b), (c), and (d).]

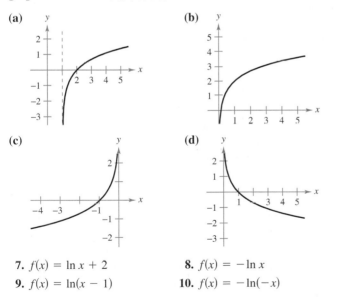

(a)   **(b)**

(c)   **(d)**

7. $f(x) = \ln x + 2$ **8.** $f(x) = -\ln x$

9. $f(x) = \ln(x - 1)$ **10.** $f(x) = -\ln(-x)$

In Exercises 11–16, sketch the graph of the function and state its domain.

11. $f(x) = 3 \ln x$ **12.** $f(x) = 2 \ln x$

13. $f(x) = \ln 2x$ **14.** $f(x) = \ln|x|$

15. $f(x) = \ln(x - 1)$ **16.** $g(x) = 2 + \ln x$

In Exercises 17 and 18, use the properties of logarithms to approximate the indicated logarithms, given that $\ln 2 \approx 0.6931$ and $\ln 3 \approx 1.0986$.

17. (a) $\ln 6$ (b) $\ln \frac{2}{3}$ (c) $\ln 81$ (d) $\ln \sqrt{3}$

18. (a) $\ln 0.25$ (b) $\ln 24$ (c) $\ln \sqrt[3]{12}$ (d) $\ln \frac{1}{72}$

In Exercises 19–28, use the properties of logarithms to expand the logarithmic expression.

19. $\ln \frac{2}{3}$ **20.** $\ln \sqrt{2^3}$

21. $\ln \frac{xy}{z}$ **22.** $\ln(xyz)$

23. $\ln \sqrt[3]{a^2 + 1}$ **24.** $\ln \sqrt{a - 1}$

25. $\ln \left(\frac{x^2 - 1}{x^3} \right)^3$ **26.** $\ln(3e^2)$

27. $\ln z(z - 1)^2$ **28.** $\ln \frac{1}{e}$

In Exercises 29–34, write the expression as a logarithm of a single quantity.

29. $\ln(x - 2) - \ln(x + 2)$

30. $3 \ln x + 2 \ln y - 4 \ln z$

31. $\frac{1}{3}[2 \ln(x + 3) + \ln x - \ln(x^2 - 1)]$

32. $2[\ln x - \ln(x + 1) - \ln(x - 1)]$

33. $2 \ln 3 - \frac{1}{2} \ln(x^2 + 1)$

34. $\frac{3}{2}[\ln(x^2 + 1) - \ln(x + 1) - \ln(x - 1)]$

In Exercises 35 and 36, show that $f = g$ by using a graphing utility to graph f and g in the same viewing window.

35. $f(x) = \ln \frac{x^2}{4}$, $x > 0$, $g(x) = 2 \ln x - \ln 4$

36. $f(x) = \ln \sqrt{x(x^2 + 1)}$, $g(x) = \frac{1}{2}[\ln x + \ln(x^2 + 1)]$

In Exercises 37–40, find the limit.

37. $\lim\limits_{x \to 3^+} \ln(x - 3)$ **38.** $\lim\limits_{x \to 6^-} \ln(6 - x)$

39. $\lim\limits_{x \to 2^-} \ln[x^2(3 - x)]$ **40.** $\lim\limits_{x \to 5^+} \ln \frac{x}{\sqrt{x - 4}}$

In Exercises 41–44, find the slope of the tangent line to the graph of the logarithmic function at the point $(1, 0)$.

41. $y = \ln x^3$ **42.** $y = \ln x^{3/2}$

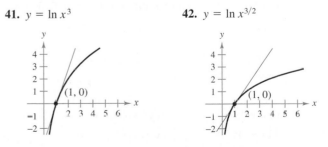

43. $y = \ln x^2$

44. $y = \ln x^{1/2}$

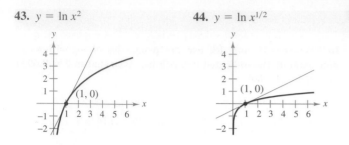

In Exercises 45–70, find the derivative of the function.

45. $g(x) = \ln x^2$

46. $h(x) = \ln(2x^2 + 1)$

47. $y = (\ln x)^4$

48. $y = x \ln x$

49. $y = \ln(x\sqrt{x^2 - 1})$

50. $y = \ln\sqrt{x^2 - 4}$

51. $f(x) = \ln\left(\dfrac{x}{x^2 + 1}\right)$

52. $f(x) = \ln\left(\dfrac{2x}{x + 3}\right)$

53. $g(t) = \dfrac{\ln t}{t^2}$

54. $h(t) = \dfrac{\ln t}{t}$

55. $y = \ln(\ln x^2)$

56. $y = \ln(\ln x)$

57. $y = \ln\sqrt{\dfrac{x + 1}{x - 1}}$

58. $y = \ln\sqrt[3]{\dfrac{x - 1}{x + 1}}$

59. $f(x) = \ln\left(\dfrac{\sqrt{4 + x^2}}{x}\right)$

60. $f(x) = \ln(x + \sqrt{4 + x^2})$

61. $y = \dfrac{-\sqrt{x^2 + 1}}{x} + \ln(x + \sqrt{x^2 + 1})$

62. $y = \dfrac{-\sqrt{x^2 + 4}}{2x^2} - \dfrac{1}{4}\ln\left(\dfrac{2 + \sqrt{x^2 + 4}}{x}\right)$

63. $y = \ln|\sin x|$

64. $y = \ln|\csc x|$

65. $y = \ln\left|\dfrac{\cos x}{\cos x - 1}\right|$

66. $y = \ln|\sec x + \tan x|$

67. $y = \ln\left|\dfrac{-1 + \sin x}{2 + \sin x}\right|$

68. $y = \ln\sqrt{1 + \sin^2 x}$

69. $f(x) = \sin 2x \ln x^2$

70. $g(x) = \displaystyle\int_1^{\ln x} (t^2 + 3)\, dt$

In Exercises 71 and 72, (a) find an equation of the tangent line to the graph of f at the indicated point, (b) use a graphing utility to graph the function and its tangent line at the point, and (c) use the *derivative* feature of a graphing utility to confirm your results.

Function	Point
71. $f(x) = 3x^2 - \ln x$	$(1, 3)$
72. $f(x) = 4 - x^2 - \ln(\tfrac{1}{2}x + 1)$	$(0, 4)$

In Exercises 73 and 74, use implicit differentiation to find dy/dx.

73. $x^2 - 3\ln y + y^2 = 10$

74. $\ln xy + 5x = 30$

In Exercises 75 and 76, show that the function is a solution of the differential equation.

Function	Differential Equation
75. $y = 2\ln x + 3$	$xy'' + y' = 0$
76. $y = x\ln x - 4x$	$x + y - xy' = 0$

In Exercises 77–82, locate any relative extrema and inflection points. Use a graphing utility to confirm your results.

77. $y = \dfrac{x^2}{2} - \ln x$

78. $y = x - \ln x$

79. $y = x \ln x$

80. $y = \dfrac{\ln x}{x}$

81. $y = \dfrac{x}{\ln x}$

82. $y = x^2 \ln \dfrac{x}{4}$

Linear and Quadratic Approximations **In Exercises 83 and 84, use a graphing utility to graph the function. Then graph**

$$P_1(x) = f(1) + f'(1)(x - 1)$$

and

$$P_2(x) = f(1) + f'(1)(x - 1) + \tfrac{1}{2}f''(1)(x - 1)^2$$

in the same viewing window. Compare the values of f, P_1, and P_2 and their first derivatives at $x = 1$.

83. $f(x) = \ln x$

84. $f(x) = x \ln x$

In Exercises 85 and 86, use Newton's Method to approximate, to three decimal places, the x-coordinate of the point of intersection of the graphs of the two equations. Use a graphing utility to verify your result.

85. $y = \ln x$
$y = -x$

86. $y = \ln x$
$y = 3 - x$

In Exercises 87–92, find dy/dx using logarithmic differentiation.

87. $y = x\sqrt{x^2 - 1}$

88. $y = \sqrt{(x - 1)(x - 2)(x - 3)}$

89. $y = \dfrac{x^2\sqrt{3x - 2}}{(x - 1)^2}$

90. $y = \sqrt{\dfrac{x^2 - 1}{x^2 + 1}}$

91. $y = \dfrac{x(x - 1)^{3/2}}{\sqrt{x + 1}}$

92. $y = \dfrac{(x + 1)(x + 2)}{(x - 1)(x - 2)}$

Getting at the Concept

93. In your own words, state the properties of the natural logarithmic function.

94. Define the base for the natural logarithmic function.

95. Explain why $\ln e^x = x$.

96. Let f be a function that is positive and differentiable on the entire real line. Let $g(x) = \ln f(x)$.

(a) If g is increasing, must f be increasing? Explain.

(b) If the graph of f is concave upward, must the graph of g be concave upward? Explain.

97. Consider the function $f(x) = x - 2\ln x$ on $[1, 3]$.

(a) Explain why Rolle's Theorem (Section 3.2) does not apply.

(b) Do you think the conclusion of Rolle's Theorem is true for f? Explain.

98. Home Mortgage The term t (in years) of a \$120,000 home mortgage at 10% interest can be approximated by

$$t = \frac{5.315}{-6.7968 + \ln x}, \quad x > 1000$$

where x is the monthly payment in dollars.

(a) Use a graphing utility to graph the model.

(b) Use the model to approximate the term of a home mortgage for which the monthly payment is \$1167.41. What is the total amount paid?

(c) Use the model to approximate the term of a home mortgage for which the monthly payment is \$1068.45. What is the total amount paid?

(d) Find the instantaneous rate of change of t with respect to x when $x = 1167.41$ and $x = 1068.45$.

(e) Write a short paragraph describing the benefit of the higher monthly payment.

99. Sound Intensity The relationship between the number of decibels β and the intensity of a sound I in watts per centimeter squared is

$$\beta = 10 \log_{10}\left(\frac{I}{10^{-16}}\right).$$

Use the properties of logarithms to write the formula in simpler form, and determine the number of decibels of a sound with an intensity of 10^{-10} watts per square centimeter.

100. Modeling Data The table shows the temperature T (°F) at which water boils at selected pressures p (pounds per square inch). *(Source: Standard Handbook of Mechanical Engineers)*

p	5	10	14.696 (1 atm)	20
T	162.24°	193.21°	212.00°	227.96°

p	30	40	60	80	100
T	250.33°	267.25°	292.71°	312.03°	327.81°

A model that approximates the data is

$$T = 87.97 + 34.96 \ln p + 7.91 \sqrt{p}.$$

(a) Use a graphing utility to plot the data and graph the model.

(b) Find the rate of change of T with respect to p when $p = 10$ and $p = 70$.

(c) Use a graphing utility to graph T'. Find

$$\lim_{p \to \infty} T'(p)$$

and interpret the result in the context of the problem.

101. Modeling Data The atmospheric pressure decreases with increasing altitude. At sea level, the average air pressure is one atmosphere (1.033227 kilograms per square centimeter). The table shows the pressure p (in atmospheres) at a given altitude h (in kilometers).

h	0	5	10	15	20	25
p	1	0.55	0.25	0.12	0.06	0.02

(a) Use a graphing utility to find a model of the form $p = a + b \ln h$ for the data. Explain why the result is an error message.

(b) Use a graphing utility to find the logarithmic model $h = a + b \ln p$ for the data.

(c) Use a graphing utility to plot the data and graph the logarithmic model.

(d) Use the model to estimate the altitude at which the pressure is 0.75 atmosphere.

(e) Use the model to estimate the pressure at an altitude of 13 kilometers.

(f) Find the rate of change of pressure when $h = 5$ and $h = 20$. Interpret the results in the context of the problem.

102. Tractrix A person walking along a dock drags a boat by a 10-meter rope. The boat travels along a path known as a *tractrix* (see figure). The equation of this path is

$$y = 10 \ln\left(\frac{10 + \sqrt{100 - x^2}}{x}\right) - \sqrt{100 - x^2}.$$

(a) Use a graphing utility to graph the function.

(b) What is the slope of this path when $x = 5$ and $x = 9$?

(c) What does the slope of the path approach as $x \to 10$?

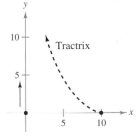

103. Conjecture Use a graphing utility to graph f and g in the same viewing window and determine which is increasing at the faster rate for "large" values of x. What can you conclude about the rate of growth of the natural logarithmic function?

(a) $f(x) = \ln x$, $\qquad g(x) = \sqrt{x}$

(b) $f(x) = \ln x$, $\qquad g(x) = \sqrt[4]{x}$

104. Prove that the natural logarithmic function is one-to-one.

True or False? **In Exercises 105 and 106, determine whether the statement is true or false. If it is false, explain why or give an example that shows it is false.**

105. $\ln(x + 25) = \ln x + \ln 25$

106. If $y = \ln \pi$, then $y' = 1/\pi$.

Section 5.2	The Natural Logarithmic Function: Integration

- Use the Log Rule for Integration to integrate a rational function.
- Integrate trigonometric functions.

Log Rule for Integration

The differentiation rules

$$\frac{d}{dx}[\ln|x|] = \frac{1}{x} \quad \text{and} \quad \frac{d}{dx}[\ln|u|] = \frac{u'}{u}$$

that you studied in the preceding section produce the following integration rule.

THEOREM 5.5   Log Rule for Integration

Let u be a differentiable function of x.

1. $\displaystyle\int \frac{1}{x}\, dx = \ln|x| + C$ **2.** $\displaystyle\int \frac{1}{u}\, du = \ln|u| + C$

Because $du = u'\, dx$, the second formula can also be written as

$$\int \frac{u'}{u}\, dx = \ln|u| + C. \qquad \text{Alternative form of Log Rule}$$

Example 1   Using the Log Rule for Integration

$$\int \frac{2}{x}\, dx = 2\int \frac{1}{x}\, dx$$
$$= 2\ln|x| + C \qquad \text{Log Rule for Integration}$$
$$= \ln(x^2) + C \qquad \text{Property of logarithms}$$

Because x^2 cannot be negative, the absolute value is unnecessary in the final form of the antiderivative.

Example 2   Using the Log Rule with a Change of Variables

Find $\displaystyle\int \frac{1}{4x - 1}\, dx$.

Solution   If you let $u = 4x - 1$, then $du = 4\, dx$.

$$\int \frac{1}{4x - 1}\, dx = \frac{1}{4}\int \left(\frac{1}{4x - 1}\right) 4\, dx \qquad \text{Multiply and divide by 4.}$$
$$= \frac{1}{4}\int \frac{1}{u}\, du \qquad \text{Substitute: } u = 4x - 1.$$
$$= \frac{1}{4}\ln|u| + C \qquad \text{Apply Log Rule.}$$
$$= \frac{1}{4}\ln|4x - 1| + C \qquad \text{Back-substitute.}$$

Example 3 uses the alternative form of the Log Rule. To apply this rule, look for quotients in which the numerator is the derivative of the denominator.

Example 3   Finding Area with the Log Rule

Find the area of the region bounded by the graph of

$$y = \frac{x}{x^2 + 1}$$

the x-axis, and the line $x = 3$.

Solution From Figure 5.8, you can see that the area of the region is given by the definite integral

$$\int_0^3 \frac{x}{x^2 + 1}\, dx.$$

If you let $u = x^2 + 1$, then $u' = 2x$. To apply the Log Rule, multiply and divide by 2 as follows.

$$\int_0^3 \frac{x}{x^2 + 1}\, dx = \frac{1}{2} \int_0^3 \frac{2x}{x^2 + 1}\, dx \qquad \text{Multiply and divide by 2.}$$

$$= \frac{1}{2}\left[\ln(x^2 + 1) \right]_0^3 \qquad \int \frac{u'}{u}\, dx = \ln|u| + C$$

$$= \frac{1}{2}(\ln 10 - \ln 1)$$

$$= \frac{1}{2}\ln 10 \qquad\qquad \ln 1 = 0$$

$$\approx 1.151$$

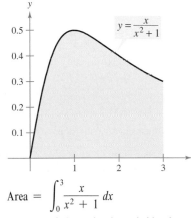

$$\text{Area} = \int_0^3 \frac{x}{x^2 + 1}\, dx$$

The area of the region bounded by the graph of y, the x-axis, and $x = 3$ is $\frac{1}{2}\ln 10$.
Figure 5.8

Example 4   Recognizing Quotient Forms of the Log Rule

a. $\displaystyle\int \frac{3x^2 + 1}{x^3 + x}\, dx = \ln|x^3 + x| + C \qquad u = x^3 + x$

b. $\displaystyle\int \frac{\sec^2 x}{\tan x}\, dx = \ln|\tan x| + C \qquad u = \tan x$

c. $\displaystyle\int \frac{x + 1}{x^2 + 2x}\, dx = \frac{1}{2}\int \frac{2x + 2}{x^2 + 2x}\, dx \qquad u = x^2 + 2x$

$$= \frac{1}{2}\ln|x^2 + 2x| + C$$

d. $\displaystyle\int \frac{1}{3x + 2}\, dx = \frac{1}{3}\int \frac{3}{3x + 2}\, dx \qquad u = 3x + 2$

$$= \frac{1}{3}\ln|3x + 2| + C$$

With antiderivatives involving logarithms, it is easy to obtain forms that look quite different but are still equivalent. For instance, which of the following are equivalent to the antiderivative listed in Example 4d?

$$\ln\left|(3x + 2)^{1/3}\right| + C, \quad \frac{1}{3}\ln\left|x + \frac{2}{3}\right| + C, \quad \ln|3x + 2|^{1/3} + C$$

Integrals to which the Log Rule can be applied often appear in disguised form. For instance, if a rational function has a *numerator of degree greater than or equal to that of the denominator,* division may reveal a form to which you can apply the Log Rule. This is illustrated in Example 5.

Example 5   Using Long Division Before Integrating

Find $\displaystyle\int \frac{x^2 + x + 1}{x^2 + 1}\, dx$.

Solution   Begin by using long division to rewrite the integrand.

$$\frac{x^2 + x + 1}{x^2 + 1} \quad \Longrightarrow \quad x^2 + 1 \overline{)\,x^2 + x + 1\,} \quad \Longrightarrow \quad 1 + \frac{x}{x^2 + 1}$$

Now, you can integrate to obtain

$$\int \frac{x^2 + x + 1}{x^2 + 1}\, dx = \int \left(1 + \frac{x}{x^2 + 1}\right) dx \qquad \text{Rewrite using long division.}$$

$$= \int dx + \frac{1}{2}\int \frac{2x}{x^2 + 1}\, dx \qquad \text{Rewrite as two integrals.}$$

$$= x + \frac{1}{2}\ln(x^2 + 1) + C. \qquad \text{Integrate.}$$

Check this result by differentiating to obtain the original integrand.

The next example gives another instance in which the use of the Log Rule is disguised. In this case, a change of variables helps you recognize the Log Rule.

Example 6   Change of Variables with the Log Rule

Find $\displaystyle\int \frac{2x}{(x + 1)^2}\, dx$.

Solution   If you let $u = x + 1$, then $du = dx$ and $x = u - 1$.

$$\int \frac{2x}{(x + 1)^2}\, dx = \int \frac{2(u - 1)}{u^2}\, du \qquad \text{Substitute.}$$

$$= 2\int \left(\frac{u}{u^2} - \frac{1}{u^2}\right) du \qquad \text{Rewrite as two fractions.}$$

$$= 2\int \frac{du}{u} - 2\int u^{-2}\, du \qquad \text{Rewrite as two integrals.}$$

$$= 2\ln|u| - 2\left(\frac{u^{-1}}{-1}\right) + C \qquad \text{Integrate.}$$

$$= 2\ln|u| + \frac{2}{u} + C \qquad \text{Simplify.}$$

$$= 2\ln|x + 1| + \frac{2}{x + 1} + C \qquad \text{Back-substitute.}$$

TECHNOLOGY   If you have access to a computer algebra system, try using it to find the indefinite integrals in Examples 5 and 6. How does the form of the antiderivative that it gives you compare with that given in Examples 5 and 6?

Check this result by differentiating to obtain the original integrand.

As you study the methods shown in Examples 5 and 6, be aware that both methods involve rewriting a disguised integrand so that it fits one or more of the basic integration formulas. Throughout the remaining sections of Chapter 5 and in Chapter 7 we will devote much time to integration techniques. To master these techniques, you must recognize the "form-fitting" nature of integration. In this sense, integration is not nearly as straightforward as differentiation. Differentiation takes the form

"Here is the question; what is the answer?"

Integration is more like

"Here is the answer; what is the question?"

We suggest the following guidelines for integration.

Guidelines for Integration

1. Learn a basic list of integration formulas. (Including those given in this section, you now have 12 formulas: the Power Rule, the Log Rule, and ten trigonometric rules. By the end of Section 5.9, this list will have expanded to 20 basic rules.)
2. Find an integration formula that resembles all or part of the integrand, and, by trial and error, find a choice of u that will make the integrand conform to the formula.
3. If you cannot find a u-substitution that works, try altering the integrand. You might try a trigonometric identity, multiplication and division by the same quantity, or addition and subtraction of the same quantity. Be creative.
4. If you have access to computer software that will find antiderivatives symbolically, use it.

STUDY TIP   Keep in mind that you can check your answer to an integration problem by differentiating the answer. For instance, in Example 7, the derivative of $y = \ln|\ln x| + C$ is $y' = 1/(x \ln x)$.

Example 7   *u*-Substitution and the Log Rule

Solve the differential equation $\dfrac{dy}{dx} = \dfrac{1}{x \ln x}$.

Solution   The solution can be written as an indefinite integral.

$$y = \int \frac{1}{x \ln x}\, dx$$

Because the integrand is a quotient whose denominator is raised to the first power, you should try the Log Rule. There are three basic choices for u. The choices $u = x$ and $u = x \ln x$ fail to fit the u'/u form of the Log Rule. However, the third choice does fit. Letting $u = \ln x$ produces $u' = 1/x$, and you obtain the following.

$$\int \frac{1}{x \ln x}\, dx = \int \frac{1/x}{\ln x}\, dx \qquad \text{Divide numerator and denominator by } x.$$

$$= \int \frac{u'}{u}\, dx \qquad \text{Substitute: } u = \ln x.$$

$$= \ln|u| + C \qquad \text{Apply Log Rule.}$$

$$= \ln|\ln x| + C \qquad \text{Back-substitute.}$$

So, the solution is $y = \ln|\ln x| + C$.

Integrals of Trigonometric Functions

In Section 4.1, you looked at six trigonometric integration rules—the six that correspond directly to differentiation rules. With the Log Rule, you can now complete the set of basic trigonometric integration formulas.

Example 8   Using a Trigonometric Identity

Find $\int \tan x \, dx$.

Solution   This integral does not seem to fit any formulas on our basic list. However, by using a trigonometric identity, you obtain the following.

$$\int \tan x \, dx = \int \frac{\sin x}{\cos x} \, dx$$

Knowing that $D_x[\cos x] = -\sin x$, you can let $u = \cos x$ and write

$$
\begin{aligned}
\int \tan x \, dx &= -\int \frac{-\sin x}{\cos x} \, dx & \text{Trigonometric identity} \\
&= -\int \frac{u'}{u} \, dx & \text{Substitute: } u = \cos x. \\
&= -\ln|u| + C & \text{Apply Log Rule.} \\
&= -\ln|\cos x| + C. & \text{Back-substitute.}
\end{aligned}
$$

Example 8 uses a trigonometric identity to derive an integration rule for the tangent function. In the next example, we take a rather unusual step (multiplying and dividing by the same quantity) to derive an integration rule for the secant function.

Example 9   Derivation of the Secant Formula

Find $\int \sec x \, dx$.

Solution   Consider the following procedure.

$$
\begin{aligned}
\int \sec x \, dx &= \int \sec x \left(\frac{\sec x + \tan x}{\sec x + \tan x} \right) dx \\
&= \int \frac{\sec^2 x + \sec x \tan x}{\sec x + \tan x} \, dx
\end{aligned}
$$

Letting u be the denominator of this quotient produces

$$u = \sec x + \tan x \quad \Longrightarrow \quad u' = \sec x \tan x + \sec^2 x.$$

Therefore, you can conclude that

$$
\begin{aligned}
\int \sec x \, dx &= \int \frac{\sec^2 x + \sec x \tan x}{\sec x + \tan x} \, dx & \text{Rewrite integrand.} \\
&= \int \frac{u'}{u} \, dx & \text{Substitute: } u = \sec x + \tan x. \\
&= \ln|u| + C & \text{Apply Log Rule.} \\
&= \ln|\sec x + \tan x| + C. & \text{Back-substitute.}
\end{aligned}
$$

With the results of Examples 8 and 9, you now have integration formulas for $\sin x$, $\cos x$, $\tan x$, and $\sec x$. All six trigonometric rules are summarized below.

NOTE   Using trigonometric identities and properties of logarithms, you could rewrite these six integration rules in other forms. For instance, you could write

$$\int \csc u\, du = \ln|\csc u - \cot u| + C.$$

(See Exercises 51–54.)

Integrals of the Six Basic Trigonometric Functions

$$\int \sin u\, du = -\cos u + C \qquad\qquad \int \cos u\, du = \sin u + C$$

$$\int \tan u\, du = -\ln|\cos u| + C \qquad\qquad \int \cot u\, du = \ln|\sin u| + C$$

$$\int \sec u\, du = \ln|\sec u + \tan u| + C \qquad \int \csc u\, du = -\ln|\csc u + \cot u| + C$$

Example 10   Integrating Trigonometric Functions

Evaluate $\displaystyle\int_0^{\pi/4} \sqrt{1 + \tan^2 x}\, dx$.

Solution   Using $1 + \tan^2 x = \sec^2 x$, you can write

$$\int_0^{\pi/4} \sqrt{1 + \tan^2 x}\, dx = \int_0^{\pi/4} \sqrt{\sec^2 x}\, dx$$

$$= \int_0^{\pi/4} \sec x\, dx \qquad\qquad \sec x \geq 0 \text{ for } 0 \leq x \leq \frac{\pi}{4}.$$

$$= \ln|\sec x + \tan x| \Big]_0^{\pi/4}$$

$$= \ln\left(\sqrt{2} + 1\right) - \ln 1$$

$$\approx 0.8814.$$

Example 11   Finding an Average Value

Find the average value of $f(x) = \tan x$ on the interval $[0, \pi/4]$.

Solution

$$\text{Average value} = \frac{1}{(\pi/4) - 0} \int_0^{\pi/4} \tan x\, dx \qquad \text{Average value} = \frac{1}{b-a} \int_a^b f(x)\, dx$$

$$= \frac{4}{\pi} \int_0^{\pi/4} \tan x\, dx \qquad\qquad \text{Simplify.}$$

$$= \frac{4}{\pi} \Big[-\ln|\cos x|\Big]_0^{\pi/4} \qquad\qquad \text{Integrate.}$$

$$= -\frac{4}{\pi}\left[\ln\left(\frac{\sqrt{2}}{2}\right) - \ln(1)\right]$$

$$= -\frac{4}{\pi} \ln\left(\frac{\sqrt{2}}{2}\right)$$

$$\approx 0.441$$

The average value is about 0.441, as indicated in Figure 5.9.

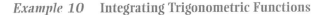

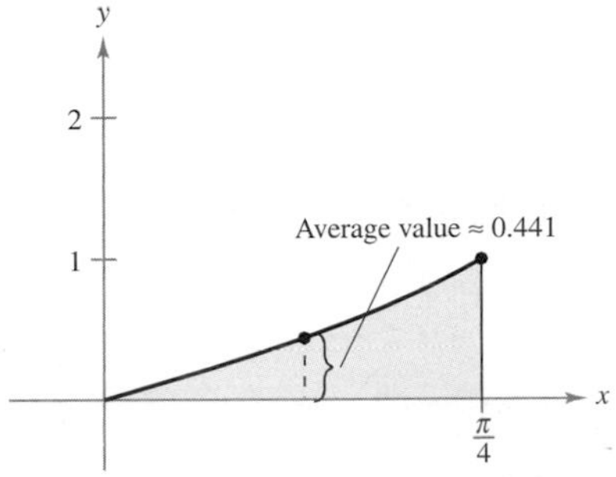

Average value ≈ 0.441

Figure 5.9

EXERCISES FOR SECTION 5.2

In Exercises 1–24, find the indefinite integral.

1. $\displaystyle\int \frac{5}{x}\,dx$

2. $\displaystyle\int \frac{10}{x}\,dx$

3. $\displaystyle\int \frac{1}{x+1}\,dx$

4. $\displaystyle\int \frac{1}{x-5}\,dx$

5. $\displaystyle\int \frac{1}{3-2x}\,dx$

6. $\displaystyle\int \frac{1}{3x+2}\,dx$

7. $\displaystyle\int \frac{x}{x^2+1}\,dx$

8. $\displaystyle\int \frac{x^2}{3-x^3}\,dx$

9. $\displaystyle\int \frac{x^2-4}{x}\,dx$

10. $\displaystyle\int \frac{x}{\sqrt{9-x^2}}\,dx$

11. $\displaystyle\int \frac{x^2+2x+3}{x^3+3x^2+9x}\,dx$

12. $\displaystyle\int \frac{x(x+2)}{x^3+3x^2-4}\,dx$

13. $\displaystyle\int \frac{x^2-3x+2}{x+1}\,dx$

14. $\displaystyle\int \frac{2x^2+7x-3}{x-2}\,dx$

15. $\displaystyle\int \frac{x^3-3x^2+5}{x-3}\,dx$

16. $\displaystyle\int \frac{x^3-6x-20}{x+5}\,dx$

17. $\displaystyle\int \frac{x^4+x-4}{x^2+2}\,dx$

18. $\displaystyle\int \frac{x^3-3x^2+4x-9}{x^2+3}\,dx$

19. $\displaystyle\int \frac{(\ln x)^2}{x}\,dx$

20. $\displaystyle\int \frac{1}{x\ln(x^3)}\,dx$

21. $\displaystyle\int \frac{1}{\sqrt{x}+1}\,dx$

22. $\displaystyle\int \frac{1}{x^{2/3}(1+x^{1/3})}\,dx$

23. $\displaystyle\int \frac{2x}{(x-1)^2}\,dx$

24. $\displaystyle\int \frac{x(x-2)}{(x-1)^3}\,dx$

In Exercises 25–28, find the indefinite integral by u-substitution. (*Hint:* Let u be the denominator of the integrand.)

25. $\displaystyle\int \frac{1}{1+\sqrt{2x}}\,dx$

26. $\displaystyle\int \frac{1}{1+\sqrt{3x}}\,dx$

27. $\displaystyle\int \frac{\sqrt{x}}{\sqrt{x}-3}\,dx$

28. $\displaystyle\int \frac{\sqrt[3]{x}}{\sqrt[3]{x}-1}\,dx$

In Exercises 29–36, find the indefinite integral.

29. $\displaystyle\int \frac{\cos\theta}{\sin\theta}\,d\theta$

30. $\displaystyle\int \tan 5\theta\,d\theta$

31. $\displaystyle\int \csc 2x\,dx$

32. $\displaystyle\int \sec\frac{x}{2}\,dx$

33. $\displaystyle\int \frac{\cos t}{1+\sin t}\,dt$

34. $\displaystyle\int \frac{\csc^2 t}{\cot t}\,dt$

35. $\displaystyle\int \frac{\sec x\tan x}{\sec x-1}\,dx$

36. $\displaystyle\int (\sec t+\tan t)\,dt$

In Exercises 37–40, solve the differential equation. Use a graphing utility to graph three solutions, one of which passes through the indicated point.

37. $\dfrac{dy}{dx}=\dfrac{3}{2-x}$, $(1,0)$

38. $\dfrac{dy}{dx}=\dfrac{2x}{x^2-9}$, $(0,4)$

39. $\dfrac{ds}{d\theta}=\tan 2\theta$, $(0,2)$

40. $\dfrac{dr}{dt}=\dfrac{\sec^2 t}{\tan t+1}$, $(\pi,4)$

Slope Fields   **In Exercises 41 and 42, a differential equation, a point, and a slope field are given. (a) Sketch two approximate solutions of the differential equation on the slope field, one of which passes through the indicated point. (b) Use integration to find the particular solution of the differential equation and use a graphing utility to graph the solution. Compare the result with the sketches in part (a). To print an enlarged copy of the graph, go to the website *www.mathgraphs.com*.**

41. $\dfrac{dy}{dx}=\dfrac{1}{x+2}$, $(0,1)$

42. $\dfrac{dy}{dx}=\dfrac{\ln x}{x}$, $(1,-2)$

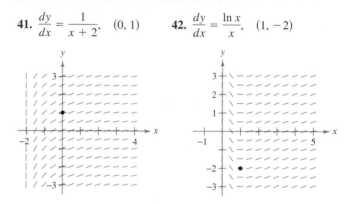

In Exercises 43–50, evaluate the definite integral. Use a graphing utility to verify your result.

43. $\displaystyle\int_0^4 \frac{5}{3x+1}\,dx$

44. $\displaystyle\int_{-1}^1 \frac{1}{x+2}\,dx$

45. $\displaystyle\int_1^e \frac{(1+\ln x)^2}{x}\,dx$

46. $\displaystyle\int_e^{e^2} \frac{1}{x\ln x}\,dx$

47. $\displaystyle\int_0^2 \frac{x^2-2}{x+1}\,dx$

48. $\displaystyle\int_0^1 \frac{x-1}{x+1}\,dx$

49. $\displaystyle\int_1^2 \frac{1-\cos\theta}{\theta-\sin\theta}\,d\theta$

50. $\displaystyle\int_{0.1}^{0.2} (\csc 2\theta-\cot 2\theta)^2\,d\theta$

In Exercises 51–54, show that the two formulas are equivalent.

51. $\displaystyle\int \tan x\,dx = -\ln|\cos x|+C$

$\displaystyle\int \tan x\,dx = \ln|\sec x|+C$

52. $\displaystyle\int \cot x\,dx = \ln|\sin x|+C$

$\displaystyle\int \cot x\,dx = -\ln|\csc x|+C$

53. $\displaystyle\int \sec x\,dx = \ln|\sec x+\tan x|+C$

$\displaystyle\int \sec x\,dx = -\ln|\sec x-\tan x|+C$

54. $\displaystyle\int \csc x\,dx = -\ln|\csc x+\cot x|+C$

$\displaystyle\int \csc x\,dx = \ln|\csc x-\cot x|+C$

In Exercises 55–60, use a computer algebra system to find or evaluate the integral.

55. $\displaystyle\int \frac{1}{1+\sqrt{x}}\,dx$

56. $\displaystyle\int \frac{1-\sqrt{x}}{1+\sqrt{x}}\,dx$

57. $\displaystyle\int \cos(1-x)\,dx$

58. $\displaystyle\int \frac{\tan^2 2x}{\sec 2x}\,dx$

59. $\displaystyle\int_{\pi/4}^{\pi/2} (\csc x - \sin x)\,dx$

60. $\displaystyle\int_{-\pi/4}^{\pi/4} \frac{\sin^2 x - \cos^2 x}{\cos x}\,dx$

In Exercises 61–64, find $F'(x)$.

61. $\displaystyle F(x) = \int_{1}^{x} \frac{1}{t}\,dt$

62. $\displaystyle F(x) = \int_{0}^{x} \tan t\,dt$

63. $\displaystyle F(x) = \int_{x}^{3x} \frac{1}{t}\,dt$

64. $\displaystyle F(x) = \int_{1}^{x^2} \frac{1}{t}\,dt$

Approximation   **In Exercises 65 and 66, determine which value best approximates the area of the region between the x-axis and the graph of the function over the given interval. (Make your selection on the basis of a sketch of the region and not by performing any calculations.)**

65. $f(x) = \sec x,\quad [0,1]$

 (a) 6   (b) -6 (c) $\frac{1}{2}$ (d) 1.25   (e) 3

66. $f(x) = \dfrac{2x}{x^2+1},\quad [0,4]$

 (a) 3   (b) 7   (c) -2 (d) 5   (e) 1

Area   **In Exercises 67–70, find the area of the region bounded by the graphs of the equations. Use a graphing utility to graph the region and verify your result.**

67. $y = \dfrac{x^2+4}{x},\ x=1,\ x=4,\ y=0$

68. $y = \dfrac{x+4}{x},\ x=1,\ x=4,\ y=0$

69. $y = 2\sec\dfrac{\pi x}{6},\ x=0,\ x=2,\ y=0$

70. $y = 2x - \tan(0.3x),\ x=1,\ x=4,\ y=0$

Getting at the Concept

In Exercises 71–74, state the integration formula you would use to perform the integration. Do not integrate.

71. $\displaystyle\int \sqrt[3]{x}\,dx$

72. $\displaystyle\int \frac{x}{(x^2+4)^3}\,dx$

73. $\displaystyle\int \frac{x}{x^2+4}\,dx$

74. $\displaystyle\int \frac{\sec^2 x}{\tan x}\,dx$

75. What is the first step when integrating
$$\int \frac{x^2}{x+1}\,dx?$$

76. Make a list of the integration formulas studied so far in the text.

In Exercises 77–80, find the average value of the function over the interval.

Function	Interval	Function	Interval
77. $f(x) = \dfrac{8}{x^2}$	$[2,4]$	**78.** $f(x) = \dfrac{4(x+1)}{x^2}$	$[2,4]$
79. $f(x) = \dfrac{\ln x}{x}$	$[1,e]$	**80.** $f(x) = \sec\dfrac{\pi x}{6}$	$[0,2]$

81. *Population Growth*   A population of bacteria is changing at a rate of
$$\frac{dP}{dt} = \frac{3000}{1+0.25t}$$
where t is the time in days. The initial population (when $t=0$) is 1000. Write an equation that gives the population at any time t, and find the population when $t=3$ days.

82. *Heat Transfer*   Find the time required for an object to cool from $300°F$ to $250°F$ by evaluating
$$t = \frac{10}{\ln 2}\int_{250}^{300} \frac{1}{T-100}\,dT$$
where t is time in minutes.

83. *Average Price*   The demand equation for a product is
$$p = \frac{90{,}000}{400+3x}.$$
Find the *average* price p on the interval $40 \le x \le 50$.

84. *Sales*   The rate of change in sales S is inversely proportional to time t ($t > 1$) measured in weeks. Find S as a function of t if sales after 2 and 4 weeks are 200 units and 300 units.

85. *Orthogonal Trajectory*

 (a) Use a graphing utility to graph the equation $2x^2 - y^2 = 8$.

 (b) Evaluate the integral to find y^2 in terms of x.
$$y^2 = e^{-\int(1/x)\,dx}$$
 For a particular value of the constant of integration, graph the result on the same screen used in part (a).

 (c) Verify that the tangents to the graphs of parts (a) and (b) are perpendicular at the points of intersection.

86. Graph the function
$$f_k(x) = \frac{x^k-1}{k}$$
for $k=1$, 0.5, and 0.1 on $[0,10]$. Find $\displaystyle\lim_{k\to 0^+} f_k(x)$.

True or False?   **In Exercises 87–90, determine whether the statement is true or false. If it is false, explain why or give an example that shows it is false.**

87. $(\ln x)^{1/2} = \frac{1}{2}(\ln x)$

88. $\int \ln x\,dx = (1/x) + C$

89. $\displaystyle\int \frac{1}{x}\,dx = \ln|cx|,\quad c \ne 0$

90. $\displaystyle\int_{-1}^{2} \frac{1}{x}\,dx = \Big[\ln|x|\Big]_{-1}^{2} = \ln 2 - \ln 1 = \ln 2$

| **Section 5.3** | **Inverse Functions** |

- Verify that one function is the inverse function of another function.
- Determine whether a function has an inverse function.
- Find the derivative of an inverse function.

Inverse Functions

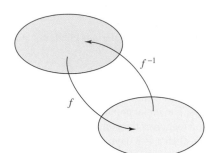

Domain of f = range of f^{-1}
Domain of f^{-1} = range of f
Figure 5.10

Recall from Section P.3 that a function can be represented by a set of ordered pairs. For instance, the function $f(x) = x + 3$ and $A = \{1, 2, 3, 4\}$ to $B = \{4, 5, 6, 7\}$ can be written as

$$f: \{(1, 4), (2, 5), (3, 6), (4, 7)\}.$$

By interchanging the first and second coordinates of each ordered pair, you can form the **inverse function** of f. This function is denoted by f^{-1}. It is a function from B to A, and can be written as

$$f^{-1}: \{(4, 1), (5, 2), (6, 3), (7, 4)\}.$$

Note that the domain of f is equal to the range of f^{-1}, and vice versa, as shown in Figure 5.10. The functions f and f^{-1} have the effect of "undoing" each other. That is, when you form the composition of f with f^{-1} or the composition of f^{-1} with f, you obtain the identity function.

$$f(f^{-1}(x)) = x \qquad \text{and} \qquad f^{-1}(f(x)) = x$$

EXPLORATION

Finding Inverse Functions Explain how to "undo" each of the following functions. Then use your explanation to write the inverse function of f.

a. $f(x) = x - 5$

b. $f(x) = 6x$

c. $f(x) = \dfrac{x}{2}$

d. $f(x) = 3x + 2$

e. $f(x) = x^3$

f. $f(x) = 4(x - 2)$

Use a graphing utility to graph each function and its inverse function in the same "square" viewing window. What observation can you make about each pair of graphs?

> **Definition of Inverse Function**
>
> A function g is the **inverse function** of the function f if
>
> $$f(g(x)) = x \quad \text{for each } x \text{ in the domain of } g$$
>
> and
>
> $$g(f(x)) = x \quad \text{for each } x \text{ in the domain of } f.$$
>
> The function g is denoted by f^{-1} (read "f inverse").

NOTE Although the notation used to denote an inverse function resembles *exponential notation*, it is a different use of -1 as a superscript. That is, in general, $f^{-1}(x) \neq 1/f(x)$.

Here are some important observations about inverse functions.

1. If g is the inverse function of f, then f is the inverse function of g.
2. The domain of f^{-1} is equal to the range of f, and the range of f^{-1} is equal to the domain of f.
3. A function need not have an inverse function, but if it does, the inverse function is unique (see Exercise 99).

You can think of f^{-1} as undoing what has been done by f. For example, subtraction can be used to undo addition, and division can be used to undo multiplication. Use the definition of an inverse function to check the following.

$$f(x) = x + c \qquad \text{and} \qquad f^{-1}(x) = x - c \quad \text{are inverse functions of each other.}$$

$$f(x) = cx \qquad \text{and} \qquad f^{-1}(x) = \frac{x}{c}, \ c \neq 0, \quad \text{are inverse functions of each other.}$$

Example 1   **Verifying Inverse Functions**

Show that the functions are inverse functions of each other.

$$f(x) = 2x^3 - 1 \quad \text{and} \quad g(x) = \sqrt[3]{\frac{x + 1}{2}}$$

Solution   Because the domains and ranges of both f and g consist of all real numbers, you can conclude that both composite functions exist for all x. The composite of f with g is given by

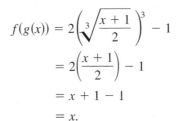

$$f(g(x)) = 2\left(\sqrt[3]{\frac{x + 1}{2}}\right)^3 - 1$$

$$= 2\left(\frac{x + 1}{2}\right) - 1$$

$$= x + 1 - 1$$

$$= x.$$

The composite of g with f is given by

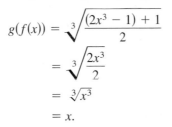

$$g(f(x)) = \sqrt[3]{\frac{(2x^3 - 1) + 1}{2}}$$

$$= \sqrt[3]{\frac{2x^3}{2}}$$

$$= \sqrt[3]{x^3}$$

$$= x.$$

Because $f(g(x)) = x$ and $g(f(x)) = x$, you can conclude that f and g are inverse functions of each other (see Figure 5.11).

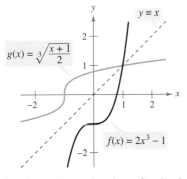

f and g are inverse functions of each other.
Figure 5.11

STUDY TIP   In Example 1, try comparing the functions f and g verbally.

For f: First cube x, then multiply by 2, then subtract 1.

For g: First add 1, then divide by 2, then take the cube root.

Do you see the "undoing pattern"?

In Figure 5.11, the graphs of f and $g = f^{-1}$ appear to be mirror images of each other with respect to the line $y = x$. The graph of f^{-1} is a **reflection** of the graph of f in the line $y = x$. This idea is generalized in the following theorem.

THEOREM 5.6   Reflective Property of Inverse Functions

The graph of f contains the point (a, b) if and only if the graph of f^{-1} contains the point (b, a).

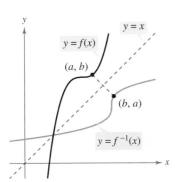

The graph of f^{-1} is a reflection of the graph of f in the line $y = x$.
Figure 5.12

Proof   If (a, b) is on the graph of f, then $f(a) = b$ and you can write

$$f^{-1}(b) = f^{-1}(f(a)) = a.$$

So, (b, a) is on the graph of f^{-1}, as shown in Figure 5.12. A similar argument will prove the theorem in the other direction.

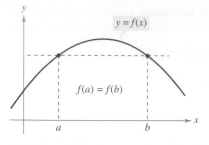

If a horizontal line intersects the graph of f twice, then f is not one-to-one.
Figure 5.13

Existence of an Inverse Function

Not every function has an inverse, and Theorem 5.6 suggests a graphical test for those that do—the **horizontal line test** for an inverse function. This test states that a function f has an inverse function if and only if every horizontal line intersects the graph of f at most once (see Figure 5.13). The following theorem formally states why the horizontal line test is valid. (Recall from Section 3.3 that a function is *strictly monotonic* if it is either increasing on its entire domain or decreasing on its entire domain.)

THEOREM 5.7   The Existence of an Inverse Function

1. A function has an inverse function if and only if it is one-to-one.

2. If f is strictly monotonic on its entire domain, then it is one-to-one and therefore has an inverse function.

Proof   To prove the second part of the theorem, recall from Section P.3 that f is one-to-one if for x_1 and x_2 in its domain

$$f(x_1) = f(x_2) \implies x_1 = x_2.$$

The *contrapositive* of this implication is logically equivalent and states that

$$x_1 \neq x_2 \implies f(x_1) \neq f(x_2).$$

Now, choose x_1 and x_2 in the domain of f. If $x_1 \neq x_2$, then, because f is strictly monotonic, it follows that either

$$f(x_1) < f(x_2) \quad \text{or} \quad f(x_1) > f(x_2).$$

In either case, $f(x_1) \neq f(x_2)$. So, f is one-to-one on the interval. The proof of the first part of the theorem is left as an exercise (see Exercise 100).

Example 2   **The Existence of an Inverse Function**

Which of the functions has an inverse function?

a. $f(x) = x^3 + x - 1$ **b.** $f(x) = x^3 - x + 1$

Solution

a. From the graph of f given in Figure 5.14a, it appears that f is increasing over its entire domain. To verify this, note that the derivative, $f'(x) = 3x^2 + 1$, is positive for all real values of x. Therefore, f is strictly monotonic and it must have an inverse function.

b. From the graph in Figure 5.14b, you can see that the function does not pass the horizontal line test. In other words, it is not one-to-one. For instance, f has the same value when $x = -1$, 0, and 1.

$$f(-1) = f(1) = f(0) = 1 \qquad \text{Not one-to-one}$$

Therefore, by Theorem 5.7, f does not have an inverse function.

NOTE   Often it is easier to prove that a function *has* an inverse function than to find the inverse function. For instance, it would be difficult algebraically to find the inverse function of the function in Example 2a.

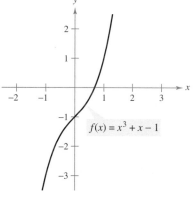

(a) Because f is increasing over its entire domain, it has an inverse function.

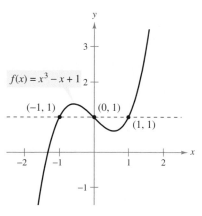

(b) Because f is not one-to-one, it does not have an inverse function.
Figure 5.14

The following guidelines suggest a procedure for finding an inverse function.

Guidelines for Finding an Inverse Function

1. Use Theorem 5.7 to determine whether the function given by $y = f(x)$ has an inverse function.
2. Solve for x as a function of y: $x = g(y) = f^{-1}(y)$.
3. Interchange x and y. The resulting equation is $y = f^{-1}(x)$.
4. Define the domain of f^{-1} to be the range of f.
5. Verify that $f(f^{-1}(x)) = x$ and $f^{-1}(f(x)) = x$.

Example 3   **Finding an Inverse Function**

Find the inverse function of $f(x) = \sqrt{2x - 3}$.

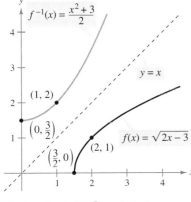

$f^{-1}(x) = \dfrac{x^2 + 3}{2}$

$y = x$

$(1, 2)$

$\left(0, \dfrac{3}{2}\right)$

$f(x) = \sqrt{2x - 3}$

$\left(\dfrac{3}{2}, 0\right)$ $(2, 1)$

The domain of f^{-1}, $[0, \infty)$, is the range of f.
Figure 5.15

Solution   The function has an inverse function because it is increasing on its entire domain (see Figure 5.15). To find an equation for the inverse function, let $y = f(x)$ and solve for x in terms of y.

$$\sqrt{2x - 3} = y \qquad\qquad \text{Let } y = f(x).$$

$$2x - 3 = y^2 \qquad\qquad \text{Square both sides.}$$

$$x = \frac{y^2 + 3}{2} \qquad\qquad \text{Solve for } x.$$

$$y = \frac{x^2 + 3}{2} \qquad\qquad \text{Interchange } x \text{ and } y.$$

$$f^{-1}(x) = \frac{x^2 + 3}{2} \qquad\qquad \text{Replace } y \text{ by } f^{-1}(x).$$

The domain of f^{-1} is the range of f, which is $[0, \infty)$. You can verify this result as follows.

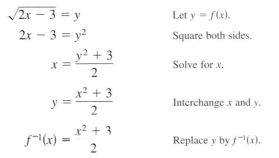

$$f(f^{-1}(x)) = \sqrt{2\left(\frac{x^2 + 3}{2}\right) - 3} = \sqrt{x^2} = x, \qquad x \ge 0$$

$$f^{-1}(f(x)) = \frac{\left(\sqrt{2x - 3}\right)^2 + 3}{2} = \frac{2x - 3 + 3}{2} = x, \qquad x \ge \frac{3}{2}$$

NOTE   Remember that any letter can be used to represent the independent variable. So,

$$f^{-1}(y) = \frac{y^2 + 3}{2}$$

$$f^{-1}(x) = \frac{x^2 + 3}{2}$$

$$f^{-1}(s) = \frac{s^2 + 3}{2}$$

all represent the same function.

Theorem 5.7 is useful in the following type of problem. Suppose you are given a function that is *not* one-to-one on its domain. By restricting the domain to an interval on which the function is strictly monotonic, you can conclude that the new function *is* one-to-one on the restricted domain.

Example 4   Testing Whether a Function Is One-to-One

Show that the sine function

$$f(x) = \sin x$$

is not one-to-one on the entire real line. Then show that $[-\pi/2, \pi/2]$ is the largest interval, centered at the origin, for which f is strictly monotonic.

Solution   It is clear that f is not one-to-one, because many different x-values yield the same y-value. For instance,

$$\sin(0) = 0 = \sin(\pi).$$

Moreover, f is increasing on the open interval $(-\pi/2, \pi/2)$, because its derivative

$$f'(x) = \cos x$$

is positive there. Finally, because the left and right endpoints correspond to relative extrema of the sine function, you can conclude that f is increasing on the closed interval $[-\pi/2, \pi/2]$ *and* that in any larger interval the function is not strictly monotonic (see Figure 5.16).

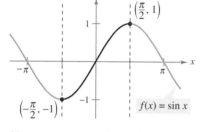

f is one-to-one on the interval $[-\pi/2, \pi/2]$.

Figure 5.16

Derivative of an Inverse Function

The next two theorems discuss the derivative of an inverse function. The reasonableness of Theorem 5.8 follows from the reflective property of inverse functions as shown in Figure 5.12. Proofs of the two theorems are given in Appendix B.

THEOREM 5.8   Continuity and Differentiability of Inverse Functions

Let f be a function whose domain is an interval I. If f has an inverse function, then the following statements are true.

1. If f is continuous on its domain, then f^{-1} is continuous on its domain.
2. If f is increasing on its domain, then f^{-1} is increasing on its domain.
3. If f is decreasing on its domain, then f^{-1} is decreasing on its domain.
4. If f is differentiable at c and $f'(c) \neq 0$, then f^{-1} is differentiable at $f(c)$.

EXPLORATION

Graph the inverse functions

$$f(x) = x^3$$

and

$$g(x) = x^{1/3}.$$

Calculate the slope of f at $(1, 1)$, $(2, 8)$, and $(3, 27)$, and the slope of g at $(1, 1)$, $(8, 2)$, and $(27, 3)$. What do you observe? What happens at $(0, 0)$?

THEOREM 5.9   The Derivative of an Inverse Function

Let f be a function that is differentiable on an interval I. If f has an inverse function g, then g is differentiable at any x for which $f'(g(x)) \neq 0$. Moreover,

$$g'(x) = \frac{1}{f'(g(x))}, \qquad f'(g(x)) \neq 0.$$

Example 5   Evaluating the Derivative of an Inverse Function

Let $f(x) = \frac{1}{4}x^3 + x - 1$.

a. What is the value of $f^{-1}(x)$ when $x = 3$?

b. What is the value of $(f^{-1})'(x)$ when $x = 3$?

Solution   Notice that f is one-to-one and hence has an inverse function.

a. Because $f(x) = 3$ when $x = 2$, you know that $f^{-1}(3) = 2$.

b. Because the function f is differentiable and has an inverse function, you can apply Theorem 5.9 (with $g = f^{-1}$) to write

$$(f^{-1})'(3) = \frac{1}{f'(f^{-1}(3))} = \frac{1}{f'(2)}.$$

Moreover, using $f'(x) = \frac{3}{4}x^2 + 1$, you can conclude that

$$(f^{-1})'(3) = \frac{1}{f'(2)} = \frac{1}{\frac{3}{4}(2^2) + 1} = \frac{1}{4}.$$

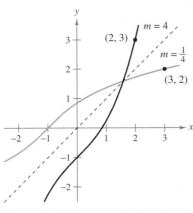

The graphs of the inverse functions f and f^{-1} have reciprocal slopes at points (a, b) and (b, a).

Figure 5.17

In Example 5, note that at the point $(2, 3)$ the slope of the graph of f is 4 and at the point $(3, 2)$ the slope of the graph of f^{-1} is $\frac{1}{4}$ (see Figure 5.17). This reciprocal relationship (which follows from Theorem 5.9) is sometimes written as

$$\frac{dy}{dx} = \frac{1}{dx/dy}.$$

Example 6   Graphs of Inverse Functions Have Reciprocal Slopes

Let $f(x) = x^2$ (for $x \geq 0$) and let $f^{-1}(x) = \sqrt{x}$. Show that the slopes of the graphs of f and f^{-1} are reciprocals at each of the following points.

a. $(2, 4)$ and $(4, 2)$

b. $(3, 9)$ and $(9, 3)$

Solution   The derivatives of f and f^{-1} are given by

$$f'(x) = 2x \qquad \text{and} \qquad (f^{-1})'(x) = \frac{1}{2\sqrt{x}}.$$

a. At $(2, 4)$, the slope of the graph of f is $f'(2) = 2(2) = 4$. At $(4, 2)$, the slope of the graph of f^{-1} is

$$(f^{-1})'(4) = \frac{1}{2\sqrt{4}} = \frac{1}{2(2)} = \frac{1}{4}.$$

b. At $(3, 9)$, the slope of the graph of f is $f'(3) = 2(3) = 6$. At $(9, 3)$, the slope of the graph of f^{-1} is

$$(f^{-1})'(9) = \frac{1}{2\sqrt{9}} = \frac{1}{2(3)} = \frac{1}{6}.$$

So, in both cases, the slopes are reciprocals, as shown in Figure 5.18.

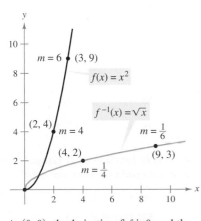

At $(0, 0)$, the derivative of f is 0, and the derivative of f^{-1} does not exist.

Figure 5.18

EXERCISES FOR SECTION 5.3

In Exercises 1–8, show that f and g are inverse functions (a) analytically and (b) graphically.

1. $f(x) = 5x + 1$, $\qquad$ $g(x) = (x - 1)/5$

2. $f(x) = 3 - 4x$, $\qquad$ $g(x) = (3 - x)/4$

3. $f(x) = x^3$, $\qquad$ $g(x) = \sqrt[3]{x}$

4. $f(x) = 1 - x^3$, $\qquad$ $g(x) = \sqrt[3]{1 - x}$

5. $f(x) = \sqrt{x - 4}$, $\qquad$ $g(x) = x^2 + 4$, $\quad x \geq 0$

6. $f(x) = 16 - x^2$, $\quad x \geq 0$, $\qquad$ $g(x) = \sqrt{16 - x}$

7. $f(x) = 1/x$, $\qquad$ $g(x) = 1/x$

8. $f(x) = \dfrac{1}{1 + x}$, $\quad x \geq 0$, $\qquad$ $g(x) = \dfrac{1 - x}{x}$, $\quad 0 < x \leq 1$

In Exercises 9–12, match the graph of the function with the graph of its inverse function. [The graphs of the inverse functions are labeled (a), (b), (c), and (d).]

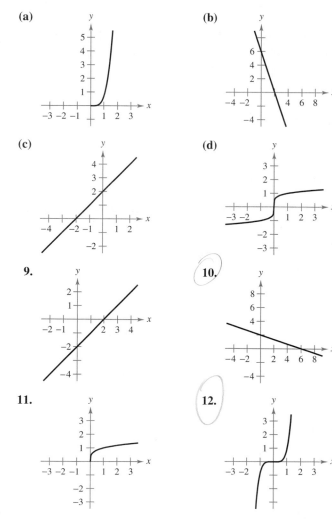

(a)

(b)

(c)

(d)

9.

10.

11.

12.

In Exercises 13–16, use the horizontal line test to determine whether the function is one-to-one on its entire domain and therefore has an inverse function. To print an enlarged copy of the graph, go to the website _www.mathgraphs.com_.

13. $f(x) = \frac{3}{4}x + 6$ $\qquad$ **14.** $f(x) = 5x - 3$

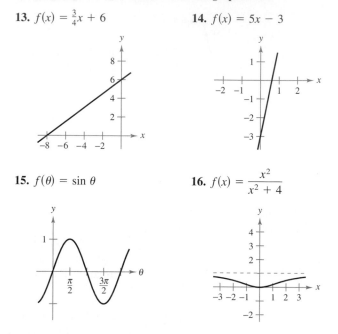

15. $f(\theta) = \sin \theta$ $\qquad$ **16.** $f(x) = \dfrac{x^2}{x^2 + 4}$

In Exercises 17–22, use a graphing utility to graph the function. Determine whether the function is one-to-one on its entire domain.

17. $h(s) = \dfrac{1}{s - 2} - 3$ $\qquad$ **18.** $g(t) = \dfrac{1}{\sqrt{t^2 + 1}}$

19. $f(x) = \ln x$ $\qquad$ **20.** $f(x) = 5x\sqrt{x - 1}$

21. $g(x) = (x + 5)^3$ $\qquad$ **22.** $h(x) = |x + 4| - |x - 4|$

In Exercises 23–28, use the derivative to determine whether the function is strictly monotonic on its entire domain and therefore has an inverse function.

23. $f(x) = (x + a)^3 + b$ $\qquad$ **24.** $f(x) = \cos \dfrac{3x}{2}$

25. $f(x) = \dfrac{x^4}{4} - 2x^2$ $\qquad$ **26.** $f(x) = x^3 - 6x^2 + 12x$

27. $f(x) = 2 - x - x^3$ $\qquad$ **28.** $f(x) = \ln(x - 3)$

In Exercises 29–36, find the inverse function of f. Graph (by hand) f and f^{-1}. Describe the relationship between the graphs.

29. $f(x) = 2x - 3$ $\qquad$ **30.** $f(x) = 3x$

31. $f(x) = x^5$ $\qquad$ **32.** $f(x) = x^3 - 1$

33. $f(x) = \sqrt{x}$ $\qquad$ **34.** $f(x) = x^2$, $\quad x \geq 0$

35. $f(x) = \sqrt{4 - x^2}$, $\quad x \geq 0$

36. $f(x) = \sqrt{x^2 - 4}$, $\quad x \geq 2$

In Exercises 37–42, find the inverse function of f. Use a graphing utility to graph f and f^{-1} in the same viewing window. Describe the relationship between the graphs.

37. $f(x) = \sqrt[3]{x - 1}$

38. $f(x) = 3\sqrt[5]{2x - 1}$

39. $f(x) = x^{2/3}, \quad x \geq 0$

40. $f(x) = x^{3/5}$

41. $f(x) = \dfrac{x}{\sqrt{x^2 + 7}}$

42. $f(x) = \dfrac{x + 2}{x}$

In Exercises 43 and 44, use the graph of the function f to complete the table and sketch the graph of f^{-1}. To print an enlarged copy of the graph, go to the website *www.mathgraphs.com*.

43.

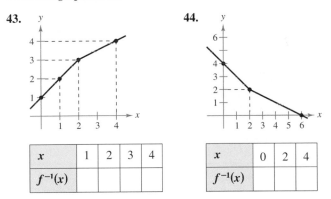

x	1	2	3	4
$f^{-1}(x)$				

44.

x	0	2	4
$f^{-1}(x)$			

45. *Cost*   Suppose you need 50 pounds of two commodities costing $1.25 and $1.60 per pound.

(a) Verify that the total cost is

$$y = 1.25x + 1.60(50 - x)$$

where x is the number of pounds of the less expensive commodity.

(b) Find the inverse function of the cost function. What does each variable represent in the inverse function?

(c) Use the context of the problem to determine the domain of the inverse function.

(d) Determine the number of pounds of the less expensive commodity purchased if the total cost is $73.

46. *Think About It*   The function $f(x) - k(2 - x - x^3)$ is one-to-one and $f^{-1}(3) = -2$. Find k.

In Exercises 47–52, show that f is strictly monotonic on the indicated interval and therefore has an inverse function on that interval.

Function	Interval
47. $f(x) = (x - 4)^2$	$[4, \infty)$
48. $f(x) = |x + 2|$	$[-2, \infty)$
49. $f(x) = \dfrac{4}{x^2}$	$(0, \infty)$
50. $f(x) = \cot x$	$(0, \pi)$
51. $f(x) = \cos x$	$[0, \pi]$
52. $f(x) = \sec x$	$\left[0, \dfrac{\pi}{2}\right)$

In Exercises 53 and 54, find the inverse function of f over the indicated interval. Use a graphing utility to graph f and f^{-1} in the same viewing window. Describe the relationship between the graphs.

Function	Interval
53. $f(x) = \dfrac{x}{x^2 - 4}$	$-2 < x < 2$
54. $f(x) = 2 - \dfrac{3}{x^2}$	$0 < x < 10$

Graphical Reasoning   In Exercises 55–58, (a) use a graphing utility to graph the function, (b) use the *drawing* feature of a graphing utility to draw the inverse of the function, and (c) determine whether the graph of the inverse relation is an inverse function. Explain your reasoning.

55. $f(x) = x^3 + x + 4$

56. $h(x) = x\sqrt{4 - x^2}$

57. $g(x) = \dfrac{3x^2}{x^2 + 1}$

58. $f(x) = \dfrac{4x}{\sqrt{x^2 + 15}}$

In Exercises 59–62, determine whether the function is one-to-one. If it is, find its inverse function.

59. $f(x) = \sqrt{x - 2}$

60. $f(x) = -3$

61. $f(x) = |x - 2|, \quad x \leq 2$

62. $f(x) = ax + b, \quad a \neq 0$

In Exercises 63–66, delete part of the domain so that the function that remains is one-to-one. Find the inverse function of the remaining function and give the domain of the inverse function. (*Note:* There is more than one correct answer.)

63. $f(x) = (x - 3)^2$

64. $f(x) = 16 - x^4$

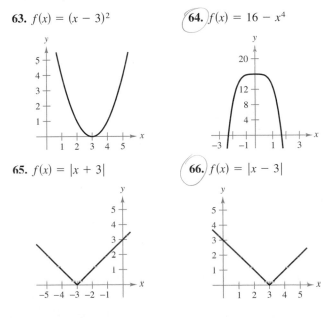

65. $f(x) = |x + 3|$

66. $f(x) = |x - 3|$

Think About It In Exercises 67–70, decide whether the function has an inverse function. If so, what is the inverse function?

67. $g(t)$ is the volume of water that has passed through a water line t minutes after a control valve is opened.

68. $h(t)$ is the height of the tide t hours after midnight, where $0 \le t < 24$.

69. $C(t)$ is the cost of a long distance call lasting t minutes.

70. $A(r)$ is the area of a circle of radius r.

In Exercises 71–76, find $(f^{-1})'(a)$ for the function f and real number a.

Function	Real Number
71. $f(x) = x^3 + 2x - 1$	$a = 2$
72. $f(x) = \frac{1}{27}(x^5 + 2x^3)$,	$a = -11$
73. $f(x) = \sin x$, $\quad -\frac{\pi}{2} \le x \le \frac{\pi}{2}$	$a = \frac{1}{2}$
74. $f(x) = \cos 2x$, $\quad 0 \le x \le \frac{\pi}{2}$	$a = 1$
75. $f(x) = x^3 - \frac{4}{x}$	$a = 6$
76. $f(x) = \sqrt{x - 4}$	$a = 2$

In Exercises 77–80, (a) find the domains of f and f^{-1}, (b) find the ranges of f and f^{-1}, (c) graph f and f^{-1}, and (d) show that the slopes of the graphs of f and f^{-1} are reciprocals at the indicated points.

Functions	Point
77. $f(x) = x^3$	$\left(\frac{1}{2}, \frac{1}{8}\right)$
$\quad f^{-1}(x) = \sqrt[3]{x}$	$\left(\frac{1}{8}, \frac{1}{2}\right)$
78. $f(x) = 3 - 4x$	$(1, -1)$
$\quad f^{-1}(x) = \dfrac{3 - x}{4}$	$(-1, 1)$
79. $f(x) = \sqrt{x - 4}$	$(5, 1)$
$\quad f^{-1}(x) = x^2 + 4$, $\quad x \ge 0$	$(1, 5)$
80. $f(x) = \dfrac{4}{1 + x^2}$, $\quad x \ge 0$	$(1, 2)$
$\quad f^{-1}(x) = \sqrt{\dfrac{4 - x}{x}}$	$(2, 1)$

In Exercises 81 and 82, find dy/dx at the indicated point for the equation.

81. $x = y^3 - 7y^2 + 2$

$\quad (-4, 1)$

82. $x = 2 \ln(y^2 - 3)$

$\quad (0, 4)$

In Exercises 83–86, use the functions $f(x) = \frac{1}{8}x - 3$ and $g(x) = x^3$ to find the indicated value.

83. $(f^{-1} \circ g^{-1})(1)$

84. $(g^{-1} \circ f^{-1})(-3)$

85. $(f^{-1} \circ f^{-1})(6)$

86. $(g^{-1} \circ g^{-1})(-4)$

In Exercises 87–90, use the functions $f(x) = x + 4$ and $g(x) = 2x - 5$ to find the indicated function.

87. $g^{-1} \circ f^{-1}$

88. $f^{-1} \circ g^{-1}$

89. $(f \circ g)^{-1}$

90. $(g \circ f)^{-1}$

Getting at the Concept

91. Describe how to find the inverse function of a one-to-one function given by an equation in x and y. Give an example.

92. Describe the relationship between the graph of a function and the graph of its inverse function.

93. Give an example of a function that does *not* have an inverse function.

94. State the theorem that gives the method for finding the derivative of an inverse function.

In Exercises 95 and 96, the derivative of the function has the same sign for all x in its domain, but the function is not one-to-one. Explain.

95. $f(x) = \tan x$

96. $f(x) = \dfrac{x}{x^2 - 4}$

97. Prove that if f and g are one-to-one functions, then $(f \circ g)^{-1}(x) = (g^{-1} \circ f^{-1})(x)$.

98. Prove that if f has an inverse function, then $(f^{-1})^{-1} = f$.

99. Prove that if a function has an inverse function, then the inverse function is unique.

100. Prove that a function has an inverse function if and only if it is one-to-one.

True or False? In Exercises 101–104, determine whether the statement is true or false. If it is false, explain why or give an example that shows it is false.

101. If f is an even function, then f^{-1} exists.

102. If the inverse function of f exists, then the y-intercept of f is an x-intercept of f^{-1}.

103. If $f(x) = x^n$ where n is odd, then f^{-1} exists.

104. There exists no function f such that $f = f^{-1}$.

105. Is the converse of the second part of Theorem 5.7 true? That is, if a function is one-to-one (and hence has an inverse function), then must the function be strictly monotonic? If so, prove it. If not, give a counterexample.

106. Let f be twice-differentiable and one-to-one on an open interval I. Show that its inverse function g satisfies

$$g''(x) = -\frac{f''(g(x))}{[f'(g(x))]^3}.$$

If f is increasing and concave downward, what is the concavity of $f^{-1} = g$?

107. If $f(x) = \displaystyle\int_2^x \frac{dt}{\sqrt{1 + t^4}}$, find $(f^{-1})'(0)$.

Exponential Functions: Differentiation and Integration

- Develop properties of the natural exponential function.
- Differentiate natural exponential functions.
- Integrate natural exponential functions.

The Natural Exponential Function

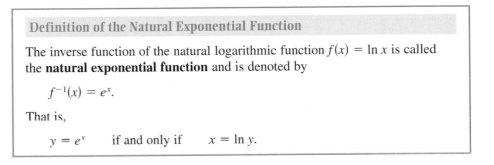

The function $f(x) = \ln x$ is increasing on its entire domain, and hence it has an inverse function f^{-1}. The domain of f^{-1} is the set of all reals, and the range is the set of positive reals, as shown in Figure 5.19. So, for any real number x,

$$f(f^{-1}(x)) = \ln[f^{-1}(x)] = x. \qquad \text{\textit{x} is any real number.}$$

If x happens to be rational, then

$$\ln(e^x) = x \ln e = x(1) = x. \qquad \text{\textit{x} is a rational number.}$$

Because the natural logarithmic function is one-to-one, you can conclude that $f^{-1}(x)$ and e^x agree for *rational* values of x. The following definition extends the meaning of e^x to include *all* real values of x.

The inverse function of the natural logarithmic function is the natural exponential function.

Figure 5.19

Definition of the Natural Exponential Function

The inverse function of the natural logarithmic function $f(x) = \ln x$ is called the **natural exponential function** and is denoted by

$$f^{-1}(x) = e^x.$$

That is,

$$y = e^x \qquad \text{if and only if} \qquad x = \ln y.$$

The inverse relationship between the natural logarithmic function and the natural exponential function can be summarized as follows.

$$\ln(e^x) = x \qquad \text{and} \qquad e^{\ln x} = x \qquad \text{Inverse relationship}$$

THE NUMBER e

The symbol e was first used by mathematician Leonhard Euler to represent the base of natural logarithms in a letter to another mathematician, Christian Goldbach, in 1731.

Example 1 **Solving Exponential Equations**

Solve $7 = e^{x+1}$.

Solution You can convert from exponential form to logarithmic form by *taking the natural log of both sides* of the equation.

$7 = e^{x+1}$	Write original equation.
$\ln 7 = \ln(e^{x+1})$	Take natural log of both sides.
$\ln 7 = x + 1$	Apply inverse property.
$-1 + \ln 7 = x$	Solve for x.
$0.946 \approx x$	Use a calculator.

Check this solution in the original equation.

Example 2   Solving a Logarithmic Equation

Solve $\ln(2x - 3) = 5$.

Solution   To convert from logarithmic form to exponential form, you can *exponentiate both sides* of the logarithmic equation.

$$\ln(2x - 3) = 5 \qquad \text{Write original equation.}$$
$$e^{\ln(2x-3)} = e^5 \qquad \text{Exponentiate both sides.}$$
$$2x - 3 = e^5 \qquad \text{Apply inverse property.}$$
$$x = \tfrac{1}{2}(e^5 + 3) \qquad \text{Solve for } x.$$
$$x \approx 75.707 \qquad \text{Use a calculator.}$$

The familiar rules for operating with rational exponents can be extended to the natural exponential function, as indicated in the following theorem.

THEOREM 5.10   Operations with Exponential Functions

Let a and b be any real numbers.

1. $e^a e^b = e^{a+b}$ **2.** $\dfrac{e^a}{e^b} = e^{a-b}$

Proof   To prove Property 1, you can write

$$\ln(e^a e^b) = \ln(e^a) + \ln(e^b)$$
$$= a + b$$
$$= \ln(e^{a+b}).$$

Because the natural log function is one-to-one, you can conclude that

$$e^a e^b = e^{a+b}.$$

The proof of the second property is left to you (see Exercise 130).

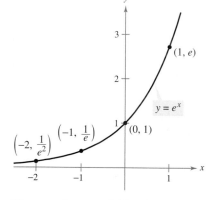

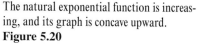

The natural exponential function is increasing, and its graph is concave upward.
Figure 5.20

In Section 5.3, you learned that an inverse function f^{-1} shares many properties with f. So, the natural exponential function inherits the following properties from the natural logarithmic function (see Figure 5.20).

Properties of the Natural Exponential Function

1. The domain of $f(x) = e^x$ is $(-\infty, \infty)$, and the range is $(0, \infty)$.

2. The function $f(x) = e^x$ is continuous, increasing, and one-to-one on its entire domain.

3. The graph of $f(x) = e^x$ is concave upward on its entire domain.

4. $\lim\limits_{x \to -\infty} e^x = 0$ and $\lim\limits_{x \to \infty} e^x = \infty$

Derivatives of Exponential Functions

One of the most intriguing (and useful) characteristics of the natural exponential function is that *it is its own derivative*. In other words, it is a solution to the differential equation $y' = y$. This result is stated in the next theorem.

FOR FURTHER INFORMATION To find out about derivatives of exponential functions of order 1/2, see the article "A Child's Garden of Fractional Derivatives" by Marcia Kleinz and Thomas J. Osler in *The College Mathematics Journal*. To view this article, go to the website *www.matharticles.com*.

> **THEOREM 5.11   The Derivative of the Natural Exponential Function**
>
> Let u be a differentiable function of x.
>
> **1.** $\dfrac{d}{dx}[e^x] = e^x$ **2.** $\dfrac{d}{dx}[e^u] = e^u \dfrac{du}{dx}$

Proof   To prove Property 1, use the fact that $\ln e^x = x$, and differentiate both sides of the equation.

$$\ln e^x = x \qquad\qquad \text{Definition of exponential function}$$

$$\frac{d}{dx}[\ln e^x] = \frac{d}{dx}[x] \qquad\qquad \text{Differentiate both sides with respect to } x.$$

$$\frac{1}{e^x}\frac{d}{dx}[e^x] = 1$$

$$\frac{d}{dx}[e^x] = e^x$$

The derivative of e^u follows from the Chain Rule.   ◢

NOTE   You can interpret this theorem geometrically by saying that the slope of the graph of $f(x) = e^x$ at any point (x, e^x) is equal to the y-coordinate of the point.

Example 3   **Differentiating Exponential Functions**

a. $\dfrac{d}{dx}[e^{2x-1}] = e^u \dfrac{du}{dx} = 2e^{2x-1}$ $u = 2x - 1$

b. $\dfrac{d}{dx}[e^{-3/x}] = e^u \dfrac{du}{dx} = \left(\dfrac{3}{x^2}\right)e^{-3/x} = \dfrac{3e^{-3/x}}{x^2}$ $u = -\dfrac{3}{x}$

Example 4   **Locating Relative Extrema**

Find the relative extrema of $f(x) = xe^x$.

Solution   The derivative of f is given by

$$f'(x) = x(e^x) + e^x(1) \qquad\qquad \text{Product Rule}$$
$$= e^x(x + 1).$$

Because e^x is never 0, the derivative is 0 only when $x = -1$. Moreover, by the First Derivative Test, you can determine that this corresponds to a relative minimum, as shown in Figure 5.21. Because the derivative $f'(x) = e^x(x + 1)$ is defined for all x, there are no other critical points.   ◢

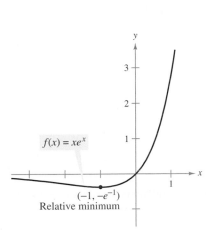

$f(x) = xe^x$

$(-1, -e^{-1})$
Relative minimum

The derivative of f changes from negative to positive at $x = -1$.
Figure 5.21

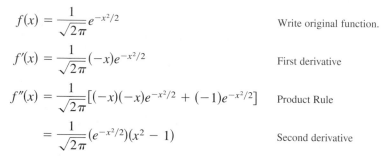

Example 5 The Normal Probability Density Function

Show that the *normal probability density function*

$$f(x) = \frac{1}{\sqrt{2\pi}} e^{-x^2/2}$$

has points of inflection when $x = \pm 1$.

Solution To locate possible points of inflection, find the x-values for which the second derivative is 0.

$$f(x) = \frac{1}{\sqrt{2\pi}} e^{-x^2/2} \qquad \text{Write original function.}$$

$$f'(x) = \frac{1}{\sqrt{2\pi}}(-x)e^{-x^2/2} \qquad \text{First derivative}$$

$$f''(x) = \frac{1}{\sqrt{2\pi}}[(-x)(-x)e^{-x^2/2} + (-1)e^{-x^2/2}] \qquad \text{Product Rule}$$

$$= \frac{1}{\sqrt{2\pi}}(e^{-x^2/2})(x^2 - 1) \qquad \text{Second derivative}$$

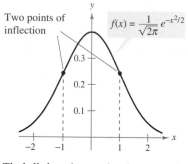

Two points of inflection

$$f(x) = \frac{1}{\sqrt{2\pi}} e^{-x^2/2}$$

The bell-shaped curve given by a normal probability density function
Figure 5.22

Therefore, $f''(x) = 0$ when $x = \pm 1$, and you can apply the techniques of Chapter 3 to conclude that these values yield the two points of inflection shown in Figure 5.22.

NOTE The general form of a normal probability density function (whose mean is 0) is given by

$$f(x) = \frac{1}{\sigma\sqrt{2\pi}} e^{-x^2/2\sigma^2}$$

where σ is the standard deviation (σ is the lowercase Greek letter sigma). This "bell-shaped curve" has points of inflection when $x = \pm\sigma$.

Example 6 M.D.s in the United States

The number y of medical doctors (in thousands) in the United States from 1980 through 1997, can be modeled by

$$y = 475,520e^{0.0271t}$$

where $t = 0$ represents 1980. At what rate was the number of M.D.s changing in 1992? *(Source: American Medical Association)*

Solution The derivative of the given model is

$$y' = (0.0271)(475,520)e^{0.0271t}$$
$$\approx 12,887e^{0.0271t}.$$

By evaluating the derivative when $t = 12$, you can conclude that the rate of change in 1992 was about

17,839 doctors per year.

The graph of this model is shown in Figure 5.23.

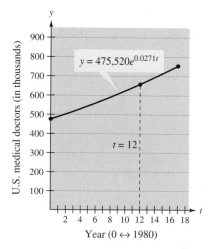

Figure 5.23

Integrals of Exponential Functions

Each differentiation formula in Theorem 5.11 has a corresponding integration formula.

> **THEOREM 5.12   Integration Rules for Exponential Functions**
>
> Let u be a differentiable function of x.
>
> **1.** $\displaystyle \int e^x\, dx = e^x + C$ **2.** $\displaystyle \int e^u\, du = e^u + C$

Example 7   **Integrating Exponential Functions**

Find $\int e^{3x+1}\, dx$.

Solution   If you let $u = 3x + 1$, then $du = 3\, dx$.

$$\int e^{3x+1} dx = \frac{1}{3}\int e^{3x+1}(3)\, dx \qquad \text{Multiply and divide by 3.}$$

$$= \frac{1}{3}\int e^u\, du \qquad \text{Substitute: } u = 3x + 1.$$

$$= \frac{1}{3}e^u + C \qquad \text{Apply Exponential Rule.}$$

$$= \frac{e^{3x+1}}{3} + C \qquad \text{Back-substitute.}$$

NOTE   In Example 7, the missing *constant* factor 3 was introduced to create $du = 3\, dx$. However, remember that you cannot introduce a missing *variable* factor in the integrand. For instance,

$$\int e^{-x^2}\, dx \neq \frac{1}{x}\int e^{-x^2}(x\, dx).$$

Example 8   **Integrating Exponential Functions**

Find $\int 5xe^{-x^2}\, dx$.

Solution   If you let $u = -x^2$, then $du = -2x\, dx$ or $x\, dx = -du/2$.

$$\int 5xe^{-x^2} dx = \int 5e^{-x^2}(x\, dx) \qquad \text{Regroup integrand.}$$

$$= \int 5e^u\left(-\frac{du}{2}\right) \qquad \text{Substitute: } u = -x^2.$$

$$= -\frac{5}{2}\int e^u\, du \qquad \text{Factor } -\tfrac{5}{2} \text{ out of integral.}$$

$$= -\frac{5}{2}e^u + C \qquad \text{Apply Exponential Rule.}$$

$$= -\frac{5}{2}e^{-x^2} + C \qquad \text{Back-substitute.}$$

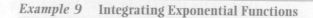

Example 9   Integrating Exponential Functions

a. $\displaystyle\int \frac{e^{1/x}}{x^2}\,dx = -\int \overbrace{e^{1/x}}^{e^u}\overbrace{\left(-\frac{1}{x^2}\right)dx}^{du} \qquad\qquad u = \frac{1}{x}$

$\qquad\qquad\quad = -e^{1/x} + C$

b. $\displaystyle\int \sin x\, e^{\cos x}\,dx = -\int \overbrace{e^{\cos x}}^{e^u}\overbrace{(-\sin x\,dx)}^{du} \qquad u = \cos x$

$\qquad\qquad\qquad\quad = -e^{\cos x} + C$

Example 10   Finding Areas Bounded by Exponential Functions

Evaluate the definite integral.

a. $\displaystyle\int_0^1 e^{-x}\,dx$ **b.** $\displaystyle\int_0^1 \frac{e^x}{1+e^x}\,dx$ **c.** $\displaystyle\int_{-1}^0 \left[e^x\cos(e^x)\right]dx$

Solution

a. $\displaystyle\int_0^1 e^{-x}\,dx = -e^{-x}\Big]_0^1 \qquad\qquad$ See Figure 5.24(a).

$\qquad\qquad\quad = -e^{-1} - (-1)$

$\qquad\qquad\quad = 1 - \dfrac{1}{e}$

$\qquad\qquad\quad \approx 0.632$

b. $\displaystyle\int_0^1 \frac{e^x}{1+e^x}\,dx = \ln(1+e^x)\Big]_0^1 \qquad$ See Figure 5.24(b).

$\qquad\qquad\qquad\quad = \ln(1+e) - \ln 2$

$\qquad\qquad\qquad\quad \approx 0.620$

c. $\displaystyle\int_{-1}^0 \left[e^x\cos(e^x)\right]dx = \sin(e^x)\Big]_{-1}^0 \qquad$ See Figure 5.24(c).

$\qquad\qquad\qquad\qquad\quad = \sin 1 - \sin(e^{-1})$

$\qquad\qquad\qquad\qquad\quad \approx 0.482$

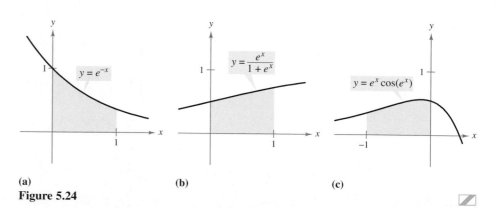

(a)   **(b)**   **(c)**

Figure 5.24

EXERCISES FOR SECTION 5.4

In Exercises 1–4, write the exponential equation as a logarithmic equation or vice versa.

1. $e^0 = 1$

2. $e^{-2} = 0.1353 \ldots$

3. $\ln 2 = 0.6931 \ldots$

4. $\ln 0.5 = -0.6931 \ldots$

In Exercises 5–18, solve for x accurate to three decimal places.

5. $e^{\ln x} = 4$

6. $e^{\ln 2x} = 12$

7. $e^x = 12$

8. $4e^x = 83$

9. $9 - 2e^x = 7$

10. $-6 + 3e^x = 8$

11. $50e^{-x} = 30$

12. $200e^{-4x} = 15$

13. $\ln x = 2$

14. $\ln x^2 = 10$

15. $\ln(x - 3) = 2$

16. $\ln 4x - 1$

17. $\ln \sqrt{x + 2} = 1$

18. $\ln(x - 2)^2 = 12$

In Exercises 19–22, sketch the graph of the function.

19. $y = e^{-x}$

20. $y = \frac{1}{2}e^x$

21. $y = e^{-x^2}$

22. $y = e^{-x/2}$

23. Use a graphing utility to graph $f(x) = e^x$ and the given function in the same viewing window. How are the two graphs related?

(a) $g(x) = e^{x-2}$ (b) $h(x) = -\frac{1}{2}e^x$ (c) $q(x) = e^{-x} + 3$

24. Use a graphing utility to graph the function. Use the graph to determine any asymptotes of the function.

(a) $f(x) = \dfrac{8}{1 + e^{-0.5x}}$ (b) $g(x) = \dfrac{8}{1 + e^{-0.5/x}}$

In Exercises 25–28, match the equation with the correct graph. Assume that a and C are positive real numbers. [The graphs are labeled (a), (b), (c), and (d).]

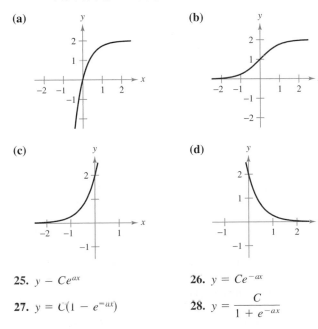

(a)

(b)

(c)

(d)

25. $y - Ce^{ax}$

26. $y = Ce^{-ax}$

27. $y = C(1 - e^{-ax})$

28. $y = \dfrac{C}{1 + e^{-ax}}$

In Exercises 29–32, illustrate that the functions are inverses of each other by graphing both functions on the same set of coordinate axes.

29. $f(x) = e^{2x}$

$g(x) = \ln \sqrt{x}$

30. $f(x) = e^{x/3}$

$g(x) = \ln x^3$

31. $f(x) = e^x - 1$

$g(x) = \ln(x + 1)$

32. $f(x) = e^{x-1}$

$g(x) = 1 + \ln x$

33. *Graphical Analysis*   Use a graphing utility to graph

$$f(x) = \left(1 + \frac{0.5}{x}\right)^x \quad \text{and} \quad g(x) = e^{0.5}$$

in the same viewing window. What is the relationship between f and g as $x \to \infty$?

34. *Conjecture*   Use the result of Exercise 33 to make a conjecture about the value of

$$\left(1 + \frac{r}{x}\right)^x$$

as $x \to \infty$.

In Exercises 35 and 36, compare the given number with the number e. Is the number less than or greater than e?

35. $\left(1 + \dfrac{1}{1,000,000}\right)^{1,000,000}$ (See Exercise 34.)

36. $1 + 1 + \dfrac{1}{2} + \dfrac{1}{6} + \dfrac{1}{24} + \dfrac{1}{120} + \dfrac{1}{720} + \dfrac{1}{5040}$

In Exercises 37 and 38, find the slope of the tangent line to the graph of the function at the point $(0, 1)$.

37. (a) $y = e^{3x}$

(b) $y = e^{-3x}$

38. (a) $y = e^{2x}$

(b) $y = e^{-2x}$

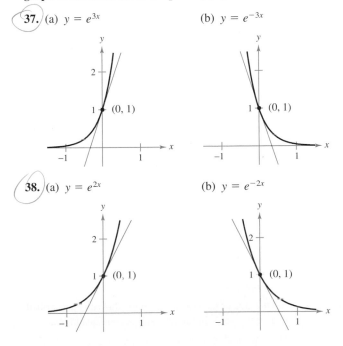

In Exercises 39–58, find the derivative of the function.

39. $f(x) = e^{2x}$

40. $f(x) = e^{1-x}$

41. $y = e^{-2x+x^2}$

42. $y = e^{-x^2}$

43. $y = e^{\sqrt{x}}$

44. $y = x^2 e^{-x}$

45. $g(t) = (e^{-t} + e^t)^3$

46. $g(t) = e^{-3/t^2}$

47. $y = \ln(e^{x^2})$

48. $y = \ln\left(\dfrac{1+e^x}{1-e^x}\right)$

49. $y = \ln(1 + e^{2x})$

50. $y = \ln\dfrac{e^x + e^{-x}}{2}$

51. $y = \dfrac{2}{e^x + e^{-x}}$

52. $y = \dfrac{e^x - e^{-x}}{2}$

53. $y = x^2 e^x - 2xe^x + 2e^x$

54. $y = xe^x - e^x$

55. $f(x) = e^{-x} \ln x$

56. $f(x) = e^3 \ln x$

57. $y = e^x(\sin x + \cos x)$

58. $y = \ln e^x$

In Exercises 59 and 60, use implicit differentiation to find dy/dx.

59. $xe^y - 10x + 3y = 0$

60. $e^{xy} + x^2 - y^2 = 10$

In Exercises 61 and 62, find the second derivative of the function.

61. $f(x) = (3 + 2x)e^{-3x}$

62. $g(x) = \sqrt{x} + e^x \ln x$

In Exercises 63 and 64, show that the function $y = f(x)$ is a solution of the differential equation.

63. $y = e^x(\cos \sqrt{2}x + \sin \sqrt{2}x)$

$y'' - 2y' + 3y = 0$

64. $y = e^x(3 \cos 2x - 4 \sin 2x)$

$y'' - 2y' + 5y = 0$

In Exercises 65–72, find the extrema and the points of inflection (if any exist) of the function. Use a graphing utility to graph the function and confirm your results.

65. $f(x) = \dfrac{e^x + e^{-x}}{2}$

66. $f(x) = \dfrac{e^x - e^{-x}}{2}$

67. $g(x) = \dfrac{1}{\sqrt{2\pi}} e^{-(x-2)^2/2}$

68. $g(x) = \dfrac{1}{\sqrt{2\pi}} e^{-(x-3)^2/2}$

69. $f(x) = x^2 e^{-x}$

70. $f(x) = xe^{-x}$

71. $g(t) = 1 + (2 + t)e^{-t}$

72. $f(x) = -2 + e^{3x}(4 - 2x)$

73. Area Find the area of the largest rectangle that can be inscribed under the curve $y = e^{-x^2}$ in the first and second quadrants.

74. Area Perform the following steps to find the maximum area of the rectangle shown in the figure.

(a) Solve for c in the equation $f(c) = f(c + x)$.

(b) Use the result in part (a) to write the area A as a function of x. [*Hint:* $A = xf(c)$]

(c) Use a graphing utility to graph the area function. Use the graph to approximate the dimensions of the rectangle of maximum area. Determine the maximum area.

(d) Use a graphing utility to graph the expression for c found in part (a). Use the graph to approximate

$$\lim_{x \to 0^+} c \qquad \text{and} \qquad \lim_{x \to \infty} c.$$

Use this result to describe the changes in dimensions and position of the rectangle for $0 < x < \infty$.

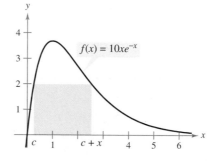

75. Verify that the function

$$y = \dfrac{L}{1 + ae^{-x/b}}, \qquad a > 0, b > 0, L > 0$$

increases at a maximum rate when $y = L/2$.

76. Find the point on the graph of $y = e^{-x}$ where the normal line to the curve passes through the origin. (Use Newton's Method or the root-finding capabilities of a graphing utility.)

77. Find, to three decimal places, the value of x such that

$$e^{-x} = x.$$

(Use Newton's Method or the root-finding capabilities of a graphing utility.)

78. Depreciation The value V of an item t years after it is purchased is

$$V = 15{,}000e^{-0.6286t}, \qquad 0 \le t \le 10.$$

(a) Use a graphing utility to graph the function.

(b) Find the rate of change of V with respect to t when $t = 1$ and $t = 5$.

(c) Use a graphing utility to graph the tangent line to the function when $t = 1$ and $t = 5$.

79. Writing Consider the function

$$f(x) = \dfrac{2}{1 + e^{1/x}}.$$

(a) Use a graphing utility to graph f.

(b) Write a short paragraph explaining why the graph has a horizontal asymptote at $y = 1$ and why the function has a nonremovable discontinuity at $x = 0$.

80. Harmonic Motion   The displacement from equilibrium of a mass oscillating on the end of a spring suspended from a ceiling is

$$y = 1.56e^{-0.22t} \cos 4.9t$$

where y is the displacement in feet and t is the time in seconds. Use a graphing utility to graph the displacement function on the interval $[0, 10]$. Find a value of t past which the displacement is less than 3 inches from equilibrium.

81. Modeling Data   A meteorologist measures the atmospheric pressure P (in kilograms per square meter) at altitude h (in kilometers). The data are shown below.

h	0	5	10	15	20
P	10,332	5583	2376	1240	517

(a) Use a graphing utility to plot the points $(h, \ln P)$. Use the regression capabilities of the graphing utility to find a linear model for the revised data points.

(b) The line in part (a) has the form $\ln P = ah + b$. Write the equation in exponential form.

(c) Use a graphing utility to plot the original data and graph the exponential model in part (b).

(d) Find the rate of change of the pressure when $h = 5$ and $h = 18$.

82. Modeling Data   A 1994 Chevrolet Camaro coupe with a 6-cylinder engine, 5-speed transmission, and air conditioning had a retail price of $17,040. A local dealership had the following guide for the approximate value of the car for the years 1995 through 2000.

Year	1994	1995	1996	1997
Value	$17,040	$14,590	$12,845	$10,995

Year	1998	1999	2000
Value	$9,220	$8,095	$6,835

In each of the following, let V represent the value of the automobile in the year t, with $t = 4$ corresponding to 1994.

(a) Use a computer algebra system to find linear and quadratic models for the data. Plot the data and graph the models.

(b) What does the slope represent in the linear model in part (a)?

(c) Use a computer algebra system to fit an exponential model to the data.

(d) Determine the horizontal asymptote of the exponential model found in part (c). Interpret its meaning in the context of the problem.

(e) Find the rate of decrease in the value of the car when $t = 5$ and $t = 9$ using the exponential model.

Linear and Quadratic Approximations   In Exercises 83 and 84, use a graphing utility to graph the function. Then graph

$$P_1(x) = f(0) + f'(0)(x - 0)$$

and

$$P_2(x) = f(0) + f'(0)(x - 0) + \tfrac{1}{2}f''(0)(x - 0)^2$$

in the same viewing window. Compare the values of f, P_1, and P_2 and their first derivatives at $x = 0$.

83. $f(x) = e^{x/2}$ **84.** $f(x) = e^{-x^2/2}$

85. Finding a Pattern   Use a graphing utility to compare the graph of the function $y = e^x$ with the graphs of each of the following functions.

(a) $y_1 = 1 + \dfrac{x}{1!}$

(b) $y_2 = 1 + \dfrac{x}{1!} + \dfrac{x^2}{2!}$

(c) $y_3 = 1 + \dfrac{x}{1!} + \dfrac{x^2}{2!} + \dfrac{x^3}{3!}$

86. Identify the pattern of successive polynomials in Exercise 85. Extend the pattern one more term and compare the graph of the resulting polynomial function with the graph of $y = e^x$. What do you think this pattern implies?

In Exercises 87–108, find or evaluate the integral.

87. $\displaystyle \int e^{5x}(5)\, dx$

88. $\displaystyle \int e^{-x^4}(-4x^3)\, dx$

89. $\displaystyle \int_0^1 e^{-2x}\, dx$

90. $\displaystyle \int_3^4 e^{3-x}\, dx$

91. $\displaystyle \int xe^{-x^2}\, dx$

92. $\displaystyle \int x^2 e^{x^3/2}\, dx$

93. $\displaystyle \int \dfrac{e^{\sqrt{x}}}{\sqrt{x}}\, dx$

94. $\displaystyle \int \dfrac{e^{1/x^2}}{x^3}\, dx$

95. $\displaystyle \int \dfrac{e^{-x}}{1 + e^{-x}}\, dx$

96. $\displaystyle \int \dfrac{e^{2x}}{1 + e^{2x}}\, dx$

97. $\displaystyle \int_1^3 \dfrac{e^{3/x}}{x^2}\, dx$

98. $\displaystyle \int_0^{\sqrt{2}} xe^{-(x^2/2)}\, dx$

99. $\displaystyle \int e^x \sqrt{1 - e^x}\, dx$

100. $\displaystyle \int \dfrac{e^x - e^{-x}}{e^x + e^{-x}}\, dx$

101. $\displaystyle \int \dfrac{e^x + e^{-x}}{e^x - e^{-x}}\, dx$

102. $\displaystyle \int \dfrac{2e^x - 2e^{-x}}{(e^x + e^{-x})^2}\, dx$

103. $\displaystyle \int \dfrac{5 - e^x}{e^{2x}}\, dx$

104. $\displaystyle \int \dfrac{e^{2x} + 2e^x + 1}{e^x}\, dx$

105. $\displaystyle \int e^{\sin \pi x} \cos \pi x\, dx$

106. $\displaystyle \int e^{\sec 2x} \sec 2x \tan 2x\, dx$

107. $\displaystyle \int e^{-x} \tan(e^{-x})\, dx$

108. $\displaystyle \int \ln(e^{2x-1})\, dx$

In Exercises 109 and 110, solve the differential equation.

109. $\dfrac{dy}{dx} = xe^{ax^2}$

110. $\dfrac{dy}{dx} = (e^x - e^{-x})^2$

In Exercises 111 and 112, find the particular solution that satisfies the initial conditions.

111. $f''(x) = \frac{1}{2}(e^x + e^{-x})$,
 $f(0) = 1, f'(0) = 0$

112. $f''(x) = \sin x + e^{2x}$,
 $f(0) = \frac{1}{4}, f'(0) = \frac{1}{2}$

Slope Fields   **In Exercises 113 and 114, a differential equation, a point, and a slope field are given. (a) Sketch two approximate solutions of the differential equation on the slope field, one of which passes through the indicated point. (b) Use integration to find the particular solution of the differential equation and use a graphing utility to graph the solution. Compare the result with the sketches in part (a). To print an enlarged copy of the graph, go to the website www.mathgraphs.com.**

113. $\dfrac{dy}{dx} = 2e^{-x/2}$, $(0, 1)$

114. $\dfrac{dy}{dx} = xe^{-0.2x^2}$, $\left(0, -\dfrac{3}{2}\right)$

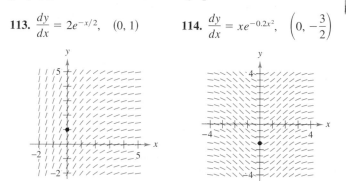

Area   **In Exercises 115–118, find the area of the region bounded by the graphs of the equations. Use a graphing utility to graph the region and verify your result.**

115. $y = e^x, y = 0, x = 0, x = 5$

116. $y = e^{-x}, y = 0, x = a, x = b$

117. $y = xe^{-x^2/4}, y = 0, x = 0, x = \sqrt{6}$

118. $y = e^{-2x} + 2, y = 0, x = 0, x = 2$

119. Given the exponential function $f(x) = e^x$, show that

(a) $f(u - v) = \dfrac{f(u)}{f(v)}$.   (b) $f(kx) = [f(x)]^k$.

120. Approximate each integral using the Midpoint Rule, the Trapezoidal Rule, and Simpson's Rule with $n = 12$. Then use the integration capabilities of a graphing utility to approximate the integrals and compare the results.

(a) $\displaystyle\int_0^4 \sqrt{x}\, e^x \, dx$ (b) $\displaystyle\int_0^2 2xe^{-x} \, dx$

121. *Probability*   A car battery has an average lifetime of 48 months with a standard deviation of 6 months. The battery lives are normally distributed. The probability that a given battery will last between 48 months and 60 months is

$$0.0665 \int_{48}^{60} e^{-0.0139(t-48)^2} \, dt.$$

Use the integration capabilities of a graphing utility to approximate the integral. Interpret the resulting probability.

122. *Probability*   The median waiting time (in minutes) for people waiting for service in a convenience store is given by the solution of the equation

$$\int_0^x 0.3e^{-0.3t} \, dt = \frac{1}{2}.$$

Solve the equation.

123. Given $e^x \geq 1$ for $x \geq 0$, it follows that

$$\int_0^x e^t \, dt \geq \int_0^x 1 \, dt.$$

Perform this integration to derive the inequality $e^x \geq 1 + x$ for $x \geq 0$.

124. *Modeling Data*   A valve on a storage tank is opened for 4 hours to release a chemical in a manufacturing process. The flow rate R (in liters per hour) at time t (in hours) is given in the table.

t	0	1	2	3	4
R	425	240	118	71	36

(a) Use the regression capabilities of a graphing utility to find a linear model for the points $(t, \ln R)$. Write the resulting equation of the form $\ln R = at + b$ in exponential form.

(b) Use a graphing utility to plot the data and graph the exponential model.

(c) Use the definite integral to approximate the number of liters of chemical released during the 4 hours.

Getting at the Concept

125. In your own words, state the properties of the natural exponential function.

126. Describe the relationship between the graph of $f(x) = \ln x$ and $g(x) = e^x$.

127. Is there a function f such that $f(x) = f'(x)$? If so, identify it.

128. Without integrating, state the integration formula you can use to integrate each of the following.

(a) $\displaystyle\int \dfrac{e^x}{e^x + 1} \, dx$ (b) $\displaystyle\int xe^{x^2} \, dx$

129. Explain why $\displaystyle\int_0^2 e^{-x} \, dx > 0$.

130. Prove that $\dfrac{e^a}{e^b} = e^{a-b}$.

131. Let $f(x) = \dfrac{\ln x}{x}$.

(a) Graph f on $(0, \infty)$ and show that f is strictly decreasing on (e, ∞).

(b) Show that if $e \leq A < B$, then $A^B > B^A$.

(c) Use part (b) to show that $e^\pi > \pi^e$.

Section 5.5   Bases Other than e and Applications

- Define exponential functions that have bases other than e.
- Differentiate and integrate exponential functions that have bases other than e.
- Use exponential functions to model compound interest and exponential growth.

Bases Other than e

The **base** of the natural exponential function is e. This "natural" base can be used to assign a meaning to a general base a.

Definition of Exponential Function to Base a

If a is a positive real number ($a \neq 1$) and x is any real number, then the **exponential function to the base a** is denoted by a^x and is defined by

$$a^x = e^{(\ln a)x}.$$

If $a = 1$, then $y = 1^x = 1$ is a constant function.

These functions obey the usual laws of exponents. For instance, here are some familiar properties.

1. $a^0 = 1$ **2.** $a^x a^y = a^{x+y}$

3. $\dfrac{a^x}{a^y} = a^{x-y}$ **4.** $(a^x)^y = a^{xy}$

When modeling the half-life of a radioactive sample, it is convenient to use $\frac{1}{2}$ as the base of the exponential model.

Example 1   **Radioactive Half-Life Model**

The half-life of carbon-14 is about 5730 years. If 1 gram of carbon-14 is present in a sample, how much will be present in 10,000 years?

Solution   Let $t = 0$ represent the present time and let y represent the amount (in grams) of carbon-14 in the sample. Using a base of $\frac{1}{2}$, you can model y by the equation

$$y = \left(\frac{1}{2}\right)^{t/5730}.$$

Notice that when $t = 5730$, the amount is reduced to half of the original amount.

$$y = \left(\frac{1}{2}\right)^{5730/5730} = \frac{1}{2} \text{ gram}$$

When $t = 11,460$, the amount is reduced to a quarter of the original amount, and so on. To find the amount of carbon-14 after 10,000 years, substitute 10,000 for t.

$$y = \left(\frac{1}{2}\right)^{10,000/5730}$$

$$\approx 0.30 \text{ gram}$$

The graph of y is shown in Figure 5.25.

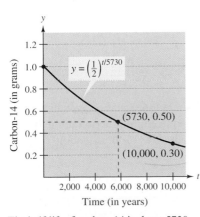

$$y = \left(\frac{1}{2}\right)^{t/5730}$$

(5730, 0.50)

(10,000, 0.30)

The half-life of carbon-14 is about 5730 years.
Figure 5.25

Logarithmic functions to bases other than e can be defined in much the same way as exponential functions to other bases are defined.

NOTE In precalculus, you learned that $\log_a x$ is the value to which a should be raised to produce x. This agrees with the definition given here because

$$a^{\log_a x} = a^{(1/\ln a)\ln x}$$
$$= (e^{\ln a})^{(1/\ln a)\ln x}$$
$$= e^{(\ln a/\ln a)\ln x}$$
$$= e^{\ln x}$$
$$= x.$$

Definition of Logarithmic Function to Base a

If a is a positive real number ($a \neq 1$) and x is any positive real number, then the **logarithmic function to the base a** is denoted by $\log_a x$ and is defined as

$$\log_a x = \frac{1}{\ln a} \ln x.$$

Logarithmic functions to the base a have properties similar to those of the natural logarithmic function given in Theorem 5.2.

1. $\log_a 1 = 0$ Log of 1
2. $\log_a xy = \log_a x + \log_a y$ Log of a product
3. $\log_a x^n = n \log_a x$ Log of a power
4. $\log_a \dfrac{x}{y} = \log_a x - \log_a y$ Log of a quotient

From the definitions of the exponential and logarithmic functions to the base a, it follows that $f(x) = a^x$ and $g(x) = \log_a x$ are inverse functions of each other.

Properties of Inverse Functions

1. $y = a^x$ if and only if $x = \log_a y$
2. $a^{\log_a x} = x, \quad$ for $x > 0$
3. $\log_a a^x = x, \quad$ for all x

The logarithmic function to the base 10 is called the **common logarithmic function.** Thus, for common logarithms, $y = 10^x$ if and only if $x = \log_{10} y$.

Example 2   **Bases Other than e**

Solve for x in each equation.

a. $3^x = \dfrac{1}{81}$ **b.** $\log_2 x = -4$

Solution

a. To solve this equation, you can apply the logarithmic function to the base 3 to both sides of the equation.

$$3^x = \frac{1}{81}$$
$$\log_3 3^x = \log_3 \frac{1}{81}$$
$$x = \log_3 3^{-4}$$
$$x = -4$$

b. To solve this equation, you can apply the exponential function to the base 2 to both sides of the equation.

$$\log_2 x = -4$$
$$2^{\log_2 x} = 2^{-4}$$
$$x = \frac{1}{2^4}$$
$$x = \frac{1}{16}$$

Differentiation and Integration

To differentiate exponential and logarithmic functions to other bases, you have three options: (1) use the definitions of a^x and $\log_a x$ and differentiate using the rules for the natural exponential and logarithmic functions, (2) use logarithmic differentiation, or (3) use the following differentiation rules for bases other than e.

THEOREM 5.13   Derivatives for Bases Other than e

Let a be a positive real number $(a \neq 1)$ and let u be a differentiable function of x.

1. $\dfrac{d}{dx}[a^x] = (\ln a)a^x$ **2.** $\dfrac{d}{dx}[a^u] = (\ln a)a^u \dfrac{du}{dx}$

3. $\dfrac{d}{dx}[\log_a x] = \dfrac{1}{(\ln a)x}$ **4.** $\dfrac{d}{dx}[\log_a u] = \dfrac{1}{(\ln a)u}\dfrac{du}{dx}$

Proof   By definition, $a^x = e^{(\ln a)x}$. Therefore, you can prove the first rule by letting $u = (\ln a)x$ and differentiating with base e to obtain

$$\frac{d}{dx}[a^x] = \frac{d}{dx}[e^{(\ln a)x}] = e^u \frac{du}{dx} = e^{(\ln a)x}(\ln a) = (\ln a)a^x.$$

To prove the third rule, you can write

$$\frac{d}{dx}[\log_a x] = \frac{d}{dx}\left[\frac{1}{\ln a}\ln x\right] = \frac{1}{\ln a}\left(\frac{1}{x}\right) = \frac{1}{(\ln a)x}.$$

The second and fourth rules are simply the Chain Rule versions of the first and third rules.

NOTE   These differentiation rules are similar to those for the natural exponential function and natural logarithmic function. In fact, they differ only by the constant factors $\ln a$ and $1/\ln a$. This points out one reason why, for calculus, e is the most convenient base.

Example 3   **Differentiating Functions to Other Bases**

Find the derivative of each of the following.

a. $y = 2^x$

b. $y = 2^{3x}$

c. $y = \log_{10} \cos x$

Solution

a. $y' = \dfrac{d}{dx}[2^x] = (\ln 2)2^x$

b. $y' = \dfrac{d}{dx}[2^{3x}] = (\ln 2)2^{3x}(3) = (3 \ln 2)2^{3x}$

Try writing 2^{3x} as 8^x and differentiating to see that you obtain the same result.

c. $y' = \dfrac{d}{dx}[\log_{10} \cos x] = \dfrac{-\sin x}{(\ln 10)\cos x} = -\dfrac{1}{\ln 10}\tan x$

Occasionally, an integrand involves an exponential function to a base other than e. When this occurs, there are two options: (1) convert to base e using the formula $a^x = e^{(\ln a)x}$ and then integrate, or (2) integrate directly, using the integration formula (which follows from Theorem 5.13)

$$\int a^x \, dx = \left(\frac{1}{\ln a}\right) a^x + C.$$

Example 4   Integrating an Exponential Function to Another Base

Find $\int 2^x \, dx$.

Solution

$$\int 2^x \, dx = \frac{1}{\ln 2} 2^x + C$$

When we introduced the Power Rule, $D_x[x^n] = nx^{n-1}$, in Chapter 2, we required the exponent n to be a rational number. We now extend the rule to cover any real value of n. Try to prove this theorem using logarithmic differentiation.

THEOREM 5.14   The Power Rule for Real Exponents

Let n be any real number and let u be a differentiable function of x.

1. $\dfrac{d}{dx}[x^n] = nx^{n-1}$ **2.** $\dfrac{d}{dx}[u^n] = nu^{n-1}\dfrac{du}{dx}$

The next example compares the derivatives of four types of functions. Each function uses a different differentiation formula, depending on whether the base and exponent are constants or variables.

Example 5   Comparing Variables and Constants

a. $\dfrac{d}{dx}[e^e] = 0$ Constant Rule

b. $\dfrac{d}{dx}[e^x] = e^x$ Exponential Rule

c. $\dfrac{d}{dx}[x^e] = ex^{e-1}$ Power Rule

NOTE   Be sure you see that there is no simple differentiation rule for calculating the derivative of $y = x^x$. In general, if $y = u(x)^{v(x)}$, you need to use logarithmic differentiation.

d. $y = x^x$ Logarithmic differentiation

$\ln y = \ln x^x$

$\ln y = x \ln x$

$\dfrac{y'}{y} = x\left(\dfrac{1}{x}\right) + (\ln x)(1) = 1 + \ln x$

$y' = y(1 + \ln x) = x^x(1 + \ln x)$

Applications of Exponential Functions

n	A
1	\$1080.00
2	\$1081.60
4	\$1082.43
12	\$1083.00
365	\$1083.28

Suppose P dollars is deposited in an account at an annual interest rate r (in decimal form). If interest accumulates in the account, what is the balance in the account at the end of 1 year? The answer depends on the number of times n the interest is compounded according to the formula

$$A = P\left(1 + \frac{r}{n}\right)^n.$$

For instance, the result for a deposit of \$1000 at 8% interest compounded n times a year is shown in the upper table at the left.

As n increases, the balance A approaches a limit. To develop this limit, we use the following theorem. To test the reasonableness of this theorem, try evaluating $[(x + 1)/x]^x$ for several values of x, as shown in the lower table at the left. (A proof of this theorem is given in Appendix B.)

x	$\left(\dfrac{x+1}{x}\right)^x$
10	2.59374
100	2.70481
1000	2.71692
10,000	2.71815
100,000	2.71827
1,000,000	2.71828

THEOREM 5.15   A Limit Involving e

$$\lim_{x \to \infty} \left(1 + \frac{1}{x}\right)^x = \lim_{x \to \infty} \left(\frac{x + 1}{x}\right)^x = e$$

Now, let's take another look at the formula for the balance A in an account in which the interest is compounded n times per year. By taking the limit as n approaches infinity, you obtain the following.

$$A = \lim_{n \to \infty} P\left(1 + \frac{r}{n}\right)^n \qquad \text{Take limit as } n \to \infty.$$

$$= P \lim_{n \to \infty} \left[\left(1 + \frac{1}{n/r}\right)^{n/r}\right]^r \qquad \text{Rewrite.}$$

$$= P\left[\lim_{x \to \infty} \left(1 + \frac{1}{x}\right)^x\right]^r \qquad \text{Let } x = n/r. \text{ Then } x \to \infty \text{ as } n \to \infty.$$

$$= Pe^r \qquad \text{Apply Theorem 5.15.}$$

This limit produces the balance after 1 year of **continuous compounding.** So, for a deposit of \$1000 at 8% interest compounded continuously, the balance at the end of 1 year would be

$$A = 1000e^{0.08} \approx \$1083.29.$$

These results are summarized as follows.

Summary of Compound Interest Formulas

Let P = amount of deposit, t = number of years, A = balance after t years, r = annual interest rate (decimal form), and n = number of compoundings per year.

1. Compounded n times per year: $A = P\left(1 + \dfrac{r}{n}\right)^{nt}$

2. Compounded continuously: $A = Pe^{rt}$

Example 6   Comparing Continuous and Quarterly Compounding

A deposit of $2500 is made in an account that pays an annual interest rate of 5%. Find the balance in the account at the end of 5 years if the interest is compounded (a) quarterly, (b) monthly, and (c) continuously.

Solution

a. $A = P\left(1 + \dfrac{r}{n}\right)^{nt} = 2500\left(1 + \dfrac{0.05}{4}\right)^{4(5)}$ Compounded quarterly

$\qquad = 2500(1.0125)^{20}$

$\qquad \approx \$3205.09$

b. $A = P\left(1 + \dfrac{r}{n}\right)^{nt} = 2500\left(1 + \dfrac{0.05}{12}\right)^{12(5)}$ Compounded monthly

$\qquad \approx 2500(1.0041667)^{60}$

$\qquad \approx \$3208.40$

c. $A = Pe^{rt} = 2500[e^{0.05(5)}]$ Compounded continuously

$\qquad = 2500e^{0.25}$

$\qquad \approx \$3210.06$

Figure 5.26 shows how the balance increases over the 5-year period. Notice that the scale used in the figure does not graphically distinguish among the three types of exponential growth in (a), (b), and (c).

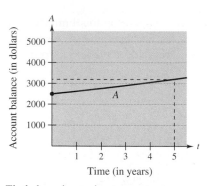

The balance in a savings account grows exponentially.
Figure 5.26

Example 7   Bacterial Culture Growth

A bacterial culture is growing according to the *logistics growth function*

$$y = \frac{1.25}{1 + 0.25e^{-0.4t}}, \qquad t \geq 0$$

where y is the weight of the culture in grams and t is the time in hours. Find the weight of the culture after (a) 0 hours, (b) 1 hour, and (c) 10 hours. (d) What is the limit as t approaches infinity?

Solution

a. When $t = 0$, $y = \dfrac{1.25}{1 + 0.25e^{-0.4(0)}}$

$\qquad = 1$ gram.

b. When $t = 1$, $y = \dfrac{1.25}{1 + 0.25e^{-0.4(1)}}$

$\qquad \approx 1.071$ grams.

c. When $t = 10$, $y = \dfrac{1.25}{1 + 0.25e^{-0.4(10)}}$

$\qquad \approx 1.244$ grams.

d. Finally, taking the limit as t approaches infinity, you obtain

$$\lim_{t \to \infty} \frac{1.25}{1 + 0.25e^{-0.4t}} = \frac{1.25}{1 + 0} = 1.25 \text{ grams.}$$

The graph of the function is shown in Figure 5.27.

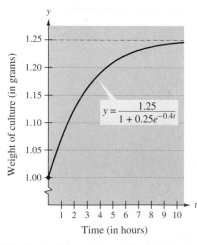

The limit of the weight of the culture as $t \to \infty$ is 1.25 grams.
Figure 5.27

EXERCISES FOR SECTION 5.5

Depreciation In Exercises 1–4, the time in which a machine depreciates to one-half its purchase price is given. Find a model that yields the fraction of the purchase price as a function of time and determine that fraction at time t_0.

Depreciation time	t_0
1. 3 years	6 years
2. 8 years	16 years
3. 7 years	10 years
4. 5 years	2 years

In Exercises 5–8, evaluate the expression without using a calculator.

5. $\log_2 \frac{1}{8}$

6. $\log_{27} 9$

7. $\log_7 1$

8. $\log_a \frac{1}{a}$

In Exercises 9–12, write the exponential equation as a logarithmic equation or vice versa.

9. (a) $2^3 = 8$

   (b) $3^{-1} = \frac{1}{3}$

10. (a) $27^{2/3} = 9$

   (b) $16^{3/4} = 8$

11. (a) $\log_{10} 0.01 = -2$

   (b) $\log_{0.5} 8 = -3$

12. (a) $\log_3 \frac{1}{9} = -2$

   (b) $49^{1/2} = 7$

In Exercises 13–18, sketch the graph of the function by hand.

13. $y = 3^x$

14. $y = 3^{x-1}$

15. $y = \left(\frac{1}{3}\right)^x$

16. $y = 2^{x^2}$

17. $h(x) = 5^{x-2}$

18. $y = 3^{-|x|}$

In Exercises 19–24, solve for x or b.

19. (a) $\log_{10} 1000 = x$

   (b) $\log_{10} 0.1 = x$

20. (a) $\log_3 \frac{1}{81} = x$

   (b) $\log_6 36 = x$

21. (a) $\log_3 x = -1$

   (b) $\log_2 x = -4$

22. (a) $\log_b 27 = 3$

   (b) $\log_b 125 = 3$

23. (a) $x^2 - x = \log_5 25$

   (b) $3x + 5 = \log_2 64$

24. (a) $\log_3 x + \log_3(x - 2) = 1$

   (b) $\log_{10}(x + 3) - \log_{10} x = 1$

In Exercises 25–34, solve the equation accurate to three decimal places.

25. $3^{2x} = 75$

26. $5^{6x} = 8320$

27. $2^{3-z} = 625$

28. $3(5^{x-1}) = 86$

29. $\left(1 + \frac{0.09}{12}\right)^{12t} = 3$

30. $\left(1 + \frac{0.10}{365}\right)^{365t} = 2$

31. $\log_2(x - 1) = 5$

32. $\log_{10}(t - 3) = 2.6$

33. $\log_3 x^2 = 4.5$

34. $\log_5 \sqrt{x - 4} = 3.2$

In Exercises 35–38, use a graphing utility to graph the function and approximate its zero(s) accurate to three decimal places.

35. $g(x) = 6(2^{1-x}) - 25$

36. $f(t) = 300(1.0075^{12t}) - 735.41$

37. $h(s) = 32 \log_{10}(s - 2) + 15$

38. $g(x) = 1 - 2 \log_{10}[x(x - 3)]$

In Exercises 39 and 40, illustrate that the functions are inverse functions of each other by sketching their graphs on the same set of coordinate axes.

39. $f(x) = 4^x$

   $g(x) = \log_4 x$

40. $f(x) = 3^x$

   $g(x) = \log_3 x$

In Exercises 41–56, find the derivative of the function.

41. $f(x) = 4^x$

42. $g(x) = 2^{-x}$

43. $y = 5^{x-2}$

44. $y = x(6^{-2x})$

45. $g(t) = t^2 2^t$

46. $f(t) = \dfrac{3^{2t}}{t}$

47. $h(\theta) = 2^{-\theta} \cos \pi\theta$

48. $g(\alpha) = 5^{-\alpha/2} \sin 2\alpha$

49. $y = \log_3 x$

50. $y = \log_{10} 2x$

51. $f(x) = \log_2 \dfrac{x^2}{x - 1}$

52. $h(x) = \log_3 \dfrac{x\sqrt{x - 1}}{2}$

53. $y = \log_5 \sqrt{x^2 - 1}$

54. $y = \log_{10} \dfrac{x^2 - 1}{x}$

55. $g(t) = \dfrac{10 \log_4 t}{t}$

56. $f(t) = t^{3/2} \log_2 \sqrt{t + 1}$

In Exercises 57–60, use logarithmic differentiation to find dy/dx.

57. $y = x^{2/x}$

58. $y = x^{x-1}$

59. $y = (x - 2)^{x+1}$

60. $y = (1 + x)^{1/x}$

In Exercises 61–68, find or evaluate the integral.

61. $\displaystyle\int 3^x \, dx$

62. $\displaystyle\int 5^{-x} \, dx$

63. $\displaystyle\int_{-1}^{2} 2^x \, dx$

64. $\displaystyle\int_{-2}^{0} (3^3 - 5^2) \, dx$

65. $\displaystyle\int x(5^{-x^2}) \, dx$

66. $\displaystyle\int (3 - x)7^{(3-x)^2} \, dx$

67. $\displaystyle\int \dfrac{3^{2x}}{1 + 3^{2x}} \, dx$

68. $\displaystyle\int 2^{\sin x} \cos x \, dx$

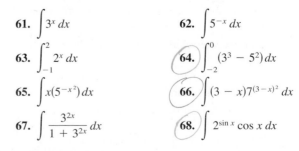

Slope Fields **In Exercises 69 and 70, a differential equation, a point, and a slope field are given. (a) Sketch two approximate solutions of the differential equation on the slope field, one of which passes through the indicated point. (b) Use integration to find the particular solution of the differential equation and use a graphing utility to graph the solution. Compare the result with the sketches in part (a). To print an enlarged copy of the graph, go to the website *www.mathgraphs.com*.**

69. $\dfrac{dy}{dx} = 0.4^{x/3}, \quad \left(0, \frac{1}{2}\right)$ **70.** $\dfrac{dy}{dx} = e^{\sin x} \cos x, \quad (\pi, 2)$

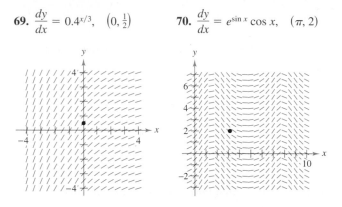

Getting at the Concept

71. List some applications of the exponential functions $f(x) = a^x$ and $g(x) = a^{-x}$.

72. Describe how to use a calculator to find the logarithm of a number if the base is not 10 or e.

73. The table of values below was obtained by evaluating a function. Determine which of the statements may be true and which must be false, and explain why.

   (a) y is an exponential function of x.

   (b) y is a logarithmic function of x.

   (c) x is an exponential function of y.

   (d) y is a linear function of x.

x	1	2	8
y	0	1	3

74. Consider the function $f(x) = \log_{10} x$.

   (a) What is the domain of f?

   (b) Find f^{-1}.

   (c) If x is a real number between 1000 and 10,000, determine the interval in which $f(x)$ will be found.

   (d) Determine the interval in which x will be found if $f(x)$ is negative.

   (e) If $f(x)$ is increased by one unit, x must have been increased by what factor?

   (f) Find the ratio of x_1 to x_2 given that $f(x_1) = 3n$ and $f(x_2) = n$.

75. *Ordering Functions* Order the functions

$$f(x) = \log_2 x, \quad g(x) = x^x, \quad h(x) = x^2, \quad \text{and} \quad k(x) = 2^x$$

from the one with the greatest rate of growth to the one with the smallest rate of growth for "large" values of x.

76. Given the exponential function $f(x) = a^x$, show that

   (a) $f(u + v) = f(u) \cdot f(v)$.

   (b) $f(2x) = [f(x)]^2$.

77. *Inflation* If the annual rate of inflation averages 5% over the next 10 years, the approximate cost C of goods or services during any year in that decade is

$$C(t) = P(1.05)^t$$

where t is the time in years and P is the present cost.

   (a) If the price of an oil change for your car is presently \$24.95, estimate the price 10 years from now.

   (b) Find the rate of change of C with respect to t when $t = 1$ and $t = 8$.

   (c) Verify that the rate of change of C is proportional to C. What is the constant of proportionality?

78. *Depreciation* After t years, the value of a car purchased for \$20,000 is

$$V(t) = 20{,}000\left(\tfrac{3}{4}\right)^t.$$

   (a) Use a graphing utility to graph the function and determine the value of the car 2 years after it was purchased.

   (b) Find the rate of change of V with respect to t when $t = 1$ and $t = 4$.

   (c) Use a graphing utility to graph $V'(t)$ and determine the horizontal asymptote of $V'(t)$. Interpret its meaning in the context of the problem.

Compound Interest **In Exercises 79–82, complete the table to determine the balance A for P dollars invested at rate r for t years and compounded n times per year.**

n	1	2	4	12	365	Continuous compounding
A						

79. $P = \$1000$
$r = 3\frac{1}{2}\%$
$t = 10$ years

80. $P = \$2500$
$r = 6\%$
$t = 20$ years

81. $P = \$1000$
$r = 5\%$
$t = 30$ years

82. $P = \$5000$
$r = 7\%$
$t = 25$ years

Compound Interest In Exercises 83–86, complete the table to determine the amount of money P (present value) that should be invested at rate r to produce a balance of $100,000 in t years.

t	1	10	20	30	40	50
P						

83. $r = 5\%$

Compounded continuously

84. $r = 6\%$

Compounded continuously

85. $r = 5\%$

Compounded monthly

86. $r = 7\%$

Compounded daily

87. ***Compound Interest*** Assume that you can earn 6% on an investment, compounded daily. Which of the following options would yield the greatest balance after 8 years?

(a) $20,000 now

(b) $30,000 after 8 years

(c) $8000 now and $20,000 after 4 years

(d) $9000 now, $9000 after 4 years, and $9000 after 8 years

88. ***Compound Interest*** Consider a deposit of $100 placed in an account for 20 years at $r\%$ compounded continuously. Use a graphing utility to graph the exponential functions giving the growth of the investment over the 20 years for each of the following interest rates. Compare the ending balances for each of the rates.

(a) $r = 3\%$ (b) $r = 5\%$ (c) $r = 6\%$

89. ***Timber Yield*** The yield V (in millions of cubic feet per acre) for a stand of timber at age t is

$$V = 6.7e^{(-48.1)/t}$$

where t is measured in years.

(a) Find the limiting volume of wood per acre as t approaches infinity.

(b) Find the rate at which the yield is changing when $t = 20$ years and $t = 60$ years.

90. ***Learning Theory*** In a group project in learning theory, a mathematical model for the proportion P of correct responses after n trials was found to be

$$P = \frac{0.86}{1 + e^{-0.25n}}.$$

(a) Find the limiting proportion of correct responses as n approaches infinity.

(b) Find the rate at which P is changing after $n = 3$ trials and $n = 10$ trials.

91. ***Forest Defoliation*** To estimate the amount of defoliation caused by the gypsy moth during a year, a forester counts the number of egg masses on $\frac{1}{40}$ of an acre the preceding fall. The percent of defoliation y is approximated by

$$y = \frac{300}{3 + 17e^{-0.0625x}}$$

where x is the number of egg masses in thousands. *(Source: USDA Forest Service)*

(a) Use a graphing utility to graph the function.

(b) Estimate the percent of defoliation if 2000 egg masses are counted.

(c) Estimate the number of egg masses that existed if you observe that approximately $\frac{2}{3}$ of a forest is defoliated.

(d) Use calculus to estimate the value of x for which y is increasing most rapidly.

92. ***Population Growth*** A lake is stocked with 500 fish, and their population increases according to the logistics curve

$$p(t) = \frac{10,000}{1 + 19e^{-t/5}}$$

where t is measured in months.

(a) Use a graphing utility to graph the function.

(b) What is the limiting size of the fish population?

(c) At what rates is the fish population changing at the end of 1 month and at the end of 10 months?

(d) After how many months is the population increasing most rapidly?

93 ***Modeling Data*** The breaking strength B (in tons) of a steel cable of diameter d (in inches) is given in the table.

d	0.50	0.75	1.00	1.25	1.50	1.75
B	9.85	21.8	38.3	59.2	84.4	114.0

(a) Use the regression capabilities of a graphing utility to fit an exponential model to the data.

(b) Use a graphing utility to plot the data and graph the model.

(c) Find the rate of growth of the model when $d = 0.8$ and $d = 1.5$.

94. ***Comparing Models*** The amount y (in billions of dollars) given to philanthropy (from individuals, foundations, corporations, and charitable bequests) in the United States for the years 1991 through 1997 is given in the table, with $x = 1$ corresponding to 1991. *(Source: AAFRC Trust for Philanthropy)*

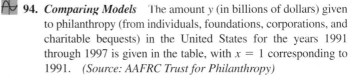

x	1	2	3	4	5	6	7
y	105.0	110.4	116.5	119.2	124.3	133.5	143.5

(a) Use the regression capabilities of a graphing utility to find the following models for the data.

$$y_1 = ax + b \qquad y_2 = a + b \ln x$$
$$y_3 = ab^x \qquad y_4 = ax^b$$

(b) Use a graphing utility to plot the data and graph each of the models. Which model do you think best fits the data?

(c) Interpret the slope of the linear model in the context of the problem.

(d) Find the rate of change of each of the models for the year 1996. Which model is increasing at the greatest rate in 1996?

95. Conjecture

(a) Use a graphing utility to approximate the integrals of the functions

$$f(t) = 4\left(\frac{3}{8}\right)^{2t/3}, \quad g(t) = 4\left(\frac{\sqrt[3]{9}}{4}\right)^{t}, \quad \text{and} \quad h(t) = 4e^{-0.653886t}$$

on the interval $[0, 4]$.

(b) Use a graphing utility to graph the three functions.

(c) Use the results in parts (a) and (b) to make a conjecture about the three functions. Could you make the conjecture using only part (a)? Explain. Prove your conjecture analytically.

96. Area Find the area of the region bounded by the graphs of $y = 3^x$, $y = 0$, $x = 0$, and $x = 3$.

97. Continuous Cash Flow The present value P of a continuous cash flow of \$2000 per year earning 6% interest compounded continuously over 10 years is

$$P = \int_{0}^{10} 2000e^{-0.06t}\, dt.$$

Find P.

98. Complete the table to demonstrate that e can also be defined as $\lim\limits_{x \to 0^+} (1 + x)^{1/x}$.

x	1	10^{-1}	10^{-2}	10^{-4}	10^{-6}
$(1 + x)^{1/x}$					

In Exercises 99 and 100, find an exponential function that fits the experimental data collected over time t.

99.

t	0	1	2	3	4
y	1200.00	720.00	432.00	259.20	155.52

100.

t	0	1	2	3	4
y	600.00	630.00	661.50	694.58	729.30

True or False? In Exercises 101–106, determine whether the statement is true or false. If it is false, explain why or give an example that shows it is false.

101. $e = 271{,}801/99{,}990$.

102. If $f(x) = \ln x$, then $f(e^{n+1}) - f(e^n) = 1$ for any value of n.

103. The functions $f(x) = 2 + e^x$ and $g(x) = \ln(x - 2)$ are inverse functions of each other.

104. The exponential function $y = Ce^x$ is a solution of the differential equation $d^n y/dx^n = y$, $n = 1, 2, 3, \ldots$.

105. The graphs of $f(x) = e^x$ and $g(x) = e^{-x}$ meet at right angles.

106. If $f(x) = g(x)e^x$, then the only zeros of f are the zeros of g.

107. Solve the logistics differential equation

$$\frac{dy}{dt} = \frac{8}{25}y\left(\frac{5}{4} - y\right), \quad y(0) = 1$$

and obtain the logistics growth function of Example 7.

$$\left[\textit{Hint:} \; \frac{1}{y\left(\frac{5}{4} - y\right)} = \frac{4}{5}\left(\frac{1}{y} + \frac{1}{\frac{5}{4} - y}\right) \right]$$

108. Find an equation of the tangent line to $y = x^{\sin x}$ at $\left(\dfrac{\pi}{2}, \dfrac{\pi}{2}\right)$.

SECTION PROJECT ▌ USING GRAPHING UTILITIES TO ESTIMATE SLOPE

Let $f(x) = \begin{cases} |x|^x, & x \neq 0 \\ 1, & x = 0. \end{cases}$

(a) Use a graphing utility to graph f in the viewing window $-3 \le x \le 3$, $-2 \le y \le 2$. What is the domain of f?

(b) Use the *zoom* and *trace* features of a graphing utility to estimate

$$\lim_{x \to 0^-} f(x).$$

(c) Write a short paragraph explaining why the function f is continuous for all real numbers.

(d) Visually estimate the slope of f at the point $(0, 1)$.

(e) Explain why the derivative of a function can be approximated by the formula

$$\frac{f(x + \Delta x) - f(x - \Delta x)}{2\Delta x}$$

for small values of Δx. Use this formula to approximate the slope of f at the point $(0, 1)$.

$$f'(0) \approx \frac{f(0 + \Delta x) - f(0 - \Delta x)}{2\Delta x} = \frac{f(\Delta x) - f(-\Delta x)}{2\Delta x}$$

What do you think the slope of the graph of f is at $(0, 1)$?

(f) Find a formula for the derivative of f and determine $f'(0)$. Write a short paragraph explaining how a graphing utility might lead you to approximate the slope of a graph incorrectly.

(g) Use your formula for the derivative of f to find the relative extrema of f. Verify your answer with a graphing utility.

FOR FURTHER INFORMATION For more information on using graphing utilities to estimate slope, see the article "Computer-Aided Delusions" by Richard L. Hall in *The College Mathematics Journal*. To view this article, go to the website *www.matharticles.com*.

Differential Equations: Growth and Decay

- Use separation of variables to solve a simple differential equation.
- Use exponential functions to model growth and decay in applied problems.

Differential Equations

Up to now in the text, you have learned to solve only two types of differential equations—those of the forms

$$y' = f(x) \quad \text{and} \quad y'' = f(x).$$

In this section, you will learn how to solve a more general type of differential equation. The strategy is to rewrite the equation so that each variable occurs on only one side of the equation. This strategy is called *separation of variables*. (You will study this strategy in detail in Section 5.7.)

Example 1   **Solving a Differential Equation**

Solve the differential equation $y' = 2x/y$.

Solution

$$y' = \frac{2x}{y} \qquad \text{Write original equation.}$$

$$yy' = 2x \qquad \text{Multiply both sides by } y.$$

$$\int yy' \, dx = \int 2x \, dx \qquad \text{Integrate with respect to } x.$$

$$\int y \, dy = \int 2x \, dx \qquad dy = y' \, dx$$

$$\frac{1}{2}y^2 = x^2 + C_1 \qquad \text{Apply Power Rule.}$$

$$y^2 - 2x^2 = C \qquad \text{Rewrite, letting } C = 2C_1.$$

So, the general solution is given by

$$y^2 - 2x^2 = C.$$

You can use implicit differentiation to check this result.   ◹

In practice, most people prefer to use Leibniz notation and differentials when applying separation of variables. Using this notation, the solution of Example 1 is as follows.

$$\frac{dy}{dx} = \frac{2x}{y}$$

$$y \, dy = 2x \, dx$$

$$\int y \, dy = \int 2x \, dx$$

$$\frac{1}{2}y^2 = x^2 + C_1$$

$$y^2 - 2x^2 = C$$

NOTE When you integrate both sides of the equation in Example 1, you don't need to add a constant of integration to both sides of the equation. If you did, you would obtain the same result as in Example 1.

$$\int y \, dy = \int 2x \, dx$$

$$\frac{1}{2}y^2 + C_2 = x^2 + C_3$$

$$\frac{1}{2}y^2 = x^2 + (C_3 - C_2)$$

$$\frac{1}{2}y^2 = x^2 + C_1$$

EXPLORATION

In Example 1, the general solution of the differential equation is

$$y^2 - 2x^2 = C.$$

Use a graphing utility to sketch several particular solutions—those given by $C = \pm 2$, $C = \pm 1$, and $C = 0$. Describe the solutions graphically. Is the following statement true of each solution?

The slope of the graph at the point (x, y) is equal to twice the ratio of x and y.

Explain your reasoning. Are all curves for which this statement is true represented by the general solution?

Growth and Decay Models

In many applications, the rate of change of a variable y is proportional to the value of y. If y is a function of time t, the proportion can be written as follows.

Rate of change of y is   proportional to y.

$$\frac{dy}{dt} = ky$$

The general solution of this differential equation is given in the following theorem.

THEOREM 5.16   **Exponential Growth and Decay Model**

If y is a differentiable function of t such that $y > 0$ and $y' = ky$, for some constant k, then

$$y = Ce^{kt}.$$

C is the **initial value** of y, and k is the **proportionality constant. Exponential growth** occurs when $k > 0$, and **exponential decay** occurs when $k < 0$.

NOTE Differentiate the function $y = Ce^{kt}$ with respect to t, and verify that $y' = ky$.

Proof

$y' = ky$	Write original equation.
$\dfrac{y'}{y} = k$	Separate variables.
$\displaystyle\int \dfrac{y'}{y}\, dt = \int k\, dt$	Integrate with respect to t.
$\displaystyle\int \dfrac{1}{y}\, dy = \int k\, dt$	$dy = y'\, dt$
$\ln y = kt + C_1$	Find antiderivative of each side.
$y = e^{kt}e^{C_1}$	Solve for y.
$y = Ce^{kt}$	Let $C = e^{C_1}$.

So, all solutions of $y' = ky$ are of the form $y = Ce^{kt}$.

Example 2   **Using an Exponential Growth Model**

The rate of change of y is proportional to y. When $t = 0$, $y = 2$. When $t = 2$, $y = 4$. What is the value of y when $t = 3$?

Solution Because $y' = ky$, you know that y and t are related by the equation $y = Ce^{kt}$. You can find the values of the constants C and k by applying the initial conditions.

$2 = Ce^0 \implies C = 2$ When $t = 0$, $y = 2$.

$4 = 2e^{2k} \implies k = \dfrac{1}{2}\ln 2 \approx 0.3466$ When $t = 2$, $y = 4$.

Therefore, the model is

$$y \approx 2e^{0.3466t}.$$

When $t = 3$, the value of y is $2e^{0.3466(3)} \approx 5.657$ (see Figure 5.28).

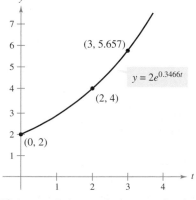

If the rate of change of y is proportional to y, then y follows an exponential model.

Figure 5.28

Radioactive decay is measured in terms of *half-life*—the number of years required for half of the atoms in a sample of radioactive material to decay. The half-lives of some common radioactive isotopes are as follows.

Uranium (^{238}U)	4,510,000,000 years
Plutonium (^{239}Pu)	24,360 years
Carbon (^{14}C)	5730 years
Radium (^{226}Ra)	1620 years
Einsteinium (^{254}Es)	270 days
Nobelium (^{257}No)	23 seconds

Example 3   **Radioactive Decay**

Suppose that 10 grams of the plutonium isotope Pu-239 was released in the Chernobyl nuclear accident. How long will it take for the 10 grams to decay to 1 gram?

Solution   Let y represent the mass (in grams) of the plutonium. Because the rate of decay is proportional to y, you know that

$$y = Ce^{kt}$$

where t is the time in years. To find the values of the constants C and k, apply the initial conditions. Using the fact that $y = 10$ when $t = 0$, you can write

$$10 = Ce^{k(0)} = Ce^0$$

which implies that $C = 10$. Next, using the fact that $y = 5$ when $t = 24,360$, you can write

$$5 = 10e^{k(24,360)}$$

$$\frac{1}{2} = e^{24,360k}$$

$$\frac{1}{24,360} \ln \frac{1}{2} = k$$

$$-2.8454 \times 10^{-5} \approx k.$$

Therefore, the model is

$$y = 10e^{\,0.000028454t}. \qquad\qquad \text{Half-life model}$$

To find the time it would take for 10 grams to decay to 1 gram, you can solve for t in the equation

$$1 = 10e^{-0.000028454t}.$$

The solution is approximately 80,923 years.

The worst nuclear accident in history happened in 1986 at the Chernobyl nuclear plant near Kiev in the Ukraine. An explosion destroyed one of the plant's four reactors, releasing large amounts of radioactive isotopes into the atmosphere.

NOTE   The exponential decay model in Example 3 could also be written as $y = 10\left(\frac{1}{2}\right)^{t/24,360}$. This model is much easier to derive, but for some applications, it is not as convenient to use.

From Example 3, notice that in an exponential growth or decay problem, it is easy to solve for C when you are given the value of y at $t = 0$. The next example demonstrates a procedure for solving for C and k when you do not know the value of y at $t = 0$.

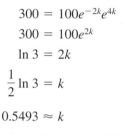

Example 4 Population Growth

Suppose an experimental population of fruit flies increases according to the law of exponential growth. There were 100 flies after the second day of the experiment and 300 flies after the fourth day. Approximately how many flies were in the original population?

Solution Let $y = Ce^{kt}$ be the number of flies at time t, where t is measured in days. Because $y = 100$ when $t = 2$ and $y = 300$ when $t = 4$, you can write

$$100 = Ce^{2k} \quad \text{and} \quad 300 = Ce^{4k}.$$

From the first equation, you know that $C = 100e^{-2k}$. Substituting this value into the second equation produces the following.

$$300 = 100e^{-2k}e^{4k}$$
$$300 = 100e^{2k}$$
$$\ln 3 = 2k$$
$$\frac{1}{2}\ln 3 = k$$
$$0.5493 \approx k$$

Therefore, the exponential growth model is

$$y = Ce^{0.5493t}.$$

To solve for C, reapply the condition $y = 100$ when $t = 2$ and obtain

$$100 = Ce^{0.5493(2)}$$
$$C = 100e^{-1.0986} \approx 33.$$

So, the original population (when $t = 0$) consisted of approximately $y = C = 33$ flies, as indicated in Figure 5.29.

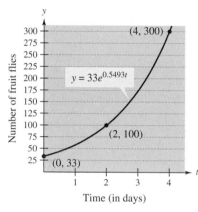

Figure 5.29

Example 5 Declining Sales

Four months after it stops advertising, a manufacturing company notices that its sales have dropped from 100,000 units per month to 80,000 units per month. If the sales follow an exponential pattern of decline, what will they be after another 2 months?

Solution Use the exponential decay model $y = Ce^{kt}$, where t is measured in months, as shown in Figure 5.30. From the initial condition ($t = 0$), you know that $C = 100,000$. Moreover, because $y = 80,000$ when $t = 4$, you have

$$80,000 = 100,000e^{4k}$$
$$0.8 = e^{4k}$$
$$\ln(0.8) = 4k$$
$$-0.0558 \approx k.$$

So, after 2 more months ($t = 6$), you can expect the monthly sales rate to be

$$y \approx 100,000e^{-0.0558(6)}$$
$$\approx 71,500 \text{ units.}$$

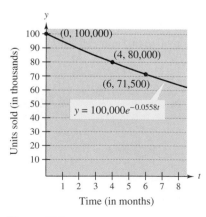

Figure 5.30

In Examples 2 through 5, you did not actually have to solve the differential equation

$$y' = ky.$$

(This was done once in the proof of Theorem 5.16.) The next example demonstrates a problem whose solution involves the separation of variables technique. The example concerns **Newton's Law of Cooling,** which states that the rate of change in the temperature of an object is proportional to the difference between the object's temperature and the temperature of the surrounding medium.

Example 6   Newton's Law of Cooling

TECHNOLOGY If you didn't read the text at the beginning of the chapter on page 312, turn back and read it now. There you can see how data collected using a *Texas Instruments Calculator-Based Laboratory (CBL) System* can be used to experimentally derive a model for Newton's Law of Cooling.

Let y represent the temperature (in °F) of an object in a room whose temperature is kept at a constant 60°. If the object cools from 100° to 90° in 10 minutes, how much longer will it take for its temperature to decrease to 80°?

Solution   From Newton's Law of Cooling, you know that the rate of change in y is proportional to the difference between y and 60. This can be written as

$$y' = k(y - 60), \qquad 80 \le y \le 100.$$

To solve this differential equation, you can use separation of variables, as follows.

$$\frac{dy}{dt} = k(y - 60) \qquad\qquad \text{Differential equation}$$

$$\left(\frac{1}{y - 60}\right) dy = k\,dt \qquad\qquad \text{Separate variables.}$$

$$\int \frac{1}{y - 60}\,dy = \int k\,dt \qquad\qquad \text{Integrate both sides.}$$

$$\ln|y - 60| = kt + C_1 \qquad\qquad \text{Find antiderivative of each side.}$$

Because $y > 60$, $|y - 60| = y - 60$, and you can omit the absolute value signs. Using exponential notation, you have

$$y - 60 = e^{kt + C_1} \quad\Longrightarrow\quad y = 60 + Ce^{kt}. \qquad C = e^{C_1}$$

Using $y = 100$ when $t = 0$, you obtain $100 = 60 + Ce^{k(0)} = 60 + C$, which implies that $C = 40$. Because $y = 90$ when $t = 10$, it follows that

$$90 = 60 + 40e^{k(10)}$$

$$30 = 40e^{10k}$$

$$k = \tfrac{1}{10} \ln \tfrac{3}{4} \approx -0.02877.$$

So, the model is

$$y = 60 + 40e^{-0.02877t} \qquad\qquad \text{Cooling model}$$

and finally, when $y = 80$, you obtain

$$80 = 60 + 40e^{-0.02877t}$$

$$20 = 40e^{-0.02877t}$$

$$\tfrac{1}{2} = e^{-0.02877t}$$

$$\ln \tfrac{1}{2} = -0.02877t$$

$$t \approx 24.09 \text{ minutes.}$$

Therefore, it will require about 14.09 *more* minutes for the object to cool to a temperature of 80° (see Figure 5.31).

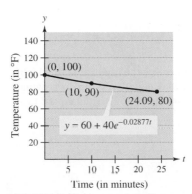

Figure 5.31

In Exercises 1–10, solve the differential equation.

1. $\dfrac{dy}{dx} = x + 2$ **2.** $\dfrac{dy}{dx} = 4 - x$

3. $\dfrac{dy}{dx} = y + 2$ **4.** $\dfrac{dy}{dx} = 4 - y$

5. $y' = \dfrac{5x}{y}$ **6.** $y' = \dfrac{\sqrt{x}}{3y}$

7. $y' = \sqrt{x}\, y$ **8.** $y' = x(1 + y)$

9. $(1 + x^2)y' - 2xy = 0$ **10.** $xy + y' = 100x$

In Exercises 11–14, write and solve the differential equation that models the verbal statement.

11. The rate of change of Q with respect to t is inversely proportional to the square of t.

12. The rate of change of P with respect to t is proportional to $10 - t$.

13. The rate of change of N with respect to s is proportional to $250 - s$.

14. The rate of change of y with respect to x varies jointly as x and $L - y$.

Slope Fields   **In Exercises 15 and 16, a differential equation, a point, and a slope field are given. (a) Sketch two approximate solutions of the differential equation on the slope field, one of which passes through the indicated point. (b) Use integration to find the particular solution of the differential equation and use a graphing utility to graph the solution. Compare the result with the sketch in part (a). To print an enlarged copy of the graph, go to the website www.mathgraphs.com.**

15. $\dfrac{dy}{dx} = x(6 - y),\quad (0, 0)$ **16.** $\dfrac{dy}{dx} = xy,\quad \left(0, \tfrac{1}{2}\right)$

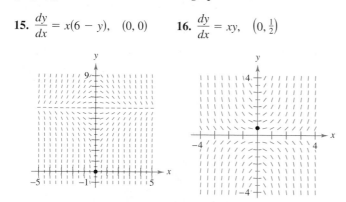

In Exercises 17–20, find the function $y = f(t)$ passing through the point $(0, 10)$ with the given first derivative. Use a graphing utility to graph the solution.

17. $\dfrac{dy}{dt} = \dfrac{1}{2}t$ **18.** $\dfrac{dy}{dt} = -\dfrac{3}{4}\sqrt{t}$

19. $\dfrac{dy}{dt} = -\dfrac{1}{2}y$ **20.** $\dfrac{dy}{dt} = \dfrac{3}{4}y$

In Exercises 21–24, write and solve the differential equation that models the verbal statement. Evaluate the solution at the specified value of the independent variable.

21. The rate of change of y is proportional to y. When $x = 0$, $y = 4$ and when $x = 3$, $y = 10$. What is the value of y when $x = 6$?

22. The rate of change of N is proportional to N. When $t = 0$, $N = 250$ and when $t = 1$, $N = 400$. What is the value of N when $t = 4$?

23. The rate of change of V is proportional to V. When $t = 0$, $V = 20{,}000$ and when $t = 4$, $V = 12{,}500$. What is the value of V when $t = 6$?

24. The rate of change of P is proportional to P. When $t = 0$, $P = 5000$ and when $t = 1$, $P = 4750$. What is the value of P when $t = 5$?

In Exercises 25–28, find the exponential function $y = Ce^{kt}$ that passes through the two given points.

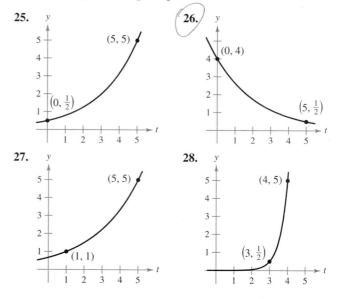

Getting at the Concept

29. In your own words, describe what is meant by a differential equation. Give an example.

30. Give the differential equation that models exponential growth and decay.

In Exercises 31 and 32, determine the quadrants in which the solution of the differential equation is an increasing function. Explain. (Do not solve the differential equation.)

31. $\dfrac{dy}{dx} = \dfrac{1}{2}xy$

32. $\dfrac{dy}{dx} = \dfrac{1}{2}x^2 y$

Radioactive Decay   In Exercises 33–40, complete the table for the radioactive isotope.

Isotope	Half-Life (in years)	Initial Quantity	Amount After 1,000 Years	Amount After 10,000 Years
33. ^{226}Ra	1620	10g		
34. ^{226}Ra	1620		1.5g	
35. ^{226}Ra	1620			0.5g
36. ^{14}C	5730			2g
37. ^{14}C	5730	5g		
38. ^{14}C	5730		3.2g	
39. ^{239}Pu	24,360		2.1g	
40. ^{239}Pu	24,360			0.4g

41. *Radioactive Decay* Radioactive radium has a half-life of approximately 1620 years. What percent of a given amount remains after 100 years?

42. *Carbon Dating* Carbon-14 dating assumes that the carbon dioxide on earth today has the same radioactive content as it did centuries ago. If this is true, the amount of ^{14}C absorbed by a tree that grew several centuries ago should be the same as the amount of ^{14}C absorbed by a tree growing today. A piece of ancient charcoal contains only 15% as much of the radioactive carbon as a piece of modern charcoal. How long ago was the tree burned to make the ancient charcoal? (The half-life of ^{14}C is 5730 years.)

Compound Interest   In Exercises 43–48, complete the table for a savings account in which interest is compounded continuously.

Initial Investment	Annual Rate	Time to Double	Amount After 10 Years
43. $1000	6%		
44. $20,000	$5\frac{1}{2}$%		
45. $750		$7\frac{3}{4}$ yr	
46. $10,000		5 yr	
47. $500			$1292.85
48. $2000			$5436.56

Compound Interest   In Exercises 49–52, find the principal P that must be invested at rate r, compounded monthly, so that $500,000 will be available for retirement in t years.

49. $r = 7\frac{1}{2}\%$, $t = 20$ **50.** $r = 6\%$, $t = 40$

51. $r = 8\%$, $t = 35$ **52.** $r = 9\%$, $t = 25$

Compound Interest   In Exercises 53–56, find the time necessary for $1000 to double if it is invested at a rate of r compounded (a) annually, (b) monthly, (c) daily, and (d) continuously.

53. $r = 7\%$ **54.** $r = 6\%$

55. $r = 8.5\%$ **56.** $r = 5.5\%$

Population   In Exercises 57–60, the population (in millions) of a country in 1999 and the continuous annual rate of change k of the population for the years 1990 through 2000 are given. Find the exponential growth model $P = Ce^{kt}$ for the population by letting $t = 0$ correspond to 2000. Use the model to predict the population of the city in 2010. (*Source: U.S. Census Bureau, "International Data Base"*)

Country	1999 Population	k
57. Bulgaria	8.2	−0.009
58. Cambodia	11.6	0.031
59. Jordan	4.6	0.036
60. Lithuania	3.6	−0.004

61. *Writing* Use the results of Exercises 57–60 and the exponential model $P = Ce^{kt}$ to discuss the relationship between the sign of k and the change in population for a given country.

62. *Modeling Data* One hundred bacteria are started in a culture and the number N of bacteria is counted each hour for 5 hours. The results are shown in the table, where t is the time in hours.

t	0	1	2	3	4	5
N	100	126	151	198	243	297

(a) Use the regression capabilities of a graphing utility to find an exponential model for the data.

(b) Use the model to estimate the time required for the population to quadruple in size.

63. *Atmospheric Pressure* Atmospheric pressure P (measured in millimeters of mercury) decreases exponentially with increasing altitude x (measured in meters). The pressure is 760 millimeters of mercury at sea level ($x = 0$) and 672.71 millimeters of mercury at an altitude of 1000 meters. Find the pressure at an altitude of 3000 meters.

64. *Revenue* Because of a slump in the economy, a company finds that its annual revenue has dropped from $742,000 in 1998 to $632,000 in 2000. If the revenue is following an exponential pattern of decline, what is the expected revenue for 2002? (Let $t = 0$ represent 1998.)

65. *Learning Curve* The management at a certain factory has found that a worker can produce at most 30 units in a day. The learning curve for the number of units N produced per day after a new employee has worked t days is

$$N = 30(1 - e^{kt}).$$

After 20 days on the job, a particular worker produces 19 units.

(a) Find the learning curve for this worker.

(b) How many days should pass before this worker is producing 25 units per day?

66. *Learning Curve* If in Exercise 65 management requires a new employee to produce at least 20 units per day after 30 days on the job, find (a) the learning curve that describes this minimum requirement and (b) the number of days before a minimal achiever is producing 25 units per day.

67. Sales The sales S (in thousands of units) of a new product after it has been on the market for t years is

$$S = Ce^{k/t}.$$

(a) Find S as a function of t if 5000 units have been sold after 1 year and the saturation point for the market is 30,000 units (that is, $\lim_{t \to \infty} S = 30$).

(b) How many units will have been sold after 5 years?

(c) Use a graphing utility to graph this sales function.

68. Sales The sales S (in thousands of units) of a new product after it has been on the market for t years is

$$S = 25(1 - e^{kt}).$$

(a) Find S as a function of t if 4000 units have been sold after 1 year.

(b) How many units will saturate this market?

(c) How many units will have been sold after 5 years?

(d) Use a graphing utility to graph this sales function.

69. Forestry The value of a tract of timber is

$$V(t) = 100,000e^{0.8\sqrt{t}}$$

where t is the time in years, with $t = 0$ corresponding to 1998. If money earns interest continuously at 10%, the present value of the timber at any time t is

$$A(t) = V(t)e^{-0.10t}.$$

Find the year in which the timber should be harvested to maximize the present value function.

70. Modeling Data The table shows the net receipts and the amounts required to service the national debt of the United States from 1990 through 1999. The monetary amounts are given in billions of dollars. *(Source: U.S. Office of Management and Budget)*

Year	1990	1991	1992	1993	1994
Receipts	1032.0	1055.0	1091.3	1154.4	1258.6
Interest	264.7	285.5	292.3	292.5	296.3

Year	1995	1996	1997	1998	1999
Receipts	1351.8	1453.1	1579.3	1721.8	1806.3
Interest	332.4	344.0	355.8	363.8	353.4

(a) Use the regression capabilities of a graphing utility to find an exponential model R for the receipts and a quartic model I for the amount required to service the debt. Let t represent the time in years, with $t = 0$ corresponding to 1990.

(b) Use a graphing utility to plot the points corresponding to the receipts, and graph the corresponding model. Based on the model, what is the continuous rate of growth of the receipts?

(c) Use a graphing utility to plot the points corresponding to the amount required to service the debt, and graph the quartic model.

(d) Find a function $P(t)$ that approximates the percent of the receipts that is required to service the national debt. Use a graphing utility to graph this function.

71. Sound Intensity The level of sound β, in decibels, with an intensity of I is

$$\beta(I) = 10 \log_{10} \frac{I}{I_0}$$

where I_0 is an intensity of 10^{-16} watts per square centimeter, corresponding roughly to the faintest sound that can be heard. Determine $\beta(I)$ for the following.

(a) $I = 10^{-14}$ watts per square centimeter (whisper)

(b) $I = 10^{-9}$ watts per square centimeter (busy street corner)

(c) $I = 10^{-6.5}$ watts per square centimeter (air hammer)

(d) $I = 10^{-4}$ watts per square centimeter (threshold of pain)

72. Noise Level With the installation of noise suppression materials, the noise level in an auditorium was reduced from 93 to 80 decibels. Use the function in Exercise 71 to find the percent decrease in the intensity level of the noise as a result of the installation of these materials.

73. Earthquake Intensity On the Richter scale, the magnitude R of an earthquake of intensity I is

$$R = \frac{\ln I - \ln I_0}{\ln 10}$$

where I_0 is the minimum intensity used for comparison. Assume that $I_0 = 1$.

(a) Find the intensity of the 1906 San Francisco earthquake ($R = 8.3$).

(b) Find the factor by which the intensity is increased if the Richter scale measurement is doubled.

(c) Find dR/dI.

74. Newton's Law of Cooling When an object is removed from a furnace and placed in an environment with a constant temperature of 80°F, its core temperature is 1500°F. One hour after it is removed, the core temperature is 1120°F. Find the core temperature 5 hours after the object is removed from the furnace.

True or False? In Exercises 75–78, determine whether the statement is true or false. If it is false, explain why or give an example that shows it is false.

75. In exponential growth the rate of growth is constant.

76. In linear growth the rate of growth is constant.

77. If prices are rising at a rate of 0.5% per month, then they are rising at a rate of 6% per year.

78. The differential equation modeling exponential growth is $dy/dx = ky$ where k is a constant.

- Use initial conditions to find particular solutions of differential equations.
- Recognize and solve differential equations that can be solved by separation of variables.
- Recognize and solve homogeneous differential equations.
- Use a differential equation to model and solve an applied problem.

General and Particular Solutions

Several times in the text, we have identified physical phenomena that can be described by differential equations. For example, in Section 5.6, you saw that problems involving radioactive decay, population growth, and Newton's Law of Cooling can be formulated in terms of differential equations.

A function $y = f(x)$ is called a **solution** of a differential equation if the equation is satisfied when y and its derivatives are replaced by $f(x)$ and its derivatives. For example, differentiation and substitution would show that $y = e^{-2x}$ is a solution of the differential equation $y' + 2y = 0$. It can be shown that every solution of this differential equation is of the form

$$y = Ce^{-2x} \qquad \text{General solution of } y' + 2y = 0$$

NOTE   First order linear differential equations are discussed in Appendix A.

where C is any real number. This solution is called the **general solution.** Some differential equations have **singular solutions** that cannot be written as special cases of the general solution. However, we will not consider such solutions in this text. The **order** of a differential equation is determined by the highest-order derivative in the equation. For instance, $y' = 4y$ is a first-order differential equation.

In Section 4.1, Example 8, you saw that the second-order differential equation $s''(t) = -32$ has the general solution

$$s(t) = -16t^2 + C_1 t + C_2 \qquad \text{General solution of } s''(t) = -32$$

which contains two arbitrary constants. It can be shown that a differential equation of order n has a general solution with n arbitrary constants.

Example 1   **Verifying Solutions**

Determine whether the given function is a solution of the differential equation $y'' - y = 0$.

a. $y = \sin x$ **b.** $y = 4e^{-x}$ **c.** $y = Ce^x$

Solution

a. Because $y = \sin x$, $y' = \cos x$, and $y'' = -\sin x$, it follows that

$$y'' - y = -\sin x - \sin x = -2\sin x \neq 0.$$

Hence, $y = \sin x$ is *not* a solution.

b. Because $y = 4e^{-x}$, $y' = -4e^{-x}$, and $y'' = 4e^{-x}$, it follows that

$$y'' - y = 4e^{-x} - 4e^{-x} = 0.$$

Hence, $y = 4e^{-x}$ *is* a solution.

c. Because $y = Ce^x$, $y' = Ce^x$, and $y'' = Ce^x$, it follows that

$$y'' - y = Ce^x - Ce^x = 0.$$

Hence, $y = Ce^x$ *is* a solution for any value of C.

Geometrically, the general solution of a first-order differential equation represents a family of curves known as **solution curves,** one for each value assigned to the arbitrary constant. For instance, you can verify that every function of the form

$$y = \frac{C}{x}$$ General solution of $xy' + y = 0$

is a solution of the differential equation $xy' + y = 0$. Figure 5.32 shows some of the solution curves corresponding to different values of C.

As discussed in Section 4.1, **particular solutions** of a differential equation are obtained from **initial conditions** that give the value of the dependent variable or one of its derivatives for a particular value of the independent variable. The term "initial condition" stems from the fact that, often in problems involving time, the value of the dependent variable or one of its derivatives is known at the *initial* time $t = 0$. For instance, the second-order differential equation $s''(t) = -32$ having the general solution

$$s(t) = -16t^2 + C_1 t + C_2$$ General solution of $s''(t) = -32$

might have the following initial conditions.

$$s(0) = 80, \qquad s'(0) = 64$$ Initial conditions

In this case, the initial conditions yield the particular solution

$$s(t) = -16t^2 + 64t + 80.$$ Particular solution

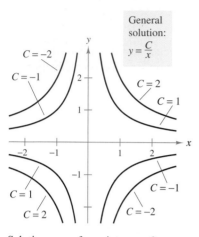

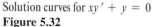

Solution curves for $xy' + y = 0$
Figure 5.32

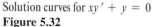 *Example 2*   **Finding a Particular Solution**

For the differential equation

$$xy' - 3y = 0$$

verify that $y = Cx^3$ is a solution, and find the particular solution determined by the initial condition $y = 2$ when $x = -3$.

Solution   You know that $y = Cx^3$ is a solution because $y' = 3Cx^2$ and

$$xy' - 3y = x(3Cx^2) - 3(Cx^3)$$
$$= 0.$$

Furthermore, the initial condition $y = 2$ when $x = -3$ yields

$$y = Cx^3$$ General solution
$$2 = C(-3)^3$$ Substitute initial condition.
$$-\frac{2}{27} = C$$ Solve for C.

and you can conclude that the particular solution is

$$y = -\frac{2x^3}{27}.$$ Particular solution

Try checking this solution by substituting for y and y' in the original differential equation.

NOTE   To determine a particular solution, the number of initial conditions must match the number of constants in the general solution.

Separation of Variables

Consider a differential equation that can be written in the form

$$M(x) + N(y)\frac{dy}{dx} = 0$$

FOR FURTHER INFORMATION For an example from engineering of a differential equation that is separable, see the article "Designing a Rose Cutter" by J. S. Hartzler in *The College Mathematics Journal.* To view this article, go to the website *www.matharticles.com.*

where M is a continuous function of x alone and N is a continuous function of y alone. As you saw in the preceding section, for this type of equation, all x terms can be collected with dx and all y terms with dy, and a solution can be obtained by integration. Such equations are said to be **separable,** and the solution procedure is called *separation of variables.* Here are some examples of differential equations that are separable.

Original Differential Equation	*Rewritten with Variables Separated*
$x^2 + 3y\dfrac{dy}{dx} = 0$	$3y\,dy = -x^2\,dx$
$\sin x\,y' - \cos x$	$dy = \cot x\,dx$
$\dfrac{xy'}{e^y + 1} = 2$	$\dfrac{1}{e^y + 1}\,dy = \dfrac{2}{x}\,dx$

Example 3   Separation of Variables

Find the general solution of $(x^2 + 4)\dfrac{dy}{dx} = xy.$

Solution   To begin, note that $y = 0$ is a solution. To find other solutions, assume that $y \neq 0$ and separate variables as follows.

$$(x^2 + 4)\,dy = xy\,dx \qquad \text{Differential form}$$

$$\frac{dy}{y} = \frac{x}{x^2 + 4}\,dx \qquad \text{Separate variables.}$$

Now, integrate to obtain

$$\int \frac{dy}{y} = \int \frac{x}{x^2 + 4}\,dx \qquad \text{Integrate.}$$

$$\ln|y| = \frac{1}{2}\ln(x^2 + 4) + C_1$$

$$\ln|y| = \ln\sqrt{x^2 + 4} + C_1$$

$$|y| = e^{C_1}\sqrt{x^2 + 4}$$

$$y = \pm e^{C_1}\sqrt{x^2 + 4}.$$

Because $y = 0$ is also a solution, you can write the general solution as

$$y = C\sqrt{x^2 + 4}. \qquad \text{General solution}$$

NOTE   We encourage you to check your solutions throughout this chapter. In Example 3 you can check the solution $y = C\sqrt{x^2 + 4}$ by differentiating and substituting into the original equation.

$$(x^2 + 4)\frac{dy}{dx} = xy \qquad \text{Original equation}$$

$$(x^2 + 4)\frac{Cx}{\sqrt{x^2 + 4}} \overset{?}{=} x\left(C\sqrt{x^2 + 4}\right) \qquad \text{Substitute.}$$

$$Cx\sqrt{x^2 + 4} = Cx\sqrt{x^2 + 4} \qquad \text{Solution checks.}$$

In some cases it is not feasible to write the general solution in the explicit form $y = f(x)$. The next example illustrates such a solution. Implicit differentiation can be used to verify this solution.

Example 4   Finding a Particular Solution

Given the initial condition $y(0) = 1$, find the particular solution of the equation

$$xy\, dx + e^{-x^2}(y^2 - 1)\, dy = 0.$$

Solution   Note that $y = 0$ is a solution of the differential equation—but this solution does not satisfy the initial condition. Hence, you can assume that $y \neq 0$. To separate variables, you must rid the first term of y and the second term of e^{-x^2}. So, you should multiply by e^{x^2}/y and obtain the following.

$$xy\, dx + e^{-x^2}(y^2 - 1)\, dy = 0$$

$$e^{-x^2}(y^2 - 1)\, dy = -xy\, dx$$

$$\int \left(y - \frac{1}{y} \right) dy = \int -xe^{x^2}\, dx$$

$$\frac{y^2}{2} - \ln y = -\frac{1}{2}e^{x^2} + C$$

From the initial condition $y(0) = 1$, you have $\frac{1}{2} - 0 = -\frac{1}{2} + C$, which implies that $C = 1$. So, the particular solution has the implicit form

$$\frac{y^2}{2} - \ln y = -\frac{1}{2}e^{x^2} + 1$$

$$y^2 - \ln y^2 + e^{x^2} = 2.$$

You can check this by differentiating and rewriting to get the original equation.

Example 5   Finding a Particular Solution Curve

Find the equation of the curve that passes through the point $(1, 3)$ and has a slope of y/x^2 at the point (x, y), as shown in Figure 5.33.

Solution   Because the slope of the curve is given by y/x^2, you have

$$\frac{dy}{dx} = \frac{y}{x^2}$$

with the initial condition $y(1) = 3$. Separating variables and integrating produces

$$\int \frac{dy}{y} = \int \frac{dx}{x^2}, \qquad\qquad y \neq 0$$

$$\ln|y| = -\frac{1}{x} + C_1$$

$$y = e^{-(1/x)+C_1} = Ce^{-1/x}.$$

Because $y = 3$ when $x = 1$, it follows that $3 = Ce^{-1}$ and $C = 3e$. Therefore, the equation of the specified curve is

$$y = (3e)e^{-1/x} = 3e^{(x-1)/x}, \qquad x > 0.$$

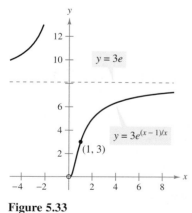

Figure 5.33

$y = 3e$

$y = 3e^{(x-1)/x}$

$(1, 3)$

Homogeneous Differential Equations

Some differential equations that are not separable in x and y can be made separable by a change of variables. This is true for differential equations of the form $y' = f(x, y)$, where f is a **homogeneous function**. The function given by $f(x, y)$ is **homogeneous of degree n** if

$$f(tx, ty) = t^n f(x, y)$$ Homogeneous function of degree n

where n is a real number.

> **NOTE**   The notation $f(x, y)$ is used to denote a function of two variables in much the same way as $f(x)$ denotes a function of one variable. You will study functions of two variables in detail in Chapter 12.

Example 6   Verifying Homogeneous Functions

a. $f(x, y) = x^2 y - 4x^3 + 3xy^2$ is a homogeneous function of degree 3 because

$$\begin{aligned}
f(tx, ty) &= (tx)^2(ty) - 4(tx)^3 + 3(tx)(ty)^2 \\
&= t^3(x^2 y) - t^3(4x^3) + t^3(3xy^2) \\
&= t^3(x^2 y - 4x^3 + 3xy^2) \\
&= t^3 f(x, y).
\end{aligned}$$

b. $f(x, y) = xe^{x/y} + y\sin(y/x)$ is a homogeneous function of degree 1 because

$$\begin{aligned}
f(tx, ty) &= txe^{tx/ty} + ty \sin\frac{ty}{tx} \\
&= t\left(xe^{x/y} + y\sin\frac{y}{x} \right) \\
&= tf(x, y).
\end{aligned}$$

c. $f(x, y) = x + y^2$ is not a homogeneous function because

$$f(tx, ty) = tx + t^2 y^2 = t(x + ty^2) \neq t^n(x + y^2).$$

d. $f(x, y) = x/y$ is a homogeneous function of degree 0 because

$$f(tx, ty) = \frac{tx}{ty} = t^0\frac{x}{y}.$$

> **Definition of Homogeneous Differential Equation**
>
> A **homogeneous differential equation** is an equation of the form
>
> $$M(x, y)\,dx + N(x, y)\,dy = 0$$
>
> where M and N are homogeneous functions of the same degree.

Example 7   Testing for Homogeneous Differential Equations

a. $(x^2 + xy)\,dx + y^2\,dy = 0$ is homogeneous of degree 2.

b. $x^3\,dx = y^3\,dy$ is homogeneous of degree 3.

c. $(x^2 + 1)\,dx + y^2\,dy = 0$ is *not* a homogeneous differential equation.

To solve a homogeneous differential equation by the method of separation of variables, we use the following change of variables theorem.

> **THEOREM 5.17   Change of Variables for Homogeneous Equations**
>
> If $M(x, y)\, dx + N(x, y)\, dy = 0$ is homogeneous, then it can be transformed into a differential equation whose variables are separable by the substitution
>
> $$y = vx$$
>
> where v is a differentiable function of x.

Example 8   Solving a Homogeneous Differential Equation

Find the general solution of

$$(x^2 - y^2)\, dx + 3xy\, dy = 0.$$

STUDY TIP   The substitution $y = vx$ will yield a differential equation that is separable with respect to the variables x and v. You must express your final solution, however, in terms of x and y.

Solution   Because $(x^2 - y^2)$ and $3xy$ are both homogeneous of degree 2, let $y = vx$ to obtain $dy = x\, dv + v\, dx$. Then, by substitution, you have

$$\overbrace{}^{dy}$$
$$(x^2 - v^2x^2)\, dx + 3x(vx)(x\, dv + v\, dx) = 0$$
$$(x^2 + 2v^2x^2)\, dx + 3x^3v\, dv = 0$$
$$x^2(1 + 2v^2)\, dx + x^2(3vx)\, dv = 0.$$

Dividing by x^2 and separating variables produces

$$(1 + 2v^2)\, dx = -3vx\, dv$$
$$\int \frac{dx}{x} = \int \frac{-3v}{1 + 2v^2}\, dv$$
$$\ln|x| = -\frac{3}{4}\ln(1 + 2v^2) + C_1$$
$$4\ln|x| = -3\ln(1 + 2v^2) + \ln|C|$$
$$\ln x^4 = \ln|C(1 + 2v^2)^{-3}|$$
$$x^4 = C(1 + 2v^2)^{-3}.$$

Substituting for v produces the following general solution.

$$x^4 = C\left[1 + 2\left(\frac{y}{x}\right)^2\right]^{-3}$$
$$\left(1 + \frac{2y^2}{x^2}\right)^3 x^4 = C$$
$$(x^2 + 2y^2)^3 = Cx^2 \qquad \text{General solution}$$

You can check this by differentiating and rewriting to get the original equation.

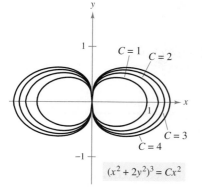

$(x^2 + 2y^2)^3 = Cx^2$

General solutions of
$(x^2 - y^2)\, dx + 3xy\, dy = 0$
Figure 5.34

TECHNOLOGY   If you have access to a graphing utility, try using it to graph several of the solutions in Example 8. For instance, Figure 5.34 shows the graphs of

$$(x^2 + 2y^2)^3 = Cx^2$$

for $C = 1, 2, 3,$ and 4.

Applications

Example 9   Wildlife Population

The rate of change of the number of coyotes $N(t)$ in a population is directly proportional to $650 - N(t)$, where t is the time in years. When $t = 0$, the population is 300, and when $t = 2$, the population has increased to 500. Find the population when $t = 3$.

Solution   Because the rate of change of the population is proportional to $650 - N(t)$, you can write the following differential equation.

$$\frac{dN}{dt} = k(650 - N)$$

You can solve this equation using separation of variables.

$$dN = k(650 - N)\, dt \qquad \text{Differential form}$$

$$\frac{dN}{650 - N} = k\, dt \qquad \text{Separate variables.}$$

$$-\ln|650 - N| = kt + C_1 \qquad \text{Integrate.}$$

$$\ln|650 - N| = -kt - C_1$$

$$650 - N = e^{-kt - C_1} \qquad \text{Assume } N < 650.$$

$$N = 650 - Ce^{-kt} \qquad \text{General solution}$$

Using $N = 300$ when $t = 0$, you can conclude that $C = 350$, which produces

$$N = 650 - 350e^{-kt}.$$

Then, using $N = 500$ when $t = 2$, it follows that

$$500 = 650 - 350e^{-2k} \quad \Longrightarrow \quad e^{-2k} = \tfrac{3}{7} \quad \Longrightarrow \quad k \approx 0.4236.$$

So, the model for the coyote population is

$$N = 650 - 350e^{-0.4236t}. \qquad \text{Model for population}$$

When $t = 3$, you can approximate the population to be

$$N = 650 - 350e^{-0.4236(3)} \approx 552 \text{ coyotes.}$$

The model for the population is shown in Figure 5.35.

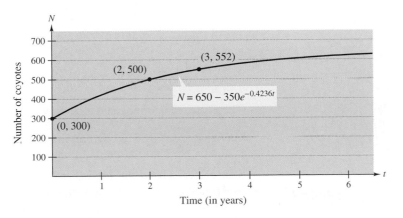

Figure 5.35

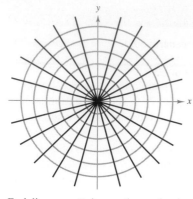

Each line $y = Kx$ is an orthogonal trajectory to the family of circles.
Figure 5.36

A common problem in electrostatics, thermodynamics, and hydrodynamics involves finding a family of curves, each of which is orthogonal to all members of a given family of curves. For example, Figure 5.36 shows a family of circles

$$x^2 + y^2 = C \qquad \text{Family of circles}$$

each of which intersects the lines in the family

$$y = Kx \qquad \text{Family of lines}$$

at right angles. Two such families of curves are said to be **mutually orthogonal,** and each curve in one of the families is called an **orthogonal trajectory** of the other family. In electrostatics, lines of force are orthogonal to the *equipotential curves.* In thermodynamics, the flow of heat across a plane surface is orthogonal to the *isothermal curves.* In hydrodynamics, the flow (stream) lines are orthogonal trajectories of the *velocity potential curves.*

Example 10   **Finding Orthogonal Trajectories**

Describe the orthogonal trajectories for the family of curves given by

$$y = \frac{C}{x}$$

for $C \neq 0$. Sketch several members of each family.

Solution   First, solve the given equation for C and write $xy = C$. Then, by differentiating implicitly with respect to x, you obtain the differential equation

$$xy' + y = 0 \qquad \text{Differential equation}$$

$$x\frac{dy}{dx} = -y$$

$$\frac{dy}{dx} = -\frac{y}{x}. \qquad \text{Slope of given family}$$

Because y' represents the slope of the given family of curves at (x, y), it follows that the orthogonal family has the negative reciprocal slope x/y, and we write

$$\frac{dy}{dx} = \frac{x}{y}. \qquad \text{Slope of orthogonal family}$$

Now you can find the orthogonal family by separating variables and integrating.

$$\int y\, dy = \int x\, dx$$

$$\frac{y^2}{2} = \frac{x^2}{2} + C_1$$

Therefore, each orthogonal trajectory is a hyperbola given by

$$\frac{y^2}{K} - \frac{x^2}{K} = 1. \qquad 2C_1 = K \neq 0$$

The centers are at the origin, and the transverse axes are vertical for $K > 0$ and horizontal for $K < 0$. Several trajectories are shown in Figure 5.37.

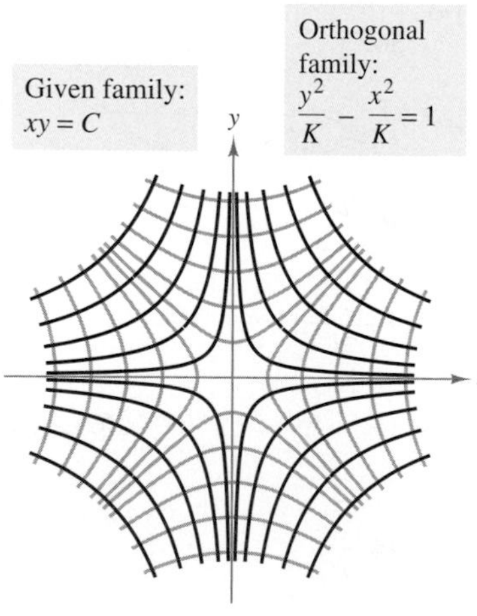

Given family:
$xy = C$

Orthogonal family:
$\dfrac{y^2}{K} - \dfrac{x^2}{K} = 1$

Orthogonal trajectories
Figure 5.37

Lab Series | LAB 7

EXERCISES FOR SECTION 5.7

In Exercises 1–6, verify the solution of the differential equation.

Solution	Differential Equation
1. $y = Ce^{4x}$	$y' = 4y$
2. $x^2 + y^2 = Cy$	$y' = 2xy/(x^2 - y^2)$
3. $y = C_1 \cos x + C_2 \sin x$	$y'' + y = 0$
4. $y = C_1 e^{-x} \cos x + C_2 e^{-x} \sin x$	$y'' + 2y' + 2y = 0$
5. $y = -\cos x \ln|\sec x + \tan x|$	$y'' + y = \tan x$
6. $y = \frac{2}{3}(e^{-2x} + e^x)$	$y'' + 2y' = 2e^x$

In Exercises 7–12, determine whether the function is a solution of the differential equation $y^{(4)} - 16y = 0$.

7. $y = 3 \cos x$ **8.** $y - 3 \cos 2x$

9. $y = e^{-2x}$ **10.** $y = 5 \ln x$

11. $y = C_1 e^{2x} + C_2 e^{-2x} + C_3 \sin 2x + C_4 \cos 2x$

12. $y = 3e^{2x} - 4 \sin 2x$

In Exercises 13–18, determine whether the function is a solution of the differential equation $xy' - 2y = x^3 e^x$.

13. $y = x^2$ **14.** $y = x^2 e^x$

15. $y = x^2(2 + e^x)$ **16.** $y = \sin x$

17. $y = \ln x$ **18.** $y = x^2 e^x - 5x^2$

19. *Think About It*   It is known that $y = Ce^{kx}$ is a solution of the differential equation $y' = 0.07y$. Is it possible to determine C or k from the information given? If so, find its value.

20. *Think About It*   It is known that $y = A \sin \omega t$ is a solution of the differential equation $y'' + 16y = 0$. Find the value of ω.

In Exercises 21 and 22, some of the curves corresponding to different values of C in the general solution of the differential equation are given. Find the particular solution that passes through the point indicated on the graph.

Solution	Differential Equation
21. $y^2 = Cx^3$	$2xy' - 3y = 0$
22. $2x^2 - y^2 = C$	$yy' - 2x = 0$

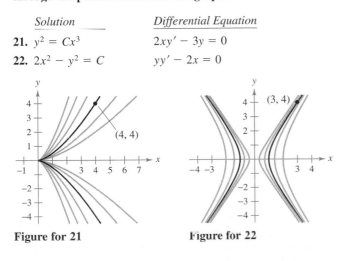

Figure for 21   **Figure for 22**

In Exercises 23 and 24, the general solution of the differential equation is given. Use a graphing utility to graph the particular solutions for the given values of C.

23. $4yy' - x = 0$
$4y^2 - x^2 = C$
$C = 0, C = \pm1, C = \pm4$

24. $yy' + x = 0$
$x^2 + y^2 = C$
$C = 0, C = 1, C = 4$

In Exercises 25–30, verify that the general solutions satisfy the differential equation. Then find the particular solution that satisfies the initial condition.

25. $y = Ce^{-2x}$
$y' + 2y = 0$
$y = 3$ when $x = 0$

26. $3x^2 + 2y^2 = C$
$3x + 2yy' = 0$
$y = 3$ when $x = 1$

27. $y = C_1 \sin 3x + C_2 \cos 3x$
$y'' + 9y = 0$
$y = 2$ when $x = \pi/6$
$y' = 1$ when $x = \pi/6$

28. $y = C_1 + C_2 \ln x$
$xy'' + y' = 0$
$y = 0$ when $x = 2$
$y' = \frac{1}{2}$ when $x = 2$

29. $y = C_1 x + C_2 x^3$
$x^2 y'' - 3xy' + 3y = 0$
$y = 0$ when $x = 2$
$y' = 4$ when $x = 2$

30. $y = e^{2x/3}(C_1 + C_2 x)$
$9y'' - 12y' + 4y = 0$
$y = 4$ when $x = 0$
$y = 0$ when $x = 3$

In Exercises 31–42, use integration to find a general solution of the differential equation.

31. $\dfrac{dy}{dx} = 3x^2$

32. $\dfrac{dy}{dx} = x^3 - 4x$

33. $\dfrac{dy}{dx} = \dfrac{x}{1 + x^2}$

34. $\dfrac{dy}{dx} = \dfrac{e^x}{1 + e^x}$

35. $\dfrac{dy}{dx} = \dfrac{x - 2}{x}$

36. $\dfrac{dy}{dx} = x \cos x^2$

37. $\dfrac{dy}{dx} = \sin 2x$

38. $\dfrac{dy}{dx} = \tan^2 x$

39. $\dfrac{dy}{dx} = x\sqrt{x - 3}$

40. $\dfrac{dy}{dx} = x\sqrt{5 - x}$

41. $\dfrac{dy}{dx} = xe^{x^2}$

42. $\dfrac{dy}{dx} = 5e^{-x/2}$

In Exercises 43–54, find the general solution of the differential equation.

43. $\dfrac{dy}{dx} = \dfrac{x}{y}$

44. $\dfrac{dy}{dx} = \dfrac{x^2 + 2}{3y^2}$

45. $\dfrac{dr}{ds} = 0.05r$

46. $\dfrac{dr}{ds} = 0.05s$

47. $(2 + x)y' = 3y$

48. $xy' = y$

49. $yy' = \sin x$

50. $yy' = 6 \cos(\pi x)$

51. $\sqrt{1 - 4x^2}\, y' - x$

52. $\sqrt{x^2 - 9}\, y' = 5x$

53. $y \ln x - xy' = 0$

54. $4yy' - 3e^x = 0$

In Exercises 55–64, find the particular solution that satisfies the initial condition.

Differential Equation	*Initial Condition*
55. $yy' - e^x = 0$	$y(0) = 4$
56. $\sqrt{x} + \sqrt{y}\,y' = 0$	$y(1) = 4$
57. $y(x + 1) + y' = 0$	$y(-2) = 1$
58. $2xy' - \ln x^2 = 0$	$y(1) = 2$
59. $y(1 + x^2)y' - x(1 + y^2) = 0$	$y(0) = \sqrt{3}$
60. $y\sqrt{1 - x^2}\,y' - x\sqrt{1 - y^2} = 0$	$y(0) = 1$
61. $\dfrac{du}{dv} = uv \sin v^2$	$u(0) = 1$
62. $\dfrac{dr}{ds} = e^{r-2s}$	$r(0) = 0$
63. $dP - kP\,dt = 0$	$P(0) = P_0$
64. $dT + k(T - 70)\,dt = 0$	$T(0) = 140$

In Exercises 65 and 66, find an equation of the graph that passes through the point and has the indicated slope.

Point	*Slope*
65. $(1, 1)$	$y' = -\dfrac{9x}{16y}$
66. $(8, 2)$	$y' = \dfrac{2y}{3x}$

In Exercises 67 and 68, find all functions f having the indicated property.

67. The tangent to the graph of f at the point (x, y) intersects the x-axis at $(x + 2, 0)$.

68. All tangents to the graph of f pass through the origin.

In Exercises 69–76, determine whether the function is homogeneous, and if it is, determine its degree.

69. $f(x, y) = x^3 - 4xy^2 + y^3$
70. $f(x, y) = x^3 + 3x^2y^2 - 2y^2$
71. $f(x, y) = \dfrac{x^2y^2}{\sqrt{x^2 + y^2}}$
72. $f(x, y) = \dfrac{xy}{\sqrt{x^2 + y^2}}$
73. $f(x, y) = 2 \ln xy$
74. $f(x, y) = \tan(x + y)$
75. $f(x, y) = 2 \ln \dfrac{x}{y}$
76. $f(x, y) = \tan \dfrac{y}{x}$

In Exercises 77–82, solve the homogeneous differential equation.

77. $y' = \dfrac{x + y}{2x}$
78. $y' = \dfrac{x^3 + y^3}{xy^2}$
79. $y' = \dfrac{x - y}{x + y}$
80. $y' = \dfrac{x^2 + y^2}{2xy}$
81. $y' = \dfrac{xy}{x^2 - y^2}$
82. $y' = \dfrac{2x + 3y}{x}$

In Exercises 83–86, find the particular solution that satisfies the initial condition.

Differential Equation	*Initial Condition*
83. $x\,dy - (2xe^{-y/x} + y)\,dx = 0$	$y(1) = 0$
84. $-y^2\,dx + x(x + y)\,dy = 0$	$y(1) = 1$
85. $\left(x \sec \dfrac{y}{x} + y\right) dx - x\,dy = 0$	$y(1) = 0$
86. $(2x^2 + y^2)\,dx + xy\,dy = 0$	$y(1) = 0$

Slope Fields In Exercises 87–90, sketch a few solutions of the differential equation on the slope field and then find the general solution analytically. To print an enlarged copy of the graph, go to the website *www.mathgraphs.com*.

87. $\dfrac{dy}{dx} = x$

88. $\dfrac{dy}{dx} = -\dfrac{x}{y}$

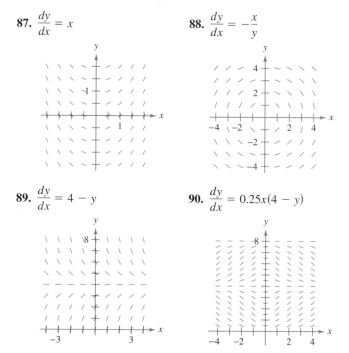

89. $\dfrac{dy}{dx} = 4 - y$

90. $\dfrac{dy}{dx} = 0.25x(4 - y)$

In Exercises 91–94, use a computer algebra system to sketch the slope field for the differential equation, and graph the solution satisfying the specified initial condition.

91. $\dfrac{dy}{dx} = 0.5y, \ y(0) = 6$

92. $\dfrac{dy}{dx} = 2 - y, \ y(0) = 4$

93. $\dfrac{dy}{dx} = 0.02y(10 - y), \ y(0) = 2$

94. $\dfrac{dy}{dx} = 0.2x(2 - y), \ y(0) = 9$

95. *Radioactive Decay* The rate of decomposition of radioactive radium is proportional to the amount present at any time. The half-life of radioactive radium is 1620 years. What percent of a present amount will remain after 25 years?

96. *Chemical Reaction* In a chemical reaction, a certain compound changes into another compound at a rate proportional to the unchanged amount. If initially there is 20 grams of the original compound, and there is 16 grams after 1 hour, when will 75 percent of the compound be changed?

Slope Fields **In Exercises 97–100, (a) write a differential equation for the statement, (b) match the differential equation with a possible slope field, and (c) verify your result by using a graphing utility to graph a slope field for the differential equation. [The slope fields are labeled (a), (b), (c), and (d).] To print an enlarged copy of the graph, go to the website www.mathgraphs.com.**

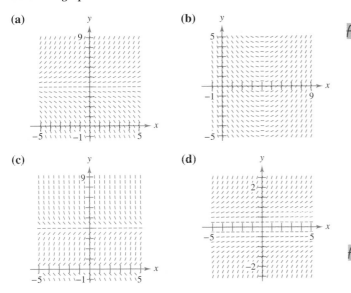

(a) **(b)** **(c)** **(d)**

97. The rate of change of y with respect to x is proportional to the difference between y and 4.

98. The rate of change of y with respect to x is proportional to the difference between x and 4.

99. The rate of change of y with respect to x is proportional to the product of y and the difference between y and 4.

100. The rate of change of y with respect to x is proportional to y^2.

101. *Weight Gain* A calf that weighs 60 pounds at birth gains weight at the rate

$$\frac{dw}{dt} = k(1200 - w)$$

where w is weight in pounds and t is time in years. Solve the differential equation.

(a) Use a computer algebra system to solve the differential equation for $k = 0.8$, $k = 0.9$, and $k = 1$. Graph the three solutions.

(b) If the animal is sold when its weight reaches 800 pounds, find the time of sale for each of the models in part (a).

(c) What is the maximum weight of the animal for each of the models?

102. *Weight Gain* A calf that weighs w_0 pounds at birth gains weight at the rate

$$\frac{dw}{dt} = 1200 - w$$

where w is weight in pounds and t is time in years. Solve the differential equation.

103. *Sailing* Ignoring resistance, a sailboat starting from rest accelerates (dv/dt) at a rate proportional to the difference between the velocities of the wind and the boat.

(a) Write the velocity as a function of time if the wind is blowing at 20 knots, and after 1 minute the boat is moving at 5 knots.

(b) Use the result in part (a) to write the distance traveled by the boat as a function of time.

104. *Radio Reception* In hilly areas, radio reception may be poor. Consider a situation where an FM transmitter is located at the point $(-1, 1)$ behind a hill modeled by the graph of $y = x - x^2$, and a radio receiver is on the opposite side of the hill. (Assume that the x-axis represents ground level at the base of the hill.)

(a) What is the closest the radio can be to the hill so that reception is unobstructed?

(b) Write the closest position x of the radio as a function of h if the transmitter is located at $(-1, h)$.

(c) Use a graphing utility to graph the function in part (b). Determine the vertical asymptote of the function and interpret the result.

In Exercises 105–110, find the orthogonal trajectories of the family. Use a graphing utility to graph several members of each family.

105. $x^2 + y^2 = C$ **106.** $x^2 - 2y^2 = C$

107. $x^2 = Cy$ **108.** $y^2 = 2Cx$

109. $y^2 = Cx^3$ **110.** $y = Ce^x$

Getting at the Concept

111. In your own words, describe the difference between a general solution of a differential equation and a particular solution.

112. When determining a particular solution, how do you determine how many initial conditions are required?

113. State the test for determining if a differential equation is homogeneous. Give an example.

114. In your own words, describe the relationship between two families of curves that are mutually orthogonal.

True or False? **In Exercises 115–118, determine whether the statement is true or false. If it is false, explain why or give an example that shows it is false.**

115. If $y = f(x)$ is a solution of a first-order differential equation, then $y = f(x) + C$ is also a solution.

116. The differential equation $y' = xy - 2y + x - 2$ can be written in separated variables form.

117. The function $f(x, y) = x^2 + xy + 2$ is homogeneous.

118. The families $x^2 + y^2 = 2Cy$ and $x^2 + y^2 = 2Kx$ are mutually orthogonal.

- Develop properties of the six inverse trigonometric functions.
- Differentiate an inverse trigonometric function.
- Review the basic differentiation formulas for elementary functions.

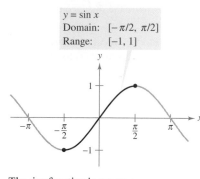

$y = \sin x$
Domain: $[-\pi/2, \pi/2]$
Range: $[-1, 1]$

The sine function is one-to-one on $[-\pi/2, \pi/2]$.

Figure 5.38

Inverse Trigonometric Functions

This section begins with a rather surprising statement: *None of the six basic trigonometric functions has an inverse function.* This statement is true because all six trigonometric functions are periodic and hence not one-to-one. In this section you will examine these six functions to see whether their domains can be redefined in such a way that they will have inverse functions on the *restricted domains*.

In Example 4 of Section 5.3, you saw that the sine function is increasing (and therefore is one-to-one) on the interval $[-\pi/2, \pi/2]$ (see Figure 5.38). On this interval you can define the inverse of the *restricted* sine function to be

$$y = \arcsin x \qquad \text{if and only if} \qquad \sin y = x$$

where $-1 \le x \le 1$ and $-\pi/2 \le \arcsin x \le \pi/2$.

Under suitable restrictions, each of the six trigonometric functions is one-to-one and so has an inverse function, as indicated in the following definition.

NOTE   The term "iff" is used to represent the phrase "if and only if."

Definition of Inverse Trigonometric Functions

Function	Domain	Range
$y = \arcsin x$ iff $\sin y = x$	$-1 \le x \le 1$	$-\dfrac{\pi}{2} \le y \le \dfrac{\pi}{2}$
$y = \arccos x$ iff $\cos y = x$	$-1 \le x \le 1$	$0 \le y \le \pi$
$y = \arctan x$ iff $\tan y = x$	$-\infty < x < \infty$	$-\dfrac{\pi}{2} < y < \dfrac{\pi}{2}$
$y = \text{arccot } x$ iff $\cot y = x$	$-\infty < x < \infty$	$0 < y < \pi$
$y = \text{arcsec } x$ iff $\sec y = x$	$|x| \ge 1$	$0 \le y \le \pi, \quad y \ne \dfrac{\pi}{2}$
$y = \text{arccsc } x$ iff $\csc y = x$	$|x| \ge 1$	$-\dfrac{\pi}{2} \le y \le \dfrac{\pi}{2}, \quad y \ne 0$

NOTE   The term "arcsin x" is read as "the arcsine of x" or sometimes "the angle whose sine is x." An alternative notation for the inverse sine function is "$\sin^{-1} x$."

EXPLORATION

The Inverse Secant Function   In the definition above, the inverse secant function is defined by restricting the domain of the secant function to the intervals $\left[0, \dfrac{\pi}{2}\right) \cup \left(\dfrac{\pi}{2}, \pi\right]$. Most other texts and reference books agree with this, but some disagree. What other domains might make sense? Explain your reasoning graphically. Most calculators do not have a key for the inverse secant function. How can you use a calculator to evaluate the inverse secant function?

The graphs of the six inverse trigonometric functions are shown in Figure 5.39.

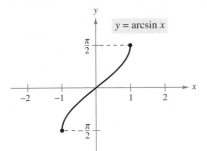

Domain: $[-1, 1]$
Range: $[-\pi/2, \pi/2]$

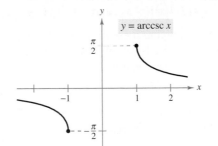

Domain: $(-\infty, -1] \cup [1, \infty)$
Range: $[-\pi/2, 0) \cup (0, \pi/2]$

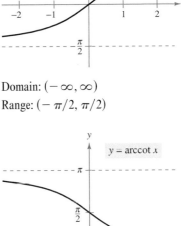

Domain: $(-\infty, \infty)$
Range: $(-\pi/2, \pi/2)$

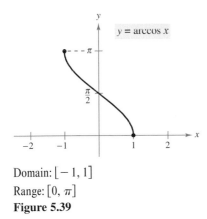

Domain: $[-1, 1]$
Range: $[0, \pi]$

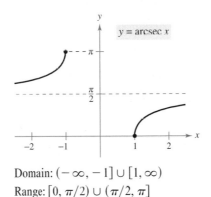

Domain: $(-\infty, -1] \cup [1, \infty)$
Range: $[0, \pi/2) \cup (\pi/2, \pi]$

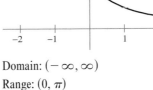

Domain: $(-\infty, \infty)$
Range: $(0, \pi)$

Figure 5.39

Example 1　　**Evaluating Inverse Trigonometric Functions**

Evaluate each of the following.

a. $\arcsin\left(-\dfrac{1}{2}\right)$　　　**b.** $\arccos 0$　　　**c.** $\arctan \sqrt{3}$　　　**d.** $\arcsin(0.3)$

NOTE　When evaluating inverse trigonometric functions, remember that they denote *angles in radian measure*.

Solution

a. By definition, $y = \arcsin\left(-\frac{1}{2}\right)$ implies that $\sin y = -\frac{1}{2}$. In the interval $[-\pi/2, \pi/2]$, the correct value of y is $-\pi/6$.

$$\arcsin\left(-\frac{1}{2}\right) = -\frac{\pi}{6}$$

b. By definition, $y = \arccos 0$ implies that $\cos y = 0$. In the interval $[0, \pi]$, you have $y = \pi/2$.

$$\arccos 0 = \frac{\pi}{2}$$

c. By definition, $y = \arctan \sqrt{3}$ implies that $\tan y = \sqrt{3}$. In the interval $(-\pi/2, \pi/2)$, you have $y = \pi/3$.

$$\arctan \sqrt{3} = \frac{\pi}{3}$$

d. Using a calculator set in *radian mode* produces

$$\arcsin(0.3) \approx 0.3047.$$

EXPLORATION

Graph $y = \arccos(\cos x)$ for $-4\pi \le x \le 4\pi$. Why isn't the graph the same as the graph of $y = x$?

Inverse functions have the properties

$$f(f^{-1}(x)) = x \quad \text{and} \quad f^{-1}(f(x)) = x.$$

When applying these properties to inverse trigonometric functions, remember that the trigonometric functions have inverse functions only in restricted domains. For x-values outside these domains, these two properties do not hold. For example, $\arcsin(\sin \pi)$ is equal to 0, not π.

Properties of Inverse Trigonometric Functions

If $-1 \le x \le 1$ and $-\pi/2 \le y \le \pi/2$, then

$$\sin(\arcsin x) = x \quad \text{and} \quad \arcsin(\sin y) = y.$$

If $-\pi/2 < y < \pi/2$, then

$$\tan(\arctan x) = x \quad \text{and} \quad \arctan(\tan y) = y.$$

If $|x| \ge 1$ and $0 \le y < \pi/2$ or $\pi/2 < y \le \pi$, then

$$\sec(\arcsec x) = x \quad \text{and} \quad \arcsec(\sec y) = y.$$

Similar properties hold for the other inverse trigonometric functions.

Example 2 **Solving an Equation**

$$\arctan(2x - 3) = \frac{\pi}{4} \qquad \text{Original equation}$$

$$\tan[\arctan(2x - 3)] = \tan\frac{\pi}{4} \qquad \text{Take tangent of both sides.}$$

$$2x - 3 = 1 \qquad \tan(\arctan x) = x$$

$$x = 2 \qquad \text{Solve for } x.$$

Some problems in calculus require that you evaluate expressions such as $\cos(\arcsin x)$, as illustrated in Example 3.

Example 3 **Using Right Triangles**

a. Given $y = \arcsin x$, where $0 < y < \pi/2$, find $\cos y$.
b. Given $y = \arcsec(\sqrt{5}/2)$, find $\tan y$.

Solution

a. Because $y = \arcsin x$, you know that $\sin y = x$. This relationship between x and y can be represented by a right triangle, as shown in Figure 5.40.

$$\cos y = \cos(\arcsin x) = \frac{\text{adj.}}{\text{hyp.}} = \sqrt{1 - x^2}$$

(This result is also valid for $-\pi/2 < y < 0$.)

b. Use the right triangle shown in Figure 5.41.

$$\tan y = \tan\left[\arcsec\left(\frac{\sqrt{5}}{2}\right)\right] = \frac{\text{opp.}}{\text{adj.}} = \frac{1}{2}$$

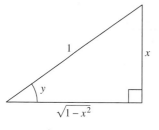

$y = \arcsin x$
Figure 5.40

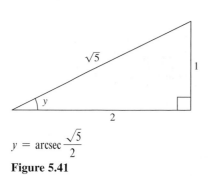

$y = \arcsec\dfrac{\sqrt{5}}{2}$
Figure 5.41

NOTE   There is no common agreement on the definition of arcsec x (or arccsc x) for negative values of x. When we defined the range of the arcsecant, we chose to preserve the reciprocal identity

$$\text{arcsec } x = \arccos \frac{1}{x}.$$

For example, to evaluate arcsec (-2), you can write

$$\text{arcsec}(-2) = \arccos(-0.5) \approx 2.09.$$

One of the consequences of the definition of the inverse secant function given in this text is that its graph has a positive slope at every x-value in its domain. (See Figure 5.39.) This accounts for the absolute value sign in the formula for the derivative of arcsec x.

Derivatives of Inverse Trigonometric Functions

In Section 5.1 you saw that the derivative of the *transcendental* function $f(x) = \ln x$ is the *algebraic* function $f'(x) = 1/x$. You will now see that the derivatives of the inverse trigonometric functions also are algebraic (even though the inverse trigonometric functions are themselves transcendental).

The following theorem lists the derivatives of the six inverse trigonometric functions. Note that the derivatives of arccos u, arccot u, and arccsc u are the *negatives* of the derivatives of arcsin u, arctan u, and arcsec u, respectively.

THEOREM 5.18   Derivatives of Inverse Trigonometric Functions

Let u be a differentiable function of x.

$$\frac{d}{dx}[\arcsin u] = \frac{u'}{\sqrt{1-u^2}} \qquad\qquad \frac{d}{dx}[\arccos u] = \frac{-u'}{\sqrt{1-u^2}}$$

$$\frac{d}{dx}[\arctan u] = \frac{u'}{1+u^2} \qquad\qquad \frac{d}{dx}[\text{arccot } u] = \frac{-u'}{1+u^2}$$

$$\frac{d}{dx}[\text{arcsec } u] = \frac{u'}{|u|\sqrt{u^2-1}} \qquad\qquad \frac{d}{dx}[\text{arccsc } u] = \frac{-u'}{|u|\sqrt{u^2-1}}$$

To derive these formulas, you can use implicit differentiation. For instance, if $y = \arcsin x$, then $\sin y = x$ and $(\cos y)y' = 1$. (See Exercise 76.)

TECHNOLOGY   If your graphing utility does not have the arcsecant function, you can obtain its graph using

$$f(x) = \text{arcsec } x = \arccos \frac{1}{x}.$$

Example 4   Differentiating Inverse Trigonometric Functions

---

a. $\dfrac{d}{dx}[\arcsin(2x)] = \dfrac{2}{\sqrt{1-(2x)^2}} = \dfrac{2}{\sqrt{1-4x^2}}$

b. $\dfrac{d}{dx}[\arctan(3x)] = \dfrac{3}{1+(3x)^2} = \dfrac{3}{1+9x^2}$

c. $\dfrac{d}{dx}[\arcsin \sqrt{x}] = \dfrac{(1/2)x^{-1/2}}{\sqrt{1-x}} = \dfrac{1}{2\sqrt{x}\sqrt{1-x}} = \dfrac{1}{2\sqrt{x-x^2}}$

d. $\dfrac{d}{dx}[\text{arcsec } e^{2x}] = \dfrac{2e^{2x}}{e^{2x}\sqrt{(e^{2x})^2-1}} = \dfrac{2e^{2x}}{e^{2x}\sqrt{e^{4x}-1}} = \dfrac{2}{\sqrt{e^{4x}-1}}$

The absolute value sign is not necessary because $e^{2x} > 0$.

NOTE   From Example 5, you can see one of the benefits of inverse trigonometric functions—they can be used to integrate common algebraic functions. For instance, from the result shown in the example, it follows that

$$\int \sqrt{1-x^2}\, dx$$
$$= \frac{1}{2}\left(\arcsin x + x\sqrt{1-x^2}\right).$$

Example 5   A Derivative That Can Be Simplified

Differentiate $y = \arcsin x + x\sqrt{1-x^2}$.

Solution

$$y' = \frac{1}{\sqrt{1-x^2}} + x\left(\frac{1}{2}\right)(-2x)(1-x^2)^{-1/2} + \sqrt{1-x^2}$$

$$= \frac{1}{\sqrt{1-x^2}} - \frac{x^2}{\sqrt{1-x^2}} + \sqrt{1-x^2}$$

$$= \sqrt{1-x^2} + \sqrt{1-x^2}$$

$$= 2\sqrt{1-x^2}$$

Example 6   Analyzing an Inverse Trigonometric Graph

Analyze the graph of $y = (\arctan x)^2$.

Solution   From the derivative

$$y' = 2(\arctan x)\left(\frac{1}{1 + x^2}\right)$$

$$= \frac{2 \arctan x}{1 + x^2}$$

you can see that the only critical number is $x = 0$. By the First Derivative Test, this value corresponds to a relative minimum. From the second derivative

$$y'' = \frac{(1 + x^2)\left(\dfrac{2}{1 + x^2}\right) - (2 \arctan x)(2x)}{(1 + x^2)^2}$$

$$= \frac{2(1 - 2x \arctan x)}{(1 + x^2)^2}$$

it follows that points of inflection occur when $2x \arctan x = 1$. Using Newton's Method, these points occur when $x \approx \pm 0.765$. Finally, because

$$\lim_{x \to \pm\infty} (\arctan x)^2 = \frac{\pi^2}{4}$$

it follows that the graph has a horizontal asymptote at $y = \pi^2/4$. The graph is shown in Figure 5.42.

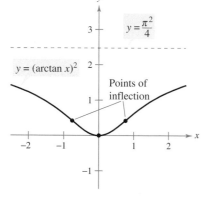

The graph of $y = (\arctan x)^2$ has a horizontal asymptote at $y = \pi^2/4$.
Figure 5.42

Example 7   Maximizing an Angle

A photographer is taking a picture of a 4-foot painting hung in an art gallery. The camera lens is 1 foot below the lower edge of the painting, as shown in Figure 5.43. How far should the camera be from the painting to maximize the angle subtended by the camera lens?

Solution   In Figure 5.43, let β be the angle to be maximized.

$$\beta = \theta - \alpha$$

$$= \text{arccot}\frac{x}{5} - \text{arccot } x$$

Differentiating produces

$$\frac{d\beta}{dx} = \frac{-1/5}{1 + (x^2/25)} - \frac{-1}{1 + x^2}$$

$$= \frac{-5}{25 + x^2} + \frac{1}{1 + x^2}$$

$$= \frac{4(5 - x^2)}{(25 + x^2)(1 + x^2)}.$$

Because $d\beta/dx = 0$ when $x = \sqrt{5}$, you can conclude from the First Derivative Test that this distance yields a maximum value of β. So, the distance is $x \approx 2.236$ feet and the angle is $\beta \approx 0.7297$ radians $\approx 41.81°$.

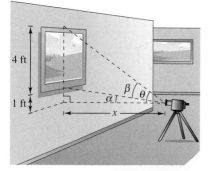

The camera should be 2.236 feet from the painting to maximize the angle β.
Figure 5.43

Review of Basic Differentiation Rules

In the 1600s, Europe was ushered into the scientific age by such great thinkers as Descartes, Galileo, Huygens, Newton, and Kepler. These men believed that nature is governed by basic laws—laws that can, for the most part, be written in terms of mathematical equations. One of the most influential publications of this period—*Dialogue on the Great World Systems*, by Galileo Galilei—has become a classic description of modern scientific thought.

As mathematics has developed during the past few hundred years, a small number of elementary functions has proven sufficient for modeling most* phenomena in physics, chemistry, biology, engineering, economics, and a variety of other fields. An **elementary function** is a function from the following list or one that can be formed as the sum, product, quotient, or composition of functions in the list.

Algebraic Functions	*Transcendental Functions*
Polynomial functions	Logarithmic functions
Rational functions	Exponential functions
Functions involving radicals	Trigonometric functions
	Inverse trigonometric functions

With the differentiation rules introduced so far in the text, you can differentiate *any* elementary function. For convenience, we summarize these differentiation rules here.

GALILEO GALILEI (1564–1642)

Galileo's approach to science departed from the accepted Aristotelian view that nature had describable *qualities*, such as "fluidity" and "potentiality." He chose to describe the physical world in terms of measurable *quantities*, such as time, distance, force, and mass.

Basic Differentiation Rules for Elementary Functions

1. $\dfrac{d}{dx}[cu] = cu'$

2. $\dfrac{d}{dx}[u \pm v] = u' \pm v'$

3. $\dfrac{d}{dx}[uv] = uv' + vu'$

4. $\dfrac{d}{dx}\left[\dfrac{u}{v}\right] = \dfrac{vu' - uv'}{v^2}$

5. $\dfrac{d}{dx}[c] = 0$

6. $\dfrac{d}{dx}[u^n] = nu^{n-1}u'$

7. $\dfrac{d}{dx}[x] = 1$

8. $\dfrac{d}{dx}[|u|] = \dfrac{u}{|u|}(u'), \quad u \neq 0$

9. $\dfrac{d}{dx}[\ln u] = \dfrac{u'}{u}$

10. $\dfrac{d}{dx}[e^u] = e^u u'$

11. $\dfrac{d}{dx}[\log_a u] = \dfrac{u'}{(\ln a)u}$

12. $\dfrac{d}{dx}[a^u] = (\ln a)a^u u'$

13. $\dfrac{d}{dx}[\sin u] = (\cos u)u'$

14. $\dfrac{d}{dx}[\cos u] = -(\sin u)u'$

15. $\dfrac{d}{dx}[\tan u] = (\sec^2 u)u'$

16. $\dfrac{d}{dx}[\cot u] = -(\csc^2 u)u'$

17. $\dfrac{d}{dx}[\sec u] = (\sec u \tan u)u'$

18. $\dfrac{d}{dx}[\csc u] = -(\csc u \cot u)u'$

19. $\dfrac{d}{dx}[\arcsin u] = \dfrac{u'}{\sqrt{1 - u^2}}$

20. $\dfrac{d}{dx}[\arccos u] = \dfrac{-u'}{\sqrt{1 - u^2}}$

21. $\dfrac{d}{dx}[\arctan u] = \dfrac{u'}{1 + u^2}$

22. $\dfrac{d}{dx}[\text{arccot } u] = \dfrac{-u'}{1 + u^2}$

23. $\dfrac{d}{dx}[\text{arcsec } u] = \dfrac{u'}{|u|\sqrt{u^2 - 1}}$

24. $\dfrac{d}{dx}[\text{arccsc } u] = \dfrac{-u'}{|u|\sqrt{u^2 - 1}}$

---

* *Some important functions used in engineering and science (such as Bessel functions and gamma functions) are not elementary functions.*

EXERCISES FOR SECTION 5.8

Numerical and Graphical Analysis In Exercises 1 and 2, (a) use a graphing utility to complete the table, (b) plot the points in the table and graph the function by hand, (c) use a graphing utility to graph the function and compare the result with your hand-drawn graph in part (b), and (d) determine any intercepts and symmetry of the graph.

x	-1	-0.8	-0.6	-0.4	-0.2	0	0.2	0.4	0.6	0.8	1
y											

1. $y = \arcsin x$

2. $y = \arccos x$

3. *True or False?* Decide whether the following statement is true or false, and explain: Because $\cos(-\pi/3) = \frac{1}{2}$, it follows that $\arccos \frac{1}{2} = -\pi/3$.

4. Determine the missing coordinates of the points on the graph of the function.

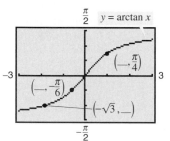

In Exercises 5–12, evaluate the expression without using a calculator.

5. $\arcsin \frac{1}{2}$

6. $\arcsin 0$

7. $\arccos \frac{1}{2}$

8. $\arccos 0$

9. $\arctan \dfrac{\sqrt{3}}{3}$

10. $\text{arccot}\left(-\sqrt{3}\right)$

11. $\text{arccsc}\left(-\sqrt{2}\right)$

12. $\arccos\left(-\dfrac{\sqrt{3}}{2}\right)$

In Exercises 13–16, use a calculator to approximate the value. Round your answer to two decimal places.

13. $\arccos(-0.8)$

14. $\arcsin(-0.39)$

15. $\text{arcsec } 1.269$

16. $\arctan(-3)$

In Exercises 17–20, evaluate the expression without using a calculator. (*Hint:* See Example 3.)

17. (a) $\sin\left(\arctan \dfrac{3}{4}\right)$

    (b) $\sec\left(\arcsin \dfrac{4}{5}\right)$

18. (a) $\tan\left(\arccos \dfrac{\sqrt{2}}{2}\right)$

    (b) $\cos\left(\arcsin \dfrac{5}{13}\right)$

19. (a) $\cot\left[\arcsin\left(-\dfrac{1}{2}\right)\right]$

    (b) $\csc\left[\arctan\left(-\dfrac{5}{12}\right)\right]$

20. (a) $\sec\left[\arctan\left(-\dfrac{3}{5}\right)\right]$

    (b) $\tan\left[\arcsin\left(-\dfrac{5}{6}\right)\right]$

In Exercises 21–28, write the expression in algebraic form.

21. $\cos(\arcsin 2x)$

22. $\sec(\arctan 4x)$

23. $\sin(\text{arcsec } x)$

24. $\cos(\text{arccot } x)$

25. $\tan\left(\text{arcsec } \dfrac{x}{3}\right)$

26. $\sec[\arcsin(x-1)]$

27. $\csc\left(\arctan \dfrac{x}{\sqrt{2}}\right)$

28. $\cos\left(\arcsin \dfrac{x-h}{r}\right)$

In Exercises 29 and 30, use a graphing utility to graph f and g in the same viewing window to verify that they are equal. Explain why they are equal. Identify any asymptotes of the graphs.

29. $f(x) = \sin(\arctan 2x)$, $g(x) = \dfrac{2x}{\sqrt{1+4x^2}}$

30. $f(x) = \tan\left(\arccos \dfrac{x}{2}\right)$, $g(x) = \dfrac{\sqrt{4-x^2}}{x}$

In Exercises 31–34, solve the equation for x.

31. $\arcsin(3x - \pi) = \dfrac{1}{2}$

32. $\arctan(2x - 5) = -1$

33. $\arcsin\sqrt{2x} = \arccos\sqrt{x}$

34. $\arccos x = \text{arcsec } x$

In Exercises 35 and 36, verify each identity.

35. (a) $\text{arccsc } x = \arcsin \dfrac{1}{x}$, $x \ge 1$

    (b) $\arctan x + \arctan \dfrac{1}{x} = \dfrac{\pi}{2}$, $x > 0$

36. (a) $\arcsin(-x) = -\arcsin x$, $|x| \le 1$

    (b) $\arccos(-x) = \pi - \arccos x$, $|x| \le 1$

In Exercises 37–40, sketch the graph of the function. Use a graphing utility to verify your graph.

37. $f(x) = \arcsin(x - 1)$

38. $f(x) = \arctan x + \dfrac{\pi}{2}$

39. $f(x) = \text{arcsec } 2x$

40. $f(x) = \arccos \dfrac{x}{4}$

In Exercises 41–60, find the derivative of the function.

41. $f(x) = 2\arcsin(x - 1)$

42. $f(t) = \arcsin t^2$

43. $g(x) = 3\arccos \dfrac{x}{2}$

44. $f(x) = \text{arcsec } 2x$

45. $f(x) = \arctan \dfrac{x}{a}$

46. $f(x) = \arctan \sqrt{x}$

47. $g(x) = \dfrac{\arcsin 3x}{x}$

48. $h(x) = x^2 \arctan x$

49. $h(t) = \sin(\arccos t)$

50. $f(x) = \arcsin x + \arccos x$

51. $y = x\arccos x - \sqrt{1 - x^2}$

52. $y = \ln(t^2 + 4) - \dfrac{1}{2}\arctan \dfrac{t}{2}$

53. $y = \dfrac{1}{2}\left(\dfrac{1}{2}\ln \dfrac{x+1}{x-1} + \arctan x\right)$

54. $y = \frac{1}{2}\left[x\sqrt{4 - x^2} + 4\arcsin\left(\frac{x}{2}\right)\right]$

55. $y = x\arcsin x + \sqrt{1 - x^2}$

56. $y = x\arctan 2x - \frac{1}{4}\ln(1 + 4x^2)$

57. $y = 8\arcsin\frac{x}{4} - \frac{x\sqrt{16 - x^2}}{2}$

58. $y = 25\arcsin\frac{x}{5} - x\sqrt{25 - x^2}$

59. $y = \arctan x + \frac{x}{1 + x^2}$ **60.** $y = \arctan\frac{x}{2} - \frac{1}{2(x^2 + 4)}$

Linear and Quadratic Approximations   **In Exercises 61 and 62, use a computer algebra system to find the linear approximation**

$$P_1(x) = f(a) + f'(a)(x - a)$$

and the quadratic approximation

$$P_2(x) = f(a) + f'(a)(x - a) + \frac{1}{2}f''(a)(x - a)^2$$

to the function f at $x = a$. Sketch the graph of the function and its linear and quadratic approximations.

61. $f(x) = \arcsin x$ **62.** $f(x) = \arctan x$
    $a = \frac{1}{2}$ $a = 1$

In Exercises 63–66, find any relative extrema of the function.

63. $f(x) = \text{arcsec } x - x$ **64.** $f(x) = \arcsin x - 2x$

65. $f(x) = \arctan x - \arctan(x - 4)$

66. $h(x) = \arcsin x - 2\arctan x$

Getting at the Concept

67. Explain why the domains of the trigonometric functions are restricted when finding the inverse trigonometric functions.

68. Explain why $\tan \pi = 0$ does not imply that $\arctan 0 = \pi$.

69. Explain how to graph $y = \text{arccot } x$ on a graphing utility that does not have the arccotangent function.

70. Are the derivatives of the inverse trigonometric functions algebraic or transcendental functions? List the derivatives of the inverse trigonometric functions.

71. *Angular Rate of Change*   An airplane flies at an altitude of 5 miles toward a point directly over an observer. Consider θ and x as shown in the figure.

(a) Write θ as a function of x.

(b) If the speed of the plane is 400 miles per hour, find $d\theta/dt$ when $x = 10$ miles and $x = 3$ miles.

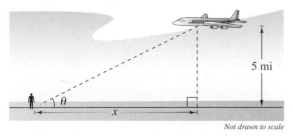

5 mi

x

Not drawn to scale

72. *Writing*   Repeat Exercise 71 if the altitude of the plane is 3 miles and describe how the altitude affects the rate of change of θ.

73. *Angular Rate of Change*   In a free-fall experiment, an object is dropped from a height of 256 feet. A camera on the ground 500 feet from the point of impact records the fall of the object.

(a) Find the position function giving the height of the object at time t assuming the object is released at time $t = 0$. At what time will the object reach ground level?

(b) Find the rate of change of the angle of elevation of the camera when $t = 1$ and $t = 2$.

74. *Angular Rate of Change*   A television camera at ground level is filming the lift-off of a space shuttle at a point 750 meters from the launch pad. Let θ be the angle of elevation of the shuttle and let s be the distance between the camera and the shuttle. Write θ as a function of s for the period of time when the shuttle is moving vertically. Differentiate the result to find $d\theta/dt$ in terms of s and ds/dt.

75. Prove that

$$\arctan x + \arctan y = \arctan\frac{x + y}{1 - xy}, \qquad xy \neq 1.$$

Use this formula to show that

$$\arctan\frac{1}{2} + \arctan\frac{1}{3} = \frac{\pi}{4}.$$

76. Verify each differentiation formula.

(a) $\frac{d}{dx}[\arcsin u] = \frac{u'}{\sqrt{1 - u^2}}$ (b) $\frac{d}{dx}[\arctan u] = \frac{u'}{1 + u^2}$

(c) $\frac{d}{dx}[\text{arcsec } u] = \frac{u'}{|u|\sqrt{u^2 - 1}}$

(d) $\frac{d}{dx}[\arccos u] = \frac{-u'}{\sqrt{1 - u^2}}$

(e) $\frac{d}{dx}[\text{arccot } u] = \frac{-u'}{1 + u^2}$ (f) $\frac{d}{dx}[\text{arccsc } u] = \frac{-u'}{|u|\sqrt{u^2 - 1}}$

77. *Existence of an Inverse*   Determine the values of k such that the function $f(x) = kx + \sin x$ has an inverse function.

78. *Think About It*   Use a graphing utility to graph

$$f(x) = \sin x \quad \text{and} \quad g(x) = \arcsin(\sin x).$$

(a) Why isn't the graph of g the line $y = x$?

(b) Determine the extrema of g.

True or False?   **In Exercises 79–82, determine whether the statement is true or false. If it is false, explain why or give an example that shows it is false.**

79. The slope of the graph of the inverse tangent function is positive for all x.

80. The range of $y = \arcsin x$ is $[0, \pi]$.

81. $\frac{d}{dx}[\arctan(\tan x)] = 1$ for all x in the domain.

82. $\arcsin^2 x + \arccos^2 x = 1$

Section 5.9	**Inverse Trigonometric Functions: Integration**

- Integrate functions whose antiderivatives involve inverse trigonometric functions.
- Use the method of completing the square to integrate a function.
- Review the basic integration formulas involving elementary functions.

Integrals Involving Inverse Trigonometric Functions

The derivatives of the six inverse trigonometric functions fall into three pairs. In each pair, the derivative of one function is the negative of the other. For example,

$$\frac{d}{dx}[\arcsin x] = \frac{1}{\sqrt{1 - x^2}}$$

and

$$\frac{d}{dx}[\arccos x] = -\frac{1}{\sqrt{1 - x^2}}.$$

When listing the *antiderivative* that corresponds to each of the inverse trigonometric functions, you need to use only one member from each pair. It is conventional to use arcsin x as the antiderivative of $1/\sqrt{1 - x^2}$, rather than $-\arccos x$. The next theorem gives one antiderivative formula for each of the three pairs. The proofs of these integration rules are left to you (see Exercise 61).

NOTE   For a proof of part 2 of Theorem 5.19, see the article "A Direct Proof of the Integral Formula for Arctangent" by Arnold J. Insel in *The College Mathematics Journal*. To view this article, go to the website *www.matharticles.com*.

> **THEOREM 5.19   Integrals Involving Inverse Trigonometric Functions**
>
> Let u be a differentiable function of x, and let $a > 0$.
>
> **1.** $\displaystyle\int \frac{du}{\sqrt{a^2 - u^2}} = \arcsin \frac{u}{a} + C$
>
> **2.** $\displaystyle\int \frac{du}{a^2 + u^2} = \frac{1}{a} \arctan \frac{u}{a} + C$
>
> **3.** $\displaystyle\int \frac{du}{u\sqrt{u^2 - a^2}} = \frac{1}{a} \text{arcsec} \frac{|u|}{a} + C$

Example 1   Integration with Inverse Trigonometric Functions

a. $\displaystyle\int \frac{dx}{\sqrt{4 - x^2}} = \arcsin \frac{x}{2} + C$

b. $\displaystyle\int \frac{dx}{2 + 9x^2} = \frac{1}{3} \int \frac{3\, dx}{\left(\sqrt{2}\right)^2 + (3x)^2}$ $u = 3x,\ a = \sqrt{2}$

$\displaystyle\qquad\qquad\quad = \frac{1}{3\sqrt{2}} \arctan \frac{3x}{\sqrt{2}} + C$

c. $\displaystyle\int \frac{dx}{x\sqrt{4x^2 - 9}} = \int \frac{2\, dx}{2x\sqrt{(2x)^2 - 3^2}}$ $u = 2x,\ a = 3$

$\displaystyle\qquad\qquad\quad = \frac{1}{3} \text{arcsec} \frac{|2x|}{3} + C$

The integrals in Example 1 are fairly straightforward applications of integration formulas. Unfortunately, this is not typical. The integration formulas for inverse trigonometric functions can be disguised in many ways.

TECHNOLOGY PITFALL　Computer software that can perform symbolic integration is useful for integrating functions such as the one in Example 2. When using such software, however, you must remember that it can fail to find an antiderivative for two reasons. First, some elementary functions simply do not have anti-derivatives that are elementary functions. Second, every symbolic integration utility has limitations—you might have entered a function that the software was not programmed to handle. You should also remember that antiderivatives involving trigonometric functions or logarithmic functions can be written in many different forms. For instance, when we used a symbolic integration utility to find the integral in Example 2, we obtained

$$\int \frac{dx}{\sqrt{e^{2x} - 1}} = \arctan \sqrt{e^{2x} - 1} + C.$$

Try showing that this antiderivative is equivalent to that obtained in Example 2.

Example 2　**Integration by Substitution**

Find $\int \dfrac{dx}{\sqrt{e^{2x} - 1}}$.

Solution　As it stands, this integral doesn't fit any of the three inverse trigonometric formulas. Using the substitution $u = e^x$, however, produces the following.

$$u = e^x \quad \Longrightarrow \quad du = e^x \, dx \quad \Longrightarrow \quad dx = \frac{du}{e^x} = \frac{du}{u}$$

With this substitution, you can integrate as follows.

$$\int \frac{dx}{\sqrt{e^{2x} - 1}} = \int \frac{dx}{\sqrt{(e^x)^2 - 1}} \qquad \text{Write } e^{2x} \text{ as } (e^x)^2.$$

$$= \int \frac{du/u}{\sqrt{u^2 - 1}} \qquad \text{Substitute.}$$

$$= \int \frac{du}{u\sqrt{u^2 - 1}} \qquad \text{Rewrite to fit Arcsecant Rule.}$$

$$= \operatorname{arcsec} \frac{|u|}{1} + C \qquad \text{Apply Arcsecant Rule.}$$

$$= \operatorname{arcsec} e^x + C \qquad \text{Back-substitute.}$$

Example 3　**Rewriting as the Sum of Two Quotients**

Find $\int \dfrac{x + 2}{\sqrt{4 - x^2}} \, dx$.

Solution　This integral does not appear to fit any of the basic integration formulas. By splitting the integrand into two parts, however, you can see that the first part can be found with the Power Rule and the second part yields an inverse sine function.

$$\int \frac{x + 2}{\sqrt{4 - x^2}} \, dx = \int \frac{x}{\sqrt{4 - x^2}} \, dx + \int \frac{2}{\sqrt{4 - x^2}} \, dx$$

$$= -\frac{1}{2} \int (4 - x^2)^{-1/2}(-2x) \, dx + 2 \int \frac{1}{\sqrt{4 - x^2}} \, dx$$

$$= -\frac{1}{2} \left[\frac{(4 - x^2)^{1/2}}{1/2} \right] + 2 \arcsin \frac{x}{2} + C$$

$$= -\sqrt{4 - x^2} + 2 \arcsin \frac{x}{2} + C$$

Completing the Square

Completing the square helps when quadratic functions are involved in the integrand. For example, the quadratic $x^2 + bx + c$ can be written as the difference of two squares by adding and subtracting $(b/2)^2$.

$$x^2 + bx + c = x^2 + bx + \left(\frac{b}{2}\right)^2 - \left(\frac{b}{2}\right)^2 + c$$

$$= \left(x + \frac{b}{2}\right)^2 - \left(\frac{b}{2}\right)^2 + c$$

Example 4   **Completing the Square**

Find $\displaystyle\int \frac{dx}{x^2 - 4x + 7}$.

Solution   You can write the denominator as the sum of two squares as follows.

$$x^2 - 4x + 7 = (x^2 - 4x + 4) - 4 + 7 = (x - 2)^2 + 3 = u^2 + a^2$$

Now, in this completed square form, let $u = x - 2$ and $a = \sqrt{3}$.

$$\int \frac{dx}{x^2 - 4x + 7} = \int \frac{dx}{(x - 2)^2 + 3} = \frac{1}{\sqrt{3}} \arctan \frac{x - 2}{\sqrt{3}} + C$$

If the leading coefficient is not 1, it helps to factor before completing the square. For instance, you can complete the square of $2x^2 - 8x + 10$ as follows.

$$2x^2 - 8x + 10 = 2(x^2 - 4x + 5)$$
$$= 2(x^2 - 4x + 4 - 4 + 5)$$
$$= 2[(x - 2)^2 + 1]$$

To complete the square when the coefficient of x^2 is negative, use the same "factoring process" illustrated above. For instance, you can complete the square for $3x - x^2$ as follows.

$$3x - x^2 = -(x^2 - 3x)$$
$$= -\left[x^2 - 3x + \left(\tfrac{3}{2}\right)^2 - \left(\tfrac{3}{2}\right)^2\right]$$
$$= \left(\tfrac{3}{2}\right)^2 - \left(x - \tfrac{3}{2}\right)^2$$

Example 5   **Completing the Square (Negative Leading Coefficient)**

Find the area of the region bounded by the graph of

$$f(x) = \frac{1}{\sqrt{3x - x^2}}$$

the x-axis, and the lines $x = \frac{3}{2}$ and $x = \frac{9}{4}$.

Solution   From Figure 5.44, you can see that the area is given by

$$\text{Area} = \int_{3/2}^{9/4} \frac{1}{\sqrt{3x - x^2}}\, dx.$$

Using the completed square form derived above, you can integrate as follows.

$$\int_{3/2}^{9/4} \frac{dx}{\sqrt{3x - x^2}} = \int_{3/2}^{9/4} \frac{dx}{\sqrt{(3/2)^2 - [x - (3/2)]^2}}$$
$$= \arcsin \frac{x - (3/2)}{3/2} \Big]_{3/2}^{9/4}$$
$$= \arcsin \frac{1}{2} - \arcsin 0$$
$$= \frac{\pi}{6}$$
$$\approx 0.524$$

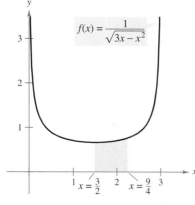

$f(x) = \dfrac{1}{\sqrt{3x - x^2}}$

The area of the region bounded by the graph of f, the x-axis, $x = \frac{3}{2}$, and $x = \frac{9}{4}$ is $\pi/6$.

Figure 5.44

TECHNOLOGY   With definite integrals such as the one given in Example 5, remember that you can resort to a numerical solution. For instance, applying Simpson's Rule (with $n = 6$) to the integral in the example, you obtain

$$\int_{3/2}^{9/4} \frac{1}{\sqrt{3x - x^2}}\, dx \approx 0.523599.$$

This differs from the exact value of the integral ($\pi/6 \approx 0.5235988$) by less than one millionth.

Review of Basic Integration Rules

You have now completed the introduction of the **basic integration rules.** To be efficient at applying these rules, you should have practiced enough so that each rule is committed to memory.

Basic Integration Rules ($a > 0$)

1. $\displaystyle\int kf(u)\,du = k\int f(u)\,du$

2. $\displaystyle\int [f(u) \pm g(u)]\,du = \int f(u)\,du \pm \int g(u)\,du$

3. $\displaystyle\int du = u + C$

4. $\displaystyle\int u^n\,du = \frac{u^{n+1}}{n+1} + C,\quad n \neq -1$

5. $\displaystyle\int \frac{du}{u} = \ln|u| + C$

6. $\displaystyle\int e^u\,du = e^u + C$

7. $\displaystyle\int a^u\,du = \left(\frac{1}{\ln a}\right)a^u + C$

8. $\displaystyle\int \sin u\,du = -\cos u + C$

9. $\displaystyle\int \cos u\,du = \sin u + C$

10. $\displaystyle\int \tan u\,du = -\ln|\cos u| + C$

11. $\displaystyle\int \cot u\,du = \ln|\sin u| + C$

12. $\displaystyle\int \sec u\,du = \ln|\sec u + \tan u| + C$

13. $\displaystyle\int \csc u\,du = -\ln|\csc u + \cot u| + C$

14. $\displaystyle\int \sec^2 u\,du = \tan u + C$

15. $\displaystyle\int \csc^2 u\,du = -\cot u + C$

16. $\displaystyle\int \sec u \tan u\,du = \sec u + C$

17. $\displaystyle\int \csc u \cot u\,du = -\csc u + C$

18. $\displaystyle\int \frac{du}{\sqrt{a^2 - u^2}} = \arcsin\frac{u}{a} + C$

19. $\displaystyle\int \frac{du}{a^2 + u^2} = \frac{1}{a}\arctan\frac{u}{a} + C$

20. $\displaystyle\int \frac{du}{u\sqrt{u^2 - a^2}} = \frac{1}{a}\operatorname{arcsec}\frac{|u|}{a} + C$

You can learn a lot about the nature of integration by comparing this list with the summary of differentiation rules given in the preceding section. For differentiation, you now have rules that allow you to differentiate *any* elementary function. For integration, this is far from true.

The integration rules listed above are primarily those that we happened on when developing differentiation rules. We do not find integration rules for the antiderivative of a general product or quotient, the natural logarithmic function, or the inverse trigonometric functions. More importantly, you cannot apply any of the rules in this list unless you can create the proper *du* corresponding to the *u* in the formula. The point is that we need to work more on integration techniques, which we will do in Chapter 7. The next two examples should give you a better feeling for the integration problems that you *can* and *cannot* do with the techniques and rules you now know.

Example 6   Comparing Integration Problems

Find as many of the following integrals as you can using the formulas and techniques you have studied so far in the text.

a. $\displaystyle\int \frac{dx}{x\sqrt{x^2 - 1}}$ **b.** $\displaystyle\int \frac{x\,dx}{\sqrt{x^2 - 1}}$ **c.** $\displaystyle\int \frac{dx}{\sqrt{x^2 - 1}}$

Solution

a. You *can* find this integral (it fits the Arcsecant Rule).

$$\int \frac{dx}{x\sqrt{x^2 - 1}} = \operatorname{arcsec}|x| + C$$

b. You *can* find this integral (it fits the Power Rule).

$$\int \frac{x\,dx}{\sqrt{x^2 - 1}} = \frac{1}{2}\int (x^2 - 1)^{-1/2}(2x)\,dx$$

$$= \frac{1}{2}\left[\frac{(x^2 - 1)^{1/2}}{1/2}\right] + C$$

$$= \sqrt{x^2 - 1} + C$$

c. You *cannot* find this integral using present techniques. (You should scan the list of basic integration rules to verify this conclusion.)

Example 7   Comparing Integration Problems

Find as many of the following integrals as you can using the formulas and techniques you have studied so far in the text.

a. $\displaystyle\int \frac{dx}{x \ln x}$ **b.** $\displaystyle\int \frac{\ln x\,dx}{x}$ **c.** $\displaystyle\int \ln x\,dx$

Solution

a. You *can* find this integral (it fits the Log Rule).

$$\int \frac{dx}{x \ln x} = \int \frac{1/x}{\ln x}\,dx$$

$$= \ln|\ln x| + C$$

b. You *can* find this integral (it fits the Power Rule).

$$\int \frac{\ln x\,dx}{x} = \int \left(\frac{1}{x}\right)(\ln x)^1\,dx$$

$$= \frac{(\ln x)^2}{2} + C$$

c. You *cannot* find this integral using present techniques.

NOTE   Note in Examples 6 and 7 that the *simplest* functions are the ones that you cannot yet integrate.

EXERCISES FOR SECTION 5.9

In Exercises 1–30, find or evaluate the integral.

1. $\displaystyle \int \frac{5}{\sqrt{9 - x^2}}\, dx$

2. $\displaystyle \int \frac{3}{\sqrt{1 - 4x^2}}\, dx$

3. $\displaystyle \int_0^{1/6} \frac{1}{\sqrt{1 - 9x^2}}\, dx$

4. $\displaystyle \int_0^1 \frac{dx}{\sqrt{4 - x^2}}$

5. $\displaystyle \int \frac{7}{16 + x^2}\, dx$

6. $\displaystyle \int \frac{4}{1 + 9x^2}\, dx$

7. $\displaystyle \int_0^{\sqrt{3}/2} \frac{1}{1 + 4x^2}\, dx$

8. $\displaystyle \int_{\sqrt{3}}^3 \frac{1}{9 + x^2}\, dx$

9. $\displaystyle \int \frac{1}{x\sqrt{4x^2 - 1}}\, dx$

10. $\displaystyle \int \frac{1}{4 + (x - 1)^2}\, dx$

11. $\displaystyle \int \frac{x^3}{x^2 + 1}\, dx$

12. $\displaystyle \int \frac{x^4 - 1}{x^2 + 1}\, dx$

13. $\displaystyle \int \frac{1}{\sqrt{1 - (x + 1)^2}}\, dx$

14. $\displaystyle \int \frac{t}{t^4 + 16}\, dt$

15. $\displaystyle \int \frac{t}{\sqrt{1 - t^4}}\, dt$

16. $\displaystyle \int \frac{1}{x\sqrt{x^4 - 4}}\, dx$

17. $\displaystyle \int_0^{1/\sqrt{2}} \frac{\arcsin x}{\sqrt{1 - x^2}}\, dx$

18. $\displaystyle \int_0^{1/\sqrt{2}} \frac{\arccos x}{\sqrt{1 - x^2}}\, dx$

19. $\displaystyle \int_{-1/2}^0 \frac{x}{\sqrt{1 - x^2}}\, dx$

20. $\displaystyle \int_{-\sqrt{3}}^0 \frac{x}{1 + x^2}\, dx$

21. $\displaystyle \int \frac{e^{2x}}{4 + e^{4x}}\, dx$

22. $\displaystyle \int_1^2 \frac{1}{3 + (x - 2)^2}\, dx$

23. $\displaystyle \int_{\pi/2}^{\pi} \frac{\sin x}{1 + \cos^2 x}\, dx$

24. $\displaystyle \int_0^{\pi/2} \frac{\cos x}{1 + \sin^2 x}\, dx$

25. $\displaystyle \int \frac{1}{\sqrt{x}\sqrt{1 - x}}\, dx$

26. $\displaystyle \int \frac{3}{2\sqrt{x}(1 + x)}\, dx$

27. $\displaystyle \int \frac{x - 3}{x^2 + 1}\, dx$

28. $\displaystyle \int \frac{4x + 3}{\sqrt{1 - x^2}}\, dx$

29. $\displaystyle \int \frac{x + 5}{\sqrt{9 - (x - 3)^2}}\, dx$

30. $\displaystyle \int \frac{x - 2}{(x + 1)^2 + 4}\, dx$

In Exercises 31–42, find or evaluate the integral. (Complete the square, if necessary.)

31. $\displaystyle \int_0^2 \frac{dx}{x^2 - 2x + 2}$

32. $\displaystyle \int_{-2}^2 \frac{dx}{x^2 + 4x + 13}$

33. $\displaystyle \int \frac{2x}{x^2 + 6x + 13}\, dx$

34. $\displaystyle \int \frac{2x - 5}{x^2 + 2x + 2}\, dx$

35. $\displaystyle \int \frac{1}{\sqrt{-x^2 - 4x}}\, dx$

36. $\displaystyle \int \frac{2}{\sqrt{-x^2 + 4x}}\, dx$

37. $\displaystyle \int \frac{x + 2}{\sqrt{-x^2 - 4x}}\, dx$

38. $\displaystyle \int \frac{x - 1}{\sqrt{x^2 - 2x}}\, dx$

39. $\displaystyle \int_2^3 \frac{2x - 3}{\sqrt{4x - x^2}}\, dx$

40. $\displaystyle \int \frac{1}{(x - 1)\sqrt{x^2 - 2x}}\, dx$

41. $\displaystyle \int \frac{x}{x^4 + 2x^2 + 2}\, dx$

42. $\displaystyle \int \frac{x}{\sqrt{9 + 8x^2 - x^4}}\, dx$

In Exercises 43 and 44, use the specified substitution to find the integral.

43. $\displaystyle \int \sqrt{e^t - 3}\, dt$
$u = \sqrt{e^t - 3}$

44. $\displaystyle \int \frac{\sqrt{x - 2}}{x + 1}\, dx$
$u = \sqrt{x - 2}$

Getting at the Concept

45. What is a perfect square trinomial?

46. What term must be added to $x^2 + 3x$ to complete the square? Explain how you found the term.

In Exercises 47–50, determine which of the integrals can be found using the basic integration formulas you have studied so far in the text.

47. (a) $\displaystyle \int \frac{1}{\sqrt{1 - x^2}}\, dx$ (b) $\displaystyle \int \frac{x}{\sqrt{1 - x^2}}\, dx$ (c) $\displaystyle \int \frac{1}{x\sqrt{1 - x^2}}\, dx$

48. (a) $\displaystyle \int e^{x^2}\, dx$ (b) $\displaystyle \int xe^{x^2}\, dx$ (c) $\displaystyle \int \frac{1}{x^2} e^{1/x}\, dx$

49. (a) $\displaystyle \int \sqrt{x - 1}\, dx$ (b) $\displaystyle \int x\sqrt{x - 1}\, dx$ (c) $\displaystyle \int \frac{x}{\sqrt{x - 1}}\, dx$

50. (a) $\displaystyle \int \frac{1}{1 + x^4}\, dx$ (b) $\displaystyle \int \frac{x}{1 + x^4}\, dx$ (c) $\displaystyle \int \frac{x^3}{1 + x^4}\, dx$

Slope Fields **In Exercises 51 and 52, a differential equation, a point, and a slope field are given. (a) Sketch two approximate solutions of the differential equation on the slope field, one of which passes through the indicated point. (b) Use integration to find the particular solution of the differential equation and use a graphing utility to graph the solution. Compare the result with the sketches in part (a). To print an enlarged copy of the graph, go to the website *www.mathgraphs.com*.**

51. $\dfrac{dy}{dx} = \dfrac{3}{1 + x^2}, \quad (0, 0)$

52. $\dfrac{dy}{dx} = x\sqrt{16 - y^2}, \quad (0, -2)$

In Exercises 53 and 54, use a computer algebra system to graph the slope field for the differential equation and graph the solution satisfying the specified initial condition.

53. $\dfrac{dy}{dx} = \dfrac{10}{x\sqrt{x^2 - 1}}$
$y(3) = 0$

54. $\dfrac{dy}{dx} = \dfrac{2y}{\sqrt{16 - x^2}}$
$y(0) = 2$

In Exercises 55 and 56, find the area of the region bounded by the graphs of the equations.

55. $y = \dfrac{1}{x^2 - 2x + 5}$, $y = 0$, $x = 1$, $x = 3$

56. $y = \dfrac{1}{\sqrt{4 - x^2}}$, $y = 0$, $x = 0$, $x = 1$

57. *Approximation*   Determine which value best approximates the area of the region between the x-axis and the function

$$f(x) = \dfrac{1}{\sqrt{1 - x^2}}$$

over the interval $[-0.5, 0.5]$. (Make your selection on the basis of a sketch of the region and *not* by performing any calculations.)

(a) 4   (b) -3 (c) 1   (d) 2   (e) 3

58. *Approximation*   Sketch the region whose area is represented by the integral

$$\int_0^1 \arcsin x \, dx$$

and use the integration capabilities of a graphing utility to approximate the area.

59. (a) Show that

$$\int_0^1 \dfrac{4}{1 + x^2} \, dx = \pi.$$

(b) Approximate the number π using Simpson's Rule (with $n = 6$) and the integral in part (a).

(c) Approximate the number π by using the integration capabilities of a graphing utility.

60. *Investigation*   Consider the function

$$F(x) = \dfrac{1}{2} \int_x^{x+2} \dfrac{2}{t^2 + 1} \, dt.$$

(a) Write a short paragraph giving a geometric interpretation of the function $F(x)$ relative to the function

$$f(x) = \dfrac{2}{x^2 + 1}.$$

Use what you have written to guess the value of x that will make F maximum.

(b) Perform the specified integration to find an alternative form of $F(x)$. Use calculus to locate the value of x that will make F maximum and compare the result with your guess in part (a).

61. Verify each rule by differentiating ($a > 0$).

(a) $\displaystyle \int \dfrac{du}{\sqrt{a^2 - u^2}} = \arcsin \dfrac{u}{a} + C$

(b) $\displaystyle \int \dfrac{du}{a^2 + u^2} = \dfrac{1}{a} \arctan \dfrac{u}{a} + C$

(c) $\displaystyle \int \dfrac{du}{u\sqrt{u^2 - a^2}} = \dfrac{1}{a} \operatorname{arcsec} \dfrac{|u|}{a} + C$

62. Consider the integral

$$\int \dfrac{1}{\sqrt{6x - x^2}} \, dx.$$

(a) Find the integral by completing the square of the radicand.

(b) Find the integral by making the substitution $u = \sqrt{x}$.

(c) The antiderivatives in parts (a) and (b) appear significantly different. Use a graphing utility to graph each in the same viewing window and determine the relationship between the two antiderivatives. Find the domain of each.

63. *Vertical Motion*   An object is projected upward from ground level with an initial velocity of 500 feet per second. In this exercise, the goal is to analyze the motion of the object during its upward flight.

(a) If air resistance is neglected, find the velocity of the object as a function of time. Use a graphing utility to graph this function.

(b) Use the result in part (a) to find the position function and determine the maximum height attained by the object.

(c) If the air resistance is proportional to the square of the velocity, you obtain the equation

$$\dfrac{dv}{dt} = -(32 + kv^2)$$

where -32 feet per second per second is the acceleration due to gravity and k is a constant. Find the velocity as a function of time by solving the equation

$$\int \dfrac{dv}{32 + kv^2} = -\int dt.$$

(d) Use a graphing utility to graph the velocity function $v(t)$ in part (c) if $k = 0.001$. Use the graph to approximate the time t_0 at which the object reaches its maximum height.

(e) Use the integration capabilities of a graphing utility to approximate the integral

$$\int_0^{t_0} v(t) \, dt$$

where $v(t)$ and t_0 are those found in part (d). This is the approximation of the maximum height of the object.

(f) Explain the difference between the results in parts (b) and (e).

FOR FURTHER INFORMATION   For more information on this topic, see "What Goes Up Must Come Down; Will Air Resistance Make It Return Sooner, or Later?" by John Lekner in *Mathematics Magazine*. To view this article, go to the website *www.matharticles.com*.

64. Graph $y_1 = \dfrac{x}{1 + x^2}$, $y_2 = \arctan x$, and $y_3 = x$ on $[0, 10]$.

Prove that $\dfrac{x}{1 + x^2} < \arctan x < x$ for $x > 0$.

- Develop properties of hyperbolic functions.
- Differentiate and integrate hyperbolic functions.
- Develop properties of inverse hyperbolic functions.
- Differentiate and integrate functions involving inverse hyperbolic functions.

Hyperbolic Functions

In this section you will look briefly at a special class of exponential functions called **hyperbolic functions**. The name *hyperbolic function* arose from comparison of the area of a semicircular region, as shown in Figure 5.45, with the area of a region under a hyperbola, as shown in Figure 5.46. The integral for the semicircular region involves an inverse trigonometric (circular) function:

$$\int_{-1}^{1} \sqrt{1 - x^2}\, dx = \frac{1}{2}\left[x\sqrt{1 - x^2} + \arcsin x \right]_{-1}^{1} = \frac{\pi}{2} \approx 1.571.$$

The integral for the hyperbolic region involves an inverse hyperbolic function:

$$\int_{-1}^{1} \sqrt{1 + x^2}\, dx = \frac{1}{2}\left[x\sqrt{1 + x^2} + \sinh^{-1} x \right]_{-1}^{1} \approx 2.296.$$

This is only one of many ways in which the hyperbolic functions are similar to the trigonometric functions.

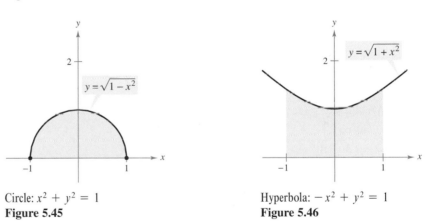

Circle: $x^2 + y^2 = 1$
Figure 5.45

Hyperbola: $-x^2 + y^2 = 1$
Figure 5.46

JOHANN HEINRICH LAMBERT (1728–1777)

The first person to publish a comprehensive study on hyperbolic functions was Johann Heinrich Lambert, a Swiss-German mathematician and colleague of Euler.

FOR FURTHER INFORMATION For more information on the development of hyperbolic functions, see the article "An Introduction to Hyperbolic Functions in Elementary Calculus" by Jerome Rosenthal in *Mathematics Teacher*. To view this article, go to the website *www.matharticles.com*.

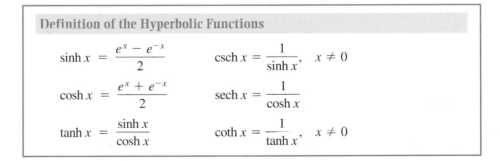

Definition of the Hyperbolic Functions

$$\sinh x = \frac{e^x - e^{-x}}{2} \qquad \operatorname{csch} x = \frac{1}{\sinh x}, \quad x \neq 0$$

$$\cosh x = \frac{e^x + e^{-x}}{2} \qquad \operatorname{sech} x = \frac{1}{\cosh x}$$

$$\tanh x = \frac{\sinh x}{\cosh x} \qquad \coth x = \frac{1}{\tanh x}, \quad x \neq 0$$

NOTE   sinh x is read as "the hyperbolic sine of x," cosh x as "the hyperbolic cosine of x," and so on.

The graphs of the six hyperbolic functions and their domains and ranges are shown in Figure 5.47. Note that the graph of sinh x can be obtained by *addition of ordinates* using the exponential functions $f(x) = \frac{1}{2}e^x$ and $g(x) = -\frac{1}{2}e^{-x}$. Likewise, the graph of cosh x can be obtained by *addition of ordinates* using the exponential functions $f(x) = \frac{1}{2}e^x$ and $h(x) = \frac{1}{2}e^{-x}$.

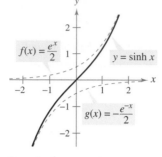

Domain: $(-\infty, \infty)$
Range: $(-\infty, \infty)$

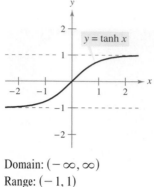

Domain: $(-\infty, \infty)$
Range: $[1, \infty)$

Domain: $(-\infty, \infty)$
Range: $(-1, 1)$

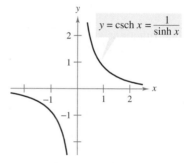

Domain: $(-\infty, 0) \cup (0, \infty)$
Range: $(-\infty, 0) \cup (0, \infty)$
Figure 5.47

Domain: $(-\infty, \infty)$
Range: $(0, 1]$

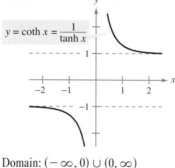

Domain: $(-\infty, 0) \cup (0, \infty)$
Range: $(-\infty, -1) \cup (1, \infty)$

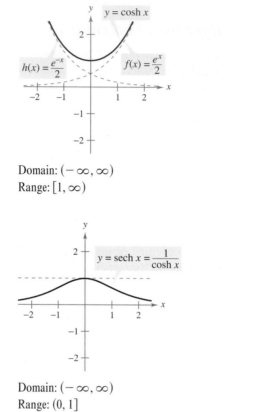

Many of the trigonometric identities have corresponding *hyperbolic identities*. For instance,

$$\cosh^2 x - \sinh^2 x = \left(\frac{e^x + e^{-x}}{2}\right)^2 - \left(\frac{e^x - e^{-x}}{2}\right)^2$$

$$= \frac{e^{2x} + 2 + e^{-2x}}{4} - \frac{e^{2x} - 2 + e^{-2x}}{4}$$

$$= \frac{4}{4}$$

$$= 1$$

FOR FURTHER INFORMATION To understand geometrically the relationship between the hyperbolic and exponential functions, see the article "A Short Proof Linking the Hyperbolic and Exponential Functions" by Michael J. Seery in *The AMATYC Review*. To view this article, go to the website *www.matharticles.com*.

and

$$2 \sinh x \cosh x = 2\left(\frac{e^x - e^{-x}}{2}\right)\left(\frac{e^x + e^{-x}}{2}\right)$$

$$= \frac{e^{2x} - e^{-2x}}{2}$$

$$= \sinh 2x.$$

Hyperbolic Identities

$$\cosh^2 x - \sinh^2 x = 1 \qquad\qquad \sinh(x + y) = \sinh x \cosh y + \cosh x \sinh y$$

$$\tanh^2 x + \operatorname{sech}^2 x = 1 \qquad\qquad \sinh(x - y) = \sinh x \cosh y - \cosh x \sinh y$$

$$\coth^2 x - \operatorname{csch}^2 x = 1 \qquad\qquad \cosh(x + y) = \cosh x \cosh y + \sinh x \sinh y$$

$$\cosh(x - y) = \cosh x \cosh y - \sinh x \sinh y$$

$$\sinh^2 x = \frac{-1 + \cosh 2x}{2} \qquad\qquad \cosh^2 x = \frac{1 + \cosh 2x}{2}$$

$$\sinh 2x = 2 \sinh x \cosh x \qquad\qquad \cosh 2x = \cosh^2 x + \sinh^2 x$$

Differentiation and Integration of Hyperbolic Functions

Because the hyperbolic functions are written in terms of e^x and e^{-x}, you can easily derive rules for their derivatives. The following theorem lists these derivatives with the corresponding integration rules.

THEOREM 5.20   Derivatives and Integrals of Hyperbolic Functions

Let u be a differentiable function of x.

$$\frac{d}{dx}[\sinh u] = (\cosh u)u' \qquad\qquad \int \cosh u\, du = \sinh u + C$$

$$\frac{d}{dx}[\cosh u] = (\sinh u)u' \qquad\qquad \int \sinh u\, du = \cosh u + C$$

$$\frac{d}{dx}[\tanh u] = (\operatorname{sech}^2 u)u' \qquad\qquad \int \operatorname{sech}^2 u\, du = \tanh u + C$$

$$\frac{d}{dx}[\coth u] = -(\operatorname{csch}^2 u)u' \qquad\qquad \int \operatorname{csch}^2 u\, du = -\coth u + C$$

$$\frac{d}{dx}[\operatorname{sech} u] = -(\operatorname{sech} u \tanh u)u' \qquad\qquad \int \operatorname{sech} u \tanh u\, du = -\operatorname{sech} u + C$$

$$\frac{d}{dx}[\operatorname{csch} u] = -(\operatorname{csch} u \coth u)u' \qquad\qquad \int \operatorname{csch} u \coth u\, du = -\operatorname{csch} u + C$$

Proof

$$\frac{d}{dx}[\sinh x] = \frac{d}{dx}\left[\frac{e^x - e^{-x}}{2}\right] = \frac{e^x + e^{-x}}{2} = \cosh x$$

$$\frac{d}{dx}[\tanh x] = \frac{d}{dx}\left[\frac{\sinh x}{\cosh x}\right] = \frac{\cosh x(\cosh x) - \sinh x(\sinh x)}{\cosh^2 x}$$

$$= \frac{1}{\cosh^2 x} = \operatorname{sech}^2 x$$

In Exercises 87 and 91, you are asked to prove some of the other differentiation rules.

Example 1   **Differentiation of Hyperbolic Functions**

a. $\dfrac{d}{dx}[\sinh(x^2 - 3)] = 2x\cosh(x^2 - 3)$

b. $\dfrac{d}{dx}[\ln(\cosh x)] = \dfrac{\sinh x}{\cosh x} = \tanh x$

c. $\dfrac{d}{dx}[x\sinh x - \cosh x] = x\cosh x + \sinh x - \sinh x = x\cosh x$

Example 2   **Finding Relative Extrema**

Find the relative extrema of $f(x) = (x - 1)\cosh x - \sinh x$.

Solution   Begin by setting the first derivative of f equal to 0.

$$f'(x) = (x - 1)\sinh x + \cosh x - \cosh x = 0$$
$$(x - 1)\sinh x = 0$$

So, the critical numbers are $x = 1$ and $x = 0$. Using the Second Derivative Test, you can verify that the point $(0, -1)$ yields a relative maximum and the point $(1, -\sinh 1)$ yields a relative minimum, as shown in Figure 5.48. Try using a graphing utility to confirm this result. If your graphing utility does not have hyperbolic functions, you can use exponential functions as follows.

$$\begin{aligned} f(x) &= (x - 1)\left(\tfrac{1}{2}\right)(e^x + e^{-x}) - \tfrac{1}{2}(e^x - e^{-x}) \\ &= \tfrac{1}{2}(xe^x + xe^{-x} - e^x - e^{-x} - e^x + e^{-x}) \\ &= \tfrac{1}{2}(xe^x + xe^{-x} - 2e^x) \end{aligned}$$

When a uniform flexible cable, such as a telephone wire, is suspended from two points, it takes the shape of a **catenary**, as discussed in Example 3.

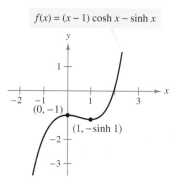

$f(x) = (x - 1)\cosh x - \sinh x$

$f''(0) < 0$, so $(0, -1)$ is a relative maximum. $f''(1) > 0$, so $(1, -\sinh 1)$ is a relative minimum.
Figure 5.48

---

iC   *Example 3*   **Hanging Power Cables**

Power cables are suspended between two towers, forming the catenary shown in Figure 5.49. The equation for this catenary is

$$y = a\cosh\frac{x}{a}.$$

The distance between the two towers is $2b$. Find the slope of the catenary at the point where the cable meets the right-hand tower.

Solution   Differentiating produces

$$y' = a\left(\frac{1}{a}\right)\sinh\frac{x}{a} = \sinh\frac{x}{a}.$$

At the point $(b, a\cosh(b/a))$, the slope (from the left) is given by

$$m = \sinh\frac{b}{a}.$$

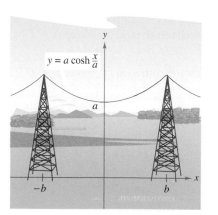

Catenary
Figure 5.49

FOR FURTHER INFORMATION In Example 3, the cable is a catenary between two supports at the same height. To learn about the shape of a cable hanging between supports of different heights, see the article "Reexamining the Catenary" by Paul Cella in *The College Mathematics Journal*. To view this article, go to the website *www.matharticles.com*.

Example 4   **Integrating a Hyperbolic Function**

Find $\int \cosh 2x \sinh^2 2x \, dx$.

Solution

$$\int \cosh 2x \sinh^2 2x \, dx = \frac{1}{2}\int (\sinh 2x)^2 (2 \cosh 2x) \, dx \qquad u = \sinh 2x$$

$$= \frac{1}{2}\left[\frac{(\sinh 2x)^3}{3} \right] + C$$

$$= \frac{\sinh^3 2x}{6} + C$$

Inverse Hyperbolic Functions

Unlike trigonometric functions, hyperbolic functions are not periodic. In fact, by looking back at Figure 5.47, you can see that four of the six hyperbolic functions are actually one-to-one (the hyperbolic sine, tangent, cosecant, and cotangent). So, you can apply Theorem 5.7 to conclude that these four functions have inverse functions. The other two (the hyperbolic cosine and secant) are one-to-one if their domains are restricted to the positive real numbers, and for this restricted domain they also have inverse functions. Because the hyperbolic functions are defined in terms of exponential functions, it is not surprising to find that the inverse hyperbolic functions can be written in terms of logarithmic functions, as shown in Theorem 5.21.

THEOREM 5.21   Inverse Hyperbolic Functions

Function	*Domain*
$\sinh^{-1} x = \ln\left(x + \sqrt{x^2 + 1}\right)$	$(-\infty, \infty)$
$\cosh^{-1} x = \ln\left(x + \sqrt{x^2 - 1}\right)$	$[1, \infty)$
$\tanh^{-1} x = \dfrac{1}{2}\ln\dfrac{1 + x}{1 - x}$	$(-1, 1)$
$\coth^{-1} x = \dfrac{1}{2}\ln\dfrac{x + 1}{x - 1}$	$(-\infty, -1) \cup (1, \infty)$
$\operatorname{sech}^{-1} x = \ln\dfrac{1 + \sqrt{1 - x^2}}{x}$	$(0, 1]$
$\operatorname{csch}^{-1} x = \ln\left(\dfrac{1}{x} + \dfrac{\sqrt{1 + x^2}}{|x|}\right)$	$(-\infty, 0) \cup (0, \infty)$

Proof   The proof of this theorem is a straightforward application of the properties of the exponential and logarithmic functions. For example, if

$$f(x) = \sinh x = \frac{e^x - e^{-x}}{2}$$

and

$$g(x) = \ln\left(x + \sqrt{x^2 + 1}\right)$$

you can show that $f(g(x)) = x$ and $g(f(x)) = x$, which implies that g is the inverse function of f.

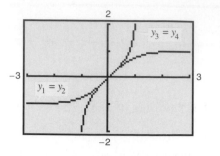

Graphs of the hyperbolic tangent function and the inverse hyperbolic tangent function
Figure 5.50

TECHNOLOGY   You can use a graphing utility to confirm graphically the results of Theorem 5.21. For instance, try sketching the graphs of the following functions.

$$y_1 = \tanh x \qquad \text{Hyperbolic tangent}$$

$$y_2 = \frac{e^x - e^{-x}}{e^x + e^{-x}} \qquad \text{Definition of hyperbolic tangent}$$

$$y_3 = \tanh^{-1} x \qquad \text{Inverse hyperbolic tangent}$$

$$y_4 = \frac{1}{2} \ln \frac{1 + x}{1 - x} \qquad \text{Definition of inverse hyperbolic tangent}$$

The resulting display is shown in Figure 5.50. As you watch the graphs being traced out, notice that $y_1 = y_2$ and $y_3 = y_4$. Also notice that the graph of y_1 is the reflection of the graph of y_3 in the line $y = x$.

The graphs of the inverse hyperbolic functions are shown in Figure 5.51.

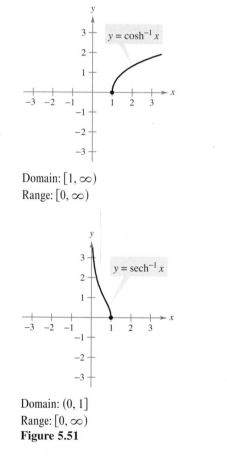

Domain: $[1, \infty)$
Range: $[0, \infty)$

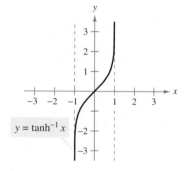

Domain: $(-1, 1)$
Range: $(-\infty, \infty)$

$y = \sinh^{-1} x$

Domain: $(-\infty, \infty)$
Range: $(-\infty, \infty)$

$y = \sech^{-1} x$

Domain: $(0, 1]$
Range: $[0, \infty)$
Figure 5.51

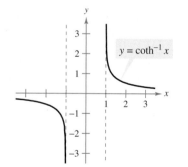

Domain: $(-\infty, -1) \cup (1, \infty)$
Range: $(-\infty, 0) \cup (0, \infty)$

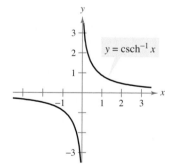

Domain: $(-\infty, 0) \cup (0, \infty)$
Range: $(-\infty, 0) \cup (0, \infty)$

The inverse hyperbolic secant can be used to define a curve called a *tractrix* or *pursuit curve*, as discussed in Example 5.

Example 5   **A Tractrix**

A person is holding a rope that is tied to a boat, as shown in Figure 5.52. As the person walks along the dock, the boat travels along a **tractrix,** given by the equation

$$y = a \operatorname{sech}^{-1} \frac{x}{a} - \sqrt{a^2 - x^2}$$

where a is the length of the rope. If $a = 20$ feet, find the distance the person must walk to bring the boat 5 feet from the dock.

Solution   In Figure 5.52, notice that the distance the person has walked is given by

$$y_1 = y + \sqrt{20^2 - x^2} = \left(20 \operatorname{sech}^{-1} \frac{x}{20} - \sqrt{20^2 - x^2}\right) + \sqrt{20^2 - x^2}$$

$$= 20 \operatorname{sech}^{-1} \frac{x}{20}.$$

When $x = 5$, this distance is

$$y_1 = 20 \operatorname{sech}^{-1} \frac{5}{20} = 20 \ln \frac{1 + \sqrt{1 - (1/4)^2}}{1/4}$$

$$= 20 \ln\left(4 + \sqrt{15}\right)$$

$$\approx 41.27 \text{ feet.}$$

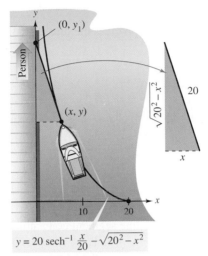

$$y = 20 \operatorname{sech}^{-1} \tfrac{x}{20} - \sqrt{20^2 - x^2}$$

A person must walk 41.27 feet to bring the boat 5 feet from the dock.
Figure 5.52

Differentiation and Integration of Inverse Hyperbolic Functions

The derivatives of the inverse hyperbolic functions, which resemble the derivatives of the inverse trigonometric functions, are listed in Theorem 5.22 with the corresponding integration formulas (in logarithmic form). You can verify each of these formulas by applying the logarithmic definitions of the inverse hyperbolic functions. (See Exercises 88–90.)

THEOREM 5.22   Differentiation and Integration Involving Inverse Hyperbolic Functions

Let u be a differentiable function of x.

$$\frac{d}{dx}[\sinh^{-1} u] = \frac{u'}{\sqrt{u^2 + 1}} \qquad \frac{d}{dx}[\cosh^{-1} u] = \frac{u'}{\sqrt{u^2 - 1}}$$

$$\frac{d}{dx}[\tanh^{-1} u] = \frac{u'}{1 - u^2} \qquad \frac{d}{dx}[\coth^{-1} u] = \frac{u'}{1 - u^2}$$

$$\frac{d}{dx}[\operatorname{sech}^{-1} u] = \frac{-u'}{u\sqrt{1 - u^2}} \qquad \frac{d}{dx}[\operatorname{csch}^{-1} u] = \frac{-u'}{|u|\sqrt{1 + u^2}}$$

$$\int \frac{du}{\sqrt{u^2 \pm a^2}} = \ln\left(u + \sqrt{u^2 \pm a^2}\right) + C$$

$$\int \frac{du}{a^2 - u^2} = \frac{1}{2a} \ln\left|\frac{a + u}{a - u}\right| + C$$

$$\int \frac{du}{u\sqrt{a^2 \pm u^2}} = -\frac{1}{a} \ln \frac{a + \sqrt{a^2 \pm u^2}}{|u|} + C$$

Example 6   **More About a Tractrix**

For the tractrix given in Example 5, show that the boat is always pointing toward the person.

Solution   For a point (x, y) on a tractrix, the slope of the graph gives the direction of the boat, as shown in Figure 5.52.

$$y' = \frac{d}{dx}\left[20 \operatorname{sech}^{-1} \frac{x}{20} - \sqrt{20^2 - x^2}\right]$$

$$= -20\left(\frac{1}{20}\right)\left[\frac{1}{(x/20)\sqrt{1 - (x/20)^2}}\right] - \left(\frac{1}{2}\right)\left(\frac{-2x}{\sqrt{20^2 - x^2}}\right)$$

$$= \frac{-20^2}{x\sqrt{20^2 - x^2}} + \frac{x}{\sqrt{20^2 - x^2}}$$

$$= -\frac{\sqrt{20^2 - x^2}}{x}$$

However, from Figure 5.52, you can see that the slope of the line segment connecting the point $(0, y_1)$ with the point (x, y) is also $m = \left(-\sqrt{20^2 - x^2}\right)/x$. Thus, the boat is always pointing toward the person. (It is because of this property that a tractrix is called a *pursuit curve*.)

Example 7   **Integration Using Inverse Hyperbolic Functions**

Find $\displaystyle\int \frac{dx}{x\sqrt{4 - 9x^2}}$.

Solution   Let $a = 2$ and $u = 3x$.

$$\int \frac{dx}{x\sqrt{4 - 9x^2}} = \int \frac{3\,dx}{(3x)\sqrt{4 - 9x^2}} \qquad\qquad \int \frac{du}{u\sqrt{a^2 - u^2}}$$

$$= -\frac{1}{2} \ln \frac{2 + \sqrt{4 - 9x^2}}{|3x|} + C \qquad -\frac{1}{a} \ln \frac{a + \sqrt{a^2 - u^2}}{|u|} + C$$

Example 8   **Integration Using Inverse Hyperbolic Functions**

Find $\displaystyle\int \frac{dx}{5 - 4x^2}$.

Solution   Let $a = \sqrt{5}$ and $u = 2x$.

$$\int \frac{dx}{5 - 4x^2} = \frac{1}{2}\int \frac{2\,dx}{(\sqrt{5})^2 - (2x)^2} \qquad\qquad \int \frac{du}{a^2 - u^2}$$

$$= \frac{1}{2}\left(\frac{1}{2\sqrt{5}} \ln \left|\frac{\sqrt{5} + 2x}{\sqrt{5} - 2x}\right|\right) + C \qquad \frac{1}{2a} \ln \left|\frac{a + u}{a - u}\right| + C$$

$$= \frac{1}{4\sqrt{5}} \ln \left|\frac{\sqrt{5} + 2x}{\sqrt{5} - 2x}\right| + C$$

EXERCISES FOR SECTION 5.10

In Exercises 1–6, evaluate the function. If the value is not a rational number, give the answer to three-decimal-place accuracy.

1. (a) $\sinh 3$

   (b) $\tanh(-2)$

3. (a) $\text{csch}(\ln 2)$

   (b) $\coth(\ln 5)$

5. (a) $\cosh^{-1} 2$

   (b) $\text{sech}^{-1} \frac{2}{3}$

2. (a) $\cosh 0$

   (b) $\text{sech } 1$

4. (a) $\sinh^{-1} 0$

   (b) $\tanh^{-1} 0$

6. (a) $\text{csch}^{-1} 2$

   (b) $\coth^{-1} 3$

In Exercises 7–12, verify the identity.

7. $\tanh^2 x + \text{sech}^2 x = 1$

8. $\cosh^2 x = \dfrac{1 + \cosh 2x}{2}$

9. $\sinh(x + y) = \sinh x \cosh y + \cosh x \sinh y$

10. $\sinh 2x = 2 \sinh x \cosh x$

11. $\sinh 3x = 3 \sinh x + 4 \sinh^3 x$

12. $\cosh x + \cosh y = 2 \cosh \dfrac{x + y}{2} \cosh \dfrac{x - y}{2}$

In Exercises 13 and 14, use the value of the given hyperbolic function to find the other hyperbolic functions.

13. $\sinh x = \frac{3}{2},$ $\cosh x = \boxed{},$ $\tanh x = \boxed{},$

   $\text{csch } x = \boxed{},$ $\text{sech } x = \boxed{},$ $\coth x = \boxed{}$

14. $\sinh x = \boxed{},$ $\cosh x = \boxed{},$ $\tanh x = \frac{1}{2},$

   $\text{csch } x = \boxed{},$ $\text{sech } x = \boxed{},$ $\coth x = \boxed{}$

In Exercises 15–28, find the derivative of the function.

15. $y = \sinh(1 - x^2)$

17. $f(x) = \ln(\sinh x)$

19. $y = \ln\left(\tanh \dfrac{x}{2}\right)$

21. $h(x) = \dfrac{1}{4} \sinh 2x - \dfrac{x}{2}$

23. $f(t) = \arctan(\sinh t)$

25. $g(x) = x^{\cosh x}$

27. $y = (\cosh x - \sinh x)^2$

16. $y = \coth 3x$

18. $g(x) = \ln(\cosh x)$

20. $y = x \cosh x - \sinh x$

22. $h(t) = t - \coth t$

24. $g(x) = \text{sech}^2 3x$

26. $f(x) = e^{\sinh x}$

28. $y = \text{sech}(x + 1)$

In Exercises 29 and 30, find any relative extrema of the function and use a graphing utility to confirm your result.

29. $f(x) = \sin x \sinh x - \cos x \cosh x,$ $-4 \le x \le 4$

30. $f(x) = x \sinh(x - 1) - \cosh(x - 1)$

In Exercises 31 and 32, use a graphing utility to graph the function and approximate any relative extrema of the function.

31. $g(x) = x \text{ sech } x$

32. $h(x) = 2 \tanh x - x$

In Exercises 33 and 34, show that the function satisfies the differential equation.

Function	*Differential Equation*
33. $y = a \sinh x$	$y''' - y' = 0$
34. $y = a \cosh x$	$y'' - y = 0$

Linear and Quadratic Approximations   In Exercises 35 and 36, use a computer algebra system to find the linear approximation

$$P_1(x) = f(a) + f'(a)(x - a)$$

and the quadratic approximation

$$P_2(x) = f(a) + f'(a)(x - a) + \tfrac{1}{2} f''(a)(x - a)^2$$

to the function f at $x = a$. Use a graphing utility to graph the function and its linear and quadratic approximations.

35. $f(x) = \tanh x$

   $a = 1$

36. $f(x) = \cosh x$

   $a = 0$

Catenary   In Exercises 37 and 38, a model for power cables suspended between two towers is given. (a) Graph the model, (b) find the height of the cable at the towers and at the midpoint between the towers, and (c) find the slope of the model at the point where the cable meets the right-hand tower.

37. $y = 10 + 15 \cosh \dfrac{x}{15},$ $-15 \le x \le 15$

38. $y = 18 + 25 \cosh \dfrac{x}{25},$ $-25 \le x \le 25$

In Exercises 39–54, find or evaluate the integral.

39. $\displaystyle\int \sinh(1 - 2x)\, dx$

40. $\displaystyle\int \dfrac{\cosh \sqrt{x}}{\sqrt{x}}\, dx$

41. $\displaystyle\int \cosh^2(x - 1) \sinh(x - 1)\, dx$

42. $\displaystyle\int \dfrac{\sinh x}{1 + \sinh^2 x}\, dx$

43. $\displaystyle\int \dfrac{\cosh x}{\sinh x}\, dx$

44. $\displaystyle\int \text{sech}^2(2x - 1)\, dx$

45. $\displaystyle\int x \,\text{csch}^2 \dfrac{x^2}{2}\, dx$

46. $\displaystyle\int \text{sech}^3 x \tanh x\, dx$

47. $\displaystyle\int \dfrac{\text{csch}(1/x) \coth(1/x)}{x^2}\, dx$

48. $\displaystyle\int \cosh^2 x\, dx$

49. $\displaystyle\int_0^4 \dfrac{1}{25 - x^2}\, dx$

50. $\displaystyle\int_0^4 \dfrac{1}{\sqrt{25 - x^2}}\, dx$

51. $\displaystyle\int_0^{\sqrt{2}/4} \dfrac{2}{\sqrt{1 - 4x^2}}\, dx$

52. $\displaystyle\int \dfrac{2}{x\sqrt{1 + 4x^2}}\, dx$

53. $\displaystyle\int \dfrac{x}{x^4 + 1}\, dx$

54. $\displaystyle\int \dfrac{\cosh x}{\sqrt{9 - \sinh^2 x}}\, dx$

In Exercises 55–62, find the derivative of the function.

55. $y = \cosh^{-1}(3x)$

56. $y = \tanh^{-1}\dfrac{x}{2}$

57. $y = \sinh^{-1}(\tan x)$

58. $y = \operatorname{sech}^{-1}(\cos 2x), \quad 0 < x < \pi/4$

59. $y = \coth^{-1}(\sin 2x)$

60. $y = (\operatorname{csch}^{-1} x)^2$

61. $y = 2x \sinh^{-1}(2x) - \sqrt{1 + 4x^2}$

62. $y = x \tanh^{-1} x + \ln\sqrt{1 - x^2}$

Getting at the Concept

63. Define the hyperbolic functions.

64. List the rules for differentiating inverse hyperbolic functions. List the corresponding integration formulas.

Tractrix **In Exercises 65 and 66, use the equation of the tractrix**

$$y = a \operatorname{sech}^{-1}\dfrac{x}{a} - \sqrt{a^2 - x^2}, \quad a > 0.$$

65. Find dy/dx.

66. Let L be the tangent line to the tractrix at the point P. If L intersects the y-axis at the point Q, show that the distance between P and Q is a.

In Exercises 67–74, find the indefinite integral using the formulas of Theorem 5.22.

67. $\displaystyle\int \dfrac{1}{\sqrt{1 + e^{2x}}}\, dx$

68. $\displaystyle\int \dfrac{x}{9 - x^4}\, dx$

69. $\displaystyle\int \dfrac{1}{\sqrt{x}\sqrt{1 + x}}\, dx$

70. $\displaystyle\int \dfrac{\sqrt{x}}{\sqrt{1 + x^3}}\, dx$

71. $\displaystyle\int \dfrac{-1}{4x - x^2}\, dx$

72. $\displaystyle\int \dfrac{dx}{(x + 2)\sqrt{x^2 + 4x + 8}}$

73. $\displaystyle\int \dfrac{1}{1 - 4x - 2x^2}\, dx$

74. $\displaystyle\int \dfrac{dx}{(x + 1)\sqrt{2x^2 + 4x + 8}}$

In Exercises 75–78, solve the differential equation.

75. $\dfrac{dy}{dx} = \dfrac{1}{\sqrt{80 + 8x - 16x^2}}$

76. $\dfrac{dy}{dx} = \dfrac{1}{(x - 1)\sqrt{-4x^2 + 8x - 1}}$

77. $\dfrac{dy}{dx} = \dfrac{x^3 - 21x}{5 + 4x - x^2}$

78. $\dfrac{dy}{dx} = \dfrac{1 - 2x}{4x - x^2}$

In Exercises 79–82, find the area of the region bounded by the graphs of the equations.

79. $y = \operatorname{sech}\dfrac{x}{2}, \quad y = 0, \quad x = -4, \quad x = 4$

80. $y = \tanh 2x, \quad y = 0, \quad x = 2$

81. $y = \dfrac{5x}{\sqrt{x^4 + 1}}, \quad y = 0, \quad x = 2$

82. $y = \dfrac{6}{\sqrt{x^2 - 4}}, \quad y = 0, \quad x = 3, \quad x = 5$

83. ***Chemical Reactions*** Suppose that chemicals A and B combine in a 3-to-1 ratio to form a compound. The amount of compound x being produced at any time t is proportional to the unchanged amounts of A and B remaining in the solution. So, if 3 kilograms of A is mixed with 2 kilograms of B, you have

$$\dfrac{dx}{dt} = k\left(3 - \dfrac{3x}{4}\right)\left(2 - \dfrac{x}{4}\right) = \dfrac{3k}{16}(x^2 - 12x + 32).$$

If 1 kilogram of the compound is formed after 10 minutes, find the amount formed after 20 minutes by solving the integral equation

$$\int \dfrac{3k}{16}\, dt = \int \dfrac{dx}{x^2 - 12x + 32}.$$

84. ***Vertical Motion*** An object is dropped from a height of 400 feet.

(a) Find the velocity of the object as a function of time (neglect air resistance on the object).

(b) Use the result in part (a) to find the position function.

(c) If the air resistance is proportional to the square of the velocity, then

$$\dfrac{dv}{dt} = -32 + kv^2$$

where -32 feet per second per second is the acceleration due to gravity and k is a constant. Show that the velocity v as a function of time is

$$v(t) = -\sqrt{\dfrac{32}{k}} \tanh\left(\sqrt{32k}\, t\right)$$

by performing the following integration and simplifying the result.

$$\int \dfrac{dv}{32 - kv^2} = -\int dt$$

(d) Use the result in part (c) to find $\lim\limits_{t \to \infty} v(t)$ and give its interpretation.

(e) Integrate the velocity function in part (c) and find the position s of the object as a function of t. Use a graphing utility to graph the position function when $k = 0.01$ and the position function in part (b) in the same viewing window. Estimate the additional time required for the object to reach ground level when air resistance is not neglected.

85. ***Writing*** Give a written description of what you believe would happen if k were increased in Exercise 84. Then test your assertion with a particular value of k.

86. Show that $\arctan(\sinh x) = \arcsin(\tanh x)$.

In Exercises 87–91, verify the differentiation formula.

87. $\dfrac{d}{dx}[\cosh x] = \sinh x$

88. $\dfrac{d}{dx}[\operatorname{sech}^{-1} x] = \dfrac{-1}{x\sqrt{1 - x^2}}$

89. $\dfrac{d}{dx}[\cosh^{-1} x] = \dfrac{1}{\sqrt{x^2 - 1}}$

90. $\dfrac{d}{dx}[\sinh^{-1} x] = \dfrac{1}{\sqrt{x^2 + 1}}$

91. $\dfrac{d}{dx}[\operatorname{sech} x] = -\operatorname{sech} x \tanh x$

SECTION PROJECT ST. LOUIS ARCH

The Gateway Arch in St. Louis, Missouri was constructed using the hyperbolic cosine function. The equation used to construct the arch was

$$y = 693.8597 - 68.7672 \cosh 0.0100333x,$$

$$-299.2239 \leq x \leq 299.2239$$

where x and y are measured in feet. Cross sections of the arch are equilateral triangles, and (x, y) traces the path of the centers of mass of the cross-sectional triangles. For each value of x, the area of the cross-sectional triangle is

$$A = 125.1406 \cosh 0.0100333x.$$

(*Source:* Owner's Manual for the Gateway Arch, *Saint Louis, MO, by William Thayer*)

(a) How high above the ground is the center of the highest triangle? (At ground level, $y = 0$.)

(b) What is the height of the arch? (*Hint:* For an equilateral triangle, $A = \sqrt{3}c^2$, where c is one-half the base of the triangle, and the center of mass of the triangle is located at two-thirds the height of the triangle.)

(c) How wide is the arch at ground level?

REVIEW EXERCISES FOR CHAPTER 5

5.1   In Exercises 1 and 2, sketch the graph of the function by hand. Identify any asymptotes of the graph.

1. $f(x) = \ln x + 3$

2. $f(x) = \ln(x - 3)$

In Exercises 3 and 4, use the properties of logarithms to expand the logarithmic function.

3. $\ln \sqrt[5]{\dfrac{4x^2 - 1}{4x^2 + 1}}$

4. $\ln[(x^2 + 1)(x - 1)]$

In Exercises 5 and 6, write the expression as the logarithm of a single quantity.

5. $\ln 3 + \frac{1}{3}\ln(4 - x^2) - \ln x$

6. $3[\ln x - 2\ln(x^2 + 1)] + 2\ln 5$

In Exercises 7 and 8, solve the equation for x.

7. $\ln \sqrt{x + 1} = 2$

8. $\ln x + \ln(x - 3) = 0$

In Exercises 9–16, find the derivative of the function.

9. $g(x) = \ln \sqrt{x}$

10. $h(x) = \ln \dfrac{x(x - 1)}{x - 2}$

11. $f(x) = x\sqrt{\ln x}$

12. $f(x) = \ln[x(x^2 - 2)^{2/3}]$

13. $y = \dfrac{1}{b^2}\left[\ln(a + bx) + \dfrac{a}{a + bx}\right]$

14. $y = \dfrac{1}{b^2}[a + bx - a\ln(a + bx)]$

15. $y = -\dfrac{1}{a}\ln\dfrac{a + bx}{x}$

16. $y = -\dfrac{1}{ax} + \dfrac{b}{a^2}\ln\dfrac{a + bx}{x}$

5.2   In Exercises 17–24, find or evaluate the integral.

17. $\displaystyle\int \frac{1}{7x - 2}\,dx$

18. $\displaystyle\int \frac{x}{x^2 - 1}\,dx$

19. $\displaystyle\int \frac{\sin x}{1 + \cos x}\,dx$

20. $\displaystyle\int \frac{\ln \sqrt{x}}{x}\,dx$

21. $\displaystyle\int_1^4 \frac{x + 1}{x}\,dx$

22. $\displaystyle\int_1^e \frac{\ln x}{x}\,dx$

23. $\displaystyle\int_0^{\pi/3} \sec \theta \, d\theta$

24. $\displaystyle\int_0^{\pi/4} \tan\left(\frac{\pi}{4} - x\right)dx$

5.3 In Exercises 25–30, (a) find the inverse of the function, (b) use a graphing utility to graph f and f^{-1} in the same viewing window, and (c) verify that $f^{-1}(f(x)) = f(f^{-1}(x)) = x$.

25. $f(x) = \frac{1}{2}x - 3$

26. $f(x) = 5x - 7$

27. $f(x) = \sqrt{x + 1}$

28. $f(x) = x^3 + 2$

29. $f(x) = \sqrt[3]{x + 1}$

30. $f(x) = x^2 - 5, \quad x \geq 0$

In Exercise 31–34, find $(f^{-1})'(a)$ for the function f and real number a.

Function	Real number
31. $f(x) = x^3 + 2$	$a = -1$
32. $f(x) = x\sqrt{x - 3}$	$a = 4$
33. $f(x) = \tan x, \quad -\frac{\pi}{4} \leq x \leq \frac{\pi}{4}$	$a = \frac{\sqrt{3}}{3}$
34. $f(x) = \ln x$	$a = 0$

5.4 In Exercises 35 and 36, (a) find the inverse function of f, (b) using a graphing utility to graph f and f^{-1} in the same viewing window, and (c) verify that $f^{-1}(f(x)) = f(f^{-1}(x)) = x$.

35. $f(x) = \ln \sqrt{x}$

36. $f(x) = e^{1-x}$

In Exercises 37 and 38, graph the function without the aid of a graphing utility.

37. $y = e^{-x/2}$

38. $y = 4e^{-x^2}$

In Exercises 39–46, find the derivative of the function.

39. $f(x) = \ln(e^{-x^2})$

40. $g(x) = \ln \dfrac{e^x}{1 + e^x}$

41. $g(t) = t^2 e^t$

42. $h(z) = e^{-z^2/2}$

43. $y = \sqrt{e^{2x} + e^{-2x}}$

44. $y = 3e^{-3/t}$

45. $g(x) = \dfrac{x^2}{e^x}$

46. $f(\theta) = \dfrac{1}{2}e^{\sin 2\theta}$

In Exercises 47 and 48, use implicit differentiation to find dy/dx.

47. $y \ln x + y^2 = 0$

48. $\cos x^2 = xe^y$

In Exercises 49–56, find the indefinite integral.

49. $\displaystyle \int xe^{-3x^2}\, dx$

50. $\displaystyle \int \frac{e^{1/x}}{x^2}\, dx$

51. $\displaystyle \int \frac{e^{4x} - e^{2x} + 1}{e^x}\, dx$

52. $\displaystyle \int \frac{e^{2x} - e^{-2x}}{e^{2x} + e^{-2x}}\, dx$

53. $\displaystyle \int xe^{1-x^2}\, dx$

54. $\displaystyle \int x^2 e^{x^3+1}\, dx$

55. $\displaystyle \int \frac{e^x}{e^x - 1}\, dx$

56. $\displaystyle \int \frac{e^{2x}}{e^{2x} + 1}\, dx$

57. Show that $y = e^x(a \cos 3x + b \sin 3x)$ satisfies the differential equation $y'' - 2y' + 10y = 0$.

58. *Depreciation* The value V of an item t years after it is purchased is

$$V = 8000e^{-0.6t}, \quad 0 \leq t \leq 5.$$

(a) Use a graphing utility to graph the function.

(b) Find the rate of change of V with respect to t when $t = 1$ and $t = 4$.

(c) Use a graphing utility to sketch the tangent line to the function when $t = 1$ and $t = 4$.

In Exercises 59 and 60, find the area of the region bounded by the graphs of the equations.

59. $y = xe^{-x^2}, \quad y = 0, \quad x = 0, \quad x = 4$

60. $y = 2e^{-x}, \quad y = 0, \quad x = 0, \quad x = 2$

5.5 In Exercises 61–64, sketch the graph of the function by hand.

61. $y = 3^{x/2}$

62. $y = 6(2^{-x^2})$

63. $y = \log_2(x - 1)$

64. $y = \log_4 x^2$

In Exercises 65–70, find the derivative of the function.

65. $f(x) = 3^{x-1}$

66. $f(x) = (4e)^x$

67. $y = x^{2x+1}$

68. $y = x(4^{-x})$

69. $g(x) = \log_3 \sqrt{1 - x}$

70. $h(x) = \log_5 \dfrac{x}{x - 1}$

In Exercises 71 and 72, find the indefinite integral.

71. $\displaystyle \int (x + 1)5^{(x+1)^2}\, dx$

72. $\displaystyle \int \frac{2^{-1/t}}{t^2}\, dt$

73. *Think About It* Find the derivative of each function, given that a is constant.

(a) $y = x^a$ (b) $y = a^x$ (c) $y = x^x$ (d) $y = a^a$

74. *Climb Rate* The time t (in minutes) for a small plane to climb to an altitude of h feet is

$$t = 50 \log_{10} \frac{18{,}000}{18{,}000 - h}$$

where 18,000 feet is the plane's absolute ceiling.

(a) Determine the domain of the function appropriate for the context of the problem.

(b) Use a graphing utility to graph the time function and identify any asymptotes.

(c) Find the time when the altitude is increasing at the greatest rate.

75. *Compound Interest* How large a deposit, at 7 percent interest compounded continuously, must be made to obtain a balance of $10,000 in 15 years?

76. *Compound Interest* A deposit earns interest at a rate of r percent compounded continuously and doubles in value in 10 years. Find r.

5.6

77. Air Pressure   Under ideal conditions, air pressure decreases continuously with height above sea level at a rate proportional to the pressure at that height. If the barometer reads 30 inches at sea level and 15 inches at 18,000 feet, find the barometric pressure at 35,000 feet.

78. Radioactive Decay   Radioactive radium has a half-life of approximately 1620 years. If the initial quantity is 5 grams, how much remains after 600 years?

79. Population Growth   A population grows continuously at the rate of 1.5%. How long will it take the population to double?

80. Fuel Economy   A certain automobile gets 28 miles per gallon of gasoline for speeds up to 50 miles per hour. Over 50 miles per hour, the number of miles per gallon drops at the rate of 12 percent for each 10 miles per hour.

(a) If s is the speed and y is the number of miles per gallon, find y as a function of s by solving the differential equation

$$\frac{dy}{ds} = -0.012y, \quad s > 50.$$

(b) Use the function in part (a) to complete the table.

Speed	50	55	60	65	70
Miles per gallon					

5.7   **In Exercises 81–86, solve the differential equation.**

81. $\dfrac{dy}{dx} = \dfrac{x^2 + 3}{x}$

82. $\dfrac{dy}{dx} = \dfrac{e^{-2x}}{1 + e^{-2x}}$

83. $y' - 2xy = 0$

84. $y' - e^y \sin x = 0$

85. $\dfrac{dy}{dx} = \dfrac{x^2 + y^2}{2xy}$

86. $\dfrac{dy}{dx} = \dfrac{3(x + y)}{x}$

87. Verify that the general solution $y = C_1 x + C_2 x^3$ satisfies the differential equation $x^2 y'' - 3xy' + 3y = 0$. Then find the particular solution that satisfies the initial condition $y = 0$ and $y' = 4$ when $x = 2$.

88. Vertical Motion   A falling object encounters air resistance that is proportional to its velocity. If the acceleration due to gravity is -9.8 meters per second per second, the net change in velocity is

$$\frac{dv}{dt} = kv - 9.8.$$

(a) Find the velocity of the object as a function of time if the initial velocity is v_0.

(b) Use the result in part (a) to find the limit of the velocity as t approaches infinity.

(c) Integrate the velocity function found in part (a) to find the position function s.

5.8   **In Exercises 89 and 90, sketch the graph of the function by hand.**

89. $f(x) = 2 \arctan(x + 3)$

90. $h(x) = -3 \arcsin 2x$

In Exercises 91 and 92, evaluate the expression without using a calculator. (Hint: Make a sketch of a right triangle.)

91. (a) $\sin\left(\arcsin \tfrac{1}{2}\right)$

(b) $\cos\left(\arcsin \tfrac{1}{2}\right)$

92. (a) $\tan(\text{arccot } 2)$

(b) $\cos\left(\text{arcsec } \sqrt{5}\right)$

In Exercises 93–98, find the derivative of the function.

93. $y = \tan(\arcsin x)$

94. $y = \arctan(x^2 - 1)$

95. $y = x \,\text{arcsec } x$

96. $y = \tfrac{1}{2} \arctan e^{2x}$

97. $y = x(\arcsin x)^2 - 2x + 2\sqrt{1 - x^2} \arcsin x$

98. $y = \sqrt{x^2 - 4} - 2\,\text{arcsec}\,\dfrac{x}{2}, \quad 2 < x < 4$

5.9   **In Exercises 99–106, find the indefinite integral.**

99. $\displaystyle\int \frac{1}{e^{2x} + e^{-2x}}\, dx$

100. $\displaystyle\int \frac{1}{3 + 25x^2}\, dx$

101. $\displaystyle\int \frac{x}{\sqrt{1 - x^4}}\, dx$

102. $\displaystyle\int \frac{1}{16 + x^2}\, dx$

103. $\displaystyle\int \frac{x}{16 + x^2}\, dx$

104. $\displaystyle\int \frac{4 - x}{\sqrt{4 - x^2}}\, dx$

105. $\displaystyle\int \frac{\arctan(x/2)}{4 + x^2}\, dx$

106. $\displaystyle\int \frac{\arcsin x}{\sqrt{1 - x^2}}\, dx$

107. Harmonic Motion   A weight of mass m is attached to a spring and oscillates with simple harmonic motion. By Hooke's Law, you can determine that

$$\int \frac{dy}{\sqrt{A^2 - y^2}} = \int \sqrt{\frac{k}{m}}\, dt$$

where A is the maximum displacement, t is the time, and k is a constant. Find y as a function of t, given that $y = 0$ when $t = 0$.

108. Think About It   Sketch the region whose area is given by $\displaystyle\int_0^1 \arcsin x \, dx$. Then find the area of the region. Explain how you arrived at your answer.

5.10   **In Exercises 109 and 110, find the derivative of the function.**

109. $y = 2x - \cosh \sqrt{x}$

110. $y = x \tanh^{-1} 2x$

In Exercises 111 and 112, find the indefinite integral.

111. $\displaystyle\int \frac{x}{\sqrt{x^4 - 1}}\, dx$

112. $\displaystyle\int x^2 \,\text{sech}^2\, x^3 \, dx$

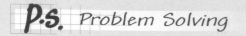

P.S. Problem Solving

1. Find the value of a that maximizes the angle θ indicated in the figure. What is the approximate measure of this angle?

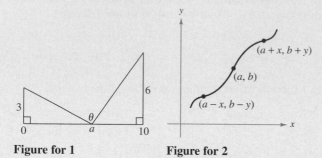

Figure for 1   **Figure for 2**

2. Recall that the graph of a function $y = f(x)$ is symmetric with respect to the origin if whenever (x, y) is a point on the graph, $(-x, -y)$ is also a point on the graph. We say that the graph of the function $y = f(x)$ is **symmetric with respect to the point** (a, b) if whenever $(a - x, b - y)$ is a point on the graph, $(a + x, b + y)$ is also a point on the graph, as indicated in the figure.

(a) Sketch the graph of $y = \sin x$ on the interval $[0, 2\pi]$. Write a short paragraph explaining how the symmetry of the graph with respect to the point $(0, \pi)$ allows you to conclude that

$$\int_0^{2\pi} \sin x \, dx = 0.$$

(b) Sketch the graph of $y = \sin x + 2$ on the interval $[0, 2\pi]$. Use the symmetry of the graph with respect to the point $(\pi, 2)$ to evaluate the integral

$$\int_0^{2\pi} (\sin x + 2) \, dx.$$

(c) Sketch the graph of $y = \arccos x$ on the interval $[-1, 1]$. Use the symmetry of the graph to evaluate the integral

$$\int_{-1}^{1} \arccos x \, dx.$$

(d) Evaluate the integral $\displaystyle\int_0^{\pi/2} \frac{1}{1 + (\tan x)^{\sqrt{2}}} \, dx.$

3. Let $f(x) = \sin(\ln x)$.

(a) Determine the domain of the function f.

(b) Find two values of x satisfying $f(x) = 1$.

(c) Find two values of x satisfying $f(x) = -1$.

(d) What is the range of the function f?

(e) Calculate $f'(x)$ and use calculus to find the maximum value of f on the interval $[1, 10]$.

(f) Use a graphing utility to graph f in the viewing window $[0, 5] \times [-2, 2]$ and estimate $\lim_{x \to 0^+} f(x)$, if it exists.

(g) Determine $\lim_{x \to 0^+} f(x)$ analytically, if it exists.

4. Graph the exponential function $y = a^x$ for $a = 0.5, 1.2,$ and 2.0. Which of these curves intersects the line $y = x$? Determine all positive numbers a for which the curve $y = a^x$ intersects the line $y = x$.

5. (a) Let $P(\cos t, \sin t)$ be a point on the unit circle $x^2 + y^2 = 1$ in the first quadrant. Show that t is equal to twice the area of the shaded circular sector AOP.

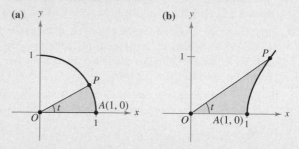

(b) Let $P(\cosh t, \sinh t)$ be a point on the unit hyperbola $x^2 - y^2 = 1$ in the first quadrant. Show that t is equal to twice the area of the shaded region AOP. Begin by showing that the area of the shaded region AOP is given by the formula

$$A(t) = \frac{1}{2} \cosh t \sinh t - \int_1^{\cosh t} \sqrt{x^2 - 1} \, dx.$$

6. Consider the three regions A, B, and C determined by the graph of $f(x) = \arcsin x$, as indicated in the figure.

(a) Calculate the areas of regions A and B.

(b) Use your answer in part (a) to evaluate the integral

$$\int_{1/2}^{\sqrt{2}/2} \arcsin x \, dx.$$

(c) Use your answer in part (a) to evaluate the integral

$$\int_1^3 \ln x \, dx.$$

(d) Use your answer in part (a) to evaluate the integral

$$\int_1^{\sqrt{3}} \arctan x \, dx.$$

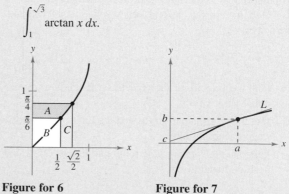

Figure for 6   **Figure for 7**

7. Let L be the tangent line to the graph of the function $y = \ln x$ at the point (a, b). Show that the distance between b and c is always equal to 1.

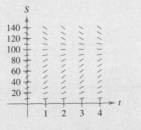

8. Let L be the tangent line to the graph of the function $y = e^x$ at the point (a, b). Show that the distance between a and c is always equal to 1.

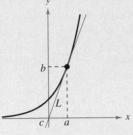

9. Use integration by substitution to find the area under the curve

$$y = \frac{1}{\sqrt{x} + x}$$

between $x = 1$ and $x = 4$.

10. Use integration by substitution to find the area under the curve

$$y = \frac{1}{\sin^2 x + 4\cos^2 x}$$

between $x = 0$ and $x = \pi/4$.

11. The differential equation $dy/dt = ky^{1+\varepsilon}$, where k and ε are positive constants, is called the **doomsday equation.**

(a) Solve the doomsday equation

$$\frac{dy}{dt} = y^{1.01}$$

given that $y(0) = 1$.

Find the time T at which $\lim\limits_{t \to T^-} y(t) = \infty$.

(b) Solve the doomsday equation $dy/dt = ky^{1+\varepsilon}$ given that $y(0) = y_0$.

Explain why this equation is called the doomsday equation.

12. The differential equation $dy/dt = ky(L - y)$, where k and L are positive constants, is called the **logistics equation.**

(a) Solve the logistics equation

$$\frac{dy}{dt} = y(1 - y)$$

given that $y(0) = \frac{1}{4}$.

$$\left(\text{Hint: } \frac{1}{y(1 - y)} = \frac{1}{y} + \frac{1}{1 - y}.\right)$$

(b) Graph the solution on the interval $-6 \le t \le 6$. Show that the rate of growth of the solution is maximum at the point of inflection.

(c) Solve the logistics equation $dy/dt = y(1 - y)$ given that $y(0) = 2$. How does this solution differ from that in part (a)?

13. A thermometer is taken from a room at 72°F to the outdoors, where the temperature is 20°F. Determine the reading on the thermometer after 5 minutes if the reading drops to 48°F after 1 minute.

14. A \$120,000 home mortgage for 35 years at $9\frac{1}{2}\%$ has a monthly payment of \$985.93. Part of the monthly payment goes for the interest charge on the unpaid balance and the remainder of the payment is used to reduce the principal. The amount that goes for interest is

$$u = M - \left(M - \frac{Pr}{12}\right)\left(1 + \frac{r}{12}\right)^{12t}$$

and the amount that goes toward reduction of the principal is

$$v = \left(M - \frac{Pr}{12}\right)\left(1 + \frac{r}{12}\right)^{12t}.$$

In these formulas, P is the size of the mortgage, r is the interest rate, M is the monthly payment, and t is the time in years.

(a) Use a graphing utility to graph each function in the same viewing window. (The viewing window should show all 35 years of mortgage payments.)

(b) In the early years of the mortgage, the larger part of the monthly payment goes for what purpose? Approximate the time when the monthly payment is evenly divided between interest and principal reduction.

(c) Use the graphs in part (a) to make a conjecture about the relationship between the slopes of the tangent lines to the two curves for a specified value of t. Give an analytical argument to verify your conjecture. Find $u'(15)$ and $v'(15)$.

(d) Repeat parts (a) and (b) for a repayment period of 20 years ($M = \$1118.56$). What can you conclude?

15. Let S represent sales of a new product (in thousands of units), let L represent the maximum level of sales (in thousands of units), and let t represent time (in months). The rate of change of S with respect to t varies jointly as the product of S and $L - S$.

(a) Write the differential equation for the sales model if $L = 100$, $S = 10$ when $t = 0$, and $S = 20$ when $t = 1$. Verify that

$$S = \frac{L}{1 + Ce^{-kt}}.$$

(b) At what time is the growth in sales increasing most rapidly?

(c) Use a graphing utility to graph the sales function.

(d) Sketch the solution in part (a) on the slope field shown in the figure below. To print an enlarged copy of the graph, go to the website *www.mathgraphs.com*.

(e) If the estimated maximum level of sales is correct, use the slope field to describe the shape of the solution curves for sales if, at some period of time, sales exceed L.

Constructing an Arch Dam

Dams were originally built to ensure water supplies during dry seasons. As technical knowledge has increased, they have begun serving other functions. Today, dams may be built to create recreational lakes, to power generators, and to prevent flooding. Every new dam creates concerns. A dam may upset an area's ecology and force the relocation of people and wildlife. Also, a poorly constructed dam endangers the entire surrounding region, creating the possibility of a massive disaster.

There are several designs used in dam construction, one of which is the arch dam. This design curves toward the water it contains, and is usually built in narrow canyons. The force of the water presses the edges of the dam against the walls of the canyon, so that the natural rock helps support the structure. This added support means that the arch dam can be built with less construction materials than its gravity-supported counterpart.

A cross section of a typical arch dam can be modeled as shown in the figure below. The model for this cross section is as follows.

$$f(x) = \begin{cases} 0.03x^2 + 7.1x + 350, & -70 \le x \le -16 \\ 389, & -16 < x < 0 \\ -6.593x + 389, & 0 \le x \le 59 \end{cases}$$

To form the arch dam, this cross section is swung through an arc, rotating it about the y-axis. The number of degrees through which it is rotated and the length of the axis of rotation vary, depending primarily on how much the water level varies. A possible configuration shows a rotation of $150°$ and an axis of rotation of 150 feet.

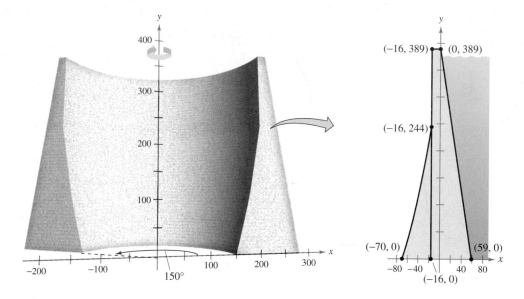

QUESTIONS

1. Find the area of a cross section of the dam.

2. Describe a strategy for estimating the volume of concrete that would be needed to build this dam.

3. Use the strategy to estimate the volume of concrete needed to build the dam described on this page.

The concepts presented here will be explored further in this chapter. For an extension of this application, see Lab 9 of the lab series that accompanies this text at college.hmco.com.

Applications of Integration

Henryk Kaiser/Leo de Wys

Hoover Dam, one of the highest concrete dams in the world, uses a gravity-arch construction. It relies on both the walls of the Black Canyon and its own mass to hold back the waters of the Colorado River.

Bettmann/Corbis

Frank Crowe calculated the winning bid of $48,890,955 for Six Companies, the private contracting firm that built Hoover Dam. Under his leadership, the dam was completed two years early.

FOR FURTHER INFORMATION To learn more about the calculus of dam design, see *Calculus, Understanding Change*, a three-part, half-hour video production by COMAP and funded by the National Science Foundation.

• Find the area of a region between two curves using integration.
• Find the area of a region between intersecting curves using integration.
• Describe integration as an accumulation process.

Area of a Region Between Two Curves

With a few modifications you can extend the application of definite integrals from the area of a region *under* a curve to the area of a region *between* two curves. Consider two functions f and g that are continuous on the interval $[a, b]$. If, as in Figure 6.1, the graphs of both f and g lie above the x-axis, and the graph of g lies below the graph of f, you can geometrically interpret the area of the region between the graphs as the area of the region under the graph of g subtracted from the area of the region under the graph of f, as shown in Figure 6.2.

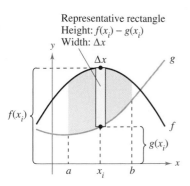

Figure 6.1

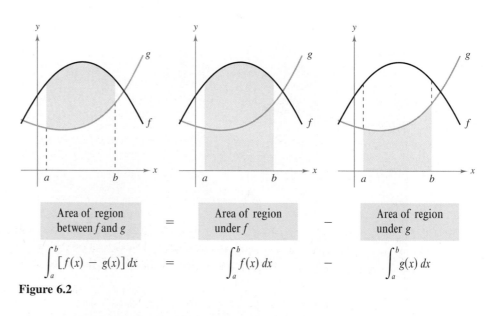

Area of region between f and g	$=$	Area of region under f	$-$	Area of region under g
$\displaystyle\int_a^b [f(x) - g(x)]\,dx$	$=$	$\displaystyle\int_a^b f(x)\,dx$	$-$	$\displaystyle\int_a^b g(x)\,dx$

Figure 6.2

To verify the reasonableness of the result shown in Figure 6.2, you can partition the interval $[a, b]$ into n subintervals, each of width Δx. Then, as shown in Figure 6.3, sketch a **representative rectangle** of width Δx and height $f(x_i) - g(x_i)$, where x_i is in the ith interval. The area of this representative rectangle is

$$\Delta A_i = (\text{height})(\text{width}) = [f(x_i) - g(x_i)]\Delta x.$$

By adding the areas of the n rectangles and taking the limit as $\|\Delta\| \to 0$ $(n \to \infty)$, you obtain

$$\lim_{n \to \infty} \sum_{i=1}^{n} [f(x_i) - g(x_i)]\Delta x.$$

Because f and g are continuous on $[a, b]$, $f - g$ is also continuous on $[a, b]$ and the limit exists. Therefore, the area of the given region is

$$\text{Area} = \lim_{n \to \infty} \sum_{i=1}^{n} [f(x_i) - g(x_i)]\Delta x$$

$$= \int_a^b [f(x) - g(x)]\,dx.$$

Representative rectangle
Height: $f(x_i) - g(x_i)$
Width: Δx

Figure 6.3

Area of a Region Between Two Curves

If f and g are continuous on $[a, b]$ and $g(x) \le f(x)$ for all x in $[a, b]$, then the area of the region bounded by the graphs of f and g and the vertical lines $x = a$ and $x = b$ is

$$A = \int_a^b [f(x) - g(x)] \, dx.$$

In Figure 6.1, the graphs of f and g are shown above the x-axis. This, however, is not necessary. The same integrand $[f(x) - g(x)]$ can be used as long as f and g are continuous and $g(x) \le f(x)$ for all x in the interval $[a, b]$. This result is summarized graphically in Figure 6.4.

NOTE   The height of a representative rectangle is $f(x) - g(x)$ regardless of the relative position of the x-axis, as shown in Figure 6.4.

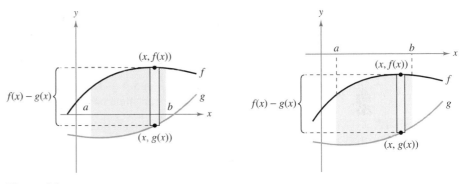

Figure 6.4

Representative rectangles are used throughout this chapter in various applications of integration. A vertical rectangle (of width Δx) implies integration with respect to x, whereas a horizontal rectangle (of width Δy) implies integration with respect to y.

Example 1   **Finding the Area of a Region Between Two Curves**

Find the area of the region bounded by the graphs of $y = x^2 + 2$, $y = -x$, $x = 0$, and $x = 1$.

Solution   Let $g(x) = -x$ and $f(x) = x^2 + 2$. Then $g(x) \le f(x)$ for all x in $[0, 1]$, as shown in Figure 6.5. Thus, the area of the representative rectangle is

$$\Delta A = [f(x) - g(x)] \, \Delta x = [(x^2 + 2) - (-x)] \, \Delta x$$

and the area of the region is

$$\begin{aligned}
A &= \int_a^b [f(x) - g(x)] \, dx = \int_0^1 [(x^2 + 2) - (-x)] \, dx \\
&= \left[\frac{x^3}{3} + \frac{x^2}{2} + 2x \right]_0^1 \\
&= \frac{1}{3} + \frac{1}{2} + 2 \\
&= \frac{17}{6}.
\end{aligned}$$

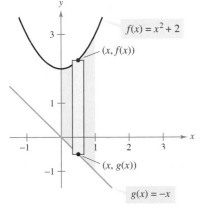

Region bounded by the graph of f, the graph of g, $x = 0$, and $x = 1$
Figure 6.5

Area of a Region Between Intersecting Curves

In Example 1, the graphs of $f(x) = x^2 + 2$ and $g(x) = -x$ do not intersect, and the values of a and b are given explicitly. A more common problem involves the area of a region bounded by two *intersecting* graphs, where the values of a and b must be calculated.

Example 2   **A Region Lying Between Two Intersecting Graphs**

Find the area of the region bounded by the graphs of $f(x) = 2 - x^2$ and $g(x) = x$.

Solution   In Figure 6.6, notice that the graphs of f and g have two points of intersection. To find the x-coordinates of these points, set $f(x)$ and $g(x)$ equal to each other and solve for x.

$$
\begin{aligned}
2 - x^2 &= x && \text{Set } f(x) \text{ equal to } g(x). \\
-x^2 - x + 2 &= 0 && \text{Write in general form.} \\
-(x + 2)(x - 1) &= 0 && \text{Factor.} \\
x = -2 &\text{ or } 1 && \text{Solve for } x.
\end{aligned}
$$

Thus, $a = -2$ and $b = 1$. Because $g(x) \leq f(x)$ for all x in the interval $[-2, 1]$, the representative rectangle has an area of

$$
\begin{aligned}
\Delta A &= [f(x) - g(x)] \Delta x \\
&= [(2 - x^2) - x] \Delta x
\end{aligned}
$$

and the area of the region is

$$
A = \int_{-2}^{1} [(2 - x^2) - x] \, dx = \left[-\frac{x^3}{3} - \frac{x^2}{2} + 2x \right]_{-2}^{1}
$$

$$
= \frac{9}{2}.
$$

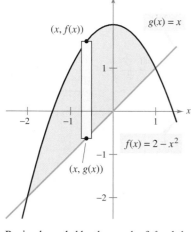

Region bounded by the graph of f and the graph of g

Figure 6.6

Example 3   **A Region Lying Between Two Intersecting Graphs**

The sine and cosine curves intersect infinitely many times, bounding regions of equal areas, as shown in Figure 6.7. Find the area of one of these regions.

Solution

$$
\begin{aligned}
\sin x &= \cos x && \text{Set } f(x) \text{ equal to } g(x). \\
\frac{\sin x}{\cos x} &= 1 && \text{Divide both sides by } \cos x. \\
\tan x &= 1 && \text{Trigonometric identity} \\
x = \frac{\pi}{4} &\text{ or } \frac{5\pi}{4}, \quad 0 \leq x \leq 2\pi && \text{Solve for } x.
\end{aligned}
$$

So, $a = \pi/4$ and $b = 5\pi/4$. Because $\sin x \geq \cos x$ for all x in the interval $[\pi/4, 5\pi/4]$, the area of the region is

$$
A = \int_{\pi/4}^{5\pi/4} [\sin x - \cos x] \, dx = \Big[-\cos x - \sin x \Big]_{\pi/4}^{5\pi/4}
$$

$$
= 2\sqrt{2}.
$$

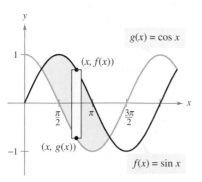

One of the regions bounded by the graphs of the sine and cosine functions

Figure 6.7

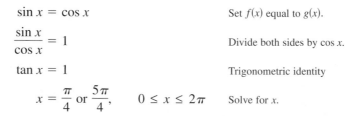

If two curves intersect at *more* than two points, then to find the area of the region between the curves, you must find all points of intersection and check to see which curve is above the other in each interval determined by these points.

 Example 4   **Curves That Intersect at More Than Two Points**

Find the area of the region between the graphs of $f(x) = 3x^3 - x^2 - 10x$ and $g(x) = -x^2 + 2x$.

Solution   Begin by setting $f(x)$ and $g(x)$ equal to each other and solving for x. This yields the x-values at each point of intersection of the two graphs.

$$3x^3 - x^2 - 10x = -x^2 + 2x \qquad \text{Set } f(x) \text{ equal to } g(x).$$
$$3x^3 - 12x = 0 \qquad \text{Write in general form.}$$
$$3x(x^2 - 4) = 0 \qquad \text{Factor.}$$
$$x = -2, 0, 2 \qquad \text{Solve for } x.$$

So, the two graphs intersect when $x = -2, 0$, and 2. In Figure 6.8, notice that $g(x) \le f(x)$ on the interval $[-2, 0]$. However, the two graphs switch at the origin, and $f(x) \le g(x)$ on the interval $[0, 2]$. Hence, you need two integrals—one for the interval $[-2, 0]$ and one for the interval $[0, 2]$.

$$A = \int_{-2}^{0} [f(x) - g(x)]\, dx + \int_{0}^{2} [g(x) - f(x)]\, dx$$
$$= \int_{-2}^{0} (3x^3 - 12x)\, dx + \int_{0}^{2} (-3x^3 + 12x)\, dx$$
$$= \left[\frac{3x^4}{4} - 6x^2 \right]_{-2}^{0} + \left[\frac{-3x^4}{4} + 6x^2 \right]_{0}^{2}$$
$$= -(12 - 24) + (-12 + 24)$$
$$= 24$$

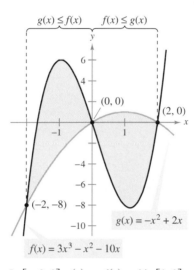

$g(x) \le f(x)$ $f(x) \le g(x)$

$(0, 0)$

$(2, 0)$

$(-2, -8)$

$g(x) = -x^2 + 2x$

$f(x) = 3x^3 - x^2 - 10x$

In $[-2, 0]$, $g(x) \le f(x)$ and in $[0, 2]$, $f(x) \le g(x)$.

Figure 6.8

NOTE   In Example 4, notice that you get an incorrect result if you integrate from -2 to 2. Such integration produces

$$\int_{-2}^{2} [f(x) - g(x)]\, dx = \int_{-2}^{2} (3x^3 - 12x)\, dx = 0.$$

If the graph of a function of y is a boundary of a region, it is often convenient to use representative rectangles that are *horizontal* and find the area by integrating with respect to y. In general, to determine the area between two curves, you can use

$$A = \int_{x_1}^{x_2} \underbrace{[(\text{top curve}) - (\text{bottom curve})]}_{\text{in variable } x}\, dx \qquad \text{Vertical rectangles}$$

$$A = \int_{y_1}^{y_2} \underbrace{[(\text{right curve}) - (\text{left curve})]}_{\text{in variable } y}\, dy \qquad \text{Horizontal rectangles}$$

where (x_1, y_1) and (x_2, y_2) are either adjacent points of intersection of the two curves involved or points on the specified boundary lines.

iC   *indicates that in the* Interactive 3.0 *CD-ROM and* Internet 3.0 *versions of this text (available at* college.hmco.com) *you will find an Open Exploration, which further explores this example using the computer algebra systems* Maple, Mathcad, Mathematica, *and* Derive.

Example 5   **Horizontal Representative Rectangles**

Find the area of the region bounded by the graphs of $x = 3 - y^2$ and $x = y + 1$.

Solution   Consider

$$g(y) = 3 - y^2 \quad \text{and} \quad f(y) = y + 1.$$

These two curves intersect when $y = -2$ and $y = 1$, as shown in Figure 6.9. Because $f(y) \le g(y)$ on this interval, you have

$$\Delta A = [g(y) - f(y)] \, \Delta y$$
$$= [(3 - y^2) - (y + 1)] \, \Delta y.$$

Hence, the area is

$$A = \int_{-2}^{1} [(3 - y^2) - (y + 1)] \, dy$$

$$= \int_{-2}^{1} (-y^2 - y + 2) \, dy$$

$$= \left[\frac{-y^3}{3} - \frac{y^2}{2} + 2y \right]_{-2}^{1}$$

$$= \left(-\frac{1}{3} - \frac{1}{2} + 2 \right) - \left(\frac{8}{3} - 2 - 4 \right) = \frac{9}{2}.$$

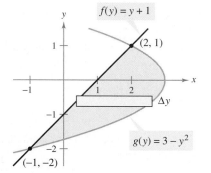

Horizontal rectangles (integration with respect to y)
Figure 6.9

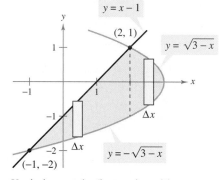

Vertical rectangles (integration with respect to x)
Figure 6.10

In Example 5, notice that by integrating with respect to y you need only one integral. If you had integrated with respect to x, you would have needed two integrals because the upper boundary changes at $x = 2$, as shown in Figure 6.10.

$$A = \int_{-1}^{2} \left[(x - 1) + \sqrt{3 - x} \right] dx + \int_{2}^{3} \left(\sqrt{3 - x} + \sqrt{3 - x} \right) dx$$

$$= \int_{-1}^{2} \left[x - 1 + (3 - x)^{1/2} \right] dx + 2 \int_{2}^{3} (3 - x)^{1/2} \, dx$$

$$= \left[\frac{x^2}{2} - x - \frac{(3 - x)^{3/2}}{3/2} \right]_{-1}^{2} - 2 \left[\frac{(3 - x)^{3/2}}{3/2} \right]_{2}^{3}$$

$$= \left(2 - 2 - \frac{2}{3} \right) - \left(\frac{1}{2} + 1 - \frac{16}{3} \right) - 2(0) + 2\left(\frac{2}{3} \right) = \frac{9}{2}$$

Integration as an Accumulation Process

In this section, we developed the integration formula for the area between two curves by using a rectangle as the *representative element*. For each new application in the remaining sections of this chapter, we will construct an appropriate representative element using precalculus formulas you already know. Each integration formula then will be obtained by summing or accumulating these representative elements.

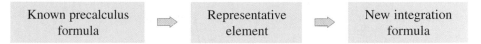

| Known precalculus formula | ⟹ | Representative element | ⟹ | New integration formula |

For example, in this section we developed the area formula as follows.

$$A = (\text{height})(\text{width}) \quad \Longrightarrow \quad \Delta A = [f(x) - g(x)]\,\Delta x \quad \Longrightarrow \quad A = \int_a^b [f(x) - g(x)]\,dx$$

Example 6   **Describing Integration as an Accumulation Process**

Find the area of the region bounded by the graphs of $y = 4 - x^2$ and the x-axis. Describe the integration as an accumulation process.

Solution   The area of the region is given by

$$A = \int_{-2}^{2} (4 - x^2)\,dx.$$

You can think of the integration as an accumulation of the areas of the rectangles formed as the representative rectangle slides from $x = -2$ to $x = 2$.

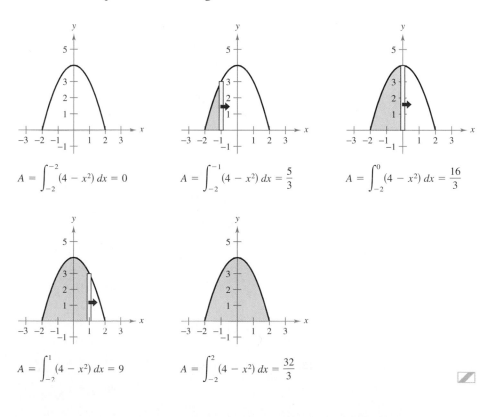

$$A = \int_{-2}^{-2} (4 - x^2)\,dx = 0$$

$$A = \int_{-2}^{-1} (4 - x^2)\,dx = \frac{5}{3}$$

$$A = \int_{-2}^{0} (4 - x^2)\,dx = \frac{16}{3}$$

$$A = \int_{-2}^{1} (4 - x^2)\,dx = 9$$

$$A = \int_{-2}^{2} (4 - x^2)\,dx = \frac{32}{3}$$

EXERCISES FOR SECTION 6.1

In Exercises 1–6, set up the definite integral that gives the area of the region.

1. $f(x) = x^2 - 6x$
   $g(x) = 0$

2. $f(x) = x^2 + 2x + 1$
   $g(x) = 2x + 5$

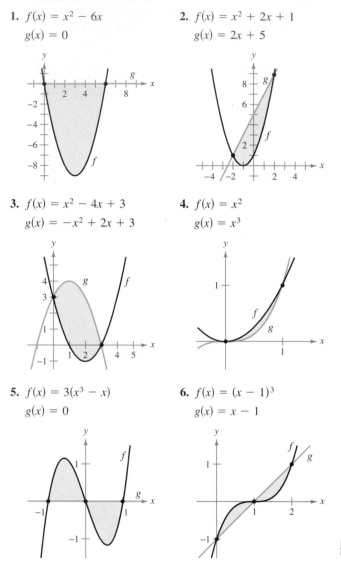

3. $f(x) = x^2 - 4x + 3$
   $g(x) = -x^2 + 2x + 3$

4. $f(x) = x^2$
   $g(x) = x^3$

5. $f(x) = 3(x^3 - x)$
   $g(x) = 0$

6. $f(x) = (x - 1)^3$
   $g(x) = x - 1$

In Exercises 7–12, the integrand of the definite integral is a difference of two functions. Sketch the graph of each function and shade the region whose area is represented by the integral.

7. $\displaystyle\int_0^4 \left[(x + 1) - \frac{x}{2} \right] dx$

8. $\displaystyle\int_{-1}^1 \left[(1 - x^2) - (x^2 - 1) \right] dx$

9. $\displaystyle\int_0^6 \left[4(2^{-x/3}) - \frac{x}{6} \right] dx$

10. $\displaystyle\int_2^3 \left[\left(\frac{x^3}{3} - x \right) - \frac{x}{3} \right] dx$

11. $\displaystyle\int_{-\pi/3}^{\pi/3} (2 - \sec x)\, dx$

12. $\displaystyle\int_{-\pi/4}^{\pi/4} (\sec^2 x - \cos x)\, dx$

Think About It   **In Exercises 13 and 14, determine which value best approximates the area of the region bounded by the graphs of f and g. (Make your selection on the basis of a sketch of the region and *not* by performing any calculations.)**

13. $f(x) = x + 1$, $g(x) = (x - 1)^2$
   (a) -2 (b) 2   (c) 10   (d) 4   (e) 8

14. $f(x) = 2 - \frac{1}{2}x$, $g(x) = 2 - \sqrt{x}$
   (a) 1   (b) 6   (c) -3 (d) 3   (e) 4

In Exercises 15–30, sketch the region bounded by the graphs of the algebraic functions and find the area of the region.

15. $y = \frac{1}{2}x^3 + 2$, $y = x + 1$, $x = 0$, $x = 2$

16. $y = -\frac{3}{8}x(x - 8)$, $y = 10 - \frac{1}{2}x$, $x = 2$, $x = 8$

17. $f(x) = x^2 - 4x$, $g(x) = 0$

18. $f(x) = -x^2 + 4x + 1$, $g(x) = x + 1$

19. $f(x) = x^2 + 2x + 1$, $g(x) = 3x + 3$

20. $f(x) = -x^2 + 4x + 2$, $g(x) = x + 2$

21. $y = x$, $y = 2 - x$, $y = 0$

22. $y = \dfrac{1}{x^2}$, $y = 0$, $x = 1$, $x = 5$

23. $f(x) = \sqrt{3x} + 1$, $g(x) = x + 1$

24. $f(x) = \sqrt[3]{x - 1}$, $g(x) = x - 1$

25. $f(y) = y^2$, $g(y) = y + 2$

26. $f(y) = y(2 - y)$, $g(y) = -y$

27. $f(y) = y^2 + 1$, $g(y) = 0$, $y = -1$, $y = 2$

28. $f(y) = \dfrac{y}{\sqrt{16 - y^2}}$, $g(y) = 0$, $y = 3$

29. $f(x) = \dfrac{10}{x}$, $x = 0$, $y = 2$, $y = 10$

30. $g(x) = \dfrac{4}{2 - x}$, $y = 4$, $x = 0$

In Exercises 31–40, use a graphing utility to graph the region bounded by the graphs of the functions, and use the integration capabilities of the graphing utility to find the area of the region.

31. $f(x) = x(x^2 - 3x + 3)$, $g(x) = x^2$

32. $f(x) = x^3 - 2x + 1$, $g(x) = -2x$, $x = 1$

33. $y = x^2 - 4x + 3$, $y = 3 + 4x - x^2$

34. $y = x^4 - 2x^2$, $y = 2x^2$

35. $f(x) = x^4 - 4x^2$, $g(x) = x^2 - 4$

36. $f(x) = x^4 - 4x^2$, $g(x) = x^3 - 4x$

37. $f(x) = 1/(1 + x^2)$, $g(x) = \frac{1}{2}x^2$

38. $f(x) = 6x/(x^2 + 1)$, $y = 0$, $0 \le x \le 3$

39. $y = \sqrt{1 + x^3}$, $y = \frac{1}{2}x + 2$, $x = 0$

40. $y = x\sqrt{\dfrac{4 - x}{4 + x}}$, $y = 0$, $x = 4$

In Exercises 41–46, sketch the region bounded by the graphs of the transcendental functions, and find the area of the region.

41. $f(x) = 2 \sin x, \quad g(x) = \tan x, \quad -\dfrac{\pi}{3} \le x \le \dfrac{\pi}{3}$

42. $f(x) = \sin x, \quad g(x) = \cos 2x, \quad -\dfrac{\pi}{2} \le x \le \dfrac{\pi}{6}$

43. $f(x) = \cos x, \quad g(x) = 2 - \cos x, \quad 0 \le x \le 2\pi$

44. $f(x) = \sec \dfrac{\pi x}{4} \tan \dfrac{\pi x}{4}, \quad g(x) = \left(\sqrt{2} - 4\right)x + 4, \quad x = 0$

45. $f(x) = xe^{-x^2}, \quad y = 0, \quad 0 \le x \le 1$

46. $f(x) = 3^x, \quad g(x) = 2x + 1$

In Exercises 47–50, use a graphing utility to graph the region bounded by the graphs of the functions, and use the integration capabilities of the graphing utility to find the area of the region.

47. $f(x) = 2 \sin x + \sin 2x, \quad y = 0, \quad 0 \le x \le \pi$

48. $f(x) = 2 \sin x + \cos 2x, \quad y = 0, \quad 0 < x \le \pi$

49. $f(x) = \dfrac{1}{x^2} e^{1/x}, \quad y = 0, \quad 1 \le x \le 3$

50. $g(x) = \dfrac{4 \ln x}{x}, \quad y = 0, \quad x = 5$

In Exercises 51 and 52, (a) use a graphing utility to graph the region bounded by the graphs of the equations. (b) Set up the integral giving the area of the region. Can you evaluate the integral by hand? (c) Use the integration capabilities of a graphing utility to approximate the area.

51. $y = \sqrt{\dfrac{x^3}{4 - x}}, \quad y = 0, \quad x = 3$

52. $y = \sqrt{x}\, e^x, \quad y = 0, \quad x = 0, \quad x = 1$

In Exercises 53–56, find the accumulation function F. Then evaluate F at each specified value of the independent variable and graphically show the area given by each value of F.

53. $F(x) = \displaystyle\int_0^x \left(\tfrac{1}{2}t + 1\right) dt$ (a) $F(0)$ (b) $F(2)$ (c) $F(6)$

54. $F(x) = \displaystyle\int_0^x \left(\tfrac{1}{2}t^2 + 2\right) dt$ (a) $F(0)$ (b) $F(4)$ (c) $F(6)$

55. $F(\alpha) = \displaystyle\int_{-1}^\alpha \cos \dfrac{\pi\theta}{2}\, d\theta$ (a) $F(-1)$ (b) $F(0)$ (c) $F(\tfrac{1}{2})$

56. $F(y) = \displaystyle\int_{-1}^y 4e^{x/2}\, dx$ (a) $F(-1)$ (b) $F(0)$ (c) $F(4)$

In Exercises 57 and 58, use integration to find the area of the triangle having the given vertices.

57. $(0, 0), (a, 0), (b, c)$

58. $(2, -3), (4, 6), (6, 1)$

In Exercises 59 and 60, set up and evaluate the definite integral that gives the area of the region bounded by the graph of the function and the tangent line to the graph at the indicated point.

59. $f(x) = x^3, \ (1, 1)$

60. $f(x) = \dfrac{1}{x^2 + 1}, \left(1, \dfrac{1}{2}\right)$

Getting at the Concept

61. Suppose horizontal representative rectangles are used when finding the area of the region between two curves. Identify the variable of integration.

62. In your own words, describe how to proceed from a precalculus formula to a new integration formula when using integration to solve applied problems.

63. The graphs of $y = x^4 - 2x^2 + 1$ and $y = 1 - x^2$ intersect at three points. However, the area between the curves *can* be found by a single integral. Explain why this is so, and write an integral for this area.

64. The area of the region bounded by the graphs of $y = x^3$ and $y = x$ *cannot* be found by the single integral

$$\int_{-1}^1 (x^3 - x)\, dx.$$

Explain why this is so. Use symmetry to write a single integral that does represent the area.

65. A college graduate has two job offers. The starting salary for each is \$32,000, and after eight years of service each will pay \$54,000. The salary increase for each offer is shown in the figure. From a strictly monetary viewpoint, which is the better offer? Explain.

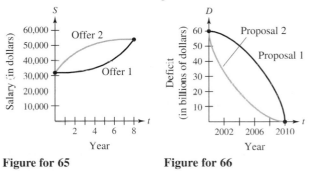

Figure for 65   Figure for 66

66. A state legislature is debating two proposals for eliminating the annual budget deficits by the year 2010. The rate of decrease of the deficits for each proposal is shown in the figure. From the viewpoint of minimizing the cumulative state deficit, which is the better proposal? Explain.

In Exercises 67 and 68, find b such that the line $y = b$ divides the region bounded by the graphs of the two equations into two regions of equal area.

67. $y = 9 - x^2, \quad y = 0$ **68.** $y = 9 - |x|, \quad y = 0$

In Exercises 69 and 70, evaluate the limit and sketch the graph of the region whose area is represented by the limit.

69. $\displaystyle\lim_{\|\Delta\|\to 0} \sum_{i=1}^n (x_i - x_i^2)\, \Delta x$

where $x_i = i/n$ and $\Delta x = 1/n$

70. $\displaystyle\lim_{\|\Delta\|\to 0} \sum_{i=1}^n (4 - x_i^2)\, \Delta x$

where $x_i = -2 + (4i/n)$ and $\Delta x = 4/n$

Revenue In Exercises 71 and 72, two models R_1 and R_2 are given for revenue (in billions of dollars per year) for a large corporation. The model R_1 gives projected annual revenues from 2000 to 2005, with $t = 0$ corresponding to 2000, and R_2 gives projected revenues if there is a decrease in the rate of growth of corporate sales over the period. Approximate the total reduction in revenue if corporate sales are actually closer to the model R_2.

71. $R_1 = 7.21 + 0.58t$

$R_2 = 7.21 + 0.45t$

72. $R_1 = 7.21 + 0.26t + 0.02t^2$

$R_2 = 7.21 + 0.1t + 0.01t^2$

73. *Modeling Data* The table shows the total receipts R and total expenditures E for the Old-Age and Survivors Insurance Trust Fund (Social Security Trust Fund) in billions of dollars. The time t is given in years, with $t = 1$ corresponding to 1991. (*Source: Social Security Administration*)

t	1	2	3	4	5
R	299.3	311.2	323.3	328.3	342.8
E	245.6	259.9	273.1	284.1	297.8

t	6	7	8	9
R	363.7	397.2	424.8	457.0
E	308.2	322.1	332.3	339.9

(a) Use a graphing utility to fit an exponential model to the data for receipts. Plot the data and graph the model.

(b) Use a graphing utility to fit an exponential model to the data for expenditures. Plot the data and graph the model.

(c) If the models are assumed true for the years 2000 through 2005, use integration to approximate the surplus revenue generated during those years.

(d) Will the models found in parts (a) and (b) intersect? Explain. Based on your answer and news reports about the fund, will these models be accurate for long-term analysis?

74. *Profit* The chief financial officer of a company reports that profits for the past fiscal year were $893,000. The officer predicts that profits for the next 5 years will grow at a continuous annual rate somewhere between $3\frac{1}{2}\%$ and 5%. Estimate the cumulative difference in total profit over the 5 years based on the predicted range of growth rates.

75. *Area* The shaded region in the figure consists of all points whose distances to the center of the square are less than the distances to the edges of the square. Find the area of the region.

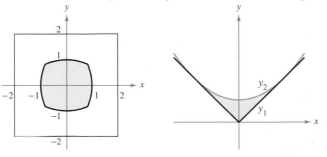

Figure for 75

76. *Mechanical Design* The surface of a machine part is the region between the graphs of $y_1 = |x|$ and $y_2 = 0.08x^2 + k$ (see figure).

(a) Find k if the parabola is tangent to the graph of y_1.

(b) Find the surface area of the machine part.

77. *Building Design* Concrete sections for a new building have the dimensions (in meters) and shape shown in the figure.

(a) Find the area of the face of the section superimposed on the rectangular coordinate system.

(b) Find the volume of concrete in one of the sections by multiplying the area in part (a) by 2 meters.

(c) One cubic meter of concrete weighs 5000 pounds. Find the weight of the section.

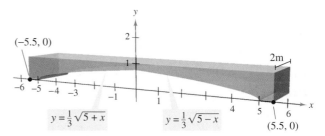

78. *Building Design* To decrease the weight and to aid in the hardening process, the concrete sections in Exercise 77 often are not solid. Rework Exercise 77 to allow for cylindrical openings such as those shown in the figure.

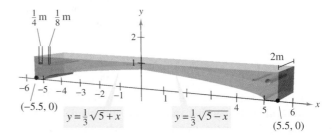

True or False? In Exercises 79–81, determine whether the statement is true or false. If it is false, explain why or give an example that shows it is false.

79. If the area of the region bounded by the graphs of f and g is 1, then the area of the region bounded by the graphs of $h(x) = f(x) + C$ and $k(x) = g(x) + C$ is also 1.

80. If $\int_a^b [f(x) - g(x)]\, dx = A$, then $\int_a^b [g(x) - f(x)]\, dx = -A$.

81. If the graphs of f and g intersect midway between $x = a$ and $x = b$, then

$$\int_a^b [f(x) - g(x)]\, dx = 0.$$

Figure for 76

- Find the volume of a solid of revolution using the disk method.
- Find the volume of a solid of revolution using the washer method.
- Find the volume of a solid with known cross sections.

The Disk Method

In Chapter 4, we mentioned that area is only *one* of the many applications of the definite integral. Another important application is its use in finding the volume of a three-dimensional solid. In this section you will study a particular type of three-dimensional solid—one whose cross sections are similar. We begin with solids of revolution. Such solids are used commonly in engineering and manufacturing. Some examples are axles, funnels, pills, bottles, and pistons, as indicated in Figure 6.11.

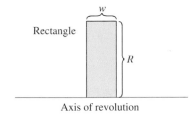

Rectangle

w

R

Axis of revolution

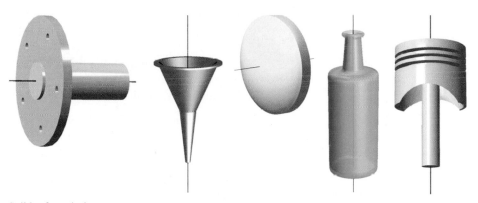

Solids of revolution
Figure 6.11

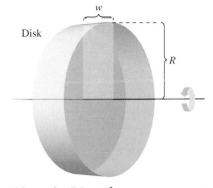

Disk

w

R

Volume of a disk: $\pi R^2 w$
Figure 6.12

If a region in the plane is revolved about a line, the resulting solid is a **solid of revolution,** and the line is called the **axis of revolution.** The simplest such solid is a right circular cylinder or **disk,** which is formed by revolving a rectangle about an axis adjacent to one side of the rectangle, as shown in Figure 6.12. The volume of such a disk is

$$\text{Volume of disk} = (\text{area of disk})(\text{width of disk})$$
$$= \pi R^2 w$$

where R is the radius of the disk and w is the width.

To see how to use the volume of a disk to find the volume of a general solid of revolution, consider a solid of revolution formed by revolving the plane region in Figure 6.13 about the indicated axis. To determine the volume of this solid, consider a representative rectangle in the plane region. When this rectangle is revolved about the axis of revolution, it generates a representative disk whose volume is

$$\Delta V = \pi R^2 \Delta x.$$

Approximating the volume of the solid by n such disks of width Δx and radius $R(x_i)$ produces

$$\text{Volume of solid} \approx \sum_{i=1}^{n} \pi [R(x_i)]^2 \Delta x$$
$$= \pi \sum_{i=1}^{n} [R(x_i)]^2 \Delta x.$$

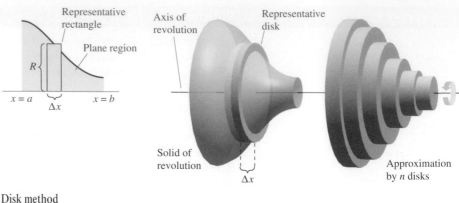

Disk method
Figure 6.13

This approximation appears to become better and better as $\|\Delta\| \to 0$ $(n \to \infty)$. Therefore, you can define the volume of the solid as

$$\text{Volume of solid} = \lim_{\|\Delta\| \to 0} \pi \sum_{i=1}^{n} [R(x_i)]^2 \, \Delta x = \pi \int_a^b [R(x)]^2 \, dx.$$

Schematically, the disk method looks like this.

Known Precalculus Formula	*Representative Element*	*New Integration Formula*
Volume of disk $V = \pi R^2 w$	$\Delta V = \pi [R(x_i)]^2 \, \Delta x$	Solid of revolution $V = \pi \int_a^b [R(x)]^2 \, dx$

A similar formula can be derived if the axis of revolution is vertical.

> **The Disk Method**
>
> To find the volume of a solid of revolution with the **disk method,** use one of the following, as indicated in Figure 6.14.
>
> | *Horizontal Axis of Revolution* | *Vertical Axis of Revolution* |
> |---|---|
> | $\text{Volume} = V = \pi \int_a^b [R(x)]^2 \, dx$ | $\text{Volume} = V = \pi \int_c^d [R(y)]^2 \, dy$ |

NOTE   In Figure 6.14, note that you can determine the variable of integration by placing a representative rectangle in the *plane* region "perpendicular" to the axis of revolution. If the width of the rectangle is Δx, integrate with respect to x, and if the width of the rectangle is Δy, integrate with respect to y.

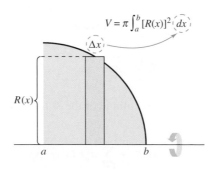

Horizontal axis of revolution
Figure 6.14

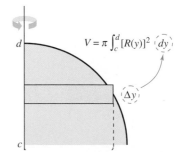

Vertical axis of revolution

The simplest application of the disk method involves a plane region bounded by the graph of f and the x-axis. If the axis of revolution is the x-axis, the radius $R(x)$ is simply $f(x)$.

Example 1   Using the Disk Method

Find the volume of the solid formed by revolving the region bounded by the graph of

$$f(x) = \sqrt{\sin x}$$

and the x-axis $(0 \le x \le \pi)$ about the x-axis.

Solution   From the representative rectangle in the upper graph in Figure 6.15, you can see that the radius of this solid is

$$R(x) = f(x)$$
$$= \sqrt{\sin x}.$$

So, the volume of the solid of revolution is

$$V = \pi \int_a^b [R(x)]^2 \, dx = \pi \int_0^\pi \left(\sqrt{\sin x} \right)^2 dx \qquad \text{Apply disk method.}$$

$$= \pi \int_0^\pi \sin x \, dx \qquad \text{Simplify.}$$

$$= \pi \left[-\cos x \right]_0^\pi \qquad \text{Integrate.}$$

$$= \pi(1 + 1)$$

$$= 2\pi.$$

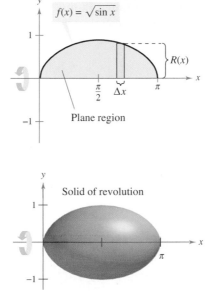

$f(x) = \sqrt{\sin x}$

Plane region

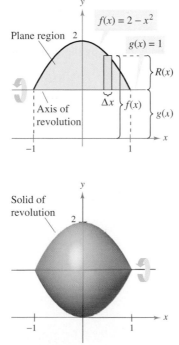

Solid of revolution

Figure 6.15

Example 2   Revolving About a Line That Is Not a Coordinate Axis

Find the volume of the solid formed by revolving the region bounded by

$$f(x) = 2 - x^2$$

and $g(x) = 1$ about the line $y = 1$, as shown in Figure 6.16.

Solution   By equating $f(x)$ and $g(x)$, you can determine that the two graphs intersect when $x = \pm 1$. To find the radius, subtract $g(x)$ from $f(x)$.

$$R(x) = f(x) - g(x)$$
$$= (2 - x^2) - 1$$
$$= 1 - x^2$$

Finally, integrate between -1 and 1 to find the volume.

$$V = \pi \int_a^b [R(x)]^2 \, dx = \pi \int_{-1}^1 (1 - x^2)^2 \, dx \qquad \text{Apply disk method.}$$

$$= \pi \int_{-1}^1 (1 - 2x^2 + x^4) \, dx \qquad \text{Simplify.}$$

$$= \pi \left[x - \frac{2x^3}{3} + \frac{x^5}{5} \right]_{-1}^1 \qquad \text{Integrate.}$$

$$= \frac{16\pi}{15}$$

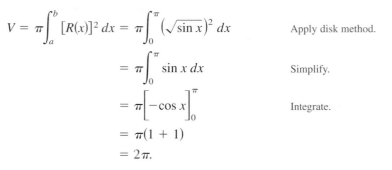

$f(x) = 2 - x^2$

Plane region

$g(x) = 1$

$R(x)$

Axis of revolution

$f(x)$

$g(x)$

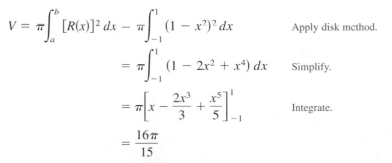

Solid of revolution

Figure 6.16

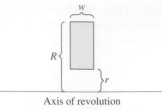

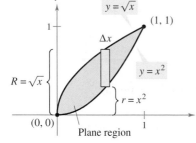

Solid of revolution

Figure 6.17

The Washer Method

The disk method can be extended to cover solids of revolution with holes by replacing the representative disk with a representative **washer.** The washer is formed by revolving a rectangle about an axis, as shown in Figure 6.17. If r and R are the inner and outer radii of the washer and w is the width of the washer, the volume is given by

Volume of washer $= \pi(R^2 - r^2)w.$

To see how this concept can be used to find the volume of a solid of revolution, consider a region bounded by an **outer radius** $R(x)$ and an **inner radius** $r(x)$, as shown in Figure 6.18. If the region is revolved about its axis of revolution, the volume of the resulting solid is given by

$$V = \pi \int_a^b ([R(x)]^2 - [r(x)]^2)\, dx. \qquad \text{Washer method}$$

Note that the integral involving the inner radius represents the volume of the hole and is *subtracted* from the integral involving the outer radius.

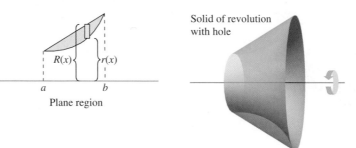

Figure 6.18

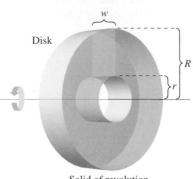

Plane region

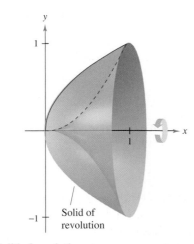

Solid of revolution
Figure 6.19

Example 3 **Using the Washer Method**

Find the volume of the solid formed by revolving the region bounded by the graphs of $y = \sqrt{x}$ and $y = x^2$ about the x-axis, as shown in Figure 6.19.

Solution In Figure 6.19, you can see that the outer and inner radii are as follows.

$$R(x) = \sqrt{x} \qquad \text{Outer radius}$$
$$r(x) = x^2 \qquad \text{Inner radius}$$

Integrating between 0 and 1 produces

$$V = \pi \int_a^b ([R(x)]^2 - [r(x)]^2)\, dx \qquad \text{Apply washer method.}$$

$$= \pi \int_0^1 \left[(\sqrt{x})^2 - (x^2)^2 \right] dx$$

$$= \pi \int_0^1 (x - x^4)\, dx \qquad \text{Simplify.}$$

$$= \pi \left[\frac{x^2}{2} - \frac{x^5}{5} \right]_0^1 \qquad \text{Integrate.}$$

$$= \frac{3\pi}{10}.$$

In each example so far, the axis of revolution has been *horizontal* and we have integrated with respect to *x*. In the next example, the axis of revolution is *vertical* and you must integrate with respect to *y*. In this example, you need two separate integrals to compute the volume.

Example 4   Integrating with Respect to *y*, Two-Integral Case

---

Find the volume of the solid formed by revolving the region bounded by the graphs of $y = x^2 + 1$, $y = 0$, $x = 0$, and $x = 1$ about the *y*-axis, as shown in Figure 6.20.

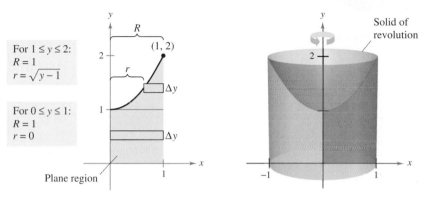

For $1 \leq y \leq 2$:
$R = 1$
$r = \sqrt{y - 1}$

For $0 \leq y \leq 1$:
$R = 1$
$r = 0$

Plane region

Solid of revolution

Figure 6.20

Solution   For the region shown in Figure 6.20, the outer radius is simply $R = 1$. There is, however, no convenient formula that represents the inner radius. When $0 \leq y \leq 1$, $r = 0$, but when $1 \leq y \leq 2$, r is determined by the equation $y = x^2 + 1$, which implies that $r = \sqrt{y - 1}$.

$$r(y) = \begin{cases} 0, & 0 \leq y \leq 1 \\ \sqrt{y - 1}, & 1 \leq y \leq 2 \end{cases}$$

Using this definition of the inner radius, you can use two integrals to find the volume.

$$V = \pi \int_0^1 (1^2 - 0^2)\, dy + \pi \int_1^2 \left[1^2 - \left(\sqrt{y - 1} \right)^2 \right] dy \qquad \text{Apply washer method.}$$

$$= \pi \int_0^1 1\, dy + \pi \int_1^2 (2 - y)\, dy \qquad \text{Simplify.}$$

$$= \pi \Big[y \Big]_0^1 + \pi \left[2y - \frac{y^2}{2} \right]_1^2 \qquad \text{Integrate.}$$

$$= \pi + \pi \left(4 - 2 - 2 + \frac{1}{2} \right) = \frac{3\pi}{2}$$

Note that the first integral $\pi \int_0^1 1\, dy$ represents the volume of a right circular cylinder of radius 1 and height 1. This portion of the volume could have been determined without using calculus.

TECHNOLOGY   Some graphing utilities have the capability to generate (or have built-in software capable of generating) a solid of revolution. If you have access to such a utility, try using it to sketch some of the solids of revolution described in this section. For instance, the solid in Example 4 might appear like that shown in Figure 6.21.

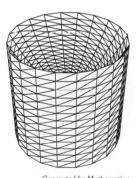

Generated by Mathematica

Figure 6.21

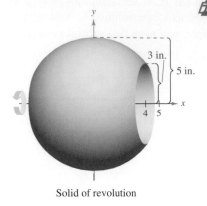

3 in.

5 in.

Solid of revolution

(a)

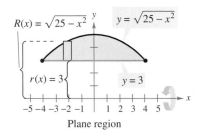

$R(x) = \sqrt{25 - x^2}$ $y = \sqrt{25 - x^2}$

$r(x) = 3$ $y = 3$

−5 −4 −3 −2 −1   1 2 3 4 5

Plane region

(b)

Figure 6.22

Example 5 Manufacturing

A manufacturer drills a hole through the center of a metal sphere of radius 5 inches, as shown in Figure 6.22(a). The hole has a radius of 3 inches. What is the volume of the resulting metal ring?

Solution You can imagine the ring to be generated by a segment of the circle whose equation is $x^2 + y^2 = 25$, as shown in Figure 6.22(b). Because the radius of the hole is 3 inches, you can let $y = 3$ and solve the equation $x^2 + y^2 = 25$ to determine that the limits of integration are $x = \pm 4$. So, the inner and outer radii are $r(x) = 3$ and $R(x) = \sqrt{25 - x^2}$ and the volume is given by

$$V = \pi \int_a^b ([R(x)]^2 - [r(x)]^2)\, dx = \pi \int_{-4}^4 \left[\left(\sqrt{25 - x^2} \right)^2 - (3)^2 \right] dx$$

$$= \pi \int_{-4}^4 (16 - x^2)\, dx$$

$$= \pi \left[16x - \frac{x^3}{3} \right]_{-4}^4$$

$$= \frac{256\pi}{3} \text{ cubic inches.}$$

Solids with Known Cross Sections

With the disk method, you can find the volume of a solid having a circular cross section whose area is $A = \pi R^2$. This method can be generalized to solids of any shape, as long as you know a formula for the area of an arbitrary cross section. Some common cross sections are squares, rectangles, triangles, semicircles, and trapezoids.

Volumes of Solids with Known Cross Sections

1. For cross sections of area $A(x)$ taken perpendicular to the x-axis,

$$\text{Volume} = \int_a^b A(x)\, dx. \qquad \text{See Figure 6.23(a).}$$

2. For cross sections of area $A(y)$ taken perpendicular to the y-axis,

$$\text{Volume} = \int_c^d A(y)\, dy. \qquad \text{See Figure 6.23(b).}$$

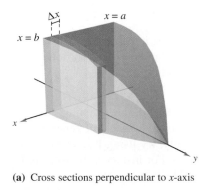

(a) Cross sections perpendicular to x-axis

(b) Cross sections perpendicular to y-axis

Figure 6.23

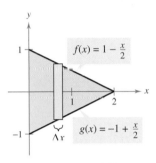

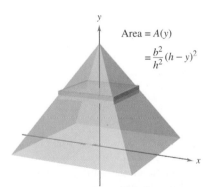

Cross sections are equilateral triangles.

Triangular base in xy-plane

Figure 6.24

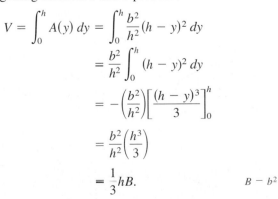

Area of base $- B = b^2$

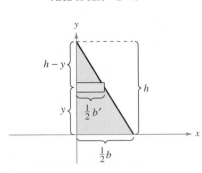

Figure 6.25

Example 6   **Triangular Cross Sections**

Find the volume of the solid shown in Figure 6.24. The base of the solid is the region bounded by the lines

$$f(x) = 1 - \frac{x}{2}, \qquad g(x) = -1 + \frac{x}{2}, \qquad \text{and} \qquad x = 0.$$

The cross sections perpendicular to the x-axis are equilateral triangles.

Solution   The base and area of each triangular cross section are as follows.

$$\text{Base} = \left(1 - \frac{x}{2}\right) - \left(-1 + \frac{x}{2}\right) = 2 - x \qquad \text{Length of base}$$

$$\text{Area} = \frac{\sqrt{3}}{4}(\text{base})^2 \qquad \text{Area of equilateral triangle}$$

$$A(x) = \frac{\sqrt{3}}{4}(2 - x)^2 \qquad \text{Area of cross section}$$

Because x ranges from 0 to 2, the volume of the solid is

$$V = \int_a^b A(x)\, dx = \int_0^2 \frac{\sqrt{3}}{4}(2 - x)^2\, dx$$

$$= -\frac{\sqrt{3}}{4}\left[\frac{(2 - x)^3}{3}\right]_0^2 = \frac{2\sqrt{3}}{3}.$$

Example 7   **An Application to Geometry**

Prove that the volume of a pyramid with a square base is $V = \frac{1}{3}hB$, where h is the height of the pyramid and B is the area of the base.

Solution   As shown in Figure 6.25, you can intersect the pyramid with a plane parallel to the base at height y to form a square cross section whose sides are of length b'. Using similar triangles, you can show that

$$\frac{b'}{b} = \frac{h - y}{h} \qquad \text{or} \qquad b' = \frac{b}{h}(h - y)$$

where b is the length of the sides of the base of the pyramid. So,

$$A(y) = (b')^2 = \frac{b^2}{h^2}(h - y)^2.$$

Integrating between 0 and h produces

$$V = \int_0^h A(y)\, dy = \int_0^h \frac{b^2}{h^2}(h - y)^2\, dy$$

$$= \frac{b^2}{h^2}\int_0^h (h - y)^2\, dy$$

$$= -\left(\frac{b^2}{h^2}\right)\left[\frac{(h - y)^3}{3}\right]_0^h$$

$$= \frac{b^2}{h^2}\left(\frac{h^3}{3}\right)$$

$$= \frac{1}{3}hB. \qquad B - b^2$$

EXERCISES FOR SECTION 6.2

In Exercises 1–6, set up and evaluate the integral that gives the volume of the solid formed by revolving the region about the x-axis.

1. $y = -x + 1$

2. $y = 4 - x^2$

3. $y = \sqrt{x}$

4. $y = \sqrt{9 - x^2}$

5. $y = x^2, \ y = x^3$

6. $y = 2, \ y = 4 - \dfrac{x^2}{4}$

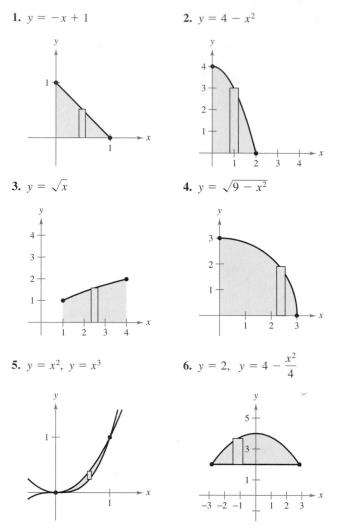

In Exercises 7–10, set up and evaluate the integral that gives the volume of the solid formed by revolving the region about the y-axis.

7. $y = x^2$

8. $y = \sqrt{16 - x^2}$

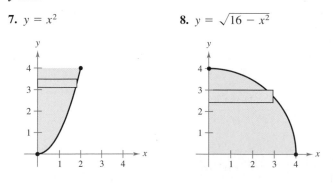

9. $y = x^{2/3}$

10. $x = -y^2 + 4y$

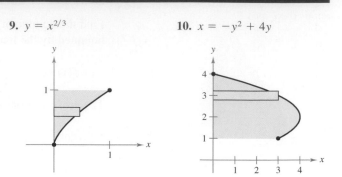

In Exercises 11–14, find the volume of the solid generated by revolving the region bounded by the graphs of the equations about the indicated lines.

11. $y = \sqrt{x}, \ y = 0, \ x = 4$
    (a) the x-axis   (b) the y-axis
    (c) the line $x = 4$ (d) the line $x = 6$

12. $y = 2x^2, \ y = 0, \ x = 2$
    (a) the y-axis   (b) the x-axis
    (c) the line $y = 8$ (d) the line $x = 2$

13. $y = x^2, \ y = 4x - x^2$
    (a) the x-axis   (b) the line $y = 6$

14. $y = 6 - 2x - x^2, \ y = x + 6$
    (a) the x-axis   (b) the line $y = 3$

In Exercises 15–18, find the volume of the solid generated by revolving the region bounded by the graphs of the equations about the line $y = 4$.

15. $y = x, \ y = 3, \ x = 0$

16. $y = \frac{1}{2}x^3, \ y = 4, \ x = 0$

17. $y = \dfrac{1}{1 + x}, \ y = 0, \ x = 0, \ x = 3$

18. $y = \sec x, \ y = 0, \ 0 \le x \le \dfrac{\pi}{3}$

In Exercises 19–22, find the volume of the solid generated by revolving the region bounded by the graphs of the equations about the line $x = 6$.

19. $y = x, \ y = 0, \ y = 4, \ x = 6$

20. $y = 6 - x, \ y = 0, \ y = 4, \ x = 0$

21. $x = y^2, \ x = 4$

22. $xy = 6, \ y = 2, \ y = 6, \ x = 6$

In Exercises 23–30, find the volume of the solid generated by revolving the region bounded by the graphs of the equations about the *x*-axis.

23. $y = \dfrac{1}{\sqrt{x+1}}$, $y = 0$, $x - 0$, $x - 3$

24. $y = x\sqrt{4 - x^2}$, $y = 0$

25. $y = \dfrac{1}{x}$, $y = 0$, $x = 1$, $x = 4$

26. $y = \dfrac{3}{x+1}$, $y = 0$, $x = 0$, $x = 8$

27. $y = e^{-x}$, $y = 0$, $x = 0$, $x = 1$

28. $y = e^{x/2}$, $y = 0$, $x = 0$, $x = 4$

29. $y = x^2 + 1$, $y = -x^2 + 2x + 5$, $x = 0$, $x = 3$

30. $y = \sqrt{x}$, $y = -\tfrac{1}{2}x + 4$, $x = 0$, $x = 8$

In Exercises 31 and 32, find the volume of the solid generated by revolving the region bounded by the graphs of the equations about the *y*-axis.

31. $y = 3(2 - x)$, $y = 0$, $x = 0$

32. $y = 9 - x^2$, $y = 0$, $x - 2$, $x - 3$

In Exercises 33–38, use the integration capabilities of a graphing utility to approximate the volume of the solid generated by revolving the region bounded by the graphs of the equations about the *x*-axis.

33. $y = \sin x$, $y = 0$, $x = 0$, $x = \pi$

34. $y = \cos x$, $y = 0$, $x = 0$, $x = \dfrac{\pi}{2}$

35. $y = e^{-x^2}$, $y = 0$, $x = 0$, $x = 2$

36. $y = \ln x$, $y = 0$, $x = 1$, $x = 3$

37. $y = e^{x/2} + e^{-x/2}$, $y = 0$, $x = -1$, $x = 2$

38. $y = 2\arctan(0.2x)$, $y = 0$, $x = 0$, $x = 5$

Think About It   In Exercises 39 and 40, determine which value best approximates the volume of the solid generated by revolving the region bounded by the graphs of the equations about the *x*-axis. (Make your selection on the basis of a sketch of the solid and *not* by performing any calculations.)

39. $y = e^{-x^2/2}$, $y = 0$, $x = 0$, $x = 2$

    (a) 3   (b) −5   (c) 10   (d) 7   (e) 20

40. $y = \arctan x$, $y = 0$, $x = 0$, $x = 1$

    (a) 10   (b) $\tfrac{3}{4}$ (c) 5   (d) −6   (e) 15

Getting at the Concept

41. Give the integration formula for finding the volumes of solids using (a) the disk method and (b) the washer method.

42. Give the integration formula for finding the volumes of solids of known cross sections.

Getting at the Concept *(continued)*

43. A region bounded by the parabola $y = 4x - x^2$ and the *x*-axis is revolved about the *x*-axis. A second region bounded by the parabola $y = 4 - x^2$ and the *x*-axis is revolved about the *x*-axis. Without integrating, how do the volumes of the two solids compare? Explain.

44. The region in the figure is revolved about the indicated axes and line. Order the volumes of the resulting solids from least to greatest. Explain your reasoning.

    (a) *x*-axis   (b) *y*-axis   (c) $x = 8$

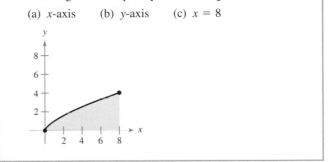

45. If the portion of the line $y = \tfrac{1}{2}x$ lying in the first quadrant is revolved about the *x*-axis, a cone is generated. Find the volume of the cone extending from $x = 0$ to $x = 6$.

46. Use the disk method to verify that the volume of a right circular cone is $\tfrac{1}{3}\pi r^2 h$, where r is the radius of the base and h is the height.

47. Use the disk method to verify that the volume of a sphere is $\tfrac{4}{3}\pi r^3$.

48. A sphere of radius r is cut by a plane h ($h < r$) units above the equator. Find the volume of the solid (spherical segment) above the plane.

49. A cone of height H with a base of radius r is cut by a plane parallel to and h units above the base. Find the volume of the solid (frustum of a cone) below the plane.

50. The region bounded by $y = \sqrt{x}$, $y = 0$, $x = 0$, and $x = 4$ is revolved about the *x*-axis.

    (a) Find the value of x in the interval $[0, 4]$ that divides the solid into two parts of equal volume.

    (b) Find the values of x in the interval $[0, 4]$ that divide the solid into three parts of equal volume.

51. *Volume of a Fuel Tank*   A tank on the wing of a jet aircraft is formed by revolving the region bounded by the graph of $y = \tfrac{1}{8}x^2\sqrt{2 - x}$ and the *x*-axis about the *x*-axis (see figure), where x and y are measured in meters. Find the tank's volume.

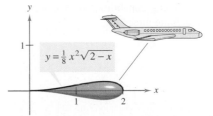
$$y = \tfrac{1}{8}x^2\sqrt{2 - x}$$

52. Volume of a Lab Glass   A glass container can be modeled by revolving the graph of

$$y = \begin{cases} \sqrt{0.1x^3 - 2.2x^2 + 10.9x + 22.2}, & 0 \le x \le 11.5 \\ 2.95, & 11.5 < x \le 15 \end{cases}$$

about the x-axis, where x and y are measured in centimeters. Use a graphing utility to graph the function and find the volume of the container.

53. Find the volume of the solid generated if the upper half of the ellipse $9x^2 + 25y^2 = 225$ is revolved about

(a) the x-axis to form a prolate spheroid (shaped like a football).

(b) the y-axis to form an oblate spheroid (shaped like half of a candy).

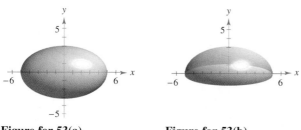

Figure for 53(a)   **Figure for 53(b)**

54. Minimum Volume   The arc of $y = 4 - (x^2/4)$ on the interval $[0, 4]$ is revolved about the line $y = b$ (see figure).

(a) Find the volume of the resulting solid as a function of b.

(b) Use a graphing utility to graph the function in part (a), and use the graph to approximate the value of b that minimizes the volume of the solid.

(c) Use calculus to find the value of b that minimizes the volume of the solid, and compare the result with the answer to part (b).

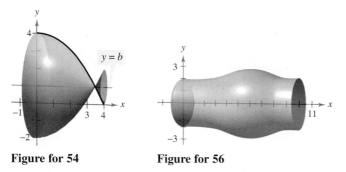

Figure for 54   **Figure for 56**

55. Water Depth in a Tank   A tank on a water tower is a sphere of radius 50 feet. Determine the depths of the water when the tank is filled to one-fourth and three-fourths of its total capacity. (*Note:* Use the root-finding capabilities of a graphing utility after evaluating the definite integral.)

56. Modeling Data   A draftsman is asked to determine the amount of material required to produce a machine part (see figure in first column). The diameters d of the part at equally spaced points x are listed in the table. The measurements are listed in centimeters.

x	0	1	2	3	4	5
d	4.2	3.8	4.2	4.7	5.2	5.7

x	6	7	8	9	10
d	5.8	5.4	4.9	4.4	4.6

(a) Use these data with Simpson's Rule to approximate the volume of the part.

(b) Use the regression capabilities of a graphing utility to find a fourth-degree polynomial through the points representing the radius of the solid. Plot the data and graph the model.

(c) Use a graphing utility to approximate the definite integral yielding the volume of the part. Compare the result with the answer to part (a).

57. Think About It   Match each integral with the solid whose volume it represents, and give the dimensions of each solid.

(a) Right circular cylinder   (b) Ellipsoid

(c) Sphere   (d) Right circular cone   (e) Torus

(i) $\pi \displaystyle\int_0^h \left(\dfrac{rx}{h}\right)^2 dx$

(ii) $\pi \displaystyle\int_0^h r^2\, dx$

(iii) $\pi \displaystyle\int_{-r}^r \left(\sqrt{r^2 - x^2}\right)^2 dx$

(iv) $\pi \displaystyle\int_{-b}^b \left(\sqrt{1 - \dfrac{x^2}{b^2}}\right)^2 dx$

(v) $\pi \displaystyle\int_{-r}^r \left[(R + \sqrt{r^2 - x^2})^2 - (R - \sqrt{r^2 - x^2})^2\right] dx$

58. Find the volume of concrete in a ramp that is 3 meters wide and whose cross sections are right triangles with base 10 meters and height 2 meters (see figure).

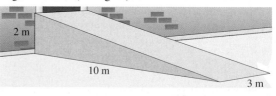

59. Find the volume of the solid whose base is bounded by the graphs of $y = x + 1$ and $y = x^2 - 1$, with the indicated cross sections taken perpendicular to the x-axis.

(a) Squares   (b) Rectangles of height 1

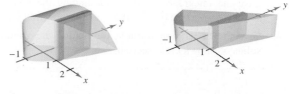

60. Find the volume of the solid whose base is bounded by the circle $x^2 + y^2 = 4$, with the indicated cross sections taken perpendicular to the x-axis.

(a) Squares
(b) Equilateral triangles

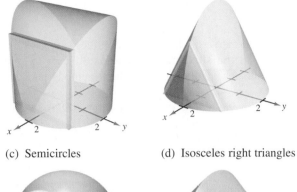

(c) Semicircles
(d) Isosceles right triangles

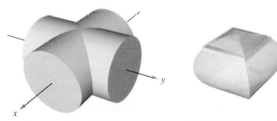

61. The base of a solid is bounded by $y = x^3$, $y = 0$, and $x = 1$. Find the volume of the solid for each of the following cross sections (taken perpendicular to the y-axis): (a) squares, (b) semicircles, (c) equilateral triangles, and (d) semiellipses whose heights are twice the lengths of their bases.

62. Find the volume of the solid of intersection (the solid common to both) of the two right circular cylinders of radius r whose axes meet at right angles (see figure).

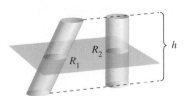

Two intersecting cylinders   Solid of intersection

FOR FURTHER INFORMATION For more information on this problem, see the article "Estimating the Volumes of Solid Figures with Curved Surfaces" by Donald Cohen in *Mathematics Teacher*. To view this article, go to the website *www.matharticles.com*.

63. *Cavalieri's Theorem* Prove that if two solids have equal altitudes and all plane sections parallel to their bases and at equal distances from their bases have equal areas, then the solids have the same volume (see figure).

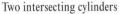

Area of R_1 = area of R_2

64. A manufacturer drills a hole through the center of a metal sphere of radius R. The hole has a radius r. Find the volume of the resulting ring.

65. For the metal sphere in Exercise 64, let $R = 5$. What value of r will produce a ring whose volume is exactly half the volume of the sphere?

66. The solid shown in the figure has cross sections bounded by the graph of

$$|x|^a + |y|^a = 1$$

where $1 \le a \le 2$.

(a) Describe the cross section when $a = 1$ and $a = 2$.

(b) Describe a procedure for approximating the volume of the solid.

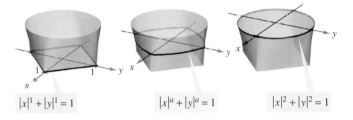

$|x|^1 + |y|^1 = 1$ $|x|^a + |y|^a = 1$ $|x|^2 + |y|^2 = 1$

67. Two planes cut a right circular cylinder to form a wedge. One plane is perpendicular to the axis of the cylinder and the second makes an angle of θ degrees with the first (see figure).

(a) Find the volume of the wedge if $\theta = 45°$.

(b) Find the volume of the wedge for an arbitrary angle θ. Assuming that the cylinder has sufficient length, how does the volume of the wedge change as θ increases from $0°$ to $90°$?

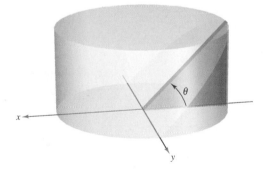

- Find the volume of a solid of revolution using the shell method.
- Compare the uses of the disk method and the shell method.

The Shell Method

In this section, you will study an alternative method for finding the volume of a solid of revolution. This method is called the **shell method** because it uses cylindrical shells. We will compare the advantages of the disk and shell methods later in this section.

To begin, consider a representative rectangle as shown in Figure 6.26, where w is the width of the rectangle, h is the height of the rectangle, and p is the distance between the axis of revolution and the *center* of the rectangle. When this rectangle is revolved about its axis of revolution, it forms a cylindrical shell (or tube) of thickness w. To find the volume of this shell, consider two cylinders. The radius of the larger cylinder corresponds to the outer radius of the shell, and the radius of the smaller cylinder corresponds to the inner radius of the shell. Because p is the average radius of the shell, you know the outer radius is $p + (w/2)$ and the inner radius is $p - (w/2)$.

$$p + \frac{w}{2} \qquad \text{Outer radius}$$

$$p - \frac{w}{2} \qquad \text{Inner radius}$$

So, the volume of the shell is

$$\text{Volume of shell} = (\text{volume of cylinder}) - (\text{volume of hole})$$
$$= \pi\left(p + \frac{w}{2}\right)^2 h - \pi\left(p - \frac{w}{2}\right)^2 h$$
$$= 2\pi phw$$
$$= 2\pi(\text{average radius})(\text{height})(\text{thickness}).$$

You can use this formula to find the volume of a solid of revolution. Assume that the plane region in Figure 6.27 is revolved about a line to form the indicated solid. If you consider a horizontal rectangle of width Δy, then, as the plane region is revolved about a line parallel to the x-axis, the rectangle generates a representative shell whose volume is

$$\Delta V = 2\pi[p(y)h(y)]\,\Delta y.$$

You can approximate the volume of the solid by n such shells of thickness Δy, height $h(y_i)$, and average radius $p(y_i)$.

$$\text{Volume of solid} \approx \sum_{i=1}^{n} 2\pi[p(y_i)h(y_i)]\,\Delta y = 2\pi\sum_{i=1}^{n}[p(y_i)h(y_i)]\,\Delta y$$

This approximation appears to become better and better as $\|\Delta\| \to 0\ (n \to \infty)$. Therefore, we define the volume of the solid to be

$$\text{Volume of solid} = \lim_{\|\Delta\| \to 0} 2\pi\sum_{i=1}^{n}[p(y_i)h(y_i)]\,\Delta y$$
$$= 2\pi\int_{c}^{d}[p(y)h(y)]\,dy.$$

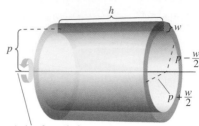

Figure 6.26

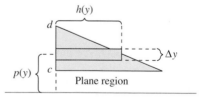
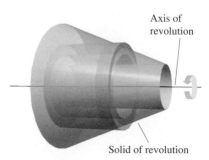

Figure 6.27

The Shell Method

To find the volume of a solid of revolution with the **shell method,** use one of the following, as shown in Figure 6.28.

Horizontal Axis of Revolution	*Vertical Axis of Revolution*
Volume $= V = 2\pi \int_{c}^{d} p(y)h(y)\,dy$	Volume $= V = 2\pi \int_{a}^{b} p(x)h(x)\,dx$

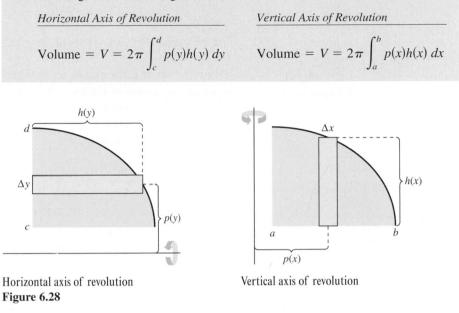

Horizontal axis of revolution

Figure 6.28

Vertical axis of revolution

Example 1   **Using the Shell Method to Find Volume**

Find the volume of the solid of revolution formed by revolving the region bounded by

$$y = x - x^3$$

and the x-axis $(0 \le x \le 1)$ about the y-axis.

Solution   Because the axis of revolution is vertical, use a vertical representative rectangle, as shown in Figure 6.29. The width Δx indicates that x is the variable of integration. The distance from the center of the rectangle to the axis of revolution is $p(x) = x$, and the height of the rectangle is

$$h(x) = x - x^3.$$

Because x ranges from 0 to 1, the volume of the solid is

$$
\begin{aligned}
V &= 2\pi \int_{a}^{b} p(x)h(x)\,dx = 2\pi \int_{0}^{1} x(x - x^3)\,dx && \text{Apply shell method.}\\
&= 2\pi \int_{0}^{1} (-x^4 + x^2)\,dx && \text{Simplify.}\\
&= 2\pi \left[-\frac{x^5}{5} + \frac{x^3}{3} \right]_{0}^{1} && \text{Integrate.}\\
&= 2\pi \left(-\frac{1}{5} + \frac{1}{3} \right)\\
&= \frac{4\pi}{15}.
\end{aligned}
$$

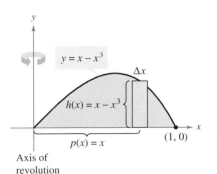

$y = x - x^3$

$h(x) = x - x^3$

$p(x) = x$

$(1, 0)$

Axis of revolution

Figure 6.29

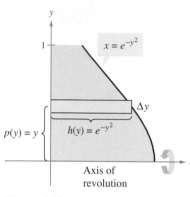

Figure 6.30

Example 2   Using the Shell Method to Find Volume

Find the volume of the solid of revolution formed by revolving the region bounded by the graph of

$$x = e^{-y^2}$$

and the y-axis $(0 \le y \le 1)$ about the x-axis.

Solution   Because the axis of revolution is horizontal, use a horizontal representative rectangle, as shown in Figure 6.30. The width Δy indicates that y is the variable of integration. The distance from the center of the rectangle to the axis of revolution is $p(y) = y$, and the height of the rectangle is $h(y) = e^{-y^2}$. Because y ranges from 0 to 1, the volume of the solid is

$$V = 2\pi \int_c^d p(y)h(y)\, dy = 2\pi \int_0^1 ye^{-y^2}\, dy \qquad \text{Apply shell method.}$$

$$= -\pi \left[e^{-y^2} \right]_0^1 \qquad \text{Integrate.}$$

$$= \pi \left(1 - \frac{1}{e} \right)$$

$$\approx 1.986.$$

NOTE   To see the advantage of using the shell method in Example 2, solve the equation $x = e^{-y^2}$ for y.

$$y = \begin{cases} 1, & 0 \le x \le 1/e \\ \sqrt{-\ln x}, & 1/e < x \le 1 \end{cases}$$

Then use this equation to find the volume using the disk method.

Comparison of Disk and Shell Methods

The disk and shell methods can be distinguished as follows. For the disk method, the representative rectangle is always *perpendicular* to the axis of revolution, whereas for the shell method, the representative rectangle is always *parallel* to the axis of revolution, as shown in Figure 6.31.

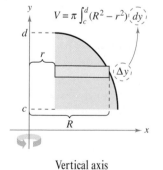

Vertical axis
of revolution

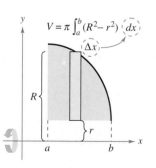

Horizontal axis
of revolution

Disk method: Representative rectangle is
perpendicular to the axis of revolution.
Figure 6.31

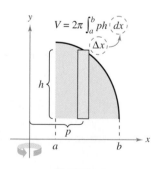

Vertical axis
of revolution

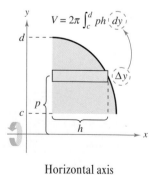

Horizontal axis
of revolution

Shell method: Representative rectangle is
parallel to the axis of revolution.

Often, one method is more convenient to use than the other. The following example illustrates a case in which the shell method is preferable.

Example 3   Shell Method Preferable

Find the volume of the solid formed by revolving the region bounded by the graphs of

$$y = x^2 + 1, \quad y = 0, \quad x = 0, \quad \text{and} \quad x = 1$$

about the y-axis.

Solution   In Example 4 in the preceding section, you saw that the disk method requires two integrals to determine the volume of this solid. See Figure 6.32(a).

$$V = \pi \int_0^1 (1^2 - 0^2)\, dy + \pi \int_1^2 \left[1^2 - \left(\sqrt{y - 1} \right)^2 \right] dy \qquad \text{Apply disk method.}$$

$$= \pi \int_0^1 1\, dy + \pi \int_1^2 (2 - y)\, dy \qquad \text{Simplify.}$$

$$= \pi \left[y \right]_0^1 + \pi \left[2y - \frac{y^2}{2} \right]_1^2 \qquad \text{Integrate.}$$

$$= \pi + \pi \left(4 - 2 - 2 + \frac{1}{2} \right)$$

$$= \frac{3\pi}{2}$$

In Figure 6.32(b), you can see that the shell method requires only one integral to find the volume.

$$V = 2\pi \int_a^b p(x)h(x)\, dx \qquad \text{Apply shell method.}$$

$$= 2\pi \int_0^1 x(x^2 + 1)\, dx$$

$$= 2\pi \left[\frac{x^4}{4} + \frac{x^2}{2} \right]_0^1 \qquad \text{Integrate.}$$

$$= 2\pi \left(\frac{3}{4} \right)$$

$$= \frac{3\pi}{2}$$

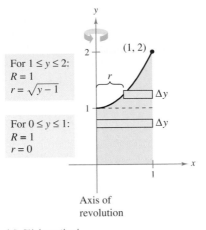

For $1 \le y \le 2$:
$R = 1$
$r = \sqrt{y - 1}$

For $0 \le y \le 1$:
$R = 1$
$r = 0$

Axis of revolution

(a) Disk method

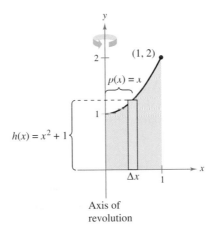

$p(x) = x$

$h(x) = x^2 + 1$

Axis of revolution

(b) Shell method

Figure 6.32

Suppose the region in Example 3 were revolved about the vertical line $x = 1$. Would the resulting solid of revolution have a greater volume or a smaller volume than the solid in Example 3? Without integrating, you should be able to reason that the resulting solid would have a smaller volume because "more" of the revolved region would be closer to the axis of revolution. To confirm this, try solving the following integral, which gives the volume of the solid.

$$V = 2\pi \int_0^1 (1 - x)(x^2 + 1)\, dx \qquad p(x) = 1 - x$$

FOR FURTHER INFORMATION To learn more about the disk and shell methods, see the article "The Disc and Shell Method" by Charles A. Cable in *The American Mathematical Monthly*. To view this article, go to the website *www.matharticles.com*.

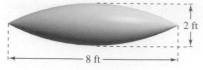

Figure 6.33

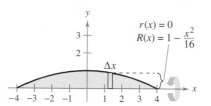

(a) Disk method

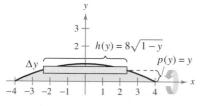

(b) Shell method

Figure 6.34

Example 4 Volume of a Pontoon

A pontoon is to be made in the shape shown in Figure 6.33. The pontoon is designed by rotating the graph of

$$y = 1 - \frac{x^2}{16}, \qquad -4 \le x \le 4$$

about the x-axis, where x and y are measured in feet. Find the volume of the pontoon.

Solution   Refer to Figure 6.34(a) and use the disk method as follows.

$$V = \pi \int_{-4}^{4} \left(1 - \frac{x^2}{16}\right)^2 dx \qquad \text{Apply disk method.}$$

$$= \pi \int_{-4}^{4} \left(1 - \frac{x^2}{8} + \frac{x^4}{256}\right) dx \qquad \text{Simplify.}$$

$$= \pi \left[x - \frac{x^3}{24} + \frac{x^5}{1280}\right]_{-4}^{4} \qquad \text{Integrate.}$$

$$= \frac{64\pi}{15} \approx 13.4 \text{ cubic feet}$$

Try using Figure 6.34(b) to set up the integral for the volume using the shell method. Does the integral seem more complicated?

For the shell method in Example 4, you would have to solve for x in terms of y in the equation

$$y = 1 - (x^2/16).$$

Sometimes, solving for x is very difficult (or even impossible). In such cases you must use a vertical rectangle (of width Δx), thus making x the variable of integration. The position (horizontal or vertical) of the axis of revolution then determines the method to be used. This is illustrated in Example 5.

Example 5 Shell Method Necessary

Find the volume of the solid formed by revolving the region bounded by the graphs of $y = x^3 + x + 1$, $y = 1$, and $x = 1$ about the line $x = 2$, as shown in Figure 6.35.

Solution   In the equation $y = x^3 + x + 1$, you cannot easily solve for x in terms of y. (See the section on Newton's Method.) Therefore, the variable of integration must be x, and you should choose a vertical representative rectangle. Because the rectangle is parallel to the axis of revolution, use the shell method and obtain

$$V = 2\pi \int_{a}^{b} p(x)h(x)\, dx = 2\pi \int_{0}^{1} (2 - x)(x^3 + x + 1 - 1)\, dx \qquad \text{Apply shell method.}$$

$$= 2\pi \int_{0}^{1} (-x^4 + 2x^3 - x^2 + 2x)\, dx \qquad \text{Simplify.}$$

$$= 2\pi \left[-\frac{x^5}{5} + \frac{x^4}{2} - \frac{x^3}{3} + x^2\right]_{0}^{1} \qquad \text{Integrate.}$$

$$= 2\pi \left(-\frac{1}{5} + \frac{1}{2} - \frac{1}{3} + 1\right)$$

$$= \frac{29\pi}{15}.$$

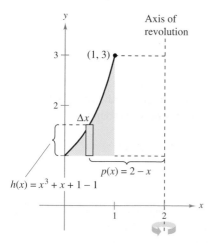

Figure 6.35

EXERCISES FOR SECTION 6.3

In Exercises 1–12, use the shell method to set up and evaluate the integral that gives the volume of the solid generated by revolving the plane region about the *y*-axis.

1. $y = x$

2. $y = 1 - x$

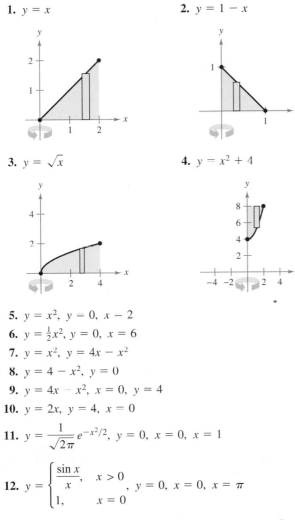

3. $y = \sqrt{x}$

4. $y = x^2 + 4$

5. $y = x^2$, $y = 0$, $x = 2$

6. $y = \frac{1}{2}x^2$, $y = 0$, $x = 6$

7. $y = x^2$, $y = 4x - x^2$

8. $y = 4 - x^2$, $y = 0$

9. $y = 4x - x^2$, $x = 0$, $y = 4$

10. $y = 2x$, $y = 4$, $x = 0$

11. $y = \dfrac{1}{\sqrt{2\pi}} e^{-x^2/2}$, $y = 0$, $x = 0$, $x = 1$

12. $y = \begin{cases} \dfrac{\sin x}{x}, & x > 0 \\ 1, & x = 0 \end{cases}$, $y = 0$, $x = 0$, $x = \pi$

In Exercises 13–16, use the shell method to set up and evaluate the integral that gives the volume of the solid generated by revolving the plane region about the *x*-axis.

13. $y = x$

14. $y = 2 - x$

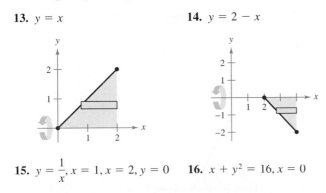

15. $y = \dfrac{1}{x}$, $x = 1$, $x = 2$, $y = 0$ **16.** $x + y^2 = 16$, $x = 0$

In Exercises 17–20, use the shell method to find the volume of the solid generated by revolving the plane region about the indicated line.

17. $y = x^2$, $y = 4x - x^2$, about the line $x = 4$

18. $y = x^2$, $y = 4x - x^2$, about the line $x = 2$

19. $y = 4x - x^2$, $y = 0$, about the line $x = 5$

20. $y = \sqrt{x}$, $y = 0$, $x = 4$, about the line $x = 6$

In Exercises 21–24, use the disk *or* the shell method to find the volume of the solid generated by revolving the region bounded by the graphs of the equations about the indicated line.

21. $y = x^3$, $y = 0$, $x = 2$

    (a) the *x*-axis   (b) the *y*-axis   (c) the line $x = 4$

22. $y = \dfrac{10}{x^2}$, $y = 0$, $x = 1$, $x = 5$

    (a) the *x*-axis   (b) the *y*-axis   (c) the line $y = 10$

23. $x^{1/2} + y^{1/2} = a^{1/2}$, $x = 0$, $y = 0$

    (a) the *x*-axis   (b) the *y*-axis   (c) the line $x = a$

24. $x^{2/3} + y^{2/3} = a^{2/3}$, $a > 0$ (hypocycloid)

    (a) the *x*-axis   (b) the *y*-axis

Getting at the Concept

25. Give the integration formula for finding the volume of a solid using the shell method.

26. The region in the figure is revolved about the indicated axes and line. Order the volumes of the resulting solids from least to greatest. Explain your reasoning.

    (a) *x*-axis   (b) *y*-axis   (c) $x = 5$

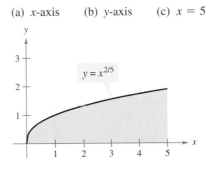

In Exercises 27 and 28, give a geometric argument that explains why the integrals have equal values.

27. $\pi \displaystyle\int_1^5 (x - 1)\, dx = 2\pi \displaystyle\int_0^2 y[5 - (y^2 + 1)]\, dy$

28. $\pi \displaystyle\int_0^2 [16 - (2y)^2]\, dy = 2\pi \displaystyle\int_0^4 x\left(\dfrac{x}{2}\right) dx$

In Exercises 29–32, (a) use a graphing utility to graph the plane region bounded by the graphs of the equations, and (b) use the integration capabilities of the graphing utility to approximate the volume of the solid generated by revolving the region about the *y*-axis.

29. $x^{4/3} + y^{4/3} = 1$, $x = 0$, $y = 0$, first quadrant

30. $y = \sqrt{1 - x^3}$, $y = 0$, $x = 0$

31. $y = \sqrt[3]{(x - 2)^2(x - 6)^2}$, $y = 0$, $x = 2$, $x = 6$

32. $y = \dfrac{2}{1 + e^{1/x}}$, $y = 0$, $x = 1$, $x = 3$

Think About It   In Exercises 33 and 34, determine which value best approximates the volume of the solid generated by revolving the region bounded by the graphs of the equations about the *y*-axis. (Make your selection on the basis of a sketch of the solid and *not* by performing any calculations.)

33. $y = 2e^{-x}$, $y = 0$, $x = 0$, $x = 2$

(a) $\frac{3}{2}$ (b) -2 (c) 4 (d) 7.5 (e) 15

34. $y = \tan x$, $y = 0$, $x = 0$, $x = \dfrac{\pi}{4}$

(a) 3.5 (b) $-\frac{9}{4}$ (c) 8 (d) 10 (e) 1

35. *Machine Part*   A solid is generated by revolving the region bounded by $y = \frac{1}{2}x^2$ and $y = 2$ about the *y*-axis. A hole, centered along the axis of revolution, is drilled through this solid so that one fourth of the volume is removed. Find the diameter of the hole.

36. *Machine Part*   A solid is generated by revolving the region bounded by $y = \sqrt{9 - x^2}$ and $y = 0$ about the *y*-axis. A hole, centered along the axis of revolution, is drilled through this solid so that one third of the volume is removed. Find the diameter of the hole.

37. *Volume of a Torus*   A torus is formed by revolving the region bounded by the circle $x^2 + y^2 = 1$ about the line $x = 2$, as shown in the figure. Find the volume of this "doughnut-shaped" solid. (*Hint:* The integral $\int_{-1}^{1} \sqrt{1 - x^2}\,dx$ represents the area of a semicircle.)

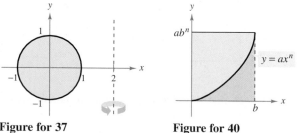

Figure for 37   **Figure for 40**

38. *Volume of a Torus*   Repeat Exercise 37 for a torus formed by revolving the region bounded by the circle $x^2 + y^2 = r^2$ about the line $x = R$, where $r < R$.

39. *Volume of a Segment of a Sphere*   Let a sphere of radius r be cut by a plane, thus forming a segment of height h. Show that the volume of this segment is $\frac{1}{3}\pi h^2(3r - h)$.

40. *Exploration*   Consider the region bounded by the graphs of $y = ax^n$, $y = ab^n$, and $x = 0$ (see figure in first column).

(a) Find the ratio $R_1(n)$ of the area of the region to the area of the circumscribed rectangle.

(b) Find $\lim\limits_{n \to \infty} R_1(n)$ and compare the result with the area of the circumscribed rectangle.

(c) Find the volume of the solid of revolution formed by revolving the region about the *y*-axis. Find the ratio $R_2(n)$ of this volume to the volume of the circumscribed right circular cylinder.

(d) Find $\lim\limits_{n \to \infty} R_2(n)$ and compare the result with the volume of the circumscribed cylinder.

(e) Use the results of parts (b) and (d) to make a conjecture about the shape of the graph of $y = ax^n$ $(0 \le x \le b)$ as $n \to \infty$.

41. *Think About It*   Match each of the integrals with the solid whose volume it represents, and give the dimensions of each solid.

(a) Right circular cone   (b) Torus   (c) Sphere

(d) Right circular cylinder   (e) Ellipsoid

(i) $2\pi \displaystyle\int_0^r hx\,dx$

(ii) $2\pi \displaystyle\int_0^r hx\left(1 - \dfrac{x}{r}\right)dx$

(iii) $2\pi \displaystyle\int_0^r 2x\sqrt{r^2 - x^2}\,dx$

(iv) $2\pi \displaystyle\int_0^b 2ax\sqrt{1 - \dfrac{x^2}{b^2}}\,dx$

(v) $2\pi \displaystyle\int_{-r}^r (R - x)\left(2\sqrt{r^2 - x^2}\right)dx$

42. *Volume of a Storage Shed*   A storage shed has a circular base of diameter 80 feet (see figure). Starting at the center, the interior height is measured every 10 feet and recorded in the table.

x	0	10	20	30	40
Height	50	45	40	20	0

(a) Use Simpson's Rule to approximate the volume of the shed.

(b) Note that the roof line consists of two line segments. Find the equations of the line segments and use integration to find the volume of the shed.

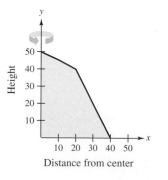

Distance from center

43. Modeling Data   A pond is approximately circular, with a diameter of 400 feet (see figure). Starting at the center, the depth of the water is measured every 25 feet and recorded in the table.

x	0	25	50	75	100	125	150	175	200
Depth	20	19	19	17	15	14	10	6	0

(a) Use Simpson's Rule to approximate the volume of water in the pond.

(b) Use the regression capabilities of a graphing utility to find a quadratic model for the depths recorded in the table. Use the graphing utility to plot the depths and graph the model.

(c) Use the integration capabilities of a graphing utility and the model in part (b) to approximate the volume of water in the pond.

(d) Use the result in part (c) to approximate the number of gallons of water in the pond if 1 cubic foot of water is approximately 7.48 gallons.

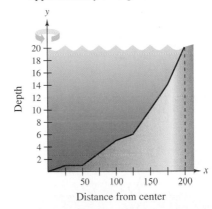

The color-enhanced photo of Saturn was taken by Voyager 1. In the photograph, the oblateness of Saturn is clearly visible.

SECTION PROJECT   SATURN

The Oblateness of Saturn   Saturn is the most oblate of the nine planets in our solar system. Its equatorial radius is 60,268 kilometers and its polar radius is 54,364 kilometers.

(a) Find the ratio of the volumes of the sphere and the oblate ellipsoid shown below.

(b) If a planet were spherical and had the same volume as Saturn, what would its radius be?

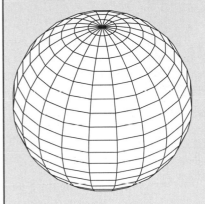

Computer model of "spherical Saturn," whose equatorial radius is equal to its polar radius. The equation of the cross section passing through the pole is

$$x^2 + y^2 = 60{,}268^2.$$

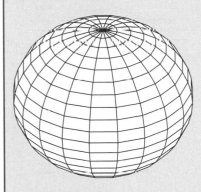

Computer model of "oblate Saturn," whose equatorial radius is greater than its polar radius. The equation of the cross section passing through the pole is

$$\frac{x^2}{60{,}268^2} + \frac{y^2}{54{,}364^2} = 1.$$

CHRISTIAN HUYGENS (1629–1695)

The Dutch mathematician Christian Huygens, who invented the pendulum clock, and James Gregory (1638–1675), a Scottish mathematician, both made early contributions to the problem of finding the length of a rectifiable curve.

Section 6.4 — Arc Length and Surfaces of Revolution

- Find the arc length of a smooth curve.
- Find the area of a surface of revolution.

Arc Length

In this section, definite integrals are used to find the arc length of a curve and the area of a surface of revolution. In both cases, we approximate an arc (a segment of a curve) by straight line segments whose lengths are given by the familiar distance formula

$$d = \sqrt{(x_2 - x_1)^2 + (y_2 - y_1)^2}.$$

A **rectifiable** curve is one that has a finite arc length. You will see that a sufficient condition for the graph of a function f to be rectifiable between $(a, f(a))$ and $(b, f(b))$ is that f' be continuous on $[a, b]$. Such a function is **continuously differentiable** on $[a, b]$, and its graph on the interval $[a, b]$ is a **smooth curve.**

Consider a function $y = f(x)$ that is continuously differentiable on the interval $[a, b]$. You can approximate the graph of f by n line segments whose endpoints are determined by the partition

$$a = x_0 < x_1 < x_2 < \cdots < x_n = b$$

as shown in Figure 6.36. By letting $\Delta x_i = x_i - x_{i-1}$ and $\Delta y_i = y_i - y_{i-1}$, you can approximate the length of the graph by

$$s \approx \sum_{i=1}^{n} \sqrt{(x_i - x_{i-1})^2 + (y_i - y_{i-1})^2}$$

$$= \sum_{i=1}^{n} \sqrt{(\Delta x_i)^2 + (\Delta y_i)^2}$$

$$= \sum_{i=1}^{n} \sqrt{(\Delta x_i)^2 + \left(\frac{\Delta y_i}{\Delta x_i}\right)^2 (\Delta x_i)^2}$$

$$= \sum_{i=1}^{n} \sqrt{1 + \left(\frac{\Delta y_i}{\Delta x_i}\right)^2} \, (\Delta x_i).$$

This approximation appears to become better and better as $\|\Delta\| \to 0 \ (n \to \infty)$. Therefore, we define the length of the graph to be

$$s = \lim_{\|\Delta\| \to 0} \sum_{i=1}^{n} \sqrt{1 + \left(\frac{\Delta y_i}{\Delta x_i}\right)^2} \, (\Delta x_i).$$

Because $f'(x)$ exists for each x in (x_{i-1}, x_i), the Mean Value Theorem guarantees the existence of c_i in (x_{i-1}, x_i) such that

$$f(x_i) - f(x_{i-1}) = f'(c_i)(x_i - x_{i-1})$$

$$\frac{\Delta y_i}{\Delta x_i} = f'(c_i).$$

Because f' is continuous on $[a, b]$, it follows that $\sqrt{1 + [f'(x)]^2}$ is also continuous (and hence integrable) on $[a, b]$, which implies that

$$s = \lim_{\|\Delta\| \to 0} \sum_{i=1}^{n} \sqrt{1 + [f'(c_i)]^2} \, (\Delta x_i)$$

$$= \int_a^b \sqrt{1 + [f'(x)]^2} \, dx.$$

We call s the **arc length** of f between a and b.

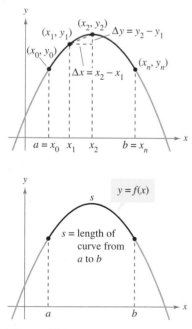

Figure 6.36

Definition of Arc Length

Let the function given by $y = f(x)$ represent a smooth curve on the interval $[a, b]$. The **arc length** of f between a and b is

$$s = \int_a^b \sqrt{1 + [f'(x)]^2}\, dx.$$

Similarly, for a smooth curve given by $x = g(y)$, the **arc length** of g between c and d is

$$s = \int_c^d \sqrt{1 + [g'(y)]^2}\, dy.$$

Because the definition of arc length can be applied to a linear function, you can check to see that this new definition agrees with the standard distance formula for the length of a line segment. This is done in Example 1.

Example 1   **The Length of a Line Segment**

Find the arc length from (x_1, y_1) to (x_2, y_2) on the graph of $f(x) = mx + b$, as shown in Figure 6.37.

Solution   Because

$$m = f'(x) = \frac{y_2 - y_1}{x_2 - x_1}$$

it follows that

$$
\begin{aligned}
s &= \int_{x_1}^{x_2} \sqrt{1 + [f'(x)]^2}\, dx && \text{Formula for arc length} \\
&= \int_{x_1}^{x_2} \sqrt{1 + \left(\frac{y_2 - y_1}{x_2 - x_1}\right)^2}\, dx \\
&= \sqrt{\frac{(x_2 - x_1)^2 + (y_2 - y_1)^2}{(x_2 - x_1)^2}}\,(x)\Bigg]_{x_1}^{x_2} && \text{Integrate and simplify.} \\
&= \sqrt{\frac{(x_2 - x_1)^2 + (y_2 - y_1)^2}{(x_2 - x_1)^2}}\,(x_2 - x_1) \\
&= \sqrt{(x_2 - x_1)^2 + (y_2 - y_1)^2}
\end{aligned}
$$

which is the formula for the distance between two points in the plane.

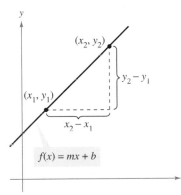

The arc length of the graph of f from (x_1, y_1) to (x_2, y_2) is the same as the standard distance formula.
Figure 6.37

TECHNOLOGY   Definite integrals representing arc length often are very difficult to evaluate. In this section we present a few examples. In the next chapter, with more advanced integration techniques, you will be able to tackle more difficult arc length problems. In the meantime, remember that you can always use a numerical integration program to approximate an arc length. For instance, try using the numerical integration feature of a graphing utility to approximate the arc lengths in Examples 2 and 3.

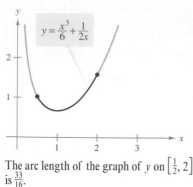

The arc length of the graph of y on $\left[\frac{1}{2}, 2\right]$ is $\frac{33}{16}$.

Figure 6.38

FOR FURTHER INFORMATION To see how arc length can be used to define trigonometric functions, see the article "Trigonometry Requires Calculus, Not Vice Versa" by Yves Nievergelt in *UMAP Modules.* To view this article, go to the website *www.matharticles.com.*

Example 2 **Finding Arc Length**

Find the arc length of the graph of

$$y = x^3/6 + 1/(2x)$$

on the interval $\left[\frac{1}{2}, 2\right]$, as shown in Figure 6.38.

Solution Using

$$\frac{dy}{dx} = \frac{3x^2}{6} - \frac{1}{2x^2} = \frac{1}{2}\left(x^2 - \frac{1}{x^2}\right)$$

yields an arc length of

$$
\begin{aligned}
s &= \int_a^b \sqrt{1 + \left(\frac{dy}{dx}\right)^2}\, dx = \int_{1/2}^2 \sqrt{1 + \left[\frac{1}{2}\left(x^2 - \frac{1}{x^2}\right)\right]^2}\, dx && \text{Formula for arc length} \\
&= \int_{1/2}^2 \sqrt{\frac{1}{4}\left(x^4 + 2 + \frac{1}{x^4}\right)}\, dx \\
&= \int_{1/2}^2 \frac{1}{2}\left(x^2 + \frac{1}{x^2}\right) dx && \text{Simplify.} \\
&= \frac{1}{2}\left[\frac{x^3}{3} - \frac{1}{x}\right]_{1/2}^2 && \text{Integrate.} \\
&= \frac{1}{2}\left(\frac{13}{6} + \frac{47}{24}\right) \\
&= \frac{33}{16}.
\end{aligned}
$$

Example 3 **Finding Arc Length**

Find the arc length of the graph of $(y - 1)^3 = x^2$ on the interval $[0, 8]$, as shown in Figure 6.39.

Solution Begin by solving for x in terms of y: $x = \pm(y - 1)^{3/2}$. Choosing the positive value of x produces

$$\frac{dx}{dy} = \frac{3}{2}(y - 1)^{1/2}.$$

The x-interval $[0, 8]$ corresponds to the y-interval $[1, 5]$, and the arc length is

$$
\begin{aligned}
s &= \int_c^d \sqrt{1 + \left(\frac{dx}{dy}\right)^2}\, dy = \int_1^5 \sqrt{1 + \left[\frac{3}{2}(y - 1)^{1/2}\right]^2}\, dy && \text{Formula for arc length} \\
&= \int_1^5 \sqrt{\frac{9}{4}y - \frac{5}{4}}\, dy \\
&= \frac{1}{2}\int_1^5 \sqrt{9y - 5}\, dy && \text{Simplify.} \\
&= \frac{1}{18}\left[\frac{(9y - 5)^{3/2}}{3/2}\right]_1^5 && \text{Integrate.} \\
&= \frac{1}{27}(40^{3/2} - 4^{3/2}) \\
&\approx 9.0734.
\end{aligned}
$$

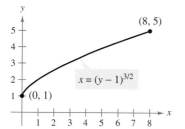

The arc length of the graph of y on $[0, 8]$ is approximately 9.0734.

Figure 6.39

 *Example 4   Finding Arc Length*

Find the arc length of the graph of $y = \ln(\cos x)$ from $x = 0$ to $x = \pi/4$, as shown in Figure 6.40.

Solution   Using

$$\frac{dy}{dx} = -\frac{\sin x}{\cos x} = -\tan x$$

yields an arc length of

$$s = \int_a^b \sqrt{1 + \left(\frac{dy}{dx}\right)^2}\, dx = \int_0^{\pi/4} \sqrt{1 + \tan^2 x}\, dx \qquad \text{Formula for arc length}$$

$$= \int_0^{\pi/4} \sqrt{\sec^2 x}\, dx \qquad \text{Trigonometric identity}$$

$$= \int_0^{\pi/4} \sec x\, dx \qquad \text{Simplify.}$$

$$= \left[\ln|\sec x + \tan x|\right]_0^{\pi/4} \qquad \text{Integrate.}$$

$$= \ln(\sqrt{2} + 1) - \ln 1$$

$$\approx 0.8814.$$

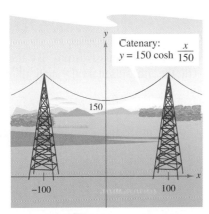

The arc length of the graph of y on $\left[0, \frac{\pi}{4}\right]$ is approximately 0.8814.
Figure 6.40

Example 5   Length of a Cable

An electric cable is hung between two towers that are 200 feet apart, as shown in Figure 6.41. The cable takes the shape of a catenary whose equation is

$$y = 75(e^{x/150} + e^{-x/150}) = 150 \cosh \frac{x}{150}.$$

Find the arc length of the cable between the two towers.

Solution   Because $y' = \frac{1}{2}(e^{x/150} - e^{-x/150})$, you can write

$$(y')^2 = \frac{1}{4}(e^{x/75} - 2 + e^{-x/75})$$

and

$$1 + (y')^2 = \frac{1}{4}(e^{x/75} + 2 + e^{-x/75}) = \left[\frac{1}{2}(e^{x/150} + e^{-x/150})\right]^2.$$

Therefore, the arc length of the cable is

$$s = \int_a^b \sqrt{1 + (y')^2}\, dx = \frac{1}{2}\int_{-100}^{100} (e^{x/150} + e^{-x/150})\, dx \qquad \text{Formula for arc length}$$

$$= 75\left[e^{x/150} - e^{-x/150}\right]_{-100}^{100} \qquad \text{Integrate.}$$

$$= 150(e^{2/3} - e^{-2/3})$$

$$\approx 215 \text{ feet.}$$

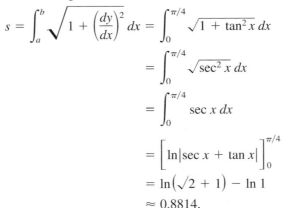

The arc length of the cable is approximately 215 feet.
Figure 6.41

Area of a Surface of Revolution

In Sections 6.2 and 6.3, integration was used to calculate the volume of a solid of revolution. We now look at a procedure for finding the area of a surface of revolution.

> **Definition of a Surface of Revolution**
>
> If the graph of a continuous function is revolved about a line, the resulting surface is a **surface of revolution.**

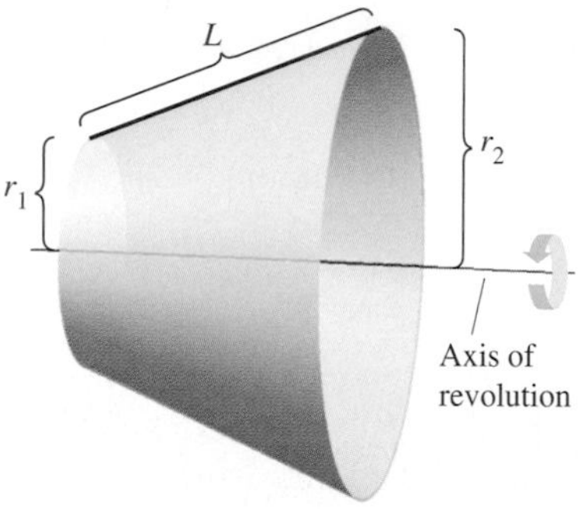

The area of a surface of revolution is derived from the formula for the lateral surface area of the frustum of a right circular cone. Consider the line segment in Figure 6.42, where L is the length of the line segment, r_1 is the radius at the left end of the line segment, and r_2 is the radius at the right end of the line segment. When the line segment is revolved about its axis of revolution, it forms a frustum of a right circular cone, with

$$S = 2\pi r L \qquad \text{Lateral surface area of frustum}$$

where

$$r = \frac{1}{2}(r_1 + r_2). \qquad \text{Average radius of frustum}$$

Figure 6.42

(In Exercise 55, you are asked to verify the formula for S.)

Suppose the graph of a function f, having a continuous derivative on the interval $[a, b]$, is revolved about the x-axis to form a surface of revolution, as shown in Figure 6.43. Let Δ be a partition of $[a, b]$, with subintervals of width Δx_i. Then the line segment of length

$$\Delta L_i = \sqrt{\Delta x_i^2 + \Delta y_i^2}$$

generates a frustum of a cone. Let r_i be the average radius of this frustum. By the Intermediate Value Theorem, a point d_i exists (in the ith subinterval) such that $r_i = f(d_i)$. The lateral surface area ΔS_i of the frustum is

$$\begin{aligned}
\Delta S_i &= 2\pi r_i \Delta L_i \\
&= 2\pi f(d_i)\sqrt{\Delta x_i^2 + \Delta y_i^2} \\
&= 2\pi f(d_i)\sqrt{1 + \left(\frac{\Delta y_i}{\Delta x_i}\right)^2}\, \Delta x_i.
\end{aligned}$$

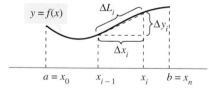

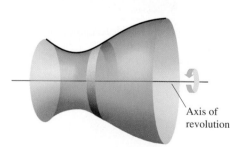

Figure 6.43

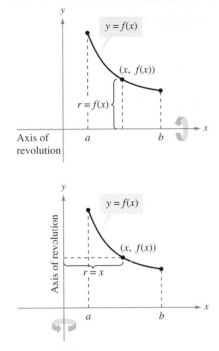

Figure 6.44

By the Mean Value Theorem, a point c_i exists in (x_{i-1}, x_i) such that

$$f'(c_i) = \frac{f(x_i) - f(x_{i-1})}{x_i - x_{i-1}} = \frac{\Delta y_i}{\Delta x_i}.$$

Therefore, $\Delta S_i = 2\pi f(d_i)\sqrt{1 + [f'(c_i)]^2}\, \Delta x_i$, and the total surface area can be approximated by

$$S \approx 2\pi \sum_{i=1}^{n} f(d_i)\sqrt{1 + [f'(c_i)]^2}\, \Delta x_i.$$

It can be shown that the limit of the right side as $\|\Delta\| \to 0$ $(n \to \infty)$, is

$$S = 2\pi \int_a^b f(x)\sqrt{1 + [f'(x)]^2}\, dx.$$

In a similar manner, if the graph of f is revolved about the y-axis, then S is

$$S = 2\pi \int_a^b x\sqrt{1 + [f'(x)]^2}\, dx.$$

In both formulas for S, you can regard the products $2\pi f(x)$ and $2\pi x$ as the circumference of the circle traced by a point (x, y) on the graph of f as it is revolved about the x- or y-axis (Figure 6.44). In one case the radius is $r = f(x)$, and in the other case the radius is $r = x$. Moreover, by appropriately adjusting r, you can generalize the formula for surface area to cover *any* horizontal or vertical axis of revolution, as indicated in the following definition.

Definition of the Area of a Surface of Revolution

Let $y = f(x)$ have a continuous derivative on the interval $[a, b]$. The area S of the surface of revolution formed by revolving the graph of f about a horizontal or vertical axis is

$$S = 2\pi \int_a^b r(x)\sqrt{1 + [f'(x)]^2}\, dx \qquad y \text{ is a function of } x.$$

where $r(x)$ is the distance between the graph of f and the axis of revolution. If $x = g(y)$ on the interval $[c, d]$, then the surface area is

$$S = 2\pi \int_c^d r(y)\sqrt{1 + [g'(y)]^2}\, dy \qquad x \text{ is a function of } y.$$

where $r(y)$ is the distance between the graph of g and the axis of revolution.

The formulas in this definition are sometimes written as

$$S = 2\pi \int_a^b r(x)\, ds \qquad y \text{ is a function of } x.$$

and

$$S = 2\pi \int_c^d r(y)\, ds \qquad x \text{ is a function of } y.$$

where $ds = \sqrt{1 + [f'(x)]^2}\, dx$ and $ds = \sqrt{1 + [g'(y)]^2}\, dy$, respectively.

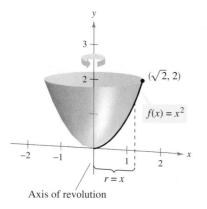

Figure 6.45

Example 6 **The Area of a Surface of Revolution**

Find the area of the surface formed by revolving the graph of

$$f(x) = x^3$$

on the interval $[0, 1]$ about the x-axis, as shown in Figure 6.45.

Solution The distance between the x-axis and the graph of f is $r(x) = f(x)$, and because $f'(x) = 3x^2$, the surface area is

$$S = 2\pi \int_a^b r(x)\sqrt{1 + [f'(x)]^2}\, dx \qquad \text{Formula for surface area}$$

$$= 2\pi \int_0^1 x^3 \sqrt{1 + (3x^2)^2}\, dx$$

$$= \frac{2\pi}{36} \int_0^1 (36x^3)(1 + 9x^4)^{1/2}\, dx \qquad \text{Simplify.}$$

$$= \frac{\pi}{18} \left[\frac{(1 + 9x^4)^{3/2}}{3/2} \right]_0^1 \qquad \text{Integrate.}$$

$$= \frac{\pi}{27} (10^{3/2} - 1)$$

$$\approx 3.563.$$

Example 7 **The Area of a Surface of Revolution**

Find the area of the surface formed by revolving the graph of

$$f(x) = x^2$$

on the interval $\left[0, \sqrt{2}\right]$ about the y-axis, as shown in Figure 6.46.

Solution In this case, the distance between the graph of f and the y-axis is $r(x) = x$. Using $f'(x) = 2x$, you can determine that the surface area is

$$S = 2\pi \int_a^b r(x)\sqrt{1 + [f'(x)]^2}\, dx \qquad \text{Formula for surface area}$$

$$= 2\pi \int_0^{\sqrt{2}} x\sqrt{1 + (2x)^2}\, dx$$

$$= \frac{2\pi}{8} \int_0^{\sqrt{2}} (1 + 4x^2)^{1/2}(8x)\, dx \qquad \text{Simplify.}$$

$$= \frac{\pi}{4} \left[\frac{(1 + 4x^2)^{3/2}}{3/2} \right]_0^{\sqrt{2}} \qquad \text{Integrate.}$$

$$= \frac{\pi}{6} \left[(1 + 8)^{3/2} - 1 \right]$$

$$= \frac{13\pi}{3}.$$

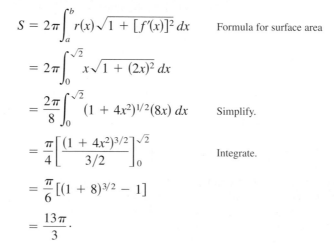

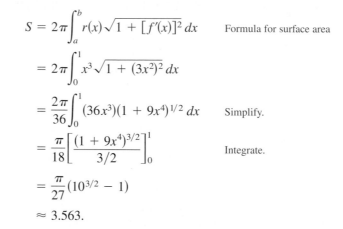

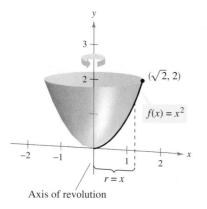

Figure 6.46

EXERCISES FOR SECTION 6.4

In Exercises 1 and 2, find the distance between the points by using (a) the Distance Formula and (b) integration.

1. $(0, 0)$, $(5, 12)$

2. $(1, 2)$, $(7, 10)$

In Exercises 3–10, find the arc length of the graph of the function over the indicated interval.

3. $y = \frac{2}{3}x^{3/2} + 1$, $[0, 1]$

4. $y = 2x^{3/2} + 3$, $[0, 9]$

5. $y = \frac{3}{2}x^{2/3}$, $[1, 8]$

6. $y = \frac{3}{2}x^{2/3} + 4$, $[1, 27]$

7. $y = \frac{x^4}{8} + \frac{1}{4x^2}$, $[1, 2]$

8. $y = \frac{x^5}{10} + \frac{1}{6x^3}$, $[1, 2]$

9. $y = \ln(\sin x)$, $\left[\frac{\pi}{4}, \frac{3\pi}{4}\right]$

10. $y = \frac{1}{2}(e^x + e^{-x})$, $[0, 2]$

In Exercises 11–20, (a) graph the function, highlighting the part indicated by the given interval, (b) find a definite integral that represents the arc length of the curve over the indicated interval and observe that the integral cannot be evaluated with the techniques studied thus far, and (c) use the integration capabilities of a graphing utility to approximate the arc length.

11. $y = 4 - x^2$, $[0, 2]$

12. $y = x^2 + x - 2$, $[-2, 1]$

13. $y = \frac{1}{x}$, $[1, 3]$

14. $y = \frac{1}{x + 1}$, $[0, 1]$

15. $y = \sin x$, $[0, \pi]$

16. $y = \cos x$, $\left[-\frac{\pi}{2}, \frac{\pi}{2}\right]$

17. $x = e^{-y}$, $[0, 2]$

18. $y = \ln x$, $[1, 5]$

19. $y = 2 \arctan x$, $[0, 1]$

20. $x = \sqrt{36 - y^2}$, $[0, 3]$

Approximation   **In Exercises 21 and 22, determine which value best approximates the length of the arc represented by the integral. (Make your selection on the basis of a sketch of the arc and *not* by performing any calculations.)**

21. $\displaystyle\int_0^2 \sqrt{1 + \left[\frac{d}{dx}\left(\frac{5}{x^2 + 1}\right)\right]^2}\, dx$

(a) 25   (b) 5   (c) 2   (d) -4 (e) 3

22. $\displaystyle\int_0^{\pi/4} \sqrt{1 + \left[\frac{d}{dx}(\tan x)\right]^2}\, dx$

(a) 3   (b) -2 (c) 4   (d) $\frac{4\pi}{3}$ (e) 1

Approximation   **In Exercises 23 and 24, approximate the arc length of the graph of the function over the interval $[0, 4]$ in four ways. (a) Use the Distance Formula to find the distance between the endpoints of the arc. (b) Use the Distance Formula to find the lengths of the four line segments connecting the points on the arc when $x = 0$, $x = 1$, $x = 2$, $x = 3$, and $x = 4$. Find the sum of the four lengths. (c) Use Simpson's Rule with $n = 10$ to approximate the integral yielding the indicated arc length. (d) Use the integration capabilities of a graphing utility to approximate the integral yielding the indicated arc length.**

23. $f(x) = x^3$

24. $f(x) = (x^2 - 4)^2$

25. *Think About It*   The figure shows the graphs of the functions $y_1 = x$, $y_2 = \frac{1}{2}x^{3/2}$, $y_3 = \frac{1}{4}x^2$, and $y_4 = \frac{1}{8}x^{5/2}$ on the interval $[0, 4]$. To print an enlarged copy of the graph, go to the website *www.mathgraphs.com*.

(a) Label the functions.

(b) List the functions in order of increasing arc length.

(c) Verify your answer in part (b) by approximating each arc length accurate to three decimal places.

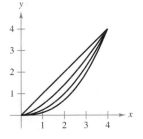

26. *Think About It*   Explain why the two integrals are equal.

$$\int_1^e \sqrt{1 + \frac{1}{x^2}}\, dx = \int_0^1 \sqrt{1 + e^{2x}}\, dx$$

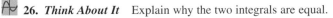

Use the integration capabilities of a graphing utility to verify that the integrals are equal.

27. *Length of Pursuit*   A fleeing object leaves the origin and moves up the y-axis (see figure). At the same time, a pursuer leaves the point $(1, 0)$ and always moves toward the fleeing object. If the pursuer's speed is twice that of the fleeing object, the equation of the path is

$$y = \frac{1}{3}(x^{3/2} - 3x^{1/2} + 2).$$

How far has the fleeing object traveled when it is caught? Show that the pursuer has traveled twice as far.

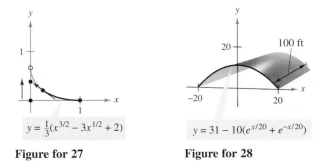

$y = \frac{1}{3}(x^{3/2} - 3x^{1/2} + 2)$ $y = 31 - 10(e^{x/20} + e^{-x/20})$

Figure for 27   **Figure for 28**

28. *Roof Area*   A barn is 100 feet long and 40 feet wide (see figure). A cross section of the roof is the inverted catenary

$$y = 31 - 10(e^{x/20} + e^{-x/20}).$$

Find the number of square feet of roofing on the barn.

29. *Length of a Catenary* Electrical wires suspended between two towers form a catenary (see figure) modeled by the equation

$$y = 20 \cosh \frac{x}{20}, \quad -20 \le x \le 20$$

where x and y are measured in meters. Find the length of the suspended cable if the towers are 40 meters apart.

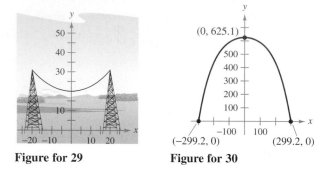

Figure for 29   **Figure for 30**

30. *Length of Gateway Arch* The Gateway Arch in St. Louis, Missouri, is modeled by

$$y = 693.8597 - 68.7672 \cosh 0.0100333x,$$
$$-299.2239 \le x \le 299.2239.$$

(See the Section Project: St. Louis Arch.) Find the length of this curve (see figure).

31. Find the arc length from $(0, 3)$ clockwise to $\left(2, \sqrt{5}\right)$ along the circle $x^2 + y^2 = 9$.

32. Find the arc length from $(-3, 4)$ clockwise to $(4, 3)$ along the circle $x^2 + y^2 = 25$. Show that the result is one-fourth the circumference of the circle.

In Exercises 33–36, set up and evaluate the definite integral for the area of the surface generated by revolving the curve about the x-axis.

33. $y = \frac{1}{3}x^3, \quad [0, 3]$ **34.** $y = 2\sqrt{x}, \quad [4, 9]$

35. $y = \frac{x^3}{6} + \frac{1}{2x}, \quad [1, 2]$ **36.** $y = \frac{x}{2}, \quad [0, 6]$

In Exercises 37 and 38, set up and evaluate the definite integral that gives the area of the surface of revolution generated by revolving the curve about the y-axis.

Function	Interval
37. $y = \sqrt[3]{x} + 2$	$[1, 8]$
38. $y = 9 - x^2$	$[0, 3]$

In Exercises 39 and 40, use the integration capabilities of a graphing utility to approximate the surface area of the solid of revolution.

Function	Interval
39. $y = \sin x$	$[0, \pi]$
revolved about the x-axis	
40. $y = \ln x$	$[1, e]$
revolved about the y-axis	

Getting at the Concept

41. Define a rectifiable curve.

42. What precalculus formula and representative element are used to develop the integration formula for arc length?

43. What precalculus formula and representative element are used to develop the integration formula for the area of a surface of revolution?

44. The graphs of the functions f_1 and f_2 on the interval $[a, b]$ are shown in the figure. The graph of each is revolved about the x-axis. Which surface of revolution has the greater surface area? Explain.

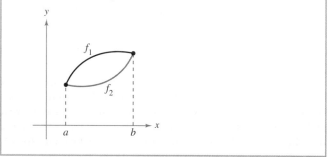

45. A right circular cone is generated by revolving the region bounded by $y = hx/r$, $y = h$, and $x = 0$ about the y-axis. Verify that the lateral surface area of the cone is

$$S = \pi r \sqrt{r^2 + h^2}.$$

46. A sphere of radius r is generated by revolving the graph of $y = \sqrt{r^2 - x^2}$ about the x-axis. Verify that the surface area of the sphere is $4\pi r^2$.

47. Find the area of the zone of a sphere formed by revolving the graph of $y = \sqrt{9 - x^2}$, $0 \le x \le 2$, about the y-axis.

48. Find the area of the zone of a sphere formed by revolving the graph of $y = \sqrt{r^2 - x^2}$, $0 \le x \le a$, about the y-axis. Assume that $a < r$.

49. *Bulb Design* An ornamental light bulb is designed by revolving the graph of

$$y = \frac{1}{3}x^{1/2} - x^{3/2}, \quad 0 \le x \le \frac{1}{3}$$

about the x-axis, where x and y are measured in feet (see figure). Find the surface area of the bulb and use the result to approximate the amount of glass needed to make the bulb. (Assume that the glass is 0.015 inch thick.)

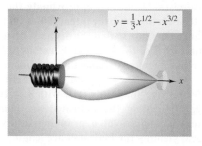

50. Modeling Data The circumference C (in inches) of a vase is measured at 3-inch intervals starting at its base. The measurements are shown in the table, where y is the vertical distance in inches from the base.

y	0	3	6	9	12	15	18
C	50	65.5	70	66	58	51	48

(a) Use the data to approximate the volume of the vase by summing the volumes of approximating disks.

(b) Use the data to approximate the outside surface area (excluding the base) of the vase by summing the outside surface areas of approximating frustums of right circular cones.

(c) Use the regression capabilities of a graphing utility to find a cubic model for the points (y, r) where $r = C/(2\pi)$. Use the graphing utility to plot the points and graph the model.

(d) Use the model in part (c) and the integration capabilities of a graphing utility to approximate the volume and outside surface area of the vase. Compare the results with your answers in parts (a) and (b).

51. Modeling Data Property bounded by two perpendicular roads and a stream is shown in the figure. All distances are measured in feet.

(a) Use the regression capabilities of a graphing utility to fit a fourth-degree polynomial to the path of the stream.

(b) Use the model of part (a) to approximate the area of the property in acres.

(c) Use the integration capabilities of a graphing utility to find the length of the stream that bounds the property.

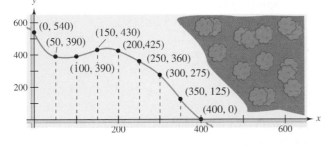

52. Individual Project Select a solid of revolution from everyday life. Measure the radius of the solid at a minimum of seven points along its axis. Use the data to approximate the volume of the solid and the surface area of the lateral sides of the solid.

53. Let R be the region bounded by $y = 1/x$, the x-axis, $x = 1$, and $x = b$, where $b > 1$. Let D be the solid formed when R is revolved about the x-axis.

(a) Find the volume V of D.

(b) Express the surface area S as an integral.

(c) Show that V approaches a finite limit as $b \to \infty$.

(d) Show that $S \to \infty$ as $b \to \infty$.

54. Think About It Consider the equation $\dfrac{x^2}{9} + \dfrac{y^2}{4} = 1$.

(a) Use a graphing utility to graph the equation.

(b) Set up the definite integral for finding the first quadrant arc length of the graph in part (a).

(c) Compare the interval of integration in part (b) and the domain of the integrand. Is it possible to evaluate the definite integral? Is it possible to use Simpson's Rule to evaluate the definite integral? Explain. (You will learn how to evaluate this type of integral in Section 7.8.)

55. (a) Given a circular sector with radius L and central angle θ (see figure), show that the area of the sector is given by

$$S = \frac{1}{2}L^2\theta.$$

(b) By joining the straight line edges of the sector in part (a), a right circular cone is formed (see figure) and the lateral surface area of the cone is the same as the area of the sector. Show that the area is

$$S = \pi r L$$

where r is the radius of the base of the cone. (*Hint:* The arc length of the sector equals the circumference of the base of the cone.)

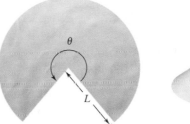

Figure for 55(a)   Figure for 55(b)

(c) Use the result in part (b) to verify that the formula for the lateral surface area of the frustum of a cone with slant height L and radii r_1 and r_2 (see figure) is

$$S = \pi(r_1 + r_2)L.$$

(*Note:* This formula was used to develop the integral for finding the surface area of a surface of revolution.)

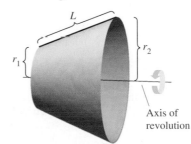

56. Writing Read the article "Arc Length, Area and the Arcsine Function" by Andrew M. Rockett in *Mathematics Magazine*. Then write a paragraph explaining how the arcsine function can be defined in terms of an arc length. (To view this article, go to the website *www.matharticles.com*.)

- Find the work done by a constant force.
- Find the work done by a variable force.

Work Done by a Constant Force

The concept of work is important to scientists and engineers for determining the energy needed to perform various jobs. For instance, it is useful to know the amount of work done when a crane lifts a steel girder, when a spring is compressed, when a rocket is propelled into the air, or when a truck pulls a load along a highway.

In general, we say that **work** is done by a force when it moves an object. If the force applied to the object is *constant*, we have the following definition of work.

Definition of Work Done by a Constant Force

If an object is moved a distance D in the direction of an applied constant force F, then the **work** W done by the force is defined as $W = FD$.

There are many types of forces—centrifugal, electromotive, and gravitational, to name a few. A **force** can be thought of as a *push* or a *pull*; a force changes the state of rest or state of motion of a body. For gravitational forces on earth, it is common to use units of measure corresponding to the weight of an object.

Example 1 **Lifting an Object**

Determine the work done in lifting a 50-pound object 4 feet.

Solution The magnitude of the required force F is the weight of the object, as shown in Figure 6.47. So, the work done in lifting the object 4 feet is

$W = FD$ Work = (force)(distance)

$= 50(4)$ Force = 50 pounds, distance = 4 feet

$= 200$ foot-pounds.

In the U.S. measurement system, work is typically expressed in foot-pounds (ft · lb), inch-pounds, or foot-tons. In the centimeter-gram-second (C-G-S) system, the basic unit of force is the **dyne**—the force required to produce an acceleration of 1 centimeter per second per second on a mass of 1 gram. In this system, work is typically expressed in dyne-centimeters (ergs) or newton-meters (joules), where 1 joule = 10^7 ergs.

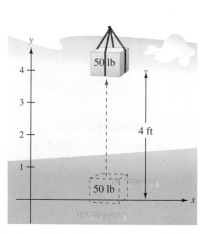

The work done in lifting a 50-pound object 4 feet is 200 foot-pounds.

Figure 6.47

EXPLORATION

How Much Work? In Example 1, 200 foot-pounds of work were needed to lift the 50-pound object 4 feet vertically off the ground. Suppose that once you lifted the object, you held it and walked a horizontal distance of 4 feet. Would this require an additional 200 foot-pounds of work? Explain your reasoning.

Work Done by a Variable Force

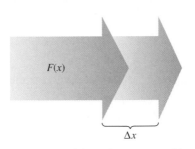

The amount of force changes as an object changes position (Δx).

Figure 6.48

In Example 1, the force involved is *constant*. If a *variable* force is applied to an object, calculus is needed to determine the work done, because the amount of force changes as the object changes position. For instance, the force required to compress a spring increases as the spring is compressed.

Suppose that an object is moved along a straight line from $x = a$ to $x = b$ by a continuously varying force $F(x)$. Let Δ be a partition that divides the interval $[a, b]$ into n subintervals determined by

$$a = x_0 < x_1 < x_2 < \cdots < x_n = b$$

and let $\Delta x_i = x_i - x_{i-1}$. For each i, choose c_i such that $x_{i-1} \le c_i \le x_i$. Then at c_i the force is given by $F(c_i)$. Because F is continuous, you can approximate the work done in moving the object through the ith subinterval by the increment

$$\Delta W_i = F(c_i)\,\Delta x_i$$

as shown in Figure 6.48. So, the total work done as the object moves from a to b is approximated by

$$W \approx \sum_{i=1}^{n} \Delta W_i$$

$$= \sum_{i=1}^{n} F(c_i)\,\Delta x_i.$$

This approximation appears to become better and better as $\|\Delta\| \to 0$ $(n \to \infty)$. Therefore, we define the work to be

$$W = \lim_{\|\Delta\| \to 0} \sum_{i=1}^{n} F(c_i)\,\Delta x_i$$

$$= \int_{a}^{b} F(x)\,dx.$$

Definition of Work Done by a Variable Force

If an object is moved along a straight line by a continuously varying force $F(x)$, then the **work** W done by the force as the object is moved from $x = a$ to $x = b$ is

$$W = \lim_{\|\Delta\| \to 0} \sum_{i=1}^{n} \Delta W_i$$

$$= \int_{a}^{b} F(x)\,dx.$$

EMILIE DE BRETEUIL (1706–1749)

Another major work by de Breteuil was the translation of Newton's "Philosophiae Naturalis Principia Mathematica" into French. Her translation and commentary greatly contributed to the acceptance of Newtonian science in Europe.

The remaining examples in this section use some well-known physical laws. The discoveries of many of these laws occurred during the same period in which calculus was being developed. In fact, during the seventeenth and eighteenth centuries, there was little difference between physicists and mathematicians. One such physicist-mathematician was Emilie de Breteuil. Breteuil was instrumental in synthesizing the work of many other scientists, including Newton, Leibniz, Huygens, Kepler, and Descartes. Her physics text *Institutions* was widely used for many years.

The following three laws of physics were developed by Robert Hooke (1635–1703), Isaac Newton (1642–1727), and Charles Coulomb (1736–1806).

1. **Hooke's Law:** The force F required to compress or stretch a spring (within its elastic limits) is proportional to the distance d that the spring is compressed or stretched from its original length. That is,

$$F = kd$$

where the constant of proportionality k (the spring constant) depends on the specific nature of the spring.

2. **Newton's Law of Universal Gravitation:** The force F of attraction between two particles of masses m_1 and m_2 is proportional to the product of the masses and inversely proportional to the square of the distance d between the two particles. That is,

$$F = k \frac{m_1 m_2}{d^2}.$$

If m_1 and m_2 are given in grams and d in centimeters, F will be in dynes for a value of $k = 6.670 \times 10^{-8}$ cubic centimeter per gram-second squared.

3. **Coulomb's Law:** The force between two charges q_1 and q_2 in a vacuum is proportional to the product of the charges and inversely proportional to the square of the distance d between the two charges. That is,

$$F = k \frac{q_1 q_2}{d^2}.$$

If q_1 and q_2 are given in electrostatic units and d in centimeters, F will be in dynes for a value of $k = 1$.

EXPLORATION

The work done in compressing the spring in Example 2 from $x = 3$ inches to $x = 6$ inches is 3375 inch-pounds. Should the work done in compressing the spring from $x = 0$ inches to $x = 3$ inches be more than, the same as, or less than this? Explain.

⚏ *Example 2* **Compressing a Spring**

A force of 750 pounds compresses a spring 3 inches from its natural length of 15 inches. Find the work done in compressing the spring an additional 3 inches.

Solution By Hooke's Law, the force $F(x)$ required to compress the spring x units (from its natural length) is $F(x) = kx$. Using the given data, it follows that $F(3) = 750 = (k)(3)$ and thus $k = 250$ and $F(x) = 250x$, as shown in Figure 6.49. To find the increment of work, assume that the force required to compress the spring over a small increment Δx is nearly constant. So, the increment of work is

$$\Delta W = (\text{force})(\text{distance increment}) = (250x)\,\Delta x.$$

Because the spring is compressed from $x = 3$ to $x = 6$ inches less than its natural length, the work required is

$$W = \int_a^b F(x)\,dx = \int_3^6 250x\,dx \qquad \text{Formula for work}$$

$$= 125x^2 \Big]_3^6 = 4500 - 1125 = 3375 \text{ inch-pounds.}$$

Note that you do *not* integrate from $x = 0$ to $x = 6$ because you were asked to determine the work done in compressing the spring an *additional* 3 inches (not including the first 3 inches). ⚏

Natural length ($F = 0$)

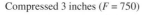

Compressed 3 inches ($F = 750$)

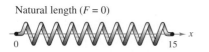

Compressed x inches ($F = 250x$)

Figure 6.49

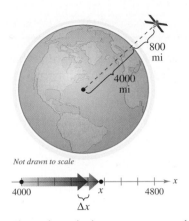

Not drawn to scale

The work required to move a space module 800 miles above earth is approximately 1.056×10^{11} foot-pounds.

Figure 6.50

Example 3   **Moving a Space Module into Orbit**

A space module weighs 15 tons on the surface of earth. How much work is done in propelling the module to a height of 800 miles above earth, as shown in Figure 6.50? (Use 4000 miles as the radius of earth. Do not consider the effect of air resistance or the weight of the propellant.)

Solution   Because the weight of a body varies inversely as the square of its distance from the center of earth, the force $F(x)$ exerted by gravity is

$$F(x) = \frac{C}{x^2}. \qquad \textit{C is the constant of proportionality.}$$

Because the module weighs 15 tons on the surface of earth and the radius of earth is approximately 4000 miles, you have

$$15 = \frac{C}{(4000)^2}$$

$$240,000,000 = C.$$

So, the increment of work is

$$\Delta W = (\text{force})(\text{distance increment})$$

$$= \frac{240,000,000}{x^2} \Delta x.$$

Finally, because the module is propelled from $x = 4000$ to $x = 4800$ miles, the total work done is

$$W = \int_a^b F(x)\, dx = \int_{4000}^{4800} \frac{240,000,000}{x^2}\, dx \qquad \text{Formula for work}$$

$$= \frac{-240,000,000}{x} \Big]_{4000}^{4800} \qquad \text{Integrate.}$$

$$= -50,000 + 60,000$$

$$= 10,000 \ \text{ mile-tons}$$

$$\approx 1.056 \times 10^{11} \ \text{ foot-pounds.}$$

In the C-G-S system, using a conversion factor of 1 foot-pound ≈ 1.35582 joules, the work done is

$$W \approx 1.432 \times 10^{11} \ \text{joules.} \qquad \blacksquare$$

The solutions to Examples 2 and 3 conform to our development of work as the summation of increments in the form

$$\Delta W = (\text{force})(\text{distance increment}) = (F)(\Delta x).$$

Another way to formulate the increment of work is

$$\Delta W = (\text{force increment})(\text{distance}) = (\Delta F)(x).$$

This second interpretation of ΔW is useful in problems involving the movement of nonrigid substances such as fluids and chains.

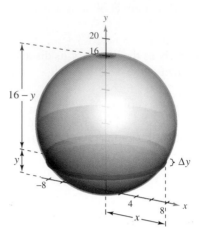

The work required to pump oil out through a hole in the top of the tank is approximately 589,782 foot-pounds.
Figure 6.51

Example 4   **Emptying a Tank of Oil**

A spherical tank of radius 8 feet is half full of oil that weighs 50 pounds per cubic foot. Find the work required to pump oil out through a hole in the top of the tank.

Solution   Consider the oil to be subdivided into disks of thickness Δy and radius x, as shown in Figure 6.51. Because the increment of force for each disk is given by its weight, you have

$$\Delta F = \text{weight}$$
$$= \left(\frac{50 \text{ pounds}}{\text{cubic foot}}\right)(\text{volume})$$
$$= 50(\pi x^2 \Delta y) \text{ pounds.}$$

For a circle of radius 8 and center at $(0, 8)$, you have

$$x^2 + (y - 8)^2 = 8^2$$
$$x^2 = 16y - y^2$$

and you can write the force increment as

$$\Delta F = 50(\pi x^2 \Delta y)$$
$$= 50\pi(16y - y^2)\,\Delta y.$$

In Figure 6.51, note that a disk y feet from the bottom of the tank must be moved a distance of $(16 - y)$ feet. Therefore, the increment of work is

$$\Delta W = \Delta F(16 - y)$$
$$= 50\pi(16y - y^2)\,\Delta y(16 - y)$$
$$= 50\pi(256y - 32y^2 + y^3)\,\Delta y.$$

Because the tank is half full, y ranges from 0 to 8, and the work required to empty the tank is

$$W = \int_0^8 50\pi(256y - 32y^2 + y^3)\,dy$$
$$= 50\pi\left[128y^2 - \frac{32}{3}y^3 + \frac{y^4}{4}\right]_0^8$$
$$= 50\pi\left(\frac{11{,}264}{3}\right)$$
$$\approx 589{,}782 \text{ foot-pounds.}$$

To estimate the reasonableness of the result in Example 4, consider that the weight of the oil in the tank is

$$\left(\frac{1}{2}\right)(\text{volume})(\text{density}) = \frac{1}{2}\left(\frac{4}{3}\pi 8^3\right)(50)$$
$$\approx 53{,}616.5 \text{ pounds.}$$

Lifting the entire half-tank of oil 8 feet would involve work of $8(53{,}616.5) \approx 428{,}932$ foot-pounds. Because the oil is actually lifted between 8 and 16 feet, it seems reasonable that the work done is 589,782 foot-pounds.

The work required to raise one end of the chain 20 feet is 1000 foot-pounds.
Figure 6.52

Example 5   **Lifting a Chain**

A 20-foot chain weighing 5 pounds per foot is lying coiled on the ground. How much work is required to raise one end of the chain to a height of 20 feet so that it is fully extended, as shown in Figure 6.52?

Solution   Imagine that the chain is divided into small sections, each of length Δy. Then the weight of each section is the increment of force

$$\Delta F = (\text{weight}) = \left(\frac{5 \text{ pounds}}{\text{foot}}\right)(\text{length}) = 5\Delta y.$$

Because a typical section (initially on the ground) is raised to a height of y, the increment of work is

$$\Delta W = (\text{force increment})(\text{distance}) = (5\,\Delta y)y = 5y\,\Delta y.$$

Because y ranges from 0 to 20, the total work is

$$W = \int_0^{20} 5y\,dy = \frac{5y^2}{2}\Big]_0^{20} = \frac{5(400)}{2} = 1000 \text{ foot-pounds.}$$

In the next example we consider a piston of radius r in a cylindrical casing, as shown in Figure 6.53. As the gas in the cylinder expands, the piston moves and work is done. If p represents the pressure of the gas (in pounds per square foot) against the piston head and V represents the volume of the gas (in cubic feet), the work increment involved in moving the piston Δx feet is

$$\Delta W = (\text{force})(\text{distance increment}) = F(\Delta x) = p(\pi r^2)\,\Delta x = p\,\Delta V.$$

So, as the volume of the gas expands from V_0 to V_1, the work done in moving the piston is

$$W = \int_{V_0}^{V_1} p\,dV.$$

Assuming the pressure of the gas to be inversely proportional to its volume, we have $p = k/V$ and the integral for work becomes

$$W = \int_{V_0}^{V_1} \frac{k}{V}\,dV.$$

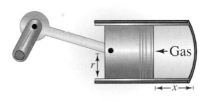

Work done by expanding gas
Figure 6.53

Example 6   **Work Done by an Expanding Gas**

A quantity of gas with an initial volume of 1 cubic foot and a pressure of 500 pounds per square foot expands to a volume of 2 cubic feet. Find the work done by the gas. (Assume that the pressure is inversely proportional to the volume.)

Solution   Because $p = k/V$ and $p = 500$ when $V = 1$, we have $k = 500$. So, the work is

$$W = \int_{V_0}^{V_1} \frac{k}{V}\,dV = \int_1^2 \frac{500}{V}\,dV = 500 \ln|V|\,\Big]_1^2 \approx 346.6 \text{ foot-pounds.}$$

EXERCISES FOR SECTION 6.5

Constant Force   **In Exercises 1–4, determine the work done by the constant force.**

1. A 100-pound bag of sugar is lifted 10 feet.

2. An electric hoist lifts a 2800-pound car 4 feet.

3. A force of 112 newtons is required to slide a cement block 4 meters in a construction project.

4. The locomotive of a freight train pulls its cars with a constant force of 9 tons a distance of one-half mile.

Getting at the Concept

5. State the definition of work done by a constant force.

6. State the definition of work done by a variable force.

7. The graphs show the force F_i (in pounds) required to move an object 9 feet along the x-axis. Order the force functions from the one that yields the least work to the one that yields the most work without doing any calculations.

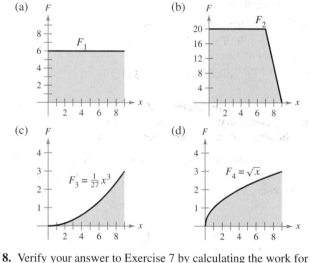

8. Verify your answer to Exercise 7 by calculating the work for each force function.

Hooke's Law   **In Exercises 9–16, use Hooke's Law to determine the variable force in the spring problem.**

9. A force of 5 pounds compresses a 15-inch spring a total of 4 inches. How much work is done in compressing the spring 7 inches?

10. How much work is done in compressing the spring in Exercise 9 from a length of 10 inches to a length of 6 inches?

11. A force of 250 newtons stretches a spring 30 centimeters. How much work is done in stretching the spring from 20 centimeters to 50 centimeters?

12. A force of 800 newtons stretches a spring 70 centimeters on a mechanical device for driving fence posts. Find the work done in stretching the spring the required 70 centimeters.

13. A force of 20 pounds stretches a spring 9 inches in an exercise machine. Find the work done in stretching the spring 1 foot from its natural position.

14. An overhead garage door has two springs, one on each side of the door. A force of 15 pounds is required to stretch each spring 1 foot. Because of the pulley system, the springs stretch only one-half the distance the door travels. Find the work done by the pair of springs if the door moves a total of 8 feet and the springs are at their natural length when the door is open.

15. Eighteen foot-pounds of work is required to stretch a spring 4 inches from its natural length. Find the work required to stretch the spring an additional 3 inches.

16. Seven and one-half foot-pounds of work is required to compress a spring 2 inches from its natural length. Find the work required to compress the spring an additional one-half inch.

17. *Propulsion*   Neglecting air resistance and the weight of the propellant, determine the work done in propelling a 5-ton satellite to a height of

   (a) 100 miles above earth.

   (b) 300 miles above earth.

18. *Propulsion*   Use the information in Exercise 17 to write the work W of the propulsion system as a function of the height h of the satellite above earth. Find the limit (if it exists) of W as h approaches infinity.

19. *Propulsion*   Neglecting air resistance and the weight of the propellant, determine the work done in propelling a 10-ton satellite to a height of

   (a) 11,000 miles above earth.

   (b) 22,000 miles above earth.

20. *Propulsion*   A lunar module weighs 12 tons on the surface of earth. How much work is done in propelling the module from the surface of the moon to a height of 50 miles? Consider the radius of the moon to be 1100 miles and its force of gravity to be one-sixth that of earth.

21. *Pumping Water*   A rectangular tank with a base 4 feet by 5 feet and a height of 4 feet is full of water (see figure). The water weighs 62.4 pounds per cubic foot. How much work is done in pumping water out over the top edge in order to empty (a) half of the tank? (b) all of the tank?

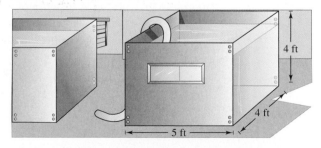

22. *Think About It*   Explain why the answer in part (b) of Exercise 21 is not twice the answer in part (a).

23. *Pumping Water* A cylindrical water tank 4 meters high with a radius of 2 meters is buried so that the top of the tank is 1 meter below ground level (see figure). How much work is done in pumping a full tank of water up to ground level? (The water weighs 9800 newtons per cubic meter.)

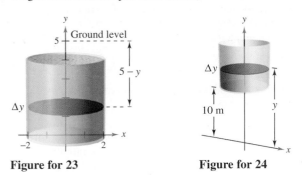

Figure for 23   **Figure for 24**

24. *Pumping Water* Suppose the tank in Exercise 23 is located on a tower so that the bottom of the tank is 10 meters above the level of a stream (see figure). How much work is done in filling the tank half full of water through a hole in the bottom, using water from the stream?

25. *Pumping Water* An open tank has the shape of a right circular cone (see figure). The tank is 8 feet across the top and 6 feet high. How much work is done in emptying the tank by pumping the water over the top edge?

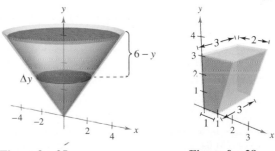

Figure for 25   **Figure for 28**

26. *Pumping Water* If water is pumped in through the bottom of the tank in Exercise 25, how much work is done to fill the tank

(a) to a depth of 2 feet?

(b) from a depth of 4 feet to a depth of 6 feet?

27. *Pumping Water* A hemispherical tank of radius 6 feet is positioned so that its base is circular. How much work is required to fill the tank with water through a hole in the base if the water source is at the base?

28. *Pumping Diesel Fuel* The fuel tank on a truck has trapezoidal cross sections with dimensions (in feet) shown in the figure. Assume that an engine is approximately 3 feet above the top of the fuel tank and that diesel fuel weighs approximately 55.6 pounds per cubic foot. Find the work done by the fuel pump in raising a full tank of fuel to the level of the engine.

Pumping Gasoline **In Exercises 29 and 30, find the work done in pumping gasoline that weighs 42 pounds per cubic foot. (*Hint:* Evaluate one integral by a geometric formula and the other by observing that the integrand is an odd function.)**

29. A cylindrical gasoline tank 3 feet in diameter and 4 feet long is carried on the back of a truck and is used to fuel tractors. The axis of the tank is horizontal. Find the work done in pumping the entire contents of the fuel tank into a tractor if the opening on the tractor tank is 5 feet above the top of the tank in the truck.

30. The top of a cylindrical storage tank for gasoline at a service station is 4 feet below ground level. The axis of the tank is horizontal and its diameter and length are 5 feet and 12 feet, respectively. Find the work done in pumping the entire contents of the full tank to a height of 3 feet above ground level.

Lifting a Chain **In Exercises 31–34, consider a 15-foot chain hanging from a winch 15 feet above ground level. Find the work done by the winch in winding up the specified amount of chain, if the chain weighs 3 pounds per foot.**

31. Wind up the entire chain.

32. Wind up one-third of the chain.

33. Run the winch until the bottom of the chain is at the 10-foot level.

34. Wind up the entire chain with a 500-pound load attached to it.

Lifting a Chain **In Exercises 35 and 36, consider a 15-foot hanging chain that weighs 3 pounds per foot. Find the work done in lifting the chain vertically to the indicated position.**

35. Take the bottom of the chain and raise it to the 15-foot level, leaving the chain doubled and still hanging vertically (see figure).

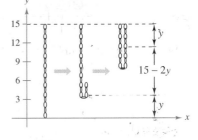

36. Repeat Exercise 35 raising the bottom of the chain to the 12-foot level.

Demolition Crane **In Exercises 37 and 38, consider a demolition crane with a 500-pound ball suspended from a 40-foot cable that weighs 1 pound per foot.**

37. Find the work required to wind up 15 feet of the apparatus.

38. Find the work required to wind up all 40 feet of the apparatus.

Boyle's Law In Exercises 39 and 40, find the work done by the gas for the given volume and pressure. Assume that the pressure is inversely proportional to the volume. (See Example 6.)

39. A quantity of gas with an initial volume of 2 cubic feet and a pressure of 1000 pounds per square foot expands to a volume of 3 cubic feet.

40. A quantity of gas with an initial volume of 1 cubic foot and a pressure of 2500 pounds per square foot expands to a volume of 3 cubic feet.

41. ***Electric Force*** Two electrons repel each other with a force that varies inversely as the square of the distance between them. If one electron is fixed at the point $(2, 4)$, find the work done in moving the second electron from $(-2, 4)$ to $(1, 4)$.

42. ***Modeling Data*** The hydraulic cylinder on a woodsplitter has a 4-inch bore (diameter) and a stroke of 2 feet. The hydraulic pump creates a maximum pressure of 2000 pounds per square inch. Therefore, the maximum force created by the cylinder is $2000(\pi 2^2) = 8000\pi$ pounds.

   (a) Find the work done through one extension of the cylinder given that the maximum force is required.

   (b) The force exerted in splitting a piece of wood is variable. Measurements of the force obtained when a piece of wood was split are shown in the table. The variable x measures the extension of the cylinder in feet, and F is the force in pounds. Use Simpson's Rule to approximate the work done in splitting the piece of wood.

x	0	$\frac{1}{3}$	$\frac{2}{3}$	1	$\frac{4}{3}$	$\frac{5}{3}$	2
$F(x)$	0	20,000	22,000	15,000	10,000	5000	0

Table for 42(b)

   (c) Use the regression capabilities of a graphing utility to find a fourth-degree polynomial model for the data. Plot the data and graph the model.

   (d) Use the model in part (c) to approximate the extension of the cylinder when the force is maximum.

   (e) Use the model in part (c) to approximate the work done in splitting the piece of wood.

Hydraulic Press In Exercises 43–46, use the integration capabilities of a graphing utility to approximate the work done by a press in a manufacturing process. A model for the variable force F (in pounds) and the distance x (in feet) the press moves is given.

Force	Interval
43. $F(x) = 1000[1.8 - \ln(x + 1)]$	$0 \le x \le 5$
44. $F(x) = \dfrac{e^{x^2} - 1}{100}$	$0 \le x \le 4$
45. $F(x) = 100x\sqrt{125 - x^3}$	$0 \le x \le 5$
46. $F(x) = 1000 \sinh x$	$0 \le x \le 2$

SECTION PROJECT **TIDAL ENERGY**

Tidal power plants use "tidal energy" to produce electrical energy. To construct a tidal power plant, a dam is built to separate a bay from the sea. Electrical energy is produced as the water flows back and forth between the bay and the sea. The amount of "natural energy" produced depends on the volume of the bay and the tidal range—the vertical distance between high and low tides. (Throughout the world, several natural bays have tidal ranges in excess of 15 feet; the Bay of Fundy in Nova Scotia has a tidal range of 47.5 feet.)

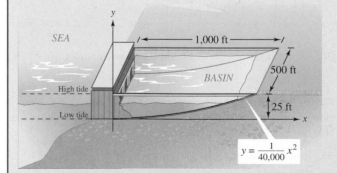

(a) Consider a basin with a rectangular base, as shown in the figure. The basin has a tidal range of 25 feet, with low tide corresponding to $y = 0$. How much water does the basin hold at high tide?

(b) The amount of energy produced during the filling (or the emptying) of the basin is proportional to the amount of work required to fill (or empty) the basin. How much work is required to fill the basin with seawater? (Use a seawater density of 64 pounds per cubic foot.)

The Bay of Fundy in Nova Scotia has an extreme tidal range, as displayed in the greatly contrasting photos above.

FOR FURTHER INFORMATION For more information on tidal power, see the article "LaRance: Six Years of Operating a Tidal Power Plant in France" by J. Cotillon in *Water Power Magazine*.

| **Section 6.6** | **Moments, Centers of Mass, and Centroids** |

- Understand the definition of mass.
- Find the center of mass in a one-dimensional system.
- Find the center of mass in a two-dimensional system.
- Find the center of mass of a planar lamina.
- Use the Theorem of Pappus to find the volume of a solid of revolution.

Mass

In this section you will study several important applications of integration that are related to **mass.** Mass is a measure of a body's resistance to changes in motion, and is independent of the particular gravitational system in which the body is located. However, because so many applications involving mass occur on earth's surface, we tend to equate an object's mass with its weight. This is not technically correct. Weight is a type of force and as such is dependent on gravity. Force and mass are related by the equation

$$\text{Force} = (\text{mass})(\text{acceleration}).$$

The table below lists some commonly used measures of mass and force, together with their conversion factors.

System of Measurement	Measure of Mass	Measure of Force
U.S.	Slug	Pound = $(\text{slug})(\text{ft/sec}^2)$
International	Kilogram	Newton = $(\text{kilogram})(\text{m/sec}^2)$
C-G-S	Gram	Dyne = $(\text{gram})(\text{cm/sec}^2)$
Conversions: 1 pound = 4.448 newtons 1 newton = 0.2248 pound 1 dyne = 0.000002247 pound 1 dyne = 0.00001 newton		1 slug = 14.59 kilograms 1 kilogram = 0.06854 slug 1 gram = 0.00006854 slug 1 meter = 0.3048 foot

---

Example 1   **Mass on the Surface of Earth**

Find the mass (in slugs) of an object whose weight at sea level is 1 pound.

Solution   Using 32 feet per second per second as the acceleration due to gravity produces

$$\text{Mass} = \frac{\text{force}}{\text{acceleration}} \qquad \text{Force} = (\text{mass})(\text{acceleration})$$

$$= \frac{1 \text{ pound}}{32 \text{ feet per second per second}}$$

$$= 0.03125 \frac{\text{pound}}{\text{foot per second per second}}$$

$$= 0.03125 \text{ slug}.$$

Because many applications involving mass occur on earth's surface, this amount of mass is called a **pound mass.**

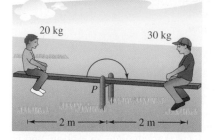

The seesaw will balance when the left and the right moments are equal.
Figure 6.54

Center of Mass in a One-Dimensional System

We will consider two types of moments of a mass—the **moment about a point** and the **moment about a line.** To define these two moments, consider an idealized situation in which a mass m is concentrated at a point. If x is the distance between this point mass and another point P, the **moment of m about the point P** is

$$\text{Moment} = mx$$

and x is the **length of the moment arm.**

The concept of moment can be demonstrated simply by a seesaw, as illustrated in Figure 6.54. Suppose a child of mass 20 kilograms sits 2 meters to the left of fulcrum P, and an older child of mass 30 kilograms sits 2 meters to the right of P. From experience, you know that the seesaw will begin to rotate clockwise, moving the larger child down. This rotation occurs because the moment produced by the child on the left is less than the moment produced by the child on the right.

$$\text{Left moment} = (20)(2) = 40 \ \text{kilogram-meters}$$
$$\text{Right moment} = (30)(2) = 60 \ \text{kilogram-meters}$$

To balance the seesaw, the two moments must be equal. For example, if the larger child moved to a position $\frac{4}{3}$ meters from the fulcrum, the seesaw would balance, because each child would produce a moment of 40 kilogram-meters.

To generalize this, you can introduce a coordinate line on which the origin corresponds to the fulcrum, as shown in Figure 6.55. Suppose several point masses are located on the x-axis. The measure of the tendency of this system to rotate about the origin is the **moment about the origin,** and it is defined as the sum of the n products $m_i x_i$.

$$M_0 = m_1 x_1 + m_2 x_2 + \cdots + m_n x_n$$

If $m_1 x_1 + m_2 x_2 + \cdots + m_n x_n = 0$, the system is in equilibrium.
Figure 6.55

If M_0 is 0, the system is said to be in **equilibrium.**

For a system that is not in equilibrium, the **center of mass** is defined as the point $\bar{x}$ at which the fulcrum could be relocated to attain equilibrium. If the system were translated $\bar{x}$ units, each coordinate x_i would become $(x_i - \bar{x})$, and because the moment of the translated system is 0, you have

$$\sum_{i=1}^{n} m_i(x_i - \bar{x}) = \sum_{i=1}^{n} m_i x_i - \sum_{i=1}^{n} m_i \bar{x} = 0.$$

Solving for $\bar{x}$ produces

$$\bar{x} = \frac{\displaystyle\sum_{i=1}^{n} m_i x_i}{\displaystyle\sum_{i=1}^{n} m_i} = \frac{\text{moment of system about origin}}{\text{total mass of system}}.$$

If $m_1 x_1 + m_2 x_2 + \cdots + m_n x_n = 0$, the system is in equilibrium.

Moments and Center of Mass: One-Dimensional System

Let the point masses $m_1, m_2, \ldots, m_n$ be located at $x_1, x_2, \ldots, x_n$.

1. The **moment about the origin** is $M_0 = m_1 x_1 + m_2 x_2 + \cdots + m_n x_n$.

2. The **center of mass** is $\bar{x} = \dfrac{M_0}{m}$, where $m = m_1 + m_2 + \cdots + m_n$ is the total mass of the system.

Example 2   **The Center of Mass of a Linear System**

Find the center of mass of the linear system shown in Figure 6.56.

Figure 6.56

Solution   The moment about the origin is

$$M_0 = m_1 x_1 + m_2 x_2 + m_3 x_3 + m_4 x_4$$
$$= 10(-5) + 15(0) + 5(4) + 10(7)$$
$$= -50 + 0 + 20 + 70$$
$$= 40.$$

Because the total mass of the system is $m = 10 + 15 + 5 + 10 = 40$, the center of mass is

$$\bar{x} = \frac{M_0}{m} = \frac{40}{40} = 1.$$

NOTE   In Example 2, where should you locate the fulcrum so that the point masses will be in equilibrium?

Rather than define the moment of a mass, you could define the moment of a *force*. In this context, the center of mass is called the **center of gravity.** Suppose that a system of point masses $m_1, m_2, \ldots, m_n$ is located at $x_1, x_2, \ldots, x_n$. Then, because force = (mass)(acceleration), the total force of the system is

$$F = m_1 a + m_2 a + \cdots + m_n a$$
$$= ma.$$

The **torque** (moment) about the origin is

$$T_0 = (m_1 a)x_1 + (m_2 a)x_2 + \cdots + (m_n a)x_n$$
$$= M_0 a$$

and the **center of gravity** is

$$\frac{T_0}{F} = \frac{M_0 a}{ma} = \frac{M_0}{m} = \bar{x}.$$

Therefore, the center of gravity and the center of mass have the same location.

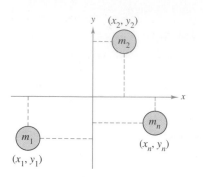

In a two-dimensional system, there is a moment about the y-axis, M_y, and a moment about the x-axis, M_x.
Figure 6.57

Center of Mass in a Two-Dimensional System

You can extend the concept of moment to two dimensions by considering a system of masses located in the xy-plane at the points $(x_1, y_1), (x_2, y_2), \ldots, (x_n, y_n)$, as shown in Figure 6.57. Rather than defining a single moment (with respect to the origin), we define two moments—one with respect to the x-axis and one with respect to the y-axis.

Moments and Center of Mass: Two-Dimensional System

Let the point masses $m_1, m_2, \ldots, m_n$ be located at $(x_1, y_1), (x_2, y_2), \ldots, (x_n, y_n)$.

1. The **moment about the y-axis** is $M_y = m_1 x_1 + m_2 x_2 + \cdots + m_n x_n$.
2. The **moment about the x-axis** is $M_x = m_1 y_1 + m_2 y_2 + \cdots + m_n y_n$.
3. The **center of mass** $(\bar{x}, \bar{y})$ (or **center of gravity**) is

$$\bar{x} = \frac{M_y}{m} \quad \text{and} \quad \bar{y} = \frac{M_x}{m}$$

where $m = m_1 + m_2 + \cdots + m_n$ is the **total mass** of the system.

The moment of a system of masses in the plane can be taken about any horizontal or vertical line. In general, the moment about a line is the sum of the product of the masses and the *directed distances* from the points to the line.

$$\text{Moment} = m_1(y_1 - b) + m_2(y_2 - b) + \cdots + m_n(y_n - b) \quad \text{Horizontal line } y = b$$

$$\text{Moment} = m_1(x_1 - a) + m_2(x_2 - a) + \cdots + m_n(x_n - a) \quad \text{Vertical line } x = a$$

Example 3 The Center of Mass of a Two-Dimensional System

Find the center of mass of a system of point masses $m_1 = 6$, $m_2 = 3$, $m_3 = 2$, and $m_4 = 9$, located at

$$(3, -2), (0, 0), (-5, 3), \text{ and } (4, 2)$$

as shown in Figure 6.58.

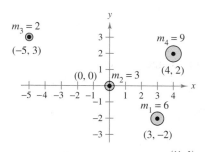

The center of mass of the system is $\left(\frac{11}{5}, \frac{3}{5}\right)$.
Figure 6.58

Solution

$$
\begin{aligned}
m &= 6 && + 3 && + 2 && + 9 && = 20 && \text{Mass} \\
M_y &= 6(3) && + 3(0) && + 2(-5) && + 9(4) && = 44 && \text{Moment about } y\text{-axis} \\
M_x &= 6(-2) && + 3(0) && + 2(3) && + 9(2) && = 12 && \text{Moment about } x\text{-axis}
\end{aligned}
$$

Therefore,

$$\bar{x} = \frac{M_y}{m} = \frac{44}{20} = \frac{11}{5}$$

and

$$\bar{y} = \frac{M_x}{m} = \frac{12}{20} = \frac{3}{5}$$

and thus the center of mass is $\left(\frac{11}{5}, \frac{3}{5}\right)$.

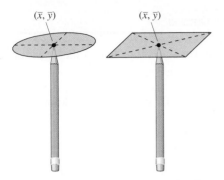

You can think of the center of mass $(\bar{x}, \bar{y})$ of a lamina as its balancing point. For a circular lamina, the center of mass is the center of the circle. For a rectangular lamina, the center of mass is the center of the rectangle.
Figure 6.59

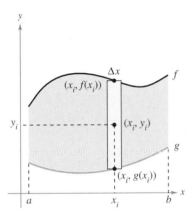

Planar lamina of uniform density ρ
Figure 6.60

Center of Mass of a Planar Lamina

So far in this section we have assumed the total mass of a system to be distributed at discrete points in a plane or on a line. We now consider a thin, flat plate of material of constant density called a **planar lamina** (see Figure 6.59). **Density** is a measure of mass per unit of volume, such as grams per cubic centimeter. For planar laminas, however, density is considered to be a measure of mass per unit of area. Density is denoted by ρ, the lowercase Greek letter rho.

Consider an irregularly shaped planar lamina of uniform density ρ, bounded by the graphs of $y = f(x)$, $y = g(x)$, and $a \le x \le b$, as shown in Figure 6.60. The mass of this region is given by

$$m = (\text{density})(\text{area})$$
$$= \rho \int_a^b [f(x) - g(x)]\, dx$$
$$= \rho A$$

where A is the area of the region. To find the center of mass of this lamina, partition the interval $[a, b]$ into n subintervals of equal width Δx. Let x_i be the center of the ith subinterval. You can approximate the portion of the lamina lying in the ith subinterval by a rectangle whose height is $h = f(x_i) - g(x_i)$. Because the density of the rectangle is ρ, its mass is

$$m_i = (\text{density})(\text{area})$$
$$= \rho \underbrace{[f(x_i) - g(x_i)]}_{\substack{| \\ \text{Density} \quad \text{Height}}} \underbrace{\Delta x}_{\text{Width}}.$$

Now, considering this mass to be located at the center (x_i, y_i) of the rectangle, the directed distance from the x-axis to (x_i, y_i) is $y_i = [f(x_i) + g(x_i)]/2$. So, the moment of m_i about the x-axis is

$$\text{Moment} = (\text{mass})(\text{distance})$$
$$= m_i y_i = \rho[f(x_i) - g(x_i)]\, \Delta x \left[\frac{f(x_i) + g(x_i)}{2}\right].$$

Summing the moments and taking the limit as $n \to \infty$ suggest the definitions below.

Moments and Center of Mass of a Planar Lamina

Let f and g be continuous functions such that $f(x) \ge g(x)$ on $[a, b]$, and consider the planar lamina of uniform density ρ bounded by the graphs of $y = f(x)$, $y = g(x)$, and $a \le x \le b$.

1. The **moments about the x- and y-axes** are

$$M_x = \rho \int_a^b \left[\frac{f(x) + g(x)}{2}\right][f(x) - g(x)]\, dx$$
$$M_y = \rho \int_a^b x[f(x) - g(x)]\, dx.$$

2. The **center of mass** $(\bar{x}, \bar{y})$ is given by $\bar{x} = \dfrac{M_y}{m}$ and $\bar{y} = \dfrac{M_x}{m}$, where $m = \rho \int_a^b [f(x) - g(x)]\, dx$ is the mass of the lamina.

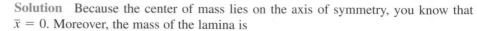

Example 4   **The Center of Mass of a Planar Lamina**

Find the center of mass of the lamina of uniform density ρ bounded by the graph of $f(x) = 4 - x^2$ and the x-axis.

Solution   Because the center of mass lies on the axis of symmetry, you know that $\bar{x} = 0$. Moreover, the mass of the lamina is

$$m = \rho \int_{-2}^{2} (4 - x^2)\, dx$$

$$= \rho \left[4x - \frac{x^3}{3} \right]_{-2}^{2}$$

$$= \frac{32\rho}{3}.$$

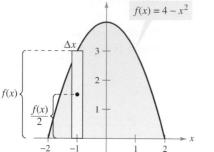

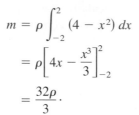

Figure 6.61

To find the moment about the x-axis, place a representative rectangle in the region, as shown in Figure 6.61. The distance from the x-axis to the center of this rectangle is

$$y_i = \frac{f(x)}{2} = \frac{4 - x^2}{2}.$$

Because the mass of the representative rectangle is

$$\rho f(x)\, \Delta x = \rho(4 - x^2)\, \Delta x$$

you have

$$M_x = \rho \int_{-2}^{2} \frac{4 - x^2}{2} (4 - x^2)\, dx$$

$$= \frac{\rho}{2} \int_{-2}^{2} (16 - 8x^2 + x^4)\, dx$$

$$= \frac{\rho}{2} \left[16x - \frac{8x^3}{3} + \frac{x^5}{5} \right]_{-2}^{2}$$

$$= \frac{256\rho}{15}$$

and $\bar{y}$ is given by

$$\bar{y} = \frac{M_x}{m} = \frac{256\rho/15}{32\rho/3} = \frac{8}{5}.$$

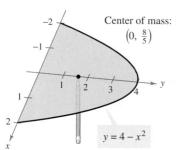

The center of mass is the balancing point.
Figure 6.62

So, the center of mass (the balancing point) of the lamina is $\left(0, \frac{8}{5}\right)$, as shown in Figure 6.62.

The density ρ in Example 4 is a common factor of both the moments and the mass, and as such cancels out of the quotients representing the coordinates of the center of mass. So, the center of mass of a lamina of *uniform* density depends only on the shape of the lamina and not on its density. For this reason, the point

$$(\bar{x}, \bar{y}) \qquad \text{Center of mass or centroid}$$

is sometimes called the center of mass of a *region* in the plane, or the **centroid** of the region. In other words, to find the centroid of a region in the plane, you simply assume that the region has a constant density of $\rho = 1$ and compute the corresponding center of mass.

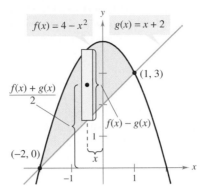

The centroid of the region is $\left(-\frac{1}{2}, \frac{12}{5}\right)$.

Figure 6.63

EXPLORATION

Cut an irregular shape from a piece of cardboard.

a. Hold a pencil vertically and move the object on the pencil point until the centroid is located.

b. Divide the object into representative elements. Make the necessary measurements and numerically approximate the centroid. Compare your result with the result in part (a).

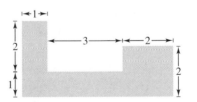

(a) Original region

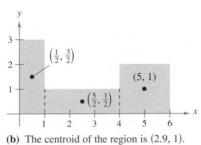

(b) The centroid of the region is (2.9, 1).

Figure 6.64

Example 5   **The Centroid of a Plane Region**

Find the centroid of the region bounded by the graphs of $f(x) = 4 - x^2$ and $g(x) = x + 2$.

Solution   The two graphs intersect at the points $(-2, 0)$ and $(1, 3)$, as shown in Figure 6.63. So, the area of the region is

$$A = \int_{-2}^{1} [f(x) - g(x)]\, dx = \int_{-2}^{1} (2 - x - x^2)\, dx = \frac{9}{2}.$$

The centroid $(\bar{x}, \bar{y})$ of the region has the following coordinates.

$$\bar{x} = \frac{1}{A} \int_{-2}^{1} x[(4 - x^2) - (x + 2)]\, dx = \frac{2}{9} \int_{-2}^{1} (-x^3 - x^2 + 2x)\, dx$$

$$= \frac{2}{9} \left[-\frac{x^4}{4} - \frac{x^3}{3} + x^2 \right]_{-2}^{1} = -\frac{1}{2}$$

$$\bar{y} = \frac{1}{A} \int_{-2}^{1} \left[\frac{(4 - x^2) + (x + 2)}{2} \right] [(4 - x^2) - (x + 2)]\, dx$$

$$= \frac{2}{9} \left(\frac{1}{2}\right) \int_{-2}^{1} (-x^2 + x + 6)(-x^2 - x + 2)\, dx$$

$$= \frac{1}{9} \int_{-2}^{1} (x^4 - 9x^2 - 4x + 12)\, dx$$

$$= \frac{1}{9} \left[\frac{x^5}{5} - 3x^3 - 2x^2 + 12x \right]_{-2}^{1} = \frac{12}{5}.$$

So, the centroid of the region is $(\bar{x}, \bar{y}) = \left(-\frac{1}{2}, \frac{12}{5}\right)$.

For simple plane regions, you may be able to find the centroid without resorting to integration.

Example 6   **The Centroid of a Simple Plane Region**

Find the centroid of the region shown in Figure 6.64(a).

Solution   By superimposing a coordinate system on the region, as shown in Figure 6.64(b), you can locate the centroids of the three rectangles at

$$\left(\frac{1}{2}, \frac{3}{2}\right), \quad \left(\frac{5}{2}, \frac{1}{2}\right), \quad \text{and} \quad (5, 1).$$

Using these three points, you can find the centroid of the region.

$$A = \text{area of region} = 3 + 3 + 4 = 10$$

$$\bar{x} = \frac{(1/2)(3) + (5/2)(3) + (5)(4)}{10} = \frac{29}{10} = 2.9$$

$$\bar{y} = \frac{(3/2)(3) + (1/2)(3) + (1)(4)}{10} = \frac{10}{10} = 1$$

So, the centroid of the region is (2.9, 1).

NOTE   In Example 6, notice that (2.9, 1) is not the "average" of $\left(\frac{1}{2}, \frac{3}{2}\right)$, $\left(\frac{5}{2}, \frac{1}{2}\right)$, and (5, 1).

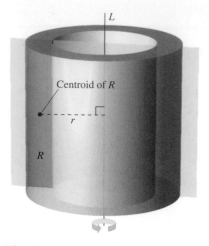

The volume V is $2\pi rA$ where A is the area of region R.

Figure 6.65

Theorem of Pappus

The final topic in this section is a useful theorem credited to Pappus of Alexandria (ca. 300 A.D.), a Greek mathematician whose eight-volume *Mathematical Collection* is a record of much of classical Greek mathematics. We delay the proof of this theorem until Section 13.4 (Exercise 54).

THEOREM 6.1   The Theorem of Pappus

Let R be a region in a plane and let L be a line in the same plane such that L does not intersect the interior of R, as shown in Figure 6.65. If r is the distance between the centroid of R and the line, then the volume V of the solid of revolution formed by revolving R about the line is

$$V = 2\pi rA$$

where A is the area of R. (Note that $2\pi r$ is the distance traveled by the centroid as the region is revolved about the line.)

The Theorem of Pappus can be used to find the volume of a torus, as shown in the following example. Recall that a torus is a doughnut-shaped solid formed by revolving a circular region about a line that lies in the same plane as the circle (but does not intersect the circle).

Example 7   **Finding Volume by the Theorem of Pappus**

Find the volume of the torus formed by revolving the circular region bounded by

$$(x - 2)^2 + y^2 = 1$$

about the y-axis, as shown in Figure 6.66(a).

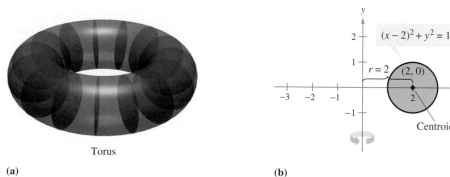

Torus

(a)   **(b)**

Figure 6.66

Solution   In Figure 6.66(b), you can see that the centroid of the circular region is $(2, 0)$. So, the distance between the centroid and the axis of revolution is $r = 2$. Because the area of the circular region is $A = \pi$, the volume of the torus is

$$
\begin{aligned}
V &= 2\pi rA \\
&= 2\pi(2)(\pi) \\
&= 4\pi^2 \\
&\approx 39.5.
\end{aligned}
$$

EXERCISES FOR SECTION 6.6

In Exercises 1–4, find the center of mass of the point masses lying on the *x*-axis.

1. $m_1 = 6, m_2 = 3, m_3 = 5$

   $x_1 = -5, x_2 = 1, x_3 = 3$

2. $m_1 = 7, m_2 = 4, m_3 = 3, m_4 = 8$

   $x_1 = -3, x_2 = -2, x_3 = 5, x_4 = 6$

3. $m_1 = 1, m_2 = 1, m_3 = 1, m_4 = 1, m_5 = 1$

   $x_1 = 7, x_2 = 8, x_3 = 12, x_4 = 15, x_5 = 18$

4. $m_1 = 12, m_2 = 1, m_3 = 6, m_4 = 3, m_5 = 11$

   $x_1 = -6, x_2 = -4, x_3 = -2, x_4 = 0, x_5 = 8$

5. *Graphical Reasoning*

   (a) Translate each point mass in Exercise 3 to the right 5 units and determine the resulting center of mass.

   (b) Translate each point mass in Exercise 4 to the left 3 units and determine the resulting center of mass.

6. *Conjecture* Use the result of Exercise 5 to make a conjecture about the change in the center of mass that results when each point mass is translated *k* units horizontally.

Statics Problems   In Exercises 7 and 8, consider a beam of length *L* with a fulcrum *x* feet from one end (see figure). If there are objects with weights W_1 and W_2 placed on opposite ends of the beam, find *x* such that the system is in equilibrium.

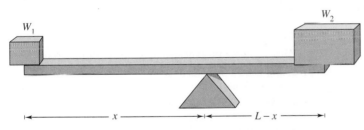

7. Two children weighing 50 pounds and 75 pounds are going to play on a seesaw that is 10 feet long.

8. In order to move a 550-pound rock, a person weighing 200 pounds wants to balance it on a beam that is 5 feet long.

In Exercise 9–12, find the center of mass of the given system of point masses.

9.

m_i	5	1	3
(x_i, y_i)	(2, 2)	(-3, 1)	(1, -4)

10.

m_i	10	2	5
(x_i, y_i)	(1, -1)	(5, 5)	(-4, 0)

11.

m_i	3	4
(x_i, y_i)	(-2, -3)	(-1, 0)

m_i	2	1	6
(x_i, y_i)	(7, 1)	(0, 0)	(-3, 0)

12.

m_i	12	6	$\frac{15}{2}$	15
(x_i, y_i)	(2, 3)	(-1, 5)	(6, 8)	(2, -2)

In Exercises 13–24, find M_x, M_y, and $(\bar{x}, \bar{y})$ for the laminas of uniform density ρ bounded by the graphs of the equations.

13. $y = \sqrt{x}, y = 0, x = 4$

14. $y = \frac{1}{2}x^2, y = 0, x = 2$

15. $y = x^2, y = x^3$

16. $y = \sqrt{x}, y = x$

17. $y = -x^2 + 4x + 2, y = x + 2$

18. $y = \sqrt{x} + 1, y = \frac{1}{3}x + 1$

19. $y = x^{2/3}, y = 0, x = 8$

20. $y = x^{2/3}, y = 4$

21. $x = 4 - y^2, x = 0$

22. $x = 2y - y^2, x = 0$

23. $x = -y, x = 2y - y^2$

24. $x = y + 2, x = y^2$

In Exercises 25–28, set up and evaluate the integrals for finding the area and moments about the *x*- and *y*-axes for the region bounded by the graphs of the equations. (Assume $\rho = 1$.)

25. $y = x^2, y = x$

26. $y = \dfrac{1}{x}, y = 0, 1 \le x < 4$

27. $y = 2x + 4, y = 0, 0 \le x \le 3$

28. $y = x^2 - 4, y = 0$

In Exercises 29–32, use a graphing utility to graph the region bounded by the graphs of the equations. Use the integration capabilities of the graphing utility to approximate the centroid of the region.

29. $y = 10x\sqrt{125 - x^3}, y = 0$

30. $y = xe^{-x/2}, y = 0, x = 0, x = 4$

31. *Prefabricated End Section of a Building*

   $y = 5\sqrt[3]{400 - x^2}, y = 0$

32. *Witch of Agnesi*

   $y = 8/(x^2 + 4), y = 0, x = -2, x = 2$

In Exercises 33–38, find and/or verify the centroid of the common region used in engineering.

33. Triangle Show that the centroid of the triangle with vertices $(-a, 0)$, $(a, 0)$, and (b, c) is the point of intersection of the medians (see figure).

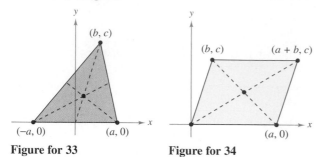

Figure for 33   **Figure for 34**

34. Parallelogram Show that the centroid of the parallelogram with vertices $(0, 0)$, $(a, 0)$, (b, c), and $(a + b, c)$ is the point of intersection of the diagonals (see figure).

35. Trapezoid Find the centroid of the trapezoid with vertices $(0, 0)$, $(0, a)$, (c, b), and $(c, 0)$. Show that it is the intersection of the line connecting the midpoints of the parallel sides and the line connecting the extended parallel sides, as shown in the figure.

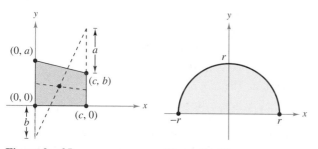

Figure for 35   **Figure for 36**

36. Semicircle Find the centroid of the region bounded by the graphs of $y = \sqrt{r^2 - x^2}$ and $y = 0$ (see figure).

37. Semiellipse Find the centroid of the region bounded by the graphs of $y = \dfrac{b}{a}\sqrt{a^2 - x^2}$ and $y = 0$ (see figure).

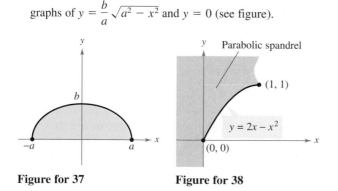

Figure for 37   **Figure for 38**

38. Parabolic Spandrel Find the centroid of the **parabolic spandrel** shown in the figure.

39. Graphical Reasoning Consider the region bounded by the graphs of $y = x^2$ and $y = b$, where $b > 0$.

(a) Sketch a graph of the region.

(b) Use the graph in part (a) to determine $\bar{x}$. Explain.

(c) Set up the integral for finding M_y. Because of the form of the integrand, the value of the integral can be obtained without integrating. What is the form of the integrand and what is the value of the integral? Compare with the result in part (b).

(d) Use the graph in part (a) to determine whether $\bar{y} > \dfrac{b}{2}$ or $\bar{y} < \dfrac{b}{2}$. Explain.

(e) Use integration to verify your answer in part (d).

40. Graphical and Numerical Reasoning Consider the region bounded by the graphs of $y = x^{2n}$ and $y = b$, where $b > 0$ and n is a positive integer.

(a) Set up the integral for finding M_y. Because of the form of the integrand, the value of the integral can be obtained without integrating. What is the form of the integrand and what is the value of the integral? Compare with the result in part (b).

(b) Is $\bar{y} > \dfrac{b}{2}$ or $\bar{y} < \dfrac{b}{2}$? Explain.

(c) Use integration to find $\bar{y}$ as a function of n.

(d) Use the result in part (c) to complete the table.

n	1	2	3	4
$\bar{y}$				

(e) Find $\lim\limits_{n \to \infty} \bar{y}$.

(f) Give a geometric explanation of the result in part (e).

41. Modeling Data The manufacturer of glass for a window in a conversion van needs to approximate its center of mass. A coordinate system is superimposed on a prototype of the glass (see figure). The measurements (in centimeters) for the right half of the symmetric piece of glass are shown in the table.

x	0	10	20	30	40
y	30	29	26	20	0

(a) Use Simpson's Rule to approximate the center of mass of the glass.

(b) Use the regression capabilities of a graphing utility to find a fourth-degree polynomial model for the data.

(c) Use the integration capabilities of a graphing utility and the model to approximate the center of mass of the glass. Compare with the result in part (a).

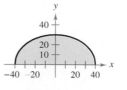

42. *Modeling Data* The manufacturer of a boat needs to approximate the center of mass of a section of the hull. A coordinate system is superimposed on a prototype (see figure). The measurements (in feet) for the right half of the symmetric prototype are listed in the table.

x	0	0.5	1.0	1.5	2
l	1.50	1.45	1.30	0.99	0
d	0.50	0.48	0.43	0.33	0

(a) Use Simpson's Rule to approximate the center of mass of the hull section.

(b) Use the regression capabilities of a graphing utility to find fourth-degree polynomial models for both curves shown in the figure. Plot the data and graph the models.

(c) Use the integration capabilities of a graphing utility and the model to approximate the center of mass of the hull section. Compare with the result in part (a).

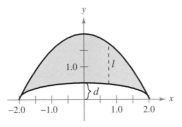

In Exercises 43–46, introduce an appropriate coordinate system and find the coordinates of the center of mass of the planar lamina. (The answer depends on the position of the coordinate system.)

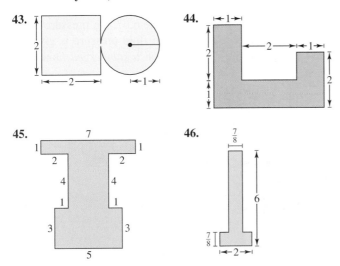

47. Find the center of mass of the lamina in Exercise 43 if the circular portion of the lamina has twice the density of the square portion of the lamina.

48. Find the center of mass of the lamina in Exercise 43 if the square portion of the lamina has twice the density of the circular portion of the lamina.

In Exercises 49–52, use the Theorem of Pappus to find the volume of the solid of revolution.

49. The torus formed by revolving the circle $(x - 5)^2 + y^2 = 16$ about the y-axis

50. The torus formed by revolving the circle $x^2 + (y - 3)^2 = 4$ about the x-axis

51. The solid formed by revolving the region bounded by the graphs of $y = x$, $y = 4$, and $x = 0$ about the x-axis

52. The solid formed by revolving the region bounded by the graphs of $y = 2\sqrt{x - 2}$, $y = 0$, and $x = 6$ about the y-axis

Getting at the Concept

53. Let the point masses $m_1, m_2, \ldots, m_n$ be located at (x_1, y_1), $(x_2, y_2), \ldots, (x_n, y_n)$. Define the center of mass $(\bar{x}, \bar{y})$.

54. What is meant by a planar lamina? Describe what is meant by the center of mass $(\bar{x}, \bar{y})$ of a planar lamina.

55. The centroid of the plane region bounded by the graphs of $y = f(x)$, $y = 0$, $x = 0$, and $x = 1$ is $\left(\frac{5}{6}, \frac{5}{18}\right)$. Is it possible to find the centroid of each of the following regions bounded by the graphs of the equations? If so, identify the centroid and explain your answer.

(a) $y = f(x) + 2$, $y = 2$, $x = 0$, and $x = 1$

(b) $y = f(x - 2)$, $y = 0$, $x = 2$, and $x = 3$

(c) $y = -f(x)$, $y = 0$, $x = 0$, and $x = 1$

(d) $y = f(x)$, $y = 0$, $x = -1$, and $x = 1$

56. State the Theorem of Pappus.

In Exercises 57 and 58, use the *Second Theorem of Pappus*, which is stated as follows. If a segment of a plane curve C is revolved about an axis that does not intersect the curve (except possibly at its endpoints), the area S of the resulting surface of revolution is given by the product of the length of C times the distance d traveled by the centroid of C.

57. A sphere is formed by revolving the graph of

$$y = \sqrt{r^2 - x^2}$$

about the x-axis. Use the formula for surface area, $S = 4\pi r^2$, to find the centroid of the semicircle $y = \sqrt{r^2 - x^2}$.

58. A torus is formed by revolving the graph of

$$(x - 1)^2 + y^2 = 1$$

about the y-axis. Find the surface area of the torus.

59. Let $n \geq 1$ be constant, and consider the region bounded by $f(x) = x^n$, the x-axis, and $x = 1$. Find the centroid of this region. As $n \to \infty$, what does the region look like, and where is its centroid?

Section 6.7	Fluid Pressure and Fluid Force

• Find fluid pressure and fluid force.

Fluid Pressure and Fluid Force

Swimmers know that the deeper an object is submerged in a fluid, the greater the pressure on the object. **Pressure** is defined as the force per unit of area over the surface of a body. For example, because a column of water that is 10 feet in height and 1 inch square weighs 4.3 pounds, the *fluid pressure* at a depth of 10 feet of water is 4.3 pounds per square inch.[*] At 20 feet, this would increase to 8.6 pounds per square inch, and in general the pressure is proportional to the depth of the object in the fluid.

Definition of Fluid Pressure

The **pressure** on an object at depth h in a liquid is

$$\text{Pressure} = P = wh$$

where w is the weight-density of the liquid per unit of volume.

Below are some common weight-densities of fluids in pounds per cubic foot.

Ethyl alcohol	49.4
Gasoline	41.0–43.0
Glycerin	78.6
Kerosene	51.2
Mercury	849.0
Seawater	64.0
Water	62.4

When calculating fluid pressure, you can use an important (and rather surprising) physical law called **Pascal's Principle**, named after the French mathematician Blaise Pascal. Pascal's Principle states that the pressure exerted by a fluid at a depth h is transmitted equally *in all directions*. For example, in Figure 6.67, the pressure at the indicated depth is the same for all three objects. Because fluid pressure is given in terms of force per unit area ($P = F/A$), the fluid force on a *submerged horizontal surface of area* A is

$$\text{Fluid force} = F = PA = (\text{pressure})(\text{area}).$$

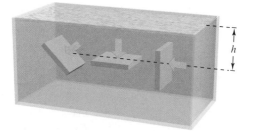

The pressure at h is the same for all three objects.
Figure 6.67

BLAISE PASCAL (1623–1662)

Pascal is well known for his work in many areas of mathematics and physics, and also for his influence on Leibniz. Although much of Pascal's work in calculus was intuitive and lacked the rigor of modern mathematics, he nevertheless anticipated many important results.

---

[*] *The total pressure on an object in 10 feet of water would also include the pressure due to earth's atmosphere. At sea level, atmospheric pressure is approximately 14.7 pounds per square inch.*

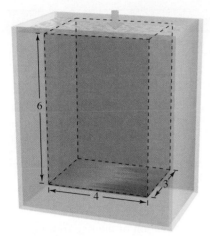

The fluid force on a horizontal metal sheet is equal to the fluid pressure times the area.
Figure 6.68

Example 1   Fluid Force on a Submerged Sheet

Find the fluid force on a rectangular metal sheet measuring 3 feet by 4 feet that is submerged in 6 feet of water, as shown in Figure 6.68.

Solution   Because the weight-density of water is 62.4 pounds per cubic foot and the sheet is submerged in 6 feet of water, the fluid pressure is

$$P = (62.4)(6) \qquad P = wh$$
$$= 374.4 \text{ pounds per square foot.}$$

Because the total area of the sheet is $A = (3)(4) = 12$ square feet, the fluid force is

$$F = PA = \left(374.4 \; \frac{\text{pounds}}{\text{square foot}}\right)(12 \text{ square feet})$$
$$= 4492.8 \text{ pounds.}$$

This result is independent of the size of the body of water. The fluid force would be the same in a swimming pool or lake.

In Example 1, the fact that the sheet is rectangular and horizontal means that you do not need the methods of calculus to solve the problem. We now look at a surface that is submerged vertically in a fluid. The problem is more difficult because the pressure is not constant over the surface.

Suppose a vertical plate is submerged in a fluid of weight-density w (per unit of volume), as shown in Figure 6.69. To determine the total force against *one side* of the region from depth c to depth d, you can subdivide the interval $[c, d]$ into n subintervals, each of width Δy. Next, consider the representative rectangle of width Δy and length $L(y_i)$, where y_i is in the ith subinterval. The force against this representative rectangle is

$$\Delta F_i = w(\text{depth})(\text{area})$$
$$= wh(y_i)L(y_i) \, \Delta y.$$

The force against n such rectangles is

$$\sum_{i=1}^{n} \Delta F_i = w \sum_{i=1}^{n} h(y_i)L(y_i) \, \Delta y.$$

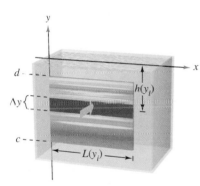

Calculus methods must be used to find the fluid force on a vertical metal plate.
Figure 6.69

Note that w is considered to be constant and is factored out of the summation. Therefore, taking the limit as $\|\Delta\| \to 0$ ($n \to \infty$) suggests the following definition.

Definition of Force Exerted by a Fluid

The **force F exerted by a fluid** of constant weight-density w (per unit of volume) against a submerged vertical plane region from $y = c$ to $y = d$ is

$$F = w \lim_{\|\Delta\| \to 0} \sum_{i=1}^{n} h(y_i)L(y_i) \, \Delta y = w \int_{c}^{d} h(y)L(y) \, dy$$

where $h(y)$ is the depth of the fluid at y and $L(y)$ is the horizontal length of the region at y.

Example 2   Fluid Force on a Vertical Surface

A vertical gate in a dam has the shape of an isosceles trapezoid 8 feet across the top and 6 feet across the bottom, with a height of 5 feet, as shown in Figure 6.70(a). What is the fluid force on the gate if the top of the gate is 4 feet below the surface of the water?

Solution   In setting up a mathematical model for this problem, you are at liberty to locate the x- and y-axis in several different ways. A convenient approach is to let the y-axis bisect the gate and place the x-axis at the surface of the water, as shown in Figure 6.70(b). So, the depth of the water at y in feet is

$$\text{Depth } = h(y) = -y.$$

To find the length $L(y)$ of the region at y, we find the equation of the line forming the right side of the gate. Because this line passes through the points $(3, -9)$ and $(4, -4)$, its equation is

$$y - (-9) = \frac{-4 - (-9)}{4 - 3}(x - 3)$$

$$y + 9 = 5(x - 3)$$

$$y = 5x - 24$$

$$x = \frac{y + 24}{5}.$$

In Figure 6.70(b) you can see that the length of the region at y is

$$\text{Length } = 2x$$

$$= \frac{2}{5}(y + 24)$$

$$= L(y).$$

Finally, by integrating from $y = -9$ to $y = -4$, you can calculate the fluid force to be

$$F = w\int_c^d h(y)L(y)\,dy$$

$$= 62.4\int_{-9}^{-4}(-y)\left(\frac{2}{5}\right)(y + 24)\,dy$$

$$= -62.4\left(\frac{2}{5}\right)\int_{-9}^{-4}(y^2 + 24y)\,dy$$

$$= -62.4\left(\frac{2}{5}\right)\left[\frac{y^3}{3} + 12y^2\right]_{-9}^{-4}$$

$$= -62.4\left(\frac{2}{5}\right)\left(\frac{-1675}{3}\right)$$

$$= 13{,}936 \text{ pounds.}$$

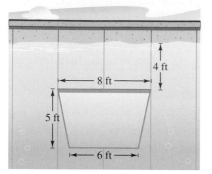

(a) Water gate in a dam

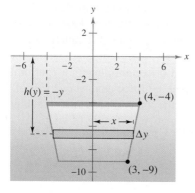

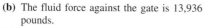

(b) The fluid force against the gate is 13,936 pounds.

Figure 6.70

NOTE   In Example 2, we let the x-axis coincide with the surface of the water. This was convenient, but arbitrary. In choosing a coordinate system to represent a physical situation, you should consider various possibilities. Often you can simplify the calculations in a problem by locating the coordinate system to take advantage of special characteristics of the problem, such as symmetry.

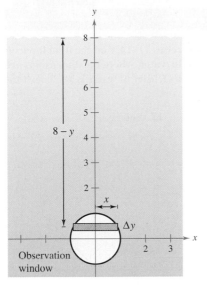

The fluid force on the window is 1608.5 pounds.
Figure 6.71

Example 3   **Fluid Force on a Vertical Surface**

A circular observation window on a marine science ship has a radius of 1 foot, and the center of the window is 8 feet below water level, as shown in Figure 6.71. What is the fluid force on the window?

Solution   To take advantage of symmetry, locate a coordinate system such that the origin coincides with the center of the window, as shown in Figure 6.71. The depth at y is then

$$\text{Depth} = h(y) = 8 - y.$$

The horizontal length of the window is $2x$, and you can use the equation for the circle, $x^2 + y^2 = 1$, to solve for x as follows.

$$\text{Length} = 2x = 2\sqrt{1 - y^2} = L(y)$$

Finally, because y ranges from -1 to 1, and using 64 pounds per cubic foot as the weight-density of seawater, you have

$$F = w \int_c^d h(y)L(y)\, dy$$
$$= 64 \int_{-1}^1 (8 - y)(2)\sqrt{1 - y^2}\, dy.$$

Initially it looks as if this integral would be difficult to solve. However, if you break the integral into two parts and apply symmetry, the solution is simple.

$$F = 64(16) \int_{-1}^1 \sqrt{1 - y^2}\, dy - 64(2) \int_{-1}^1 y\sqrt{1 - y^2}\, dy$$

The second integral is 0 (because the integrand is odd and the limits of integration are symmetric to the origin). Moreover, by recognizing that the first integral represents the area of a semicircle of radius 1, you obtain

$$F = 64(16)\left(\frac{\pi}{2}\right) - 64(2)(0)$$
$$= 512\pi$$
$$\approx 1608.5 \text{ pounds.}$$

So, the fluid force on the window is 1608.5 pounds.

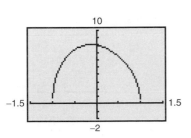

f is not differentiable at $x = \pm 1$.
Figure 6.72

TECHNOLOGY   To confirm the result obtained in Example 3, you might have considered using Simpson's Rule to approximate the value of

$$128 \int_{-1}^1 (8 - x)\sqrt{1 - x^2}\, dx.$$

From the graph of

$$f(x) = (8 - x)\sqrt{1 - x^2}$$

however, you can see that f is not differentiable when $x = \pm 1$ (see Figure 6.72). This means that you cannot apply Theorem 4.19 from Section 4.6 to determine the potential error in Simpson's Rule. Without knowing the potential error, the approximation is of little value. Try using a graphing utility to approximate the integral.

EXERCISES FOR SECTION 6.7

Force on a Submerged Sheet   In Exercises 1 and 2, the area of the top side of a piece of sheet metal is given. The sheet metal is submerged horizontally in 5 feet of water. Find the fluid force on the top side.

1. 3 square feet

2. 16 square feet

Buoyant Force   In Exercises 3 and 4, find the buoyant force of a rectangular solid of the given dimensions submerged in water so that the top side is parallel to the surface of the water. The buoyant force is the difference between the fluid forces on the top and bottom sides of the solid.

3.

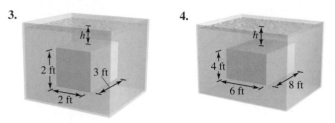

4.

Fluid Force on a Tank Wall   In Exercises 5–10, find the fluid force on the vertical side of the tank, where the dimensions are given in feet. Assume that the tank is full of water.

5. Rectangle

6. Triangle

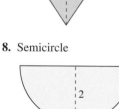

7. Trapezoid

8. Semicircle

9. Parabola, $y = x^2$

10. Semiellipse,
$$y = -\frac{1}{2}\sqrt{36 - 9x^2}$$

Fluid Force of Water   In Exercises 11–14, find the fluid force on the vertical plate submerged in water, where the dimensions are given in meters and the weight-density of water is 9800 newtons per cubic meter.

11. Square

12. Square

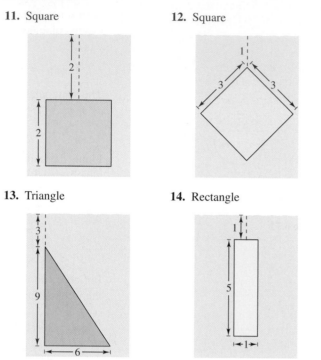

13. Triangle

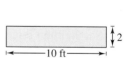

14. Rectangle

Force on a Concrete Form   In Exercises 15–18, the figure is the vertical side of a form for poured concrete that weighs 140.7 pounds per cubic foot. Determine the force on this part of the concrete form.

15. Rectangle

16. Semiellipse,
$$y = -\frac{3}{4}\sqrt{16 - x^2}$$

17. Rectangle

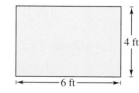

18. Triangle

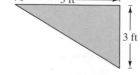

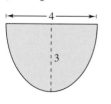

19. *Fluid Force of Gasoline*   A cylindrical gasoline tank is placed so that the axis of the cylinder is horizontal. Find the fluid force on a circular end of the tank if the tank is half full, assuming that the diameter is 3 feet and the gasoline weighs 42 pounds per cubic foot.

20. Fluid Force of Gasoline Repeat Exercise 19 for a tank that is full. (Evaluate one integral by a geometric formula and the other by observing that the integrand is an odd function.)

21. Fluid Force on a Circular Plate A circular plate of radius r feet is submerged vertically in a tank of fluid that weighs w pounds per cubic foot. The center of the circle is k $(k > r)$ feet below the surface of the fluid. Show that the fluid force on the surface of the plate is

$$F = wk(\pi r^2).$$

(Evaluate one integral by a geometric formula and the other by observing that the integrand is an odd function.)

22. Fluid Force on a Circular Plate Use the result of Exercise 21 to find the fluid force on each of the circular plates shown in the figure. Assume the plates are in the wall of a tank filled with water and the measurements are given in feet.

(a)   (b)

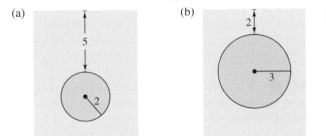

23. Fluid Force on a Rectangular Plate A rectangular plate of height h feet and base b feet is submerged vertically in a tank of fluid that weighs w pounds per cubic foot. The center is k feet below the surface of the fluid, where $h \le k/2$. Show that the fluid force on the surface of the plate is

$$F = wkhb.$$

24. Fluid Force on a Rectangular Plate Use the result of Exercise 23 to find the fluid force on each of the rectangular plates shown in the figure. Assume the plates are in the wall of a tank filled with water and the measurements are given in feet.

(a)   (b)

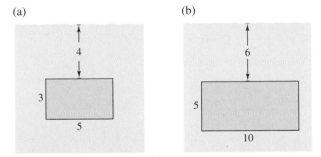

25. Submarine Porthole A porthole on a vertical side of a submarine (submerged in seawater) is 1 foot square. Find the fluid force on the porthole, assuming that the center of the square is 15 feet below the surface.

26. Submarine Porthole Repeat Exercise 25 for a circular porthole that has a diameter of 1 foot. The center is 15 feet below the surface.

27. Modeling Data The vertical stern of a boat with a superimposed coordinate system is shown in the figure. The table shows the width w of the stern at indicated values of y. Find the fluid force against the stern if the measurements are given in feet.

y	0	$\frac{1}{2}$	1	$\frac{3}{2}$	2	$\frac{5}{2}$	3	$\frac{7}{2}$	4
w	0	3	5	8	9	10	10.25	10.5	10.5

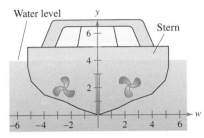

28. Irrigation Canal Gate The vertical cross section of an irrigation canal is modeled by

$$f(x) = \frac{5x^2}{x^2 + 4}$$

where x is measured in feet and $x = 0$ corresponds to the center of the canal. Use the integration capabilities of a graphing utility to approximate the fluid force against a vertical gate used to stop the flow of water if the water is 3 feet deep.

In Exercises 29 and 30, use the integration capabilities of a graphing utility to approximate the fluid force on the vertical plate bounded by the x-axis and the top half of the graph of the equation. Assume that the base of the plate is 12 feet beneath the surface of the water.

29. $x^{2/3} + y^{2/3} = 4^{2/3}$ **30.** $\dfrac{x^2}{28} + \dfrac{y^2}{16} = 1$

31. Think About It

(a) Approximate the depth of the water in the tank in Exercise 5 if the fluid force is one-half as great as when the tank is full.

(b) Explain why the answer in part (a) is not $\frac{3}{2}$.

Getting at the Concept

32. Define fluid pressure.

33. Define fluid force against a submerged vertical plane region.

34. Two identical semicircular windows are placed at the same depth in the vertical wall of an aquarium (see figure). Which has the greater fluid force? Explain.

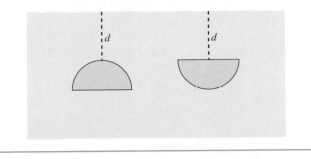

REVIEW EXERCISES FOR CHAPTER 6

6.1   *Area*   In Exercises 1–10, sketch the region bounded by the graphs of the equations, and determine the area of the region.

1. $y = \dfrac{1}{x^2}$, $y = 0$, $x = 1$, $x = 5$ **2.** $y = \dfrac{1}{x^2}$, $y = 4$, $x = 5$

3. $y = \dfrac{1}{x^2 + 1}$, $y = 0$, $x = -1$, $x = 1$

4. $x = y^2 - 2y$, $x = -1$, $y = 0$

5. $y = x$, $y = x^3$

6. $x = y^2 + 1$, $x = y + 3$

7. $y = e^x$, $y = e^2$, $x = 0$

8. $y = \csc x$, $y = 2$ (one region)

9. $y = \sin x$, $y = \cos x$, $\dfrac{\pi}{4} \le x \le \dfrac{5\pi}{4}$

10. $x = \cos y$, $x = \dfrac{1}{2}$, $\dfrac{\pi}{3} \le y \le \dfrac{7\pi}{3}$

In Exercises 11–14, use a graphing utility to graph the region bounded by the graphs of the functions, and use the integration capabilities of the graphing utility to find the area of the region.

11. $y = x^2 - 8x + 3$, $y = 3 + 8x - x^2$

12. $y = x^2 - 4x + 3$, $y = x^3$, $x = 0$

13. $\sqrt{x} + \sqrt{y} = 1$, $y = 0$, $x = 0$

14. $y = x^4 - 2x^2$, $y = 2x^2$

In Exercises 15–18, use vertical and horizontal representative rectangles to set up integrals for finding the area of the region bounded by the graphs of the equations. Find the area of the region by evaluating the easier of the two integrals.

15. $x = y^2 - 2y$, $x = 0$

16. $y = \sqrt{x - 1}$, $y = \dfrac{x - 1}{2}$

17. $y = 1 - \dfrac{x}{2}$, $y = x - 2$, $y = 1$

18. $y = \sqrt{x - 1}$, $y = 2$, $y = 0$, $x = 0$

19. *Think About It*   A person has two job offers. The starting salary for each is \$30,000, and after 10 years of service each will pay \$56,000. The salary increases for each offer are shown in the figure. From a strictly monetary viewpoint, which is the better offer? Explain.

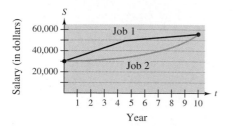

20. *Modeling Data*   The table shows the annual service revenue R_1 in billions of dollars for the cellular telephone industry for the years 1992 through 1998. *(Source: Cellular Telecommunications Industry Association)*

Year	1992	1993	1994	1995	1996	1997	1998
R_1	7.8	10.9	14.2	19.1	23.6	27.5	33.1

(a) Use the regression capabilities of a graphing utility to fit an exponential model to the data. Let t be time in years, with $t = 2$ corresponding to 1992. Use the graphing utility to plot the data and graph the model.

(b) A financial consultant believes that a model for service revenue for the years 2000 through 2005 is

$$R_2 = 10 + 5.28e^{0.2t}.$$

What is the difference in total service revenue between the two models for the years 2000 through 2005?

6.2, 6.3   In Exercises 21–28, find the volume of the solid generated by revolving the plane region bounded by the equations about the indicated lines.

21. $y = x$, $y = 0$, $x = 4$

   (a) the x-axis   (b) the y-axis

   (c) the line $x = 4$ (d) the line $x = 6$

22. $y = \sqrt{x}$, $y = 2$, $x = 0$

   (a) the x-axis   (b) the line $y = 2$

   (c) the y-axis   (d) the line $x = -1$

23. $\dfrac{x^2}{16} + \dfrac{y^2}{9} = 1$ (a) the y-axis (oblate spheroid)

                         (b) the x-axis (prolate spheroid)

24. $\dfrac{x^2}{a^2} + \dfrac{y^2}{b^2} = 1$ (a) the y-axis (oblate spheroid)

                         (b) the x-axis (prolate spheroid)

25. $y = \dfrac{1}{x^4 + 1}$, $y = 0$, $x = 0$, $x = 1$

   revolved about the y-axis

26. $y = \dfrac{1}{\sqrt{1 + x^2}}$, $y = 0$, $x = -1$, $x = 1$

   revolved about the x-axis

27. $y = 1/(1 + \sqrt{x - 2})$, $y = 0$, $x = 2$, $x = 6$

   revolved about the y-axis

28. $y = e^{-x}$, $y = 0$, $x = 0$, $x = 1$

   revolved about the x-axis

In Exercises 29 and 30, consider the region bounded by the graphs of the equations $y = x\sqrt{x + 1}$ and $y = 0$.

29. *Area*   Find the area of the region.

30. *Volume*   Find the volume of the solid generated by revolving the region about (a) the x-axis and (b) the y-axis.

31. Depth of Gasoline in a Tank   A gasoline tank is an oblate spheroid generated by revolving the region bounded by the graph of $(x^2/16) + (y^2/9) = 1$ about the y-axis, where x and y are measured in feet. Find the depth of the gasoline in the tank when it is filled to one-fourth its capacity.

32. Magnitude of a Base   The base of a solid is a circle of radius a, and its vertical cross sections are equilateral triangles. Find the radius of the circle if the volume of the solid is 10 cubic meters.

6.4   Arc Length   In Exercises 33 and 34, find the arc length of the graph of the function over the indicated interval.

33. $f(x) = \dfrac{4}{5}x^{5/4}$, $[0, 4]$ **34.** $y = \dfrac{1}{6}x^3 + \dfrac{1}{2x}$, $[1, 3]$

35. Length of a Catenary   A cable of a suspension bridge forms a catenary modeled by the equation

$$y = 300 \cosh\left(\frac{x}{2000}\right) - 280, \quad -2000 \leq x \leq 2000$$

where x and y are measured in feet. Use a graphing utility to approximate the length of the cable.

36. Approximation   Determine which value best approximates the length of the arc represented by the integral

$$\int_0^{\pi/4} \sqrt{1 + (\sec^2 x)^2}\, dx.$$

(Make your selection on the basis of a sketch of the arc and *not* by performing any calculations.)

(a) -2 (b) 1   (c) π (d) 4   (e) 3

37. Surface Area   Use integration to find the lateral surface area of a right circular cone of height 4 and radius 3.

38. Surface Area   The region bounded by the graphs of $y = 2\sqrt{x}$, $y = 0$, and $x = 3$ is revolved about the x-axis. Find the surface area of the solid generated.

6.5

39. Work   Find the work done in stretching a spring from its natural length of 10 inches to a length of 15 inches, if a force of 4 pounds is needed to stretch it 1 inch from its natural position.

40. Work   Find the work done in stretching a spring from its natural length of 9 inches to double that length. The force required to stretch the spring is 50 pounds.

41. Work   A water well has an 8-inch casing (diameter) and is 175 feet deep. If the water is 25 feet from the top of the well, determine the amount of work done in pumping it dry, assuming that no water enters the well while it is being pumped.

42. Work   Repeat Exercise 41, assuming that water enters the well at a rate of 4 gallons per minute and the pump works at a rate of 12 gallons per minute. How many gallons are pumped in this case?

43. Work   A chain 10 feet long weighs 5 pounds per foot and is hung from a platform 20 feet above the ground. How much work is required to raise the entire chain to the 20-foot level?

44. Work   A windlass, 200 feet above ground level on the top of a building, uses a cable weighing 4 pounds per foot. Find the work done in winding up the cable if

(a) one end is at ground level.

(b) there is a 300-pound load attached to the end of the cable.

45. Work   The work done by a variable force in a press is 80 foot-pounds. The press moves a distance of 4 feet and the force is a quadratic of the form $F = ax^2$. Find a.

46. Work   Find the work done by the force F shown in the figure.

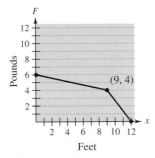

6.6   In Exercises 47–50, find the centroid of the region bounded by the graphs of the equations.

47. $\sqrt{x} + \sqrt{y} = \sqrt{a}$, $x = 0$, $y = 0$ **48.** $y = x^2$, $y = 2x + 3$

49. $y = a^2 - x^2$, $y = 0$ **50.** $y = x^{2/3}$, $y = \dfrac{1}{2}x$

51. Centroid   A blade on an industrial fan has the configuration of a semicircle attached to a trapezoid (see figure). Find the centroid of the blade.

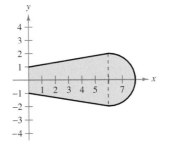

6.7

52. Fluid Force   A swimming pool is 5 feet deep at one end and 10 feet deep at the other, and the bottom is an inclined plane. The length and width of the pool are 40 feet and 20 feet. If the pool is full of water, what is the fluid force on each of the vertical walls?

53. Fluid Force   Show that the fluid force against any vertical region in a liquid is the product of the weight per cubic volume of the liquid, the area of the region, and the depth of the centroid of the region.

54. Fluid Force   Using the result of Exercise 53, find the fluid force on one side of a vertical circular plate of radius 4 feet that is submerged in water so that its center is 5 feet below the surface.

P.S. Problem Solving

1. Let R be the area of the region in the first quadrant bounded by the parabola $y = x^2$ and the line $y = cx$, $c > 0$. Let T be the area of the triangle AOB. Calculate the limit

$$\lim_{c \to 0^+} \frac{T}{R}.$$

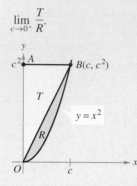

2. Let R be the region bounded by the parabola $y = x - x^2$ and the x-axis. Find the equation of the line $y = mx$ that divides this region into two regions of equal area.

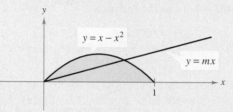

3. (a) A torus is formed by revolving the region bounded by the circle

$$(x - 2)^2 + y^2 = 1$$

about the y-axis (see figure). Use the disk method to calculate the volume of the torus.

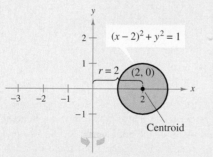

(b) Use the disk method to find the volume of the general torus if the circle has radius r and its center is R units from the axis of rotation.

4. Graph the curve $8y^2 = x^2(1 - x^2)$. Use a computer algebra system to find the surface area of the solid of revolution obtained by revolving the curve about the x-axis.

5. A hole is cut through the center of a sphere of radius r (see figure). The height of the remaining spherical ring is h. Find the volume of the ring and show that it is independent of the radius of the sphere.

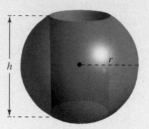

6. A rectangle R of length l and width w is revolved about the line L (see figure). Find the volume of the resulting solid of revolution.

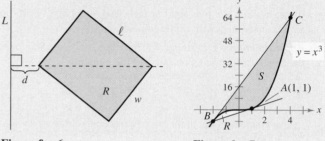

Figure for 6   **Figure for 7**

7. (a) The tangent line to the curve $y = x^3$ at the point $A(1, 1)$ intersects the curve at another point B. Let R be the area of the region bounded by the curve and the tangent line. The tangent line at B intersects the curve at another point C (see figure). Let S be the area of the region bounded by the curve and this second tangent line. How are the areas R and S related?

(b) Repeat the above construction by selecting an arbitrary point A on the curve $y = x^3$. Show that the two areas R and S are always related in the same way.

8. The graph of $y = f(x)$ passes through the origin. The arc length of the curve from $(0, 0)$ to $(x, f(x))$ is given by

$$s(x) = \int_0^x \sqrt{1 + e^t}\, dt.$$

Identify the function f.

9. Let f be rectifiable on the interval $[a, b]$, and let

$$s(x) = \int_a^x \sqrt{1 + [f'(t)]^2}\, dt.$$

(a) Find $\dfrac{ds}{dx}$.

(b) Find ds and $(ds)^2$.

(c) If $f(t) = t^{3/2}$, find $s(x)$ on $[1, 3]$.

(d) Calculate $s(2)$ and describe what it signifies.

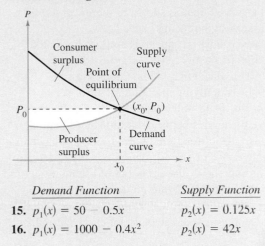

10. The Archimedes Principle states that the upward or buoyant force on an object within a fluid is equal to the weight of the fluid that the object displaces. For a partially submerged object, you can obtain information about the relative densities of the floating object and the fluid by observing how much of the object is above and below the surface. You can also determine the size of a floating object if you know the amount that is above the surface and the relative densities. Suppose you can see the top of a floating iceberg. The density of ocean water is 1.03×10^3 kg/m^3, and that of ice is 0.92×10^3 kg/m^3. What percent of the total iceberg is below the surface?

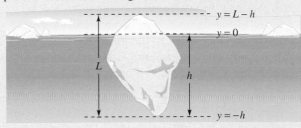

11. Sketch the region bounded on the left by $x = 1$, bounded above by $y = 1/x^3$, and bounded below by $y = -1/x^3$.

(a) Find the centroid of the region for $1 \le x \le 6$.

(b) Find the centroid of the region for $1 \le x \le b$.

(c) Where is the centroid as $b \to \infty$?

12. Sketch the region to the right of the y-axis, bounded above by $y = 1/x^4$ and bounded below by $y = -1/x^4$.

(a) Find the centroid of the region for $1 \le x \le 6$.

(b) Find the centroid of the region for $1 \le x \le b$.

(c) Where is the centroid as $b \to \infty$?

13. Find the work done by each force F.

(a)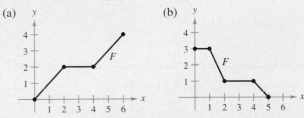

(b)

14. To estimate the surface area of a pond, a surveyor takes several measurements, as shown in the figure. Estimate the surface area of the pond using (a) the Trapezoidal Rule and (b) Simpson's Rule.

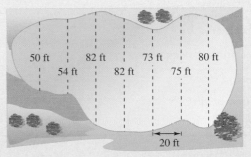

In Exercises 15 and 16, find the consumer surplus and producer surplus for the supply and demand curves. The consumer surplus and producer surplus are represented by the areas shown in the figure.

Demand Function	Supply Function
15. $p_1(x) = 50 - 0.5x$	$p_2(x) = 0.125x$
16. $p_1(x) = 1000 - 0.4x^2$	$p_2(x) = 42x$

17. A swimming pool is 20 feet wide, 40 feet long, 4 feet deep at one end, and 8 feet deep at the other end (see figure). The bottom is an inclined plane. Find the fluid force on each of the vertical walls.

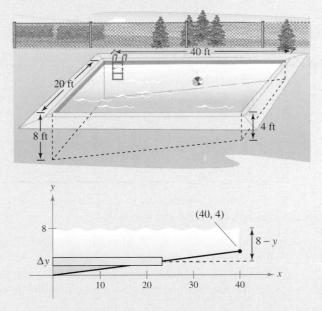

Making a Mercator Map

When flying or sailing, pilots expect to be given a steady compass course to follow. On a standard flat map, this is difficult because a steady compass course results in a curved line, as shown in the lower left and middle figures on the facing page.

For curved lines to appear as straight lines on a flat map, Flemish geographer Gerardus Mercator (1512–1594) realized that latitude lines must be stretched horizontally by a scaling factor of $\sec \phi$, where ϕ is the angle of the latitude line. For the map to preserve the angles between latitude and longitude lines, the lengths of longitude lines are also stretched by a scaling factor of $\sec \phi$ at latitude ϕ. The Mercator map has latitude lines that are not equidistant, as shown in the lower left figure on the facing page.

To calculate these vertical lengths, imagine a globe with latitude lines marked at angles of every $\Delta \phi$ radians,

with $\Delta \phi = \phi_i - \phi_{i-1}$. The arc length of consecutive latitude lines is $R\Delta \phi$. On the Mercator map, the vertical distance between the equator and the first latitude line is $R\Delta \phi \sec \phi_1$. The vertical distance between the first and second latitude lines is $R\Delta \phi \sec \phi_2$. The vertical distance between the second and third latitude lines is $R\Delta \phi \sec \phi_3$, and so on, as shown in the figure on the right below.

On a globe, the angle between consecutive latitude lines is $\Delta \phi$, and the arc length between them is $R\Delta \phi$ (see the left-hand figure below). On a Mercator map, the vertical distance between the ith and $(i - 1)$st latitude lines is $R\Delta \phi \sec \phi_i$, and the distance from the equator to the ith latitude line is approximately

$$R\Delta \phi \sec \phi_1 + R\Delta \phi \sec \phi_2 + \cdots + R\Delta \phi \sec \phi_i$$

(see right-hand figure below).

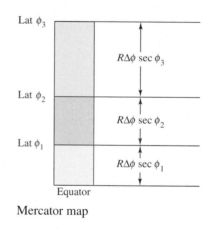

Globe

Mercator map

QUESTIONS

1. Use summation notation to write an expression to calculate how far from the equator to draw the line representing latitude ϕ_n.

2. In the calculations above, Mercator realized that the smaller the value used for $\Delta \phi$, the better the map became (better in the sense that straight lines could be used to plot steady compass courses). From your knowledge of calculus, how could you use Mercator's observation to calculate the total vertical distance of a latitude line from the equator?

3. Use the result of Question 2 to find how far from the equator to place latitude lines whose angles are 10°, 20°, 30°, 40°, and 50°. (Use a globe radius of $R = 6$ inches.)

4. What problem do you encounter when you attempt to calculate how far from the equator to place the North Pole?

The concepts presented here will be explored further in this chapter. For an extension of this application, see Lab 10 in the lab series that accompanies this text at college.hmco.com.

Integration Techniques, L'Hôpital's Rule, and Improper Integrals

7

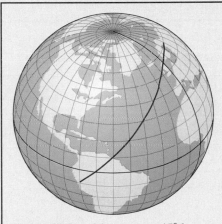

Globe: flight with constant 45° bearing

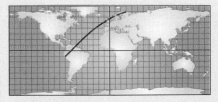

Standard flat map: flight with constant 45° bearing

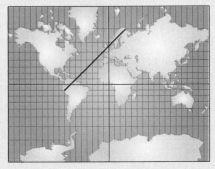

Mercator map: flight with constant 45° bearing

Richard Pasley/Stock Boston

Gerardus Mercator was known as one of the best geographers of the Renaissance. He was also the first to refer to a collection of maps as an "atlas."

Bettmann/Corbis

Section 7.1	Basic Integration Rules

• Review procedures for fitting an integrand to one of the basic integration rules.

Fitting Integrands to Basic Rules

In this chapter, you will study several integration techniques that greatly expand the set of integrals to which the basic integration rules can be applied. The formulas are reviewed on page 484 and on the inside front cover. A major step in solving any integration problem is recognizing the proper basic integration rule to be used. This is not easy. As demonstrated in Example 1, slight differences in the integrand can lead to very different solution techniques.

---

iC   *Example 1*   **A Comparison of Three Similar Integrals**

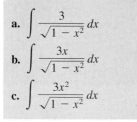

EXPLORATION

A Comparison of Three Similar Integrals   Which, if any, of the following integrals can be evaluated using the 20 basic integration rules? For any that can be evaluated, do so. For any that can't, explain why.

a. $\int \dfrac{3}{\sqrt{1-x^2}}\,dx$

b. $\int \dfrac{3x}{\sqrt{1-x^2}}\,dx$

c. $\int \dfrac{3x^2}{\sqrt{1-x^2}}\,dx$

Evaluate each of the integrals.

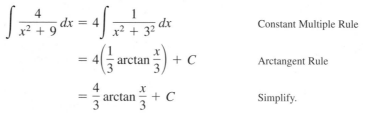

a. $\int \dfrac{4}{x^2+9}\,dx$ **b.** $\int \dfrac{4x}{x^2+9}\,dx$ **c.** $\int \dfrac{4x^2}{x^2+9}\,dx$

Solution

a. Use the Arctangent Rule and let $u = x$ and $a = 3$.

$$\int \frac{4}{x^2+9}\,dx = 4\int \frac{1}{x^2+3^2}\,dx \qquad \text{Constant Multiple Rule}$$

$$= 4\left(\frac{1}{3}\arctan\frac{x}{3}\right) + C \qquad \text{Arctangent Rule}$$

$$= \frac{4}{3}\arctan\frac{x}{3} + C \qquad \text{Simplify.}$$

b. Here the Arctangent Rule does not apply because the numerator contains a factor of x. Consider the Log Rule and let $u = x^2 + 9$. Then $du = 2x\,dx$, and you have

$$\int \frac{4x}{x^2+9}\,dx = 2\int \frac{2x\,dx}{x^2+9} \qquad \text{Constant Multiple Rule}$$

$$= 2\int \frac{du}{u} \qquad \text{Substitution: } u = x^2 + 9$$

$$= 2\ln|u| + C \qquad \text{Log Rule}$$

$$= 2\ln(x^2+9) + C. \qquad \text{Rewrite as a function of } x.$$

c. Because the degree of the numerator is equal to the degree of the denominator, you should first use division to rewrite the improper rational function as the sum of a polynomial and a proper rational function.

$$\int \frac{4x^2}{x^2+9}\,dx = \int \left(4 - \frac{36}{x^2+9}\right)\,dx \qquad \text{Rewrite using long division.}$$

$$= \int 4\,dx - 36\int \frac{1}{x^2+9}\,dx \qquad \text{Write as two integrals.}$$

$$= 4x - 36\left(\frac{1}{3}\arctan\frac{x}{3}\right) + C \qquad \text{Integrate.}$$

$$= 4x - 12\arctan\frac{x}{3} + C \qquad \text{Simplify.}$$

NOTE Notice in Example 1c that some preliminary algebra was required before applying the rules for integration, and that subsequently more than one rule was needed to evaluate the resulting integral.

---

iC   indicates that in the Interactive 3.0 *CD-ROM and* Internet 3.0 *versions of this text (available at* college.hmco.com) *you will find an Open Exploration, which further explores this example using the computer algebra systems* Maple, Mathcad, Mathematica, *and* Derive.

Example 2   Using Two Basic Rules to Solve a Single Integral

Evaluate $\int_0^1 \dfrac{x+3}{\sqrt{4-x^2}}\,dx$.

Solution   Begin by writing the integral as the sum of two integrals. Then apply the Power Rule and the Arcsine Rule as follows.

$$\int_0^1 \frac{x+3}{\sqrt{4-x^2}}\,dx = \int_0^1 \frac{x}{\sqrt{4-x^2}}\,dx + \int_0^1 \frac{3}{\sqrt{4-x^2}}\,dx$$

$$= -\frac{1}{2}\int_0^1 (4-x^2)^{-1/2}(-2x)\,dx + 3\int_0^1 \frac{1}{\sqrt{2^2-x^2}}\,dx$$

$$= \left[-(4-x^2)^{1/2} + 3\arcsin\frac{x}{2}\right]_0^1$$

$$= \left(-\sqrt{3} + \frac{1}{2}\pi\right) - (-2 + 0)$$

$$\approx 1.839$$

(See Figure 7.1.)

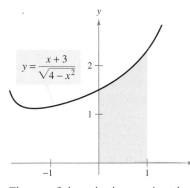

$y = \dfrac{x+3}{\sqrt{4-x^2}}$

The area of the region is approximately 1.839.

Figure 7.1

TECHNOLOGY   Simpson's Rule can be used to give a good approximation of the value of the integral in Example 2 (for $n = 10$, the approximation is 1.839). When using numerical integration, however, you should be aware that Simpson's Rule does not give good approximations when one or both of the limits of integration are near a vertical asymptote. For instance, using the Fundamental Theorem of Calculus, you can obtain

$$\int_0^{1.99} \frac{x+3}{\sqrt{4-x^2}}\,dx \approx 6.213.$$

Applying Simpson's Rule (with $n = 10$) to this integral produces an approximation of 6.889.

Example 3   A Substitution Involving $a^2 - u^2$

Evaluate $\int \dfrac{x^2}{\sqrt{16-x^6}}\,dx$.

Solution   Because the radical in the denominator can be written in the form

$$\sqrt{a^2 - u^2} = \sqrt{4^2 - (x^3)^2}$$

you can try the substitution $u = x^3$. Then $du = 3x^2\,dx$, and you have

$$\int \frac{x^2}{\sqrt{16-x^6}}\,dx = \frac{1}{3}\int \frac{3x^2\,dx}{\sqrt{16-(x^3)^2}} \qquad \text{Rewrite integral.}$$

$$= \frac{1}{3}\int \frac{du}{\sqrt{4^2-u^2}} \qquad \text{Substitution: } u = x^3$$

$$= \frac{1}{3}\arcsin\frac{u}{4} + C \qquad \text{Arcsine Rule}$$

$$= \frac{1}{3}\arcsin\frac{x^3}{4} + C. \qquad \text{Rewrite as a function of } x.$$

STUDY TIP   Rules 18, 19, and 20 of the basic integration rules (see page 484) all have expressions involving the sum or difference of two squares:

$$a^2 - u^2$$

$$a^2 + u^2$$

$$u^2 - a^2$$

With such an expression, consider the substitution $u = f(x)$, as in Example 3.

Surprisingly, two of the most commonly overlooked integration rules are the Log Rule and the Power Rule. Notice in the next two examples how these two integration rules can be disguised.

Example 4   A Disguised Form of the Log Rule

Evaluate $\displaystyle\int \frac{1}{1 + e^x}\, dx$.

Solution   The integral does not appear to fit any of the basic rules. However, the quotient form suggests the Log Rule. If you let $u = 1 + e^x$, then $du = e^x\, dx$. You can obtain the required du by adding and subtracting e^x in the numerator, as follows.

$$\int \frac{1}{1 + e^x}\, dx = \int \frac{1 + e^x - e^x}{1 + e^x}\, dx \qquad \text{Add and subtract } e^x \text{ in numerator.}$$

$$= \int \left(\frac{1 + e^x}{1 + e^x} - \frac{e^x}{1 + e^x} \right) dx \qquad \text{Rewrite as two fractions.}$$

$$= \int dx - \int \frac{e^x\, dx}{1 + e^x} \qquad \text{Rewrite as two integrals.}$$

$$= x - \ln(1 + e^x) + C \qquad \text{Integrate.}$$

NOTE   There is usually more than one way to solve an integration problem. For instance, in Example 4, try integrating by multiplying the numerator and denominator by e^{-x} to obtain an integral of the form $-\int du/u$. See if you can get the same answer by this procedure. (Be careful: the answer will appear in a different form.)

Example 5   A Disguised Form of the Power Rule

Evaluate $\int (\cot x)[\ln(\sin x)]\, dx$.

Solution   Again, the integral does not appear to fit any of the basic rules. However, considering the two primary choices for u ($u = \cot x$ and $u = \ln \sin x$), you can see that the second choice is the appropriate one because

$$u = \ln \sin x \qquad \text{and} \qquad du = \frac{\cos x}{\sin x}\, dx = \cot x\, dx.$$

So, you have

$$\int (\cot x)[\ln(\sin x)]\, dx = \int u\, du \qquad \text{Substitution: } u = \ln \sin x$$

$$= \frac{u^2}{2} + C \qquad \text{Integrate.}$$

$$= \frac{1}{2}[\ln(\sin x)]^2 + C. \qquad \text{Rewrite as a function of } x.$$

NOTE   In Example 5, try *checking* that the derivative of

$$\frac{1}{2}[\ln(\sin x)]^2 + C$$

is the integrand of the original integral.

Review of Basic Integration Rules ($a > 0$)

1. $\displaystyle\int kf(u)\, du = k \int f(u)\, du$

2. $\displaystyle\int [f(u) \pm g(u)]\, du =$
$$\int f(u)\, du \pm \int g(u)\, du$$

3. $\displaystyle\int du = u + C$

4. $\displaystyle\int u^n\, du = \frac{u^{n+1}}{n + 1} + C, \ n \neq -1$

5. $\displaystyle\int \frac{du}{u} = \ln|u| + C$

6. $\displaystyle\int e^u\, du = e^u + C$

7. $\displaystyle\int a^u\, du = \left(\frac{1}{\ln a} \right) a^u + C$

8. $\displaystyle\int \sin u\, du = -\cos u + C$

9. $\displaystyle\int \cos u\, du = \sin u + C$

10. $\displaystyle\int \tan u\, du = -\ln|\cos u| + C$

11. $\displaystyle\int \cot u\, du = \ln|\sin u| + C$

12. $\displaystyle\int \sec u\, du =$
$$\ln|\sec u + \tan u| + C$$

13. $\displaystyle\int \csc u\, du =$
$$-\ln|\csc u + \cot u| + C$$

14. $\displaystyle\int \sec^2 u\, du = \tan u + C$

15. $\displaystyle\int \csc^2 u\, du = -\cot u + C$

16. $\displaystyle\int \sec u \tan u\, du = \sec u + C$

17. $\displaystyle\int \csc u \cot u\, du = -\csc u + C$

18. $\displaystyle\int \frac{du}{\sqrt{a^2 - u^2}} = \arcsin \frac{u}{a} + C$

19. $\displaystyle\int \frac{du}{a^2 + u^2} = \frac{1}{a} \arctan \frac{u}{a} + C$

20. $\displaystyle\int \frac{du}{u\sqrt{u^2 - a^2}} = \frac{1}{a} \operatorname{arcsec} \frac{|u|}{a} + C$

Trigonometric identities can often be used to fit integrals to one of the basic integration rules.

Example 6   Using Trigonometric Identities

Evaluate $\int \tan^2 2x \, dx$.

Solution   Note that $\tan^2 u$ is not in the list of basic integration rules. However, $\sec^2 u$ is in the list. This suggests the trigonometric identity $\tan^2 u = \sec^2 u - 1$. If you let $u = 2x$, then $du = 2 \, dx$ and

$$\int \tan^2 2x \, dx = \frac{1}{2} \int \tan^2 u \, du \qquad \text{Substitution: } u = 2x$$

$$= \frac{1}{2} \int (\sec^2 u - 1) \, du \qquad \text{Trigonometric identity}$$

$$= \frac{1}{2} \int \sec^2 u \, du - \frac{1}{2} \int du \qquad \text{Rewrite as two integrals.}$$

$$= \frac{1}{2} \tan u - \frac{u}{2} + C \qquad \text{Integrate.}$$

$$= \frac{1}{2} \tan 2x - x + C. \qquad \text{Rewrite as a function of } x.$$

> **TECHNOLOGY**   If you have access to a computer algebra system, try using it to evaluate the integrals in this section. Compare the *form* of the antiderivative given by the software with the form obtained by hand. Sometimes the forms will be the same, but often they will differ. For instance, why is the antiderivative $\ln 2x + C$ equivalent to the antiderivative $\ln x + C$?

We conclude this section with a summary of the common procedures for fitting integrands to the basic integration rules.

Procedures for Fitting Integrands to Basic Rules

Technique	Example
Expand (numerator).	$(1 + e^x)^2 = 1 + 2e^x + e^{2x}$
Separate numerator.	$\dfrac{1 + x}{x^2 + 1} = \dfrac{1}{x^2 + 1} + \dfrac{x}{x^2 + 1}$
Complete the square.	$\dfrac{1}{\sqrt{2x - x^2}} = \dfrac{1}{\sqrt{1 - (x - 1)^2}}$
Divide improper rational function.	$\dfrac{x^2}{x^2 + 1} = 1 - \dfrac{1}{x^2 + 1}$
Add and subtract terms in numerator.	$\dfrac{2x}{x^2 + 2x + 1} = \dfrac{2x + 2 - 2}{x^2 + 2x + 1} = \dfrac{2x + 2}{x^2 + 2x + 1} - \dfrac{2}{(x + 1)^2}$
Use trigonometric identities.	$\cot^2 x = \csc^2 x - 1$
Multiply and divide by Pythagorean conjugate.	$\dfrac{1}{1 + \sin x} = \left(\dfrac{1}{1 + \sin x}\right)\left(\dfrac{1 - \sin x}{1 - \sin x}\right) = \dfrac{1 - \sin x}{1 - \sin^2 x}$
	$= \dfrac{1 - \sin x}{\cos^2 x} = \sec^2 x - \dfrac{\sin x}{\cos^2 x}$

NOTE   Remember that you can separate numerators but not denominators. Watch out for this common error when fitting integrands to basic rules.

$$\frac{1}{x^2 + 1} \ne \frac{1}{x^2} + \frac{1}{1} \qquad \text{Do not separate denominators.}$$

EXERCISES FOR SECTION 7.1

In Exercises 1–4, select the correct antiderivative.

1. $\dfrac{dy}{dx} = \dfrac{x}{\sqrt{x^2 + 1}}$

  (a) $2\sqrt{x^2 + 1} + C$ (b) $\sqrt{x^2 + 1} + C$

  (c) $\frac{1}{2}\sqrt{x^2 + 1} + C$ (d) $\ln(x^2 + 1) + C$

2. $\dfrac{dy}{dx} = \dfrac{x}{x^2 + 1}$

  (a) $\ln\sqrt{x^2 + 1} + C$ (b) $\dfrac{2x}{(x^2 + 1)^2} + C$

  (c) $\arctan x + C$ (d) $\ln(x^2 + 1) + C$

3. $\dfrac{dy}{dx} = \dfrac{1}{x^2 + 1}$

  (a) $\ln\sqrt{x^2 + 1} + C$ (b) $\dfrac{2x}{(x^2 + 1)^2} + C$

  (c) $\arctan x + C$ (d) $\ln(x^2 + 1) + C$

4. $\dfrac{dy}{dx} = x \cos(x^2 + 1)$

  (a) $2x \sin(x^2 + 1) + C$ (b) $-\frac{1}{2}\sin(x^2 + 1) + C$

  (c) $\frac{1}{2}\sin(x^2 + 1) + C$ (d) $-2x \sin(x^2 + 1) + C$

In Exercises 5–14, select the basic integration formula you can use to evaluate the integral, and identify u and a when appropriate.

5. $\displaystyle\int (3x - 2)^4 \, dx$

6. $\displaystyle\int \dfrac{2t - 1}{t^2 - t + 2} \, dt$

7. $\displaystyle\int \dfrac{1}{\sqrt{x}\left(1 - 2\sqrt{x}\right)} \, dx$

8. $\displaystyle\int \dfrac{2}{(2t - 1)^2 + 4} \, dt$

9. $\displaystyle\int \dfrac{3}{\sqrt{1 - t^2}} \, dt$

10. $\displaystyle\int \dfrac{-2x}{\sqrt{x^2 - 4}} \, dx$

11. $\displaystyle\int t \sin t^2 \, dt$

12. $\displaystyle\int \sec 3x \tan 3x \, dx$

13. $\displaystyle\int \cos x e^{\sin x} \, dx$

14. $\displaystyle\int \dfrac{1}{x\sqrt{x^2 - 4}} \, dx$

In Exercises 15–54, evaluate the indefinite integral.

15. $\displaystyle\int (-2x + 5)^{3/2} \, dx$

16. $\displaystyle\int 6(x - 4)^5 \, dx$

17. $\displaystyle\int \dfrac{5}{(z - 4)^5} \, dz$

18. $\displaystyle\int \dfrac{2}{(t - 9)^2} \, dt$

19. $\displaystyle\int t^2 \sqrt[3]{t^3 - 1} \, dt$

20. $\displaystyle\int x\sqrt{4 - 2x^2} \, dx$

21. $\displaystyle\int \left[v + \dfrac{1}{(3v - 1)^3}\right] dv$

22. $\displaystyle\int \left[x - \dfrac{3}{(2x + 3)^2}\right] dx$

23. $\displaystyle\int \dfrac{t^2 - 3}{-t^3 + 9t + 1} \, dt$

24. $\displaystyle\int \dfrac{x + 1}{\sqrt{x^2 + 2x - 4}} \, dx$

25. $\displaystyle\int \dfrac{x^2}{x - 1} \, dx$

26. $\displaystyle\int \dfrac{2x}{x - 4} \, dx$

27. $\displaystyle\int \dfrac{e^x}{1 + e^x} \, dx$

28. $\displaystyle\int \left(\dfrac{1}{3x - 1} - \dfrac{1}{3x + 1}\right) dx$

29. $\displaystyle\int (1 + 2x^2)^2 \, dx$

30. $\displaystyle\int x\left(1 + \dfrac{1}{x}\right)^3 \, dx$

31. $\displaystyle\int x \cos 2\pi x^2 \, dx$

32. $\displaystyle\int \sec 4x \, dx$

33. $\displaystyle\int \csc \pi x \cot \pi x \, dx$

34. $\displaystyle\int \dfrac{\sin x}{\sqrt{\cos x}} \, dx$

35. $\displaystyle\int e^{5x} \, dx$

36. $\displaystyle\int \csc^2 x e^{\cot x} \, dx$

37. $\displaystyle\int \dfrac{2}{e^{-x} + 1} \, dx$

38. $\displaystyle\int \dfrac{5}{3e^x - 2} \, dx$

39. $\displaystyle\int \dfrac{\ln x^2}{x} \, dx$

40. $\displaystyle\int (\tan x)[\ln(\cos x)] \, dx$

41. $\displaystyle\int \dfrac{1 + \sin x}{\cos x} \, dx$

42. $\displaystyle\int \dfrac{1 + \cos \alpha}{\sin \alpha} \, d\alpha$

43. $\displaystyle\int \dfrac{1}{\cos \theta - 1} \, d\theta$

44. $\displaystyle\int \dfrac{2}{3(\sec x - 1)} \, dx$

45. $\displaystyle\int \dfrac{3z + 2}{z^2 + 9} \, dz$

46. $\displaystyle\int \dfrac{3}{t^2 + 1} \, dt$

47. $\displaystyle\int \dfrac{-1}{\sqrt{1 - (2t - 1)^2}} \, dt$

48. $\displaystyle\int \dfrac{1}{4 + 3x^2} \, dx$

49. $\displaystyle\int \dfrac{\tan(2/t)}{t^2} \, dt$

50. $\displaystyle\int \dfrac{e^{1/t}}{t^2} \, dt$

51. $\displaystyle\int \dfrac{3}{\sqrt{6x - x^2}} \, dx$

52. $\displaystyle\int \dfrac{1}{(x - 1)\sqrt{4x^2 - 8x + 3}} \, dx$

53. $\displaystyle\int \dfrac{4}{4x^2 + 4x + 65} \, dx$

54. $\displaystyle\int \dfrac{1}{\sqrt{1 - 4x - x^2}} \, dx$

Slope Fields **In Exercises 55 and 56, a differential equation, a point, and a slope field are given. (a) Sketch two approximate solutions of the differential equation on the slope field, one of which passes through the indicated point. (b) Use integration to find the particular solution of the differential equation and use a graphing utility to graph the solution. Compare the result with the sketches in part (a). To print an enlarged copy of the graph, go the the website *www.mathgraphs.com*.**

55. $\dfrac{ds}{dt} = \dfrac{t}{\sqrt{1 - t^4}}, \quad \left(0, -\frac{1}{2}\right)$ **56.** $\dfrac{dy}{dx} = \tan^2(2x), \quad (0, 0)$

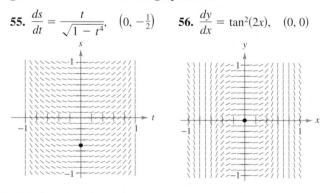

In Exercises 57 and 58, use a computer algebra system to sketch the slope field for the differential equation and graph the solution through the specified initial condition.

57. $\dfrac{dy}{dx} = 0.2y, \quad y(0) = 3$ **58.** $\dfrac{dy}{dx} = 5 - y, \quad y(0) = 1$

In Exercises 59–62, solve the differential equation.

59. $\dfrac{dy}{dx} = (1 + e^x)^2$

60. $\dfrac{dr}{dt} = \dfrac{(1 + e^t)^2}{e^t}$

61. $(4 + \tan^2 x)\, y' = \sec^2 x$

62. $y' = \dfrac{1}{x\sqrt{4x^2 - 1}}$

In Exercises 63–70, evaluate the definite integral. Use the integration capabilities of a graphing utility to verify your result.

63. $\displaystyle\int_0^{\pi/4} \cos 2x\, dx$

64. $\displaystyle\int_0^{\pi} \sin^2 t \cos t\, dt$

65. $\displaystyle\int_0^1 xe^{-x^2}\, dx$

66. $\displaystyle\int_1^e \dfrac{1 - \ln x}{x}\, dx$

67. $\displaystyle\int_0^4 \dfrac{2x}{\sqrt{x^2 + 9}}\, dx$

68. $\displaystyle\int_1^2 \dfrac{x - 2}{x}\, dx$

69. $\displaystyle\int_0^{2/\sqrt{3}} \dfrac{1}{4 + 9x^2}\, dx$

70. $\displaystyle\int_0^4 \dfrac{1}{\sqrt{25 - x^2}}\, dx$

In Exercises 71–74, use a computer algebra system to evaluate the integral. Use the computer algebra system to graph two antiderivatives. Describe the relationship between the two graphs of the antiderivatives.

71. $\displaystyle\int \dfrac{1}{x^2 + 4x + 13}\, dx$

72. $\displaystyle\int \dfrac{x - 2}{x^2 + 4x + 13}\, dx$

73. $\displaystyle\int \dfrac{1}{1 + \sin \theta}\, d\theta$

74. $\displaystyle\int \left(\dfrac{e^x + e^{-x}}{2}\right)^3 dx$

Getting at the Concept

In Exercises 75–78, state the integration formula you would use to perform the integration. Do not integrate.

75. $\displaystyle\int x(x^2 + 1)^3\, dx$

76. $\displaystyle\int x \sec(x^2 + 1) \tan(x^2 + 1)\, dx$

77. $\displaystyle\int \dfrac{x}{x^2 + 1}\, dx$

78. $\displaystyle\int \dfrac{1}{x^2 + 1}\, dx$

79. Explain why the antiderivative $y_1 = e^{x + C_1}$ is equivalent to the antiderivative $y_2 = Ce^x$.

80. Explain why the antiderivative $y_1 = \sec^2 x + C_1$ is equivalent to the antiderivative $y_2 = \tan^2 x + C$.

81. Determine the constants a and b such that

$$\sin x + \cos x = a \sin(x + b).$$

Use this result to integrate $\displaystyle\int \dfrac{dx}{\sin x + \cos x}$.

82. *Think About It*   Use a graphing utility to graph the function $f(x) = \frac{1}{5}(x^3 - 7x^2 + 10x)$. Use the graph to determine whether

$$\int_0^5 f(x)\, dx$$

is positive or negative. Explain.

Approximation   **In Exercises 83 and 84, determine which value best approximates the area of the region between the x-axis and the function over the given interval. (Make your selection on the basis of a sketch of the region and *not* by integrating.)**

83. $f(x) = \dfrac{4x}{x^2 + 1}$, $[0, 2]$

   (a) 3   (b) 1   (c) -8 (d) 8   (e) 10

84. $f(x) = \dfrac{4}{x^2 + 1}$, $[0, 2]$

   (a) 3   (b) 1   (c) -4 (d) 4   (e) 10

Area   **In Exercises 85 and 86, find the area of the region bounded by the graph(s) of the equation(s).**

85. $y^2 = x^2(1 - x^2)$

86. $y = \sin 2x$, $y = 0$, $x - 0$, $x = \pi/2$

87. *Area*   The graphs of $f(x) = x$ and $g(x) = ax^2$ intersect at the points $(0, 0)$ and $(1/a, 1/a)$. Find a $(a > 0)$ such that the area of the region bounded by the graphs of these two functions is $\frac{2}{3}$.

88. *Interpreting an Integral*   You are given the integral

$$\int_0^2 2\pi x^2\, dx$$

but are not told what it represents. (There is more than one correct answer for each part.)

   (a) Sketch the region whose area is given by the integral.

   (b) Sketch the solid whose volume is given by the integral if the disk method is used.

   (c) Sketch the solid whose volume is given by the integral if the shell method is used.

89. *Volume*   The region bounded by $y = e^{-x^2}$, $y = 0$, $x = 0$, and $x = b$ $(b > 0)$ is revolved about the y-axis.

   (a) Find the volume of the solid generated if $b = 1$.

   (b) Find b such that the volume of the generated solid is $\frac{4}{3}$ cubic units.

90. *Average Value*   Compute the average value of each of the functions over the indicated interval.

   (a) $f(x) - \sin nx$, $0 \le x \le \pi/n$, n is a positive integer

   (b) $f(x) = \dfrac{1}{1 + x^2}$, $-3 \le x \le 3$

91. *Centroid*   Find the x-coordinate of the centroid of the region bounded by the graphs of

$$y = \dfrac{5}{\sqrt{25 - x^2}}, \quad y = 0, \quad x = 0, \quad \text{and} \quad x = 4.$$

92. *Surface Area*   Find the area of the surface formed by revolving the graph of $y = 2\sqrt{x}$ on the interval $[0, 9]$ about the x-axis.

Arc Length   **In Exercises 93 and 94, use the integration capabilities of a graphing utility to approximate the arc length of the curve over the indicated interval.**

93. $y = \tan \pi x$, $\left[0, \frac{1}{4}\right]$

94. $y = x^{2/3}$, $[1, 8]$

Section 7.2	Integration by Parts

- Find an antiderivative using integration by parts.
- Use a tabular method to perform integration by parts.

Integration by Parts

In this section you will study an important integration technique called **integration by parts.** This technique can be applied to a wide variety of functions and is particularly useful for integrands involving *products* of algebraic and transcendental functions. For instance, integration by parts works well with integrals such as

$$\int x \ln x \, dx, \quad \int x^2 e^x \, dx, \quad \text{and} \quad \int e^x \sin x \, dx.$$

Integration by parts is based on the formula for the derivative of a product

$$\frac{d}{dx}[uv] = u\frac{dv}{dx} + v\frac{du}{dx}$$

$$= uv' + vu'$$

where both u and v are differentiable functions of x. If u' and v' are continuous, you can integrate both sides of this equation to obtain

$$uv = \int uv' \, dx + \int vu' \, dx$$

$$= \int u \, dv + \int v \, du.$$

By rewriting this equation, you obtain the following theorem.

THEOREM 7.1   Integration by Parts

If u and v are functions of x and have continuous derivatives, then

$$\int u \, dv = uv - \int v \, du.$$

This formula expresses the original integral in terms of another integral. Depending on the choices of u and dv, it may be easier to evaluate the second integral than the original one. Because the choices of u and dv are critical in the integration by parts process, the following guidelines are provided.

Guidelines for Integration by Parts

1. Try letting dv be the most complicated portion of the integrand that fits a basic integration rule. Then u will be the remaining factor(s) of the integrand.

2. Try letting u be the portion of the integrand whose derivative is a function simpler than u. Then dv will be the remaining factor(s) of the integrand.

EXPLORATION

Proof Without Words   Here is a different approach to proving the formula for integration by parts. Exercise taken from "Proof Without Words: Integration by Parts" by Roger B. Nelsen, *Mathematics Magazine*, April 1991. Used by permission of the author.

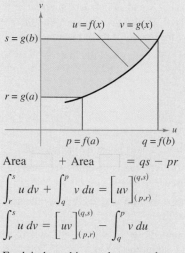

Area ☐ + Area ☐ = $qs - pr$

$$\int_r^s u \, dv + \int_q^p v \, du = \left[uv \right]_{(p,r)}^{(q,s)}$$

$$\int_r^s u \, dv = \left[uv \right]_{(p,r)}^{(q,s)} - \int_q^p v \, du$$

Explain how this graph proves the theorem. Which notation in this proof is unfamiliar? What do you think it means?

Example 1   Integration by Parts

Evaluate $\int xe^x \, dx$.

Solution To apply integration by parts, you need to write the integral in the form $\int u \, dv$. There are several ways to do this.

$$\int \underbrace{(x)}_{u} \underbrace{(e^x \, dx)}_{dv}, \quad \int \underbrace{(e^x)}_{u} \underbrace{(x \, dx)}_{dv}, \quad \int \underbrace{(1)}_{u} \underbrace{(xe^x \, dx)}_{dv}, \quad \int \underbrace{(xe^x)}_{u} \underbrace{(dx)}_{dv}$$

The guidelines on page 488 suggest choosing the first option because the derivative of $u = x$ is simpler than x, and $dv = e^x \, dx$ is the most complicated portion of the integrand that fits a basic integration formula.

$$dv = e^x \, dx \quad \Longrightarrow \quad v = \int dv = \int e^x \, dx = e^x$$

$$u = x \quad \Longrightarrow \quad du = dx$$

Now, integration by parts produces the following.

$$\int u \, dv = uv - \int v \, du \qquad \text{Integration by parts formula}$$

$$\int xe^x \, dx = xe^x - \int e^x \, dx \qquad \text{Substitute.}$$

$$= xe^x - e^x + C \qquad \text{Integrate.}$$

To check this, differentiate $xe^x - e^x + C$ to see that you obtain the original integrand.

NOTE In Example 1, note that it is not necessary to include a constant of integration when solving

$$v = \int e^x \, dx = e^x + C_1.$$

To illustrate this, replace $v = e^x$ by $v = e^x + C_1$ and apply integration by parts to see that you obtain the same result.

Example 2   Integration by Parts

Evaluate $\int x^2 \ln x \, dx$.

Solution In this case, x^2 is more easily integrated than $\ln x$. Furthermore, the derivative of $\ln x$ is simpler than $\ln x$. Therefore, you should let $dv = x^2 \, dx$.

$$dv = x^2 \, dx \quad \Longrightarrow \quad v = \int x^2 \, dx = \frac{x^3}{3}$$

$$u = \ln x \quad \Longrightarrow \quad du = \frac{1}{x} \, dx$$

Integration by parts produces the following.

$$\int u \, dv = uv - \int v \, du \qquad \text{Integration by parts formula}$$

$$\int x^2 \ln x \, dx = \frac{x^3}{3} \ln x - \int \left(\frac{x^3}{3}\right)\left(\frac{1}{x}\right) dx \qquad \text{Substitute.}$$

$$= \frac{x^3}{3} \ln x - \frac{1}{3} \int x^2 \, dx \qquad \text{Simplify.}$$

$$= \frac{x^3}{3} \ln x - \frac{x^3}{9} + C \qquad \text{Integrate.}$$

You can check this result by differentiating.

$$\frac{d}{dx}\left[\frac{x^3}{3} \ln x - \frac{x^3}{9}\right] = \frac{x^3}{3}\left(\frac{1}{x}\right) + (\ln x)(x^2) - \frac{x^2}{3} = x^2 \ln x$$

FOR FURTHER INFORMATION To see how integration by parts is used to prove Sterling's approximation

$$\ln(n!) = n \ln n - n,$$

see the article "The Validity of Stirling's Approximation: A Physical Chemistry Project" by A. S. Wallner and K. A. Brandt in *Journal of Chemical Education*. To view this article, go to the website *www.matharticles.com*.

TECHNOLOGY Try graphing

$$\int x^2 \ln x \, dx \quad \text{and} \quad \frac{x^3}{3} \ln x - \frac{x^3}{9}$$

on your graphing utility. Do you get the same graph? (This will take a while, so be patient.)

One surprising application of integration by parts involves integrands consisting of a single factor, such as $\int \ln x \, dx$ or $\int \arcsin x \, dx$. In such cases, you should let $dv = dx$, as illustrated in the next example.

Example 3   An Integrand with a Single Term

Evaluate $\displaystyle\int_0^1 \arcsin x \, dx$.

Solution   Let $dv = dx$.

$$dv = dx \quad \Longrightarrow \quad v = \int dx = x$$

$$u = \arcsin x \quad \Longrightarrow \quad du = \frac{1}{\sqrt{1 - x^2}} \, dx$$

Integration by parts now produces the following.

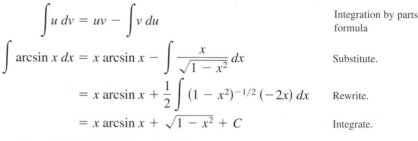

$$\int u \, dv = uv - \int v \, du \qquad \text{Integration by parts formula}$$

$$\int \arcsin x \, dx = x \arcsin x - \int \frac{x}{\sqrt{1 - x^2}} \, dx \qquad \text{Substitute.}$$

$$= x \arcsin x + \frac{1}{2} \int (1 - x^2)^{-1/2} \, (-2x) \, dx \qquad \text{Rewrite.}$$

$$= x \arcsin x + \sqrt{1 - x^2} + C \qquad \text{Integrate.}$$

Using this antiderivative, you can evaluate the definite integral as follows.

$$\int_0^1 \arcsin x \, dx = \left[x \arcsin x + \sqrt{1 - x^2} \right]_0^1$$

$$= \frac{\pi}{2} - 1$$

$$\approx 0.571$$

The area represented by this definite integral is shown in Figure 7.2.

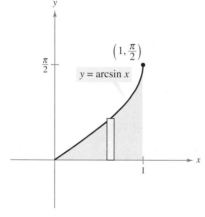

The area of the region is approximately 0.571.

Figure 7.2

TECHNOLOGY   Remember that there are two ways to use technology to evaluate a definite integral: (1) you can use a numerical approximation such as the Trapezoidal Rule or Simpson's Rule, or (2) you can use a computer algebra system to find the antiderivative and then apply the Fundamental Theorem of Calculus. Both methods have shortcomings. To find the possible error when using a numerical method, the integrand must have a second derivative (Trapezoidal Rule) or a fourth derivative (Simpson's Rule) in the interval of integration: the integrand in Example 3 fails to meet this requirement. To apply the Fundamental Theorem of Calculus, the symbolic integration utility must be able to find the antiderivative.

Which method would you use to evaluate

$$\int_0^1 \arctan x \, dx?$$

Which method would you use to evaluate

$$\int_0^1 \arctan x^2 \, dx?$$

Some integrals require repeated use of the integration by parts formula.

Example 4   **Repeated Use of Integration by Parts**

Evaluate $\int x^2 \sin x \, dx$.

Solution   The factors x^2 and $\sin x$ are equally easy to integrate. However, the derivative of x^2 becomes simpler, whereas the derivative of $\sin x$ does not. Therefore, you should let $u = x^2$.

$$dv = \sin x \, dx \quad \Longrightarrow \quad v = \int \sin x \, dx = -\cos x$$

$$u = x^2 \quad \Longrightarrow \quad du = 2x \, dx$$

Now, integration by parts produces the following.

$$\int x^2 \sin x \, dx = -x^2 \cos x + \int 2x \cos x \, dx \qquad \text{First use of integration by parts}$$

This first use of integration by parts has succeeded in simplifying the original integral, but the integral on the right still doesn't fit a basic integration rule. To evaluate that integral, you can apply integration by parts again. This time, let $u = 2x$.

$$dv = \cos x \, dx \quad \Longrightarrow \quad v = \int \cos x \, dx = \sin x$$

$$u = 2x \quad \Longrightarrow \quad du = 2 \, dx$$

Now, integration by parts produces

$$\int 2x \cos x \, dx = 2x \sin x - \int 2 \sin x \, dx \qquad \text{Second use of integration by parts}$$

$$= 2x \sin x + 2 \cos x + C.$$

Combining these two results, you can write

$$\int x^2 \sin x \, dx = -x^2 \cos x + 2x \sin x + 2 \cos x + C.$$

When making repeated applications of integration by parts, you need to be careful not to interchange the substitutions in successive applications. For instance, in Example 4, the first substitution was $u = x^2$ and $dv = \sin x \, dx$. If, in the second application, you had switched the substitution to $u = \cos x$ and $dv = 2x$, you would have obtained

$$\int x^2 \sin x \, dx = -x^2 \cos x + \int 2x \cos x \, dx$$

$$= -x^2 \cos x + x^2 \cos x + \int x^2 \sin x \, dx$$

$$= \int x^2 \sin x \, dx$$

thus undoing the previous integration and returning to the *original* integral. When making repeated applications of integration by parts, you should also watch for the appearance of a *constant multiple* of the original integral. For instance, this occurs when you use integration by parts to evaluate $\int e^x \cos 2x \, dx$, and also occurs in the next example.

EXPLORATION

Try to evaluate

$$\int e^x \cos 2x \, dx$$

by letting $u = \cos 2x$ and $dv = e^x \, dx$ in the first substitution. For the second substitution, let $u = \sin 2x$ and $dv = e^x \, dx$.

NOTE The integral in Example 5 is an important one. In Section 7.4 (Example 5), you will see that it is used to find the arc length of a parabolic segment.

STUDY TIP The trigonometric identities

$$\sin^2 x = \frac{1 - \cos 2x}{2}$$

$$\cos^2 x = \frac{1 + \cos 2x}{2}$$

play an important role in this chapter.

Example 5 Integration by Parts

Evaluate $\int \sec^3 x\, dx$.

Solution The most complicated portion of the integrand that can be easily integrated is $\sec^2 x$, so you should let $dv = \sec^2 x\, dx$ and $u = \sec x$.

$$dv = \sec^2 x\, dx \implies v = \int \sec^2 x\, dx = \tan x$$

$$u = \sec x \implies du = \sec x \tan x\, dx$$

Integration by parts produces the following.

$$\int u\, dv = uv - \int v\, du \qquad \text{Integration by parts formula}$$

$$\int \sec^3 x\, dx = \sec x \tan x - \int \sec x \tan^2 x\, dx \qquad \text{Substitute.}$$

$$\int \sec^3 x\, dx = \sec x \tan x - \int \sec x (\sec^2 x - 1)\, dx \qquad \text{Trigonometric identity}$$

$$\int \sec^3 x\, dx = \sec x \tan x - \int \sec^3 x\, dx + \int \sec x\, dx \qquad \text{Rewrite.}$$

$$2 \int \sec^3 x\, dx = \sec x \tan x + \int \sec x\, dx \qquad \text{Collect like integrals.}$$

$$\int \sec^3 x\, dx = \frac{1}{2} \sec x \tan x + \frac{1}{2} \ln|\sec x + \tan x| + C \qquad \text{Integrate and divide by 2.}$$

Example 6 Finding a Centroid

A machine part is modeled by the region bounded by the graph of $y = \sin x$ and the x-axis, $0 \le x \le \pi/2$, as shown in Figure 7.3. Find the centroid of this region.

Solution Begin by finding the area of the region.

$$A = \int_0^{\pi/2} \sin x\, dx = \Big[-\cos x \Big]_0^{\pi/2} = 1$$

Now, you can find the coordinates of the centroid as follows.

$$\bar{y} = \frac{1}{A} \int_0^{\pi/2} \frac{\sin x}{2}(\sin x)\, dx = \frac{1}{4} \int_0^{\pi/2} (1 - \cos 2x)\, dx = \frac{1}{4}\Big[x - \frac{\sin 2x}{2} \Big]_0^{\pi/2} = \frac{\pi}{8}$$

You can evaluate the integral for $\bar{x}$, $(1/A) \int_0^{\pi/2} x \sin x\, dx$, with integration by parts. To do this, let $dv = \sin x\, dx$ and $u = x$. This produces $v = -\cos x$ and $du = dx$, and you can write

$$\int x \sin x\, dx = -x \cos x + \int \cos x\, dx$$

$$= -x \cos x + \sin x + C.$$

Finally, you can determine $\bar{x}$ to be

$$\bar{x} = \frac{1}{A} \int_0^{\pi/2} x \sin x\, dx = \Big[-x \cos x + \sin x \Big]_0^{\pi/2} = 1.$$

Therefore, the centroid of the region is $(1, \pi/8)$.

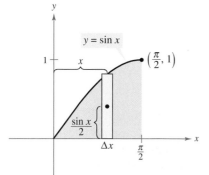

Figure 7.3

As you gain experience in using integration by parts, your skill in determining u and dv will increase. The following summary lists several common integrals with suggestions for the choices of u and dv.

Summary of Common Integrals Using Integration by Parts

1. For integrals of the form

$$\int x^n e^{ax}\, dx, \qquad \int x^n \sin ax\, dx, \qquad \text{or} \qquad \int x^n \cos ax\, dx$$

let $u = x^n$ and let $dv = e^{ax}\, dx$, $\sin ax\, dx$, or $\cos ax\, dx$.

2. For integrals of the form

$$\int x^n \ln x\, dx, \qquad \int x^n \arcsin ax\, dx, \qquad \text{or} \qquad \int x^n \arctan ax\, dx$$

let $u = \ln x$, $\arcsin ax$, or $\arctan ax$ and let $dv = x^n\, dx$.

3. For integrals of the form

$$\int e^{ax} \sin bx\, dx \qquad \text{or} \qquad \int e^{ax} \cos bx\, dx$$

let $u = \sin bx$ or $\cos bx$ and let $dv = e^{ax}\, dx$.

STUDY TIP You can use the acronym LIATE as a guideline for choosing u in integration by parts. In order, check the integrand for the following.

Is there a **L**ogarithmic part?

Is there an **I**nverse trigonometric part?

Is there an **A**lgebraic part?

Is there a **T**rigonometric part?

Is there an **E**xponential part?

Tabular Method

In problems involving repeated applications of integration by parts, a tabular method, illustrated in Example 7, can help to organize the work. This method works well for integrals of the form $\int x^n \sin ax\, dx$, $\int x^n \cos ax\, dx$, and $\int x^n e^{ax}\, dx$.

Example 7   Using the Tabular Method

Evaluate $\int x^2 \sin 4x\, dx$.

Solution Begin as usual by letting $u = x^2$ and $dv = v'\, dx = \sin 4x\, dx$. Next, create a table consisting of three columns, as follows.

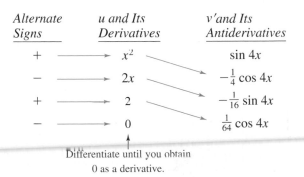

Alternate Signs	u and Its Derivatives	v' and Its Antiderivatives
$+$	x^2	$\sin 4x$
$-$	$2x$	$-\frac{1}{4}\cos 4x$
$+$	2	$-\frac{1}{16}\sin 4x$
$-$	0	$\frac{1}{64}\cos 4x$

Differentiate until you obtain
0 as a derivative.

FOR FURTHER INFORMATION For more information on the tabular method, see the article "Tabular Integration by Parts" by David Horowitz in *The College Mathematics Journal*, and the article "More on Tabular Integration by Parts" by Leonard Gillman in *The College Mathematics Journal*. To view these articles, go to the website *www.matharticles.com*.

The solution is obtained by adding the signed products of the diagonal entries:

$$\int x^2 \sin 4x\, dx = -\frac{1}{4} x^2 \cos 4x + \frac{1}{8} x \sin 4x + \frac{1}{32} \cos 4x + C.$$

EXERCISES FOR SECTION 7.2

In Exercises 1–4, match the antiderivative with the correct integral. [Integrals are labeled (a), (b), (c), and (d).]

(a) $\int \ln x \, dx$ **(b)** $\int x \sin x \, dx$

(c) $\int x^2 e^x \, dx$ **(d)** $\int x^2 \cos x \, dx$

1. $y = \sin x - x \cos x$

2. $y = x^2 \sin x + 2x \cos x - 2 \sin x$

3. $y = x^2 e^x - 2xe^x + 2e^x$

4. $y = -x + x \ln x$

In Exercises 5–10, identify u and dv for evaluating the integral using integration by parts. (Do not evaluate the integral.)

5. $\displaystyle\int xe^{2x} \, dx$ **6.** $\displaystyle\int x^2 e^{2x} \, dx$

7. $\displaystyle\int (\ln x)^2 \, dx$ **8.** $\displaystyle\int \ln 3x \, dx$

9. $\displaystyle\int x \sec^2 x \, dx$ **10.** $\displaystyle\int x^2 \cos x \, dx$

In Exercises 11–36, evaluate the integral. (*Note:* Solve by the simplest method—not all require integration by parts.)

11. $\displaystyle\int xe^{-2x} \, dx$ **12.** $\displaystyle\int \frac{2x}{e^x} \, dx$

13. $\displaystyle\int x^3 e^x \, dx$ **14.** $\displaystyle\int \frac{e^{1/t}}{t^2} \, dt$

15. $\displaystyle\int x^2 e^{x^3} \, dx$ **16.** $\displaystyle\int x^4 \ln x \, dx$

17. $\displaystyle\int t \ln(t + 1) \, dt$ **18.** $\displaystyle\int \frac{1}{x(\ln x)^3} \, dx$

19. $\displaystyle\int \frac{(\ln x)^2}{x} \, dx$ **20.** $\displaystyle\int \frac{\ln x}{x^2} \, dx$

21. $\displaystyle\int \frac{xe^{2x}}{(2x + 1)^2} \, dx$ **22.** $\displaystyle\int \frac{x^3 e^{x^2}}{(x^2 + 1)^2} \, dx$

23. $\displaystyle\int (x^2 - 1)e^x \, dx$ **24.** $\displaystyle\int \frac{\ln 2x}{x^2} \, dx$

25. $\displaystyle\int x\sqrt{x - 1} \, dx$ **26.** $\displaystyle\int \frac{x}{\sqrt{2 + 3x}} \, dx$

27. $\displaystyle\int x \cos x \, dx$ **28.** $\displaystyle\int x \sin x \, dx$

29. $\displaystyle\int x^3 \sin x \, dx$ **30.** $\displaystyle\int x^2 \cos x \, dx$

31. $\displaystyle\int t \csc t \cot t \, dt$ **32.** $\displaystyle\int \theta \sec \theta \tan \theta \, d\theta$

33. $\displaystyle\int \arctan x \, dx$ **34.** $\displaystyle\int 4 \arccos x \, dx$

35. $\displaystyle\int e^{2x} \sin x \, dx$ **36.** $\displaystyle\int e^x \cos 2x \, dx$

In Exercises 37–42, solve the differential equation.

37. $y' = xe^{x^2}$ **38.** $y' = \ln x$

39. $\dfrac{dy}{dt} = \dfrac{t^2}{\sqrt{2 + 3t}}$ **40.** $\dfrac{dy}{dx} = x^2\sqrt{x - 1}$

41. $(\cos y)y' = 2x$ **42.** $y' = \arctan \dfrac{x}{2}$

Slope Fields   In Exercises 43 and 44, a differential equation, a point, and a slope field are given. (a) Sketch two approximate solutions of the differential equation on the slope field, one of which passes through the indicated point. (b) Use integration to find the particular solution of the differential equation and use a graphing utility to graph the solution. Compare the result with the sketches in part (a). To print an enlarged copy of the graph, go the the website *www.mathgraphs.com*.

43. $\dfrac{dy}{dx} = x\sqrt{y} \cos x$, $(0, 4)$ **44.** $\dfrac{dy}{dx} = e^{-x/3} \sin 2x$, $\left(0, -\frac{18}{37}\right)$

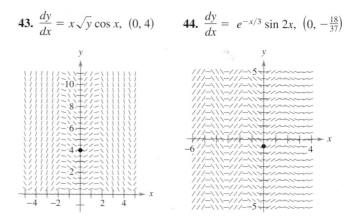

In Exercises 45 and 46, use a computer algebra system to sketch the slope field for the differential equation and graph the solution through the specified initial condition.

45. $\dfrac{dy}{dx} = \dfrac{x}{y}e^{x/8}$

$y(0) = 2$

46. $\dfrac{dy}{dx} = \dfrac{x}{y} \sin x$

$y(0) = 4$

In Exercises 47–58, evaluate the definite integral. Use a graphing utility to confirm your result.

47. $\displaystyle\int_0^4 xe^{-x/2} \, dx$ **48.** $\displaystyle\int_0^1 x^2 e^x \, dx$

49. $\displaystyle\int_0^{\pi/2} x \cos x \, dx$ **50.** $\displaystyle\int_0^\pi x \sin 2x \, dx$

51. $\displaystyle\int_0^{1/2} \arccos x \, dx$ **52.** $\displaystyle\int_0^1 x \arcsin x^2 \, dx$

53. $\displaystyle\int_0^1 e^x \sin x \, dx$ **54.** $\displaystyle\int_0^2 e^{-x} \cos x \, dx$

55. $\displaystyle\int_1^2 x^2 \ln x \, dx$ **56.** $\displaystyle\int_0^1 \ln(1 + x^2) \, dx$

57. $\displaystyle\int_2^4 x \, \text{arcsec} \, x \, dx$ **58.** $\displaystyle\int_0^{\pi/4} x \sec^2 x \, dx$

In Exercises 59–64, use the tabular method to evaluate the integral.

59. $\displaystyle\int x^2 e^{2x}\, dx$

60. $\displaystyle\int x^3 e^{-2x}\, dx$

61. $\displaystyle\int x^3 \sin x\, dx$

62. $\displaystyle\int x^3 \cos 2x\, dx$

63. $\displaystyle\int x \sec^2 x\, dx$

64. $\displaystyle\int x^2 (x - 2)^{3/2}\, dx$

Getting at the Concept

65. Integration by parts is based on what differentiation rule?

66. In your own words, state guidelines for integration by parts.

In Exercises 67–72, state whether you would use integration by parts to evaluate the integral. If so, identify what you would use for u and dv.

67. $\displaystyle\int \frac{\ln x}{x}\, dx$

68. $\displaystyle\int x \ln x\, dx$

69. $\displaystyle\int x^2 e^{2x}\, dx$

70. $\displaystyle\int 2x\, e^{x^2}\, dx$

71. $\displaystyle\int \frac{x}{\sqrt{x + 1}}\, dx$

72. $\displaystyle\int \frac{x}{\sqrt{x^2 + 1}}\, dx$

In Exercises 73–76, use a computer algebra system to evaluate the integral.

73. $\displaystyle\int t^3\, e^{-4t}\, dt$

74. $\displaystyle\int \alpha^4 \sin \pi\alpha\, d\alpha$

75. $\displaystyle\int_0^{\pi/2} e^{-2x} \sin 3x\, dx$

76. $\displaystyle\int_0^5 x^4 (25 - x^2)^{3/2}\, dx$

77. Integrate $\displaystyle\int 2x\sqrt{2x - 3}\, dx$

    (a) by parts, letting $dv = \sqrt{2x - 3}\, dx$.

    (b) by substitution, letting $u = 2x - 3$.

78. Integrate $\displaystyle\int x\sqrt{4 + x}\, dx$

    (a) by parts, letting $dv = \sqrt{4 + x}\, dx$.

    (b) by substitution, letting $u = 4 + x$.

79. Integrate $\displaystyle\int \frac{x^3}{\sqrt{4 + x^2}}\, dx$

    (a) by parts, letting $dv = \left(x/\sqrt{4 + x^2}\right) dx$.

    (b) by substitution, letting $u = 4 + x^2$.

80. Integrate $\displaystyle\int x\sqrt{4 - x}\, dx$

    (a) by parts, letting $dv = \sqrt{4 - x}\, dx$.

    (b) by substitution, letting $u = 4 - x$.

In Exercises 81 and 82, use a computer algebra system to evaluate the integral for $n = 0, 1, 2,$ and 3. Use the result to obtain a general rule for the integral for any positive integer n and test your results for $n = 4$.

81. $\displaystyle\int x^n \ln x\, dx$

82. $\displaystyle\int x^n e^x\, dx$

In Exercises 83–88, use integration by parts to verify the formula. (For Exercises 83–86, assume that n is a positive integer.)

83. $\displaystyle\int x^n \sin x\, dx = -x^n \cos x + n \int x^{n-1} \cos x\, dx$

84. $\displaystyle\int x^n \cos x\, dx = x^n \sin x - n \int x^{n-1} \sin x\, dx$

85. $\displaystyle\int x^n \ln x\, dx = \frac{x^{n+1}}{(n+1)^2}\left[-1 + (n+1)\ln x\right] + C$

86. $\displaystyle\int x^n e^{ax}\, dx = \frac{x^n e^{ax}}{a} - \frac{n}{a} \int x^{n-1} e^{ax}\, dx$

87. $\displaystyle\int e^{ax} \sin bx\, dx = \frac{e^{ax}(a \sin bx - b \cos bx)}{a^2 + b^2} + C$

88. $\displaystyle\int e^{ax} \cos bx\, dx = \frac{e^{ax}(a \cos bx + b \sin bx)}{a^2 + b^2} + C$

In Exercises 89–92, evaluate the integral by using the appropriate formula from Exercises 83–88.

89. $\displaystyle\int x^3 \ln x\, dx$

90. $\displaystyle\int x^2 \cos x\, dx$

91. $\displaystyle\int e^{2x} \cos 3x\, dx$

92. $\displaystyle\int x^3 e^{2x}\, dx$

Area **In Exercises 93–96, use a graphing utility to sketch the region bounded by the graphs of the equations, and find the area of the region.**

93. $y = xe^{-x}, \; y = 0, \; x = 4$

94. $y = \frac{1}{9} xe^{-x/3}, \; y = 0, \; x = 0, \; x = 3$

95. $y = e^{-x} \sin \pi x, \; y = 0, \; x = 0, \; x = 1$

96. $y = x \sin x, \; y = 0, \; x = 0, \; x = \pi$

97. *Area, Volume, and Centroid* Given the region bounded by the graphs of $y = \ln x, \; y = 0,$ and $x = e,$ find

    (a) the area of the region.

    (b) the volume of the solid generated by revolving the region about the x-axis.

    (c) the volume of the solid generated by revolving the region about the y-axis.

    (d) the centroid of the region.

98. *Centroid* Find the centroid of the region bounded by the graphs of $y = \arcsin x, \; x = 0,$ and $y = \pi/2.$ How is this problem related to Example 6 in this section?

99. Average Displacement A damping force affects the vibration of a spring so that the displacement of the spring is

$$y = e^{-4t}(\cos 2t + 5 \sin 2t).$$

Find the average value of y on the interval from $t = 0$ to $t = \pi$.

100. Memory Model A model for the ability M of a child to memorize, measured on a scale from 0 to 10, is

$$M = 1 + 1.6t \ln t, \quad 0 < t \le 4$$

where t is the child's age in years. Find the average value of this function

(a) between the child's first and second birthdays.

(b) between the child's third and fourth birthdays.

Present Value In Exercises 101 and 102, find the present value P of a continuous income flow of $c(t)$ dollars per year if

$$P = \int_0^{t_1} c(t)e^{-rt}\, dt$$

where t_1 is the time in years and r is the annual interest rate compounded continuously.

101. $c(t) = 100,000 + 4000t,\ r = 5\%,\ t_1 = 10$

102. $c(t) = 30,000 + 500t,\ r = 7\%,\ t_1 = 5$

Integrals Used to Find Fourier Coefficients In Exercises 103 and 104, verify the value of the definite integral, where n is a positive integer.

103. $\displaystyle \int_{-\pi}^{\pi} x \sin nx\, dx = \begin{cases} \dfrac{2\pi}{n}, & n \text{ is odd} \\[2mm] -\dfrac{2\pi}{n}, & n \text{ is even} \end{cases}$

104. $\displaystyle \int_{-\pi}^{\pi} x^2 \cos nx\, dx = \dfrac{(-1)^n\, 4\pi}{n^2}$

105. Vibrating String A string stretched between the two points $(0, 0)$ and $(2, 0)$ is plucked by displacing the string h units at its midpoint. The motion of the string is modeled by a **Fourier Sine Series** whose coefficients are given by

$$b_n = h \int_0^1 x \sin \frac{n\pi x}{2}\, dx + h \int_1^2 (-x + 2) \sin \frac{n\pi x}{2}\, dx.$$

Find b_n.

106. Find the fallacy in the following argument that $0 = 1$.

$$dv = dx \quad \Longrightarrow \quad v = \int dx = x$$

$$u = \frac{1}{x} \quad \Longrightarrow \quad du = -\frac{1}{x^2}\, dx$$

$$0 + \int \frac{dx}{x} = \left(\frac{1}{x}\right)(x) - \int \left(-\frac{1}{x^2}\right)(x)\, dx = 1 + \int \frac{dx}{x}$$

So, $0 = 1$.

107. Let $y = f(x)$ be positive and strictly increasing on the interval $0 < a \le x \le b$. Consider the region R bounded by the graphs of $y = f(x)$, $y = 0$, $x = a$, and $x = b$. If R is revolved about the y-axis, show that the disk method and shell method yield the same volume.

108. Think About It Explain why

$$\int_0^{\pi/2} x \sin x\, dx \le \int_0^{\pi/2} x\, dx.$$

Evaluate the integrals to verify the inequality.

109. Consider the differential equation $f'(x) = xe^{-x}$ with the initial condition $f(0) = 0$.

(a) Use integration to solve the differential equation.

(b) Use a graphing utility to graph the solution of the differential equation.

(c) *Euler's Method* From the definition of the derivative it follows that for "small" Δx

$$f'(x) \approx \frac{f(x + \Delta x) - f(x)}{\Delta x}$$

$$f(x + \Delta x) \approx f(x) + [f'(x)]\, \Delta x.$$

Consider points of the form

$$(x_n, y_n) = (n\, \Delta x,\ y_{n-1} + f'(x_{n-1})\, \Delta x)$$

where $(x_0, y_0) = (0, 0)$. Starting with $n = 0$, use the recursive capabilities of a graphing utility to generate the next 80 points of this form when $\Delta x = 0.05$. Use the graphing utility to plot the points and compare the result with the graph in part (b).

(d) Starting with $n = 0$, repeat part (c) by generating the next 40 points when $\Delta x = 0.1$.

(e) Give a geometric explanation of the process described in part (c). Why do you think the result in part (c) is a better approximation of the solution than the result in part (d)?

110. Euler's Method Consider the differential equation

$$f'(x) = \cos \sqrt{x}$$

with the initial condition $f(0) = 2$.

(a) Try solving the differential equation by integration. Can you perform the integration?

(b) Starting with $n = 0$, use the recursive capabilities of a graphing utility to generate 80 points of the form shown in part (c) of Exercise 109 when $\Delta x = 0.05$. Plot the points for an approximation of the graph of the solution of the differential equation.

Section 7.3   Trigonometric Integrals

- Solve trigonometric integrals involving powers of sine and cosine.
- Solve trigonometric integrals involving powers of secant and tangent.
- Solve trigonometric integrals involving sine-cosine products with different angles.

Integrals Involving Powers of Sine and Cosine

In this section you will study techniques for evaluating integrals of the form

$$\int \sin^m x \cos^n x \, dx \qquad \text{and} \qquad \int \sec^m x \tan^n x \, dx$$

where either m or n is a positive integer. To find antiderivatives for these forms, try to break them into combinations of trigonometric integrals to which you can apply the Power Rule.

For instance, you can evaluate $\int \sin^5 x \cos x \, dx$ with the Power Rule by letting $u = \sin x$. Then, $du = \cos x \, dx$ and you have

$$\int \sin^5 x \cos x \, dx = \int u^5 \, du = \frac{u^6}{6} + C = \frac{\sin^6 x}{6} + C.$$

To break up $\int \sin^m x \cos^n x \, dx$ into forms to which you can apply the Power Rule, use the following identities.

$$\sin^2 x + \cos^2 x = 1 \qquad \text{Pythagorean identity}$$

$$\sin^2 x = \frac{1 - \cos 2x}{2} \qquad \text{Half-angle identity for } \sin^2 x$$

$$\cos^2 x = \frac{1 + \cos 2x}{2} \qquad \text{Half-angle identity for } \cos^2 x$$

SHEILA SCOTT MACINTYRE (1910–1960)

Sheila Scott Macintyre published her first paper on the asymptotic periods of integral functions in 1935. She completed her doctorate work at Aberdeen University, where she taught. In 1958 she accepted a visiting research fellowship at the University of Cincinnati.

Guidelines for Evaluating Integrals Involving Sine and Cosine

1. If the power of the sine is odd and positive, save one sine factor and convert the remaining factors to cosines. Then, expand and integrate.

$$\int \underbrace{\sin^{2k+1} x}_{\text{Odd}} \cos^n x \, dx = \int \underbrace{(\sin^2 x)^k}_{\text{Convert to cosines}} \cos^n x \underbrace{\sin x \, dx}_{\text{Save for } du} = \int (1 - \cos^2 x)^k \cos^n x \sin x \, dx$$

2. If the power of the cosine is odd and positive, save one cosine factor and convert the remaining factors to sines. Then, expand and integrate.

$$\int \sin^m x \underbrace{\cos^{2k+1} x}_{\text{Odd}} \, dx = \int \sin^m x \underbrace{(\cos^2 x)^k}_{\text{Convert to sines}} \underbrace{\cos x \, dx}_{\text{Save for } du} = \int \sin^m x (1 - \sin^2 x)^k \cos x \, dx$$

3. If the powers of both the sine and cosine are even and nonnegative, make repeated use of the identities

$$\sin^2 x = \frac{1 - \cos 2x}{2} \qquad \text{and} \qquad \cos^2 x = \frac{1 + \cos 2x}{2}$$

to convert the integrand to odd powers of the cosine. Then proceed as in guideline 2.

TECHNOLOGY
Try using a computer algebra system to evaluate the integral in Example 1. When we did this, we obtained

$$\int \sin^3 x \cos^4 x \, dx =$$
$$-\cos^5 x\left(\frac{1}{7}\sin^2 x + \frac{2}{35}\right) + C.$$

Is this equivalent to the result obtained in Example 1?

Example 1 **Power of Sine Is Odd and Positive**

Evaluate $\int \sin^3 x \cos^4 x \, dx$.

Solution Because you expect to use the Power Rule with $u = \cos x$, *save one sine factor* to form du and convert the remaining sine factors to cosines.

$$\int \sin^3 x \cos^4 x \, dx = \int \sin^2 x \cos^4 x (\sin x) \, dx \qquad \text{Rewrite.}$$

$$= \int (1 - \cos^2 x)\cos^4 x \sin x \, dx \qquad \text{Trigonometric identity}$$

$$= \int (\cos^4 x - \cos^6 x)\sin x \, dx \qquad \text{Multiply.}$$

$$= \int \cos^4 x \sin x \, dx - \int \cos^6 x \sin x \, dx$$

$$= -\int \cos^4 x(-\sin x) \, dx + \int \cos^6 x(-\sin x) \, dx$$

$$= -\frac{\cos^5 x}{5} + \frac{\cos^7 x}{7} + C \qquad \text{Integrate.}$$

In Example 1, *both* of the powers m and n happened to be positive integers. However, the same strategy will work as long as either m or n is odd and positive. For instance, in the next example the power of the cosine is 3, but the power of the sine is $-\frac{1}{2}$.

Example 2 **Power of Cosine Is Odd and Positive**

Evaluate $\int_{\pi/6}^{\pi/3} \frac{\cos^3 x}{\sqrt{\sin x}} \, dx$.

Solution Because you expect to use the Power Rule with $u = \sin x$, *save one cosine factor* to form du and convert the remaining cosine factors to sines.

$$\int_{\pi/6}^{\pi/3} \frac{\cos^3 x}{\sqrt{\sin x}} \, dx = \int_{\pi/6}^{\pi/3} \frac{\cos^2 x \cos x}{\sqrt{\sin x}} \, dx$$

$$= \int_{\pi/6}^{\pi/3} \frac{(1 - \sin^2 x)(\cos x)}{\sqrt{\sin x}} \, dx$$

$$= \int_{\pi/6}^{\pi/3} [(\sin x)^{-1/2} \cos x - (\sin x)^{3/2} \cos x] \, dx$$

$$= \left[\frac{(\sin x)^{1/2}}{1/2} - \frac{(\sin x)^{5/2}}{5/2}\right]_{\pi/6}^{\pi/3}$$

$$= 2\left(\frac{\sqrt{3}}{2}\right)^{1/2} - \frac{2}{5}\left(\frac{\sqrt{3}}{2}\right)^{5/2} - \sqrt{2} + \frac{\sqrt{32}}{80}$$

$$\approx 0.239$$

Figure 7.4 shows the region whose area is represented by this integral.

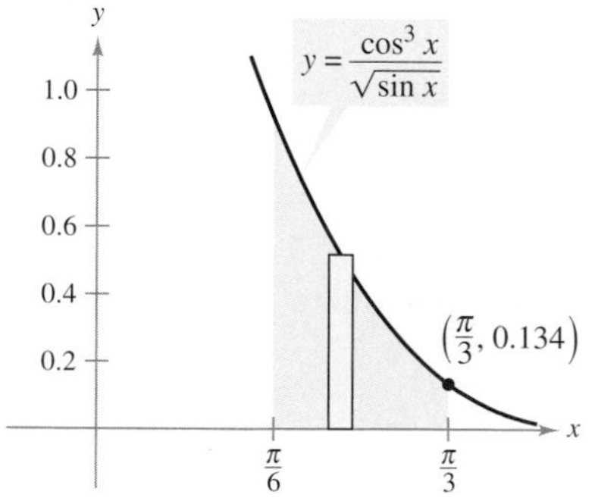

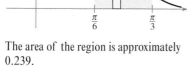

$$y = \frac{\cos^3 x}{\sqrt{\sin x}}$$

$\left(\frac{\pi}{3}, 0.134\right)$

The area of the region is approximately 0.239.
Figure 7.4

Bettmann/Corbis

JOHN WALLIS (1616–1703)

Wallis did much of his work in calculus prior to Newton and Leibniz, and he influenced the thinking of both of these men. Wallis is also credited with introducing the present symbol (∞) for infinity.

Example 3   **Power of Cosine Is Even and Nonnegative**

Evaluate $\int \cos^4 x \, dx$.

Solution   Because m and n are both even and nonnegative ($m = 0$), you can replace $\cos^4 x$ by $[(1 + \cos 2x)/2]^2$.

$$\int \cos^4 x \, dx = \int \left(\frac{1 + \cos 2x}{2} \right)^2 dx$$

$$= \int \left(\frac{1}{4} + \frac{\cos 2x}{2} + \frac{\cos^2 2x}{4} \right) dx$$

$$= \int \left[\frac{1}{4} + \frac{\cos 2x}{2} + \frac{1}{4} \left(\frac{1 + \cos 4x}{2} \right) \right] dx$$

$$= \frac{3}{8} \int dx + \frac{1}{4} \int 2 \cos 2x \, dx + \frac{1}{32} \int 4 \cos 4x \, dx$$

$$= \frac{3x}{8} + \frac{\sin 2x}{4} + \frac{\sin 4x}{32} + C$$

Try using a symbolic differentiation utility to verify this. Can you simplify the derivative to obtain the original integrand?

In Example 3, if you were to evaluate the definite integral from 0 to $\pi/2$, you would obtain

$$\int_0^{\pi/2} \cos^4 x \, dx = \left[\frac{3x}{8} + \frac{\sin 2x}{4} + \frac{\sin 4x}{32} \right]_0^{\pi/2}$$

$$= \left(\frac{3\pi}{16} + 0 + 0 \right) - (0 + 0 + 0)$$

$$= \frac{3\pi}{16}.$$

Note that the only term that contributes to the solution is $3x/8$. This observation is generalized in the following formulas developed by John Wallis.

Wallis's Formulas

1. If n is odd ($n \geq 3$), then

$$\int_0^{\pi/2} \cos^n x \, dx = \left(\frac{2}{3} \right)\left(\frac{4}{5} \right)\left(\frac{6}{7} \right) \cdots \left(\frac{n-1}{n} \right).$$

2. If n is even ($n \geq 2$), then

$$\int_0^{\pi/2} \cos^n x \, dx = \left(\frac{1}{2} \right)\left(\frac{3}{4} \right)\left(\frac{5}{6} \right) \cdots \left(\frac{n-1}{n} \right)\left(\frac{\pi}{2} \right).$$

These formulas are also valid if $\cos^n x$ is replaced by $\sin^n x$. (You are asked to prove both formulas in Exercise 96.)

Integrals Involving Powers of Secant and Tangent

The following guidelines can help you evaluate integrals of the form $\int \sec^m x \tan^n x \, dx$.

Guidelines for Evaluating Integrals Involving Secant and Tangent

1. If the power of the secant is even and positive, save a secant-squared factor and convert the remaining factors to tangents. Then expand and integrate.

$$\int \overset{\text{Even}}{\overbrace{\sec^{2k} x}} \tan^n x \, dx = \int \overset{\text{Convert to tangents}}{\overbrace{(\sec^2 x)^{k-1}}} \tan^n x \overset{\text{Save for } du}{\overbrace{\sec^2 x \, dx}} = \int (1 + \tan^2 x)^{k-1} \tan^n x \sec^2 x \, dx$$

2. If the power of the tangent is odd and positive, save a secant-tangent factor and convert the remaining factors to secants. Then expand and integrate.

$$\int \sec^m x \overset{\text{Odd}}{\overbrace{\tan^{2k+1} x}} \, dx = \int \sec^{m-1} x \overset{\text{Convert to secants}}{\overbrace{(\tan^2 x)^k}} \overset{\text{Save for } du}{\overbrace{\sec x \tan x \, dx}} = \int \sec^{m-1} x (\sec^2 x - 1)^k \sec x \tan x \, dx$$

3. If there are no secant factors and the power of the tangent is even and positive, convert a tangent-squared factor to a secant-squared factor, then expand and repeat if necessary.

$$\int \tan^n x \, dx = \int \tan^{n-2} x \overset{\text{Convert to secants}}{\overbrace{(\tan^2 x)}} \, dx = \int \tan^{n-2} x (\sec^2 x - 1) \, dx$$

4. If the integral is of the form $\int \sec^m x \, dx$, where m is odd and positive, use integration by parts, as illustrated in Example 5 in the preceding section.

5. If none of the first four guidelines applies, try converting to sines and cosines.

Example 4   **Power of Tangent Is Odd and Positive**

Evaluate $\displaystyle \int \frac{\tan^3 x}{\sqrt{\sec x}} \, dx$.

Solution   Because you expect to use the Power Rule with $u = \sec x$, *save a factor of* $(\sec x \tan x)$ to form du and convert the remaining tangent factors to secants.

$$\int \frac{\tan^3 x}{\sqrt{\sec x}} \, dx = \int (\sec x)^{-1/2} \tan^3 x \, dx$$

$$= \int (\sec x)^{-3/2} (\tan^2 x)(\sec x \tan x) \, dx$$

$$= \int (\sec x)^{-3/2} (\sec^2 x - 1)(\sec x \tan x) \, dx$$

$$= \int [(\sec x)^{1/2} - (\sec x)^{-3/2}](\sec x \tan x) \, dx$$

$$= \frac{2}{3}(\sec x)^{3/2} + 2(\sec x)^{-1/2} + C$$

NOTE   In Example 5, the power of the tangent is odd and positive. So, you could also evaluate the integral with the procedure described in guideline 2 on page 500. In Exercise 81, you are asked to show that the results obtained by these two procedures differ only by a constant.

Example 5   **Power of Secant Is Even and Positive**

Evaluate $\int \sec^4 3x \tan^3 3x \, dx$.

Solution   Let $u - \tan 3x$, then $du = 3 \sec^2 3x \, dx$ and you can write

$$\int \sec^4 3x \tan^3 3x \, dx = \int \sec^2 3x \tan^3 3x(\sec^2 3x) \, dx$$

$$= \int (1 + \tan^2 3x) \tan^3 3x \, (\sec^2 3x) \, dx$$

$$= \frac{1}{3}\int (\tan^3 3x + \tan^5 3x)(3 \sec^2 3x) \, dx$$

$$= \frac{1}{3}\left(\frac{\tan^4 3x}{4} + \frac{\tan^6 3x}{6} \right) + C$$

$$= \frac{\tan^4 3x}{12} + \frac{\tan^6 3x}{18} + C.$$

Example 6   **Power of Tangent Is Even**

Evaluate $\int_0^{\pi/4} \tan^4 x \, dx$.

Solution   Because there are no secant factors, you can begin by converting a tangent-squared factor to a secant-squared factor.

$$\int \tan^4 x \, dx = \int \tan^2 x(\tan^2 x) \, dx$$

$$= \int \tan^2 x(\sec^2 x - 1) \, dx$$

$$= \int \tan^2 x \sec^2 x \, dx - \int \tan^2 x \, dx$$

$$= \int \tan^2 x \sec^2 x \, dx - \int (\sec^2 x - 1) \, dx$$

$$= \frac{\tan^3 x}{3} - \tan x + x + C$$

You can evaluate the definite integral as follows.

$$\int_0^{\pi/4} \tan^4 x \, dx = \left[\frac{\tan^3 x}{3} - \tan x + x \right]_0^{\pi/4}$$

$$= \frac{\pi}{4} - \frac{2}{3}$$

$$\approx 0.119$$

The area represented by the definite integral is shown in Figure 7.5. Try using Simpson's Rule to approximate this integral. With $n = 10$, you should obtain an approximation that is within 0.00001 of the actual value.

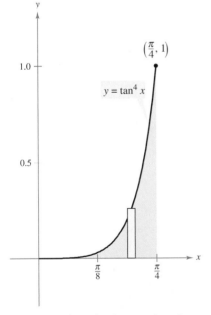

The area of the region is approximately 0.119.

Figure 7.5

For integrals involving powers of cotangents and cosecants, you can follow a strategy similar to that used for powers of tangents and secants. Also, when integrating trigonometric functions, remember that it sometimes helps to convert the entire integrand to powers of sines and cosines.

Example 7   **Converting to Sines and Cosines**

Evaluate $\displaystyle\int \frac{\sec x}{\tan^2 x}\, dx$.

Solution   Because the first four guidelines on page 500 do not apply, try converting the integrand to sines and cosines. In this case, you are able to integrate the resulting powers of sine and cosine as follows.

$$\int \frac{\sec x}{\tan^2 x}\, dx = \int \left(\frac{1}{\cos x}\right)\left(\frac{\cos x}{\sin x}\right)^2 dx$$

$$= \int (\sin x)^{-2}(\cos x)\, dx$$

$$= -(\sin x)^{-1} + C$$

$$= -\csc x + C$$

Integrals Involving Sine-Cosine Products with Different Angles

Integrals involving the products of sines and cosines of two *different* angles occur in many applications. In such instances you can use the following product-to-sum identities.

$$\sin mx \sin nx = \frac{1}{2}(\cos[(m - n)x] - \cos[(m + n)x])$$

$$\sin mx \cos nx = \frac{1}{2}(\sin[(m - n)x] + \sin[(m + n)x])$$

$$\cos mx \cos nx = \frac{1}{2}(\cos[(m - n)x] + \cos[(m + n)x])$$

Example 8   **Using Product-to-Sum Identities**

Evaluate $\int \sin 5x \cos 4x \, dx$.

Solution   Considering the second product-to-sum identity above, you can write the following.

$$\int \sin 5x \cos 4x \, dx = \frac{1}{2}\int (\sin x + \sin 9x)\, dx$$

$$= \frac{1}{2}\left(-\cos x - \frac{\cos 9x}{9}\right) + C$$

$$= -\frac{\cos x}{2} - \frac{\cos 9x}{18} + C$$

FOR FURTHER INFORMATION   To learn more about integrals involving sine-cosine products with different angles, see the article "Integrals of Products of Sine and Cosine with Different Arguments" by Sherrie J. Nicol in *The College Mathematics Journal*. To view this article, go to the website *www.matharticles.com*.

EXERCISES FOR SECTION 7.3

1. Consider the function $f(x) = \sin^4 x + \cos^4 x$.

(a) Use the power-reducing formulas to write the function in terms of the first power of the cosine.

(b) Determine another way of rewriting the function. Use a graphing utility to verify your result.

(c) Determine a trigonometric expression to add to the function so that it becomes a perfect square trinomial. Rewrite the function as a perfect square trinomial minus the term that you added. Use a graphing utility to verify your result.

(d) Rewrite the result in part (c) in terms of the sine of a double angle. Use a graphing utility to verify your result.

(e) In how many ways have you rewritten the trigonometric function? When rewriting a trigonometric expression, your result may not be the same as another person's result. Does this mean that one of you is wrong? Explain.

2. Match the antiderivative in the left column with the correct integral in the right column.

(a) $y = \sec x$ (i) $\int \sin x \tan^2 x \, dx$

(b) $y = \cos x + \sec x$ (ii) $8 \int \cos^4 x \, dx$

(c) $y = x - \tan x + \frac{1}{3} \tan^3 x$ (iii) $\int \sin x \sec^2 x \, dx$

(d) $y = 3x + 2 \sin x \cos^3 x +$ (iv) $\int \tan^4 x \, dx$
$\quad\quad 3 \sin x \cos x$

In Exercises 3–16, evaluate the integral.

3. $\int \cos^3 x \sin x \, dx$ **4.** $\int \cos^3 x \sin^4 x \, dx$

5. $\int \sin^5 2x \cos 2x \, dx$ **6.** $\int \sin^3 x \, dx$

7. $\int \sin^5 x \cos^2 x \, dx$ **8.** $\int \cos^3 \frac{x}{3} \, dx$

9. $\int \cos^3 \theta \sqrt{\sin \theta} \, d\theta$ **10.** $\int \frac{\sin^5 t}{\sqrt{\cos t}} \, dt$

11. $\int \cos^2 3x \, dx$ **12.** $\int \sin^2 2x \, dx$

13. $\int \sin^2 \alpha \cos^2 \alpha \, d\alpha$ **14.** $\int \sin^4 2\theta \, d\theta$

15. $\int x \sin^2 x \, dx$ **16.** $\int x^2 \sin^2 x \, dx$

In Exercises 17–20, verify Wallis's Formulas by evaluating the integral.

17. $\int_0^{\pi/2} \cos^3 x \, dx = \frac{2}{3}$ **18.** $\int_0^{\pi/2} \cos^5 x \, dx = \frac{8}{15}$

19. $\int_0^{\pi/2} \cos^7 x \, dx = \frac{16}{35}$ **20.** $\int_0^{\pi/2} \sin^2 x \, dx = \frac{\pi}{4}$

In Exercises 21–38, evaluate the integral involving secant and tangent.

21. $\int \sec 3x \, dx$ **22.** $\int \sec^2(2x - 1) \, dx$

23. $\int \sec^4 5x \, dx$ **24.** $\int \sec^6 3x \, dx$

25. $\int \sec^3 \pi x \, dx$ **26.** $\int \tan^2 x \, dx$

27. $\int \tan^5 \frac{x}{4} \, dx$ **28.** $\int \tan^3 \frac{\pi x}{2} \sec^2 \frac{\pi x}{2} \, dx$

29. $\int \sec^2 x \tan x \, dx$ **30.** $\int \tan^3 2t \sec^3 2t \, dt$

31. $\int \tan^2 x \sec^2 x \, dx$ **32.** $\int \tan^5 2x \sec^2 2x \, dx$

33. $\int \sec^6 4x \tan 4x \, dx$ **34.** $\int \sec^2 \frac{x}{2} \tan \frac{x}{2} \, dx$

35. $\int \sec^3 x \tan x \, dx$ **36.** $\int \tan^3 3x \, dx$

37. $\int \frac{\tan^2 x}{\sec x} \, dx$ **38.** $\int \frac{\tan^2 x}{\sec^5 x} \, dx$

In Exercises 39–42, solve the differential equation.

39. $\dfrac{dr}{d\theta} = \sin^4 \pi \theta$ **40.** $\dfrac{ds}{d\alpha} = \sin^2 \dfrac{\alpha}{2} \cos^2 \dfrac{\alpha}{2}$

41. $y' = \tan^3 3x \sec 3x$ **42.** $y' = \sqrt{\tan x} \sec^4 x$

Slope Fields In Exercises 43 and 44, a differential equation, a point, and a slope field are given. (a) Sketch two approximate solutions of the differential equation on the slope field, one of which passes through the indicated point. (b) Use integration to find the particular solution of the differential equation and use a graphing utility to graph the solution. Compare the result with the sketches in part (a). To print an enlarged copy of the graph, go to the website *www.mathgraphs.com*.

43. $\dfrac{dy}{dx} = \sin^2 x, \ (0, 0)$ **44.** $\dfrac{dy}{dx} = \sec^2 x \tan^2 x, \ \left(0, -\dfrac{1}{4}\right)$

In Exercises 45 and 46, use a computer algebra system to sketch the slope field for the differential equation, and graph the solution through the specified initial condition.

45. $\dfrac{dy}{dx} = \dfrac{3 \sin x}{y}$, $y(0) = 2$ **46.** $\dfrac{dy}{dx} = 3\sqrt{y} \tan^2 x$, $y(0) = 3$

In Exercises 47–50, evaluate the integral.

47. $\displaystyle\int \sin 3x \cos 2x \, dx$ **48.** $\displaystyle\int \cos 4\theta \cos(-3\theta) \, d\theta$

49. $\displaystyle\int \sin\theta \sin 3\theta \, d\theta$ **50.** $\displaystyle\int \sin(-4x) \cos 3x \, dx$

In Exercises 51–60, evaluate the integral. Use a computer algebra system to confirm your result.

51. $\displaystyle\int \cot^3 2x \, dx$ **52.** $\displaystyle\int \tan^4 \dfrac{x}{2} \sec^4 \dfrac{x}{2} \, dx$

53. $\displaystyle\int \csc^4 \theta \, d\theta$ **54.** $\displaystyle\int \csc^2 3x \cot 3x \, dx$

55. $\displaystyle\int \dfrac{\cot^2 t}{\csc t} \, dt$ **56.** $\displaystyle\int \dfrac{\cot^3 t}{\csc t} \, dt$

57. $\displaystyle\int \dfrac{1}{\sec x \tan x} \, dx$ **58.** $\displaystyle\int \dfrac{\sin^2 x - \cos^2 x}{\cos x} \, dx$

59. $\displaystyle\int (\tan^4 t - \sec^4 t) \, dt$ **60.** $\displaystyle\int \dfrac{1 - \sec t}{\cos t - 1} \, dt$

In Exercises 61–68, evaluate the definite integral.

61. $\displaystyle\int_{-\pi}^{\pi} \sin^2 x \, dx$ **62.** $\displaystyle\int_0^{\pi/3} \tan^2 x \, dx$

63. $\displaystyle\int_0^{\pi/4} \tan^3 x \, dx$ **64.** $\displaystyle\int_0^{\pi/4} \sec^2 t \sqrt{\tan t} \, dt$

65. $\displaystyle\int_0^{\pi/2} \dfrac{\cos t}{1 + \sin t} \, dt$ **66.** $\displaystyle\int_{-\pi}^{\pi} \sin 3\theta \cos\theta \, d\theta$

67. $\displaystyle\int_{-\pi/2}^{\pi/2} \cos^3 x \, dx$ **68.** $\displaystyle\int_{-\pi/2}^{\pi/2} (\sin^2 x + 1) \, dx$

In Exercises 69–74, use a computer algebra system to evaluate the integral. Graph the antiderivatives for two different values of the constant of integration.

69. $\displaystyle\int \cos^4 \dfrac{x}{2} \, dx$ **70.** $\displaystyle\int \sin^2 x \cos^2 x \, dx$

71. $\displaystyle\int \sec^5 \pi x \, dx$ **72.** $\displaystyle\int \tan^3(1 - x) \, dx$

73. $\displaystyle\int \sec^5 \pi x \tan \pi x \, dx$ **74.** $\displaystyle\int \sec^4(1 - x) \tan(1 - x) \, dx$

In Exercises 75–78, use a computer algebra system to evaluate the definite integral.

75. $\displaystyle\int_0^{\pi/4} \sin 2\theta \sin 3\theta \, d\theta$ **76.** $\displaystyle\int_0^{\pi/2} (1 - \cos\theta)^2 \, d\theta$

77. $\displaystyle\int_0^{\pi/2} \sin^4 x \, dx$ **78.** $\displaystyle\int_0^{\pi/2} \sin^6 x \, dx$

Getting at the Concept

79. In your own words, describe how you would integrate $\int \sin^m x \cos^n x \, dx$ for each of the following.

(a) m is positive and odd.

(b) n is positive and odd.

(c) m and n are both positive and even.

80. In your own words, describe how you would integrate $\int \sec^m x \tan^n x \, dx$ for each of the following.

(a) m is positive and even.

(b) n is positive and odd.

(c) n is positive and even, and there are no secant factors.

(d) m is positive and odd, and there are no tangent factors.

In Exercises 81 and 82, (a) find the indefinite integral in two different ways. (b) Use a graphing utility to graph the anti-derivative (without the constant of integration) obtained by each method to show that the results differ only by a constant. (c) Verify analytically that the results differ only by a constant.

81. $\displaystyle\int \sec^4 3x \tan^3 3x \, dx$ **82.** $\displaystyle\int \sec^2 x \tan x \, dx$

83. *Area*   Find the area of the region bounded by the graphs of the equations $y = \sin^2 \pi x$, $y = 0$, $x = 0$, and $x = 1$.

84. *Volume*   Find the volume of the solid generated by revolving the region bounded by the graphs of the equations $y = \tan x$, $y = 0$, $x = -\pi/4$, and $x = \pi/4$ about the x-axis.

Volume and Centroid   **In Exercises 85 and 86, for the region bounded by the graphs of the equations, find (a) the volume of the solid formed by revolving the region about the x-axis and (b) the centroid of the region.**

85. $y = \sin x$, $y = 0$, $x = 0$, $x = \pi$

86. $y = \cos x$, $y = 0$, $x = 0$, $x = \dfrac{\pi}{2}$

In Exercises 87–90, use integration by parts to verify the reduction formula.

87. $\displaystyle\int \sin^n x \, dx = -\dfrac{\sin^{n-1} x \cos x}{n} + \dfrac{n-1}{n} \int \sin^{n-2} x \, dx$

88. $\displaystyle\int \cos^n x \, dx = \dfrac{\cos^{n-1} x \sin x}{n} + \dfrac{n-1}{n} \int \cos^{n-2} x \, dx$

89. $\displaystyle\int \cos^m x \sin^n x \, dx = -\dfrac{\cos^{m+1} x \sin^{n-1} x}{m+n} + \dfrac{n-1}{m+n} \int \cos^m x \sin^{n-2} x \, dx$

90. $\displaystyle\int \sec^n x \, dx = \dfrac{1}{n-1} \sec^{n-2} x \tan x + \dfrac{n-2}{n-1} \int \sec^{n-2} x \, dx$

In Exercises 91–94, use the results of Exercises 87–90 to evaluate the integral.

91. $\displaystyle\int \sin^5 x \, dx$

92. $\displaystyle\int \cos^4 x \, dx$

93. $\displaystyle\int \sec^4 \frac{2\pi x}{5} \, dx$

94. $\displaystyle\int \sin^4 x \cos^2 x \, dx$

95. *Modeling Data*   The table shows the normal maximum (high) and minimum (low) temperatures for Erie, Pennsylvania for each month of a year.   *(Source: NOAA)*

Month	Jan	Feb	Mar	Apr	May	Jun
Max	30.9	32.2	41.1	53.7	64.6	74.0
Min	18.0	17.7	25.8	36.1	45.4	55.2

Month	Jul	Aug	Sep	Oct	Nov	Dec
Max	78.2	77.0	71.0	60.1	47.1	35.7
Min	59.9	59.4	53.1	43.2	34.3	24.2

The maximum and minimum temperatures can be modeled by

$$f(t) = a_0 + a_1 \cos \frac{\pi t}{6} + b_1 \sin \frac{\pi t}{6}$$

where a_0, a_1, and b_1 are as follows.

$$a_0 = \frac{1}{12}\int_0^{12} f(t)\, dt$$

$$a_1 = \frac{1}{6}\int_0^{12} f(t) \cos \frac{\pi t}{6}\, dt$$

$$b_1 = \frac{1}{6}\int_0^{12} f(t) \sin \frac{\pi t}{6}\, dt$$

(a) Approximate the model $H(t)$ for the maximum temperatures. Let $t = 0$ correspond to January. (*Hint:* Use Simpson's Rule to approximate the integrals and use the January data twice.)

(b) Repeat part (a) for a model $L(t)$ for the minimum temperature data.

(c) Use a graphing utility to compare each model with the actual data. During what part of the year is the difference between the maximum and minimum temperatures greatest?

96. *Wallis's Formulas*   Use the result of Exercise 88 to prove the following versions of Wallis's Formulas.

(a) If n is odd $(n \geq 3)$, then

$$\int_0^{\pi/2} \cos^n x \, dx = \left(\frac{2}{3}\right)\left(\frac{4}{5}\right)\left(\frac{6}{7}\right) \cdots \left(\frac{n-1}{n}\right).$$

(b) If n is even $(n \geq 2)$, then

$$\int_0^{\pi/2} \cos^n x \, dx = \left(\frac{1}{2}\right)\left(\frac{3}{4}\right)\left(\frac{5}{6}\right) \cdots \left(\frac{n-1}{n}\right)\left(\frac{\pi}{2}\right).$$

97. The **inner product** of two functions f and g on $[a, b]$ is given by $\langle f, g \rangle = \int_a^b f(x)g(x)\, dx$. Two distinct functions f and g are said to be **orthogonal** if $\langle f, g \rangle = 0$. Show that the following set of functions is orthogonal on $[-\pi, \pi]$.

$$\{\sin x, \sin 2x, \sin 3x, \ldots, \cos x, \cos 2x, \cos 3x, \ldots\}$$

SECTION PROJECT   **POWER LINES**

Power lines are constructed by stringing wire between supports and adjusting the tension on each span. The wire hangs between supports in the shape of a catenary, as shown in the figure.

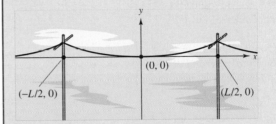

Let T be the tension (in pounds) on a span of wire, let u be the density (in pounds per foot), let $g \approx 32.2$ be the acceleration due to gravity (in feet per second per second), and let L be the distance (in feet) between the supports. Then the equation of the catenary is

$$y = \frac{T}{ug}\left(\cosh \frac{ugx}{T} - 1\right)$$

where x and y are measured in feet.

(a) Find the length of the wire between two spans.

(b) To measure the tension in a span, power line workers use the *return wave method*. The wire is struck at one support, creating a wave in the line, and the time t (in seconds) it takes for the wave to make a round trip is measured. The velocity v (in feet per second) is given by $v = \sqrt{T/u}$. How long does it take the wave to make a round trip between supports?

(c) The sag s (in inches) can be obtained by evaluating y when $x = L/2$ in the equation for the catenary (and multiplying by 12). In practice, however, power line workers use the "lineman's equation" given by $s \approx 12.075t^2$. Use the fact that $[\cosh(ugL/2T) + 1] \approx 2$ to derive this equation.

FOR FURTHER INFORMATION To learn more about the mathematics of power lines, see the article "Constructing Power Lines" by Thomas O'Neil in *The UMAP Journal*. To view this article, go to the website *www.matharticles.com*.

Trigonometric Substitution

- Use trigonometric substitution to solve an integral.
- Use integrals to model and solve real-life applications.

Trigonometric Substitution

Now that you can evaluate integrals involving powers of trigonometric functions, you can use **trigonometric substitution** to evaluate integrals involving the radicals

$$\sqrt{a^2 - u^2}, \qquad \sqrt{a^2 + u^2}, \qquad \text{and} \qquad \sqrt{u^2 - a^2}.$$

The objective with trigonometric substitution is to eliminate the radical in the integrand. You do this with the Pythagorean identities

$$\cos^2 \theta = 1 - \sin^2 \theta, \quad \sec^2 \theta = 1 + \tan^2 \theta, \quad \text{and} \quad \tan^2 \theta = \sec^2 \theta - 1.$$

For example, if $a > 0$, let $u = a \sin \theta$, where $-\pi/2 \le \theta \le \pi/2$. Then

$$\begin{aligned}
\sqrt{a^2 - u^2} &= \sqrt{a^2 - a^2 \sin^2 \theta} \\
&= \sqrt{a^2(1 - \sin^2 \theta)} \\
&= \sqrt{a^2 \cos^2 \theta} \\
&= a \cos \theta.
\end{aligned}$$

Note that $\cos \theta \ge 0$, because $-\pi/2 \le \theta \le \pi/2$.

Trigonometric Substitution ($a > 0$)

1. For integrals involving $\sqrt{a^2 - u^2}$, let

$$u = a \sin \theta.$$

Then $\sqrt{a^2 - u^2} = a \cos \theta$, where $-\pi/2 \le \theta \le \pi/2$.

2. For integrals involving $\sqrt{a^2 + u^2}$, let

$$u = a \tan \theta.$$

Then $\sqrt{a^2 + u^2} = a \sec \theta$, where $-\pi/2 < \theta < \pi/2$.

3. For integrals involving $\sqrt{u^2 - a^2}$, let

$$u = a \sec \theta.$$

Then $\sqrt{u^2 - a^2} = \pm a \tan \theta$, where $0 \le \theta < \pi/2$ or $\pi/2 < \theta \le \pi$. Use the positive value if $u > a$ and the negative value if $u < -a$.

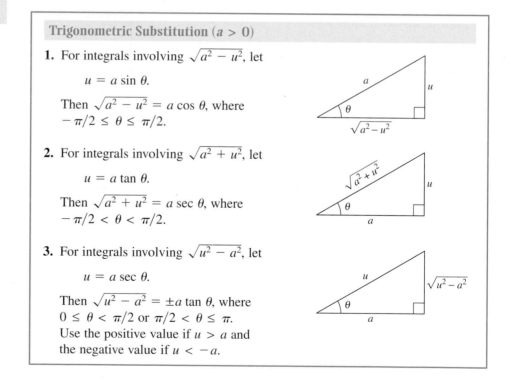

NOTE The restrictions on θ ensure that the function that defines the substitution is one-to-one. In fact, these are the same intervals over which the arcsine, arctangent, and arcsecant are defined.

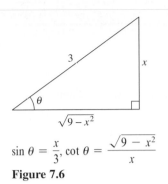

$$\sin \theta = \frac{x}{3}, \ \cot \theta = \frac{\sqrt{9 - x^2}}{x}$$

Figure 7.6

Example 1   Trigonometric Substitution: $u = a \sin \theta$

Evaluate $\displaystyle\int \frac{dx}{x^2 \sqrt{9 - x^2}}$.

Solution First, note that none of the basic integration rules applies. To use trigonometric substitution, you should observe that $\sqrt{9 - x^2}$ is of the form $\sqrt{a^2 - u^2}$. So, you can use the substitution

$$x = a \sin \theta = 3 \sin \theta.$$

Using differentiation and the triangle shown in Figure 7.6, you obtain

$$dx = 3 \cos \theta \, d\theta, \qquad \sqrt{9 - x^2} = 3 \cos \theta, \qquad \text{and} \qquad x^2 = 9 \sin^2 \theta.$$

Therefore, trigonometric substitution yields the following.

$$
\begin{aligned}
\int \frac{dx}{x^2 \sqrt{9 - x^2}} &= \int \frac{3 \cos \theta \, d\theta}{(9 \sin^2 \theta)(3 \cos \theta)} && \text{Substitute.} \\
&= \frac{1}{9} \int \frac{d\theta}{\sin^2 \theta} && \text{Simplify.} \\
&= \frac{1}{9} \int \csc^2 \theta \, d\theta && \text{Trigonometric identity} \\
&= -\frac{1}{9} \cot \theta + C && \text{Apply Cosecant Rule.} \\
&= -\frac{1}{9} \left(\frac{\sqrt{9 - x^2}}{x} \right) + C && \text{Substitute for } \cot \theta. \\
&= -\frac{\sqrt{9 - x^2}}{9x} + C
\end{aligned}
$$

Note that the triangle in Figure 7.6 can be used to convert the θ's back to x's as follows.

$$\cot \theta = \frac{\text{adj.}}{\text{opp.}} = \frac{\sqrt{9 - x^2}}{x}$$

TECHNOLOGY Use a computer algebra system to integrate each of the following.

$$\int \frac{dx}{\sqrt{9 - x^2}} \qquad \int \frac{dx}{x\sqrt{9 - x^2}} \qquad \int \frac{dx}{x^2\sqrt{9 - x^2}} \qquad \int \frac{dx}{x^3\sqrt{9 - x^2}}$$

Then use trigonometric substitution to duplicate the results obtained with the computer algebra system.

In an earlier chapter, you saw how the inverse hyperbolic functions can be used to evaluate the integrals

$$\int \frac{du}{\sqrt{u^2 \pm a^2}}, \qquad \int \frac{du}{a^2 - u^2}, \qquad \text{and} \qquad \int \frac{du}{u\sqrt{a^2 \pm u^2}}.$$

You can also evaluate these integrals using trigonometric substitution. This is illustrated in the next example.

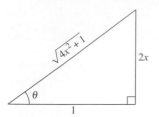

$\tan \theta = 2x, \ \sec \theta = \sqrt{4x^2 + 1}$
Figure 7.7

Example 2 **Trigonometric Substitution:** $u = a \tan \theta$

Evaluate $\displaystyle\int \frac{dx}{\sqrt{4x^2 + 1}}$.

Solution Let $u = 2x$, $a = 1$, and $2x = \tan \theta$, as shown in Figure 7.7. Then,

$$dx = \frac{1}{2} \sec^2 \theta \, d\theta \qquad \text{and} \qquad \sqrt{4x^2 + 1} = \sec \theta.$$

Trigonometric substitution produces the following.

$$\int \frac{1}{\sqrt{4x^2 + 1}} \, dx = \frac{1}{2} \int \frac{\sec^2 \theta \, d\theta}{\sec \theta} \qquad\qquad \text{Substitute.}$$

$$= \frac{1}{2} \int \sec \theta \, d\theta \qquad\qquad \text{Simplify.}$$

$$= \frac{1}{2} \ln |\sec \theta + \tan \theta| + C \qquad\qquad \text{Apply Secant Rule.}$$

$$= \frac{1}{2} \ln \left| \sqrt{4x^2 + 1} + 2x \right| + C \qquad\qquad \text{Back-substitute.}$$

Try checking this result with a computer algebra system. Is the result given in this form or in the form of an inverse hyperbolic function?

You can extend the use of trigonometric substitution to cover integrals involving expressions such as $(a^2 - u^2)^{n/2}$ by writing the expression as

$$(a^2 - u^2)^{n/2} = \left(\sqrt{a^2 - u^2} \right)^n.$$

Example 3 **Trigonometric Substitution: Rational Powers**

Evaluate $\displaystyle\int \frac{dx}{(x^2 + 1)^{3/2}}$.

Solution Begin by writing $(x^2 + 1)^{3/2}$ as $\left(\sqrt{x^2 + 1} \right)^3$. Then, let $a = 1$ and $u = x = \tan \theta$, as shown in Figure 7.8. Using

$$dx = \sec^2 \theta \, d\theta \qquad \text{and} \qquad \sqrt{x^2 + 1} = \sec \theta$$

you can apply trigonometric substitution as follows.

$$\int \frac{dx}{(x^2 + 1)^{3/2}} = \int \frac{dx}{\left(\sqrt{x^2 + 1} \right)^3} \qquad\qquad \text{Rewrite denominator.}$$

$$= \int \frac{\sec^2 \theta \, d\theta}{\sec^3 \theta} \qquad\qquad \text{Substitute.}$$

$$= \int \frac{d\theta}{\sec \theta} \qquad\qquad \text{Simplify.}$$

$$= \int \cos \theta \, d\theta \qquad\qquad \text{Trigonometric identity}$$

$$= \sin \theta + C \qquad\qquad \text{Apply Cosine Rule.}$$

$$= \frac{x}{\sqrt{x^2 + 1}} + C \qquad\qquad \text{Back-substitute.}$$

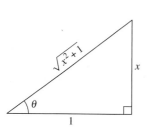

$\tan \theta = x, \ \sin \theta = \dfrac{x}{\sqrt{x^2 + 1}}$

Figure 7.8

For definite integrals, it is often convenient to determine the integration limits for θ that avoid converting back to x. You might want to review this procedure in Section 4.5, Examples 8 and 9.

Example 4 **Converting the Limits of Integration**

Evaluate $\displaystyle\int_{\sqrt{3}}^{2} \frac{\sqrt{x^2 - 3}}{x}\, dx$.

Solution Because $\sqrt{x^2 - 3}$ has the form $\sqrt{u^2 - a^2}$, you can consider

$$u = x, \quad a = \sqrt{3}, \quad \text{and} \quad x = \sqrt{3}\sec\theta$$

as shown in Figure 7.9. Then,

$$dx = \sqrt{3}\sec\theta\tan\theta\, d\theta \quad\text{and}\quad \sqrt{x^2 - 3} = \sqrt{3}\tan\theta.$$

To determine the upper and lower limits of integration, use the substitution $x = \sqrt{3}\sec\theta$, as follows.

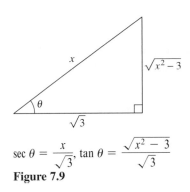

$\sec\theta = \dfrac{x}{\sqrt{3}}$, $\tan\theta = \dfrac{\sqrt{x^2 - 3}}{\sqrt{3}}$

Figure 7.9

Lower Limit	*Upper Limit*
When $x = \sqrt{3}$, $\sec\theta = 1$ and $\theta = 0$.	When $x = 2$, $\sec\theta = \dfrac{2}{\sqrt{3}}$ and $\theta = \dfrac{\pi}{6}$.

Therefore, you have

Integration limits for x $\downarrow$

Integration limits for θ $\downarrow$

$$\int_{\sqrt{3}}^{2} \frac{\sqrt{x^2 - 3}}{x}\, dx = \int_{0}^{\pi/6} \frac{\left(\sqrt{3}\tan\theta\right)\left(\sqrt{3}\sec\theta\tan\theta\right) d\theta}{\sqrt{3}\sec\theta}$$

$$= \int_{0}^{\pi/6} \sqrt{3}\tan^2\theta\, d\theta$$

$$= \sqrt{3}\int_{0}^{\pi/6} (\sec^2\theta - 1)\, d\theta$$

$$= \sqrt{3}\left[\tan\theta - \theta\right]_{0}^{\pi/6}$$

$$= \sqrt{3}\left(\frac{1}{\sqrt{3}} - \frac{\pi}{6}\right)$$

$$= 1 - \frac{\sqrt{3}\,\pi}{6}$$

$$\approx 0.0931.$$

In Example 4, try converting back to the variable x and evaluating the antiderivative at the original limits of integration. You should obtain

$$\int_{\sqrt{3}}^{2} \frac{\sqrt{x^2 - 3}}{x}\, dx = \sqrt{3}\left. \frac{\sqrt{x^2 - 3}}{\sqrt{3}} - \text{arcsec}\frac{x}{\sqrt{3}}\right]_{\sqrt{3}}^{2}.$$

When using trigonometric substitution to evaluate definite integrals, you must be careful to check that the values of θ lie in the intervals discussed at the beginning of this section. For instance, if in Example 4 you had been asked to evaluate the definite integral

$$\int_{-2}^{-\sqrt{3}} \frac{\sqrt{x^2-3}}{x}\, dx$$

then using $u = x$ and $a = \sqrt{3}$ in the interval $\left[-2, -\sqrt{3}\right]$ would imply that $u < -a$. So, when determining the upper and lower limits of integration, you would have to choose θ such that $\pi/2 < \theta \le \pi$. In this case the integral would be evaluated as follows.

$$\int_{-2}^{-\sqrt{3}} \frac{\sqrt{x^2-3}}{x}\, dx = \int_{5\pi/6}^{\pi} \frac{\left(-\sqrt{3}\tan\theta\right)\left(\sqrt{3}\sec\theta\tan\theta\right)d\theta}{\sqrt{3}\sec\theta}$$

$$= \int_{5\pi/6}^{\pi} -\sqrt{3}\tan^2\theta\, d\theta$$

$$= -\sqrt{3}\int_{5\pi/6}^{\pi} (\sec^2\theta - 1)\, d\theta$$

$$= -\sqrt{3}\left[\tan\theta - \theta\right]_{5\pi/6}^{\pi}$$

$$= -\sqrt{3}\left[(0 - \pi) - \left(-\frac{1}{\sqrt{3}} - \frac{5\pi}{6}\right)\right]$$

$$= -1 + \frac{\sqrt{3}\pi}{6}$$

$$\approx -0.0931$$

Trigonometric substitution can be used with completing the square. For instance, try evaluating the following integral.

$$\int \sqrt{x^2 - 2x}\, dx$$

To begin, you could complete the square and write the integral as

$$\int \sqrt{(x-1)^2 - 1^2}\, dx.$$

Trigonometric substitution can be used to evaluate the three integrals listed in the following theorem. These integrals will be encountered several times in the remainder of the text. When this happens, we will simply refer to this theorem. (In Exercise 81, you are asked to verify the formulas given in the theorem.)

THEOREM 7.2   Special Integration Formulas ($a > 0$)

1. $\displaystyle\int \sqrt{a^2 - u^2}\, du = \frac{1}{2}\left(a^2\arcsin\frac{u}{a} + u\sqrt{a^2 - u^2}\right) + C$

2. $\displaystyle\int \sqrt{u^2 - a^2}\, du = \frac{1}{2}(u\sqrt{u^2 - a^2} - a^2\ln|u + \sqrt{u^2 + a^2}|) + C,\ u > a$

3. $\displaystyle\int \sqrt{u^2 + a^2}\, du = \frac{1}{2}(u\sqrt{u^2 + a^2} + a^2\ln|u + \sqrt{u^2 + a^2}|) + C$

Applications

Example 5   Finding Arc Length

Find the arc length of the graph of $f(x) = \frac{1}{2}x^2$ from $x = 0$ to $x = 1$ (see Figure 7.10).

Solution   (Refer to the arc length formula in Section 6.4.)

$$
\begin{aligned}
s &= \int_0^1 \sqrt{1 + [f'(x)]^2}\, dx && \text{Formula for arc length} \\
&= \int_0^1 \sqrt{1 + x^2}\, dx && f'(x) = x \\
&= \int_0^{\pi/4} \sec^3 \theta\, d\theta && \text{Let } a = 1 \text{ and } x = \tan \theta. \\
&= \frac{1}{2}\left[\sec \theta \tan \theta + \ln|\sec \theta + \tan \theta| \right]_0^{\pi/4} && \text{Example 5, Section 7.2} \\
&= \frac{1}{2}\left[\sqrt{2} + \ln\left(\sqrt{2} + 1 \right) \right] \approx 1.148
\end{aligned}
$$

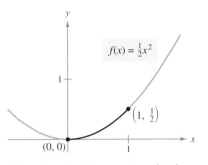

The arc length of the curve from $(0, 0)$ to $\left(1, \frac{1}{2}\right)$ is approximately 1.148.

Figure 7.10

Example 6   Comparing Two Fluid Forces

A sealed barrel of oil (weighing 48 pounds per cubic foot) is floating in seawater (weighing 64 pounds per cubic foot), as shown in Figures 7.11 and 7.12. (The barrel is not completely full of oil—on its side, the top 0.2 foot of the barrel is empty.) Compare the fluid forces against one end of the barrel from the inside and from the outside.

Solution   In Figure 7.12, locate the coordinate system with the origin at the center of the circle given by $x^2 + y^2 - 1$. To find the fluid force against an end of the barrel *from the inside*, integrate between -1 and 0.8 (using a weight of $w = 48$).

$$
\begin{aligned}
F &= w \int_c^d h(y)L(y)\, dy && \text{General equation (see Section 6.7)} \\
F_{\text{inside}} &= 48 \int_{-1}^{0.8} (0.8 - y)(2)\sqrt{1 - y^2}\, dy \\
&= 76.8 \int_{-1}^{0.8} \sqrt{1 - y^2}\, dy - 96 \int_{-1}^{0.8} y\sqrt{1 - y^2}\, dy
\end{aligned}
$$

To find the fluid force *from the outside*, integrate between -1 and 0.4 (using a weight of $w = 64$).

$$
\begin{aligned}
F_{\text{outside}} &= 64 \int_{-1}^{0.4} (0.4 - y)(2)\sqrt{1 - y^2}\, dy \\
&= 51.2 \int_{-1}^{0.4} \sqrt{1 - y^2}\, dy - 128 \int_{-1}^{0.4} y\sqrt{1 - y^2}\, dy
\end{aligned}
$$

We leave the details of integration for you to complete in Exercise 74. Intuitively, would you say that the force from the oil (the inside) or the force from the seawater (the outside) is greater? By evaluating these two integrals, you can determine that

$$F_{\text{inside}} \approx 121.3 \text{ pounds} \qquad \text{and} \qquad F_{\text{outside}} \approx 93.0 \text{ pounds.}$$

The barrel is not quite full of oil—the top 0.2 foot of the barrel is empty.

Figure 7.11

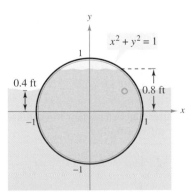

Figure 7.12

EXERCISES FOR SECTION 7.4

In Exercises 1–4, match the antiderivative with the correct integral. [Integrals are labeled (a), (b), (c), and (d).]

(a) $\displaystyle\int \frac{x^2}{\sqrt{16 - x^2}}\, dx$

(b) $\displaystyle\int \frac{\sqrt{x^2 + 16}}{x}\, dx$

(c) $\displaystyle\int \sqrt{7 + 6x - x^2}\, dx$

(d) $\displaystyle\int \frac{x^2}{\sqrt{x^2 - 16}}\, dx$

1. $4 \ln\left| \dfrac{\sqrt{x^2 + 16} - 4}{x} \right| + \sqrt{x^2 + 16} + C$

2. $8 \ln\left| \sqrt{x^2 - 16} + x \right| + \dfrac{x\sqrt{x^2 - 16}}{2} + C$

3. $8 \arcsin \dfrac{x}{4} - \dfrac{x\sqrt{16 - x^2}}{2} + C$

4. $8 \arcsin \dfrac{x - 3}{4} + \dfrac{(x - 3)\sqrt{7 + 6x - x^2}}{2} + C$

In Exercises 5–8, evaluate the indefinite integral using the substitution $x = 5 \sin \theta$.

5. $\displaystyle\int \frac{1}{(25 - x^2)^{3/2}}\, dx$

6. $\displaystyle\int \frac{10}{x^2\sqrt{25 - x^2}}\, dx$

7. $\displaystyle\int \frac{\sqrt{25 - x^2}}{x}\, dx$

8. $\displaystyle\int \frac{x^2}{\sqrt{25 - x^2}}\, dx$

In Exercises 9–12, evaluate the indefinite integral using the substitution $x = 2 \sec \theta$.

9. $\displaystyle\int \frac{1}{\sqrt{x^2 - 4}}\, dx$

10. $\displaystyle\int \frac{\sqrt{x^2 - 4}}{x}\, dx$

11. $\displaystyle\int x^3\sqrt{x^2 - 4}\, dx$

12. $\displaystyle\int \frac{x^3}{\sqrt{x^2 - 4}}\, dx$

In Exercises 13–16, evaluate the indefinite integral using the substitution $x = \tan \theta$.

13. $\displaystyle\int x\sqrt{1 + x^2}\, dx$

14. $\displaystyle\int \frac{9x^3}{\sqrt{1 + x^2}}\, dx$

15. $\displaystyle\int \frac{1}{(1 + x^2)^2}\, dx$

16. $\displaystyle\int \frac{x^2}{(1 + x^2)^2}\, dx$

In Exercises 17 and 18, use Theorem 7.2 to evaluate the integral.

17. $\displaystyle\int \sqrt{4 + 9x^2}\, dx$

18. $\displaystyle\int \sqrt{1 + x^2}\, dx$

In Exercises 19–40, evaluate the integral.

19. $\displaystyle\int \frac{x}{\sqrt{x^2 + 9}}\, dx$

20. $\displaystyle\int \frac{x}{\sqrt{9 - x^2}}\, dx$

21. $\displaystyle\int \frac{1}{\sqrt{16 - x^2}}\, dx$

22. $\displaystyle\int \frac{1}{\sqrt{25 - x^2}}\, dx$

23. $\displaystyle\int \sqrt{16 - 4x^2}\, dx$

24. $\displaystyle\int x\sqrt{16 - 4x^2}\, dx$

25. $\displaystyle\int \frac{1}{\sqrt{x^2 - 9}}\, dx$

26. $\displaystyle\int \frac{t}{(1 - t^2)^{3/2}}\, dt$

27. $\displaystyle\int \frac{\sqrt{1 - x^2}}{x^4}\, dx$

28. $\displaystyle\int \frac{\sqrt{4x^2 + 9}}{x^4}\, dx$

29. $\displaystyle\int \frac{1}{x\sqrt{4x^2 + 9}}\, dx$

30. $\displaystyle\int \frac{1}{x\sqrt{4x^2 + 16}}\, dx$

31. $\displaystyle\int \frac{-5x}{(x^2 + 5)^{3/2}}\, dx$

32. $\displaystyle\int \frac{1}{(x^2 + 3)^{3/2}}\, dx$

33. $\displaystyle\int e^{2x}\sqrt{1 + e^{2x}}\, dx$

34. $\displaystyle\int (x + 1)\sqrt{x^2 + 2x + 2}\, dx$

35. $\displaystyle\int e^x\sqrt{1 - e^{2x}}\, dx$

36. $\displaystyle\int \frac{\sqrt{1 - x}}{\sqrt{x}}\, dx$

37. $\displaystyle\int \frac{1}{4 + 4x^2 + x^4}\, dx$

38. $\displaystyle\int \frac{x^3 + x + 1}{x^4 + 2x^2 + 1}\, dx$

39. $\displaystyle\int \operatorname{arcsec} 2x\, dx, \ \ x > \tfrac{1}{2}$

40. $\displaystyle\int x \arcsin x\, dx$

In Exercises 41–44, complete the square and evaluate the integral.

41. $\displaystyle\int \frac{1}{\sqrt{4x - x^2}}\, dx$

42. $\displaystyle\int \frac{x^2}{\sqrt{2x - x^2}}\, dx$

43. $\displaystyle\int \frac{x}{\sqrt{x^2 + 4x + 8}}\, dx$

44. $\displaystyle\int \frac{x}{\sqrt{x^2 - 6x + 5}}\, dx$

In Exercises 45–50, evaluate the integral using (a) the given integration limits and (b) the limits obtained by trigonometric substitution.

45. $\displaystyle\int_0^{\sqrt{3}/2} \frac{t^2}{(1 - t^2)^{3/2}}\, dt$

46. $\displaystyle\int_0^{\sqrt{3}/2} \frac{1}{(1 - t^2)^{5/2}}\, dt$

47. $\displaystyle\int_0^3 \frac{x^3}{\sqrt{x^2 + 9}}\, dx$

48. $\displaystyle\int_0^{3/5} \sqrt{9 - 25x^2}\, dx$

49. $\displaystyle\int_4^6 \frac{x^2}{\sqrt{x^2 - 9}}\, dx$

50. $\displaystyle\int_3^6 \frac{\sqrt{x^2 - 9}}{x^2}\, dx$

In Exercises 51–54, use a computer algebra system to evaluate the integral. Verify the result by differentiation.

51. $\displaystyle\int \frac{x^2}{\sqrt{x^2 + 10x + 9}}\, dx$

52. $\displaystyle\int (x^2 + 2x + 11)^{3/2}\, dx$

53. $\displaystyle\int \frac{x^2}{\sqrt{x^2 - 1}}\, dx$

54. $\displaystyle\int x^2\sqrt{x^2 - 4}\, dx$

Getting at the Concept

55. State the substitution you would make if you used trigonometric substitution and the integral involved the given radical, where $a > 0$.

(a) $\sqrt{a^2 - u^2}$ (b) $\sqrt{a^2 + u^2}$

(c) $\sqrt{u^2 - a^2}$

56. State the method of integration you would use to perform the integration. Do not integrate.

(a) $\displaystyle\int x\sqrt{x^2 + 1}\, dx$ (b) $\displaystyle\int x^2\sqrt{x^2 - 1}\, dx$

---

57. *Area* Find the area enclosed by the ellipse shown in the figure.

$$\frac{x^2}{a^2} + \frac{y^2}{b^2} = 1$$

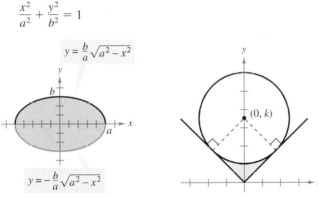

Figure for 57   **Figure for 58**

58. *Mechanical Design* The surface of a machine part is the region between the graphs of $y = |x|$ and $x^2 + (y - k)^2 = 25$ (see figure).

(a) Find k if the circle is tangent to the graph of $y = |x|$.

(b) Find the area of the surface of the machine part.

(c) Find the area of the surface of the machine part as a function of the radius of the circle r.

59. *Area* Find the area of the shaded region of the circle of radius a, if the chord is h units $(0 < h < a)$ from the center of the circle (see figure).

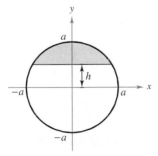

 60. *Volume* The axis of a storage tank in the form of a right circular cylinder is horizontal (see figure). The radius and length of the tank are 1 meter and 3 meters.

(a) Determine the volume of fluid in the tank as a function of its depth d.

(b) Use a graphing utility to graph the function in part (a).

(c) Design a dip stick for the tank with markings of $\frac{1}{4}$, $\frac{1}{2}$, and $\frac{3}{4}$.

(d) If fluid is entering the tank at a rate of $\frac{1}{4}$ cubic meter per minute, determine the rate of change of depth of the fluid as a function of its depth d.

(e) Use a graphing utility to graph the function in part (d). When will the rate of change of depth be minimum? Does this agree with your intuition? Explain.

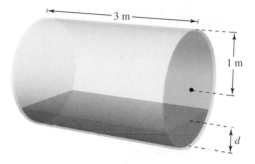

Volume of a Torus In Exercises 61 and 62, find the volume of the torus generated by revolving the region bounded by the graph of the circle about the y-axis.

61. $(x - 3)^2 + y^2 = 1$ (see figure)

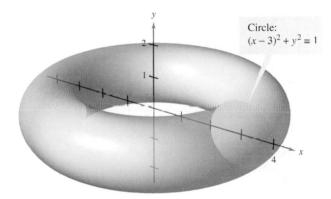

62. $(x - h)^2 + y^2 = r^2, \ h > r$

Arc Length In Exercises 63 and 64, find the arc length of the curve over the indicated interval.

Function	Interval
63. $y = \ln x$	$[1, 5]$
64. $y = \frac{1}{2}x^2$	$[0, 4]$

65. *Arc Length* Show that the length of one arch of the sine curve is equal to the length of one arch of the cosine curve.

66. *Conjecture*

(a) Find formulas for the distance between $(0, 0)$ and (a, a^2) along the line and along the parabola $y = x^2$.

(b) Use the formulas from part (a) to find the distances for $a = 1$ and $a = 10$.

(c) Make a conjecture about the difference between the two distances as a increases.

Projectile Motion   In Exercises 67 and 68, (a) use a graphing utility to graph the path of a projectile that follows the path given by the graph of the equation, (b) determine the range of the projectile, and (c) use the integration capabilities of a graphing utility to determine the distance the projectile travels.

67. $y = x - 0.005x^2$ **68.** $y = x - \dfrac{x^2}{72}$

Centroid   In Exercises 69 and 70, find the centroid of the region determined by the graphs of the inequalities.

69. $y \le 3/\sqrt{x^2 + 9}$, $y \ge 0$, $x \ge -4$, $x \le 4$

70. $y \le \frac{1}{4}x^2$, $(x - 4)^2 + y^2 \le 16$, $y \ge 0$

71. *Surface Area*   Find the surface area of the solid generated by revolving the region bounded by the graphs of $y = x^2$, $y = 0$, $x = 0$, and $x = \sqrt{2}$ about the x-axis.

72. *Average Field Strength*   The field strength H of a magnet of length $2L$ on a particle r units from the center of the magnet is

$$H = \frac{2mL}{(r^2 + L^2)^{3/2}}$$

where $\pm m$ are the poles of the magnet (see figure). Find the average field strength as the particle moves from 0 to R units from the center by evaluating the integral

$$\frac{1}{R}\int_0^R \frac{2mL}{(r^2 + L^2)^{3/2}}\, dr.$$

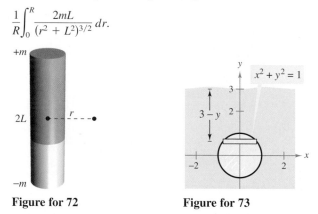

Figure for 72   **Figure for 73**

73. *Fluid Force*   Find the fluid force on a circular observation window of radius 1 foot in a vertical wall of a large water-filled tank at a fish hatchery for each of the indicated depths (see figure). Use trigonometric substitution to evaluate the one integral. (Recall that in Section 6.7 in a similar problem, you evaluated one integral by a geometric formula and the other by observing that the integrand was odd.)

(a) The center of the window is 3 feet below the water's surface.

(b) The center of the window is d feet below the water's surface $(d > 1)$.

74. *Fluid Force*   Evaluate the following two integrals, which yield the fluid forces given in Example 6.

(a) $F_{\text{inside}} = 48\displaystyle\int_{-1}^{0.8} (0.8 - y)(2)\sqrt{1 - y^2}\, dy$

(b) $F_{\text{outside}} = 64\displaystyle\int_{-1}^{0.4} (0.4 - y)(2)\sqrt{1 - y^2}\, dy$

75. *Tractrix*   A person moves from the origin along the positive y-axis pulling a weight at the end of a 12-meter rope (see figure). Initially, the weight is located at the point $(12, 0)$.

(a) Show that the slope of the tangent line of the path of the weight is

$$\frac{dy}{dx} = -\frac{\sqrt{144 - x^2}}{x}.$$

(b) Use the result in part (a) to find the equation of the path of the weight. Use a graphing utility to graph the path and compare it with the figure.

(c) Find any vertical asymptotes of the graph in part (b).

(d) When the person has reached the point $(0, 12)$, how far has the weight moved?

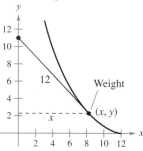

76. *Modeling Data*   For the years 1990 through 1997, the average size S (in thousands of dollars) of ordinary life insurance policies in force in the United States is given in the table. *(Source: American Council of Life Insurance)*

Year	1990	1991	1992	1993	1994	1995	1996	1997
S	37.9	41.5	43.0	45.8	45.9	49.1	52.3	56.0

A model for these data is

$$S = \sqrt{1520.4 + 111.2t + 15.8t^2}$$

where t is the time in years, with $t = 0$ corresponding to 1990. Use a graphing utility to answer each of the following.

(a) Graph the model for $0 \le t \le 7$.

(b) Find the rate of increase in S when $t = 5$.

(c) Use the model and integration to predict the average value of S for the years 2000 through 2002.

True or False?   In Exercises 77–80, determine whether the statement is true or false. If it is false, explain why or give an example that shows it is false.

77. If $x = \sin \theta$, then $\displaystyle\int \frac{dx}{\sqrt{1 - x^2}} = \int d\theta$.

78. If $x = \sec \theta$, then $\displaystyle\int \frac{\sqrt{x^2 - 1}}{x}\, dx = \int \sec \theta \tan \theta\, d\theta$.

79. If $x = \tan \theta$, then $\displaystyle\int_0^{\sqrt{3}} \frac{dx}{(1 + x^2)^{3/2}} = \int_0^{4\pi/3} \cos \theta\, d\theta$.

80. If $x = \sin \theta$, then $\displaystyle\int_{-1}^{1} x^2\sqrt{1 - x^2}\, dx = 2\int_0^{\pi/2} \sin^2 \theta \cos^2 \theta\, d\theta$.

81. Use trigonometric substitution to verify the integration formulas given in Theorem 7.2.

Section 7.5   Partial Fractions

- Understand the concept of a partial fraction decomposition.
- Use partial fraction decomposition with linear factors to integrate rational functions.
- Use partial fraction decomposition with quadratic factors to integrate rational functions.

Partial Fractions

This section examines a procedure for decomposing a rational function into simpler rational functions to which you can apply the basic integration formulas. This procedure is called the **method of partial fractions.** To see the benefit of the method of partial fractions, consider the integral

$$\int \frac{1}{x^2 - 5x + 6}\, dx.$$

To evaluate this integral *without* partial fractions, you can complete the square and use trigonometric substitution (see Figure 7.13) to obtain the following.

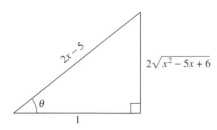

$\sec \theta = 2x - 5$
Figure 7.13

$$\int \frac{1}{x^2 - 5x + 6}\, dx = \int \frac{dx}{(x - 5/2)^2 - (1/2)^2} \qquad a = \tfrac{1}{2}, x - \tfrac{5}{2} = \tfrac{1}{2} \sec \theta$$

$$= \int \frac{(1/2)\sec \theta \tan \theta\, d\theta}{(1/4)\tan^2 \theta} \qquad dx = \tfrac{1}{2} \sec \theta \tan \theta\, d\theta$$

$$= 2 \int \csc \theta\, d\theta$$

$$= 2 \ln|\csc \theta - \cot \theta| + C$$

$$= 2 \ln\left| \frac{2x - 5}{2\sqrt{x^2 - 5x + 6}} - \frac{1}{2\sqrt{x^2 - 5x + 6}} \right| + C$$

$$- 2 \ln\left| \frac{x - 3}{\sqrt{x^2 - 5x + 6}} \right| + C$$

$$= 2 \ln\left| \frac{\sqrt{x - 3}}{\sqrt{x - 2}} \right| + C$$

$$= \ln\left| \frac{x - 3}{x - 2} \right| + C$$

$$= \ln|x - 3| - \ln|x - 2| + C$$

Now, suppose you had observed that

$$\frac{1}{x^2 - 5x + 6} = \frac{1}{x - 3} - \frac{1}{x - 2}. \qquad \text{Partial fraction decomposition}$$

Then you could evaluate the integral easily, as follows.

$$\int \frac{1}{x^2 - 5x + 6}\, dx = \int \left(\frac{1}{x - 3} - \frac{1}{x - 2} \right) dx$$

$$= \ln|x - 3| - \ln|x - 2| + C$$

This method is clearly preferable to trigonometric substitution. However, its use depends on the ability to factor the denominator, $x^2 - 5x + 6$, and to find the **partial fractions**

$$\frac{1}{x - 3} \qquad \text{and} \qquad -\frac{1}{x - 2}.$$

In this section, you will study techniques for finding partial fraction decompositions.

JOHN BERNOULLI (1667–1748)

The method of partial fractions was introduced by John Bernoulli, a Swiss mathematician who was instrumental in the early development of calculus. John Bernoulli was a professor at the University of Basel and taught many outstanding students, the most famous of whom was Leonhard Euler.

STUDY TIP In precalculus you learned how to combine functions such as

$$\frac{1}{x-2} + \frac{-1}{x+3} = \frac{5}{(x-2)(x+3)}.$$

The method of partial fractions shows you how to reverse this process.

$$\frac{5}{(x-2)(x+3)} = \frac{?}{x-2} + \frac{?}{x+3}$$

Recall from algebra that every polynomial with real coefficients can be factored into linear and irreducible quadratic factors.* For instance, the polynomial

$$x^5 + x^4 - x - 1$$

can be written as

$$
\begin{aligned}
x^5 + x^4 - x - 1 &= x^4(x+1) - (x+1) \\
&= (x^4 - 1)(x+1) \\
&= (x^2 + 1)(x^2 - 1)(x+1) \\
&= (x^2 + 1)(x+1)(x-1)(x+1) \\
&= (x-1)(x+1)^2(x^2+1)
\end{aligned}
$$

where $(x-1)$ is a linear factor, $(x+1)^2$ is a repeated linear factor, and (x^2+1) is an irreducible quadratic factor. Using this factorization, you can write the partial fraction decomposition of the rational expression

$$\frac{N(x)}{x^5 + x^4 - x - 1}$$

where $N(x)$ is a polynomial of degree less than 5, as follows.

$$\frac{N(x)}{(x-1)(x+1)^2(x^2+1)} = \frac{A}{x-1} + \frac{B}{x+1} + \frac{C}{(x+1)^2} + \frac{Dx+E}{x^2+1}$$

Decomposition of $N(x)/D(x)$ into Partial Fractions

1. **Divide if improper:** If $N(x)/D(x)$ is an improper fraction (that is, if the degree of the numerator is greater than or equal to the degree of the denominator), divide the denominator into the numerator to obtain

$$\frac{N(x)}{D(x)} = (\text{a polynomial}) + \frac{N_1(x)}{D(x)}$$

where the degree of $N_1(x)$ is less than the degree of $D(x)$. Then apply steps 2, 3, and 4 to the proper rational expression $N_1(x)/D(x)$.

2. **Factor denominator:** Completely factor the denominator into factors of the form

$$(px+q)^m \quad \text{and} \quad (ax^2 + bx + c)^n$$

where $ax^2 + bx + c$ is irreducible.

3. **Linear factors:** For each factor of the form $(px+q)^m$, the partial fraction decomposition must include the following sum of m fractions.

$$\frac{A_1}{(px+q)} + \frac{A_2}{(px+q)^2} + \cdots + \frac{A_m}{(px+q)^m}$$

4. **Quadratic factors:** For each factor of the form $(ax^2 + bx + c)^n$, the partial fraction decomposition must include the following sum of n fractions.

$$\frac{B_1 x + C_1}{ax^2 + bx + c} + \frac{B_2 x + C_2}{(ax^2 + bx + c)^2} + \cdots + \frac{B_n x + C_n}{(ax^2 + bx + c)^n}$$

---

*For a review of factorization techniques, see Precalculus, *5th edition,* by Larson and Hostetler or Precalculus: A Graphing Approach, *3rd edition, by Larson, Hostetler, and Edwards (Boston, Massachusetts: Houghton Mifflin, 2001).*

Linear Factors

Algebraic techniques for determining the constants in the numerators of a partial decomposition with linear or repeated linear factors are demonstrated in Examples 1 and 2.

Example 1   **Distinct Linear Factors**

Write the partial fraction decomposition for $\dfrac{1}{x^2 - 5x + 6}$.

Solution   Because $x^2 - 5x + 6 = (x - 3)(x - 2)$, you should include one partial fraction for each factor and write

$$\frac{1}{x^2 - 5x + 6} = \frac{A}{x - 3} + \frac{B}{x - 2}$$

where A and B are to be determined. Multiplying this equation by the least common denominator $(x - 3)(x - 2)$ yields the **basic equation**

$$1 = A(x - 2) + B(x - 3). \qquad \text{Basic equation}$$

Because this equation is to be true for all x, you can substitute any *convenient* values for x to obtain equations in A and B. The most convenient values are the ones that make particular factors equal to 0.

NOTE   Note that the substitutions for x in Example 1 are chosen for their convenience in determining values for A and B; $x = 2$ is chosen to eliminate the term $A(x - 2)$, and $x = 3$ is chosen to eliminate the term $B(x - 3)$. The goal is to make *convenient* substitutions whenever possible.

To solve for A, let $x = 3$ and obtain

$$1 = A(3 - 2) + B(3 - 3) \qquad \text{Let } x = 3 \text{ in basic equation.}$$
$$1 = A(1) + B(0)$$
$$A = 1.$$

To solve for B, let $x = 2$ and obtain

$$1 = A(2 - 2) + B(2 - 3) \qquad \text{Let } x = 2 \text{ in basic equation.}$$
$$1 = A(0) + B(-1)$$
$$B = -1.$$

Therefore, the decomposition is

$$\frac{1}{x^2 - 5x + 6} = \frac{1}{x - 3} - \frac{1}{x - 2}$$

as indicated at the beginning of this section.

FOR FURTHER INFORMATION   To learn a different method for finding the partial fraction decomposition, called the Heavyside Method, see the article "Calculus to Algebra Connections in Partial Fraction Decomposition" by Joseph Wiener and Will Watkins in *The AMATYC Review*. To view this article, go to the website *www.matharticles.com.*

Be sure you see that the method of partial fractions is practical only for integrals of rational functions whose denominators factor "nicely." For instance, if the denominator in Example 1 were changed to $x^2 - 5x + 5$, its factorization as

$$x^2 - 5x + 5 = \left[x + \frac{5 + \sqrt{5}}{2} \right]\left[x - \frac{5 - \sqrt{5}}{2} \right]$$

would be too cumbersome to use with partial fractions. In such cases, you should use completing the square or a computer algebra system to perform the integration. If you do this, you should obtain

$$\int \frac{1}{x^2 - 5x + 5}\, dx = \frac{\sqrt{5}}{5} \ln \left| 2x - \sqrt{5} - 5 \right| - \frac{\sqrt{5}}{5} \ln \left| 2x + \sqrt{5} - 5 \right| + C.$$

Example 2 **Repeated Linear Factors**

Evaluate $\int \dfrac{5x^2 + 20x + 6}{x^3 + 2x^2 + x}\, dx$.

Solution Because

$$x^3 + 2x^2 + x = x(x^2 + 2x + 1)$$
$$= x(x + 1)^2$$

you should include one fraction for *each power* of x and $(x + 1)$ and write

$$\frac{5x^2 + 20x + 6}{x(x + 1)^2} = \frac{A}{x} + \frac{B}{x + 1} + \frac{C}{(x + 1)^2}.$$

FOR FURTHER INFORMATION For an alternative approach to using partial fractions, see the article " A Shortcut in Partial Fractions" by Xun-Cheng Huang in *The College Mathematics Journal.* To view this article, go to the website *www.matharticles.com.*

Multiplying by the least common denominator $x(x + 1)^2$ yields the *basic equation*

$$5x^2 + 20x + 6 = A(x + 1)^2 + Bx(x + 1) + Cx. \qquad \text{Basic equation}$$

To solve for A, let $x = 0$. This eliminates the B and C terms and yields

$$6 = A(1) + 0 + 0$$
$$A = 6.$$

To solve for C, let $x = -1$. This eliminates the A and B terms and yields

$$5 - 20 + 6 = 0 + 0 - C$$
$$C = 9.$$

The most convenient choices for x have been used, so to find the value of B, you can use *any other value* of x along with the calculated values of A and C. Using $x = 1$, $A = 6$, and $C = 9$ produces

$$5 + 20 + 6 = A(4) + B(2) + C$$
$$31 = 6(4) + 2B + 9$$
$$-2 = 2B$$
$$B = -1.$$

Therefore, it follows that

$$\int \frac{5x^2 + 20x + 6}{x(x + 1)^2}\, dx = \int \left(\frac{6}{x} - \frac{1}{x + 1} + \frac{9}{(x + 1)^2} \right) dx$$

$$= 6\ln|x| - \ln|x + 1| + 9\frac{(x + 1)^{-1}}{-1} + C$$

$$= \ln\left| \frac{x^6}{x + 1} \right| - \frac{9}{x + 1} + C.$$

TECHNOLOGY Most computer algebra systems, such as *Derive, Maple, Mathcad, Mathematica,* and the *TI-89,* can be used to convert a rational function to its partial fraction decomposition. For instance, using *Maple,* you obtain the following.

> convert$\left(\dfrac{5x^2 + 20x + 6}{x^3 + 2x^2 + x}, \text{parfrac}, x \right)$

$\dfrac{6}{x} + \dfrac{9}{(x + 1)^2} - \dfrac{1}{x + 1}$

Try checking this result by differentiating. Include algebra in your check, simplifying the derivative until you have obtained the original integrand. ⬚

NOTE It is necessary to make as many substitutions for x as there are unknowns $(A, B, C, \ldots)$ to be determined. For instance, in Example 2, we made three substitutions $(x = -1, x = 0, \text{and } x = 1)$ to solve for C, A, and B.

Quadratic Factors

When using the method of partial fractions with *linear* factors, a convenient choice of x immediately yields a value for one of the coefficients. With *quadratic* factors, a system of linear equations usually has to be solved, regardless of the choice of x.

Example 3 **Distinct Linear and Quadratic Factors**

Evaluate $\displaystyle \int \frac{2x^3 - 4x - 8}{(x^2 - x)(x^2 + 4)}\, dx$.

Solution Because

$$(x^2 - x)(x^2 + 4) = x(x - 1)(x^2 + 4)$$

you should include one partial fraction for each factor and write

$$\frac{2x^3 - 4x - 8}{x(x - 1)(x^2 + 4)} = \frac{A}{x} + \frac{B}{x - 1} + \frac{Cx + D}{x^2 + 4}.$$

Multiplying by the least common denominator $x(x - 1)(x^2 + 4)$ yields the *basic equation*

$$2x^3 - 4x - 8 = A(x - 1)(x^2 + 4) + Bx(x^2 + 4) + (Cx + D)(x)(x - 1).$$

To solve for A, let $x = 0$ and obtain

$$8 = A(-1)(4) + 0 + 0 \quad \Longrightarrow \quad 2 = A.$$

To solve for B, let $x = 1$ and obtain

$$-10 = 0 + B(5) + 0 \quad \Longrightarrow \quad -2 = B.$$

At this point, C and D are yet to be determined. You can find these remaining constants by choosing two other values for x and solving the resulting system of linear equations. If $x = -1$, then, using $A = 2$ and $B = -2$, you can write

$$-6 = (2)(-2)(5) + (-2)(-1)(5) + (-C + D)(-1)(-2)$$
$$2 = -C + D.$$

If $x = 2$, you have

$$0 = (2)(1)(8) + (-2)(2)(8) + (2C + D)(2)(1)$$
$$8 = 2C + D.$$

Solving the linear system by subtracting the first equation from the second

$$-C + D = 2$$
$$2C + D = 8$$

yields $C = 2$. Consequently, $D = 4$, and it follows that

$$\int \frac{2x^3 - 4x - 8}{x(x - 1)(x^2 + 4)}\, dx =$$

$$\int \left(\frac{2}{x} - \frac{2}{x - 1} + \frac{2x}{x^2 + 4} + \frac{4}{x^2 + 4} \right) dx =$$

$$2 \ln|x| - 2 \ln|x - 1| + \ln(x^2 + 4) + 2 \arctan \frac{x}{2} + C.$$

In Examples 1, 2, and 3, we began the solution of the basic equation by substituting values of x that made the linear factors equal to 0. This method works well when the partial fraction decomposition involves linear factors. However, if the decomposition involves only quadratic factors, an alternative procedure is often more convenient.

Example 4   **Repeated Quadratic Factors**

Evaluate $\displaystyle\int \frac{8x^3 + 13x}{(x^2 + 2)^2}\, dx.$

Solution   Include one partial fraction for each power of $(x^2 + 2)$ and write

$$\frac{8x^3 + 13x}{(x^2 + 2)^2} = \frac{Ax + B}{x^2 + 2} + \frac{Cx + D}{(x^2 + 2)^2}.$$

Multiplying by the least common denominator $(x^2 + 2)^2$ yields the *basic equation*

$$8x^3 + 13x = (Ax + B)(x^2 + 2) + Cx + D.$$

Expanding the basic equation and collecting like terms produces

$$8x^3 + 13x = Ax^3 + 2Ax + Bx^2 + 2B + Cx + D$$
$$8x^3 + 13x = Ax^3 + Bx^2 + (2A + C)x + (2B + D).$$

Now, you can equate the coefficients of like terms on opposite sides of the equation.

$$8 = A \qquad\qquad\qquad 0 = 2B + D$$

$$8x^3 + 0x^2 + 13x + 0 = Ax^3 + Bx^2 + (2A + C)x + (2B + D)$$

$$0 = B$$

$$13 = 2A + C$$

Using the known values $A = 8$ and $B = 0$, you can write the following.

$$13 = 2A + C = 2(8) + C \quad\Longrightarrow\quad C = -3$$
$$0 = 2B + D = 2(0) + D \quad\Longrightarrow\quad D = 0$$

Finally, you can conclude that

$$\int \frac{8x^3 + 13x}{(x^2 + 2)^2}\, dx = \int \left(\frac{8x}{x^2 + 2} + \frac{-3x}{(x^2 + 2)^2} \right) dx$$

$$= 4 \ln(x^2 + 2) + \frac{3}{2(x^2 + 2)} + C.$$

TECHNOLOGY   Use a computer algebra system to evaluate the integral in Example 4—you might find that the form of the antiderivative is different. For instance, when you use a computer algebra system to work Example 4, you obtain

$$\int \frac{8x^3 + 13x}{(x^2 + 2)^2}\, dx = \ln(x^8 + 8x^6 + 24x^4 + 32x^2 + 16) + \frac{3}{2(x^2 + 2)} + C.$$

Is this result equivalent to that obtained in Example 4?

When integrating rational expressions, keep in mind that for *improper* rational expressions such as

$$\frac{N(x)}{D(x)} = \frac{2x^3 + x^2 - 7x + 7}{x^2 + x - 2}$$

you must first divide to obtain

$$\frac{N(x)}{D(x)} = 2x - 1 + \frac{-2x + 5}{x^2 + x - 2}.$$

The proper rational expression is then decomposed into its partial fractions by the usual methods. Here are some guidelines for solving the basic equation that is obtained in a partial fraction decomposition.

Guidelines for Solving the Basic Equation

Linear Factors

1. Substitute the roots of the distinct linear factors into the basic equation.
2. For repeated linear factors, use the coefficients determined in guideline 1 to rewrite the basic equation. Then substitute other convenient values of x and solve for the remaining coefficients.

Quadratic Factors

1. Expand the basic equation.
2. Collect terms according to powers of x.
3. Equate the coefficients of like powers to obtain a system of linear equations involving A, B, C, and so on.
4. Solve the system of linear equations.

Before concluding this section, here are a few things you should remember. First, it is not necessary to use the partial fractions technique on all rational functions. For instance, the following integral is evaluated more easily by the Log Rule.

$$\int \frac{x^2 + 1}{x^3 + 3x - 4} \, dx = \frac{1}{3} \int \frac{3x^2 + 3}{x^3 + 3x - 4} \, dx = \frac{1}{3} \ln|x^3 + 3x - 4| + C$$

Second, if the integrand is not in reduced form, reducing it may eliminate the need for partial fractions, as shown in the following integral.

$$\int \frac{x^2 - x - 2}{x^3 - 2x - 4} \, dx = \int \frac{(x + 1)(x - 2)}{(x - 2)(x^2 + 2x + 2)} \, dx$$

$$= \int \frac{x + 1}{x^2 + 2x + 2} \, dx$$

$$= \frac{1}{2} \ln|x^2 + 2x + 2| + C$$

Finally, partial fractions can be used with some quotients involving transcendental functions. For instance, the substitution $u = \sin x$ allows you to write

$$\int \frac{\cos x}{\sin x(\sin x - 1)} \, dx = \int \frac{du}{u(u - 1)}. \qquad u = \sin x, \, du = \cos x \, dx$$

EXERCISES FOR SECTION 7.5

In Exercises 1–6, give the form of the partial fraction decomposition of the rational expression. Do not solve for the constants.

1. $\dfrac{5}{x^2 - 10x}$

2. $\dfrac{4x^2 + 3}{(x - 5)^3}$

3. $\dfrac{2x - 3}{x^3 + 10x}$

4. $\dfrac{x - 2}{x^2 + 4x + 3}$

5. $\dfrac{16x}{x^3 - 10x^2}$

6. $\dfrac{2x - 1}{x(x^2 + 1)^2}$

In Exercises 7–28, use partial fractions to evaluate the integral.

7. $\displaystyle\int \dfrac{1}{x^2 - 1}\,dx$

8. $\displaystyle\int \dfrac{1}{4x^2 - 9}\,dx$

9. $\displaystyle\int \dfrac{3}{x^2 + x - 2}\,dx$

10. $\displaystyle\int \dfrac{x + 1}{x^2 + 4x + 3}\,dx$

11. $\displaystyle\int \dfrac{5 - x}{2x^2 + x - 1}\,dx$

12. $\displaystyle\int \dfrac{5x^2 - 12x - 12}{x^3 - 4x}\,dx$

13. $\displaystyle\int \dfrac{x^2 + 12x + 12}{x^3 - 4x}\,dx$

14. $\displaystyle\int \dfrac{x^3 - x + 3}{x^2 + x - 2}\,dx$

15. $\displaystyle\int \dfrac{2x^3 - 4x^2 - 15x + 5}{x^2 - 2x - 8}\,dx$

16. $\displaystyle\int \dfrac{x + 2}{x^2 - 4x}\,dx$

17. $\displaystyle\int \dfrac{4x^2 + 2x - 1}{x^3 + x^2}\,dx$

18. $\displaystyle\int \dfrac{2x - 3}{(x - 1)^2}\,dx$

19. $\displaystyle\int \dfrac{x^2 + 3x - 4}{x^3 - 4x^2 + 4x}\,dx$

20. $\displaystyle\int \dfrac{4x^2}{x^3 + x^2 - x - 1}\,dx$

21. $\displaystyle\int \dfrac{x^2 - 1}{x^3 + x}\,dx$

22. $\displaystyle\int \dfrac{6x}{x^3 - 8}\,dx$

23. $\displaystyle\int \dfrac{x^2}{x^4 - 2x^2 - 8}\,dx$

24. $\displaystyle\int \dfrac{x^2 - x + 9}{(x^2 + 9)^2}\,dx$

25. $\displaystyle\int \dfrac{x}{16x^4 - 1}\,dx$

26. $\displaystyle\int \dfrac{x^2 - 4x + 7}{x^3 - x^2 + x + 3}\,dx$

27. $\displaystyle\int \dfrac{x^2 + 5}{x^3 - x^2 + x + 3}\,dx$

28. $\displaystyle\int \dfrac{x^2 + x + 3}{x^4 + 6x^2 + 9}\,dx$

In Exercises 29–32, evaluate the definite integral. Use a graphing utility to verify your result.

29. $\displaystyle\int_0^1 \dfrac{3}{2x^2 + 5x + 2}\,dx$

30. $\displaystyle\int_1^5 \dfrac{x - 1}{x^2(x + 1)}\,dx$

31. $\displaystyle\int_1^2 \dfrac{x + 1}{x(x^2 + 1)}\,dx$

32. $\displaystyle\int_0^1 \dfrac{x^2 - x}{x^2 + x + 1}\,dx$

In Exercises 33–40, use a computer algebra system to determine the antiderivative that passes through the indicated point. Use the system to graph the resulting antiderivative.

33. $\displaystyle\int \dfrac{3x}{x^2 - 6x + 9}\,dx,\ (4, 0)$

34. $\displaystyle\int \dfrac{6x^2 + 1}{x^2(x - 1)^3}\,dx,\ (2, 1)$

35. $\displaystyle\int \dfrac{x^2 + x + 2}{(x^2 + 2)^2}\,dx,\ (0, 1)$

36. $\displaystyle\int \dfrac{x^3}{(x^2 - 4)^2}\,dx,\ (3, 4)$

37. $\displaystyle\int \dfrac{2x^2 - 2x + 3}{x^3 - x^2 - x - 2}\,dx,\ (3, 10)$

38. $\displaystyle\int \dfrac{x(2x - 9)}{x^3 - 6x^2 + 12x - 8}\,dx,\ (3, 2)$

39. $\displaystyle\int \dfrac{1}{x^2 - 4}\,dx,\ (6, 4)$

40. $\displaystyle\int \dfrac{x^2 - x + 2}{x^3 - x^2 + x - 1}\,dx,\ (2, 6)$

In Exercises 41–46, use substitution to evaluate the integral.

41. $\displaystyle\int \dfrac{\sin x}{\cos x(\cos x - 1)}\,dx$

42. $\displaystyle\int \dfrac{\sin x}{\cos x + \cos^2 x}\,dx$

43. $\displaystyle\int \dfrac{3 \cos x}{\sin^2 x + \sin x - 2}\,dx$

44. $\displaystyle\int \dfrac{\sec^2 x}{\tan x(\tan x + 1)}\,dx$

45. $\displaystyle\int \dfrac{e^x}{(e^x - 1)(e^x + 4)}\,dx$

46. $\displaystyle\int \dfrac{e^x}{(e^{2x} + 1)(e^x - 1)}\,dx$

In Exercises 47–50, use the method of partial fractions to verify the integration formula.

47. $\displaystyle\int \dfrac{1}{x(a + bx)}\,dx = \dfrac{1}{a}\ln\left|\dfrac{x}{a + bx}\right| + C$

48. $\displaystyle\int \dfrac{1}{a^2 - x^2}\,dx = \dfrac{1}{2a}\ln\left|\dfrac{a + x}{a - x}\right| + C$

49. $\displaystyle\int \dfrac{x}{(a + bx)^2}\,dx = \dfrac{1}{b^2}\left(\dfrac{a}{a + bx} + \ln|a + bx|\right) + C$

50. $\displaystyle\int \dfrac{1}{x^2(a + bx)}\,dx = -\dfrac{1}{ax} - \dfrac{b}{a^2}\ln\left|\dfrac{x}{a + bx}\right| + C$

In Exercises 51 and 52, use a computer algebra system to sketch the slope field for the differential equation, and graph the solution through the specified initial condition.

51. $\dfrac{dy}{dx} = \dfrac{6}{4 - x^2}$

$y(0) = 3$

52. $\dfrac{dy}{dx} = \dfrac{4}{x^2 - 2x - 3}$

$y(0) = 5$

Getting at the Concept

53. What is the first step when integrating $\displaystyle\int \dfrac{x^3}{x - 5}\,dx$?

54. Describe the decomposition of the proper rational function $N(x)/D(x)$ (a) if $D(x) = (px + q)^m$, and (b) if $D(x) = (ax^2 + bx + c)^n$ where $ax^2 + bx + c$ is irreducible.

55. State the method you would use to evaluate each integral. Do not integrate.

(a) $\displaystyle\int \dfrac{x + 1}{x^2 + 2x - 8}\,dx$

(b) $\displaystyle\int \dfrac{7x + 4}{x^2 + 2x - 8}\,dx$

(c) $\displaystyle\int \dfrac{4}{x^2 + 2x + 5}\,dx$

56. Area   Find the area of the region bounded by the graphs of $y = 7/(16 - x^2)$ and $y = 1$.

57. Modeling Data   The predicted cost C (in 100,000s of dollars) for a company to remove $p\%$ of a chemical from its waste water is shown in the table.

p	0	10	20	30	40	50	60	70	80	90
C	0	0.7	1.0	1.3	1.7	2.0	2.7	3.6	5.5	11.2

A model for the data is

$$C = \frac{124p}{(10 + p)(100 - p)}, \quad 0 \le p < 100.$$

Use the model to find the average cost for removing between 75% and 80% of the chemical.

58. Logistics Growth   In Chapter 5, the exponential growth equation was derived from the assumption that the rate of growth was proportional to the existing quantity. In practice, there often exists some upper limit L past which growth cannot occur. In such cases, we assume the rate of growth to be proportional not only to the existing quantity, but also to the difference between the existing quantity y and the upper limit L. That is,

$$\frac{dy}{dt} = ky(L - y).$$

In integral form, we can express this relationship as

$$\int \frac{dy}{y(L - y)} = \int k \, dt.$$

(a) A slope field for the differential equation $dy/dt = y(3 - y)$ is shown. Draw a possible solution to the differential equation if $y(0) = 5$, and another if $y(0) = \frac{1}{2}$. To print an enlarged copy of the graph, go to the website *www.mathgraphs.com*.

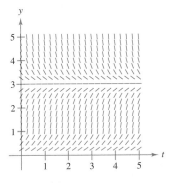

(b) Where $y(0)$ is greater than 3, what is the sign of the slope of the solution?

(c) For $y > 0$, find $\lim_{t \to \infty} y(t)$.

(d) Evaluate the two integrals above and solve for y as a function of t, where y_0 is the initial quantity.

(e) Use the result in part (d) to find and graph the solutions in part (a). Use a graphing utility to graph the solutions and compare the results with the solutions in part (a).

(f) The graph of the function y is called a **logistics curve.** Show that the rate of growth is maximum at the point of inflection, and that this occurs when $y = L/2$.

59. Approximation   Determine which value best approximates the area of the region between the x-axis and the graph of the function $10/[x(x^2 + 1)]$ over the interval $[1, 3]$. Make your selection on the basis of a sketch of the region and not by performing any calculations.

(a) -6 (b) 6   (c) 3   (d) 5   (e) 8

60. Volume and Centroid   Consider the region bounded by the graphs of

$y = 2x/(x^2 + 1)$, $y = 0$, $x = 0$, and $x = 3$.

(a) Find the volume of the solid generated by revolving the region about the x-axis.

(b) Find the centroid of the region.

61. Epidemic Model   A single infected individual enters a community of n susceptible individuals. Let x be the number of newly infected individuals at time t. The common epidemic model assumes that the disease spreads at a rate proportional to the product of the total number infected and the number not yet infected. So

$$\frac{dx}{dt} = k(x + 1)(n - x)$$

and you obtain

$$\int \frac{1}{(x + 1)(n - x)} \, dx = \int k \, dt.$$

Solve for x as a function of t.

62. Chemical Reactions   In a chemical reaction, one unit of compound Y and one unit of compound Z are converted into a single unit of compound X. If x is the amount of compound X formed, and the rate of formation of X is proportional to the product of the amounts of unconverted compounds Y and Z, then

$$\frac{dx}{dt} = k(y_0 - x)(z_0 - x)$$

where y_0 and z_0 are the initial amounts of compounds Y and Z. From the above equation you obtain

$$\int \frac{1}{(y_0 - x)(z_0 - x)} \, dx = \int k \, dt.$$

(a) Perform the two integrations and solve for x in terms of t.

(b) Use the result in part (a) to find x as $t \to \infty$ if (1) $y_0 < z_0$, (2) $y_0 > z_0$, and (3) $y_0 = z_0$.

63. Evaluate

$$\int_0^1 \frac{x}{1 + x^4} \, dx$$

in two different ways, one of which is partial fractions.

Section 7.6	Integration by Tables and Other Integration Techniques

- Evaluate an indefinite integral using a table of integrals.
- Evaluate an indefinite integral using reduction formulas.
- Evaluate an indefinite integral involving rational functions of sine and cosine.

Integration by Tables

So far in this chapter you have studied several integration techniques that can be used with the basic integration rules. But merely knowing *how* to use the various techniques is not enough. You also need to know *when* to use them. Integration is first and foremost a problem of recognition. That is, you must recognize which rule or technique to apply to obtain an antiderivative. Frequently, a slight alteration of an integrand will require a different integration technique (or produce a function whose antiderivative is not an elementary function), as shown below.

$$\int x \ln x \, dx = \frac{x^2}{2} \ln x - \frac{x^2}{4} + C \qquad \text{Integration by parts}$$

$$\int \frac{\ln x}{x} \, dx = \frac{(\ln x)^2}{2} + C \qquad \text{Power Rule}$$

$$\int \frac{1}{x \ln x} \, dx = \ln|\ln x| + C \qquad \text{Log Rule}$$

$$\int \frac{x}{\ln x} \, dx = ? \qquad \text{Not an elementary function}$$

TECHNOLOGY   A computer algebra system consists, in part, of a database of integration formulas. The primary difference between using a computer algebra system and using tables of integrals is that with a computer algebra system the computer searches through the database to find a fit. With integration tables, *you* must do the searching.

Many people find tables of integrals to be a valuable supplement to the integration techniques discussed in this chapter. Tables of common integrals can be found in Appendix C. **Integration by tables** is not a "cure-all" for all of the difficulties that can accompany integration—using tables of integrals requires considerable thought and insight and often involves substitution.

Each integration formula in Appendix C can be developed using one or more of the techniques in this chapter. You should try to verify several of the formulas. For instance, Formula 4.

$$\int \frac{u}{(a + bu)^2} \, du = \frac{1}{b^2} \left(\frac{a}{a + bu} + \ln|a + bu| \right) + C \qquad \text{Formula 4}$$

can be verified using the method of partial fractions, and Formula 19

$$\int \frac{\sqrt{a + bu}}{u} \, du = 2\sqrt{a + bu} + a \int \frac{du}{u\sqrt{a + bu}} \qquad \text{Formula 19}$$

can be verified using integration by parts. Note that the integrals in Appendix C are classified according to forms involving the following.

u^n	$(a + bu)$
$(a + bu + cu^2)$	$\sqrt{a + bu}$
$(a^2 \pm u^2)$	$\sqrt{u^2 \pm a^2}$
$\sqrt{a^2 - u^2}$	Trigonometric functions
Inverse trigonometric functions	Exponential functions
Logarithmic functions	

Example 1   **Integration by Tables**

Evaluate $\displaystyle\int \frac{dx}{x\sqrt{x - 1}}$.

Solution   Because the expression inside the radical is linear, you should consider forms involving $\sqrt{a + bu}$.

$$\int \frac{du}{u\sqrt{a + bu}} = \frac{2}{\sqrt{-a}} \arctan \sqrt{\frac{a + bu}{-a}} + C \qquad \text{Formula 17 } (a < 0)$$

Let $a = -1$, $b = 1$, and $u = x$. Then $du = dx$, and you can write

$$\int \frac{dx}{x\sqrt{x - 1}} = 2 \arctan \sqrt{x - 1} + C.$$

 *Example 2*   **Integration by Tables**

Evaluate $\int x\sqrt{x^4 - 9}\,dx$.

Solution   Because the radical has the form $\sqrt{u^2 - a^2}$, you should consider Formula 26.

$$\int \sqrt{u^2 - a^2}\,du = \frac{1}{2}\left(u\sqrt{u^2 - a^2} - a^2 \ln\left|u + \sqrt{u^2 - a^2}\right|\right) + C$$

Let $u = x^2$ and $a = 3$. Then $du = 2x\,dx$, and you have

$$\int x\sqrt{x^4 - 9}\,dx = \frac{1}{2}\int \sqrt{(x^2)^2 - 3^2}\,(2x)\,dx$$

$$= \frac{1}{4}\left(x^2\sqrt{x^4 - 9} - 9\ln\left|x^2 + \sqrt{x^4 - 9}\right|\right) + C.$$

Example 3   **Integration by Tables**

Evaluate $\displaystyle\int \frac{x}{1 + e^{-x^2}}\,dx$.

Solution   Of the forms involving e^u, consider the following.

$$\int \frac{du}{1 + e^u} = u - \ln(1 + e^u) + C \qquad \text{Formula 84}$$

Let $u = -x^2$. Then $du = -2x\,dx$, and you have

$$\int \frac{x}{1 + e^{-x^2}}\,dx = -\frac{1}{2}\int \frac{-2x\,dx}{1 + e^{-x^2}}$$

$$= -\frac{1}{2}\left[-x^2 - \ln(1 + e^{-x^2})\right] + C$$

$$= \frac{1}{2}\left[x^2 + \ln(1 + e^{-x^2})\right] + C.$$

Reduction Formulas

Several of the integrals in the integration tables have the form $\int f(x)\, dx = g(x) + \int h(x)\, dx$. Such integration formulas are called **reduction formulas** because they reduce a given integral to the sum of a function and a simpler integral.

Example 4   Using a Reduction Formula

Evaluate $\int x^3 \sin x\, dx$.

Solution   Consider the following three formulas.

$$\int u \sin u\, du = \sin u - u \cos u + C \qquad \text{Formula 52}$$

$$\int u^n \sin u\, du = -u^n \cos u + n \int u^{n-1} \cos u\, du \qquad \text{Formula 54}$$

$$\int u^n \cos u\, du = u^n \sin u - n \int u^{n-1} \sin u\, du \qquad \text{Formula 55}$$

Using Formula 54, Formula 55, and then Formula 52 produces

$$\int x^3 \sin x\, dx = -x^3 \cos x + 3 \int x^2 \cos x\, dx$$

$$= -x^3 \cos x + 3 \left(x^2 \sin x - 2 \int x \sin x\, dx \right)$$

$$= -x^3 \cos x + 3x^2 \sin x + 6x \cos x - 6 \sin x + C.$$

TECHNOLOGY   Sometimes when you use computer algebra systems you obtain results that look very different, but are actually equivalent. We used several to evaluate the integral in Example 5, as follows.

Maple

$$\sqrt{3 - 5x} -$$
$$\sqrt{3} \operatorname{arctanh}\left(\tfrac{1}{3} \sqrt{3 - 5x} \sqrt{3} \right)$$

Derive

$$\sqrt{3} \ln \left[\frac{\sqrt{(3 - 5x)} - \sqrt{3}}{\sqrt{x}} \right] +$$
$$\sqrt{(3 - 5x)}$$

Mathematica

$$\text{Sqrt}[3 - 5x] -$$
$$\text{Sqrt}[3]\, \text{ArcTanh}\left[\frac{\text{Sqrt}[3 - 5x]}{\text{Sqrt}[3]} \right]$$

Mathcad

$$\sqrt{3 - 5x} +$$
$$\tfrac{1}{2}\sqrt{3} \ln \left[-\tfrac{1}{5} \frac{(-6 + 5x + 2\sqrt{3}\sqrt{3 - 5x})}{x} \right]$$

Notice that computer algebra systems do not include a constant of integration.

Example 5   Using a Reduction Formula

Evaluate $\int \dfrac{\sqrt{3 - 5x}}{2x}\, dx$.

Solution   Consider the following two formulas.

$$\int \frac{du}{u\sqrt{a + bu}} = \frac{1}{\sqrt{a}} \ln \left| \frac{\sqrt{a + bu} - \sqrt{a}}{\sqrt{a + bu} + \sqrt{a}} \right| + C \qquad \text{Formula 17 } (a > 0)$$

$$\int \frac{\sqrt{a + bu}}{u}\, du = 2\sqrt{a + bu} + a \int \frac{du}{u\sqrt{a + bu}} \qquad \text{Formula 19}$$

Using Formula 19, with $a = 3$, $b = -5$, and $u = x$, produces

$$\frac{1}{2} \int \frac{\sqrt{3 - 5x}}{x}\, dx = \frac{1}{2} \left(2\sqrt{3 - 5x} + 3 \int \frac{dx}{x\sqrt{3 - 5x}} \right)$$

$$= \sqrt{3 - 5x} + \frac{3}{2} \int \frac{dx}{x\sqrt{3 - 5x}}.$$

Using Formula 17, with $a = 3$, $b = -5$, and $u = x$, produces

$$\int \frac{\sqrt{3 - 5x}}{2x}\, dx = \sqrt{3 - 5x} + \frac{3}{2} \left(\frac{1}{\sqrt{3}} \ln \left| \frac{\sqrt{3 - 5x} - \sqrt{3}}{\sqrt{3 - 5x} + \sqrt{3}} \right| \right) + C$$

$$= \sqrt{3 - 5x} + \frac{\sqrt{3}}{2} \ln \left| \frac{\sqrt{3 - 5x} - \sqrt{3}}{\sqrt{3 - 5x} + \sqrt{3}} \right| + C.$$

Rational Functions of Sine and Cosine

Example 6   Integration by Tables

Evaluate $\displaystyle \int \frac{\sin 2x}{2 + \cos x}\, dx$.

Solution   Substituting $2 \sin x \cos x$ for $\sin 2x$ produces

$$\int \frac{\sin 2x}{2 + \cos x}\, dx = 2 \int \frac{\sin x \cos x}{2 + \cos x}\, dx.$$

A check of the forms involving $\sin u$ or $\cos u$ in Appendix C shows that none of those listed applies. Therefore, you can consider forms involving $a + bu$. For example,

$$\int \frac{u\, du}{a + bu} = \frac{1}{b^2}\,(bu - a\,\ln|a + bu|) + C. \qquad \text{Formula 3}$$

Let $a = 2$, $b = 1$, and $u = \cos x$. Then $du = -\sin x\, dx$, and you have

$$2 \int \frac{\sin x \cos x}{2 + \cos x}\, dx = -2 \int \frac{\cos x(-\sin x\, dx)}{2 + \cos x}$$

$$= -2(\cos x - 2\,\ln|2 + \cos x|) + C$$

$$= -2 \cos x + 4\,\ln|2 + \cos x| + C.$$ ▨

Example 6 involves a rational expression of $\sin x$ and $\cos x$. If you are unable to find an integral of this form in the integration tables, try using the following special substitution to convert the trigonometric expression to a standard rational expression.

Substitution for Rational Functions of Sine and Cosine

For integrals involving rational functions of sine and cosine, the substitution

$$u = \frac{\sin x}{1 + \cos x} = \tan \frac{x}{2}$$

yields

$$\cos x = \frac{1 - u^2}{1 + u^2}, \quad \sin x = \frac{2u}{1 + u^2}, \quad \text{and} \quad dx = \frac{2\, du}{1 + u^2}.$$

Proof   From the substitution for u, it follows that

$$u^2 = \frac{\sin^2 x}{(1 + \cos x)^2} = \frac{1 - \cos^2 x}{(1 + \cos x)^2} = \frac{1 - \cos x}{1 + \cos x}.$$

Solving for $\cos x$ produces $\cos x = (1 - u^2)/(1 + u^2)$. To find $\sin x$, write $u = \sin x/(1 + \cos x)$ as

$$\sin x = u(1 + \cos x) = u\left(1 + \frac{1 - u^2}{1 + u^2}\right) = \frac{2u}{1 + u^2}.$$

Finally, to find dx, consider $u = \tan(x/2)$. Then you have $\arctan u = x/2$ and $dx = (2\, du)/(1 + u^2)$. ▨

EXERCISES FOR SECTION 7.6

In Exercises 1 and 2, use a table of integrals with forms involving $a + bu$ to evaluate the integral.

1. $\displaystyle\int \frac{x^2}{1 + x}\,dx$

2. $\displaystyle\int \frac{2}{3x^2(2x - 5)^2}\,dx$

In Exercises 3 and 4, use a table of integrals with forms involving $\sqrt{u^2 \pm a^2}$ to evaluate the integral.

3. $\displaystyle\int e^x \sqrt{1 + e^{2x}}\,dx$

4. $\displaystyle\int \frac{\sqrt{x^2 - 9}}{3x}\,dx$

In Exercises 5 and 6, use a table of integrals with forms involving $\sqrt{a^2 - u^2}$ to evaluate the integral.

5. $\displaystyle\int \frac{1}{x^2\sqrt{1 - x^2}}\,dx$

6. $\displaystyle\int \frac{x}{\sqrt{9 - x^4}}\,dx$

In Exercises 7–10, use a table of integrals with forms involving the trigonometric functions to evaluate the integral.

7. $\displaystyle\int \sin^4 2x\,dx$

8. $\displaystyle\int \frac{\cos^3 \sqrt{x}}{\sqrt{x}}\,dx$

9. $\displaystyle\int \frac{1}{\sqrt{x}\left(1 - \cos\sqrt{x}\right)}\,dx$

10. $\displaystyle\int \frac{1}{1 - \tan 5x}\,dx$

In Exercises 11 and 12, use a table of integrals with forms involving e^u to evaluate the integral.

11. $\displaystyle\int \frac{1}{1 + e^{2x}}\,dx$

12. $\displaystyle\int e^{-x/2}\sin 2x\,dx$

In Exercises 13 and 14, use a table of integrals with forms involving $\ln u$ to evaluate the integral.

13. $\displaystyle\int x^3 \ln x\,dx$

14. $\displaystyle\int (\ln x)^3\,dx$

In Exercises 15–18, find the indefinite integral (a) using integration tables and (b) using the indicated method.

Integral	Method
15. $\displaystyle\int x^2 e^x\,dx$	Integration by parts
16. $\displaystyle\int x^4 \ln x\,dx$	Integration by parts
17. $\displaystyle\int \frac{1}{x^2(x + 1)}\,dx$	Partial fractions
18. $\displaystyle\int \frac{1}{x^2 - 75}\,dx$	Partial fractions

In Exercises 19–50, use integration tables to evaluate the integral.

19. $\displaystyle\int xe^{x^2}\,dx$

20. $\displaystyle\int \frac{x}{\sqrt{1 + x}}\,dx$

21. $\displaystyle\int x\,\mathrm{arcsec}(x^2 + 1)\,dx$

22. $\displaystyle\int \mathrm{arcsec}\,2x\,dx$

23. $\displaystyle\int x^2 \ln x\,dx$

24. $\displaystyle\int x \sin x\,dx$

25. $\displaystyle\int \frac{1}{x^2\sqrt{x^2 - 4}}\,dx$

26. $\displaystyle\int \frac{x^2}{(3x - 5)^2}\,dx$

27. $\displaystyle\int \frac{2x}{(1 - 3x)^2}\,dx$

28. $\displaystyle\int \frac{1}{x^2 + 2x + 2}\,dx$

29. $\displaystyle\int e^x \arccos e^x\,dx$

30. $\displaystyle\int \frac{\theta^2}{1 - \sin \theta^3}\,d\theta$

31. $\displaystyle\int \frac{x}{1 - \sec x^2}\,dx$

32. $\displaystyle\int \frac{e^x}{1 - \tan e^x}\,dx$

33. $\displaystyle\int \frac{\cos x}{1 + \sin^2 x}\,dx$

34. $\displaystyle\int \frac{1}{t[1 + (\ln t)^2]}\,dt$

35. $\displaystyle\int \frac{\cos \theta}{3 + 2\sin \theta + \sin^2 \theta}\,d\theta$

36. $\displaystyle\int \sqrt{3 + x^2}\,dx$

37. $\displaystyle\int \frac{1}{x^2\sqrt{2 + 9x^2}}\,dx$

38. $\displaystyle\int x^2\sqrt{2 + 9x^2}\,dx$

39. $\displaystyle\int t^3 \cos t\,dt$

40. $\displaystyle\int \sqrt{x}\,\arctan x^{3/2}\,dx$

41. $\displaystyle\int \frac{\ln x}{x(3 + 2\ln x)}\,dx$

42. $\displaystyle\int \frac{e^x}{(1 - e^{2x})^{3/2}}\,dx$

43. $\displaystyle\int \frac{x}{(x^2 - 6x + 10)^2}\,dx$

44. $\displaystyle\int (2x - 3)^2\sqrt{(2x - 3)^2 + 4}\,dx$

45. $\displaystyle\int \frac{x}{\sqrt{x^4 - 6x^2 + 5}}\,dx$

46. $\displaystyle\int \frac{\cos x}{\sqrt{\sin^2 x + 1}}\,dx$

47. $\displaystyle\int \frac{x^3}{\sqrt{4 - x^2}}\,dx$

48. $\displaystyle\int \sqrt{\frac{3 - x}{3 + x}}\,dx$

49. $\displaystyle\int \frac{e^{3x}}{(1 + e^x)^3}\,dx$

50. $\displaystyle\int \tan^3 \theta\,d\theta$

In Exercises 51–56, verify the integration formula.

51. $\displaystyle\int \frac{u^2}{(a + bu)^2}\,du = \frac{1}{b^3}\left(bu - \frac{a^2}{a + bu} - 2a \ln|a + bu|\right) + C$

52. $\displaystyle\int \frac{u^n}{\sqrt{a + bu}}\,du = \frac{2}{(2n + 1)b}\left(u^n\sqrt{a + bu} - na\int \frac{u^{n-1}}{\sqrt{a + bu}}\,du\right)$

53. $\displaystyle\int \frac{1}{(u^2 \pm a^2)^{3/2}}\,du = \frac{\pm u}{a^2\sqrt{u^2 \pm a^2}} + C$

54. $\displaystyle\int u^n \cos u\,du = u^n \sin u - n\int u^{n-1}\sin u\,du$

55. $\displaystyle\int \arctan u\,du = u \arctan u - \ln\sqrt{1 + u^2} + C$

56. $\displaystyle\int (\ln u)^n\,du = u(\ln u)^n - n\int (\ln u)^{n-1}\,du$

In Exercises 57–62, use a computer algebra system to determine the antiderivative that passes through the indicated point. Use the system to graph the resulting antiderivative.

57. $\int \dfrac{1}{x^{3/2}\sqrt{1-x}}\,dx,\ \left(\frac{1}{2}, 5\right)$

58. $\int x\sqrt{x^2 + 2x}\,dx,\ (0, 0)$

59. $\int \dfrac{1}{(x^2 - 6x + 10)^2}\,dx,\ (3, 0)$

60. $\int \dfrac{\sqrt{2 - 2x - x^2}}{x + 1}\,dx,\ \left(0, \sqrt{2}\right)$

61. $\int \dfrac{1}{\sin\theta\tan\theta}\,d\theta,\ \left(\frac{\pi}{4}, 2\right)$

62. $\int \dfrac{\sin\theta}{(\cos\theta)(1 + \sin\theta)}\,d\theta,\ (0, 1)$

In Exercises 63–70, evaluate the integral.

63. $\int \dfrac{1}{2 - 3\sin\theta}\,d\theta$

64. $\int \dfrac{\sin\theta}{1 + \cos^2\theta}\,d\theta$

65. $\int_0^{\pi/2} \dfrac{1}{1 + \sin\theta + \cos\theta}\,d\theta$

66. $\int_0^{\pi/2} \dfrac{1}{3 - 2\cos\theta}\,d\theta$

67. $\int \dfrac{\sin\theta}{3 - 2\cos\theta}\,d\theta$

68. $\int \dfrac{\cos\theta}{1 + \cos\theta}\,d\theta$

69. $\int \dfrac{\cos\sqrt{\theta}}{\sqrt{\theta}}\,d\theta$

70. $\int \dfrac{1}{\sec\theta - \tan\theta}\,d\theta$

Area   In Exercises 71 and 72, find the area of the region bounded by the graphs of the equations.

71. $y = \dfrac{x}{\sqrt{x + 1}},\ y = 0, x = 8$ **72.** $y = \dfrac{x}{1 + e^{x^2}},\ y = 0, x = 2$

Getting at the Concept

In Exercises 73–78, state (if possible) the method or integration formula you would use to find the antiderivative. Do not integrate.

73. $\int \dfrac{e^x}{e^{2x} + 1}\,dx$

74. $\int \dfrac{e^x}{e^x + 1}\,dx$

75. $\int x\,e^{x^2}\,dx$

76. $\int x\,e^x\,dx$

77. $\int e^{x^2}\,dx$

78. $\int e^{2x}\sqrt{e^{2x} + 1}\,dx$

---

79. Generate four integration problems that can be integrated from a table of integrals after an appropriate substitution. Use four different integration formulas from the table in the text.

80. Describe what is meant by a reduction formula. Give an example.

81. *Work*   A hydraulic cylinder on an industrial machine pushes a steel block a distance of x feet $(0 \le x \le 5)$, where the variable force required is

$$F(x) = 2000xe^{-x} \text{ pounds.}$$

Find the work done in pushing the block the full 5 feet through the machine.

82. *Work*   Repeat Exercise 81, using a force of

$$F(x) = \frac{500x}{\sqrt{26 - x^2}} \text{ pounds.}$$

83. *Building Design*   The cross section of a precast concrete beam for a building is bounded by the graphs of the equations

$$x = \frac{2}{\sqrt{1 + y^2}},\ x = \frac{-2}{\sqrt{1 + y^2}}, y = 0, \text{ and } y = 3$$

where x and y are measured in feet. The length of the beam is 20 feet (see figure).

(a) Find the volume V and the weight W of the beam. Assume the concrete weighs 148 pounds per cubic foot.

(b) Find the centroid of a cross section of the beam.

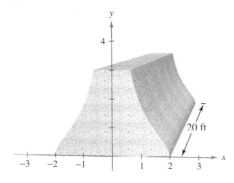

84. *Average Population Size*   A population is growing according to the logistics model

$$N = \frac{5000}{1 + e^{4.8 - 1.9t}}$$

where t is the time in days. Find the average population over the interval $[0, 2]$.

In Exercises 85 and 86, use a graphing utility to (a) solve the integral equation for the constant k and (b) graph the region whose area is given by the integral.

85. $\int_0^4 \dfrac{k}{2 + 3x}\,dx = 10$ **86.** $\int_0^k 6x^2\,e^{-x/2}\,dx = 50$

True or False   In Exercises 87 and 88, determine whether the statement is true or false. If it is false, explain why or give an example that shows it is false.

87. To use a table of integrals, the integral you are evaluating must appear in the table.

88. When using a table of integrals, you may have to make substitutions to rewrite your integral in the form in which it appears in the table.

Indeterminate Forms and L'Hôpital's Rule

- Recognize limits that produce indeterminate forms.
- Apply L'Hôpital's Rule to evaluate a limit.

Indeterminate Forms

Recall from Chapters 1 and 3 that the forms $0/0$ and ∞/∞ are called *indeterminate* because they do not guarantee that a limit exists, nor do they indicate what the limit is, if one does exist. When you encountered one of these indeterminate forms earlier in the text, you attempted to rewrite the expression by using various algebraic techniques.

Indeterminate Form	Limit	Algebraic Technique
$\dfrac{0}{0}$	$\displaystyle\lim_{x \to -1} \frac{2x^2 - 2}{x + 1} = \lim_{x \to -1} 2(x - 1)$ $= -4$	Divide numerator and denominator by $(x + 1)$.
$\dfrac{\infty}{\infty}$	$\displaystyle\lim_{x \to \infty} \frac{3x^2 - 1}{2x^2 + 1} = \lim_{x \to \infty} \frac{3 - (1/x^2)}{2 + (1/x^2)}$ $= \dfrac{3}{2}$	Divide numerator and denominator by x^2.

Occasionally, you can extend these algebraic techniques to find limits of transcendental functions. For instance, the limit

$$\lim_{x \to 0} \frac{e^{2x} - 1}{e^x - 1}$$

produces the indeterminate form $0/0$. Factoring and then dividing produces

$$\lim_{x \to 0} \frac{e^{2x} - 1}{e^x - 1} = \lim_{x \to 0} \frac{(e^x + 1)(e^x - 1)}{e^x - 1} = \lim_{x \to 0} (e^x + 1) = 2.$$

However, not all indeterminate forms can be evaluated by algebraic manipulation. This is particularly true when *both* algebraic and transcendental functions are involved. For instance, the limit

$$\lim_{x \to 0} \frac{e^{2x} - 1}{x}$$

produces the indeterminate form $0/0$. Rewriting the expression to obtain

$$\lim_{x \to 0} \left(\frac{e^{2x}}{x} - \frac{1}{x} \right)$$

merely produces another indeterminate form, $\infty - \infty$. Of course, you could use technology to estimate the limit, as shown in the table and in Figure 7.14. From the table and the graph, the limit appears to be 2. (This limit will be verified in Example 1.)

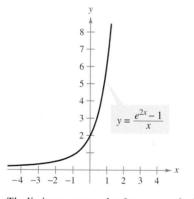

$$y = \frac{e^{2x} - 1}{x}$$

The limit as x approaches 0 appears to be 2.

Figure 7.14

x	-1	-0.1	-0.01	-0.001	0	0.001	0.01	0.1	1
$\dfrac{e^{2x} - 1}{x}$	0.865	1.813	1.980	1.998	?	2.002	2.020	2.214	6.389

L'Hôpital's Rule

To find the limit illustrated in Figure 7.14, you can use a theorem called **L'Hôpital's Rule.** This theorem states that under certain conditions the limit of the quotient $f(x)/g(x)$ is determined by the limit of the quotient of the derivatives

$$\frac{f'(x)}{g'(x)}.$$

To prove this theorem, you can use a more general result called the **Extended Mean Value Theorem.**

THEOREM 7.3   The Extended Mean Value Theorem

If f and g are differentiable on an open interval (a, b) and continuous on $[a, b]$ such that $g'(x) \neq 0$ for any x in (a, b), then there exists a point c in (a, b) such that

$$\frac{f'(c)}{g'(c)} = \frac{f(b) - f(a)}{g(b) - g(a)}.$$

NOTE   To see why this is called the Extended Mean Value Theorem, consider the special case in which $g(x) = x$. For this case, you obtain the "standard" Mean Value Theorem as presented in Section 3.2.

The Extended Mean Value Theorem and L'Hôpital's Rule are both proved in Appendix B.

THEOREM 7.4   L'Hôpital's Rule

Let f and g be functions that are differentiable on an open interval (a, b) containing c, except possibly at c itself. Assume that $g'(x) \neq 0$ for all x in (a, b), except possibly at c itself. If the limit of $f(x)/g(x)$ as x approaches c produces the indeterminate form $0/0$, then

$$\lim_{x \to c} \frac{f(x)}{g(x)} = \lim_{x \to c} \frac{f'(x)}{g'(x)}$$

provided the limit on the right exists (or is infinite). This result also applies if the limit of $f(x)/g(x)$ as x approaches c produces any one of the indeterminate forms ∞/∞, $(-\infty)/\infty$, $\infty/(-\infty)$, or $(-\infty)/(-\infty)$.

NOTE   People occasionally use L'Hôpital's Rule incorrectly by applying the Quotient Rule to $f(x)/g(x)$. Be sure you see that the rule involves $f'(x)/g'(x)$, not the derivative of $f(x)/g(x)$.

L'Hôpital's Rule can also be applied to one-sided limits. For instance, if the limit of $f(x)/g(x)$ as x approaches c *from the right* produces the indeterminate form $0/0$, then

$$\lim_{x \to c^+} \frac{f(x)}{g(x)} = \lim_{x \to c^+} \frac{f'(x)}{g'(x)}$$

provided the limit exists (or is infinite).

GUILLAUME L'HÔPITAL (1661–1704)

L'Hôpital's Rule is named after the French mathematician Guillaume François Antoine de L'Hôpital. L'Hôpital is credited with writing the first text on differential calculus (in 1696) in which the rule publicly appeared. It was recently discovered that the rule and its proof were written in a letter from John Bernoulli to L'Hôpital. "... I acknowledge that I owe very much to the bright minds of the Bernoulli brothers. ... I have made free use of their discoveries ...," said L'Hôpital.

The Granger Collection

FOR FURTHER INFORMATION
To further understand the necessity of the restriction that $g'(x)$ be nonzero for all x in (a, b), except possibly at c, see the article "Counterexamples to L'Hôpital's Rule" by R. P. Boas in *The American Mathematical Monthly.* To view this article, go to the website *www.matharticles.com.*

Example 1 Indeterminate Form 0/0

Evaluate $\lim_{x \to 0} \dfrac{e^{2x} - 1}{x}$.

Solution Because direct substitution results in the indeterminate form 0/0

$$\lim_{x \to 0} \frac{e^{2x} - 1}{x} \quad \nearrow \quad \lim_{x \to 0} (e^{2x} - 1) = 0$$
$$\searrow \quad \lim_{x \to 0} x = 0$$

you can apply L'Hôpital's Rule as follows.

$$\lim_{x \to 0} \frac{e^{2x} - 1}{x} = \lim_{x \to 0} \frac{\dfrac{d}{dx}[e^{2x} - 1]}{\dfrac{d}{dx}[x]} \qquad \text{Apply L'Hôpital's Rule.}$$

$$= \lim_{x \to 0} \frac{2e^{2x}}{1} \qquad \text{Differentiate numerator and denominator.}$$

$$= 2 \qquad \text{Evaluate the limit.}$$

NOTE In writing the string of equations in Example 1, you actually do not know that the first limit is equal to the second until you have shown that the second limit exists. In other words, if the second limit had not existed, it would not have been permissible to apply L'Hôpital's Rule.

Another form of L'Hôpital's Rule states that if the limit of $f(x)/g(x)$ as x approaches ∞ (or $-\infty$) produces the indeterminate form 0/0 or ∞/∞, then

$$\lim_{x \to \infty} \frac{f(x)}{g(x)} = \lim_{x \to \infty} \frac{f'(x)}{g'(x)}$$

provided the limit on the right exists.

Example 2 Indeterminate Form ∞/∞

Evaluate $\lim_{x \to \infty} \dfrac{\ln x}{x}$.

Solution Because direct substitution results in the indeterminate form ∞/∞, you can apply L'Hôpital's Rule to obtain

$$\lim_{x \to \infty} \frac{\ln x}{x} = \lim_{x \to \infty} \frac{\dfrac{d}{dx}[\ln x]}{\dfrac{d}{dx}[x]} \qquad \text{Apply L'Hôpital's Rule.}$$

$$= \lim_{x \to \infty} \frac{1}{x} \qquad \text{Differentiate numerator and denominator.}$$

$$= 0. \qquad \text{Evaluate the limit.}$$

NOTE Try graphing $y_1 = \ln x$ and $y_2 = x$ in the same viewing window. Which function grows faster as x approaches ∞? How is this observation related to Example 2?

Occasionally it is necessary to apply L'Hôpital's Rule more than once to remove an indeterminate form, as illustrated in Example 3.

Example 3   Applying L'Hôpital's Rule More than Once

Evaluate $\lim\limits_{x \to -\infty} \dfrac{x^2}{e^{-x}}$.

Solution   Because direct substitution results in the indeterminate form ∞/∞, you can apply L'Hôpital's Rule.

$$\lim_{x \to -\infty} \frac{x^2}{e^{-x}} = \lim_{x \to -\infty} \frac{\dfrac{d}{dx}[x^2]}{\dfrac{d}{dx}[e^{-x}]} = \lim_{x \to -\infty} \frac{2x}{-e^{-x}}$$

This limit yields the indeterminate form $(-\infty)/(-\infty)$, so you can apply L'Hôpital's Rule again to obtain

$$\lim_{x \to -\infty} \frac{2x}{-e^{-x}} = \lim_{x \to -\infty} \frac{\dfrac{d}{dx}[2x]}{\dfrac{d}{dx}[-e^{-x}]} = \lim_{x \to -\infty} \frac{2}{e^{-x}} = 0.$$

In addition to the forms $0/0$ and ∞/∞, there are other indeterminate forms such as $0 \cdot \infty$, 1^∞, ∞^0, 0^0, and $\infty - \infty$. For example, consider the following four limits that lead to the indeterminate form $0 \cdot \infty$.

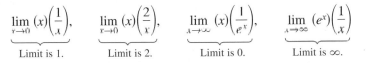

$$\underbrace{\lim_{x \to 0} (x)\left(\frac{1}{x}\right)}_{\text{Limit is } 1.}, \qquad \underbrace{\lim_{x \to 0} (x)\left(\frac{2}{x}\right)}_{\text{Limit is } 2.}, \qquad \underbrace{\lim_{x \to \infty} (x)\left(\frac{1}{e^x}\right)}_{\text{Limit is } 0.}, \qquad \underbrace{\lim_{x \to \infty} (e^x)\left(\frac{1}{x}\right)}_{\text{Limit is } \infty.}$$

Because each limit is different, it is clear that the form $0 \cdot \infty$ is indeterminate in the sense that it does not determine the value (or even the existence) of the limit. The following examples indicate methods for evaluating these forms. Basically, you attempt to convert each of these forms to $0/0$ or ∞/∞ so that L'Hôpital's Rule can be applied.

Example 4   Indeterminate Form $0 \cdot \infty$

Evaluate $\lim\limits_{x \to \infty} e^{-x}\sqrt{x}$.

Solution   Because direct substitution produces the indeterminate form $0 \cdot \infty$, you should try to rewrite the limit to fit the form $0/0$ or ∞/∞. In this case, you can rewrite the limit to fit the second form.

$$\lim_{x \to \infty} e^{-x}\sqrt{x} = \lim_{x \to \infty} \frac{\sqrt{x}}{e^x}$$

Now, by L'Hôpital's Rule, you have

$$\lim_{x \to \infty} \frac{\sqrt{x}}{e^x} = \lim_{x \to \infty} \frac{1/(2\sqrt{x})}{e^x} = \lim_{x \to \infty} \frac{1}{2\sqrt{x}\,e^x} = 0.$$

If rewriting a limit in one of the forms $0/0$ or ∞/∞ does not seem to work, try the other form. For instance, in Example 4 you can write the limit as

$$\lim_{x \to \infty} e^{-x}\sqrt{x} = \lim_{x \to \infty} \frac{e^{-x}}{x^{-1/2}}$$

which yields the indeterminate form $0/0$. As it happens, applying L'Hôpital's Rule to this limit produces

$$\lim_{x \to \infty} \frac{e^{-x}}{x^{-1/2}} = \lim_{x \to \infty} \frac{-e^{-x}}{-1/(2x^{3/2})}$$

which also yields the indeterminate form $0/0$.

The indeterminate forms 1^{∞}, ∞^{0}, and 0^{0} arise from limits of functions that have variable bases and variable exponents. When we previously encountered this type of function, we used logarithmic differentiation to find the derivative. You can use a similar procedure when taking limits, as indicated in the next example.

Example 5 Indeterminate Form 1^{∞}

Evaluate $\displaystyle\lim_{x \to \infty} \left(1 + \frac{1}{x}\right)^{x}$.

Solution Because direct substitution yields the indeterminate form 1^{∞}, you can proceed as follows. To begin, assume that the limit exists and is equal to y.

$$y = \lim_{x \to \infty} \left(1 + \frac{1}{x}\right)^{x}$$

Taking the natural logarithm of both sides produces

$$\ln y = \ln\left[\lim_{x \to \infty} \left(1 + \frac{1}{x}\right)^{x}\right].$$

Because the natural logarithmic function is continuous, you can write the following.

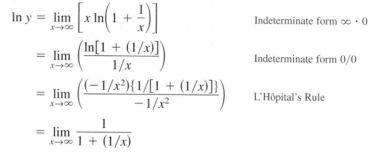

$$\ln y = \lim_{x \to \infty} \left[x \ln\left(1 + \frac{1}{x}\right)\right] \qquad \text{Indeterminate form } \infty \cdot 0$$

$$= \lim_{x \to \infty} \left(\frac{\ln[1 + (1/x)]}{1/x}\right) \qquad \text{Indeterminate form } 0/0$$

$$= \lim_{x \to \infty} \left(\frac{(-1/x^2)\{1/[1 + (1/x)]\}}{-1/x^2}\right) \qquad \text{L'Hôpital's Rule}$$

$$= \lim_{x \to \infty} \frac{1}{1 + (1/x)}$$

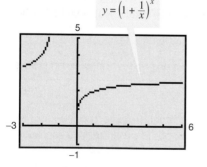

$y = \left(1 + \frac{1}{x}\right)^{x}$

The limit of $[1 + (1/x)]^{x}$ as x approaches infinity is e.

Figure 7.15

Now, because you have shown that $\ln y = 1$, you can conclude that $y = e$ and obtain

$$\lim_{x \to \infty} \left(1 + \frac{1}{x}\right)^{x} = e.$$

You can use a graphing utility to confirm this result, as shown in Figure 7.15.

L'Hôpital's Rule can also be applied to one-sided limits, as demonstrated in Examples 6 and 7.

Example 6   Indeterminate Form 0^0

Evaluate $\lim\limits_{x \to 0^+} (\sin x)^x$.

Solution   Because direct substitution produces the indeterminate form 0^0, you can proceed as follows. To begin, assume that the limit exists and is equal to y.

$$y = \lim_{x \to 0^+} (\sin x)^x \qquad \text{Indeterminate form } 0^0$$

$$\ln y = \ln \left[\lim_{x \to 0^+} (\sin x)^x \right] \qquad \text{Take natural log of both sides.}$$

$$= \lim_{x \to 0^+} \left[\ln(\sin x)^x \right] \qquad \text{Continuity}$$

$$= \lim_{x \to 0^+} \left[x \ln(\sin x) \right] \qquad \text{Indeterminate form } 0 \cdot (-\infty)$$

$$= \lim_{x \to 0^+} \frac{\ln(\sin x)}{1/x} \qquad \text{Indeterminate form } -\infty/\infty$$

$$= \lim_{x \to 0^+} \frac{\cot x}{-1/x^2} \qquad \text{L'Hôpital's Rule}$$

$$= \lim_{x \to 0^+} \frac{-x^2}{\tan x} \qquad \text{Indeterminate form } 0/0$$

$$= \lim_{x \to 0^+} \frac{-2x}{\sec^2 x} = 0 \qquad \text{L'Hôpital's Rule}$$

Now, because $\ln y = 0$, you can conclude that $y = e^0 = 1$, and it follows that

$$\lim_{x \to 0^+} (\sin x)^x = 1.$$

TECHNOLOGY   When evaluating complicated limits such as the one in Example 6, it is helpful to check the reasonableness of the solution with a computer or with a graphing utility. For instance, the calculations in the following table and the graph in Figure 7.16 are consistent with the conclusion that $(\sin x)^x$ approaches 1 as x approaches 0 from the right.

x	1.0	0.1	0.01	0.001	0.0001	0.00001
$(\sin x)^x$	0.8415	0.7942	0.9550	0.9931	0.9991	0.9999

Try using a computer or graphing utility to estimate the following limits.

$$\lim_{x \to 0} (1 - \cos x)^x$$

and

$$\lim_{x \to 0^+} (\tan x)^x$$

Then see if you can verify your estimates analytically.

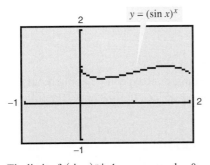

$y = (\sin x)^x$

The limit of $(\sin x)^x$ is 1 as x approaches 0 from the right.
Figure 7.16

STUDY TIP   In each of the examples presented in this section, L'Hôpital's Rule is used to find a limit that exists. It can also be used to conclude that a limit is infinite. For instance, try using L'Hôpital's Rule to show that

$$\lim_{x \to \infty} \frac{e^x}{x} = \infty.$$

Example 7   **Indeterminate Form** $\infty - \infty$

Evaluate $\displaystyle\lim_{x \to 1^+} \left(\frac{1}{\ln x} - \frac{1}{x - 1} \right)$.

Solution   Because direct substitution yields the indeterminate form $\infty - \infty$, you should try to rewrite the expression to produce a form to which you can apply L'Hôpital's Rule. In this case, you can combine the two fractions to obtain

$$\lim_{x \to 1^+} \left(\frac{1}{\ln x} - \frac{1}{x - 1} \right) = \lim_{x \to 1^+} \left[\frac{x - 1 - \ln x}{(x - 1) \ln x} \right].$$

Now, because direct substitution produces the indeterminate form $0/0$, you can apply L'Hôpital's Rule to obtain

$$\lim_{x \to 1^+} \left(\frac{1}{\ln x} - \frac{1}{x - 1} \right) = \lim_{x \to 1^+} \frac{\dfrac{d}{dx}[x - 1 - \ln x]}{\dfrac{d}{dx}[(x - 1) \ln x]}$$

$$= \lim_{x \to 1^+} \left[\frac{1 - (1/x)}{(x - 1)(1/x) + \ln x} \right]$$

$$= \lim_{x \to 1^+} \left(\frac{x - 1}{x - 1 + x \ln x} \right).$$

This limit also yields the indeterminate form $0/0$, so you can apply L'Hôpital's Rule again to obtain

$$\lim_{x \to 1^+} \left(\frac{1}{\ln x} - \frac{1}{x - 1} \right) = \lim_{x \to 1^+} \left[\frac{1}{1 + x(1/x) + \ln x} \right]$$

$$= \frac{1}{2}.$$

We have identified the forms $0/0$, ∞/∞, $\infty - \infty$, $0 \cdot \infty$, 0^0, 1^∞, and ∞^0 as *indeterminate*. There are similar forms that you should recognize as "determinate."

$$\infty + \infty \to \infty \qquad \text{Limit is positive infinity.}$$
$$-\infty - \infty \to -\infty \qquad \text{Limit is negative infinity.}$$
$$0^\infty \to 0 \qquad \text{Limit is zero.}$$
$$0^{-\infty} \to \infty \qquad \text{Limit is positive infinity.}$$

(You are asked to verify two of these in Exercises 95 and 96.)

As a final comment, we remind you that L'Hôpital's Rule can be applied only to quotients leading to the indeterminate forms $0/0$ and ∞/∞. For instance, the following application of L'Hôpital's Rule is *incorrect*.

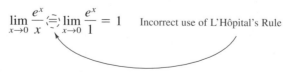

$$\lim_{x \to 0} \frac{e^x}{x} \overset{?}{=} \lim_{x \to 0} \frac{e^x}{1} = 1 \qquad \text{Incorrect use of L'Hôpital's Rule}$$

The reason this application is incorrect is that, even though the limit of the denominator is 0, the limit of the numerator is 1, which means that the hypotheses of L'Hôpital's Rule have not been satisfied.

EXERCISES FOR SECTION 7.7

Numerical and Graphical Analysis   **In Exercises 1–4, complete the table and use the result to estimate the limit. Use a graphing utility to graph the function to support your result.**

1. $\lim\limits_{x\to 0} \dfrac{\sin 5x}{\sin 2x}$

x	-0.1	-0.01	-0.001	0.001	0.01	0.1
$f(x)$						

2. $\lim\limits_{x\to 0} \dfrac{1 - e^x}{x}$

x	-0.1	-0.01	-0.001	0.001	0.01	0.1
$f(x)$						

3. $\lim\limits_{x\to\infty} x^5 e^{-x/100}$

x	1	10	10^2	10^3	10^4	10^5
$f(x)$						

4. $\lim\limits_{x\to\infty} \dfrac{6x}{\sqrt{3x^2 - 2x}}$

x	1	10	10^2	10^3	10^4	10^5
$f(x)$						

In Exercises 5–10, evaluate the limit (a) using techniques from Chapters 1 and 3 and (b) using L'Hôpital's Rule.

5. $\lim\limits_{x\to 3} \dfrac{2(x-3)}{x^2 - 9}$

6. $\lim\limits_{x\to -1} \dfrac{2x^2 - x - 3}{x + 1}$

7. $\lim\limits_{x\to 3} \dfrac{\sqrt{x+1} - 2}{x - 3}$

8. $\lim\limits_{x\to 0} \dfrac{\sin 4x}{2x}$

9. $\lim\limits_{x\to\infty} \dfrac{5x^2 - 3x + 1}{3x^2 - 5}$

10. $\lim\limits_{x\to\infty} \dfrac{2x + 1}{4x^2 + x}$

In Exercises 11–36, evaluate the limit, using L'Hôpital's Rule if necessary. (In Exercise 17, n is a positive integer.)

11. $\lim\limits_{x\to 2} \dfrac{x^2 - x - 2}{x - 2}$

12. $\lim\limits_{x\to -1} \dfrac{x^2 - x - 2}{x + 1}$

13. $\lim\limits_{x\to 0} \dfrac{\sqrt{4 - x^2} - 2}{x}$

14. $\lim\limits_{x\to 2^-} \dfrac{\sqrt{4 - x^2}}{x - 2}$

15. $\lim\limits_{x\to 0} \dfrac{e^x - (1 - x)}{x}$

16. $\lim\limits_{x\to 0^+} \dfrac{e^x - (1 + x)}{x^3}$

17. $\lim\limits_{x\to 0^+} \dfrac{e^x - (1 + x)}{x^n}$

18. $\lim\limits_{x\to 1} \dfrac{\ln x^2}{x^2 - 1}$

19. $\lim\limits_{x\to 0} \dfrac{\sin 2x}{\sin 3x}$

20. $\lim\limits_{x\to 0} \dfrac{\sin ax}{\sin bx}$

21. $\lim\limits_{x\to 0} \dfrac{\arcsin x}{x}$

22. $\lim\limits_{x\to 1} \dfrac{\arctan x - (\pi/4)}{x - 1}$

23. $\lim\limits_{x\to\infty} \dfrac{3x^2 - 2x + 1}{2x^2 + 3}$

24. $\lim\limits_{x\to\infty} \dfrac{x - 1}{x^2 + 2x + 3}$

25. $\lim\limits_{x\to\infty} \dfrac{x^2 + 2x + 3}{x - 1}$

26. $\lim\limits_{x\to\infty} \dfrac{x^3}{x + 2}$

27. $\lim\limits_{x\to 0} \dfrac{x^3}{e^{x/2}}$

28. $\lim\limits_{x\to\infty} \dfrac{x^2}{e^x}$

29. $\lim\limits_{x\to\infty} \dfrac{x}{\sqrt{x^2 + 1}}$

30. $\lim\limits_{x\to\infty} \dfrac{x^2}{\sqrt{x^2 + 1}}$

31. $\lim\limits_{x\to\infty} \dfrac{\cos x}{x}$

32. $\lim\limits_{x\to\infty} \dfrac{\sin x}{x - \pi}$

33. $\lim\limits_{x\to\infty} \dfrac{\ln x}{x^2}$

34. $\lim\limits_{x\to\infty} \dfrac{\ln x^4}{x^3}$

35. $\lim\limits_{x\to\infty} \dfrac{e^x}{x^2}$

36. $\lim\limits_{x\to\infty} \dfrac{e^{x/2}}{x}$

In Exercises 37–54, (a) describe the type of indeterminate form (if any) that is obtained by direct substitution. (b) Evaluate the limit, using L'Hôpital's Rule if necessary. (c) Use a graphing utility to graph the function and verify the result in part (b). (For a geometric approach to Exercise 37, see the article "A Geometric Proof of $\lim\limits_{d\to 0^+} (-d \ln d) = 0$" by John H. Mathews in *The College Mathematics Journal*. To view this article, go to the website *www.matharticles.com*.)

37. $\lim\limits_{x\to 0^+} (-x \ln x)$

38. $\lim\limits_{x\to 0^+} x^3 \cot x$

39. $\lim\limits_{x\to 0} \left(x \sin \dfrac{1}{x}\right)$

40. $\lim\limits_{x\to\infty} x \tan \dfrac{1}{x}$

41. $\lim\limits_{x\to 0^+} x^{1/x}$

42. $\lim\limits_{x\to 0^+} (e^x + x)^{2/x}$

43. $\lim\limits_{x\to\infty} x^{1/x}$

44. $\lim\limits_{x\to\infty} \left(1 + \dfrac{1}{x}\right)^x$

45. $\lim\limits_{x\to 0^+} (1 + x)^{1/x}$

46. $\lim\limits_{x\to\infty} (1 + x)^{1/x}$

47. $\lim\limits_{x\to 0^+} \left[3(x)^{x/2}\right]$

48. $\lim\limits_{x\to 4^+} \left[3(x - 4)\right]^{x-4}$

49. $\lim\limits_{x\to 1^+} (\ln x)^{x-1}$

50. $\lim\limits_{x\to 0^+} \left[\cos\left(\dfrac{\pi}{2} - x\right)\right]^x$

51. $\lim\limits_{x\to 2^+} \left(\dfrac{8}{x^2 - 4} - \dfrac{x}{x - 2}\right)$

52. $\lim\limits_{x\to 2^+} \left(\dfrac{1}{x^2 - 4} - \dfrac{\sqrt{x - 1}}{x^2 - 4}\right)$

53. $\lim\limits_{x\to 1^+} \left(\dfrac{3}{\ln x} - \dfrac{2}{x - 1}\right)$

54. $\lim\limits_{x\to 0^+} \left(\dfrac{10}{x} - \dfrac{3}{x^2}\right)$

In Exercises 55–58, use a graphing utility to (a) graph the function and (b) find the required limit (if it exists).

55. $\lim\limits_{x\to 3} \dfrac{x - 3}{\ln(2x - 5)}$

56. $\lim\limits_{x\to 0^+} (\sin x)^x$

57. $\lim\limits_{x\to\infty} \left(\sqrt{x^2 + 5x + 2} - x\right)$

58. $\lim\limits_{x\to\infty} \dfrac{x^3}{e^{2x}}$

Getting at the Concept

59. List six different indeterminate forms.

60. State L'Hôpital's Rule.

61. Find the differentiable functions f and g that satisfy the specified condition such that $\lim\limits_{x \to 5} f(x) = 0$ and $\lim\limits_{x \to 5} g(x) = 0$.
(*Note:* There are many correct answers.)

(a) $\lim\limits_{x \to 5} \dfrac{f(x)}{g(x)} = 10$ (b) $\lim\limits_{x \to 5} \dfrac{f(x)}{g(x)} = 0$ (c) $\lim\limits_{x \to 5} \dfrac{f(x)}{g(x)} = \infty$

62. Find differentiable functions f and g such that

$$\lim_{x \to \infty} f(x) = \lim_{x \to \infty} g(x) = \infty$$

$$\lim_{x \to \infty} [f(x) - g(x)] = 25.$$

(*Note:* There are many correct answers.)

Comparing Functions In Exercises 63–68, use L'Hôpital's Rule to determine the comparative rates of increase of the functions

$$f(x) = x^m, \quad g(x) = e^{nx}, \quad \text{and} \quad h(x) = (\ln x)^n$$

where $n > 0, m > 0$, and $x \to \infty$.

63. $\lim\limits_{x \to \infty} \dfrac{x^2}{e^{5x}}$

64. $\lim\limits_{x \to \infty} \dfrac{x^3}{e^{2x}}$

65. $\lim\limits_{x \to \infty} \dfrac{(\ln x)^3}{x}$

66. $\lim\limits_{x \to \infty} \dfrac{(\ln x)^2}{x^3}$

67. $\lim\limits_{x \to \infty} \dfrac{(\ln x)^n}{x^m}$

68. $\lim\limits_{x \to \infty} \dfrac{x^m}{e^{nx}}$

69. *Numerical Approach* Complete the table to show that x eventually "overpowers" $(\ln x)^4$.

x	10	10^2	10^4	10^6	10^8	10^{10}
$\dfrac{(\ln x)^4}{x}$						

70. *Numerical Approach* Complete the table to show that e^x eventually "overpowers" x^5.

x	1	5	10	20	30	40	50	100
$\dfrac{e^x}{x^5}$								

In Exercises 71–74, find any asymptotes and relative extrema that may exist and use a graphing utility to graph the function. (*Hint:* Some of the limits required in finding asymptotes have been found in preceding exercises.)

71. $y = x^{1/x}, \quad x > 0$

72. $y = x^x, \quad x > 0$

73. $y = 2xe^{-x}$

74. $y = \dfrac{\ln x}{x}$

Think About It In Exercises 75–78, L'Hôpital's Rule is used incorrectly. Describe the error.

75. $\lim\limits_{x \to 0} \dfrac{e^{2x} - 1}{e^x} = \lim\limits_{x \to 0} \dfrac{2e^{2x}}{e^x} = \lim\limits_{x \to 0} 2e^x = 2$ ✗

76. $\lim\limits_{x \to 0} \dfrac{\sin \pi x - 1}{x} = \lim\limits_{x \to 0} \dfrac{\pi \cos \pi x}{1} = \pi$ ✗

77. $\lim\limits_{x \to \infty} x \cos \dfrac{1}{x} = \lim\limits_{x \to \infty} \dfrac{\cos(1/x)}{1/x}$ ✗

$= \lim\limits_{x \to \infty} \dfrac{[-\sin(1/x)](1/x^2)}{-1/x^2}$

$= 0$

78. $\lim\limits_{x \to \infty} \dfrac{e^{-x}}{1 + e^{-x}} = \lim\limits_{x \to \infty} \dfrac{-e^{-x}}{-e^{-x}} = \lim\limits_{x \to \infty} 1 = 1$ ✗

79. *Analytical Approach* Consider $\lim\limits_{x \to \infty} \dfrac{x}{\sqrt{x^2 + 1}}$.

(a) Find the limit analytically without trying to use L'Hôpital's Rule.

(b) Show that L'Hôpital's Rule fails.

(c) Use a graphing utility to graph the function and approximate the limit from the graph. Compare the result with that in part (a).

80. *Compound Interest* The formula for the amount A in a savings account compounded n times per year for t years at an interest rate r and an initial deposit of P is

$$A = P\left(1 + \dfrac{r}{n}\right)^{nt}.$$

Use L'Hôpital's Rule to show that the limiting formula as the number of compoundings per year becomes infinite is

$$A = Pe^{rt}.$$

81. *Velocity in a Resisting Medium* The velocity of an object falling through a resisting medium such as air or water is

$$v = \dfrac{32}{k}\left(1 - e^{-kt} + \dfrac{v_0 k e^{-kt}}{32}\right)$$

where v_0 is the initial velocity, t is the time in seconds, and k is the resistance constant of the medium. Use L'Hôpital's Rule to find the formula for the velocity of a falling body in a vacuum by fixing v_0 and t and letting k approach zero. (Assume that the downward direction is positive.)

82. *The Gamma Function* The Gamma Function $\Gamma(n)$ is defined in terms of the integral of the function

$$f(x) = x^{n-1}e^{-x}, \quad n > 0.$$

Show that for any fixed value of n, the limit of $f(x)$ as x approaches infinity is zero.

83. *Area*   Find the limit, as x approaches 0, of the ratio of the area of the triangle to the total shaded area in the figure.

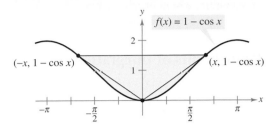

84. Use a graphing utility to graph

$$f(x) = \frac{x^k - 1}{k}$$

for $k = 1, 0.1,$ and 0.01. Then evaluate the limit

$$\lim_{k \to 0^+} \frac{x^k - 1}{k}.$$

In Exercises 85–88, apply the Extended Mean Value Theorem to the functions f and g on the indicated interval. Find all values c in the interval (a, b) such that

$$\frac{f'(c)}{g'(c)} = \frac{f(b) - f(a)}{g(b) - g(a)}.$$

Functions	Interval
85. $f(x) = x^3$	$[0, 1]$
$g(x) = x^2 + 1$	
86. $f(x) = \dfrac{1}{x}$	$[1, 2]$
$g(x) = x^2 - 4$	
87. $f(x) = \sin x$	$\left[0, \dfrac{\pi}{2}\right]$
$g(x) = \cos x$	
88. $f(x) = \ln x$	$[1, 4]$
$g(x) = x^3$	

True or False?   **In Exercises 89–92, determine whether the statement is true or false. If it is false, explain why or give an example that shows it is false.**

89. $\lim\limits_{x \to 0} \left[\dfrac{x^2 + x + 1}{x}\right] = \lim\limits_{x \to 0} \left[\dfrac{2x + 1}{1}\right] = 1$

90. If $y = e^x/x^2$, then $y' = e^x/2x$.

91. If $p(x)$ is a polynomial, then $\lim\limits_{x \to \infty} \left[p(x)/e^x\right] = 0$.

92. If $\lim\limits_{x \to \infty} \dfrac{f(x)}{g(x)} = 1$, then $\lim\limits_{x \to \infty} \left[f(x) - g(x)\right] = 0$.

93. In Chapter 1 we used a geometric argument (see figure) to prove that

$$\lim_{\theta \to 0} \frac{\sin \theta}{\theta} = 1.$$

(a) Express the area of the triangle $\triangle ABD$ in terms of θ.

(b) Express the area of the shaded region in terms of θ.

(c) Express the ratio R of the area of $\triangle ABD$ to that of the shaded region.

(d) Find $\lim\limits_{\theta \to 0} R$.

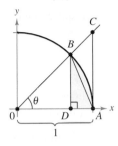

94. Sketch the graph of

$$g(x) = \begin{cases} e^{-1/x^2}, & x \neq 0 \\ 0, & x = 0 \end{cases}$$

and determine $g'(0)$.

95. Prove that if $f(x) \geq 0$, $\lim\limits_{x \to a} f(x) = 0$, and $\lim\limits_{x \to a} g(x) = \infty$, then

$$\lim_{x \to a} f(x)^{g(x)} = 0.$$

96. Prove that if $f(x) \geq 0$, $\lim\limits_{x \to a} f(x) = 0$, and $\lim\limits_{x \to a} g(x) = -\infty$, then

$$\lim_{x \to a} f(x)^{g(x)} = \infty.$$

97. Prove the following generalization of the Mean Value Theorem. If f is twice differentiable on the closed interval $[a, b]$, then

$$f(b) - f(a) = f'(a)(b - a) - \int_a^b f''(t)(t - b)\, dt.$$

98. Show that the indeterminate form 0^0 is not always equal to 1 by evaluating

$$\lim_{x \to 0^+} x^{\ln 2/(1 + \ln x)}.$$

Improper Integrals

- Evaluate an improper integral that has an infinite limit of integration.
- Evaluate an improper integral that has an infinite discontinuity.

Improper Integrals with Infinite Limits of Integration

The definition of a definite integral

$$\int_a^b f(x)\,dx$$

requires that the interval $[a, b]$ be finite. Furthermore, the Fundamental Theorem of Calculus, by which you have been evaluating definite integrals, requires that f be continuous on $[a, b]$. In this section you will study a procedure for evaluating integrals that do not satisfy these requirements—usually because either one or both of the limits of integration are infinite, or f has a finite number of infinite discontinuities in the interval $[a, b]$. Integrals that possess either property are **improper integrals.** Note that a function f is said to have an **infinite discontinuity** at c if, *from the right or left,*

$$\lim_{x \to c} f(x) = \infty \quad \text{or} \quad \lim_{x \to c} f(x) = -\infty.$$

To get an idea of how to evaluate an improper integral, consider the integral

$$\int_1^b \frac{dx}{x^2} = -\frac{1}{x}\Big]_1^b = -\frac{1}{b} + 1 = 1 - \frac{1}{b}$$

which can be interpreted as the area of the shaded region shown in Figure 7.17. Taking the limit as $b \to \infty$ produces

$$\int_1^\infty \frac{dx}{x^2} = \lim_{b \to \infty} \left(\int_1^b \frac{dx}{x^2} \right) = \lim_{b \to \infty} \left(1 - \frac{1}{b} \right) = 1.$$

This improper integral can be interpreted as the area of the *unbounded* region between the graph of $f(x) = 1/x^2$ and the x-axis (to the right of $x = 1$).

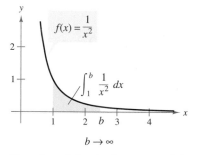

$f(x) = \dfrac{1}{x^2}$

$\displaystyle\int_1^b \frac{1}{x^2}\,dx$

$b \to \infty$

The unbounded region has an area of 1.

Figure 7.17

Definition of Improper Integrals with Infinite Integration Limits

1. If f is continuous on the interval $[a, \infty)$, then

$$\int_a^\infty f(x)\,dx = \lim_{b \to \infty} \int_a^b f(x)\,dx.$$

2. If f is continuous on the interval $(-\infty, b]$, then

$$\int_{-\infty}^b f(x)\,dx = \lim_{a \to -\infty} \int_a^b f(x)\,dx.$$

3. If f is continuous on the interval $(-\infty, \infty)$, then

$$\int_{-\infty}^\infty f(x)\,dx = \int_{-\infty}^c f(x)\,dx + \int_c^\infty f(x)\,dx$$

where c is any real number.

In the first two cases, the improper integral **converges** if the limit exists—otherwise, the improper integral **diverges.** In the third case, the improper integral on the left diverges if either of the improper integrals on the right diverges.

Example 1   **An Improper Integral That Diverges**

Evaluate $\displaystyle\int_1^\infty \frac{dx}{x}$.

Solution

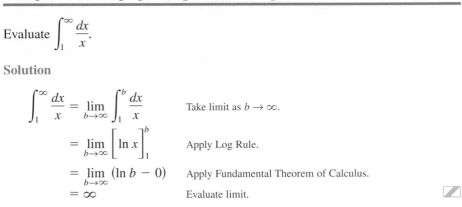

$$\int_1^\infty \frac{dx}{x} = \lim_{b\to\infty} \int_1^b \frac{dx}{x} \qquad \text{Take limit as } b \to \infty.$$

$$= \lim_{b\to\infty} \left[\ln x\right]_1^b \qquad \text{Apply Log Rule.}$$

$$= \lim_{b\to\infty} (\ln b - 0) \qquad \text{Apply Fundamental Theorem of Calculus.}$$

$$= \infty \qquad \text{Evaluate limit.}$$

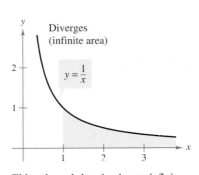

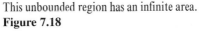

This unbounded region has an infinite area.
Figure 7.18

NOTE   Try comparing the regions shown in Figures 7.17 and 7.18. They look similar, yet the region in Figure 7.17 has a finite area of 1 and the region in Figure 7.18 has an infinite area.

Example 2   **Improper Integrals That Converge**

Evaluate each of the improper integrals.

a. $\displaystyle\int_0^\infty e^{-x}\, dx$
b. $\displaystyle\int_0^\infty \frac{1}{x^2 + 1}\, dx$

Solution

a. $\displaystyle\int_0^\infty e^{-x}\, dx = \lim_{b\to\infty} \int_0^b e^{-x}\, dx$

$$= \lim_{b\to\infty} \left[-e^{-x}\right]_0^b$$

$$= \lim_{b\to\infty} (-e^{-b} + 1)$$

$$= 1$$

(See Figure 7.19.)

b. $\displaystyle\int_0^\infty \frac{1}{x^2 + 1}\, dx = \lim_{b\to\infty} \int_0^b \frac{1}{x^2 + 1}\, dx$

$$= \lim_{b\to\infty} \left[\arctan x\right]_0^b$$

$$= \lim_{b\to\infty} \arctan b$$

$$= \frac{\pi}{2}$$

(See Figure 7.20.)

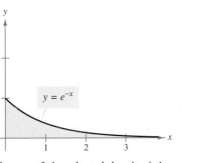

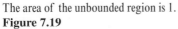

The area of the unbounded region is 1.
Figure 7.19

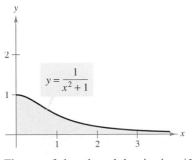

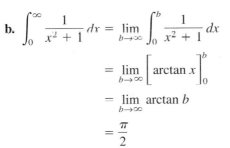

The area of the unbounded region is $\pi/2$.
Figure 7.20

In the following example, note how L'Hôpital's Rule can be used to evaluate an improper integral.

Example 3   Using L'Hôpital's Rule with an Improper Integral

Evaluate $\displaystyle\int_{1}^{\infty} (1 - x)e^{-x}\,dx$.

Solution   Use integration by parts, with $dv = e^{-x}\,dx$ and $u = (1 - x)$.

$$\int (1 - x)e^{-x}\,dx = -e^{-x}(1 - x) - \int e^{-x}\,dx$$

$$= -e^{-x} + xe^{-x} + e^{-x} + C$$

$$= xe^{-x} + C$$

Now, apply the definition of an improper integral.

$$\int_{1}^{\infty} (1 - x)e^{-x}\,dx = \lim_{b \to \infty} \left[xe^{-x} \right]_{1}^{b} = \left(\lim_{b \to \infty} \frac{b}{e^{b}} \right) - \frac{1}{e}$$

Finally, using L'Hôpital's Rule on the right-hand limit produces

$$\lim_{b \to \infty} \frac{b}{e^{b}} = \lim_{b \to \infty} \frac{1}{e^{b}} = 0$$

from which you can conclude that

$$\int_{1}^{\infty} (1 - x)e^{-x}\,dx = -\frac{1}{e}.$$

(See Figure 7.21.)

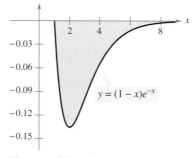

The area of the unbounded region is $\left| -1/e \right|$.

Figure 7.21

$y = (1 - x)e^{-x}$

Example 4   Infinite Upper and Lower Limits of Integration

Evaluate $\displaystyle\int_{-\infty}^{\infty} \frac{e^{x}}{1 + e^{2x}}\,dx$.

Solution   Note that the integrand is continuous on $(-\infty, \infty)$. To evaluate the integral, you can break it into two parts, choosing $c = 0$ as a convenient value.

$$\int_{-\infty}^{\infty} \frac{e^{x}}{1 + e^{2x}}\,dx = \int_{-\infty}^{0} \frac{e^{x}}{1 + e^{2x}}\,dx + \int_{0}^{\infty} \frac{e^{x}}{1 + e^{2x}}\,dx$$

$$= \lim_{b \to -\infty} \left[\arctan e^{x} \right]_{b}^{0} + \lim_{b \to \infty} \left[\arctan e^{x} \right]_{0}^{b}$$

$$= \lim_{b \to -\infty} \left(\frac{\pi}{4} - \arctan e^{b} \right) + \lim_{b \to \infty} \left(\arctan e^{b} - \frac{\pi}{4} \right)$$

$$= \frac{\pi}{4} - 0 + \frac{\pi}{2} - \frac{\pi}{4}$$

$$= \frac{\pi}{2}$$

(See Figure 7.22.)

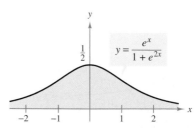

$y = \dfrac{e^{x}}{1 + e^{2x}}$

The area of the unbounded region is $\pi/2$.

Figure 7.22

Example 5   **Sending a Space Module into Orbit**

In Example 3 of Section 6.5, you found that it would require 10,000 mile-tons of work to propel a 15-ton space module to a height of 800 miles above earth. How much work is required to propel the module an unlimited distance away from earth's surface?

Solution   At first you might think that an infinite amount of work would be required. But if this were the case, it would be impossible to send rockets into outer space. Because this has been done, the work required must be finite. You can determine the work in the following manner. Using the integral of Example 3, Section 6.5, replace the upper bound of 4800 miles by ∞ and write

$$W = \int_{4000}^{\infty} \frac{240,000,000}{x^2}\, dx$$

$$= \lim_{b \to \infty} \left[-\frac{240,000,000}{x} \right]_{4000}^{b}$$

$$= \lim_{b \to \infty} \left(-\frac{240,000,000}{b} + \frac{240,000,000}{4000} \right)$$

$$= 60,000 \text{ mile-tons}$$

$$= 6.336 \times 10^{11} \text{ foot-pounds.}$$

(See Figure 7.23.)

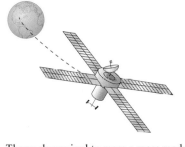

The work required to move a space module an unlimited distance away from earth is approximately 6.336×10^{11} foot-pounds.
Figure 7.23

Improper Integrals with Infinite Discontinuities

The second basic type of improper integral is one that has an infinite discontinuity *at or between* the limits of integration.

Definition of Improper Integrals with Infinite Discontinuities

1. If f is continuous on the interval $[a, b)$ and has an infinite discontinuity at b, then

$$\int_{a}^{b} f(x)\, dx = \lim_{c \to b^-} \int_{a}^{c} f(x)\, dx.$$

2. If f is continuous on the interval $(a, b]$ and has an infinite discontinuity at a, then

$$\int_{a}^{b} f(x)\, dx = \lim_{c \to a^+} \int_{c}^{b} f(x)\, dx.$$

3. If f is continuous on the interval $[a, b]$, except for some c in (a, b) at which f has an infinite discontinuity, then

$$\int_{a}^{b} f(x)\, dx = \int_{a}^{c} f(x)\, dx + \int_{c}^{b} f(x)\, dx.$$

In the first two cases, the improper integral **converges** if the limit exists—otherwise, the improper integral **diverges.** In the third case, the improper integral on the left diverges if either of the improper integrals on the right diverges.

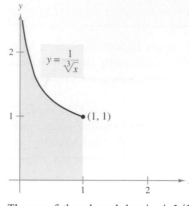

The area of the unbounded region is 3/2.
Figure 7.24

Example 6   An Improper Integral with an Infinite Discontinuity

Evaluate $\int_0^1 \dfrac{dx}{\sqrt[3]{x}}$.

Solution   The integrand has an infinite discontinuity at $x = 0$, as shown in Figure 7.24. You can evaluate this integral as follows.

$$\int_0^1 x^{-1/3}\, dx = \lim_{b \to 0^+} \left[\frac{x^{2/3}}{2/3} \right]_b^1$$

$$= \lim_{b \to 0^+} \frac{3}{2}(1 - b^{2/3})$$

$$= \frac{3}{2}$$

Example 7   An Improper Integral That Diverges

Evaluate $\int_0^2 \dfrac{dx}{x^3}$.

Solution   Because the integrand has an infinite discontinuity at $x = 0$, you can write the following.

$$\int_0^2 \frac{dx}{x^3} = \lim_{b \to 0^+} \left[-\frac{1}{2x^2} \right]_b^2 = \lim_{b \to 0^+} \left(-\frac{1}{8} + \frac{1}{2b^2} \right) = \infty$$

So, you can conclude that the improper integral diverges.

Example 8   An Improper Integral with an Interior Discontinuity

Evaluate $\int_{-1}^2 \dfrac{dx}{x^3}$.

Solution   This integral is improper because the integrand has an infinite discontinuity at the interior point $x = 0$, as shown in Figure 7.25. So, you can write the following.

$$\int_{-1}^2 \frac{dx}{x^3} = \int_{-1}^0 \frac{dx}{x^3} + \int_0^2 \frac{dx}{x^3}$$

From Example 7 you know that the second integral diverges. Therefore, the original improper integral also diverges.

NOTE   Remember to check for infinite discontinuities at interior points as well as endpoints when determining whether an integral is improper. For instance, if you had not recognized that the integral in Example 8 was improper, you would have obtained the *incorrect* result

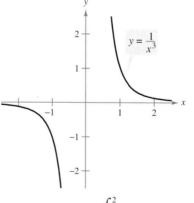

The improper integral $\int_{-1}^2 1/x^3\, dx$ diverges.
Figure 7.25

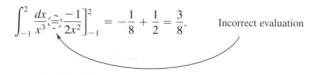

$$\int_{-1}^2 \frac{dx}{x^3} = \left[\frac{-1}{2x^2} \right]_{-1}^2 = -\frac{1}{8} + \frac{1}{2} = \frac{3}{8}. \qquad \text{Incorrect evaluation}$$

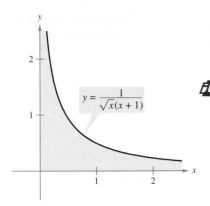

$$y = \frac{1}{\sqrt{x}(x+1)}$$

The area of the unbounded region is π.

Figure 7.26

The integral in the next example is improper for *two* reasons. One limit of integration is infinite, and the integrand has an infinite discontinuity at the outer limit of integration, as shown in Figure 7.26.

Example 9   A Doubly-Improper Integral

Evaluate $\displaystyle\int_0^\infty \frac{dx}{\sqrt{x}(x+1)}$.

Solution   To evaluate this integral, split it at a convenient point (say, $x = 1$) and write

$$\int_0^\infty \frac{dx}{\sqrt{x}(x+1)} = \int_0^1 \frac{dx}{\sqrt{x}(x+1)} + \int_1^\infty \frac{dx}{\sqrt{x}(x+1)}$$

$$= \lim_{b \to 0^+} \left[2 \arctan \sqrt{x} \right]_b^1 + \lim_{c \to \infty} \left[2 \arctan \sqrt{x} \right]_1^c$$

$$= 2\left(\frac{\pi}{4}\right) - 0 + 2\left(\frac{\pi}{2}\right) - 2\left(\frac{\pi}{4}\right)$$

$$= \pi.$$

Example 10   An Application Involving Arc Length

Use the formula for arc length to show that the circumference of the circle $x^2 + y^2 = 1$ is 2π.

Solution   To simplify the work, consider the quarter circle given by $y = \sqrt{1 - x^2}$, where $0 \le x \le 1$. The function y is differentiable for any x in this interval except $x = 1$. Therefore, the arc length of the quarter circle is given by the improper integral

$$s = \int_0^1 \sqrt{1 + (y')^2}\, dx$$

$$= \int_0^1 \sqrt{1 + \left(\frac{-x}{\sqrt{1-x^2}}\right)^2}\, dx$$

$$= \int_0^1 \frac{dx}{\sqrt{1-x^2}}.$$

This integral is improper because it has an infinite discontinuity at $x = 1$. So, you can write

$$s = \int_0^1 \frac{dx}{\sqrt{1-x^2}}$$

$$= \lim_{b \to 1^-} \left[\arcsin x \right]_0^b$$

$$= \frac{\pi}{2} - 0$$

$$= \frac{\pi}{2}.$$

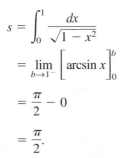

$$y = \sqrt{1-x^2}$$

The circumference of the circle is 2π.

Figure 7.27

Finally, multiplying by 4, you can conclude that the circumference of the circle is $4s = 2\pi$, as shown in Figure 7.27.

We conclude this section with a useful theorem describing the convergence or divergence of a common type of improper integral. The proof of this theorem is left as an exercise (see Exercise 43).

THEOREM 7.5   A Special Type of Improper Integral

$$\int_1^\infty \frac{dx}{x^p} = \begin{cases} \dfrac{1}{p-1}, & \text{if } p > 1 \\[2mm] \text{diverges}, & \text{if } p \le 1 \end{cases}$$

Example 11   **An Application Involving A Solid of Revolution**

The solid formed by revolving (about the x-axis) the *unbounded* region lying between the graph of $f(x) = 1/x$ and the x-axis ($x \ge 1$) is called **Gabriel's Horn.** (See Figure 7.28.) Show that this solid has a finite volume and an infinite surface area.

Solution   Using the disk method and Theorem 7.5, you can determine the volume to be

$$V = \pi \int_1^\infty \left(\frac{1}{x}\right)^2 dx \qquad \text{Theorem 7.5, } p = 2 > 1$$

$$= \pi \left(\frac{1}{2-1}\right)$$

$$= \pi.$$

The surface area is given by

$$S = 2\pi \int_1^\infty f(x)\sqrt{1 + [f'(x)]^2}\, dx = 2\pi \int_1^\infty \frac{1}{x}\sqrt{1 + \frac{1}{x^4}}\, dx.$$

Because

$$\sqrt{1 + \frac{1}{x^4}} > 1$$

on the interval $[1, \infty)$, and the improper integral

$$\int_1^\infty \frac{1}{x}\, dx$$

diverges, you can conclude that the improper integral

$$\int_1^\infty \frac{1}{x}\sqrt{1 + \frac{1}{x^4}}\, dx$$

also diverges. (See Exercise 46.) So, the surface area is infinite.

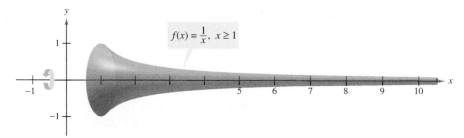

Gabriel's Horn has a finite volume and an infinite surface area.
Figure 7.28

FOR FURTHER INFORMATION To further investigate solids that have finite volumes and infinite surface areas, see the article "Supersolids: Solids Having Finite Volume and Infinite Surfaces" by William P. Love in *Mathematics Teacher*. To view this article, go to the website *www.matharticles.com*.

FOR FURTHER INFORMATION To learn about another function that has a finite volume and an infinite surface area, see the article "Gabriel's Wedding Cake" by Julian F. Fleron in *The College Mathematics Journal*. To view this article, go to the website *www.matharticles.com*.

EXERCISES FOR SECTION 7.8

In Exercises 1–6, explain why the integral is improper and determine whether it diverges or converges. Evaluate the integral if it converges.

1. $\displaystyle\int_0^4 \frac{1}{\sqrt{x}}\, dx$

2. $\displaystyle\int_3^4 \frac{1}{(x-3)^{3/2}}\, dx$

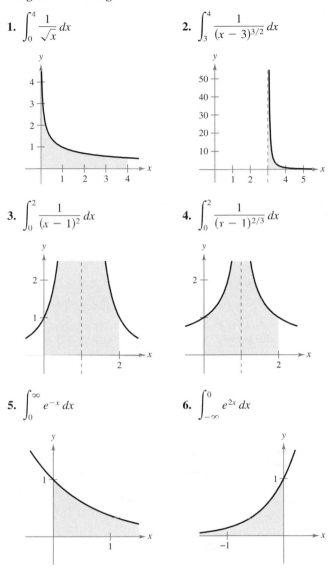

3. $\displaystyle\int_0^2 \frac{1}{(x-1)^2}\, dx$

4. $\displaystyle\int_0^2 \frac{1}{(x-1)^{2/3}}\, dx$

5. $\displaystyle\int_0^\infty e^{-x}\, dx$

6. $\displaystyle\int_{-\infty}^0 e^{2x}\, dx$

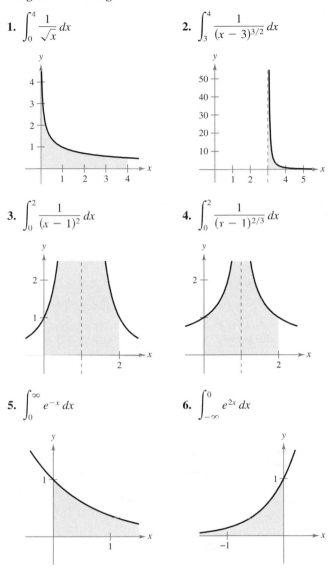

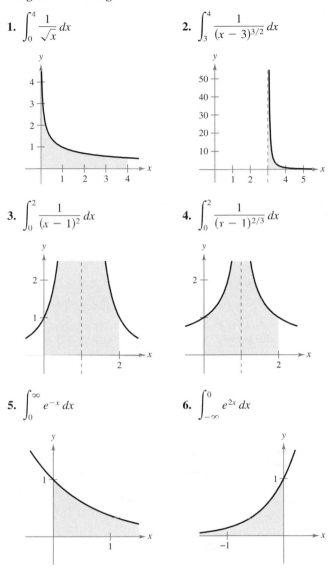 *Writing*   In Exercises 7 and 8, explain why the evaluation of the integral is *incorrect*. Use the integration capabilities of a graphing utility to attempt to evaluate the integral. Determine whether the utility gives the correct answer.

7. $\displaystyle\int_{-1}^1 \frac{1}{x^2}\, dx = -2$ ✗

8. $\displaystyle\int_0^\infty e^{-x}\, dx = 0$ ✗

In Exercises 9–26, determine whether the improper integral diverges or converges. Evaluate the integral if it converges.

9. $\displaystyle\int_1^\infty \frac{1}{x^2}\, dx$

10. $\displaystyle\int_1^\infty \frac{5}{x^3}\, dx$

11. $\displaystyle\int_1^\infty \frac{3}{\sqrt[3]{x}}\, dx$

12. $\displaystyle\int_1^\infty \frac{4}{\sqrt[4]{x}}\, dx$

13. $\displaystyle\int_{-\infty}^0 xe^{-2x}\, dx$

14. $\displaystyle\int_0^\infty xe^{-x/2}\, dx$

15. $\displaystyle\int_0^\infty x^2 e^{-x}\, dx$

16. $\displaystyle\int_0^\infty (x-1)e^{-x}\, dx$

17. $\displaystyle\int_0^\infty e^{-x}\cos x\, dx$

18. $\displaystyle\int_0^\infty e^{-ax}\sin bx\, dx, \quad a > 0$

19. $\displaystyle\int_4^\infty \frac{1}{x(\ln x)^3}\, dx$

20. $\displaystyle\int_1^\infty \frac{\ln x}{x}\, dx$

21. $\displaystyle\int_{-\infty}^\infty \frac{2}{4+x^2}\, dx$

22. $\displaystyle\int_0^\infty \frac{x^3}{(x^2+1)^2}\, dx$

23. $\displaystyle\int_0^\infty \frac{1}{e^x + e^{-x}}\, dx$

24. $\displaystyle\int_0^\infty \frac{e^x}{1+e^x}\, dx$

25. $\displaystyle\int_0^\infty \cos \pi x\, dx$

26. $\displaystyle\int_0^\infty \sin \frac{x}{2}\, dx$

 In Exercises 27–42, determine whether the improper integral diverges or converges. Evaluate the integral if it converges, and check your results with the results obtained by using the integration capabilities of a graphing utility.

27. $\displaystyle\int_0^1 \frac{1}{x^2}\, dx$

28. $\displaystyle\int_0^4 \frac{8}{x}\, dx$

29. $\displaystyle\int_0^8 \frac{1}{\sqrt[3]{8-x}}\, dx$

30. $\displaystyle\int_0^6 \frac{4}{\sqrt{6-x}}\, dx$

31. $\displaystyle\int_0^1 x \ln x\, dx$

32. $\displaystyle\int_0^e \ln x^2\, dx$

33. $\displaystyle\int_0^{\pi/2} \tan \theta\, d\theta$

34. $\displaystyle\int_0^{\pi/2} \sec \theta\, d\theta$

35. $\displaystyle\int_2^4 \frac{2}{x\sqrt{x^2-4}}\, dx$

36. $\displaystyle\int_0^2 \frac{1}{\sqrt{4-x^2}}\, dx$

37. $\displaystyle\int_2^4 \frac{1}{\sqrt{x^2-4}}\, dx$

38. $\displaystyle\int_0^2 \frac{1}{4-x^2}\, dx$

39. $\displaystyle\int_0^2 \frac{1}{\sqrt[3]{x-1}}\, dx$

40. $\displaystyle\int_1^3 \frac{2}{(x-2)^{8/3}}\, dx$

41. $\displaystyle\int_0^\infty \frac{4}{\sqrt{x}(x+6)}\, dx$

42. $\displaystyle\int_1^\infty \frac{1}{x \ln x}\, dx$

In Exercises 43 and 44, determine all values of p for which the improper integral converges.

43. $\displaystyle\int_1^\infty \frac{1}{x^p}\, dx$

44. $\displaystyle\int_0^1 \frac{1}{x^p}\, dx$

45. Use mathematical induction to verify that the following integral converges for any positive integer n.

$$\int_0^\infty x^n e^{-x}\, dx$$

46. Given continuous functions f and g such that $0 \leq f(x) \leq g(x)$ on the interval $[a, \infty)$, prove the following.

(a) If $\int_a^\infty g(x)\, dx$ converges, then $\int_a^\infty f(x)\, dx$ converges.

(b) If $\int_a^\infty f(x)\, dx$ diverges, then $\int_a^\infty g(x)\, dx$ diverges.

In Exercises 47–56, use the results of Exercises 43–46 to determine whether the improper integral converges or diverges.

47. $\displaystyle\int_0^1 \frac{1}{x^3}\, dx$

48. $\displaystyle\int_0^1 \frac{1}{\sqrt[3]{x}}\, dx$

49. $\displaystyle\int_1^\infty \frac{1}{x^3}\, dx$

50. $\displaystyle\int_0^\infty x^4 e^{-x}\, dx$

51. $\displaystyle\int_1^\infty \frac{1}{x^2 + 5}\, dx$

52. $\displaystyle\int_2^\infty \frac{1}{\sqrt{x-1}}\, dx$

53. $\displaystyle\int_2^\infty \frac{1}{\sqrt[3]{x(x-1)}}\, dx$

54. $\displaystyle\int_1^\infty \frac{1}{\sqrt{x}(x+1)}\, dx$

55. $\displaystyle\int_0^\infty e^{-x^2}\, dx$

56. $\displaystyle\int_2^\infty \frac{1}{\sqrt{x}\ln x}\, dx$

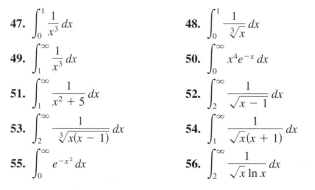

Getting at the Concept

57. List the different types of improper integrals.

58. Define the terms *converges* and *diverges* when working with improper integrals.

59. Explain why $\displaystyle\int_{-1}^1 \frac{1}{x^3}\, dx \neq 0$.

60. Give examples of an improper integral with infinite limits that (a) converges and (b) diverges.

Laplace Transforms   Let $f(t)$ be a function defined for all positive values of t. The Laplace Transform of $f(t)$ is defined by

$$F(s) = \int_0^\infty e^{-st} f(t)\, dt$$

if the improper integral exists. Laplace Transforms are used to solve differential equations. In Exercises 61–68, find the Laplace Transform of the function.

61. $f(t) = 1$

62. $f(t) = t$

63. $f(t) = t^2$

64. $f(t) = e^{at}$

65. $f(t) = \cos at$

66. $f(t) = \sin at$

67. $f(t) = \cosh at$

68. $f(t) = \sinh at$

Area and Volume   In Exercises 69 and 70, consider the region satisfying the inequalities. **(a) Find the area of the region. (b) Find the volume of the solid generated by revolving the region about the x-axis. (c) Find the volume of the solid generated by revolving the region about the y-axis.**

69. $y \leq e^{-x}, \ y \geq 0, \ x \geq 0$

70. $y \leq 1/x^2, \ y \geq 0, \ x \geq 1$

71. *Arc Length*   Sketch the graph of the hypocycloid of four cusps

$$x^{2/3} + y^{2/3} = 4$$

and find its perimeter.

72. *Surface Area*   The region bounded by

$$(x-2)^2 + y^2 = 1$$

is revolved about the y-axis to form a torus. Find the surface area of the torus.

73. *The Gamma Function*   The Gamma Function $\Gamma(n)$ is defined by

$$\Gamma(n) = \int_0^\infty x^{n-1} e^{-x}\, dx, \quad n > 0.$$

(a) Find $\Gamma(1)$, $\Gamma(2)$, and $\Gamma(3)$.

(b) Use integration by parts to show that $\Gamma(n+1) = n\Gamma(n)$.

(c) Express $\Gamma(n)$ in terms of factorial notation where n is a positive integer.

74. *Work*   A 5-ton rocket is fired from the surface of earth into outer space.

(a) How much work is required to overcome earth's gravitational force?

(b) How far has the rocket traveled when half the total work has occurred?

Probability   A nonnegative function f is called a *probability density function* if

$$\int_{-\infty}^\infty f(t)\, dt = 1.$$

The probability that x lies between a and b is given by

$$P(a \leq x \leq b) = \int_a^b f(t)\, dt.$$

The expected value of x is given by

$$E(x) = \int_{-\infty}^\infty t f(t)\, dt.$$

In Exercises 75 and 76, (a) show that the nonnegative function is a probability density function, (b) find $P(0 \leq x \leq 4)$, and (c) find $E(x)$.

75. $f(t) = \begin{cases} \frac{1}{7}e^{-t/7}, & t \geq 0 \\ 0, & t < 0 \end{cases}$

76. $f(t) = \begin{cases} \frac{2}{5}e^{-2t/5}, & t \geq 0 \\ 0, & t < 0 \end{cases}$

Capitalized Cost In Exercises 77 and 78, find the capitalized cost C of an asset (a) for $n = 5$ years, (b) for $n = 10$ years, and (c) forever. The capitalized cost is given by

$$C = C_0 + \int_0^n c(t)e^{-rt}\,dt$$

where C_0 is the original investment, t is the time in years, r is the annual interest rate compounded continuously, and $c(t)$ is the annual cost of maintenance.

77. $C_0 = \$650{,}000$
    $c(t) = \$25{,}000$
    $r = 0.06$

78. $C_0 = \$650{,}000$
    $c(t) = \$25{,}000(1 + 0.08t)$
    $r = 0.06$

79. *Electromagnetic Theory* Find the value of the following integral used in electromagnetic theory.

$$P = k\int_1^\infty \frac{1}{(a^2 + x^2)^{3/2}}\,dx$$

80. *Writing*

(a) The improper integrals

$$\int_1^\infty \frac{1}{x}\,dx \qquad \text{and} \qquad \int_1^\infty \frac{1}{x^2}\,dx$$

diverge and converge, respectively. Describe the essential differences between the integrands that cause one integral to converge and the other to diverge.

(b) Sketch a graph of the function $y = \sin x / x$ over the interval $(1, \infty)$. Use your knowledge of the definite integral to make an inference as to whether or not the integral

$$\int_1^\infty \frac{\sin x}{x}\,dx$$

converges. Give reasons for your answer.

(c) Use one iteration of integration by parts on the integral in part (b) to determine its divergence or convergence.

81. *Think About It* Consider the integral

$$\int_0^3 \frac{10}{x^2 - 2x}\,dx.$$

To determine the convergence or divergence of the integral, how many improper integrals must be analyzed? What must be true of each of these integrals if the given integral converges?

82. *Exploration* Consider the integral

$$\int_0^{\pi/2} \frac{4}{1 + (\tan x)^n}\,dx$$

where n is a positive integer.

(a) Is the integral improper? Explain.

(b) Use a graphing utility to graph the integrand for $n = 2, 4, 8,$ and 12.

(c) Use the graphs to approximate the integral as $n \to \infty$.

(d) Use a computer algebra system to evaluate the integral for the values of n in part (b). Make a conjecture about the value of the integral for any positive integer n. Compare the results with your answer in part (c).

83. Let $I_n = \displaystyle\int_0^\infty \frac{x^{2n-1}}{(x^2 + 1)^{n+3}}\,dx, \quad n \geq 1.$

Prove that $I_n = \left(\dfrac{n-1}{n+2}\right)I_{n-1}$

and then evaluate each of the following.

(a) $\displaystyle\int_0^\infty \frac{x}{(x^2 + 1)^4}\,dx$

(b) $\displaystyle\int_0^\infty \frac{x^3}{(x^2 + 1)^5}\,dx$

(c) $\displaystyle\int_0^\infty \frac{x^5}{(x^2 + 1)^6}\,dx$

84. *Normal Probability* The mean height of American men between 18 and 24 years old is 70 inches, and the standard deviation is 3 inches. If an 18- to 24-year-old man is chosen at random from the population, the probability that he is 6 feet tall or taller is

$$P(72 \leq x < \infty) = \int_{72}^\infty \frac{1}{3\sqrt{2\pi}}e^{-(x-70)^2/18}\,dx.$$

(Source: National Center for Health Statistics)

(a) Use a graphing utility to graph the integrand. Use the graphing utility to convince yourself that the area between the x-axis and the integrand is 1.

(b) Use a graphing utility to approximate $P(72 \leq x < \infty)$.

(c) Approximate $0.5 - P(70 \leq x \leq 72)$ using a graphing utility. Use the graph in part (a) to explain why this result is the same as the answer in part (b).

True or False? In Exercises 85–88, determine whether the statement is true or false. If it is false, explain why or give an example that shows it is false.

85. If f is continuous on $[0, \infty)$ and $\lim_{x\to\infty} f(x) = 0$, then $\int_0^\infty f(x)\,dx$ converges.

86. If f is continuous on $[0, \infty)$ and $\int_0^\infty f(x)\,dx$ diverges, then $\lim_{x\to\infty} f(x) \neq 0$.

87. If f' is continuous on $[0, \infty)$ and $\lim_{x\to\infty} f(x) = 0$, then $\int_0^\infty f'(x)\,dx = -f(0)$.

88. If the graph of f is symmetric with respect to the origin or the y-axis, then $\int_0^\infty f(x)\,dx$ converges if and only if $\int_{-\infty}^\infty f(x)\,dx$ converges.

REVIEW EXERCISES FOR CHAPTER 7

7.1 In Exercises 1–8, use the basic integration rules to evaluate the integral.

1. $\int x\sqrt{x^2 - 1}\, dx$

2. $\int xe^{x^2 - 1}\, dx$

3. $\int \dfrac{x}{x^2 - 1}\, dx$

4. $\int \dfrac{x}{\sqrt{1 - x^2}}\, dx$

5. $\int \dfrac{\ln(2x)}{x}\, dx$

6. $\int 2x\sqrt{2x - 3}\, dx$

7. $\int \dfrac{16}{\sqrt{16 - x^2}}\, dx$

8. $\int \dfrac{x^4 + 2x^2 + x + 1}{(x^2 + 1)^2}\, dx$

7.2 In Exercises 9–16, use integration by parts to evaluate the integral.

9. $\int e^{2x} \sin 3x\, dx$

10. $\int (x^2 - 1)e^x\, dx$

11. $\int x\sqrt{x - 5}\, dx$

12. $\int \arctan 2x\, dx$

13. $\int x^2 \sin 2x\, dx$

14. $\int \ln\sqrt{x^2 - 1}\, dx$

15. $\int x \arcsin 2x\, dx$

16. $\int e^x \arctan e^x\, dx$

7.3 In Exercises 17–22, evaluate the trigonometric integral.

17. $\int \cos^3(\pi x - 1)\, dx$

18. $\int \sin^2 \dfrac{\pi x}{2}\, dx$

19. $\int \sec^4 \dfrac{x}{2}\, dx$

20. $\int \tan \theta \sec^4 \theta\, d\theta$

21. $\int \dfrac{1}{1 - \sin \theta}\, d\theta$

22. $\int \cos 2\theta(\sin \theta + \cos \theta)^2\, d\theta$

7.4 In Exercises 23–28, use trigonometric substitution to evaluate the integral.

23. $\int \dfrac{-12}{x^2\sqrt{4 - x^2}}\, dx$

24. $\int \dfrac{\sqrt{x^2 - 9}}{x}\, dx, \quad x > 3$

25. $\int \dfrac{x^3}{\sqrt{4 + x^2}}\, dx$

26. $\int \sqrt{9 - 4x^2}\, dx$

27. $\int \sqrt{4 - x^2}\, dx$

28. $\int \dfrac{\sin \theta}{1 + 2\cos^2 \theta}\, d\theta$

In Exercises 29 and 30, evaluate the integral using the indicated methods.

29. $\int \dfrac{x^3}{\sqrt{4 + x^2}}\, dx$

(a) Trigonometric substitution

(b) Substitution: $u^2 = 4 + x^2$

(c) Integration by parts: $dv = \left(x/\sqrt{4 + x^2}\right) dx$

30. $\int x\sqrt{4 + x}\, dx$

(a) Trigonometric substitution

(b) Substitution: $u^2 = 4 + x$

(c) Substitution: $u = 4 + x$

(d) Integration by parts: $dv = \sqrt{4 + x}\, dx$

7.5 In Exercises 31–36, use partial fractions to evaluate the integral.

31. $\int \dfrac{x - 28}{x^2 - x - 6}\, dx$

32. $\int \dfrac{2x^3 - 5x^2 + 4x - 4}{x^2 - x}\, dx$

33. $\int \dfrac{x^2 + 2x}{x^3 - x^2 + x - 1}\, dx$

34. $\int \dfrac{4x - 2}{3(x - 1)^2}\, dx$

35. $\int \dfrac{x^2}{x^2 + 2x - 15}\, dx$

36. $\int \dfrac{\sec^2 \theta}{\tan \theta(\tan \theta - 1)}\, d\theta$

7.6 In Exercises 37–44, use integration tables to evaluate the integral.

37. $\int \dfrac{x}{(2 + 3x)^2}\, dx$

38. $\int \dfrac{x}{\sqrt{2 + 3x}}\, dx$

39. $\int \dfrac{x}{1 + \sin x^2}\, dx$

40. $\int \dfrac{x}{1 + e^{x^2}}\, dx$

41. $\int \dfrac{x}{x^2 + 4x + 8}\, dx$

42. $\int \dfrac{3}{2x\sqrt{9x^2 - 1}}\, dx, \quad x > \dfrac{1}{3}$

43. $\int \dfrac{1}{\sin \pi x \cos \pi x}\, dx$

44. $\int \dfrac{1}{1 + \tan \pi x}\, dx$

45. Verify the reduction formula

$$\int (\ln x)^n\, dx = x(\ln x)^n - n\int (\ln x)^{n-1}\, dx.$$

46. Verify the reduction formula

$$\int \tan^n x\, dx = \dfrac{1}{n - 1}\tan^{n-1} x - \int \tan^{n-2} x\, dx.$$

In Exercises 47–54, evaluate the integral using any method.

47. $\int \theta \sin \theta \cos \theta\, d\theta$

48. $\int \dfrac{\csc \sqrt{2x}}{\sqrt{x}}\, dx$

49. $\int \dfrac{x^{1/4}}{1 + x^{1/2}}\, dx$

50. $\int \sqrt{1 + \sqrt{x}}\, dx$

51. $\int \sqrt{1 + \cos x}\, dx$

52. $\int \dfrac{3x^3 + 4x}{(x^2 + 1)^2}\, dx$

53. $\int \cos x \ln(\sin x)\, dx$

54. $\int (\sin \theta + \cos \theta)^2\, d\theta$

In Exercises 55–58, solve the differential equation using any method.

55. $\dfrac{dy}{dx} = \dfrac{9}{x^2 - 9}$

56. $\dfrac{dy}{dx} = \dfrac{\sqrt{4 - x^2}}{2x}$

57. $y' = \ln(x^2 + x)$

58. $y' = \sqrt{1 - \cos\theta}$

In Exercises 59–64, evaluate the definite integral using any method. Use a graphing utility to verify your result.

59. $\displaystyle\int_2^{\sqrt5} x(x^2 - 4)^{3/2}\, dx$

60. $\displaystyle\int_0^1 \dfrac{x}{(x - 2)(x - 4)}\, dx$

61. $\displaystyle\int_1^4 \dfrac{\ln x}{x}\, dx$

62. $\displaystyle\int_0^2 xe^{3x}\, dx$

63. $\displaystyle\int_0^{\pi} x\sin x\, dx$

64. $\displaystyle\int_0^3 \dfrac{x}{\sqrt{1 + x}}\, dx$

Area In Exercises 65 and 66, find the area of the region bounded by the graphs of the equations.

65. $y = x\sqrt{4 - x}, \quad y = 0$

66. $y = \dfrac{1}{25 - x^2}, \quad y = 0, x = 0, x = 4$

In Exercises 67 and 68, find the centroid of the region bounded by the graphs of the equations.

67. $y = \sqrt{1 - x^2}, \quad y = 0$

68. $(x - 1)^2 + y^2 = 1, \quad (x - 4)^2 + y^2 = 4$

Arc Length In Exercises 69 and 70, approximate to two decimal places the arc length of the curve over the given interval.

Function	Interval
69. $y = \sin x$	$[0, \pi]$
70. $y = \sin^2 x$	$[0, \pi]$

7.7 In Exercises 71–78, use L'Hôpital's Rule to evaluate the limit.

71. $\displaystyle\lim_{x\to1} \dfrac{(\ln x)^2}{x - 1}$

72. $\displaystyle\lim_{x\to0} \dfrac{\sin\pi x}{\sin 2\pi x}$

73. $\displaystyle\lim_{x\to\infty} \dfrac{e^{2x}}{x^2}$

74. $\displaystyle\lim_{x\to\infty} xe^{-x^2}$

75. $\displaystyle\lim_{x\to\infty} (\ln x)^{2/x}$

76. $\displaystyle\lim_{x\to1^+} (x - 1)^{\ln x}$

77. $\displaystyle\lim_{n\to\infty} 1000\left(1 + \dfrac{0.09}{n}\right)^n$

78. $\displaystyle\lim_{x\to1^+} \left(\dfrac{2}{\ln x} - \dfrac{2}{x - 1}\right)$

7.8 In Exercises 79–82, determine whether the improper integral converges or diverges. Evaluate the integral if it converges.

79. $\displaystyle\int_0^{16} \dfrac{1}{\sqrt[4]{x}}\, dx$

80. $\displaystyle\int_0^1 \dfrac{6}{x - 1}\, dx$

81. $\displaystyle\int_1^{\infty} x^2 \ln x\, dx$

82. $\displaystyle\int_0^{\infty} \dfrac{e^{-1/x}}{x^2}\, dx$

83. *Present Value* The board of directors of a corporation is calculating the price to pay for a business that is forecast to yield a continuous flow of profit of \$500,000 per year. If money will earn a nominal rate of 5% per year compounded continuously, what is the present value of the business

(a) for 20 years?

(b) forever (in perpetuity)?

(*Note:* The present value for t_0 years is $\int_0^{t_0} 500{,}000e^{-0.05t}\, dt$.)

84. *Volume* Find the volume of the solid generated by revolving the region bounded by the graphs of $y = xe^{-x}$, $y = 0$, and $x = 0$ about the *x*-axis.

85. *Probability* The average lengths (from beak to tail) of different species of warblers in the eastern United States are approximately normally distributed with a mean of 12.9 centimeters and a standard deviation of 0.95 centimeter (see figure). The probability that a randomly selected warbler has a length between a and b centimeters is

$$P(a \le x \le b) = \dfrac{1}{0.95\sqrt{2\pi}} \int_a^b e^{-(x - 12.9)^2/2(0.95)^2}\, dx.$$

Use a graphing utility to approximate the probability that a randomly selected warbler has a length of (a) 13 centimeters or greater and (b) 15 centimeters or greater. (*Source: Peterson's Field Guide: Eastern Birds*)

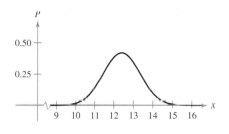

86. Using the inequality

$$\dfrac{1}{x^5} + \dfrac{1}{x^{10}} + \dfrac{1}{x^{15}} < \dfrac{1}{x^5 - 1} < \dfrac{1}{x^5} + \dfrac{1}{x^{10}} + \dfrac{2}{x^{15}}$$

for $x \ge 2$, approximate $\displaystyle\int_2^{\infty} \dfrac{1}{x^5 - 1}\, dx$.

P.S. Problem Solving

1. (a) Evaluate the integrals

$$\int_{-1}^{1} (1 - x^2)\, dx \quad \text{and} \quad \int_{-1}^{1} (1 - x^2)^2\, dx.$$

(b) Use Wallis's Formulas to prove that

$$\int_{-1}^{1} (1 - x^2)^n\, dx = \frac{2^{2n+1}(n!)^2}{(2n + 1)!}$$

for all positive integers n.

2. (a) Evaluate the integrals

$$\int_{0}^{1} \ln x\, dx \quad \text{and} \quad \int_{0}^{1} (\ln x)^2\, dx.$$

(b) Prove that

$$\int_{0}^{1} (\ln x)^n\, dx = (-1)^n\, n!$$

for all positive integers n.

3. Find the value of the positive constant c such that

$$\lim_{x \to \infty} \left(\frac{x + c}{x - c} \right)^x = 9.$$

4. Find the value of the positive constant c such that

$$\lim_{x \to \infty} \left(\frac{x - c}{x + c} \right)^x = \frac{1}{4}.$$

5. In the figure, the line $x = 1$ is tangent to the unit circle at A. The length of segment QA equals the length of the circular arc $\overset{\frown}{PA}$. Show that the length of segment OR approaches 2 as P approaches A.

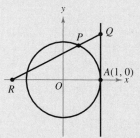

6. In the figure, the segment BD is the height of triangle $\triangle OAB$. Let R be the ratio of the area of $\triangle DAB$ to that of the shaded region formed by deleting $\triangle OAB$ from the circular sector subtended by angle θ. Find $\lim_{\theta \to 0^+} R$.

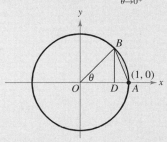

7. Consider the problem of finding the area of the region bounded by the curve

$$y = \frac{x^2}{[x^2 + 9]^{3/2}},$$

the x-axis and $x = 4$.

(a) Use a graphing utility to graph the region and approximate its area.

(b) Use an appropriate trigonometric substitution to find the exact area.

(c) Use the substitution $x = 3 \sinh u$ to find the exact area and verify that you obtain the same answer as in part (b).

8. Use the substitution $u = \tan \dfrac{x}{2}$ to find the area of the shaded region under the graph of $y = \dfrac{1}{2 + \cos x}$, $0 \le x \le \pi/2$.

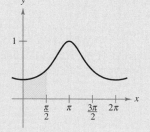

9. Find the arc length of the graph of the function

$$y = \ln(1 - x^2)$$

on the interval $0 \le x \le \frac{1}{2}$.

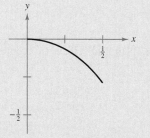

10. Find the centroid of the region above the x-axis and bounded above by the curve $y = e^{-c^2 x^2}$, where c is a positive constant.

$$\left(\textit{Hint:} \text{ Show that } \int_{0}^{\infty} e^{-c^2 x^2}\, dx = \frac{1}{c} \int_{0}^{\infty} e^{-x^2}\, dx. \right)$$

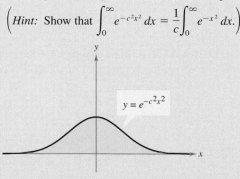

$y = e^{-c^2 x^2}$

11. Some elementary functions, such as $f(x) = \sin(x^2)$, do not have antiderivatives that are elementary functions. Joseph Liouville proved that

$$\int \frac{e^x}{x}\, dx$$

does not have an elementary antiderivative. Use this fact to prove that

$$\int \frac{1}{\ln x}\, dx$$

is not elementary.

12. (a) Let $y = f^{-1}(x)$ be the inverse of f. Use integration by parts to derive the formula

$$\int f^{-1}(x)\, dx = xf^{-1}(x) - \int f(y)\, dy.$$

(b) Use the formula in part (a) to evaluate the integral

$$\int \arcsin x\, dx.$$

(c) Use the formula in part (a) to find the area under the graph of $y = \ln x$, $1 \le x \le e$.

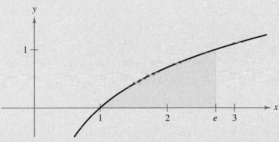

13. Factor the polynomial $p(x) = x^4 + 1$ and then find the area under the graph of $y = \dfrac{1}{x^4 + 1}$, $0 \le x \le 1$.

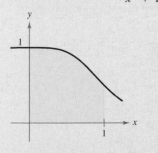

14. (a) Use the substitution $u = \dfrac{\pi}{2} - x$ to evaluate the integral

$$\int_0^{\pi/2} \frac{\sin x}{\cos x + \sin x}\, dx.$$

(b) Let n be a positive integer. Evaluate the integral

$$\int_0^{\pi/2} \frac{\sin^n x}{\cos^n x + \sin^n x}\, dx.$$

15. Use a graphing utility to estimate each limit. Then calculate each limit using L'Hôpital's Rule. What can you conclude about the indeterminate form $0 \cdot \infty$?

(a) $\displaystyle \lim_{x \to 0^+} \left(\cot x + \frac{1}{x} \right)$

(b) $\displaystyle \lim_{x \to 0^+} \left(\cot x - \frac{1}{x} \right)$

(c) $\displaystyle \lim_{x \to 0^+} \left[\left(\cot x + \frac{1}{x} \right)\left(\cot x - \frac{1}{x} \right) \right]$

16. Suppose the denominator of a rational function can be factored into distinct linear factors

$$D(x) = (x - c_1)(x - c_2) \cdots (x - c_n)$$

for a positive integer n and distinct real numbers $c_1, c_2, \ldots, c_n$. If N is a polynomial of degree less than n, show that

$$\frac{N(x)}{D(x)} = \frac{P_1}{x - c_1} + \frac{P_2}{x - c_2} + \cdots + \frac{P_n}{x - c_n}$$

where $P_k = N(c_k)/D'(c_k)$ for $k = 1, 2, \ldots, n$. Note that this is the partial fraction decomposition of $N(x)/D(x)$.

17. Use the results of Exercise 16 to find the partial fraction decomposition of

$$\frac{x^3 - 3x^2 + 1}{x^4 - 13x^2 + 12x}.$$

18. The velocity (in feet per second) of a rocket whose initial mass (including fuel) is m is

$$v = gt + u \ln \frac{m}{m - rt}, \quad t < \frac{m}{r}$$

where u is the expulsion speed of the fuel, r is the rate at which the fuel is consumed, and $g = -32$ feet per second per second is the acceleration due to gravity. Find the position equation for a rocket for which $m = 50{,}000$ pounds, $u = 12{,}000$ feet per second, and $r = 400$ pounds per second. What is the height of the rocket when $t = 100$ seconds? (Assume that the rocket was fired from ground level and is moving straight up.)

19. Suppose that $f(a) = f(b) = g(a) = g(b) = 0$ and the second derivatives of f and g are continuous on the closed interval $[a, b]$. Prove that

$$\int_a^b f(x)g''(x)\, dx = \int_a^b f''(x)g(x)\, dx.$$

20. Suppose that $f(a) = f(b) = 0$ and the second derivatives of f exist on the closed interval $[a, b]$. Prove that

$$\int_a^b (x - a)(x - b)f''(x)\, dx = 2 \int_a^b f(x)\, dx.$$

The Koch Snowflake: Infinite Perimeter?

Why is geometry often described as "cold" and "dry"? One reason lies in its inability to describe the shape of a cloud, a mountain, a coastline, or a tree. Clouds are not spheres, mountains are not cones, coastlines are not circles, and bark is not smooth, nor does lightening travel in a straight line. ... Nature exhibits not simply a higher degree but an altogether different level of complexity.

Benoit Mandelbrot (1924–)

To meet the challenge of creating a geometry capable of describing nature, Mandelbrot developed *fractal* geometry. Fractal sets come in diverse forms. Some are curves, others are disconnected "dust" and still others are such odd forms that there are no existing geometric terms to describe them.

One of the "classic" fractals is the Koch snowflake, named after the Swedish mathematician Helge von Koch (1870–1924). It is sometimes classified as a "coastline curve" because of the way a coastline appears increasingly more complex with magnification. To describe the Koch snowflake, Mandelbrot coined the term *teragon*, which translates literally from the Greek words for "monster curve."

The construction of the Koch snowflake begins with an equilateral triangle whose sides are one unit long. In the first iteration, a triangle with sides one-third unit long is added in the center of each side of the original. In the second iteration, a triangle with sides one-ninth unit long is added in the center of each side. Successive iterations continue this process—without stopping.

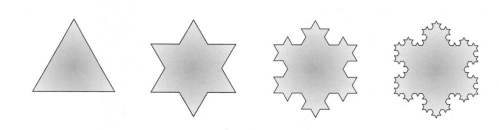

QUESTIONS

1. Write a formula that describes the side length of the triangles that will be added in the nth iteration.

2. Make a table of the perimeter of the original triangle and of the teragon in the first three iterations, as shown above. Write an expression describing the perimeter of the teragon after the nth iteration. What do you expect will happen to the perimeter as n approaches infinity?

3. Make a table of the area of the teragon in the first four iterations. Write an expression describing the area after the nth iteration. What do you expect will happen to the area as n approaches infinity?

4. Is it possible for a closed and bounded region in the plane to have a finite area and an infinite perimeter? Explain your reasoning.

The concepts presented here will be explored further in this chapter. For an extension of this application, see Lab 11 in the lab series that accompanies this text at college.hmco.com.

Infinite Series 8

Eric Haines generated the sphere-like fractal.

The sphereflake fractal is a three-dimensional version of the Koch snowflake. You are asked to prove that its surface area is infinite in Exercise 84 on page 575.

After developing some of the first computer graphics programs, Benoit Mandelbrot was able to share some of the most beautiful fractals with the world and create a growing interest in this new area of fractal geometry.

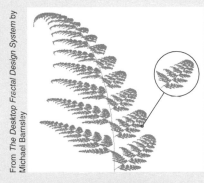

Fractals are **self-similar**, as seen in the fern fractal. When magnifying a small portion of a fractal image, you see an image similar to the original fractal.

Section 8.1   Sequences

- List the terms of a sequence.
- Determine whether a sequence converges or diverges.
- Write a formula for the nth term of a sequence.
- Use properties of monotonic sequences and bounded sequences.

Sequences

In mathematics, the word "sequence" is used in much the same way as in ordinary English. To say that a collection of objects or events is *in sequence* usually means that the collection is ordered so that it has an identified first member, second member, third member, and so on.

Mathematically, a **sequence** is defined as a function whose domain is the set of positive integers. Although a sequence is a function, it is common to represent sequences by subscript notation rather than by the standard function notation. For instance, in the sequence

$$1, \quad 2, \quad 3, \quad 4, \quad \ldots, \quad n, \quad \ldots$$
$$\downarrow \quad \downarrow \quad \downarrow \quad \downarrow \quad \quad \downarrow \qquad \qquad \text{Sequence}$$
$$a_1, \quad a_2, \quad a_3, \quad a_4, \quad \ldots, \quad a_n, \quad \ldots$$

1 is mapped onto a_1, 2 is mapped onto a_2, and so on. The numbers $a_1, a_2, a_3, \ldots, a_n$, $\ldots$ are the **terms** of the sequence. The number a_n is the **nth term** of the sequence, and the entire sequence is denoted by $\{a_n\}$.

NOTE   Occasionally, it is convenient to begin a sequence with a_0, so that the terms of the sequence become

$$a_0, a_1, a_2, a_3, \ldots, a_n, \ldots$$

EXPLORATION

Finding Patterns   Describe a pattern for each of the following sequences. Then use your description to write a formula for the nth term of each sequence. As n increases, do the terms appear to be approaching a limit? Explain your reasoning.

a. $1, \frac{1}{2}, \frac{1}{4}, \frac{1}{8}, \frac{1}{16}, \ldots$

b. $1, \frac{1}{2}, \frac{1}{6}, \frac{1}{24}, \frac{1}{120}, \ldots$

c. $10, \frac{10}{3}, \frac{10}{6}, \frac{10}{10}, \frac{10}{15}, \ldots$

d. $\frac{1}{4}, \frac{4}{9}, \frac{9}{16}, \frac{16}{25}, \frac{25}{36}, \ldots$

e. $\frac{3}{7}, \frac{5}{10}, \frac{7}{13}, \frac{9}{16}, \frac{11}{19}, \ldots$

Example 1   Listing the Terms of a Sequence

a. The terms of the sequence $\{a_n\} = \{3 + (-1)^n\}$ are

$$3 + (-1)^1, \ 3 + (-1)^2, \ 3 + (-1)^3, \ 3 + (-1)^4, \ \ldots$$
$$2, \qquad\qquad 4, \qquad\qquad 2, \qquad\qquad 4, \qquad\qquad \ldots\ .$$

b. The terms of the sequence $\{b_n\} = \left\{ \dfrac{n}{1 - 2n} \right\}$ are

$$\frac{1}{1 - 2 \cdot 1}, \ \frac{2}{1 - 2 \cdot 2}, \ \frac{3}{1 - 2 \cdot 3}, \ \frac{4}{1 - 2 \cdot 4}, \ \ldots$$
$$-1, \qquad -\frac{2}{3}, \qquad -\frac{3}{5}, \qquad -\frac{4}{7}, \qquad \ldots\ .$$

c. The terms of the sequence $\{c_n\} = \left\{ \dfrac{n^2}{2^n - 1} \right\}$ are

$$\frac{1^2}{2^1 - 1}, \ \frac{2^2}{2^2 - 1}, \ \frac{3^2}{2^3 - 1}, \ \frac{4^2}{2^4 - 1}, \ \ldots$$
$$\frac{1}{1}, \qquad \frac{4}{3}, \qquad \frac{9}{7}, \qquad \frac{16}{15}, \qquad \ldots\ .$$

Limit of a Sequence

The primary focus of this chapter concerns sequences whose terms approach limiting values. Such sequences are said to **converge.** For instance, the sequence $\{1/2^n\}$

$$\frac{1}{2}, \frac{1}{4}, \frac{1}{8}, \frac{1}{16}, \frac{1}{32}, \ldots$$

converges to 0, as indicated in the following definition.

Definition of the Limit of a Sequence

Let L be a real number. The **limit** of a sequence $\{a_n\}$ is L, written as

$$\lim_{n \to \infty} a_n = L$$

if for each $\varepsilon > 0$, there exists $M > 0$ such that $|a_n - L| < \varepsilon$ whenever $n > M$. Sequences that have limits **converge,** whereas sequences that do not have limits **diverge.**

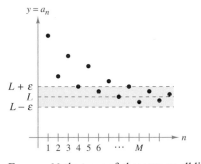

$y = a_n$

$L + \varepsilon$
L
$L - \varepsilon$

1 2 3 4 5 6 $\cdots$ M $\quad n$

For $n > M$, the terms of the sequence all lie within ε units of L.
Figure 8.1

Graphically, this definition says that eventually (for $n > M$) the terms of a sequence that converges to L will lie within the band between the lines $y = L + \varepsilon$ and $y = L - \varepsilon$, as illustrated in Figure 8.1.

If a sequence $\{a_n\}$ agrees with a function f at every positive integer, and if $f(x)$ approaches a limit L as $x \to \infty$, the sequence must converge to the same limit L.

THEOREM 8.1   Limit of a Sequence

Let L be a real number. Let f be a function of a real variable such that

$$\lim_{x \to \infty} f(x) = L.$$

If $\{a_n\}$ is a sequence such that $f(n) = a_n$ for every positive integer n, then

$$\lim_{n \to \infty} a_n = L.$$

Example 2   **Finding the Limit of a Sequence**

NOTE   There are different ways in which a sequence can fail to have a limit. One way is that the terms of the sequence increase without bound or decrease without bound. These cases are written symbolically as follows.

Terms increase without bound:

$$\lim_{n \to \infty} a_n = \infty$$

Terms decrease without bound:

$$\lim_{n \to \infty} a_n = -\infty$$

Find the limit of the sequence whose nth term is

$$a_n = \left(1 + \frac{1}{n}\right)^n.$$

Solution   Previously you learned that

$$\lim_{x \to \infty} \left(1 + \frac{1}{x}\right)^x = e.$$

Therefore, you can apply Theorem 8.1 to conclude that

$$\lim_{n \to \infty} a_n = \lim_{n \to \infty} \left(1 + \frac{1}{n}\right)^n$$

$$= e.$$

The following properties of limits of sequences parallel those given for limits of functions of a real variable in Section 1.3.

THEOREM 8.2   Properties of Limits of Sequences

Let $\lim\limits_{n\to\infty} a_n = L$ and $\lim\limits_{n\to\infty} b_n = K$.

1. $\lim\limits_{n\to\infty} (a_n \pm b_n) = L \pm K$ **2.** $\lim\limits_{n\to\infty} ca_n = cL$, c is any real number

3. $\lim\limits_{n\to\infty} (a_n b_n) = LK$ **4.** $\lim\limits_{n\to\infty} \dfrac{a_n}{b_n} = \dfrac{L}{K}$, $b_n \neq 0$ and $K \neq 0$

Example 3   Determining Convergence or Divergence

a. Because the sequence $\{a_n\} = \{3 + (-1)^n\}$ has terms

$$2, 4, 2, 4, \ldots$$ See Example 1a, page 556.

that alternate between 2 and 4, the limit

$$\lim_{n\to\infty} a_n$$

does not exist. So, the sequence diverges.

b. For $\{b_n\} = \left\{\dfrac{n}{1 - 2n}\right\}$, you can divide the numerator and denominator by n to obtain

$$\lim_{n\to\infty} \frac{n}{1 - 2n} = \lim_{n\to\infty} \frac{1}{(1/n) - 2} = -\frac{1}{2}$$ See Example 1b, page 556.

which implies that the sequence converges to $-\frac{1}{2}$.

Example 4   Using L'Hôpital's Rule to Determine Convergence

Show that the sequence whose nth term is $a_n = \dfrac{n^2}{2^n - 1}$ converges.

Solution   Consider the function of a real variable

$$f(x) = \frac{x^2}{2^x - 1}.$$

Applying L'Hôpital's Rule twice produces

$$\lim_{x\to\infty} \frac{x^2}{2^x - 1} = \lim_{x\to\infty} \frac{2x}{(\ln 2)2^x} = \lim_{x\to\infty} \frac{2}{(\ln 2)^2 2^x} = 0.$$

Because $f(n) = a_n$ for every positive integer, you can apply Theorem 8.1 to conclude that

$$\lim_{n\to\infty} \frac{n^2}{2^n - 1} = 0.$$ See Example 1c, page 556.

So, the sequence converges to 0.

TECHNOLOGY   Use a graphing utility to graph the function in Example 4. Notice that as x approaches infinity, the value of the function gets closer and closer to 0. If you have access to a graphing utility that can generate terms of a sequence, try using it to calculate the first 20 terms of the sequence in Example 4. Then view the terms to observe numerically that the sequence converges to 0.

---

indicates that in the Interactive 3.0 *CD-ROM and* Internet 3.0 *versions of this text* (*available at* college.hmco.com) *you will find an Open Exploration, which further explores this example using the computer algebra systems* Maple, Mathcad, Mathematica, *and* Derive.

To simplify some of the formulas developed in this chapter, we use the symbol $n!$ (read "n factorial"). Let n be a positive integer; then **n factorial** is given by

$$n! = 1 \cdot 2 \cdot 3 \cdot 4 \cdots (n-1) \cdot n.$$

Zero factorial is given by $0! = 1$. From this definition, you can see that $1! = 1$, $2! = 1 \cdot 2 = 2$, $3! = 1 \cdot 2 \cdot 3 = 6$, and so on. Factorials follow the same conventions for order of operations as exponents. That is, just as $2x^3$ and $(2x)^3$ imply different orders of operations, $2n!$ and $(2n)!$ imply the following orders.

$$2n! = 2(n!) = 2(1 \cdot 2 \cdot 3 \cdot 4 \cdots n)$$

and

$$(2n)! = 1 \cdot 2 \cdot 3 \cdot 4 \cdots n \cdot (n+1) \cdots 2n$$

Another useful limit theorem that can be rewritten for sequences is the Squeeze Theorem from Section 1.3.

THEOREM 8.3   Squeeze Theorem for Sequences

If

$$\lim_{n \to \infty} a_n = L = \lim_{n \to \infty} b_n$$

and there exists an integer N such that $a_n \le c_n \le b_n$ for all $n > N$, then

$$\lim_{n \to \infty} c_n = L.$$

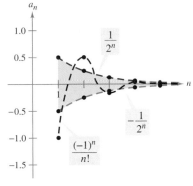

a_n

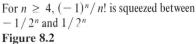

For $n \ge 4$, $(-1)^n / n!$ is squeezed between $-1/2^n$ and $1/2^n$
Figure 8.2

Example 5   Using the Squeeze Theorem

Show that the sequence $\{c_n\} = \left\{(-1)^n \dfrac{1}{n!}\right\}$ converges, and find its limit.

Solution   To apply the Squeeze Theorem, you must find two convergent sequences that can be related to the given sequence. Two possibilities are $a_n = -1/2^n$ and $b_n = 1/2^n$, both of which converge to 0. By comparing the term $n!$ with 2^n, you can see that

$$n! = 1 \cdot 2 \cdot 3 \cdot 4 \cdot 5 \cdot 6 \cdots n = 24 \cdot \underbrace{5 \cdot 6 \cdots n}_{n-4 \text{ factors}} \qquad (n \ge 4)$$

and

$$2^n = 2 \cdot 2 \cdot 2 \cdot 2 \cdot 2 \cdot 2 \cdots 2 = 16 \cdot \underbrace{2 \cdot 2 \cdots 2}_{n-4 \text{ factors}}. \qquad (n \ge 4)$$

This implies that for $n \ge 4$, $2^n < n!$, and you have

$$\frac{-1}{2^n} \le (-1)^n \frac{1}{n!} \le \frac{1}{2^n}, \qquad n \ge 4$$

as illustrated in Figure 8.2. Therefore, by the Squeeze Theorem it follows that

$$\lim_{n \to \infty} (-1)^n \frac{1}{n!} = 0.$$

NOTE   Example 5 suggests something about the rate at which $n!$ increases as $n \to \infty$. As Figure 8.2 suggests, both $1/2^n$ and $1/n!$ approach 0 as $n \to \infty$. Yet $1/n!$ approaches 0 so much faster than $1/2^n$ does that

$$\lim_{n \to \infty} \frac{1/n!}{1/2^n} = \lim_{n \to \infty} \frac{2^n}{n!} = 0.$$

In fact, it can be shown that for any fixed number k,

$$\lim_{n \to \infty} \frac{k^n}{n!} = 0.$$

This means that *the factorial function grows faster than any exponential function.*

In Example 5, the sequence $\{c_n\}$ has both positive and negative terms. For this sequence, it happens that the sequence of absolute values, $\{|c_n|\}$, also converges to 0. You can show this by the Squeeze Theorem using the inequality

$$0 \le \frac{1}{n!} \le \frac{1}{2^n}, \qquad n \ge 4.$$

In such cases, it is often convenient to consider the sequence of absolute values—and then apply Theorem 8.4, which states that if the absolute value sequence converges to 0, the original signed sequence also converges to 0.

THEOREM 8.4   Absolute Value Theorem

For the sequence $\{a_n\}$, if

$$\lim_{n\to\infty} |a_n| = 0 \qquad \text{then} \qquad \lim_{n\to\infty} a_n = 0.$$

Proof   Consider the two sequences $\{|a_n|\}$ and $\{-|a_n|\}$. Because both of these sequences converge to 0 and

$$-|a_n| \le a_n \le |a_n|$$

you can use the Squeeze Theorem to conclude that $\{a_n\}$ converges to 0.   ▨

Pattern Recognition for Sequences

Sometimes the terms of a sequence are generated by some rule that does not explicitly identify the nth term of the sequence. In such cases, you may be required to discover a *pattern* in the sequence and to describe the nth term. Once the nth term has been specified, you can investigate the convergence or divergence of the sequence.

Example 6   **Finding the nth Term of a Sequence**

Find a sequence $\{a_n\}$ whose first five terms are

$$\frac{2}{1}, \frac{4}{3}, \frac{8}{5}, \frac{16}{7}, \frac{32}{9}, \ldots$$

and then determine whether the particular sequence you have chosen converges or diverges.

Solution   First, note that the numerators are successive powers of 2, and the denominators form the sequence of positive odd integers. By comparing a_n with n, you have the following pattern.

$$\frac{2^1}{1}, \frac{2^2}{3}, \frac{2^3}{5}, \frac{2^4}{7}, \frac{2^5}{9}, \ldots, \frac{2^n}{2n-1}$$

Using L'Hôpital's Rule to evaluate the limit of $f(x) = 2^x/(2x-1)$, you obtain

$$\lim_{x\to\infty} \frac{2^x}{2x-1} = \lim_{x\to\infty} \frac{2^x(\ln 2)}{2} = \infty \qquad \Longrightarrow \qquad \lim_{n\to\infty} \frac{2^n}{2n-1} = \infty.$$

Hence, the sequence *diverges*.   ▨

Without a specific rule for generating the terms of a sequence or some knowledge of the context in which the terms of the sequence are obtained, it is not possible to determine the convergence or divergence of the sequence merely from its first several terms. For instance, although the first three terms of the following four sequences are identical, the first two sequences converge to 0, the third sequence converges to $\frac{1}{9}$, and the fourth sequence diverges.

$$\{a_n\}: \frac{1}{2}, \frac{1}{4}, \frac{1}{8}, \frac{1}{16}, \cdots, \frac{1}{2^n}, \cdots$$

$$\{b_n\}: \frac{1}{2}, \frac{1}{4}, \frac{1}{8}, \frac{1}{15}, \cdots, \frac{6}{(n+1)(n^2-n+6)}, \cdots$$

$$\{c_n\}: \frac{1}{2}, \frac{1}{4}, \frac{1}{8}, \frac{7}{62}, \cdots, \frac{n^2-3n+3}{9n^2-25n+18}, \cdots$$

$$\{d_n\}: \frac{1}{2}, \frac{1}{4}, \frac{1}{8}, 0, \cdots, \frac{-n(n+1)(n-4)}{6(n^2+3n-2)}, \cdots$$

The process of determining an nth term from the pattern observed in the first several terms of a sequence is an example of *inductive reasoning*.

Example 7   **Finding the nth Term of a Sequence**

Determine an nth term for a sequence whose first five terms are

$$-\frac{2}{1}, \frac{8}{2}, -\frac{26}{6}, \frac{80}{24}, -\frac{242}{120}, \cdots$$

and then decide whether the sequence converges or diverges.

Solution   Note that the numerators are 1 less than 3^n. Hence, you can reason that the numerators are given by the rule $3^n - 1$. Factoring the denominators produces

$$1 = 1$$
$$2 = 1 \cdot 2$$
$$6 = 1 \cdot 2 \cdot 3$$
$$24 = 1 \cdot 2 \cdot 3 \cdot 4$$
$$120 = 1 \cdot 2 \cdot 3 \cdot 4 \cdot 5 \cdots.$$

This suggests that the denominators are represented by $n!$. Finally, because the signs alternate, you can write the nth term as

$$a_n = (-1)^n \left(\frac{3^n - 1}{n!} \right).$$

From the discussion about the growth of $n!$, it follows that

$$\lim_{n \to \infty} |a_n| = \lim_{n \to \infty} \frac{3^n - 1}{n!} = 0.$$

Applying Theorem 8.4, you can conclude that

$$\lim_{n \to \infty} a_n = 0.$$

So, the sequence $\{a_n\}$ converges to 0.

Monotonic Sequences and Bounded Sequences

So far you have determined the convergence of a sequence by finding its limit. Even if you cannot determine the limit of a particular sequence, it still may be useful to know whether the sequence converges. Theorem 8.5 identifies a test for convergence of sequences without determining the limit. First, we give some preliminary definitions.

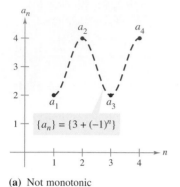

(a) Not monotonic

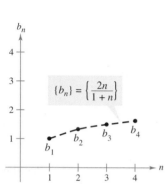

(b) Monotonic

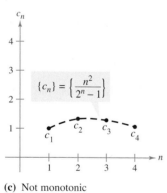

(c) Not monotonic

Figure 8.3

Definition of a Monotonic Sequence

A sequence $\{a_n\}$ is **monotonic** if its terms are nondecreasing

$$a_1 \leq a_2 \leq a_3 \leq \cdots \leq a_n \leq \cdots$$

or if its terms are nonincreasing

$$a_1 \geq a_2 \geq a_3 \geq \cdots \geq a_n \geq \cdots.$$

Example 8   Determining Whether a Sequence Is Monotonic

Determine whether each sequence having the given nth term is monotonic.

a. $a_n = 3 + (-1)^n$ **b.** $b_n = \dfrac{2n}{1 + n}$ **c.** $c_n = \dfrac{n^2}{2^n - 1}$

Solution

a. This sequence alternates between 2 and 4. Therefore, it is not monotonic.

b. This sequence is monotonic because each successive term is larger than its predecessor. To see this, compare the terms b_n and b_{n+1}. [Note that, because n is positive, you can multiply both sides of the inequality by $(1 + n)$ and $(2 + n)$ without reversing the inequality sign.]

$$b_n = \frac{2n}{1 + n} \overset{?}{<} \frac{2(n + 1)}{1 + (n + 1)} = b_{n+1}$$

$$2n(2 + n) \overset{?}{<} (1 + n)(2n + 2)$$

$$4n + 2n^2 \overset{?}{<} 2 + 4n + 2n^2$$

$$0 < 2$$

Starting with the final inequality, which is valid, you can reverse the steps to conclude that the original inequality is also valid.

c. This sequence is not monotonic, because the second term is larger than the first term, and larger than the third. (Note that if we drop the first term, the remaining sequence $c_2, c_3, c_4, \ldots$ is monotonic.)

Figure 8.3 graphically illustrates these three sequences.

NOTE   In Example 8b, another way to see that the sequence is monotonic is to argue that the derivative of the corresponding differentiable function $f(x) = 2x/(1 + x)$ is positive for all x. This implies that f is increasing, which in turn implies that $\{a_n\}$ is increasing.

NOTE   All three sequences shown in Figure 8.3 are bounded. To see this, consider the following.

$$2 \le a_n \le 4$$

$$1 \le b_n \le 2$$

$$0 \le c_n \le \frac{4}{3}$$

Definition of a Bounded Sequence

1. A sequence $\{a_n\}$ is **bounded above** if there is a real number M such that $a_n \le M$ for all n. The number M is called an **upper bound** of the sequence.

2. A sequence $\{a_n\}$ is **bounded below** if there is a real number N such that $N \le a_n$ for all n. The number N is called a **lower bound** of the sequence.

3. A sequence $\{a_n\}$ is **bounded** if it is bounded above and bounded below.

One important property of the real numbers is that they are **complete.** Informally, this means that there are no holes or gaps on the real number line. (The set of rational numbers does not have the completeness property.) The completeness axiom for real numbers can be used to conclude that if a sequence has an upper bound, it must have a **least upper bound** (an upper bound that is smaller than all other upper bounds for the sequence). For example, the least upper bound of the sequence $\{a_n\} = \{n/(n + 1)\}$,

$$\frac{1}{2}, \frac{2}{3}, \frac{3}{4}, \frac{4}{5}, \cdots, \frac{n}{n + 1}, \cdots$$

is 1. We use the completeness axiom in the proof of Theorem 8.5.

THEOREM 8.5   Bounded Monotonic Sequences

If a sequence $\{a_n\}$ is bounded and monotonic, then it converges.

Proof   Assume that the sequence is nondecreasing, as shown in Figure 8.4. For the sake of simplicity, also assume that each term in the sequence is positive. Because the sequence is bounded, there must exist an upper bound M such that

$$a_1 \le a_2 \le a_3 \le \cdots \le a_n \le \cdots \le M.$$

From the completeness axiom, it follows that there is a least upper bound L such that

$$a_1 \le a_2 \le a_3 \le \cdots \le a_n \le \cdots \le L.$$

For $\varepsilon > 0$, it follows that $L - \varepsilon < L$, and therefore $L - \varepsilon$ cannot be an upper bound for the sequence. Consequently, at least one term of $\{a_n\}$ is greater than $L - \varepsilon$. That is, $L - \varepsilon < a_N$ for some positive integer N. Because the terms of $\{a_n\}$ are nondecreasing, it follows that $a_N \le a_n$ for $n > N$. You now know that $L - \varepsilon < a_N \le a_n \le L < L + \varepsilon$, for every $n > N$. It follows that $|a_n - L| < \varepsilon$ for $n > N$, which by definition means that $\{a_n\}$ converges to L. The proof for a nonincreasing sequence is similar.

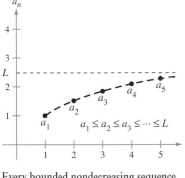

Every bounded nondecreasing sequence converges.

Figure 8.4

Example 9   **Bounded and Monotonic Sequences**

a. The sequence $\{a_n\} = \{1/n\}$ is both bounded and monotonic and so, by Theorem 8.5, must converge.

b. The divergent sequence $\{b_n\} = \{n^2/(n + 1)\}$ is monotonic, but not bounded. (It *is* bounded below.)

c. The divergent sequence $\{c_n\} = \{(-1)^n\}$ is bounded, but not monotonic.

EXERCISES FOR SECTION 8.1

In Exercises 1–12, write the first five terms of the sequence.

1. $a_n = 2^n$

2. $a_n = \dfrac{2n}{n+3}$

3. $a_n = \left(-\dfrac{1}{2}\right)^n$

4. $a_n = \left(-\dfrac{2}{3}\right)^n$

5. $a_n = \sin\dfrac{n\pi}{2}$

6. $a_n = \cos\dfrac{n\pi}{2}$

7. $a_n = \dfrac{(-1)^{n(n+1)/2}}{n^2}$

8. $a_n = (-1)^{n+1}\left(\dfrac{2}{n}\right)$

9. $a_n = 5 - \dfrac{1}{n} + \dfrac{1}{n^2}$

10. $a_n = 10 + \dfrac{2}{n} + \dfrac{6}{n^2}$

11. $a_n = \dfrac{3^n}{n!}$

12. $a_n = \dfrac{3n!}{(n-1)!}$

In Exercises 13–16, write the first five terms of the recursively defined sequence.

13. $a_1 = 3,\ a_{k+1} = 2(a_k - 1)$

14. $a_1 = 4,\ a_{k+1} = \left(\dfrac{k+1}{2}\right)a_k$

15. $a_1 = 32,\ a_{k+1} = \dfrac{1}{2}a_k$

16. $a_1 = 6,\ a_{k+1} = \dfrac{1}{3}a_k^2$

In Exercises 17–20, match the sequence with its graph. [The graphs are labeled (a), (b), (c), and (d).]

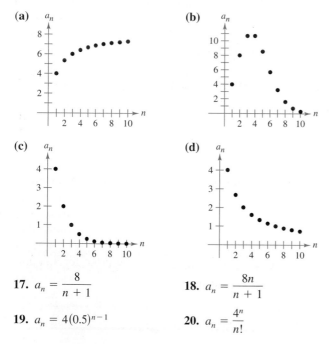

(a)

(b)

(c)

(d)

17. $a_n = \dfrac{8}{n+1}$

18. $a_n = \dfrac{8n}{n+1}$

19. $a_n = 4(0.5)^{n-1}$

20. $a_n = \dfrac{4^n}{n!}$

In Exercises 21–26, use a graphing utility to graph the first ten terms of the sequence.

21. $a_n = \dfrac{2}{3}n$

22. $a_n = 2 - \dfrac{4}{n}$

23. $a_n = 16(-0.5)^{n-1}$

24. $a_n = 8(0.75)^{n-1}$

25. $a_n = \dfrac{2n}{n+1}$

26. $a_n = \dfrac{3n^2}{n^2+1}$

In Exercises 27–30, write the next two *apparent* terms of the sequence. Describe the pattern you used to find these terms.

27. $2, 5, 8, 11, \ldots$

28. $\dfrac{7}{2}, 4, \dfrac{9}{2}, 5, \ldots$

29. $3, -\dfrac{3}{2}, \dfrac{3}{4}, -\dfrac{3}{8}, \ldots$

30. $5, 10, 20, 40, \ldots$

In Exercises 31–36, simplify the ratio of factorials.

31. $\dfrac{10!}{8!}$

32. $\dfrac{25!}{23!}$

33. $\dfrac{(n+1)!}{n!}$

34. $\dfrac{(n+2)!}{n!}$

35. $\dfrac{(2n-1)!}{(2n+1)!}$

36. $\dfrac{(2n+2)!}{(2n)!}$

In Exercises 37–42, find the limit (if possible) of the sequence.

37. $a_n = \dfrac{5n^2}{n^2+2}$

38. $a_n = 5 - \dfrac{1}{n^2}$

39. $a_n = \dfrac{2n}{\sqrt{n^2+1}}$

40. $a_n = \dfrac{5n}{\sqrt{n^2+4}}$

41. $a_n = \sin\dfrac{1}{n}$

42. $a_n = \cos\dfrac{2}{n}$

In Exercises 43–46, use a graphing utility to graph the first ten terms of the sequence. Use the graph to make an inference about the convergence or divergence of the sequence. Verify your inference analytically and, if the sequence converges, find its limit.

43. $a_n = \dfrac{n+1}{n}$

44. $a_n = \dfrac{1}{n^{3/2}}$

45. $a_n = \cos\dfrac{n\pi}{2}$

46. $a_n = 3 - \dfrac{1}{2^n}$

In Exercises 47–66, determine the convergence or divergence of the sequence with the given nth term. If the sequence converges, find its limit.

47. $a_n = (-1)^n\left(\dfrac{n}{n+1}\right)$

48. $a_n = 1 + (-1)^n$

49. $a_n = \dfrac{3n^2 - n + 4}{2n^2 + 1}$

50. $a_n = \dfrac{\sqrt[3]{n}}{\sqrt[3]{n}+1}$

51. $a_n = \dfrac{1 + (-1)^n}{n}$

52. $a_n = \dfrac{1 + (-1)^n}{n^2}$

53. $a_n = \dfrac{\ln(n^3)}{2n}$

54. $a_n = \dfrac{\ln\sqrt{n}}{n}$

55. $a_n = \dfrac{3^n}{4^n}$

56. $a_n = (0.5)^n$

57. $a_n = \dfrac{(n+1)!}{n!}$

58. $a_n = \dfrac{(n-2)!}{n!}$

59. $a_n = \dfrac{n-1}{n} - \dfrac{n}{n-1},\ n \geq 2$

60. $a_n = \dfrac{n^2}{2n+1} - \dfrac{n^2}{2n-1}$

61. $a_n = \dfrac{n^p}{e^n}$, $p > 0$

62. $a_n = n \sin \dfrac{1}{n}$

63. $a_n = \left(1 + \dfrac{k}{n}\right)^n$

64. $a_n = 2^{1/n}$

65. $a_n = \dfrac{\sin n}{n}$

66. $a_n = \dfrac{\cos \pi n}{n^2}$

In Exercises 67–80, write an expression for the nth term of the sequence. (There is more than one correct answer.)

67. $1, 4, 7, 10, \ldots$

68. $3, 7, 11, 15, \ldots$

69. $-1, 2, 7, 14, 23, \ldots$

70. $1, -\frac{1}{4}, \frac{1}{9}, -\frac{1}{16}, \ldots$

71. $\frac{2}{3}, \frac{3}{4}, \frac{4}{5}, \frac{5}{6}, \ldots$

72. $\frac{3}{2}, \frac{4}{5}, \frac{5}{8}, \frac{6}{11}, \frac{1}{2}, \ldots$

73. $2, -1, \frac{1}{2}, -\frac{1}{4}, \frac{1}{8}, \ldots$

74. $-\frac{1}{3}, \frac{1}{2}, -\frac{3}{4}, \frac{9}{8}, -\frac{27}{16}, \ldots$

75. $2, 1 + \frac{1}{2}, 1 + \frac{1}{3}, 1 + \frac{1}{4}, 1 + \frac{1}{5}, \ldots$

76. $1 + \frac{1}{2}, 1 + \frac{3}{4}, 1 + \frac{7}{8}, 1 + \frac{15}{16}, 1 + \frac{31}{32}, \ldots$

77. $\dfrac{1}{2 \cdot 3}, \dfrac{2}{3 \cdot 4}, \dfrac{3}{4 \cdot 5}, \dfrac{4}{5 \cdot 6}, \ldots$

78. $1, \frac{1}{2}, \frac{1}{6}, \frac{1}{24}, \frac{1}{120}, \ldots$

79. $1, -\dfrac{1}{1 \cdot 3}, \dfrac{1}{1 \cdot 3 \cdot 5}, -\dfrac{1}{1 \cdot 3 \cdot 5 \cdot 7}, \ldots$

80. $1, x, \dfrac{x^2}{2}, \dfrac{x^3}{6}, \dfrac{x^4}{24}, \dfrac{x^5}{120}, \ldots$

In Exercises 81–90, determine whether the sequence with the given nth term is monotonic. Discuss the boundedness of the sequence. Use a graphing utility to confirm your results.

81. $a_n = 4 - \dfrac{1}{n}$

82. $a_n = \dfrac{3n}{n + 2}$

83. $a_n = \dfrac{n}{2^{n+2}}$

84. $a_n = ne^{-n/2}$

85. $a_n = (-1)^n \left(\dfrac{1}{n}\right)$

86. $a_n = \left(-\dfrac{2}{3}\right)^n$

87. $a_n = \left(\dfrac{2}{3}\right)^n$

88. $a_n = \left(\dfrac{3}{2}\right)^n$

89. $a_n = \sin \dfrac{n\pi}{6}$

90. $a_n = \dfrac{\cos n}{n}$

In Exercises 91–94, (a) use Theorem 8.5 to show that the sequence with the given nth term converges and (b) use a graphing utility to graph the first ten terms of the sequence and find its limit.

91. $a_n = 5 + \dfrac{1}{n}$

92. $a_n = 4 - \dfrac{3}{n}$

93. $a_n = \dfrac{1}{3}\left(1 - \dfrac{1}{3^n}\right)$

94. $a_n = 4 + \dfrac{1}{2^n}$

95. *Compound Interest* Consider the sequence $\{A_n\}$ whose nth term is given by

$$A_n = P\left(1 + \dfrac{r}{12}\right)^n$$

where P is the principal, A_n is the account balance after n months, and r is the interest rate compounded annually.

(a) Is $\{A_n\}$ a convergent sequence? Explain.

(b) Find the first ten terms of the sequence if $P = \$9000$ and $r = 0.115$.

96. *Investment* A deposit of $100 is made at the beginning of each month in an account at an annual interest rate of 12% compounded monthly. The balance in the account after n months is

$$A_n = 100(101)[(1.01)^n - 1].$$

(a) Compute the first six terms of the sequence $\{A_n\}$.

(b) Find the balance after 5 years by computing the 60th term of the sequence.

(c) Find the balance after 20 years by computing the 240th term of the sequence.

Getting at the Concept

97. In your own words, define each of the following.

(a) Sequence

(b) Convergence of a sequence

(c) Bounded monotonic sequence

98. The graphs of two sequences are given in the figures. Which graph represents the sequence with alternating signs? Explain.

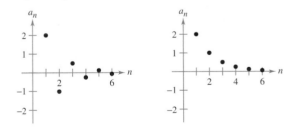

In Exercises 99–102, give an example of a sequence satisfying the condition or explain why no such sequence exists. (Examples are not unique.)

99. A monotonically increasing sequence that converges to 10

100. A monotonically increasing bounded sequence that does not converge

101. A sequence that converges to $\frac{3}{4}$

102. An unbounded sequence that converges to 100

103. *Government Expenditures*   A government program that currently costs taxpayers $2.5 billion per year is cut back by 20 percent per year.

   (a) Write an expression for the amount budgeted for this program after n years.

   (b) Compute the budgets for the first 4 years.

   (c) Determine the convergence or divergence of the sequence of reduced budgets. If the sequence converges, find its limit.

104. *Inflation*   If the rate of inflation is $4\frac{1}{2}\%$ per year and the average price of a car is currently $16,000, the average price after n years is

$$P_n = \$16{,}000(1.045)^n.$$

Compute the average price for the next 5 years.

105. *Modeling Data*   The average cost per day for a hospital room from 1990 through 1997 is shown in the table, where a_n is the average cost in dollars and n is the year, with $n = 0$ corresponding to 1990.   *(Source: American Hospital Association)*

n	0	1	2	3	4	5	6	7
a_n	687	752	820	881	931	968	1006	1033

   (a) Use the regression capabilities of a graphing utility to find a model of the form

$$a_n = bn^2 + cn + d, \qquad n = 0, 1, 2, 3, 4, 5, 6, 7$$

   for the data. Use the graphing utility to plot the points and graph the model.

   (b) Use the model to predict the cost in the year 2004.

106. *Modeling Data*   The annual sales a_n (in millions of dollars) of H. J. Heinz Company from 1990 through 1999 are given below as ordered pairs of the form (n, a_n), where n is the year, with $n = 0$ corresponding to 1990.   *(Source: 1999 H. J. Heinz Report)*

$(0, 6086),\ (1, 6647),\ (2, 6582),\ (3, 7103),\ (4, 7047),$

$(5, 8087),\ (6, 9112),\ (7, 9357),\ (8, 9209),\ (9, 9300)$

   (a) Use the regression capabilities of a graphing utility to find a model of the form

$$a_n = bn + c, \quad n = 0, 1, \ldots, 9$$

   for the data. Graphically compare the points and the model.

   (b) Use the model to predict sales in the year 2004.

107. *Comparing Exponential and Factorial Growth*   Consider the sequence $a_n = 10^n/n!$.

   (a) Find two consecutive terms that are equal in magnitude.

   (b) Are the terms following those found in part (a) increasing or decreasing?

   (c) In Section 7.7, Exercises 63–68, it was shown that for "large" values of the independent variable an exponential function increases more rapidly than a polynomial function. From the result in part (b), what inference can you make about the rate of growth of an exponential function versus a factorial function for "large" integer values of n?

108. Compute the first six terms of the sequence

$$\{a_n\} = \{(1 + 1/n)^n\}.$$

If the sequence converges, find its limit.

109. Compute the first six terms of the sequence $\{a_n\} = \left\{\sqrt[n]{n}\right\}$. If the sequence converges, find its limit.

110. Prove that if $\{s_n\}$ converges to L and $L > 0$, then there exists a number N such that $s_n > 0$ for $n > N$.

111. *Fibonacci Sequence*   In a study of the progeny of rabbits, Fibonacci (ca. 1175–ca. 1250) encountered the sequence now bearing his name. It is defined recursively by

$$a_{n+2} = a_n + a_{n+1}, \qquad \text{where} \quad a_1 = 1 \text{ and } a_2 = 1.$$

   (a) Write the first 12 terms of the sequence.

   (b) Write the first ten terms of the sequence defined by

$$b_n = \frac{a_{n+1}}{a_n}, \ n \geq 1.$$

   (c) Using the definition in part (b), show that

$$b_n = 1 + \frac{1}{b_{n-1}}.$$

   (d) The **golden ratio** ρ can be defined by $\lim_{n \to \infty} b_n = \rho$. Show that

$$\rho = 1 + 1/\rho$$

   and solve this equation for ρ.

112. Complete the proof of Theorem 8.5.

True or False?   In Exercises 113–116, determine whether the statement is true or false. If it is false, explain why or give an example that shows it is false.

113. If $\{a_n\}$ converges to 3 and $\{b_n\}$ converges to 2, then $\{a_n + b_n\}$ converges to 5.

114. If $\{a_n\}$ converges, then $\lim_{n \to \infty} (a_n - a_{n+1}) = 0$.

115. If $n > 1$, then $n! = n(n - 1)!$.

116. If $\{a_n\}$ converges, then $\{a_n/n\}$ converges to 0.

117. Consider the sequence

$$\sqrt{2}, \ \sqrt{2 + \sqrt{2}}, \ \sqrt{2 + \sqrt{2 + \sqrt{2}}}, \ldots$$

where $a_n = \sqrt{2 + a_{n-1}}$ for $n \geq 2$. Compute the first five terms of this sequence. Find $\lim_{n \to \infty} a_n$.

118. *Conjecture*   Let $x_0 = 1$ and consider the sequence x_n given by the formula

$$x_n = \frac{1}{2}x_{n-1} + \frac{1}{x_{n-1}}, \qquad n = 1, 2, \ldots.$$

Use a graphing utility to compute the first ten terms of the sequence and make a conjecture about the limit of the sequence.

- Understand the definition of a convergent infinite series.
- Use properties of infinite geometric series.
- Use the nth-Term Test for Divergence of an infinite series.

Infinite Series

INFINITE SERIES

The study of infinite series was considered a novelty in the fourteenth century. Logician Richard Suiseth, whose nickname was Calculator, solved this problem.

If throughout the first half of a given time interval a variation continues at a certain intensity, throughout the next quarter of the interval at double the intensity, throughout the following eighth at triple the intensity and so ad infinitum, then the average intensity for the whole interval will be the intensity of the variation during the second subinterval (or double the intensity).

This is the same as saying that the sum of the infinite series

$$\frac{1}{2} + \frac{2}{4} + \frac{3}{8} + \cdots + \frac{n}{2^n} + \cdots$$

is 2.

One important application of infinite sequences is in representing "infinite summations." Informally, if $\{a_n\}$ is an infinite sequence, then

$$\sum_{n=1}^{\infty} a_n = a_1 + a_2 + a_3 + \cdots + a_n + \cdots \qquad \text{Infinite series}$$

is an **infinite series** (or simply a **series**). The numbers a_1, a_2, a_3, are the **terms** of the series. For some series it is convenient to begin the index at $n = 0$ (or some other integer). As a typesetting convention, it is common to represent an infinite series as simply $\Sigma\, a_n$. In such cases, the starting value for the index must be taken from the context of the statement.

To find the sum of an infinite series, consider the following **sequence of partial sums.**

$$S_1 = a_1$$
$$S_2 = a_1 + a_2$$
$$S_3 = a_1 + a_2 + a_3$$
$$\vdots$$
$$S_n = a_1 + a_2 + a_3 + \cdots + a_n$$

If this sequence of partial sums converges, the series is said to converge and has the sum indicated in the following definition.

Definition of Convergent and Divergent Series

For the infinite series $\Sigma\, a_n$, the **nth partial sum** is given by

$$S_n = a_1 + a_2 + \cdots + a_n.$$

If the sequence of partial sums $\{S_n\}$ converges to S, then the series $\Sigma\, a_n$ **converges.** The limit S is called the **sum of the series.**

$$S = a_1 + a_2 + \cdots + a_n + \cdots$$

If $\{S_n\}$ diverges, then the series **diverges.**

STUDY TIP   As you study this chapter, you will see that there are two basic questions involving infinite series. Does a series converge or does it diverge? If a series converges, what is its sum? These questions are not always easy to answer, especially the second one.

EXPLORATION

Finding the Sum of an Infinite Series   Find the sum of each infinite series. Explain your reasoning.

a. $0.1 + 0.01 + 0.001 + 0.0001 + \cdots$ **b.** $\frac{3}{10} + \frac{3}{100} + \frac{3}{1000} + \frac{3}{10,000} + \cdots$

c. $1 + \frac{1}{2} + \frac{1}{4} + \frac{1}{8} + \frac{1}{16} + \cdots$ **d.** $\frac{15}{100} + \frac{15}{10,000} + \frac{15}{1,000,000} + \cdots$

TECHNOLOGY Figure 8.5 shows the first 15 partial sums of the infinite series in Example 1a. Notice how the values appear to approach the line $y = 1$.

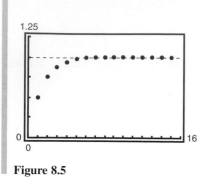

Figure 8.5

NOTE You can geometrically determine the partial sums of the series in Example 1a using Figure 8.6.

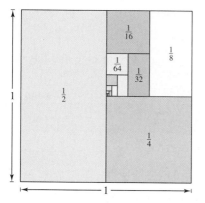

Figure 8.6

FOR FURTHER INFORMATION To learn more about the partial sums of infinite series, see the article "Six Ways to Sum a Series" by Dan Kalman in *The College Mathematics Journal*. To view this article, go to the website *www.matharticles.com*.

Example 1 **Convergent and Divergent Series**

a. The series

$$\sum_{n=1}^{\infty} \frac{1}{2^n} = \frac{1}{2} + \frac{1}{4} + \frac{1}{8} + \frac{1}{16} + \cdots$$

has the following partial sums.

$$S_1 = \frac{1}{2}$$

$$S_2 = \frac{1}{2} + \frac{1}{4} = \frac{3}{4}$$

$$S_3 = \frac{1}{2} + \frac{1}{4} + \frac{1}{8} = \frac{7}{8}$$

$$\vdots$$

$$S_n = \frac{1}{2} + \frac{1}{4} + \frac{1}{8} + \cdots + \frac{1}{2^n} = \frac{2^n - 1}{2^n}$$

Because

$$\lim_{x \to \infty} \frac{2^n - 1}{2^n} = 1$$

it follows that the series converges and its sum is 1.

b. The nth partial sum of the series

$$\sum_{n=1}^{\infty} \left(\frac{1}{n} - \frac{1}{n+1} \right) = \left(1 - \frac{1}{2} \right) + \left(\frac{1}{2} - \frac{1}{3} \right) + \left(\frac{1}{3} - \frac{1}{4} \right) + \cdots$$

is given by

$$S_n = 1 - \frac{1}{n+1}.$$

Because the limit of S_n is 1, the series converges and its sum is 1.

c. The series

$$\sum_{n=1}^{\infty} 1 = 1 + 1 + 1 + 1 + \cdots$$

diverges because $S_n = n$ and the sequence of partial sums diverges.

The series in Example 1b is a **telescoping series.** That is, it is of the form

$$(b_1 - b_2) + (b_2 - b_3) + (b_3 - b_4) + (b_4 - b_5) + \cdots. \qquad \text{Telescoping series}$$

Note that b_2 is canceled by the second term, b_3 is canceled by the third term, and so on. Because the nth partial sum of this series is

$$S_n = b_1 - b_{n+1}$$

it follows that a telescoping series will converge if and only if b_n approaches a finite number as $n \to \infty$. Moreover, if the series converges, its sum is

$$S = b_1 - \lim_{n \to \infty} b_{n+1}.$$

Example 2   Writing a Series in Telescoping Form

Find the sum of the series $\displaystyle\sum_{n=1}^{\infty} \frac{2}{4n^2 - 1}$.

Solution   Using partial fractions, you can write

$$a_n = \frac{2}{4n^2 - 1} = \frac{2}{(2n - 1)(2n + 1)} = \frac{1}{2n - 1} - \frac{1}{2n + 1}.$$

From this telescoping form, you can see that the nth partial sum is

$$S_n = \left(\frac{1}{1} - \frac{1}{3}\right) + \left(\frac{1}{3} - \frac{1}{5}\right) + \cdots + \left(\frac{1}{2n - 1} - \frac{1}{2n + 1}\right) = 1 - \frac{1}{2n + 1}.$$

So, the series converges and its sum is 1. That is,

$$\sum_{n=1}^{\infty} \frac{2}{4n^2 - 1} = \lim_{n \to \infty} S_n = \lim_{n \to \infty} \left(1 - \frac{1}{2n + 1}\right) = 1.$$

Geometric Series

The series given in Example 1a is a **geometric series.** In general, the series given by

$$\sum_{n=0}^{\infty} ar^n = a + ar + ar^2 + \cdots + ar^n + \cdots, \quad a \neq 0 \qquad \text{Geometric series}$$

is a **geometric series** with ratio r.

THEOREM 8.6   Convergence of a Geometric Series

A geometric series with ratio r diverges if $|r| \geq 1$. If $0 < |r| < 1$, then the series converges to the sum

$$\sum_{n=0}^{\infty} ar^n = \frac{a}{1 - r}, \qquad 0 < |r| < 1.$$

Proof   It is easy to see that the series diverges if $r = \pm 1$. If $r \neq \pm 1$, then $S_n = a + ar + ar^2 + \cdots + ar^{n-1}$. Multiplication by r yields

$$rS_n = ar + ar^2 + ar^3 + \cdots + ar^n.$$

Subtracting the second equation from the first produces $S_n - rS_n = a - ar^n$. Therefore, $S_n(1 - r) = a(1 - r^n)$, and the nth partial sum is

$$S_n = \frac{a}{1 - r}(1 - r^n).$$

If $0 < |r| < 1$, it follows that $r^n \to 0$ as $n \to \infty$, and you obtain

$$\lim_{n \to \infty} S_n = \lim_{n \to \infty} \left[\frac{a}{1 - r}(1 - r^n)\right] = \frac{a}{1 - r}\left[\lim_{n \to \infty}(1 - r^n)\right] = \frac{a}{1 - r}$$

which means that the series *converges* and its sum is $a/(1 - r)$. We leave it to you to show that the series diverges if $|r| > 1$.

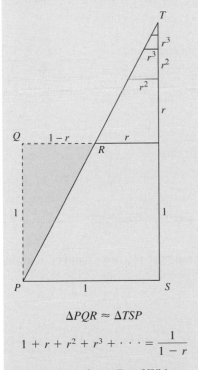

Example 3 **Convergent and Divergent Geometric Series**

TECHNOLOGY Try using a graphing utility or writing a computer program to compute the sum of the first 20 terms of the sequence in Example 3a. You should obtain a sum of about 5.999994.

a. The geometric series

$$\sum_{n=0}^{\infty} \frac{3}{2^n} = \sum_{n=0}^{\infty} 3\left(\frac{1}{2}\right)^n$$

$$= 3(1) + 3\left(\frac{1}{2}\right) + 3\left(\frac{1}{2}\right)^2 + \cdots$$

has a ratio of $r = \frac{1}{2}$ with $a = 3$. Because $0 < |r| < 1$, the series converges and its sum is

$$S = \frac{a}{1 - r} = \frac{3}{1 - (1/2)} = 6.$$

b. The geometric series

$$\sum_{n=0}^{\infty} \left(\frac{3}{2}\right)^n = 1 + \frac{3}{2} + \frac{9}{4} + \frac{27}{8} + \cdots$$

has a ratio of $r = \frac{3}{2}$. Because $|r| \geq 1$, the series diverges.

The formula for the sum of a geometric series can be used to write a repeating decimal as the ratio of two integers, as demonstrated in the next example.

Example 4 **A Geometric Series for a Repeating Decimal**

Use a geometric series to express 0.080808 as the ratio of two integers.

Solution For the repeating decimal $0.08\overline{08}$, you can write

$$0.080808\ldots = \frac{8}{10^2} + \frac{8}{10^4} + \frac{8}{10^6} + \frac{8}{10^8} + \cdots$$

$$= \sum_{n=0}^{\infty} \left(\frac{8}{10^2}\right)\left(\frac{1}{10^2}\right)^n.$$

For this series, you have $a = 8/10^2$ and $r = 1/10^2$. So,

$$0.080808\ldots = \frac{a}{1 - r} = \frac{8/10^2}{1 - (1/10^2)} = \frac{8}{99}.$$

Try dividing 8 by 99 on a calculator to see that it produces $0.08\overline{08}$.

The convergence of a series is not affected by removal of a finite number of terms from the beginning of the series. For instance, the geometric series

$$\sum_{n=4}^{\infty} \left(\frac{1}{2}\right)^n \qquad \text{and} \qquad \sum_{n=0}^{\infty} \left(\frac{1}{2}\right)^n$$

both converge. Furthermore, because the sum of the second series is $a/(1 - r) = 2$, you can conclude that the sum of the first series is

$$S = 2 - \left[\left(\frac{1}{2}\right)^0 + \left(\frac{1}{2}\right)^1 + \left(\frac{1}{2}\right)^2 + \left(\frac{1}{2}\right)^3\right]$$

$$= 2 - \frac{15}{8}$$

$$= \frac{1}{8}.$$

The following properties are direct consequences of the corresponding properties of limits of sequences.

THEOREM 8.7   Properties of Infinite Series

If $\Sigma\, a_n = A$, $\Sigma\, b_n = B$, and c is a real number, then the following series converge to the indicated sums.

1. $\displaystyle\sum_{n=1}^{\infty} ca_n = cA$

2. $\displaystyle\sum_{n=1}^{\infty} (a_n + b_n) = A + B$

3. $\displaystyle\sum_{n=1}^{\infty} (a_n - b_n) = A - B$

---

nth-Term Test for Divergence

The following theorem states that if a series converges, the limit of its nth term must be 0.

THEOREM 8.8   Limit of nth Term of a Convergent Series

If $\displaystyle\sum_{n=1}^{\infty} a_n$ converges, then $\displaystyle\lim_{n\to\infty} a_n = 0$.

Proof Assume that

$$\sum_{n=1}^{\infty} a_n = \lim_{n\to\infty} S_n = L.$$

Then, because $S_n = S_{n-1} + a_n$ and

$$\lim_{n\to\infty} S_n = \lim_{n\to\infty} S_{n-1} = L$$

it follows that

$$
\begin{aligned}
L = \lim_{n\to\infty} S_n &= \lim_{n\to\infty} (S_{n-1} + a_n) \\
&= \lim_{n\to\infty} S_{n-1} + \lim_{n\to\infty} a_n \\
&= L + \lim_{n\to\infty} a_n
\end{aligned}
$$

which implies that $\{a_n\}$ converges to 0.

The contrapositive of Theorem 8.8 provides a useful test for *divergence*. This **nth-Term Test for Divergence** states that if the limit of the nth term of a series does *not* converge to 0, the series must diverge.

THEOREM 8.9 nth-Term Test for Divergence

If $\displaystyle\lim_{n\to\infty} a_n \neq 0$, then $\displaystyle\sum_{n=1}^{\infty} a_n$ diverges.

Example 5   Using the *n*th-Term Test for Divergence

a. For the series $\displaystyle\sum_{n=0}^{\infty} 2^n$, you have

$$\lim_{n\to\infty} 2^n = \infty.$$

So, the limit of the *n*th term is not 0, and the series diverges.

b. For the series $\displaystyle\sum_{n=1}^{\infty} \frac{n!}{2n! + 1}$, you have

$$\lim_{n\to\infty} \frac{n!}{2n! + 1} = \frac{1}{2}.$$

So, the limit of the *n*th term is not 0, and the series diverges.

c. For the series $\displaystyle\sum_{n=1}^{\infty} \frac{1}{n}$, you have

$$\lim_{n\to\infty} \frac{1}{n} = 0.$$

Because the limit of the *n*th term is 0, the *n*th-Term Test for Divergence does *not* apply and you can draw no conclusions about convergence or divergence. (In the next section, you will see that this particular series diverges.)

STUDY TIP   The series in Example 5c will play an important role in this chapter.

$$\sum_{n=1}^{\infty} \frac{1}{n} = 1 + \frac{1}{2} + \frac{1}{3} + \frac{1}{4} + \cdots$$

You will see that this series diverges even though the *n*th term approaches 0 as *n* approaches ∞.

Example 6   Bouncing Ball Problem

A ball is dropped from a height of 6 feet and begins bouncing, as shown in Figure 8.7. The height of each bounce is three-fourths the height of the previous bounce. Find the total vertical distance traveled by the ball.

Solution   When the ball hits the ground for the first time, it has traveled a distance of $D_1 = 6$ feet. For subsequent bounces, let D_i be the distance traveled up *and* down. For example, D_2 and D_3 are as follows.

$$D_2 = \underbrace{6\left(\tfrac{3}{4}\right)}_{\text{Up}} + \underbrace{6\left(\tfrac{3}{4}\right)}_{\text{Down}} = 12\left(\tfrac{3}{4}\right)$$

$$D_3 = \underbrace{6\left(\tfrac{3}{4}\right)\left(\tfrac{3}{4}\right)}_{\text{Up}} + \underbrace{6\left(\tfrac{3}{4}\right)\left(\tfrac{3}{4}\right)}_{\text{Down}} = 12\left(\tfrac{3}{4}\right)^2$$

By continuing this process, it can be determined that the total vertical distance is

$$
\begin{aligned}
D &= 6 + 12\left(\tfrac{3}{4}\right) + 12\left(\tfrac{3}{4}\right)^2 + 12\left(\tfrac{3}{4}\right)^3 + \cdots \\
&= 6 + 12\sum_{n=0}^{\infty} \left(\tfrac{3}{4}\right)^{n+1} \\
&= 6 + 12\left(\tfrac{3}{4}\right)\sum_{n=0}^{\infty} \left(\tfrac{3}{4}\right)^n \\
&= 6 + 9\left(\frac{1}{1 - \tfrac{3}{4}}\right) \\
&= 6 + 9(4) \\
&= 42 \text{ feet.}
\end{aligned}
$$

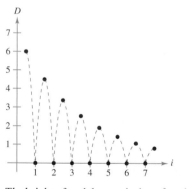

The height of each bounce is three-fourths the height of the previous bounce.
Figure 8.7

Lab Series | LAB 11

EXERCISES FOR SECTION 8.2

In Exercises 1–6, find the first five terms of the sequence of partial sums.

1. $1 + \frac{1}{4} + \frac{1}{9} + \frac{1}{16} + \frac{1}{25} + \cdots$

2. $\frac{1}{2 \cdot 3} + \frac{2}{3 \cdot 4} + \frac{3}{4 \cdot 5} + \frac{4}{5 \cdot 6} + \frac{5}{6 \cdot 7} + \cdots$

3. $3 - \frac{9}{2} + \frac{27}{4} - \frac{81}{8} + \frac{243}{16} - \cdots$

4. $\frac{1}{1} + \frac{1}{3} + \frac{1}{5} + \frac{1}{7} + \frac{1}{9} + \frac{1}{11} + \cdots$

5. $\sum_{n=1}^{\infty} \frac{3}{2^{n-1}}$

6. $\sum_{n=1}^{\infty} \frac{(-1)^{n+1}}{n!}$

In Exercises 7–16, verify that the infinite series diverges.

7. $\sum_{n=0}^{\infty} 3\left(\frac{3}{2}\right)^n$

8. $\sum_{n=0}^{\infty} \left(\frac{4}{3}\right)^n$

9. $\sum_{n=0}^{\infty} 1000(1.055)^n$

10. $\sum_{n=0}^{\infty} 2(-1.03)^n$

11. $\sum_{n=1}^{\infty} \frac{n}{n+1}$

12. $\sum_{n=1}^{\infty} \frac{n}{2n+3}$

13. $\sum_{n=1}^{\infty} \frac{n^2}{n^2+1}$

14. $\sum_{n=1}^{\infty} \frac{n}{\sqrt{n^2+1}}$

15. $\sum_{n=1}^{\infty} \frac{2^n+1}{2^{n+1}}$

16. $\sum_{n=1}^{\infty} \frac{n!}{2^n}$

In Exercises 17–20, match the series with the graph of its sequence of partial sums. [The graphs are labeled (a), (b), (c), and (d).] Use the graph to estimate the sum of the series. Confirm your answer analytically.

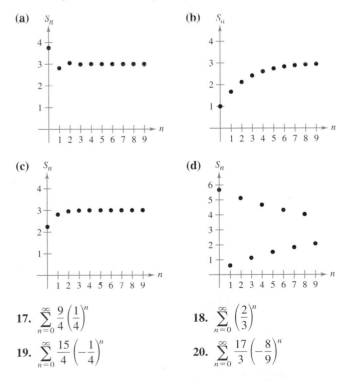

(a) S_n

(b) S_n

(c) S_n

(d) S_n

17. $\sum_{n=0}^{\infty} \frac{9}{4}\left(\frac{1}{4}\right)^n$

18. $\sum_{n=0}^{\infty} \left(\frac{2}{3}\right)^n$

19. $\sum_{n=0}^{\infty} \frac{15}{4}\left(-\frac{1}{4}\right)^n$

20. $\sum_{n=0}^{\infty} \frac{17}{3}\left(-\frac{8}{9}\right)^n$

In Exercises 21–26, verify that the infinite series converges.

21. $\sum_{n=1}^{\infty} \frac{1}{n(n+1)}$ (Use partial fractions.)

22. $\sum_{n=1}^{\infty} \frac{1}{n(n+2)}$ (Use partial fractions.)

23. $\sum_{n=0}^{\infty} 2\left(\frac{3}{4}\right)^n$

24. $\sum_{n=1}^{\infty} 2\left(-\frac{1}{2}\right)^n$

25. $\sum_{n=0}^{\infty} (0.9)^n = 1 + 0.9 + 0.81 + 0.729 + \cdots$

26. $\sum_{n=0}^{\infty} (-0.6)^n = 1 - 0.6 + 0.36 - 0.216 + \cdots$

Numerical, Graphical, and Analytic Analysis **In Exercises 27–32, (a) find the sum of the series, (b) use a graphing utility to find the indicated partial sum S_n and complete the table, (c) use a graphing utility to graph the first ten terms of the sequence of partial sums and a horizontal line representing the sum, and (d) explain the relationship between the magnitude of the terms of the series and the rate at which the sequence of partial sums approaches the sum of the series.**

n	5	10	20	50	100
S_n					

27. $\sum_{n=1}^{\infty} \frac{6}{n(n+3)}$

28. $\sum_{n=1}^{\infty} \frac{4}{n(n+4)}$

29. $\sum_{n=1}^{\infty} 2(0.9)^{n-1}$

30. $\sum_{n=1}^{\infty} 3(0.85)^{n-1}$

31. $\sum_{n=1}^{\infty} 10(0.25)^{n-1}$

32. $\sum_{n=1}^{\infty} 5\left(-\frac{1}{3}\right)^{n-1}$

In Exercises 33–46, find the sum of the convergent series.

33. $\sum_{n=2}^{\infty} \frac{1}{n^2-1}$

34. $\sum_{n=1}^{\infty} \frac{4}{n(n+2)}$

35. $\sum_{n=1}^{\infty} \frac{8}{(n+1)(n+2)}$

36. $\sum_{n=1}^{\infty} \frac{1}{(2n+1)(2n+3)}$

37. $\sum_{n=0}^{\infty} \left(\frac{1}{2}\right)^n$

38. $\sum_{n=0}^{\infty} 6\left(\frac{4}{5}\right)^n$

39. $\sum_{n=0}^{\infty} \left(-\frac{1}{2}\right)^n$

40. $\sum_{n=0}^{\infty} 2\left(-\frac{2}{3}\right)^n$

41. $1 + 0.1 + 0.01 + 0.001 + \cdots$

42. $8 + 6 + \frac{9}{2} + \frac{27}{8} + \cdots$

43. $3 - 1 + \frac{1}{3} - \frac{1}{9} + \cdots$

44. $4 - 2 + 1 - \frac{1}{2} + \cdots$

45. $\sum_{n=0}^{\infty} \left(\frac{1}{2^n} - \frac{1}{3^n}\right)$

46. $\sum_{n=1}^{\infty} \left[(0.7)^n + (0.9)^n\right]$

In Exercises 47–50, express the repeating decimal as a geometric series, and write its sum as the ratio of two integers.

47. $0.\overline{4}$

48. $0.81\overline{81}$

49. $0.075\overline{75}$

50. $0.215\overline{15}$

In Exercises 51–62, determine the convergence or divergence of the series.

51. $\displaystyle\sum_{n=1}^{\infty} \frac{n+10}{10n+1}$

52. $\displaystyle\sum_{n=1}^{\infty} \frac{n+1}{2n-1}$

53. $\displaystyle\sum_{n=1}^{\infty} \left(\frac{1}{n} - \frac{1}{n+2}\right)$

54. $\displaystyle\sum_{n=1}^{\infty} \frac{1}{n(n+3)}$

55. $\displaystyle\sum_{n=1}^{\infty} \frac{3n-1}{2n+1}$

56. $\displaystyle\sum_{n=1}^{\infty} \frac{3^n}{n^3}$

57. $\displaystyle\sum_{n=0}^{\infty} \frac{4}{2^n}$

58. $\displaystyle\sum_{n=0}^{\infty} \frac{1}{4^n}$

59. $\displaystyle\sum_{n=0}^{\infty} (1.075)^n$

60. $\displaystyle\sum_{n=1}^{\infty} \frac{2^n}{100}$

61. $\displaystyle\sum_{n=2}^{\infty} \frac{n}{\ln n}$

62. $\displaystyle\sum_{n=1}^{\infty} \left(1 + \frac{k}{n}\right)^n$

Getting at the Concept

63. State the definition of convergent and divergent series.

64. Describe the difference between $\displaystyle\lim_{n\to\infty} a_n = 5$ and $\displaystyle\sum_{n=1}^{\infty} a_n = 5$.

65. Define a geometric series, state when it converges, and give the formula for the sum of a convergent geometric series.

66. State the nth-Term Test for Divergence.

In Exercises 67 and 68, (a) find the common ratio of the geometric series, (b) write the function that gives the sum of the series, and (c) use a graphing utility to graph the function and the partial sum S_2.

67. $1 + x + x^2 + x^3 + \cdots$

68. $1 - \dfrac{x}{2} + \dfrac{x^2}{4} - \dfrac{x^3}{8} + \cdots$

In Exercises 69 and 70, use a graphing utility to graph the function. Identify the horizontal asymptote of the graph and determine its relationship to the sum of the series.

Function	Series
69. $f(x) = 3\left[\dfrac{1 - (0.5)^x}{1 - 0.5}\right]$	$\displaystyle\sum_{n=0}^{\infty} 3\left(\frac{1}{2}\right)^n$
70. $f(x) = 2\left[\dfrac{1 - (0.8)^x}{1 - 0.8}\right]$	$\displaystyle\sum_{n=0}^{\infty} 2\left(\frac{4}{5}\right)^n$

Writing **In Exercises 71 and 72, use a graphing utility to determine the first term that is less than 0.0001 in each of the convergent series. Note that the answers are very different. Explain how this will affect the rate at which the series converges.**

71. $\displaystyle\sum_{n=1}^{\infty} \frac{1}{n(n+1)}$, $\displaystyle\sum_{n=1}^{\infty} \left(\frac{1}{8}\right)^n$

72. $\displaystyle\sum_{n=1}^{\infty} \frac{1}{2^n}$, $\displaystyle\sum_{n=1}^{\infty} (0.01)^n$

73. *Marketing* A company producing a new product estimates the annual sales to be 8000 units. Each year 10% of the units that have been sold will become inoperative. So, 8000 units will be in use after 1 year, $[8000 + 0.9(8000)]$ units will be in use after 2 years, and so on. How many units will be in use after n years?

74. *Depreciation* A company buys a machine for \$225,000 that depreciates at a rate of 30% per year. Find a formula for the value of the machine after n years. What is its value after 5 years?

75. *Multiplier Effect* The annual spending by tourists in a resort city is \$100 million. Approximately 75% of that revenue is again spent in the resort city, and of that amount approximately 75% is again spent in the same city, and so on. Write the geometric series that gives the total amount of spending generated by the \$100 million and find the sum of the series.

76. *Multiplier Effect* Repeat Exercise 75 if the percent of the revenue that is spent again in the city decreases to 60%.

77. *Distance* A ball is dropped from a height of 16 feet. Each time it drops h feet, it rebounds $0.81h$ feet. Find the total distance traveled by the ball.

78. *Time* The ball in Exercise 77 takes the following times for each fall.

$s_1 = -16t^2 + 16,$	$s_1 = 0$ if $t = 1$
$s_2 = -16t^2 + 16(0.81),$	$s_2 = 0$ if $t = 0.9$
$s_3 = -16t^2 + 16(0.81)^2,$	$s_3 = 0$ if $t = (0.9)^2$
$s_4 = -16t^2 + 16(0.81)^3,$	$s_4 = 0$ if $t = (0.9)^3$
$\vdots$	$\vdots$
$s_n = -16t^2 + 16(0.81)^{n-1},$	$s_n = 0$ if $t = (0.9)^{n-1}$

Beginning with s_2, the ball takes the same amount of time to bounce up as it does to fall, and thus the total time elapsed before it comes to rest is

$$t = 1 + 2\sum_{n=1}^{\infty} (0.9)^n.$$

Find this total time.

Probability **In Exercises 79 and 80, the random variable n represents the number of units of a certain product sold per day in a store. The probability distribution of n is given by $P(n)$. Find the probability that two units are sold in a given day $[P(2)]$ and show that $P(1) + P(2) + P(3) + \cdots = 1$.**

79. $P(n) = \dfrac{1}{2}\left(\dfrac{1}{2}\right)^n$

80. $P(n) = \dfrac{1}{3}\left(\dfrac{2}{3}\right)^n$

81. *Probability* If a fair coin is tossed repeatedly, the probability that the first head occurs on the nth toss is given by $P(n) = \left(\frac{1}{2}\right)^n$, where $n \geq 1$.

(a) Show that $\displaystyle\sum_{n=1}^{\infty} \left(\frac{1}{2}\right)^n = 1$.

(b) The expected number of tosses required until the first head occurs in the experiment is given by

$$\sum_{n=1}^{\infty} n\left(\frac{1}{2}\right)^n.$$

Is this series geometric?

(c) Use a computer algebra system to find the sum in part (b).

82. *Area* The sides of a square are 16 inches in length. A new square is formed by connecting the midpoints of the sides of the original square, and two of the triangles outside the second square are shaded (see figure). Determine the area of the shaded region (a) if this process is continued five more times and (b) if this pattern of shading is continued infinitely.

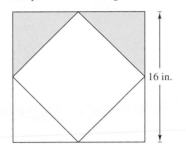

16 in.

In Exercises 83–86, use the formula for the *n*th partial sum of a geometric series

$$\sum_{i=0}^{n-1} ar^i = \frac{a(1-r^n)}{1-r}.$$

83. *Present Value* The winner of a $1,000,000 sweepstakes will be paid $50,000 per year for 20 years. If the money earns 6% interest per year, the present value of the winnings is

$$\sum_{n=1}^{19} 50,000\left(\frac{1}{1.06}\right)^n.$$

Compute the present value and interpret its meaning.

84. *Sphereflake* The sphereflake shown on page 554 is a computer-generated fractal that was created by Eric Haines, 3D/Eye Inc. The radius of the large sphere is 1. To the large sphere, nine spheres of radius $\frac{1}{3}$ are attached. To each of these, nine spheres of radius $\frac{1}{9}$ are attached. This process is continued infinitely. Prove that the sphereflake has an infinite surface area.

85. *Income* Suppose you go to work at a company that pays $0.01 for the first day, $0.02 for the second day, $0.04 for the third day, and so on. If the daily wage keeps doubling, what would your total income be for working (a) 29 days, (b) 30 days, and (c) 31 days?

86. *Annuities* When an employee receives a paycheck at the end of each month, P dollars is invested in a retirement account. These deposits are made each month for t years and the account earns interest at the annual percentage rate r. If the interest is compounded monthly, the amount A in the account at the end of t years is

$$A = P + P\left(1 + \frac{r}{12}\right) + \cdots + P\left(1 + \frac{r}{12}\right)^{12t-1}$$

$$= P\left(\frac{12}{r}\right)\left[\left(1 + \frac{r}{12}\right)^{12t} - 1\right].$$

If the interest is compounded continuously, the amount A in the account after t years is

$$A = P + Pe^{r/12} + Pe^{2r/12} + \cdots + Pe^{(12t-1)r/12}$$

$$= \frac{P(e^{rt} - 1)}{e^{r/12} - 1}.$$

Verify the formulas for the sums given above.

Annuities In Exercises 87–90, consider making monthly deposits of P dollars in a savings account at an annual interest rate r. Use the results of Exercise 86 to find the balance A after t years if the interest is compounded (a) monthly and (b) continuously.

87. $P = \$50$, $r = 3\%$, $t = 20$ years

88. $P = \$75$, $r = 5\%$, $t = 25$ years

89. $P = \$100$, $r = 4\%$, $t = 40$ years

90. $P = \$20$, $r = 6\%$, $t = 50$ years

91. *Modeling Data* The annual sales a_n (in millions of dollars) of H. J. Heinz Company from 1990 through 1999 are given below as ordered pairs of the form (n, a_n), where n is the year, with $n = 0$ corresponding to 1990. *(Source: 1999 H. J. Heinz Report)*

$(0, 6086)$, $(1, 6647)$, $(2, 6582)$, $(3, 7103)$, $(4, 7047)$,

$(5, 8087)$, $(6, 9112)$, $(7, 9357)$, $(8, 9209)$, $(9, 9300)$,

(a) Use the regression capabilities of a graphing utility to find a model of the form

$$a_n = ce^{kn}, \quad n = 0, 1, \ldots, 9$$

for the data. Graphically compare the points and the model.

(b) Use the data to find the total sales for the 10-year period.

(c) Approximate the total sales for the 10-year period using the formula for the sum of a geometric series. Compare the result with that in part (b).

92. *Salary* You accept a job that pays a salary of $40,000 for the first year. Suppose that during the next 39 years you receive a 4% raise each year. What would be your total compensation over the 40-year period?

93. Prove that $0.75 = 0.749999 \ldots$.

94. Prove that every decimal with a repeating pattern of digits is a rational number.

95. Show that the series

$$\sum_{n=1}^{\infty} a_n$$

can be written in the telescoping form

$$\sum_{n=1}^{\infty} [(c - S_{n-1}) - (c - S_n)]$$

where $S_0 = 0$ and S_n is the nth partial sum.

96. Let Σa_n be a convergent series, and let

$$R_N = a_{N+1} + a_{N+2} + \cdots$$

be the remainder of the series after the first N terms. Prove that

$$\lim_{N \to \infty} R_N = 0.$$

97. Find two divergent series Σa_n and Σb_n such that $\Sigma(a_n + b_n)$ converges.

98. Given two infinite series Σa_n and Σb_n such that Σa_n converges and Σb_n diverges, prove that $\Sigma(a_n + b_n)$ diverges.

True or False? **In Exercises 99–102, determine whether the statement is true or false. If it is false, explain why or give an example that shows it is false.**

99. If $\lim_{n \to \infty} a_n = 0$, then $\sum_{n=1}^{\infty} a_n$ converges.

100. If $\sum_{n=1}^{\infty} a_n = L$, then $\sum_{n=0}^{\infty} a_n = L + a_0$.

101. If $|r| < 1$, then $\sum_{n=1}^{\infty} ar^n = a/(1 - r)$.

102. The series $\sum_{n=1}^{\infty} \dfrac{n}{1000(n + 1)}$ diverges.

103. ***Writing*** Read the article "The Exponential-Decay Law Applied to Medical Dosages" by Gerald M. Armstrong and Calvin P. Midgley in *Mathematics Teacher*. (To view this article, go to the website *www.matharticles.com.*) Then write a paragraph on how a geometric sequence can be used to find the total amount of a drug that remains in a patient's system after n equal dosages have been administered (at equal time intervals).

104. Prove that

$$\frac{1}{r} + \frac{1}{r^2} + \frac{1}{r^3} + \cdots = \frac{1}{r - 1}$$

for $|r| > 1$.

SECTION PROJECT **CANTOR'S DISAPPEARING TABLE**

The following procedure shows how to make a table disappear by removing only half of the table!

(a) Original table has a length of L.

(b) Remove $\frac{1}{4}$ of the table centered at the midpoint. Each remaining piece has a length that is less than $\frac{1}{2}L$.

(c) Remove $\frac{1}{8}$ of the table by taking sections of length $\frac{1}{16}L$ from the centers of each of the two remaining pieces. Now, you have removed $\frac{1}{4} + \frac{1}{8}$ of the table. Each remaining piece has a length that is less than $\frac{1}{4}L$.

(d) Remove $\frac{1}{16}$ of the table by taking sections of length $\frac{1}{64}L$ from the centers of each of the four remaining pieces. Now, you have removed $\frac{1}{4} + \frac{1}{8} + \frac{1}{16}$ of the table. Each remaining piece has a length that is less than $\frac{1}{8}L$.

Will continuing this process cause the table to disappear, even though you have only removed half of the table? Why?

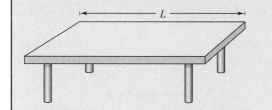

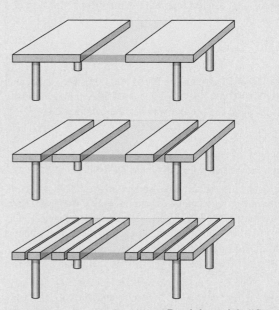

FOR FURTHER INFORMATION Read the article "Cantor's Disappearing Table" by Larry E. Knop in *The College Mathematics Journal*. To view this article, go to the website *www.matharticles.com.*

The Integral Test and p-Series

- Use the Integral Test to determine whether an infinite series converges or diverges.
- Use properties of p-series and harmonic series.

The Integral Test

In this and the following section, you will study several convergence tests that apply to series with *positive* terms.

THEOREM 8.10   The Integral Test

If f is positive, continuous, and decreasing for $x \geq 1$ and $a_n = f(n)$, then

$$\sum_{n=1}^{\infty} a_n \quad \text{and} \quad \int_1^{\infty} f(x)\, dx$$

either both converge or both diverge.

Proof   Begin by partitioning the interval $[1, n]$ into $n - 1$ unit intervals, as shown in Figure 8.8. The total areas of the inscribed rectangles and the circumscribed rectangles are as follows.

$$\sum_{i=2}^{n} f(i) = f(2) + f(3) + \cdots + f(n) \qquad \text{Inscribed area}$$

$$\sum_{i=1}^{n-1} f(i) = f(1) + f(2) + \cdots + f(n - 1) \qquad \text{Circumscribed area}$$

The exact area under the graph of f from $x = 1$ to $x = n$ lies between the inscribed and circumscribed areas.

$$\sum_{i=2}^{n} f(i) \leq \int_1^{n} f(x)\, dx \leq \sum_{i=1}^{n-1} f(i)$$

Using the nth partial sum, $S_n = f(1) + f(2) + \cdots + f(n)$, you can write this inequality as

$$S_n - f(1) \leq \int_1^{n} f(x)\, dx \leq S_{n-1}.$$

Now, assuming that $\int_1^{\infty} f(x)\, dx$ converges to L, it follows that for $n \geq 1$

$$S_n - f(1) \leq L \quad \implies \quad S_n \leq L + f(1).$$

Consequently, $\{S_n\}$ is bounded and monotonic, and by Theorem 8.5 it converges. So, $\Sigma\, a_n$ converges. For the other direction of the proof, assume that the improper integral diverges. Then $\int_1^{n} f(x)\, dx$ approaches infinity as $n \to \infty$, and the inequality $S_{n-1} \geq \int_1^{n} f(x)\, dx$ implies that $\{S_n\}$ diverges. So, $\Sigma\, a_n$ diverges.

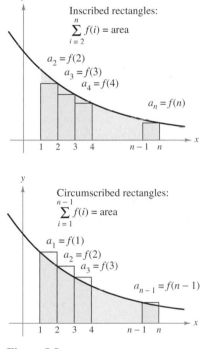

Inscribed rectangles:
$$\sum_{i=2}^{n} f(i) = \text{area}$$
$a_2 = f(2)$
$a_3 = f(3)$
$a_4 = f(4)$
$a_n = f(n)$

Circumscribed rectangles:
$$\sum_{i=1}^{n-1} f(i) = \text{area}$$
$a_1 = f(1)$
$a_2 = f(2)$
$a_3 = f(3)$
$a_{n-1} = f(n-1)$

Figure 8.8

NOTE   Remember that the convergence or divergence of $\Sigma\, a_n$ is not affected by deleting the first N terms. Similarly, if the conditions for the Integral Test are satisfied for all $x \geq N > 1$, you can simply use the integral $\int_N^{\infty} f(x)\, dx$ to test for convergence or divergence. (This is illustrated in Example 4.)

Example 1   **Using the Integral Test**

Apply the Integral Test to the series $\displaystyle\sum_{n=1}^{\infty} \frac{n}{n^2 + 1}$.

Solution   Because $f(x) = x/(x^2 + 1)$ satisfies the conditions for the Integral Test (check this), you can integrate to obtain

$$\int_1^{\infty} \frac{x}{x^2 + 1}\,dx = \frac{1}{2}\int_1^{\infty} \frac{2x}{x^2 + 1}\,dx$$

$$= \frac{1}{2} \lim_{b\to\infty} \int_1^b \frac{2x}{x^2 + 1}\,dx$$

$$= \frac{1}{2} \lim_{b\to\infty} \left[\ln(x^2 + 1)\right]_1^b$$

$$= \frac{1}{2} \lim_{b\to\infty} \left[\ln(b^2 + 1) - \ln 2\right]$$

$$= \infty.$$

So, the series *diverges*.

Example 2   **Using the Integral Test**

Apply the Integral Test to the series $\displaystyle\sum_{n=1}^{\infty} \frac{1}{n^2 + 1}$.

Solution   Because $f(x) = 1/(x^2 + 1)$ satisfies the conditions for the Integral Test, you can integrate to obtain

$$\int_1^{\infty} \frac{1}{x^2 + 1}\,dx = \lim_{b\to\infty} \int_1^b \frac{1}{x^2 + 1}\,dx$$

$$= \lim_{b\to\infty} \left[\arctan x\right]_1^b$$

$$= \lim_{b\to\infty} (\arctan b - \arctan 1)$$

$$= \frac{\pi}{2} - \frac{\pi}{4}$$

$$= \frac{\pi}{4}.$$

So, the series *converges* (see Figure 8.9).

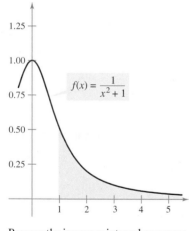

$f(x) = \dfrac{1}{x^2 + 1}$

Because the improper integral converges, the infinite series also converges.
Figure 8.9

TECHNOLOGY   In Example 2, the fact that the improper integral converges to $\pi/4$ does not imply that the infinite series converges to $\pi/4$. To approximate the sum of the series, you can use the inequality

$$\sum_{n=1}^{N} \frac{1}{n^2 + 1} \le \sum_{n=1}^{\infty} \frac{1}{n^2 + 1} \le \sum_{n=1}^{N} \frac{1}{n^2 + 1} + \int_N^{\infty} \frac{1}{x^2 + 1}\,dx.$$

(See Exercise 36.) The larger the value of N, the better the approximation. For instance, using $N = 200$ produces $1.072 \le \Sigma 1/(n^2 + 1) \le 1.077$.

p-Series and Harmonic Series

In the remainder of this section, we investigate a second type of series that has a simple arithmetic test for convergence or divergence. A series of the form

$$\sum_{n=1}^{\infty} \frac{1}{n^p} = \frac{1}{1^p} + \frac{1}{2^p} + \frac{1}{3^p} + \cdots \qquad \text{\textit{p}-series}$$

is a ***p*-series,** where p is a positive constant. For $p = 1$, the series

$$\sum_{n=1}^{\infty} \frac{1}{n} = 1 + \frac{1}{2} + \frac{1}{3} + \cdots \qquad \text{Harmonic series}$$

is the **harmonic series.** A **general harmonic series** is of the form $\Sigma 1/(an + b)$. In music, strings of the same material, diameter, and tension, whose lengths form a harmonic series, produce harmonic tones.

The Integral Test is convenient for establishing the convergence or divergence of *p*-series. This is shown in the proof of Theorem 8.11.

THEOREM 8.11   Convergence of *p*-Series

The *p*-series

$$\sum_{n=1}^{\infty} \frac{1}{n^p} = \frac{1}{1^p} + \frac{1}{2^p} + \frac{1}{3^p} + \frac{1}{4^p} + \cdots$$

1. converges if $p > 1$, and

2. diverges if $0 < p \leq 1$.

Proof The proof follows from the Integral Test and from Theorem 7.5, which states that

$$\int_{1}^{\infty} \frac{1}{x^p} \, dx$$

converges if $p > 1$ and diverges if $0 < p \leq 1$.

Example 3   **Convergent and Divergent *p*-Series**

Discuss the convergence or divergence of (a) the harmonic series and (b) the *p*-series with $p = 2$.

Solution

a. From Theorem 8.11, it follows that the harmonic series

$$\sum_{n=1}^{\infty} \frac{1}{n} = \frac{1}{1} + \frac{1}{2} + \frac{1}{3} + \cdots \qquad p = 1$$

diverges.

b. From Theorem 8.11, it follows that the *p*-series

$$\sum_{n=1}^{\infty} \frac{1}{n^2} = \frac{1}{1^2} + \frac{1}{2^2} + \frac{1}{3^2} + \cdots \qquad p = 2$$

converges.

NOTE The sum of the series in Example 3b can be shown to be $\pi^2/6$. (This was proved by Leonhard Euler, but the proof is too difficult to present here.) Be sure you see that the Integral Test does not tell you that the sum of the series is equal to the value of the integral. For instance, the sum of the series in Example 3b is

$$\sum_{n=1}^{\infty} \frac{1}{n^2} = \frac{\pi^2}{6} \approx 1.645$$

but the value of the corresponding improper integral is

$$\int_{1}^{\infty} \frac{1}{x^2} \, dx = 1.$$

Example 4   Testing a Series for Convergence

Determine whether the following series converges or diverges.

$$\sum_{n=2}^{\infty} \frac{1}{n \ln n}$$

Solution   This series is similar to the divergent harmonic series. If its terms were larger than those of the harmonic series, you would expect it to diverge. However, because its terms are smaller, you are not sure what to expect. Using the Integral Test with

$$f(x) = \frac{1}{x \ln x}$$

you can see that the series diverges.

$$\int_{2}^{\infty} \frac{1}{x \ln x} \, dx = \int_{2}^{\infty} \frac{1/x}{\ln x} \, dx$$

$$= \lim_{b \to \infty} \left[\ln(\ln x) \right]_{2}^{b}$$

$$= \lim_{b \to \infty} \left[\ln(\ln b) - \ln(\ln 2) \right]$$

$$= \infty$$

NOTE   The infinite series in Example 4 diverges very slowly. For instance, the sum of the first ten terms is approximately 1.6878196, whereas the sum of the first 100 terms is just slightly larger, 2.3250871. In fact, the sum of the first 10,000 terms is approximately 3.015021704. You can see that although the infinite series "adds up to infinity," it does so very slowly!

EXERCISES FOR SECTION 8.3

In Exercises 1–10, use the Integral Test to determine the convergence or divergence of the series.

1. $\displaystyle\sum_{n=1}^{\infty} \frac{1}{n+1}$

2. $\displaystyle\sum_{n=1}^{\infty} \frac{2}{3n+5}$

3. $\displaystyle\sum_{n=1}^{\infty} e^{-n}$

4. $\displaystyle\sum_{n=1}^{\infty} n e^{-n/2}$

5. $\dfrac{1}{2} + \dfrac{1}{5} + \dfrac{1}{10} + \dfrac{1}{17} + \dfrac{1}{26} + \cdots$

6. $\dfrac{1}{3} + \dfrac{1}{5} + \dfrac{1}{7} + \dfrac{1}{9} + \dfrac{1}{11} + \cdots$

7. $\dfrac{\ln 2}{2} + \dfrac{\ln 3}{3} + \dfrac{\ln 4}{4} + \dfrac{\ln 5}{5} + \dfrac{\ln 6}{6} + \cdots$

8. $\dfrac{1}{4} + \dfrac{2}{7} + \dfrac{3}{12} + \cdots + \dfrac{n}{n^2 + 3} + \cdots$

9. $\displaystyle\sum_{n=1}^{\infty} \frac{n^{k-1}}{n^k + c}$, k is a positive integer

10. $\displaystyle\sum_{n=1}^{\infty} n^k e^{-n}$, k is a positive integer

In Exercises 11 and 12, use the Integral Test to determine the convergence or divergence of the *p*-series.

11. $\displaystyle\sum_{n=1}^{\infty} \frac{1}{n^3}$

12. $\displaystyle\sum_{n=1}^{\infty} \frac{1}{n^{1/3}}$

In Exercises 13–20, use Theorem 8.11 to determine the convergence or divergence of the *p*-series.

13. $\displaystyle\sum_{n=1}^{\infty} \frac{1}{\sqrt[5]{n}}$

14. $\displaystyle\sum_{n=1}^{\infty} \frac{3}{n^{5/3}}$

15. $1 + \dfrac{1}{\sqrt{2}} + \dfrac{1}{\sqrt{3}} + \dfrac{1}{\sqrt{4}} + \cdots$

16. $1 + \dfrac{1}{4} + \dfrac{1}{9} + \dfrac{1}{16} + \dfrac{1}{25} + \cdots$

17. $1 + \dfrac{1}{2\sqrt{2}} + \dfrac{1}{3\sqrt{3}} + \dfrac{1}{4\sqrt{4}} + \dfrac{1}{5\sqrt{5}} + \cdots$

18. $1 + \dfrac{1}{\sqrt[3]{4}} + \dfrac{1}{\sqrt[3]{9}} + \dfrac{1}{\sqrt[3]{16}} + \dfrac{1}{\sqrt[3]{25}} + \cdots$

19. $\displaystyle\sum_{n=1}^{\infty} \frac{1}{n^{1.04}}$

20. $\displaystyle\sum_{n=1}^{\infty} \frac{1}{n^{\pi}}$

In Exercises 21–24, match the series with the graph of its sequence of partial sums. [The graphs are labeled (a), (b), (c), and (d).] Determine the convergence or divergence of the series.

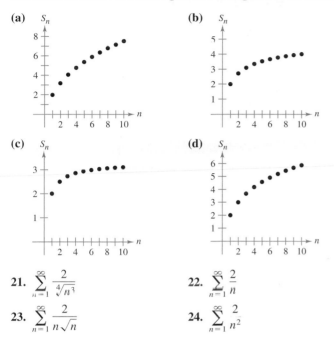

(a) S_n

(b) S_n

(c) S_n

(d) S_n

21. $\displaystyle\sum_{n=1}^{\infty} \frac{2}{\sqrt[4]{n^3}}$

22. $\displaystyle\sum_{n=1}^{\infty} \frac{2}{n}$

23. $\displaystyle\sum_{n=1}^{\infty} \frac{2}{n\sqrt{n}}$

24. $\displaystyle\sum_{n=1}^{\infty} \frac{2}{n^2}$

25. *Writing*   In Exercises 21–24, $\lim_{n\to\infty} a_n = 0$ for each series but they do not all converge. Is this a contradiction of Theorem 8.9? Why do you think some converge and others diverge?

26. *Numerical and Graphical Analysis*   (a) Use a graphing utility to find the indicated partial sum S_n and complete the table. (b) Use a graphing utility to graph the first ten terms of the sequence of partial sums. (c) Compare the rate at which the sequence of partial sums approaches the sum of the series for each series.

n	5	10	20	50	100
S_n					

(a) $\displaystyle\sum_{n=1}^{\infty} 3\left(\frac{1}{5}\right)^{n-1} = \frac{15}{4}$ (b) $\displaystyle\sum_{n=1}^{\infty} \frac{1}{n^2} = \frac{\pi^2}{6}$

27. *Numerical Reasoning*   Because the harmonic series diverges, it follows that for any positive real number M there exists a positive integer N such that the partial sum

$$\sum_{n=1}^{N} \frac{1}{n} > M.$$

(a) Use a graphing utility to complete the table.

M	2	4	6	8
N				

(b) As the real number M increases in equal increments, does the number N increase in equal increments? Explain.

28. The **Riemann zeta function** for real numbers is defined for all x for which the series

$$\zeta(x) = \sum_{n=1}^{\infty} n^{-x}$$

converges. Find the domain of the function.

In Exercises 29 and 30, find the positive values of p for which the series converges.

29. $\displaystyle\sum_{n=2}^{\infty} \frac{1}{n(\ln n)^p}$

30. $\displaystyle\sum_{n=2}^{\infty} \frac{\ln n}{n^p}$

Getting at the Concept

31. State the Integral Test and give an example of its use.

32. Define a *p*-series and state the requirements for its convergence.

33. A friend in your calculus class tells you that the following series converges because the terms are very small and approach 0 rapidly. Is your friend correct? Explain.

$$\frac{1}{10,000} + \frac{1}{10,001} + \frac{1}{10,002} + \cdots$$

34. Find a series such that the *n*th term goes to 0, but the series diverges.

35. Let f be a positive, continuous, and decreasing function for $x \ge 1$, such that $a_n = f(n)$. Prove that if the series

$$\sum_{n=1}^{\infty} a_n$$

converges to S, then the remainder $R_N = S - S_N$ is bounded by

$$0 \le R_N \le \int_N^{\infty} f(x)\, dx.$$

36. Show that the result of Exercise 35 can be written as

$$\sum_{n=1}^{N} a_n \le \sum_{n=1}^{\infty} a_n \le \sum_{n=1}^{N} a_n + \int_N^{\infty} f(x)\, dx.$$

In Exercises 37–42, use the result of Exercise 35 to approximate the sum of the convergent series using the indicated number of terms. Include an estimate of the maximum error for your approximation.

37. $\displaystyle\sum_{n=1}^{\infty} \frac{1}{n^4}$

Six terms

38. $\displaystyle\sum_{n=1}^{\infty} \frac{1}{n^5}$

Four terms

39. $\displaystyle\sum_{n=1}^{\infty} \frac{1}{n^2 + 1}$

Ten terms

40. $\displaystyle\sum_{n=1}^{\infty} \frac{1}{(n+1)[\ln(n+1)]^3}$

Ten terms

41. $\displaystyle\sum_{n=1}^{\infty} n e^{-n^2}$

Four terms

42. $\displaystyle\sum_{n=1}^{\infty} e^{-n}$

Four terms

In Exercises 43–48, use the result of Exercise 35 to find N such that $R_N \le 0.001$ for the convergent series.

43. $\displaystyle\sum_{n=1}^{\infty} \frac{1}{n^4}$

44. $\displaystyle\sum_{n=1}^{\infty} \frac{1}{n^{3/2}}$

45. $\displaystyle\sum_{n=1}^{\infty} e^{-5n}$

46. $\displaystyle\sum_{n=1}^{\infty} e^{-n/2}$

47. $\displaystyle\sum_{n=1}^{\infty} \frac{1}{n^2 + 1}$

48. $\displaystyle\sum_{n=1}^{\infty} \frac{2}{n^2 + 5}$

49. (a) Show that $\displaystyle\sum_{n=2}^{\infty} \frac{1}{n^{1.1}}$ converges and $\displaystyle\sum_{n=2}^{\infty} \frac{1}{n \ln n}$ diverges.

(b) Compare the first five terms of each series in part (a).

(c) Find $n > 3$ such that

$$\frac{1}{n^{1.1}} < \frac{1}{n \ln n}.$$

50. Ten terms are used to approximate a convergent p-series. Therefore, the remainder is a function of p and is

$$0 \le R_{10}(p) \le \int_{10}^{\infty} \frac{1}{x^p}\, dx, \qquad p > 1.$$

(a) Perform the integration in the inequality.

(b) Use a graphing utility to represent the inequality graphically.

(c) Identify any asymptotes of the error function and interpret their meaning.

51. *Euler's Constant* Let

$$S_n = \sum_{k=1}^{n} \frac{1}{k} = 1 + \frac{1}{2} + \cdots + \frac{1}{n}.$$

(a) Show that $\ln(n+1) \le S_n \le 1 + \ln n$.

(b) Show that the sequence $\{a_n\} = \{S_n - \ln n\}$ is bounded.

(c) Show that the sequence $\{a_n\}$ is decreasing.

(d) Show that a_n converges to a limit γ (called Euler's constant).

(e) Approximate γ using a_{100}.

52. Find the sum of the series $\displaystyle\sum_{n=2}^{\infty} \ln\left(1 - \frac{1}{n^2}\right)$.

Review **In Exercises 53–64, determine the convergence or divergence of the series.**

53. $\displaystyle\sum_{n=1}^{\infty} \frac{1}{2n - 1}$

54. $\displaystyle\sum_{n=2}^{\infty} \frac{1}{n\sqrt{n^2 - 1}}$

55. $\displaystyle\sum_{n=1}^{\infty} \frac{1}{n\sqrt[4]{n}}$

56. $3\displaystyle\sum_{n=1}^{\infty} \frac{1}{n^{0.95}}$

57. $\displaystyle\sum_{n=0}^{\infty} \left(\frac{2}{3}\right)^n$

58. $\displaystyle\sum_{n=0}^{\infty} (1.075)^n$

59. $\displaystyle\sum_{n=1}^{\infty} \frac{n}{\sqrt{n^2 + 1}}$

60. $\displaystyle\sum_{n=1}^{\infty} \left(\frac{1}{n^2} - \frac{1}{n^3}\right)$

61. $\displaystyle\sum_{n=1}^{\infty} \left(1 + \frac{1}{n}\right)^n$

62. $\displaystyle\sum_{n=2}^{\infty} \ln n$

63. $\displaystyle\sum_{n=2}^{\infty} \frac{1}{n(\ln n)^3}$

64. $\displaystyle\sum_{n=2}^{\infty} \frac{\ln n}{n^3}$

SECTION PROJECT **THE HARMONIC SERIES**

The harmonic series

$$\sum_{n=1}^{\infty} \frac{1}{n} = 1 + \frac{1}{2} + \frac{1}{3} + \frac{1}{4} + \cdots + \frac{1}{n} + \cdots$$

is one of the most important series in this chapter. Even though its terms tend to zero as n increases,

$$\lim_{n\to\infty} \frac{1}{n} = 0,$$

the harmonic series diverges. In other words, even though the terms are getting smaller and smaller, the sum "adds up to infinity."

(a) One way to show that the harmonic series diverges is attributed to J. Bernoulli. He grouped the terms of the harmonic series as follows:

$$1 + \underbrace{\frac{1}{2}}_{} + \underbrace{\frac{1}{3} + \frac{1}{4}}_{> \frac{1}{2}} + \underbrace{\frac{1}{5} + \cdots + \frac{1}{8}}_{> \frac{1}{2}} + \underbrace{\frac{1}{9} + \cdots + \frac{1}{16}}_{> \frac{1}{2}} +$$

$$\underbrace{\frac{1}{17} + \cdots + \frac{1}{32}}_{> \frac{1}{2}} + \cdots$$

Write a short paragraph explaining how you can use this grouping to show that the harmonic series diverges.

(b) Use the proof of the Integral Test, Theorem 8.10, to show that

$$\ln(n+1) \le 1 + \frac{1}{2} + \frac{1}{3} + \frac{1}{4} + \cdots + \frac{1}{n} \le 1 + \ln n.$$

(c) Use part (b) to determine how many terms M you would need so that

$$\sum_{n=1}^{M} \frac{1}{n} > 50.$$

(d) Show that the sum of the first million terms of the harmonic series is less than 15.

(e) Show that the following inequalities are valid.

$$\ln \frac{21}{10} \le \frac{1}{10} + \frac{1}{11} + \cdots + \frac{1}{20} \le \ln \frac{20}{9}$$

$$\ln \frac{201}{100} \le \frac{1}{100} + \frac{1}{101} + \cdots + \frac{1}{200} \le \ln \frac{200}{99}$$

(f) Use the ideas in part (e) to find the limit

$$\lim_{m\to\infty} \sum_{n=m}^{2m} \frac{1}{n}.$$

- Use the Direct Comparison Test to determine whether a series converges or diverges.
- Use the Limit Comparison Test to determine whether a series converges or diverges.

Direct Comparison Test

For the convergence tests developed so far, the terms of the series had to be fairly simple and the series had to have special characteristics in order for the convergence tests to be applied. A slight deviation from these special characteristics can make a test nonapplicable. For example, in the following pairs, the second series cannot be tested by the same convergence test as the first series even though it is similar to the first.

1. $\displaystyle\sum_{n=0}^{\infty} \frac{1}{2^n}$ is geometric, but $\displaystyle\sum_{n=0}^{\infty} \frac{n}{2^n}$ is not.

2. $\displaystyle\sum_{n=1}^{\infty} \frac{1}{n^3}$ is a p-series, but $\displaystyle\sum_{n=1}^{\infty} \frac{1}{n^3 + 1}$ is not.

3. $a_n = \dfrac{n}{(n^2 + 3)^2}$ is easily integrated, but $b_n = \dfrac{n^2}{(n^2 + 3)^2}$ is not.

In this section you will study two additional tests for positive-term series. These two tests greatly expand the variety of series you are able to test for convergence or divergence. They allow you to *compare* a series having complicated terms with a simpler series whose convergence or divergence is known.

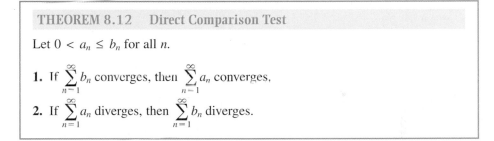

THEOREM 8.12   Direct Comparison Test

Let $0 < a_n \leq b_n$ for all n.

1. If $\displaystyle\sum_{n=1}^{\infty} b_n$ converges, then $\displaystyle\sum_{n=1}^{\infty} a_n$ converges.

2. If $\displaystyle\sum_{n=1}^{\infty} a_n$ diverges, then $\displaystyle\sum_{n=1}^{\infty} b_n$ diverges.

Proof   To prove the first property, let $L = \displaystyle\sum_{n=1}^{\infty} b_n$ and let

$$S_n = a_1 + a_2 + \cdots + a_n.$$

Because $0 < a_n \leq b_n$, the sequence $S_1, S_2, S_3, \ldots$ is nondecreasing and bounded above by L; so, it must converge. Because

$$\lim_{n \to \infty} S_n = \sum_{n=1}^{\infty} a_n$$

it follows that $\Sigma \, a_n$ converges. The second property is logically equivalent to the first.

NOTE   As stated, the Direct Comparison Test requires that $0 < a_n \leq b_n$ for all n. Because the convergence of a series is not dependent on its first several terms, you could modify the test to require only that $0 < a_n \leq b_n$ for all n greater than some integer N.

Example 1   **Using the Direct Comparison Test**

Determine the convergence or divergence of

$$\sum_{n=1}^{\infty} \frac{1}{2 + 3^n}.$$

Solution   This series resembles

$$\sum_{n=1}^{\infty} \frac{1}{3^n}.$$ Convergent geometric series

Term-by-term comparison yields

$$a_n = \frac{1}{2 + 3^n} < \frac{1}{3^n} = b_n, \qquad n \geq 1.$$

So, by the Direct Comparison Test, the series converges.

 Example 2   **Using the Direct Comparison Test**

Determine the convergence or divergence of

$$\sum_{n=1}^{\infty} \frac{1}{2 + \sqrt{n}}.$$

Solution   This series resembles

$$\sum_{n=1}^{\infty} \frac{1}{n^{1/2}}.$$ Divergent *p*-series

Term-by-term comparison yields

$$\frac{1}{2 + \sqrt{n}} \leq \frac{1}{\sqrt{n}}, \qquad n \geq 1$$

which *does not* meet the requirements for divergence. (Remember that if term-by-term comparison reveals a series that is *smaller* than a divergent series, the Direct Comparison Test tells you nothing.) Still expecting the series to diverge, you can compare the given series with

$$\sum_{n=1}^{\infty} \frac{1}{n}.$$ Divergent harmonic series

In this case, term-by-term comparison yields

$$a_n = \frac{1}{n} \leq \frac{1}{2 + \sqrt{n}} = b_n, \qquad n \geq 4$$

and, by the Direct Comparison Test, the given series diverges.

NOTE   To verify the last inequality in Example 2, try showing that $2 + \sqrt{n} \leq n$ whenever $n \geq 4$.

Remember that both parts of the Direct Comparison Test require that $0 < a_n \leq b_n$. Informally, the test says the following about the two series with nonnegative terms.

1. If the "larger" series converges, the "smaller" series must also converge.

2. If the "smaller" series diverges, the "larger" series must also diverge.

Limit Comparison Test

Often a given series closely resembles a p-series or a geometric series, yet you cannot establish the term-by-term comparison necessary to apply the Direct Comparison Test. Under these circumstances you may be able to apply a second comparison test, called the **Limit Comparison Test.**

THEOREM 8.13   Limit Comparison Test

Suppose that $a_n > 0$, $b_n > 0$, and

$$\lim_{n \to \infty} \left(\frac{a_n}{b_n} \right) = L$$

where L is *finite and positive*. Then the two series $\Sigma \, a_n$ and $\Sigma \, b_n$ either both converge or both diverge.

NOTE As with the Direct Comparison Test, the Limit Comparison Test could be modified to require only that a_n and b_n be positive for all n greater than some integer N.

Proof Because $a_n > 0$, $b_n > 0$, and

$$\lim_{n \to \infty} \frac{a_n}{b_n} = L$$

there exists $N > 0$ such that

$$0 < \frac{a_n}{b_n} < L + 1, \quad \text{for } n \geq N.$$

This implies that

$$0 < a_n < (L + 1)b_n.$$

Hence, by the Direct Comparison Test, the convergence of $\Sigma \, b_n$ implies the convergence of $\Sigma \, a_n$. Similarly, the fact that

$$\lim_{n \to \infty} \left(\frac{b_n}{a_n} \right) = \frac{1}{L}$$

can be used to show that the convergence of $\Sigma \, a_n$ implies the convergence of $\Sigma \, b_n$.

Example 3   **Using the Limit Comparison Test**

Show that the following general harmonic series diverges.

$$\sum_{n=1}^{\infty} \frac{1}{an + b}, \qquad a > 0, \qquad b > 0$$

Solution By comparison with

$$\sum_{n=1}^{\infty} \frac{1}{n} \qquad \text{Divergent harmonic series}$$

you have

$$\lim_{n \to \infty} \frac{1/(an + b)}{1/n} = \lim_{n \to \infty} \frac{n}{an + b} = \frac{1}{a}.$$

Because this limit is greater than 0, you can conclude from the Limit Comparison Test that the given series diverges.

The Limit Comparison Test works well for comparing a "messy" algebraic series with a p-series. In choosing an appropriate p-series, you must choose one with an nth term of the same magnitude as the nth term of the given series.

Given Series	Comparison Series	Conclusion
$\displaystyle\sum_{n=1}^{\infty} \frac{1}{3n^2 - 4n + 5}$	$\displaystyle\sum_{n=1}^{\infty} \frac{1}{n^2}$	Both series converge.
$\displaystyle\sum_{n=1}^{\infty} \frac{1}{\sqrt{3n - 2}}$	$\displaystyle\sum_{n=1}^{\infty} \frac{1}{\sqrt{n}}$	Both series diverge.
$\displaystyle\sum_{n=1}^{\infty} \frac{n^2 - 10}{4n^5 + n^3}$	$\displaystyle\sum_{n=1}^{\infty} \frac{n^2}{n^5} = \sum_{n=1}^{\infty} \frac{1}{n^3}$	Both series converge.

In other words, when choosing a series for comparison, you can disregard all but the *highest powers of n* in both the numerator and the denominator.

Example 4 Using the Limit Comparison Test

Determine the convergence or divergence of

$$\sum_{n=1}^{\infty} \frac{\sqrt{n}}{n^2 + 1}.$$

Solution Disregarding all but the highest powers of n in the numerator and the denominator, you can compare the series with

$$\sum_{n=1}^{\infty} \frac{\sqrt{n}}{n^2} = \sum_{n=1}^{\infty} \frac{1}{n^{3/2}}. \qquad \text{Convergent } p\text{-series}$$

Because

$$\lim_{n\to\infty} \frac{a_n}{b_n} = \lim_{n\to\infty} \left(\frac{\sqrt{n}}{n^2 + 1}\right)\left(\frac{n^{3/2}}{1}\right)$$

$$= \lim_{n\to\infty} \frac{n^2}{n^2 + 1} = 1$$

you can conclude by the Limit Comparison Test that the given series converges.

Example 5 Using the Limit Comparison Test

Determine the convergence or divergence of

$$\sum_{n=1}^{\infty} \frac{n2^n}{4n^3 + 1}.$$

Solution A reasonable comparison would be with the series

$$\sum_{n=1}^{\infty} \frac{2^n}{n^2}. \qquad \text{Divergent series}$$

Note that this series diverges by the nth-Term Test. From the limit

$$\lim_{n\to\infty} \frac{a_n}{b_n} = \lim_{n\to\infty} \left(\frac{n2^n}{4n^3 + 1}\right)\left(\frac{n^2}{2^n}\right)$$

$$= \lim_{n\to\infty} \frac{1}{4 + (1/n^3)} = \frac{1}{4}$$

you can conclude that the given series diverges.

EXERCISES FOR SECTION 8.4

1. Graphical Analysis   The figures show the graphs of the first ten terms, and the graphs of the first ten terms of the sequence of partial sums, of each series.

$$\sum_{n=1}^{\infty} \frac{6}{n^{3/2}}, \quad \sum_{n=1}^{\infty} \frac{6}{n^{3/2}+3}, \quad \text{and} \quad \sum_{n=1}^{\infty} \frac{6}{n\sqrt{n^2+0.5}}$$

(a) Identify the series in each figure.

(b) Which series is a *p*-series? Does it converge or diverge?

(c) For the series that are not *p*-series, how do the magnitudes of the terms compare with the magnitudes of the terms of the *p*-series? What conclusion can you draw about the convergence or divergence of the series?

(d) Explain the relationship between the magnitudes of the terms of the series and the magnitudes of the terms of the partial sums.

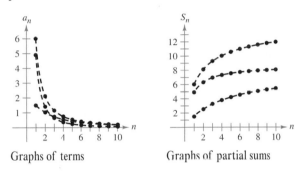

Graphs of terms   Graphs of partial sums

2. Graphical Analysis   The figures show the graphs of the first ten terms, and the graphs of the first ten terms of the sequence of partial sums, of each series.

$$\sum_{n=1}^{\infty} \frac{2}{\sqrt{n}}, \quad \sum_{n=1}^{\infty} \frac{2}{\sqrt{n}-0.5}, \quad \text{and} \quad \sum_{n=1}^{\infty} \frac{4}{\sqrt{n}+0.5}$$

(a) Identify the series in each figure.

(b) Which series is a *p*-series? Does it converge or diverge?

(c) For the series that are not *p*-series, how do the magnitudes of the terms compare with the magnitudes of the terms of the *p*-series? What conclusion can you draw about the convergence or divergence of the series?

(d) Explain the relationship between the magnitudes of the terms of the series and the magnitudes of the terms of the partial sums.

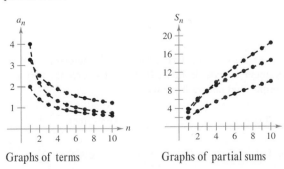

Graphs of terms   Graphs of partial sums

In Exercises 3–14, use the Direct Comparison Test to determine the convergence or divergence of the series.

3. $\displaystyle\sum_{n=1}^{\infty} \frac{1}{n^2+1}$

4. $\displaystyle\sum_{n=1}^{\infty} \frac{1}{3n^2+2}$

5. $\displaystyle\sum_{n=2}^{\infty} \frac{1}{n-1}$

6. $\displaystyle\sum_{n=2}^{\infty} \frac{1}{\sqrt{n}-1}$

7. $\displaystyle\sum_{n=0}^{\infty} \frac{1}{3^n+1}$

8. $\displaystyle\sum_{n=0}^{\infty} \frac{3^n}{4^n+5}$

9. $\displaystyle\sum_{n=2}^{\infty} \frac{\ln n}{n+1}$

10. $\displaystyle\sum_{n=1}^{\infty} \frac{1}{\sqrt{n^3+1}}$

11. $\displaystyle\sum_{n=0}^{\infty} \frac{1}{n!}$

12. $\displaystyle\sum_{n=1}^{\infty} \frac{1}{4\sqrt[3]{n}-1}$

13. $\displaystyle\sum_{n=0}^{\infty} e^{-n^2}$

14. $\displaystyle\sum_{n=1}^{\infty} \frac{4^n}{3^n-1}$

In Exercises 15–28, use the Limit Comparison Test to determine the convergence or divergence of the series.

15. $\displaystyle\sum_{n=1}^{\infty} \frac{n}{n^2+1}$

16. $\displaystyle\sum_{n=1}^{\infty} \frac{2}{3^n-5}$

17. $\displaystyle\sum_{n=0}^{\infty} \frac{1}{\sqrt{n^2+1}}$

18. $\displaystyle\sum_{n=3}^{\infty} \frac{3}{\sqrt{n^2-4}}$

19. $\displaystyle\sum_{n=1}^{\infty} \frac{2n^2-1}{3n^5+2n+1}$

20. $\displaystyle\sum_{n=1}^{\infty} \frac{5n-3}{n^2-2n+5}$

21. $\displaystyle\sum_{n=1}^{\infty} \frac{n+3}{n(n+2)}$

22. $\displaystyle\sum_{n=1}^{\infty} \frac{1}{n(n^2+1)}$

23. $\displaystyle\sum_{n=1}^{\infty} \frac{1}{n\sqrt{n^2+1}}$

24. $\displaystyle\sum_{n=1}^{\infty} \frac{n}{(n+1)2^{n-1}}$

25. $\displaystyle\sum_{n=1}^{\infty} \frac{n^{k-1}}{n^k+1}, \quad k > 2$

26. $\displaystyle\sum_{n=1}^{\infty} \frac{5}{n+\sqrt{n^2+4}}$

27. $\displaystyle\sum_{n=1}^{\infty} \sin\frac{1}{n}$

28. $\displaystyle\sum_{n=1}^{\infty} \tan\frac{1}{n}$

In Exercises 29–36, test for convergence or divergence, using each test at least once. Identify the test used.

(a) **nth-Term Test**   (b) **Geometric Series Test**

(c) **p-Series Test**   (d) **Telescoping Series Test**

(e) **Integral Test**   (f) **Direct Comparison Test**

(g) **Limit Comparison Test**

29. $\displaystyle\sum_{n=1}^{\infty} \frac{\sqrt{n}}{n}$

30. $\displaystyle\sum_{n=0}^{\infty} 5\left(-\frac{1}{5}\right)^n$

31. $\displaystyle\sum_{n=1}^{\infty} \frac{1}{3^n+2}$

32. $\displaystyle\sum_{n=4}^{\infty} \frac{1}{3n^2-2n-15}$

33. $\displaystyle\sum_{n=1}^{\infty} \frac{n}{2n+3}$

34. $\displaystyle\sum_{n=1}^{\infty} \left(\frac{1}{n+1}-\frac{1}{n+2}\right)$

35. $\displaystyle\sum_{n=1}^{\infty} \frac{n}{(n^2+1)^2}$

36. $\displaystyle\sum_{n=1}^{\infty} \frac{3}{n(n+3)}$

37. Use the Limit Comparison Test with the harmonic series to show that the series $\Sigma\, a_n$ (where $0 < a_n < a_{n-1}$) diverges if

$$\lim_{n\to\infty} na_n$$

is finite and nonzero.

38. Prove that, if $P(n)$ and $Q(n)$ are polynomials of degree j and k, respectively, then the series

$$\sum_{n=1}^{\infty} \frac{P(n)}{Q(n)}$$

converges if $j < k - 1$ and diverges if $j \geq k - 1$.

In Exercises 39–42, use the polynomial test given in Exercise 38 to determine whether the series converges or diverges.

39. $\frac{1}{2} + \frac{2}{5} + \frac{3}{10} + \frac{4}{17} + \frac{5}{26} + \cdots$

40. $\frac{1}{3} + \frac{1}{8} + \frac{1}{15} + \frac{1}{24} + \frac{1}{35} + \cdots$

41. $\displaystyle\sum_{n=1}^{\infty} \frac{1}{n^3 + 1}$

42. $\displaystyle\sum_{n=1}^{\infty} \frac{n^2}{n^3 + 1}$

In Exercises 43 and 44, use the divergence test given in Exercise 37 to show that the series diverges.

43. $\displaystyle\sum_{n=1}^{\infty} \frac{n^3}{5n^4 + 3}$

44. $\displaystyle\sum_{n=2}^{\infty} \frac{1}{\ln n}$

Getting at the Concept

45. State the Direct Comparison Test and give an example of its use.

46. State the Limit Comparison Test and give an example of its use.

47. The figure shows the first 20 terms of the convergent series

$$\sum_{n=1}^{\infty} a_n$$

and the first 20 terms of the series

$$\sum_{n=1}^{\infty} a_n^2.$$

Identify the two series and explain your reasoning in making the selection.

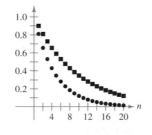

Getting at the Concept *(continued)*

48. It appears that the terms of the series

$$\frac{1}{1000} + \frac{1}{1001} + \frac{1}{1002} + \frac{1}{1003} + \cdots$$

are less than the corresponding terms of the convergent series

$$1 + \frac{1}{4} + \frac{1}{9} + \frac{1}{16} + \cdots.$$

If the statement above is correct, the first series converges. Is this correct? Why or why not? Make a statement about how the divergence or convergence of a series is affected by inclusion or exclusion of the first finite number of terms.

In Exercises 49–52, determine the convergence or divergence of the series.

49. $\frac{1}{200} + \frac{1}{400} + \frac{1}{600} + \frac{1}{800} + \cdots$

50. $\frac{1}{200} + \frac{1}{210} + \frac{1}{220} + \frac{1}{230} + \cdots$

51. $\frac{1}{201} + \frac{1}{204} + \frac{1}{209} + \frac{1}{216} + \cdots$

52. $\frac{1}{201} + \frac{1}{208} + \frac{1}{227} + \frac{1}{264} + \cdots$

53. *Think About It* Review the results of Exercises 49–52. Explain why careful analysis is required to determine the convergence or divergence of a series and why only considering the magnitudes of the terms of a series could be misleading.

54. Consider the following series and its sum.

$$\sum_{n=1}^{\infty} \frac{1}{(2n - 1)^2} = \frac{\pi^2}{8}$$

(a) Verify that the series converges.

(b) Use a graphing utility to complete the table.

n	5	10	20	50	100
S_n					

(c) Find the sum of the series

$$\sum_{n=3}^{\infty} \frac{1}{(2n - 1)^2}$$

by hand. Describe how you found the sum.

(d) Use a graphing utility to find the sum of the series

$$\sum_{n=10}^{\infty} \frac{1}{(2n - 1)^2}.$$

True or False? **In Exercises 55–58, determine whether the statement is true or false. If it is false, explain why or give an example that shows it is false.**

55. If $0 < a_n \leq b_n$ and $\displaystyle\sum_{n=1}^{\infty} a_n$ converges, then $\displaystyle\sum_{n=1}^{\infty} b_n$ diverges.

56. If $0 < a_{n+10} \leq b_n$ and $\displaystyle\sum_{n=1}^{\infty} b_n$ converges, then $\displaystyle\sum_{n=1}^{\infty} a_n$ converges.

57. If $a_n + b_n \le c_n$ and $\displaystyle\sum_{n=1}^{\infty} c_n$ converges, then the series $\displaystyle\sum_{n=1}^{\infty} a_n$ and $\displaystyle\sum_{n=1}^{\infty} b_n$ both converge. (Assume that the terms of all three series are positive.)

58. If $a_n \le b_n + c_n$ and $\displaystyle\sum_{n=1}^{\infty} a_n$ diverges, then the series $\displaystyle\sum_{n=1}^{\infty} b_n$ and $\displaystyle\sum_{n=1}^{\infty} c_n$ both diverge. (Assume that the terms of all three series are positive.)

59. Prove that if the nonnegative series

$$\sum_{n=1}^{\infty} a_n \qquad \text{and} \qquad \sum_{n=1}^{\infty} b_n$$

converge, then so does the series

$$\sum_{n=1}^{\infty} a_n b_n.$$

60. Use the result of Exercise 59 to prove that if the nonnegative series

$$\sum_{n=1}^{\infty} a_n$$

converges, then so does the series

$$\sum_{n=1}^{\infty} a_n^2.$$

61. Find two series that demonstrate the result of Exercise 59.

62. Find two series that demonstrate the result of Exercise 60.

63. Suppose that $\Sigma\, a_n$ and $\Sigma\, b_n$ are series with positive terms. Prove the following.

(a) If $\displaystyle\lim_{n \to \infty} \frac{a_n}{b_n} = 0$ and $\Sigma\, b_n$ converges, then $\Sigma\, a_n$ also converges.

(b) If $\displaystyle\lim_{n \to \infty} \frac{a_n}{b_n} = \infty$ and $\Sigma\, b_n$ diverges, then $\Sigma\, a_n$ also diverges.

64. Find two series that demonstrate the results of Exercise 63.

65. *Investigation*   Consider an equilateral triangle with sides of length 9. Center equilateral triangles with sides of length 3 on each side of the first triangle. Center equilateral triangles with sides of length 1 on each side of the second set of triangles. Continue this process of centering equilateral triangles on the previous set of triangles where the length of the sides of each set is $\frac{1}{3}$ that of the previous set. This forms the **Koch snowflake** as described on page 554. Use infinite series to find (if possible) the area and the perimeter of the figure below.

SECTION PROJECT   SOLERA METHOD

Most wines are produced entirely from grapes grown in a single year. Sherry, however, is a complex mixture of older wines with new wines. This is done with a sequence of barrels (called a solera) stacked on top of each other, as shown in the photo.

Everton/The Image Works

The oldest wine is in the bottom tier of barrels, and the newest is in the top tier. Each year, half of each barrel in the bottom tier is bottled as sherry. The bottom barrels are then refilled with the

wine from the barrels above. This process is repeated throughout the solera, with new wine being added to the top barrels. A mathematical model for the amount of n-year-old wine that is removed from a solera (with k tiers) each year is

$$f(n, k) = \left(\frac{n-1}{k-1}\right)\left(\frac{1}{2}\right)^{n+1}, \qquad k \le n.$$

(a) Consider a solera that has five tiers, numbered $k = 1, 2, 3, 4,$ and 5. In 1980 ($n = 0$), half of each barrel in the top tier (tier 1) was refilled with new wine. How much of this wine was removed from the solera in 1981? In 1982? In 1983? . . . In 1995? During which year(s) was the greatest amount of the 1980 wine removed from the solera?

(b) In part (a), let a_n be the amount of 1980 wine that is removed from the solera in year n. Evaluate

$$\sum_{n=0}^{\infty} a_n.$$

FOR FURTHER INFORMATION See the article "Finding Vintage Concentrations in a Sherry Solera" by Rhodes Peele and John T. MacQueen in the *UMAP Modules*. To view this article, go to the website *www.matharticles.com*.

- Use the Alternating Series Test to determine whether an infinite series converges.
- Use the Alternating Series Remainder to approximate the sum of an alternating series.
- Classify a convergent series as absolutely or conditionally convergent.
- Rearrange an infinite series to obtain a different sum.

Alternating Series

So far, most series we have dealt with have had positive terms. In this section and the following section, you will study series that contain both positive and negative terms. The simplest such series is an **alternating series,** whose terms alternate in sign. For example, the geometric series

$$\sum_{n=0}^{\infty} \left(-\frac{1}{2}\right)^n = \sum_{n=0}^{\infty} (-1)^n \frac{1}{2^n}$$

$$= 1 - \frac{1}{2} + \frac{1}{4} - \frac{1}{8} + \frac{1}{16} - \cdots$$

is an *alternating geometric series* with $r = -\frac{1}{2}$. Alternating series occur in two ways: either the odd terms are negative or the even terms are negative.

THEOREM 8.14 Alternating Series Test

Let $a_n > 0$. The alternating series

$$\sum_{n=1}^{\infty} (-1)^n a_n \quad \text{and} \quad \sum_{n=1}^{\infty} (-1)^{n+1} a_n$$

converge if the following two conditions are met.

1. $\lim_{n \to \infty} a_n = 0$ **2.** $a_{n+1} \leq a_n$, for all n

---

Proof Consider the alternating series $\sum (-1)^{n+1} a_n$. For this series, the partial sum (where $2n$ is even)

$$S_{2n} = (a_1 - a_2) + (a_3 - a_4) + (a_5 - a_6) + \cdots + (a_{2n-1} - a_{2n})$$

has all nonnegative terms, and therefore $\{S_{2n}\}$ is a nondecreasing sequence. But you can also write

$$S_{2n} = a_1 - (a_2 - a_3) - (a_4 - a_5) - \cdots - (a_{2n-2} - a_{2n-1}) - a_{2n}$$

which implies that $S_{2n} \leq a_1$ for every integer n. Thus $\{S_{2n}\}$ is a bounded, nondecreasing sequence that converges to some value L. Because $S_{2n-1} - a_{2n} = S_{2n}$ and $a_{2n} \to 0$, you have

$$\lim_{n \to \infty} S_{2n-1} = \lim_{n \to \infty} S_{2n} + \lim_{n \to \infty} a_{2n}$$

$$= L + \lim_{n \to \infty} a_{2n}$$

$$= L.$$

Because both S_{2n} and S_{2n-1} converge to the same limit L, it follows that $\{S_n\}$ also converges to L. Consequently, the given alternating series converges. ◼

NOTE The second condition in the Alternating Series Test can be modified to require only that $0 < a_{n+1} \leq a_n$ for all n greater than some integer N.

NOTE The series in Example 1 is called the *alternating harmonic series*—more is said about this series in Example 7.

Example 1 Using the Alternating Series Test

Determine the convergence or divergence of $\displaystyle\sum_{n=1}^{\infty} (-1)^{n+1} \frac{1}{n}$.

Solution Because

$$\frac{1}{n+1} \le \frac{1}{n}$$

for all n and the limit as $n \to \infty$ of $1/n$ is 0, you can apply the Alternating Series Test to conclude that the series converges.

Example 2 Using the Alternating Series Test

Determine the convergence or divergence of $\displaystyle\sum_{n=1}^{\infty} \frac{n}{(-2)^{n-1}}$.

Solution To apply the Alternating Series Test, note that, for $n \ge 1$,

$$\frac{1}{2} \le \frac{n}{n+1}$$

$$\frac{2^{n-1}}{2^n} \le \frac{n}{n+1}$$

$$(n+1)2^{n-1} \le n2^n$$

$$\frac{n+1}{2^n} \le \frac{n}{2^{n-1}}.$$

Hence, $a_{n+1} = (n+1)/2^n \le n/2^{n-1} = a_n$ for all n. Furthermore, by L'Hôpital's Rule,

$$\lim_{x \to \infty} \frac{x}{2^{x-1}} = \lim_{x \to \infty} \frac{1}{2^{x-1}(\ln 2)} = 0 \quad\Longrightarrow\quad \lim_{n \to \infty} \frac{n}{2^{n-1}} = 0.$$

Therefore, by the Alternating Series Test, the series converges.

Example 3 Cases for Which the Alternating Series Test Fails

NOTE In Example 3a, remember that whenever a series does not pass the first condition of the Alternating Series Test, you can use the nth-Term Test for Divergence to conclude that the series diverges.

a. The alternating series

$$\sum_{n=1}^{\infty} \frac{(-1)^{n+1}(n+1)}{n} = \frac{2}{1} - \frac{3}{2} + \frac{4}{3} - \frac{5}{4} + \frac{6}{5} - \cdots$$

passes the second condition of the Alternating Series Test because $a_{n+1} \le a_n$ for all n. You cannot apply the Alternating Series Test, however, because the series does not pass the first condition. In fact, the series diverges.

b. The alternating series

$$\frac{2}{1} - \frac{1}{1} + \frac{2}{2} - \frac{1}{2} + \frac{2}{3} - \frac{1}{3} + \frac{2}{4} - \frac{1}{4} + \cdots$$

passes the first condition because a_n approaches 0 as $n \to \infty$. You cannot apply the Alternating Series Test, however, because the series does not pass the second condition. To conclude that the series diverges, you can argue that S_{2N} equals the Nth partial sum of the divergent harmonic series. This implies that the sequence of partial sums diverges. Hence, the series diverges.

Alternating Series Remainder

For a convergent alternating series, the partial sum S_N can be a useful approximation for the sum S of the series. Just how close S_N is to S is stated in the following theorem.

> **THEOREM 8.15** **Alternating Series Remainder**
>
> If a convergent alternating series satisfies the condition $a_{n+1} \le a_n$, then the absolute value of the remainder R_N involved in approximating the sum S by S_N is less than (or equal to) the first neglected term. That is,
>
> $$|S - S_N| = |R_N| \le a_{N+1}.$$

Proof The series obtained by deleting the first N terms of the given series satisfies the conditions of the Alternating Series Test and has a sum of R_N.

$$
\begin{aligned}
R_N = S - S_N &= \sum_{n=1}^{\infty} (-1)^{n+1} a_n - \sum_{n=1}^{N} (-1)^{n+1} a_n \\
&= (-1)^N a_{N+1} + (-1)^{N+1} a_{N+2} + (-1)^{N+2} a_{N+3} + \cdots \\
&= (-1)^N (a_{N+1} - a_{N+2} + a_{N+3} - \cdots) \\
|R_N| &= a_{N+1} - a_{N+2} + a_{N+3} - a_{N+4} + a_{N+5} - \cdots \\
&= a_{N+1} - (a_{N+2} - a_{N+3}) - (a_{N+4} - a_{N+5}) - \cdots \le a_{N+1}
\end{aligned}
$$

Consequently, $|S - S_N| = |R_N| \le a_{N+1}$, which establishes the theorem.

🖙 *Example 4* **Approximating the Sum of an Alternating Series**

Approximate the sum of the following series by its first six terms.

$$\sum_{n=1}^{\infty} (-1)^{n+1} \left(\frac{1}{n!}\right) = \frac{1}{1!} - \frac{1}{2!} + \frac{1}{3!} - \frac{1}{4!} + \frac{1}{5!} - \frac{1}{6!} + \cdots$$

Solution The series converges by the Alternating Series Test because

$$\frac{1}{(n+1)!} \le \frac{1}{n!} \quad \text{and} \quad \lim_{n \to \infty} \frac{1}{n!} = 0.$$

The sum of the first six terms is

$$S_6 = 1 - \frac{1}{2} + \frac{1}{6} - \frac{1}{24} + \frac{1}{120} - \frac{1}{720} \approx 0.63194$$

and, by the Alternating Series Remainder, you have

$$|S - S_6| = |R_6| \le a_7 = \frac{1}{5040} \approx 0.0002.$$

Therefore, the sum S lies between $0.63194 - 0.0002$ and $0.63194 + 0.0002$, and you have

$$0.63174 \le S \le 0.63214.$$

TECHNOLOGY Later, in Section 8.10, you will be able to show that the series in Example 4 converges to

$$\frac{e - 1}{e} \approx 0.63212.$$

For now, try using a computer to obtain an approximation of the sum of the series. How many terms do you need to obtain an approximation that is within 0.00001 unit of the actual sum?

Absolute and Conditional Convergence

Occasionally, a series may have both positive and negative terms and not be an alternating series. For instance, the series

$$\sum_{n=1}^{\infty} \frac{\sin n}{n^2} = \frac{\sin 1}{1} + \frac{\sin 2}{4} + \frac{\sin 3}{9} + \cdots$$

has both positive and negative terms, yet it is not an alternating series. One way to obtain some information about the convergence of this series is to investigate the convergence of the series $\sum_{n=1}^{\infty} \left| \frac{\sin n}{n^2} \right|$. By direct comparison, you have $|\sin n| \leq 1$ for all n, so $\left| \frac{\sin n}{n^2} \right| \leq \frac{1}{n^2}$, $n \geq 1$. Therefore, by the Direct Comparison Test, the series $\sum \left| \frac{\sin n}{n^2} \right|$ converges. The next theorem tells you that the original series also converges. A proof is given in Appendix B.

THEOREM 8.16   Absolute Convergence

If the series $\sum |a_n|$ converges, then the series $\sum a_n$ also converges.

The converse of Theorem 8.16 is not true. For instance, the **alternating harmonic series**

$$\sum_{n=1}^{\infty} \frac{(-1)^{n+1}}{n} = \frac{1}{1} - \frac{1}{2} + \frac{1}{3} - \frac{1}{4} + \cdots$$

converges by the Alternating Series Test. Yet the harmonic series diverges. This type of convergence is called **conditional.**

Definition of Absolute and Conditional Convergence

1. $\sum a_n$ is **absolutely convergent** if $\sum |a_n|$ converges.
2. $\sum a_n$ is **conditionally convergent** if $\sum a_n$ converges but $\sum |a_n|$ diverges.

Example 5   Absolute and Conditional Convergence

Determine whether each of the series is convergent or divergent. Classify any convergent series as absolutely or conditionally convergent.

a. $\displaystyle\sum_{n=0}^{\infty} \frac{(-1)^n \, n!}{2^n} = \frac{0!}{2^0} - \frac{1!}{2^1} + \frac{2!}{2^2} - \frac{3!}{2^3} + \cdots$

b. $\displaystyle\sum_{n=1}^{\infty} \frac{(-1)^n}{\sqrt{n}} = -\frac{1}{\sqrt{1}} + \frac{1}{\sqrt{2}} - \frac{1}{\sqrt{3}} + \frac{1}{\sqrt{4}} - \cdots$

Solution

a. By the nth-Term Test for Divergence, you can conclude that this series diverges.

b. The given series can be shown to be convergent by the Alternating Series Test. Moreover, because the p-series

$$\sum_{n=1}^{\infty} \left| \frac{(-1)^n}{\sqrt{n}} \right| = \frac{1}{\sqrt{1}} + \frac{1}{\sqrt{2}} + \frac{1}{\sqrt{3}} + \frac{1}{\sqrt{4}} + \cdots$$

diverges, the given series is *conditionally* convergent.

Example 6   Absolute and Conditional Convergence

Determine whether each of the series is convergent or divergent. Classify any convergent series as absolutely or conditionally convergent.

a. $\displaystyle\sum_{n=1}^{\infty} \frac{(-1)^{n(n+1)/2}}{3^n} = -\frac{1}{3} - \frac{1}{9} + \frac{1}{27} + \frac{1}{81} - \cdots$

b. $\displaystyle\sum_{n=1}^{\infty} \frac{(-1)^n}{\ln(n+1)} = -\frac{1}{\ln 2} + \frac{1}{\ln 3} - \frac{1}{\ln 4} + \frac{1}{\ln 5} - \cdots$

Solution

a. This is *not* an alternating series. However, because

$$\sum_{n=1}^{\infty} \left| \frac{(-1)^{n(n+1)/2}}{3^n} \right| = \sum_{n=1}^{\infty} \frac{1}{3^n}$$

is a convergent geometric series, you can apply Theorem 8.16 to conclude that the given series is *absolutely* convergent (and hence convergent).

b. In this case, the Alternating Series Test indicates that the given series converges. However, the series

$$\sum_{n=1}^{\infty} \left| \frac{(-1)^n}{\ln(n+1)} \right| = \frac{1}{\ln 2} + \frac{1}{\ln 3} + \frac{1}{\ln 4} + \cdots$$

diverges by direct comparison with the terms of the harmonic series. Therefore, the given series is *conditionally* convergent.

Rearrangement of Series

A finite sum such as $(1 + 3 - 2 + 5 - 4)$ can be rearranged without changing the value of the sum. This is not necessarily true of an infinite series—it depends on whether the series is absolutely convergent (every rearrangement has the same sum) or conditionally convergent.

Example 7   Rearrangement of a Series

FOR FURTHER INFORMATION Georg Friedrich Riemann (1826–1866) proved that if $\Sigma\, a_n$ is conditionally convergent and S is any real number, the terms of the series can be rearranged to converge to S. For more on this topic, see the article "Riemann's Rearrangement Theorem" by Stewart Galanor in *Mathematics Teacher*. To view this article, go to the website *www.matharticles.com*.

The alternating harmonic series converges to ln 2. That is,

$$\sum_{n=1}^{\infty} (-1)^{n+1} \frac{1}{n} = \frac{1}{1} - \frac{1}{2} + \frac{1}{3} - \frac{1}{4} + \cdots = \ln 2. \qquad \text{(See Exercise 47, Section 8.10.)}$$

Rearrange the series to produce a different sum.

Solution   Consider the following rearrangement.

$$1 - \frac{1}{2} - \frac{1}{4} + \frac{1}{3} - \frac{1}{6} - \frac{1}{8} + \frac{1}{5} - \frac{1}{10} - \frac{1}{12} + \frac{1}{7} - \frac{1}{14} - \cdots$$

$$= \left(1 - \frac{1}{2}\right) - \frac{1}{4} + \left(\frac{1}{3} - \frac{1}{6}\right) - \frac{1}{8} + \left(\frac{1}{5} - \frac{1}{10}\right) - \frac{1}{12} + \left(\frac{1}{7} - \frac{1}{14}\right) - \cdots$$

$$= \frac{1}{2} - \frac{1}{4} + \frac{1}{6} - \frac{1}{8} + \frac{1}{10} - \frac{1}{12} + \frac{1}{14} - \cdots$$

$$= \frac{1}{2}\left(1 - \frac{1}{2} + \frac{1}{3} - \frac{1}{4} + \frac{1}{5} - \frac{1}{6} + \frac{1}{7} - \cdots\right) = \frac{1}{2}(\ln 2)$$

By rearranging the terms, you obtain a sum that is half the original sum.

EXERCISES FOR SECTION 8.5

In Exercises 1–4, match the series with the graph of its sequence of partial sums. [The graphs are labeled (a), (b), (c), and (d).]

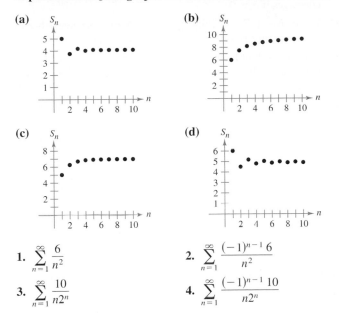

(a)   (b)   (c)   (d)

1. $\displaystyle\sum_{n=1}^{\infty} \frac{6}{n^2}$

2. $\displaystyle\sum_{n=1}^{\infty} \frac{(-1)^{n-1} 6}{n^2}$

3. $\displaystyle\sum_{n=1}^{\infty} \frac{10}{n2^n}$

4. $\displaystyle\sum_{n=1}^{\infty} \frac{(-1)^{n-1} 10}{n2^n}$

Numerical and Graphical Analysis   In Exercises 5–8, explore the Alternating Series Remainder.

(a) Use a graphing utility to find the indicated partial sum S_n and complete the table.

(b) Use a graphing utility to graph the first ten terms of the sequence of partial sums and a horizontal line representing the sum.

(c) What pattern exists between the plot of the successive points in part (b) relative to the horizontal line representing the sum of the series? Do the distances between the successive points and the horizontal line increase or decrease?

(d) Discuss the relationship between the answers in part (c) and the Alternating Series Remainder as given in Theorem 8.15.

n	1	2	3	4	5	6	7	8	9	10
S_n										

5. $\displaystyle\sum_{n=1}^{\infty} \frac{(-1)^{n-1}}{2n-1} = \frac{\pi}{4}$

6. $\displaystyle\sum_{n=1}^{\infty} \frac{(-1)^{n-1}}{(n-1)!} = \frac{1}{e}$

7. $\displaystyle\sum_{n=1}^{\infty} \frac{(-1)^{n-1}}{n^2} = \frac{\pi^2}{12}$

8. $\displaystyle\sum_{n=1}^{\infty} \frac{(-1)^{n-1}}{(2n-1)!} = \sin 1$

In Exercises 9–28, determine the convergence or divergence of the series.

9. $\displaystyle\sum_{n=1}^{\infty} \frac{(-1)^{n+1}}{n}$

10. $\displaystyle\sum_{n=1}^{\infty} \frac{(-1)^{n+1}n}{2n-1}$

11. $\displaystyle\sum_{n=1}^{\infty} \frac{(-1)^{n+1}}{2n-1}$

12. $\displaystyle\sum_{n=1}^{\infty} \frac{(-1)^n}{\ln(n+1)}$

13. $\displaystyle\sum_{n=1}^{\infty} \frac{(-1)^n n^2}{n^2+1}$

14. $\displaystyle\sum_{n=1}^{\infty} \frac{(-1)^{n+1} n}{n^2+1}$

15. $\displaystyle\sum_{n=1}^{\infty} \frac{(-1)^n}{\sqrt{n}}$

16. $\displaystyle\sum_{n=1}^{\infty} \frac{(-1)^{n+1} n^2}{n^2+5}$

17. $\displaystyle\sum_{n=1}^{\infty} \frac{(-1)^{n+1}(n+1)}{\ln(n+1)}$

18. $\displaystyle\sum_{n=1}^{\infty} \frac{(-1)^{n+1}\ln(n+1)}{n+1}$

19. $\displaystyle\sum_{n=1}^{\infty} \sin \frac{(2n-1)\pi}{2}$

20. $\displaystyle\sum_{n=1}^{\infty} \frac{1}{n}\sin \frac{(2n-1)\pi}{2}$

21. $\displaystyle\sum_{n=1}^{\infty} \cos n\pi$

22. $\displaystyle\sum_{n=1}^{\infty} \frac{1}{n}\cos n\pi$

23. $\displaystyle\sum_{n=0}^{\infty} \frac{(-1)^n}{n!}$

24. $\displaystyle\sum_{n=0}^{\infty} \frac{(-1)^n}{(2n+1)!}$

25. $\displaystyle\sum_{n=1}^{\infty} \frac{(-1)^{n+1}\sqrt{n}}{n+2}$

26. $\displaystyle\sum_{n=1}^{\infty} \frac{(-1)^{n+1}\sqrt{n}}{\sqrt[3]{n}}$

27. $\displaystyle\sum_{n=1}^{\infty} \frac{2(-1)^{n+1}}{e^n - e^{-n}} = \sum_{n=1}^{\infty} (-1)^{n+1}\operatorname{csch} n$

28. $\displaystyle\sum_{n=1}^{\infty} \frac{2(-1)^{n+1}}{e^n + e^{-n}} = \sum_{n=1}^{\infty} (-1)^{n+1}\operatorname{sech} n$

In Exercises 29–32, approximate the sum of the series by using the first six terms. (See Example 4.)

29. $\displaystyle\sum_{n=1}^{\infty} \frac{(-1)^{n+1} 3}{n^2}$

30. $\displaystyle\sum_{n=1}^{\infty} \frac{(-1)^{n+1} 4}{\ln(n+1)}$

31. $\displaystyle\sum_{n=0}^{\infty} \frac{(-1)^n 2}{n!}$

32. $\displaystyle\sum_{n=1}^{\infty} \frac{(-1)^{n+1} n}{2^n}$

In Exercises 33–38, (a) use Theorem 8.15 to determine the number of terms required to approximate the sum of the convergent series with an error of less than 0.001, and (b) use a graphing utility to approximate the sum of the series with an error of less than 0.001.

33. $\displaystyle\sum_{n=0}^{\infty} \frac{(-1)^n}{n!} = \frac{1}{e}$

34. $\displaystyle\sum_{n=0}^{\infty} \frac{(-1)^n}{2^n n!} = \frac{1}{\sqrt{e}}$

35. $\displaystyle\sum_{n=0}^{\infty} \frac{(-1)^n}{(2n+1)!} = \sin 1$

36. $\displaystyle\sum_{n=0}^{\infty} \frac{(-1)^n}{(2n)!} = \cos 1$

37. $\displaystyle\sum_{n=1}^{\infty} \frac{(-1)^{n+1}}{n} = \ln 2$

38. $\displaystyle\sum_{n=1}^{\infty} \frac{(-1)^{n+1}}{n4^n} = \ln \frac{5}{4}$

In Exercises 39 and 40, use Theorem 8.15 to determine the number of terms required to approximate the sum of the series with an error of less than 0.001.

39. $\displaystyle\sum_{n=1}^{\infty} \frac{(-1)^{n+1}}{2n^3 - 1}$

40. $\displaystyle\sum_{n=1}^{\infty} \frac{(-1)^{n+1}}{n^4}$

In Exercises 41–56, determine whether the series converges conditionally or absolutely, or diverges.

41. $\displaystyle\sum_{n=1}^{\infty} \frac{(-1)^{n+1}}{(n+1)^2}$

42. $\displaystyle\sum_{n=1}^{\infty} \frac{(-1)^{n+1}}{n+1}$

43. $\displaystyle\sum_{n=1}^{\infty} \frac{(-1)^{n+1}}{\sqrt{n}}$

44. $\displaystyle\sum_{n=1}^{\infty} \frac{(-1)^{n+1}}{n\sqrt{n}}$

45. $\displaystyle\sum_{n=1}^{\infty} \frac{(-1)^{n+1} n^2}{(n+1)^2}$

46. $\displaystyle\sum_{n=1}^{\infty} \frac{(-1)^{n+1}(2n+3)}{n+10}$

47. $\displaystyle\sum_{n=2}^{\infty} \frac{(-1)^n}{\ln n}$

48. $\displaystyle\sum_{n=0}^{\infty} (-1)^n e^{-n^2}$

49. $\displaystyle\sum_{n=2}^{\infty} \frac{(-1)^n n}{n^3 - 1}$

50. $\displaystyle\sum_{n=1}^{\infty} \frac{(-1)^{n+1}}{n^{1.5}}$

51. $\displaystyle\sum_{n=0}^{\infty} \frac{(-1)^n}{(2n+1)!}$

52. $\displaystyle\sum_{n=0}^{\infty} \frac{(-1)^n}{\sqrt{n+4}}$

53. $\displaystyle\sum_{n=0}^{\infty} \frac{\cos n\pi}{n+1}$

54. $\displaystyle\sum_{n=1}^{\infty} (-1)^{n+1} \arctan n$

55. $\displaystyle\sum_{n=1}^{\infty} \frac{\cos n\pi}{n^2}$

56. $\displaystyle\sum_{n=1}^{\infty} \frac{\sin[(2n-1)\pi/2]}{n}$

Getting at the Concept

57. Define an alternating series and state the Alternating Series Test.

58. Give the remainder after N terms of a convergent alternating series.

59. In your own words, state the difference between absolute and conditional convergence of an alternating series.

60. Give an example of an alternating series that converges while the series of its absolute values diverges.

61. The graphs of the sequences of partial sums of two series are shown in the figures. Which graph represents the partial sums of an alternating series? Explain.

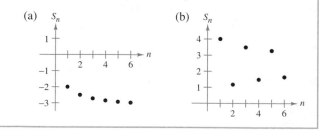

62. Prove that the alternating p-series

$$\sum_{n=1}^{\infty} (-1)^n \left(\frac{1}{n^p}\right)$$

converges if $p > 0$.

63. Prove that if $\Sigma |a_n|$ converges, then Σa_n^2 converges. Is the converse true? If not, give an example that shows it is false.

64. Use the result of Exercise 62 to give an example of an alternating p-series that converges, but whose corresponding p-series diverges.

65. Give an example of a series that demonstrates the statement you proved in Exercise 63.

66. Find all values of x for which the series $\Sigma (x^n/n)$ (a) converges absolutely and (b) converges conditionally.

True or False? In Exercises 67 and 68, determine whether the statement is true or false. If false, explain why.

67. If both Σa_n and $\Sigma (-a_n)$ converge, then $\Sigma |a_n|$ converges.

68. If Σa_n does not converge, then $\Sigma |a_n|$ does not converge.

In Exercises 69–78, test for convergence or divergence and identify the test used.

69. $\displaystyle\sum_{n=1}^{\infty} \frac{10}{n^{3/2}}$

70. $\displaystyle\sum_{n=1}^{\infty} \frac{3}{n^2 + 5}$

71. $\displaystyle\sum_{n=1}^{\infty} \frac{3^n}{n^2}$

72. $\displaystyle\sum_{n=1}^{\infty} \frac{1}{2^n + 1}$

73. $\displaystyle\sum_{n=0}^{\infty} 5\left(\frac{7}{8}\right)^n$

74. $\displaystyle\sum_{n=1}^{\infty} \frac{3n^2}{2n^2 + 1}$

75. $\displaystyle\sum_{n=1}^{\infty} 100e^{-n/2}$

76. $\displaystyle\sum_{n=0}^{\infty} \frac{(-1)^n}{n+4}$

77. $\displaystyle\sum_{n=1}^{\infty} \frac{(-1)^{n+1} 4}{3n^2 - 1}$

78. $\displaystyle\sum_{n=2}^{\infty} \frac{\ln n}{n}$

79. The following argument, that $0 = 1$, is *incorrect*. Describe the error.

$$\begin{aligned}
0 &= 0 + 0 + 0 + \cdots \\
&= (1 - 1) + (1 - 1) + (1 - 1) + \cdots \\
&= 1 + (-1 + 1) + (-1 + 1) + \cdots \\
&= 1 + 0 + 0 + \cdots \\
&= 1
\end{aligned}$$

Section 8.6   The Ratio and Root Tests

- Use the Ratio Test to determine whether a series converges or diverges.
- Use the Root Test to determine whether a series converges or diverges.
- Review the tests for convergence and divergence of an infinite series.

The Ratio Test

This section begins with a test for absolute convergence—the **Ratio Test.**

THEOREM 8.17   Ratio Test

Let $\Sigma\, a_n$ be a series with nonzero terms.

1. $\Sigma\, a_n$ converges absolutely if $\lim\limits_{n\to\infty} \left|\dfrac{a_{n+1}}{a_n}\right| < 1$.

2. $\Sigma\, a_n$ diverges if $\lim\limits_{n\to\infty} \left|\dfrac{a_{n+1}}{a_n}\right| > 1$ or $\lim\limits_{n\to\infty} \left|\dfrac{a_{n+1}}{a_n}\right| = \infty$.

3. The Ratio Test is inconclusive if $\lim\limits_{n\to\infty} \left|\dfrac{a_{n+1}}{a_n}\right| = 1$.

Proof   To prove Property 1, assume that

$$\lim_{n\to\infty} \left|\frac{a_{n+1}}{a_n}\right| = r < 1$$

and choose R such that $0 \le r < R < 1$. By the definition of the limit of a sequence, there exists some $N > 0$ such that $|a_{n+1}/a_n| < R$ for all $n > N$. Therefore, you can write the following inequalities.

$$|a_{N+1}| < |a_N|R$$
$$|a_{N+2}| < |a_{N+1}|R < |a_N|R^2$$
$$|a_{N+3}| < |a_{N+2}|R < |a_{N+1}|R^2 < |a_N|R^3$$
$$\vdots$$

The geometric series $\Sigma\, |a_N|R^n = |a_N|R + |a_N|R^2 + \cdots + |a_N|R^n + \cdots$ converges, and so, by the Direct Comparison Test, the series

$$\sum_{n=1}^{\infty} |a_{N+n}| = |a_{N+1}| + |a_{N+2}| + \cdots + |a_{N+n}| + \cdots$$

also converges. This in turn implies that the series $\Sigma\, |a_n|$ converges, because discarding a finite number of terms ($n = N - 1$) does not affect convergence. Consequently, by Theorem 8.16, the series $\Sigma\, a_n$ converges absolutely. The proof of Property 2 is similar and is left as an exercise (see Exercise 74).   ◻

NOTE   The fact that the Ratio Test is inconclusive when $|a_{n+1}/a_n| \to 1$ can be seen by comparing the two series $\Sigma\,(1/n)$ and $\Sigma\,(1/n^2)$. The first series diverges and the second one converges, but in both cases

$$\lim_{n\to\infty} \left|\frac{a_{n+1}}{a_n}\right| = 1.$$

EXPLORATION

Writing a Series   One of the following conditions guarantees that a series will diverge, two conditions guarantee that a series will converge, and one has no guarantee—the series can either converge or diverge. Which is which? Explain your reasoning.

a. $\lim\limits_{n\to\infty} \left|\dfrac{a_{n+1}}{a_n}\right| = 0$

b. $\lim\limits_{n\to\infty} \left|\dfrac{a_{n+1}}{a_n}\right| = \dfrac{1}{2}$

c. $\lim\limits_{n\to\infty} \left|\dfrac{a_{n+1}}{a_n}\right| = 1$

d. $\lim\limits_{n\to\infty} \left|\dfrac{a_{n+1}}{a_n}\right| = 2$

Although the Ratio Test is not a cure for all ills related to tests for convergence, it is particularly useful for series that *converge rapidly*. Series involving factorials or exponentials are frequently of this type.

Example 1   Using the Ratio Test

Determine the convergence or divergence of

$$\sum_{n=0}^{\infty} \frac{2^n}{n!}.$$

Solution   Because $a_n = 2^n/n!$, you can write the following.

$$\lim_{n \to \infty} \left| \frac{a_{n+1}}{a_n} \right| = \lim_{n \to \infty} \left[\frac{2^{n+1}}{(n+1)!} \div \frac{2^n}{n!} \right]$$

$$= \lim_{n \to \infty} \left[\frac{2^{n+1}}{(n+1)!} \cdot \frac{n!}{2^n} \right]$$

$$= \lim_{n \to \infty} \frac{2}{n+1}$$

$$= 0$$

STUDY TIP   A step frequently used in applications of the Ratio Test involves simplifying quotients of factorials. In Example 1, for instance, notice that

$$\frac{n!}{(n+1)!} = \frac{n!}{(n+1)n!} = \frac{1}{n+1}.$$

Therefore, the series converges.

Example 2   Using the Ratio Test

Determine whether each series converges or diverges.

a. $\displaystyle \sum_{n=0}^{\infty} \frac{n^2 2^{n+1}}{3^n}$ **b.** $\displaystyle \sum_{n=1}^{\infty} \frac{n^n}{n!}$

Solution
a. This series converges because the limit of $|a_{n+1}/a_n|$ is less than 1.

$$\lim_{n \to \infty} \left| \frac{a_{n+1}}{a_n} \right| = \lim_{n \to \infty} \left[(n+1)^2 \left(\frac{2^{n+2}}{3^{n+1}} \right) \left(\frac{3^n}{n^2 2^{n+1}} \right) \right]$$

$$= \lim_{n \to \infty} \frac{2(n+1)^2}{3n^2}$$

$$= \frac{2}{3} < 1$$

b. This series diverges because the limit of $|a_{n+1}/a_n|$ is greater than 1.

$$\lim_{n \to \infty} \left| \frac{a_{n+1}}{a_n} \right| = \lim_{n \to \infty} \left[\frac{(n+1)^{n+1}}{(n+1)!} \left(\frac{n!}{n^n} \right) \right]$$

$$= \lim_{n \to \infty} \left[\frac{(n+1)^{n+1}}{(n+1)} \left(\frac{1}{n^n} \right) \right]$$

$$= \lim_{n \to \infty} \frac{(n+1)^n}{n^n}$$

$$= \lim_{n \to \infty} \left(1 + \frac{1}{n} \right)^n$$

$$= e > 1$$

Example 3   A Failure of the Ratio Test

Determine the convergence or divergence of $\displaystyle\sum_{n=1}^{\infty} (-1)^n \frac{\sqrt{n}}{n+1}$.

Solution   The limit of $|a_{n+1}/a_n|$ is equal to 1.

$$\lim_{n\to\infty} \left| \frac{a_{n+1}}{a_n} \right| = \lim_{n\to\infty} \left[\left(\frac{\sqrt{n+1}}{n+2} \right)\left(\frac{n+1}{\sqrt{n}} \right) \right]$$

$$= \lim_{n\to\infty} \left[\sqrt{\frac{n+1}{n}} \left(\frac{n+1}{n+2} \right) \right]$$

$$= \sqrt{1}(1)$$

$$= 1$$

So, the Ratio Test is inconclusive. To determine whether the series converges, you need to try a different test. In this case, you can apply the Alternating Series Test. To show that $a_{n+1} \le a_n$, let

$$f(x) = \frac{\sqrt{x}}{x+1}.$$

Then the derivative is

$$f'(x) = \frac{-x+1}{2\sqrt{x}(x+1)^2}.$$

Because the derivative is negative for $x > 1$, you know that f is a decreasing function. Also, by L'Hôpital's Rule,

$$\lim_{x\to\infty} \frac{\sqrt{x}}{x+1} = \lim_{x\to\infty} \frac{1/(2\sqrt{x})}{1}$$

$$= \lim_{x\to\infty} \frac{1}{2\sqrt{x}}$$

$$= 0.$$

Therefore, by the Alternating Series Test, the series converges.

The series in Example 3 is *conditionally convergent*. This follows from the fact that the series

$$\sum_{n=1}^{\infty} |a_n|$$

diverges $\left(\text{by the Limit Comparison Test with } \Sigma\, 1/\sqrt{n}\right)$, but the series

$$\sum_{n=1}^{\infty} a_n$$

converges.

TECHNOLOGY   A computer or programmable calculator can reinforce the conclusion that the series in Example 3 converges *conditionally*. By adding the first 100 terms of the series, you obtain a sum of about -0.2. (The sum of the first 100 terms of the series $\Sigma\, |a_n|$ is about 17.)

The Root Test

The next test for convergence or divergence of series works especially well for series involving nth powers. The proof of this theorem is similar to that given for the Ratio Test, and we leave it as an exercise (see Exercise 75).

THEOREM 8.18 Root Test

Let $\Sigma\, a_n$ be a series.

1. $\Sigma\, a_n$ converges absolutely if $\lim\limits_{n\to\infty} \sqrt[n]{|a_n|} < 1$.

2. $\Sigma\, a_n$ diverges if $\lim\limits_{n\to\infty} \sqrt[n]{|a_n|} > 1$ or $\lim\limits_{n\to\infty} \sqrt[n]{|a_n|} = \infty$.

3. The Root Test is inconclusive if $\lim\limits_{n\to\infty} \sqrt[n]{|a_n|} = 1$.

Example 4 **Using the Root Test**

Determine the convergence or divergence of

$$\sum_{n=1}^{\infty} \frac{e^{2n}}{n^n}.$$

Solution You can apply the Root Test as follows.

$$\lim_{n\to\infty} \sqrt[n]{|a_n|} = \lim_{n\to\infty} \sqrt[n]{\frac{e^{2n}}{n^n}}$$

$$= \lim_{n\to\infty} \frac{e^{2n/n}}{n^{n/n}}$$

$$= \lim_{n\to\infty} \frac{e^2}{n}$$

$$= 0 < 1$$

Because this limit is less than 1, you can conclude that the series converges absolutely (and hence converges).

FOR FURTHER INFORMATION For more information on the usefulness of the Root Test, see the article *"N! and the Root Test"* by Charles C. Mumma II in *The American Mathematical Monthly*. To view this article, go to the website *www.matharticles.com*.

To see the usefulness of the Root Test for the series in Example 4, try applying the Ratio Test to that series. When you do this, you obtain the following.

$$\lim_{n\to\infty} \left| \frac{a_{n+1}}{a_n} \right| = \lim_{n\to\infty} \left[\frac{e^{2(n+1)}}{(n+1)^{n+1}} \div \frac{e^{2n}}{n^n} \right]$$

$$= \lim_{n\to\infty} e^2 \frac{n^n}{(n+1)^{n+1}}$$

$$= \lim_{n\to\infty} e^2 \left(\frac{n}{n+1} \right)^n \left(\frac{1}{n+1} \right)$$

$$= 0$$

Note that this limit is not as easily evaluated as the limit obtained by the Root Test in Example 4.

Strategies for Testing Series

You have now studied ten tests for determining the convergence or divergence of an infinite series. (See the summary in the table on page 602.) Skill in choosing and applying the various tests will come only with practice. Below is a set of guidelines for choosing an appropriate test.

> **Guidelines for Testing a Series for Convergence or Divergence**
>
> 1. Does the nth term approach 0? If not, the series diverges.
> 2. Is the series one of the special types—geometric, p-series, telescoping, or alternating?
> 3. Can the Integral Test, the Root Test, or the Ratio Test be applied?
> 4. Can the series be compared favorably to one of the special types?

In some instances, more than one test is applicable. However, your objective should be to learn to choose the most efficient test.

Example 5 **Applying the Strategies for Testing Series**

Determine the convergence or divergence of each series.

a. $\displaystyle\sum_{n=1}^{\infty} \frac{n+1}{3n+1}$
  b. $\displaystyle\sum_{n=1}^{\infty} \left(\frac{\pi}{6}\right)^n$
  c. $\displaystyle\sum_{n=1}^{\infty} ne^{-n^2}$

d. $\displaystyle\sum_{n=1}^{\infty} \frac{1}{3n+1}$
  e. $\displaystyle\sum_{n=1}^{\infty} (-1)^n \frac{3}{4n+1}$
  f. $\displaystyle\sum_{n=1}^{\infty} \frac{n!}{10^n}$

g. $\displaystyle\sum_{n=1}^{\infty} \left(\frac{n+1}{2n+1}\right)^n$

Solution

a. For this series, the limit of the nth term is not 0 $\left(a_n \to \frac{1}{3}\text{ as }n\to\infty\right)$. So, by the nth-Term Test, the series diverges.

b. This series is geometric. Moreover, because the common ratio of the terms is less than 1 in absolute value $(r = \pi/6)$, you can conclude that the series converges.

c. Because the function $f(x) = xe^{-x^2}$ is easily integrated, you can use the Integral Test to conclude that the series converges.

d. The nth term of this series can be compared to the nth term of the harmonic series. After using the Limit Comparison Test, you can conclude that the series diverges.

e. This is an alternating series whose nth term approaches 0. Because $a_{n+1} \le a_n$, you can use the Alternating Series Test to conclude that the series converges.

f. The nth term of this series involves a factorial, which indicates that the Ratio Test may work well. After applying the Ratio Test, you can conclude that the series diverges.

g. The nth term of this series involves a variable that is raised to the nth power, which indicates that the Root Test may work well. After applying the Root Test, you can conclude that the series converges.

Summary of Tests for Series

Test	Series	Condition(s) of Convergence	Condition(s) of Divergence	Comment
nth-Term	$\displaystyle\sum_{n=1}^{\infty} a_n$		$\displaystyle\lim_{n\to\infty} a_n \neq 0$	This test cannot be used to show convergence.
Geometric Series	$\displaystyle\sum_{n=0}^{\infty} ar^n$	$\lvert r \rvert < 1$	$\lvert r \rvert \geq 1$	Sum: $S = \dfrac{a}{1-r}$
Telescoping Series	$\displaystyle\sum_{n=1}^{\infty} (b_n - b_{n+1})$	$\displaystyle\lim_{n\to\infty} b_n = L$		Sum: $S = b_1 - L$
p-Series	$\displaystyle\sum_{n=1}^{\infty} \dfrac{1}{n^p}$	$p > 1$	$p \leq 1$	
Alternating Series	$\displaystyle\sum_{n=1}^{\infty} (-1)^{n-1} a_n$	$0 < a_{n+1} \leq a_n$ and $\displaystyle\lim_{n\to\infty} a_n = 0$		Remainder: $\lvert R_N \rvert \leq a_{N+1}$
Integral (f is continuous, positive, and decreasing)	$\displaystyle\sum_{n=1}^{\infty} a_n,$ $a_n = f(n) \geq 0$	$\displaystyle\int_1^{\infty} f(x)\,dx$ converges	$\displaystyle\int_1^{\infty} f(x)\,dx$ diverges	Remainder: $0 < R_N < \displaystyle\int_N^{\infty} f(x)\,dx$
Root	$\displaystyle\sum_{n=1}^{\infty} a_n$	$\displaystyle\lim_{n\to\infty} \sqrt[n]{\lvert a_n \rvert} < 1$	$\displaystyle\lim_{n\to\infty} \sqrt[n]{\lvert a_n \rvert} > 1$	Test is inconclusive if $\displaystyle\lim_{n\to\infty} \sqrt[n]{\lvert a_n \rvert} = 1$.
Ratio	$\displaystyle\sum_{n=1}^{\infty} a_n$	$\displaystyle\lim_{n\to\infty} \left\lvert \dfrac{a_{n+1}}{a_n} \right\rvert < 1$	$\displaystyle\lim_{n\to\infty} \left\lvert \dfrac{a_{n+1}}{a_n} \right\rvert > 1$	Test is inconclusive if $\displaystyle\lim_{n\to\infty} \left\lvert \dfrac{a_{n+1}}{a_n} \right\rvert = 1$.
Direct Comparison ($a_n, b_n > 0$)	$\displaystyle\sum_{n=1}^{\infty} a_n$	$0 < a_n \leq b_n$ and $\displaystyle\sum_{n=1}^{\infty} b_n$ converges	$0 < b_n \leq a_n$ and $\displaystyle\sum_{n=1}^{\infty} b_n$ diverges	
Limit Comparison ($a_n, b_n > 0$)	$\displaystyle\sum_{n=1}^{\infty} a_n$	$\displaystyle\lim_{n\to\infty} \dfrac{a_n}{b_n} = L > 0$ and $\displaystyle\sum_{n=1}^{\infty} b_n$ converges	$\displaystyle\lim_{n\to\infty} \dfrac{a_n}{b_n} = L > 0$ and $\displaystyle\sum_{n=1}^{\infty} b_n$ diverges	

EXERCISES FOR SECTION 8.6

In Exercises 1–4, verify the formula.

1. $\dfrac{(n+1)!}{(n-2)!} = (n+1)(n)(n-1)$

2. $\dfrac{(2k-2)!}{(2k)!} = \dfrac{1}{(2k)(2k-1)}$

3. $1 \cdot 3 \cdot 5 \cdots (2k-1) = \dfrac{(2k)!}{2^k k!}$

4. $\dfrac{1}{1 \cdot 3 \cdot 5 \cdots (2k-5)} = \dfrac{2^k k!(2k-3)(2k-1)}{(2k)!}, \quad k \geq 3$

In Exercises 5–10, match the series with the graph of its sequence of partial sums. [The graphs are labeled (a), (b), (c), (d), (e), and (f).]

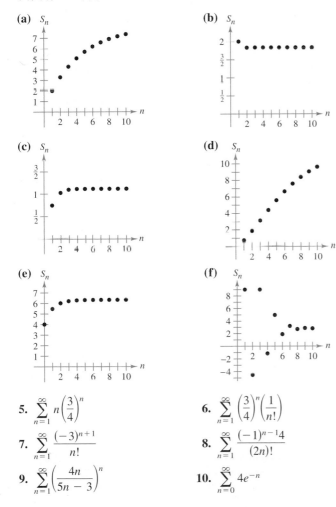

(a) S_n

(b) S_n

(c) S_n

(d) S_n

(e) S_n

(f) S_n

5. $\displaystyle\sum_{n=1}^{\infty} n\left(\dfrac{3}{4}\right)^n$

6. $\displaystyle\sum_{n=1}^{\infty} \left(\dfrac{3}{4}\right)^n\left(\dfrac{1}{n!}\right)$

7. $\displaystyle\sum_{n=1}^{\infty} \dfrac{(-3)^{n+1}}{n!}$

8. $\displaystyle\sum_{n=1}^{\infty} \dfrac{(-1)^{n-1}4}{(2n)!}$

9. $\displaystyle\sum_{n=1}^{\infty} \left(\dfrac{4n}{5n-3}\right)^n$

10. $\displaystyle\sum_{n=0}^{\infty} 4e^{-n}$

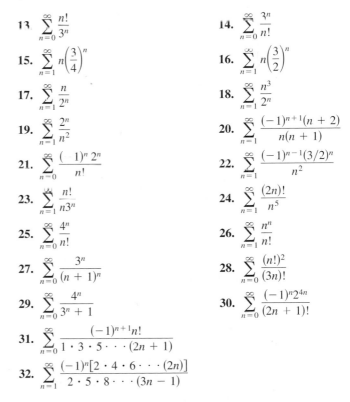

📈 *Numerical, Graphical, and Analytic Analysis*   In Exercises 11 and 12, (a) verify that the series converges. (b) Use a graphing utility to find the indicated partial sum S_n and complete the table. (c) Use a graphing utility to graph the first ten terms of the sequence of partial sums. (d) Use the table to estimate the sum of the series. (e) Explain the relationship between the magnitude of the terms of the series and the rate at which the sequence of partial sums approaches the sum of the series.

n	5	10	15	20	25
S_n					

11. $\displaystyle\sum_{n=1}^{\infty} n^2\left(\dfrac{5}{8}\right)^n$

12. $\displaystyle\sum_{n=1}^{\infty} \dfrac{n^2+1}{n!}$

In Exercises 13–32, use the Ratio Test to determine the convergence or divergence of the series.

13. $\displaystyle\sum_{n=0}^{\infty} \dfrac{n!}{3^n}$

14. $\displaystyle\sum_{n=0}^{\infty} \dfrac{3^n}{n!}$

15. $\displaystyle\sum_{n=1}^{\infty} n\left(\dfrac{3}{4}\right)^n$

16. $\displaystyle\sum_{n=1}^{\infty} n\left(\dfrac{3}{2}\right)^n$

17. $\displaystyle\sum_{n=1}^{\infty} \dfrac{n}{2^n}$

18. $\displaystyle\sum_{n=1}^{\infty} \dfrac{n^3}{2^n}$

19. $\displaystyle\sum_{n=1}^{\infty} \dfrac{2^n}{n^2}$

20. $\displaystyle\sum_{n=1}^{\infty} \dfrac{(-1)^{n+1}(n+2)}{n(n+1)}$

21. $\displaystyle\sum_{n=0}^{\infty} \dfrac{(-1)^n 2^n}{n!}$

22. $\displaystyle\sum_{n=1}^{\infty} \dfrac{(-1)^{n-1}(3/2)^n}{n^2}$

23. $\displaystyle\sum_{n=1}^{\infty} \dfrac{n!}{n3^n}$

24. $\displaystyle\sum_{n=1}^{\infty} \dfrac{(2n)!}{n^5}$

25. $\displaystyle\sum_{n=0}^{\infty} \dfrac{4^n}{n!}$

26. $\displaystyle\sum_{n=1}^{\infty} \dfrac{n^n}{n!}$

27. $\displaystyle\sum_{n=0}^{\infty} \dfrac{3^n}{(n+1)^n}$

28. $\displaystyle\sum_{n=0}^{\infty} \dfrac{(n!)^2}{(3n)!}$

29. $\displaystyle\sum_{n=0}^{\infty} \dfrac{4^n}{3^n+1}$

30. $\displaystyle\sum_{n=0}^{\infty} \dfrac{(-1)^n 2^{4n}}{(2n+1)!}$

31. $\displaystyle\sum_{n=0}^{\infty} \dfrac{(-1)^{n+1}n!}{1 \cdot 3 \cdot 5 \cdots (2n+1)}$

32. $\displaystyle\sum_{n=1}^{\infty} \dfrac{(-1)^n[2 \cdot 4 \cdot 6 \cdots (2n)]}{2 \cdot 5 \cdot 8 \cdots (3n-1)}$

In Exercises 33 and 34, verify that the Ratio Test is inconclusive for the p-series.

33. (a) $\displaystyle\sum_{n=1}^{\infty} \dfrac{1}{n^{3/2}}$ (b) $\displaystyle\sum_{n=1}^{\infty} \dfrac{1}{n^{1/2}}$

34. (a) $\displaystyle\sum_{n=1}^{\infty} \dfrac{1}{n^4}$ (b) $\displaystyle\sum_{n=1}^{\infty} \dfrac{1}{n^p}$

In Exercises 35–42, use the Root Test to determine the convergence or divergence of the series.

35. $\displaystyle\sum_{n=1}^{\infty} \left(\frac{n}{2n+1}\right)^n$

36. $\displaystyle\sum_{n=1}^{\infty} \left(\frac{2n}{n+1}\right)^n$

37. $\displaystyle\sum_{n=2}^{\infty} \frac{(-1)^n}{(\ln n)^n}$

38. $\displaystyle\sum_{n=1}^{\infty} \left(\frac{-3n}{2n+1}\right)^{3n}$

39. $\displaystyle\sum_{n=1}^{\infty} \left(2\sqrt[n]{n}+1\right)^n$

40. $\displaystyle\sum_{n=0}^{\infty} e^{-n}$

41. $\dfrac{1}{(\ln 3)^3} + \dfrac{1}{(\ln 4)^4} + \dfrac{1}{(\ln 5)^5} + \dfrac{1}{(\ln 6)^6} + \cdots$

42. $1 + \dfrac{2}{3} + \dfrac{3}{3^2} + \dfrac{4}{3^3} + \dfrac{5}{3^4} + \dfrac{6}{3^5} + \cdots$

In Exercises 43–60, determine the convergence or divergence of the series using any appropriate test from this chapter. Identify the test used.

43. $\displaystyle\sum_{n=1}^{\infty} \frac{(-1)^{n+1}5}{n}$

44. $\displaystyle\sum_{n=1}^{\infty} \frac{5}{n}$

45. $\displaystyle\sum_{n=1}^{\infty} \frac{3}{n\sqrt{n}}$

46. $\displaystyle\sum_{n=1}^{\infty} \left(\frac{\pi}{4}\right)^n$

47. $\displaystyle\sum_{n=1}^{\infty} \frac{2n}{n+1}$

48. $\displaystyle\sum_{n=1}^{\infty} \frac{n}{2n^2+1}$

49. $\displaystyle\sum_{n=1}^{\infty} \frac{(-1)^n 3^{n-2}}{2^n}$

50. $\displaystyle\sum_{n=1}^{\infty} \frac{10}{3\sqrt{n^3}}$

51. $\displaystyle\sum_{n=1}^{\infty} \frac{10n+3}{n2^n}$

52. $\displaystyle\sum_{n=1}^{\infty} \frac{2^n}{4n^2-1}$

53. $\displaystyle\sum_{n=1}^{\infty} \frac{\cos n}{2^n}$

54. $\displaystyle\sum_{n=2}^{\infty} \frac{(-1)^n}{n \ln n}$

55. $\displaystyle\sum_{n=1}^{\infty} \frac{n7^n}{n!}$

56. $\displaystyle\sum_{n=1}^{\infty} \frac{\ln n}{n^2}$

57. $\displaystyle\sum_{n=1}^{\infty} \frac{(-1)^n 3^{n-1}}{n!}$

58. $\displaystyle\sum_{n=1}^{\infty} \frac{(-1)^n 3^n}{n2^n}$

59. $\displaystyle\sum_{n=1}^{\infty} \frac{(-3)^n}{3 \cdot 5 \cdot 7 \cdots (2n+1)}$

60. $\displaystyle\sum_{n=1}^{\infty} \frac{3 \cdot 5 \cdot 7 \cdots (2n+1)}{18^n(2n-1)n!}$

In Exercises 61–64, identify the two series that are the same.

61. (a) $\displaystyle\sum_{n=1}^{\infty} \frac{n5^n}{n!}$

(b) $\displaystyle\sum_{n=0}^{\infty} \frac{n5^n}{n!}$

(c) $\displaystyle\sum_{n=0}^{\infty} \frac{(n+1)5^{n+1}}{(n+1)!}$

62. (a) $\displaystyle\sum_{n=4}^{\infty} n\left(\frac{3}{4}\right)^n$

(b) $\displaystyle\sum_{n=0}^{\infty} (n+1)\left(\frac{3}{4}\right)^n$

(c) $\displaystyle\sum_{n=1}^{\infty} n\left(\frac{3}{4}\right)^{n-1}$

63. (a) $\displaystyle\sum_{n=0}^{\infty} \frac{(-1)^n}{(2n+1)!}$

(b) $\displaystyle\sum_{n=1}^{\infty} \frac{(-1)^{n-1}}{(2n-1)!}$

(c) $\displaystyle\sum_{n=1}^{\infty} \frac{(-1)^{n-1}}{(2n+1)!}$

64. (a) $\displaystyle\sum_{n=2}^{\infty} \frac{(-1)^n}{(n-1)2^{n-1}}$

(b) $\displaystyle\sum_{n=1}^{\infty} \frac{(-1)^{n+1}}{n2^n}$

(c) $\displaystyle\sum_{n=0}^{\infty} \frac{(-1)^{n+1}}{(n+1)2^n}$

In Exercises 65 and 66, write an equivalent series with the index of summation beginning at $n = 0$.

65. $\displaystyle\sum_{n=1}^{\infty} \frac{n}{4^n}$

66. $\displaystyle\sum_{n=2}^{\infty} \frac{2^n}{(n-2)!}$

In Exercises 67 and 68, (a) determine the number of terms required to approximate the sum of the series with an error less than 0.0001, and (b) use a graphing utility to approximate the sum of the series with an error less than 0.0001.

67. $\displaystyle\sum_{k=1}^{\infty} \frac{(-3)^k}{2^k k!}$

68. $\displaystyle\sum_{k=0}^{\infty} \frac{(-3)^k}{1 \cdot 3 \cdot 5 \cdots (2k+1)}$

Getting at the Concept

69. State the Ratio Test.

70. State the Root Test.

71. You are told that the terms of a positive series appear to approach zero rapidly as n approaches infinity. In fact, $a_7 \le 0.0001$. Given no other information, does this imply that the series converges? Support your conclusion with examples.

72. The graph shows the first ten terms of the sequence of partial sums of the convergent series

$$\sum_{n=1}^{\infty} \left(\frac{2n}{3n+2}\right)^n.$$

Find a series such that the terms of its sequence of partial sums are less than the corresponding terms of the sequence in the figure, but such that the series diverges.

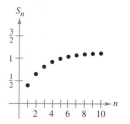

73. Using the Ratio Test, it is determined that an alternating series converges. Does the series converge conditionally or absolutely?

74. Prove Property 2 of Theorem 8.17.

75. Prove Theorem 8.18. (*Hint for Property 1:* If the limit equals $r < 1$, choose a real number R such that $r < R < 1$. By the definition of the limit, there exists some $N > 0$ such that

$$\sqrt[n]{|a_n|} < R \text{ for } n > N.)$$

76. *Writing* Read the article "A Differentiation Test for Absolute Convergence" by Yaser S. Abu-Mostafa in *Mathematics Magazine*. (To view this article, go to the website *www.matharticles.com*.) Then write a paragraph that describes the test. Include examples of series that converge and examples of series that diverge.

| Section 8.7 | **Taylor Polynomials and Approximations** |

- Find polynomial approximations of elementary functions and compare them with the elementary function.
- Find Taylor and Maclaurin polynomial approximations of elementary functions.
- Use the remainder of a Taylor polynomial.

Polynomial Approximations of Elementary Functions

The goal of this section is to show how polynomial functions can be used as approximations for other elementary functions. To find a polynomial function P that approximates another function f, begin by choosing a number c in the domain of f at which f and P have the same value. That is,

$$P(c) = f(c).   \text{Graphs of } f \text{ and } P \text{ pass through } (c, f(c)).$$

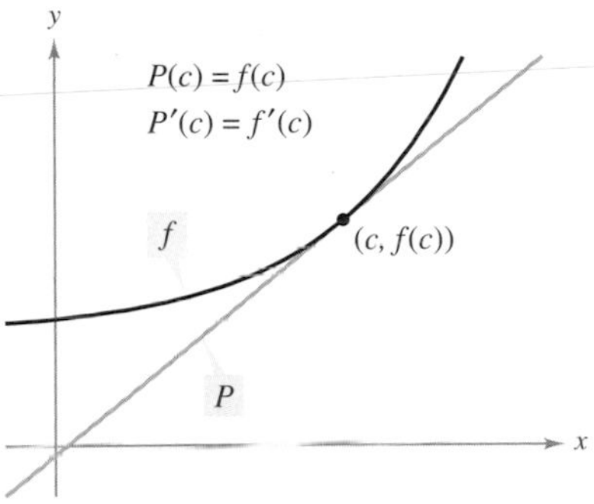

The approximating polynomial is said to be **expanded about c** or **centered at c.** Geometrically, the requirement that $P(c) = f(c)$ means that the graph of P passes through the point $(c, f(c))$. Of course, there are many polynomials whose graphs pass through the point $(c, f(c))$. Your task is to find a polynomial whose graph resembles the graph of f near this point. One way to do this is to impose the additional requirement that the slope of the polynomial function be the same as the slope of the graph of f at the point $(c, f(c))$.

$$P'(c) = f'(c)   \text{Graphs of } f \text{ and } P \text{ have the same slope at } (c, f(c)).$$

With these two requirements, you can obtain a simple linear approximation of f, as shown in Figure 8.10.

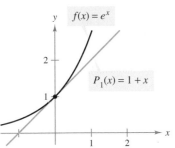

$P(c) = f(c)$
$P'(c) = f'(c)$

Near $(c, f(c))$, the graph of P can be used to approximate the graph of f.
Figure 8.10

Example 1   **First-Degree Polynomial Approximation of $f(x) = e^x$**

For the function $f(x) = e^x$, find a first-degree polynomial function

$$P_1(x) = a_0 + a_1 x$$

whose value and slope agree with the value and slope of f at $x = 0$.

Solution   Because $f(x) = e^x$ and $f'(x) = e^x$, the value and the slope of f, at $x = 0$, are given by

$$f(0) = e^0 = 1$$

and

$$f'(0) = e^0 = 1.$$

Because $P_1(x) = a_0 + a_1 x$, you can use the condition that $P_1(0) = f(0)$ to conclude that $a_0 = 1$. Moreover, because $P_1'(x) = a_1$, you can use the condition that $P_1'(0) = f'(0)$ to conclude that $a_1 = 1$. Therefore,

$$P_1(x) = 1 + x.$$

Figure 8.11 shows the graphs of $P_1(x) = 1 + x$ and $f(x) = e^x$.   ◢

$f(x) = e^x$

$P_1(x) = 1 + x$

P_1 is the first-degree polynomial approximation of $f(x) = e^x$.
Figure 8.11

NOTE   Example 1 isn't the first time you have used a linear function to approximate another function. The same procedure was used as the basis for Newton's Method.

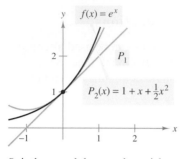

P_2 is the second-degree polynomial approximation of $f(x) = e^x$.
Figure 8.12

In Figure 8.12 you can see that, at points near $(0, 1)$, the graph of

$$P_1(x) = 1 + x \qquad \text{1st-degree approximation}$$

is reasonably close to the graph of $f(x) = e^x$. However, as you move away from $(0, 1)$, the graphs move farther from each other and the accuracy of the approximation decreases. To improve the approximation, you can impose yet another requirement— that the values of the second derivatives of P and f agree when $x = 0$. The polynomial, P_2, of least degree that satisfies all three requirements $P_2(0) = f(0)$, $P_2{}'(0) = f'(0)$, and $P_2{}''(0) = f''(0)$ can be shown to be

$$P_2(x) = 1 + x + \frac{1}{2}x^2. \qquad \text{2nd-degree approximation}$$

Moreover, in Figure 8.12, you can see that P_2 is a better approximation of f than P_1. If you continue this pattern, requiring that the values of $P_n(x)$ and its first n derivatives match those of $f(x) = e^x$ at $x = 0$, you obtain the following.

$$P_n(x) = 1 + x + \frac{1}{2}x^2 + \frac{1}{3!}x^3 + \cdots + \frac{1}{n!}x^n \qquad \text{nth-degree approximation}$$
$$\approx e^x$$

Example 2 Third-Degree Polynomial Approximation of $f(x) = e^x$

Construct a table comparing the values of the polynomial

$$P_3(x) = 1 + x + \frac{1}{2}x^2 + \frac{1}{3!}x^3 \qquad \text{3rd-degree approximation}$$

with $f(x) = e^x$ for several values of x near 0.

Solution Using a calculator or a computer, you can obtain the results shown in the table below. Note that for $x = 0$, the two functions have the same value, but that as x moves farther away from 0, the accuracy of the approximating polynomial $P_3(x)$ decreases.

x	-1.0	-0.2	-0.1	0.0	0.1	0.2	1.0
e^x	0.3679	0.81873	0.904837	1	1.105171	1.22140	2.7183
$P_3(x)$	0.3333	0.81867	0.904833	1	1.105167	1.22133	2.6667

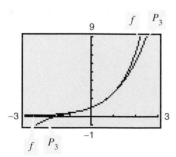

P_3 is the third-degree polynomial approximation of $f(x) = e^x$.
Figure 8.13

TECHNOLOGY A graphing utility can be used to compare the graph of the approximating polynomial with the graph of the function f. For instance, in Figure 8.13, the graph of

$$P_3(x) = 1 + x + \tfrac{1}{2}x^2 + \tfrac{1}{6}x^3 \qquad \text{3rd-degree approximation}$$

is compared with the graph of $f(x) = e^x$. If you have access to a graphing utility, try comparing the graphs of

$$P_4(x) = 1 + x + \tfrac{1}{2}x^2 + \tfrac{1}{6}x^3 + \tfrac{1}{24}x^4 \qquad \text{4th-degree approximation}$$
$$P_5(x) = 1 + x + \tfrac{1}{2}x^2 + \tfrac{1}{6}x^3 + \tfrac{1}{24}x^4 + \tfrac{1}{120}x^5 \qquad \text{5th-degree approximation}$$
$$P_6(x) = 1 + x + \tfrac{1}{2}x^2 + \tfrac{1}{6}x^3 + \tfrac{1}{24}x^4 + \tfrac{1}{120}x^5 + \tfrac{1}{720}x^6 \qquad \text{6th-degree approximation}$$

with the graph of f. What do you notice?

Taylor and Maclaurin Polynomials

The polynomial approximation of $f(x) = e^x$ given in Example 2 is expanded about $c = 0$. For expansions about an arbitrary value of c, it is convenient to write the polynomial in the form

$$P_n(x) = a_0 + a_1(x - c) + a_2(x - c)^2 + a_3(x - c)^3 + \cdots + a_n(x - c)^n.$$

In this form, repeated differentiation produces

$$P_n{}'(x) = a_1 + 2a_2(x - c) + 3a_3(x - c)^2 + \cdots + na_n(x - c)^{n-1}$$
$$P_n{}''(x) = 2a_2 + 2(3a_3)(x - c) + \cdots + n(n - 1)a_n(x - c)^{n-2}$$
$$P_n{}'''(x) = 2(3a_3) + \cdots + n(n - 1)(n - 2)a_n(x - c)^{n-3}$$
$$\vdots$$
$$P_n^{(n)}(x) = n(n - 1)(n - 2) \cdots (2)(1)a_n.$$

Letting $x = c$, you then obtain

$$P_n(c) = a_0, \qquad P_n{}'(c) = a_1, \qquad P_n{}''(c) = 2a_2, \quad \ldots, \qquad P_n^{(n)}(c) = n!a_n$$

and because the value of f and its first n derivatives must agree with the value of P_n and its first n derivatives at $x = c$, it follows that

$$f(c) = a_0, \qquad f'(c) = a_1, \qquad \frac{f''(c)}{2!} = a_2, \quad \ldots, \qquad \frac{f^{(n)}(c)}{n!} = a_n.$$

With these coefficients, you can obtain the following definition of **Taylor polynomials,** named after the English mathematician Brook Taylor, and **Maclaurin polynomials,** named after the English mathematician Colin Maclaurin (1698–1746).

BROOK TAYLOR (1685–1731)

Although Taylor was not the first to seek polynomial approximations of transcendental functions, his account published in 1715 was one of the first comprehensive works on the subject.

NOTE   Maclaurin polynomials are special types of Taylor polynomials for which $c = 0$.

Definition of nth Taylor Polynomial and nth Maclaurin Polynomial

If f has n derivatives at c, then the polynomial

$$P_n(x) = f(c) + f'(c)(x - c) + \frac{f''(c)}{2!}(x - c)^2 + \cdots + \frac{f^{(n)}(c)}{n!}(x - c)^n$$

is called the **nth Taylor polynomial for f at c.** If $c = 0$, then

$$P_n(x) = f(0) + f'(0)x + \frac{f''(0)}{2!}x^2 + \frac{f'''(0)}{3!}x^3 + \cdots + \frac{f^{(n)}(0)}{n!}x^n$$

is also called the **nth Maclaurin polynomial for f.**

Example 3   A Maclaurin Polynomial for $f(x) = e^x$

Find the nth Maclaurin polynomial for $f(x) = e^x$.

Solution   From the discussion on page 606, the nth Maclaurin polynomial for

$$f(x) = e^x$$

is given by

$$P_n(x) = 1 + x + \frac{1}{2!}x^2 + \frac{1}{3!}x^3 + \cdots + \frac{1}{n!}x^n.$$

FOR FURTHER INFORMATION To see how to use series to obtain other approximations to e, see the article "Novel Series-based Approximations to e" by John Knox and Harlan J. Brothers in *The College Mathematics Journal.* To view this article, go to the website *www.matharticles.com.*

Example 4   Finding Taylor Polynomials for ln x

Find the Taylor polynomials P_0, P_1, P_2, P_3, and P_4 for $f(x) = \ln x$ centered at $c = 1$.

Solution   Expanding about $c = 1$ yields the following.

$$f(x) = \ln x \qquad\qquad f(1) = \ln 1 = 0$$

$$f'(x) = \frac{1}{x} \qquad\qquad f'(1) = \frac{1}{1} = 1$$

$$f''(x) = -\frac{1}{x^2} \qquad\qquad f''(1) = -\frac{1}{1^2} = -1$$

$$f'''(x) = \frac{2!}{x^3} \qquad\qquad f'''(1) = \frac{2!}{1^3} = 2$$

$$f^{(4)}(x) = -\frac{3!}{x^4} \qquad\qquad f^{(4)}(1) = -\frac{3!}{1^4} = -6$$

Therefore, the Taylor polynomials are as follows.

$$P_0(x) = f(1) = 0$$

$$P_1(x) = f(1) + f'(1)(x - 1) = (x - 1)$$

$$P_2(x) = f(1) + f'(1)(x - 1) + \frac{f''(1)}{2!}(x - 1)^2$$

$$= (x - 1) - \frac{1}{2}(x - 1)^2$$

$$P_3(x) = f(1) + f'(1)(x - 1) + \frac{f''(1)}{2!}(x - 1)^2 + \frac{f'''(1)}{3!}(x - 1)^3$$

$$= (x - 1) - \frac{1}{2}(x - 1)^2 + \frac{1}{3}(x - 1)^3$$

$$P_4(x) = f(1) + f'(1)(x - 1) + \frac{f''(1)}{2!}(x - 1)^2 + \frac{f'''(1)}{3!}(x - 1)^3$$

$$+ \frac{f^{(4)}(1)}{4!}(x - 1)^4$$

$$= (x - 1) - \frac{1}{2}(x - 1)^2 + \frac{1}{3}(x - 1)^3 - \frac{1}{4}(x - 1)^4$$

Figure 8.14 compares the graphs of P_1, P_2, P_3, and P_4 with the graph of $f(x) = \ln x$. Note that near $x = 1$ the graphs are nearly indistinguishable. For instance, $P_4(0.9) \approx -0.105358$ and $\ln(0.9) \approx -0.105361$.

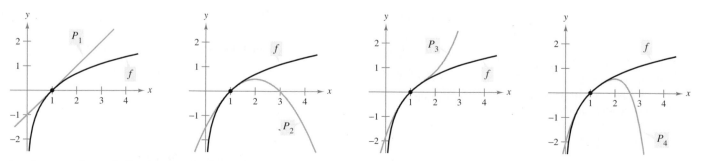

As n increases, the graph of P_n becomes a better and better approximation of the graph of $f(x) = \ln x$ near $x = 1$.
Figure 8.14

Example 5   Finding Maclaurin Polynomials for cos x

Find the Maclaurin polynomials P_0, P_2, P_4, and P_6 for $f(x) = \cos x$. Use $P_6(x)$ to approximate the value of $\cos(0.1)$.

Solution   Expanding about $c = 0$ yields the following.

$$f(x) = \cos x \qquad\qquad f(0) = \cos 0 = 1$$
$$f'(x) = -\sin x \qquad\qquad f'(0) = -\sin 0 = 0$$
$$f''(x) = -\cos x \qquad\qquad f''(0) = -\cos 0 = -1$$
$$f'''(x) = \sin x \qquad\qquad f'''(0) = \sin 0 = 0$$

Through repeated differentiation, you can see that the pattern $1, 0, -1, 0$ continues, and you obtain the following Maclaurin polynomials.

$$P_0(x) = 1$$
$$P_2(x) = 1 - \frac{1}{2!}x^2$$
$$P_4(x) = 1 - \frac{1}{2!}x^2 + \frac{1}{4!}x^4$$
$$P_6(x) = 1 - \frac{1}{2!}x^2 + \frac{1}{4!}x^4 - \frac{1}{6!}x^6$$

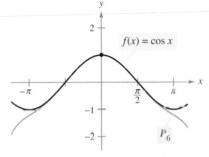

Near $(0, 1)$, the graph of P_6 can be used to approximate the graph of $f(x) = \cos x$.
Figure 8.15

Using $P_6(x)$, you obtain the approximation $\cos(0.1) \approx 0.995004165$, which coincides with the calculator value to nine decimal places. Figure 8.15 compares the graphs of $f(x) = \cos x$ and P_6.

Note in Example 5 that the Maclaurin polynomials for cos x have only even powers of x. Similarly, the Maclaurin polynomials for sin x have only odd powers of x (see Exercise 17). This is not generally true of the Taylor polynomials for sin x and cos x expanded about $c \neq 0$, as you can see in the next example.

Example 6   Finding a Taylor Polynomial for sin x

Find the third Taylor polynomial for $f(x) = \sin x$, expanded about $c = \pi/6$.

Solution   Expanding about $c = \pi/6$ yields the following.

$$f(x) = \sin x \qquad\qquad f\left(\frac{\pi}{6}\right) = \sin\frac{\pi}{6} = \frac{1}{2}$$

$$f'(x) = \cos x \qquad\qquad f'\left(\frac{\pi}{6}\right) = \cos\frac{\pi}{6} = \frac{\sqrt{3}}{2}$$

$$f''(x) = -\sin x \qquad\qquad f''\left(\frac{\pi}{6}\right) = -\sin\frac{\pi}{6} = -\frac{1}{2}$$

$$f'''(x) = -\cos x \qquad\qquad f'''\left(\frac{\pi}{6}\right) = -\cos\frac{\pi}{6} = -\frac{\sqrt{3}}{2}$$

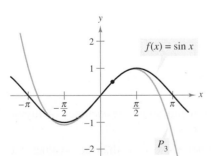

Near $(\pi/6, 1/2)$, the graph of P_3 can be used to approximate the graph of $f(x) = \sin x$.
Figure 8.16

So, the third Taylor polynomial for $f(x) = \sin x$, expanded about $c = \pi/6$, is

$$P_3(x) = f\left(\frac{\pi}{6}\right) + f'\left(\frac{\pi}{6}\right)\left(x - \frac{\pi}{6}\right) + \frac{f''\left(\frac{\pi}{6}\right)}{2!}\left(x - \frac{\pi}{6}\right)^2 + \frac{f'''\left(\frac{\pi}{6}\right)}{3!}\left(x - \frac{\pi}{6}\right)^3$$

$$= \frac{1}{2} + \frac{\sqrt{3}}{2}\left(x - \frac{\pi}{6}\right) - \frac{1}{2(2!)}\left(x - \frac{\pi}{6}\right)^2 - \frac{\sqrt{3}}{2(3!)}\left(x - \frac{\pi}{6}\right)^3.$$

Figure 8.16 compares the graphs of $f(x) = \sin x$ and P_3.

Taylor polynomials and Maclaurin polynomials can be used to approximate the value of a function at a specific point. For instance, to approximate the value of $\ln(1.1)$, you can use Taylor polynomials for $f(x) = \ln x$ expanded about $c = 1$, as shown in Example 4, or you can use Maclaurin polynomials, as shown in Example 7.

Example 7   **Approximation Using Maclaurin Polynomials**

Use a fourth Maclaurin polynomial to approximate the value of $\ln(1.1)$.

Solution   Because 1.1 is closer to 1 than to 0, you should consider Maclaurin polynomials for the function $g(x) = \ln(1 + x)$.

$$
\begin{aligned}
g(x) &= \ln(1 + x) & g(0) &= \ln(1 + 0) = 0 \\
g'(x) &= (1 + x)^{-1} & g'(0) &= (1 + 0)^{-1} = 1 \\
g''(x) &= -(1 + x)^{-2} & g''(0) &= -(1 + 0)^{-2} = -1 \\
g'''(x) &= 2(1 + x)^{-3} & g'''(0) &= 2(1 + 0)^{-3} = 2 \\
g^{(4)}(x) &= -6(1 + x)^{-4} & g^{(4)}(0) &= -6(1 + 0)^{-4} = -6
\end{aligned}
$$

Note that you obtain the same coefficients as in Example 4. Therefore, the fourth Maclaurin polynomial for $g(x) = \ln(1 + x)$ is

$$
P_4(x) = g(0) + g'(0)x + \frac{g''(0)}{2!}x^2 + \frac{g'''(0)}{3!}x^3 + \frac{g^{(4)}(0)}{4!}x^4
$$

$$
= x - \frac{1}{2}x^2 + \frac{1}{3}x^3 - \frac{1}{4}x^4.
$$

Consequently,

$$
\ln(1.1) = \ln(1 + 0.1) \approx P_4(0.1) \approx 0.0953083.
$$

Check to see that the fourth Taylor polynomial (from Example 4), evaluated at $x = 1.1$, yields the same result.   ▨

n	$P_n(0.1)$
1	0.1000000
2	0.0950000
3	0.0953333
4	0.0953083

The table at the left illustrates the accuracy of the Taylor polynomial approximation of the calculator value of $\ln(1.1)$. You can see that as n becomes larger, $P_n(0.1)$ approaches the calculator value of 0.0953102.

On the other hand, the table below illustrates that as you move away from the expansion point $c = 1$, the accuracy of the approximation decreases.

Fourth Taylor Polynomial Approximation of $\ln(1 + x)$

x	0.0	0.1	0.5	0.75	1.0
$\ln(1 + x)$	0.0000000	0.0953102	0.4054651	0.5596158	0.6931472
$P_4(x)$	0.0000000	0.0953083	0.4010417	0.5302734	0.5833333

These two tables illustrate two very important points about the accuracy of Taylor (or Maclaurin) polynomials for use in approximations.

1. The approximation is usually better at x-values close to c than at x-values far from c.

2. The approximation is usually better for higher-degree Taylor (or Maclaurin) polynomials than for those of lower degree.

Remainder of a Taylor Polynomial

An approximation technique is of little value without some idea of its accuracy. To measure the accuracy of approximating a function value $f(x)$ by the Taylor polynomial $P_n(x)$, you can use the concept of a **remainder** $R_n(x)$, defined as follows.

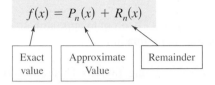

$$f(x) = P_n(x) + R_n(x)$$

Exact value   Approximate Value   Remainder

So, $R_n(x) = f(x) - P_n(x)$. The absolute value of $R_n(x)$ is called the **error** associated with the approximation. That is,

$$\text{Error} = |R_n(x)| = |f(x) - P_n(x)|.$$

The next theorem gives a general procedure for estimating the remainder associated with a Taylor polynomial. This important theorem is called **Taylor's Theorem,** and the remainder given in the theorem is called the **Lagrange form of the remainder.** (The proof of the theorem is lengthy, and is given in Appendix B.)

THEOREM 8.19   Taylor's Theorem

If a function f is differentiable through order $n + 1$ in an interval I containing c, then, for each x in I, there exists z between x and c such that

$$f(x) = f(c) + f'(c)(x - c) + \frac{f''(c)}{2!}(x - c)^2 + \cdots + \frac{f^{(n)}(c)}{n!}(x - c)^n + R_n(x)$$

where

$$R_n(x) = \frac{f^{(n+1)}(z)}{(n+1)!}(x - c)^{n+1}.$$

NOTE   One useful consequence of Taylor's Theorem is that

$$|R_n(x)| \le \frac{|x - c|^{n+1}}{(n+1)!} \max |f^{(n+1)}(z)|$$

where $\max|f^{(n+1)}(z)|$ is the maximum value of $f^{(n+1)}(z)$ between x and c.

For $n = 0$, Taylor's Theorem states that if f is differentiable in an interval I containing c, then, for each x in I, there exists z between x and c such that

$$f(x) = f(c) + f'(z)(x - c) \quad \text{or} \quad f'(z) = \frac{f(x) - f(c)}{x - c}.$$

Do you recognize this special case of Taylor's Theorem? (It is the Mean Value Theorem.)

When applying Taylor's Theorem, you should not expect to be able to find the exact value of z. (If you could do this, an approximation would not be necessary.) Rather, you try to find bounds for $f^{(n+1)}(z)$ from which you are able to tell how large the remainder $R_n(x)$ is.

Example 8   Determining the Accuracy of an Approximation

The third Maclaurin polynomial for $\sin x$ is given by

$$P_3(x) = x - \frac{x^3}{3!}.$$

Use Taylor's Theorem to approximate $\sin(0.1)$ by $P_3(0.1)$ and determine the accuracy of the approximation.

Solution   Using Taylor's Theorem, you have

$$\sin x = x - \frac{x^3}{3!} + R_3(x) = x - \frac{x^3}{3!} + \frac{f^{(4)}(z)}{4!}x^4$$

where $0 < z < 0.1$. Therefore,

$$\sin(0.1) \approx 0.1 - \frac{(0.1)^3}{3!} \approx 0.1 - 0.000167 = 0.099833.$$

Because $f^{(4)}(z) = \sin z$, it follows that the error $|R_3(0.1)|$ can be bounded as follows.

$$0 < R_3(0.1) = \frac{\sin z}{4!}(0.1)^4 < \frac{0.0001}{4!} \approx 0.000004$$

This implies that

$$0.099833 < \sin(0.1) = 0.099833 + R_3(x) < 0.099833 + 0.000004$$
$$0.099833 < \sin(0.1) < 0.099837.$$

NOTE   Try using a calculator to verify the results obtained in Examples 8 and 9. For Example 8, you obtain

$$\sin(0.1) \approx 0.0998334.$$

For Example 9, you obtain

$$P_3(1.2) \approx 0.1827$$

and

$$\ln(1.2) \approx 0.1823.$$

Example 9   Approximating a Value to a Desired Accuracy

Determine the degree of the Taylor polynomial $P_n(x)$ expanded about $c = 1$ that should be used to approximate $\ln(1.2)$ so that the error is less than 0.001.

Solution   Following the pattern of Example 4, you can see that the $(n + 1)$st derivative of $f(x) = \ln x$ is given by

$$f^{(n+1)}(x) = (-1)^n \frac{n!}{x^{n+1}}.$$

Using Taylor's Theorem, you know that the error $|R_n(1.2)|$ is given by

$$|R_n(1.2)| = \left| \frac{f^{(n+1)}(z)}{(n+1)!}(1.2 - 1)^{n+1} \right| = \frac{n!}{z^{n+1}}\left[\frac{1}{(n+1)!} \right](0.2)^{n+1}$$

$$= \frac{(0.2)^{n+1}}{z^{n+1}(n+1)}$$

where $1 < z < 1.2$. In this interval, $(0.2)^{n+1}/z^{n+1}(n+1)$ is less than $(0.2)^{n+1}/(n+1)$. So, you are seeking a value of n such that

$$\frac{(0.2)^{n+1}}{(n+1)}0.001 \quad \Longrightarrow \quad 1000 < (n+1)5^{n+1}.$$

By trial and error, you can determine that the smallest value of n that satisfies this inequality is $n = 3$. So, you would need the third Taylor polynomial to achieve the desired accuracy in approximating $\ln(1.2)$.

EXERCISES FOR SECTION 8.7

In Exercises 1–4, match the Taylor polynomial approximation of the function $f(x) = e^{-x^2/2}$ with the correct graph. [The graphs are labeled (a), (b), (c), and (d).]

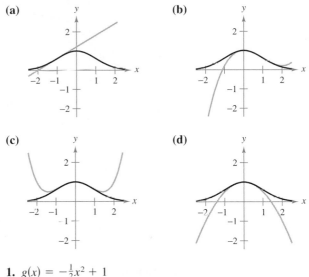

(a) **(b)**

(c) **(d)**

1. $g(x) = -\frac{1}{2}x^2 + 1$

2. $g(x) = \frac{1}{8}x^4 - \frac{1}{2}x^2 + 1$

3. $g(x) = e^{-1/2}[(x + 1) + 1]$

4. $g(x) = e^{-1/2}\left[\frac{1}{3}(x - 1)^3 - (x - 1) + 1\right]$

In Exercises 5–8, find a first-degree polynomial function P_1 whose value and slope agree with the value and slope of f at $x = c$. Use a graphing utility to graph f and P_1. What is P_1 called?

5. $f(x) = \dfrac{4}{\sqrt{x}}, \quad c = 1$

6. $f(x) = \dfrac{4}{\sqrt[3]{x}}, \quad c = 8$

7. $f(x) = \sec x, \quad c = \dfrac{\pi}{4}$

8. $f(x) = \tan x, \quad c = \dfrac{\pi}{4}$

Graphical and Numerical Analysis **In Exercises 9 and 10, use a graphing utility to graph f and its second-degree polynomial approximation P_2 at $x = c$. Complete the table comparing the values of f and P_2.**

9. $f(x) = \dfrac{4}{\sqrt{x}}, \quad c = 1$

$P_2(x) = 4 - 2(x - 1) + \frac{3}{2}(x - 1)^2$

x	0	0.8	0.9	1	1.1	1.2	2
$f(x)$							
$P_2(x)$							

10. $f(x) = \sec x, \quad c = \dfrac{\pi}{4}$

$P_2(x) = \sqrt{2} + \sqrt{2}\left(x - \dfrac{\pi}{4}\right) + \dfrac{3}{2}\sqrt{2}\left(x - \dfrac{\pi}{4}\right)^2$

x	-2.15	0.585	0.685	$\dfrac{\pi}{4}$	0.885	0.985	1.785
$f(x)$							
$P_2(x)$							

11. *Conjecture* Consider the function $f(x) = \cos x$ and its Maclaurin polynomials P_2, P_4, and P_6 (see Example 5).

   (a) Use a graphing utility to graph f and the indicated polynomial approximations.

   (b) Evaluate and compare the values of $f^{(n)}(0)$ and $P_n^{(n)}(0)$ for $n = 2$, 4, and 6.

   (c) Use the results in part (b) to make a conjecture about $f^{(n)}(0)$ and $P_n^{(n)}(0)$.

12. *Conjecture* Consider the function $f(x) = x^2 e^x$.

   (a) Find the Maclaurin polynomials P_2, P_3, and P_4 for f.

   (b) Use a graphing utility to graph f, P_2, P_3, and P_4.

   (c) Evaluate and compare the values of $f^{(n)}(0)$ and $P_n^{(n)}(0)$ for $n = 2$, 3, and 4.

   (d) Use the results in part (c) to make a conjecture about $f^{(n)}(0)$ and $P_n^{(n)}(0)$.

In Exercises 13–24, find the Maclaurin polynomial of degree n for the function.

13. $f(x) = e^{-x}, \quad n = 3$

14. $f(x) = e^{-x}, \quad n = 5$

15. $f(x) = e^{2x}, \quad n = 4$

16. $f(x) = e^{3x}, \quad n = 4$

17. $f(x) = \sin x, \quad n = 5$

18. $f(x) = \sin \pi x, \quad n = 3$

19. $f(x) = xe^x, \quad n = 4$

20. $f(x) = x^2 e^{-x}, \quad n = 4$

21. $f(x) = \dfrac{1}{x + 1}, \quad n = 4$

22. $f(x) = \dfrac{x}{x + 1}, \quad n = 4$

23. $f(x) = \sec x, \quad n = 2$

24. $f(x) = \tan x, \quad n = 3$

In Exercises 25–30, find the nth Taylor polynomial centered at c.

25. $f(x) = \dfrac{1}{x}, \quad n = 4, \quad c = 1$

26. $f(x) = \dfrac{2}{x^2}, \quad n = 4, \quad c = 2$

27. $f(x) = \sqrt{x}, \quad n = 4, \quad c = 1$

28. $f(x) = \sqrt[3]{x}, \quad n = 3, \quad c = 8$

29. $f(x) = \ln x, \quad n = 4, \quad c = 1$

30. $f(x) = x^2 \cos x, \quad n = 2, \quad c = \pi$

In Exercises 31 and 32, use a computer algebra system to find the indicated Taylor polynomials for the function f. Graph the function and the Taylor polynomials.

31. $f(x) = \tan x$

(a) $n = 3$, $c = 0$

(b) $n = 5$, $c = 0$

(c) $n = 3$, $c = \pi/4$

32. $f(x) = \dfrac{1}{x^2 + 1}$

(a) $n = 2$, $c = 0$

(b) $n = 4$, $c = 0$

(c) $n = 4$, $c = 1$

33. *Numerical and Graphical Approximations*

(a) Use the Maclaurin polynomials $P_1(x)$, $P_3(x)$, $P_5(x)$, and $P_7(x)$ for $f(x) = \sin x$ to complete the table.

x	0	0.25	0.50	0.75	1.00
$\sin x$	0	0.2474	0.4794	0.6816	0.8415
$P_1(x)$					
$P_3(x)$					
$P_5(x)$					
$P_7(x)$					

(b) Use a graphing utility to graph $f(x) = \sin x$ and the Maclaurin polynomials in part (a).

(c) Describe the change in accuracy of a polynomial approximation as the distance from the point where the polynomial is centered increases.

34. *Numerical and Graphical Approximations*

(a) Use the Taylor polynomials $P_1(x)$ and $P_4(x)$ for $f(x) = \ln x$ centered at $c = 1$ to complete the table.

x	1.00	1.25	1.50	1.75	2.00
$\ln x$	0	0.2231	0.4055	0.5596	0.6931
$P_1(x)$					
$P_4(x)$					

(b) Use a graphing utility to graph $f(x) = \ln x$ and the Taylor polynomials in part (a).

(c) Describe the change in accuracy of polynomial approximations as the degree increases.

Numerical and Graphical Approximations In Exercises 35 and 36, (a) find the Maclaurin polynomial $P_3(x)$ for $f(x)$, (b) complete the table for $f(x)$ and $P_3(x)$, and (c) sketch the graphs of $f(x)$ and $P_3(x)$ on the same set of coordinate axes.

x	-0.75	-0.50	-0.25	0	0.25	0.50	0.75
$f(x)$							
$P_3(x)$							

35. $f(x) = \arcsin x$

36. $f(x) = \arctan x$

In Exercises 37–40, the graph of $y = f(x)$ is shown with four of its Maclaurin polynomials. Identify the Maclaurin polynomials and use a graphing utility to confirm your results.

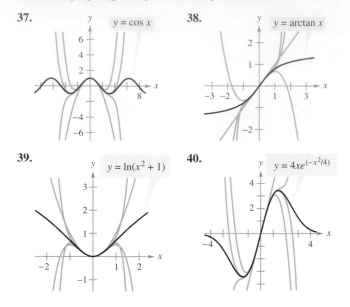

37. $y = \cos x$

38. $y = \arctan x$

39. $y = \ln(x^2 + 1)$

40. $y = 4xe^{(-x^2/4)}$

In Exercises 41–44, approximate the function at the given value of x, using the polynomial found in the indicated exercise.

41. $f(x) = e^{-x}$, $\quad f\left(\frac{1}{2}\right)$, Exercise 13

42. $f(x) = x^2 e^{-x}$, $\quad f\left(\frac{1}{5}\right)$, Exercise 20

43. $f(x) = \ln x$, $\quad f(1.2)$, Exercise 29

44. $f(x) = x^2 \cos x$, $\quad f\left(\dfrac{7\pi}{8}\right)$, Exercise 30

In Exercises 45–48, use Taylor's Theorem to obtain an upper bound for the error of the approximation. Then calculate the exact value of the error.

45. $\cos(0.3) \approx 1 - \dfrac{(0.3)^2}{2!} + \dfrac{(0.3)^4}{4!}$

46. $e \approx 1 + 1 + \dfrac{1^2}{2!} + \dfrac{1^3}{3!} + \dfrac{1^4}{4!} + \dfrac{1^5}{5!}$

47. $\arcsin(0.4) \approx 0.4 + \dfrac{(0.4)^3}{2 \cdot 3}$ **48.** $\arctan(0.4) \approx 0.4 - \dfrac{(0.4)^3}{3}$

In Exercises 49 and 50, determine the degree of the Maclaurin polynomial required for the error in the approximation of the function at the indicated value of x to be less than 0.001.

49. $\sin(0.3)$

50. $e^{0.6}$

In Exercises 51 and 52, determine the degree of the Maclaurin polynomial required for the error in the approximation of the function at the indicated value of x to be less than 0.0001. Use a computer algebra system to obtain and evaluate the required derivatives.

51. $f(x) = \ln(x + 1)$, approximate $f(0.5)$.

52. $f(x) = \cos(\pi x^2)$, approximate $f(0.6)$.

In Exercises 53 and 54, determine the values of x for which the function can be replaced by the Taylor polynomial if the error cannot exceed 0.001.

53. $f(x) = e^x \approx 1 + x + \dfrac{x^2}{2!} + \dfrac{x^3}{3!},$ $x < 0$

54. $f(x) = \sin x \approx x - \dfrac{x^3}{3!}$

Getting at the Concept

55. An elementary function is approximated by a polynomial. In your own words, describe what is meant by saying that the polynomial is *expanded about c* or *centered at c*.

56. When an elementary function f is approximated by a second-degree polynomial P_2 centered at c, what is known about f and P_2 at c?

57. State the definition of an nth-degree Taylor polynomial of f centered at c.

58. Describe the accuracy of the nth-degree Taylor polynomial of f centered at c as the distance between c and x increases.

59. In general, how does the accuracy of a Taylor polynomial change as the degree of the polynomial is increased?

60. The graphs show first-, second-, and third-degree polynomial approximations P_1, P_2, and P_3 of a function f. Label the graphs of P_1, P_2, and P_3. To print an enlarged copy of the graph, go to the website *www.mathgraphs.com*.

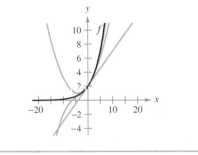

61. *Comparing Maclaurin Polynomials*

(a) Compare the Maclaurin polynomials of degree 4 and degree 5, respectively, for the functions

$f(x) = e^x$ and $g(x) = xe^x$.

What is the relationship between them?

(b) Use the result in part (a) and the Maclaurin polynomial of degree 5 for $f(x) = \sin x$ to find a Maclaurin polynomial of degree 6 for the function $g(x) = x \sin x$.

(c) Use the result in part (a) and the Maclaurin polynomial of degree 5 for $f(x) = \sin x$ to find a Maclaurin polynomial of degree 4 for the function $g(x) = (\sin x)/x$.

62. *Differentiating Maclaurin Polynomials*

(a) Differentiate the Maclaurin polynomial of degree 5 for $f(x) = \sin x$ and compare the result with the Maclaurin polynomial of degree 4 for $g(x) = \cos x$.

(b) Differentiate the Maclaurin polynomial of degree 6 for $f(x) = \cos x$ and compare the result with the Maclaurin polynomial of degree 5 for $g(x) = \sin x$.

(c) Differentiate the Maclaurin polynomial of degree 4 for $f(x) = e^x$. Describe the relationship between the two series.

63. *Graphical Reasoning*   The figure shows the graph of the function

$$f(x) = \sin\left(\frac{\pi x}{4}\right)$$

and the second-degree Taylor polynomial

$$P_2(x) = 1 - \frac{\pi^2}{32}(x - 2)^2$$

centered at $x = 2$.

(a) Use the symmetry of the graph of f to write the second-degree Taylor polynomial for f centered at $x = -2$.

(b) Use a horizontal translation of the result in part (a) to find the second-degree Taylor polynomial for f centered at $x = 6$.

(c) Is it possible to use a horizontal translation of the result in part (a) to write a second-degree Taylor polynomial for f centered at $x = 4$? Explain.

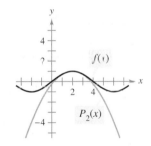

64. Prove that if f is an odd function, then its nth Maclaurin polynomial contains only terms with odd powers of x.

65. Prove that if f is an even function, then its nth Maclaurin polynomial contains only terms with even powers of x.

66. Let $P_n(x)$ be the nth Taylor polynomial for f at c. Prove that $P_n(c) = f(c)$ and $P^{(k)}(c) = f^{(k)}(c)$ for $1 \le k \le n$. (See Exercises 9 and 10.)

67. *Writing*   The proof in Exercise 66 guarantees that the Taylor polynomial and its derivatives agree with the function and its derivatives at $x = c$. Use the graphs and tables in Exercises 33–36 to discuss what happens to the accuracy of the Taylor polynomial as you move away from $x = c$.

- Understand the definition of a power series.
- Find the radius and interval of convergence of a power series.
- Determine the endpoint convergence of a power series.
- Differentiate and integrate a power series.

Power Series

In Section 8.7, we introduced the concept of approximating functions by Taylor polynomials. For instance, the function $f(x) = e^x$ can be *approximated* by its Maclaurin polynomials as follows.

$$e^x \approx 1 + x \qquad \text{1st-degree polynomial}$$

$$e^x \approx 1 + x + \frac{x^2}{2!} \qquad \text{2nd-degree polynomial}$$

$$e^x \approx 1 + x + \frac{x^2}{2!} + \frac{x^3}{3!} \qquad \text{3rd-degree polynomial}$$

$$e^x \approx 1 + x + \frac{x^2}{2!} + \frac{x^3}{3!} + \frac{x^4}{4!} \qquad \text{4th-degree polynomial}$$

$$e^x \approx 1 + x + \frac{x^2}{2!} + \frac{x^3}{3!} + \frac{x^4}{4!} + \frac{x^5}{5!} \qquad \text{5th-degree polynomial}$$

In that section, you saw that the higher the degree of the approximating polynomial, the better the approximation becomes.

In this and the next two sections, you will see that several important types of functions, including

$$f(x) = e^x$$

can be represented *exactly* by an infinite series called a **power series.** For example, the power series representation for e^x is

$$e^x = 1 + x + \frac{x^2}{2!} + \frac{x^3}{3!} + \cdots + \frac{x^n}{n!} + \cdots.$$

For each real number x, it can be shown that the infinite series on the right converges to the number e^x. Before doing this, however, we will discuss some preliminary results dealing with power series—beginning with the following definition.

EXPLORATION

Graphical Reasoning Use a graphing utility to approximate the graphs of the following power series near $x = 0$. (Use the first several terms of each series.) Each series represents a well-known function. What is the function?

a. $\displaystyle\sum_{n=0}^{\infty} \frac{(-1)^n x^n}{n!}$

b. $\displaystyle\sum_{n=0}^{\infty} \frac{(-1)^n x^{2n}}{(2n)!}$

c. $\displaystyle\sum_{n=0}^{\infty} \frac{(-1)^n x^{2n+1}}{(2n+1)!}$

d. $\displaystyle\sum_{n=0}^{\infty} \frac{(-1)^n x^{2n+1}}{2n+1}$

e. $\displaystyle\sum_{n=0}^{\infty} \frac{2^n x^n}{n!}$

Definition of Power Series

If x is a variable, then an infinite series of the form

$$\sum_{n=0}^{\infty} a_n x^n = a_0 + a_1 x + a_2 x^2 + a_3 x^3 + \cdots + a_n x^n + \cdots$$

is called a **power series.** More generally, series of the form

$$\sum_{n=0}^{\infty} a_n (x - c)^n = a_0 + a_1(x - c) + a_2(x - c)^2 + \cdots + a_n(x - c)^n + \cdots$$

is called a **power series centered at c,** where c is a constant.

NOTE   To simplify the notation for power series, we agree that $(x - c)^0 = 1$, even if $x = c$.

Example 1   **Power Series**

a. The following power series is centered at 0.

$$\sum_{n=0}^{\infty} \frac{x^n}{n!} = 1 + x + \frac{x^2}{2} + \frac{x^3}{3!} + \cdots$$

b. The following power series is centered at -1.

$$\sum_{n=0}^{\infty} (-1)^n (x + 1)^n = 1 - (x + 1) + (x + 1)^2 - (x + 1)^3 + \cdots$$

c. The following power series is centered at 1.

$$\sum_{n=1}^{\infty} \frac{1}{n} (x - 1)^n = (x - 1) + \frac{1}{2} (x - 1)^2 + \frac{1}{3} (x - 1)^3 + \cdots$$

Radius and Interval of Convergence

A power series in x can be viewed as a function of x

$$f(x) = \sum_{n=0}^{\infty} a_n (x - c)^n$$

where the *domain of f* is the set of all x for which the power series converges. Determination of the domain of a power series is the primary concern in this section. Of course, every power series converges at its center c because

$$f(c) = \sum_{n=0}^{\infty} a_n (c - c)^n$$
$$= a_0(1) + 0 + 0 + \cdots + 0 + \cdots$$
$$= a_0.$$

So, c always lies in the domain of f. The following important theorem states that the domain of a power series can take three basic forms: a single point, an interval centered at c, or the entire real line, as shown in Figure 8.17. A proof is given in Appendix B.

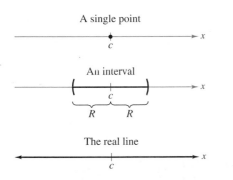

A single point

An interval

R R

The real line

The domain of a power series has only three basic forms: a single point, an interval centered at c, or the entire real line.
Figure 8.17

> **THEOREM 8.20   Convergence of a Power Series**
>
> For a power series centered at c, precisely one of the following is true.
>
> **1.** The series converges only at c.
> **2.** There exists a real number $R > 0$ such that the series converges absolutely for $|x - c| < R$, and diverges for $|x - c| > R$.
> **3.** The series converges absolutely for all x.
>
> The number R is the **radius of convergence** of the power series. If the series converges only at c, the radius of convergence is $R = 0$, and if the series converges for all x, the radius of convergence is $R = \infty$. The set of all values of x for which the power series converges is the **interval of convergence** of the power series.

Example 2 Finding the Radius of Convergence

Find the radius of convergence of $\displaystyle\sum_{n=0}^{\infty} n!x^n$.

Solution For $x = 0$, you obtain

$$f(0) = \sum_{n=0}^{\infty} n!0^n = 1 + 0 + 0 + \cdots = 1.$$

For any fixed value of x such that $|x| > 0$, let $u_n = n!x^n$. Then

$$\lim_{n\to\infty} \left|\frac{u_{n+1}}{u_n}\right| = \lim_{n\to\infty} \left|\frac{(n+1)!x^{n+1}}{n!x^n}\right|$$
$$= |x| \lim_{n\to\infty} (n+1)$$
$$= \infty.$$

Therefore, by the Ratio Test, the series diverges for $|x| > 0$ and converges only at its center, 0. Hence, the radius of convergence is $R = 0$.

Example 3 Finding the Radius of Convergence

Find the radius of convergence of

$$\sum_{n=0}^{\infty} 3(x-2)^n.$$

Solution For $x \neq 2$, let $u_n = 3(x-2)^n$. Then

$$\lim_{n\to\infty} \left|\frac{u_{n+1}}{u_n}\right| = \lim_{n\to\infty} \left|\frac{3(x-2)^{n+1}}{3(x-2)^n}\right|$$
$$= \lim_{n\to\infty} |x-2|$$
$$= |x-2|.$$

By the Ratio Test, the series converges if $|x-2| < 1$ and diverges if $|x-2| > 1$. Therefore, the radius of convergence of the series is $R = 1$.

Example 4 Finding the Radius of Convergence

Find the radius of convergence of

$$\sum_{n=0}^{\infty} \frac{(-1)^n x^{2n+1}}{(2n+1)!}.$$

Solution Let $u_n = (-1)^n x^{2n+1}/(2n+1)!$. Then

$$\lim_{n\to\infty} \left|\frac{u_{n+1}}{u_n}\right| = \lim_{n\to\infty} \left|\frac{[(-1)^{n+1} x^{2n+3}]/(2n+3)!}{[(-1)^n x^{2n+1}]/(2n+1)!}\right|$$
$$= \lim_{n\to\infty} \frac{x^2}{(2n+3)(2n+2)}.$$

For any *fixed* value of x, this limit is 0. So, by the Ratio Test, the series converges for all x. Therefore, the radius of convergence is $R = \infty$.

Endpoint Convergence

Note that for a power series whose radius of convergence is a finite number R, Theorem 8.20 says nothing about the convergence at the *endpoints* of the interval of convergence. Each endpoint must be tested separately for convergence or divergence. As a result, the interval of convergence of a power series can take any one of the six forms shown in Figure 8.18.

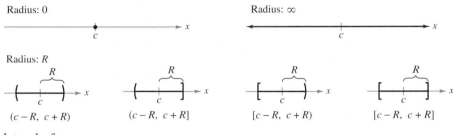

Intervals of convergence
Figure 8.18

Example 5   **Finding the Interval of Convergence**

Find the interval of convergence of $\displaystyle\sum_{n=1}^{\infty} \frac{x^n}{n}$.

Solution   Letting $u_n = x^n/n$ produces

$$\lim_{n\to\infty} \left| \frac{u_{n+1}}{u_n} \right| = \lim_{n\to\infty} \left| \frac{x^{n+1}/(n+1)}{x^n/n} \right|$$

$$= \lim_{n\to\infty} \left| \frac{nx}{n+1} \right|$$

$$= |x|.$$

Therefore, by the Ratio Test, the radius of convergence is $R = 1$. Moreover, because the series is centered at 0, it converges in the interval $(-1, 1)$. This interval, however, is not necessarily the *interval of convergence*. To determine this, you must test for convergence at each endpoint. When $x = 1$, you obtain the *divergent* harmonic series

$$\sum_{n=1}^{\infty} \frac{1}{n} = \frac{1}{1} + \frac{1}{2} + \frac{1}{3} + \cdots. \qquad \text{Diverges when } x = 1$$

When $x = -1$, you obtain the *convergent* alternating harmonic series

$$\sum_{n=1}^{\infty} \frac{(-1)^n}{n} = -1 + \frac{1}{2} - \frac{1}{3} + \frac{1}{4} - \cdots. \qquad \text{Converges when } x = -1$$

Therefore, the interval of convergence for the series is $[-1, 1)$, as shown in Figure 8.19.

Interval: $[-1, 1)$
Radius: $R = 1$

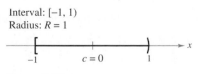

Figure 8.19

Example 6   Finding the Interval of Convergence

Find the interval of convergence of

$$\sum_{n=0}^{\infty} \frac{(-1)^n(x+1)^n}{2^n}.$$

Solution   Letting $u_n = (-1)^n(x+1)^n/2^n$ produces

$$\lim_{n\to\infty}\left|\frac{u_{n+1}}{n}\right| = \lim_{n\to\infty}\left|\frac{(-1)^{n+1}(x+1)^{n+1}/2^{n+1}}{(-1)^n(x+1)^n/2^n}\right|$$

$$= \lim_{n\to\infty}\left|\frac{2^n(x+1)}{2^{n+1}}\right|$$

$$= \left|\frac{x+1}{2}\right|.$$

By the Ratio Test, the series converges if $|(x+1)/2| < 1$ or $|x+1| < 2$. So, the radius of convergence is $R = 2$. Because the series is centered at $x = -1$, it will converge in the interval $(-3, 1)$. Furthermore, at the endpoints you have

$$\sum_{n=0}^{\infty} \frac{(-1)^n(-2)^n}{2^n} = \sum_{n=0}^{\infty}\frac{2^n}{2^n} = \sum_{n=0}^{\infty} 1 \qquad\text{Diverges when } x = -3$$

and

$$\sum_{n=0}^{\infty} \frac{(-1)^n(2)^n}{2^n} = \sum_{n=0}^{\infty}(-1)^n \qquad\text{Diverges when } x = 1$$

both of which diverge. So, the interval of convergence is $(-3, 1)$, as shown in Figure 8.20.

Interval: $(-3, 1)$
Radius: $R = 2$

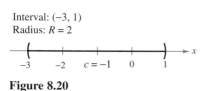

Figure 8.20

Example 7   Finding the Interval of Convergence

Find the interval of convergence of

$$\sum_{n=1}^{\infty} \frac{x^n}{n^2}.$$

Solution   Letting $u_n = x^n/n^2$ produces

$$\lim_{n\to\infty}\left|\frac{u_{n+1}}{u_n}\right| = \lim_{n\to\infty}\left|\frac{x^{n+1}/(n+1)^2}{x^n/n^2}\right|$$

$$= \lim_{n\to\infty}\left|\frac{n^2 x}{(n+1)^2}\right| = |x|.$$

So, the radius of convergence is $R = 1$. Because the series is centered at $x = 0$, it converges in the interval $(-1, 1)$. When $x = 1$, you obtain the *convergent p*-series

$$\sum_{n=1}^{\infty} \frac{1}{n^2} = \frac{1}{1^2} + \frac{1}{2^2} + \frac{1}{3^2} + \frac{1}{4^2} + \cdots. \qquad\text{Converges when } x = 1$$

When $x = -1$, you obtain the *convergent* alternating series

$$\sum_{n=1}^{\infty} \frac{(-1)^n}{n^2} = -\frac{1}{1^2} + \frac{1}{2^2} - \frac{1}{3^2} + \frac{1}{4^2} - \cdots. \qquad\text{Converges when } x = -1$$

Therefore, the interval of convergence for the given series is $[-1, 1]$.

Differentiation and Integration of Power Series

Power series representation of functions has played an important role in the development of calculus. In fact, much of Newton's work with differentiation and integration was done in the context of power series—especially his work with complicated algebraic functions and transcendental functions. Euler, Lagrange, Leibniz, and the Bernoullis all used power series extensively in calculus.

Once you have defined a function with a power series, it is natural to wonder how you can determine the characteristics of the function. Is it continuous? Differentiable? Theorem 8.21, which we state without proof, answers these questions.

JAMES GREGORY (1638–1675)

One of the earliest mathematicians to work with power series was a Scotsman, James Gregory. He developed a power series method for interpolating table values—a method that was later used by Brook Taylor in the development of Taylor polynomials and Taylor series.

THEOREM 8.21   Properties of Functions Defined by Power Series

If the function given by

$$f(x) = \sum_{n=0}^{\infty} a_n(x - c)^n$$
$$= a_0 + a_1(x - c) + a_2(x - c)^2 + a_3(x - c)^3 + \cdots$$

has a radius of convergence of $R > 0$, then, on the interval $(c - R, c + R)$, f is differentiable (and therefore continuous). Moreover, the derivative and antiderivative of f are as follows.

1. $f'(x) = \sum_{n=1}^{\infty} na_n(x - c)^{n-1}$

$$= a_1 + 2a_2(x - c) + 3a_3(x - c)^2 + \cdots$$

2. $\int f(x)\, dx = C + \sum_{n=0}^{\infty} a_n \frac{(x - c)^{n+1}}{n + 1}$

$$= C + a_0(x - c) + a_1 \frac{(x - c)^2}{2} + a_2 \frac{(x - c)^3}{3} + \cdots$$

The *radius of convergence* of the series obtained by differentiating or integrating a power series is the same as that of the original power series. The *interval of convergence*, however, may differ as a result of the behavior at the endpoints.

Theorem 8.21 states that, in many ways, a function defined by a power series behaves like a polynomial. It is continuous in its interval of convergence, and both its derivative and its antiderivative can be determined by differentiating and integrating each term of the given power series. For instance, the derivative of the power series

$$f(x) = \sum_{n=0}^{\infty} \frac{x^n}{n!}$$
$$= 1 + x + \frac{x^2}{2} + \frac{x^3}{3!} + \frac{x^4}{4!} + \cdots$$

is

$$f'(x) = 1 + (2)\frac{x}{2} + (3)\frac{x^2}{3!} + (4)\frac{x^3}{4!} + \cdots$$
$$= 1 + x + \frac{x^2}{2} + \frac{x^3}{3!} + \frac{x^4}{4!} + \cdots$$
$$= f(x).$$

Notice that $f'(x) = f(x)$. Do you recognize this function?

Example 8　Intervals of Convergence for $f(x)$, $f'(x)$, and $\int f(x)\,dx$

Consider the function given by

$$f(x) = \sum_{n=1}^{\infty} \frac{x^n}{n} = x + \frac{x^2}{2} + \frac{x^3}{3} + \cdots.$$

Find the intervals of convergence for each of the following.

a. $\int f(x)\,dx$　　　**b.** $f(x)$　　　**c.** $f'(x)$

Solution　By Theorem 8.21, you have

$$f'(x) = \sum_{n=1}^{\infty} x^{n-1}$$

$$= 1 + x + x^2 + x^3 + \cdots$$

and

$$\int f(x)\,dx = C + \sum_{n=1}^{\infty} \frac{x^{n+1}}{n(n+1)}$$

$$= C + \frac{x^2}{1\cdot 2} + \frac{x^3}{2\cdot 3} + \frac{x^4}{3\cdot 4} + \cdots.$$

By the Ratio Test, you can show that each series has a radius of convergence of $R = 1$. Considering the interval $(-1, 1)$, you have the following.

a. For $\int f(x)\,dx$, the series

$$\sum_{n=1}^{\infty} \frac{x^{n+1}}{n(n+1)} \qquad \text{Interval of convergence: } [-1, 1]$$

converges for $x = \pm 1$, and its interval of convergence is $[-1, 1]$. See Figure 8.21(a).

b. For $f(x)$, the series

$$\sum_{n=1}^{\infty} \frac{x^n}{n} \qquad \text{Interval of convergence: } [-1, 1)$$

converges for $x = -1$ and diverges for $x = 1$. Hence, its interval of convergence is $[-1, 1)$. See Figure 8.21(b).

c. For $f'(x)$, the series

$$\sum_{n=1}^{\infty} x^{n-1} \qquad \text{Interval of convergence: } (-1, 1)$$

diverges for $x = \pm 1$, and its interval of convergence is $(-1, 1)$. See Figure 8.21(c).

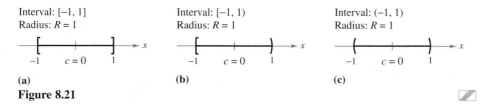

Interval: $[-1, 1]$　　　　　Interval: $[-1, 1)$　　　　　Interval: $(-1, 1)$
Radius: $R = 1$　　　　　　Radius: $R = 1$　　　　　　Radius: $R = 1$

(a)　　　　　　　　　　　(b)　　　　　　　　　　　(c)

Figure 8.21

From Example 8, it appears that of the three series, the one for the derivative, $f'(x)$, is the least likely to converge at the endpoints. In fact, it can be shown that if the series for $f'(x)$ converges at the endpoints $x = c \pm R$, the series for $f(x)$ will also converge there.

Lab Series | LAB 12

EXERCISES FOR SECTION 8.8

In Exercises 1–4, state where the power series is centered.

1. $\displaystyle\sum_{n=0}^{\infty} n x^n$

2. $\displaystyle\sum_{n=1}^{\infty} \frac{(-1)^n 1 \cdot 3 \cdots (2n-1)}{2^n n!} x^n$

3. $\displaystyle\sum_{n=1}^{\infty} \frac{(x-2)^n}{n^3}$

4. $\displaystyle\sum_{n=0}^{\infty} \frac{(-1)^n (x-\pi)^{2n}}{(2n)!}$

In Exercises 5–10, find the radius of convergence of the power series.

5. $\displaystyle\sum_{n=0}^{\infty} (-1)^n \frac{x^n}{n+1}$

6. $\displaystyle\sum_{n=0}^{\infty} (2x)^n$

7. $\displaystyle\sum_{n=1}^{\infty} \frac{(2x)^n}{n^2}$

8. $\displaystyle\sum_{n=0}^{\infty} \frac{(-1)^n x^n}{2^n}$

9. $\displaystyle\sum_{n=0}^{\infty} \frac{(2x)^{2n}}{(2n)!}$

10. $\displaystyle\sum_{n=0}^{\infty} \frac{(2n)! x^{2n}}{n!}$

In Exercises 11–34, find the interval of convergence of the power series. (Be sure to include a check for convergence at the endpoints of the interval.)

11. $\displaystyle\sum_{n=0}^{\infty} \left(\frac{x}{2}\right)^n$

12. $\displaystyle\sum_{n=0}^{\infty} \left(\frac{x}{k}\right)^n, \quad k>0$

13. $\displaystyle\sum_{n=1}^{\infty} \frac{(-1)^n x^n}{n}$

14. $\displaystyle\sum_{n=0}^{\infty} (-1)^{n+1}(n+1)x^n$

15. $\displaystyle\sum_{n=0}^{\infty} \frac{x^n}{n!}$

16. $\displaystyle\sum_{n=0}^{\infty} \frac{(3x)^n}{(2n)!}$

17. $\displaystyle\sum_{n=0}^{\infty} (2n)! \left(\frac{x}{2}\right)^n$

18. $\displaystyle\sum_{n=0}^{\infty} \frac{(-1)^n x^n}{(n+1)(n+2)}$

19. $\displaystyle\sum_{n=1}^{\infty} \frac{(-1)^{n+1} x^n}{4^n}$

20. $\displaystyle\sum_{n=0}^{\infty} \frac{(-1)^n n!(x-4)^n}{3^n}$

21. $\displaystyle\sum_{n=1}^{\infty} \frac{(-1)^{n+1}(x-5)^n}{n5^n}$

22. $\displaystyle\sum_{n=0}^{\infty} \frac{(x-2)^{n+1}}{(n+1)4^{n+1}}$

23. $\displaystyle\sum_{n=0}^{\infty} \frac{(-1)^{n+1}(x-1)^{n+1}}{n+1}$

24. $\displaystyle\sum_{n=1}^{\infty} \frac{(-1)^{n+1}(x-c)^n}{nc^n}$

25. $\displaystyle\sum_{n=1}^{\infty} \frac{(x-c)^{n-1}}{c^{n-1}}, \quad c>0$

26. $\displaystyle\sum_{n=0}^{\infty} \frac{(-1)^n x^{2n+1}}{2n+1}$

27. $\displaystyle\sum_{n=1}^{\infty} \frac{n}{n+1}(-2x)^{n-1}$

28. $\displaystyle\sum_{n=0}^{\infty} \frac{(-1)^n x^{2n}}{n!}$

29. $\displaystyle\sum_{n=0}^{\infty} \frac{x^{2n+1}}{(2n+1)!}$

30. $\displaystyle\sum_{n=1}^{\infty} \frac{n! x^n}{(2n)!}$

31. $\displaystyle\sum_{n=1}^{\infty} \frac{k(k+1)(k+2)\cdots(k+n-1)x^n}{n!}, \quad k\geq 1$

32. $\displaystyle\sum_{n=1}^{\infty} \left[\frac{2\cdot4\cdot6\cdots2n}{3\cdot5\cdot7\cdots(2n+1)}\right] x^{2n+1}$

33. $\displaystyle\sum_{n=1}^{\infty} \frac{(-1)^{n+1} 3\cdot7\cdot11\cdots(4n-1)(x-3)^n}{4^n}$

34. $\displaystyle\sum_{n=1}^{\infty} \frac{n!(x-c)^n}{1\cdot3\cdot5\cdots(2n-1)}$

In Exercises 35–38, find the intervals of convergence of (a) $f(x)$, (b) $f'(x)$, (c) $f''(x)$, and (d) $\int f(x)\,dx$. Include a check for convergence at the endpoints.

35. $\displaystyle f(x) = \sum_{n=0}^{\infty} \left(\frac{x}{2}\right)^n$

36. $\displaystyle f(x) = \sum_{n=1}^{\infty} \frac{(-1)^{n+1}(x-5)^n}{n5^n}$

37. $\displaystyle f(x) = \sum_{n=0}^{\infty} \frac{(-1)^{n+1}(x-1)^{n+1}}{n+1}$

38. $\displaystyle f(x) = \sum_{n=1}^{\infty} \frac{(-1)^{n+1}(x-2)^n}{n}$

Writing In Exercises 39–42, match the graph of the first ten terms of the sequence of partial sums of the series

$$g(x) = \sum_{n=0}^{\infty} \left(\frac{x}{3}\right)^n$$

with the indicated value of the function. [The graphs are labeled (a), (b), (c), and (d).] Explain how you made your choice.

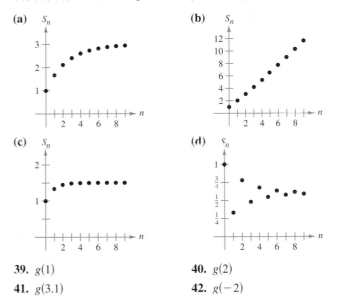

(a) S_n

(b) S_n

(c) S_n

(d) S_n

39. $g(1)$

40. $g(2)$

41. $g(3.1)$

42. $g(-2)$

Getting at the Concept

43. Define a power series centered at c.

44. What is the radius of convergence of a power series? What is the interval of convergence of a power series?

45. What are the three basic forms of the domain of a power series?

46. Describe how to differentiate and integrate a power series with a radius of convergence R. Will the series resulting from the operations of differentiation and integration have a different radius of convergence? Explain.

47. Let $f(x) = \displaystyle\sum_{n=0}^{\infty} \frac{(-1)^n x^{2n+1}}{(2n+1)!}$ and $g(x) = \displaystyle\sum_{n=0}^{\infty} \frac{(-1)^n x^{2n}}{(2n)!}$.

    (a) Find the intervals of convergence of f and g.

    (b) Show that $f'(x) = g(x)$.

    (c) Show that $g'(x) = -f(x)$.

    (d) Identify the functions f and g.

48. Let $f(x) = \displaystyle\sum_{n=0}^{\infty} \frac{x^n}{n!}$.

    (a) Find the interval of convergence of f.

    (b) Show that $f'(x) = f(x)$.

    (c) Show that $f(0) = 1$.

    (d) Identify the function f.

In Exercises 49 and 50, show that the function represented by the power series is a solution of the differential equation.

49. $y = \displaystyle\sum_{n=0}^{\infty} \frac{x^{2n}}{2^n n!}$, $\quad y'' - xy' - y = 0$

50. $y = 1 + \displaystyle\sum_{n=1}^{\infty} \frac{(-1)^n x^{4n}}{2^{2n} n! \cdot 3 \cdot 7 \cdot 11 \cdots (4n-1)}$, $\quad y'' + x^2 y = 0$

51. Bessel Function The Bessel function of order 0 is

$$J_0(x) = \sum_{k=0}^{\infty} \frac{(-1)^k x^{2k}}{2^{2k} (k!)^2}.$$

    (a) Show that the series converges for all x.

    (b) Show that the series is a solution of the differential equation $x^2 J_0'' + x J_0' + x^2 J_0 = 0$.

    (c) Use a graphing utility to graph the polynomial composed of the first four terms of J_0.

    (d) Approximate $\int_0^1 J_0\, dx$ accurate to two decimal places.

52. Bessel Function The Bessel function of order 1 is

$$J_1(x) = x \sum_{k=0}^{\infty} \frac{(-1)^k x^{2k}}{2^{2k+1} k!(k+1)!}.$$

    (a) Show that the series converges for all x.

    (b) Show that the series is a solution of the differential equation $x^2 J_1'' + x J_1' + (x^2 - 1) J_1 = 0$.

    (c) Use a graphing utility to graph the polynomial composed of the first four terms of J_1.

    (d) Show that $J_0'(x) = -J_1(x)$.

In Exercises 53–56, the series represents a well-known function. Use a computer algebra system to graph the partial sum S_{10} and identify the function from the graph.

53. $f(x) = \displaystyle\sum_{n=0}^{\infty} (-1)^n \frac{x^{2n}}{(2n)!}$

54. $f(x) = \displaystyle\sum_{n=0}^{\infty} (-1)^n \frac{x^{2n+1}}{(2n+1)!}$

55. $f(x) = \displaystyle\sum_{n=0}^{\infty} (-1)^n x^n, \quad -1 < x < 1$

56. $f(x) = \displaystyle\sum_{n=0}^{\infty} (-1)^n \frac{x^{2n+1}}{2n+1}, \quad -1 \le x \le 1$

57. Investigation In Exercise 11 you found that the interval of convergence of the geometric series

$$\sum_{n=0}^{\infty} \left(\frac{x}{2}\right)^n$$

is $(-2, 2)$.

    (a) Find the sum of the series when $x = \frac{3}{4}$. Use a graphing utility to graph the first six terms of the sequence of partial sums and the horizontal line representing the sum of the series.

    (b) Repeat part (a) for $x = -\frac{3}{4}$.

    (c) Write a short paragraph comparing the rate of convergence of the partial sums with the sum of the series in parts (a) and (b). How do the plots of the partial sums differ as they converge toward the sum of the series?

    (d) Given any positive real number M, there exists a positive integer N such that the partial sum

$$\sum_{n=0}^{N} \left(\frac{3}{2}\right)^n > M.$$

    Use a graphing utility to complete the table.

M	10	100	1000	10,000
N				

58. Write a series equivalent to

$$\sum_{n=0}^{\infty} \frac{x^{2n+1}}{(2n+1)!}$$

where the index of summation has been adjusted to begin at $n = 1$.

True or False? **In Exercises 59–62, determine whether the statement is true or false. If it is false, explain why or give an example that shows it is false.**

59. If the power series $\displaystyle\sum_{n=0}^{\infty} a_n x^n$ converges for $x = 2$, then it also converges for $x = -2$.

60. If the power series $\displaystyle\sum_{n=0}^{\infty} a_n x^n$ converges for $x = 2$, then it also converges for $x = -1$.

61. If the interval of convergence for $\displaystyle\sum_{a=0}^{\infty} a_n x^n$ is $(-1, 1)$, then the interval of convergence for $\displaystyle\sum_{n=0}^{\infty} a_n (x-1)^2$ is $(0, 2)$.

62. If $f(x) = \displaystyle\sum_{n=0}^{\infty} a_n x^n$ converges for $|x| < 2$, then $\displaystyle\int_0^1 f(x)\, dx = \sum_{n=0}^{\infty} \frac{a_n}{n+1}$.

Representation of Functions by Power Series

- Find a geometric power series that represents a function.
- Construct a power series using series operations.

Geometric Power Series

In this section and the next, you will study several techniques for finding a power series that represents a given function.

Consider the function given by $f(x) = 1/(1 - x)$. The form of f closely resembles the sum of a geometric series

$$\sum_{n=0}^{\infty} ar^n = \frac{a}{1 - r}, \quad |r| < 1.$$

In other words, if you let $a = 1$ and $r = x$, a power series representation for $1/(1 - x)$, centered at 0, is

$$\frac{1}{1 - x} = \sum_{n=0}^{\infty} x^n$$

$$= 1 + x + x^2 + x^3 + \cdots, \quad |x| < 1.$$

Of course, this series represents $f(x) = 1/(1 - x)$ only on the interval $(-1, 1)$, whereas f is defined for all $x \neq 1$, as shown in Figure 8.22. To represent f in another interval, you must develop a different series. For instance, to obtain the power series centered at -1, you could write

$$\frac{1}{1 - x} = \frac{1}{2 - (x + 1)} = \frac{1/2}{1 - [(x + 1)/2]} = \frac{a}{1 - r}$$

which implies that $a = \frac{1}{2}$ and $r = (x + 1)/2$. So, for $|x + 1| < 2$, you have

$$\frac{1}{1 - x} = \sum_{n=0}^{\infty} \left(\frac{1}{2}\right)\left(\frac{x + 1}{2}\right)^n$$

$$= \frac{1}{2}\left[1 + \frac{(x + 1)}{2} + \frac{(x + 1)^2}{4} + \frac{(x + 1)^3}{8} + \cdots\right], \quad |x + 1| < 2$$

which converges on the interval $(-3, 1)$.

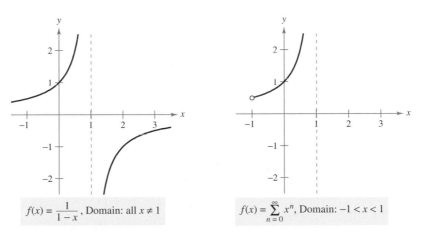

$f(x) = \dfrac{1}{1 - x}$, Domain: all $x \neq 1$

$f(x) = \displaystyle\sum_{n=0}^{\infty} x^n$, Domain: $-1 < x < 1$

Figure 8.22

JOSEPH FOURIER (1768–1830)

Some of the early work in representing functions by power series was done by the French mathematician Joseph Fourier. Fourier's work is important in the history of calculus, partly because it forced eighteenth century mathematicians to question the then-prevailing narrow concept of a function. Both Cauchy and Dirichlet were motivated by Fourier's work with series, and in 1837 Dirichlet published the general definition of a function that is used today.

Example 1 Finding a Geometric Power Series Centered at 0

Find a power series for $f(x) = \dfrac{4}{x+2}$, centered at 0.

Solution Writing $f(x)$ in the form $a/(1-r)$ produces

$$\frac{4}{2+x} = \frac{2}{1-(-x/2)} = \frac{a}{1-r}$$

which implies that $a = 2$ and $r = -x/2$. So, the power series for $f(x)$ is

$$\frac{4}{x+2} = \sum_{n=0}^{\infty} ar^n$$

$$= \sum_{n=0}^{\infty} 2\left(-\frac{x}{2}\right)^n$$

$$= 2\left(1 - \frac{x}{2} + \frac{x^2}{4} - \frac{x^3}{8} + \cdots\right).$$

This power series converges when

$$\left|-\frac{x}{2}\right| < 1$$

which implies that the interval of convergence is $(-2, 2)$.

Long Division

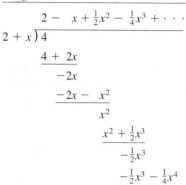

Another way to determine a power series for a rational function such as the one in Example 1 is to use long division. For instance, by dividing $2 + x$ into 4, you obtain the result shown at the left.

Example 2 Finding a Geometric Power Series Centered at 1

Find a power series for $f(x) = \dfrac{1}{x}$, centered at 1.

Solution Writing $f(x)$ in the form $a/(1-r)$ produces

$$\frac{1}{x} = \frac{1}{1-(-x+1)} = \frac{a}{1-r}$$

which implies that $a = 1$ and $r = 1 - x = -(x-1)$. So, the power series for $f(x)$ is

$$\frac{1}{x} = \sum_{n=0}^{\infty} ar^n$$

$$= \sum_{n=0}^{\infty} [-(x-1)]^n$$

$$= \sum_{n=0}^{\infty} (-1)^n (x-1)^n$$

$$= 1 - (x-1) + (x-1)^2 - (x-1)^3 + \cdots.$$

This power series converges when

$$|x-1| < 1$$

which implies that the interval of convergence is $(0, 2)$.

Operations with Power Series

The versatility of geometric power series will be shown later in this section, following a discussion of power series operations. These operations, used with differentiation and integration, provide a means of developing power series for a variety of elementary functions. (For simplicity, the following properties are stated for a series centered at 0.)

Operations with Power Series

Let $f(x) = \Sigma a_n x^n$ and $g(x) = \Sigma b_n x^n$.

1. $f(kx) = \displaystyle\sum_{n=0}^{\infty} a_n k^n x^n$

2. $f(x^N) = \displaystyle\sum_{n=0}^{\infty} a_n x^{nN}$

3. $f(x) \pm g(x) = \displaystyle\sum_{n=0}^{\infty} (a_n \pm b_n)x^n$

The operations described above can change the interval of convergence for the resulting series. For example, in the following addition, the interval of convergence for the sum is the *intersection* of the intervals of convergence of the two original series.

$$\underbrace{\sum_{n=0}^{\infty} x^n}_{(-1,1)} + \underbrace{\sum_{n=0}^{\infty} \left(\frac{x}{2}\right)^n}_{(-2,2)} = \underbrace{\sum_{n=0}^{\infty} \left(1 + \frac{1}{2^n}\right)x^n}_{(-1,1)}$$

Example 3   Adding Two Power Series

Find a power series, centered at 0, for $f(x) = \dfrac{3x - 1}{x^2 - 1}$.

Solution   Using partial fractions, you can write $f(x)$ as

$$\frac{3x - 1}{x^2 - 1} = \frac{2}{x + 1} + \frac{1}{x - 1}.$$

By adding the two geometric power series

$$\frac{2}{x + 1} = \frac{2}{1 - (-x)} = \sum_{n=0}^{\infty} 2(-1)^n x^n, \quad |x| < 1$$

and

$$\frac{1}{x - 1} = \frac{-1}{1 - x} = -\sum_{n=0}^{\infty} x^n, \quad |x| < 1$$

you obtain the following power series.

$$\frac{3x - 1}{x^2 - 1} = \sum_{n=0}^{\infty} [2(-1)^n - 1]x^n = 1 - 3x + x^2 - 3x^3 + x^4 - \cdots$$

The interval of convergence for this power series is $(-1, 1)$.

Example 4 **Finding a Power Series by Integration**

Find a power series for $f(x) = \ln x$, centered at 1.

Solution From Example 2, you know that

$$\frac{1}{x} = \sum_{n=0}^{\infty} (-1)^n (x-1)^n. \qquad \text{Interval of convergence: } (0, 2)$$

Integrating this series produces

$$\ln x = \int \frac{1}{x}\, dx + C$$

$$= C + \sum_{n=0}^{\infty} (-1)^n \frac{(x-1)^{n+1}}{n+1}.$$

By letting $x = 1$, you can conclude that $C = 0$. Therefore,

$$\ln x = \sum_{n=0}^{\infty} (-1)^n \frac{(x-1)^{n+1}}{n+1}$$

$$= \frac{(x-1)}{1} - \frac{(x-1)^2}{2} + \frac{(x-1)^3}{3} - \frac{(x-1)^4}{4} + \cdots. \qquad \text{Interval of convergence: } (0, 2]$$

Note that the series converges at $x = 2$. This is consistent with the observation in the preceding section that integration of a power series may alter the convergence at the endpoints of the interval of convergence.

TECHNOLOGY In Section 8.7, the fourth-degree Taylor polynomial for the natural logarithmic function

$$\ln x \approx (x-1) - \frac{(x-1)^2}{2} + \frac{(x-1)^3}{3} - \frac{(x-1)^4}{4}$$

was used to approximate $\ln(1.1)$.

$$\ln(1.1) \approx (0.1) - \frac{1}{2}(0.1)^2 + \frac{1}{3}(0.1)^3 - \frac{1}{4}(0.1)^4$$

$$\approx 0.0953083$$

You now know from Example 4 that this polynomial represents the first four terms of the power series for $\ln x$. Moreover, using the Alternating Series Remainder, you can determine that the error in this approximation is less than

$$|R_4| \le |a_5|$$

$$= \frac{1}{5}(0.1)^5$$

$$= 0.000002.$$

During the seventeenth and eighteenth centuries, mathematical tables for logarithms and values of other transcendental functions were computed in this manner. Such numerical techniques are far from outdated, because it is precisely by such means that many modern calculating devices are programmed to evaluate transcendental functions.

🖥 *Example 5   Finding a Power Series by Integration*

Find a power series for $g(x) = \arctan x$, centered at 0.

Solution   Because $D_x[\arctan x] = 1/(1 + x^2)$, you can use the series

$$f(x) = \frac{1}{1 + x} = \sum_{n=0}^{\infty} (-1)^n x^n. \qquad \text{Interval of convergence: } (-1, 1)$$

Substituting x^2 for x produces

$$f(x^2) = \frac{1}{1 + x^2} = \sum_{n=0}^{\infty} (-1)^n x^{2n}.$$

Finally, by integrating, you obtain

$$\arctan x = \int \frac{1}{1 + x^2} \, dx + C$$

$$= C + \sum_{n=0}^{\infty} (-1)^n \frac{x^{2n+1}}{2n + 1}$$

$$= \sum_{n=0}^{\infty} (-1)^n \frac{x^{2n+1}}{2n + 1} \qquad \text{Let } x = 0, \text{ then } C = 0.$$

$$= x - \frac{x^3}{3} + \frac{x^5}{5} - \frac{x^7}{7} + \cdots. \qquad \text{Interval of convergence: } (-1, 1)$$ ▨

It can be shown that the power series developed for $\arctan x$ in Example 5 also converges (to $\arctan x$) for $x = \pm 1$. For instance, when $x = 1$, you can write

$$\arctan 1 = 1 - \frac{1}{3} + \frac{1}{5} - \frac{1}{7} + \cdots$$

$$= \frac{\pi}{4}.$$

However, this series (developed by James Gregory in 1671) does not give us a practical way of approximating π because it converges so slowly that hundreds of terms would have to be used to obtain reasonable accuracy. Example 6 shows how to use *two* different arctangent series to obtain a very good approximation of π using only a few terms. This approximation was developed by John Machin in 1706.

Example 6   Approximating π with a Series

Use the trigonometric identity

$$4 \arctan \frac{1}{5} - \arctan \frac{1}{239} = \frac{\pi}{4}$$

to approximate the number π [see Exercise 48(b)].

Solution   By using only five terms from each of the series for $\arctan(1/5)$ and $\arctan(1/239)$, you obtain

$$4\left(4 \arctan \frac{1}{5} - \arctan \frac{1}{239} \right) \approx 3.1415926$$

which agrees with the decimal representation of π with an error of less than 0.0000001. ▨

SRINIVASA RAMANUJAN (1887–1920)

Series that can be used to approximate π have interested mathematicians for the past 300 years. An amazing series for approximating $1/\pi$ was discovered by the Indian mathematician Srinivasa Ramanujan in 1914. Each successive term of Ramanujan's series adds roughly eight more correct digits to the value of $1/\pi$. For more information about Ramanujan's work, see the article "Ramanujan and Pi" by Jonathan M. Borwein and Peter B. Borwein in *Scientific American*. (To view this article, go to the website *www.matharticles.com*.)

EXERCISES FOR SECTION 8.9

In Exercises 1–4, find a geometric power series for the function, centered at 0, (a) by the technique shown in Examples 1 and 2 and (b) by long division.

1. $f(x) = \dfrac{1}{2 - x}$

2. $f(x) = \dfrac{4}{5 - x}$

3. $f(x) = \dfrac{1}{2 + x}$

4. $f(x) = \dfrac{1}{1 + x}$

In Exercises 5–16, find a power series for the function, centered at c, and determine the interval of convergence.

5. $f(x) = \dfrac{1}{2 - x}, \quad c = 5$

6. $f(x) = \dfrac{4}{5 - x}, \quad c = -2$

7. $f(x) = \dfrac{3}{2x - 1}, \quad c = 0$

8. $f(x) = \dfrac{3}{2x - 1}, \quad c = 2$

9. $g(x) = \dfrac{1}{2x - 5}, \quad c = -3$

10. $h(x) = \dfrac{1}{2x - 5}, \quad c = 0$

11. $f(x) = \dfrac{3}{x + 2}, \quad c = 0$

12. $f(x) = \dfrac{4}{3x + 2}, \quad c = 2$

13. $g(x) = \dfrac{3x}{x^2 + x - 2}, \quad c = 0$

14. $g(x) = \dfrac{4x - 7}{2x^2 + 3x - 2}, \quad c = 0$

15. $f(x) = \dfrac{2}{1 - x^2}, \quad c = 0$

16. $f(x) = \dfrac{4}{4 + x^2}, \quad c = 0$

In Exercises 17–26, use the power series

$$\frac{1}{1 + x} = \sum_{n=0}^{\infty} (-1)^n x^n$$

to determine a power series, centered at 0, for the function. Identify the interval of convergence.

17. $h(x) = \dfrac{-2}{x^2 - 1} = \dfrac{1}{1 + x} + \dfrac{1}{1 - x}$

18. $h(x) = \dfrac{x}{x^2 - 1} = \dfrac{1}{2(1 + x)} - \dfrac{1}{2(1 - x)}$

19. $f(x) = -\dfrac{1}{(x + 1)^2} = \dfrac{d}{dx}\left[\dfrac{1}{x + 1}\right]$

20. $f(x) = \dfrac{2}{(x + 1)^3} = \dfrac{d^2}{dx^2}\left[\dfrac{1}{x + 1}\right]$

21. $f(x) = \ln(x + 1) = \displaystyle\int \dfrac{1}{x + 1}\, dx$

22. $f(x) = \ln(1 - x^2) = \displaystyle\int \dfrac{1}{1 + x}\, dx - \int \dfrac{1}{1 - x}\, dx$

23. $g(x) = \dfrac{1}{x^2 + 1}$

24. $f(x) = \ln(x^2 + 1)$

25. $h(x) = \dfrac{1}{4x^2 + 1}$

26. $f(x) = \arctan 2x$

Graphical and Numerical Analysis In Exercises 27 and 28, let

$$S_n = x - \frac{x^2}{2} + \frac{x^3}{3} - \frac{x^4}{4} + \cdots \pm \frac{x^n}{n}.$$

Use a graphing utility to confirm the inequality graphically. Then complete the table to confirm the inequality numerically.

x	0.0	0.2	0.4	0.6	0.8	1.0
S_n						
$\ln(x + 1)$						
S_{n+1}						

27. $S_2 \leq \ln(x + 1) \leq S_3$

28. $S_4 \leq \ln(x + 1) \leq S_5$

In Exercises 29–32, match the polynomial approximation of the function $f(x) = \arctan x$ with the correct graph. [The graphs are labeled (a), (b), (c), and (d).]

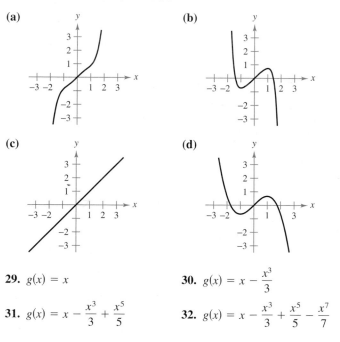

(a)

(b)

(c)

(d)

29. $g(x) = x$

30. $g(x) = x - \dfrac{x^3}{3}$

31. $g(x) = x - \dfrac{x^3}{3} + \dfrac{x^5}{5}$

32. $g(x) = x - \dfrac{x^3}{3} + \dfrac{x^5}{5} - \dfrac{x^7}{7}$

33. *Think About It* Use the results of Exercises 29–32 to make a geometric argument for why the series approximations of $f(x) = \arctan x$ have only odd powers of x.

34. *Conjecture* Use the results of Exercises 29–32 to make a conjecture about the degree of series approximations of $f(x) = \arctan x$ that have relative extrema.

In Exercises 35–38, use the series for $f(x) = \arctan x$ to approximate the value, using $R_N \le 0.001$.

35. $\arctan \dfrac{1}{4}$

36. $\displaystyle\int_0^{3/4} \arctan x^2 \, dx$

37. $\displaystyle\int_0^{1/2} \dfrac{\arctan x^2}{x} \, dx$

38. $\displaystyle\int_0^{1/2} x^2 \arctan x \, dx$

In Exercises 39–42, use the power series

$$\dfrac{1}{1-x} = \sum_{n=0}^{\infty} x^n, \quad |x| < 1.$$

39. Find the series representation of the function and determine its interval of convergence.

(a) $f(x) = \dfrac{1}{(1-x)^2}$

(b) $f(x) = \dfrac{x}{(1-x)^2}$

(c) $f(x) = \dfrac{1+x}{(1-x)^2}$

(d) $f(x) = \dfrac{x(1+x)}{(1-x)^2}$

40. Adjust the index of summation for the series found in Exercise 39(a) to begin with $n = 0$.

41. *Probability* If a fair coin is tossed repeatedly, the probability that the first head occurs on the nth toss is

$$P(n) = \left(\dfrac{1}{2}\right)^n.$$

When this game is repeated many times, the average number of tosses required until the first head occurs is

$$E(n) = \sum_{n=1}^{\infty} nP(n).$$

(This value is called the *expected value* of n.) Use the results of Exercises 39 and 40 to find $E(n)$. Is the answer what you expected? Why or why not?

42. Use the results of Exercises 39 and 40 to find the sum of each of the following series.

(a) $\dfrac{1}{3} \displaystyle\sum_{n=1}^{\infty} n\left(\dfrac{2}{3}\right)^n$

(b) $\dfrac{1}{10} \displaystyle\sum_{n=1}^{\infty} n\left(\dfrac{9}{10}\right)^n$

Getting at the Concept

In Exercises 43–46, explain how to use the geometric series

$$g(x) = \dfrac{1}{1-x} = \sum_{n=0}^{\infty} x^n, \quad |x| < 1$$

to find the series for the function. Do not find the series.

43. $f(x) = \dfrac{1}{1+x}$

44. $f(x) = \dfrac{1}{1-x^2}$

45. $f(x) = \dfrac{5}{1+x}$

46. $f(x) = \ln(1-x)$

47. Prove that

$$\arctan x + \arctan y = \arctan \dfrac{x+y}{1-xy} \quad \text{for } xy \ne 1$$

provided the value of the left side of the equation is between $-\pi/2$ and $\pi/2$.

48. Use the result of Exercise 47 to verify the identity.

(a) $\arctan \dfrac{120}{119} - \arctan \dfrac{1}{239} = \dfrac{\pi}{4}$

(b) $4 \arctan \dfrac{1}{5} - \arctan \dfrac{1}{239} = \dfrac{\pi}{4}$

[*Hint:* Use Exercise 47 twice to find $4 \arctan \frac{1}{5}$. Then use part (a).]

In Exercises 49 and 50, (a) verify the given equation and (b) use the equation and the series for the arctangent to approximate π to two-decimal-place accuracy.

49. $2 \arctan \dfrac{1}{2} - \arctan \dfrac{1}{7} = \dfrac{\pi}{4}$

50. $\arctan \dfrac{1}{2} + \arctan \dfrac{1}{3} = \dfrac{\pi}{4}$

In Exercises 51–56, find the sum of the convergent series by using a well-known function. Identify the function and explain how you obtained the sum.

51. $\displaystyle\sum_{n=1}^{\infty} (-1)^{n+1} \dfrac{1}{2^n n}$

52. $\displaystyle\sum_{n=1}^{\infty} (-1)^{n+1} \dfrac{1}{3^n n}$

53. $\displaystyle\sum_{n=1}^{\infty} (-1)^{n+1} \dfrac{2^n}{5^n n}$

54. $\displaystyle\sum_{n=0}^{\infty} (-1)^n \dfrac{1}{2n+1}$

55. $\displaystyle\sum_{n=0}^{\infty} (-1)^n \dfrac{1}{2^{2n+1}(2n+1)}$

56. $\displaystyle\sum_{n=1}^{\infty} (-1)^{n+1} \dfrac{1}{3^{2n-1}(2n-1)}$

57. *Writing* One of the series in Exercises 51–56 converges to its sum at a much slower rate than the other five series. Which is it? Explain why this series converges so slowly. Use a graphing utility to illustrate the rate of convergence.

58. Prove that $\displaystyle\sum_{n=0}^{\infty} \dfrac{(-1)^n}{3^n(2n+1)} = \dfrac{\pi}{2\sqrt{3}}$.

59. Use a graphing utility and 50 terms of the series

$$f(x) = \sum_{n=1}^{\infty} \dfrac{(-1)^{n+1}(x-1)^n}{n}, \quad 0 < x \le 2$$

to approximate $f(0.5)$. (The actual sum is $\ln 0.5$.)

- Find a Taylor or Maclaurin series for a function.
- Find a binomial series.
- Use a basic list of Taylor series to find other Taylor series.

Taylor Series and Maclaurin Series

In Section 8.9, you derived power series for several functions using geometric series with term-by-term differentiation or integration. In this section you will study a *general* procedure for deriving the power series for a function that has derivatives of all orders. The following theorem gives the form that *every* convergent power series must take.

THEOREM 8.22 **The Form of a Convergent Power Series**

If f is represented by a power series $f(x) = \sum a_n(x - c)^n$ for all x in an open interval I containing c, then $a_n = f^{(n)}(c)/n!$ and

$$f(x) = f(c) + f'(c)(x - c) + \frac{f''(c)}{2!}(x - c)^2 + \cdots + \frac{f^{(n)}(c)}{n!}(x - c)^n + \cdots.$$

Proof Suppose the power series $\sum a_n(x - c)^n$ has a radius of convergence R. Then, by Theorem 8.21, you know that the nth derivative of f exists for $|x - c| < R$, and by successive differentiation you obtain the following.

$$f^{(0)}(x) = a_0 + a_1(x - c) + a_2(x - c)^2 + a_3(x - c)^3 + a_4(x - c)^4 + \cdots$$
$$f^{(1)}(x) = a_1 + 2a_2(x - c) + 3a_3(x - c)^2 + 4a_4(x - c)^3 + \cdots$$
$$f^{(2)}(x) = 2a_2 + 3!a_3(x - c) + 4 \cdot 3a_4(x - c)^2 + \cdots$$
$$f^{(3)}(x) = 3!a_3 + 4!a_4(x - c) + \cdots$$
$$\vdots$$
$$f^{(n)}(x) = n!a_n + (n + 1)!a_{n+1}(x - c) + \cdots$$

Evaluating each of these derivatives at $x = c$ yields

$$f^{(0)}(c) = 0!a_0$$
$$f^{(1)}(c) = 1!a_1$$
$$f^{(2)}(c) = 2!a_2$$
$$f^{(3)}(c) = 3!a_3$$

and, in general, $f^{(n)}(c) = n!a_n$. By solving for a_n, you find that the coefficients of the power series representation of $f(x)$ are

$$a_n = \frac{f^{(n)}(c)}{n!}.$$

Notice that the coefficients of the power series in Theorem 8.22 are precisely the coefficients of the Taylor polynomials for $f(x)$ at c as defined in Section 8.7. For this reason, the series is called the **Taylor series** for $f(x)$ at c.

COLIN MACLAURIN (1698–1746)

The development of power series to represent functions is credited to the combined work of many seventeenth and eighteenth century mathematicians. Gregory, Newton, John and James Bernoulli, Leibniz, Euler, Lagrange, Wallis, and Fourier all contributed to this work. However, the two names that are most commonly associated with power series are Brook Taylor (1685-1731) and Colin Maclaurin.

NOTE Be sure you understand Theorem 8.22. The theorem says that *if a power series converges to $f(x)$, the series must be a Taylor series*. The theorem does *not* say that every series formed with the Taylor coefficients $a_n = f^{(n)}(c)/n!$ will converge to $f(x)$.

Definition of Taylor and Maclaurin Series

If a function f has derivatives of all orders at $x = c$, then the series

$$\sum_{n=0}^{\infty} \frac{f^{(n)}(c)}{n!} (x - c)^n = f(c) + f'(c)(x - c) + \cdots + \frac{f^{(n)}(c)}{n!}(x - c)^n + \cdots$$

is called the **Taylor series for $f(x)$ at c.** Moreover, if $c = 0$, then the series is the **Maclaurin series for f.**

If you know the pattern for the coefficients of the Taylor polynomials for a function, you can extend the pattern easily to form the corresponding Taylor series. For instance, in Example 4 of Section 8.7, you found the fourth Taylor polynomial for $\ln x$, centered at 1, to be

$$P_4(x) = (x - 1) - \frac{1}{2}(x - 1)^2 + \frac{1}{3}(x - 1)^3 - \frac{1}{4}(x - 1)^4.$$

From this pattern, you can obtain the Taylor series for $\ln x$ centered at $c = 1$,

$$(x - 1) - \frac{1}{2}(x - 1)^2 + \cdots + \frac{(-1)^{n+1}}{n}(x - 1)^n + \cdots.$$

Example 1   Forming a Power Series

Use the function $f(x) = \sin x$ to form the Maclaurin series

$$\sum_{n=0}^{\infty} \frac{f^{(n)}(0)}{n!} x^n = f(0) + f'(0)x + \frac{f''(0)}{2!}x^2 + \frac{f^{(3)}(0)}{3!}x^3 + \frac{f^{(4)}(0)}{4!}x^4 + \cdots$$

and determine the interval of convergence.

Solution   Successive differentiation of $f(x)$ yields

$$f(x) = \sin x \qquad\qquad f(0) = \sin 0 = 0$$
$$f'(x) = \cos x \qquad\qquad f'(0) = \cos 0 = 1$$
$$f''(x) = -\sin x \qquad\qquad f''(0) = -\sin 0 = 0$$
$$f^{(3)}(x) = -\cos x \qquad\qquad f^{(3)}(0) = -\cos 0 = -1$$
$$f^{(4)}(x) = \sin x \qquad\qquad f^{(4)}(0) = \sin 0 = 0$$
$$f^{(5)}(x) = \cos x \qquad\qquad f^{(5)}(0) = \cos 0 = 1$$

and so on. The pattern repeats after the third derivative. Hence, the power series is as follows.

$$\sum_{n=0}^{\infty} \frac{f^{(n)}(0)}{n!} x^n = f(0) + f'(0)x + \frac{f''(0)}{2!}x^2 + \frac{f^{(3)}(0)}{3!}x^3 + \frac{f^{(4)}(0)}{4!}x^4 + \cdots$$

$$\sum_{n=0}^{\infty} \frac{(-1)^n x^{2n+1}}{(2n + 1)!} = 0 + (1)x + \frac{0}{2!}x^2 + \frac{(-1)}{3!}x^3 + \frac{0}{4!}x^4 + \frac{1}{5!}x^5 + \frac{0}{6!}x^6$$

$$+ \frac{(-1)}{7!}x^7 + \cdots$$

$$= x - \frac{x^3}{3!} + \frac{x^5}{5!} - \frac{x^7}{7!} + \cdots$$

By the Ratio Test, you can conclude that this series converges for all x.

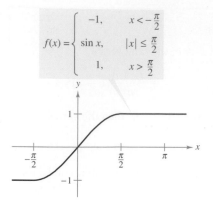

$$f(x) = \begin{cases} -1, & x < -\frac{\pi}{2} \\ \sin x, & |x| \le \frac{\pi}{2} \\ 1, & x > \frac{\pi}{2} \end{cases}$$

Figure 8.23

Notice that in Example 1 we do not conclude that the power series converges to $\sin x$ for all x. We simply conclude that the power series converges to some function, but we are not sure what function it is. This is a subtle, but important, point in dealing with Taylor or Maclaurin series. To persuade yourself that the series

$$f(c) + f'(c)(x - c) + \frac{f''(c)}{2!}(x - c)^2 + \cdots + \frac{f^{(n)}(c)}{n!}(x - c)^n + \cdots$$

might converge to a function other than f, remember that the derivatives are being evaluated at a single point. It can easily happen that another function will agree with the values of $f^{(n)}(x)$ when $x = c$ and disagree at other x-values. For instance, if you formed the power series (centered at 0) for the function shown in Figure 8.23, you would obtain the same series as in Example 1. You know that the series converges for all x, and yet it obviously cannot converge to both $f(x)$ and $\sin x$ for all x.

Let f have derivatives of all orders in an open interval I centered at c. The Taylor series for f may fail to converge for some x in I. Or, even if it is convergent, it may fail to have $f(x)$ as its sum. Nevertheless, Theorem 8.19 tells us that for each n,

$$f(x) = f(c) + f'(c)(x - c) + \frac{f''(c)}{2!}(x - c)^2 + \cdots + \frac{f^{(n)}(c)}{n!}(x - c)^n + R_n(x),$$

where

$$R_n(x) = \frac{f^{(n+1)}(z)}{(n+1)!}(x - c)^{n+1}.$$

Note that in this remainder formula the particular value of z that makes the remainder formula true depends on the values of x and n. If $R_n \to 0$ then the following theorem tells us that the Taylor series for f actually converges to $f(x)$ for all x in I.

THEOREM 8.23 **Convergence of Taylor Series**

If $\lim_{n \to \infty} R_n = 0$ for all x in the interval I, then the Taylor series for f converges and equals $f(x)$,

$$f(x) = \sum_{n=0}^{\infty} \frac{f^{(n)}(c)}{n!}(x - c)^n.$$

Proof For a Taylor series, the nth partial sum coincides with the nth Taylor polynomial. That is, $S_n(x) = P_n(x)$. Moreover, because

$$P_n(x) = f(x) - R_n(x)$$

it follows that

$$\lim_{n \to \infty} S_n(x) = \lim_{n \to \infty} P_n(x)$$
$$= \lim_{n \to \infty} [f(x) - R_n(x)]$$
$$= f(x) - \lim_{n \to \infty} R_n(x).$$

Hence, for a given x, the Taylor series (the sequence of partial sums) converges to $f(x)$ if and only if $R_n(x) \to 0$ as $n \to \infty$.

NOTE Stated another way, Theorem 8.23 says that a power series formed with Taylor coefficients $a_n = f^{(n)}(c)/n!$ converges to the function from which it was derived at precisely those values for which the remainder approaches 0 as $n \to \infty$.

In Example 1, you derived the power series from the sine function and you also concluded that the series converges to some function on the entire real line. In Example 2, you will see that the series actually converges to $\sin x$. The key observation is that although the value of z is not known, it is possible to obtain an upper bound for $\left|f^{(n+1)}(z)\right|$.

Example 2   A Convergent Maclaurin Series

Show that the Maclaurin series for $f(x) = \sin x$ converges to $\sin x$ for all x.

Solution   Using the result in Example 1, you need to show that

$$\sin x = x - \frac{x^3}{3!} + \frac{x^5}{5!} - \frac{x^7}{7!} + \cdots + \frac{(-1)^n x^{2n+1}}{(2n+1)!} + \cdots$$

is true for all x. Because

$$f^{(n+1)}(x) = \pm\sin x$$

or

$$f^{(n+1)}(x) = \pm\cos x$$

you know that $\left|f^{(n+1)}(z)\right| \le 1$ for every real number z. Therefore, for any fixed x, you can apply Taylor's Theorem (Theorem 8.19) to conclude that

$$0 \le \left|R_n(x)\right| = \left|\frac{f^{(n+1)}(z)}{(n+1)!}x^{n+1}\right| \le \frac{|x|^{n+1}}{(n+1)!}.$$

From the discussion in Section 8.1 regarding the relative rates of convergence of exponential and factorial sequences, it follows that for a fixed x

$$\lim_{n\to\infty} \frac{|x|^{n+1}}{(n+1)!} = 0.$$

Finally, by the Squeeze Theorem, it follows that for all x, $R_n(x) \to 0$ as $n \to \infty$. Hence, by Theorem 8.23, the Maclaurin series for $\sin x$ converges to $\sin x$ for all x.

Figure 8.24 visually illustrates the convergence of the Maclaurin series for $\sin x$ by comparing the graphs of the Maclaurin polynomials $P_1(x)$, $P_3(x)$, $P_5(x)$, and $P_7(x)$ with the graph of the sine function. Notice that as the degree of the polynomial increases, its graph more closely resembles that of the sine function.

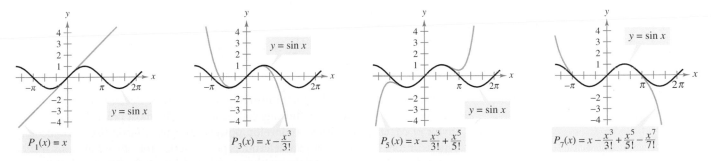

As n increases, the graph of P_n more closely resembles the sine function.
Figure 8.24

The guidelines for finding a Taylor series for $f(x)$ at c are summarized below.

> **Guidelines for Finding a Taylor Series**
>
> 1. Differentiate $f(x)$ several times and evaluate each derivative at c.
>
> $$f(c), f'(c), f''(c), f'''(c), \cdots, f^{(n)}(c), \cdots$$
>
> Try to recognize a pattern in these numbers.
>
> 2. Use the sequence developed in the first step to form the Taylor coefficients $a_n = f^{(n)}(c)/n!$, and determine the interval of convergence for the resulting power series
>
> $$f(c) + f'(c)(x - c) + \frac{f''(c)}{2!}(x - c)^2 + \cdots + \frac{f^{(n)}(c)}{n!}(x - c)^n + \cdots.$$
>
> 3. Within this interval of convergence, determine whether or not the series converges to $f(x)$.

The direct determination of Taylor or Maclaurin coefficients using successive differentiation can be difficult, and the next example illustrates a shortcut for finding the coefficients indirectly—using the coefficients of a known Taylor or Maclaurin series.

Example 3   **Maclaurin Series for a Composite Function**

Find the Maclaurin series for $f(x) = \sin x^2$.

Solution   To find the coefficients for this Maclaurin series *directly*, you must calculate successive derivatives of $f(x) = \sin x^2$. By calculating just the first two,

$$f'(x) = 2x \cos x^2 \quad \text{and} \quad f''(x) = -4x^2 \sin x^2 + 2 \cos x^2$$

you can see that this task would be quite cumbersome. Fortunately, there is an alternative. Suppose you first consider the Maclaurin series for $\sin x$ found in Example 1.

$$g(x) = \sin x$$
$$= x - \frac{x^3}{3!} + \frac{x^5}{5!} - \frac{x^7}{7!} + \cdots$$

Now, because $\sin x^2 = g(x^2)$, you can substitute x^2 for x in the series for $\sin x$ to obtain

$$\sin x^2 = g(x^2)$$
$$= x^2 - \frac{x^6}{3!} + \frac{x^{10}}{5!} - \frac{x^{14}}{7!} + \cdots.$$

Be sure to understand the point illustrated in Example 3. Because direct computation of Taylor or Maclaurin coefficients can be tedious, the most practical way to find a Taylor or Maclaurin series is to develop power series for a *basic list* of elementary functions. From this list, you can determine power series for other functions by the operations of addition, subtraction, multiplication, division, differentiation, integration, or composition with known power series.

Binomial Series

Before presenting the basic list for elementary functions, we develop one more series—for a function of the form $f(x) = (1 + x)^k$. This produces the **binomial series.**

Example 4   **Binomial Series**

Find the Maclaurin series for $f(x) = (1 + x)^k$ and determine its radius of convergence. Assume that R is not a positive integer.

Solution   By successive differentiation, you have

$$f(x) = (1 + x)^k \qquad\qquad\qquad f(0) = 1$$
$$f'(x) = k(1 + x)^{k-1} \qquad\qquad\qquad f'(0) = k$$
$$f''(x) = k(k - 1)(1 + x)^{k-2} \qquad\qquad f''(0) = k(k - 1)$$
$$f'''(x) = k(k - 1)(k - 2)(1 + x)^{k-3} \qquad f'''(0) = k(k - 1)(k - 2)$$
$$\vdots \qquad\qquad\qquad\qquad\qquad \vdots$$
$$f^{(n)}(x) = k \cdots (k - n + 1)(1 + x)^{k-n} \qquad f^{(n)}(0) = k(k - 1) \cdots (k - n + 1)$$

which produces the series

$$1 + kx + \frac{k(k - 1)x^2}{2} + \cdots + \frac{k(k - 1) \cdots (k - n + 1)x^n}{n!} + \cdots.$$

Because $a_{n+1}/a_n \to 1$, you can apply the Ratio Test to conclude that the radius of convergence is $R = 1$. So, the series converges to some function in the interval $(-1, 1)$.

Note that in Example 4 we showed that the Taylor series for $(1 + x)^k$ converges to *some* function in the interval $(-1, 1)$. However, we did not show that the series actually converges to $(1 + x)^k$. To do this, you could show that the remainder $R_n(x)$ converges to 0, as illustrated in Example 2.

Example 5   **Finding a Binomial Series**

Find the power series for $f(x) = \sqrt[3]{1 + x}$.

Solution   Using the binomial series

$$(1 + x)^k = 1 + kx + \frac{k(k - 1)x^2}{2!} + \frac{k(k - 1)(k - 2)x^3}{3!} + \cdots$$

let $k = \frac{1}{3}$ and write

$$(1 + x)^{1/3} = 1 + \frac{x}{3} - \frac{2x^2}{3^2 2!} + \frac{2 \cdot 5x^3}{3^3 3!} - \frac{2 \cdot 5 \cdot 8x^4}{3^4 4!} + \cdots$$

which converges for $-1 \le x \le 1$.

> **TECHNOLOGY**   Try using a graphing utility to confirm the result in Example 5. When you graph the functions
>
> $$f(x) = (1 + x)^{1/3} \quad \text{and} \quad P_4(x) = 1 + \frac{x}{3} - \frac{x^2}{9} + \frac{5x^3}{81} - \frac{10x^4}{243}$$
>
> in the same viewing window, you should obtain the result shown in Figure 8.25.

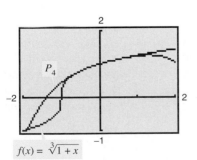

$f(x) = \sqrt[3]{1 + x}$

Figure 8.25

Deriving Taylor Series from a Basic List

In the following list, we provide the power series for several elementary functions with the corresponding intervals of convergence.

Power Series for Elementary Functions

Function	Interval of Convergence
$\dfrac{1}{x} = 1 - (x - 1) + (x - 1)^2 - (x - 1)^3 + (x - 1)^4 - \cdots + (-1)^n (x - 1)^n + \cdots$	$0 < x < 2$
$\dfrac{1}{1 + x} = 1 - x + x^2 - x^3 + x^4 - x^5 + \cdots + (-1)^n x^n + \cdots$	$-1 < x < 1$
$\ln x = (x - 1) - \dfrac{(x - 1)^2}{2} + \dfrac{(x - 1)^3}{3} - \dfrac{(x - 1)^4}{4} + \cdots + \dfrac{(-1)^{n-1}(x - 1)^n}{n} + \cdots$	$0 < x \leq 2$
$e^x = 1 + x + \dfrac{x^2}{2!} + \dfrac{x^3}{3!} + \dfrac{x^4}{4!} + \dfrac{x^5}{5!} + \cdots + \dfrac{x^n}{n!} + \cdots$	$-\infty < x < \infty$
$\sin x = x - \dfrac{x^3}{3!} + \dfrac{x^5}{5!} - \dfrac{x^7}{7!} + \dfrac{x^9}{9!} - \cdots + \dfrac{(-1)^n x^{2n+1}}{(2n + 1)!} + \cdots$	$-\infty < x < \infty$
$\cos x = 1 - \dfrac{x^2}{2!} + \dfrac{x^4}{4!} - \dfrac{x^6}{6!} + \dfrac{x^8}{8!} - \cdots + \dfrac{(-1)^n x^{2n}}{(2n)!} + \cdots$	$-\infty < x < \infty$
$\arctan x = x - \dfrac{x^3}{3} + \dfrac{x^5}{5} - \dfrac{x^7}{7} + \dfrac{x^9}{9} - \cdots + \dfrac{(-1)^n x^{2n+1}}{2n + 1} + \cdots$	$-1 \leq x \leq 1$
$\arcsin x = x + \dfrac{x^3}{2 \cdot 3} + \dfrac{1 \cdot 3 x^5}{2 \cdot 4 \cdot 5} + \dfrac{1 \cdot 3 \cdot 5 x^7}{2 \cdot 4 \cdot 6 \cdot 7} + \cdots + \dfrac{(2n)! x^{2n+1}}{(2^n n!)^2 (2n + 1)} + \cdots$	$-1 \leq x \leq 1$
$(1 + x)^k = 1 + kx + \dfrac{k(k - 1)x^2}{2!} + \dfrac{k(k - 1)(k - 2)x^3}{3!} + \dfrac{k(k - 1)(k - 2)(k - 3)x^4}{4!} + \cdots$	$-1 < x < 1*$

The convergence at $x = \pm1$ depends on the value of k.

NOTE   The binomial series is valid for noninteger values of k. Moreover, if k happens to be a positive integer, the binomial series reduces to a simple binomial expansion.

Example 6   **Deriving a Power Series from a Basic List**

Find the power series for $f(x) = \cos \sqrt{x}$.

Solution   Using the power series

$$\cos x = 1 - \frac{x^2}{2!} + \frac{x^4}{4!} - \frac{x^6}{6!} + \frac{x^8}{8!} - \cdots$$

you can replace x by $\sqrt{x}$ to obtain the series

$$\cos \sqrt{x} = 1 - \frac{x}{2!} + \frac{x^2}{4!} - \frac{x^3}{6!} + \frac{x^4}{8!} - \cdots.$$

This series converges for all x in the domain of $\cos \sqrt{x}$—that is, for $x \geq 0$.

Power series can be multiplied and divided like polynomials. After finding the first few terms of the product (or quotient), you may be able to recognize a pattern.

Example 7   Multiplication and Division of Power Series

Find the first three nonzero terms in each of the Maclaurin series.

a. $e^x \arctan x$ **b.** $\tan x$

Solution

a. Using the Maclaurin series for e^x and $\arctan x$ in the table, you have

$$e^x \arctan x = \left(1 + \frac{x}{1!} + \frac{x^2}{2!} + \frac{x^3}{3!} + \frac{x^4}{4!} + \cdots\right)\left(x - \frac{x^3}{3} + \frac{x^5}{5} - \cdots\right).$$

Multiply these expressions and collect like terms as you would for multiplying polynomials.

$$
\begin{array}{l}
1 + x + \frac{1}{2}x^2 + \frac{1}{6}x^3 + \frac{1}{24}x^4 + \cdots \\[4pt]
\underline{\quad x \qquad\qquad - \frac{1}{3}x^3 \qquad\quad + \frac{1}{5}x^5 - \cdots} \\[4pt]
x + \ x^2 + \frac{1}{2}x^3 + \ \frac{1}{6}x^4 + \ \frac{1}{24}x^5 + \cdots \\[4pt]
\qquad\qquad -\frac{1}{3}x^3 - \ \frac{1}{3}x^4 - \ \frac{1}{6}x^5 - \cdots \\[4pt]
\underline{\qquad\qquad\qquad\qquad\qquad + \ \frac{1}{5}x^5 + \cdots} \\[4pt]
x + \ x^2 + \frac{1}{6}x^3 - \ \frac{1}{6}x^4 + \ \frac{3}{40}x^5 + \cdots
\end{array}
$$

So, $e^x \arctan x = x + x^2 + \frac{1}{6}x^3 + \cdots.$

b. Using the Maclaurin series for $\sin x$ and $\cos x$ in the table, you have

$$\tan x = \frac{\sin x}{\cos x} = \frac{x - \dfrac{x^3}{3!} + \dfrac{x^5}{5!} - \cdots}{1 - \dfrac{x^2}{2!} + \dfrac{x^4}{4!} - \cdots}.$$

Divide using long division.

$$
\begin{array}{r}
x + \dfrac{1}{3}x^3 + \dfrac{2}{15}x^5 + \cdots \\[4pt]
1 - \dfrac{1}{2}x^2 + \dfrac{1}{24}x^4 - \cdots \overline{\smash{\big)}\, x - \dfrac{1}{6}x^3 + \dfrac{1}{120}x^5 - \cdots} \\[4pt]
\underline{x - \dfrac{1}{2}x^3 + \dfrac{1}{24}x^5 - \cdots} \\[4pt]
\dfrac{1}{3}x^3 - \dfrac{1}{30}x^5 + \cdots \\[4pt]
\underline{\dfrac{1}{3}x^3 - \dfrac{1}{6}x^5 + \cdots} \\[4pt]
\dfrac{2}{15}x^5 + \cdots
\end{array}
$$

So, $\tan x = x + \frac{1}{3}x^3 + \frac{2}{15}x^5 + \cdots.$

Example 8   A Power Series for sin² *x*

Find the power series for $f(x) = \sin^2 x$.

Solution   Consider rewriting $\sin^2 x$ as follows.

$$\sin^2 x = \frac{1 - \cos 2x}{2} = \frac{1}{2} - \frac{\cos 2x}{2}$$

Now, use the series for cos *x*.

$$\cos x = 1 - \frac{x^2}{2!} + \frac{x^4}{4!} - \frac{x^6}{6!} + \frac{x^8}{8!} - \cdots$$

$$\cos 2x = 1 - \frac{2^2}{2!}x^2 + \frac{2^4}{4!}x^4 - \frac{2^6}{6!}x^6 + \frac{2^8}{8!}x^8 - \cdots$$

$$-\frac{1}{2}\cos 2x = -\frac{1}{2} + \frac{2}{2!}x^2 - \frac{2^3}{4!}x^4 + \frac{2^5}{6!}x^6 - \frac{2^7}{8!}x^8 + \cdots$$

$$\sin^2 x = \frac{1}{2} - \frac{1}{2}\cos 2x = \frac{1}{2} - \frac{1}{2} + \frac{2}{2!}x^2 - \frac{2^3}{4!}x^4 + \frac{2^5}{6!}x^6 - \frac{2^7}{8!}x^8 + \cdots$$

$$= \frac{2}{2!}x^2 - \frac{2^3}{4!}x^4 + \frac{2^5}{6!}x^6 - \frac{2^7}{8!}x^8 + \cdots$$

This series converges for $-\infty < x < \infty$.   ▨

   As mentioned in the preceding section, power series can be used to obtain tables of values of transcendental functions. They are also useful for estimating the values of definite integrals for which antiderivatives cannot be found. The next example demonstrates this use.

⏪ *Example 9*   **Power Series Approximation of a Definite Integral**

Use a power series to approximate

$$\int_0^1 e^{-x^2}\, dx$$

with an error of less than 0.01.

Solution   Replacing *x* with $-x^2$ in the series for e^x produces the following.

$$e^{-x^2} = 1 - x^2 + \frac{x^4}{2!} - \frac{x^6}{3!} + \frac{x^8}{4!} - \cdots$$

$$\int_0^1 e^{-x^2}\, dx = \left[x - \frac{x^3}{3} + \frac{x^5}{5 \cdot 2!} - \frac{x^7}{7 \cdot 3!} + \frac{x^9}{9 \cdot 4!} - \cdots \right]_0^1$$

$$= 1 - \frac{1}{3} + \frac{1}{10} - \frac{1}{42} + \frac{1}{216} - \cdots$$

Summing the first *four* terms, you have

$$\int_0^1 e^{-x^2}\, dx \approx 0.74$$

which, by the Alternating Series Test, has an error of less than $\frac{1}{216} \approx 0.005$.   ▨

EXERCISES FOR SECTION 8.10

In Exercises 1–10, use the definition to find the Taylor series (centered at *c*) for the function.

1. $f(x) = e^{2x}, \quad c = 0$

2. $f(x) = e^{3x}, \quad c = 0$

3. $f(x) = \cos x, \quad c = \dfrac{\pi}{4}$

4. $f(x) = \sin x, \quad c = \dfrac{\pi}{4}$

5. $f(x) = \ln x, \quad c = 1$

6. $f(x) = e^x, \quad c = 1$

7. $f(x) = \sin 2x, \quad c = 0$

8. $f(x) = \ln(x^2 + 1), \quad c = 0$

9. $f(x) = \sec x, \quad c = 0$ (first three nonzero terms)

10. $f(x) = \tan x, \quad c = 0$ (first three nonzero terms)

In Exercises 11 and 12, prove that the Maclaurin series for the function converges to the function for all *x*.

11. $f(x) = \cos x$

12. $f(x) = e^{-2x}$

In Exercises 13–18, use the binomial series to find the Maclaurin series for the function.

13. $f(x) = \dfrac{1}{(1 + x)^2}$

14. $f(x) = \dfrac{1}{\sqrt{1 - x}}$

15. $f(x) = \dfrac{1}{\sqrt{4 + x^2}}$

16. $f(x) = \sqrt[3]{1 + x}$

17. $f(x) = \sqrt{1 + x^2}$

18. $f(x) = \sqrt{1 + x^3}$

In Exercises 19–28, find the Maclaurin series for the function. (Use the table of power series for elementary functions.)

19. $f(x) = e^{x^2/2}$

20. $g(x) = e^{-3x}$

21. $g(x) = \sin 2x$

22. $f(x) = \cos 4x$

23. $f(x) = \cos x^{3/2}$

24. $g(x) = 2 \sin x^3$

25. $f(x) = \frac{1}{2}(e^x - e^{-x}) = \sinh x$

26. $f(x) = e^x + e^{-x} = 2 \cosh x$

27. $f(x) = \cos^2 x$

[*Hint:* $\cos^2 x = \frac{1}{2}(1 + \cos 2x)$]

28. $f(x) = \sinh^{-1} x = \ln\left(x + \sqrt{x^2 + 1}\right)$

$\left(\text{\textit{Hint:} Integrate the series for } \dfrac{1}{\sqrt{x^2 + 1}}.\right)$

In Exercises 29–32, find the Maclaurin series for the function. (See Example 7.)

29. $f(x) = x \sin x$

30. $h(x) = x \cos x$

31. $g(x) = \begin{cases} \dfrac{\sin x}{x}, & x \neq 0 \\ 1, & x = 0 \end{cases}$

32. $f(x) = \begin{cases} \dfrac{\arcsin x}{x}, & x \neq 0 \\ 1, & x = 0 \end{cases}$

In Exercises 33 and 34, use a power series and the fact that $i^2 = -1$ to verify the formula.

33. $g(x) = \dfrac{1}{2i}(e^{ix} - e^{-ix}) = \sin x$

34. $g(x) = \frac{1}{2}(e^{ix} + e^{-ix}) = \cos x$

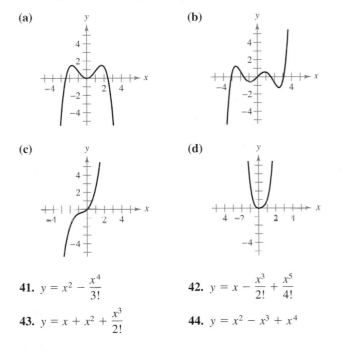

In Exercises 35–40, find the first four nonzero terms of the Maclaurin series for the function by multiplying or dividing the appropriate power series. Use the table of power series for elementary functions on page 638. Use a graphing utility to obtain a graph of the function and its corresponding polynomial approximation.

35. $f(x) = e^x \sin x$

36. $g(x) = e^x \cos x$

37. $h(x) = \cos x \ln(1 + x)$

38. $f(x) = e^x \ln(1 + x)$

39. $g(x) = \dfrac{\sin x}{1 + x}$

40. $f(x) = \dfrac{e^x}{1 + x}$

In Exercises 41–44, match the polynomial with its graph. [The graphs are labeled (a), (b), (c), and (d).] Factor a common factor from each polynomial and identify the function approximated by the remaining Taylor polynomial.

(a)

(b)

(c)

(d)

41. $y = x^2 - \dfrac{x^4}{3!}$

42. $y = x - \dfrac{x^3}{2!} + \dfrac{x^5}{4!}$

43. $y = x + x^2 + \dfrac{x^3}{2!}$

44. $y = x^2 - x^3 + x^4$

In Exercises 45 and 46, find a Maclaurin series for $f(x)$.

45. $f(x) = \displaystyle\int_0^x (e^{-t^2} - 1)\, dt$

46. $f(x) = \displaystyle\int_0^x \sqrt{1 + t^3}\, dt$

In Exercises 47–50, verify the sum. Then use a graphing utility to approximate the sum with an error of less than 0.0001.

47. $\displaystyle\sum_{n=1}^{\infty} (-1)^{n+1} \frac{1}{n} = \ln 2$

48. $\displaystyle\sum_{n=0}^{\infty} (-1)^n \left[\frac{1}{(2n + 1)!}\right] = \sin 1$

49. $\displaystyle\sum_{n=0}^{\infty} \frac{2^n}{n!} = e^2$

50. $\displaystyle\sum_{n=1}^{\infty} (-1)^{n-1} \left(\frac{1}{n!}\right) = \frac{e - 1}{e}$

In Exercises 51 and 52, use the series representation of the function f to find $\lim\limits_{x \to 0} f(x)$ (if it exists).

51. $f(x) = \dfrac{1 - \cos x}{x}$

52. $f(x) = \dfrac{\sin x}{x}$

In Exercises 53–58, use power series to approximate the value of the integral with an error of less than 0.0001. (In Exercises 53 and 54, assume that the integrand is defined as 1 when $x = 0$.)

53. $\displaystyle\int_0^1 \dfrac{\sin x}{x}\, dx$

54. $\displaystyle\int_0^{1/2} \dfrac{\arctan x}{x}\, dx$

55. $\displaystyle\int_0^{\pi/2} \sqrt{x}\cos x\, dx$

56. $\displaystyle\int_{0.5}^1 \cos \sqrt{x}\, dx$

57. $\displaystyle\int_{0.1}^{0.3} \sqrt{1 + x^3}\, dx$

58. $\displaystyle\int_0^{1/4} x \ln(x + 1)\, dx$

Probability **In Exercises 59 and 60, approximate the normal probability with an error of less than 0.0001, where the probability is given by**

$$P(a < x < b) = \dfrac{1}{\sqrt{2\pi}} \int_a^b e^{-x^2/2}\, dx.$$

59. $P(0 < x < 1)$

60. $P(1 < x < 2)$

In Exercises 61–64, use a computer algebra system to find the fifth-degree Taylor polynomial (centered at c) for the function. Graph the function and the polynomial. Use the graph to determine the largest interval on which the polynomial is a reasonable approximation of the function.

61. $f(x) = x \cos 2x, \quad c = 0$

62. $f(x) = \sin \dfrac{x}{2} \ln(1 + x), \quad c = 0$

63. $g(x) = \sqrt{x} \ln x, \quad c = 1$

64. $h(x) = \sqrt[3]{x} \arctan x, \quad c = 1$

Getting at the Concept

65. State the guidelines for finding a Taylor series.

66. If f is an even function, what must be true about the coefficients a_n in the Maclaurin series

$$f(x) = \sum_{n=0}^{\infty} a_n x^n?$$

67. Explain how to use the series $g(x) = e^x = \displaystyle\sum_{n=0}^{\infty} \dfrac{x^n}{n!}$ to find the series for the functions. Do not find the series.

(a) $f(x) = e^{-x}$

(b) $f(x) = e^{3x}$

(c) $f(x) = xe^x$

(d) $f(x) = e^{2x} + e^{-2x}$

68. Summarize the use of power series in approximating elementary functions.

---

69. *Projectile Motion* A projectile fired from the ground follows the trajectory given by

$$y = \left(\tan \theta - \dfrac{g}{kv_0 \cos \theta}\right) x - \dfrac{g}{k^2} \ln\left(1 - \dfrac{kx}{v_0 \cos \theta}\right)$$

where v_0 is the initial speed, θ is the angle of projection, g is the acceleration due to gravity, and k is the drag factor caused by air resistance. Using the power series representation

$$\ln(1 + x) = x - \dfrac{x^2}{2} + \dfrac{x^3}{3} - \dfrac{x^4}{4} + \cdots, \quad -1 < x < 1$$

verify that the trajectory can be rewritten as

$$y = (\tan \theta)x + \dfrac{gx^2}{2v_0^2 \cos^2 \theta} + \dfrac{kgx^3}{3v_0 \cos^3 \theta} + \dfrac{k^2 gx^4}{4v_0 \cos^4 \theta} + \cdots.$$

70. *Projectile Motion* Use the result of Exercise 69 to determine the series for the path of a projectile projected from ground level at an angle of $\theta = 60°$, with an initial speed of $v_0 = 64$ feet per second and a drag factor of $k = \frac{1}{16}$.

71. *Investigation* Consider the function f defined by

$$f(x) = \begin{cases} e^{-1/x^2}, & x \neq 0 \\ 0, & x = 0. \end{cases}$$

(a) Sketch a graph of the function.

(b) Use the alternative form of the definition of the derivative (Section 2.1) and L'Hôpital's Rule to show that $f'(0) = 0$. [By continuing this process, it can be shown that $f^{(n)}(0) = 0$ for $n > 1$.]

(c) Using the result in part (b), find the Maclaurin series for f. Does the series converge to f?

72. *Investigation*

(a) Find the power series centered at 0 for the function

$$f(x) = \dfrac{\ln(x^2 + 1)}{x^2}.$$

(b) Use a graphing utility to graph f and the eighth-degree Taylor polynomial $P_8(x)$ for f.

(c) Complete the following table, where

$$F(x) = \int_0^x \dfrac{\ln(t^2 + 1)}{t^2}\, dt \quad \text{and} \quad G(x) = \int_0^x P_8(t)\, dt.$$

x	0.25	0.50	0.75	1.00	1.50	2.00
$F(x)$						
$G(x)$						

(d) Describe the relationship between the graphs of f and P_8 and the results given in the table in part (c).

73. Prove that $\displaystyle\lim_{n \to 0} \dfrac{x^n}{n!} = 0$ for any real x.

74. Prove that e is irrational. $\left[$ *Hint:* Assume that $e = p/q$ is rational (p, q integers) and consider

$$e = 1 + 1 + \dfrac{1}{2!} + \cdots + \dfrac{1}{n!} + \cdots. \right]$$

REVIEW EXERCISES FOR CHAPTER 8

8.1 In Exercises 1 and 2, write an expression for the *n*th term of the sequence.

1. $1, \dfrac{1}{2}, \dfrac{1}{6}, \dfrac{1}{24}, \dfrac{1}{120}, \cdots$

2. $\dfrac{1}{2}, \dfrac{2}{5}, \dfrac{3}{10}, \dfrac{4}{17}, \cdots$

In Exercises 3–6, match the sequence with its graph. [The graphs arc labeled (a), (b), (c), and (d).]

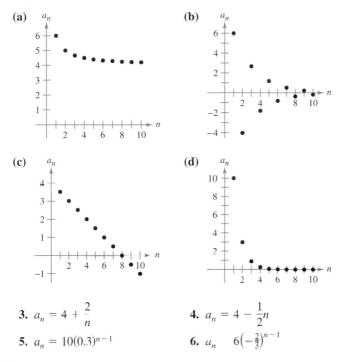

(a)   **(b)**

(c)   **(d)**

3. $a_n = 4 + \dfrac{2}{n}$

4. $a_n = 4 - \dfrac{1}{2}n$

5. $a_n = 10(0.3)^{n-1}$

6. $a_n = 6\left(-\dfrac{2}{3}\right)^{n-1}$

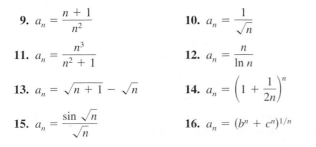
In Exercises 7 and 8, use a graphing utility to graph the first ten terms of the sequence. Use the graph to make an inference about the convergence or divergence of the sequence. Verify your inference analytically, and if the sequence converges, find its limit.

7. $a_n = \dfrac{5n + 2}{n}$

8. $a_n = \sin\dfrac{n\pi}{2}$

In Exercises 9–16, determine the convergence or divergence of the sequence with the given *n*th term. (*b* and *c* are positive real numbers.)

9. $a_n = \dfrac{n + 1}{n^2}$

10. $a_n = \dfrac{1}{\sqrt{n}}$

11. $a_n = \dfrac{n^3}{n^2 + 1}$

12. $a_n = \dfrac{n}{\ln n}$

13. $a_n = \sqrt{n + 1} - \sqrt{n}$

14. $a_n = \left(1 + \dfrac{1}{2n}\right)^n$

15. $a_n = \dfrac{\sin\sqrt{n}}{\sqrt{n}}$

16. $a_n = (b^n + c^n)^{1/n}$

17. *Compound Interest* A deposit of $5000 is made in an account that earns 5% interest compounded quarterly. The balance in the account after *n* quarters is

$$A_n = 5000\left(1 + \dfrac{0.05}{4}\right)^n, \quad n = 1, 2, 3, \cdots.$$

(a) Compute the first eight terms of the sequence.

(b) Find the balance in the account after 10 years by computing the 40th term of the sequence.

18. *Depreciation* A company buys a machine for $120,000. During the next 5 years the machine will depreciate at a rate of 30% per year. (That is, at the end of each year, the depreciated valued will be 70% of what it was at the beginning of the year.)

(a) Find a formula for the *n*th term of the sequence that gives the value *V* of the machine *t* full years after it was purchased.

(b) Find the depreciated value of the machine at the end of 5 full years.

8.2 *Numerical, Graphical, and Analytic Analysis* In Exercises 19–22, (a) use a graphing utility to find the indicated partial sum S_k and complete the table, and (b) use the graphing utility to graph the first ten terms of the sequence of partial sums.

k	5	10	15	20	25
S_k					

19. $\displaystyle\sum_{n=1}^{\infty} \left(\dfrac{3}{2}\right)^{n-1}$

20. $\displaystyle\sum_{n=1}^{\infty} \dfrac{(-1)^{n+1}}{2n}$

21. $\displaystyle\sum_{n=1}^{\infty} \dfrac{(-1)^{n+1}}{(2n)!}$

22. $\displaystyle\sum_{n=1}^{\infty} \dfrac{1}{n(n + 1)}$

In Exercises 23–26, determine the convergence or divergence of the series.

23. $\displaystyle\sum_{n=0}^{\infty} (0.82)^n$

24. $\displaystyle\sum_{n=0}^{\infty} (1.82)^n$

25. $\displaystyle\sum_{n=1}^{\infty} \dfrac{(-1)^n n}{\ln n}$

26. $\displaystyle\sum_{n=0}^{\infty} \dfrac{2n + 1}{3n + 2}$

In Exercises 27–30, find the sum of the series.

27. $\displaystyle\sum_{n=0}^{\infty} \left(\dfrac{2}{3}\right)^n$

28. $\displaystyle\sum_{n=0}^{\infty} \dfrac{2^{n+2}}{3^n}$

29. $\displaystyle\sum_{n=0}^{\infty} \left(\dfrac{1}{2^n} - \dfrac{1}{3^n}\right)$

30. $\displaystyle\sum_{n=0}^{\infty} \left[\left(\dfrac{2}{3}\right)^n - \dfrac{1}{(n + 1)(n + 2)}\right]$

In Exercises 31 and 32, express the repeating decimal as a geometric series and write its sum as the ratio of two integers.

31. $0.\overline{09}$

32. $0.\overline{923076}$

33. *Bouncing Ball*   A ball is dropped from a height of 8 meters. Each time it drops h meters, it rebounds $0.7h$ meters. Find the total distance traveled by the ball.

34. *Total Compensation*   Suppose you accept a job that pays a salary of \$32,000 the first year. During the next 39 years, you will receive a 5.5% raise each year. Find your total salary over the 40-year period.

35. *Compound Interest*   A deposit of \$200 is made at the end of each month for 2 years in an account that pays 6% interest, compounded continuously. Determine the balance in the account at the end of 2 years.

36. *Compound Interest*   A deposit of \$100 is made at the end of each month for 10 years in an account that pays 6.5%, compounded monthly. Determine the balance in the account at the end of 10 years.

8.3   **In Exercises 37–40, determine the convergence or divergence of the series.**

37. $\displaystyle\sum_{n=1}^{\infty} \frac{\ln n}{n^4}$

38. $\displaystyle\sum_{n=1}^{\infty} \frac{1}{\sqrt[4]{n^3}}$

39. $\displaystyle\sum_{n=1}^{\infty} \left(\frac{1}{n^2} - \frac{1}{n}\right)$

40. $\displaystyle\sum_{n=1}^{\infty} \left(\frac{1}{n^2} - \frac{1}{2^n}\right)$

8.4   **In Exercises 41–44, determine the convergence or divergence of the series.**

41. $\displaystyle\sum_{n=1}^{\infty} \frac{1}{\sqrt{n^3 + 2n}}$

42. $\displaystyle\sum_{n=1}^{\infty} \frac{n+1}{n(n+2)}$

43. $\displaystyle\sum_{n=1}^{\infty} \frac{1 \cdot 3 \cdot 5 \cdots (2n-1)}{2 \cdot 4 \cdot 6 \cdots (2n)}$

44. $\displaystyle\sum_{n=1}^{\infty} \frac{1}{3^n - 5}$

8.5   **In Exercises 45–48, determine the convergence or divergence of the series.**

45. $\displaystyle\sum_{n=2}^{\infty} \frac{(-1)^n n}{n^2 - 3}$

46. $\displaystyle\sum_{n=1}^{\infty} \frac{(-1)^n \sqrt{n}}{n+1}$

47. $\displaystyle\sum_{n=4}^{\infty} \frac{(-1)^n n}{n-3}$

48. $\displaystyle\sum_{n=2}^{\infty} \frac{(-1)^n \ln n^3}{n}$

8.6   **In Exercises 49–52, determine the convergence or divergence of the series.**

49. $\displaystyle\sum_{n=1}^{\infty} \frac{n}{e^{n^2}}$

50. $\displaystyle\sum_{n=1}^{\infty} \frac{n!}{e^n}$

51. $\displaystyle\sum_{n=1}^{\infty} \frac{2^n}{n^3}$

52. $\displaystyle\sum_{n=1}^{\infty} \frac{1 \cdot 3 \cdot 5 \cdots (2n-1)}{2 \cdot 5 \cdot 8 \cdots (3n-1)}$

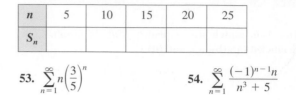

Numerical, Graphical, and Analytic Analysis   **In Exercises 53 and 54, (a) verify that the series converges, (b) use a graphing utility to find the indicated partial sum S_n and complete the table, (c) use the graphing utility to graph the first ten terms of the sequence of partial sums, and (d) use the table to estimate the sum of the series.**

n	5	10	15	20	25
S_n					

53. $\displaystyle\sum_{n=1}^{\infty} n\left(\frac{3}{5}\right)^n$

54. $\displaystyle\sum_{n=1}^{\infty} \frac{(-1)^{n-1}n}{n^3 + 5}$

55. *Writing*   Use a graphing utility to complete the table for (a) $p = 2$ and (b) $p = 5$. Write a short paragraph describing and comparing the entries in the table.

N	5	10	20	30	40
$\displaystyle\sum_{n=1}^{N} \frac{1}{n^P}$					
$\displaystyle\int_{N}^{\infty} \frac{1}{x^P}\,dx$					

56. *Writing*   You are told that the terms of a positive series appear to approach zero very slowly as n approaches infinity. (In fact, $a_{75} = 0.7$.) If you are given no other information, can you conclude that the series diverges? Support your answer with an example.

8.7   **In Exercises 57 and 58, use the definition of Taylor polynomial to find the third-degree Taylor polynomial centered at c.**

57. $f(x) = e^{-x/2}, \quad c = 0$

58. $f(x) = \tan x, \quad c = -\dfrac{\pi}{4}$

In Exercises 59–62, use a Taylor polynomial to approximate the function with an error of less than 0.001.

59. $\sin 95°$

60. $\cos(0.75)$

61. $\ln(1.75)$

62. $e^{-0.25}$

63. A Taylor polynomial centered at 0 will be used to approximate the cosine function. Find the degree of the polynomial required to obtain the desired accuracy over the indicated interval.

	Maximum Error	*Interval*
(a)	0.001	$[-0.5, 0.5]$
(b)	0.001	$[-1, 1]$
(c)	0.0001	$[-0.5, 0.5]$
(d)	0.0001	$[-2, 2]$

64. Use a graphing utility to graph the cosine function and the Taylor polynomials in Exercise 63.

8.8 In Exercises 65–70, find the interval of convergence of the power series.

65. $\displaystyle\sum_{n=0}^{\infty} \left(\frac{x}{10}\right)^n$

66. $\displaystyle\sum_{n=0}^{\infty} (2x)^n$

67. $\displaystyle\sum_{n=0}^{\infty} \frac{(-1)^n (x-2)^n}{(n+1)^2}$

68. $\displaystyle\sum_{n=1}^{\infty} \frac{3^n (x-2)^n}{n}$

69. $\displaystyle\sum_{n=0}^{\infty} n!(x-2)^n$

70. $\displaystyle\sum_{n=0}^{\infty} \frac{(x-2)^n}{2^n}$

In Exercises 71 and 72, show that the function defined by the series is a solution of the differential equation.

71. $y = \displaystyle\sum_{n=0}^{\infty} (-1)^n \frac{x^{2n}}{4^n (n!)^2}$

$x^2 y'' + xy' + x^2 y = 0$

72. $y = \displaystyle\sum_{n=0}^{\infty} \frac{(-3)^n x^{2n}}{2^n n!}$

$y'' + 3xy' + 3y = 0$

8.9 In Exercises 73 and 74, find the geometric power series centered at 0 for the function.

73. $g(x) = \dfrac{2}{3-x}$

74. $h(x) = \dfrac{3}{2+x}$

75. Find the power series for the derivative of the function in Exercise 73.

76. Find the power series for the integral of the function in Exercise 74.

In Exercises 77 and 78, find a function represented by the series and give the domain of the function.

77. $1 + \dfrac{2}{3}x + \dfrac{4}{9}x^2 + \dfrac{8}{27}x^3 + \cdots$

78. $8 - 2(x-3) + \dfrac{1}{2}(x-3)^2 - \dfrac{1}{8}(x-3)^3 + \cdots$

In Exercises 79–86, find the power series for the function centered at c.

79. $f(x) = \sin x, \quad c - \dfrac{3\pi}{4}$

80. $f(x) = \cos x, \quad c = -\dfrac{\pi}{4}$

81. $f(x) = 3^x, \quad c = 0$

82. $f(x) = \csc x, \quad c = \dfrac{\pi}{2}$

(first three terms)

83. $f(x) = \dfrac{1}{x}, \quad c = -1$

84. $f(x) = \sqrt{x}, \quad c = 4$

85. $g(x) = \sqrt[5]{1+x}, \quad c = 0$

86. $h(x) = \dfrac{1}{(1+x)^3}, \quad c = 0$

In Exercises 87–92, find the sum of the convergent series. Explain how you obtained the sum. (*Hint:* Use the power series for elementary functions.)

87. $\displaystyle\sum_{n=1}^{\infty} (-1)^{n+1} \frac{1}{4^n n}$

88. $\displaystyle\sum_{n=1}^{\infty} (-1)^{n+1} \frac{1}{5^n n}$

89. $\displaystyle\sum_{n=0}^{\infty} \frac{1}{2^n n!}$

90. $\displaystyle\sum_{n=0}^{\infty} \frac{2^n}{3^n n!}$

91. $\displaystyle\sum_{n=0}^{\infty} (-1)^n \frac{2^{2n}}{3^{2n} (2n)!}$

92. $\displaystyle\sum_{n=0}^{\infty} (-1)^n \frac{1}{3^{2n+1} (2n+1)!}$

8.10

93. *Writing* One of the series in Exercises 41 and 49 converges to its sum at a much slower rate than the other series. Which is it? Explain why this series converges so slowly. Use a graphing utility to illustrate the rate of convergence.

94. Find the Maclaurin series for $f(x) = xe^x$. Integrate the series term-by-term over the closed interval $[0, 1]$, and show that

$$\sum_{n=0}^{\infty} \frac{1}{(n+2)n!} = 1.$$

95. *Forming Maclaurin Series* Determine the first four terms of the Maclaurin series for e^{2x}

(a) by using the definition of the Maclaurin series and the formula for the coefficient of the nth term, $a_n = f^{(n)}(0)/n!$.

(b) by replacing x by $2x$ in the series for e^x.

(c) by multiplying the series for e^x by itself, because $e^{2x} = e^x \cdot e^x$.

96. *Forming Maclaurin Series* Follow the pattern of Exercise 95 to find the first four terms of the series for $\sin 2x$. (*Hint:* $\sin 2x - 2 \sin x \cos x$.)

In Exercises 97–100, find the series representation of the function defined by the integral.

97. $\displaystyle\int_0^x \frac{\sin t}{t} \, dt$

98. $\displaystyle\int_0^x \cos \frac{\sqrt{t}}{2} \, dt$

99. $\displaystyle\int_0^x \frac{\ln(t+1)}{t} \, dt$

100. $\displaystyle\int_0^x \frac{e^t - 1}{t} \, dt$

In Exercises 101 and 102, use power series to find the limit (if it exists). Verify the result by using L'Hôpital's Rule.

101. $\displaystyle\lim_{x \to 0} \frac{\arctan x}{\sqrt{x}}$

102. $\displaystyle\lim_{x \to 0} \frac{\arcsin x}{x}$

P.S. Problem Solving

1. The Cantor set (Georg Cantor, 1845–1918) is a subset of the unit interval $[0, 1]$. To construct the Cantor set, first remove the middle third $\left(\frac{1}{3}, \frac{2}{3}\right)$ of the interval, leaving two line segments. For the second step, remove the middle third of each of the two remaining segments, leaving four line segments. Continue this procedure indefinitely, as indicated in the figure. The Cantor set consists of all numbers in the unit interval $[0, 1]$ that still remain.

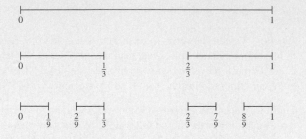

(a) Find the total length of all the line segments that are removed.

(b) Write down three numbers that are in the Cantor set.

(c) Let C_n denote the total length of the remaining line segments after n steps. Find $\lim\limits_{n \to \infty} C_n$.

GEORG CANTOR (1845–1918)

Cantor was a German mathematician known for his work on the development of set theory, which is the basis of modern mathematical analysis. This theory extends to the concept of infinite (or transfinite) numbers.

2. It can be shown that

$$\sum_{n=1}^{\infty} \frac{1}{n^2} = \frac{\pi^2}{6} \text{ (see Example 3, Section 8.3)}.$$

Use this fact to show that $\displaystyle\sum_{n=1}^{\infty} \frac{1}{(2n-1)^2} = \frac{\pi^2}{8}$.

3. Let T be an equilateral triangle with sides of length 1. Let a_n be the number of circles that can be packed tightly in n rows inside the triangle. For example, $a_1 = 1$, $a_2 = 3$, and $a_3 = 6$, as shown in the figure. Let A_n be the combined area of the a_n circles. Find $\lim\limits_{n \to \infty} A_n$.

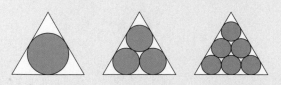

4. Identical blocks of unit length are stacked on top of each other at the edge of a table. The center of gravity of the top block must lie over the block below it, the center of gravity of the top two blocks must lie over the block below them, and so on.

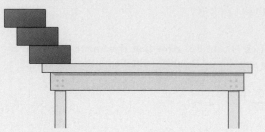

(a) If there are three blocks, show that it is possible to stack them so that the left edge of the top block extends $\frac{11}{12}$ unit beyond the edge of the table.

(b) Is it possible to stack the blocks so that the right edge of the top block extends beyond the edge of the table?

(c) How far beyond the table can the blocks be stacked?

5. (a) Consider the power series

$$\sum_{n=0}^{\infty} a_n x^n = 1 + 2x + 3x^2 + x^3 + 2x^4 + 3x^5 + x^6 + \cdots$$

in which the coefficients $a_n = 1, 2, 3, 1, 2, 3, 1, \ldots$ are periodic of period $p = 3$. Find the radius of convergence and the sum of this power series.

(b) Consider a power series

$$\sum_{n=0}^{\infty} a_n x^n$$

in which the coefficients are periodic, $a_{n+p} = a_p$. Find the radius of convergence and the sum of this power series.

6. For what values of the positive constants a and b does the following series converge absolutely? For what values does it converge conditionally?

$$a - \frac{b}{2} + \frac{a}{3} - \frac{b}{4} + \frac{a}{5} - \frac{b}{6} + \frac{a}{7} - \frac{b}{8} + \cdots$$

7. Find a power series for the function

$$f(x) = xe^x$$

centered at 0. Use this representation to find the sum of the infinite series

$$\sum_{n=1}^{\infty} \frac{1}{n!(n+2)}.$$

8. Find $f^{(12)}(0)$ if

$$f(x) = e^{x^2}.$$

(*Hint:* Do not calculate 12 derivatives!)

9. The graph of the function

$$f(x) = \begin{cases} 1, & x = 0 \\ \dfrac{\sin x}{x}, & x > 0 \end{cases}$$

is shown below. Use the Alternating Series Test to show that the improper integral $\int_1^\infty f(x)\,dx$ converges.

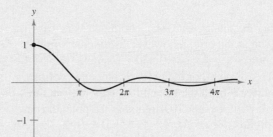

10. (a) Prove that $\displaystyle\int_2^\infty \frac{1}{x(\ln x)^p}\,dx$ converges if and only if $p > 1$.

(b) Determine the convergence or divergence of the series

$$\sum_{n=4}^\infty \frac{1}{n\,\ln(n^2)}.$$

11. (a) Consider the following sequence of numbers defined recursively.

$$a_1 = 3$$
$$a_2 = \sqrt{3}$$
$$a_3 = \sqrt{3 + \sqrt{3}}$$
$$\vdots$$
$$a_{n+1} = \sqrt{3 + a_n}$$

Write the decimal approximations for the first six terms of this sequence. Prove that the sequence converges and find its limit.

(b) Consider the following sequence defined recursively by $a_1 = \sqrt{a}$ and $a_{n+1} = \sqrt{a + a_n}$, where $a > 2$.

$$\sqrt{a},\ \sqrt{a + \sqrt{a}},\ \sqrt{a + \sqrt{a + \sqrt{a}}},\ \ldots$$

Prove that this sequence converges and find its limit.

12. Let $\{a_n\}$ be a sequence of positive numbers satisfying

$$\lim_{n\to\infty} (a_n)^{1/n} = L < \frac{1}{r},\ r > 0.$$ Prove that the series $\displaystyle\sum_{n=1}^\infty a_n r^n$ converges.

13. Consider the infinite series $\displaystyle\sum_{n=1}^\infty \frac{1}{2^{n+(-1)^n}}$.

(a) Find the first five terms of the sequence of partial sums.

(b) Show that the Ratio Test is inconclusive for this series.

(c) Use the Root Test to test for the convergence or divergence of this series.

14. Derive each identity using the appropriate geometric series.

(a) $\dfrac{1}{0.99} = 1.01010101 \ldots$

(b) $\dfrac{1}{0.98} = 1.0204081632 \ldots$

15. Consider an idealized population with the characteristic that each member of the population produces one offspring at the end of every time period. If each member has a life span of three time periods and the population begins with ten newborn members, then the following table gives the population during the first five time periods.

	Time Period				
Age Bracket	**1**	**2**	**3**	**4**	**5**
0–1	10	10	20	40	70
1–2		10	10	20	40
2–3			10	10	20
Total	10	20	40	70	130

The sequence for the total population has the property that

$$S_n = S_{n-1} + S_{n-2} + S_{n-3}, \qquad n > 3.$$

Find the total population during the next five time periods.

16. Imagine you are stacking an infinite number of spheres of decreasing radii on top of each other, as indicated in the figure. The radii of the spheres are 1 m, $1/\sqrt{2}$ m, $1/\sqrt{3}$ m, etc. The spheres are made of a material that weighs 1 newton per cubic meter.

(a) How high is this infinite stack of spheres?

(b) What is the total surface area of all the spheres in the stack?

(c) Show that the weight of the stack is finite.

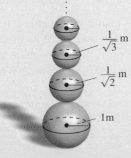

Exploring New Planets

Planets outside our own solar system are difficult to find because they are so dim compared with their parent stars. To discover these planets, astronomers rely on the influence that the planet may have on the star. An orbiting planet's gravitational pull drags the star back and forth as the planet rotates around it. This wobbling results in a subtle red-blue shift in the color of the star's light, known as the Doppler effect. Using a spectrometer, astronomers can monitor a star's Doppler variations, and use the results to calculate details pertaining to the orbiting body.

It was this technique that allowed Geoffrey Marcy and Paul Butler, of San Francisco State University, to identify a body rotating around the star 70 Virginis. They theorize that it is a large planet, 6.6 times as massive as Jupiter, although there is a small probability that it is a brown dwarf star. Marcy and Butler have calculated that the planet, named 70 Vir B, completes an orbit once every 116.6 days.

According to the astronomers, the planet's orbit is an ellipse with an eccentricity of 0.4, and a major axis length of 0.86 AU. (An astronomical unit, or AU, is the mean distance from the earth to the sun, about 93 million miles.) Placed on a rectangular coordinate system and centered at the origin, the equation for this ellipse is

$$\frac{x^2}{0.1849} + \frac{y^2}{0.1553} = 1$$

as shown in the graph.

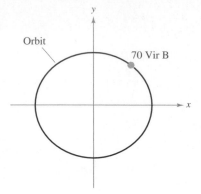

Rather than using Cartesian coordinates and centering the orbit at the origin, however, astronomers find it convenient to use polar coordinates. Using the sun as the main reference point, or the pole, each point is defined by its distance r from the sun and its angle θ from the horizontal. With the star 70 Virginis as the pole, the new planet's orbit is

$$r = \frac{0.3612}{1 - 0.4 \cos \theta}. \qquad \text{Polar equation for orbit of 70 Vir B}$$

Kepler's second law of planetary motion allows you to set up the proportion

$$\frac{t}{\text{period}} = \frac{\text{area of segment}}{\text{area of ellipse}}$$

$$= \frac{\frac{1}{2}\int_{\alpha}^{\beta}\left(\frac{0.3612}{1 - 0.4 \cos \theta}\right)^2 d\theta}{0.5324}$$

which you can solve to find the time t (in days) that it takes this particular planet to move in its orbit from $\theta = \alpha$ to $\theta = \beta$.

QUESTIONS

1. Set your graphing utility to polar mode and enter the polar equation for the orbit of 70 Vir B. Graph the equation using a window with θ varying from 0 to π. Then graph the equation again with θ varying from 0 to 2π, and again with θ varying from 0 to 4π. What do you observe?

2. When θ varies from 0 to π, the planet moves through half of its orbit. Starting with $\theta = 0$, what value of θ corresponds to one-fourth of the orbit? Explain.

3. Use the result of Question 2 to estimate the time it takes the planet to travel from $\theta = 0$ through one-quarter of its orbit. Then estimate the time it takes to travel through the second quarter of its orbit. Are these times the same? Describe the motion of this planet. When does it have a maximum speed? When does it have a minimum speed?

The concepts presented here will be explored further in this chapter. For an extension of this application, see Lab 13 in the lab series that accompanies this text at college.hmco.com.

Conics, Parametric Equations, and Polar Coordinates

9

In April 2001, Geoffrey Marcy and Paul Butler were awarded the Henry Draper Medal by the National Academy of Sciences "for their pioneering investigations of planets orbiting other stars via high-precision radial velocities." With their colleagues, Marcy and Butler have found 38 of 53 known extra-solar planets since 1995.

Geoffrey Marcy, left, and Paul Butler, right, used a technique known as the Doppler effect to identify the new planet 70 Vir B.

FOR FURTHER INFORMATION For more information on the discovery of the new planet 70 Vir B, see the article "Searching for Other Worlds" in *Time*. To view this article, go to the website *www.matharticles.com*.

Section 9.1	Conics and Calculus

- Understand the definition of a conic section.
- Analyze and write equations of parabolas using properties of parabolas.
- Analyze and write equations of ellipses using properties of ellipses.
- Analyze and write equations of hyperbolas using properties of hyperbolas.

Conic Sections

Each **conic section** (or simply **conic**) can be described as the intersection of a plane and a double-napped cone. Notice in Figure 9.1 that for the four basic conics, the intersecting plane does not pass through the vertex of the cone. When the plane passes through the vertex, the resulting figure is a **degenerate conic,** as shown in Figure 9.2.

Circle   Parabola   Ellipse   Hyperbola
Conic sections
Figure 9.1

 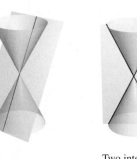

Point   Line   Two intersecting lines
Degenerate conics
Figure 9.2

There are several ways to study conics. You could begin as the Greeks did by defining the conics in terms of the intersections of planes and cones, or you could define them algebraically in terms of the general second-degree equation

$$Ax^2 + Bxy + Cy^2 + Dx + Ey + F = 0. \qquad \text{General second-degree equation}$$

However, a third approach, in which each of the conics is defined as a **locus** (collection) of points satisfying a certain geometric property, suits our needs best. For example, a circle can be defined as the collection of all points (x, y) that are equidistant from a fixed point (h, k). This locus definition easily produces the standard equation of a circle,

$$(x - h)^2 + (y - k)^2 = r^2. \qquad \text{Standard equation of a circle}$$

FOR FURTHER INFORMATION To learn more about the mathematical activities of Hypatia, see the article "Hypatia and Her Mathematics" by Michael A. B. Deakin in *The American Mathematical Monthly.* To view this article, go to the website *www.matharticles.com.*

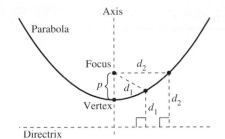

Figure 9.3

Parabolas

A **parabola** is the set of all points (x, y) that are equidistant from a fixed line called the **directrix** and a fixed point called the **focus** not on the line. The midpoint between the focus and the directrix is the **vertex,** and the line passing through the focus and the vertex is the **axis** of the parabola. Note in Figure 9.3 that a parabola is symmetric with respect to its axis.

THEOREM 9.1   Standard Equation of a Parabola

The **standard form** of the equation of a parabola with vertex (h, k) and directrix $y = k - p$ is

$$(x - h)^2 = 4p(y - k). \qquad \text{Vertical axis}$$

For directrix $x = h - p$, the equation is

$$(y - k)^2 = 4p(x - h). \qquad \text{Horizontal axis}$$

The focus lies on the axis p units (*directed distance*) from the vertex. The coordinates of the focus are as follows.

$$(h, k + p) \qquad \text{Vertical axis}$$
$$(h + p, k) \qquad \text{Horizontal axis}$$

Example 1   **Finding the Focus of a Parabola**

Find the focus of the parabola given by $y = -\frac{1}{2}x^2 - x + \frac{1}{2}$.

Solution   To find the focus, convert to standard form by completing the square.

$$y = \frac{1}{2} - x - \frac{1}{2}x^2 \qquad \text{Write original equation.}$$

$$y = \frac{1}{2}(1 - 2x - x^2) \qquad \text{Factor out } \tfrac{1}{2}.$$

$$2y = 1 - 2x - x^2 \qquad \text{Multiply each side by 2.}$$
$$2y = 1 - (x^2 + 2x) \qquad \text{Group terms.}$$
$$2y = 2 - (x^2 + 2x + 1) \qquad \text{Add and subtract 1 on right side.}$$
$$x^2 + 2x + 1 = -2y + 2$$
$$(x + 1)^2 = -2(y - 1) \qquad \text{Standard form}$$

Comparing this equation with $(x - h)^2 = 4p(y - k)$, you can conclude that

$$h = -1, \qquad k = 1, \qquad \text{and} \qquad p = -\frac{1}{2}.$$

Because p is negative, the parabola opens downward, as shown in Figure 9.4. Therefore, the focus of the parabola is p units from the vertex, or

$$(h, k + p) = \left(-1, \frac{1}{2}\right). \qquad \text{Focus}$$

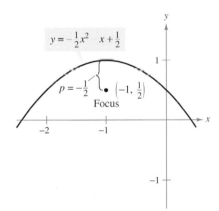

Parabola with a vertical axis, $p < 0$
Figure 9.4

A line segment that passes through the focus of a parabola and has endpoints on the parabola is called a **focal chord.** The specific focal chord perpendicular to the axis of the parabola is the **latus rectum.** The next example shows how to determine the length of the latus rectum and the length of the corresponding intercepted arc.

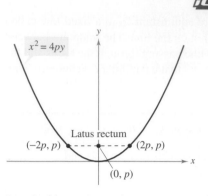

$x^2 = 4py$

Latus rectum

$(-2p, p)$ $(2p, p)$

$(0, p)$

Length of latus rectum: $4p$
Arc length: $4.59p$
Figure 9.5

i⊂ *Example 2   Focal Chord Length and Arc Length*

Find the length of the latus rectum of the parabola given by

$$x^2 = 4py.$$

Then find the length of the parabolic arc intercepted by the latus rectum.

Solution   Because the latus rectum passes through the focus $(0, p)$ and is perpendicular to the y-axis, the coordinates of its endpoints are $(-x, p)$ and (x, p). Substituting p for y in the equation of the parabola produces

$$x^2 = 4p(p) \quad \Longrightarrow \quad x = \pm 2p.$$

So, the endpoints of the latus rectum are $(-2p, p)$ and $(2p, p)$, and you can conclude that its length is $4p$, as shown in Figure 9.5. In contrast, the length of the intercepted arc is given by the following.

$$s = \int_{-2p}^{2p} \sqrt{1 + (y')^2}\, dx \qquad \text{Use arc length formula.}$$

$$= 2\int_{0}^{2p} \sqrt{1 + \left(\frac{x}{2p}\right)^2}\, dx \qquad y = \frac{x^2}{4p} \implies y' = \frac{x}{2p}$$

$$= \frac{1}{p}\int_{0}^{2p} \sqrt{4p^2 + x^2}\, dx \qquad \text{Simplify.}$$

$$= \frac{1}{2p}\left[x\sqrt{4p^2 + x^2} + 4p^2 \ln\left|x + \sqrt{4p^2 + x^2}\right| \right]_{0}^{2p} \qquad \text{Theorem 7.2}$$

$$= \frac{1}{2p}\left[2p\sqrt{8p^2} + 4p^2 \ln\left(2p + \sqrt{8p^2}\right) - 4p^2 \ln(2p) \right]$$

$$= 2p\left[\sqrt{2} + \ln\left(1 + \sqrt{2}\right) \right]$$

$$\approx 4.59p$$

One widely used property of a parabola is its reflective property. In physics, a surface is called **reflective** if the tangent line at any point on the surface makes equal angles with an incoming ray and the resulting outgoing ray. The angle corresponding to the incoming ray is the **angle of incidence,** and the angle corresponding to the outgoing ray is the **angle of reflection.** One example of a reflective surface is a flat mirror.

Another type of reflective surface is that formed by revolving a parabola about its axis. A special property of parabolic reflectors is that they allow us to direct all incoming rays parallel to the axis through the focus of the parabola—this is the principle behind the design of the parabolic mirrors used in reflecting telescopes. Conversely, all light rays emanating from the focus of a parabolic reflector used in a flashlight are parallel, as shown in Figure 9.6.

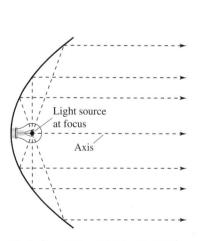

Light source
at focus

Axis

Parabolic reflector: light is reflected in
parallel rays.
Figure 9.6

> **THEOREM 9.2   Reflective Property of a Parabola**
>
> Let P be a point on a parabola. The tangent line to the parabola at the point P makes equal angles with the following two lines.
>
> **1.** The line passing through P and the focus
> **2.** The line passing through P parallel to the axis of the parabola

i⊂ *indicates that in the* Interactive 3.0 *CD-ROM and* Internet 3.0 *versions of this text* (*available at* college.hmco.com) *you will find an Open Exploration, which further explores this example using the computer algebra systems* Maple, Mathcad, Mathematica, *and* Derive.

Ellipses

More than a thousand years after the close of the Alexandrian period of Greek mathematics, Western civilization finally began a Renaissance of mathematical and scientific discovery. One of the principal figures in this rebirth was the Polish astronomer Nicolaus Copernicus. In his work *On the Revolutions of the Heavenly Spheres*, Copernicus claimed that all of the planets, including earth, revolved about the sun in circular orbits. Although some of Copernicus's claims were invalid, the controversy set off by his heliocentric theory motivated astronomers to search for a mathematical model to explain the observed movements of the sun and planets. The first to find the correct model was the German astronomer Johannes Kepler (1571–1630). Kepler discovered that the planets move about the sun in elliptical orbits, with the sun not as the center but as a focal point of the orbit.

The use of ellipses to explain the movement of the planets is only one of many practical and aesthetic uses. As with parabolas, we begin our study of this second type of conic by defining it as a locus of points. Now, however, we use *two* focal points rather than one.

An **ellipse** is the set of all points (x, y) the sum of whose distances from two distinct fixed points called **foci** is constant. (See Figure 9.7.) The line through the foci intersects the ellipse at two points, called the **vertices.** The chord joining the vertices is the **major axis,** and its midpoint is the **center** of the ellipse. The chord perpendicular to the major axis at the center is the **minor axis** of the ellipse.

NICOLAUS COPERNICUS (1473–1543)

Copernicus began to study planetary motion when asked to revise the calendar. At that time, the exact length of the year could not be accurately predicted using the theory that earth was the center of the universe.

FOR FURTHER INFORMATION To learn about how an ellipse may be "exploded" into a parabola, see the article "Exploding the Ellipse" by Arnold Good in *Mathematics Teacher*. To view this article, go to the website *www.matharticles.com*.

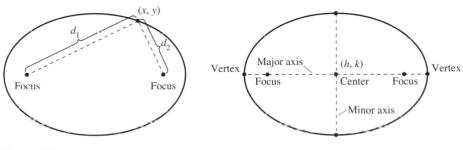

Figure 9.7

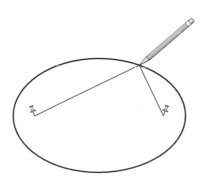

Figure 9.8

THEOREM 9.3　Standard Equation of an Ellipse

The standard form of the equation of an ellipse with center (h, k) and major and minor axes of lengths $2a$ and $2b$, where $a > b$, is

$$\frac{(x - h)^2}{a^2} + \frac{(y - k)^2}{b^2} = 1 \quad \text{Major axis is horizontal.}$$

or

$$\frac{(x - h)^2}{b^2} + \frac{(y - k)^2}{a^2} = 1. \quad \text{Major axis is vertical.}$$

The foci lie on the major axis, c units from the center, with $c^2 = a^2 - b^2$.

NOTE　You can visualize the definition of an ellipse by imagining two thumbtacks placed at the foci, as shown in Figure 9.8. If the ends of a fixed length of string are fastened to the thumbtacks and the string is drawn taut with a pencil, the path traced by the pencil will be an ellipse.

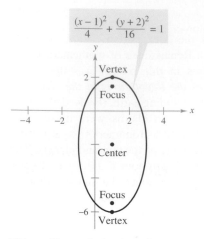

$$\frac{(x-1)^2}{4} + \frac{(y+2)^2}{16} = 1$$

Ellipse with a vertical major axis
Figure 9.9

Example 3   Completing the Square

Find the center, vertices, and foci of the ellipse given by

$$4x^2 + y^2 - 8x + 4y - 8 = 0.$$

Solution   By completing the square, you can write the given equation in standard form.

$$4x^2 + y^2 - 8x + 4y - 8 = 0 \qquad \text{Write original equation.}$$
$$4x^2 - 8x + y^2 + 4y = 8$$
$$4(x^2 - 2x + 1) + (y^2 + 4y + 4) = 8 + 4 + 4$$
$$4(x - 1)^2 + (y + 2)^2 = 16$$
$$\frac{(x-1)^2}{4} + \frac{(y+2)^2}{16} = 1 \qquad \text{Standard form}$$

So, the major axis is parallel to the y-axis, where $h = 1$, $k = -2$, $a = 4$, $b = 2$, and $c = \sqrt{16 - 4} = 2\sqrt{3}$. Therefore, you obtain the following.

Center:	$(1, -2)$	(h, k)
Vertices:	$(1, -6)$ and $(1, 2)$	$(h, k \pm a)$
Foci:	$\left(1, -2 - 2\sqrt{3}\right)$ and $\left(1, -2 + 2\sqrt{3}\right)$	$(h, k \pm c)$

The graph of the ellipse is shown in Figure 9.9.

NOTE   If the constant term $F = -8$ in the equation in Example 3 had been greater than or equal to 8, you would have obtained one of the following degenerate cases.

1. $F = 8$, single point, $(1, -2)$: $\dfrac{(x-1)^2}{4} + \dfrac{(y+2)^2}{16} = 0$

2. $F > 8$, no solution points: $\dfrac{(x-1)^2}{4} + \dfrac{(y+2)^2}{16} < 0$

Example 4   The Orbit of the Moon

The moon orbits earth in an elliptical path with the center of earth at one focus, as shown in Figure 9.10. The major and minor axes of the orbit have lengths of 768,806 kilometers and 767,746 kilometers. Find the greatest and least distances (the apogee and perigee) from earth's center to the moon's center.

Solution   Begin by solving for a and b.

$2a = 768,806$	Length of major axis
$a = 384,403$	Solve for a.
$2b = 767,746$	Length of minor axis
$b = 383,873$	Solve for b.

Now, using these values, you can solve for c as follows.

$$c = \sqrt{a^2 - b^2} \approx 20,179$$

The greatest distance between the center of earth and the center of the moon is $a + c \approx 404,582$ kilometers, and the least distance is $a - c \approx 364,224$ kilometers.

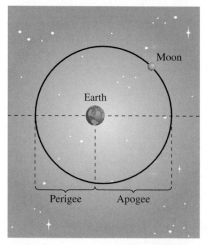

Figure 9.10

FOR FURTHER INFORMATION For more information on some uses of the reflective properties of conics, see the article "Parabolic Mirrors, Elliptic and Hyperbolic Lenses" by Mohsen Maesumi in *The American Mathematical Monthly.* Also see the article "The Geometry of Microwave Antennas" by William R. Parzynski in *Mathematics Teacher.* To view these articles, go to the website *www.matharticles.com.*

Theorem 9.2 presented a reflective property of parabolas. Ellipses have a similar reflective property. You are asked to prove the following theorem in Exercise 110.

THEOREM 9.4   Reflective Property of an Ellipse

Let P be a point on an ellipse. The tangent line to the ellipse at point P makes equal angles with the lines through P and the foci.

One of the reasons that astronomers had difficulty in detecting that the orbits of the planets are ellipses is that the foci of the planetary orbits are relatively close to the center of the sun, making the orbits nearly circular. To measure the ovalness of an ellipse, we use the concept of **eccentricity.**

Definition of Eccentricity of an Ellipse

The **eccentricity** e of an ellipse is given by the ratio

$$e = \frac{c}{a}.$$

To see how this ratio is used to describe the shape of an ellipse, note that because the foci of an ellipse are located along the major axis between the vertices and the center, it follows that

$$0 < c < a.$$

For an ellipse that is nearly circular, the foci are close to the center and the ratio c/a is small, and for an elongated ellipse, the foci are close to the vertices and the ratio is close to 1, as shown in Figure 9.11. Note that $0 < e < 1$ for every ellipse.

The orbit of the moon has an eccentricity of $e = 0.0549$, and the eccentricities of the nine planetary orbits are as follows.

Mercury:	$e = 0.2056$	Saturn:	$e = 0.0543$
Venus:	$e = 0.0068$	Uranus:	$e = 0.0460$
Earth:	$e = 0.0167$	Neptune:	$e = 0.0082$
Mars:	$e = 0.0934$	Pluto:	$e = 0.2481$
Jupiter:	$e = 0.0484$		

You can use integration to show that the area of an ellipse is $A = \pi ab$. For instance, the area of the ellipse

$$\frac{x^2}{a^2} + \frac{y^2}{b^2} = 1$$

is given by

$$A = 4\int_0^a \frac{b}{a}\sqrt{a^2 - x^2}\, dx$$
$$= \frac{4b}{a}\int_0^{\pi/2} a^2\cos^2\theta\, d\theta. \quad \text{Trigonometric substitution } x = a\sin\theta.$$

However, it is not so simple to find the *circumference* of an ellipse. The next example shows how to use eccentricity to set up an "elliptic integral" for the circumference of an ellipse.

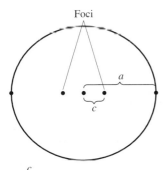

Foci

(a) $\dfrac{c}{a}$ is small.

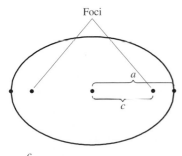

Foci

(b) $\dfrac{c}{a}$ is close to 1.

Eccentricity is the ratio $\dfrac{c}{a}$.
Figure 9.11

iC *Example 5* **Finding the Circumference of an Ellipse**

Show that the circumference of the ellipse $(x^2/a^2) + (y^2/b^2) = 1$ is

$$4a \int_0^{\pi/2} \sqrt{1 - e^2 \sin^2 \theta} \, d\theta. \qquad e = \frac{c}{a}$$

Solution Because the given ellipse is symmetric with respect to both the x-axis and the y-axis, you know that its circumference C is four times the arc length of $y = (b/a)\sqrt{a^2 - x^2}$ in the first quadrant. The function y is differentiable for all x in the interval $[0, a]$ except at $x = a$. So, the circumference is given by the improper integral

$$C = \lim_{d \to a} 4 \int_0^d \sqrt{1 + (y')^2} \, dx = 4 \int_0^a \sqrt{1 + (y')^2} \, dx = 4 \int_0^a \sqrt{1 + \frac{b^2 x^2}{a^2(a^2 - x^2)}} \, dx.$$

Using the trigonometric substitution $x = a \sin \theta$, you obtain

$$C = 4 \int_0^{\pi/2} \sqrt{1 + \frac{b^2 \sin^2 \theta}{a^2 \cos^2 \theta}} \, (a \cos \theta) \, d\theta$$

$$= 4 \int_0^{\pi/2} \sqrt{a^2 \cos^2 \theta + b^2 \sin^2 \theta} \, d\theta$$

$$= 4 \int_0^{\pi/2} \sqrt{a^2(1 - \sin^2 \theta) + b^2 \sin^2 \theta} \, d\theta$$

$$= 4 \int_0^{\pi/2} \sqrt{a^2 - (a^2 - b^2)\sin^2 \theta} \, d\theta.$$

Because $e^2 = c^2/a^2 = (a^2 - b^2)/a^2$, you can rewrite this integral as

$$C = 4a \int_0^{\pi/2} \sqrt{1 - e^2 \sin^2 \theta} \, d\theta.$$

AREA AND CIRCUMFERENCE OF AN ELLIPSE

In his work with elliptic orbits in the early 1600's, Johannes Kepler successfully developed a formula for the area of an ellipse, $A = \pi ab$. He was less successful in developing a formula for the circumference of an ellipse, however; the best he could do was to give the approximate formula $C = \pi(a + b)$.

A great deal of time has been devoted to the study of elliptic integrals. Such integrals generally do not have elementary antiderivatives. To find the circumference of an ellipse, you must usually resort to an approximation technique.

Example 6 **Approximating the Value of an Elliptic Integral**

Use the elliptic integral in Example 5 to approximate the circumference of the ellipse

$$\frac{x^2}{25} + \frac{y^2}{16} = 1.$$

Solution Because $e^2 = c^2/a^2 = (a^2 - b^2)/a^2 = 9/25$, you have

$$C = (4)(5) \int_0^{\pi/2} \sqrt{1 - \frac{9 \sin^2 \theta}{25}} \, d\theta.$$

Applying Simpson's Rule with $n = 4$ produces

$$C \approx 20 \left(\frac{\pi}{6}\right)\left(\frac{1}{4}\right)[1 + 4(0.9733) + 2(0.9055) + 4(0.8323) + 0.8]$$

$$\approx 28.36.$$

So, the ellipse has a circumference of about 28.36 units, as shown in Figure 9.12.

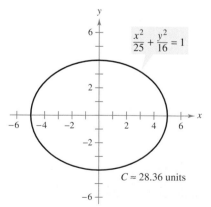

$C \approx 28.36$ units

Figure 9.12

Hyperbolas

The definition of a hyperbola is similar to that of an ellipse. For an ellipse, the *sum* of the distances between the foci and a point on the ellipse is fixed, whereas for a hyperbola, the absolute value of the *difference* between these distances is fixed.

A **hyperbola** is the set of all points (x, y) for which the absolute value of the difference between the distances from two distinct fixed points called **foci** is constant. (See Figure 9.13.) The line through the two foci intersects a hyperbola at two points called the **vertices.** The line segment connecting the vertices is the **transverse axis,** and the midpoint of the transverse axis is the **center** of the hyperbola. One distinguishing feature of a hyperbola is that its graph has two separate *branches*.

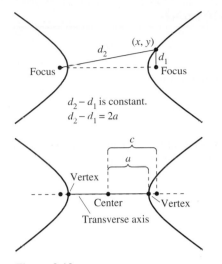

Figure 9.13

THEOREM 9.5   Standard Equation of a Hyperbola

The standard form of the equation of a hyperbola with center at (h, k) is

$$\frac{(x - h)^2}{a^2} - \frac{(y - k)^2}{b^2} = 1 \qquad \text{Transverse axis is horizontal.}$$

or

$$\frac{(y - k)^2}{a^2} - \frac{(x - h)^2}{b^2} = 1. \qquad \text{Transverse axis is vertical.}$$

The vertices are a units from the center, and the foci are c units from the center. Moreover, $c^2 = a^2 + b^2$.

NOTE   The constants a, b, and c do not have the same relationship for hyperbolas as they do for ellipses. For hyperbolas, $c^2 = a^2 + b^2$, but for ellipses, $c^2 = a^2 - b^2$.

An important aid in sketching the graph of a hyperbola is the determination of its **asymptotes,** as shown in Figure 9.14. Each hyperbola has two asymptotes that intersect at the center of the hyperbola. The asymptotes pass through the vertices of a rectangle of dimensions $2a$ by $2b$, with its center at (h, k). The line segment of length $2b$ joining $(h, k + b)$ and $(h, k - b)$ is referred to as the **conjugate axis** of the hyperbola.

THEOREM 9.6   Asymptotes of a Hyperbola

For a *horizontal* transverse axis, the equations of the asymptotes are

$$y = k + \frac{b}{a}(x - h) \qquad \text{and} \qquad y = k - \frac{b}{a}(x - h).$$

For a *vertical* transverse axis, the equations of the asymptotes are

$$y = k + \frac{a}{b}(x - h) \qquad \text{and} \qquad y = k - \frac{a}{b}(x - h).$$

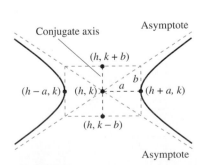

Figure 9.14

In Figure 9.14 you can see that the asymptotes coincide with the diagonals of the rectangle with dimensions $2a$ and $2b$, centered at (h, k). This provides you with a quick means of sketching the asymptotes, which in turn aids in sketching the hyperbola.

Example 7 Using Asymptotes to Sketch a Hyperbola

Sketch the graph of the hyperbola whose equation is $4x^2 - y^2 = 16$.

Solution Begin by rewriting the equation in standard form.

$$\frac{x^2}{4} - \frac{y^2}{16} = 1$$

The transverse axis is horizontal and the vertices occur at $(-2, 0)$ and $(2, 0)$. The ends of the conjugate axis occur at $(0, -4)$ and $(0, 4)$. Using these four points, you can sketch the rectangle shown in Figure 9.15(a). By drawing the asymptotes through the corners of this rectangle, you can complete the sketch as shown in Figure 9.15(b).

TECHNOLOGY You can use a graphing utility to verify the graph obtained in Example 7 by solving the original equation for y and graphing the following.

$$y_1 = \sqrt{4x^2 - 16}$$
$$y_2 = -\sqrt{4x^2 - 16}$$

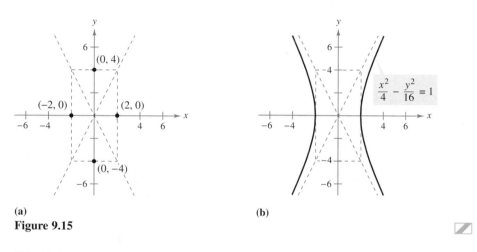

(a)

(b)

Figure 9.15

Definition of Eccentricity of a Hyperbola

The **eccentricity** e of a hyperbola is given by the ratio

$$e = \frac{c}{a}.$$

As with an ellipse, the **eccentricity** of a hyperbola is $e = c/a$. Because $c > a$ for hyperbolas, it follows that $e > 1$ for hyperbolas. If the eccentricity is large, the branches of the hyperbola are nearly flat. If the eccentricity is close to 1, the branches of the hyperbola are more pointed, as shown in Figure 9.16.

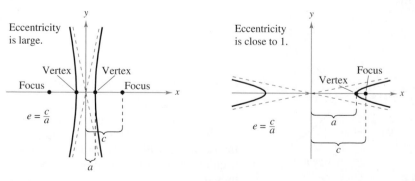

Eccentricity is large.

Eccentricity is close to 1.

Figure 9.16

The following application was developed during World War II. It shows how the properties of hyperbolas can be used in radar and other detection systems.

Example 8   A Hyperbolic Detection System

Two microphones, 1 mile apart, record an explosion. Microphone A receives the sound 2 seconds before microphone B. Where was the explosion?

Solution   Assuming that sound travels at 1100 feet per second, you know that the explosion took place 2200 feet farther from B than from A, as shown in Figure 9.17. The locus of all points that are 2200 feet closer to A than to B is one branch of the hyperbola $(x^2/a^2) - (y^2/b^2) = 1$, where

$$c = \frac{1 \text{ mile}}{2} = \frac{5280 \text{ ft}}{2} = 2640 \text{ ft}$$

and

$$a = \frac{2200 \text{ ft}}{2} = 1100 \text{ ft.}$$

Because $c^2 = a^2 + b^2$, it follows that

$$b^2 = c^2 - a^2$$
$$= 5,759,600$$

and you can conclude that the explosion occurred somewhere on the right branch of the hyperbola given by

$$\frac{x^2}{1,210,000} - \frac{y^2}{5,759,600} = 1.$$

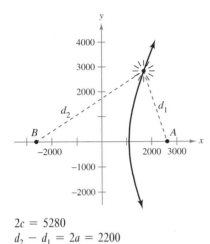

$2c = 5280$
$d_2 - d_1 = 2a = 2200$

Figure 9.17

In Example 8, you were able to determine only the hyperbola on which the explosion occurred, but not the exact location of the explosion. If, however, you had received the sound at a third position C, then two other hyperbolas would be determined. The exact location of the explosion would be the point at which these three hyperbolas intersect.

Another interesting application of conics involves the orbits of comets in our solar system. Of the 610 comets identified prior to 1970, 245 have elliptical orbits, 295 have parabolic orbits, and 70 have hyperbolic orbits. The center of the sun is a focus of each orbit, and each orbit has a vertex at the point at which the comet is closest to the sun. Undoubtedly, many comets with parabolic or hyperbolic orbits have not been identified—such comets pass through our solar system once. Only comets with elliptical orbits such as Halley's comet remain in our solar system.

The type of orbit for a comet can be determined as follows.

1. Ellipse: $v < \sqrt{2GM/p}$
2. Parabola: $v = \sqrt{2GM/p}$
3. Hyperbola: $v > \sqrt{2GM/p}$

In these three formulas, p is the distance between one vertex and one focus of the comet's orbit (in meters), v is the velocity of the comet at the vertex (in meters per second), $M \approx 1.991 \times 10^{30}$ kilograms is the mass of the sun, and $G \approx 6.67 \times 10^{-11}$ cubic meters per kilogram-second squared is the gravitational constant.

CAROLINE HERSCHEL (1750–1848)

The first woman to be credited with detecting a new comet was the English astronomer Caroline Herschel. During her life, Caroline Herschel discovered a total of eight new comets.

EXERCISES FOR SECTION 9.1

In Exercises 1–8, match the equation with its graph. [The graphs are labeled (a), (b), (c), (d), (e), (f), (g), and (h).]

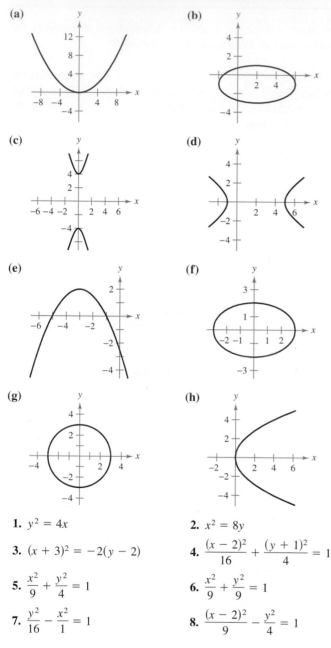

(a)

(b)

(c)

(d)

(e)

(f)

(g)

(h)

1. $y^2 = 4x$

2. $x^2 = 8y$

3. $(x + 3)^2 = -2(y - 2)$

4. $\dfrac{(x - 2)^2}{16} + \dfrac{(y + 1)^2}{4} = 1$

5. $\dfrac{x^2}{9} + \dfrac{y^2}{4} = 1$

6. $\dfrac{x^2}{9} + \dfrac{y^2}{9} = 1$

7. $\dfrac{y^2}{16} - \dfrac{x^2}{1} = 1$

8. $\dfrac{(x - 2)^2}{9} - \dfrac{y^2}{4} = 1$

In Exercises 9–16, find the vertex, focus, and directrix of the parabola, and sketch its graph.

9. $y^2 = -6x$

10. $x^2 + 8y = 0$

11. $(x + 3) + (y - 2)^2 = 0$

12. $(x - 1)^2 + 8(y + 2) = 0$

13. $y^2 - 4y - 4x = 0$

14. $y^2 + 6y + 8x + 25 = 0$

15. $x^2 + 4x + 4y - 4 = 0$

16. $y^2 + 4y + 8x - 12 = 0$

In Exercises 17–20, find the vertex, focus, and directrix of the parabola. Then use a graphing utility to graph the parabola.

17. $y^2 + x + y = 0$

18. $y = -\frac{1}{6}(x^2 - 8x + 6)$

19. $y^2 - 4x - 4 = 0$

20. $x^2 - 2x + 8y + 9 = 0$

In Exercises 21–28, find an equation of the parabola.

21. Vertex: $(3, 2)$
    Focus: $(1, 2)$

22. Vertex: $(-1, 2)$
    Focus: $(-1, 0)$

23. Vertex: $(0, 4)$
    Directrix: $y = -2$

24. Focus: $(2, 2)$
    Directrix: $x = -2$

25.

26.

27. Axis is parallel to y-axis; graph passes through $(0, 3)$, $(3, 4)$, and $(4, 11)$.

28. Directrix: $y = -2$; endpoints of latus rectum are $(0, 2)$ and $(8, 2)$.

In Exercises 29–34, find the center, foci, vertices, and eccentricity of the ellipse, and sketch its graph.

29. $x^2 + 4y^2 = 4$

30. $5x^2 + 7y^2 = 70$

31. $\dfrac{(x - 1)^2}{9} + \dfrac{(y - 5)^2}{25} = 1$

32. $(x + 2)^2 + \dfrac{(y + 4)^2}{1/4} = 1$

33. $9x^2 + 4y^2 + 36x - 24y + 36 = 0$

34. $16x^2 + 25y^2 - 64x + 150y + 279 = 0$

In Exercises 35–38, find the center, foci, and vertices of the ellipse. Use a graphing utility to graph the ellipse.

35. $12x^2 + 20y^2 - 12x + 40y - 37 = 0$

36. $36x^2 + 9y^2 + 48x - 36y + 43 = 0$

37. $x^2 + 2y^2 - 3x + 4y + 0.25 = 0$

38. $2x^2 + y^2 + 4.8x - 6.4y + 3.12 = 0$

In Exercises 39–44, find an equation of the ellipse.

39. Center: $(0, 0)$
    Focus: $(2, 0)$
    Vertex: $(3, 0)$

40. Vertices: $(0, 2)$, $(4, 2)$
    Eccentricity: $\frac{1}{2}$

41. Vertices: $(3, 1), (3, 9)$

Minor axis length: 6

42. Foci: $(0, \pm 5)$

Major axis length: 14

43. Center: $(0, 0)$

Major axis: horizontal

Points on the ellipse:

$(3, 1), (4, 0)$

44. Center: $(1, 2)$

Major axis: vertical

Points on the ellipse:

$(1, 6), (3, 2)$

70. $25x^2 - 10x - 200y - 119 = 0$

71. $4x^2 + 4y^2 - 16y + 15 = 0$

72. $y^2 - 4y = x + 5$

73. $9x^2 + 9y^2 - 36x + 6y + 34 = 0$

74. $2x(x - y) = y(3 - y - 2x)$

75. $3(x - 1)^2 = 6 + 2(y + 1)^2$

76. $9(x + 3)^2 = 36 - 4(y - 2)^2$

In Exercises 45–52, find the center, foci, and vertices of the hyperbola, and sketch its graph using asymptotes as an aid.

45. $y^2 - \dfrac{x^2}{4} = 1$

46. $\dfrac{x^2}{25} - \dfrac{y^2}{9} = 1$

47. $\dfrac{(x - 1)^2}{4} - \dfrac{(y + 2)^2}{1} = 1$

48. $\dfrac{(y + 1)^2}{144} - \dfrac{(x - 4)^2}{25} = 1$

49. $9x^2 - y^2 - 36x - 6y + 18 = 0$

50. $y^2 - 9x^2 + 36x - 72 = 0$

51. $x^2 - 9y^2 + 2x - 54y - 80 = 0$

52. $9x^2 - 4y^2 + 54x + 8y + 78 = 0$

In Exercises 53–56, find the center, foci, and vertices of the hyperbola. Use a graphing utility to graph the hyperbola and its asymptotes.

53. $9y^2 - x^2 + 2x + 54y + 62 = 0$

54. $9x^2 - y^2 + 54x + 10y + 55 = 0$

55. $3x^2 - 2y^2 - 6x - 12y - 27 = 0$

56. $3y^2 - x^2 + 6x - 12y = 0$

In Exercises 57–64, find an equation of the hyperbola.

57. Vertices: $(\pm 1, 0)$

Asymptotes: $y = \pm 3x$

58. Vertices: $(0, \pm 3)$

Asymptotes: $y = \pm 3x$

59. Vertices: $(2, \pm 3)$

Point on graph: $(0, 5)$

60. Vertices: $(2, \pm 3)$

Foci: $(2, \pm 5)$

61. Center: $(0, 0)$

Vertex: $(0, 2)$

Focus: $(0, 4)$

62. Center: $(0, 0)$

Vertex: $(3, 0)$

Focus: $(5, 0)$

63. Vertices: $(0, 2), (6, 2)$

Asymptotes: $y = \frac{2}{3}x$

$y = 4 - \frac{2}{3}x$

64. Focus: $(10, 0)$

Asymptotes: $y = \pm \frac{3}{4}x$

In Exercises 65 and 66, find equations for (a) the tangent and (b) the normal lines to the hyperbola for the given value of x.

65. $\dfrac{x^2}{9} - y^2 = 1, \quad x = 6$

66. $\dfrac{y^2}{4} - \dfrac{x^2}{2} = 1, \quad x = 4$

In Exercises 67–76, classify the graph of the equation as a circle, a parabola, an ellipse, or a hyperbola.

67. $x^2 + 4y^2 - 6x + 16y + 21 = 0$

68. $4x^2 - y^2 - 4x - 3 = 0$

69. $y^2 - 4y - 4x = 0$

Getting at the Concept

77. (a) Give the definition of a parabola.

(b) Give the standard forms of a parabola with vertex at (h, k).

(c) In your own words, state the reflective property of a parabola.

78. (a) Give the definition of an ellipse.

(b) Give the standard forms of an ellipse with center at (h, k).

79. (a) Give the definition of a hyperbola.

(b) Give the standard forms of a hyperbola with center at (h, k).

(c) Write equations for the asymptotes of a hyperbola.

80. Define the eccentricity of an ellipse. In your own words, describe how changes in the eccentricity affect the ellipse.

81. *Solar Collector* A solar collector for heating water is constructed with a sheet of stainless steel that is formed into the shape of a parabola (see figure). The water will flow through a pipe that is located at the focus of the parabola. At what distance from the vertex is the pipe?

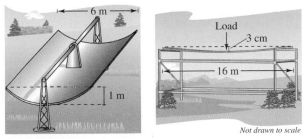

Figure for 81   **Figure for 82**

82. *Beam Deflection* A simply supported beam that is 16 meters long has a load concentrated at the center (see figure). The deflection of the beam at its center is 3 centimeters. Assume that the shape of the deflected beam is parabolic.

(a) Find an equation of the parabola. (Assume that the origin is at the center of the beam.)

(b) How far from the center of the beam is the deflection 1 centimeter?

83. Find an equation of the tangent line to the parabola $y = ax^2$ at $x = x_0$. Prove that the x-intercept of this tangent line is $(x_0/2, 0)$.

84. (a) Prove that any two distinct tangent lines to a parabola intersect.

(b) Demonstrate the result in part (a) by finding the point of intersection of the tangent lines to the parabola $x^2 - 4x - 4y = 0$ at the points $(0, 0)$ and $(6, 3)$.

85. (a) Prove that if any two tangent lines to a parabola intersect at right angles, their point of intersection must lie on the directrix.

(b) Demonstrate the result in part (a) by proving that the tangent lines to the parabola $x^2 - 4x - 4y + 8 = 0$ at the points $(-2, 5)$ and $\left(3, \frac{5}{4}\right)$ intersect at right angles, and that the point of intersection lies on the directrix.

86. Find the point on the graph of $x^2 = 8y$ that is closest to the focus of the parabola.

87. *Radio and Television Reception*   In mountainous areas, reception of radio and television is sometimes poor. Consider an idealized case where a hill is represented by the graph of the parabola $y = x - x^2$, a transmitter is located at the point $(-1, 1)$, and a receiver is located on the other side of the hill at the point $(x_0, 0)$. What is the closest the receiver can be to the hill so that the reception is unobstructed?

88. *Modeling Data*   The per capita consumption C (in pounds) of commercially produced fruits in the United States for selected years is given in the table. *(Source: U.S. Department of Agriculture)*

Year	1980	1985	1990	1995	1996	1997
C	262.4	269.4	273.5	285.4	289.8	294.7

(a) Use the regression capabilities of a graphing utility to find a quadratic model for the data, where t is the time in years, with $t = 0$ corresponding to 1980.

(b) Use a graphing utility to plot the data and graph the model.

(c) Find dC/dt and sketch its graph for $0 \le t \le 17$. What information about the consumption of fruits is given by the graph of the derivative?

89. *Architecture*   A church window is bounded on top by a parabola and below by the arc of a circle (see figure). Find the surface area of the window.

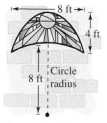

Figure for 89

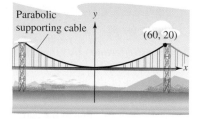

Figure for 91

90. *Arc Length*   Find the arc length of the parabola $4x - y^2 = 0$ over the interval $0 \le y \le 4$.

91. *Bridge Design*   A cable of a suspension bridge is suspended (in the shape of a parabola) between two towers that are 120 meters apart and 20 meters above the roadway (see figure). The cables touch the roadway midway between the towers.

(a) Find an equation for the parabolic shape of each cable.

(b) Find the length of the parabolic supporting cable.

92. *Surface Area*   A satellite-signal receiving dish is formed by revolving the parabola given by the graph of

$$x^2 = 20y$$

about the y-axis. If the radius of the dish is r feet, verify that the surface area of the dish is given by

$$2\pi \int_0^r x \sqrt{1 + \left(\frac{x}{10}\right)^2}\, dx = \frac{\pi}{15}[(100 + r^2)^{3/2} - 1000].$$

93. *Investigation*   Sketch the graphs of $x^2 = 4py$ for $p = \frac{1}{4}, \frac{1}{2}, 1, \frac{3}{2}$, and 2 on the same coordinate axes. Discuss the change in the graphs as p increases.

94. *Area*   Find a formula for the area of the shaded region in the figure.

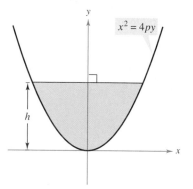

95. Sketch the ellipse that consists of all points (x, y) such that the sum of the distances between (x, y) and two fixed points is 16 units, and the foci are located at the centers of the two sets of concentric circles in the figure. To print an enlarged copy of the graph, go to the website *www.mathgraphs.com*.

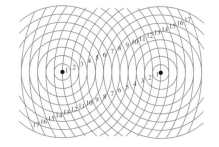

96. *Writing*   On page 653, it was noted that an ellipse can be drawn using two thumbtacks, a string of fixed length (greater than the distance between the tacks), and a pencil. If the ends of the string are fastened at the tacks and the string is drawn taut with a pencil, the path traced by the pencil will be an ellipse.

(a) What is the length of the string in terms of a?

(b) Explain why the path is an ellipse.

97. Construction of a Semielliptical Arch A fireplace arch is to be constructed in the shape of a semiellipse. The opening is to have a height of 2 feet at the center and a width of 5 feet along the base (see figure). The contractor draws the outline of the ellipse by the method shown in Exercise 96. Where should the tacks be placed and what should be the length of the piece of string?

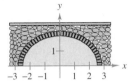

98. Orbit of the Earth Earth moves in an elliptical orbit with the sun at one of the foci. The length of half of the major axis is 149,570,000 kilometers, and the eccentricity is 0.0167. Find the minimum distance (*perihelion*) and the maximum distance (*aphelion*) of earth from the sun.

99. Satellite Orbit If the apogee and the perigee of an elliptical orbit of an earth satellite are given by A and P, show that the eccentricity of the orbit is

$$e = \frac{A - P}{A + P}.$$

100. Explorer 18 On November 26, 1963, the United States launched Explorer 18. Its low and high points above the surface of earth were 119 miles and 122,000 miles. Find the eccentricity of its elliptical orbit.

101. Halley's Comet Probably the most famous of all comets, Halley's comet, has an elliptical orbit with the sun at the focus. Its maximum distance from the sun is approximately 35.34 AU (astronomical unit $\approx 92,956 \times 10^6$ miles), and its minimum distance is approximately 0.59 AU. Find the eccentricity of the orbit.

102. The equation of an ellipse with its center at the origin can be written as

$$\frac{x^2}{a^2} + \frac{y^2}{a^2(1 - e^2)} = 1.$$

Show that as $e \to 0$, with a remaining fixed, the ellipse approaches a circle.

103. Consider a particle traveling clockwise on the elliptical path $x^2/100 + y^2/25 = 1$. The particle leaves the orbit at the point $(-8, 3)$ and travels in a straight line tangent to the ellipse. At what point will the particle cross the y-axis?

104. Volume The water tank on a fire truck is 16 feet long, and its cross sections are ellipses. Find the volume of water in the partially filled tank as shown in the figure.

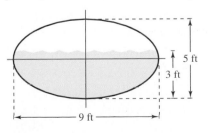

In Exercises 105 and 106, determine the points at which dy/dx is zero or does not exist to locate the endpoints of the major and minor axes of the ellipse.

105. $16x^2 + 9y^2 + 96x + 36y + 36 = 0$

106. $9x^2 + 4y^2 + 36x - 24y + 36 = 0$

Area and Volume In Exercises 107 and 108, find (a) the area of the region bounded by the ellipse, (b) the volume and surface area of the solid generated by revolving the region about its major axis (prolate spheroid), and (c) the volume and surface area of the solid generated by revolving the region about its minor axis (oblate spheroid).

107. $\dfrac{x^2}{4} + \dfrac{y^2}{1} = 1$ **108.** $\dfrac{x^2}{16} + \dfrac{y^2}{9} = 1$

109. Arc Length Use the integration capabilities of a graphing utility to approximate to two decimal-place accuracy the elliptical integral representing the circumference of the ellipse.

$$\frac{x^2}{25} + \frac{y^2}{49} = 1$$

110. Prove that the tangent line to an ellipse at a point P makes equal angles with lines through P and the foci (see figure). [*Hint:* (1) Find the slope of the tangent line at P, (2) find the slopes of the lines through P and each focus, and (3) use the formula for the tangent of the angle between two lines.]

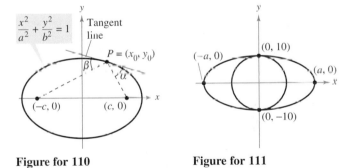

Figure for 110   Figure for 111

111. Geometry The area of the ellipse in the figure is twice the area of the circle. What is the length of the major axis?

112. Conjecture

(a) Show that the equation of an ellipse can be written as

$$\frac{(x - h)^2}{a^2} + \frac{(y - k)^2}{a^2(1 - e^2)} = 1.$$

(b) Use a graphing utility to graph the ellipse

$$\frac{(x - 2)^2}{4} + \frac{(y - 3)^2}{4(1 - e^2)} = 1$$

for $e = 0.95$, $e = 0.75$, $e = 0.5$, $e = 0.25$, and $e = 0$.

(c) Use the results in part (b) to make a conjecture about the change in the shape of the ellipse as e approaches 0.

113. Find an equation of the hyperbola such that for any point on the hyperbola, the difference between its distance from the points $(2, 2)$ and $(10, 2)$ is 6.

114. Find an equation of the hyperbola such that for any point on the hyperbola, the difference between its distances from the points $(-3, 0)$ and $(-3, 3)$ is 2.

115. Sketch the hyperbola that consists of all points (x, y) such that the difference of the distances between (x, y) and two fixed points is 10 units, and the foci are located at the centers of the two sets of concentric circles in the figure. To print an enlarged copy of the graph, go to the website *www.mathgraphs.com*.

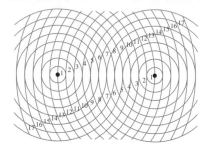

116. Consider a hyperbola centered at the origin with a horizontal transverse axis. Use the definition of a hyperbola to derive its standard form:

$$\frac{x^2}{a^2} - \frac{y^2}{b^2} = 1.$$

117. *Sound Location* A rifle positioned at point $(-c, 0)$ is fired at a target positioned at point $(c, 0)$. A person hears the sound of the rifle and the sound of the bullet hitting the target at the same time. Prove that the person is positioned on one branch of the hyperbola given by

$$\frac{x^2}{c^2 v_s^2 / v_m^2} - \frac{y^2}{c^2(v_m^2 - v_s^2)/v_m^2} = 1$$

where v_m is the muzzle velocity of the rifle and v_s is the speed of sound, which is about 1100 feet per second.

118. *Navigation* LORAN (long distance radio navigation) for aircraft and ships uses synchronized pulses transmitted by widely separated transmitting stations. These pulses travel at the speed of light (186,000 miles per second). The difference in the times of arrival of these pulses at an aircraft or ship is constant on a hyperbola having the transmitting stations as foci. Assume that two stations, 300 miles apart, are positioned on the rectangular coordinate system at $(-150, 0)$ and $(150, 0)$ and that a ship is traveling on a path with coordinates $(x, 75)$ (see figure). Find the x-coordinate of the position of the ship if the time difference between the pulses from the transmitting stations is 1000 microseconds (0.001 second).

119. *Hyperbolic Mirror* A hyperbolic mirror (used in some telescopes) has the property that a light ray directed at the focus will be reflected to the other focus. The mirror in the figure has the equation $(x^2/36) - (y^2/64) = 1$. At which point on the mirror will light from the point $(0, 10)$ be reflected to the other focus?

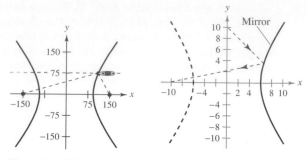

Figure for 118   **Figure for 119**

120. Show that the equation of the tangent line to

$$\frac{x^2}{a^2} - \frac{y^2}{b^2} = 1$$

at the point (x_0, y_0) is $(x_0/a^2)x - (y_0/b^2)y = 1$.

121. Show that the graphs of the equations intersect at right angles:

$$\frac{x^2}{a^2} + \frac{2y^2}{b^2} = 1 \quad \text{and} \quad \frac{x^2}{a^2 - b^2} - \frac{2y^2}{b^2} = 1.$$

122. Prove that the graph of the equation

$$Ax^2 + Cy^2 + Dx + Ey + F = 0$$

is one of the following (except in degenerate cases).

Conic	Condition
(a) Circle	$A = C$
(b) Parabola	$A = 0$ or $C = 0$ (but not both)
(c) Ellipse	$AC > 0$
(d) Hyperbola	$AC < 0$

True or False? In Exercises 123–129, determine whether the statement is true or false. If it is false, explain why or give an example that shows it is false.

123. It is possible for a parabola to intersect its directrix.

124. The point on a parabola closest to its focus is its vertex.

125. If C is the circumference of the ellipse

$$\frac{x^2}{a^2} + \frac{y^2}{b^2} = 1, \quad b < a,$$

then $2\pi b \le C \le 2\pi a$.

126. The graph of $(x^2/4) + y^4 = 1$ is an ellipse.

127. If $D \ne 0$ or $E \ne 0$, then the graph of

$$y^2 - x^2 + Dx + Ey = 0$$

is a hyperbola.

128. If the asymptotes of the hyperbola $(x^2/a^2) - (y^2/b^2) = 1$ intersect at right angles, then $a = b$.

129. Every tangent line to a hyperbola intersects the hyperbola only at the point of tangency.

- Sketch the graph of a curve given by a set of parametric equations.
- Eliminate the parameter in a set of parametric equations.
- Find a set of parametric equations to represent a curve.
- Understand two classic calculus problems, the tautochrone and brachistochrone problems.

Plane Curves and Parametric Equations

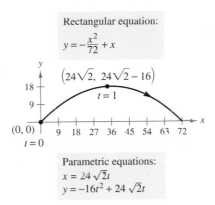

Rectangular equation:

$$y = -\frac{x^2}{72} + x$$

$\left(24\sqrt{2},\ 24\sqrt{2} - 16\right)$

$t = 1$

$(0, 0)$

$t = 0$

Parametric equations:

$x = 24\sqrt{2}\,t$

$y = -16t^2 + 24\sqrt{2}\,t$

Curvilinear motion: two variables for position, one variable for time
Figure 9.18

Until now, we have been representing a graph by a single equation involving *two* variables. In this section you will study situations in which *three* variables are used to represent a curve in the plane.

Consider the path followed by an object that is propelled into the air at an angle of 45°. If the initial velocity of the object is 48 feet per second, the object travels the parabolic path given by

$$y = -\frac{x^2}{72} + x \qquad \text{Rectangular equation}$$

as shown in Figure 9.18. However, this equation does not tell the whole story. Although it does tell you *where* the object has been, it doesn't tell you *when* the object was at a given point (x, y). To determine this time, you can introduce a third variable t, called a **parameter.** By writing both x and y as functions of t, you obtain the **parametric equations**

$$x = 24\sqrt{2}\,t \qquad \text{Parametric equation for } x$$

and

$$y = -16t^2 + 24\sqrt{2}\,t. \qquad \text{Parametric equation for } y$$

From this set of equations, you can determine that at time $t = 0$, the object is at the point $(0, 0)$. Similarly, at time $t = 1$, the object is at the point $\left(24\sqrt{2},\ 24\sqrt{2} - 16\right)$, and so on. (We will discuss a method for determining this particular set of parametric equations—the equations of motion—later, in Section 11.3.)

For this particular motion problem, x and y are continuous functions of t, and the resulting path is called a **plane curve.**

Definition of a Plane Curve

If f and g are continuous functions of t on an interval I, then the equations

$$x = f(t) \quad \text{and} \quad y = g(t)$$

are called **parametric equations** and t is called the **parameter.** The set of points (x, y) obtained as t varies over the interval I is called the **graph** of the parametric equations. Taken together, the parametric equations and the graph are called a **plane curve,** denoted by C.

NOTE   At times it is important to distinguish between a graph (the set of points) and a curve (the points together with their defining parametric equations). When it is important, we will make the distinction explicit. When it is not important, we will use C to represent the graph or the curve.

When sketching (by hand) a curve represented by a pair of parametric equations, you can plot points in the xy-plane. Each set of coordinates (x, y) is determined from a value chosen for the parameter t. By plotting the resulting points in order of *increasing* values of t, the curve is traced out in a specific direction. This is called the **orientation** of the curve.

Example 1   **Sketching a Curve**

Sketch the curve described by the parametric equations

$$x = t^2 - 4 \quad \text{and} \quad y = \frac{t}{2}, \quad -2 \leq t \leq 3.$$

Solution   For values of t on the given interval, the parametric equations yield the points (x, y) shown in the table.

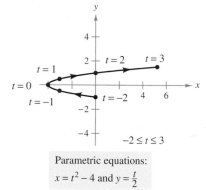

t	-2	-1	0	1	2	3
x	0	-3	-4	-3	0	5
y	-1	$-\frac{1}{2}$	0	$\frac{1}{2}$	1	$\frac{3}{2}$

Parametric equations:
$x = t^2 - 4$ and $y = \frac{t}{2}$

Figure 9.19

By plotting these points in order of increasing t and using the continuity of f and g, you obtain the curve C shown in Figure 9.19. Note that the arrows on the curve indicate its orientation as t increases from -2 to 3.

NOTE   From the vertical line test, you can see that the graph shown in Figure 9.19 does not define y as a function of x. This points out one benefit of parametric equations—they can be used to represent graphs that are more general than graphs of functions.

It often happens that two different sets of parametric equations have the same graph. For example, the set of parametric equations

$$x = 4t^2 - 4 \quad \text{and} \quad y = t, \quad -1 \leq t \leq \frac{3}{2}$$

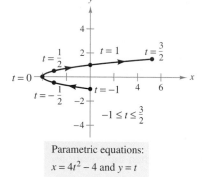

Parametric equations:
$x = 4t^2 - 4$ and $y = t$

Figure 9.20

has the same graph as the set given in Example 1. However, comparing the values of t in Figures 9.19 and 9.20, you can see that the second graph is traced out more *rapidly* (considering t as time) than the first graph. So, in applications, different parametric representations can be used to represent various *speeds* at which objects travel along a given path.

TECHNOLOGY   Most graphing utilities have a parametric graphing mode. If you have access to such a utility, try using it to confirm the graphs shown in Figures 9.19 and 9.20. Does the curve given by

$$x = 4t^2 - 8t \quad \text{and} \quad y = 1 - t, \quad -\tfrac{1}{2} \leq t \leq 2$$

represent the same graph as that shown in Figures 9.19 and 9.20? What do you notice about the *orientation* of this curve?

Eliminating the Parameter

Finding a rectangular equation that represents the graph of a set of parametric equations is called **eliminating the parameter**. For instance, you can eliminate the parameter from the set of parametric equations in Example 1 as follows.

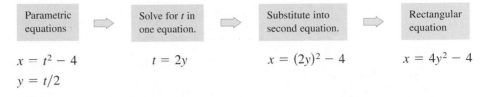

| Parametric equations | Solve for t in one equation. | Substitute into second equation. | Rectangular equation |

$$x = t^2 - 4 \qquad\qquad t = 2y \qquad\qquad x = (2y)^2 - 4 \qquad\qquad x = 4y^2 - 4$$
$$y = t/2$$

Once you have eliminated the parameter, you can recognize that the equation $x = 4y^2 - 4$ represents a parabola with a horizontal axis and vertex at $(-4, 0)$, as shown in Figure 9.19.

The range of x and y implied by the parametric equations may be altered by the change to rectangular form. In such instances the domain of the rectangular equation must be adjusted so that its graph matches the graph of the parametric equations. Such a situation is demonstrated in the next example.

Example 2   Adjusting the Domain After Eliminating the Parameter

Sketch the curve represented by the equations

$$x = \frac{1}{\sqrt{t+1}} \quad \text{and} \quad y = \frac{t}{t+1}, \quad t > -1$$

by eliminating the parameter and adjusting the domain of the resulting rectangular equation.

Solution   Begin by solving one of the parametric equations for t. For instance, you can solve the first equation for t as follows.

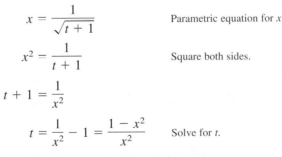

$$x = \frac{1}{\sqrt{t+1}} \qquad\qquad \text{Parametric equation for } x$$

$$x^2 = \frac{1}{t+1} \qquad\qquad \text{Square both sides.}$$

$$t + 1 = \frac{1}{x^2}$$

$$t = \frac{1}{x^2} - 1 = \frac{1 - x^2}{x^2} \qquad \text{Solve for } t.$$

Now, substituting into the parametric equation for y produces the following.

$$y = \frac{t}{t+1} \qquad\qquad \text{Parametric equation for } y$$

$$y = \frac{(1 - x^2)/x^2}{[(1 - x^2)/x^2] + 1} \qquad \text{Substitute } (1 - x^2)/x^2 \text{ for } t.$$

$$y = 1 - x^2 \qquad\qquad \text{Simplify.}$$

The rectangular equation, $y = 1 - x^2$, is defined for all values of x, but from the parametric equation for x you can see that the curve is defined only when $t > -1$. This implies that you should restrict the domain of x to positive values, as shown in Figure 9.21.

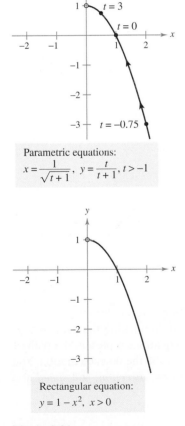

Parametric equations:
$$x = \frac{1}{\sqrt{t+1}}, \; y = \frac{t}{t+1}, t > -1$$

Rectangular equation:
$$y = 1 - x^2, \; x > 0$$

Figure 9.21

It is not necessary for the parameter in a set of parametric equations to represent time. The next example uses an *angle* as the parameter.

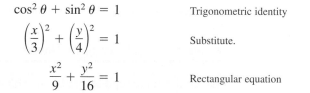

Example 3 **Using Trigonometry to Eliminate a Parameter**

Sketch the curve represented by

$$x = 3 \cos \theta \quad \text{and} \quad y = 4 \sin \theta, \quad 0 \le \theta \le 2\pi$$

by eliminating the parameter and finding the corresponding rectangular equation.

Solution Begin by solving for $\cos \theta$ and $\sin \theta$ in the given equations.

$$\cos \theta = \frac{x}{3} \quad \text{and} \quad \sin \theta = \frac{y}{4} \qquad \text{Solve for } \cos \theta \text{ and } \sin \theta.$$

Next, make use of the identity $\sin^2 \theta + \cos^2 \theta = 1$ to form an equation involving only x and y.

$$\cos^2 \theta + \sin^2 \theta = 1 \qquad \text{Trigonometric identity}$$

$$\left(\frac{x}{3}\right)^2 + \left(\frac{y}{4}\right)^2 = 1 \qquad \text{Substitute.}$$

$$\frac{x^2}{9} + \frac{y^2}{16} = 1 \qquad \text{Rectangular equation}$$

From this rectangular equation you can see that the graph is an ellipse centered at $(0, 0)$, with vertices at $(0, 4)$ and $(0, -4)$ and minor axis of length $2b = 6$, as shown in Figure 9.22. Note that the elliptic curve is traced out *counterclockwise* as θ varies from 0 to 2π.

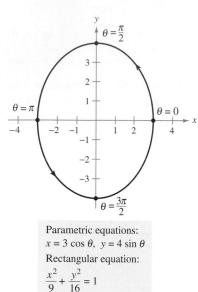

Parametric equations:
$x = 3 \cos \theta, \ y = 4 \sin \theta$
Rectangular equation:
$\dfrac{x^2}{9} + \dfrac{y^2}{16} = 1$

Figure 9.22

Using the technique shown in Example 3, you can conclude that the graph of the parametric equations

$$x = h + a \cos \theta \quad \text{and} \quad y = k + b \sin \theta, \quad 0 \le \theta \le 2\pi$$

is the ellipse (traced counterclockwise) given by

$$\frac{(x - h)^2}{a^2} + \frac{(y - k)^2}{b^2} = 1.$$

The graph of the parametric equations

$$x = h + a \sin \theta \quad \text{and} \quad y = k + b \cos \theta, \quad 0 \le \theta \le 2\pi$$

is also the ellipse (traced clockwise) given by

$$\frac{(x - h)^2}{a^2} + \frac{(y - k)^2}{b^2} = 1.$$

Try using a graphing utility in parametric mode to sketch several ellipses.

In Examples 2 and 3, it is important to realize that eliminating the parameter is primarily an *aid to curve sketching*. If the parametric equations represent the path of a moving object, the graph alone is not sufficient to describe the object's motion. You still need the parametric equations to tell you the *position*, *direction*, and *speed* at a given time.

Finding Parametric Equations

The first three examples in this section illustrated techniques for sketching the graph represented by a set of parametric equations. We now look at the reverse problem. How can you determine a set of parametric equations for a given graph or a given physical description? From the discussion following Example 1, you know that such a representation is not unique. This is demonstrated further in the following example, which finds two different parametric representations for a given graph.

Example 4   Finding Parametric Equations for a Given Graph

Find a set of parametric equations to represent the graph of $y = 1 - x^2$, using each of the following parameters.

a. $t = x$ **b.** the slope $m = \dfrac{dy}{dx}$ at the point (x, y)

Solution

a. Letting $x = t$ produces the parametric equations

$$x = t \quad \text{and} \quad y = 1 - x^2 = 1 - t^2.$$

b. To express x and y in terms of the parameter m, you can proceed as follows.

$$m = \frac{dy}{dx} = -2x \qquad \text{Differentiate } y = 1 - x^2.$$

$$x = -\frac{m}{2} \qquad \text{Solve for } x.$$

This produces a parametric equation for x. To obtain a parametric equation for y, substitute $-m/2$ for x in the original equation.

$$y = 1 - x^2 \qquad \text{Write original rectangular equation.}$$

$$y = 1 - \left(-\frac{m}{2}\right)^2 \qquad \text{Substitute } -m/2 \text{ for } x.$$

$$y = 1 - \frac{m^2}{4} \qquad \text{Simplify.}$$

So, the parametric equations are

$$x = -\frac{m}{2} \quad \text{and} \quad y = 1 - \frac{m^2}{4}.$$

In Figure 9.23, note that the resulting curve has a right-to-left orientation as determined by the direction of increasing values of slope m. For part (a), the curve would have the opposite orientation. ◢

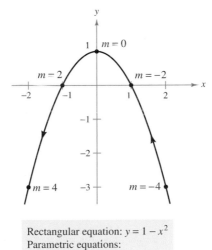

Rectangular equation: $y = 1 - x^2$
Parametric equations:

$$x = -\frac{m}{2}, y = 1 - \frac{m^2}{4}$$

Figure 9.23

TECHNOLOGY   To be efficient at using a graphing utility, it is important that you develop skill in representing a graph by a set of parametric equations. The reason for this is that many graphing utilities have only three graphing modes—(1) functions, (2) parametric equations, and (3) polar equations. Most graphing utilities are not programmed to sketch the graph of a general equation. For instance, suppose you want to sketch the graph of the hyperbola $x^2 - y^2 = 1$. To sketch the graph in function mode, you need two equations: $y = \sqrt{x^2 - 1}$ and $y = -\sqrt{x^2 - 1}$. In parametric mode, you can represent the graph by $x = \sec t$ and $y = \tan t$.

CYCLOIDS

Galileo first called attention to the cycloid, once recommending that it be used for the arches of bridges. Pascal once spent 8 days attempting to solve many of the problems of cycloids, such as finding the area under one arch, and the volume of the solid of revolution formed by revolving the curve about a line. The cycloid has so many interesting properties and has caused so many quarrels among mathematicians that it has been called "the Helen of geometry" and "the apple of discord."

FOR FURTHER INFORMATION For more information on cycloids, see the article "The Geometry of Rolling Curves" by John Bloom and Lee Whitt in *The American Mathematical Monthly.* To view this article, go to the website *www.matharticles.com.*

Example 5 Parametric Equations for a Cycloid

Determine the curve traced by a point P on the circumference of a circle of radius a rolling along a straight line in a plane. Such a curve is called a **cycloid**.

Solution Let the parameter θ be the measure of the circle's rotation, and let the point $P = (x, y)$ begin at the origin. When $\theta = 0$, P is at the origin. When $\theta = \pi$, P is at a maximum point $(\pi a, 2a)$. When $\theta = 2\pi$, P is back on the x-axis at $(2\pi a, 0)$. From Figure 9.24, you can see that $\angle APC = 180° - \theta$. Hence,

$$\sin \theta = \sin(180° - \theta) = \sin(\angle APC) = \frac{AC}{a} = \frac{BD}{a}$$

$$\cos \theta = -\cos(180° - \theta) = -\cos(\angle APC) = \frac{AP}{-a}$$

which implies that

$$AP = -a \cos \theta \quad \text{and} \quad BD = a \sin \theta.$$

Because the circle rolls along the x-axis, you know that $OD = \overset{\frown}{PD} = a\theta$. Furthermore, because $BA = DC = a$, you have

$$x = OD - BD = a\theta - a \sin \theta$$
$$y = BA + AP = a - a \cos \theta.$$

Therefore, the parametric equations are

$$x = a(\theta - \sin \theta) \quad \text{and} \quad y = a(1 - \cos \theta).$$

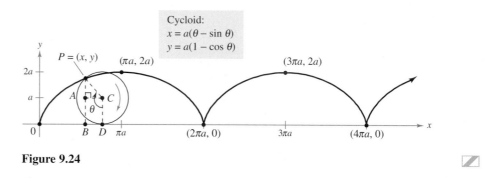

Cycloid:
$x = a(\theta - \sin \theta)$
$y = a(1 - \cos \theta)$

Figure 9.24

TECHNOLOGY Some graphing utilities allow you to simulate the motion of an object that is moving in the plane or in space. If you have access to such a utility, try using it to trace out the path of the cycloid shown in Figure 9.24.

The cycloid in Figure 9.24 has sharp corners at the values $x = 2n\pi a$. Notice that the derivatives $x'(\theta)$ and $y'(\theta)$ are both zero at the points for which $\theta = 2n\pi$.

$$x(\theta) = a(\theta - \sin \theta) \qquad y(\theta) = a(1 - \cos \theta)$$
$$x'(\theta) = a - a \cos \theta \qquad y'(\theta) = a \sin \theta$$
$$x'(2n\pi) = 0 \qquad y'(2n\pi) = 0$$

Between these points, the cycloid is called **smooth.**

Definition of a Smooth Curve

A curve C represented by $x = f(t)$ and $y = g(t)$ on an interval I is called **smooth** if f' and g' are continuous on I and not simultaneously 0, except possibly at the endpoints of I. The curve C is called **piecewise smooth** if it is smooth on each subinterval of some partition of I.

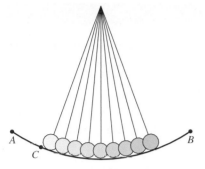

The time required to complete a full swing of the pendulum when starting from point C is only approximately the same as when starting from point A.
Figure 9.25

JAMES BERNOULLI (1654–1705)

James Bernoulli, also called Jacques, was the older brother of John. He was one of several accomplished mathematicians of the Swiss Bernoulli family. James's mathematical accomplishments have given him a prominent place in the early development of calculus.

The Tautochrone and Brachistochrone Problems

The type of curve described in Example 5 is related to one of the most famous pairs of problems in the history of calculus. The first problem (called the **tautochrone problem**) began with Galileo's discovery that the time required to complete a full swing of a given pendulum is *approximately* the same whether it makes a large movement at high speeds or a small movement at lower speeds (see Figure 9.25). Late in his life, Galileo (1564–1642) realized that he could use this principle to construct a clock. However, he was not able to conquer the mechanics of actual construction. Christian Huygens (1629–1695) was the first to design and construct a working model. In his work with pendulums, Huygens realized that a pendulum does not take *exactly* the same time to complete swings of varying lengths. (This doesn't affect a pendulum clock, because the length of the circular arc is kept constant by giving the pendulum a slight boost each time it passes its lowest point.) But, in studying the problem, Huygens discovered that a ball rolling back and forth on an inverted cycloid does complete each cycle in exactly the same time.

An inverted cycloid is the path down which a ball will roll in the shortest time.
Figure 9.26

The second problem, posed by John Bernoulli in 1696, is called the **brachistochrone problem**—in Greek, *brachys* means *short* and *chronos* means *time*. The problem was to determine the path down which a particle will slide from point A to point B in the *shortest time*. Several mathematicians took up the challenge, and the following year the problem was solved by Newton, Leibniz, L'Hôpital, John Bernoulli, and James Bernoulli. As it turns out, the solution is not a straight line from A to B, but an inverted cycloid passing through the points A and B, as shown in Figure 9.26. The amazing part of the solution is that a particle starting at rest at *any* other point C of the cycloid between A and B will take exactly the same time to reach B, as indicated in Figure 9.27.

A ball starting at point C takes the same time to reach point B as one that starts at point A.
Figure 9.27

FOR FURTHER INFORMATION To see a proof of the famous brachistochrone problem, see the article "A New Minimization Proof for the Brachistochrone" by Gary Lawlor in *The American Mathematical Monthly*. To view this article, go to the website *www.matharticles.com*.

EXERCISES FOR SECTION 9.2

1. Consider the parametric equations $x = \sqrt{t}$ and $y = 1 - t$.

(a) Complete the table.

t	0	1	2	3	4
x					
y					

(b) Plot the points (x, y) generated in the table, and sketch a graph of the parametric equations. Indicate the orientation of the graph.

(c) Use a graphing utility to confirm your graph in part (b).

(d) Find the rectangular equation by eliminating the parameter. Compare the graph in part (b) with the graph of the rectangular equation.

2. Consider the parametric equations $x = 4 \cos^2 \theta$ and $y = 2 \sin \theta$.

(a) Complete the table.

θ	$-\dfrac{\pi}{2}$	$-\dfrac{\pi}{4}$	0	$\dfrac{\pi}{4}$	$\dfrac{\pi}{2}$
x					
y					

(b) Plot the points (x, y) generated in the table, and sketch a graph of the parametric equations. Indicate the orientation of the graph.

(c) Use a graphing utility to confirm your graph in part (b).

(d) Find the rectangular equation by eliminating the parameter. Compare the graph in part (b) with the graph of the rectangular equation.

(e) If values of θ were selected from the interval $[\pi/2, 3\pi/2]$ for the table in part (a), would the graph in part (b) be different? Explain.

In Exercises 3–20, sketch the curve represented by the parametric equations (indicate the orientation of the curve), and write the corresponding rectangular equation by eliminating the parameter.

3. $x = 3t - 1, \quad y = 2t + 1$

4. $x = 3 - 2t, \quad y = 2 + 3t$

5. $x = t + 1, \quad y = t^2$

6. $x = 2t^2, \quad y = t^4 + 1$

7. $x = t^3, \quad y = \dfrac{t^2}{2}$

8. $x = t^2 + t, \quad y = t^2 - t$

9. $x = \sqrt{t}, \quad y = t - 2$

10. $x = \sqrt[4]{t}, \quad y = 3 - t$

11. $x = t - 1, \quad y = \dfrac{t}{t - 1}$

12. $x = 1 + \dfrac{1}{t}, \quad y = t - 1$

13. $x = 2t, \quad y = |t - 2|$

14. $x = |t - 1|, \quad y = t + 2$

15. $x = e^t, \quad y = e^{3t} + 1$

16. $x = e^{-t}, \quad y = e^{2t} - 1$

17. $x = \sec \theta, \quad y = \cos \theta, \quad 0 \le \theta < \pi/2, \quad \pi/2 < \theta \le \pi$

18. $x = \tan^2 \theta, \quad y = \sec^2 \theta$

19. $x = 3 \cos \theta, \quad y = 3 \sin \theta$

20. $x = 2 \cos \theta, \quad y = 6 \sin \theta$

In Exercises 21–32, use a graphing utility to sketch the curve represented by the parametric equations (indicate the orientation of the curve). Eliminate the parameter and write the corresponding rectangular equation.

21. $x = 4 \sin 2\theta, y = 2 \cos 2\theta$

22. $x = \cos \theta, y = 2 \sin 2\theta$

23. $x = 4 + 2 \cos \theta$
$y = -1 + \sin \theta$

24. $x = 4 + 2 \cos \theta$
$y = -1 + 2 \sin \theta$

25. $x = 4 + 2 \cos \theta$
$y = -1 + 4 \sin \theta$

26. $x = \sec \theta$
$y = \tan \theta$

27. $x = 4 \sec \theta, \quad y = 3 \tan \theta$

28. $x = \cos^3 \theta, \quad y = \sin^3 \theta$

29. $x = t^3, \quad y = 3 \ln t$

30. $x = \ln 2t, \quad y = t^2$

31. $x = e^{-t}, \quad y = e^{3t}$

32. $x = e^{2t}, \quad y = e^t$

Comparing Plane Curves In Exercises 33–36, determine any differences between the curves of the parametric equations. Are the graphs the same? Are the orientations the same? Are the curves smooth?

33. (a) $x = t$
$y = 2t + 1$

(b) $x = \cos \theta$
$y = 2 \cos \theta + 1$

(c) $x = e^{-t}$
$y = 2e^{-t} + 1$

(d) $x = e^t$
$y = 2e^t + 1$

34. (a) $x = 2 \cos \theta$
$y = 2 \sin \theta$

(b) $x = \sqrt{4t^2 - 1}/|t|$
$y = 1/t$

(c) $x = \sqrt{t}$
$y = \sqrt{4 - t}$

(d) $x = -\sqrt{4 - e^{2t}}$
$y = e^t$

35. (a) $x = \cos \theta$
$y = 2 \sin^2 \theta$
$0 < \theta < \pi$

(b) $x = \cos(-\theta)$
$y = 2 \sin^2(-\theta)$
$0 < \theta < \pi$

36. (a) $x = t + 1, y = t^3$

(b) $x = -t + 1, y = (-t)^3$

37. *Conjecture*

(a) Use a graphing utility to sketch the curves represented by the two sets of parametric equations.

$x = 4 \cos t \qquad x = 4 \cos(-t)$
$y = 3 \sin t \qquad y = 3 \sin(-t)$

(b) Describe the change in the graph when the sign of the parameter is changed.

(c) Make a conjecture about the change in the graph of parametric equations when the sign of the parameter is changed.

(d) Test your conjecture with another set of parametric equations.

38. *Writing* Review Exercises 33–36 and write a short paragraph describing how the graphs of curves represented by different sets of parametric equations can differ even though eliminating the parameter from each yields the same rectangular equation.

In Exercises 39–42, eliminate the parameter and obtain the standard form of the rectangular equation.

39. Line through (x_1, y_1) and (x_2, y_2):

$x = x_1 + t(x_2 - x_1)$, $y = y_1 + t(y_2 - y_1)$

40. Circle: $x = h + r \cos \theta$, $y = k + r \sin \theta$

41. Ellipse: $x = h + a \cos \theta$, $y = k + b \sin \theta$

42. Hyperbola: $x = h + a \sec \theta$, $y = k + b \tan \theta$

In Exercises 43–50, use the results of Exercises 39–42 to find a set of parametric equations for the line or conic.

43. Line: Passes through $(0, 0)$ and $(5, -2)$

44. Line: Passes through $(1, 4)$ and $(5, -2)$

45. Circle: Center: $(2, 1)$; Radius: 4

46. Circle: Center: $(-3, 1)$; Radius: 3

47. Ellipse: Vertices: $(\pm 5, 0)$; Foci: $(\pm 4, 0)$

48. Ellipse: Vertices: $(4, 7)$, $(4, -3)$; Foci: $(4, 5)$, $(4, -1)$

49. Hyperbola: Vertices: $(\pm 4, 0)$; Foci: $(\pm 5, 0)$

50. Hyperbola: Vertices: $(0, \pm 1)$; Foci: $(0, \pm 2)$

In Exercises 51–54, find two different sets of parametric equations for the given rectangular equation.

51. $y = 3x - 2$

52. $y = \dfrac{2}{x - 1}$

53. $y = x^3$

54. $y = x^2$

In Exercises 55–62, use a graphing utility to graph the curve represented by the parametric equations. Indicate the direction of the curve. Identify any points at which the curve is not smooth.

55. Cycloid: $x = 2(\theta - \sin \theta)$, $y = 2(1 - \cos \theta)$

56. Cycloid: $x = \theta + \sin \theta$, $y = 1 - \cos \theta$

57. Prolate cycloid: $x = \theta - \frac{3}{2}\sin \theta$, $y = 1 - \frac{3}{2}\cos \theta$

58. Prolate cycloid: $x = 2\theta - 4 \sin \theta$, $y = 2 - 4 \cos \theta$

59. Hypocycloid: $x = 3 \cos^3 \theta$, $y = 3 \sin^3 \theta$

60. Curtate cycloid: $x = 2\theta - \sin \theta$, $y = 2 - \cos \theta$

61. Witch of Agnesi: $x = 2 \cot \theta$, $y = 2 \sin^2 \theta$

62. Folium of Descartes: $x = 3t/(1 + t^3)$, $y = 3t^2/(1 + t^3)$

Getting at the Concept

63. State the definition of a plane curve given by parametric equations.

64. Explain the process of sketching a plane curve given by parametric equations. What is meant by the orientation of the curve?

65. State the definition of a smooth curve.

Getting at the Concept *(continued)*

66. Match each graph with a set of parametric equations. Explain your reasoning.

(i) $x = t^2 - 1$

$y = t + 2$

(ii) $x = \sin^2 \theta - 1$

$y = \sin \theta + 2$

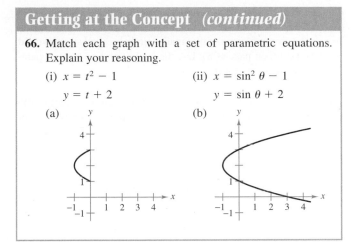

In Exercises 67–70, match the set of parametric equations with the correct graph. [The graphs are labeled (a), (b), (c), and (d).]

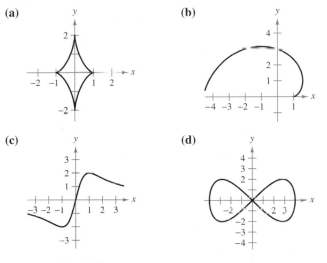

67. Lissajous curve: $x = 4 \cos \theta$, $y = 2 \sin 2\theta$

68. Evolute of ellipse: $x = \cos^3 \theta$, $y = 2 \sin^3 \theta$

69. Involute of circle: $x = \cos \theta + \theta \sin \theta$, $y = \sin \theta - \theta \cos \theta$

70. Serpentine curve: $x = \cot \theta$, $y = 4 \sin \theta \cos \theta$

71. *Curtate Cycloid* A wheel of radius a rolls along a line without slipping. The curve traced by a point P that is b units from the center $(b < a)$ is called a **curtate cycloid** (see figure). Use the angle θ to find a set of parametric equations for this curve.

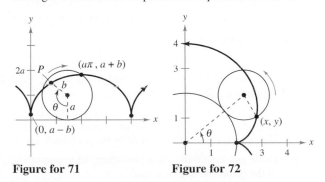

Figure for 71   Figure for 72

72. *Epicycloid* A circle of radius 1 rolls around the outside of a circle of radius 2 without slipping. The curve traced by a point on the circumference of the smaller circle is called an **epicycloid** (see figure on page 673). Use the angle θ to find a set of parametric equations for this curve.

True or False? **In Exercises 73 and 74, determine whether the statement is true or false. If it is false, explain why or give an example that shows it is false.**

73. The graph of the parametric equations $x = t^2$ and $y = t^2$ is the line $y = x$.

74. If y is a function of t and x is a function of t, then y is a function of x.

Projectile Motion **In Exercises 75 and 76, consider a projectile launched at a height h feet above the ground and at an angle θ with the horizontal. If the initial velocity is v_0 feet per second, the path of the projectile is modeled by the parametric equations**

$$x = (v_0 \cos \theta)t \quad \text{and} \quad y = h + (v_0 \sin \theta)t - 16t^2.$$

75. *Baseball* The center field fence in a ballpark is 10 feet high and 400 feet from home plate. The ball is hit 3 feet above the ground. It leaves the bat at an angle of θ degrees with the horizontal at a speed of 100 miles per hour (see figure).

(a) Write a set of parametric equations for the path of the ball.

(b) Use a graphing utility to graph the path of the ball if $\theta = 15°$. Is the hit a home run?

(c) Use a graphing utility to graph the path of the ball if $\theta = 23°$. Is the hit a home run?

(d) Find the minimum angle for the ball to leave the bat in order for the hit to be a home run.

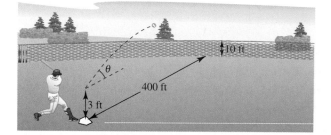

76. A rectangular equation for the path of a projectile is

$$y = 5 + x - 0.005x^2.$$

(a) Eliminate the parameter t from the position function for the motion of a projectile to show that the rectangular equation is

$$y = -\frac{16 \sec^2 \theta}{v_0^2}x^2 + (\tan \theta)\, x + h.$$

(b) Use the result in part (a) to find h, v_0, and θ. Find the parametric equations of the path.

(c) Use a graphing utility to graph the rectangular equation for the path of the projectile. Confirm your answer in part (b) by sketching the curve represented by the parametric equations.

(d) Use a graphing utility to approximate the maximum height of the projectile and its range.

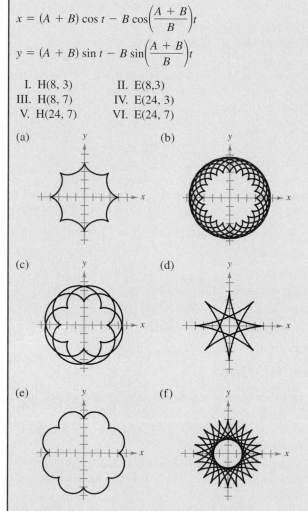

SECTION PROJECT **CYCLOIDS**

In Greek, the word *cycloid* means *wheel*, the word *hypocycloid* means *under the wheel*, and the word *epicycloid* means *upon the wheel*. Match the hypocycloid or epicycloid with its graph. [The graphs are labeled (a), (b), (c), (d), (e), and (f).]

Hypocycloid, H(A, B)

Path traced by a fixed point on a circle of radius B as it rolls around the *inside* of a circle of radius A.

$$x = (A - B)\cos t + B \cos\left(\frac{A - B}{B}\right)t$$

$$y = (A - B)\sin t - B \sin\left(\frac{A - B}{B}\right)t$$

Epicycloid, E(A, B)

Path traced by a fixed point on a circle of radius B as it rolls around the *outside* of a circle of radius A.

$$x = (A + B)\cos t - B \cos\left(\frac{A + B}{B}\right)t$$

$$y = (A + B)\sin t - B \sin\left(\frac{A + B}{B}\right)t$$

  I. H(8, 3)   II. E(8, 3)
III. H(8, 7)   IV. E(24, 3)
 V. H(24, 7)   VI. E(24, 7)

Exercises based on "Mathematical Discovery via Computer Graphics: Hypocycloids and Epicycloids" by Florence S. Gordon and Sheldon P. Gordon, *The College Mathematics Journal*, November 1984, p. 441. Used by permission of the authors.

* Find the slope of a tangent line to a curve given by a set of parametric equations.
* Find the arc length of a curve given by a set of parametric equations.
* Find the area of a surface of revolution (parametric form).

Slope and Tangent Lines

Now that you can represent a graph in the plane by a set of parametric equations, it is natural to ask how to use calculus to study plane curves. To begin, let's take another look at the projectile represented by the parametric equations

$$x = 24\sqrt{2}t \quad \text{and} \quad y = -16t^2 + 24\sqrt{2}t$$

as shown in Figure 9.28. From Section 9.2, you know that these equations enable you to locate the position of the projectile at a given time. You also know that the object is initially projected at an angle of 45°. But how can you find the angle θ representing the object's direction at some other time t? The following theorem answers this question by giving a formula for the slope of the tangent line as a function of t.

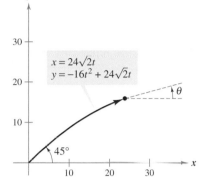

$$x = 24\sqrt{2}t$$
$$y = -16t^2 + 24\sqrt{2}t$$

At time t, the angle of elevation of the projectile is θ, the slope of the tangent line at that point.
Figure 9.28

THEOREM 9.7   Parametric Form of the Derivative

If a smooth curve C is given by the equations $x = f(t)$ and $y = g(t)$, then the slope of C at (x, y) is

$$\frac{dy}{dx} = \frac{dy/dt}{dx/dt}, \qquad \frac{dx}{dt} \neq 0.$$

Proof In Figure 9.29, consider $\Delta t > 0$ and let

$$\Delta y = g(t + \Delta t) - g(t) \quad \text{and} \quad \Delta x = f(t + \Delta t) - f(t).$$

Because $\Delta x \to 0$ as $\Delta t \to 0$, you can write

$$\frac{dy}{dx} = \lim_{\Delta x \to 0} \frac{\Delta y}{\Delta x}$$

$$= \lim_{\Delta t \to 0} \frac{g(t + \Delta t) - g(t)}{f(t + \Delta t) - f(t)}.$$

Dividing both the numerator and denominator by Δt, you can use the differentiability of f and g to conclude that

$$\frac{dy}{dx} = \lim_{\Delta t \to 0} \frac{[g(t + \Delta t) - g(t)]/\Delta t}{[f(t + \Delta t) - f(t)]/\Delta t}$$

$$= \frac{\displaystyle\lim_{\Delta t \to 0} \frac{g(t + \Delta t) - g(t)}{\Delta t}}{\displaystyle\lim_{\Delta t \to 0} \frac{f(t + \Delta t) - f(t)}{\Delta t}}$$

$$= \frac{g'(t)}{f'(t)}$$

$$= \frac{dy/dt}{dx/dt}.$$

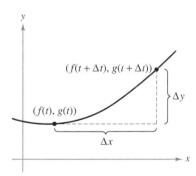

The slope of the secant line through the points $(f(t), g(t))$ and $(f(t + \Delta t), g(t + \Delta t))$ is $\Delta y / \Delta x$.
Figure 9.29

Example 1 Differentiation and Parametric Form

Find dy/dx for the curve given by $x = \sin t$ and $y = \cos t$.

STUDY TIP The curve traced out in Example 1 is a circle. Use the formula

$$\frac{dy}{dx} = -\tan t$$

to find the slope at the points $(1, 0)$ and $(0, 1)$.

Solution

$$\frac{dy}{dx} = \frac{dy/dt}{dx/dt} = \frac{-\sin t}{\cos t} = -\tan t$$

Because dy/dx is a function of t, you can use Theorem 9.7 repeatedly to find *higher-order* derivatives. For instance,

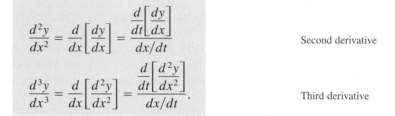

$$\frac{d^2y}{dx^2} = \frac{d}{dx}\left[\frac{dy}{dx}\right] = \frac{\frac{d}{dt}\left[\frac{dy}{dx}\right]}{dx/dt} \qquad \text{Second derivative}$$

$$\frac{d^3y}{dx^3} = \frac{d}{dx}\left[\frac{d^2y}{dx^2}\right] = \frac{\frac{d}{dt}\left[\frac{d^2y}{dx^2}\right]}{dx/dt}. \qquad \text{Third derivative}$$

Example 2 Finding Slope and Concavity

For the curve given by

$$x = \sqrt{t} \qquad \text{and} \qquad y = \frac{1}{4}(t^2 - 4), \qquad t \geq 0$$

find the slope and concavity at the point $(2, 3)$.

Solution Because

$$\frac{dy}{dx} = \frac{dy/dt}{dx/dt} = \frac{(1/2)t}{(1/2)t^{-1/2}} = t^{3/2} \qquad \text{Parametric form of first derivative}$$

you can find the second derivative to be

$$\frac{d^2y}{dx^2} = \frac{\frac{d}{dt}[dy/dx]}{dx/dt} = \frac{\frac{d}{dt}[t^{3/2}]}{dx/dt} = \frac{(3/2)t^{1/2}}{(1/2)t^{-1/2}} = 3t. \qquad \text{Parametric form of second derivative}$$

At $(x, y) = (2, 3)$, it follows that $t = 4$, and the slope is

$$\frac{dy}{dx} = (4)^{3/2} = 8.$$

Moreover, when $t = 4$, the second derivative is

$$\frac{d^2y}{dx^2} = 3(4) = 12 > 0$$

and you can conclude that the graph is concave upward at $(2, 3)$, as shown in Figure 9.30.

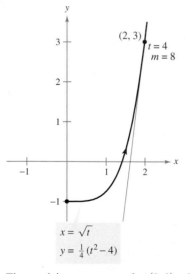

The graph is concave upward at $(2, 3)$, when $t = 4$.
Figure 9.30

Because the parametric equations $x = f(t)$ and $y = g(t)$ need not define y as a function of x, it follows that a plane curve can loop and cross itself. At such points the curve may have more than one tangent line, as shown in the next example.

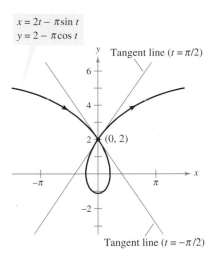

$x = 2t - \pi \sin t$
$y = 2 - \pi \cos t$

Tangent line ($t = \pi/2$)

(0, 2)

Tangent line ($t = -\pi/2$)

This prolate cycloid has two tangent lines at the point (0, 2).
Figure 9.31

Example 3   A Curve with Two Tangent Lines at a Point

The **prolate cycloid** given by

$$x = 2t - \pi \sin t \qquad \text{and} \qquad y = 2 - \pi \cos t$$

crosses itself at the point (0, 2), as shown in Figure 9.31. Find the equations of both tangent lines at this point.

Solution   Because $x = 0$ and $y = 2$ when $t = \pm \pi/2$, and

$$\frac{dy}{dx} = \frac{dy/dt}{dx/dt} = \frac{\pi \sin t}{2 - \pi \cos t}$$

you have $dy/dx = -\pi/2$ when $t = -\pi/2$ and $dy/dx = \pi/2$ when $t = \pi/2$. Therefore, the two tangent lines at (0, 2) are

$$y - 2 = -\left(\frac{\pi}{2}\right)x \qquad \text{Tangent line when } t = -\frac{\pi}{2}$$

$$y - 2 = \left(\frac{\pi}{2}\right)x. \qquad \text{Tangent line when } t = \frac{\pi}{2}$$

If $dy/dt = 0$ and $dx/dt \neq 0$ when $t = t_0$, the curve represented by $x = f(t)$ and $y = g(t)$ has a *horizontal* tangent at $(f(t_0), g(t_0))$. For instance, in Example 3, the given curve has a horizontal tangent at the point $(0, 2 - \pi)$ (when $t = 0$). Similarly, if $dx/dt = 0$ and $dy/dt \neq 0$ when $t = t_0$, the curve represented by $x = f(t)$ and $y = g(t)$ has a *vertical* tangent at $(f(t_0), g(t_0))$.

Arc Length

You have seen how parametric equations can be used to describe the path of a particle moving in the plane. We now develop a formula for determining the *distance* traveled by the particle along its path.

Recall from Section 6.4 that the formula for the arc length of a curve C given by $y = h(x)$ over the interval $[x_0, x_1]$ is

$$s = \int_{x_0}^{x_1} \sqrt{1 + [h'(x)]^2}\, dx$$

$$= \int_{x_0}^{x_1} \sqrt{1 + \left(\frac{dy}{dx}\right)^2}\, dx.$$

If C is represented by the parametric equations $x = f(t)$ and $y = g(t)$, $a \leq t \leq b$, and if $dx/dt = f'(t) > 0$, you can write

$$s = \int_{x_0}^{x_1} \sqrt{1 + \left(\frac{dy}{dx}\right)^2}\, dx = \int_{x_0}^{x_1} \sqrt{1 + \left(\frac{dy/dt}{dx/dt}\right)^2}\, dx$$

$$= \int_a^b \sqrt{\frac{(dx/dt)^2 + (dy/dt)^2}{(dx/dt)^2}}\, \frac{dx}{dt}\, dt$$

$$= \int_a^b \sqrt{\left(\frac{dx}{dt}\right)^2 + \left(\frac{dy}{dt}\right)^2}\, dt$$

$$= \int_a^b \sqrt{[f'(t)]^2 + [g'(t)]^2}\, dt.$$

NOTE   When applying the arc length formula to a curve, be sure that the curve is traced out only once on the interval of integration. For instance, the circle given by $x = \cos t$ and $y = \sin t$ is traced out once on the interval $0 \leq t \leq 2\pi$, but is traced out twice on the interval $0 \leq t \leq 4\pi$.

THEOREM 9.8   Arc Length in Parametric Form

If a smooth curve C is given by $x = f(t)$ and $y = g(t)$ such that C does not intersect itself on the interval $a \leq t \leq b$ (except possibly at the endpoints), then the arc length of C over the interval is given by

$$s = \int_a^b \sqrt{\left(\frac{dx}{dt}\right)^2 + \left(\frac{dy}{dt}\right)^2}\, dt = \int_a^b \sqrt{[f'(t)]^2 + [g'(t)]^2}\, dt.$$

In the preceding section you saw that if a circle rolls along a line, a point on its circumference will trace a path called a cycloid. If the circle rolls around the circumference of another circle, the path of the point is an **epicycloid.** The next example shows how to find the arc length of an epicycloid.

ARCH OF A CYCLOID

The arc length of an arch of a cycloid was first calculated in 1658 by British architect and mathematician Christopher Wren, famous for rebuilding many buildings and churches in London, including St. Paul's Cathedral.

Example 4   **Finding Arc Length**

A circle of radius 1 rolls around the circumference of a larger circle of radius 4, as shown in Figure 9.32. The epicycloid traced by a point on the circumference of the smaller circle is given by

$$x = 5 \cos t - \cos 5t$$

and

$$y = 5 \sin t - \sin 5t.$$

Find the distance traveled by the point in one complete trip about the larger circle.

Solution   Before applying Theorem 9.8, note in Figure 9.32 that the curve has sharp points when $t = 0$ and $t = \pi/2$. Between these two points, dx/dt and dy/dt are not simultaneously 0. So, the portion of the curve generated from $t = 0$ to $t = \pi/2$ is smooth. To find the total distance traveled by the point, you can find the arc length of that portion lying in the first quadrant and multiply by 4.

$$\begin{aligned}
s &= 4\int_0^{\pi/2} \sqrt{\left(\frac{dx}{dt}\right)^2 + \left(\frac{dy}{dt}\right)^2}\, dt && \text{Parametric form for arc length}\\
&= 4\int_0^{\pi/2} \sqrt{(-5\sin t + 5\sin 5t)^2 + (5\cos t - 5\cos 5t)^2}\, dt\\
&= 20\int_0^{\pi/2} \sqrt{2 - 2\sin t \sin 5t - 2\cos t \cos 5t}\, dt\\
&= 20\int_0^{\pi/2} \sqrt{2 - 2\cos 4t}\, dt\\
&= 20\int_0^{\pi/2} \sqrt{4\sin^2 2t}\, dt && \text{Trigonometric identity}\\
&= 40\int_0^{\pi/2} \sin 2t\, dt\\
&= -20\Big[\cos 2t\Big]_0^{\pi/2}\\
&= 40
\end{aligned}$$

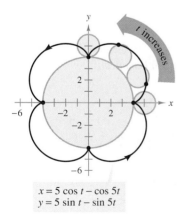

$x = 5 \cos t - \cos 5t$
$y = 5 \sin t - \sin 5t$

An epicycloid is traced by a point on the smaller circle as it rolls around the larger circle.
Figure 9.32

For the epicycloid shown in Figure 9.32, an arc length of 40 seems about right because the circumference of a circle of radius 6 is $2\pi r = 12\pi \approx 37.7$.

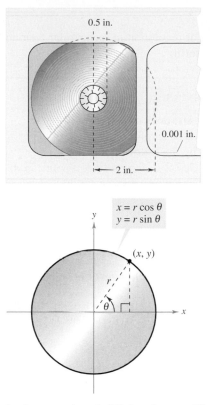

It takes approximately 982 feet of tape to fill the reel.

Figure 9.33

Example 5   **Length of a Recording Tape**

A recording tape 0.001 inch thick is wound around a reel whose inner radius is 0.5 inch and outer radius is 2 inches, as shown in Figure 9.33. How much tape is required to fill the reel?

Solution   To create a model for this problem, assume that as the tape is wound around the reel its distance r from the center increases linearly at a rate of 0.001 inch per revolution, or

$$r = (0.001)\frac{\theta}{2\pi} = \frac{\theta}{2000\pi}, \qquad 1000\pi \le \theta \le 4000\pi,$$

where θ is measured in radians. You can determine the coordinates of the point (x, y) corresponding to a given radius to be

$$x = r \cos \theta$$

and

$$y = r \sin \theta.$$

Substituting for r, you obtain the parametric equations

$$x = \left(\frac{\theta}{2000\pi}\right) \cos \theta \qquad \text{and} \qquad y = \left(\frac{\theta}{2000\pi}\right) \sin \theta.$$

You can use the arc length formula to determine the total length of the tape to be

$$
\begin{aligned}
s &= \int_{1000\pi}^{4000\pi} \sqrt{\left(\frac{dx}{d\theta}\right)^2 + \left(\frac{dy}{d\theta}\right)^2} \, d\theta \\
&= \frac{1}{2000\pi} \int_{1000\pi}^{4000\pi} \sqrt{(-\theta \sin \theta + \cos \theta)^2 + (\theta \cos \theta + \sin \theta)^2} \, d\theta \\
&= \frac{1}{2000\pi} \int_{1000\pi}^{4000\pi} \sqrt{\theta^2 + 1} \, d\theta \\
&= \frac{1}{2000\pi}\left(\frac{1}{2}\right)\left[\theta\sqrt{\theta^2 + 1} + \ln\left|\theta + \sqrt{\theta^2 + 1}\right|\right]_{1000\pi}^{4000\pi} \quad \begin{array}{l}\text{Integration tables}\\ \text{(Appendix C), Formula 26}\end{array} \\
&\approx 11{,}781 \text{ in.} \\
&\approx 982 \text{ ft.}
\end{aligned}
$$

FOR FURTHER INFORMATION   For more information on the mathematics of recording tape, see "Tape Counters" by Richard L. Roth in *The American Mathematical Monthly*. To view this article, go to the website *www.matharticles.com*.

NOTE   The graph of $r = a\theta$ is called the **spiral of Archimedes**. The graph of $r = \theta/2000\pi$ (in Example 5) is of this form.

The length of the tape in Example 5 can be approximated by adding the circumferences of circular pieces of tape. The smallest circle has a radius of 0.501 and the largest has a radius of 2.

$$
\begin{aligned}
s &\approx 2\pi(0.501) + 2\pi(0.502) + 2\pi(0.503) + \cdots + 2\pi(2.000) \\
&= \sum_{i=1}^{1500} 2\pi(0.5 + 0.001i) \\
&= 2\pi[1500(0.5) + 0.001(1500)(1501)/2] \\
&\approx 11{,}786 \text{ in.}
\end{aligned}
$$

Area of a Surface of Revolution

You can use the formula for the area of a surface of revolution in rectangular form to develop a formula for surface area in parametric form.

THEOREM 9.9   Area of a Surface of Revolution

If a smooth curve C given by $x = f(t)$ and $y = g(t)$ does not cross itself on an interval $a \le t \le b$, then the area S of the surface of revolution formed by revolving C about the coordinate axes is given by the following.

1. $S = 2\pi \displaystyle\int_a^b g(t) \sqrt{\left(\dfrac{dx}{dt}\right)^2 + \left(\dfrac{dy}{dt}\right)^2}\, dt$ Revolution about the x-axis: $g(t) \ge 0$

2. $S = 2\pi \displaystyle\int_a^b f(t) \sqrt{\left(\dfrac{dx}{dt}\right)^2 + \left(\dfrac{dy}{dt}\right)^2}\, dt$ Revolution about the y-axis: $f(t) \ge 0$

These formulas are easy to remember if you think of the differential of arc length as

$$ds = \sqrt{\left(\frac{dx}{dt}\right)^2 + \left(\frac{dy}{dt}\right)^2}\, dt.$$

Then the formulas are written as follows.

1. $S = 2\pi \displaystyle\int_a^b g(t)\, ds$ **2.** $S = 2\pi \displaystyle\int_a^b f(t)\, ds$

Example 6   Finding the Area of a Surface of Revolution

Let C be the arc of the circle

$$x^2 + y^2 = 9$$

from $(3, 0)$ to $\left(3/2, 3\sqrt{3}/2\right)$, as shown in Figure 9.34. Find the area of the surface formed by revolving C about the x-axis.

Solution   You can represent C parametrically by the equations

$$x = 3\cos t \quad \text{and} \quad y = 3\sin t, \quad 0 \le t \le \pi/3.$$

(Note that you can determine the interval for t by observing that $t = 0$ when $x = 3$ and $t = \pi/3$ when $x = 3/2$.) On this interval, C is smooth and y is nonnegative, and you can apply Theorem 9.9 to obtain a surface area of

$$S = 2\pi \int_0^{\pi/3} (3\sin t)\sqrt{(-3\sin t)^2 + (3\cos t)^2}\, dt \qquad \text{Formula for area of a surface of revolution}$$

$$= 6\pi \int_0^{\pi/3} \sin t\sqrt{9(\sin^2 t + \cos^2 t)}\, dt$$

$$= 6\pi \int_0^{\pi/3} 3\sin t\, dt \qquad \text{Trigonometric identity}$$

$$= -18\pi\Big[\cos t\Big]_0^{\pi/3}$$

$$= -18\pi\left(\frac{1}{2} - 1\right)$$

$$= 9\pi.$$

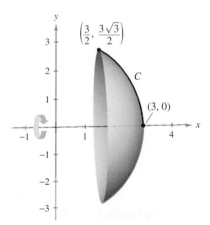

This surface of revolution has a surface area of 9π.

Figure 9.34

EXERCISES FOR SECTION 9.3

In Exercises 1–4, find dy/dx.

1. $x = t^2$, $y = 5 - 4t$

2. $x = \sqrt[3]{t}$, $y = 4 - t$

3. $x = \sin^2 \theta$, $y = \cos^2 \theta$

4. $x = 2e^\theta$, $y = e^{-\theta/2}$

In Exercises 5–14, find dy/dx and d^2y/dx^2, and find the slope and concavity (if possible) at the indicated value of the parameter.

Parametric Equations	Point
5. $x = 2t$, $y = 3t - 1$	$t = 3$
6. $x = \sqrt{t}$, $y = 3t - 1$	$t = 1$
7. $x = t + 1$, $y = t^2 + 3t$	$t = -1$
8. $x = t^2 + 3t + 2$, $y = 2t$	$t = 0$
9. $x = 2 \cos \theta$, $y = 2 \sin \theta$	$\theta = \dfrac{\pi}{4}$
10. $x = \cos \theta$, $y = 3 \sin \theta$	$\theta = 0$
11. $x = 2 + \sec \theta$, $y = 1 + 2 \tan \theta$	$\theta = \dfrac{\pi}{6}$
12. $x = \sqrt{t}$, $y = \sqrt{t - 1}$	$t = 2$
13. $x = \cos^3 \theta$, $y = \sin^3 \theta$	$\theta = \dfrac{\pi}{4}$
14. $x = \theta - \sin \theta$, $y = 1 - \cos \theta$	$\theta = \pi$

In Exercises 15 and 16, find an equation of the tangent line at the indicated points on the curve.

15. $x = 2 \cot \theta$
$y = 2 \sin^2 \theta$

16. $x = 2 - 3 \cos \theta$
$y = 3 + 2 \sin \theta$

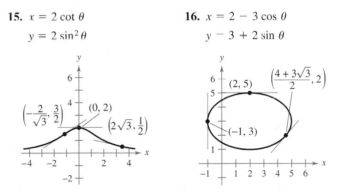

In Exercises 17–20, (a) use a graphing utility to graph the curve represented by the parametric equations, (b) use a graphing utility to find dx/dt, dy/dt, and dy/dx at the indicated value of the parameter, (c) find an equation of the tangent line to the curve at the indicated value of the parameter, and (d) confirm the result in part (c) by using a graphing utility to graph the tangent line.

Parametric Equations	Parameter
17. $x = 2t$, $y = t^2 - 1$	$t = 2$
18. $x = t - 1$, $y = \dfrac{1}{t} + 1$	$t = 1$
19. $x = t^2 - t + 2$, $y = t^3 - 3t$	$t = -1$
20. $x = 4 \cos \theta$, $y = 3 \sin \theta$	$\theta = \dfrac{3\pi}{4}$

In Exercises 21 and 22, find the equations of the tangent lines at the point where the curve crosses itself.

21. $x = 2 \sin 2t$, $y = 3 \sin t$

22. $x = t^2 - t$, $y = t^3 - 3t - 1$

In Exercises 23 and 24, find all points (if any) of horizontal and vertical tangency to the portion of the curve shown.

23. Involute of a circle:

$x = \cos \theta + \theta \sin \theta$

$y = \sin \theta - \theta \cos \theta$

24. $x = 2\theta$

$y = 2(1 - \cos \theta)$

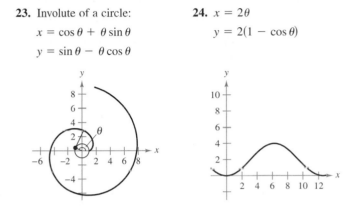

In Exercises 25–34, find all points (if any) of horizontal and vertical tangency to the curve. Use a graphing utility to confirm your results.

25. $x = 1 - t$, $y = t^2$

26. $x = t + 1$, $y = t^2 + 3t$

27. $x = 1 - t$, $y = t^3 - 3t$

28. $x = t^2 - t + 2$, $y = t^3 - 3t$

29. $x = 3 \cos \theta$, $y = 3 \sin \theta$

30. $x = \cos \theta$, $y = 2 \sin 2\theta$

31. $x = 4 + 2 \cos \theta$, $y = -1 + \sin \theta$

32. $x = 4 \cos^2 \theta$, $y = 2 \sin \theta$

33. $x = \sec \theta$, $y = \tan \theta$

34. $x = \cos^2 \theta$, $y = \cos \theta$

Arc Length   **In Exercises 35–40, find the arc length of the given curve on the indicated interval.**

Parametric Equations	Interval
35. $x = t^2$, $y = 2t$	$0 \leq t \leq 2$
36. $x = t^2 + 1$, $y = 4t^3 + 3$	$-1 \leq t \leq 0$
37. $x = e^{-t} \cos t$, $y = e^{-t} \sin t$	$0 \leq t \leq \dfrac{\pi}{2}$
38. $x = \arcsin t$, $y = \ln \sqrt{1 - t^2}$	$0 \leq t \leq \dfrac{1}{2}$
39. $x = \sqrt{t}$, $y = 3t - 1$	$0 \leq t \leq 1$
40. $x = t$, $y = \dfrac{t^5}{10} + \dfrac{1}{6t^3}$	$1 \leq t \leq 2$

Arc Length **In Exercises 41–44, find the arc length of the curve on the interval $[0, 2\pi]$.**

41. Hypocycloid perimeter: $x = a \cos^3 \theta, y = a \sin^3 \theta$

42. Circle circumference: $x = a \cos \theta, y = a \sin \theta$

43. Cycloid arch: $x = a(\theta - \sin \theta), y = a(1 - \cos \theta)$

44. Involute of a circle: $x = \cos \theta + \theta \sin \theta, y = \sin \theta - \theta \cos \theta$

45. *Path of a Projectile* The path of a projectile is modeled by the parametric equations

$$x = (90 \cos 30°)t \quad \text{and} \quad y = (90 \sin 30°)t - 16t^2$$

where x and y are measured in feet. Use a graphing utility to perform the following.

(a) Graph the path of the projectile.

(b) Approximate the range of the projectile.

(c) Use the integration capabilities of the graphing utility to approximate the arc length of the path. Compare this result with the range of the projectile.

(d) If the projectile is launched at an angle θ with the horizontal, its parametric equations are

$$x = (90 \cos \theta)t \quad \text{and} \quad y = (90 \sin \theta)t - 16t^2.$$

What angle maximizes its range? What angle maximizes the arc length of the trajectory?

46. *Folium of Descartes* Given the parametric equations

$$x = \frac{4t}{1 + t^3} \quad \text{and} \quad y = \frac{4t^2}{1 + t^3}$$

use a graphing utility to perform the following.

(a) Sketch the curve described by the parametric equations.

(b) Find the points of horizontal tangency to the curve.

(c) Use the integration capabilities of the graphing utility to approximate the arc length of the closed loop. (*Hint:* Use symmetry and integrate over the interval $0 \le t \le 1$.)

47. *Writing*

(a) Use a graphing utility to graph each set of parametric equations.

$$x = t - \sin t \qquad x = 2t - \sin(2t)$$
$$y = 1 - \cos t \qquad y = 1 - \cos(2t)$$
$$0 \le t \le 2\pi \qquad 0 \le t \le \pi$$

(b) Compare the graphs of the two sets of parametric equations in part (a). If the curve represents the motion of a particle and t is time, what can you infer about the average speed of the particle on the paths represented by the two sets of parametric equations?

(c) Without graphing the curve, determine the time required for a particle to traverse the same path as in parts (a) and (b) if the path is modeled by

$$x = \tfrac{1}{2}t - \sin\left(\tfrac{1}{2}t\right) \quad \text{and} \quad y = 1 - \cos\left(\tfrac{1}{2}t\right).$$

48. *Circumference of an Ellipse* Use the integration capabilities of a graphing utility to approximate the circumference of the ellipse given by the parametric equations $x = 3 \cos \theta$ and $y = 4 \sin \theta$.

Surface Area **In Exercises 49–54, find the area of the surface generated by revolving the curve about the given axis.**

49. $x = t, y = 2t, \quad 0 \le t \le 4,$ (a) x-axis   (b) y-axis

50. $x = t, y = 4 - 2t, \quad 0 \le t \le 2,$ (a) x-axis   (b) y-axis

51. $x = 4 \cos \theta, y = 4 \sin \theta, \quad 0 \le \theta \le \dfrac{\pi}{2}, \quad y$-axis

52. $x = \tfrac{1}{3}t^3, y = t + 1, \quad 1 \le t \le 2, \quad y$-axis

53. $x = a \cos^3 \theta, y = a \sin^3 \theta, \quad 0 \le \theta \le \pi, \quad x$-axis

54. $x = a \cos \theta, y = b \sin \theta, \quad 0 \le \theta \le 2\pi,$

    (a) x-axis   (b) y-axis

Getting at the Concept

55. Give the parametric form of the derivative.

56. Mentally determine dy/dx.

  (a) $x = t$ (b) $x = t$

     $y = 4$ $y = 4t - 3$

57. Sketch a graph of a curve defined by the parametric equations $x = g(t)$ and $y = f(t)$ such that $dx/dt > 0$ and $dy/dt < 0$ for all real numbers t.

58. Sketch a graph of a curve defined by the parametric equations $x = g(t)$ and $y = f(t)$ such that $dx/dt < 0$ and $dy/dt < 0$ for all real numbers t.

59. Give the integral formula for arc length in parametric form.

60. Give the integral formulas for the area of a surface of revolution formed when a smooth curve C is revolved about (a) the x-axis and (b) the y-axis.

---

61. *Surface Area* A portion of a sphere of radius r is removed by cutting out a circular cone with its vertex at the center of the sphere. Find the surface area removed from the sphere if the vertex of the cone forms an angle of 2θ.

62. Use integration by substitution to show that if y is a continuous function of x on the interval $a \le x \le b$, where $x = f(t)$ and $y = g(t)$, then

$$\int_a^b y \, dx = \int_{t_1}^{t_2} g(t) f'(t) \, dt,$$

where $f(t_1) = a, f(t_2) = b$, and both g and f' are continuous on $[t_1, t_2]$.

Centroid **In Exercises 63 and 64, find the centroid of the region bounded by the graph of the parametric equations and the coordinate axes. (Use the result in Exercise 62.)**

63. $x = \sqrt{t}, y = 4 - t$ **64.** $x = \sqrt{4 - t}, y = \sqrt{t}$

Volume **In Exercises 65 and 66, find the volume of the solid formed by revolving the region bounded by the graphs of the given equations about the *x*-axis. (Use the result in Exercise 62.)**

65. $x = 3 \cos \theta, y = 3 \sin \theta$ **66.** $x = \cos \theta, y = 3 \sin \theta$

Area **In Exercises 67 and 68, find the area of the region. (Use the result in Exercise 62.)**

67. $x = 2 \sin^2 \theta$
$y = 2 \sin^2 \theta \tan \theta$
$0 \le \theta < \dfrac{\pi}{2}$

68. $x = 2 \cot \theta$
$y = 2 \sin^2 \theta$
$0 < \theta < \pi$

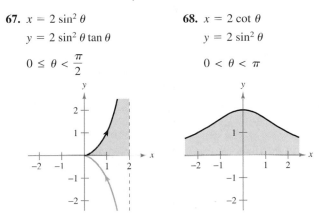

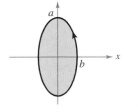 ***Areas of Simple Closed Curves*** **In Exercises 69–74, use a computer algebra system and the result in Exercise 62 to match the closed curve with its area. (These exercises were adapted from the article "The Surveyor's Area Formula" by Bart Braden in the September 1986 issue of *The College Mathematics Journal*. Used by permission of the author.)**

(a) $\frac{8}{3}ab$ (b) $\frac{3}{8}\pi a^2$ (c) $2\pi a^2$

(d) πab (e) $2\pi ab$ (f) $6\pi a^2$

69. Ellipse: $(0 \le t \le 2\pi)$
$x = b \cos t$
$y = a \sin t$

70. Asteroid: $(0 \le t \le 2\pi)$
$x = a \cos^3 t$
$y = a \sin^3 t$

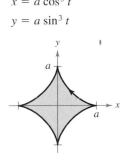

71. Cardioid: $(0 \le t \le 2\pi)$
$x = 2a \cos t - a \cos 2t$
$y = 2a \sin t - a \sin 2t$

72. Deltoid: $(0 \le t \le 2\pi)$
$x = 2a \cos t + a \cos 2t$
$y = 2a \sin t - a \sin 2t$

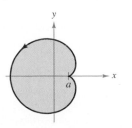

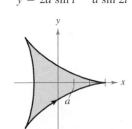

73. Hourglass: $(0 \le t \le 2\pi)$
$x = a \sin 2t$
$y = b \sin t$

74. Teardrop: $(0 \le t \le 2\pi)$
$x = 2a \cos t - a \sin 2t$
$y = b \sin t$

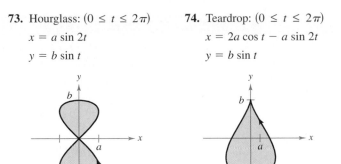

75. Use a graphing utility to graph the curve given by
$$x = \frac{1 - t^2}{1 + t^2}, \quad y = \frac{2t}{1 + t^2}, \quad -20 \le t \le 20.$$

(a) Describe the graph and confirm your result analytically.

(b) Discuss the speed at which the curve is traced as *t* increases from -20 to 20.

76. ***Tractrix***   A person moves from the origin along the positive *y*-axis pulling a weight at the end of a 12-meter rope. Initially, the weight is located at the point $(12, 0)$.

(a) In Exercise 75 of Section 7.4, it was shown that the path of the weight is modeled by the rectangular equation
$$y = -12 \ln\left(\frac{12 - \sqrt{144 - x^2}}{x}\right) - \sqrt{144 - x^2}$$

where $0 < x \le 12$. Use a graphing utility to graph the rectangular equation.

(b) Use a graphing utility to graph the parametric equations
$$x = 12 \operatorname{sech} \frac{t}{12} \quad \text{and} \quad y = t - 12 \tanh \frac{t}{12}$$

where $t \ge 0$. How does this graph compare with the graph in part (a)? Which graph (if either) do you think is a better representation of the path?

(c) Use the parametric equations for the tractrix to verify that the distance from the *y*-intercept of the tangent line to the point of tangency is independent of the location of the point of tangency.

True or False?   **In Exercises 77 and 78, determine whether the statement is true or false. If it is false, explain why or give an example that shows it is false.**

77. If $x = f(t)$ and $y = g(t)$, then $d^2y/dx^2 = g''(t)/f''(t)$.

78. The curve given by $x = t^3$, $y = t^2$ has a horizontal tangent at the origin because $dy/dt = 0$ when $t = 0$.

- Understand the polar coordinate system.
- Rewrite rectangular equations in polar form and vice versa.
- Sketch the graph of an equation given in polar form.
- Find the slope of a tangent line to a polar graph.
- Identify several types of special polar graphs.

Polar Coordinates

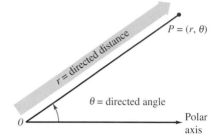

Polar coordinates
Figure 9.35

So far, we have been representing graphs as collections of points (x, y) on the rectangular coordinate system. The corresponding equations for these graphs have been in either rectangular or parametric form. In this section we introduce a coordinate system called the **polar coordinate system.**

To form the polar coordinate system in the plane, we fix a point O, called the **pole** (or **origin**), and construct from O an initial ray called the **polar axis,** as shown in Figure 9.35. Then each point P in the plane can be assigned **polar coordinates** (r, θ), as follows.

$r = $ *directed distance* from O to P

$\theta = $ *directed angle*, counterclockwise from polar axis to segment $\overline{OP}$

Figure 9.36 shows three points on the polar coordinate system. Notice that in this system, it is convenient to locate points with respect to a grid of concentric circles intersected by **radial lines** through the pole.

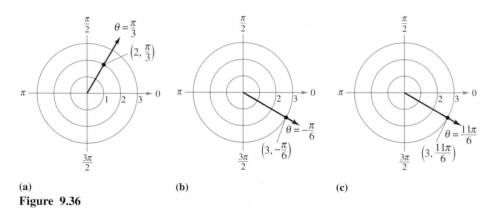

(a)   **(b)**   **(c)**

Figure 9.36

With rectangular coordinates, each point (x, y) has a unique representation. This is not true with polar coordinates. For instance, the coordinates (r, θ) and $(r, 2\pi + \theta)$ represent the same point [see parts (b) and (c) in Figure 9.36]. Also, because r is a *directed distance*, the coordinates (r, θ) and $(-r, \theta + \pi)$ represent the same point. In general, the point (r, θ) can be written as

$$(r, \theta) = (r, \theta + 2n\pi)$$

or

$$(r, \theta) = (-r, \theta + (2n + 1)\pi)$$

where n is any integer. Moreover, the pole is represented by $(0, \theta)$, where θ is any angle.

POLAR COORDINATES

The mathematician credited with first using polar coordinates was James Bernoulli, who introduced them in 1691. However, there is some evidence that it may have been Isaac Newton who first used them.

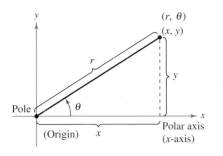

Relating polar and rectangular coordinates
Figure 9.37

Coordinate Conversion

To establish the relationship between polar and rectangular coordinates, let the polar axis coincide with the positive x-axis and the pole with the origin, as shown in Figure 9.37. Because (x, y) lies on a circle of radius r, it follows that $r^2 = x^2 + y^2$. Moreover, for $r > 0$, the definition of the trigonometric functions implies that

$$\tan \theta = \frac{y}{x}, \qquad \cos \theta = \frac{x}{r}, \qquad \text{and} \qquad \sin \theta = \frac{y}{r}.$$

If $r < 0$, you can show that the same relationships hold.

THEOREM 9.10   Coordinate Conversion

The polar coordinates (r, θ) of a point are related to the rectangular coordinates (x, y) of the point as follows.

1. $x = r \cos \theta$ **2.** $\tan \theta = \dfrac{y}{x}$

     $y = r \sin \theta$ $r^2 = x^2 + y^2$

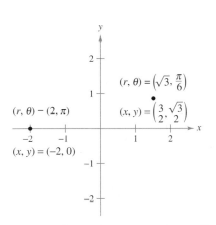

To convert from polar to rectangular coordinates, let $x = r \cos \theta$ and $y = r \sin \theta$.
Figure 9.38

Example 1   **Polar-to-Rectangular Conversion**

a. For the point $(r, \theta) = (2, \pi)$,

$$x = r \cos \theta = 2 \cos \pi = -2 \qquad \text{and} \qquad y = r \sin \theta = 2 \sin \pi = 0.$$

So, the rectangular coordinates are $(x, y) = (-2, 0)$.

b. For the point $(r, \theta) = \left(\sqrt{3}, \pi/6\right)$,

$$x = \sqrt{3} \cos \frac{\pi}{6} = \frac{3}{2} \qquad \text{and} \qquad y = \sqrt{3} \sin \frac{\pi}{6} = \frac{\sqrt{3}}{2}$$

So, the rectangular coordinates are $(x, y) = \left(3/2, \sqrt{3}/2\right)$.

(See Figure 9.38.)

Example 2   **Rectangular-to-Polar Conversion**

a. For the second quadrant point $(x, y) = (-1, 1)$,

$$\tan \theta = \frac{y}{x} = -1 \qquad \Longrightarrow \qquad \theta = \frac{3\pi}{4}.$$

Because θ was chosen to be in the same quadrant as (x, y), you should use a positive value of r.

$$r = \sqrt{x^2 + y^2}$$
$$= \sqrt{(-1)^2 + (1)^2}$$
$$= \sqrt{2}$$

This implies that *one* set of polar coordinates is $(r, \theta) = \left(\sqrt{2}, 3\pi/4\right)$.

b. Because the point $(x, y) = (0, 2)$ lies on the positive y-axis, we choose $\theta = \pi/2$ and $r = 2$, and one set of polar coordinates is $(r, \theta) = (2, \pi/2)$.

(See Figure 9.39.)

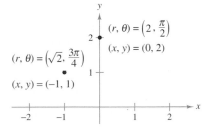

To convert from rectangular to polar coordinates, let $\tan \theta = y/x$ and $r = \sqrt{x^2 + y^2}$.
Figure 9.39

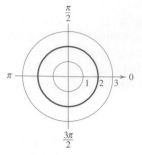

(a) Circle: $r = 2$

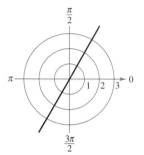

(b) Radial line: $\theta = \dfrac{\pi}{3}$

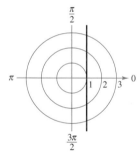

(c) Vertical line: $r = \sec \theta$

Figure 9.40

Polar Graphs

One way to sketch the graph of a polar equation is to convert to rectangular coordinates and then sketch the graph of the rectangular equation.

Example 3 Graphing Polar Equations

Describe the graph of each polar equation. Confirm each description by converting to a rectangular equation.

a. $r = 2$ **b.** $\theta = \dfrac{\pi}{3}$ **c.** $r = \sec \theta$

Solution

a. The graph of the polar equation $r = 2$ consists of all points that are two units from the pole. In other words, this graph is a circle centered at the origin with a radius of 2. (See Figure 9.40a.) You can confirm this by using the relationship $r^2 = x^2 + y^2$ to obtain the rectangular equation

$$x^2 + y^2 = 2^2. \qquad \text{Rectangular equation}$$

b. The graph of the polar equation $\theta = \pi/3$ consists of all points on the line that makes an angle of $\pi/3$ with the positive x-axis. (See Figure 9.40b.) You can confirm this by using the relationship $\tan \theta = y/x$ to obtain the rectangular equation

$$y = \sqrt{3}\, x. \qquad \text{Rectangular equation}$$

c. The graph of the polar equation $r = \sec \theta$ is not evident by simple inspection, so you can begin by converting to rectangular form using the relationship $r \cos \theta = x$.

$$r = \sec \theta \qquad \text{Polar equation}$$
$$r \cos \theta = 1$$
$$x = 1 \qquad \text{Rectangular equation}$$

From the rectangular equation, you can see that the graph is a vertical line. (See Figure 9.40c.) ◻

TECHNOLOGY Sketching the graphs of complicated polar equations *by hand* can be tedious. With technology, however, the task is not difficult. If your graphing utility has a polar mode, try using it to sketch the graphs in the exercise set. If your graphing utility doesn't have a polar mode, but does have a parametric mode, you can sketch the graph of $r = f(\theta)$ by writing the equation as

$$x = f(\theta) \cos \theta$$
$$y = f(\theta) \sin \theta.$$

For instance, the graph of $r = \frac{1}{2}\theta$ shown in Figure 9.41 was produced with a graphing calculator in parametric mode. To sketch the graph, we entered the parametric equations

$$x = \frac{1}{2}\theta \cos \theta$$

$$y = \frac{1}{2}\theta \sin \theta$$

and let the values of θ vary from -4π to 4π. This curve is of the form $r = a\theta$ and is called a **spiral of Archimedes.**

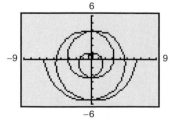

Spiral of Archimedes
Figure 9.41

Example 4   Sketching a Polar Graph

Sketch the graph of $r = 2 \cos 3\theta$.

NOTE   One way to sketch the graph of $r = 2 \cos 3\theta$ by hand is to make a table of values.

θ	0	$\dfrac{\pi}{6}$	$\dfrac{\pi}{3}$	$\dfrac{\pi}{2}$	$\dfrac{2\pi}{3}$
r	2	0	-2	0	2

By extending the table and plotting the points, you will obtain the curve shown in Example 4.

Solution   Begin by writing the polar equation in parametric form.

$$x = 2 \cos 3\theta \cos \theta \qquad \text{and} \qquad y = 2 \cos 3\theta \sin \theta$$

After some experimentation, you will find that the entire curve, which is called a **rose curve,** can be sketched by letting θ vary from 0 to π, as shown in Figure 9.42. If you try duplicating this graph with a graphing utility, you will find that by letting θ vary from 0 to 2π, you will actually trace the entire curve *twice*.

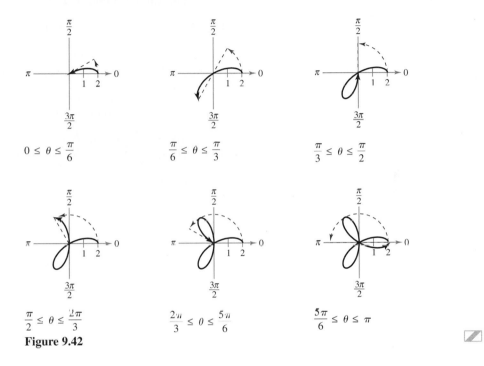

Figure 9.42

Try using a graphing utility to experiment with other rose curves (they are of the form $r = a \cos n\theta$ or $r = a \sin n\theta$). For instance, Figure 9.43 shows the graphs of two other rose curves.

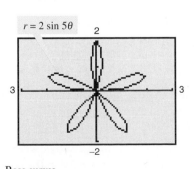

$r = 2 \sin 5\theta$

Rose curves
Figure 9.43

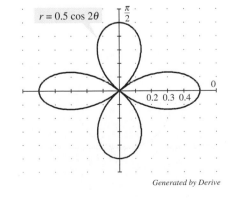

$r = 0.5 \cos 2\theta$

Generated by Derive

Slope and Tangent Lines

To find the slope of a tangent line to a polar graph, consider a differentiable function given by $r = f(\theta)$. To find the slope in polar form, use the parametric equations

$$x = r \cos \theta = f(\theta) \cos \theta \qquad \text{and} \qquad y = r \sin \theta = f(\theta) \sin \theta.$$

Using the parametric form of dy/dx given in Theorem 9.7, you have

$$\frac{dy}{dx} = \frac{dy/d\theta}{dx/d\theta}$$

$$= \frac{f(\theta) \cos \theta + f'(\theta) \sin \theta}{-f(\theta) \sin \theta + f'(\theta) \cos \theta}$$

which establishes the following theorem.

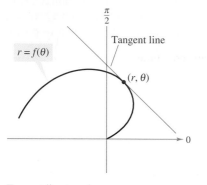

$r = f(\theta)$

Tangent line

(r, θ)

Tangent line to polar curve
Figure 9.44

THEOREM 9.11   Slope in Polar Form

If f is a differentiable function of θ, then the *slope* of the tangent line to the graph of $r = f(\theta)$ at the point (r, θ) is

$$\frac{dy}{dx} = \frac{dy/d\theta}{dx/d\theta} = \frac{f(\theta) \cos \theta + f'(\theta) \sin \theta}{-f(\theta) \sin \theta + f'(\theta) \cos \theta}$$

provided that $dx/d\theta \neq 0$ at (r, θ). (See Figure 9.44.)

From Theorem 9.11, you can make the following observations.

1. Solutions to $\dfrac{dy}{d\theta} = 0$ yield horizontal tangents, provided that $\dfrac{dx}{d\theta} \neq 0$.

2. Solutions to $\dfrac{dx}{d\theta} = 0$ yield vertical tangents, provided that $\dfrac{dy}{d\theta} \neq 0$.

If $dy/d\theta$ and $dx/d\theta$ are *simultaneously* 0, no conclusion can be drawn about tangent lines.

Example 5   **Finding Horizontal and Vertical Tangent Lines**

Find the horizontal and vertical tangent lines of $r = \sin \theta$, $0 \leq \theta \leq \pi$.

Solution   Begin by writing the equation in parametric form.

$$x = r \cos \theta = \sin \theta \cos \theta$$

and

$$y = r \sin \theta = \sin \theta \sin \theta = \sin^2 \theta$$

Next, differentiate x and y with respect to θ and set each derivative equal to 0.

$$\frac{dx}{d\theta} = \cos^2 \theta - \sin^2 \theta = \cos 2\theta = 0 \quad \Longrightarrow \quad \theta = \frac{\pi}{4}, \frac{3\pi}{4}$$

$$\frac{dy}{d\theta} = 2 \sin \theta \cos \theta = \sin 2\theta = 0 \quad \Longrightarrow \quad \theta = 0, \frac{\pi}{2}$$

So, the graph has vertical tangent lines at $\left(\sqrt{2}/2, \pi/4\right)$ and $\left(\sqrt{2}/2, 3\pi/4\right)$, and it has horizontal tangent lines at $(0, 0)$ and $(1, \pi/2)$, as shown in Figure 9.45.

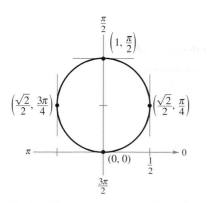

Horizontal and vertical tangent lines of
$r = \sin \theta$
Figure 9.45

Example 6   Finding Horizontal and Vertical Tangent Lines

Find the horizontal and vertical tangents to the graph of $r = 2(1 - \cos \theta)$.

Solution   Using $y = r \sin \theta$, differentiate and set $dy/d\theta$ equal to 0.

$$y = r \sin \theta = 2(1 - \cos \theta) \sin \theta$$

$$\frac{dy}{d\theta} = 2[(1 - \cos \theta)(\cos \theta) + \sin \theta(\sin \theta)]$$

$$= -2(2 \cos \theta + 1)(\cos \theta - 1) = 0$$

So, $\cos \theta = -\frac{1}{2}$ and $\cos \theta = 1$, and you can conclude that $dy/d\theta = 0$ when $\theta = 2\pi/3, 4\pi/3,$ and 0. Similarly, using $x = r \cos \theta$, you have

$$x = r \cos \theta = 2 \cos \theta - 2 \cos^2 \theta$$

$$\frac{dx}{d\theta} = -2 \sin \theta + 4 \cos \theta \sin \theta = 2 \sin \theta(2 \cos \theta - 1) = 0.$$

So, $\sin \theta = 0$ or $\cos \theta = \frac{1}{2}$, and you can conclude that $dx/d\theta = 0$ when $\theta = 0$, π, $\pi/3$, and $5\pi/3$. From these results, and from the graph shown in Figure 9.46, you can conclude that the graph has horizontal tangents at $(3, 2\pi/3)$ and $(3, 4\pi/3)$, and has vertical tangents at $(1, \pi/3)$, $(1, 5\pi/3)$, and $(4, \pi)$. This graph is called a **cardioid.** Note that both derivatives ($dy/d\theta$ and $dx/d\theta$) are 0 when $\theta = 0$. Using this information alone, you don't know whether the graph has a horizontal or vertical tangent line at the pole. From Figure 9.46, however, you can see that the graph has a cusp at the pole.   ◢

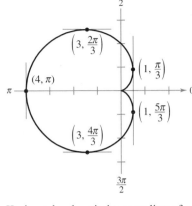

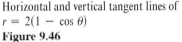

Horizontal and vertical tangent lines of $r = 2(1 - \cos \theta)$
Figure 9.46

Theorem 9.11 has an important consequence. Suppose the graph of $r = f(\theta)$ passes through the pole when $\theta = \alpha$ and $f'(\alpha) \neq 0$. Then the formula for dy/dx simplifies as follows.

$$\frac{dy}{dx} = \frac{f'(\alpha) \sin \alpha + f(\alpha) \cos \alpha}{f'(\alpha) \cos \alpha - f(\alpha) \sin \alpha} = \frac{f'(\alpha) \sin \alpha + 0}{f'(\alpha) \cos \alpha - 0} = \frac{\sin \alpha}{\cos \alpha} = \tan \alpha$$

So, the line $\theta = \alpha$ is tangent to the graph at the pole, $(0, \alpha)$.

THEOREM 9.12   Tangent Lines at the Pole

If $f(\alpha) = 0$ and $f'(\alpha) \neq 0$, then the line $\theta = \alpha$ is tangent at the pole to the graph of $r = f(\theta)$.

Theorem 9.12 is useful because it states that the zeros of $r = f(\theta)$ can be used to find the tangent lines at the pole. Note that because a polar curve can cross the pole more than once, it can have more than one tangent line at the pole. For example, the rose curve

$$f(\theta) = 2 \cos 3\theta$$

has three tangent lines at the pole, as shown in Figure 9.47. For this curve, $f(\theta) = 2 \cos 3\theta$ is 0 when θ is $\pi/6$, $\pi/2$, and $5\pi/6$. Moreover, the derivative $f'(\theta) = -6 \sin 3\theta$ is not 0 for these values of θ.

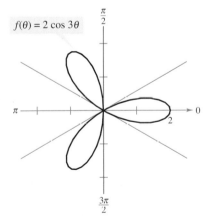

$f(\theta) = 2 \cos 3\theta$

This rose curve has three tangent lines ($\theta = \pi/6, \theta = \pi/2,$ and $\theta = 5\pi/6$) at the pole.
Figure 9.47

Special Polar Graphs

Several important types of graphs have equations that are simpler in polar form than in rectangular form. For example, the polar equation of a circle having a radius of a and centered at the origin is simply $r = a$. Later in the text you will come to appreciate this benefit. For now, we summarize some other types of graphs that have simpler equations in polar form. (Conics are considered in Section 9.6.)

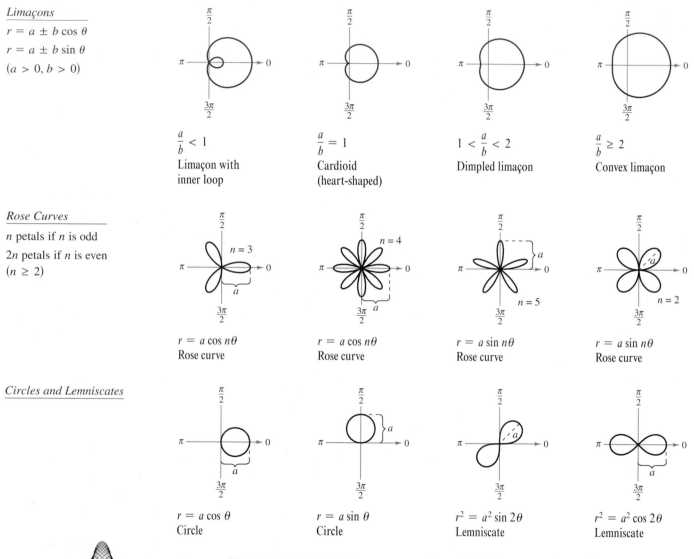

Limaçons

$r = a \pm b \cos \theta$

$r = a \pm b \sin \theta$

$(a > 0, b > 0)$

$\dfrac{a}{b} < 1$

Limaçon with inner loop

$\dfrac{a}{b} = 1$

Cardioid (heart-shaped)

$1 < \dfrac{a}{b} < 2$

Dimpled limaçon

$\dfrac{a}{b} \geq 2$

Convex limaçon

Rose Curves

n petals if n is odd

$2n$ petals if n is even

$(n \geq 2)$

$r = a \cos n\theta$

Rose curve

$r = a \cos n\theta$

Rose curve

$r = a \sin n\theta$

Rose curve

$r = a \sin n\theta$

Rose curve

Circles and Lemniscates

$r = a \cos \theta$

Circle

$r = a \sin \theta$

Circle

$r^2 = a^2 \sin 2\theta$

Lemniscate

$r^2 = a^2 \cos 2\theta$

Lemniscate

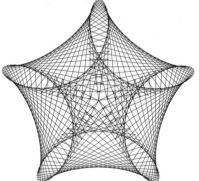

TECHNOLOGY   The rose curves described above are of the form $r = a \cos n\theta$ or $r = a \sin n\theta$, where n is a positive integer that is greater than or equal to 2. Try using a graphing utility to sketch the graph of $r = a \cos n\theta$ or $r = a \sin n\theta$ for some noninteger values of n. Are these graphs also rose curves? For example, try sketching the graph of $r = \cos \frac{2}{3}\theta$, $0 \leq \theta \leq 6\pi$.

FOR FURTHER INFORMATION   For more information on rose curves and related curves, see the article "A Rose is a Rose . . ." by Peter M. Maurer in *The American Mathematical Monthly*. To view this article, go to the website *www.matharticles.com*. (The computer-generated graph at the left is the result of an algorithm that Maurer calls "The Rose.")

EXERCISES FOR SECTION 9.4

In Exercises 1–6, plot the point in polar coordinates and find the corresponding rectangular coordinates for the point.

1. $(4, 3\pi/6)$

2. $(-2, 7\pi/4)$

3. $(-4, -\pi/3)$

4. $(0, -7\pi/6)$

5. $(\sqrt{2}, 2.36)$

6. $(-3, -1.57)$

In Exercises 7–10, use the *angle* feature of a graphing utility to find the rectangular coordinates for the point given in polar coordinates. Plot the point.

7. $(5, 3\pi/4)$

8. $(-2, 11\pi/6)$

9. $(-3.5, 2.5)$

10. $(8.25, 1.3)$

In Exercises 11–14, the rectangular coordinates of a point are given. Plot the point and find *two* sets of polar coordinates for the point for $0 \le \theta < 2\pi$.

11. $(1, 1)$

12. $(0, -5)$

13. $(-3, 4)$

14. $(4, \quad 2)$

In Exercises 15–18, use the *angle* feature of a graphing utility to find one set of polar coordinates for the point given in rectangular coordinates.

15. $(3, -2)$

16. $(3\sqrt{2}, 3\sqrt{2})$

17. $\left(\frac{5}{2}, \frac{4}{3}\right)$

18. $(0, -5)$

19. Plot the point $(4, 3.5)$ if the point is given in (a) rectangular coordinates and (b) polar coordinates.

20. *Graphical Reasoning*

(a) Set the window format of a graphing utility to rectangular coordinates and locate the cursor at any position off the coordinate axes. Move the cursor horizontally and describe any changes in the displayed coordinates of the points. Repeat the process moving the cursor vertically.

(b) Set the window format of a graphing utility to polar coordinates and locate the cursor at any position off the coordinate axes. Move the cursor horizontally and describe any changes in the displayed coordinates of the points. Repeat the process moving the cursor vertically.

(c) Why are the results in parts (a) and (b) different?

In Exercises 21–28, convert the rectangular equation to polar form and sketch its graph.

21. $x^2 + y^2 = a^2$

22. $x^2 + y^2 - 2ax = 0$

23. $y = 4$

24. $x = 10$

25. $3x - y + 2 = 0$

26. $xy = 4$

27. $y^2 = 9x$

28. $(x^2 + y^2)^2 - 9(x^2 - y^2) = 0$

In Exercises 29–36, convert the polar equation to rectangular form and sketch its graph.

29. $r = 3$

30. $r = -2$

31. $r = \sin \theta$

32. $r = 5 \cos \theta$

33. $r = \theta$

34. $\theta = \dfrac{5\pi}{6}$

35. $r = 3 \sec \theta$

36. $r = 2 \csc \theta$

In Exercises 37–46, use a graphing utility to graph the polar equation. Find an interval for θ over which the graph is traced *only once.*

37. $r = 3 - 4 \cos \theta$

38. $r = 5(1 - 2 \sin \theta)$

39. $r = 2 + \sin \theta$

40. $r = 4 + 3 \cos \theta$

41. $r = \dfrac{2}{1 + \cos \theta}$

42. $r = \dfrac{2}{4 - 3 \sin \theta}$

43. $r = 2 \cos\left(\dfrac{3\theta}{2}\right)$

44. $r = 3 \sin\left(\dfrac{5\theta}{2}\right)$

45. $r^2 = 4 \sin 2\theta$

46. $r^2 = \dfrac{1}{\theta}$

47. Convert the equation

$$r = 2(h \cos \theta + k \sin \theta)$$

to rectangular form and verify that it is the equation of a circle. Find the radius and the rectangular coordinates of the center of the circle.

48. *Distance Formula*

(a) Verify that the Distance Formula for the distance between the two points (r_1, θ_1) and (r_2, θ_2) in polar coordinates is

$$d = \sqrt{r_1^2 + r_2^2 - 2r_1 r_2 \cos(\theta_1 - \theta_2)}.$$

(b) Describe the position of the points relative to each other if $\theta_1 = \theta_2$. Simplify the Distance Formula for this case. Is the simplification what you expected? Explain.

(c) Simplify the Distance Formula if $\theta_1 - \theta_2 = 90°$. Is the simplification what you expected? Explain.

(d) Choose two points on the polar coordinate system and find the distance between them. Then choose different polar representations of the same two points and apply the Distance Formula again. Discuss the result.

In Exercises 49–52, use the result of Exercise 48 to approximate the distance between the two points in polar coordinates.

49. $\left(4, \dfrac{2\pi}{3}\right), \left(2, \dfrac{\pi}{6}\right)$

50. $\left(10, \dfrac{7\pi}{6}\right), (3, \pi)$

51. $(2, 0.5), (7, 1.2)$

52. $(4, 2.5), (12, 1)$

In Exercises 53 and 54, find dy/dx **and the slope of the tangent lines shown on the graph of the polar equation.**

53. $r = 2 + 3 \sin \theta$ **54.** $r = 2(1 - \sin \theta)$

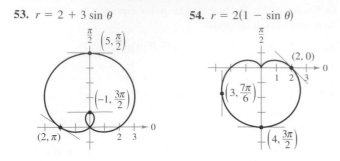

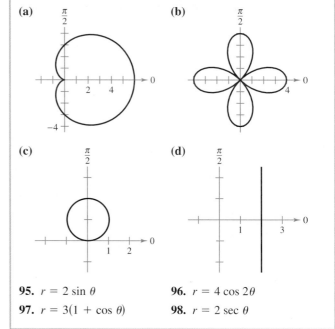

In Exercises 55–58, use a graphing utility to (a) graph the polar equation, (b) draw the tangent line at the given value of θ, and (c) find dy/dx at the given value of θ. (*Hint:* Let the increment between the values of θ equal $\pi/24$.)

55. $r = 3(1 - \cos \theta)$, $\theta = \dfrac{\pi}{2}$ **56.** $r = 3 - 2 \cos \theta$, $\theta = 0$

57. $r = 3 \sin \theta$, $\theta = \dfrac{\pi}{3}$ **58.** $r = 4$, $\theta = \dfrac{\pi}{4}$

In Exercises 59 and 60, find the points of horizontal and vertical tangency (if any) to the polar curve.

59. $r = 1 - \sin \theta$ **60.** $r = a \sin \theta$

In Exercises 61 and 62, find the points of horizontal tangency (if any) to the polar curve.

61. $r = 2 \csc \theta + 3$ **62.** $r = a \sin \theta \cos^2 \theta$

In Exercises 63–66, use a graphing utility to graph the polar equation and find all points of horizontal tangency.

63. $r = 4 \sin \theta \cos^2 \theta$ **64.** $r = 3 \cos 2\theta \sec \theta$

65. $r = 2 \csc \theta + 5$ **66.** $r = 2 \cos(3\theta - 2)$

In Exercises 67–74, sketch the graph of the polar equation and find the tangents at the pole.

67. $r = 3 \sin \theta$ **68.** $r = 3 \cos \theta$

69. $r = 2(1 - \sin \theta)$ **70.** $r = 3(1 - \cos \theta)$

71. $r = 2 \cos 3\theta$ **72.** $r = -\sin 5\theta$

73. $r = 3 \sin 2\theta$ **74.** $r = 3 \cos 2\theta$

In Exercises 75–86, sketch the graph of the polar equation.

75. $r = 5$ **76.** $r = 2$

77. $r = 4(1 + \cos \theta)$ **78.** $r = 1 + \sin \theta$

79. $r = 3 - 2 \cos \theta$ **80.** $r = 5 - 4 \sin \theta$

81. $r = 3 \csc \theta$ **82.** $r = \dfrac{6}{2 \sin \theta - 3 \cos \theta}$

83. $r = 2\theta$ **84.** $r = \dfrac{1}{\theta}$

85. $r^2 = 4 \cos 2\theta$ **86.** $r^2 = 4 \sin \theta$

In Exercises 87–90, use a graphing utility to graph the equation and show that the indicated line is an asymptote of the graph.

Name of Graph	Polar Equation	Asymptote
87. Conchoid	$r = 2 - \sec \theta$	$x = -1$
88. Conchoid	$r = 2 + \csc \theta$	$y = 1$
89. Hyperbolic spiral	$r = 2/\theta$	$y = 2$
90. Strophoid	$r = 2 \cos 2\theta \sec \theta$	$x = -2$

Getting at the Concept

91. In your own words, describe the differences between the rectangular coordinate system and the polar coordinate system.

92. Give the equations for the coordinate conversion from rectangular to polar coordinates and vice versa.

93. For constants a and b, describe the graphs of the equations $r = a$ and $\theta = b$ in polar coordinates.

94. How are the slopes of tangent lines determined in polar coordinates? What are tangent lines at the pole and how are they determined?

In Exercises 95–98, match the graph with its polar equation. [The graphs are labeled (a), (b), (c), and (d).]

(a)

(b)

(c)

(d)

95. $r = 2 \sin \theta$ **96.** $r = 4 \cos 2\theta$

97. $r = 3(1 + \cos \theta)$ **98.** $r = 2 \sec \theta$

99. Sketch the graph of $r = 4 \sin \theta$ over each interval.

(a) $0 \le \theta \le \dfrac{\pi}{2}$ (b) $\dfrac{\pi}{2} \le \theta \le \pi$ (c) $-\dfrac{\pi}{2} \le \theta \le \dfrac{\pi}{2}$

100. *Think About It*   Use a graphing utility to graph the polar equation $r = 6[1 + \cos(\theta - \phi)]$ for (a) $\phi = 0$, (b) $\phi = \pi/4$, and (c) $\phi = \pi/2$. Use the graphs to describe the effect of the angle ϕ. Write the equation as a function of $\sin \theta$ for part (c).

101. Verify that if the curve whose polar equation is $r = f(\theta)$ is rotated about the pole through an angle ϕ, then an equation for the rotated curve is $r = f(\theta - \phi)$.

102. The polar form of an equation for a curve is $r = f(\sin \theta)$. Show that the form becomes

  (a) $r = f(-\cos \theta)$ if the curve is rotated counterclockwise $\pi/2$ radians about the pole.

  (b) $r = f(-\sin \theta)$ if the curve is rotated counterclockwise π radians about the pole.

  (c) $r = f(\cos \theta)$ if the curve is rotated counterclockwise $3\pi/2$ radians about the pole.

In Exercises 103–106, use the results of Exercises 101 and 102.

103. Write an equation for the limaçon $r = 2 - \sin \theta$ after it has been rotated by the given amount. Verify the results by using a graphing utility to graph the rotated limaçon.

  (a) $\dfrac{\pi}{4}$ (b) $\dfrac{\pi}{2}$ (c) π (d) $\dfrac{3\pi}{2}$

104. Write an equation for the rose curve $r = 2 \sin 2\theta$ after it has been rotated by the given amount. Verify the results by using a graphing utility to graph the rotated rose curve.

  (a) $\dfrac{\pi}{6}$ (b) $\dfrac{\pi}{2}$ (c) $\dfrac{2\pi}{3}$ (d) π

105. Sketch the graph of each equation.

  (a) $r = 1 - \sin \theta$ (b) $r = 1 - \sin\left(\theta - \dfrac{\pi}{4}\right)$

106. Prove that the tangent of the angle $\psi\ (0 \le \psi \le \pi/2)$ between the radial line and the tangent line at the point (r, θ) on the graph of $r = f(\theta)$ (see figure) is given by $\tan \psi = \left| r/(dr/d\theta) \right|$.

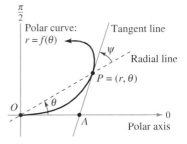

In Exercises 107–112, use the result of Exercise 106 to find the angle ψ between the radial and tangent lines to the graph for the indicated value of θ. Use a graphing utility to graph the polar equation, the radial line, and the tangent line for the indicated value of θ. Identify the angle ψ.

Polar Equation	Value of θ
107. $r = 2(1 - \cos \theta)$	$\theta = \pi$
108. $r = 3(1 - \cos \theta)$	$\theta = 3\pi/4$
109. $r = 2 \cos 3\theta$	$\theta = \pi/6$
110. $r = 4 \sin 2\theta$	$\theta = \pi/6$
111. $r = \dfrac{6}{1 - \cos \theta}$	$\theta = 2\pi/3$
112. $r = 5$	$\theta = \pi/6$

True or False? **In Exercises 113–116, determine whether the statement is true or false. If it is false, explain why or give an example that shows it is false.**

113. If (r_1, θ_1) and (r_2, θ_2) represent the same point on the polar coordinate system, then $|r_1| = |r_2|$.

114. If (r, θ_1) and (r, θ_2) represent the same point on the polar coordinate system, then $\theta_1 = \theta_2 + 2\pi n$ for some integer n.

115. If $x > 0$, then the point (x, y) on the rectangular coordinate system can be represented by (r, θ) on the polar coordinate system, where $r = \sqrt{x^2 + y^2}$ and $\theta = \arctan(y/x)$.

116. The polar equations $r = \sin 2\theta$ and $r = -\sin 2\theta$ have the same graph.

SECTION PROJECT **ANAMORPHIC ART**

Use the anamorphic transformations

$$r = y + 16 \quad \text{and} \quad \theta = -\frac{\pi}{8}x, \quad -\frac{3\pi}{4} \le \theta \le \frac{3\pi}{4}$$

to sketch the transformed polar image of the rectangular graph. When the reflection (in a cylindrical mirror centered at the pole) of each polar image is viewed from the polar axis, the viewer will see the original rectangular image.

  (a) $y = 3$ (b) $x = 2$

  (c) $y = x + 5$ (d) $x^2 + (y - 5)^2 = 5^2$

Museum of Science and Industry in Manchester, England

This example of anamorphic art is from the Museum of Science and Industry in Manchester, England. When the reflection of the transformed "polar painting" is viewed in the mirror, the viewer sees faces.

FOR FURTHER INFORMATION For more information on anamorphic art, see the article "Anamorphisms" by Philip Hickin in the *Mathematical Gazette*. To view this article, go to the website *www.matharticles.com*.

- Find the area of a region bounded by a polar graph.
- Find the points of intersection of two polar graphs.
- Find the arc length of a polar graph.
- Find the area of a surface of revolution (polar form).

Area of a Polar Region

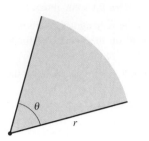

The area of a sector of a circle is $A = \frac{1}{2}\theta r^2$.
Figure 9.48

The development of a formula for the area of a polar region parallels that for the area of a region on the rectangular coordinate system, but uses *sectors* of a circle instead of rectangles as the basic element of area. In Figure 9.48, note that the area of a circular sector of radius r is given by $\frac{1}{2}\theta r^2$, provided θ is measured in radians.

Consider the function given by $r = f(\theta)$, where f is continuous and nonnegative in the interval given by $\alpha \le \theta \le \beta$. The region bounded by the graph of f and the radial lines $\theta = \alpha$ and $\theta = \beta$ is shown in Figure 9.49. To find the area of this region, partition the interval $[a, \beta]$ into n equal subintervals,

$$\alpha = \theta_0 < \theta_1 < \theta_2 < \cdots < \theta_{n-1} < \theta_n = \beta.$$

Then, approximate the area of the region by the sum of the areas of the n sectors.

$$\text{Radius of } i\text{th sector} = f(\theta_i)$$

$$\text{Central angle of } i\text{th sector} = \frac{\beta - \alpha}{n} = \Delta\theta$$

$$A \approx \sum_{i=1}^{n} \left(\frac{1}{2}\right)\Delta\theta[f(\theta_i)]^2$$

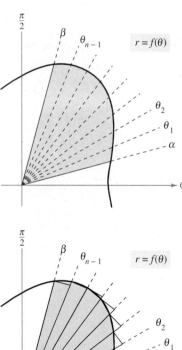

Taking the limit as $n \to \infty$ produces

$$A = \lim_{n \to \infty} \frac{1}{2}\sum_{i=1}^{n}[f(\theta_i)]^2\,\Delta\theta$$

$$= \frac{1}{2}\int_{\alpha}^{\beta}[f(\theta)]^2\,d\theta$$

which leads to the following theorem.

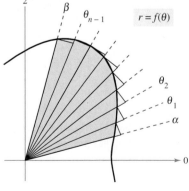

Figure 9.49

THEOREM 9.13   Area in Polar Coordinates

If f is continuous and nonnegative on the interval $[\alpha, \beta]$, $0 < \beta - \alpha \le 2\pi$, then the area of the region bounded by the graph of $r = f(\theta)$ between the radial lines $\theta = \alpha$ and $\theta = \beta$ is given by

$$A = \frac{1}{2}\int_{\alpha}^{\beta}[f(\theta)]^2\,d\theta$$

$$= \frac{1}{2}\int_{\alpha}^{\beta} r^2\,d\theta. \qquad 0 < \beta - \alpha \le 2\pi$$

NOTE   You can use the same formula to find the area of a region bounded by the graph of a continuous *nonpositive* function. However, the formula is not necessarily valid if f takes on both positive *and* negative values in the interval $[\alpha, \beta]$.

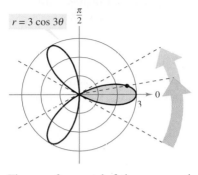

$r = 3 \cos 3\theta$

The area of one petal of the rose curve that lies between the radial lines $\theta = -\pi/6$ and $\theta = \pi/6$ is $3\pi/4$.
Figure 9.50

NOTE: To find the area of the region lying inside all three petals of the rose curve in Example 1, you could not simply integrate between 0 and 2π. In doing this you would obtain $9\pi/2$, which is twice the area of the three petals—the duplication occurs because the rose curve is traced *twice* as θ increases from 0 to 2π.

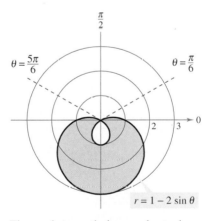

$\theta = \dfrac{5\pi}{6}$ $\theta = \dfrac{\pi}{6}$

$r = 1 - 2 \sin \theta$

The area between the inner and outer loops is approximately 8.34.
Figure 9.51

Example 1   Finding the Area of a Polar Region

Find the area of *one petal* of the rose curve given by $r = 3 \cos 3\theta$.

Solution   In Figure 9.50, you can see that the right petal is traced as θ increases from $-\pi/6$ to $\pi/6$. So, the area is

$$A = \frac{1}{2}\int_{\alpha}^{\beta} r^2 \, d\theta = \frac{1}{2}\int_{-\pi/6}^{\pi/6} (3 \cos 3\theta)^2 \, d\theta \qquad \text{Formula for area in polar coordinates}$$

$$= \frac{9}{2}\int_{-\pi/6}^{\pi/6} \frac{1 + \cos 6\theta}{2} \, d\theta \qquad \text{Trigonometric identity}$$

$$= \frac{9}{4}\left[\theta + \frac{\sin 6\theta}{6} \right]_{-\pi/6}^{\pi/6}$$

$$= \frac{9}{4}\left(\frac{\pi}{6} + \frac{\pi}{6} \right)$$

$$= \frac{3\pi}{4}.$$

Example 2   Finding the Area Bounded by a Single Curve

Find the area of the region lying between the inner and outer loops of the limaçon $r = 1 - 2 \sin \theta$.

Solution   In Figure 9.51, note that the inner loop is traced as θ increases from $\pi/6$ to $5\pi/6$. So, the area inside the *inner loop* is

$$A_1 = \frac{1}{2}\int_{\alpha}^{\beta} r^2 \, d\theta = \frac{1}{2}\int_{\pi/6}^{5\pi/6} (1 - 2 \sin \theta)^2 \, d\theta \qquad \text{Formula for area in polar coordinates}$$

$$= \frac{1}{2}\int_{\pi/6}^{5\pi/6} (1 - 4 \sin \theta + 4 \sin^2 \theta) \, d\theta$$

$$= \frac{1}{2}\int_{\pi/6}^{5\pi/6} \left[1 - 4 \sin \theta + 4\left(\frac{1 - \cos 2\theta}{2} \right) \right] d\theta \qquad \text{Trigonometric identity}$$

$$= \frac{1}{2}\int_{\pi/6}^{5\pi/6} (3 - 4 \sin \theta - 2 \cos 2\theta) \, d\theta \qquad \text{Simplify.}$$

$$= \frac{1}{2}\left[3\theta + 4 \cos \theta - \sin 2\theta \right]_{\pi/6}^{5\pi/6}$$

$$= \frac{1}{2}\left(2\pi - 3\sqrt{3} \right)$$

$$= \pi - \frac{3\sqrt{3}}{2}.$$

In a similar way, you can integrate from $5\pi/6$ to $13\pi/6$ to find that the area of the region lying inside the *outer loop* is $A_2 = 2\pi + \left(3\sqrt{3}/2 \right)$. The area of the region lying between the two loops is the difference of A_2 and A_1.

$$A = A_2 - A_1 = \left(2\pi + \frac{3\sqrt{3}}{2} \right) - \left(\pi - \frac{3\sqrt{3}}{2} \right) = \pi + 3\sqrt{3} \approx 8.34$$

Points of Intersection of Polar Graphs

Because a point may be represented in different ways in polar coordinates, care must be taken in determining the points of intersection of two polar graphs. For example, consider the points of intersection of the graphs of

$$r = 1 - 2\cos\theta \quad \text{and} \quad r = 1$$

as shown in Figure 9.52. If, as with rectangular equations, you attempted to find the points of intersection by solving the two equations simultaneously, you would obtain the following.

$$r = 1 - 2\cos\theta \qquad \text{First equation}$$

$$1 = 1 - 2\cos\theta \qquad \text{Substitute } r = 1 \text{ from 2nd equation into 1st equation.}$$

$$\cos\theta = 0 \qquad \text{Simplify.}$$

$$\theta = \frac{\pi}{2}, \frac{3\pi}{2} \qquad \text{Solve for } \theta.$$

FOR FURTHER INFORMATION For more information on using technology to find points of intersection, see the article "Finding Points of Intersection of Polar-Coordinate Graphs" by Warren W. Esty in *Mathematics Teacher*. To view this article, go to the website *www.matharticles.com*.

The corresponding points of intersection are $(1, \pi/2)$ and $(1, 3\pi/2)$. However, from Figure 9.52 you can see that there is a *third* point of intersection that did not show up when the two polar equations were solved simultaneously. (This is one reason we stress sketching a graph when finding the area of a polar region.) The reason the third point was not found is that it does not occur with the same coordinates in the two graphs. On the graph of $r = 1$, the point occurs with coordinates $(1, \pi)$, but on the graph of $r = 1 - 2\cos\theta$, the point occurs with coordinates $(-1, 0)$.

You can compare the problem of finding points of intersection of two polar graphs with that of finding collision points of two satellites in intersecting orbits about earth, as shown in Figure 9.53. The satellites will not collide as long as they reach the points of intersection at different times (θ-values). A collision will occur only at the points of intersection that are "simultaneous points"—those reached at the same time (θ-value).

NOTE   Because the pole can be represented by $(0, \theta)$, where θ is *any* angle, you should check separately for the pole when hunting for points of intersection.

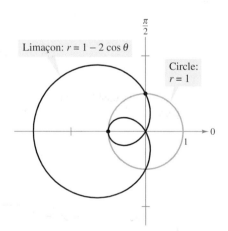

Three points of intersection: $(1, \pi/2)$, $(-1, 0), (1, 3\pi/2)$

Figure 9.52

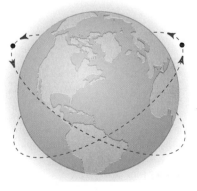

The paths of satellites can cross without causing a collision.

Figure 9.53

Example 3   **Finding the Area of a Region Between Two Curves**

Find the area of the region common to the two regions bounded by the following curves.

$$r = -6 \cos \theta \qquad \text{Circle}$$
$$r = 2 - 2 \cos \theta \qquad \text{Cardioid}$$

Solution   Because both curves are symmetric with respect to the x-axis, you can work with the upper half-plane, as shown in Figure 9.54. The gray shaded region lies between the circle and the radial line $\theta = 2\pi/3$. Because the circle has coordinates $(0, \pi/2)$ at the pole, you can integrate between $\pi/2$ and $2\pi/3$ to obtain the area of this region. The region that is shaded red is bounded by the radial lines $\theta = 2\pi/3$ and $\theta = \pi$ and the cardioid. So, you can find the area of this second region by integrating between $2\pi/3$ and π. The sum of these two integrals gives the area of the common region lying *above* the radial line $\theta = \pi$.

$$
\frac{A}{2} = \underbrace{\frac{1}{2} \int_{\pi/2}^{2\pi/3} (-6 \cos \theta)^2 \, d\theta}_{\substack{\text{Region between circle} \\ \text{and radial line } \theta = 2\pi/3}} + \underbrace{\frac{1}{2} \int_{2\pi/3}^{\pi} (2 - 2 \cos \theta)^2 \, d\theta}_{\substack{\text{Region between cardioid and} \\ \text{radial lines } \theta = 2\pi/3 \text{ and } \theta = \pi}}
$$

$$
= 18 \int_{\pi/2}^{2\pi/3} \cos^2 \theta \, d\theta + \frac{1}{2} \int_{2\pi/3}^{\pi} (4 - 8 \cos \theta + 4 \cos^2 \theta) \, d\theta
$$

$$
= 9 \int_{\pi/2}^{2\pi/3} (1 + \cos 2\theta) \, d\theta + \int_{2\pi/3}^{\pi} (3 - 4 \cos \theta + \cos 2\theta) \, d\theta
$$

$$
= 9 \left[\theta + \frac{\sin 2\theta}{2} \right]_{\pi/2}^{2\pi/3} + \left[3\theta - 4 \sin \theta + \frac{\sin 2\theta}{2} \right]_{2\pi/3}^{\pi}
$$

$$
= 9 \left(\frac{2\pi}{3} - \frac{\sqrt{3}}{4} - \frac{\pi}{2} \right) + \left(3\pi - 2\pi + 2\sqrt{3} + \frac{\sqrt{3}}{4} \right)
$$

$$
= \frac{5\pi}{2}
$$

$$
\approx 7.85
$$

Finally, multiplying by 2, you can conclude that the total area is 5π.

NOTE   To check the reasonableness of the result obtained in Example 3, note that the area of the circular region is $\pi r^2 = 9\pi$. So, it seems reasonable that the area of the region lying inside the circle and the cardioid is 5π.

To see the benefit of polar coordinates for finding the area in Example 3, consider the following integral, which gives the comparable area in rectangular coordinates.

$$
\frac{A}{2} = \int_{-4}^{-3/2} \sqrt{2\sqrt{1 - 2x} - x^2 - 2x + 2} \, dx + \int_{-3/2}^{0} \sqrt{-x^2 - 6x} \, dx
$$

Try using the integration capabilities of a graphing utility to show that you obtain the same area as that found in Example 3.

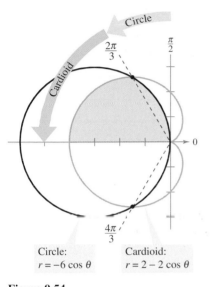

Circle:
$r = -6 \cos \theta$

Cardioid:
$r = 2 - 2 \cos \theta$

Figure 9.54

Arc Length in Polar Form

NOTE When applying the arc length formula to a polar curve, be sure that the curve is traced out only once on the interval of integration. For instance, the rose curve given by $r = \cos 3\theta$ is traced out once on the interval $0 \le \theta \le \pi$, but is traced out twice on the interval $0 \le \theta \le 2\pi$.

The formula for the length of a polar arc can be obtained from the arc length formula for a curve described by parametric equations. (See Exercise 65.)

THEOREM 9.14   Arc Length of a Polar Curve

Let f be a function whose derivative is continuous on an interval $\alpha \le \theta \le \beta$. The length of the graph of $r = f(\theta)$ from $\theta = \alpha$ to $\theta = \beta$ is

$$s = \int_\alpha^\beta \sqrt{[f(\theta)]^2 + [f'(\theta)]^2}\, d\theta = \int_\alpha^\beta \sqrt{r^2 + \left(\frac{dr}{d\theta}\right)^2}\, d\theta.$$

Example 4   **Finding the Length of a Polar Curve**

Find the length of the arc from $\theta = 0$ to $\theta = 2\pi$ for the cardioid

$$r = f(\theta) = 2 - 2\cos\theta$$

as shown in Figure 9.55.

Solution   Because $f'(\theta) = 2\sin\theta$, you can find the arc length as follows.

$$
\begin{aligned}
s &= \int_\alpha^\beta \sqrt{[f(\theta)]^2 + [f'(\theta)]^2}\, d\theta && \text{Formula for arc length of a polar curve}\\
&= \int_0^{2\pi} \sqrt{(2 - 2\cos\theta)^2 + (2\sin\theta)^2}\, d\theta \\
&= 2\sqrt{2}\int_0^{2\pi} \sqrt{1 - \cos\theta}\, d\theta && \text{Simplify.}\\
&= 2\sqrt{2}\int_0^{2\pi} \sqrt{2\sin^2\frac{\theta}{2}}\, d\theta && \text{Trigonometric identity}\\
&= 4\int_0^{2\pi} \sin\frac{\theta}{2}\, d\theta && \sin\frac{\theta}{2} \ge 0 \text{ for } 0 \le \theta \le 2\pi\\
&= 8\left[-\cos\frac{\theta}{2}\right]_0^{2\pi} \\
&= 8(1 + 1) \\
&= 16
\end{aligned}
$$

In the fifth step of the solution, it is legitimate to write

$$\sqrt{2\sin^2(\theta/2)} = \sqrt{2}\sin(\theta/2)$$

rather than

$$\sqrt{2\sin^2(\theta/2)} = \sqrt{2}\,|\sin(\theta/2)|$$

because $\sin(\theta/2) \ge 0$ for $0 \le \theta \le 2\pi$.

NOTE   Using Figure 9.55, you can determine the reasonableness of this answer by comparing it with the circumference of a circle. For example, a circle of radius $\frac{5}{2}$ has a circumference of $5\pi \approx 15.7$.

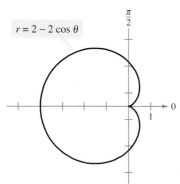

$r = 2 - 2\cos\theta$

The arc length of this cardioid is 16.
Figure 9.55

Area of a Surface of Revolution

The polar coordinate version of the formulas for the area of a surface of revolution can be obtained from the parametric versions given in Theorem 9.9, using the equations $x = r \cos \theta$ and $y = r \sin \theta$.

> **THEOREM 9.15   Area of a Surface of Revolution**
>
> Let f be a function whose derivative is continuous on an interval $\alpha \leq \theta \leq \beta$. The area of the surface formed by revolving the graph of $r = f(\theta)$ from $\theta = \alpha$ to $\theta = \beta$ about the indicated line is as follows.
>
> **1.** $S = 2\pi \displaystyle\int_{\alpha}^{\beta} f(\theta) \sin \theta \sqrt{[f(\theta)]^2 + [f'(\theta)]^2} \, d\theta$ About the polar axis
>
> **2.** $S = 2\pi \displaystyle\int_{\alpha}^{\beta} f(\theta) \cos \theta \sqrt{[f(\theta)]^2 + [f'(\theta)]^2} \, d\theta$ About the line $\theta = \dfrac{\pi}{2}$

NOTE   When using Theorem 9.15, check to see that the graph of $r = f(\theta)$ is traced only once on the interval $\alpha \leq \theta \leq \beta$. For example, the circle given by $r = \cos \theta$ is traced once on the interval $0 \leq \theta \leq \pi$.

Example 5   **Finding the Area of a Surface of Revolution**

Find the area of the surface formed by revolving the circle $r = f(\theta) = \cos \theta$ about the line $\theta = \pi/2$, as shown in Figure 9.56.

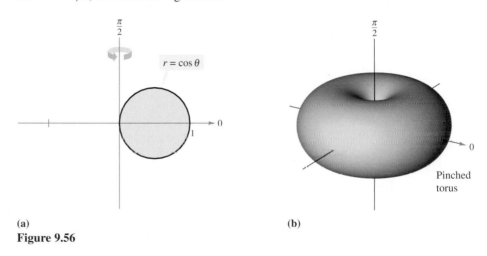

$r = \cos \theta$

Pinched torus

(a)   **(b)**

Figure 9.56

Solution   You can use the second formula given in Theorem 9.15 with $f'(\theta) = -\sin \theta$. Because the circle is traced once as θ increases from 0 to π, we have

$$S = 2\pi \int_{\alpha}^{\beta} f(\theta) \cos \theta \sqrt{[f(\theta)]^2 + [f'(\theta)]^2} \, d\theta \qquad \text{Formula for area of a surface of revolution}$$

$$= 2\pi \int_{0}^{\pi} \cos \theta (\cos \theta) \sqrt{\cos^2 \theta + \sin^2 \theta} \, d\theta$$

$$= 2\pi \int_{0}^{\pi} \cos^2 \theta \, d\theta \qquad \text{Trigonometric identity}$$

$$= \pi \int_{0}^{\pi} (1 + \cos 2\theta) \, d\theta \qquad \text{Trigonometric identity}$$

$$= \pi \left[\theta + \frac{\sin 2\theta}{2} \right]_{0}^{\pi} = \pi^2.$$

EXERCISES FOR SECTION 9.5

In Exercises 1 and 2, find the area of the region bounded by the graph of the polar equation using (a) a geometric formula, and (b) integration.

1. $r = 8 \sin \theta$
2. $r = 3 \cos \theta$

In Exercises 3–8, find the area of the region.

3. One petal of $r = 2 \cos 3\theta$
4. One petal of $r = 6 \sin 2\theta$
5. One petal of $r = \cos 2\theta$
6. One petal of $r = \cos 5\theta$
7. Interior of $r = 1 - \sin \theta$
8. Interior of $r = 1 - \sin \theta$ (above the polar axis)

In Exercises 9–12, use a graphing utility to graph the polar equation and find the area of the indicated region.

9. Inner loop of $r = 1 + 2 \cos \theta$
10. Inner loop of $r = 4 - 6 \sin \theta$
11. Between the loops of $r = 1 + 2 \cos \theta$
12. Between the loops of $r = 2(1 + 2 \sin \theta)$

In Exercises 13–22, find the points of intersection of the graphs of the equations.

13. $r = 1 + \cos \theta$
$r = 1 - \cos \theta$

14. $r = 3(1 + \sin \theta)$
$r = 3(1 - \sin \theta)$

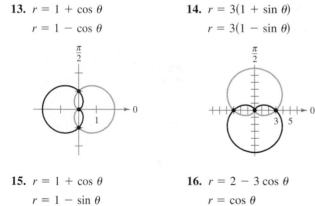

15. $r = 1 + \cos \theta$
$r = 1 - \sin \theta$

16. $r = 2 - 3 \cos \theta$
$r = \cos \theta$

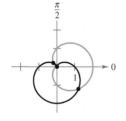
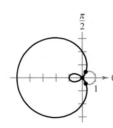

17. $r = 4 - 5 \sin \theta$
$r = 3 \sin \theta$

18. $r = 1 + \cos \theta$
$r = 3 \cos \theta$

19. $r = \dfrac{\theta}{2}$
$r = 2$

20. $\theta = \dfrac{\pi}{4}$
$r = 2$

21. $r = 4 \sin 2\theta$
$r = 2$

22. $r = 3 + \sin \theta$
$r = 2 \csc \theta$

In Exercises 23 and 24, use a graphing utility to approximate the points of intersection of the graphs of the polar equations. Confirm your results analytically.

23. $r = 2 + 3 \cos \theta$
$r = \dfrac{\sec \theta}{2}$

24. $r = 3(1 - \cos \theta)$
$r = \dfrac{6}{1 - \cos \theta}$

Writing **In Exercises 25 and 26, use a graphing utility to find the points of intersection of the graphs of the polar equations. Watch the graphs as they are traced in the viewing window. Explain why the pole is not a point of intersection obtained by solving the equations simultaneously.**

25. $r = \cos \theta$
$r = 2 - 3 \sin \theta$

26. $r = 4 \sin \theta$
$r = 2(1 + \sin \theta)$

In Exercises 27–32, use a graphing utility to graph the polar equations and find the area of the indicated region.

27. Common interior of $r = 4 \sin 2\theta$ and $r = 2$
28. Common interior of $r = 3(1 + \sin \theta)$ and $r = 3(1 - \sin \theta)$
29. Common interior of $r = 3 - 2 \sin \theta$ and $r = -3 + 2 \sin \theta$
30. Common interior of $r = 5 - 3 \sin \theta$ and $r = 5 - 3 \cos \theta$
31. Common interior of $r = 4 \sin \theta$ and $r = 2$
32. Inside $r = 3 \sin \theta$ and outside $r = 2 - \sin \theta$

In Exercises 33–36, find the area of the region.

33. Inside $r = a(1 + \cos \theta)$ and outside $r = a \cos \theta$
34. Inside $r = 2a \cos \theta$ and outside $r = a$
35. Common interior of $r = a(1 + \cos \theta)$ and $r = a \sin \theta$
36. Common interior of $r = a \cos \theta$ and $r = a \sin \theta$ where $a > 0$.

37. *Antenna Radiation* The radiation from a transmitting antenna is not uniform in all directions. The intensity from a particular antenna is modeled by

$r = a \cos^2 \theta$.

(a) Convert the polar equation to rectangular form.

(b) Use a graphing utility to graph the model for $a = 4$ and $a = 6$.

(c) Find the area of the geographical region between the two curves in part (b).

38. *Area* The area inside one or more of the three interlocking circles

$r = 2a \cos \theta, \quad r = 2a \sin \theta, \quad \text{and} \quad r = a$

is divided into seven regions. Find the area of each region.

39. *Conjecture* Find the area of the region enclosed by $r = a \cos(n\theta)$ for $n = 1, 2, 3, \ldots$. Use the results to make a conjecture about the area enclosed by the function if n is even and if n is odd.

40. *Area* Sketch the strophoid

$$r = \sec\theta - 2\cos\theta, \quad -\frac{\pi}{2} < \theta < \frac{\pi}{2}.$$

Convert this equation to rectangular coordinates. Find the area enclosed by the loop.

In Exercises 41–44, find the length of the curve over the indicated interval.

Polar Equation	Interval
41. $r = a$	$0 \le \theta \le 2\pi$
42. $r = 2a\cos\theta$	$-\dfrac{\pi}{2} \le \theta \le \dfrac{\pi}{2}$
43. $r = 1 + \sin\theta$	$0 \le \theta \le 2\pi$
44. $r = 8(1 + \cos\theta)$	$0 \le \theta \le 2\pi$

In Exercises 45–50, use a graphing utility to graph the polar equation over the indicated interval. Use the integration capabilities of the graphing utility to approximate the length of the curve accurate to two decimal places.

Polar Equation	Interval
45. $r = 2\theta$	$0 \le \theta \le \dfrac{\pi}{2}$
46. $r = \sec\theta$	$0 \le \theta \le \dfrac{\pi}{3}$
47. $r = \dfrac{1}{\theta}$	$\pi \le \theta \le 2\pi$
48. $r = e^{\theta}$	$0 \le \theta \le \pi$
49. $r = \sin(3\cos\theta)$	$0 \le \theta \le \pi$
50. $r = 2\sin(2\cos\theta)$	$0 \le \theta \le \pi$

In Exercises 51–54, find the area of the surface formed by revolving the curve about the given line.

Polar Equation	Interval	Axis of Revolution
51. $r = 6\cos\theta$	$0 \le \theta \le \dfrac{\pi}{2}$	Polar axis
52. $r = a\cos\theta$	$0 \le \theta \le \dfrac{\pi}{2}$	$\theta = \dfrac{\pi}{2}$
53. $r = e^{a\theta}$	$0 \le \theta \le \dfrac{\pi}{2}$	$\theta = \dfrac{\pi}{2}$
54. $r = a(1 + \cos\theta)$	$0 \le \theta \le \pi$	Polar axis

In Exercises 55 and 56, use the integration capabilities of a graphing utility to approximate to two decimal places the area of the surface formed by revolving the curve about the polar axis.

Polar Equation	Interval
55. $r = 4\cos 2\theta$	$0 \le \theta \le \dfrac{\pi}{4}$
56. $r = \theta$	$0 \le \theta \le \pi$

Getting at the Concept

57. Give the integral formulas for area and arc length in polar coordinates.

58. Explain why finding points of intersection of polar graphs may require further analysis beyond solving two equations simultaneously.

59. Which integral yields the arc length of $r = 3(1 - \cos 2\theta)$? State why the other integrals are incorrect.

(a) $3\displaystyle\int_0^{2\pi} \sqrt{(1 - \cos 2\theta)^2 + 4\sin^2 2\theta}\, d\theta$

(b) $12\displaystyle\int_0^{\pi/4} \sqrt{(1 - \cos 2\theta)^2 + 4\sin^2 2\theta}\, d\theta$

(c) $3\displaystyle\int_0^{\pi} \sqrt{(1 - \cos 2\theta)^2 + 4\sin^2 2\theta}\, d\theta$

(d) $6\displaystyle\int_0^{\pi/2} \sqrt{(1 - \cos 2\theta)^2 + 4\sin^2 2\theta}\, d\theta$

60. Give the integral formulas for the area of the surface of revolution formed when the graph of $r = f(\theta)$ is revolved about (a) the x-axis and (b) the y-axis.

61. *Surface Area of a Torus* Find the surface area of the torus generated by revolving the circle given by $r = a$ about the line $r = b\sec\theta$, where $0 < a < b$.

62. *Approximating Area* Consider the circle $r = 8\cos\theta$.

(a) Find the area of the circle.

(b) Complete the table giving the areas A of the sectors of the circle between $\theta = 0$ and the values of θ in the table.

θ	0.2	0.4	0.6	0.8	1.0	1.2	1.4
A							

(c) Use the table in part (b) to approximate the values of θ for which the sector of the circle composes $\frac{1}{4}$, $\frac{1}{2}$, and $\frac{3}{4}$ of the total area of the circle.

(d) Use a graphing utility to approximate to two-decimal-place accuracy the angles θ for which the sector of the circle composes $\frac{1}{4}$, $\frac{1}{2}$, and $\frac{3}{4}$ of the total area of the circle.

(e) Do the results in part (d) depend on the radius of the circle? Explain.

True or False? **In Exercises 63 and 64, determine whether the statement is true or false. If it is false, explain why or give an example that shows it is false.**

63. If $f(\theta) > 0$ for all θ and $g(\theta) < 0$ for all θ, then the graphs of $r = f(\theta)$ and $r = g(\theta)$ do not intersect.

64. If $f(\theta) = g(\theta)$ for $\theta = 0$, $\pi/2$, and $3\pi/2$, then the graphs of $r = f(\theta)$ and $r = g(\theta)$ have at least four points of intersection.

65. Use the formula for the arc length of a curve in parametric form to derive the formula for the arc length of a polar curve.

Section 9.6	Polar Equations of Conics and Kepler's Laws

- Analyze and write polar equations of conics.
- Understand and use Kepler's Laws of planetary motion.

Polar Equations of Conics

EXPLORATION

Graphing Conics   Set a graphing utility to polar mode and enter polar equations of the form

$$r = \frac{a}{1 \pm b \cos \theta}$$

or

$$r = \frac{a}{1 \pm b \sin \theta}.$$

As long as $a \neq 0$, the graph should be a conic. Describe the values of a and b that produce parabolas. What values produce ellipses? What values produce hyperbolas?

In this chapter you have seen that the rectangular equations of ellipses and hyperbolas take simple forms when the origin lies at their *centers*. As it happens, there are many important applications of conics in which it is more convenient to use one of the *foci* as the reference point (the origin) for the coordinate system. For example, the sun lies at a focus of earth's orbit. Similarly, the light source of a parabolic reflector lies at its focus. In this section you will see that polar equations of conics take simple forms if one of the foci lies at the pole.

The following theorem uses the concept of *eccentricity*, as defined in Section 9.1, to classify the three basic types of conics. A proof of this theorem is given in Appendix B.

THEOREM 9.16   Classification of Conics by Eccentricity

Let F be a fixed point (*focus*) and D be a fixed line (*directrix*) in the plane. Let P be another point in the plane and let e (*eccentricity*) be the ratio of the distance between P and F to the distance between P and D. The collection of all points P with a given eccentricity is a conic.

1. The conic is an ellipse if $0 < e < 1$.
2. The conic is a parabola if $e = 1$.
3. The conic is a hyperbola if $e > 1$.

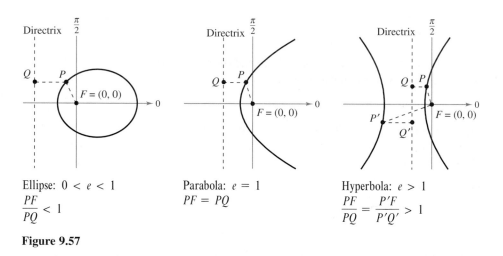

Ellipse: $0 < e < 1$
$$\frac{PF}{PQ} < 1$$

Parabola: $e = 1$
$$PF = PQ$$

Hyperbola: $e > 1$
$$\frac{PF}{PQ} = \frac{P'F}{P'Q'} > 1$$

Figure 9.57

In Figure 9.57, note that for each type of conic the pole corresponds to the fixed point (focus) given in the definition. The benefit of this location can be seen in the proof of the following theorem.

> **THEOREM 9.17   Polar Equations of Conics**
>
> The graph of a polar equation of the form
>
> $$r = \frac{ed}{1 \pm e \cos \theta} \qquad \text{or} \qquad r = \frac{ed}{1 \pm e \sin \theta}$$
>
> is a conic, where $e > 0$ is the eccentricity and $|d|$ is the distance between the focus at the pole and its corresponding directrix.

Proof We give a proof for $r = ed/(1 + e \cos \theta)$ with $d > 0$. In Figure 9.58, consider a vertical directrix d units to the right of the focus $F = (0, 0)$. If $P = (r, \theta)$ is a point on the graph of $r = ed/(1 + e \cos \theta)$, the distance between P and the directrix can be shown to be

$$PQ = |d - x| = |d - r \cos \theta| = \left| \frac{r(1 + e \cos \theta)}{e} - r \cos \theta \right| = \left| \frac{r}{e} \right|.$$

Because the distance between P and the pole is simply $PF = |r|$, the ratio of PF to PQ is $PF/PQ = |r|/|r/e| = |e| = e$ and, by Theorem 9.16, the graph of the equation must be a conic. The proofs of the other cases are similar.   ◾

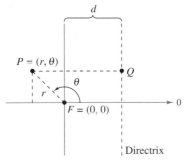

Figure 9.58

The four types of equations indicated in Theorem 9.17 can be classified as follows, where $d > 0$.

a. Horizontal directrix above the pole: $r = \dfrac{ed}{1 + e \sin \theta}$

b. Horizontal directrix below the pole: $r = \dfrac{ed}{1 - e \sin \theta}$

c. Vertical directrix to the right of the pole: $r = \dfrac{ed}{1 + e \cos \theta}$

d. Vertical directrix to the left of the pole: $r = \dfrac{ed}{1 - e \cos \theta}$

Figure 9.59 illustrates these four possibilities for a parabola.

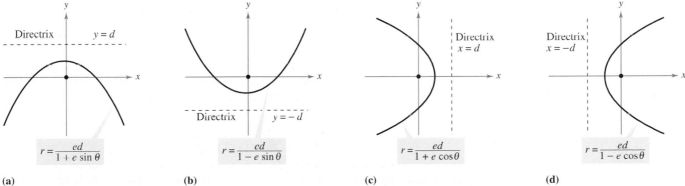

(a)   **(b)**   **(c)**   **(d)**

The four types of polar equations for a parabola
Figure 9.59

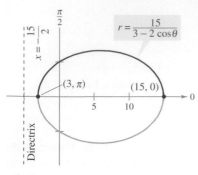

The graph of the conic is an ellipse with $e = \frac{2}{3}$.

Figure 9.60

Example 1 Determining a Conic from Its Equation

Sketch the graph of the conic given by $r = \dfrac{15}{3 - 2 \cos \theta}$.

Solution To determine the type of conic, rewrite the equation as

$$r = \frac{15}{3 - 2 \cos \theta}$$

$$= \frac{5}{1 - (2/3) \cos \theta}. \qquad \text{Divide numerator and denominator by 3.}$$

So, the graph is an ellipse with $e = \frac{2}{3}$. You can sketch the upper half of the ellipse by plotting points from $\theta = 0$ to $\theta = \pi$, as shown in Figure 9.60. Then, using symmetry with respect to the polar axis, you can sketch the lower half. ◢

For the ellipse in Figure 9.60, the major axis is horizontal and the vertices lie at $(15, 0)$ and $(3, \pi)$. So, the length of the *major* axis is $2a = 18$. To find the length of the *minor* axis, you can use the equations $e = c/a$ and $b^2 = a^2 - c^2$ to conclude

$$b^2 = a^2 - c^2 = a^2 - (ea)^2 = a^2(1 - e^2). \qquad \text{Ellipse}$$

Because $e = \frac{2}{3}$, you have

$$b^2 = 9^2\left[1 - \left(\tfrac{2}{3}\right)^2\right] = 45$$

which implies that $b = \sqrt{45} = 3\sqrt{5}$. So, the length of the minor axis is $2b = 6\sqrt{5}$. A similar analysis for hyperbolas yields

$$b^2 = c^2 - a^2 = (ea)^2 - a^2 = a^2(e^2 - 1). \qquad \text{Hyperbola}$$

Example 2 Sketching a Conic from Its Polar Equation

Sketch the graph of the polar equation $r = \dfrac{32}{3 + 5 \sin \theta}$.

Solution Dividing the numerator and denominator by 3 produces

$$r = \frac{32/3}{1 + (5/3) \sin \theta}.$$

Because $e = \frac{5}{3} > 1$, the graph is a hyperbola. Because $d = \frac{32}{5}$, the directrix is the line $y = \frac{32}{5}$. The transverse axis of the hyperbola lies on the line $\theta = \pi/2$, and the vertices occur at

$$(r, \theta) = \left(4, \frac{\pi}{2}\right) \quad \text{and} \quad (r, \theta) = \left(-16, \frac{3\pi}{2}\right).$$

Because the length of the transverse axis is 12, you can see that $a = 6$. To find b, write

$$b^2 = a^2(e^2 - 1) = 6^2\left[\left(\frac{5}{3}\right)^2 - 1\right] = 64.$$

Therefore, $b = 8$. Finally, you can use a and b to determine the asymptotes of the hyperbola and obtain the sketch shown in Figure 9.61. ◢

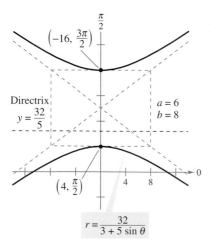

The graph of the conic is a hyperbola with $e = \frac{5}{3}$.

Figure 9.61

JOHANNES KEPLER (1571–1630)

Kepler formulated his three laws from the extensive data recorded by Danish astronomer Tycho Brahe, and from direct observation of the orbit of Mars.

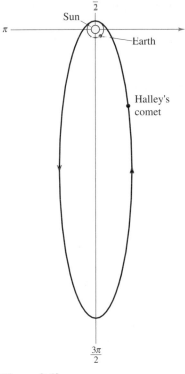

Figure 9.62

Kepler's Laws

Kepler's Laws, named after the German astronomer Johannes Kepler, can be used to describe the orbits of the planets about the sun.

1. Each planet moves in an elliptical orbit with the sun as a focus.
2. The ray from the sun to the planet sweeps out equal areas of the ellipse in equal times.
3. The square of the period is proportional to the cube of the mean distance between the planet and the sun.[*]

Although Kepler derived these laws empirically, they were later validated by Newton. In fact, Newton was able to show that each law can be deduced from a set of universal laws of motion and gravitation that govern the movement of all heavenly bodies, including comets and satellites. This is illustrated in the next example, involving the comet named after the English mathematician and physicist Edmund Halley (1656–1742).

Example 3   Halley's Comet

Halley's comet has an elliptical orbit with an eccentricity of $e \approx 0.97$. The length of the major axis of the orbit is approximately 36.18 astronomical units. (An astronomical unit is defined to be the mean distance between earth and the sun, 93 million miles.) Find a polar equation for the orbit. How close does Halley's comet come to the sun?

Solution   Using a vertical axis, you can choose an equation of the form

$$r = \frac{ed}{(1 + e \sin \theta)}.$$

Because the vertices of the ellipse occur when $\theta = \pi/2$ and $\theta = 3\pi/2$, you can determine the length of the major axis to be the sum of the r-values of the vertices, as shown in Figure 9.62. That is,

$$2a = \frac{0.97d}{1 + 0.97} + \frac{0.97d}{1 - 0.97}$$

$$36.18 \approx 32.83d. \qquad 2a \approx 36.18$$

So, $d \approx 1.102$ and $ed \approx (0.97)(1.102) \approx 1.069$. Using this value in the equation produces

$$r = \frac{1.069}{1 + 0.97 \sin \theta}$$

where r is measured in astronomical units. To find the closest point to the sun (the focus), you can write $c = ea \approx (0.97)(18.09) \approx 17.55$. Because c is the distance between the focus and the center, the closest point is

$$a - c \approx 18.09 - 17.55$$

$$\approx 0.54 \text{ AU}$$

$$\approx 50{,}000{,}000 \text{ miles}$$

---

[*] *If earth is used as a reference with a period of 1 year and a distance of 1 astronomical unit, the proportionality constant is 1. For example, because Mars has a mean distance to the sun of $D = 1.523$ AU, its period P is given by $D^3 = P^2$. So, the period for Mars is $P = 1.88$.*

Kepler's Second Law states that as a planet moves about the sun, a ray from the sun to the planet sweeps out equal areas in equal times. This law can also be applied to comets or asteroids with elliptical orbits. For example, Figure 9.63 shows the orbit of the asteroid Apollo about the sun. Applying Kepler's Second Law to this asteroid, you know that the closer it is to the sun, the greater its velocity, because a short ray must be moving quickly to sweep out as much area as a long ray.

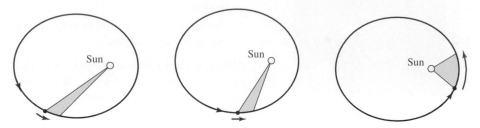

A ray from the sun to the asteroid sweeps out equal areas in equal times.
Figure 9.63

Example 4   **The Asteroid Apollo**

The asteroid Apollo has a period of 478 earth days, and its orbit is approximated by the ellipse

$$r = \frac{1}{1 + (5/9) \cos \theta} = \frac{9}{9 + 5 \cos \theta}$$

where r is measured in astronomical units. How long does it take Apollo to move from the position given by $\theta = -\pi/2$ to $\theta = \pi/2$, as shown in Figure 9.64?

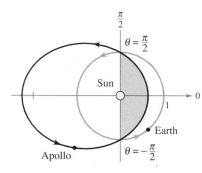

Figure 9.64

Solution   Begin by finding the area swept out as θ increases from $-\pi/2$ to $\pi/2$.

$$A = \frac{1}{2} \int_{\alpha}^{\beta} r^2 \, d\theta \qquad \text{Formula for area of a polar graph}$$

$$= \frac{1}{2} \int_{-\pi/2}^{\pi/2} \left(\frac{9}{9 + 5 \cos \theta} \right)^2 d\theta$$

Using the substitution $u = \tan(\theta/2)$, as discussed in Section 7.6, you obtain

$$A = \frac{81}{112} \left[\frac{-5 \sin \theta}{9 + 5 \cos \theta} + \frac{18}{\sqrt{56}} \arctan \frac{\sqrt{56} \tan(\theta/2)}{14} \right]_{-\pi/2}^{\pi/2} \approx 0.90429.$$

Because the major axis of the ellipse has length $2a = 81/28$ and the eccentricity is $e = 5/9$, you can determine that $b = a\sqrt{1 - e^2} = 9/\sqrt{56}$. So, the area of the ellipse is

$$\text{Area of ellipse} \ = \pi ab = \pi \left(\frac{81}{56} \right) \left(\frac{9}{\sqrt{56}} \right) \approx 5.46507.$$

Because the time required to complete the orbit is 478 days, you can apply Kepler's Second Law to conclude that the time t required to move from the position $\theta = -\pi/2$ to $\theta = \pi/2$ is given by

$$\frac{t}{478} = \frac{\text{area of elliptical segment}}{\text{area of ellipse}} \approx \frac{0.90429}{5.46507}$$

which implies that $t \approx 79$ days.

Lab Series | LAB 13

EXERCISES FOR SECTION 9.6

Graphical Reasoning In Exercises 1–4, use a graphing utility to graph the polar equation when (a) $e = 1$, (b) $e = 0.5$, and (c) $e = 1.5$. Identify the conic.

1. $r = \dfrac{2e}{1 + e \cos \theta}$

2. $r = \dfrac{2e}{1 - e \cos \theta}$

3. $r = \dfrac{2e}{1 - e \sin \theta}$

4. $r = \dfrac{2e}{1 + e \sin \theta}$

5. Consider the polar equation

$$r = \frac{4}{1 + e \sin \theta}.$$

(a) Use a graphing utility to graph the equation for $e = 0.1$, $e = 0.25$, $e = 0.5$, $e = 0.75$, and $e = 0.9$. Identify the conic and discuss the change in its shape as $e \to 1^-$ and $e \to 0^+$.

(b) Use a graphing utility to graph the equation for $e = 1$. Identify the conic.

(c) Use a graphing utility to graph the equation for $e = 1.1$, $e = 1.5$, and $e = 2$. Identify the conic and discuss the change in its shape as $e \to 1^+$ and $e \to \infty$.

6. Consider the polar equation

$$r = \frac{4}{1 - 0.4 \cos \theta}.$$

(a) Identify the conic without graphing the equation.

(b) Without graphing the following polar equations, describe how each differs from the polar equation above.

$$r = \frac{4}{1 + 0.4 \cos \theta}, \quad r = \frac{4}{1 - 0.4 \sin \theta}$$

(c) Verify the results in part (b) graphically.

In Exercises 7–12, match the polar equation with the correct graph. [The graphs are labeled (a), (b), (c), (d), (e), and (f).]

(a)
(b)
(c)
(d)

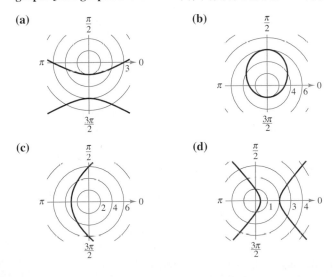

(e)
(f)

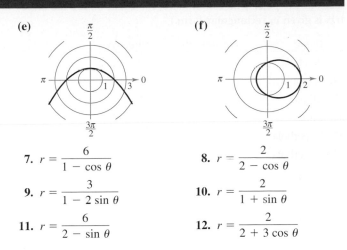

7. $r = \dfrac{6}{1 - \cos \theta}$

8. $r = \dfrac{2}{2 - \cos \theta}$

9. $r = \dfrac{3}{1 - 2 \sin \theta}$

10. $r = \dfrac{2}{1 + \sin \theta}$

11. $r = \dfrac{6}{2 - \sin \theta}$

12. $r = \dfrac{2}{2 + 3 \cos \theta}$

In Exercises 13–22, sketch and identify the graph. Use a graphing utility to confirm your results.

13. $r = \dfrac{-1}{1 - \sin \theta}$

14. $r = \dfrac{6}{1 + \cos \theta}$

15. $r = \dfrac{6}{2 + \cos \theta}$

16. $r = \dfrac{5}{5 + 3 \sin \theta}$

17. $r(2 + \sin \theta) = 4$

18. $r(3 - 2 \cos \theta) = 6$

19. $r = \dfrac{5}{-1 + 2 \cos \theta}$

20. $r = \dfrac{-6}{3 + 7 \sin \theta}$

21. $r = \dfrac{3}{2 + 6 \sin \theta}$

22. $r = \dfrac{4}{1 + 2 \cos \theta}$

In Exercises 23–26, use a graphing utility to graph the polar equation. Identify the graph.

23. $r = \dfrac{3}{-4 + 2 \sin \theta}$

24. $r = \dfrac{-3}{2 + 4 \sin \theta}$

25. $r = \dfrac{-1}{1 - \cos \theta}$

26. $r = \dfrac{2}{2 + 3 \sin \theta}$

In Exercises 27–30, use a graphing utility to graph the conic. Describe how the graph differs from that in the indicated exercise.

27. $r = \dfrac{-1}{1 - \sin(\theta - \pi/4)}$ (See Exercise 13.)

28. $r = \dfrac{6}{1 + \cos(\theta - \pi/3)}$ (See Exercise 14.)

29. $r = \dfrac{6}{2 + \cos(\theta + \pi/6)}$ (See Exercise 15.)

30. $r = \dfrac{-6}{3 + 7 \sin(\theta + 2\pi/3)}$ (See Exercise 20.)

31. Write the equation for the ellipse rotated $\pi/4$ radians clockwise from the ellipse $r = 5/(5 + 3 \cos\theta)$.

32. Write the equation for the parabola rotated $\pi/6$ radians counterclockwise from the parabola $r = 2/(1 + \sin \theta)$.

In Exercises 33–44, find a polar equation for the conic with its focus at the pole. (For convenience, the equation for the directrix is given in rectangular form.)

	Conic	Eccentricity	Directrix
33.	Parabola	$e = 1$	$x = -1$
34.	Parabola	$e = 1$	$y = 1$
35.	Ellipse	$e = \frac{1}{2}$	$y = 1$
36.	Ellipse	$e = \frac{3}{4}$	$y = -2$
37.	Hyperbola	$e = 2$	$x = 1$
38.	Hyperbola	$e = \frac{3}{2}$	$x = -1$

	Conic	Vertex or Vertices
39.	Parabola	$(1, -\pi/2)$
40.	Parabola	$(5, \pi)$
41.	Ellipse	$(2, 0), \ (8, \pi)$
42.	Ellipse	$(2, \pi/2), \ (4, 3\pi/2)$
43.	Hyperbola	$(1, 3\pi/2), \ (9, 3\pi/2)$
44.	Hyperbola	$(2, 0), \ (10, 0)$

Getting at the Concept

45. Classify the conics by their eccentricities.

46. Explain how the graph of each conic differs from the graph of $r = \dfrac{4}{1 + \sin \theta}$.

(a) $r = \dfrac{4}{1 - \cos \theta}$ (b) $r = \dfrac{4}{1 - \sin \theta}$

(c) $r = \dfrac{4}{1 + \cos \theta}$ (d) $r = \dfrac{4}{1 - \sin(\theta - \pi/4)}$

47. Identify the conic.

(a) $r = \dfrac{5}{1 - 2 \cos \theta}$ (b) $r = \dfrac{5}{10 - \sin \theta}$

(c) $r = \dfrac{5}{3 - 3 \cos \theta}$ (d) $r = \dfrac{5}{1 - 3 \sin(\theta - \pi/4)}$

48. (a) Show that the polar equation for $(x^2/a^2) + (y^2/b^2) = 1$ is

$$r^2 = \frac{b^2}{1 - e^2 \cos^2 \theta}. \qquad \text{Ellipse}$$

(b) Show that the polar equation for $(x^2/a^2) - (y^2/b^2) = 1$ is

$$r^2 = \frac{-b^2}{1 - e^2 \cos^2 \theta}. \qquad \text{Hyperbola}$$

In Exercises 49–52, use the results of Exercise 48 to write the polar form of the equation of the conic.

49. Ellipse: Focus at $(4, 0)$; Vertices at $(5, 0), \ (5, \pi)$

50. Hyperbola: Focus at $(5, 0)$; Vertices at $(4, 0), \ (4, \pi)$

51. $\dfrac{x^2}{9} - \dfrac{y^2}{16} = 1$ **52.** $\dfrac{x^2}{4} + y^2 = 1$

In Exercises 53 and 54, use the integration capabilities of a graphing utility to approximate to two decimal places the area of the region bounded by the graph of the polar equation.

53. $r = \dfrac{3}{2 - \cos \theta}$ **54.** $r = \dfrac{2}{3 - 2 \sin \theta}$

55. *Explorer 18* On November 26, 1963, the United States launched Explorer 18. Its low and high points above the surface of earth were 119 miles and 122,000 miles (see figure). The center of earth is the focus of the orbit. Find the polar equation for the orbit and find the distance between the surface of earth and the satellite when $\theta = 60°$. (Assume that the radius of earth is 4000 miles.)

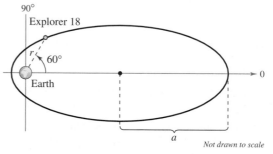

Not drawn to scale

56. *Planetary Motion* The planets travel in elliptical orbits with the sun as a focus, as shown in the figure.

(a) Show that the polar equation of the orbit is given by

$$r = \frac{(1 - e^2)a}{1 - e \cos \theta}$$

where e is the eccentricity.

(b) Show that the minimum distance (*perihelion distance*) from the sun to the planet is $r = a(1 - e)$ and the maximum distance (*aphelion distance*) is $r = a(1 + e)$.

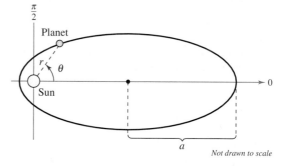

Not drawn to scale

In Exercises 57–60, use Exercise 56 to find the polar equation of the elliptical orbit of the planet, and the perihelion and aphelion distances.

57. Earth $a = 92.957 \times 10^6$ miles
                  $e = 0.0167$

58. Saturn $a = 1.427 \times 10^9$ kilometers
                  $e = 0.0543$

59. Pluto $a = 5.900 \times 10^9$ kilometers
                  $e = 0.2481$

60. Mercury $a = 36.0 \times 10^6$ miles
                  $e = 0.206$

61. *Planetary Motion*   In Exercise 59, the polar equation for the elliptical orbit of Pluto was found. Use the equation and a computer algebra system to perform each of the following.

(a) Approximate the area swept out by a ray from the sun to the planet as θ increases from 0 to $\pi/9$. Use this result to determine the number of years for the planet to move through this arc if the period of one revolution around the sun is 248 years.

(b) By trial and error, approximate the angle α such that the area swept out by a ray from the sun to the planet as θ increases from π to α equals the area found in part (a) (see figure). Does the ray sweep through a larger or smaller angle than in part (a) to generate the same area? Why is this the case?

(c) Approximate the distances the planet traveled in parts (a) and (b). Use these distances to approximate the average number of kilometers per year the planet traveled in the two cases.

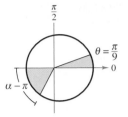

Figure for 61

62. What conic section does the following polar equation represent?

$$r = a \sin \theta + b \cos \theta$$

63. Show that the graphs of the following equations intersect at right angles.

$$r = \frac{ed}{1 + \sin \theta} \quad \text{and} \quad r = \frac{ed}{1 - \sin \theta}$$

REVIEW EXERCISES FOR CHAPTER 9

9.1   **In Exercises 1–4, match the equation with the correct graph. [The graphs are labeled (a), (b), (c), and (d).]**

(a)

(b)

(c)

(d)

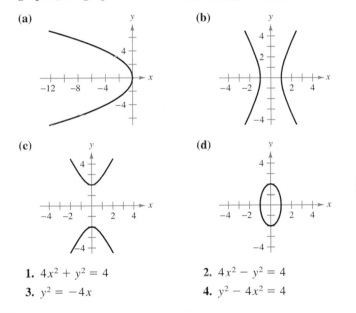

1. $4x^2 + y^2 = 4$

2. $4x^2 - y^2 = 4$

3. $y^2 = -4x$

4. $y^2 - 4x^2 = 4$

In Exercises 5–10, analyze each equation and sketch its graph. Use a graphing utility to confirm your results.

5. $16x^2 + 16y^2 - 16x + 24y - 3 = 0$

6. $y^2 - 12y - 8x + 20 = 0$

7. $3x^2 - 2y^2 + 24x + 12y + 24 = 0$

8. $4x^2 + y^2 - 16x + 15 = 0$

9. $3x^2 + 2y^2 - 12x + 12y + 29 = 0$

10. $4x^2 - 4y^2 - 4x + 8y - 11 = 0$

In Exercises 11 and 12, find an equation of the parabola.

11. Vertex: $(0, 2)$; Directrix: $x = -3$

12. Vertex: $(4, 2)$; Focus: $(4, 0)$

In Exercises 13 and 14, find an equation of the ellipse.

13. Vertices: $(-3, 0)$, $(7, 0)$; Foci: $(0, 0)$, $(4, 0)$

14. Center: $(0, 0)$; Solution points: $(1, 2)$, $(2, 0)$

In Exercises 15 and 16, find an equation of the hyperbola.

15. Vertices: $(\pm 4, 0)$; Foci: $(\pm 6, 0)$

16. Foci: $(0, \pm 8)$; Asymptotes: $y = \pm 4x$

In Exercises 17 and 18, use a graphing utility to approximate the perimeter of the ellipse.

17. $\dfrac{x^2}{9} + \dfrac{y^2}{4} = 1$

18. $\dfrac{x^2}{4} + \dfrac{y^2}{25} = 1$

19. A line is tangent to the parabola $y = x^2 - 2x + 2$ and perpendicular to the line $y = x - 2$. Find the equation of the line.

20. *Satellite Antenna*   A cross section of a large parabolic antenna is modeled by the graph of $y = x^2/200$, $-100 \le x \le 100$. The receiving and transmitting equipment is positioned at the focus.

(a) Find the coordinates of the focus.

(b) Find the surface area of the antenna.

21. Consider a fire truck with a water tank 16 feet long whose vertical cross sections are ellipses modeled by the equation $x^2/16 + y^2/9 = 1$.

(a) Find the volume of the tank.

(b) Find the force on the end of the tank when it is full of water. (The density of water is 62.4 pounds per cubic foot.)

(c) Find the depth of the water in the tank if it is $\frac{3}{4}$ full (by volume) and the truck is on level ground.

(d) Approximate the tank's surface area.

22. Consider the region bounded by the ellipse

$$\frac{x^2}{a^2} + \frac{y^2}{b^2} = 1,$$

eccentricity $e = c/a$.

(a) Show that the area of the region is πab.

(b) Show that the solid (oblate spheroid) generated by revolving the region about the minor axis of the ellipse has a volume of $V = 4\pi a^2 b/3$ and a surface area of

$$S = 2\pi a^2 + \pi\left(\frac{b^2}{e}\right)\ln\left(\frac{1+e}{1-e}\right).$$

(c) Show that the solid (prolate spheroid) generated by revolving the region about the major axis of the ellipse has a volume of $V = 4\pi ab^2/3$ and a surface area of

$$S = 2\pi b^2 + 2\pi\left(\frac{ab}{e}\right)\arcsin e.$$

9.2 In Exercises 23–28, sketch the curve represented by the parametric equations (indicate the orientation of the curve), and write the corresponding rectangular equation by eliminating the parameter.

23. $x = 1 + 4t,\ y = 2 - 3t$

24. $x = t + 4,\ y = t^2$

25. $x = 6\cos\theta,\ y = 6\sin\theta$

26. $x = 3 + 3\cos\theta,\ y = 2 + 5\sin\theta$

27. $x = 2 + \sec\theta,\ y = 3 + \tan\theta$

28. $x = 5\sin^3\theta,\ y = 5\cos^3\theta$

In Exercises 29–32, find a parametric representation of the line or conic.

29. Line: Passes through $(-2, 6)$ and $(3, 2)$

30. Circle: Center at $(5, 3)$; Radius 2

31. Ellipse: Center at $(-3, 4)$; Horizontal major axis of length 8 and minor axis of length 6

32. Hyperbola: Vertices at $(0, \pm 4)$; Foci at $(0, \pm 5)$

33. *Rotary Engine* The rotary engine was developed by Felix Wankel in the 1950s (see Chapter 4). It features a rotor, which is a modified equilateral triangle. The rotor moves in a chamber that, in two dimensions, is an epitrochoid. Use a graphing utility to graph the chamber modeled by the parametric equations

$$x = \cos 3\theta + 5\cos\theta$$

and

$$y = \sin 3\theta + 5\sin\theta.$$

34. *Hypocycloids* A hypocycloid has the parametric equations

$$x = (a - b)\cos t + b\cos\left(\frac{a-b}{b}t\right)\quad\text{and}$$

$$y = (a - b)\sin t - b\sin\left(\frac{a-b}{b}t\right).$$

Use a graphing utility to graph the hypocycloid for each of the following values of a and b.

(a) $a = 2,\ b = 1$ (b) $a = 3,\ b = 1$ (c) $a = 4,\ b = 1$

(d) $a = 10,\ b = 1$ (e) $a = 3,\ b = 2$ (f) $a = 4,\ b = 3$

35. *Serpentine Curve* Consider the parametric equations $x = 2\cot\theta$ and $y = 4\sin\theta\cos\theta$, $0 < \theta < \pi$.

(a) Use a graphing utility to sketch the curve.

(b) Eliminate the parameter to show that the rectangular equation of the serpentine curve is $(4 + x^2)y = 8x$.

36. *Involute of a Circle* The involute of a circle is described by the endpoint P of a string that is held taut as it is unwound from a spool that does not turn (see figure). Show that a parametric representation of the involute is

$$x = r(\cos\theta + \theta\sin\theta)\quad\text{and}\quad y = r(\sin\theta - \theta\cos\theta).$$

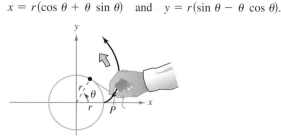

9.3 In Exercises 37–46, (a) find dy/dx and all points of horizontal tangency, (b) eliminate the parameter where possible, and (c) sketch the curve represented by the parametric equations.

37. $x = 1 + 4t,\quad y = 2 - 3t$

38. $x = t + 4,\quad y = t^2$

39. $x = \dfrac{1}{t},\quad y = 2t + 3$

40. $x = \dfrac{1}{t},\quad y = t^2$

41. $x = \dfrac{1}{2t + 1}$

  $y = \dfrac{1}{t^2 - 2t}$

42. $x = 2t - 1$

  $y = \dfrac{1}{t^2 - 2t}$

43. $x = 3 + 2\cos\theta$

  $y = 2 + 5\sin\theta$

44. $x = 6\cos\theta$

  $y = 6\sin\theta$

45. $x = \cos^3\theta$

  $y = 4\sin^3\theta$

46. $x = e^t$

  $y = e^{-t}$

In Exercises 47 and 48, (a) use a graphing utility to sketch the curve represented by the parametric equations, (b) use a graphing utility to find $dx/d\theta$, $dy/d\theta$, and dy/dx for $\theta = \pi/6$, and (c) use a graphing utility to graph the tangent line to the curve when $\theta = \pi/6$.

47. $x = \cot\theta$

  $y = \sin 2\theta$

48. $x = 2\theta - \sin\theta$

  $y = 2 - \cos\theta$

In Exercises 49 and 50, find the length of the curve represented by the parametric equations over the given interval.

49. $x = r(\cos\theta + \theta\sin\theta)$
$y = r(\sin\theta - \theta\cos\theta)$
$0 \le \theta \le \pi$

50. $x = 6\cos\theta$
$y = 6\sin\theta$
$0 \le \theta \le \pi$

9.4   **In Exercises 51 and 52, the rectangular coordinates of a point are given. Plot the point and find two sets of polar coordinates for the point for $0 \le \theta \le 2\pi$.**

51. $(4, -4)$

52. $(-1, 3)$

In Exercises 53–60, convert the polar equation to rectangular form.

53. $r = 3\cos\theta$

54. $r = 10$

55. $r = -2(1 + \cos\theta)$

56. $r = \dfrac{1}{2 - \cos\theta}$

57. $r^2 = \cos 2\theta$

58. $r = 4\sec\left(\theta - \dfrac{\pi}{3}\right)$

59. $r = 4\cos 2\theta \sec\theta$

60. $\theta = \dfrac{3\pi}{4}$

In Exercises 61–64, convert the rectangular equation to polar form.

61. $(x^2 + y^2)^2 = ax^2y$

62. $x^2 + y^2 - 4x = 0$

63. $x^2 + y^2 = a^2\left(\arctan\dfrac{y}{x}\right)^2$

64. $(x^2 + y^2)\left(\arctan\dfrac{y}{x}\right)^2 = a^2$

In Exercises 65–76, sketch a graph of the polar equation.

65. $r = 4$

66. $\theta = \dfrac{\pi}{12}$

67. $r = -\sec\theta$

68. $r = 3\csc\theta$

69. $r = -2(1 + \cos\theta)$

70. $r = 3 - 4\cos\theta$

71. $r = 4 - 3\cos\theta$

72. $r = 2\theta$

73. $r = -3\cos 2\theta$

74. $r = \cos 5\theta$

75. $r^2 = 4\sin^2 2\theta$

76. $r^2 = \cos 2\theta$

In Exercises 77–80, use a graphing utility to graph the polar equation.

77. $r = \dfrac{3}{\cos(\theta - \pi/4)}$

78. $r = 2\sin\theta\cos^2\theta$

79. $r = 4\cos 2\theta \sec\theta$

80. $r = 4(\sec\theta - \cos\theta)$

In Exercises 81 and 82, (a) find the tangents at the pole, (b) find all points of horizontal and vertical tangency, and (c) use a graphing utility to graph the polar equation and draw a tangent line to the graph for $\theta = \pi/6$.

81. $r = 1 - 2\cos\theta$

82. $r^2 = 4\sin 2\theta$

83. Find the angle between the circle $r = 3\sin\theta$ and the limaçon $r = 4 - 5\sin\theta$ at the point of intersection $(3/2, \pi/6)$.

84. *True or False?*   There is a unique polar coordinate representation for each point in the plane. Explain.

9.5   **In Exercises 85 and 86, show that the graphs of the polar equations are orthogonal at the points of intersection. Use a graphing utility to confirm your results graphically.**

85. $r = 1 + \cos\theta$
$r = 1 - \cos\theta$

86. $r = a\sin\theta$
$r = a\cos\theta$

In Exercises 87–94, use a graphing utility to graph the polar equation. Set up an integral for finding the area of the indicated region and use the integration capabilities of a graphing utility to approximate the integral accurate to two decimal places.

87. Interior of $r = 2 + \cos\theta$

88. Interior of $r = 5(1 - \sin\theta)$

89. Interior of $r = \sin\theta\cos^2\theta$

90. Interior of $r = 4\sin 3\theta$

91. Interior of $r^2 = 4\sin 2\theta$

92. Common interior of $r = 3$ and $r^2 = 18\sin 2\theta$

93. Common interior of $r = 4\cos\theta$ and $r = 2$

94. Region bounded by the polar axis and $r = e^\theta$ for $0 \le \theta \le \pi$

In Exercises 95 and 96, find the perimeter of the curve.

95. $r = a(1 - \cos\theta)$

96. $r = a\cos 2\theta$

9.6   **In Exercises 97–102, sketch and identify the graph. Use a graphing utility to confirm your results.**

97. $r = \dfrac{2}{1 - \sin\theta}$

98. $r = \dfrac{2}{1 + \cos\theta}$

99. $r = \dfrac{6}{3 + 2\cos\theta}$

100. $r = \dfrac{4}{5 - 3\sin\theta}$

101. $r = \dfrac{4}{2 - 3\sin\theta}$

102. $r = \dfrac{8}{2 - 5\cos\theta}$

In Exercises 103–108, find a polar equation for the line or conic.

103. Circle
Center: $(5, \pi/2)$
Solution point: $(0, 0)$

104. Line
Solution point: $(0, 0)$
Slope: $\sqrt{3}$

105. Parabola
Vertex: $(2, \pi)$
Focus: $(0, 0)$

106. Parabola
Vertex: $(2, \pi/2)$
Focus: $(0, 0)$

107. Ellipse
Vertices: $(5, 0)$, $(1, \pi)$
One focus: $(0, 0)$

108. Hyperbola
Vertices: $(1, 0)$, $(7, 0)$
One focus: $(0, 0)$

P.S. Problem Solving

1. Consider the parabola $x^2 = 4y$ and the focal chord $y = \frac{3}{4}x + 1$.

(a) Sketch the graph of the parabola and the focal chord.

(b) Show that the tangent lines to the parabola at the endpoints of the focal chord intersect at right angles.

(c) Show that the tangent lines to the parabola at the endpoints of the focal chord intersect on the directrix of the parabola.

2. Consider the parabola $x^2 = 4py$ and one of its focal chords.

(a) Show that the tangent lines to the parabola at the endpoints of the focal chord intersect at right angles.

(b) Show that the tangent lines to the parabola at the endpoints of the focal chord intersect on the directrix of the parabola.

3. Prove Theorem 9.2, the Reflective Property of a Parabola, as illustrated in the figure.

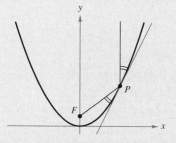

4. Consider the hyperbola

$$\frac{x^2}{a^2} - \frac{y^2}{b^2} = 1$$

with foci F_1 and F_2, as indicated in the figure. Let T be the tangent line at a point M on the hyperbola. Show that incoming rays of light aimed at one focus are reflected by a hyperbolic mirror toward the other focus.

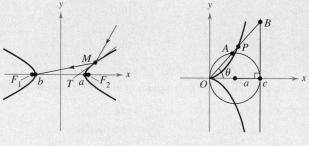

Figure for 4   **Figure for 5**

5. Consider a circle of radius a tangent to the y-axis and the line $x = 2a$, as indicated in the figure. Let A be the point where the segment OB intersects the circle. The **cissoid of Diocles** consists of all points P such that $OP = AB$.

(a) Find a polar equation of the cissoid.

(b) Find a set of parametric equations for the cissoid that does not contain trigonometric functions.

(c) Find a rectangular equation of the cissoid.

6. The curve given by the parametric equations

$$x(t) = \frac{1 - t^2}{1 + t^2} \quad \text{and} \quad y(t) = \frac{t(1 - t^2)}{1 + t^2}$$

is called a **strophoid.**

(a) Find a rectangular equation of the strophoid.

(b) Find a polar equation of the strophoid.

(c) Sketch a graph of the strophoid.

(d) Find the equations of the two tangent lines at the origin.

(e) Find the points on the graph where the tangent lines are horizontal.

7. Find the rectangular equation of the portion of the cycloid given by the parametric equations $x = a(\theta - \sin \theta)$ and $y = a(1 - \cos \theta)$, $0 \le \theta \le \pi$, as indicated in the figure.

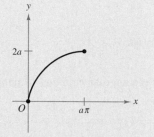

8. Consider the **cornu spiral** given by

$$x(t) = \int_0^t \cos\left(\frac{\pi u^2}{2}\right) du \quad \text{and} \quad y(t) = \int_0^t \sin\left(\frac{\pi u^2}{2}\right) du.$$

(a) Use a graphing utility to graph the spiral over the interval $-\pi \le t \le \pi$.

(b) Show that the cornu spiral is symmetric with respect to the origin.

(c) Find the length of the cornu spiral from $t = 0$ to $t = a$. What is the length of the spiral from $t = -\pi$ to $t = \pi$?

9. A particle is moving along the path described by the parametric equations

$$x = \frac{1}{t} \quad \text{and} \quad y = \frac{\sin t}{t}, \quad 1 \le t < \infty,$$

as indicated in the figure. Find the length of this path.

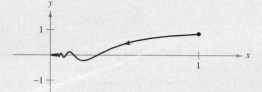

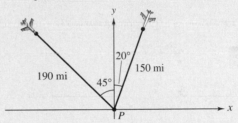

10. Let a and b be positive constants. Find the area of the region in the first quadrant bounded by the graph of the polar equation

$$r = \frac{ab}{(a \sin \theta + b \cos \theta)}, \quad 0 \le \theta \le \frac{\pi}{2}.$$

11. Consider the right triangle in the figure.

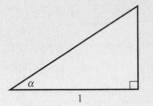

  (a) Show that the area of the triangle is

$$A(\alpha) = \frac{1}{2} \int_0^\alpha \sec^2 \theta \, d\theta.$$

  (b) Show that $\tan \alpha = \int_0^\alpha \sec^2 \theta \, d\theta.$

  (c) Use part (b) to derive the formula for the derivative of the tangent function.

12. Determine the polar equation of the set of all points (r, θ), the product of whose distances from the points $(1, 0)$ and $(-1, 0)$ is equal to 1, as indicated in the figure.

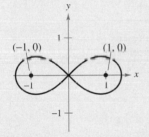

13. Four dogs are located at the corners of a square with sides of length d. The dogs all move counterclockwise at the same speed directly toward the next dog, as indicated in the figure. Find the polar equation of a dog's path as it spirals toward the center of the square.

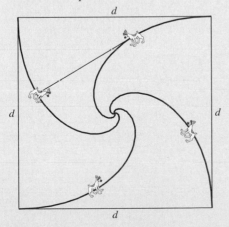

14. Use a graphing utility to graph the polar equation $r = 2 + k \cos \theta$ for $k = 0, 1, 2,$ and 3. Identify each graph.

15. A controller spots two planes at the same altitude flying toward each other (see figure). Their flight paths are S 20° W and S 45° E. One plane is 150 miles from point P with a speed of 375 miles per hour. The other is 190 miles from point P with a speed of 450 miles per hour.

  (a) Find parametric equations for the path of each plane where t is the time in hours, with $t = 0$ corresponding to the time at which the air traffic controller spots the planes.

  (b) Use the result in part (a) to write the distance between the planes as a function of t.

  (c) Use a graphing utility to graph the function in part (b). When will the distance between the planes be minimum? If the planes must keep a separation of at least 3 miles, is the requirement met?

16. Use a graphing utility to produce the curve shown below. The curve is given by

$$r = e^{\cos \theta} - 2 \cos 4\theta + \sin^5 \frac{\theta}{12}.$$

Over what interval must θ vary to produce the curve?

FOR FURTHER INFORMATION For more information on this curve, see the article "A Study in Step Size" by Temple H. Fay in *Mathematics Magazine*. To view this article, go to the website *www.matharticles.com*.

17. Use a graphing utility to graph the polar equation

$$r = \cos 5\theta + n \cos \theta$$

for $0 \le \theta < \pi$ for the integers $n = -5$ to $n = 5$. What values of n produce the "heart" portion of the curve? What values of n produce the "bell" portion? (This curve, created by Michael W. Chamberlin, appeared in *The College Mathematics Journal*.)

Appendices

The remaining appendices are located on the website that accompanies this text at *college.hmco.com*.

Additional Topics in Differential Equations A

- Use a slope field to sketch solutions of a differential equation.
- Use Euler's Method to approximate a solution of a differential equation.
- Solve a first-order linear differential equation.

Slope Fields

In this appendix, you will study two techniques for approximating solutions of differential equations of the form $y' = F(x, y)$. The first technique is a graphical approach that uses **slope fields,** or *direction fields*. The second technique is a numerical approach and is called *Euler's method*.

Consider a differential equation of the form

$$y' = F(x, y). \qquad \text{Differential equation}$$

You can interpret this differential equation graphically to mean that the slope of the graph of each solution at the point (x, y) is y'. You can use a slope field to visualize the family of solutions. To sketch a slope field, pick several points (x, y) and draw short line segments with slope $F(x, y)$. The slope field shows the general shape of all the solutions. An initial condition is needed to sketch a particular solution, as shown in Example 1.

Example 1   **Sketching a Solution Using a Slope Field**

Sketch a slope field for the differential equation

$$y' = 2x + y.$$

Use the slope field to sketch the solution that passes through the point $(1, 1)$.

Solution   Make a table showing the slope at several points. The table shown is a small sample. The slope at many other points should be calculated to get a representative slope field. Next draw line segments at the points with their respective slopes, as shown in Figure A.1.

x	-2	-2	-1	-1	0	0	1	1	2	2
y	-1	1	-1	1	-1	1	-1	1	-1	1
$y' = 2x + y$	-5	-3	-3	-1	-1	1	1	3	3	5

After the slope field is drawn, start at the initial point $(1, 1)$ and move to the right in the direction of the line segment. Continue to draw the solution curve so that it moves parallel to the line segments. Do the same to the left of $(1, 1)$. The resulting solution is shown in Figure A.2.

NOTE   Drawing a slope field by hand is tedious. In practice, slope fields are usually drawn using a graphing utility.

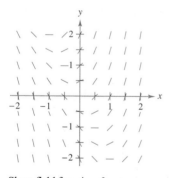

Slope field for $y' = 2x + y$
Figure A.1

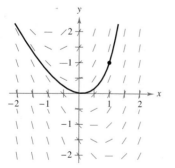

Particular solution for $y' = 2x + y$ passing through $(1, 1)$
Figure A.2

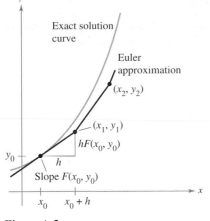

Figure A.3

Euler's Method

Euler's Method is a numerical approach to approximate the particular solution of the differential equation $y' = F(x, y)$ that passes through the point (x_0, y_0). From the given information, you know that the graph of the solution passes through the point (x_0, y_0) and has a slope of $F(x_0, y_0)$ at this point. This gives you a "starting point" for approximating the solution.

From this starting point, you can proceed in the direction indicated by the slope. Using a small step h, move along the tangent line until you arrive at the point (x_1, y_1), where

$$x_1 = x_0 + h \quad \text{and} \quad y_1 = y_0 + hF(x_0, y_0)$$

as shown in Figure A.3. If you think of (x_1, y_1) as a new starting point, you can repeat the process to obtain a second point (x_2, y_2). The values of x_i and y_i are as follows.

$$x_1 = x_0 + h \qquad\qquad y_1 = y_0 + hF(x_0, y_0)$$
$$x_2 = x_1 + h \qquad\qquad y_2 = y_1 + hF(x_1, y_1)$$
$$\vdots \qquad\qquad\qquad \vdots$$
$$x_n = x_{n-1} + h \qquad\qquad y_n = y_{n-1} + hF(x_{n-1}, y_{n-1})$$

NOTE   You can obtain better approximations to the exact solution by choosing smaller and smaller step sizes.

Example 2   Approximating a Solution Using Euler's Method

Use Euler's Method to approximate the particular solution of the differential equation

$$y' = x - y$$

passing through $(0, 1)$. Use a step of $h = 0.1$.

Solution   Using $h = 0.1$, $x_0 = 0$, $y_0 = 1$, and $F(x, y) = x - y$, you have $x_0 = 0$, $x_1 = 0.1$, $x_2 = 0.2$, $x_3 = 0.3, \ldots,$ and

$$y_1 = y_0 + hF(x_0, y_0) = 1 + (0.1)(0 - 1) = 0.9$$
$$y_2 = y_1 + hF(x_1, y_1) = 0.9 + (0.1)(0.1 - 0.9) = 0.82$$
$$y_3 = y_2 + hF(x_2, y_2) = 0.82 + (0.1)(0.2 - 0.82) = 0.758.$$

The first ten approximations are shown in the table. You can plot these values to see a graph of the approximate solution, as shown in Figure A.4.

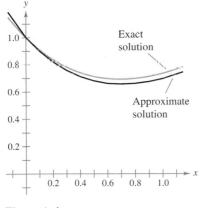

Figure A.4

n	0	1	2	3	4	5	6	7	8	9	10
x_n	0	0.1	0.2	0.3	0.4	0.5	0.6	0.7	0.8	0.9	1.0
y_n	1	0.900	0.820	0.758	0.712	0.681	0.663	0.657	0.661	0.675	0.697

NOTE   For the differential equation in Example 2, you can find the exact solution to be $y = x - 1 + 2e^{-x}$. Figure A.4 compares this exact solution with the approximate solution obtained in Example 2.

First-Order Linear Differential Equations

As a final topic in this appendix, you will learn how to solve a very important class of first-order differential equations—first-order *linear* differential equations.

Definition of a First-Order Linear Differential Equation

A first-order linear differential equation is an equation of the form

$$\frac{dy}{dx} + P(x)y = Q(x)$$

where P and Q are continuous functions of x. This first-order linear differential equation is said to be in **standard form.**

To solve a first-order linear differential equation, you can use an *integrating factor* $u(x)$, which converts the left side into the derivative of the product $u(x)y$. That is, you need a factor $u(x)$ such that

$$u(x)\frac{dy}{dx} + u(x)P(x)y = \frac{d[u(x)y]}{dx}$$

$$u(x)y' + u(x)P(x)y = u(x)y' + yu'(x)$$

$$u(x)P(x)y = yu'(x)$$

$$P(x) = \frac{u'(x)}{u(x)}$$

$$\ln|u(x)| = \int P(x)\,dx + C_1$$

$$u(x) = Ce^{\int P(x)\,dx}.$$

Because you don't need the most general integrating factor, let $C = 1$. Multiplying the original equation $y' + P(x)y = Q(x)$ by $u(x) = e^{\int P(x)dx}$ produces

$$y'e^{\int P(x)\,dx} + yP(x)e^{\int P(x)\,dx} = Q(x)e^{\int P(x)\,dx}$$

$$\frac{d}{dx}\left[ye^{\int P(x)\,dx}\right] = Q(x)e^{\int P(x)\,dx}.$$

The general solution is given by

$$ye^{\int P(x)\,dx} = \int Q(x)e^{\int P(x)\,dx}\,dx + C.$$

THEOREM A.1   Solution of a First-Order Linear Differential Equation

An integrating factor for the first-order linear differential equation

$$y' + P(x)y = Q(x)$$

is $u(x) = e^{\int P(x)\,dx}$. The solution of the differential equation is

$$ye^{\int P(x)\,dx} = \int Q(x)e^{\int P(x)\,dx}\,dx + C.$$

STUDY TIP Rather than memorizing this formula, just remember that multiplication by the integrating factor $e^{\int P(x)\,dx}$ converts the left side of the differential equation into the derivative of the product $ye^{\int P(x)\,dx}$.

Example 3   **Solving a First-Order Linear Differential Equation**

Find the general solution of $xy' - 2y = x^2$.

Solution   The *standard form* of the given equation is

$$y' + P(x)y = Q(x)$$

$$y' - \left(\frac{2}{x}\right)y = x. \qquad \text{Standard form}$$

So, $P(x) = -2/x$, and you have

$$\int P(x)\, dx = -\int \frac{2}{x}\, dx = -\ln x^2$$

$$e^{\int P(x)\, dx} = e^{-\ln x^2} = \frac{1}{x^2}. \qquad \text{Integrating factor}$$

Therefore, multiplying both sides of the standard form by $1/x^2$ yields

$$\frac{y'}{x^2} - \frac{2y}{x^3} = \frac{1}{x}$$

$$\frac{d}{dx}\left[\frac{y}{x^2}\right] = \frac{1}{x}$$

$$\frac{y}{x^2} = \int \frac{1}{x}\, dx$$

$$\frac{y}{x^2} = \ln|x| + C$$

$$y = x^2(\ln|x| + C). \qquad \text{General solution}$$

Several solution curves (for $C = -2, -1, 0, 1, 2, 3,$ and 4) are shown in Figure A.5.

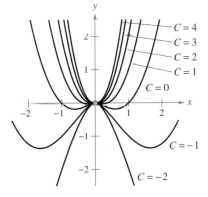

Figure A.5

Example 4   **Solving a First-Order Linear Differential Equation**

Find the general solution of $y' - y\tan t = 1$, $-\dfrac{\pi}{2} < t < \dfrac{\pi}{2}$.

Solution   The equation is already in the standard form $y' + P(t)y = Q(t)$. So,

$$\int P(t)\, dt = -\int \tan t\, dt = \ln|\cos t|$$

which implies that the integrating factor is $e^{\int P(t)\, dt} = e^{\ln|\cos t|} = |\cos t|$.

A quick check shows that $\cos t$ is also an integrating factor. So, multiplying $y' - y\tan t = 1$ by $\cos t$ produces

$$\frac{d}{dt}[y\cos t] = \cos t$$

$$y\cos t = \int \cos t\, dt$$

$$y\cos t = \sin t + C$$

$$y = \tan t + C\sec t. \qquad \text{General solution}$$

Several solution curves are shown in Figure A.6.

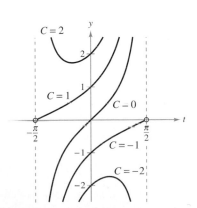

Figure A.6

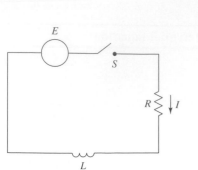

Figure A.7

Application

A simple electrical circuit consists of electric current I (in amperes), a resistance R (in ohms), an inductance L (in henrys), and a constant electromotive force E (in volts), as shown in Figure A.7. According to Kirchhoff's Second Law, if the switch S is closed when $t = 0$, the applied electromotive force (voltage) is equal to the sum of the voltage drops in the rest of the circuit. This in turn means that the current I satisfies the differential equation

$$L\frac{dI}{dt} + RI = E.$$

Example 5   An Electric Circuit Problem

Find the current I as a function of time t (in seconds), given that I satisfies the differential equation $L(dI/dt) + RI = \sin 2t$, where R and L are nonzero constants.

Solution   In standard form, the given linear equation is

$$\frac{dI}{dt} + \frac{R}{L}I = \frac{1}{L}\sin 2t.$$

Let $P(t) = R/L$, so that $e^{\int P(t)\,dt} = e^{(R/L)t}$, and, by Theorem A.1,

$$Ie^{(R/L)t} = \frac{1}{L}\int e^{(R/L)t}\sin 2t\,dt = \frac{1}{4L^2 + R^2}e^{(R/L)t}(R\sin 2t - 2L\cos 2t) + C.$$

So, the general solution is

$$I = e^{-(R/L)t}\left[\frac{1}{4L^2 + R^2}e^{(R/L)t}(R\sin 2t - 2L\cos 2t) + C\right]$$

$$I = \frac{1}{4L^2 + R^2}(R\sin 2t - 2L\cos 2t) + Ce^{-(R/L)t}.$$

EXERCISES FOR APPENDIX A

In Exercises 1 and 2, a differential equation and its slope field are given. Determine the slope (if possible) in the slope field at the points given in the table.

x	-4	-2	0	2	4	8
y	2	0	4	4	6	8
dy/dx						

1. $\dfrac{dy}{dx} = \dfrac{x}{y}$

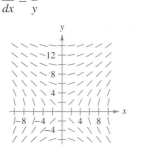

2. $\dfrac{dy}{dx} = x\cos\dfrac{\pi y}{8}$

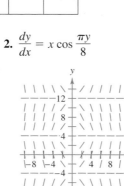

In Exercises 3–6, (a) sketch an approximate solution of the differential equation satisfying the initial condition by hand on the slope field, (b) find the particular solution that satisfies the initial condition, and (c) use a graphing utility to graph the particular solution. Compare the graph with the hand-drawn graph of part (a).

Differential Equation	*Initial Condition*
3. $\dfrac{dy}{dx} = e^x - y$	$(0, 1)$
4. $y' + 2y = \sin x$	$(0, 4)$

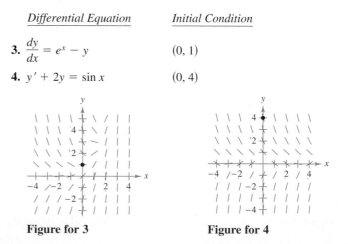

Figure for 3   Figure for 4

Differential Equation   *Initial Condition*

5. $y' = \csc x + y \cot x$ $(1, 1)$

6. $y' = \csc x - y \cot x$ $(1, 2)$

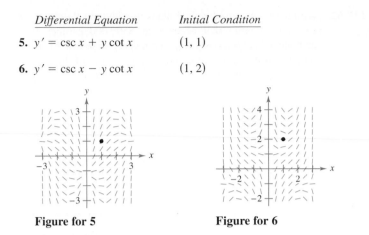

Figure for 5   Figure for 6

In Exercises 7 and 8, use a computer algebra system to sketch the slope field for the differential equation and graph the solution satisfying the specified initial condition.

7. $\dfrac{dy}{dx} = 0.4y(3 - x), \quad y(0) = 1$

8. $\dfrac{dy}{dx} = \dfrac{1}{2}e^{-x/8} \sin \dfrac{\pi y}{4}, \quad y(0) = 2$

Euler's Method   In Exercise 9–14, use Euler's method to make a table of values for the approximate solution of the differential equation with the specified initial value. Use *n* steps of size *h*.

9. $y' = x + y, \quad y(0) = 2, \quad n = 10, \quad h = 0.1$

10. $y' = x + y, \quad y(0) = 2, \quad n = 20, \quad h = 0.05$

11. $y' = 3x - 2y, \quad y(0) = 3, \quad n = 10, \quad h = 0.05$

12. $y' = 0.5x(3 - y), \quad y(0) = 1, \quad n = 5, \quad h = 0.4$

13. $y' = e^{xy}, \quad y(0) = 1, \quad n = 10, \quad h = 0.1$

14. $y' = \cos x + \sin y, \quad y(0) = 5, \quad n = 10, \quad h = 0.1$

True or False?   In Exercises 15 and 16, determine whether the statement is true or false. If it is false, explain why or give an example that shows it is false.

15. $y' + x\sqrt{y} = x^2$ is a first-order linear differential equation.

16. $y' + xy = e^x y$ is a first-order linear differential equation.

In Exercises 17–32, solve the first-order linear differential equation.

17. $\dfrac{dy}{dx} + \left(\dfrac{1}{x}\right)y = 3x + 4$

18. $\dfrac{dy}{dx} + \left(\dfrac{2}{x}\right)y = 3x + 1$

19. $\dfrac{dy}{dx} - 3x^2 y = e^{x^3}$

20. $\dfrac{dy}{dx} - \dfrac{3y}{x^2} = \dfrac{1}{x^2}$

21. $y' - y = \cos x$

22. $y' + 2xy = 2x$

23. $(x + y)\,dx - x\,dy = 0$

24. $(2y - e^x)\,dx + x\,dy = 0$

25. $(3y + \sin 2x)\,dx - dy = 0$

26. $(y - 1)\sin x\,dx - dy = 0$

27. $(x - 1)y' + y = x^2 - 1$

28. $y' + 5y = e^{5x}$

29. $dy = (y \tan x + 2e^x)\,dx$

30. $xy' + y = \sin x$

31. $xy' - ay = bx^4$

32. $y' = y + 2x(y - e^x)$

In Exercises 33–40, find the particular solution of the differential equation that satisfies the boundary condition.

Differential Equation   *Boundary Condition*

33. $y' \cos^2 x + y - 1 = 0$ $y(0) = 5$

34. $x^3 y' + 2y = e^{1/x^2}$ $y(1) = e$

35. $y' + y \tan x = \sec x + \cos x$ $y(0) = 1$

36. $y' + y \sec x = \sec x$ $y(0) = 4$

37. $y' + \left(\dfrac{1}{x}\right)y = 0$ $y(2) = 2$

38. $y' + (2x - 1)y = 0$ $y(1) = 2$

39. $x\,dy = (x + y + 2)\,dx$ $y(1) = 10$

40. $2xy' - y = x^3 - x$ $y(4) = 2$

In Exercises 41 and 42, (a) use a graphing utility to graph the slope field for the differential equation, (b) find the particular solutions of the differential equation passing through the specified points, and (c) use a graphing utility to graph the particular solutions on the slope field.

Differential Equation   *Points*

41. $\dfrac{dy}{dx} - \dfrac{1}{x}y = x^2$ $(-2, 4), \ (2, 8)$

42. $\dfrac{dy}{dx} + (\cot x)y = x$ $(1, 1), \ (3, -1)$

Electrical Circuits   In Exercises 43–46, use the differential equation for electrical circuits given by

$$L\dfrac{dI}{dt} + RI = E.$$

In this equation, I is the current, R is the resistance, L is the inductance, and E is the electromotive force (voltage).

43. Solve the differential equation given a constant voltage E_0.

44. Use the result of Exercise 43 to find the equation for the current if $I(0) = 0$, $E_0 = 110$ volts, $R = 550$ ohms, and $L = 4$ henrys. When does the current reach 90% of its limiting value?

45. Solve the differential equation given a periodic electromotive force $E_0 \sin \omega t$.

46. Verify that the solution of Exercise 45 can be written in the form

$$I = ce^{-(R/L)t} + \frac{E_0}{\sqrt{R^2 + \omega^2 L^2}} \sin(\omega t + \phi)$$

where ϕ, the phase angle, is given by $\arctan(-\omega L/R)$. (Note that the exponential term approaches 0 as $t \to \infty$. This implies that the current approaches a periodic function.)

47. *Population Growth* When predicting population growth, demographers must consider birth and death rates as well as the net change caused by the difference between the rates of immigration and emigration. Let P be the population at time t and let N be the net increase per unit time resulting from the difference between immigration and emigration. So, the rate of growth of the population is given by

$$\frac{dP}{dt} = kP + N, \qquad N \text{ is constant.}$$

Solve this differential equation to find P as a function of time if at time $t = 0$ the size of the population is P_0.

48. *Investment Growth* A large corporation starts at time $t = 0$ to continuously invest part of its receipts at a rate of P dollars per year in a fund for future corporate expansion. Assume that the fund earns r percent interest per year compounded continuously. So, the rate of growth of the amount A in the fund is given by

$$\frac{dA}{dt} = rA + P$$

where $A = 0$ when $t = 0$. Solve this differential equation for A as a function of t.

Investment Growth In Exercises 49 and 50, use the result of Exercise 48.

49. Find A for the following.

(a) $P = \$100,000$, $r = 6\%$, and $t = 5$ years

(b) $P = \$250,000$, $r = 5\%$, and $t = 10$ years

50. Find t if the corporation needs $\$800,000$ and it can invest $\$75,000$ per year in a fund earning 8% interest compounded continuously.

51. *Investment* Let $A(t)$ be the amount in a fund earning interest at an annual rate r compounded continuously. If a continuous cash flow of P dollars per year is withdrawn from the fund, the rate of change of A is given by the differential equation

$$\frac{dA}{dt} = rA - P$$

where $A = A_0$ when $t = 0$. Solve this differential equation for A as a function of t.

52. *Investment* A retired couple plans to withdraw P dollars per year from a retirement account of $\$500,000$ earning 10% compounded continuously. Use the result of Exercise 51 and a graphing utility to graph the function A for each of the following continuous annual cash flows. Use the graphs to describe what happens to the balance in the fund for each of the cases.

(a) $P = \$40,000$

(b) $P = \$50,000$

(c) $P = \$60,000$

53. *Intravenous Feeding* Glucose is added intravenously to the bloodstream at the rate of q units per minute, and the body removes glucose from the bloodstream at a rate proportional to the amount present. Assume $Q(t)$ is the amount of glucose in the bloodstream at time t.

(a) Determine the differential equation describing the rate of change with respect to time of glucose in the bloodstream.

(b) Solve the differential equation from part (a), letting $Q = Q_0$ when $t = 0$.

(c) Find the limit of $Q(t)$ as $t \to \infty$.

54. *Learning Curve* The management at a certain factory has found that the maximum number of units a worker can produce in a day is 30. The rate of increase in the number of units N produced with respect to time t in days by a new employee is proportional to $30 - N$.

(a) Determine the differential equation describing the rate of change of performance with respect to time.

(b) Solve the differential equation from part (a).

(c) Find the particular solution for a new employee who produced ten units on the first day at the factory and 19 units on the twentieth day.

In Exercises 55–58, match the differential equation with its solution.

Differential Equation	Solution
55. $y' - 2x = 0$	(a) $y = Ce^{x^2}$
56. $y' - 2y = 0$	(b) $y = -\frac{1}{2} + Ce^{x^2}$
57. $y' - 2xy = 0$	(c) $y = x^2 + C$
58. $y' - 2xy = x$	(d) $y = Ce^{2x}$

Proofs of Selected Theorems

B

> **THEOREM 1.2   Properties of Limits (Properties 2, 3, 4, and 5) (page 57)**
>
> Let b and c be real numbers, let n be a positive integer, and let f and g be functions with the following limits.
>
> $$\lim_{x \to c} f(x) = L \quad \text{and} \quad \lim_{x \to c} g(x) = K$$
>
> **2.** Sum or difference:　　$\lim\limits_{x \to c} [f(x) \pm g(x)] = L \pm K$
>
> **3.** Product:　　　　　　　$\lim\limits_{x \to c} [f(x)g(x)] = LK$
>
> **4.** Quotient:　　　　　　　$\lim\limits_{x \to c} \dfrac{f(x)}{g(x)} = \dfrac{L}{K}, \quad$ provided $K \neq 0$
>
> **5.** Power:　　　　　　　　$\lim\limits_{x \to c} [f(x)]^n = L^n$

Proof   To prove Property 2, choose $\varepsilon > 0$. Because $\varepsilon/2 > 0$, you know that there exists $\delta_1 > 0$ such that $0 < |x - c| < \delta_1$ implies $|f(x) - L| < \varepsilon/2$. You also know that there exists $\delta_2 > 0$ such that $0 < |x - c| < \delta_2$ implies $|g(x) - K| < \varepsilon/2$. Let δ be the smaller of δ_1 and δ_2; then $0 < |x - c| < \delta$ implies that

$$|f(x) - L| < \frac{\varepsilon}{2} \quad \text{and} \quad |g(x) - K| < \frac{\varepsilon}{2}.$$

So, you can apply the Triangle Inequality to conclude that

$$|[f(x) + g(x)] - (L + K)| \leq |f(x) - L| + |g(x) - K| < \frac{\varepsilon}{2} + \frac{\varepsilon}{2} = \varepsilon$$

which implies that

$$\lim_{x \to c} [f(x) + g(x)] = L + K = \lim_{x \to c} f(x) + \lim_{x \to c} g(x).$$

The proof that

$$\lim_{x \to c} [f(x) - g(x)] = L - K$$

is similar.

To prove Property 3, given that

$$\lim_{x \to c} f(x) = L \quad \text{and} \quad \lim_{x \to c} g(x) = K$$

you can write

$$f(x)g(x) = [f(x) - L][g(x) - K] + [Lg(x) + Kf(x)] - LK.$$

Because the limit of $f(x)$ is L, and the limit of $g(x)$ is K, you have

$$\lim_{x \to c} [f(x) - L] = 0 \quad \text{and} \quad \lim_{x \to c} [g(x) - K] = 0.$$

Let $0 < \varepsilon < 1$. Then there exists $\delta > 0$ such that if $0 < |x - c| < \delta$, then

$$|f(x) - L - 0| < \varepsilon \quad \text{and} \quad |g(x) - K - 0| < \varepsilon$$

which implies that

$$|[f(x) - L][g(x) - K] - 0| = |f(x) - L| \, |g(x) - K| < \varepsilon\varepsilon < \varepsilon.$$

Hence,

$$\lim_{x \to c} [f(x) - L][g(x) - K] = 0.$$

Furthermore, by Property 1, you have

$$\lim_{x \to c} Lg(x) = LK \quad \text{and} \quad \lim_{x \to c} Kf(x) = KL.$$

Finally, by Property 2, you obtain

$$\lim_{x \to c} f(x)g(x) = \lim_{x \to c} [f(x) - L][g(x) - K] + \lim_{x \to c} Lg(x) + \lim_{x \to c} Kf(x) - \lim_{x \to c} LK$$

$$= 0 + LK + KL - LK$$

$$= LK.$$

To prove Property 4, note that it is sufficient to prove that

$$\lim_{x \to c} \frac{1}{g(x)} = \frac{1}{K}.$$

Then you can use Property 3 to write

$$\lim_{x \to c} \frac{f(x)}{g(x)} = \lim_{x \to c} f(x) \frac{1}{g(x)} = \lim_{x \to c} f(x) \cdot \lim_{x \to c} \frac{1}{g(x)} = \frac{L}{K}.$$

Let $\varepsilon > 0$. Because $\lim_{x \to c} g(x) = K$, there exists $\delta_1 > 0$ such that if

$$0 < |x - c| < \delta_1, \text{ then } |g(x) - K| < \frac{|K|}{2}$$

which implies that

$$|K| = |g(x) + [|K| - g(x)]| \le |g(x)| + ||K| - g(x)| < |g(x)| + \frac{|K|}{2}.$$

That is, for $0 < |x - c| < \delta_1$,

$$\frac{|K|}{2} < |g(x)| \quad \text{or} \quad \frac{1}{|g(x)|} < \frac{2}{|K|}.$$

Similarly, there exists a $\delta_2 > 0$ such that if $0 < |x - c| < \delta_2$, then

$$|g(x) - K| < \frac{|K|^2}{2} \varepsilon.$$

Let δ be the smaller of δ_1 and δ_2. For $0 < |x - c| < \delta$, you have

$$\left| \frac{1}{g(x)} - \frac{1}{K} \right| = \left| \frac{K - g(x)}{g(x)K} \right| = \frac{1}{|K|} \cdot \frac{1}{|g(x)|} |K - g(x)| < \frac{1}{|K|} \cdot \frac{2}{|K|} \frac{|K|^2}{2} \varepsilon = \varepsilon.$$

So, $\lim_{x \to c} \frac{1}{g(x)} = \frac{1}{K}.$

Finally, the proof of Property 5 can be obtained by a straightforward application of mathematical induction coupled with Property 3.   ◿

THEOREM 1.4   The Limit of a Function Involving a Radical (page 58)

Let n be a positive integer. The following limit is valid for all c if n is odd, and is valid for $c > 0$ if n is even.

$$\lim_{x \to c} \sqrt[n]{x} = \sqrt[n]{c}.$$

Proof   Consider the case for which $c > 0$ and n is any positive integer. For a given $\varepsilon > 0$, you need to find $\delta > 0$ such that

$$\left| \sqrt[n]{x} - \sqrt[n]{c} \right| < \varepsilon \quad \text{whenever} \quad 0 < |x - c| < \delta$$

which is the same as saying

$$-\varepsilon < \sqrt[n]{x} - \sqrt[n]{c} < \varepsilon \quad \text{whenever} \quad -\delta < x - c < \delta.$$

Assume $\varepsilon < \sqrt[n]{c}$, which implies that $0 < \sqrt[n]{c} - \varepsilon < \sqrt[n]{c}$. Now, let δ be the smaller of the two numbers.

$$c - \left(\sqrt[n]{c} - \varepsilon \right)^n \quad \text{and} \quad \left(\sqrt[n]{c} + \varepsilon \right)^n - c$$

Then you have

$$-\delta < x - c \qquad\qquad < \delta$$
$$-\left[c - \left(\sqrt[n]{c} - \varepsilon \right)^n \right] < x - c \qquad < \left(\sqrt[n]{c} + \varepsilon \right)^n - c$$
$$\left(\sqrt[n]{c} - \varepsilon \right)^n - c < x - c \qquad < \left(\sqrt[n]{c} + \varepsilon \right)^n - c$$
$$\left(\sqrt[n]{c} - \varepsilon \right)^n < x \qquad\qquad < \left(\sqrt[n]{c} + \varepsilon \right)^n$$
$$\sqrt[n]{c} - \varepsilon < \sqrt[n]{x} \qquad\qquad < \sqrt[n]{c} + \varepsilon$$
$$-\varepsilon < \sqrt[n]{x} - \sqrt[n]{c} < \varepsilon.$$

THEOREM 1.5   The Limit of a Composite Function (page 59)

If f and g are functions such that $\lim_{x \to c} g(x) = L$ and $\lim_{x \to L} f(x) = f(L)$, then

$$\lim_{x \to c} f(g(x)) = f\left(\lim_{x \to c} g(x) \right) = f(L).$$

Proof   For a given $\varepsilon > 0$, you must find $\delta > 0$ such that

$$|f(g(x)) - f(L)| < \varepsilon \quad \text{whenever} \quad 0 < |x - c| < \delta.$$

Because the limit of $f(x)$ as $x \to L$ is $f(L)$, you know there exists $\delta_1 > 0$ such that

$$|f(u) - f(L)| < \varepsilon \quad \text{whenever} \quad |u - L| < \delta_1.$$

Moreover, because the limit of $g(x)$ as $x \to c$ is L, you know there exists $\delta > 0$ such that

$$|g(x) - L| < \delta_1 \quad \text{whenever} \quad 0 < |x - c| < \delta.$$

Finally, letting $u = g(x)$, you have

$$|f(g(x)) - f(L)| < \varepsilon \quad \text{whenever} \quad 0 < |x - c| < \delta.$$

> **THEOREM 1.7   Functions That Agree at All But One Point (page 60)**
>
> Let c be a real number and let $f(x) = g(x)$ for all $x \neq c$ in an open interval containing c. If the limit of $g(x)$ as x approaches c exists, then the limit of $f(x)$ also exists and
>
> $$\lim_{x \to c} f(x) = \lim_{x \to c} g(x).$$

Proof   Let L be the limit of $g(x)$ as $x \to c$. Then, for each $\varepsilon > 0$ there exists a $\delta > 0$ such that $f(x) = g(x)$ in the open intervals $(c - \delta, c)$ and $(c, c + \delta)$, and

$$|g(x) - L| < \varepsilon \quad \text{whenever} \quad 0 < |x - c| < \delta.$$

Because $f(x) = g(x)$ for all x in the open interval other than $x = c$, it follows that

$$|f(x) - L| < \varepsilon \quad \text{whenever} \quad 0 < |x - c| < \delta.$$

So, the limit of $f(x)$ as $x \to c$ is also L.   ◢

> **THEOREM 1.8   The Squeeze Theorem (page 63)**
>
> If $h(x) \leq f(x) \leq g(x)$ for all x in an open interval containing c, except possibly at c itself, and if
>
> $$\lim_{x \to c} h(x) = L = \lim_{x \to c} g(x)$$
>
> then $\lim_{x \to c} f(x)$ exists and is equal to L.

Proof   For $\varepsilon > 0$ there exist δ_1 and δ_2 such that

$$|h(x) - L| < \varepsilon \quad \text{whenever} \quad 0 < |x - c| < \delta_1$$

and

$$|g(x) - L| < \varepsilon \quad \text{whenever} \quad 0 < |x - c| < \delta_2.$$

Because $h(x) \leq f(x) \leq g(x)$ for all x in an open interval containing c, except possibly at c itself, there exists $\delta_3 > 0$ such that $h(x) \leq f(x) \leq g(x)$ for $0 < |x - c| < \delta_3$. Let δ be the smallest of δ_1, δ_2, and δ_3. Then, if $0 < |x - c| < \delta$, it follows that $|h(x) - L| < \varepsilon$ and $|g(x) - L| < \varepsilon$, which implies that

$$-\varepsilon < h(x) - L < \varepsilon \quad \text{and} \quad -\varepsilon < g(x) - L < \varepsilon$$

$$L - \varepsilon < h(x) \quad \text{and} \quad g(x) < L + \varepsilon.$$

Now, because $h(x) \leq f(x) \leq g(x)$, it follows that $L - \varepsilon < f(x) < L + \varepsilon$, which implies that $|f(x) - L| < \varepsilon$. Therefore,

$$\lim_{x \to c} f(x) = L.$$ ◢

> **THEOREM 1.14   Vertical Asymptotes (page 82)**
>
> Let f and g be continuous on an open interval containing c. If $f(c) \neq 0$, $g(c) = 0$, and there exists an open interval containing c such that $g(x) \neq 0$ for all $x \neq c$ in the interval, then the graph of the function given by
>
> $$h(x) = \frac{f(x)}{g(x)}$$
>
> has a vertical asymptote at $x = c$.

Proof   Consider the case for which $f(c) > 0$, and there exists $b > c$ such that $c < x < b$ implies $g(x) > 0$. Then for $M > 0$, choose δ_1 such that

$$0 < x - c < \delta_1 \quad \text{implies that} \quad \frac{f(c)}{2} < f(x) < \frac{3f(c)}{2}$$

and δ_2 such that

$$0 < x - c < \delta_2 \quad \text{implies that} \quad 0 < g(x) < \frac{f(c)}{2M}.$$

Now let δ be the smaller of δ_1 and δ_2. Then it follows that

$$0 < x - c < \delta \quad \text{implies that} \quad \frac{f(x)}{g(x)} > \frac{f(c)}{2}\left[\frac{2M}{f(c)}\right] = M.$$

Therefore, it follows that

$$\lim_{x \to c^+} \frac{f(x)}{g(x)} = \infty$$

and the line $x = c$ is a vertical asymptote of the graph of h.

> **Alternative Form of the Derivative (page 99)**
>
> The derivative of f at c is given by
>
> $$f'(c) = \lim_{x \to c} \frac{f(x) - f(c)}{x - c}$$
>
> provided this limit exists.

Proof   The derivative of f at c is given by

$$f'(c) = \lim_{\Delta x \to 0} \frac{f(c + \Delta x) - f(c)}{\Delta x}.$$

Let $x = c + \Delta x$. Then $x \to c$ as $\Delta x \to 0$. So, replacing $c + \Delta x$ by x, you have

$$f'(c) = \lim_{\Delta x \to 0} \frac{f(c + \Delta x) - f(c)}{\Delta x} = \lim_{x \to c} \frac{f(x) - f(c)}{x - c}.$$

THEOREM 2.10   The Chain Rule (page 128)

If $y = f(u)$ is a differentiable function of u, and $u = g(x)$ is a differentiable function of x, then $y = f(g(x))$ is a differentiable function of x and

$$\frac{dy}{dx} = \frac{dy}{du} \cdot \frac{du}{dx} \quad \text{or, equivalently,} \quad \frac{d}{dx}[f(g(x))] = f'(g(x))g'(x).$$

Proof   In Section 2.4, we let $h(x) = f(g(x))$ and used the alternative form of the derivative to show that $h'(c) = f'(g(c))g'(c)$, provided $g(x) \neq g(c)$ for values of x other than c. Now consider a more general proof. Begin by considering the derivative of f.

$$f'(x) = \lim_{\Delta x \to 0} \frac{f(x + \Delta x) - f(x)}{\Delta x} = \lim_{\Delta x \to 0} \frac{\Delta y}{\Delta x}$$

For a fixed value of x, define a function η such that

$$\eta(\Delta x) = \begin{cases} 0, & \Delta x = 0 \\ \dfrac{\Delta y}{\Delta x} - f'(x), & \Delta x \neq 0. \end{cases}$$

Because the limit of $\eta(\Delta x)$ as $\Delta x \to 0$ doesn't depend on the value of $\eta(0)$, you have

$$\lim_{\Delta x \to 0} \eta(\Delta x) = \lim_{\Delta x \to 0} \left[\frac{\Delta y}{\Delta x} - f'(x) \right] = 0$$

and you can conclude that η is continuous at 0. Moreover, because $\Delta y = 0$ when $\Delta x = 0$, the equation

$$\Delta y = \Delta x \eta(\Delta x) + \Delta x f'(x)$$

is valid whether Δx is zero or not. Now, by letting $\Delta u = g(x + \Delta x) - g(x)$, you can use the continuity of g to conclude that

$$\lim_{\Delta x \to 0} \Delta u = \lim_{\Delta x \to 0} [g(x + \Delta x) - g(x)] = 0$$

which implies that

$$\lim_{\Delta x \to 0} \eta(\Delta u) = 0.$$

Finally,

$$\Delta y = \Delta u \eta(\Delta u) + \Delta u f'(u) \to \frac{\Delta y}{\Delta x} = \frac{\Delta u}{\Delta x} \eta(\Delta u) + \frac{\Delta u}{\Delta x} f'(u), \quad \Delta x \neq 0$$

and taking the limit as $\Delta x \to 0$, you have

$$\frac{dy}{dx} = \frac{du}{dx} \left[\lim_{\Delta x \to 0} \eta(\Delta u) \right] + \frac{du}{dx} f'(u) = \frac{dy}{dx}(0) + \frac{du}{dx} f'(u)$$

$$= \frac{du}{dx} f'(u) = \frac{du}{dx} \cdot \frac{dy}{du}.$$

Concavity Interpretation (page 184)

1. Let f be differentiable on an open interval I. If the graph of f is concave *upward* on I, then the graph of f lies *above* all of its tangent lines on I.

2. Let f be differentiable on an open interval I. If the graph of f is concave *downward* on I, then the graph of f lies *below* all of its tangent lines on I.

Proof Assume that f is concave upward on $I = (a, b)$. Then, f' is increasing on (a, b). Let c be a point in the interval $I = (a, b)$. The equation of the tangent line to the graph of f at c is given by

$$g(x) = f(c) + f'(c)(x - c).$$

If x is in the open interval (c, b), then the directed distance from point $(x, f(x))$ (on the graph of f) to the point $(x, g(x))$ (on the tangent line) is given by

$$d = f(x) - [f(c) + f'(c)(x - c)]$$
$$= f(x) - f(c) - f'(c)(x - c).$$

Moreover, by the Mean Value Theorem there exists a number z in (c, x) such that

$$f'(z) = \frac{f(x) - f(c)}{x - c}.$$

So, you have

$$d = f(x) - f(c) - f'(c)(x - c)$$
$$= f'(z)(x - c) - f'(c)(x - c)$$
$$= [f'(z) - f'(c)](x - c).$$

The second factor $(x - c)$ is positive because $c < x$. Moreover, because f' is increasing, it follows that the first factor $[f'(z) - f'(c)]$ is also positive. Therefore, $d > 0$ and you can conclude that the graph of f lies above the tangent line at x. If x is in the open interval (a, c), a similar argument can be given. This proves the first statement. The proof of the second statement is similar.

THEOREM 3.10   Limits at Infinity (page 193)

If r is a positive rational number, and c is any real number, then

$$\lim_{x \to \infty} \frac{c}{x^r} = 0.$$

Furthermore, if x^r is defined when $x < 0$, then $\displaystyle\lim_{x \to -\infty} \frac{c}{x^r} = 0.$

Proof Begin by proving that

$$\lim_{x \to \infty} \frac{1}{x} = 0.$$

For $\varepsilon > 0$, let $M = 1/\varepsilon$. Then, for $x > M$, you have

$$x > M = \frac{1}{\varepsilon} \quad \Longrightarrow \quad \frac{1}{x} < \varepsilon \quad \Longrightarrow \quad \left| \frac{1}{x} - 0 \right| < \varepsilon.$$

Therefore, by the definition of a limit at infinity, you can conclude that the limit of $1/x$ as $x \to \infty$ is 0. Now, using this result, and letting $r = m/n$, you can write the following.

$$\lim_{x \to \infty} \frac{c}{x^r} = \lim_{x \to \infty} \frac{c}{x^{m/n}}$$

$$= c \left[\lim_{x \to \infty} \left(\frac{1}{\sqrt[n]{x}} \right)^m \right]$$

$$= c \left(\lim_{x \to \infty} \sqrt[n]{\frac{1}{x}} \right)^m$$

$$= c \left(\sqrt[n]{\lim_{x \to \infty} \frac{1}{x}} \right)^m$$

$$= c \left(\sqrt[n]{0} \right)^m$$

$$= 0$$

The proof of the second part of the theorem is similar.

> **THEOREM 4.2**   **Summation Formulas (page 254)**
>
> **1.** $\displaystyle\sum_{i=1}^{n} c = cn$ **2.** $\displaystyle\sum_{i=1}^{n} i = \frac{n(n+1)}{2}$
>
> **3.** $\displaystyle\sum_{i=1}^{n} i^2 = \frac{n(n+1)(2n+1)}{6}$ **4.** $\displaystyle\sum_{i=1}^{n} i^3 = \frac{n^2(n+1)^2}{4}$

Proof   The proof of Property 1 is straightforward. By adding c to itself n times, you obtain a sum of cn.

To prove Property 2, write the sum in increasing and decreasing order and add corresponding terms as follows.

$$\sum_{i=1}^{n} i = \quad 1 \quad + \quad 2 \quad + \quad 3 \quad + \cdots + (n-1) + \quad n$$

$$\downarrow \qquad\qquad \downarrow \qquad\qquad \downarrow \qquad\qquad \downarrow$$

$$\sum_{i=1}^{n} i = \quad n \quad + (n-1) + (n-2) + \cdots + \quad 2 \quad + \quad 1$$

$$\downarrow \qquad \downarrow \qquad \downarrow \qquad\qquad \downarrow \qquad \downarrow$$

$$2\sum_{i=1}^{n} i = (n+1) + (n+1) + (n+1) + \cdots + (n+1) + (n+1)$$

$$\underbrace{\hspace{8cm}}_{n \text{ terms}}$$

Therefore,

$$\sum_{i=1}^{n} i = \frac{n(n+1)}{2}.$$

To prove Property 3, use mathematical induction. First, if $n = 1$, the result is true because

$$\sum_{i=1}^{1} i^2 = 1^2 = 1 = \frac{1(1+1)(2+1)}{6}.$$

Now, assuming the result is true for $n = k$, you can show that it is true for $n = k + 1$, as follows.

$$\sum_{i=1}^{k+1} i^2 = \sum_{i=1}^{k} i^2 + (k + 1)^2$$

$$= \frac{k(k + 1)(2k + 1)}{6} + (k + 1)^2$$

$$= \frac{k + 1}{6}(2k^2 + k + 6k + 6)$$

$$= \frac{k + 1}{6}[(2k + 3)(k + 2)]$$

$$= \frac{(k + 1)(k + 2)[2(k + 1) + 1]}{6}$$

Property 4 can be proved using a similar argument with mathematical induction. ◪

THEOREM 4.8   Preservation of Inequality (page 272)

1. If f is integrable and nonnegative on the closed interval $[a, b]$, then

$$0 \leq \int_a^b f(x)\, dx.$$

2. If f and g are integrable on the closed interval $[a, b]$, and $f(x) \leq g(x)$ for every x in $[a, b]$, then

$$\int_a^b f(x)\, dx \leq \int_a^b g(x)\, dx.$$

Proof   To prove Property 1, suppose, on the contrary, that

$$\int_a^b f(x)\, dx = I < 0.$$

Then, let $a = x_0 < x_1 < x_2 < \cdots < x_n = b$ be a partition of $[a, b]$, and let

$$R = \sum_{i=1}^{n} f(c_i)\, \Delta x_i$$

be a Riemann sum. Because $f(x) \geq 0$, it follows that $R \geq 0$. Now, for $\|\Delta\|$ sufficiently small, you have $|R - I| < -I/2$, which implies that

$$\sum_{i=1}^{n} f(c_i)\, \Delta x_i = R < I - \frac{I}{2} < 0$$

which is not possible. From this contradiction, you can conclude that

$$0 \leq \int_a^b f(x)\, dx.$$

To prove Property 2 of the theorem, note that $f(x) \le g(x)$ implies that $g(x) - f(x) \ge 0$. Hence, you can apply the result of Property 1 to conclude that

$$0 \le \int_a^b [g(x) - f(x)] \, dx$$

$$0 \le \int_a^b g(x) \, dx - \int_a^b f(x) \, dx$$

$$\int_a^b f(x) \, dx \le \int_a^b g(x) \, dx.$$

◨

Properties of the Natural Logarithmic Function (page 315)

$$\lim_{x \to 0^+} \ln x = -\infty \qquad \text{and} \qquad \lim_{x \to \infty} \ln x = \infty$$

Proof   To begin, show that $\ln 2 \ge \frac{1}{2}$. From the Mean Value Theorem for Integrals, you can write

$$\ln 2 = \int_1^2 \frac{1}{x} \, dx = (2 - 1)\frac{1}{c} = \frac{1}{c}$$

where c is in $[1, 2]$. This implies that

$$1 \le \quad c \quad \le 2$$

$$1 \ge \quad \frac{1}{c} \quad \ge \frac{1}{2}$$

$$1 \ge \ln 2 \ge \frac{1}{2}.$$

Now, let N be any positive (large) number. Because $\ln x$ is increasing, it follows that if $x > 2^{2N}$, then

$$\ln x > \ln 2^{2N} = 2N \ln 2.$$

However, because $\ln 2 \ge \frac{1}{2}$, it follows that

$$\ln x > 2N \ln 2 \ge 2N\left(\frac{1}{2}\right) = N.$$

This verifies the second limit. To verify the first limit, let $z = 1/x$. Then, $z \to \infty$ as $x \to 0^+$, and you can write

$$\lim_{x \to 0^+} \ln x = \lim_{x \to 0^+} \left(-\ln \frac{1}{x}\right)$$

$$= \lim_{z \to \infty} (-\ln z)$$

$$= -\lim_{z \to \infty} \ln z$$

$$= -\infty$$

◨

> ### THEOREM 5.8   Continuity and Differentiability of Inverse Functions
> (page 336)
>
> Let f be a function whose domain is an interval I. If f has an inverse function, then the following statements are true.
>
> **1.** If f is continuous on its domain, then f^{-1} is continuous on its domain.
> **2.** If f is increasing on its domain, then f^{-1} is increasing on its domain.
> **3.** If f is decreasing on its domain, then f^{-1} is decreasing on its domain.
> **4.** If f is differentiable at c and $f'(c) \neq 0$, then f^{-1} is differentiable at $f(c)$.

Proof   To prove Property 1, first show that if f is continuous on I and has an inverse function, then f is strictly monotonic on I. Suppose that f were not strictly monotonic. Then there would exist numbers x_1, x_2, x_3 in I such that $x_1 < x_2 < x_3$, but $f(x_2)$ is not between $f(x_1)$ and $f(x_3)$. Without loss of generality, assume $f(x_1) < f(x_3) < f(x_2)$. By the Intermediate Value Theorem, there exists a number x_0 between x_1 and x_2 such that $f(x_0) = f(x_3)$. So, f is not one-to-one and cannot have an inverse function. Hence, f must be strictly monotonic.

Because f is continuous, the Intermediate Value Theorem implies that the set of values of f,

$$\{f(x) : x \in I\},$$

forms an interval J. Assume that a is an interior point of J. From the previous argument, $f^{-1}(a)$ is an interior point of I. Let $\varepsilon > 0$. There exists $0 < \varepsilon_1 < \varepsilon$ such that

$$I_1 = (f^{-1}(a) - \varepsilon_1, f^{-1}(a) + \varepsilon_1) \subseteq I.$$

Because f is strictly monotonic on I_1, the set of values $\{f(x) : x \in I_1\}$ forms an interval $J_1 \subseteq J$. Let $\delta > 0$ such that $(a - \delta, a + \delta) \subseteq J_1$. Finally, if

$$|y - a| < \delta, \text{ then } |f^{-1}(y) - f^{-1}(a)| < \varepsilon_1 < \varepsilon.$$

Hence, f^{-1} is continuous at a. A similar proof can be given if a is an endpoint.

To prove Property 2, let y_1 and y_2 be in the domain of f^{-1}, with $y_1 < y_2$. Then, there exist x_1 and x_2 in the domain of f such that

$$f(x_1) = y_1 < y_2 = f(x_2).$$

Because f is increasing, $f(x_1) < f(x_2)$ holds precisely when $x_1 < x_2$. Therefore,

$$f^{-1}(y_1) = x_1 < x_2 = f^{-1}(y_2),$$

which implies that f^{-1} is increasing. (Property 3 can be proved in a similar way.)

Finally, to prove Property 4, consider the limit

$$(f^{-1})'(a) = \lim_{y \to a} \frac{f^{-1}(y) - f^{-1}(a)}{y - a}$$

where a is in the domain of f^{-1} and $f^{-1}(a) = c$. Because f is differentiable at c, f is continuous at c, and so is f^{-1} at a. So, $y \to a$ implies that $x \to c$, and you have

$$(f^{-1})'(a) = \lim_{x \to c} \frac{x - c}{f(x) - f(c)}$$

$$= \lim_{x \to c} \frac{1}{\left(\dfrac{f(x) - f(c)}{x - c} \right)}$$

$$= \frac{1}{\lim\limits_{x \to c} \dfrac{f(x) - f(c)}{x - c}}$$

$$= \frac{1}{f'(c)}.$$

Hence, $(f^{-1})'(a)$ exists, and f^{-1} is differentiable at $f(c)$.

THEOREM 5.9   The Derivative of an Inverse Function (page 336)

Let f be a function that is differentiable on an interval I. If f has an inverse function g, then g is differentiable at any x for which $f'(g(x)) \neq 0$. Moreover,

$$g'(x) = \frac{1}{f'(g(x))}, \qquad f'(g(x)) \neq 0.$$

Proof From the proof of Theorem 5.8, letting $a = x$, you know that g is differentiable. Using the Chain Rule, differentiate both sides of the equation $x = f(g(x))$ to obtain

$$1 = f'(g(x)) \frac{d}{dx}[g(x)].$$

Because $f'(g(x)) \neq 0$, you can divide by this quantity to obtain

$$\frac{d}{dx}[g(x)] = \frac{1}{f'(g(x))}.$$

THEOREM 5.15   A Limit Involving e (page 355)

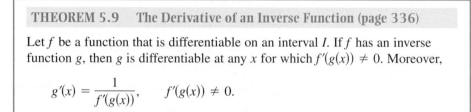

$$\lim_{x \to \infty} \left(1 + \frac{1}{x}\right)^x = \lim_{x \to \infty} \left(\frac{x + 1}{x}\right)^x = e$$

Proof Let $y = \lim\limits_{x \to \infty} \left(1 + \dfrac{1}{x}\right)^x$. Taking the natural logs of both sides, you have

$$\ln y = \ln\left[\lim_{x \to \infty} \left(1 + \frac{1}{x}\right)^x \right].$$

Because the natural logarithmic function is continuous, you can write

$$\ln y = \lim_{x \to \infty} \left[x \ln\left(1 + \frac{1}{x}\right) \right] = \lim_{x \to \infty} \left\{ \frac{\ln[1 + (1/x)]}{1/x} \right\}.$$

Letting $x = \frac{1}{t}$, you have

$$\ln y = \lim_{t \to 0^+} \frac{\ln(1 + t)}{t}$$

$$= \lim_{t \to 0^+} \frac{\ln(1 + t) - \ln 1}{t}$$

$$= \frac{d}{dx} \ln x \text{ at } x = 1$$

$$= \frac{1}{x} \text{ at } x = 1$$

$$= 1.$$

Finally, because $\ln y = 1$, you know that $y = e$, and you can conclude that

$$\lim_{x \to \infty} \left(1 + \frac{1}{x}\right)^x = e.$$

> **THEOREM 7.3   The Extended Mean Value Theorem (page 531)**
>
> If f and g are differentiable on an open interval (a, b) and continuous on $[a, b]$ such that $g'(x) \neq 0$ for any x in (a, b), then there exists a point c in (a, b) such that
>
> $$\frac{f'(c)}{g'(c)} = \frac{f(b) - f(a)}{g(b) - g(a)}.$$

Proof   You can assume that $g(a) \neq g(b)$, because otherwise, by Rolle's Theorem, it would follow that $g'(x) = 0$ for some x in (a, b). Now, define $h(x)$ to be

$$h(x) = f(x) - \left[\frac{f(b) - f(a)}{g(b) - g(a)}\right] g(x).$$

Then

$$h(a) = f(a) - \left[\frac{f(b) - f(a)}{g(b) - g(a)}\right] g(a) = \frac{f(a)g(b) - f(b)g(a)}{g(b) - g(a)}$$

and

$$h(b) = f(b) - \left[\frac{f(b) - f(a)}{g(b) - g(a)}\right] g(b) = \frac{f(a)g(b) - f(b)g(a)}{g(b) - g(a)}$$

and by Rolle's Theorem there exists a point c in (a, b) such that

$$h'(c) = f'(c) - \frac{f(b) - f(a)}{g(b) - g(a)} g'(c) = 0$$

which implies that

$$\frac{f'(c)}{g'(c)} = \frac{f(b) - f(a)}{g(b) - g(a)}.$$

THEOREM 7.4   L'Hôpital's Rule (page 531)

Let f and g be functions that are differentiable on an open interval (a, b) containing c, except possibly at c itself. Assume that $g'(x) \neq 0$ for all x in (a, b), except possibly at c itself. If the limit of $f(x)/g(x)$ as x approaches c produces the indeterminate form $0/0$, then

$$\lim_{x \to c} \frac{f(x)}{g(x)} = \lim_{x \to c} \frac{f'(x)}{g'(x)}$$

provided the limit on the right exists (or is infinite). This result also applies if the limit of $f(x)/g(x)$ as x approaches c produces any one of the indeterminate forms ∞/∞, $(-\infty)/\infty$, $\infty/(-\infty)$, or $(-\infty)/(-\infty)$.

You can use the Extended Mean Value Theorem to prove L'Hôpital's Rule. Of the several different cases of this rule, the proof of only one case is illustrated. The remaining cases where $x \to c^-$ and $x \to c$ are left for you to prove.

Proof   Consider the case for which

$$\lim_{x \to c^+} f(x) = 0 \quad \text{and} \quad \lim_{x \to c^+} g(x) = 0.$$

Define the following new functions:

$$F(x) = \begin{cases} f(x), & x \neq c \\ 0, & x = c \end{cases} \quad \text{and} \quad G(x) = \begin{cases} g(x), & x \neq c \\ 0, & x = c \end{cases}.$$

For any $x, c < x < b$, F and G are differentiable on $(c, x]$ and continuous on $[c, x]$. You can apply the Extended Mean Value Theorem to conclude that there exists a number z in (c, x) such that

$$\frac{F'(z)}{G'(z)} = \frac{F(x) - F(c)}{G(x) - G(c)} = \frac{F(x)}{G(x)} = \frac{f'(z)}{g'(z)} = \frac{f(x)}{g(x)}.$$

Finally, by letting x approach c from the right, $x \to c^+$, we have $z \to c^+$ because $c < z < x$, and

$$\lim_{x \to c^+} \frac{f(x)}{g(x)} = \lim_{x \to c^+} \frac{f'(z)}{g'(z)} = \lim_{z \to c^+} \frac{f'(z)}{g'(z)} = \lim_{x \to c^+} \frac{f'(x)}{g'(x)}. \qquad \blacksquare$$

THEOREM 8.16   Absolute Convergence (page 593)

If the series $\Sigma |a_n|$ converges, then the series Σa_n also converges.

Proof   Because $0 \leq a_n + |a_n| \leq 2|a_n|$ for all n, the series

$$\sum_{n=1}^{\infty} (a_n + |a_n|)$$

converges by comparison with the convergent series

$$\sum_{n=1}^{\infty} 2|a_n|.$$

Furthermore, because $a_n = (a_n + |a_n|) - |a_n|$, you can write

$$\sum_{n=1}^{\infty} a_n = \sum_{n=1}^{\infty} (a_n + |a_n|) - \sum_{n=1}^{\infty} |a_n|$$

where both series on the right converge. Hence it follows that Σa_n converges. $\qquad \blacksquare$

THEOREM 8.19   Taylor's Theorem (page 611)

If a function f is differentiable through order $n + 1$ in an interval I containing c, then, for each x in I, there exists z between x and c such that

$$f(x) = f(c) + f'(c)(x - c) + \frac{f''(c)}{2!}(x - c)^2 + \cdots + \frac{f^{(n)}(c)}{n!}(x - c)^n + R_n(x)$$

where

$$R_n(x) = \frac{f^{(n+1)}(z)}{(n+1)!}(x - c)^{n+1}.$$

Proof To find $R_n(x)$, fix x in I ($x \neq c$) and write

$$R_n(x) = f(x) - P_n(x)$$

where $P_n(x)$ is the nth Taylor polynomial for $f(x)$. Then let g be a function of t defined by

$$g(t) = f(x) - f(t) - f'(t)(x - t) - \cdots - \frac{f^{(n)}(t)}{n!}(x - t)^n - R_n(x)\frac{(x - t)^{n+1}}{(x - c)^{n+1}}.$$

The reason for defining g in this way is that differentiation with respect to t has a telescoping effect. For example, you have

$$\frac{d}{dt}[-f(t) - f'(t)(x - t)] = -f'(t) + f'(t) - f''(t)(x - t)$$

$$= -f''(t)(x - t).$$

The result is that the derivative $g'(t)$ simplifies to

$$g'(t) = -\frac{f^{(n+1)}(t)}{n!}(x - t)^n + (n + 1)R_n(x)\frac{(x - t)^n}{(x - c)^{n+1}}$$

for all t between c and x. Moreover, for a fixed x,

$$g(c) = f(x) - [P_n(x) + R_n(x)] = f(x) - f(x) = 0$$

and

$$g(x) = f(x) - f(x) - 0 - \cdots - 0 = f(x) - f(x) = 0.$$

Therefore, g satisfies the conditions of Rolle's Theorem, and it follows that there is a number z between c and x such that $g'(z) = 0$. Substituting z for t in the equation for $g'(t)$ and then solving for $R_n(x)$, you obtain

$$g'(z) = -\frac{f^{(n+1)}(z)}{n!}(x - z)^n + (n + 1)R_n(x)\frac{(x - z)^n}{(x - c)^{n+1}} = 0$$

$$R_n(x) = \frac{f^{(n+1)}(z)}{(n+1)!}(x - c)^{n+1}.$$

Finally, because $g(c) = 0$, you have

$$0 = f(x) - f(c) - f'(c)(x - c) - \cdots - \frac{f^{(n)}(c)}{n!}(x - c)^n - R_n(x)$$

$$f(x) = f(c) + f'(c)(x - c) + \cdots + \frac{f^{(n)}(c)}{n!}(x - c)^n + R_n(x).$$

> **THEOREM 8.20** **Convergence of a Power Series (page 617)**
>
> For a power series centered at c, precisely one of the following is true.
>
> **1.** The series converges only at c.
> **2.** There exists a real number $R > 0$ such that the series converges absolutely for $|x - c| < R$, and diverges for $|x - c| > R$.
> **3.** The series converges absolutely for all x.
>
> The number R is the **radius of convergence** of the power series. If the series converges only at c, the radius of convergence is $R = 0$, and if the series converges for all x, the radius of convergence is $R = \infty$. The set of all values of x for which the power series converges is the **interval of convergence** of the power series.

Proof In order to simplify the notation, we will prove the theorem for the power series $\Sigma\, a_n x^n$ centered at $x = 0$. The proof for a power series centered at $x = c$ follows easily. A key step in this proof uses the Completeness Property of the set of real numbers: If a nonempty set S of real numbers has an upper bound, then it must have a least upper bound (see page 563).

We must show that if a power series $\Sigma\, a_n x^n$ converges at $x = d$, $d \ne 0$ then it converges for all b satisfying $|b| < |d|$. Because $\Sigma\, a_n x^n$ converges, $\lim\limits_{x \to \infty} a_n d^n = 0$. Hence, there exists $N > 0$ such that $a_n d^n < 1$ for all $n \ge N$. Then for $n \ge N$,

$$\left| a_n b^n \right| = \left| a_n b^n \frac{d^n}{d^n} \right| = \left| a_n d^n \right| \left| \frac{b^n}{d^n} \right| < \left| \frac{b^n}{d^n} \right|.$$

So, for $|b| < |d|$, $\left| \dfrac{b}{d} \right| < 1$ which implies that

$$\Sigma \left| \frac{b^n}{d^n} \right|$$

is a convergent geometric series. By the Comparison Test, the series $\Sigma\, a_n b^n$ converges.

Similarly, if the power series $\Sigma\, a_n x^n$ diverges at $x = b$, where $b \ne 0$, then it diverges for all d satisfying $|d| > |b|$. If $\Sigma\, a_n d^n$ converged, then the above argument would imply that $\Sigma\, a_n b^n$ converged as well.

Finally, to prove the theorem, suppose that neither case 1 nor case 3 is true. Then there exist points b and d such that $\Sigma\, a_n x^n$ converges to b and diverges at d. Let $S = \{x : \Sigma\, a_n x^n \text{ converges}\}$. S is nonempty because $b \in S$. If $b \in S$ then $|x| \le |d|$, which shows that $|d|$ is an upper bound for the nonempty set S. By the Completeness Property, S has a least upper bound, R.

Now, if $|x| > R$, then $x \ne S$ so $\Sigma\, a_n x^n$ diverges. And if $|x| < R$, then $|x|$ is not an upper bound for S, so there exists b in S satisfying $|b| > |x|$. Since $b \in S$, $\Sigma\, a_n b^n$ converges, which implies that $\Sigma\, a_n x^n$ converges.

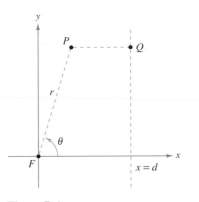

Figure B.1

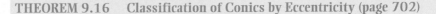

> **THEOREM 9.16   Classification of Conics by Eccentricity (page 702)**
>
> Let F be the fixed point (*focus*) and D be a fixed line (*directrix*) in the plane. Let P be another point in the plane and let e (*eccentricity*) be the ratio of the distance between P and F to the distance between P and D. The collection of all points P with a given eccentricity is a conic.
>
> 1. The conic is an ellipse if $0 < e < 1$.
> 2. The conic is a parabola if $e = 1$.
> 3. The conic is a hyperbola if $e > 1$.

Proof If $e = 1$, then, by definition, the conic must be a parabola. If $e \neq 1$, then you can consider the focus F to lie at the origin and the directrix $x = d$ to lie to the right of the origin, as shown in Figure B.1. For the point $P = (r, \theta) = (x, y)$, you have $|PF| = r$ and $|PQ| = d - r \cos \theta$. Given that $e = |PF|/|PQ|$, it follows that

$$|PF| = |PQ|e \quad \Longrightarrow \quad r = e(d - r \cos \theta).$$

By converting to rectangular coordinates and squaring both sides, you obtain

$$x^2 + y^2 = e^2(d - x)^2 = e^2(d^2 - 2\, dx + x^2).$$

Completing the square produces

$$\left(x + \frac{e^2 d}{1 - e^2}\right)^2 + \frac{y^2}{1 - e^2} = \frac{e^2 d^2}{(1 - e^2)^2}.$$

If $e < 1$, this equation represents an ellipse. If $e > 1$, then $1 - e^2 < 0$, and the equation represents a hyperbola. ◢

> **THEOREM 12.4   Sufficient Condition for Differentiability (page 870)**
>
> If f is a function of x and y, where f_x and f_y are continuous in an open region R, then f is differentiable on R.

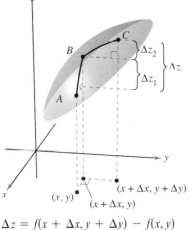

$\Delta z = f(x + \Delta x, y + \Delta y) - f(x, y)$

Figure B.2

Proof Let S be the surface defined by $z = f(x, y)$, where $f, f_x,$ and f_y are continuous at (x, y). Let A, B, and C be points on surface S, as shown in Figure B.2. From this figure, you can see that the change in f from point A to point C is given by

$$\begin{aligned} \Delta z &= f(x + \Delta x, y + \Delta y) - f(x, y) \\ &= [f(x + \Delta x, y) - f(x, y)] + [f(x + \Delta x, y + \Delta y) - f(x + \Delta x, y)] \\ &= \Delta z_1 + \Delta z_2. \end{aligned}$$

Between A and B, y is fixed and x changes. Hence, by the Mean Value Theorem, there is a value x_1 between x and $x + \Delta x$ such that

$$\Delta z_1 = f(x + \Delta x, y) - f(x, y) = f_x(x_1, y)\, \Delta x.$$

Similarly, between B and C, x is fixed and y changes, and there is a value y_1 between y and $y + \Delta y$ such that

$$\Delta z_2 = f(x + \Delta x, y + \Delta y) - f(x + \Delta x, y) = f_y(x + \Delta x, y_1)\, \Delta y.$$

By combining these two results, you can write

$$\Delta z = \Delta z_1 + \Delta z_2 = f_x(x_1, y)\Delta x + f_y(x + \Delta x, y_1)\, \Delta y.$$

If you define ε_1 and ε_2 as

$$\varepsilon_1 = f_x(x_1, y) - f_x(x, y) \qquad \text{and} \qquad \varepsilon_2 = f_y(x + \Delta x, y_1) - f_y(x, y)$$

it follows that

$$\Delta z = \Delta z_1 + \Delta z_2 = [\varepsilon_1 + f_x(x, y)]\,\Delta x + [\varepsilon_2 + f_y(x, y)]\,\Delta y$$
$$= [f_x(x, y)\,\Delta x + f_y(x, y)\,\Delta y] + \varepsilon_1 \Delta x + \varepsilon_2 \Delta y.$$

By the continuity of f_x and f_y and the fact that $x \le x_1 \le x + \Delta x$ and $y \le y_1 \le y + \Delta y$, it follows that $\varepsilon_1 \to 0$ and $\varepsilon_2 \to 0$ as $\Delta x \to 0$ and $\Delta y \to 0$. Therefore, by definition, f is differentiable.   ◼

THEOREM 12.6   Chain Rule: One Independent Variable (page 876)

Let $w = f(x, y)$, where f is a differentiable function of x and y. If $x = g(t)$ and $y = h(t)$, where g and h are differentiable functions of t, then w is a differentiable function of t, and

$$\frac{dw}{dt} = \frac{\partial w}{\partial x}\frac{dx}{dt} + \frac{\partial w}{\partial y}\frac{dy}{dt}.$$

Proof   Because g and h are differentiable functions of t, you know that both Δx and Δy approach zero as Δt approaches zero. Moreover, because f is a differentiable function of x and y, you know that

$$\Delta w = \frac{\partial w}{\partial x}\,\Delta x + \frac{\partial w}{\partial y}\,\Delta y + \varepsilon_1 \Delta x + \varepsilon_2 \Delta y$$

where both ε_1 and $\varepsilon_2 \to 0$ as $(\Delta x, \Delta y) \to (0, 0)$. So, for $\Delta t \ne 0$, we have

$$\frac{\Delta w}{\Delta t} = \frac{\partial w}{\partial x}\frac{\Delta x}{\Delta t} + \frac{\partial w}{\partial y}\frac{\Delta y}{\Delta t} + \varepsilon_1 \frac{\Delta x}{\Delta t} + \varepsilon_2 \frac{\Delta y}{\Delta t}$$

from which it follows that

$$\frac{dw}{dt} = \lim_{\Delta t \to 0} \frac{\Delta w}{\Delta t} = \frac{\partial w}{\partial x}\frac{dx}{dt} + \frac{\partial w}{\partial y}\frac{dy}{dt} + 0\left(\frac{dx}{dt}\right) + 0\left(\frac{dy}{dt}\right)$$
$$= \frac{\partial w}{\partial x}\frac{dx}{dt} + \frac{\partial w}{\partial y}\frac{dy}{dt}.$$ ◼

Integration Tables C

Forms Involving u^n

1. $\displaystyle \int u^n \, du = \frac{u^{n+1}}{n+1} + C, \ n \neq -1$

2. $\displaystyle \int \frac{1}{u} \, du = \ln|u| + C$

Forms Involving $a + bu$

3. $\displaystyle \int \frac{u}{a+bu} \, du = \frac{1}{b^2}\big(bu - a\ln|a+bu|\big) + C$

4. $\displaystyle \int \frac{u}{(a+bu)^2} \, du = \frac{1}{b^2}\left(\frac{a}{a+bu} + \ln|a+bu|\right) + C$

5. $\displaystyle \int \frac{u}{(a+bu)^n} \, du = \frac{1}{b^2}\left[\frac{-1}{(n-2)(a+bu)^{n-2}} + \frac{a}{(n-1)(a+bu)^{n-1}}\right] + C, \ n \neq 1, 2$

6. $\displaystyle \int \frac{u^2}{a+bu} \, du = \frac{1}{b^3}\left[-\frac{bu}{2}(2a-bu) + a^2\ln|a+bu|\right] + C$

7. $\displaystyle \int \frac{u^2}{(a+bu)^2} \, du = \frac{1}{b^3}\left(bu - \frac{a^2}{a+bu} - 2a\ln|a+bu|\right) + C$

8. $\displaystyle \int \frac{u^2}{(a+bu)^3} \, du = \frac{1}{b^3}\left[\frac{2a}{a+bu} - \frac{a^2}{2(a+bu)^2} + \ln|a+bu|\right] + C$

9. $\displaystyle \int \frac{u^2}{(a+bu)^n} \, du = \frac{1}{b^3}\left[\frac{-1}{(n-3)(a+bu)^{n-3}} + \frac{2a}{(n-2)(a+bu)^{n-2}} - \frac{a^2}{(n-1)(a+bu)^{n-1}}\right] + C, \ n \neq 1, 2, 3$

10. $\displaystyle \int \frac{1}{u(a+bu)} \, du = \frac{1}{a}\ln\left|\frac{u}{a+bu}\right| + C$

11. $\displaystyle \int \frac{1}{u(a+bu)^2} \, du = \frac{1}{a}\left(\frac{1}{a+bu} + \frac{1}{a}\ln\left|\frac{u}{a+bu}\right|\right) + C$

12. $\displaystyle \int \frac{1}{u^2(a+bu)} \, du = -\frac{1}{a}\left(\frac{1}{u} + \frac{b}{a}\ln\left|\frac{u}{a+bu}\right|\right) + C$

13. $\displaystyle \int \frac{1}{u^2(a+bu)^2} \, du = -\frac{1}{a^2}\left[\frac{a+2bu}{u(a+bu)} + \frac{2b}{a}\ln\left|\frac{u}{a+bu}\right|\right] + C$

Forms Involving $a + bu + cu^2,\ b^2 \neq 4ac$

14. $\displaystyle \int \frac{1}{a + bu + cu^2}\, du = \begin{cases} \dfrac{2}{\sqrt{4ac - b^2}} \arctan \dfrac{2cu + b}{\sqrt{4ac - b^2}} + C, & b^2 < 4ac \\[3mm] \dfrac{1}{\sqrt{b^2 - 4ac}} \ln\left|\dfrac{2cu + b - \sqrt{b^2 - 4ac}}{2cu + b + \sqrt{b^2 - 4ac}}\right| + C, & b^2 > 4ac \end{cases}$

15. $\displaystyle \int \frac{u}{a + bu + cu^2}\, du = \frac{1}{2c}\left(\ln|a + bu + cu^2| - b \int \frac{1}{a + bu + cu^2}\, du\right)$

Forms Involving $\sqrt{a + bu}$

16. $\displaystyle \int u^n \sqrt{a + bu}\, du = \frac{2}{b(2n + 3)}\left[u^n(a + bu)^{3/2} - na \int u^{n-1}\sqrt{a + bu}\, du\right]$

17. $\displaystyle \int \frac{1}{u\sqrt{a + bu}}\, du = \begin{cases} \dfrac{1}{\sqrt{a}} \ln\left|\dfrac{\sqrt{a + bu} - \sqrt{a}}{\sqrt{a + bu} + \sqrt{a}}\right| + C, & a > 0 \\[3mm] \dfrac{2}{\sqrt{-a}} \arctan \sqrt{\dfrac{a + bu}{-a}} + C, & a < 0 \end{cases}$

18. $\displaystyle \int \frac{1}{u^n \sqrt{a + bu}}\, du = \frac{-1}{a(n - 1)}\left[\frac{\sqrt{a + bu}}{u^{n-1}} + \frac{(2n - 3)b}{2} \int \frac{1}{u^{n-1}\sqrt{a + bu}}\, du\right],\ n \neq 1$

19. $\displaystyle \int \frac{\sqrt{a + bu}}{u}\, du = 2\sqrt{a + bu} + a \int \frac{1}{u\sqrt{a + bu}}\, du$

20. $\displaystyle \int \frac{\sqrt{a + bu}}{u^n}\, du = \frac{-1}{a(n - 1)}\left[\frac{(a + bu)^{3/2}}{u^{n-1}} + \frac{(2n - 5)b}{2} \int \frac{\sqrt{a + bu}}{u^{n-1}}\, du\right],\ n \neq 1$

21. $\displaystyle \int \frac{u}{\sqrt{a + bu}}\, du = \frac{-2(2a - bu)}{3b^2}\sqrt{a + bu} + C$

22. $\displaystyle \int \frac{u^n}{\sqrt{a + bu}}\, du = \frac{2}{(2n + 1)b}\left(u^n \sqrt{a + bu} - na \int \frac{u^{n-1}}{\sqrt{a + bu}}\, du\right)$

Forms Involving $a^2 \pm u^2,\ a > 0$

23. $\displaystyle \int \frac{1}{a^2 + u^2}\, du = \frac{1}{a} \arctan \frac{u}{a} + C$

24. $\displaystyle \int \frac{1}{u^2 - a^2}\, du = -\int \frac{1}{a^2 - u^2}\, du = \frac{1}{2a} \ln\left|\frac{u - a}{u + a}\right| + C$

25. $\displaystyle \int \frac{1}{(a^2 \pm u^2)^n}\, du = \frac{1}{2a^2(n - 1)}\left[\frac{u}{(a^2 \pm u^2)^{n-1}} + (2n - 3) \int \frac{1}{(a^2 \pm u^2)^{n-1}}\, du\right],\ n \neq 1$

Forms Involving $\sqrt{u^2 \pm a^2},\ a > 0$

26. $\displaystyle \int \sqrt{u^2 \pm a^2}\, du = \frac{1}{2}\left(u\sqrt{u^2 \pm a^2} \pm a^2 \ln\left|u + \sqrt{u^2 \pm a^2}\right|\right) + C$

27. $\displaystyle \int u^2 \sqrt{u^2 \pm a^2}\, du = \frac{1}{8}\left[u(2u^2 \pm a^2)\sqrt{u^2 \pm a^2} - a^4 \ln\left|u + \sqrt{u^2 \pm a^2}\right|\right] + C$

28. $\displaystyle \int \frac{\sqrt{u^2 + a^2}}{u}\, du = \sqrt{u^2 + a^2} - a \ln\left|\frac{a + \sqrt{u^2 + a^2}}{u}\right| + C$

29. $\displaystyle\int \frac{\sqrt{u^2 - a^2}}{u}\, du = \sqrt{u^2 - a^2} - a \operatorname{arcsec} \frac{|u|}{a} + C$

30. $\displaystyle\int \frac{\sqrt{u^2 \pm a^2}}{u^2}\, du = \frac{-\sqrt{u^2 \pm a^2}}{u} + \ln\left|u + \sqrt{u^2 \pm a^2}\right| + C$

31. $\displaystyle\int \frac{1}{\sqrt{u^2 \pm a^2}}\, du = \ln\left|u + \sqrt{u^2 \pm a^2}\right| + C$

32. $\displaystyle\int \frac{1}{u\sqrt{u^2 + a^2}}\, du = \frac{-1}{a} \ln\left|\frac{a + \sqrt{u^2 + a^2}}{u}\right| + C$

33. $\displaystyle\int \frac{1}{u\sqrt{u^2 - a^2}}\, du = \frac{1}{a} \operatorname{arcsec} \frac{|u|}{a} + C$

34. $\displaystyle\int \frac{u^2}{\sqrt{u^2 \pm a^2}}\, du = \frac{1}{2}\left(u\sqrt{u^2 \pm a^2} \mp a^2 \ln\left|u + \sqrt{u^2 \pm a^2}\right|\right) + C$

35. $\displaystyle\int \frac{1}{u^2\sqrt{u^2 \pm a^2}}\, du = \mp \frac{\sqrt{u^2 \pm a^2}}{a^2 u} + C$

36. $\displaystyle\int \frac{1}{(u^2 \pm a^2)^{3/2}}\, du = \frac{\pm u}{a^2\sqrt{u^2 \pm a^2}} + C$

Forms Involving $\sqrt{a^2 - u^2},\ a > 0$

37. $\displaystyle\int \sqrt{a^2 - u^2}\, du - \frac{1}{2}\left(u\sqrt{a^2 - u^2} + a^2 \arcsin \frac{u}{a}\right) + C$

38. $\displaystyle\int u^2\sqrt{a^2 - u^2}\, du = \frac{1}{8}\left[u(2u^2 - a^2)\sqrt{a^2 - u^2} + a^4 \arcsin \frac{u}{a}\right] + C$

39. $\displaystyle\int \frac{\sqrt{a^2 - u^2}}{u}\, du = \sqrt{a^2 - u^2} - a \ln\left|\frac{a + \sqrt{a^2 - u^2}}{u}\right| + C$

40. $\displaystyle\int \frac{\sqrt{a^2 - u^2}}{u^2}\, du = \frac{-\sqrt{a^2 - u^2}}{u} - \arcsin \frac{u}{a} + C$

41. $\displaystyle\int \frac{1}{\sqrt{a^2 - u^2}}\, du = \arcsin \frac{u}{a} + C$

42. $\displaystyle\int \frac{1}{u\sqrt{a^2 - u^2}}\, du = \frac{-1}{a} \ln\left|\frac{a + \sqrt{a^2 - u^2}}{u}\right| + C$

43. $\displaystyle\int \frac{u^2}{\sqrt{a^2 - u^2}}\, du = \frac{1}{2}\left(-u\sqrt{a^2 - u^2} + a^2 \arcsin \frac{u}{a}\right) + C$

44. $\displaystyle\int \frac{1}{u^2\sqrt{a^2 - u^2}}\, du = \frac{-\sqrt{a^2 - u^2}}{a^2 u} + C$

45. $\displaystyle\int \frac{1}{(a^2 - u^2)^{3/2}}\, du = \frac{u}{a^2\sqrt{a^2 - u^2}} + C$

Forms Involving $\sin u$ or $\cos u$

46. $\displaystyle\int \sin u \, du = -\cos u + C$

47. $\displaystyle\int \cos u \, du = \sin u + C$

48. $\displaystyle\int \sin^2 u \, du = \frac{1}{2}(u - \sin u \cos u) + C$

49. $\displaystyle\int \cos^2 u \, du = \frac{1}{2}(u + \sin u \cos u) + C$

50. $\displaystyle\int \sin^n u \, du = -\frac{\sin^{n-1} u \cos u}{n} + \frac{n-1}{n}\int \sin^{n-2} u \, du$

51. $\displaystyle\int \cos^n u \, du = \frac{\cos^{n-1} u \sin u}{n} + \frac{n-1}{n}\int \cos^{n-2} u \, du$

52. $\displaystyle\int u \sin u \, du = \sin u - u \cos u + C$

53. $\displaystyle\int u \cos u \, du = \cos u + u \sin u + C$

54. $\displaystyle\int u^n \sin u \, du = -u^n \cos u + n\int u^{n-1} \cos u \, du$

55. $\displaystyle\int u^n \cos u \, du = u^n \sin u - n\int u^{n-1} \sin u \, du$

56. $\displaystyle\int \frac{1}{1 \pm \sin u} \, du = \tan u \mp \sec u + C$

57. $\displaystyle\int \frac{1}{1 \pm \cos u} \, du = -\cot u \pm \csc u + C$

58. $\displaystyle\int \frac{1}{\sin u \cos u} \, du = \ln|\tan u| + C$

Forms Involving $\tan u$, $\cot u$, $\sec u$, $\csc u$

59. $\displaystyle\int \tan u \, du = -\ln|\cos u| + C$

60. $\displaystyle\int \cot u \, du = \ln|\sin u| + C$

61. $\displaystyle\int \sec u \, du = \ln|\sec u + \tan u| + C$

62. $\displaystyle\int \csc u \, du = \ln|\csc u - \cot u| + C$

63. $\displaystyle\int \tan^2 u \; du = -u + \tan u + C$

64. $\displaystyle\int \cot^2 u \; du = -u - \cot u + C$

65. $\displaystyle\int \sec^2 u \; du = \tan u + C$

66. $\displaystyle\int \csc^2 u \; du = -\cot u + C$

67. $\displaystyle\int \tan^n u \; du = \frac{\tan^{n-1} u}{n - 1} - \int \tan^{n-2} u \; du, \; n \neq 1$

68. $\displaystyle\int \cot^n u \; du = -\frac{\cot^{n-1} u}{n - 1} - \int (\cot^{n-2} u) \; du, \; n \neq 1$

69. $\displaystyle\int \sec^n u \; du = \frac{\sec^{n-2} u \tan u}{n - 1} + \frac{n - 2}{n - 1} \int \sec^{n-2} u \; du, \; n \neq 1$

70. $\displaystyle\int \csc^n u \; du = -\frac{\csc^{n-2} u \cot u}{n - 1} + \frac{n - 2}{n - 1} \int \csc^{n-2} u \; du, \; n \neq 1$

71. $\displaystyle\int \frac{1}{1 \pm \tan u} \; du = \frac{1}{2}\big(u \pm \ln|\cos u \pm \sin u|\big) + C$

72. $\displaystyle\int \frac{1}{1 \pm \cot u} \; du = \frac{1}{2}\big(u \mp \ln|\sin u \pm \cos u|\big) + C$

73. $\displaystyle\int \frac{1}{1 \pm \sec u} \; du = u + \cot u \mp \csc u + C$

74. $\displaystyle\int \frac{1}{1 \pm \csc u} \; du = u - \tan u + \sec u + C$

Forms Involving Inverse Trigonometric Functions

75. $\displaystyle\int \arcsin u \; du = u \arcsin u + \sqrt{1 - u^2} + C$

76. $\displaystyle\int \arccos u \; du = u \arccos u - \sqrt{1 - u^2} + C$

77. $\displaystyle\int \arctan u \; du = u \arctan u - \ln\sqrt{1 + u^2} + C$

78. $\displaystyle\int \text{arccot } u \; du = u \text{ arccot } u + \ln\sqrt{1 + u^2} + C$

79. $\displaystyle\int \text{arcsec } u \; du = u \text{ arcsec } u - \ln\left|u + \sqrt{u^2 - 1}\right| + C$

80. $\displaystyle\int \text{arccsc } u \; du = u \text{ arccsc } u + \ln\left|u + \sqrt{u^2 - 1}\right| + C$

Forms Involving e^u

81. $\displaystyle\int e^u \, du = e^u + C$

82. $\displaystyle\int u e^u \, du = (u - 1)e^u + C$

83. $\displaystyle\int u^n e^u \, du = u^n e^u - n \int u^{n-1} e^u \, du$

84. $\displaystyle\int \frac{1}{1 + e^u} \, du = u - \ln(1 + e^u) + C$

85. $\displaystyle\int e^{au} \sin bu \, du = \frac{e^{au}}{a^2 + b^2}(a \sin bu - b \cos bu) + C$

86. $\displaystyle\int e^{au} \cos bu \, du = \frac{e^{au}}{a^2 + b^2}(a \cos bu + b \sin bu) + C$

Forms Involving $\ln u$

87. $\displaystyle\int \ln u \, du = u(-1 + \ln u) + C$

88. $\displaystyle\int u \ln u \, du = \frac{u^2}{4}(-1 + 2 \ln u) + C$

89. $\displaystyle\int u^n \ln u \, du = \frac{u^{n+1}}{(n+1)^2}[-1 + (n+1) \ln u] + C, \; n \neq -1$

90. $\displaystyle\int (\ln u)^2 \, du = u\left[2 - 2 \ln u + (\ln u)^2\right] + C$

91. $\displaystyle\int (\ln u)^n \, du = u(\ln u)^n - n \int (\ln u)^{n-1} \, du$

Answers to Odd-Numbered Exercises

Chapter P

Section P.1 (page 8)

1. b **2.** d **3.** a **4.** c

5. Answers may vary.

x	-4	-2	0	2	4
y	-5	-2	1	4	7

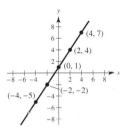

7. Answers may vary.

x	-3	-2	0	2	3
y	-5	0	4	0	-5

9. Answers may vary.

x	-5	-4	-3	-2	-1	0	1
y	3	2	1	0	1	2	3

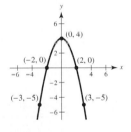

11. Answers may vary.

x	0	1	4	9	16
y	-4	-3	-2	-1	0

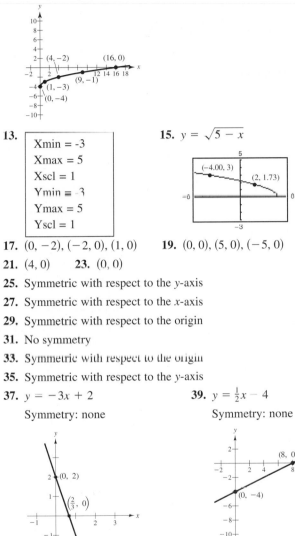

13.

```
Xmin = -3
Xmax = 5
Xscl = 1
Ymin = -3
Ymax = 5
Yscl = 1
```

15. $y = \sqrt{5 - x}$

17. $(0, -2), (-2, 0), (1, 0)$ **19.** $(0, 0), (5, 0), (-5, 0)$

21. $(4, 0)$ **23.** $(0, 0)$

25. Symmetric with respect to the y-axis

27. Symmetric with respect to the x-axis

29. Symmetric with respect to the origin

31. No symmetry

33. Symmetric with respect to the origin

35. Symmetric with respect to the y-axis

37. $y = -3x + 2$
Symmetry: none

39. $y = \frac{1}{2}x - 4$
Symmetry: none

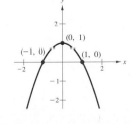

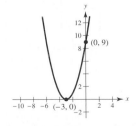

41. $y = 1 - x^2$
Symmetry: y-axis

43. $y = (x + 3)^2$
Symmetry: none

45. $y = x^3 + 2$

Symmetry: none

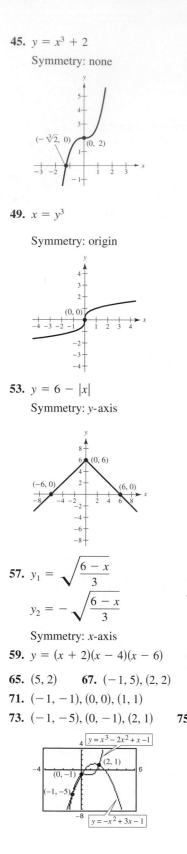

47. $y = x\sqrt{x + 2}$

Symmetry: none

49. $x = y^3$

Symmetry: origin

51. $y = \dfrac{1}{x}$

Symmetry: origin

53. $y = 6 - |x|$

Symmetry: y-axis

55. $y_1 = \sqrt{x + 9}$

$y_2 = -\sqrt{x + 9}$

Symmetry: x-axis

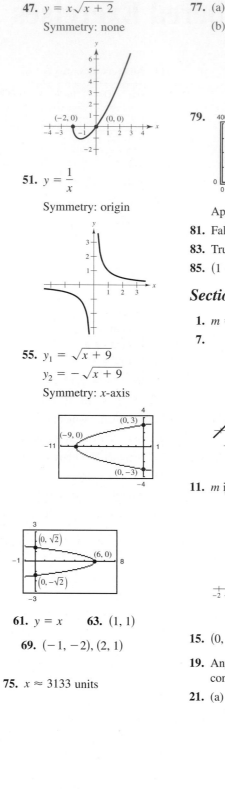

57. $y_1 = \sqrt{\dfrac{6 - x}{3}}$

$y_2 = -\sqrt{\dfrac{6 - x}{3}}$

Symmetry: x-axis

59. $y = (x + 2)(x - 4)(x - 6)$ **61.** $y = x$ **63.** $(1, 1)$

65. $(5, 2)$ **67.** $(-1, 5), (2, 2)$ **69.** $(-1, -2), (2, 1)$

71. $(-1, -1), (0, 0), (1, 1)$

73. $(-1, -5), (0, -1), (2, 1)$ **75.** $x \approx 3133$ units

77. (a) $y = -0.0153t^2 + 4.9971t + 34.9405$

(b)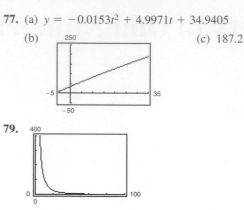

(c) 187.2

79.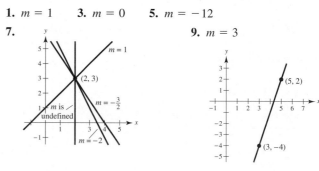

Approximately $\tfrac{1}{4}$

81. False: $(-1, -2)$ is not a point on the graph of $x = \tfrac{1}{4}y^2$.

83. True

85. $(1 - K^2)x^2 + (1 - K^2)y^2 + 4K^2x - 4K^2 = 0$

Section P.2 *(page 16)*

1. $m = 1$ **3.** $m = 0$ **5.** $m = -12$

7.   **9.** $m = 3$

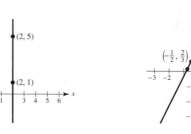

11. m is undefined.   **13.** $m = 2$

15. $(0, 1), (1, 1), (3, 1)$ **17.** $(0, 10), (2, 4), (3, 1)$

19. Any two points can be used because the rate of change remains constant.

21. (a) 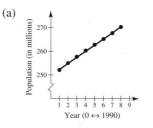   (b) Population increased most rapidly: 1991–1992

23. $m = -\tfrac{1}{5}, (0, 4)$ **25.** m is undefined, no y-intercept

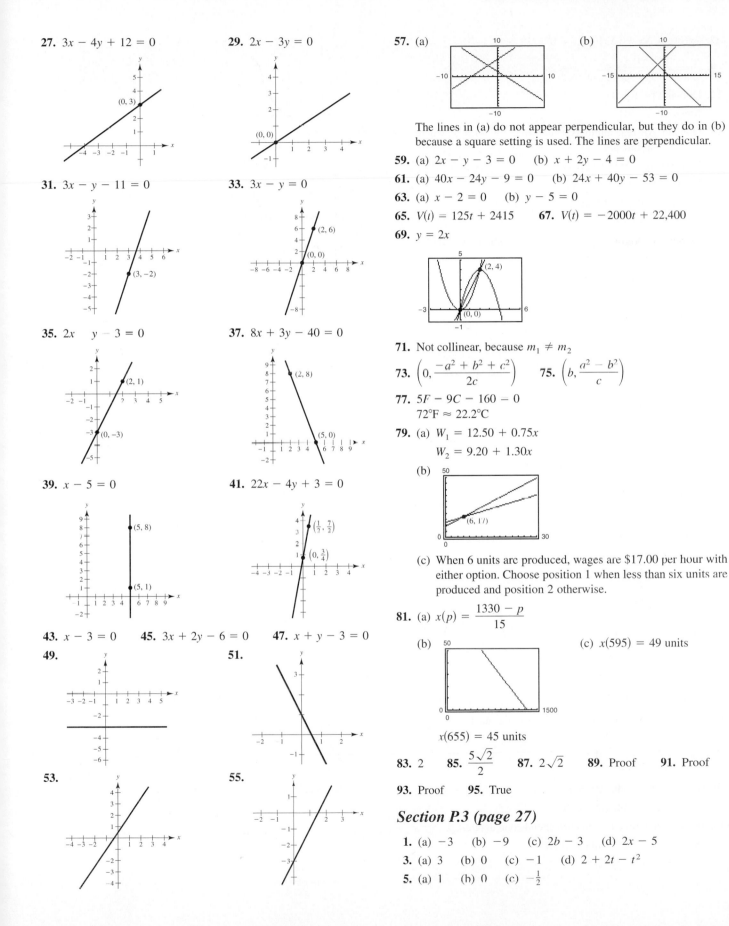

27. $3x - 4y + 12 = 0$

29. $2x - 3y = 0$

31. $3x - y - 11 = 0$

33. $3x - y = 0$

35. $2x \quad y - 3 = 0$

37. $8x + 3y - 40 = 0$

39. $x - 5 = 0$

41. $22x - 4y + 3 = 0$

43. $x - 3 = 0$ **45.** $3x + 2y - 6 = 0$ **47.** $x + y - 3 = 0$

49.

51.

53.

55.

57. (a)

(b)

The lines in (a) do not appear perpendicular, but they do in (b) because a square setting is used. The lines are perpendicular.

59. (a) $2x - y - 3 = 0$ (b) $x + 2y - 4 = 0$

61. (a) $40x - 24y - 9 = 0$ (b) $24x + 40y - 53 = 0$

63. (a) $x - 2 = 0$ (b) $y - 5 = 0$

65. $V(t) = 125t + 2415$ **67.** $V(t) = -2000t + 22,400$

69. $y = 2x$

71. Not collinear, because $m_1 \neq m_2$

73. $\left(0, \dfrac{-a^2 + b^2 + c^2}{2c}\right)$ **75.** $\left(b, \dfrac{a^2 - b^2}{c}\right)$

77. $5F - 9C - 160 - 0$
$72°F \approx 22.2°C$

79. (a) $W_1 = 12.50 + 0.75x$
$W_2 = 9.20 + 1.30x$

(b)

(c) When 6 units are produced, wages are $17.00 per hour with either option. Choose position 1 when less than six units are produced and position 2 otherwise.

81. (a) $x(p) = \dfrac{1330 - p}{15}$

(b)

(c) $x(595) = 49$ units

$x(655) = 45$ units

83. 2   **85.** $\dfrac{5\sqrt{2}}{2}$ **87.** $2\sqrt{2}$ **89.** Proof   **91.** Proof

93. Proof   **95.** True

Section P.3 (page 27)

1. (a) -3 (b) -9 (c) $2b - 3$ (d) $2x - 5$

3. (a) 3 (b) 0 (c) -1 (d) $2 + 2t - t^2$

5. (a) 1 (b) 0 (c) $-\frac{1}{2}$

7. $3x^2 + 3x\,\Delta x + (\Delta x)^2$, $\Delta x \neq 0$

9. $\dfrac{-1}{\sqrt{x-1}\,(1 + \sqrt{x-1})}$, $x \neq 2$

11. Domain: $[-3, \infty)$

    Range: $(-\infty, 0]$

13. Domain: All real numbers t such that $t \neq 4n + 2$, where n is
      an integer.

    Range: $(-\infty, -1] \cup [1, \infty)$

15. Domain: $(-\infty, 0) \cup (0, \infty)$

    Range: $(-\infty, 0) \cup (0, \infty)$

17. (a) -1 (b) 2 (c) 6 (d) $2t^2 + 4$

    Domain: $(-\infty, \infty)$

    Range: $(-\infty, 1) \cup [2, \infty)$

19. (a) 4 (b) 0 (c) -2 (d) $-b^2$

    Domain: $(-\infty, \infty)$

    Range: $(-\infty, 0] \cup [1, \infty)$

21. $f(x) = 4 - x$ **23.** $h(x) = \sqrt{x-1}$

    Domain: $(-\infty, \infty)$ Domain: $[1, \infty)$

    Range: $(-\infty, \infty)$ Range: $[0, \infty)$

25. $f(x) = \sqrt{9 - x^2}$ **27.** $g(t) = 2 \sin \pi t$

    Domain: $[-3, 3]$ Domain: $(-\infty, \infty)$

    Range: $[0, 3]$ Range: $[-2, 2]$

29. y is not a function of x.   **31.** y is a function of x.

33. y is not a function of x.   **35.** y is not a function of x.

37. $f(x) = \begin{cases} -2x + 2, & x < 0 \\ 2, & 0 \le x < 2 \\ 2x - 2, & x \ge 2 \end{cases}$

39. ii, $c = -2$ **40.** i, $c = \frac{1}{4}$

41. iv, $c = 32$ **42.** iii, $c = 3$

43. (a) For each time t there corresponds one depth d.

    (b) Domain: $[0, 5]$; Range: $[0, 30]$

(c)

45.

47. (a)

(b)

(c)

(d)

(e)

(f)

49. (a) Vertical translation   (b) Reflection about the x-axis

(c) Horizontal translation

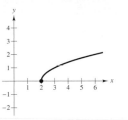

51. (a) $T(4) = 16$, $T(15) = 23$

(b) The changes in temperature would occur 1 hour later.

(c) The temperatures are 1° lower.

53. $(f \circ g)(x) = x$

Domain: $[0, \infty)$

$(g \circ f)(x) = |x|$

Domain: $(-\infty, \infty)$

No, their domains are different.

55. $(f \circ g)(x) = \dfrac{3}{x^2 - 1}$

Domain: $(-\infty, -1) \cup (-1, 1) \cup (1, \infty)$

$(g \circ f)(x) = \dfrac{9}{x^2} - 1$

Domain: $(-\infty, 0) \cup (0, \infty)$

No

57. $(A \circ r)(t) = 0.36\pi t^2$

$A \circ r$ represents the area of the circle at time t.

59. Even   **61.** Odd   **63.** (a) $\left(\frac{3}{2}, 4\right)$ (b) $\left(\frac{3}{2}, -4\right)$

65. Proof   **67.** Proof

69. (a) $f(x) = x^2(4 - x^2)$ (b) $f(x) = x(1 - x^2)$

71. (a)

Height, x	Length and Width	Volume, V
1	$24 - 2(1)$	$1[24 - 2(1)]^2 = 484$
2	$24 - 2(2)$	$2[24 - 2(2)]^2 = 800$
3	$24 - 2(3)$	$3[24 - 2(3)]^2 = 972$
4	$24 - 2(4)$	$4[24 - 2(4)]^2 = 1024$
5	$24 - 2(5)$	$5[24 - 2(5)]^2 = 980$
6	$24 - 2(6)$	$6[24 - 2(6)]^2 = 864$

Guess of maximum volume: 1024 cubic centimeters

(b)

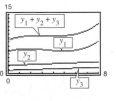

(c) $V = 4x(12 - x)^2$

Domain: $(0, 12)$

V is a function of x.

(d) $4 \times 16 \times 16$ centimeters

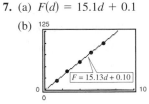

73. False: if $f(x) = x^2$, then $f(-1) = f(1)$.   **75.** True

Section P.4 (page 33)

1. Quadratic   **3.** Linear

5. (a) and (b)

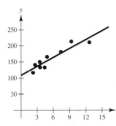

Approximately linear

(c) 136

7. (a) $F(d) = 15.1d + 0.1$

(b)

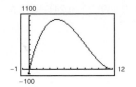

$F = 15.13d + 0.10$

Answers will vary.

(c) 3.6 centimeters

9. (a) $y = 0.08x + 5.0$

$r \approx 0.705$

(b)

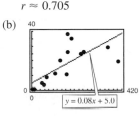

$y = 0.08x + 5.0$

(c) Greater per capita energy usage by a country tends to relate to greater per capita gross national product of the country. Japan, Denmark, Canada

(d) $y = 0.10x + 1.1$

$r \approx 0.9202$

11. (a) $y_1 = 0.03434t^3 - 0.3451t^2 + 0.884t + 5.61$

$y_2 = 0.110t + 2.07$

$y_3 = 0.092t + 0.79$

(b) $y_1 + y_2 + y_3 = 0.03434t^3 - 0.3451t^2 + 1.086t + 8.47$

$y_1 + y_2 + y_3$

y_1

y_2

y_3

31.1 cents per mile

13. (a) Linear: $y_1 = 4.04t + 29.0$

Cubic: $y_2 = -0.010t^3 + 0.549t^2 + 0.24t + 33.1$

(b)

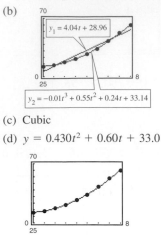

$y_1 = 4.04t + 28.96$

$y_2 = -0.01t^3 + 0.55t^2 + 0.24t + 33.14$

(c) Cubic

(d) $y = 0.430t^2 + 0.60t + 33.0$

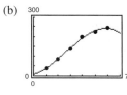

(e) The slope represents the average increase per year in the number of people receiving care in HMOs.

(f) Linear: 69.4 million

Cubic: 80.4 million

15. (a) $y = -1.81x^3 + 14.58x^2 + 16.39x + 10$

(b)

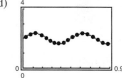

(c) 214

17. (a) Yes. At time t there is one and only one displacement y.

(b) Amplitude: 0.35; Period: 0.5

(c) $y = 0.35 \sin(4\pi t) + 2$

(d)

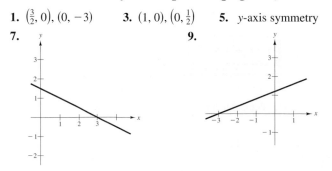

19. Answers will vary.

Review Exercises for Chapter P (page 36)

1. $\left(\frac{3}{2}, 0\right), (0, -3)$ **3.** $(1, 0), \left(0, \frac{1}{2}\right)$ **5.** y-axis symmetry

7.

9.

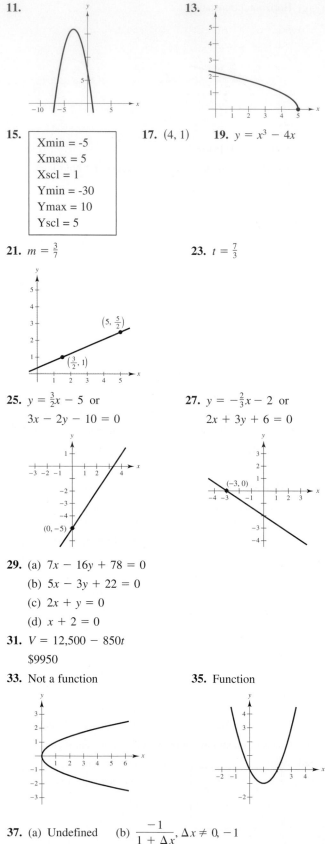

11.

13.

15.

Xmin = -5
Xmax = 5
Xscl = 1
Ymin = -30
Ymax = 10
Yscl = 5

17. $(4, 1)$ **19.** $y = x^3 - 4x$

21. $m = \frac{3}{7}$ **23.** $t = \frac{7}{3}$

25. $y = \frac{3}{2}x - 5$ or

$3x - 2y - 10 = 0$

27. $y = -\frac{2}{3}x - 2$ or

$2x + 3y + 6 = 0$

29. (a) $7x - 16y + 78 = 0$

(b) $5x - 3y + 22 = 0$

(c) $2x + y = 0$

(d) $x + 2 = 0$

31. $V = 12,500 - 850t$

$9950

33. Not a function   **35.** Function

37. (a) Undefined   (b) $\dfrac{-1}{1 + \Delta x}$, $\Delta x \neq 0, -1$

39. (a) $D: [-6, 6];$ $R: [0, 6]$

(b) $D: (-\infty, 5) \cup (5, \infty);$ $R: (-\infty, 0) \cup (0, \infty)$

(c) $D: (-\infty, \infty);$ $R: (-\infty, \infty)$

41. (a) (b)

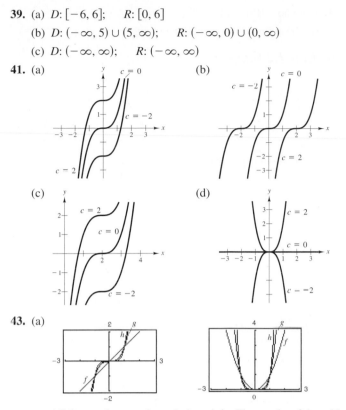

(c) (d)

43. (a)

All the graphs pass through the origin. The graphs of the odd powers of x are symmetric with respect to the origin and the graphs of the even powers are symmetric with respect to the y-axis. As the powers increase, the graphs become flatter in the interval $-1 < x < 1$. Graphs of these equations with odd powers pass through Quadrants I and III. Graphs of these equations with even powers pass through Quadrants I and II.

(b) The graph of $y = x^7$ should pass through the origin and Quadrants I and III. It should be symmetric with respect to the origin and be fairly flat in the interval $(-1, 1)$. The graph of $y = x^8$ should pass through the origin and Quadrants I and II. It should be symmetric with respect to the y-axis and be fairly flat in the interval $(-1, 1)$.

45. (a) $A = x(12 - x)$

(b) Domain: $(0, 12)$

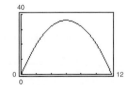

(c) Maximum area: 36; 6×6 inches

47. (a) Minimum degree: 3; Leading coefficient: negative

(b) Minimum degree: 4; Leading coefficient: positive

(c) Minimum degree: 2; Leading coefficient: negative

(d) Minimum degree: 5; Leading coefficient: positive

49. (a) Yes. For each time t there corresponds one and only one displacement y.

(b) Amplitude: 0.25; Period: 1.1

(c) $y = \frac{1}{4}\cos(5.7t)$

(d)

P.S. Problem Solving (page 38)

1. (a) Center: $(3, 4);$ Radius: 5

(b) $y = -\frac{3}{4}x$ (c) $y = \frac{3}{4}x - \frac{9}{2}$ (d) $\left(3, -\frac{9}{4}\right)$

3.

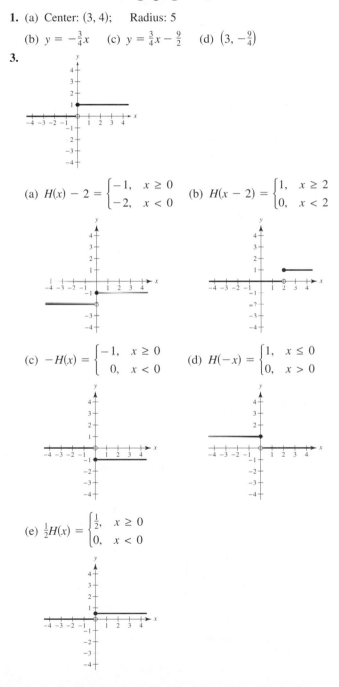

(a) $H(x) - 2 = \begin{cases} -1, & x \geq 0 \\ -2, & x < 0 \end{cases}$ (b) $H(x - 2) = \begin{cases} 1, & x \geq 2 \\ 0, & x < 2 \end{cases}$

(c) $-H(x) = \begin{cases} -1, & x \geq 0 \\ 0, & x < 0 \end{cases}$ (d) $H(-x) = \begin{cases} 1, & x \leq 0 \\ 0, & x > 0 \end{cases}$

(e) $\frac{1}{2}H(x) = \begin{cases} \frac{1}{2}, & x \geq 0 \\ 0, & x < 0 \end{cases}$

(f) $-H(x-2) + 2 = \begin{cases} 1, & x \geq 2 \\ 2, & x < 2 \end{cases}$

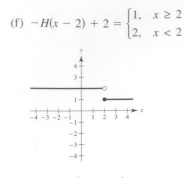

5. (a) $A(x) = x\left(\dfrac{100 - x}{2}\right)$;   Domain: (0, 100)

(b)

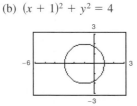

Dimensions 50 m × 25 m yield maximum area of 1250 square meters.

(c) 50 m × 25 m; Area = 1250 square meters

7. $T(x) = \dfrac{2\sqrt{4 + x^2} + \sqrt{(3 - x)^2 + 1}}{4}$

9. (a) 5, less   (b) 3, greater   (c) 4.1, less   (d) $4 + h$
(e) 4; Answers will vary.

11. (a) $x = 1, x = -3$ 13. Answers will vary.
(b) $(x + 1)^2 + y^2 = 4$

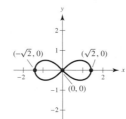

Chapter 1

Section 1.1 (page 47)

1. Precalculus: 300 feet

3. Calculus: Slope of the tangent line at $x = 2$ is 0.16.

5. Precalculus: $\frac{15}{2}$ square units

7. Precalculus: 24 cubic units

9. (a)

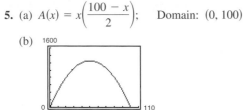

(b) The graphs of y_2 approach the tangent line to y_1 at $x = 1$.

(c) 2; Use numbers increasingly closer to zero such as 0.2, 0.01, 0.001, ...

---

11. (a) 5.66   (b) 6.11
(c) Increase the number of line segments.

Section 1.2 (page 54)

1.

x	1.9	1.99	1.999	2.001	2.01	2.1
$f(x)$	0.3448	0.3344	0.3334	0.3332	0.3322	0.3226

$\lim\limits_{x \to 2} \dfrac{x - 2}{x^2 - x - 2} \approx 0.3333$ $\left(\text{Actual limit is } \dfrac{1}{3}.\right)$

3.

x	-0.1	-0.01	-0.001	0.001	0.01	0.1
$f(x)$	0.2911	0.2889	0.2887	0.2887	0.2884	0.2863

$\lim\limits_{x \to 0} \dfrac{\sqrt{x + 3} - \sqrt{3}}{x} \approx 0.2887$ $\left(\text{Actual limit is } \dfrac{1}{2\sqrt{3}}.\right)$

5.

x	2.9	2.99	2.999
$f(x)$	-0.0641	-0.0627	-0.0625

x	3.001	3.01	3.1
$f(x)$	-0.0625	-0.0623	-0.0610

$\lim\limits_{x \to 3} \dfrac{[1/(x + 1)] - (1/4)}{x - 3} \approx -0.0625$ $\left(\text{Actual limit is } -\dfrac{1}{16}.\right)$

7.

x	-0.1	-0.01	-0.001	0.001	0.01	0.1
$f(x)$	0.9983	0.99998	1.0000	1.0000	0.99998	0.9983

$\lim\limits_{x \to 0} \dfrac{\sin x}{x} \approx 1.0000$ (Actual limit is 1.)

9. 1   11. 2

13. Limit does not exist. The function approaches 1 from the right side of 5 but it approaches -1 from the left side of 5.

15. Limit does not exist. The function increases without bound as x approaches $\dfrac{\pi}{2}$ from the left and decreases without bound as x approaches $\dfrac{\pi}{2}$ from the right.

17. Limit does not exist. The function oscillates between 1 and -1 as x approaches 0.

19. (a)

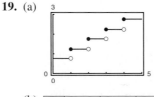

(b)

t	3	3.3	3.4	3.5	3.6	3.7	4
C	1.75	2.25	2.25	2.25	2.25	2.25	2.25

$\lim\limits_{t \to 3.5} C(t) = 2.25$

(c)

t	2	2.5	2.9	3	3.1	3.5	4
C	1.25	1.75	1.75	1.75	2.25	2.25	2.25

The limit does not exist, because the limits from the right and left are not equal.

21. $\delta = \dfrac{1}{11} \approx 0.91$ **23.** $L = 8$. Let $\delta = \dfrac{0.01}{3} \approx 0.0033$.

25. $L = 1$. Assume $1 < x < 3$ and let $\delta = \dfrac{0.01}{5} = 0.002$.

27. 5   **29.** -3 **31.** 3   **33.** 0   **35.** 4   **37.** 2

39. Answers will vary.   **41.** Answers will vary.

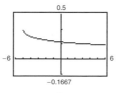

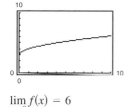

$\lim\limits_{x\to4} f(x) = \frac{1}{6}$

Domain: $[-5, 4) \cup (4, \infty)$

The graph has a hole at $x = 4$.

$\lim\limits_{x\to9} f(x) = 6$

Domain: $[0, 9) \cup (9, \infty)$

The graph has a hole at $x - 9$.

43. Answers will vary. Sample answer: As x approaches 8 from either side, $f(x)$ becomes arbitrarily close to 25.

45. Examples will vary.

Type 1: $f(x)$ approaches a different number from the right of c than it approaches from the left

Type 2: $f(x)$ increases or decreases without bound as x approaches c.

$\lim\limits_{x\to0} \dfrac{|2x|}{x}$

$\lim\limits_{x\to1} \left(\dfrac{1}{x-1}\right)^2$

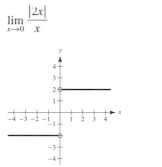

Type 3: $f(x)$ oscillates between two fixed values as x approaches c.

$\lim\limits_{x\to0} \left(2\cos\left(\dfrac{\pi}{x}\right)\right)$

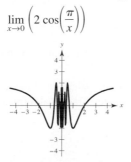

47.

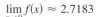

x	-0.001	-0.0001	-0.00001
f(x)	2.7196	2.7184	2.7183

x	0.00001	0.0001	0.001
f(x)	2.7183	2.7181	2.7169

$\lim\limits_{x\to0} f(x) \approx 2.7183$

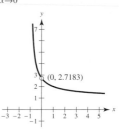

(0, 2.7183)

49. False: the existence or nonexistence of $f(x)$ at $x = c$ has no bearing on the existence of the limit of $f(x)$ as $x \to c$.

51. False: see Exercise 11.

53. Answers will vary.   **55.** Proof   **57.** Proof

Section 1.3 (page 65)

1.

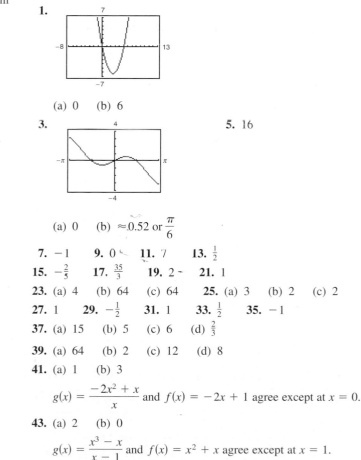

(a) 0   (b) 6

3.

(a) 0   (b) ≈ 0.52 or $\dfrac{\pi}{6}$

5. 16

7. -1 **9.** 0   **11.** 7   **13.** $\frac{1}{2}$

15. $-\frac{2}{5}$ **17.** $\frac{35}{3}$ **19.** 2   **21.** 1

23. (a) 4   (b) 64   (c) 64   **25.** (a) 3   (b) 2   (c) 2

27. 1   **29.** $-\frac{1}{2}$ **31.** 1   **33.** $\frac{1}{2}$ **35.** -1

37. (a) 15   (b) 5   (c) 6   (d) $\frac{2}{3}$

39. (a) 64   (b) 2   (c) 12   (d) 8

41. (a) 1   (b) 3

$g(x) = \dfrac{-2x^2 + x}{x}$ and $f(x) = -2x + 1$ agree except at $x = 0$.

43. (a) 2   (b) 0

$g(x) = \dfrac{x^3 - x}{x - 1}$ and $f(x) = x^2 + x$ agree except at $x = 1$.

45. -2

$f(x) = \dfrac{x^2 - 1}{x + 1}$ and $g(x) = x - 1$ agree except at $x = -1$.

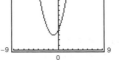

The graph has a hole at $x = -1$.

47. 12

$f(x) = \dfrac{x^3 - 8}{x - 2}$ and $g(x) = x^2 + 2x + 4$ agree except at $x = 2$.

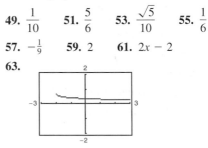

The graph has a hole at $x = 2$.

49. $\dfrac{1}{10}$ **51.** $\dfrac{5}{6}$ **53.** $\dfrac{\sqrt{5}}{10}$ **55.** $\dfrac{1}{6}$

57. $-\dfrac{1}{9}$ **59.** 2 **61.** $2x - 2$

63.

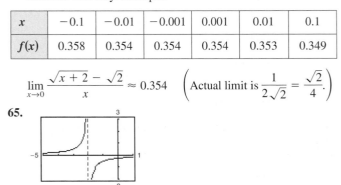

The graph has a hole at $x = 0$.

Answers will vary. Example:

x	-0.1	-0.01	-0.001	0.001	0.01	0.1
$f(x)$	0.358	0.354	0.354	0.354	0.353	0.349

$\displaystyle\lim_{x \to 0} \dfrac{\sqrt{x + 2} - \sqrt{2}}{x} \approx 0.354$ $\left(\text{Actual limit is } \dfrac{1}{2\sqrt{2}} = \dfrac{\sqrt{2}}{4}.\right)$

65.

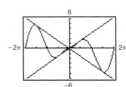

The graph has a hole at $x = 0$.

Answers will vary. Example:

x	-0.1	-0.01	-0.001	0.001	0.01	0.1
$f(x)$	-0.263	-0.251	-0.250	-0.250	-0.249	-0.238

$\displaystyle\lim_{x \to 0} \dfrac{[1/(2 + x)] - (1/2)}{x} \approx -0.250$ $\left(\text{Actual limit is } -\dfrac{1}{4}.\right)$

67. $\dfrac{1}{5}$ **69.** 0 **71.** 0 **73.** 0 **75.** 1 **77.** $\dfrac{3}{2}$

79.

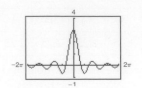

The graph has a hole at $t = 0$.

Answers will vary. Example:

t	-0.1	-0.01	0	0.01	0.1
$f(t)$	2.96	2.9996	$?$	2.9996	2.96

$\displaystyle\lim_{t \to 0} \dfrac{\sin 3t}{t} = 3$

81.

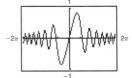

The graph has a hole at $x = 0$.

Answers will vary. Example:

x	-0.1	-0.01	-0.001	0	0.001	0.01	0.1
$f(x)$	-0.1	-0.01	-0.001	$?$	0.001	0.01	0.1

$\displaystyle\lim_{x \to 0} \dfrac{\sin x^2}{x} = 0$

83. 2 **85.** $-\dfrac{4}{x^2}$ **87.** 4

89. 0 **91.** 0

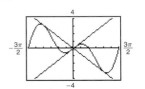 

93. 0

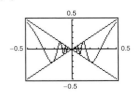

The graph has a hole at $x = 0$.

95. f and g agree at all but one point if c is a real number such that $f(x) = g(x)$ for all $x \neq c$.

97. An indeterminate form is obtained when evaluating a limit using direct substitution produces a meaningless fractional form, such as $\dfrac{0}{0}$.

99.

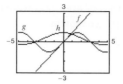

The magnitudes of $f(x)$ and $g(x)$ are approximately equal when x is "close to" 0. Therefore, their ratio is approximately 1.

101. 160 feet per second   **103.** -29.4 meters per second

105. Let $f(x) = \dfrac{1}{x}$ and $g(x) = -\dfrac{1}{x}$.

$\lim\limits_{x \to 0} f(x)$ and $\lim\limits_{x \to 0} g(x)$ do not exist.

However, $\lim\limits_{x \to 0} [f(x) + g(x)] = \lim\limits_{x \to 0} \left[\dfrac{1}{x} + \left(-\dfrac{1}{x} \right) \right] = \lim\limits_{x \to 0} 0 = 0$

and therefore does exist.

107. Proof   **109.** Proof   **111.** Proof

113. False. The limit does not exist because the function approaches 1 from the right side of 0 and approaches -1 from the left side of 0. (See graph below.)

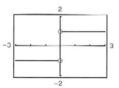

115. True. Theorem 1.7

117. False. The limit does not exist because $f(x)$ approaches 3 from the left side of 2 and approaches 0 from the right side of 2. (See graph below.)

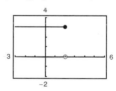

119. Let $f(x) = \begin{cases} 4, & \text{if } x \geq 0 \\ -4, & \text{if } x < 0 \end{cases}$

$\lim\limits_{x \to 0} |f(x)| = \lim\limits_{x \to 0} 4 = 4$

$\lim\limits_{x \to 0} f(x)$ does not exist because for $x < 0$, $f(x) = -4$ and for $x \geq 0$, $f(x) = 4$.

121. $\lim\limits_{x \to 0} f(x)$ does not exist because $f(x)$ oscillates between two fixed values as x approaches 0.

$\lim\limits_{x \to 0} g(x) = 0$ because, as x gets increasingly closer to 0, the values of $g(x)$ become increasingly closer to 0.

123. (a) $\dfrac{1}{2}$

(b) Because $\dfrac{1 - \cos x}{x^2} \approx \dfrac{1}{2}$, it follows that

$1 - \cos x \approx \dfrac{1}{2}x^2$

$\cos x \approx 1 - \dfrac{1}{2}x^2$ when $x \approx 0$.

(c) 0.995

(d) Calculator: $\cos(0.1) \approx .9950$

Section 1.4 *(page 76)*

1. (a) 1   (b) 1   (c) 1

$f(x)$ is continuous on $(-\infty, \infty)$.

3. (a) 0   (b) 0   (c) 0

Discontinuity at $x = 3$

5. (a) 2   (b) -2 (c) Limit does not exist.

Discontinuity at $x = 4$

7. $\dfrac{1}{10}$

9. Limit does not exist. The function decreases without bound as x approaches -3 from the left.

11. -1 **13.** $-\dfrac{1}{x^2}$ **15.** $\dfrac{5}{2}$ **17.** 2

19. Limit does not exist. The function decreases without bound as x approaches π from the left and increases without bound as x approaches π from the right.

21. 4

23. Limit does not exist. The function approaches 5 from the left side of 3 but approaches 6 from the right side of 3.

25. Discontinuous at $x = -2$ and $x = 2$

27. Discontinuous at every integer

29. Continuous on $[-5, 5]$ **31.** Continuous on $[-1, 4]$

33. Continuous for all real x **35.** Continuous for all real x

37. Nonremovable discontinuity at $x = 1$

Removable discontinuity at $x = 0$

39. Continuous for all real x

41. Removable discontinuity at $x = -2$

Nonremovable discontinuity at $x = 5$

43. Nonremovable discontinuity at $x = -2$

45. Continuous for all real x

47. Nonremovable discontinuity at $x = 2$

49. Continuous for all real x

51. Nonremovable discontinuities at integer multiples of $\dfrac{\pi}{2}$

53. Nonremovable discontinuity at each integer

55.

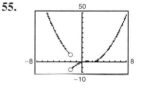

$\lim\limits_{x \to 0^+} f(x) = 0$

$\lim\limits_{x \to 0^-} f(x) = 0$

Discontinuity at $x = -2$

57. $a = 2$ **59.** $a = -1$, $b = 1$

61. Continuous for all real x

63. Nonremovable discontinuities at $x = 1$ and $x = -1$

65. Nonremovable discontinuity at each integer

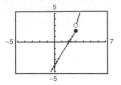

67. Discontinuous at $x = 3$

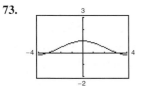

69. Continuous on $(-\infty, \infty)$

71. $f(x)$ is continuous on the open intervals
. . . $(-6, -2), (-2, 2), (2, 6), \ldots$.

73.

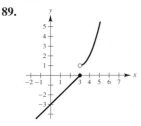

The graph has a hole at $x = 0$.

The graph appears continuous but the function is not continuous on $[-4, 4]$.

It is not obvious from the graph that the function has a discontinuity at $x = 0$.

75. Because $f(x)$ is continuous on the interval $[1, 2]$ and $f(1) = 2.0625$ and $f(2) = -4$, by the Intermediate Value Theorem there exists a real number c in $[1, 2]$ such that $f(c) = 0$.

77. Because $f(x)$ is continuous on the interval $[0, \pi]$ and $f(0) = -3$ and $f(\pi) \approx 8.87$, by the Intermediate Value Theorem there exists a real number c in $[0, \pi]$ such that $f(c) = 0$.

79. $0.68, 0.6823$ **81.** $0.56, 0.5636$

83. $f(3) = 11$ **85.** $f(2) = 4$

87. (a) The limit does not exist at $x = c$.

(b) The function is not defined at $x = c$.

(c) The limit exists, but it is not equal to the value of the function at $x = c$.

(d) The limit does not exist at $x = c$.

89.

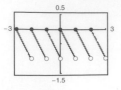

Not continuous because $\lim_{x \to 3} f(x)$ does not exist.

91. $g(x) = f(x)$ where x is an integer, but $g(x) = f(x) + 1$ elsewhere.

93. The function is discontinuous at every even positive integer. The company must replenish every two months.

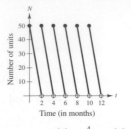

95. Because $V(1) = \frac{4}{3}\pi$, $V(5) = 523.6$, and V is continuous, there is at least one real number r, $1 \le r \le 5$, such that $V(r) = 275$.

97. If c is an element of the real numbers, then $\lim_{x \to c} f(x)$ does not exist since there are both rational and irrational numbers arbitrarily close to c. Therefore, f is not continuous at c.

99.

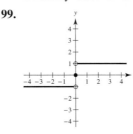

(a) -1

(b) 1

(c) Limit does not exist.

101. True

103. False. $f(x)$ is not defined at $x = 1$.

105. (a) $f(x) = \begin{cases} 0, & 0 \le x < b \\ b, & b < x \le 2b \end{cases}$

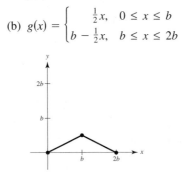

$f(x)$ is not continuous. There is a discontinuity at $x = b$.

(b) $g(x) = \begin{cases} \frac{1}{2}x, & 0 \le x \le b \\ b - \frac{1}{2}x, & b \le x \le 2b \end{cases}$

$g(x)$ is continuous on $[0, 2b]$ because $g(x)$ is continuous on $[0, b]$ and on $[b, 2b]$, and $\lim_{x \to b} g(x) = g(b)$.

107. Domain: $[-c^2, 0) \cup (0, \infty)$; Let $f(0) = \dfrac{1}{2c}$.

109. $h(x)$ has a nonremovable discontinuity at every integer except 0.

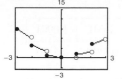

57.

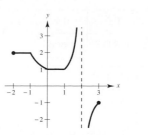

59. ∞

Section 1.5 (page 85)

1. $\lim\limits_{x \to -2^+} 2\left|\dfrac{x}{x^2 - 4}\right| = \infty$ $\lim\limits_{x \to -2^-} 2\left|\dfrac{x}{x^2 - 4}\right| = \infty$

3. $\lim\limits_{x \to -2^+} \tan \dfrac{\pi x}{4} = -\infty$ $\lim\limits_{x \to -2^-} \tan \dfrac{\pi x}{4} = \infty$

5.

x	-3.5	-3.1	-3.01	-3.001
$f(x)$	0.31	1.64	16.6	167

x	-2.999	-2.99	-2.9	-2.5
$f(x)$	-167	-16.6	-1.7	-0.36

$\lim\limits_{x \to -3^+} f(x) = -\infty$ $\lim\limits_{x \to -3^-} f(x) = \infty$

7.

x	3.5	-3.1	-3.01	-3.001
$f(x)$	3.8	16	151	1501

x	-2.999	-2.99	-2.9	-2.5
$f(x)$	-1499	-149	-14	-2.3

$\lim\limits_{x \to -3^+} f(x) = -\infty$ $\lim\limits_{x \to -3^-} f(x) = \infty$

9. $x = 0$ **11.** $x = 2, \quad x = -1$ **13.** $x = \pm 2$

15. No vertical asymptote

17. $x = \dfrac{\pi}{4} + \dfrac{n\pi}{2}$, n is an integer.   **19.** $t = 0$

21. $x = -2, \quad x = 1$ **23.** No vertical asymptote

25. No vertical asymptote   **27.** $t = n\pi$, n is a nonzero integer.

29. Removable discontinuity at $x = -1$

31. Vertical asymptote at $x = -1$

33. $-\infty$ **35.** ∞ **37.** $\frac{4}{5}$ **39.** $\frac{1}{2}$

41. $-\infty$ **43.** ∞ **45.** 0   **47.** Does not exist

49.

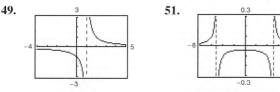

51.

$\lim\limits_{x \to 1^+} f(x) = \infty$ $\lim\limits_{x \to 5^-} f(x) = -\infty$

53. Answers will vary.

55. Answers will vary. Example: $f(x) = \dfrac{x - 3}{x^2 - 4x - 12}$

61. (a) \$176 million   (b) \$528 million   (c) \$1584 million
(d) ∞; As the percentage of drugs seized increases and approaches 100%, the cost to the government increases without bound.

63. (a) $\frac{7}{12}$ foot per second   (b) $\frac{3}{2}$ feet per second   (c) ∞

65. (a)

x	1	0.5	0.2	0.1
$f(x)$	0.1585	0.0411	0.0067	0.0017

x	0.01	0.001	0.0001
$f(x)$	1.7×10^{-5}	1.7×10^{-7}	1.7×10^{-9}

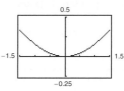

The graph has a hole at $x = 0$.

(b)

x	1	0.5	0.2	0.1
$f(x)$	0.1585	0.0823	0.0333	0.0167

x	0.01	0.001	0.0001
$f(x)$	0.0017	1.7×10^{-4}	1.7×10^{-5}

The graph has a hole at $x = 0$.

(c)

x	1	0.5	0.2	0.1
$f(x)$	0.1585	0.1646	0.1663	0.1666

x	0.01	0.001	0.0001
$f(x)$	0.1667	0.1667	0.1667

The graph has a hole at $x = 0$.

(d)

x	1	0.5	0.2	0.1
f(x)	0.1585	0.3292	0.8317	1.666

x	0.01	0.001	0.0001
f(x)	16.67	166.7	1667

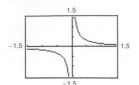

The value of the limit when the power on x in the denominator is greater than 3 is ∞.

67. (a) 850 revolutions per minute

(b) Reverse direction

(c) $L = 60 \cot \phi + 30(\pi + 2\phi)$

Domain: $\left(0, \dfrac{\pi}{2}\right)$

(d)

ϕ	0.3	0.6	0.9	1.2	1.5
L	306.2	217.9	195.9	189.6	188.5

(e)

(f) $60\pi \approx 188.5$

(g) ∞

69. False: let $f(x) = \dfrac{1}{x^2 + 1}$.   **71.** False: let $f(x) = \begin{cases} \dfrac{1}{x}, & x \neq 0 \\ 0, & x = 0 \end{cases}$

73. Proof   **75.** Proof

Review Exercises for Chapter 1 (page 88)

1. Calculus

Estimate: 8.261

3.

x	−0.1	−0.01	−0.001	0.001
f(x)	−1.0526	−1.0050	−1.0005	−0.9995

x	0.01	0.1
f(x)	−0.9950	−0.9524

The estimate of the limit of f(x), as x approaches zero, is −1.00.

5. (a) −2   (b) −3

7. 2; Proof   **9.** 1; Proof   **11.** $\sqrt{6} \approx 2.45$ **13.** $-\frac{1}{4}$

15. $\dfrac{1}{4}$ **17.** −1   **19.** 75   **21.** 0   **23.** $\dfrac{\sqrt{3}}{2}$ **25.** $-\dfrac{1}{2}$

27. (a)

x	1.1	1.01	1.001	1.0001
f(x)	0.5680	0.5764	0.5773	0.5773

(b)

The graph has a hole at x = 1.

(c) $\displaystyle\lim_{x \to 1^+} \dfrac{\sqrt{2x + 1} - \sqrt{3}}{x - 1} \approx 0.577$ $\left(\text{Actual limit is } \dfrac{\sqrt{3}}{3}.\right)$

29. −39.2 meters per second   **31.** −1   **33.** 0

35. Limit does not exist. The limit as t approaches 1 from the left is 2 whereas the limit as t approaches 1 from the right is 1.

37. Nonremovable discontinuity at each integer
Continuous on $(k, k + 1)$ for all integers k

39. Removable discontinuity at x = 1
Continuous on $(-\infty, 1) \cup (1, \infty)$

41. Nonremovable discontinuity at x = 2
Continuous on $(-\infty, 2) \cup (2, \infty)$

43. Nonremovable discontinuity at x = −1
Continuous on $(-\infty, -1) \cup (-1, \infty)$

45. Nonremovable discontinuity at each even integer
Continuous on $(2k, 2k + 2)$ for all integers k

47. $c = -\frac{1}{2}$ **49.** Proof

51. (a) −4   (b) 4   (c) Limit does not exist.

53. x = 0   **55.** x = 10   **57.** $-\infty$ **59.** $\frac{1}{3}$

61. $-\infty$ **63.** $-\infty$ **65.** $\frac{4}{5}$ **67.** ∞

69. (a) \$14,117.65   (b) \$80,000.00   (c) \$720,000.00   (d) ∞

P.S. Problem Solving (page 90)

1. (a) Perimeter $\triangle PAO = 1 + \sqrt{(x^2 - 1)^2 + x^2} + \sqrt{x^4 + x^2}$
Perimeter $\triangle PBO = 1 + \sqrt{x^4 + (x - 1)^2} + \sqrt{x^4 + x^2}$

(b)

x	4	2	1
Perimeter $\triangle PAO$	33.0166	9.0777	3.4142
Perimeter $\triangle PBO$	33.7712	9.5952	3.4142
r(x)	0.9777	0.9461	1.0000

x	0.1	0.01
Perimeter $\triangle PAO$	2.0955	2.0100
Perimeter $\triangle PBO$	2.0006	2.0000
r(x)	1.0475	1.005

(c) 1

3. (a) Area (hexagon) $= \dfrac{3\sqrt{3}}{2} \approx 2.5981$

Area (circle) $= \pi \approx 3.1416$

Area (circle) $-$ Area (hexagon) ≈ 0.5435

(b) $A_n = \dfrac{n}{2} \sin\left(\dfrac{2\pi}{n}\right)$

(c)

n	6	12	24	48	96
A_n	2.5981	3.0000	3.1058	3.1326	3.1394

(d) 3.1416 or π

5. (a) $m = -\dfrac{12}{5}$ (b) $y = \dfrac{5}{12}x - \dfrac{169}{12}$

(c) $m_x = \dfrac{-\sqrt{169 - x^2} + 12}{x - 5}$

(d) $\dfrac{5}{12}$; It is the same as the slope of the tangent line found in (b).

7. (a) Domain: $[-27, 1) \cup (1, \infty)$

(b)

The graph has a hole at $x = 1$.

(c) $\dfrac{1}{14}$ (d) $\dfrac{1}{12}$

9. (a) g_1, g_4 (b) g_1 (c) g_1, g_3, g_4

11.

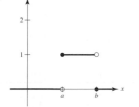

The graph jumps at every integer.

(a) $f(1) = 0$, $f(0) = 0$, $f\left(\dfrac{1}{2}\right) = -1$, $f(-2.7) = -1$

(b) $\lim\limits_{x \to 1^-} f(x) = -1$, $\lim\limits_{x \to 1^+} f(x) = -1$, $\lim\limits_{x \to 1/2} f(x) = -1$

(c) There is a discontinuity at each integer.

13. (a)

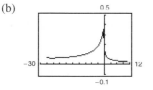

(b) (i) $\lim\limits_{x \to a^+} P_{a,b}(x) = 1$ (ii) $\lim\limits_{x \to a^-} P_{a,b}(x) = 0$

(iii) $\lim\limits_{x \to b^+} P_{a,b}(x) = 0$ (iv) $\lim\limits_{x \to b^-} P_{a,b}(x) = 1$

(c) Continuous for all positive real numbers except a and b

(d) The area under the curve gives a value of 1.

Chapter 2

Section 2.1 (page 101)

1. (a) $m = 0$ (b) $m = -3$

3.

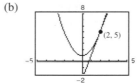

$y = \dfrac{f(4) - f(1)}{4 - 1}(x - 1) + f(1) = x + 1$

5. $m = -2$ **7.** $m = 2$ **9.** $m = 3$

11. $f'(x) = 0$ **13.** $f'(x) = -5$ **15.** $h'(s) = \dfrac{2}{3}$

17. $f'(x) = 4x + 1$ **19.** $f'(x) = 3x^2 - 12$

21. $f'(x) = \dfrac{-1}{(x - 1)^2}$ **23.** $f'(x) = \dfrac{1}{2\sqrt{x + 1}}$

25. (a) Tangent line: $y = 4x - 3$

(b)

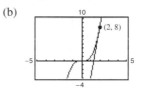

27. (a) Tangent line: $y = 12x - 16$

(b)

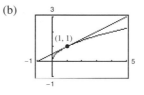

29. (a) Tangent line: $y = \dfrac{1}{2}x + \dfrac{1}{2}$

(b)

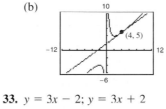

31. (a) Tangent line: $y = \dfrac{3}{4}x + 2$

(b)

33. $y = 3x - 2$; $y = 3x + 2$

35. $y = -\dfrac{1}{2}x + \dfrac{3}{2}$

37. $g(5) = 2; g'(5) = -\frac{1}{2}$

39. b   **40.** d   **41.** a   **42.** c

43. Answers will vary. Sample answer: $y = -x$

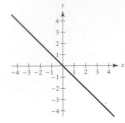

45. (a) $f'(-c) = 3$ (b) $f'(-c) = -3$

47. $y = 2x + 1; y = -2x + 9$

49. (a) -3

(b) 0

(c) The graph is moving downward to the right when $x = 1$.

(d) The graph is moving upward to the right when $x = -4$.

(e) Positive. Because $g'(x) > 0$ on $[3, 6]$, the graph of g is moving upward to the right.

(f) No. Knowing only $g'(2)$ is not sufficient information. $g'(2)$ remains the same for any vertical translation of g.

51.

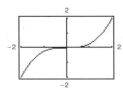

x	-2	-1.5	-1	-0.5	0	0.5	1	1.5	2
$f(x)$	-2	$-\frac{27}{32}$	$-\frac{1}{4}$	$-\frac{1}{32}$	0	$\frac{1}{32}$	$\frac{1}{4}$	$\frac{27}{32}$	2
$f'(x)$	3	$\frac{27}{16}$	$\frac{3}{4}$	$\frac{3}{16}$	0	$\frac{3}{16}$	$\frac{3}{4}$	$\frac{27}{16}$	3

53.

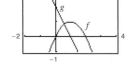

$g(x) \approx f'(x)$

55. $f(2) = 4; f(2.1) = 3.99; f'(2) \approx -0.1;$ Exact $f'(2) = 0$

57.

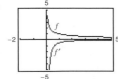

As x approaches infinity, the graph of f approaches a line of slope 0. Thus $f'(x)$ approaches 0.

59. (a)

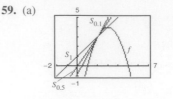

(b) The graphs of S for decreasing values of Δx are secant lines approaching the tangent line to the graph of f at the point $(2, f(2))$.

61. 4   **63.** 4

65. $g(x)$ is not differentiable at $x = 0$.

67. $f(x)$ is not differentiable at $x = 6$.

69. $h(x)$ is not differentiable at $x = -5$.

71. $(-\infty, -3) \cup (-3, \infty)$ **73.** $(-\infty, -1) \cup (-1, \infty)$

75. $(-\infty, 3) \cup (3, \infty)$ **77.** $(1, \infty)$ **79.** $(-\infty, 0) \cup (0, \infty)$

81. The derivative from the left is -1 and from the right is 1, so f is not differentiable at $x = 1$.

83. The derivative from both the right and left is 0, so $f'(1) = 0$.

85. f is differentiable at $x = 2$.

87. (a) $d = \dfrac{3|m + 1|}{\sqrt{m^2 + 1}}$

(b)

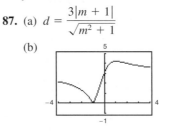

Not differentiable at $m = -1$

89. False. It is $\lim\limits_{\Delta x \to 0} \dfrac{f(2 + \Delta x) - f(2)}{\Delta x}$.

91. False. For example: $f(x) = |x|$. The derivative from the left and the derivative from the right both exist but are not equal.

93. Proof

Section 2.2 (page 113)

1. (a) $\frac{1}{2}$ (b) $\frac{3}{2}$ (c) 2   (d) 3

3. 0   **5.** $6x^5$ **7.** $-\dfrac{7}{x^8}$ **9.** $\dfrac{1}{5x^{4/5}}$ **11.** 1

13. $-4t + 3$ **15.** $2x + 12x^2$ **17.** $3t^2 - 2$

19. $\dfrac{\pi}{2} \cos \theta + \sin \theta$ **21.** $2x + \dfrac{1}{2} \sin x$ **23.** $-\dfrac{1}{x^2} - 3 \cos x$

	Function	Rewrite	Derivative	Simplify
25.	$y = \dfrac{5}{2x^2}$	$y = \dfrac{5}{2}x^{-2}$	$y' = -5x^{-3}$	$y' = -\dfrac{5}{x^3}$
27.	$y = \dfrac{3}{(2x)^3}$	$y = \dfrac{3}{8}x^{-3}$	$y' = -\dfrac{9}{8}x^{-4}$	$y' = -\dfrac{9}{8x^4}$
29.	$y = \dfrac{\sqrt{x}}{x}$	$y = x^{-1/2}$	$y' = -\dfrac{1}{2}x^{-3/2}$	$y' = -\dfrac{1}{2x^{3/2}}$

31. -6 **33.** 0   **35.** 4   **37.** 3   **39.** $2x + \dfrac{6}{x^3}$

41. $2t + \dfrac{12}{t^4}$ **43.** $\dfrac{x^3 - 8}{x^3}$ **45.** $3x^2 + 1$

47. $\dfrac{1}{2\sqrt{x}} - \dfrac{2}{x^{2/3}}$ **49.** $\dfrac{4}{5s^{1/5}} - \dfrac{2}{3s^{1/3}}$ **51.** $\dfrac{3}{\sqrt{x}} - 5\sin x$

53. (a) $2x + y - 2 = 0$ **55.** (a) $3x + 2y - 7 = 0$

(b)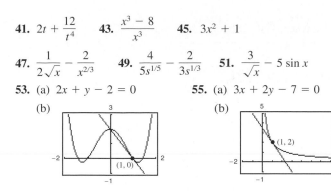

57. $(0, 2), (-2, -14), (2, -14)$ **59.** No horizontal tangents

61. (π, π) **63.** $k = 2, k = -10$ **65.** $k = 3$

67. (a) A and B (b) Greater

(c)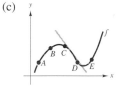

69. $g'(x) = f'(x)$

71.

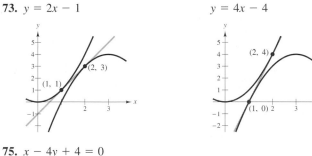

The rate of change of f is constant and therefore f' is a constant function.

73. $y = 2x - 1$ $y = 4x - 4$

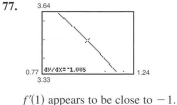

75. $x - 4y + 4 = 0$

77.

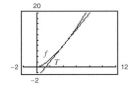

$f'(1)$ appears to be close to -1.
$f'(1) = -1$

79. (a)

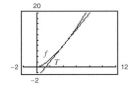

$(3.9, 7.7019)$, $S(x) = 2.981x - 3.924$

(b) $T(x) = 3(x - 4) + 8 = 3x - 4$

The slope (and equation) of the secant line approaches that of the tangent line at $(4, 8)$ as you choose points closer to $(4, 8)$.

(c) It becomes less accurate.

(d)

Δx	-3	-2	-1	-0.5	-0.1	0
$f(4 + \Delta x)$	1	2.828	5.196	6.458	7.702	8
$T(4 + \Delta x)$	-1	2	5	6.5	7.7	8

Δx	0.1	0.5	1	2	3
$f(4 + \Delta x)$	8.302	9.546	11.180	14.697	18.520
$T(4 + \Delta x)$	8.3	9.5	11	14	17

81. False: let $f(x) = x$ and $g(x) = x + 1$.

83. False: $dy/dx = 0$.   **85.** True

87. Average rate: 2 **89.** Average rate: $\frac{1}{2}$

Instantaneous rates: Instantaneous rates:

$f'(1) = 2$ $f'(1) = 1$

$f'(2) = 2$ $f'(2) = \frac{1}{4}$

91. (a) $s(t) = -16t^2 + 1362$

$v(t) = -32t$

(b) -48 feet per second

(c) $s'(1) = -32$ feet per second

$s'(2) = -64$ feet per second

(d) $t = \dfrac{\sqrt{1362}}{4} \approx 9.226$ seconds

(e) -295.242 feet per second

93. $v(5) = 71$ meters per second

$v(10) = 22$ meters per second

95.   **97.**

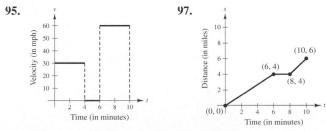

99. (a) $R(v) = 0.167v - 0.02$

    (b) $B(v) = 0.006v^2 - 0.024v + 0.460$

    (c) $T(v) = 0.006v^2 + 0.143v + 0.440$

    (d)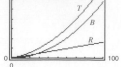

    (e) $T'(v) = 0.012v + 0.143$

        $T'(40) = 0.623$

        $T'(80) = 1.103$

        $T'(100) = 1.343$

    (f) Stopping distance increases at an increasing rate.

101. 8 square meters per meter change in s

103. $-\$1.91,\ -\$1.93$

105. (a) The rate of change of gallons of gasoline sold when the price is \$1.479

    (b) In general, the rate of change when $p = 1.479$ should be negative.

107. $y = 2x^2 - 3x + 1$ **109.** $y = -9x,\ y = -\frac{9}{4}x - \frac{27}{4}$

111. $a = \frac{1}{3},\ b = -\frac{4}{3}$ **113.** Proof

Section 2.3 (page 124)

1. $2(2x^3 - 3x^2 + x - 1)$ **3.** $\dfrac{7t^2 + 4}{3t^{2/3}}$

5. $x^2(3\cos x - x\sin x)$ **7.** $\dfrac{1 - x^2}{(x^2 + 1)^2}$

9. $\dfrac{1 - 8x^3}{3x^{2/3}(x^3 + 1)^2}$ **11.** $\dfrac{x\cos x - 2\sin x}{x^3}$

13. $f'(x) = (x^3 - 3x)(4x + 3) + (2x^2 + 3x + 5)(3x^2 - 3)$

        $= 10x^4 + 12x^3 - 3x^2 - 18x - 15$

    $f'(0) = -15$

15. $f'(x) = \dfrac{x^2 - 6x + 4}{(x - 3)^2}$ **17.** $f'(x) = \cos x - x\sin x$

    $f'(1) = -\dfrac{1}{4}$ $f'\!\left(\dfrac{\pi}{4}\right) = \dfrac{\sqrt{2}}{8}(4 - \pi)$

Function	Rewrite	Differentiate	Simplify
19. $y = \dfrac{x^2 + 2x}{3}$	$y = \dfrac{1}{3}(x^2 + 2x)$	$y' = \dfrac{1}{3}(2x + 2)$	$y' = \dfrac{2(x + 1)}{3}$
21. $y = \dfrac{7}{3x^3}$	$y = \dfrac{7}{3}x^{-3}$	$y' = -7x^{-4}$	$y' = -\dfrac{7}{x^4}$
23. $y = \dfrac{4x^{3/2}}{x}$	$y = 4x^{1/2},$ $x > 0$	$y' = 2x^{-1/2}$	$y' = \dfrac{2}{\sqrt{x}},$ $x > 0$

25. $\dfrac{(x^2 - 1)(-2 - 2x) - (3 - 2x - x^2)(2x)}{(x^2 - 1)^2} = \dfrac{2}{(x + 1)^2},\ x \neq 1$

27. $1 - \dfrac{12}{(x + 3)^2} = \dfrac{x^2 + 6x - 3}{(x + 3)^2}$

29. $\dfrac{\sqrt{x}(2) - (2x + 5)\dfrac{1}{2\sqrt{x}}}{x} = \dfrac{2x - 5}{2x^{3/2}}$

31. $6s^2(s^3 - 2)$ **33.** $-\dfrac{2x^2 - 2x + 3}{x^2(x - 3)^2}$

35. $(3x^3 + 4x)[(x - 5) \cdot 1 + (x + 1) \cdot 1]$

        $+ [(x - 5)(x + 1)](9x^2 + 4)$

        $= 15x^4 - 48x^3 - 33x^2 - 32x - 20$

37. $\dfrac{(x^2 - c^2)(2x) - (x^2 + c^2)(2x)}{(x^2 - c^2)^2} = -\dfrac{4xc^2}{(x^2 - c^2)^2}$

39. $t(t\cos t + 2\sin t)$ **41.** $-\dfrac{t\sin t + \cos t}{t^2}$

43. $-1 + \sec^2 x = \tan^2 x$ **45.** $\dfrac{1}{4t^{3/4}} + 8\sec t\tan t$

47. $\dfrac{-6\cos^2 x + 6\sin x - 6\sin^2 x}{4\cos^2 x} = \dfrac{3}{2}(-1 + \tan x\sec x - \tan^2 x)$

                                   $= \dfrac{3}{2}\sec x(\tan x - \sec x)$

49. $\csc x\cot x - \cos x = \cos x\cot^2 x$ **51.** $x(x\sec^2 x + 2\tan x)$

53. $2x\cos x + 2\sin x - x^2\sin x + 2x\cos x$

    $= 4x\cos x + (2 - x^2)\sin x$

55. $\left(\dfrac{x + 1}{x + 2}\right)(2) + (2x - 5)\left[\dfrac{(x + 2)(1) - (x + 1)(1)}{(x + 2)^2}\right]$

    $= \dfrac{2x^2 + 8x - 1}{(x + 2)^2}$

57. $\dfrac{1 - \sin\theta + \theta\cos\theta}{(1 - \sin\theta)^2}$ **59.** $y' = \dfrac{-2\csc x\cot x}{(1 - \csc x)^2},\ -4\sqrt{3}$

61. $h'(t) = \dfrac{\sec t(t\tan t - 1)}{t^2},\ \dfrac{1}{\pi^2}$

63. (a) $y = -x - 2$ **65.** (a) $y = -x + 4$

    (b) 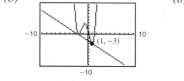   (b)

67. (a) $4x - 2y - \pi + 2 = 0$

    (b)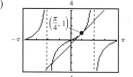

69. $(0, 0),\ (2, 4)$ **71.** $f(x) + 2 = g(x)$

73. $n = 1,\ f'(x) = x\cos x + \sin x$

    $n = 2,\ f'(x) = x^2\cos x + 2x\sin x$

    $n = 3,\ f'(x) = x^3\cos x + 3x^2\sin x$

    $n = 4,\ f'(x) = x^4\cos x + 4x^3\sin x$

    $f'(x) = x^n\cos x + nx^{n-1}\sin x$

75. $\dfrac{6t + 1}{2\sqrt{t}}$ square centimeters per second

77. (a) $-\$38.13$ (b) $-\$10.37$ (c) $-\$3.80$

The costs decrease with increasing order size.

79. 31.55 bacteria per hour   **81.** Proof   **83.** $\dfrac{3}{\sqrt{x}}$

85. $\dfrac{2}{(x - 1)^3}$ **87.** $-3\sin x$ **89.** $2x$ **91.** $\dfrac{1}{\sqrt{x}}$

93. Answers will vary. For example: $(x - 2)^2$

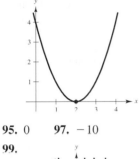

95. 0   **97.** -10

99.

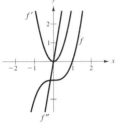

101. $v(3) = 27$ meters per second

$a(3) = -6$ meters per second per second

The speed of the object is decreasing, but the rate of that decrease is increasing.

103. (a) 2.4 ft/sec^2 (b) 1.2 ft/sec^2 (c) 0.5 ft/sec^2

105. (a) $f''(x) = g(x)h''(x) + 2g'(x)h'(x) + g''(x)h(x)$

$f'''(x) = g(x)h'''(x) + 3g'(x)h''(x) +$
$\qquad 3g''(x)h'(x) + g'''(x)h(x)$

$f^{(4)}(x) = g(x)h^{(4)}(x) + 4g'(x)h'''(x) + 6g''(x)h''(x) +$
$\qquad 4g'''(x)h'(x) + g^{(4)}(x)h(x)$

(b) $f^{(n)}(x) = g(x)h^{(n)}(x) + \dfrac{n!}{1!(n - 1)!}g'(x)h^{(n-1)}(x) +$

$\qquad \dfrac{n!}{2!(n - 2)!}g''(x)h^{(n-2)}(x) + \cdots +$

$\qquad \dfrac{n!}{(n - 1)!1!}g^{(n-1)}(x)h'(x) + g^{(n)}(x)h(x)$

107. (a) $P_1(x) = -\dfrac{\sqrt{3}}{2}\left(x - \dfrac{\pi}{3}\right) + \dfrac{1}{2}$

$P_2(x) = -\dfrac{1}{4}\left(x - \dfrac{\pi}{3}\right)^2 - \dfrac{\sqrt{3}}{2}\left(x - \dfrac{\pi}{3}\right) + \dfrac{1}{2}$

(b)

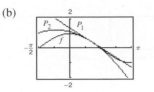

(c) P_2

(d) P_1 and P_2 become less accurate as you move farther from $x = a$.

109. False: $dy/dx = f(x)g'(x) + g(x)f'(x)$ **111.** True

113. True   **115.** $f'(x) = 2|x|$; $f''(0)$ does not exist.

Section 2.4 (page 133)

$y = f(g(x))$	$u = g(x)$	$y = f(u)$
1. $y = (6x - 5)^4$	$u = 6x - 5$	$y = u^4$
3. $y = \sqrt{x^2 - 1}$	$u = x^2 - 1$	$y = \sqrt{u}$
5. $y = \csc^3 x$	$u = \csc x$	$y = u^3$

7. $6(2x - 7)^2$ **9.** $-108(4 - 9x)^3$

11. $\dfrac{2}{3}(9 - x^2)^{-1/3}(-2x) = -\dfrac{4x}{3(9 - x^2)^{1/3}}$

13. $\dfrac{1}{2}(1 - t)^{-1/2}(-1) = -\dfrac{1}{2\sqrt{1 - t}}$

15. $\dfrac{1}{3}(9x^2 + 4)^{-2/3}(18x) = \dfrac{6x}{(9x^2 + 4)^{2/3}}$

17. $\dfrac{1}{2}(4 - x^2)^{-3/4}(-2x) = \dfrac{-x}{\sqrt[4]{(4 - x^2)^3}}$

19. $-\dfrac{1}{(x - 2)^2}$ **21.** $-2(t - 3)^{-3}(1) = -\dfrac{2}{(t - 3)^3}$

23. $-\dfrac{1}{2(x + 2)^{3/2}}$

25. $x^2[4(x - 2)^3(1)] + (x - 2)^4(2x) = 2x(x - 2)^3(3x - 2)$

27. $x\left(\dfrac{1}{2}\right)(1 - x^2)^{-1/2}(-2x) + (1 - x^2)^{1/2}(1) = \dfrac{1 - 2x^2}{\sqrt{1 - x^2}}$

29. $\dfrac{(x^2 + 1)^{1/2}(1) - x(1/2)(x^2 + 1)^{-1/2}(2x)}{x^2 + 1} = \dfrac{1}{(x^2 + 1)^{3/2}}$

31. $\dfrac{-2(x + 5)(x^2 + 10x - 2)}{(x^2 + 2)^3}$ **33.** $\dfrac{-9(2v - 1)^2}{(v + 1)^4}$

35. $\dfrac{1 - 3x^2 - 4x^{3/2}}{2\sqrt{x}(x^2 + 1)^2}$

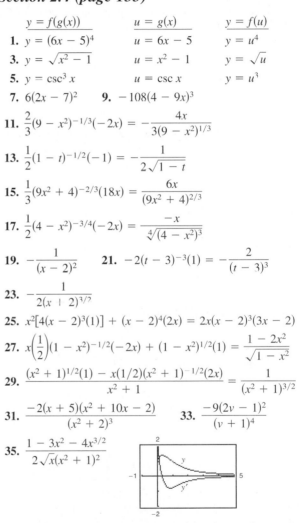

The zero of y' corresponds to the point on the graph of the function where the tangent line is horizontal.

37. $\dfrac{3t(t^2 + 3t - 2)}{(t^2 + 2t - 1)^{3/2}}$

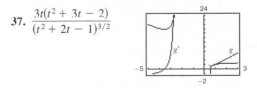

The zeros of $g'(t)$ correspond to the points on the graph of the function where the tangent line is horizontal.

39. $-\dfrac{\sqrt{\dfrac{x+1}{x}}}{2x(x+1)}$

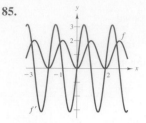

y' has no zeros.

41. $\dfrac{t}{\sqrt{1+t}}$

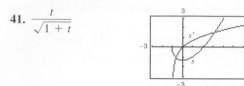

The zero of $s'(t)$ corresponds to the point on the graph of the function where the tangent line is horizontal.

43. $-\dfrac{\pi x \sin \pi x + \cos \pi x + 1}{x^2}$

The zeros of y' correspond to the points on the graph of the function where the tangent lines are horizontal.

45. (a) 1   (b) 2; The slope of $\sin ax$ at the origin is a.

47. $-3 \sin 3x$ **49.** $12 \sec^2 4x$ **51.** $2\pi^2 x \cos(\pi x)^2$

53. $2 \cos(4x)$ **55.** $\dfrac{-1 - \cos^2 x}{\sin^3 x}$

57. $8 \sec^2 x \tan x = \dfrac{8 \sin x}{\cos^3 x}$ **59.** $\sin 2\theta \cos 2\theta = \dfrac{1}{2} \sin 4\theta$

61. $\dfrac{6\pi \sin(\pi t - 1)}{\cos^3(\pi t - 1)}$ **63.** $\dfrac{1}{2\sqrt{x}} + 2x \cos(2x)^2$

65. $-\sin x \cos(\cos x)$ **67.** $s'(t) = \dfrac{t+1}{\sqrt{t^2 + 2t + 8}}, \dfrac{3}{4}$

69. $f'(x) = \dfrac{-9x^2}{(x^3 - 4)^2}, -\dfrac{9}{25}$ **71.** $f'(t) = \dfrac{-5}{(t-1)^2}, -5$

73. $y' = -6 \sec^3(2x) \tan(2x), 0$

75. (a) $9x - 5y - 2 = 0$ **77.** (a) $2x - y - 2\pi = 0$

(b)   (b)

79. $12(5x^2 - 1)(x^2 - 1)$ **81.** $2(\cos x^2 - 2x^2 \sin x^2)$

83.

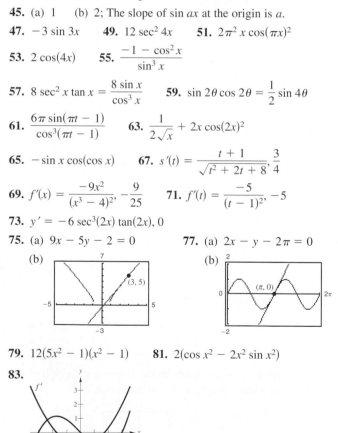

The zeros of f' correspond to the points where the graph of f has horizontal tangents.

85.

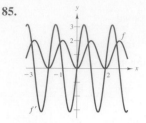

The zeros of f' correspond to the points where the graph of f has horizontal tangents.

87. The rate of change of g will be three times as fast as the rate of change of f.

89. (a) 24   (b) Not possible because $g'(h(5))$ is not known.

(c) $\dfrac{4}{3}$ (d) 162

91. (a) 1.461   (b) -1.016

93. 0.2 radian, 1.45 radians per second   **95.** 0.04224

97. (a) $x = -1.637t^3 + 19.31t^2 - 0.5t - 1$

(b) $\dfrac{dC}{dt} = -294.66t^2 + 2317.2t - 30$

(c) Because x, the number of units produced in t hours, is not a linear function, and therefore the cost with respect to time t is not linear.

99. (a) $f'(x) = \beta \cos \beta x$

$f''(x) = -\beta^2 \sin \beta x$

$f'''(x) = -\beta^3 \cos \beta x$

$f^{(4)}(x) = \beta^4 \sin \beta x$

(b) $f''(x) + \beta^2 f(x) = -\beta^2 \sin \beta x + \beta^2 \sin \beta x = 0$

(c) $f^{(2k)}(x) = (-1)^k \beta^{2k} \sin \beta x$

$f^{(2k-1)}(x) = (-1)^{k+1} \beta^{2k-1} \cos \beta x$

101. (a) 0   (b) $\dfrac{5}{8}$ **103.** Proof   **105.** $\dfrac{2(2x-3)}{|2x-3|}, x \neq \dfrac{3}{2}$

107. $-|x| \sin x + \dfrac{x}{|x|} \cos x, \quad x \neq 0$

109. (a) $P_1(x) = \dfrac{\pi}{2}(x - 1) + 1$

$P_2(x) = \dfrac{\pi^2}{8}(x - 1)^2 + \dfrac{\pi}{2}(x - 1) + 1$

(b)   (c) P_2

(d) P_1 and P_2 become less accurate as you move farther from $x = 1$.

111. False. $y' = -\dfrac{1}{2}(1 - x)^{-1/2}$ **113.** True

Section 2.5 (page 142)

1. $-\dfrac{x}{y}$ **3.** $-\sqrt{\dfrac{y}{x}}$ **5.** $\dfrac{y - 3x^2}{2y - x}$ **7.** $\dfrac{1 - 3x^2y^3}{3x^3y^2 - 1}$

9. $\dfrac{6xy - 3x^2 - 2y^2}{4xy - 3x^2}$ **11.** $\dfrac{\cos x}{4 \sin 2y}$

13. $\dfrac{\cos x - \tan y - 1}{x \sec^2 y}$ **15.** $\dfrac{y \cos(xy)}{1 - x \cos(xy)}$

17. (a) $y_1 = \sqrt{16 - x^2}$

$y_2 = -\sqrt{16 - x^2}$

(b)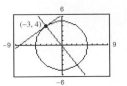

(c) $y' = \mp \dfrac{x}{\sqrt{16 - x^2}} = -\dfrac{x}{y}$

(d) $y' = -\dfrac{x}{y}$

19. (a) $y_1 = \dfrac{3}{4}\sqrt{16 - x^2}$

$y_2 = -\dfrac{3}{4}\sqrt{16 - x^2}$

(b)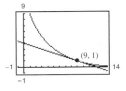

(c) $y' = \mp \dfrac{3x}{4\sqrt{16 - x^2}} = -\dfrac{9x}{16y}$

(d) $y' = -\dfrac{9x}{16y}$

21. $-\dfrac{y}{x}, \ -\dfrac{1}{4}$ **23.** $\dfrac{8x}{y(x^2 + 4)^2}$, Undefined

25. $-\sqrt[3]{\dfrac{y}{x}}, \ -\dfrac{1}{2}$ **27.** $-\sin^2(x + y)$ or $-\dfrac{x^2}{x^2 + 1}, \ 0$

29. $-\dfrac{1}{2}$ **31.** 0 **33.** $\cos^2 y, \ -\dfrac{\pi}{2} < y < \dfrac{\pi}{2}, \ \dfrac{1}{1 + x^2}$

35. $-\dfrac{36}{y^3}$ **37.** $-\dfrac{16}{y^3}$ **39.** $\dfrac{3x}{4y}$

41. $x + 3y - 12 = 0$

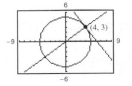

43. At $(4, 3)$:

Tangent line: $4x + 3y - 25 = 0$

Normal line: $3x - 4y = 0$

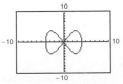

At $(-3, 4)$:

Tangent line: $3x - 4y + 25 = 0$

Normal line: $4x + 3y = 0$

(continued)

45. $x^2 + y^2 = r^2 \Rightarrow y' = -\dfrac{x}{y} \Rightarrow \dfrac{y}{x} = $ slope of normal line. Then for (x_0, y_0) on the circle, $x_0 \neq 0$, an equation of the normal line is $y = \dfrac{y_0}{x_0}x$, which passes through the origin. If $x_0 = 0$, the normal line is vertical and passes through the origin.

47. Horizontal tangents: $(-4, 0), (-4, 10)$

Vertical tangents: $(0, 5), (-8, 5)$

49.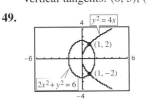

At $(1, 2)$:

Slope of ellipse: -1

Slope of parabola: 1

At $(1, -2)$:

Slope of ellipse: 1

Slope of parabola: -1

51.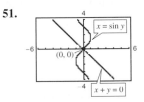

At $(0, 0)$:

Slope of line: -1

Slope of sine curve: 1

53.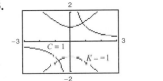

Derivatives: $\dfrac{dy}{dx} = -\dfrac{y}{x}, \ \dfrac{dy}{dx} = \dfrac{x}{y}$

55. (a) $4y\dfrac{dy}{dx} - 12x^3 = 0$ (b) $4y\dfrac{dy}{dt} - 12x^3\dfrac{dx}{dt} = 0$

57. (a) $-\pi \sin \pi y \left(\dfrac{dy}{dx}\right) - 3\pi \cos \pi x = 0$

(b) $-\pi \sin \pi y \left(\dfrac{dy}{dt}\right) - 3\pi \cos \pi x \left(\dfrac{dx}{dt}\right) = 0$

59. Answers will vary. In the explicit form of a function, the variable is explicitly written as a function of x. In an implicit equation, the function is only implied by an equation. An example of an implicit function is $x^2 + xy = 5$. In explicit form it would be $y = \dfrac{5 - x^2}{x}$.

61. (a) $x^4 = 4(4x^2 - y^2)$

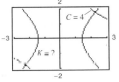

(b)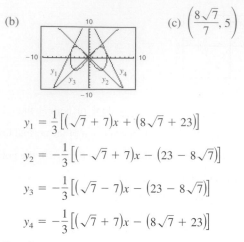

(c) $\left(\dfrac{8\sqrt{7}}{7}, 5\right)$

$$y_1 = \frac{1}{3}\left[(\sqrt{7} + 7)x + (8\sqrt{7} + 23)\right]$$

$$y_2 = -\frac{1}{3}\left[(-\sqrt{7} + 7)x - (23 - 8\sqrt{7})\right]$$

$$y_3 = -\frac{1}{3}\left[(\sqrt{7} - 7)x - (23 - 8\sqrt{7})\right]$$

$$y_4 = -\frac{1}{3}\left[(\sqrt{7} + 7)x - (8\sqrt{7} + 23)\right]$$

63. Proof

Section 2.6 (page 149)

1. (a) $\frac{3}{4}$ (b) 20   **3.** (a) $-\frac{5}{8}$ (b) $\frac{3}{2}$

5. (a) -4 centimeters per second

    (b) 0 centimeter per second

    (c) 4 centimeters per second

7. (a) 8 centimeters per second

    (b) 4 centimeters per second

    (c) 2 centimeters per second

9. (a) Positive   (b) Negative

11. In a linear function, if x changes at a constant rate, so does y. However, unless $a = 1$, y does not change at the same rate as x.

13. $\dfrac{2(2x^3 + 3x)}{\sqrt{x^4 + 3x^2 + 1}}$

15. (a) 36π square centimeters per minute

    (b) 144π square centimeters per minute

17. (a) Proof

    (b) When $\theta = \dfrac{\pi}{6}$, $\dfrac{dA}{dt} = \dfrac{\sqrt{3}}{8}s^2$.

     When $\theta = \dfrac{\pi}{3}$, $\dfrac{dA}{dt} = \dfrac{1}{8}s^2$.

    (c) If s and $d\theta/dt$ are constant, dA/dt is proportional to $\cos\theta$.

19. (a) $\dfrac{2}{9\pi}$ centimeter per minute

    (b) $\dfrac{1}{18\pi}$ centimeter per minute

21. (a) 36 square centimeters per second

    (b) 360 square centimeters per second

23. $\dfrac{8}{405\pi}$ foot per minute

25. (a) 12.5%   (b) $\frac{1}{144}$ meter per minute

27. (a) $-\frac{7}{12}$ foot per second; $-\frac{3}{2}$ feet per second;

     $-\frac{48}{7}$ feet per second

    (b) $\frac{527}{24}$ square feet per second   (c) $\frac{1}{12}$ radian per second

29. Rate of vertical change: $\frac{1}{5}$ meter per second

    Rate of horizontal change: $-\dfrac{\sqrt{3}}{15}$ meter per second

31. (a) -750 miles per hour   (b) 20 minutes

33. $-\dfrac{28}{\sqrt{10}} \approx -8.85$ feet per second

35. (a) $\frac{25}{3}$ feet per second   (b) $\frac{10}{3}$ feet per second

37. (a) 12 seconds

    (b) $\dfrac{1}{2}\sqrt{3}$ meter

    (c) $\dfrac{\sqrt{5}\,\pi}{120}$ meter per second

39. Evaporation rate proportional to

$$S \Longrightarrow \frac{dV}{dt} = k(4\pi r^2)$$

$$V = \left(\frac{4}{3}\right)\pi r^3 \Longrightarrow \frac{dV}{dt} = 4\pi r^2 \frac{dr}{dt}$$

    So $k = \dfrac{dr}{dt}$.

41. $V^{0.3}\left(1.3p\dfrac{dV}{dt} + V\dfrac{dp}{dt}\right) = 0$ **43.** $\dfrac{1}{20}$ radian per second

45. (a) $\frac{1}{2}$ radian per minute   (b) $\frac{3}{2}$ radians per minute

    (c) 1.87 radians per minute

47. (a) $\dfrac{dx}{dt} = -600\pi \sin\theta$

    (b)

    (c) $\theta = 90° + n \cdot 180°$; $\theta = 0° + n \cdot 180°$

    (d) -300π centimeters per second;

     $-300\sqrt{3}\pi$ centimeters per second

49. $\dfrac{1}{25}\cos^2\theta$, $-\dfrac{\pi}{4} \le \theta \le \dfrac{\pi}{4}$

51. -0.1808 foot per second per second

53. (a) $m(s) = -0.881s^2 + 29.10s - 206.2$

    (b) $(-1.762s + 29.10)\dfrac{ds}{dt}$ (c) 2.15 million

Review Exercises for Chapter 2 (page 153)

1. $f'(x) = 2x - 2$ **3.** $f'(x) = \dfrac{1}{2\sqrt{x}} = \dfrac{\sqrt{x}}{2x}$

5. f is differentiable at all $x \ne -1$.

7.

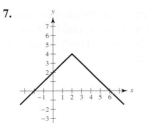

(a) Yes

(b) No, because the derivatives from the left and right are not equal.

9. $-\frac{3}{2}$

11. (a) $y = 3x + 1$ **13.** 8

(b)

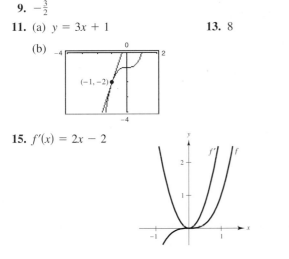

15. $f'(x) = 2x - 2$

$f' > 0$ where the slopes of tangent lines to the graph of f are positive.

17. 0   **19.** $8x^7$ **21.** $12t^3$ **23.** $3x(x - 2)$

25. $\dfrac{3}{\sqrt{x}} + \dfrac{1}{x^{2/3}}$ **27.** $-\dfrac{4}{3t^3}$ **29.** $2 - 3\cos\theta$

31. $-3\sin\theta - \dfrac{\cos\theta}{4}$

33. (a) 50 vibrations per second per pound

    (b) 33.33 vibrations per second per pound

35. 414.74 meters or 1,354 feet

37. (a)

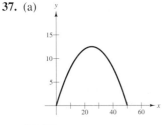

    (b) 50

    (c) $x - 25$

    (d) $y' = 1 - 0.04x$

x	0	10	25	30	50
y'	1	0.6	0	-0.2	-1

    (e) $y'(25) = 0$

39. (a) $x'(t) = 2t - 3$ (b) $(-\infty, 1.5)$ (c) $x = -\frac{1}{4}$ (d) 1

41. $2(6x^3 - 9x^2 + 16x - 7)$ **43.** $\sqrt{x}\cos x + \dfrac{1}{2\sqrt{x}}\sin x$

45. $2 + \dfrac{2}{x^3}$ **47.** $-\dfrac{x^2 + 1}{(x^2 - 1)^2}$ **49.** $\dfrac{6x}{(4 - 3x^2)^2}$

51. $\dfrac{2x\cos x + x^2\sin x}{\cos^2 x}$ **53.** $3x^2\sec x\tan x + 6x\sec x$

55. $-x\sec^2 x - \tan x$ **57.** $-x\sin x$ **59.** $6t$

61. $6\sec^2\theta\tan\theta$

63. $y'' + y = -(2\sin x + 3\cos x) + (2\sin x + 3\cos x) = 0$

65. $\dfrac{-3x^2}{2\sqrt{1 - x^3}}$ **67.** $\dfrac{2(x - 3)(-x^2 + 6x + 1)}{(x^2 + 1)^3}$

69. $s(s^2 - 1)^{3/2}(8s^3 - 3s + 25)$ **71.** $-9\sin(3x + 1)$

73. $-\csc 2x\cot 2x$ **75.** $\frac{1}{2}(1 - \cos 2x) = \sin^2 x$

77. $\sin^{1/2} x\cos x - \sin^{5/2} x\cos x = \cos^3 x\sqrt{\sin x}$

79. $\dfrac{(x + 2)[\pi\cos(\pi x)] - \sin(\pi x)}{(x + 2)^2}$

81. $t(t - 1)^4(7t - 2)$

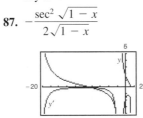

The zeros of f' correspond to the points on the graph of the function where the tangent line is horizontal.

83. $\dfrac{x + 2}{(x + 1)^{3/2}}$ **85.** $\dfrac{5}{6(t + 1)^{1/6}}$

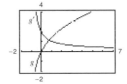   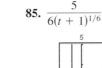

g' is not equal to zero for any x. f' has no zeros.

87. $-\dfrac{\sec^2\sqrt{1 - x}}{2\sqrt{1 - x}}$

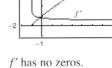

y' has no zeros.

89. $4 - 4\sin 2x$ **91.** $2\csc^2 x\cot x$

93. $\dfrac{2(t + 2)}{(1 - t)^4}$ **95.** $18\sec^2(3\theta)\tan(3\theta) + \sin(\theta - 1)$

97. (a) -18.667 degrees per hour

    (b) -7.284 degrees per hour

    (c) -3.240 degrees per hour

    (d) -0.747 degree per hour

99. $-\dfrac{2x+3y}{3(x+y^2)}$ **101.** $\dfrac{2y\sqrt{x}-y\sqrt{y}}{2x\sqrt{y}-x\sqrt{x}}$ **103.** $\dfrac{y\sin x+\sin y}{\cos x-x\cos y}$

105. Tangent line: $x+2y-10=0$

Normal line: $2x-y=0$

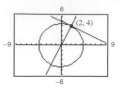

107. (a) $2\sqrt{2}$ units/sec   (b) 4 units/sec   (c) 8 units/sec

109. $\frac{2}{25}$ meter per minute   **111.** -38.34 meters per second

P.S. Problem Solving (page 156)

1. (a) $r=\frac{1}{2}$ (b) Center: $\left(0,\frac{5}{4}\right)$

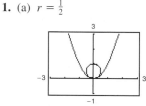

3. (a) $P_1(x)=1$

(b) $P_2(x)=1-\frac{1}{2}x^2$

(c)

x	-1.0	-0.1	-0.001	0	0.001
$\cos x$	0.5403	0.9950	1.000	1	1
$P_2(x)$	0.5000	0.9950	1.000	1	1

x	0.1	1.0
$\cos x$	0.9950	0.5403
$P_2(x)$	0.9950	0.5000

$P_2(x)$ is a good approximation of $f(x)=\cos x$ when x is very close to 0.

(d) $P_3(x)=x-\frac{1}{6}x^3$

5. $p(x)=2x^3+4x^2-5$

7. (a) Graph $\begin{cases} y_1=\dfrac{1}{a}\sqrt{x^2(a^2-x^2)} \\ y_2=-\dfrac{1}{a}\sqrt{x^2(a^2-x^2)} \end{cases}$ as separate equations.

(b) Answers will vary.

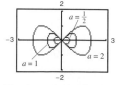

The intercepts will always be $(0,0)$, $(a,0)$, and $(-a,0)$, and the maximum and minimum y-values appear to be $\pm\frac{1}{2}a$.

(c) $\left(\dfrac{a\sqrt{2}}{2},\dfrac{a}{2}\right)$, $\left(\dfrac{a\sqrt{2}}{2},-\dfrac{a}{2}\right)$, $\left(-\dfrac{a\sqrt{2}}{2},\dfrac{a}{2}\right)$, $\left(-\dfrac{a\sqrt{2}}{2},-\dfrac{a}{2}\right)$

9. (a) When the man is 90 feet from the light, the tip of his shadow is $112\frac{1}{2}$ feet from the light. The tip of the child's shadow is $111\frac{1}{9}$ feet from the light, so the man's shadow extends $1\frac{7}{18}$ feet beyond the child's shadow.

(b) When the man is 60 feet from the light, the tip of his shadow is 75 feet from the light. The tip of the child's shadow is $77\frac{7}{9}$ feet from the light, so the child's shadow extends $2\frac{7}{9}$ feet beyond the man's shadow.

(c) $d=80$ feet

(d) Let x be the distance of the man from the light and s be the distance from the light to the tip of the shadow.

If $0<x<80$, $\dfrac{ds}{dt}=-\dfrac{50}{9}$.

If $x>80$, $\dfrac{ds}{dt}=-\dfrac{25}{4}$.

There is a discontinuity at $x=80$.

11. Proof. The graph of L is a line passing through the origin $(0,0)$.

13. (a)

$z°$	0.1	0.01	0.0001
$\dfrac{\sin z}{z}$	0.01745241	0.0174532924	0.0174532925

(b) $\dfrac{\pi}{180}$ (c) $\dfrac{\pi}{180}\cos z$ (d) $\dfrac{\pi}{180}C(z)$

(e) Answers will vary.

15. (a) j would be the rate of change of acceleration/deceleration.

(b) $j=0$. Deceleration is constant, so there is no change in deceleration.

Chapter 3

Section 3.1 (page 165)

1. $f'(0)=0$ **3.** $f'(3)=0$

5. $f'(-2)$ is undefined.   **7.** 2, absolute maximum

9. 1, absolute maximum; 2, absolute minimum; 3, absolute maximum

11. $x=0,\ x=2$ **13.** $t=\dfrac{8}{3}$ **15.** $x=\dfrac{\pi}{3},\pi,\dfrac{5\pi}{3}$

17. Minimum: $(2,2)$
Maximum: $(-1,8)$

19. Minimum: $(0,0)$ and $(3,0)$
Maximum: $\left(\frac{3}{2},\frac{9}{4}\right)$

21. Minimum: $\left(-1,-\frac{5}{2}\right)$
Maximum: $(2,2)$

23. Minimum: $(0,0)$
Maximum: $(-1,5)$

25. Minimum: $(0,0)$
Maximum: $\left(-1,\frac{1}{4}\right)$ and $\left(1,\frac{1}{4}\right)$

27. Minimum: $(1,-1)$
Maximum: $\left(0,-\frac{1}{2}\right)$

29. Minimum: $\left(\dfrac{1}{6},\dfrac{\sqrt{3}}{2}\right)$
Maximum: $(0,1)$

31. Minimum: $(2,3)$
Maximum: $\left(1,\sqrt{2}+3\right)$

33. (a) Minimum: $(0,-3)$;
Maximum: $(2,1)$
(b) Minimum: $(0,-3)$
(c) Maximum: $(2,1)$
(d) No extrema

35. (a) Minimum: $(1,-1)$;
Maximum: $(-1,3)$
(b) Minimum: $(3,3)$
(c) Minimum: $(1,-1)$
(d) Minimum: $(1,-1)$

37.

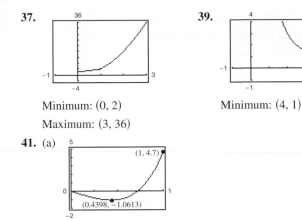

Minimum: (0, 2)

Maximum: (3, 36)

39.

Minimum: (4, 1)

41. (a)

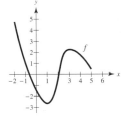

(b) Minimum: (0.4398, −1.0613)

43. Maximum: $\left|f''\left(\sqrt[3]{-10 + \sqrt{108}}\right)\right| = f''\left(\sqrt{3} - 1\right) \approx 1.47$

45. Maximum: $\left|f^{(4)}(0)\right| = \frac{56}{81}$

47. Because f is continuous on $\left[0, \frac{\pi}{4}\right]$, but not continuous on $[0, \pi]$.

49. Answers will vary. Example:

51. (a) Yes (b) No **53.** (a) No (b) Yes

55. Maximum: $P(12) = 72$

No. P is decreasing for $t \geq 12$.

57. 0.9553 radian

59. (a) $y = \frac{3}{40,000}x^2 - \frac{3}{200}x + \frac{75}{4}$

(b)

x	−500	−400	−300	−200	−100	0
d	0	0.75	3	6.75	12	18.75

x	100	200	300	400	500
d	12	6.75	3	0.75	0

(c) Lowest point ≈ (100, 18); No

61. True **63.** True

Section 3.2 (page 172)

1. $f(0) = f(2) = 0$; f is not differentiable on $(0, 2)$.

3. $(2, 0), (-1, 0); f'\left(\frac{1}{2}\right) = 0$ **5.** $(0, 0), (-4, 0); f'\left(-\frac{8}{3}\right) = 0$

7. $f'(1) = 0$ **9.** $f'\left(\frac{6 - \sqrt{3}}{3}\right) = 0; f'\left(\frac{6 + \sqrt{3}}{3}\right) = 0$

11. Not differentiable at $x = 0$ **13.** $f'\left(-2 + \sqrt{5}\right) = 0$

15. $f'\left(\frac{\pi}{2}\right) = 0; f'\left(\frac{3\pi}{2}\right) = 0$ **17.** $f'(0.249) \approx 0$

19. Not continuous on $[0, \pi]$

21.

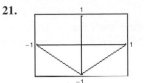

Rolle's Theorem does not apply.

23.

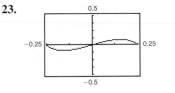

$f'(\pm 0.1533) = 0$

25. (a) $f(1) = f(2) = 64$

(b) Velocity $= 0$ for some t in $(-1, 2)$; $t = \frac{3}{2}$ seconds

27.

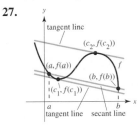

29. The function is discontinuous on $[0, 6]$.

31. $f'\left(-\frac{1}{2}\right) = -1$ **33.** $f'\left(\frac{8}{27}\right) = 1$

35. $f'\left(-\frac{1}{4}\right) = -\frac{1}{3}$ **37.** $f'\left(\frac{\pi}{2}\right) = 0$

39. Secant line: $2x - 3y - 2 = 0$

Tangent line: $c = \frac{-2 + \sqrt{6}}{2}$, $2x - 3y + 5 - 2\sqrt{6} = 0$

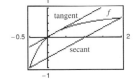

41. Secant line: $x - 4y + 3 = 0$

Tangent line: $c = 4$, $x - 4y + 4 = 0$

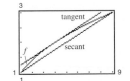

43. (a) −14.7 meters per second (b) 1.5 seconds

45. No. Let $f(x) = x^2$ on $[-1, 2]$.

47. By the Mean Value Theorem, there is a time when the speed of the plane must equal the average speed of 454.5 miles per hour. The speed was 400 miles per hour when the plane was accelerating to 454.5 miles per hour and decelerating from 454.5 miles per hour.

49. (a) f is continuous and changes signs in $[-10, 4]$ (Intermediate Value Theorem).

(b) There exist real numbers a and b such that $-10 < a < b < 4$ and $f(a) = f(b) = 2$. Therefore, f' has a zero in the interval by Rolle's Theorem.

(c) 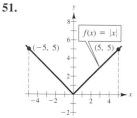 (d)

(e) No, by Theorem 2.1.

51.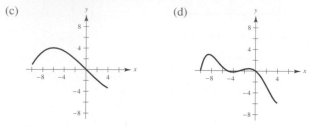

53. False. f is not continuous on $[-1, 1]$.

55. True   **57.** Proof   **59.** Proof   **61.** Proof

Section 3.3 (page 181)

1. Increasing on $(3, \infty)$; Decreasing on $(-\infty, 3)$

3. Increasing on $(-\infty, -2)$ and $(2, \infty)$; Decreasing on $(-2, 2)$

5. Increasing on $(-\infty, 0)$; Decreasing on $(0, \infty)$

7. Increasing on $(1, \infty)$; Decreasing on $(-\infty, 1)$

9. Increasing on $\left(-2\sqrt{2}, 2\sqrt{2}\right)$

Decreasing on $\left(-4, -2\sqrt{2}\right), \left(2\sqrt{2}, 4\right)$

11. Critical number: $x = 3$

Increasing on $(3, \infty)$

Decreasing on $(-\infty, 3)$

Relative minimum: $(3, -9)$

13. Critical number: $x = 1$

Increasing on $(-\infty, 1)$

Decreasing on $(1, \infty)$

Relative maximum: $(1, 5)$

15. Critical numbers: $x = -2, 1$

Increasing on $(-\infty, -2)$ and $(1, \infty)$

Decreasing on $(-2, 1)$

Relative maximum: $(-2, 20)$

Relative minimum: $(1, -7)$

17. Critical numbers: $x = 0, 2$

Increasing on $(0, 2)$

Decreasing on $(-\infty, 0), (2, \infty)$

Relative maximum: $(2, 4)$

Relative minimum: $(0, 0)$

19. Critical numbers: $x = -1, 1$

Increasing on $(-\infty, -1)$ and $(1, \infty)$

Decreasing on $(-1, 1)$

Relative maximum: $\left(-1, \frac{4}{5}\right)$

Relative minimum: $\left(1, -\frac{4}{5}\right)$

21. Critical number: $x = 0$

Increasing on $(-\infty, \infty)$

No relative extrema

23. Critical number: $x = 1$

Increasing on $(1, \infty)$

Decreasing on $(-\infty, 1)$

Relative minimum: $(1, 0)$

25. Critical number: $x = 5$

Increasing on $(-\infty, 5)$

Increasing on $(-\infty, 5)$

Relative maximum: $(5, 5)$

27. Critical numbers: $x = -1, 1$

Discontinuity: $x = 0$

Increasing on $(-\infty, -1)$ and $(1, \infty)$

Decreasing on $(-1, 0)$ and $(0, 1)$

Relative maximum: $(-1, -2)$

Relative minimum: $(1, 2)$

29. Critical number: $x = 0$

Discontinuities: $x = -3, 3$

Increasing on $(-\infty, -3)$ and $(-3, 0)$

Decreasing on $(0, 3)$ and $(3, \infty)$

Relative maximum: $(0, 0)$

31. Critical numbers: $x = -3, 1$

Discontinuity: $x = -1$

Increasing on $(-\infty, -3)$ and $(1, \infty)$

Decreasing on $(-3, -1)$ and $(-1, 1)$

Relative maximum: $(-3, -8)$

Relative minimum: $(1, 0)$

33. Critical numbers: $x = \dfrac{\pi}{6}, \dfrac{5\pi}{6}$

Increasing on $\left(0, \dfrac{\pi}{6}\right), \left(\dfrac{5\pi}{6}, 2\pi\right)$

Decreasing on $\left(\dfrac{\pi}{6}, \dfrac{5\pi}{6}\right)$

Relative maximum: $\left(\dfrac{\pi}{6}, \dfrac{\left[\pi + 6\sqrt{3}\right]}{12}\right)$

Relative minimum: $\left(\dfrac{5\pi}{6}, \dfrac{\left[5\pi - 6\sqrt{3}\right]}{12}\right)$

35. Critical numbers: $x = \dfrac{\pi}{2}, \dfrac{7\pi}{6}, \dfrac{3\pi}{2}, \dfrac{11\pi}{6}$

Increasing on $\left(0, \dfrac{\pi}{2}\right), \left(\dfrac{7\pi}{6}, \dfrac{3\pi}{2}\right), \left(\dfrac{11\pi}{6}, 2\pi\right)$

Decreasing on $\left(\dfrac{\pi}{2}, \dfrac{7\pi}{6}\right), \left(\dfrac{3\pi}{2}, \dfrac{11\pi}{6}\right)$

Relative maxima: $\left(\dfrac{\pi}{2}, 2\right), \left(\dfrac{3\pi}{2}, 0\right)$

Relative minima: $\left(\dfrac{7\pi}{6}, -\dfrac{1}{4}\right), \left(\dfrac{11\pi}{6}, -\dfrac{1}{4}\right)$

37. (a) $f'(x) = \dfrac{2(9 - 2x^2)}{\sqrt{9 - x^2}}$

(b)

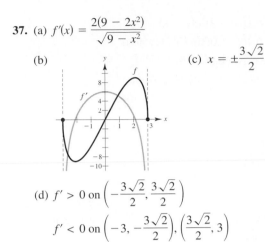

(c) $x = \pm\dfrac{3\sqrt{2}}{2}$

(d) $f' > 0$ on $\left(-\dfrac{3\sqrt{2}}{2}, \dfrac{3\sqrt{2}}{2}\right)$

$f' < 0$ on $\left(-3, -\dfrac{3\sqrt{2}}{2}\right), \left(\dfrac{3\sqrt{2}}{2}, 3\right)$

39. (a) $f'(t) = t(t \cos t + 2 \sin t)$

(b)

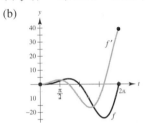

(c) Critical numbers: $t = 2.2889, 5.0870$

(d) $f' > 0$ on $(0, 2.2889), (5.0870, 2\pi)$

$f' < 0$ on $(2.2889, 5.0870)$

41. $f(x)$ is symmetric with respect to the origin.

Zeros: $(0, 0), (\pm\sqrt{3}, 0)$

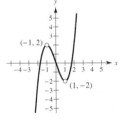

$g(x)$ is continuous on $(-\infty, \infty)$ and $f(x)$ has holes at $x = 1$ and $x = -1$.

43.

45.

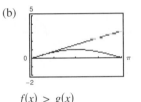

47.

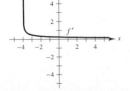

49. $g'(0) < 0$ **51.** $g'(-6) < 0$ **53.** $g'(0) > 0$

55.

57.

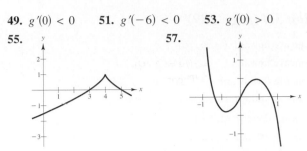

Minimum at the approximate critical number $x = -0.40$

Maximum at the approximate critical number $x = 0.48$

59. (a)

x	0.5	1	1.5	2	2.5	3
$f(x)$	0.5	1	1.5	2	2.5	3
$g(x)$	0.48	0.84	1.00	0.91	0.60	0.14

$f(x) > g(x)$

(b)

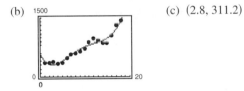

(c) Proof

$f(x) > g(x)$

61. $r = \dfrac{2R}{3}$ **63.** Maximum when $R_2 = R_1$

65. (a) $B = 0.11980t^4 - 4.4879t^3 + 56.991t^2 - 223.02t + 580.0$

(b)

(c) $(2.8, 311.2)$

67. (a) 3

(b) $a_3(0)^3 + a_2(0)^2 + a_1(0) + a_0 = 0$

$a_3(2)^3 + a_2(2)^2 + a_1(2) + a_0 = 2$

$3a_3(0)^2 + 2a_2(0) + a_1 = 0$

$3a_3(2)^2 + 2a_2(2) + a_1 = 0$

(c) $f(x) = -\dfrac{1}{2}x^3 + \dfrac{3}{2}x^2$

69. (a) 4

(b) $a_4(0)^4 + a_3(0)^3 + a_2(0)^2 + a_1(0) + a_0 = 0$

$a_4(2)^4 + a_3(2)^3 + a_2(2)^2 + a_1(2) + a_0 = 4$

$a_4(4)^4 + a_3(4)^3 + a_2(4)^2 + a_1(4) + a_0 = 0$

$4a_4(0)^3 + 3a_3(0)^2 + 2a_2(0) + a_1 = 0$

$4a_4(2)^3 + 3a_3(2)^2 + 2a_2(2) + a_1 = 0$

(c) $f(x) = \dfrac{1}{4}x^4 - 2x^3 + 4x^2$

71. True   **73.** False. Let $f(x) = x^3$.

75. False. Let $f(x) = x^3$. There is a critical number at $x = 0$, but not a relative extremum.

77. Proof   **79.** Proof

Section 3.4 (page 189)

1. Concave upward: $(-\infty, \infty)$

3. Concave upward: $(-\infty, -2), (2, \infty)$

Concave downward: $(-2, 2)$

5. Concave upward: $(-\infty, -1), (1, \infty)$

Concave downward: $(-1, 1)$

7. Concave upward: $(-\infty, 1)$

Concave downward: $(1, \infty)$

9. Concave upward: $\left(-\dfrac{\pi}{2}, 0\right)$

Concave downward: $\left(0, \dfrac{\pi}{2}\right)$

11. Point of inflection: $(2, 8)$

Concave downward: $(-\infty, 2)$

Concave upward: $(2, \infty)$

13. Points of inflection: $\left(\pm\dfrac{2}{\sqrt{3}}, -\dfrac{20}{9}\right)$

Concave upward: $\left(-\infty, -\dfrac{2}{\sqrt{3}}\right), \left(\dfrac{2}{\sqrt{3}}, \infty\right)$

Concave downward: $\left(-\dfrac{2}{\sqrt{3}}, \dfrac{2}{\sqrt{3}}\right)$

15. Points of inflection: $(2, -16), (4, 0)$

Concave upward: $(-\infty, 2), (4, \infty)$

Concave downward: $(2, 4)$

17. Concave upward: $(-3, \infty)$

19. Points of inflection: $\left(-\sqrt{3}, -\dfrac{\sqrt{3}}{4}\right), (0, 0), \left(\sqrt{3}, \dfrac{\sqrt{3}}{4}\right)$

Concave upward: $\left(-\sqrt{3}, 0\right), \left(\sqrt{3}, \infty\right)$

Concave downward: $\left(-\infty, -\sqrt{3}\right), \left(0, \sqrt{3}\right)$

21. Point of inflection: $(2\pi, 0)$

Concave upward: $(2\pi, 4\pi)$

Concave downward: $(0, 2\pi)$

23. Concave upward: $(0, \pi), (2\pi, 3\pi)$

Concave downward: $(\pi, 2\pi), (3\pi, 4\pi)$

25. Points of inflection: $(\pi, 0), (1.823, 1.452), (4.46, -1.452)$

Concave upward: $(1.823, \pi), (4.46, 2\pi)$

Concave downward: $(0, 1.823), (\pi, 4.46)$

27. Relative minimum: $(3, -25)$

29. Relative minimum: $(5, 0)$

31. Relative maximum: $(0, 3)$

Relative minimum: $(2, -1)$

33. Relative maximum: $(2.4, 268.74)$

Relative minimum: $(0, 0)$

35. Relative minimum: $(0, -3)$

37. Relative maximum: $(-2, -4)$

Relative minimum: $(2, 4)$

39. No relative extrema, because f is nonincreasing.

41. (a) $f'(x) = 0.2x(x - 3)^2(5x - 6)$

$f''(x) = 0.4(x - 3)(10x^2 - 24x + 9)$

(b) Relative maximum: $(0, 0)$

Relative minimum: $(1.2, -1.6796)$

Points of inflection: $(0.4652, -0.7048),$

$(1.9348, -0.9048), (3, 0)$

(c)

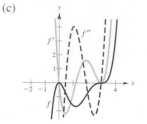

f is increasing when f' is positive, and decreasing when f' is negative. f is concave upward when f'' is positive, and concave downward when f'' is negative.

43. (a) $f'(x) = \cos x - \cos 3x + \cos 5x$

$f''(x) = -\sin x + 3 \sin 3x - 5 \sin 5x$

(b) Relative maximum: $(\pi/2, 1.53333)$

Points of inflection:

$(0.5236, 0.2667), (1.1731, 0.9637),$

$(1.9685, 0.9637), (2.6180, 0.2667)$

(c)

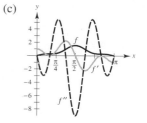

f is increasing when f' is positive, and decreasing when f' is negative. f is concave upward when f'' is positive, and concave downward when f'' is negative.

45. (a)   (b)

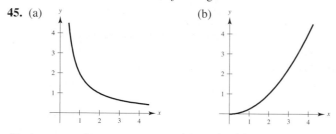

47. Answers will vary. Example: $f(x) = x^4$ $f''(0) = 0$, but $(0, 0)$ is not point of inflection.

49.

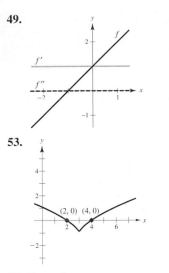

51.

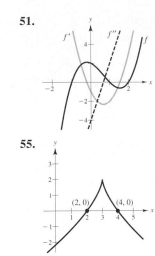

53.

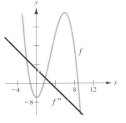

55.

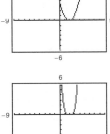

57. Example:

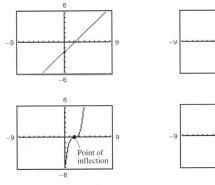

59. (a) $f(x) = (x - 2)^n$ has a point of inflection at $(2, 0)$ if n is odd and $n \geq 3$.

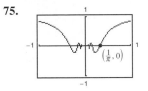

(b) Proof

61. $f(x) = \frac{1}{2}x^3 - 6x^2 + \frac{45}{2}x - 24$

63. (a) $f(x) = \frac{1}{32}x^3 + \frac{3}{16}x^2$ (b) Two miles from touchdown

65. $x = \left(\dfrac{15 - \sqrt{33}}{16}\right)L \approx 0.578L$ **67.** $x = 100$ units

69. $t = \sqrt{\frac{8}{3}} \approx 1.633$ years

71. $P_1(x) = 2\sqrt{2}$

$P_2(x) = 2\sqrt{2} - \sqrt{2}\left(x - \dfrac{\pi}{4}\right)^2$

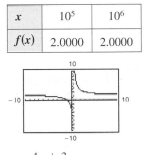

The values of f, P_1, and P_2 and their first derivatives are equal when $x = \pi/4$. The approximations worsen as you move away from $x = \pi/4$.

73. $P_1(x) = 1 - \dfrac{x}{2}$

$P_2(x) = 1 - \dfrac{x}{2} - \dfrac{x^2}{8}$

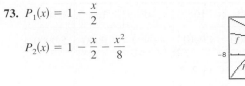

The values of f, P_1, and P_2 and their first derivatives are equal when $x = 0$. The approximations worsen as you move away from $x = 0$.

75.

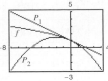

77. Proof   **79.** True

81. False. The maximum value is $\sqrt{13} \approx 3.60555$.

83. False. f is concave upward at $x = c$ if $f''(c) > 0$.

Section 3.5 (page 199)

1. f   **2.** c   **3.** d   **4.** a   **5.** b   **6.** e

7.

x	10^0	10^1	10^2	10^3	10^4
$f(x)$	7	2.2632	2.0251	2.0025	2.0003

x	10^5	10^6
$f(x)$	2.0000	2.0000

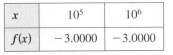

$$\lim_{x \to \infty} \frac{4x + 3}{2x - 1} = 2$$

9.

x	10^0	10^1	10^2	10^3	10^4
$f(x)$	-2	-2.9814	-2.9998	-3.0000	-3.0000

x	10^5	10^6
$f(x)$	-3.0000	-3.0000

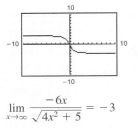

$$\lim_{x \to \infty} \frac{-6x}{\sqrt{4x^2 + 5}} = -3$$

11.

x	10^0	10^1	10^2	10^3	10^4
$f(x)$	4.5000	4.9901	4.9999	5.0000	5.0000

x	10^5	10^6
$f(x)$	5.0000	5.0000

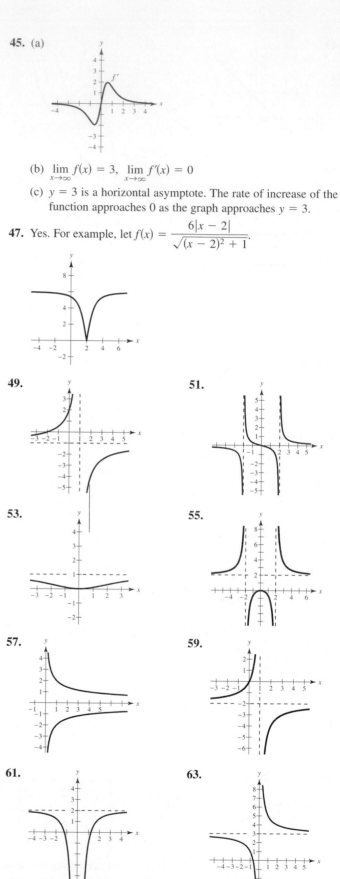

$$\lim_{x\to\infty}\left(5-\frac{1}{x^2+1}\right)=5$$

13. (a) ∞ (b) 5   (c) 0   **15.** (a) 0   (b) 1   (c) ∞

17. (a) 0   (b) $-\frac{2}{3}$ (c) $-\infty$

19. $\frac{2}{3}$ **21.** 0   **23.** $-\infty$

25. -1 **27.** -2 **29.** 0   **31.** 0

33.

35. 1   **37.** 0   **39.** $-\frac{1}{2}$

41.

x	10^0	10^1	10^2	10^3	10^4	10^5	10^6
$f(x)$	1.000	0.513	0.501	0.500	0.500	0.500	0.500

$$\lim_{x\to\infty}\left[x-\sqrt{x(x-1)}\right]=\frac{1}{2}$$

43.

x	10^0	10^1	10^2	10^3	10^4	10^5	10^6
$f(x)$	0.479	0.500	0.500	0.500	0.500	0.500	0.500

The graph has a hole at $x=0$.

$$\lim_{x\to\infty} x\sin\frac{1}{2x}=\frac{1}{2}$$

45. (a)

(b) $\lim\limits_{x\to\infty} f(x)=3,\ \lim\limits_{x\to\infty} f'(x)=0$

(c) $y=3$ is a horizontal asymptote. The rate of increase of the function approaches 0 as the graph approaches $y=3$.

47. Yes. For example, let $f(x)=\dfrac{6|x-2|}{\sqrt{(x-2)^2+1}}$.

49.

51.

53.

55.

57.

59.

61.

63.

65.

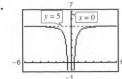

67.

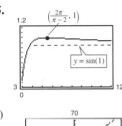

69.

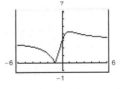

71.

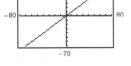

73.

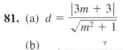

75.

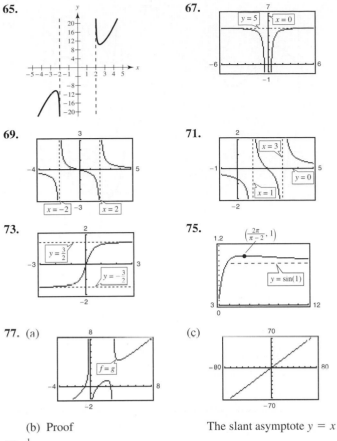

77. (a)

(c)

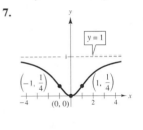

(b) Proof

The slant asymptote $y = x$

79. $\frac{1}{2}$

81. (a) $d = \dfrac{|3m + 3|}{\sqrt{m^2 + 1}}$

(b)

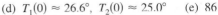

(c) $\displaystyle\lim_{m \to \infty} d(m) = 3$; $\displaystyle\lim_{m \to -\infty} d(m) = 3$

As $m \to \infty$, the line approaches the vertical line $x = 0$. Therefore, the distance approaches 3.

83. (a) $T_1 = -0.003t^2 + 0.68t + 26.6$

(b)

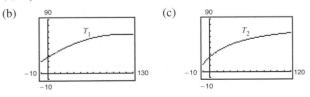

(c)

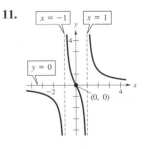

(d) $T_1(0) \approx 26.6°$, $T_2(0) \approx 25.0°$ (e) 86

(f) The limiting temperature is 86°.

No. T_1 has no horizontal asymptote.

85. Answers will vary. See "Guidelines for Finding Limits at Infinity of Rational Functions" on page 195. Examples:

(a) $\displaystyle\lim_{x \to \infty} \left(\dfrac{5 - 2x}{3x^2 - 4} \right) = 0$ since the degree of the numerator is less than the degree of the denominator.

(b) $\displaystyle\lim_{x \to \infty} \left(\dfrac{2x - 1}{3x + 2} \right) = \dfrac{2}{3}$ since the degree of the numerator is equal to the degree of the denominator.

(c) $\displaystyle\lim_{x \to \infty} \left(\dfrac{x^2 + 2}{x - 1} \right) = \infty$ since the degree of the numerator is greater than the degree of the denominator.

87. False. Let $f(x) = \dfrac{2x}{\sqrt{x^2 + 2}}$.

$f'(x) > 0$ for all real numbers.

Section 3.6 (page 208)

1. d **2.** c **3.** a **4.** b

5. (a) $f'(x) = 0$ for $x = \pm 2$

$f'(x) > 0$ for $(-\infty, -2), (2, \infty)$

$f'(x) < 0$ for $(-2, 2)$

(b) $f''(x) = 0$ for $x = 0$

$f''(x) > 0$ for $(0, \infty)$

$f''(x) < 0$ for $(-\infty, 0)$

(c) $(0, \infty)$

(d) f' is minimum for $x = 0$.

f is decreasing at the fastest rate.

7.

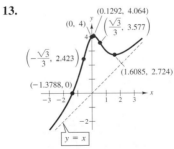

9.

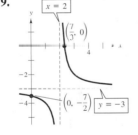

11.

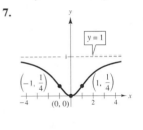

13.

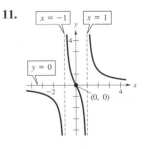

15.

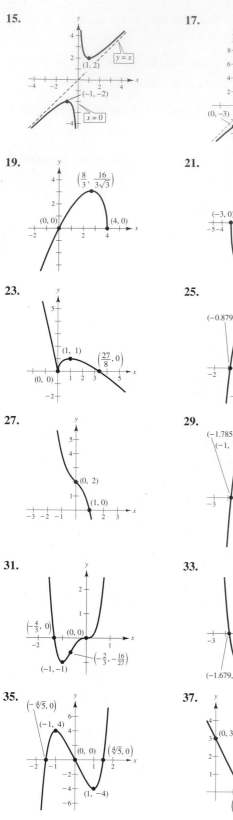

17.

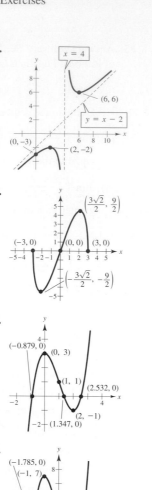

19.

21.

23.

25.

27.

29.

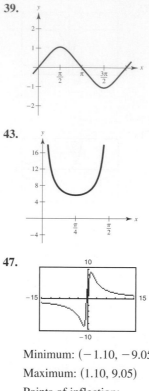

31.

33.

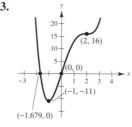

35.

37.

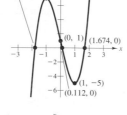

39.

41.

43.

45.

47.

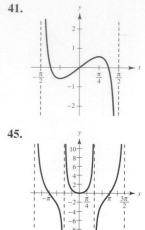

Minimum: $(-1.10, -9.05)$

Maximum: $(1.10, 9.05)$

Points of inflection:

$(-1.84, -7.86), (1.84, 7.86)$

Vertical asymptote: $x = 0$

Horizontal asymptote: $y = 0$

49.

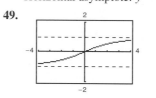

Point of inflection: $(0, 0)$

Horizontal asymptotes: $y = \pm 1$

51.

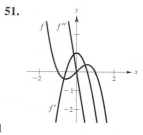

53.

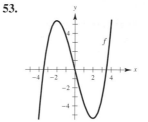

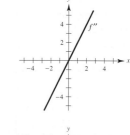

55.

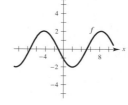

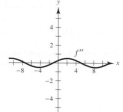

57. f is decreasing on $(2, 8)$ and therefore $f(3) > f(5)$.

59.

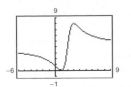

The graph crosses the horizontal asymptote $y = 4$. The graph of f does not cross its vertical asymptote $x = c$ because $f(c)$ does not exist.

61.

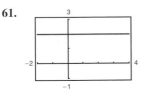

The graph has a hole at $x = 3$. The rational function is not reduced to lowest terms.

63.

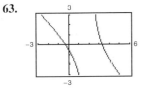

The graph appears to approach the line $y = -x + 1$, which is the slant asymptote.

65. (a)

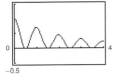

The graph has a hole at $x = 0$ and at $x = 1$.

Visual approximate critical numbers: $\dfrac{1}{2}, 1, \dfrac{3}{2}, 2, \dfrac{5}{2}, 3, \dfrac{7}{2}$

(b) $f'(x) = \dfrac{-\cos^2(\pi x)}{(x^2 + 1)^{3/2}} - \dfrac{2\pi \sin(\pi x)\cos(\pi x)}{\sqrt{x^2 + 1}}$

Approximate critical numbers: $\dfrac{1}{2}, 0.97, \dfrac{3}{2}, 1.98, \dfrac{5}{2}, 2.98, \dfrac{7}{2}$

The critical numbers where maxima occur appear to be integers in part (a), but approximating them using f' you see that they are not integers.

67. Answers will vary. Example: $y = \dfrac{1}{x - 5}$

69. Answers will vary. Example: $y = \dfrac{3x^2 - 13x - 9}{x - 5}$

71. (a) Rate of change of f changes as a varies. If the sign of a is changed, the graph is reflected through the x-axis.

(b) The locations of the vertical asymptote and the minimum (if $a > 0$) or maximum (if $a < 0$) are changed.

73. (a) If n is even, f is symmetric with respect to the y-axis.

If n is odd, f is symmetric with respect to the origin.

(b) $n = 0, 1, 2, 3$ (c) $n = 4$

(d) When $n = 5$, the slant asymptote is $y = 3x$.

(e)

n	0	1	2	3	4	5
M	1	2	3	2	1	0
N	2	3	4	5	2	3

75. (a)

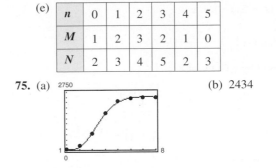

(b) 2434

(c) The number of bacteria reached its maximum early on the seventh day.

(d) The rate of increase in the number of bacteria was greatest approximately in the middle of the third day.

(e) $\dfrac{13{,}250}{7}$

Section 3.7 (page 216)

1. (a) and (b)

First Number, x	Second Number	Product P
10	$110 - 10$	$10(110 - 10) = 1000$
20	$110 - 20$	$20(110 - 20) = 1800$
30	$110 - 30$	$30(110 - 30) = 2400$
40	$110 - 40$	$40(110 - 40) = 2800$
50	$110 - 50$	$50(110 - 50) = 3000$
60	$110 - 60$	$60(110 - 60) = 3000$
70	$110 - 70$	$70(110 - 70) = 2800$
80	$110 - 80$	$80(110 - 80) = 2400$
90	$110 - 90$	$90(110 - 90) = 1800$
100	$110 - 100$	$100(110 - 100) = 1000$

(c) $P = x(110 - x)$

(d)

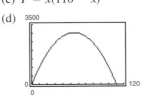

(e) 55 and 55

3. $\sqrt{192}$ and $\sqrt{192}$ **5.** 1 and 1   **7.** $l = w = 25$ meters

9. $l = w = 8$ feet   **11.** $\left(\dfrac{7}{2}, \sqrt{\dfrac{7}{2}}\right)$ **13.** $(1, 1)$

15. $x = \dfrac{Q_0}{2}$ **17.** 600×300 meters

19. (a) Proof

(b) $V_1 = 99$ cubic inches

$V_2 = 125$ cubic inches

$V_3 = 117$ cubic inches

(c) $5 \times 5 \times 5$ inches

21. (a) $V = x(s - 2x)^2, 0 < x < \dfrac{s}{2}$

Maximum: $V\left(\dfrac{s}{6}\right) = \dfrac{2s^3}{27}$

(b) Increased by a factor of 8

23. Rectangular portion: $\dfrac{16}{\pi + 4} \times \dfrac{32}{\pi + 4}$ feet

25. (a) $L = \sqrt{x^2 + 4 + \dfrac{8}{x - 1} + \dfrac{4}{(x - 1)^2}}, \quad x > 1$

(b)

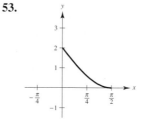

(2.587, 4.162)

Minimum when $x \approx 2.587$

(c) $(0, 0), (2, 0), (0, 4)$

27. Width: $\dfrac{5\sqrt{2}}{2}$; Length: $5\sqrt{2}$

29. Dimensions of page: $\left(2 + \sqrt{30}\right)$ inches $\times \left(2 + \sqrt{30}\right)$ inches

31. (a) and (b)

Radius, r	Height	Surface Area, S
0.2	$\dfrac{22}{\pi(0.2)^2}$	$2\pi(0.2)\left[0.2 + \dfrac{22}{\pi(0.2)^2}\right] \approx 220.3$
0.4	$\dfrac{22}{\pi(0.4)^2}$	$2\pi(0.4)\left[0.4 + \dfrac{22}{\pi(0.4)^2}\right] \approx 111.0$
0.6	$\dfrac{22}{\pi(0.6)^2}$	$2\pi(0.6)\left[0.6 + \dfrac{22}{\pi(0.6)^2}\right] \approx 75.6$
0.8	$\dfrac{22}{\pi(0.8)^2}$	$2\pi(0.8)\left[0.8 + \dfrac{22}{\pi(0.8)^2}\right] \approx 59.0$
1.0	$\dfrac{22}{\pi(1.0)^2}$	$2\pi(1.0)\left[1.0 + \dfrac{22}{\pi(1.0)^2}\right] \approx 50.3$
1.2	$\dfrac{22}{\pi(1.2)^2}$	$2\pi(1.2)\left[1.2 + \dfrac{22}{\pi(1.2)^2}\right] \approx 45.7$
1.4	$\dfrac{22}{\pi(1.4)^2}$	$2\pi(1.4)\left[1.4 + \dfrac{22}{\pi(1.4)^2}\right] \approx 43.7$
1.6	$\dfrac{22}{\pi(1.6)^2}$	$2\pi(1.6)\left[1.6 + \dfrac{22}{\pi(1.6)^2}\right] \approx 43.6$
1.8	$\dfrac{22}{\pi(1.8)^2}$	$2\pi(1.8)\left[1.8 + \dfrac{22}{\pi(1.8)^2}\right] \approx 44.8$
2.0	$\dfrac{22}{\pi(2.0)^2}$	$2\pi(2.0)\left[2.0 + \dfrac{22}{\pi(2.0)^2}\right] \approx 47.1$

(c) $S = 2\pi r\left(r + \dfrac{22}{\pi r^2}\right)$

(d)

(1.52, 43.46)

(e) $r = \sqrt[3]{\dfrac{11}{\pi}}, h = 2r$

Minimum area of 43.46 square inches when $r = 1.52$

33. $18 \times 18 \times 36$ inches

35. $\dfrac{32\pi r^3}{81}$

37. Answers will vary. If area is expressed as a function of either length or width, the feasible domain is the interval $(0, 10)$. No dimensions will yield a minimum area because the second derivative on this open interval is always negative.

39. $r = \sqrt[3]{\dfrac{9}{\pi}} \approx 1.42$ cm

41. Side of square: $\dfrac{10\sqrt{3}}{9 + 4\sqrt{3}}$

Side of triangle: $\dfrac{30}{9 + 4\sqrt{3}}$

43. $w = 8\sqrt{3}$ inches, $h = 8\sqrt{6}$ inches   **45.** $\theta = \dfrac{\pi}{4}$

47. $h = \sqrt{2}$ feet

49. One mile from the nearest point on the coast

51. Proof

53.

(a) Origin to y-intercept: 2

Origin to x-intercept: $\dfrac{\pi}{2}$

(b) $d = \sqrt{x^2 + (2 - 2\sin x)^2}$

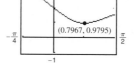

(0.7967, 0.9795)

(c) Minimum distance is 0.9795 when $x \approx 0.7967$.

55. $F = \dfrac{kW}{\sqrt{k^2 + 1}}; \quad \theta = \arctan k$

57. (a)

Base 1	Base 2	Altitude	Area
8	$8 + 16\cos 10°$	$8\sin 10°$	≈ 22.1
8	$8 + 16\cos 20°$	$8\sin 20°$	≈ 42.5
8	$8 + 16\cos 30°$	$8\sin 30°$	≈ 59.7
8	$8 + 16\cos 40°$	$8\sin 40°$	≈ 72.7
8	$8 + 16\cos 50°$	$8\sin 50°$	≈ 80.5
8	$8 + 16\cos 60°$	$8\sin 60°$	≈ 83.1

(b)

Base 1	Base 2	Altitude	Area
8	$8 + 16 \cos 10°$	$8 \sin 10°$	≈ 22.1
8	$8 + 16 \cos 20°$	$8 \sin 20°$	≈ 42.5
8	$8 + 16 \cos 30°$	$8 \sin 30°$	≈ 59.7
8	$8 + 16 \cos 40°$	$8 \sin 40°$	≈ 72.7
8	$8 + 16 \cos 50°$	$8 \sin 50°$	≈ 80.5
8	$8 + 16 \cos 60°$	$8 \sin 60°$	≈ 83.1
8	$8 + 16 \cos 70°$	$8 \sin 70°$	≈ 80.7
8	$8 + 16 \cos 80°$	$8 \sin 80°$	≈ 74.0
8	$8 + 16 \cos 90°$	$8 \sin 90°$	≈ 64.0

The maximum cross-sectional area is approximately 83.1 square feet.

(c) $A = (a + b)\dfrac{h}{2}$

$\quad = [8 + (8 + 16 \cos \theta)]\dfrac{8 \sin \theta}{2}$

$\quad = 64(1 + \cos \theta)\sin \theta, \, 0° < \theta < 90°$

(d) $\dfrac{dA}{d\theta} = 64(1 + \cos \theta)\cos \theta + (-64 \sin \theta)\sin \theta$

$\quad = 64(\cos \theta + \cos^2 \theta - \sin^2 \theta)$

$\quad = 64(2 \cos^2 \theta + \cos \theta - 1)$

$\quad = 64(2 \cos \theta - 1)(\cos \theta + 1)$

$\quad = 0$ when $\theta = 60°, 180°, 300°$

The maximum occurs when $\theta = 60°$.

(e)

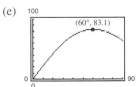

59. 4045 units

61. $y = \frac{64}{141}x$; $S_1 = 6.1$ miles

63. $y = \frac{3}{10}x$; $S_3 = 4.50$ miles

Section 3.8 (page 226)

1.

n	x_n	$f(x_n)$	$f'(x_n)$	$\dfrac{f(x_n)}{f'(x_n)}$	$x_n - \dfrac{f(x_n)}{f'(x_n)}$
1	1.7000	-0.1100	3.4000	-0.0324	1.7324
2	1.7324	0.0012	3.4648	0.0003	1.7321

3.

n	x_n	$f(x_n)$	$f'(x_n)$	$\dfrac{f(x_n)}{f'(x_n)}$	$x_n - \dfrac{f(x_n)}{f'(x_n)}$
1	3	0.1411	-0.9900	-0.1425	3.1425
2	3.1425	-0.0009	-1.0000	0.0009	3.1416

5. 0.682   **7.** 1.146, 7.854   **9.** -1.442

11. 0.900, 1.100, 1.900   **13.** -0.489 **15.** 0.569

17. 4.493   **19.** $x_{i+1} = \dfrac{x_i^2 + a}{2x_i}$ **21.** 2.646

23. 1.565   **25.** 3.141   **27.** $f'(x_1) = 0$

29. $2 = x_1 = x_3 = \ldots$

$\quad 1 = x_2 = x_4 = \ldots$

31. If f is a function continuous on $[a, b]$ and differentiable on (a, b), where $c \in [a, b]$ and $f(c) = 0$, Newton's Method uses tangent lines to approximate c. First, estimate an initial x_1 close to c. (See graph.) Then determine x_2 by $x_2 = x_1 - \dfrac{f(x_1)}{f'(x_1)}$. Calculate a third estimate by $x_3 = x_2 - \dfrac{f(x_2)}{f'(x_2)}$. Continue this process until $|x_n - x_{n+1}|$ is within the desired accuracy and let x_{n+1} be the final approximation of c.

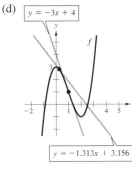

33. 0.74

35. (a)

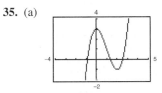

(b) 1.347

(c) 2.532

(d)

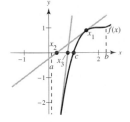

x-intercept of $y = -3x + 4$ is $\frac{4}{3}$.

x-intercept of $y = -1.313x + 3.156$ is approximately 2.404.

(e) If the initial estimate $x = x_1$ is not sufficiently close to the desired zero of a function, the x-intercept of the corresponding tangent line to the function may approximate a second zero of the function.

37. Proof

39. 0.860

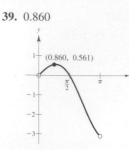

41. (1.939, 0.240)   **43.** $x \approx 1.563$ miles   **45.** \$384,356

47. False: let $f(x) = \dfrac{x^2 - 1}{x - 1}$.   **49.** True   **51.** $x \approx 11.803$

Section 3.9 (page 233)

1. $T(x) = 4x - 4$

x	1.9	1.99	2	2.01	2.1
$f(x)$	3.610	3.960	4	4.040	4.410
$T(x)$	3.600	3.960	4	4.040	4.400

3. $T(x) = 80x - 128$

x	1.9	1.99	2	2.01	2.1
$f(x)$	24.761	31.208	32	32.808	40.841
$T(x)$	24.000	31.200	32	32.800	40.000

5. $T(x) = (\cos 2)(x - 2) + \sin 2$

x	1.9	1.99	2	2.01	2.1
$f(x)$	0.946	0.913	0.909	0.905	0.863
$T(x)$	0.951	0.913	0.909	0.905	0.868

7. $\Delta y = 0.6305$; $dy = 0.6000$ **9.** $\Delta y = -0.039$; $dy = -0.040$

11. $6x\,dx$ **13.** $-\dfrac{3}{(2x - 1)^2}\,dx$ **15.** $\dfrac{1 - 2x^2}{\sqrt{1 - x^2}}\,dx$

17. $(2 + 2\cot x + 2\cot^3 x)\,dx$ **19.** $-\pi \sin\left(\dfrac{6\pi x - 1}{2}\right)dx$

21. (a) 0.9   (b) 1.04   **23.** (a) 1.05   (b) 0.98

25. (a) 8.035   (b) 7.95   **27.** (a) 8   (b) 8

29. $\pm\frac{3}{8}$ square inch   **31.** $\pm 7\pi$ square inches

33. (a) $\frac{2}{3}\%$ (b) 1.25%

35. (a) $\pm 2.88\pi$ cubic inches   (b) $\pm 0.96\pi$ square inches
   (c) 1%, $\frac{2}{3}\%$

37. 80π cubic centimeters

39. (a) $\frac{1}{4}\%$ (b) 216 seconds = 3.6 minutes

41. (a) 0.87%   (b) 2.16%   **43.** 4961 feet

45. $f(x) = \sqrt{x}$, $dy = \dfrac{1}{2\sqrt{x}}\,dx$

$$f(99.4) \approx \sqrt{100} + \dfrac{1}{2\sqrt{100}}(-0.6) = 9.97$$

Calculator: 9.97

47. $f(x) = \sqrt[4]{x}$, $dy = \dfrac{1}{4x^{3/4}}\,dx$

$$f(624) \approx \sqrt[4]{625} + \dfrac{1}{4(625)^{3/4}}(-1) = 4.998$$

Calculator: 4.998

49. $f(x) = \sqrt{x}$; $dy = \dfrac{1}{2\sqrt{x}}\,dx$

$$f(4.02) \approx \sqrt{4} + \dfrac{1}{2\sqrt{4}}(0.02) = 2 + \dfrac{1}{4}(0.02)$$

51. The value of dy becomes closer to the value of Δy as Δx decreases.

53. True   **55.** True

Review Exercises for Chapter 3 (page 235)

1. Let f be defined at c. If $f'(c) = 0$ or if f' is undefined at c, then c is a critical number of f.

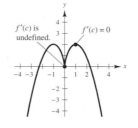

3. Maximum: $(2\pi, 17.57)$
   Minimum: $(2.73, 0.88)$

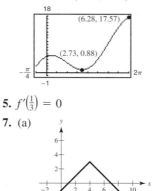

5. $f'\left(\frac{1}{3}\right) = 0$

7. (a)

   (b) f is not differentiable at $x = 4$.

9. $f'\left(\dfrac{2744}{729}\right) = \dfrac{3}{7}$ **11.** $f'(0) = 1$ **13.** $c = \dfrac{x_1 + x_2}{2}$

15. Critical numbers: $x = 1, \frac{7}{3}$
   Increasing on $(-\infty, 1)$, $\left(\frac{7}{3}, \infty\right)$
   Decreasing on $\left(1, \frac{7}{3}\right)$

17. Critical number: $x = 1$
   Increasing on $(1, \infty)$
   Decreasing on $(0, 1)$

19. Minimum: $(2, -12)$

21. (a) $y = \frac{1}{4}$ inch; $v = 4$ inches per second

(b) Proof

(c) Period: $\frac{\pi}{6}$; Frequency: $\frac{6}{\pi}$

23. $\left(\frac{\pi}{2}, \frac{\pi}{2}\right), \left(\frac{3\pi}{2}, \frac{3\pi}{2}\right)$

25.

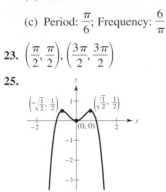

Relative maxima: $\left(\frac{\sqrt{2}}{2}, \frac{1}{2}\right), \left(-\frac{\sqrt{2}}{2}, \frac{1}{2}\right)$

Relative minimum: $(0, 0)$

27.

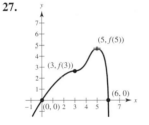

29. Increasing and concave down

31. (a) $D = 0.00340t^4 - 0.2352t^3 + 4.942t^2 - 20.86t + 94.4$

(b)

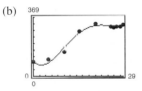

(c) Maximum occurs in 1991; Minimum occurs in 1972.

(d) 1979

33. $\frac{2}{3}$ **35.** 0

37. Vertical asymptote: $x = 4$

Horizontal asymptote: $y = 2$

39. Vertical asymptote: $x = 0$

Horizontal asymptote: $y = -2$

41.

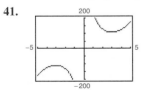

Vertical asymptote: $x = 0$

Relative minimum: $(3, 108)$

Relative maximum: $(-3, -108)$

43.

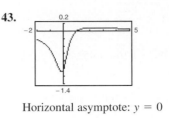

Horizontal asymptote: $y = 0$

Relative minimum: $(-0.155, -1.077)$

Relative maximum: $(2.155, 0.077)$

45.

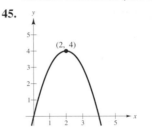

47.

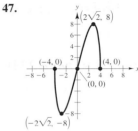

49.

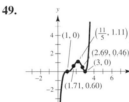

51.

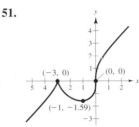

53.

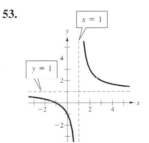

55.

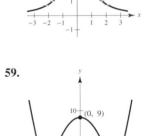

57.

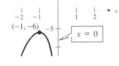

59.

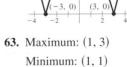

61.

63. Maximum: $(1, 3)$

Minimum: $(1, 1)$

65. $t \approx 4.92 \approx 4\!:\!55$ P.M.; $d \approx 64$ kilometers

67. $(0, 0), (5, 0), (0, 10)$ **69.** Proof   **71.** 14.05 feet

73. $3(3^{2/3} + 2^{2/3})^{3/2} \approx 21.07$ feet

75. $v \approx 54.77$ miles per hour

77. $-1.532, -0.347, 1.879$ **79.** $-1.164, 1.453$

81. $dy = (1 - \cos x + x \sin x)\, dx$

83. $dS = \pm 1.8\pi$ square centimeters, $\dfrac{dS}{S} \times 100 \approx \pm 0.56\%$

     $dV = \pm 8.1\pi$ cubic centimeters, $\dfrac{dV}{V} \times 100 \approx \pm 0.83\%$

P.S. Problem Solving (page 238)

1. Proof

3. (a)

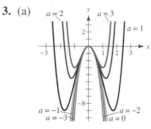

Relative maximum for all a at $(0, 0)$

Two relative minima for $a = 1, 2, 3$

(b) $p = ax^4 - 6x^2$

$p' = 4ax^3 - 12x$ has critical points at $x = 0$ and

$x = \pm\sqrt{3/a},\ a > 0$

$p'' = 12ax^2 - 12,\ p''(0) = -12$

Therefore, by the Second Derivative Test, p has a relative maximum for all a at $x = 0$.

(c) $p''\!\left(\pm\sqrt{3/a}\right) = 24$. Therefore, by the Second Derivative Test, p has a relative minimum when $x = \pm\sqrt{3/a},\ a > 0$.

(d) Relative extrema of p occur at $x = 0,\ \pm\sqrt{3/a},\ a > 0$. If $x = 0$, $p(x) = 0$ and $(0, 0)$ also lies on the graph of $y = -3x^2$. If $x = \pm\sqrt{3/a}$, $p(x) = -9/a$ and $\left(\pm\sqrt{3/a}, -9/a\right)$ also lies on the graph of $y = -3x^2$.

5. Choices of a may vary.

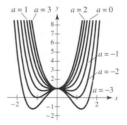

(a) One relative minimum at $(0, 1)$ for $a \geq 0$

(b) One relative maximum at $(0, 1)$ for $a < 0$

(c) Two relative minima for $a < 0$ when $x = \pm\sqrt{-\dfrac{a}{2}}$

(d) If $a < 0$, there are three critical points; if $a \geq 0$, there is only one critical point.

7. All c where c is a real number

9. Proof   **11.** $\phi \approx 42.1°$ or 0.736 radians

13. $\theta = \dfrac{\pi}{2} + 2n\pi$ and $\theta = \dfrac{3\pi}{2} + 2n\pi$, where n is an integer.

15. Rectangle: $\frac{3}{2} \times 2$

     Circle: $r = 1$

     Semicircle: $r = \frac{12}{7}$

     Calculus was helpful for the rectangle.

17. Greatest slope at $\left(-\dfrac{\sqrt{3}}{3}, \dfrac{3}{4}\right)$; Least slope at $\left(\dfrac{\sqrt{3}}{3}, \dfrac{3}{4}\right)$

19. (a) Proof   (b) Proof

Chapter 4

Section 4.1 (page 249)

1. Proof   **3.** Proof   **5.** $y = t^3 + C$

7. $y = \frac{2}{5}x^{5/2} + C$

Original Integral	Rewrite	Integrate	Simplify
9. $\displaystyle\int \sqrt[3]{x}\, dx$	$\displaystyle\int x^{1/3}\, dx$	$\dfrac{x^{4/3}}{4/3} + C$	$\dfrac{3}{4}x^{4/3} + C$
11. $\displaystyle\int \dfrac{1}{x\sqrt{x}}\, dx$	$\displaystyle\int x^{-3/2}\, dx$	$\dfrac{x^{-1/2}}{-1/2} + C$	$-\dfrac{2}{\sqrt{x}} + C$
13. $\displaystyle\int \dfrac{1}{2x^3}\, dx$	$\dfrac{1}{2}\displaystyle\int x^{-3}\, dx$	$\dfrac{1}{2}\!\left(\dfrac{x^{-2}}{-2}\right) + C$	$-\dfrac{1}{4x^2} + C$

15. $\frac{1}{2}x^2 + 3x + C$ **17.** $x^2 - x^3 + C$ **19.** $\frac{1}{4}x^4 + 2x + C$

21. $\frac{2}{5}x^{5/2} + x^2 + x + C$ **23.** $\frac{3}{5}x^{5/3} + C$ **25.** $-\dfrac{1}{2x^2} + C$

27. $\frac{2}{15}x^{1/2}(3x^2 + 5x + 15) + C$ **29.** $x^3 + \frac{1}{2}x^2 - 2x + C$

31. $\frac{2}{7}y^{7/2} + C$ **33.** $x + C$ **35.** $-2\cos x + 3\sin x + C$

37. $t + \csc t + C$ **39.** $\tan\theta + \cos\theta + C$ **41.** $\tan y + C$

43.

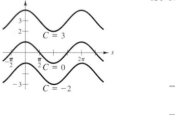

45. Answers will vary. Example:

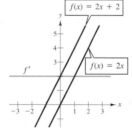

47. Answers will vary. Example:

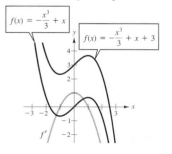

49. $y = x^2 - x + 1$ **51.** $y = \sin x + 4$

53. (a) Answers will vary.   (b) $y = \frac{1}{4}x^2 - x + 2$
Example:

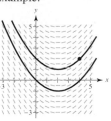 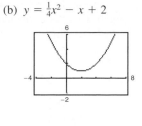

55. $f(x) = 2x^2 + 6$ **57.** $h(t) = 2t^4 + 5t - 11$

59. $f(x) = x^2 + x + 4$ **61.** $f(x) = -4\sqrt{x} + 3x$

63. (a) $h(t) = \frac{3}{4}t^2 + 5t + 12$ (b) 69 centimeters

65. (a) -1; $f'(4)$ represents the slope of f at $x = 4$.

    (b) No. The slope of the tangent lines are greater than 2 on $[0, 2]$. Therefore, f must increase more than four units on $[0, 2]$.

    (c) No. The function is decreasing on $[4, 5]$.

    (d) 3.5; $f'(3.5) \approx 0$

    (e) Concave upward: $(-\infty, 1), (5, \infty)$
Concave downward: $(1, 5)$
Points of inflection at $x \approx 1$ and $x \approx 5$

    (f) 3

    (g)

67. 62.25 feet   **69.** $v_0 \approx 187.617$ feet per second

71. $v(t) = -9.8t + C_1 = -9.8t + v_0$ **73.** 7.1 meters
$f(t) = -4.9t^2 + v_0t + C_2 = -4.9t^2 + v_0t + s_0$

75. 320 meters; -32 meters per second

77. (a) $v(t) = 3t^2 - 12t + 9$; $a(t) = 6t - 12$

    (b) $(0, 1), (3, 5)$ (c) -3

79. $a(t) = \dfrac{-1}{2t^{3/2}}$; $s(t) = 2\sqrt{t} + 2$

81. (a) 1.18 meters per second per second   (b) 190 meters

83. (a) 300 feet   (b) 60 feet per second ≈ 41 miles per hour

85. (a)

t	0	5	10	15	20	25	30
v_1	0	3.67	10.27	23.47	42.53	66.00	95.33
v_2	0	30.80	55.73	74.80	88.00	93.87	95.33

    (b) $v_1(t) = 0.1068t^2 - 0.042t + 0.37$
$v_2(t) = -0.1208t^2 + 6.799t - 0.07$

    (c) Distance of car 1 ≈ 953 feet
Distance of car 2 ≈ 1970 feet
Car 2 traveled farther because it accelerated faster for about the first 15 seconds.

87. 7.45 feet per second per second

89. True   **91.** True

93. False. Let $f(x) = x$ and $g(x) = x + 1$.

95. $f(x) = \begin{cases} x + 2, & 0 \le x < 2 \\ \frac{3}{2}x^2 - 2, & 2 \le x < 5 \end{cases}$

    f is not differentiable at $x = 2$ because the left- and right-hand derivatives at $x = 2$ do not agree.

Section 4.2 (page 261)

1. 35   **3.** $\dfrac{158}{85}$ **5.** $4c$ **7.** $\displaystyle\sum_{i=1}^{9} \frac{1}{3i}$ **9.** $\displaystyle\sum_{j=1}^{8} \left[5\left(\frac{j}{8}\right) + 3 \right]$

11. $\dfrac{2}{n} \displaystyle\sum_{i=1}^{n} \left[\left(\frac{2i}{n}\right)^3 - \left(\frac{2i}{n}\right) \right]$ **13.** $\dfrac{3}{n} \displaystyle\sum_{i=1}^{n} \left[2\left(1 + \frac{3i}{n}\right)^2 \right]$

15. 420   **17.** 2470   **19.** 12,040   **21.** 2930

23. The area of the shaded region falls between 12.5 square units and 16.5 square units.

25. The area of the shaded region falls between 7 square units and 11 square units.

27. $A \approx S \approx 0.768$ **29.** $A \approx S \approx 0.746$
$A \approx s \approx 0.518$ $A \approx s \approx 0.646$

31. $\dfrac{81}{4}$ **33.** 9

35. $\dfrac{n + 2}{n}$

    $n = 10$; $S = 1.2$
$n = 100$; $S = 1.02$
$n = 1000$; $S = 1.002$
$n = 10{,}000$; $S = 1.0002$

37. $\dfrac{2(n + 1)(n - 1)}{n^2}$

    $n = 10$; $S = 1.98$
$n = 100$; $S = 1.9998$
$n = 1000$; $S = 1.999998$
$n = 10{,}000$; $S = 1.99999998$

39. $\displaystyle\lim_{n \to \infty} \left[8\left(\frac{n^2 + n}{n^2}\right) \right] = 8$ **41.** $\displaystyle\lim_{n \to \infty} \frac{1}{6}\left(\frac{2n^3 - 3n^2 + n}{n^3}\right) = \frac{1}{3}$

43. $\displaystyle\lim_{n \to \infty} \left(\frac{3n + 1}{n}\right) = 3$

45. (a)

    (b) $\Delta x = \dfrac{2 - 0}{n} = \dfrac{2}{n}$

    (c) $s(n) = \displaystyle\sum_{i=1}^{n} f(x_{i-1})\, \Delta x$
$= \displaystyle\sum_{i=1}^{n} \left[(i - 1)\left(\frac{2}{n}\right) \right]\left(\frac{2}{n}\right)$

    (d) $S(n) = \displaystyle\sum_{i=1}^{n} f(x_i)\, \Delta x$
$= \displaystyle\sum_{i=1}^{n} \left[i\left(\frac{2}{n}\right) \right]\left(\frac{2}{n}\right)$

    (e)

n	5	10	50	100
$s(n)$	1.6	1.8	1.96	1.98
$S(n)$	2.4	2.2	2.04	2.02

    (f) $\displaystyle\lim_{n \to \infty} \sum_{i=1}^{n} \left[(i - 1)\left(\frac{2}{n}\right) \right]\left(\frac{2}{n}\right) = 2$
$\displaystyle\lim_{n \to \infty} \sum_{i=1}^{n} \left[i\left(\frac{2}{n}\right) \right]\left(\frac{2}{n}\right) = 2$

47. $A = 2$ **49.** $A = \frac{7}{3}$

51. $A = \frac{70}{3}$ **53.** $A = \frac{513}{4}$

55. $A = \frac{2}{3}$ **57.** $A = 6$

59. $A = 9$ **61.** $A = \frac{44}{3}$

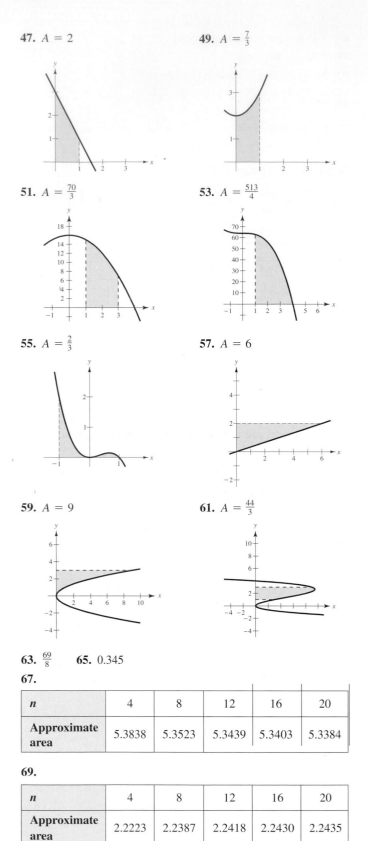

63. $\frac{69}{8}$ **65.** 0.345

67.

n	4	8	12	16	20
Approximate area	5.3838	5.3523	5.3439	5.3403	5.3384

69.

n	4	8	12	16	20
Approximate area	2.2223	2.2387	2.2418	2.2430	2.2435

71. We can use the line $y = x$ bounded by $x = a$ and $x = b$. The sum of the areas of the inscribed rectangles in the figure below is the lower sum.

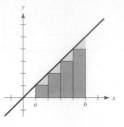

The sum of the areas of the circumscribed rectangles in the figure below is the upper sum.

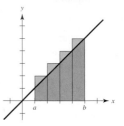

The rectangles in the first graph do not contain all of the area of the region, and the rectangles in the second graph cover more than the area of the region. The exact value of the area lies between these two sums.

73. (a)   (b)

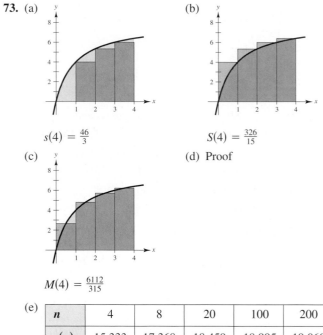

$s(4) = \frac{46}{3}$ $S(4) = \frac{326}{15}$

(c)   (d) Proof

$M(4) = \frac{6112}{315}$

(e)

n	4	8	20	100	200
$s(n)$	15.333	17.368	18.459	18.995	19.060
$S(n)$	21.733	20.568	19.739	19.251	19.188
$M(n)$	19.403	19.201	19.137	19.125	19.125

(f) f is an increasing function.

75. b   **77.** True   **79.** Answers will vary.

81. Suppose there are n rows in the figure. The stars on the left total $1 + 2 + \cdots + n$, as do the stars on the right. There are $n(n + 1)$ stars in total. So,

$$2[1 + 2 + \cdots + n] = n(n + 1)$$

$$1 + 2 + \cdots + n = \frac{n(n + 1)}{2}.$$

83. (a) $y = (-4.09 \times 10^{-5})x^3 + 0.016x^2 - 2.67x + 452.9$

(b) 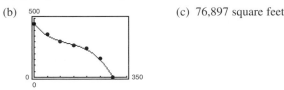 (c) 76,897 square feet

Section 4.3 (page 272)

1. $2\sqrt{3} \approx 3.464$ **3.** 36 **5.** 0 **7.** $\frac{10}{3}$

9. $\int_{-1}^{5} (3x + 10) \, dx$ **11.** $\int_{0}^{3} \sqrt{x^2 + 4} \, dx$

13. $\int_{0}^{5} 3 \, dx$ **15.** $\int_{-4}^{4} \left(4 - |x|\right) dx$ **17.** $\int_{-2}^{2} (4 - x^2) \, dx$

19. $\int_{0}^{\pi} \sin x \, dx$ **21.** $\int_{0}^{2} y^3 \, dy$

23. $A = 12$ **25.** $A = 8$

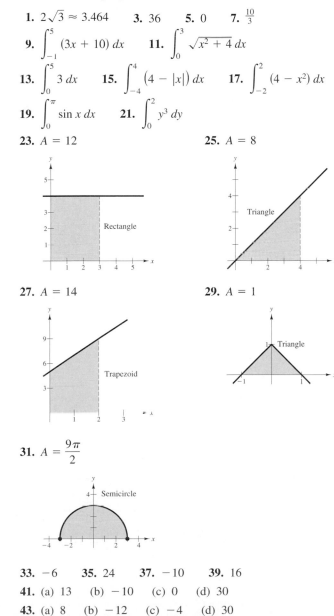

27. $A = 14$ **29.** $A = 1$

31. $A = \dfrac{9\pi}{2}$

33. -6 **35.** 24 **37.** -10 **39.** 16

41. (a) 13 (b) -10 (c) 0 (d) 30

43. (a) 8 (b) -12 (c) -4 (d) 30

45. (a) $-\pi$ (b) 4 (c) $-(1 + 2\pi)$ (d) $3 - 2\pi$

(e) $5 + 2\pi$ (f) $23 - 2\pi$

47. $\sum_{i=1}^{n} f(x_i) \Delta x > \int_{4}^{5} f(x) \, dx$ **49.** $\sum_{i=1}^{n} f(x_i) \Delta x < \int_{1}^{5} f(x) \, dx$

51. No. There is a discontinuity at $x = 4$.

53. a **55.** d

57.

n	4	8	12	16	20
$L(n)$	3.6830	3.9956	4.0707	4.1016	4.1177
$M(n)$	4.3082	4.2076	4.1838	4.1740	4.1690
$R(n)$	3.6830	3.9956	4.0707	4.1016	4.1177

59.

n	4	8	12	16	20
$L(n)$	0.5890	0.6872	0.7199	0.7363	0.7461
$M(n)$	0.7854	0.7854	0.7854	0.7854	0.7854
$R(n)$	0.9817	0.8836	0.8508	0.8345	0.8247

61. True **63.** True

65. False: $\int_{0}^{2} (-x) \, dx = -2$ **67.** 272

69. No. No matter how small the subintervals, the number of both rational and irrational numbers within each subinterval is infinite and $f(c_i) = 0$ or $f(c_i) = 1$.

71. $\frac{1}{3}$

Section 4.4 (page 284)

1. 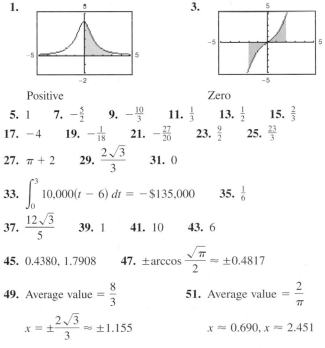 **3.**

Positive Zero

5. 1 **7.** $-\frac{5}{2}$ **9.** $-\frac{10}{3}$ **11.** $\frac{1}{3}$ **13.** $\frac{1}{2}$ **15.** $\frac{2}{3}$

17. -4 **19.** $-\frac{1}{18}$ **21.** $-\frac{27}{20}$ **23.** $\frac{9}{2}$ **25.** $\frac{23}{3}$

27. $\pi + 2$ **29.** $\dfrac{2\sqrt{3}}{3}$ **31.** 0

33. $\int_{0}^{3} 10,000(t - 6) \, dt = -\$135,000$ **35.** $\frac{1}{6}$

37. $\dfrac{12\sqrt{3}}{5}$ **39.** 1 **41.** 10 **43.** 6

45. 0.4380, 1.7908 **47.** $\pm \arccos \dfrac{\sqrt{\pi}}{2} \approx \pm 0.4817$

49. Average value $= \dfrac{8}{3}$ **51.** Average value $= \dfrac{2}{\pi}$

$x = \pm \dfrac{2\sqrt{3}}{3} \approx \pm 1.155$ $x \approx 0.690, x \approx 2.451$

53. The Fundamental Theorem of Calculus states that if a function f is continuous on $[a, b]$ and F is an antiderivative of f on $[a, b]$, then $\int_{a}^{b} f(x) \, dx = F(b) - F(a)$.

55. -1.5 **57.** 6.5 **59.** 15.5

61. (a) $F(x) = 500 \sec^2 x$ (b) 827 newtons

63. ≈ 0.5318 liter

65. (a)

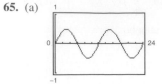

The average value of $f(t)$ over the interval $0 \le t \le 24$ is represented by $\displaystyle\int_0^{24} 0.5 \sin\!\left(\frac{\pi t}{6}\right) = 0$.

(b)

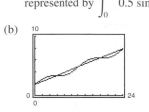

Even though the average value of $f(t) = 0$, the trend represented by g increases over the interval $0 \le t \le 24$ as does $S(t)$.

67. (a) $v = -0.00086t^3 + 0.0782t^2 - 0.208t + 0.10$

(b)

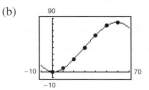

(c) 2475.6 meters

69. $F(x) = \dfrac{1}{2}x^2 - 5x$

$F(2) = -8$

$F(5) = -12\frac{1}{2}$

$F(8) = -8$

71. $F(x) = -\dfrac{10}{x} + 10$

$F(2) = 5$

$F(5) = 8$

$F(8) = 8\frac{3}{4}$

73. $F(x) = \sin x - \sin 1$

$F(2) = \sin 2 - \sin 1 \approx 0.0678$

$F(5) = \sin 5 - \sin 1 \approx -1.8004$

$F(8) = \sin 8 - \sin 1 \approx 0.1479$

75. $\frac{1}{2}x^2 + 2x$ **77.** $\frac{3}{4}x^{4/3} - 12$ **79.** $\tan x - 1$

81. $x^2 - 2x$ **83.** $\sqrt{x^4 + 1}$ **85.** $x \cos x$ **87.** 8

89. $\cos x \sqrt{\sin x}$ **91.** $3x^2 \sin x^6$

93.

An extremum of g occurs at $x = 2$.

95. (a) $C(x) = 1000(12x^{5/4} + 125)$

(b) $C(1) = \$137{,}000$

$C(5) = \$214{,}721$

$C(10) = \$338{,}394$

97. True

99. False: $f(x) = x^{-2}$ has a nonremovable discontinuity at $x = 0$.

101. $f'(x) = \dfrac{1}{(1/x)^2 + 1}\left(-\dfrac{1}{x^2}\right) + \dfrac{1}{x^2 + 1} = 0$

Since $f'(x) = 0$, $f(x)$ is constant.

103. 28 units   **105.** 2 units

Section 4.5 (page 297)

$\displaystyle\int f(g(x))g'(x)\,dx$	$u = g(x)$	$du = g'(x)\,dx$
1. $\displaystyle\int (5x^2 + 1)^2(10x)\,dx$	$5x^2 + 1$	$10x\,dx$
3. $\displaystyle\int \dfrac{x}{\sqrt{x^2 + 1}}\,dx$	$x^2 + 1$	$2x\,dx$
5. $\displaystyle\int \tan^2 x \sec^2 x\,dx$	$\tan x$	$\sec^2 x\,dx$

7. $\dfrac{(1 + 2x)^5}{5} + C$ **9.** $\dfrac{2}{3}(9 - x^2)^{3/2} + C$

11. $\dfrac{(x^4 + 3)^3}{12} + C$ **13.** $\dfrac{(x^3 - 1)^5}{15} + C$

15. $\dfrac{(t^2 + 2)^{3/2}}{3} + C$ **17.** $-\dfrac{15}{8}(1 - x^2)^{4/3} + C$

19. $\dfrac{1}{4(1 - x^2)^2} + C$ **21.** $-\dfrac{1}{3(1 + x^3)} + C$

23. $-\sqrt{1 - x^2} + C$ **25.** $-\dfrac{1}{4}\left(1 + \dfrac{1}{t}\right)^4 + C$

27. $\sqrt{2x} + C$

29. $\frac{2}{5}x^{5/2} + 2x^{3/2} + 14x^{1/2} + C = \frac{2}{5}\sqrt{x}(x^2 + 5x + 35) + C$

31. $\frac{1}{4}t^4 - t^2 + C$

33. $6y^{3/2} - \frac{2}{5}y^{5/2} + C = \frac{2}{5}y^{3/2}(15 - y) + C$

35. $2x^2 - 4\sqrt{16 - x^2} + C$ **37.** $-\dfrac{1}{2(x^2 + 2x - 3)} + C$

39. (a) Answers will vary.   (b) $y = -\frac{1}{3}(4 - x^2)^{3/2} + 2$

Example:

41. $-\cos(\pi x) + C$ **43.** $-\dfrac{1}{2}\cos 2x + C$ **45.** $-\sin\dfrac{1}{\theta} + C$

47. $\frac{1}{4}\sin^2 2x + C_1$ or $-\frac{1}{4}\cos^2 2x + C_2$ or $-\frac{1}{8}\cos 4x + C_3$

49. $\frac{1}{5}\tan^5 x + C$ **51.** $\frac{1}{2}\tan^2 x + C$ or $\frac{1}{2}\sec^2 x + C_1$

53. $\cot x - x + C$ **55.** $f(x) = 2\sin\dfrac{x}{2} + 3$

57. $\frac{2}{15}(x + 2)^{3/2}(3x - 4) + C$

59. $-\frac{2}{105}(1 - x)^{3/2}(15x^2 + 12x + 8) + C$

61. $\dfrac{\sqrt{2x - 1}}{15}(3x^2 + 2x - 13) + C$

63. $-x - 1 - 2\sqrt{x + 1} + C$ or $-(x + 2\sqrt{x + 1}) + C_1$

65. 0 **67.** $12 - \dfrac{8\sqrt{2}}{9}$ **69.** 2 **71.** $\dfrac{1}{2}$ **73.** $\dfrac{4}{15}$

75. $\dfrac{3\sqrt{3}}{4}$ **77.** $\dfrac{1209}{28}$ **79.** 4 **81.** $2(\sqrt{3} - 1)$

83. $\frac{10}{3}$ **85.** $\frac{144}{5}$

87. 7.38

89. $\frac{1}{6}(2x - 1)^3 + C_1 = \frac{4}{3}x^3 - 2x^2 + x - \frac{1}{6} + C_1$
or $\frac{4}{3}x^3 - 2x^2 + x + C_2$

Answers differ by a constant: $C_2 = C_1 - \frac{1}{6}$

91. $\frac{272}{15}$ **93.** 0

95. (a) $\frac{8}{3}$ (b) $\frac{16}{3}$ (c) $-\frac{8}{3}$ (d) 8

97. $2\displaystyle\int_0^4 (6x^2 - 3)\,dx = 232$

99. Answers will vary. See "Guidelines for Making a Change of Variables" on page 292.

101. It is an odd function.

103. $V(t) = \dfrac{200,000}{t + 1} + 300,000$

$\$340,000$

105. (a) 102.352 thousand units   (b) 102.352 thousand units
(c) 74.5 thousand units

107. (a) 1.273 amperes   (b) 1.382 amperes
(c) 0 amperes

109. False. $\displaystyle\int (2x + 1)^2\,dx = \frac{1}{6}(2x + 1)^3 + C$

111. True   **113.** True   **115.** Proof

Section 4.6 (page 305)

	Trapezoidal	Simpson's	Exact
1.	2.7500	2.6667	2.6667
3.	4.2500	4.0000	4.0000
5.	4.0625	4.0000	4.0000
7.	12.6640	12.6667	12.6667
9.	0.1676	0.1667	0.1667

	Trapezoidal	Simpson's	Graphing utility
11.	3.2833	3.2396	3.2413
13.	0.3415	0.3720	0.3927
15.	0.9567	0.9778	0.9775
17.	0.0891	0.0888	0.0891
19.	0.1940	0.1860	0.1858

21. The Trapezoidal Rule will yield a result greater than $\int_a^b f(x)\,dx$ if f is concave upward on $[a, b]$ because the graph of f will lie within the trapezoids.

23. (a) 0.500   (b) 0.000   **25.** (a) $n = 366$ (b) $n = 26$

27. (a) $n = 130$ (b) $n = 12$

29. (a) $n = 643$ (b) $n = 48$ **31.** Proof

33.

n	$L(n)$	$M(n)$	$R(n)$	$T(n)$	$S(n)$
4	12.7771	15.3965	18.4340	15.6055	15.4845
8	14.0868	15.4480	16.9152	15.5010	15.4662
10	14.3569	15.4544	16.6197	15.4883	15.4658
12	14.5386	15.4578	16.4242	15.4814	15.4657
16	14.7674	15.4613	16.1816	15.4745	15.4657
20	14.9056	15.4628	16.0370	15.4713	15.4657

35.

n	$L(n)$	$M(n)$	$R(n)$	$T(n)$	$S(n)$
4	2.8163	3.5456	3.7256	3.2709	3.3996
8	3.1809	3.5053	3.6356	3.4083	3.4541
10	3.2478	3.4990	3.6115	3.4296	3.4624
12	3.2909	3.4952	3.5940	3.4425	3.4674
16	3.3431	3.4910	3.5704	3.4568	3.4730
20	3.3734	3.4888	3.5552	3.4643	3.4759

37. 0.701

39. 10,233.58 foot-pounds   **41.** 3.1416

43. 89,250 square meters   **45.** 2.477

Review Exercises for Chapter 4 (page 307)

1.

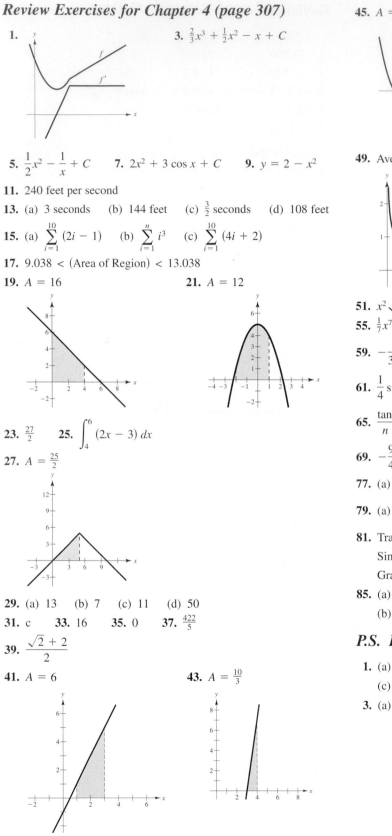

3. $\frac{2}{3}x^3 + \frac{1}{2}x^2 - x + C$

5. $\frac{1}{2}x^2 - \frac{1}{x} + C$ **7.** $2x^2 + 3\cos x + C$ **9.** $y = 2 - x^2$

11. 240 feet per second

13. (a) 3 seconds   (b) 144 feet   (c) $\frac{3}{2}$ seconds   (d) 108 feet

15. (a) $\sum_{i=1}^{10} (2i - 1)$ (b) $\sum_{i=1}^{n} i^3$ (c) $\sum_{i=1}^{10} (4i + 2)$

17. $9.038 < (\text{Area of Region}) < 13.038$

19. $A = 16$ **21.** $A = 12$

23. $\frac{27}{2}$ **25.** $\int_{4}^{6} (2x - 3)\, dx$

27. $A = \frac{25}{2}$

29. (a) 13   (b) 7   (c) 11   (d) 50

31. c   **33.** 16   **35.** 0   **37.** $\frac{422}{5}$

39. $\dfrac{\sqrt{2} + 2}{2}$

41. $A = 6$ **43.** $A = \frac{10}{3}$

45. $A = \frac{1}{4}$ **47.** $A = 16$

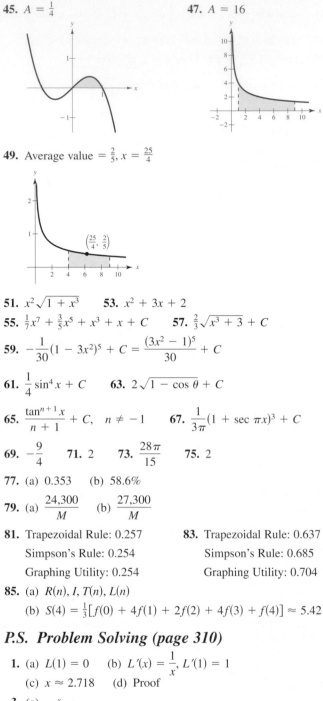

49. Average value $= \frac{2}{5}, x = \frac{25}{4}$

$\left(\frac{25}{4}, \frac{2}{5}\right)$

51. $x^2\sqrt{1 + x^3}$ **53.** $x^2 + 3x + 2$

55. $\frac{1}{7}x^7 + \frac{3}{5}x^5 + x^3 + x + C$ **57.** $\frac{2}{3}\sqrt{x^3 + 3} + C$

59. $-\dfrac{1}{30}(1 - 3x^2)^5 + C = \dfrac{(3x^2 - 1)^5}{30} + C$

61. $\dfrac{1}{4}\sin^4 x + C$ **63.** $2\sqrt{1 - \cos\theta} + C$

65. $\dfrac{\tan^{n+1} x}{n + 1} + C, \quad n \neq -1$ **67.** $\dfrac{1}{3\pi}(1 + \sec\pi x)^3 + C$

69. $-\dfrac{9}{4}$ **71.** 2   **73.** $\dfrac{28\pi}{15}$ **75.** 2

77. (a) 0.353   (b) 58.6%

79. (a) $\dfrac{24{,}300}{M}$ (b) $\dfrac{27{,}300}{M}$

81. Trapezoidal Rule: 0.257   **83.** Trapezoidal Rule: 0.637
      Simpson's Rule: 0.254   Simpson's Rule: 0.685
      Graphing Utility: 0.254   Graphing Utility: 0.704

85. (a) $R(n), I, T(n), L(n)$

     (b) $S(4) = \frac{1}{3}[f(0) + 4f(1) + 2f(2) + 4f(3) + f(4)] \approx 5.42$

P.S. Problem Solving (page 310)

1. (a) $L(1) = 0$ (b) $L'(x) = \dfrac{1}{x}, L'(1) = 1$

     (c) $x \approx 2.718$ (d) Proof

3. (a)

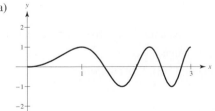

(b)

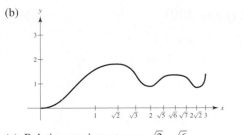

(c) Relative maxima at $x = \sqrt{2}, \sqrt{6}$
    Relative minima at $x = 2, 2\sqrt{2}$

(d) Points of inflection at $x = 1, \sqrt{3}, \sqrt{5}, \sqrt{7}$

5. (a)

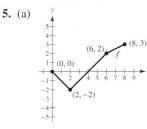

(b)

x	0	1	2	3	4	5	6	7	8
$F(x)$	0	$-\frac{1}{2}$	-2	$-\frac{7}{2}$	-4	$-\frac{7}{2}$	-2	$\frac{1}{4}$	3

(c) $x = 4, 8$ (d) $x = 2$

7. (a) 1.6758; Error of approximation ≈ 0.0071

(b) $\frac{3}{2}$ (c) Proof

9. Proof

11. $\lim\limits_{n\to\infty} \sum\limits_{i=1}^{n} \left(\dfrac{i}{n}\right)^5 \left(\dfrac{1}{n}\right) = \dfrac{1}{6}$ **13.** $1 \le \displaystyle\int_0^1 \sqrt{1 + x^4}\, dx \le \sqrt{2}$

15. Proof   **17.** 100,000 pounds

Chapter 5

Section 5.1 (page 321)

1.

x	0.5	1.5	2	2.5	3
$\int_1^x (1/t)\, dt$	-0.6932	0.4055	0.6932	0.9163	1.0987

x	3.5	4
$\int_1^x (1/t)\, dt$	1.2529	1.3865

3. (a) 3.8067   (b) $\ln 45 = \displaystyle\int_1^{45} \dfrac{1}{t}\, dt \approx 3.8067$

5. (a) -0.2231 (b) $\ln 0.8 = \displaystyle\int_1^{0.8} \dfrac{1}{t}\, dt \approx -0.2231$

7. b   **8.** d   **9.** a   **10.** c

11. Domain: $x > 0$

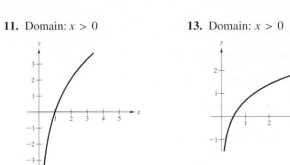

13. Domain: $x > 0$

15. Domain: $x > 1$

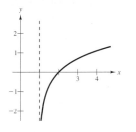

17. (a) 1.7917   (b) -0.4055 (c) 4.3944   (d) 0.5493

19. $\ln 2 - \ln 3$ **21.** $\ln x + \ln y - \ln z$ **23.** $\frac{1}{3}\ln(a^2 + 1)$

25. $3[\ln(x + 1) + \ln(x - 1) - 3 \ln x]$ **27.** $\ln z + 2 \ln(z - 1)$

29. $\ln \dfrac{x - 2}{x + 2}$ **31.** $\ln \sqrt[3]{\dfrac{x(x + 3)^2}{x^2 - 1}}$ **33.** $\ln \dfrac{9}{\sqrt{x^2 + 1}}$

35.

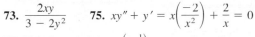

37. $-\infty$

39. $\ln 4$ **41.** 3   **43.** 2   **45.** $\dfrac{2}{x}$ **47.** $\dfrac{4(\ln x)^3}{x}$

49. $\dfrac{2x^2 - 1}{x(x^2 - 1)}$ **51.** $\dfrac{1 - x^2}{x(x^2 + 1)}$ **53.** $\dfrac{1 - 2 \ln t}{t^3}$

55. $\dfrac{2}{x \ln x^2} = \dfrac{1}{x \ln x}$ **57.** $\dfrac{1}{1 - x^2}$ **59.** $\dfrac{-4}{x(x^2 + 4)}$

61. $\dfrac{\sqrt{x^2 + 1}}{x^2}$ **63.** $\cot x$ **65.** $-\tan x + \dfrac{\sin x}{\cos x - 1}$

67. $\dfrac{3 \cos x}{(\sin x - 1)(\sin x + 2)}$ **69.** $\dfrac{2}{x}(\sin 2x + x \cos 2x \ln x^2)$

71. (a) $5x - y - 2 = 0$

(b)

73. $\dfrac{2xy}{3 - 2y^2}$ **75.** $xy'' + y' = x\left(\dfrac{-2}{x^2}\right) + \dfrac{2}{x} = 0$

77. Relative minimum: $\left(1, \frac{1}{2}\right)$

79. Relative minimum: $(e^{-1}, -e^{-1})$

81. Relative minimum: (e, e)

Point of inflection: $\left(e^2, \dfrac{e^2}{2}\right)$

83. $P_1 = x - 1$; $P_2 = x - 1 - \frac{1}{2}(x - 1)^2$

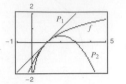

The values of f, P_1, and P_2 and their first derivatives agree at $x = 1$.

85. $x \approx 0.567$ **87.** $\dfrac{2x^2 - 1}{\sqrt{x^2 - 1}}$

89. $\dfrac{3x^3 - 15x^2 + 8x}{2(x - 1)^3 \sqrt{3x - 2}}$ **91.** $\dfrac{(2x^2 + 2x - 1)\sqrt{x - 1}}{(x + 1)^{3/2}}$

93. The domain of the natural logarithmic function is $(0, \infty)$ and the range is $(-\infty, \infty)$. The function is continuous, increasing, and one-to-one and its graph is concave downward. In addition, if a and b are positive numbers and n is rational, then $\ln(1) = 0$, $\ln(a \cdot b) = \ln a + \ln b$, $\ln(a^n) = n \ln a$, and $\ln\left(\dfrac{a}{b}\right) = \ln a - \ln b$.

95. Using properties of logarithms, $\ln e^x$ can be rewritten as $x \ln e$. Then, since $\ln e = 1$ by the definition of e, $x \ln e = x(1) = x$.

97. (a) Rolle's Theorem does not apply because $f(1) \neq f(3)$.
(b) Yes. $f'(2) = 0$ and $2 \in [1, 3]$.

99. $\beta = 160 + 10 \log_{10} I$; $\beta = 60$ decibels

101. (a) $h = 0$ is not in the domain of the function.
(b) $h = 0.86 - 6.447 \ln p$

(c)

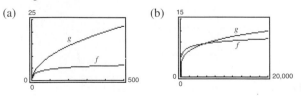

(d) 2.7 kilometers
(e) 0.15 atmosphere
(f) $h = 5$: $\dfrac{dp}{dh} = -0.085$

$h = 20$: $\dfrac{dp}{dh} = -0.009$

As the altitude increases, the pressure decreases at a slower rate.

103. For large values of x, g increases at a faster rate than f in both cases. The natural logarithmic function increases very slowly for large values of x.

(a) (b)

105. False: $\ln x + \ln 25 = \ln 25x$.

Section 5.2 (page 330)

1. $5 \ln|x| + C$ **3.** $\ln|x + 1| + C$

5. $-\frac{1}{2}\ln|3 - 2x| + C$ **7.** $\ln\sqrt{x^2 + 1} + C$

9. $\dfrac{x^2}{2} - \ln(x^4) + C$ **11.** $\frac{1}{3}\ln|x^3 + 3x^2 + 9x| + C$

13. $\dfrac{x^2}{2} - 4x + 6\ln|x + 1| + C$ **15.** $\dfrac{x^3}{3} + 5\ln|x - 3| + C$

17. $\dfrac{x^3}{3} - 2x + \ln\sqrt{x^2 + 2} + C$ **19.** $\frac{1}{3}(\ln x)^3 + C$

21. $2\sqrt{x + 1} + C$ **23.** $2\ln|x - 1| - \dfrac{2}{x - 1} + C$

25. $\sqrt{2x} - \ln\left|1 + \sqrt{2x}\right| + C$

27. $x + 6\sqrt{x} + 18\ln\left|\sqrt{x} - 3\right| + C$ **29.** $\ln|\sin \theta| + C$

31. $-\frac{1}{2}\ln|\csc 2x + \cot 2x| + C$ **33.** $\ln|1 + \sin t| + C$

35. $\ln|\sec x - 1| + C$

37. $y = -3\ln|2 - x| + C$ **39.** $y = -\frac{1}{2}\ln|\cos 2\theta| + C$

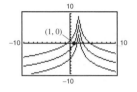 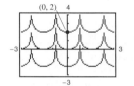

The graph has a hole at $x = 2$.

41. (a)

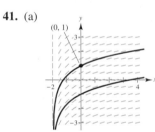

(b) $y = \ln\left|\dfrac{x + 2}{2}\right| + 1$

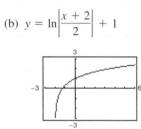

43. $\frac{5}{3}\ln 13 \approx 4.275$ **45.** $\frac{7}{3}$ **47.** $-\ln 3 \approx -1.099$

49. $\ln\left|\dfrac{2 - \sin 2}{1 - \sin 1}\right| \approx 1.929$

51. $-\ln|\cos x| + C = \ln\left|\dfrac{1}{\cos x}\right| + C = \ln|\sec x| + C$

53. $\ln|\sec x + \tan x| + C = \ln\left|\dfrac{\sec^2 x - \tan^2 x}{\sec x - \tan x}\right| + C$

$= -\ln|\sec x - \tan x| + C$

55. $2\left[\sqrt{x} - \ln\left(1 + \sqrt{x}\right)\right] + C$ **57.** $-\sin(1 - x) + C$

59. $\ln\left(\sqrt{2} + 1\right) - \dfrac{\sqrt{2}}{2} \approx 0.174$ **61.** $\dfrac{1}{x}$ **63.** 0 **65.** d

67. $\frac{15}{2} + 8\ln 2 \approx 13.045$

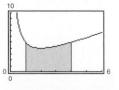

69. $\dfrac{12}{\pi}\left[2\ln(\sqrt{3}+1)-\ln 2\right] \approx 5.03$

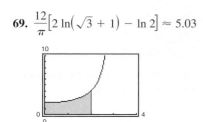

71. Power Rule   **73.** Log rule

75. Use long division to rewrite the integrand.

77. 1   **79.** $\dfrac{1}{2(e-1)} \approx 0.291$

81. $P(t) = 1000(12\ln|1+0.25t|+1)$; $P(3) \approx 7715$

83. \$168.27

85. (a)

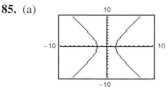

(b) Answers will vary. Example: $y^2 - e^{-\ln x + \ln 4} = \dfrac{4}{x}$

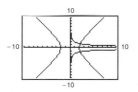

87. False. $\frac{1}{2}(\ln x) = \ln x^{1/2}$ **89.** True

Section 5.3 (page 338)

1. (a) $f(g(x)) = 5\left(\dfrac{x-1}{5}\right) + 1 = x$

$g(f(x)) = \dfrac{(5x+1)-1}{5} = x$

(b)

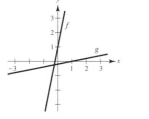

3. (a) $f(g(x)) = \left(\sqrt[3]{x}\right)^3 = x$; $g(f(x)) = \sqrt[3]{x^3} = x$

(b)

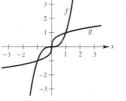

5. (a) $f(g(x)) = \sqrt{x^2+4-4} = x$;

$g(f(x)) = \left(\sqrt{x-4}\right)^2 + 4 = x$

(b)

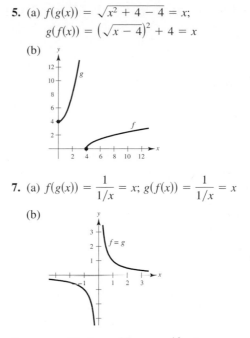

7. (a) $f(g(x)) = \dfrac{1}{1/x} = x$; $g(f(x)) = \dfrac{1}{1/x} = x$

(b)

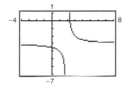

9.   c   **10.** b   **11.** a   **12.** d

13. Inverse exists.   **15.** Inverse does not exist.

17. One-to-one   **19.** One-to-one

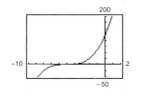

21. One-to-one

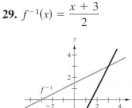

23. Inverse exists.   **25.** Inverse does not exist.

27. Inverse exists.

29. $f^{-1}(x) = \dfrac{x+3}{2}$ **31.** $f^{-1}(x) = x^{1/5}$

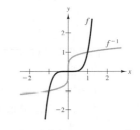

f and f^{-1} are symmetric f and f^{-1} are symmetric
about $y = x$.   about $y = x$.

33. $f^{-1}(x) = x^2, \; x \geq 0$

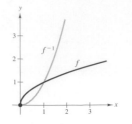

f and f^{-1} are symmetric about $y = x$.

35. $f^{-1}(x) = \sqrt{4 - x^2}, \; 0 \leq x \leq 2$

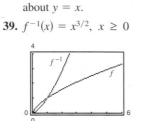

f and f^{-1} are symmetric about $y = x$.

37. $f^{-1}(x) = x^3 + 1$

f and f^{-1} are symmetric about $y = x$.

39. $f^{-1}(x) = x^{3/2}, \; x \geq 0$

f and f^{-1} are symmetric about $y = x$.

41. $f^{-1}(x) = \dfrac{\sqrt{7}x}{\sqrt{1 - x^2}}, \; -1 < x < 1$

f and f^{-1} are symmetric about $y = x$.

43.

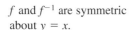

x	1	2	3	4
$f^{-1}(x)$	0	1	2	4

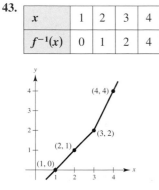

45. (a) Proof

(b) $y = \frac{20}{7}(80 - x)$

x: total cost

y: number of pounds of the less expensive commodity

(c) $[62.5, 80]$

(d) 20 pounds

47. $f'(x) = 2(x - 4) > 0$ on $(4, \infty)$

49. $f'(x) = -\dfrac{8}{x^3} < 0$ on $(0, \infty)$

51. $f'(x) = -\sin x < 0$ on $(0, \pi)$

53. $f^{-1}(x) = \begin{cases} \dfrac{1 - \sqrt{1 + 16x^2}}{2x}, & \text{if } x \neq 0 \\ 0, & \text{if } x = 0 \end{cases}$

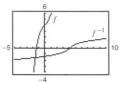

The graph of f^{-1} is a reflection of the graph of f in the line $y = x$.

55. (a) and (b)

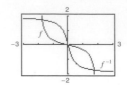

(c) f is one-to-one and has an inverse function.

57. (a) and (b)

(c) g is not one-to-one and does not have an inverse function.

59. One-to-one

$f^{-1}(x) = x^2 + 2, \; x \geq 0$

61. One-to-one

$f^{-1}(x) = 2 - x, \; x \geq 0$

63. $f^{-1}(x) = \sqrt{x} + 3, \; x \geq 0$

(Answer is not unique.)

65. $f^{-1}(x) = x - 3, \; x \geq 0$

(Answer is not unique.)

67. Inverse exists. Volume is an increasing function, therefore one-to-one. The inverse function gives the time t corresponding to the volume V.

69. Inverse does not exist.   **71.** $\dfrac{1}{5}$ **73.** $\dfrac{2\sqrt{3}}{3}$ **75.** $\dfrac{1}{13}$

77. (a) Domain of f: $(-\infty, \infty)$

Domain of f^{-1}: $(-\infty, \infty)$

(b) Range of f: $(-\infty, \infty)$

Range of f^{-1}: $(-\infty, \infty)$

(c)

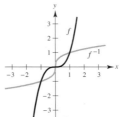

(d) $f'(\frac{1}{2}) = \frac{3}{4}, \; (f^{-1})'(\frac{1}{8}) = \frac{4}{3}$

79. (a) Domain of f: $[4, \infty)$

Domain of f^{-1}: $[0, \infty)$

(b) Range of f: $[0, \infty)$

Range of f^{-1}: $[4, \infty)$

(c)

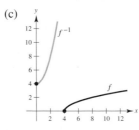

(d) $f'(5) = \frac{1}{2}, \; (f^{-1})'(1) = 2$

81. $-\frac{1}{11}$ **83.** 32   **85.** 600   **87.** $(g^{-1} \circ f^{-1})(x) = \dfrac{x+1}{2}$

89. $(f \circ g)^{-1}(x) = \dfrac{x+1}{2}$

91. Let $y = f(x)$ be one-to-one. Solve for x as a function of y. Interchange x and y to get $y = f^{-1}(x)$. Let the domain of f^{-1} be the range of f. Verify that $f(f^{-1}(x)) = x$ and $f^{-1}(f(x)) = x$.

Example: $f(x) = x^3$

$$y = x^3$$
$$x = \sqrt[3]{y}$$
$$y = \sqrt[3]{x}$$
$$f^{-1}(x) = \sqrt[3]{x}$$

93. Answers will vary. Example: $y = x^4 - 2x^3$

95. Many x-values yield the same y-value.

For example, $f(\pi) = 0 = f(0)$.

The graph is not continuous at $x = \dfrac{(2n-1)\pi}{2}$, where n is an integer.

97. Proof   **99.** Proof

101. False. Let $f(x) = x^2$.   **103.** True

105. No. Let $f(x) = \begin{cases} x, & 0 \le x \le 1 \\ 1 - x, & 1 \le x \le 2 \end{cases}$.   **107.** $\sqrt{17}$

Section 5.4 (page 347)

1. $\ln 1 = 0$ **3.** $e^{0.6931 \cdots} = 2$ **5.** $x = 4$

7. $x \approx 2.485$ **9.** $x = 0$ **11.** $x \approx 0.511$

13. $x \approx 7.389$ **15.** $x \approx 10.389$ **17.** $x \approx 5.389$

19.   **21.**

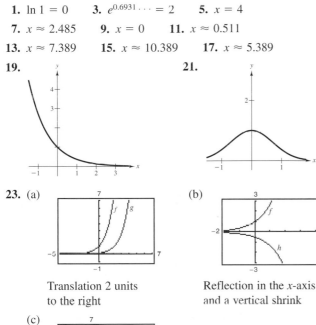

23. (a)   (b)

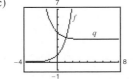

Translation 2 units   Reflection in the x-axis
to the right   and a vertical shrink

(c)

Reflection in the y-axis and
a translation 3 units upward

25. c   **26.** d   **27.** a   **28.** b

29.   **31.**

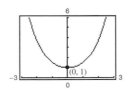

33.

$$\lim_{x \to \infty} f(x) = \lim_{x \to \infty} g(x) = e^{0.5}$$

35. $2.7182805 < e$ **37.** (a) 3   (b) -3

39. $2e^{2x}$ **41.** $2(x-1)e^{-2x+x^2}$ **43.** $\dfrac{e^{\sqrt{x}}}{2\sqrt{x}}$

45. $3(e^{-t} + e^t)^2(e^t - e^{-t})$ **47.** $2x$ **49.** $\dfrac{2e^{2x}}{1 + e^{2x}}$

51. $\dfrac{-2(e^x - e^{-x})}{(e^x + e^{-x})^2}$ **53.** $x^2 e^x$ **55.** $e^{-x}\left(\dfrac{1}{x} - \ln x\right)$

57. $2e^x \cos x$ **59.** $\dfrac{10 - e^y}{xe^y + 3}$ **61.** $3(6x + 5)e^{-3x}$

63. $y'' - 2y' + 3y - 0$

$e^x\left[-\cos\sqrt{2}x - \sin\sqrt{2}x - 2\sqrt{2}\sin\sqrt{2}x + 2\sqrt{2}\cos\sqrt{2}x\right] -$
$2e^x\left[-\sqrt{2}\sin\sqrt{2}x + \sqrt{2}\cos\sqrt{2}x + \cos\sqrt{2}x + \sin\sqrt{2}x\right] +$
$3e^x\left[\cos\sqrt{2}x + \sin\sqrt{2}x\right]$

$$= 0$$
$$0 = 0$$

65. Relative minimum: $(0, 1)$

67. Relative maximum: $\left(2, \dfrac{1}{\sqrt{2\pi}}\right)$

Points of inflection: $\left(1, \dfrac{e^{-0.5}}{\sqrt{2\pi}}\right), \left(3, \dfrac{e^{-0.5}}{\sqrt{2\pi}}\right)$

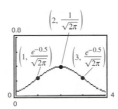

69. Relative minimum: $(0, 0)$

Relative maximum: $(2, 4e^{-2})$

Points of inflection: $\left(2 \pm \sqrt{2}, \left(6 \pm 4\sqrt{2}\right)e^{-(2 \pm \sqrt{2})}\right)$

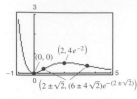

71. Relative maximum: $(-1, 1 + e)$

Point of inflection: $(0, 3)$

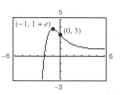

73. $A = \sqrt{2}e^{-1/2}$ **75.** Proof   **77.** 0.567

79. (a)

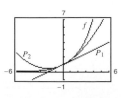

(b) When x increases without bound, $1/x$ approaches zero and $e^{1/x}$ approaches 1. Therefore, $f(x)$ approaches $\frac{2}{1+1} = 1$. Thus, $f(x)$ has a horizontal asymptote at $y = 1$. As x approaches zero from the right, $1/x$ approaches ∞, $e^{1/x}$ approaches ∞, and $f(x)$ approaches 0. As x approaches zero from the left, $1/x$ approaches $-\infty$, $e^{1/x}$ approaches 0, and $f(x)$ approaches 2. The limit does not exist, because the limit from the left does not equal the limit from the right. Therefore, $x = 0$ is a nonremovable discontinuity.

81. (a) $\ln P = -0.1499h + 9.3018$ (b) $P = 10,957.7e^{-0.1499h}$

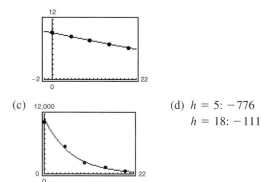

(c)
 (d) $h = 5: -776$
 $h = 18: -111$

83. $P_1 = 1 + \dfrac{x}{2}$; $P_2 = 1 + \dfrac{x}{2} + \dfrac{x^2}{8}$

The values of f, P_1, and P_2 and their first derivatives agree at $x = 0$.

85. (a)   (b)

(c)

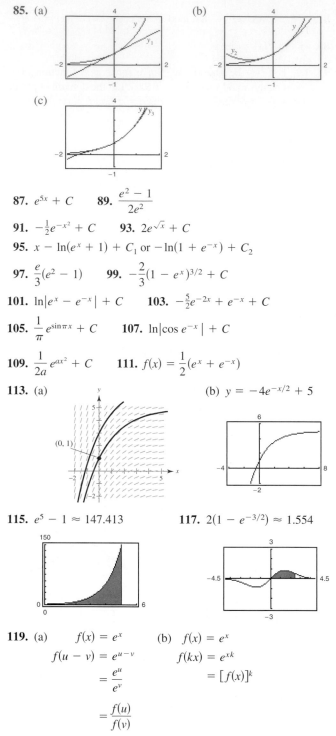

87. $e^{5x} + C$ **89.** $\dfrac{e^2 - 1}{2e^2}$

91. $-\frac{1}{2}e^{-x^2} + C$ **93.** $2e^{\sqrt{x}} + C$

95. $x - \ln(e^x + 1) + C_1$ or $-\ln(1 + e^{-x}) + C_2$

97. $\dfrac{e}{3}(e^2 - 1)$ **99.** $-\dfrac{2}{3}(1 - e^x)^{3/2} + C$

101. $\ln|e^x - e^{-x}| + C$ **103.** $-\frac{5}{2}e^{-2x} + e^{-x} + C$

105. $\dfrac{1}{\pi}e^{\sin \pi x} + C$ **107.** $\ln|\cos e^{-x}| + C$

109. $\dfrac{1}{2a}e^{ax^2} + C$ **111.** $f(x) = \frac{1}{2}(e^x + e^{-x})$

113. (a)   (b) $y = -4e^{-x/2} + 5$

115. $e^5 - 1 \approx 147.413$ **117.** $2(1 - e^{-3/2}) \approx 1.554$

119. (a) $f(x) = e^x$ (b) $f(x) = e^x$

$f(u - v) = e^{u-v}$ $f(kx) = e^{xk}$

$= \dfrac{e^u}{e^v}$ $= [f(x)]^k$

$= \dfrac{f(u)}{f(v)}$

121. The probability that a given battery will last between 48 months and 60 months is approximately 47.72%.

123. $\displaystyle\int_0^x e^t \, dt \geq \int_0^x 1 \, dt$; $e^x - 1 \geq x$; $e^x > x + 1$ for $x \geq 0$

125. $f(x) = e^x$

The domain of $f(x)$ is $(-\infty, \infty)$ and the range of $f(x)$ is $(0, \infty)$. $f(x)$ is continuous, increasing, one-to-one, and concave upward on its entire domain.

$$\lim_{x \to -\infty} e^x = 0 \text{ and } \lim_{x \to \infty} e^x = \infty$$

127. $f(x) = e^x = f'(x)$

129. $e^{-x} > 0$ implies $\displaystyle\int_0^2 e^{-x}\,dx > 0$

131. (a) $f'(x) = \dfrac{1 - \ln x}{x^2} = 0$ when $x = e$.

On $(0, e), f'(x) > 0 \Rightarrow f$ is increasing.

On $(e, \infty), f'(x) < 0 \Rightarrow f$ is decreasing.

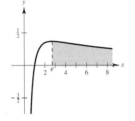

(b) For $e \le A < B$, we have:

$$\frac{\ln A}{A} > \frac{\ln B}{B}$$

$$B \ln A > A \ln B$$

$$\ln A^B > \ln B^A$$

$$A^B > B^A.$$

(c) Since $e < \pi$, from part (b) we have $e^\pi > \pi^e$.

Section 5.5 (page 357)

1. $y(t) = \left(\frac{1}{2}\right)^{t/3}, \frac{1}{4}$ **3.** $y(t) = \left(\frac{1}{2}\right)^{t/7}, \left(\frac{1}{2}\right)^{10/7} \approx 0.371$

5. -3 **7.** 0 **9.** (a) $\log_2 8 = 3$ (b) $\log_3(1/3) = -1$

11. (a) $10^{-2} = 0.01$ (b) $\left(\frac{1}{2}\right)^{-3} = 8$

13.   **15.**

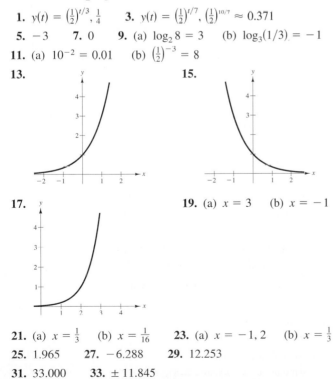

17.   **19.** (a) $x = 3$ (b) $x = -1$

21. (a) $x = \frac{1}{3}$ (b) $x = \frac{1}{16}$ **23.** (a) $x = -1, 2$ (b) $x = \frac{1}{3}$

25. 1.965 **27.** -6.288 **29.** 12.253

31. 33.000 **33.** ± 11.845

35. 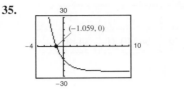   **37.**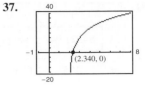

39.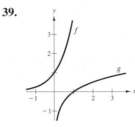

41. $(\ln 4)4^x$ **43.** $(\ln 5)5^{x-2}$

45. $t2^t (t \ln 2 + 2)$ **47.** $-2^{-\theta}[(\ln 2)\cos \pi\theta + \pi \sin \pi\theta]$

49. $\dfrac{1}{x(\ln 3)}$ **51.** $\dfrac{x - 2}{(\ln 2)x(x - 1)}$ **53.** $\dfrac{x}{(\ln 5)(x^2 - 1)}$

55. $\dfrac{5}{(\ln 2)t^2}(1 - \ln t)$ **57.** $2(1 - \ln x)x^{(2/x)-2}$

59. $(x - 2)^{x+1}\left[\dfrac{x + 1}{x - 2} + \ln(x - 2)\right]$

61. $\dfrac{3^x}{\ln 3} + C$ **63.** $\dfrac{7}{\ln 4}$ **65.** $-\dfrac{1}{2\ln 5}(5^{-x^2}) + C$

67. $\dfrac{\ln(3^{2x} + 1)}{2 \ln 3} + C$

69. (a)   (b) $y = \dfrac{3(1 - 0.4^{x/3})}{\ln 2.5} + \dfrac{1}{2}$

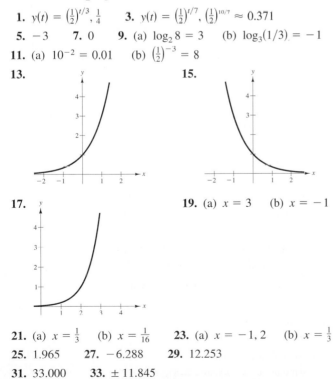

71. Answers will vary. Example: Growth and decay problems

73. (a) False. $y = a^x \Rightarrow 0 = a' \Rightarrow a = 0$, but exponential functions are not defined for $a = 0$.

(b) True: $y = \log_2 x$

(c) True: $2^y = x$

(d) False. $(1, 0)$, $(2, 1)$, and $(8, 3)$ are not collinear.

75. $g(x) = x^x, k(x) = 2^x, h(x) = x^2, f(x) = \log_2 x$

77. (a) $\$40.64$ (b) $C'(1) \approx 0.051P$, $C'(8) \approx 0.072P$

(c) $\ln 1.05$

79.

n	1	2	4	12
A	$\$1410.60$	$\$1414.78$	$\$1416.91$	$\$1418.34$

n	365	Continuous
A	$\$1419.04$	$\$1419.07$

81.

n	1	2	4	12
A	$4321.94	$4399.79	$4440.21	$4467.74

n	365	Continuous
A	$4481.23	$4481.69

83.

t	1	10	20	30
P	$95,122.94	$60,653.07	$36,787.94	$22,313.02

t	40	50
P	$13,533.53	$8208.50

85.

t	1	10	20	30
P	$95,132.82	$60,716.10	$36,864.45	$22,382.66

t	40	50
P	$13,589.88	$8251.24

87. c

89. (a) 6.7 million cubic feet per acre

(b) $t = 20: \dfrac{dV}{dt} = 0.073$

$t = 60: \dfrac{dV}{dt} = 0.040$

91. (a)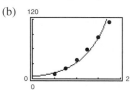

(b) 16.7%

(c) $x \approx 38.8$ or 38,800 egg masses

(d) $x \approx 2.78$ or 27,800 egg masses

93. (a) $B = 4.75(6.774)^d$

(b)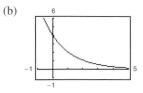

(c) When $d = 0.8$, the rate of growth is 41.99.

When $d = 1.5$, the rate of growth is 160.21.

95. (a) 5.67

(b)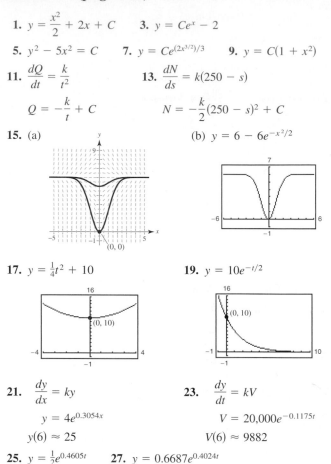

(c) $f(t) = g(t) = h(t)$. No, because the definite integrals of two functions over a given interval may be equal even though the functions are not equal.

97. $15,039.61   **99.** $y = 1200(0.6^t)$

101. False: e is an irrational number.

103. True   **105.** True   **107.** Proof

Section 5.6 (page 366)

1. $y = \dfrac{x^2}{2} + 2x + C$ **3.** $y = Ce^x - 2$

5. $y^2 - 5x^2 = C$ **7.** $y = Ce^{(2x^{3/2})/3}$ **9.** $y = C(1 + x^2)$

11. $\dfrac{dQ}{dt} = \dfrac{k}{t^2}$ **13.** $\dfrac{dN}{ds} = k(250 - s)$

$Q = -\dfrac{k}{t} + C$ $N = -\dfrac{k}{2}(250 - s)^2 + C$

15. (a)   (b) $y = 6 - 6e^{-x^2/2}$

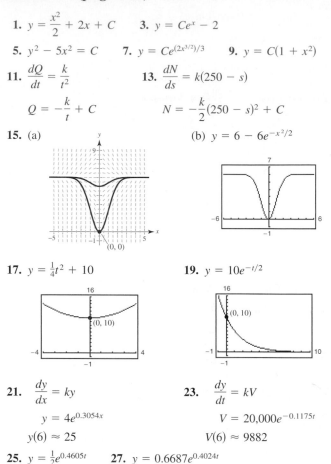

17. $y = \frac{1}{4}t^2 + 10$ **19.** $y = 10e^{-t/2}$

21. $\dfrac{dy}{dx} = ky$ **23.** $\dfrac{dy}{dt} = kV$

$y = 4e^{0.3054x}$ $V = 20{,}000e^{-0.1175t}$

$y(6) \approx 25$ $V(6) \approx 9882$

25. $y = \frac{1}{2}e^{0.4605t}$ **27.** $y = 0.6687e^{0.4024t}$

29. A differential equation in x and y is an equation that involves x, y, and derivatives of y.

Example: $y' = \dfrac{3x}{y}$

31. Quadrants I and III; dy/dx is positive when both x and y are positive (Quadrant I) or when both x and y are negative (Quadrant III).

33. Amount after 1000 years: 6.52 grams
Amount after 10,000 years: 0.14 gram

35. Initial quantity: 36.07 grams
Amount after 1000 years: 23.65 grams

37. Amount after 1000 years: 4.43 grams
Amount after 10,000 years: 1.49 grams

39. Initial quantity: 2.16 grams
Amount after 10,000 years: 1.63 grams

41. 95.81%

43. Time to double: 11.55 years
Amount after 10 years: $1822.12

45. Annual rate: 8.94%
Amount after 10 years: $1833.67

47. Annual rate: 9.50%

Time to double: 7.30 years

49. $112,087.09   **51.** $30,688.87

53. (a) 10.24 years   **55.** (a) 8.50 years

   (b) 9.93 years   (b) 8.18 years

   (c) 9.90 years   (c) 8.16 years

   (d) 9.90 years   (d) 8.15 years

57. 7.43 million   **59.** 6.83 million

61. When $k > 0$, the population is increasing.

When $k < 0$, the population is decreasing.

63. 527.06 millimeters of mercury

65. (a) $N \approx 30(1 - e^{-0.0502t})$ (b) 36 days

67. (a) $S \approx 30e^{-1.7918/t}$ (b) 20,965 units

(c)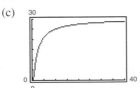

69. 2014 $(t = 16)$

71. (a) 20 decibels   (b) 70 decibels

   (c) 95 decibels   (d) 120 decibels

73. (a) $10^{8.3} \approx 199{,}526{,}231.5$ (b) 10^R (c) $\dfrac{1}{I \ln 10}$

75. False. The rate of growth $\dfrac{dy}{dx}$ is proportional to y.

77. True

Section 5.7 (page 377)

1. Proof   **3.** Proof   **5.** Proof   **7.** Not a solution

9. Solution   **11.** Solution   **13.** Not a solution

15. Solution   **17.** Not a solution   **19.** $k = 0.07$

21. $4y^2 = x^3$

23.
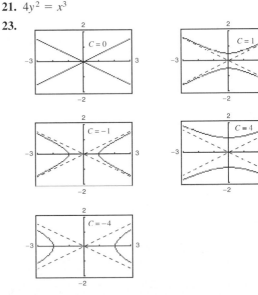

25. $y = 3e^{-2x}$ **27.** $y = 2 \sin 3x - \frac{1}{3}\cos 3x$

29. $y = -2x + \frac{1}{2}x^3$ **31.** $y = x^3 + C$

33. $y = \frac{1}{2}\ln(1 + x^2) + C$

35. $y = x - \ln x^2 + C$ **37.** $y = -\frac{1}{2}\cos 2x + C$

39. $y = \frac{2}{5}(x - 3)^{5/2} + 2(x - 3)^{3/2} + C$ **41.** $y = \frac{1}{2}e^{x^2} + C$

43. $y^2 - x^2 = C$ **45.** $r = Ce^{0.05s}$ **47.** $y = C(x + 2)^3$

49. $y^2 = C - 2\cos x$ **51.** $y = -\frac{1}{4}\sqrt{1 - 4x^2} + C$

53. $y = Ce^{(\ln x)^2/2}$ **55.** $y^2 = 2e^x + 14$ **57.** $y = e^{-(x^2+2x)/2}$

59. $y^2 = 4x^2 + 3$ **61.** $u = e^{(1-\cos v^2)/2}$ **63.** $P = P_0 e^{kt}$

65. $9x^2 + 16y^2 = 25$ **67.** $f(x) = Ce^{-x/2}$

69. Homogeneous of degree 3   **71.** Homogeneous of degree 3

73. Not homogeneous   **75.** Homogeneous of degree 0

77. $|x| = C(x - y)^2$ **79.** $|y^2 + 2xy - x^2| = C$

81. $y = Ce^{-x^2/2y^2}$ **83.** $e^{y/x} = 1 + \ln x^2$ **85.** $x = e^{\sin(y/x)}$

87.

$y = \frac{1}{2}x^2 + C$

89.

$y = 4 + Ce^{-x}$

91.

93.

95. 98.9% of the original amount

97. (a) $\dfrac{dy}{dx} = k(y - 4)$ (b) a   (c) Proof

98. (a) $\dfrac{dy}{dx} = k(x - 4)$ (b) b   (c) Proof

99. (a) $\dfrac{dy}{dx} = ky(y - 4)$ (b) c   (c) Proof

100. (a) $\dfrac{dy}{dx} = ky^2$ (b) d   (c) Proof

101. (a) $w = 1200 - 1140e^{-0.8t}$ $w = 1200 - 1140e^{-0.9t}$

$w = 1200 - 1140e^{-t}$

(b) 1.31 years; 1.16 years; 1.05 years

(c) 1200 pounds

103. (a) $\dfrac{dv}{dt} = k(W - v); \; v = 20(1 - e^{-0.2877t})$

    (b) $s = 20t + 69.5(e^{-0.2877t} - 1)$

105. Circles: $x^2 + y^2 = C$

    Lines: $y = Kx$

    Graphs will vary.

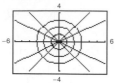

107. Parabolas: $x^2 = Cy$

    Ellipses: $x^2 + 2y^2 = K$

    Graphs will vary.

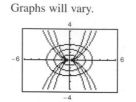

109. Curves: $y^2 = Cx^3$

    Ellipses: $2x^2 + 3y^2 = K$

    Graphs will vary.

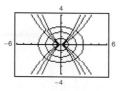

111. The general solution is a family of curves that satisfies the differential equation. A particular solution is one member of the family that satisfies given conditions.

    Example: $(x^3 + y^3)\,dx - (xy^2)\,dy = 0$

113. A homogeneous differential equation is an equation of the form $M(x, y)\,dx + N(x, y)\,dy = 0$, where M and N are homogeneous functions of the same degree.

115. False: $y = x^3$ is a solution to $xy' - 3y = 0$, but $y = x^3 + 1$ is not a solution.

117. False: $f(tx, ty) \ne t^n f(x, y)$.

Section 5.8 (page 386)

1. (a)

x	-1	-0.8	-0.6	-0.4	-0.2	0
y	-1.57	-0.93	-0.64	-0.41	-0.20	0

x	0.2	0.4	0.6	0.8	1
y	0.20	0.41	0.64	0.93	1.57

(b)   (c)

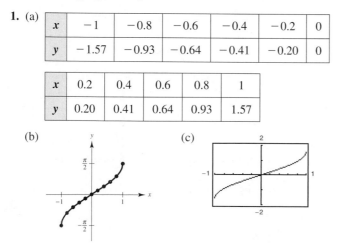

(d) Intercept: $(0, 0)$; Symmetry: origin

3. False: the range of $y = \arccos x$ is $[0, \pi]$.   5. $\dfrac{\pi}{6}$ 7. $\dfrac{\pi}{3}$

9. $\dfrac{\pi}{6}$ 11. $-\dfrac{\pi}{4}$ 13. 2.50   15. $\arccos\!\left(\dfrac{1}{1.269}\right) \approx 0.66$

17. (a) $\dfrac{3}{5}$ (b) $\dfrac{5}{3}$ 19. (a) $-\sqrt{3}$ (b) $-\dfrac{13}{5}$

21. $\sqrt{1 - 4x^2}$ 23. $\dfrac{\sqrt{x^2 - 1}}{|x|}$

25. $\dfrac{\sqrt{x^2 - 9}}{3}$ 27. $\dfrac{\sqrt{x^2 + 2}}{x}$

29.

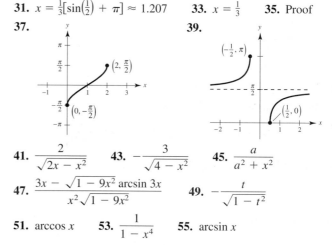

g is the algebraic form of f.

Horizontal asymptotes: $y = -1, y = 1$

31. $x = \frac{1}{3}\left[\sin\!\left(\frac{1}{2}\right) + \pi\right] \approx 1.207$ 33. $x = \frac{1}{3}$ 35. Proof

37.   39.

41. $\dfrac{2}{\sqrt{2x - x^2}}$ 43. $-\dfrac{3}{\sqrt{4 - x^2}}$ 45. $\dfrac{a}{a^2 + x^2}$

47. $\dfrac{3x - \sqrt{1 - 9x^2}\,\arcsin 3x}{x^2\sqrt{1 - 9x^2}}$ 49. $-\dfrac{t}{\sqrt{1 - t^2}}$

51. $\arccos x$ 53. $\dfrac{1}{1 - x^4}$ 55. $\arcsin x$

57. $\dfrac{x^2}{\sqrt{16 - x^2}}$ 59. $\dfrac{2}{(1 + x^2)^2}$

61. $P_1(x) = \dfrac{\pi}{6} + \dfrac{2\sqrt{3}}{3}\left(x - \dfrac{1}{2}\right)$

    $P_2(x) = \dfrac{\pi}{6} + \dfrac{2\sqrt{3}}{3}\left(x - \dfrac{1}{2}\right) + \dfrac{2\sqrt{3}}{9}\left(x - \dfrac{1}{2}\right)^2$

63. Relative maximum: $(1.272, -0.606)$

    Relative minimum: $(-1.272, 3.747)$

65. Relative minimum: $(2, 2.214)$

67. If the domains were not restricted, then the trigonometric functions would not be one-to-one and hence would have no inverses.

69. If $x > 0$, $y = \text{arccot } x = \arctan \dfrac{1}{x}$; If $x < 0$, $y = \arctan \dfrac{1}{x} + \pi$.

71. (a) $\theta = \text{arccot } \dfrac{x}{5}$

  (b) $x = 10$: 16 radians per hour

  $x = 3$: 58.824 radians per hour

73. (a) $h(t) = -16t^2 + 256$

  $t = 4$ seconds

  (b) $t = 1$: -0.0520 radian per second

  $t = 2$: -0.1116 radian per second

75. Proof   **77.** $k \leq -1$ or $k \geq 1$ **79.** True   **81.** True

Section 5.9 (page 393)

1. $5 \arcsin \dfrac{x}{3} + C$ **3.** $\dfrac{\pi}{18}$ **5.** $\dfrac{7}{4} \arctan \dfrac{x}{4} + C$ **7.** $\dfrac{\pi}{6}$

9. $\text{arcsec} |2x| + C$ **11.** $\dfrac{1}{2}x^2 - \dfrac{1}{2} \ln(x^2 + 1) + C$

13. $\arcsin(x + 1) + C$ **15.** $\dfrac{1}{2} \arcsin t^2 + C$

17. $\dfrac{\pi^2}{32} \approx 0.308$ **19.** $\dfrac{\sqrt{3} - 2}{2} \approx -0.134$

21. $\dfrac{1}{4} \arctan \dfrac{e^{2x}}{2} + C$ **23.** $\dfrac{\pi}{4}$ **25.** $2 \arcsin \sqrt{x} + C$

27. $\ln \sqrt{x^2 + 1} - 3 \arctan x + C$

29. $8 \arcsin \left(\dfrac{x - 3}{3} \right) - \sqrt{6x - x^2} + C$

31. $\dfrac{\pi}{2}$ **33.** $\ln |x^2 + 6x + 13| - 3 \arctan \left(\dfrac{x + 3}{2} \right) + C$

35. $\arcsin \left(\dfrac{x + 2}{2} \right) + C$ **37.** $-\sqrt{-x^2 - 4x} + C$

39. $4 - 2\sqrt{3} + \dfrac{\pi}{6} \approx 1.059$ **41.** $\dfrac{1}{2} \arctan(x^2 + 1) + C$

43. $2\sqrt{e^t - 3} - 2\sqrt{3} \arctan \left(\dfrac{\sqrt{e^t - 3}}{\sqrt{3}} \right) + C$

45. A trinomial of the form $x^2 + 2bx + b^2$

47. a and b   **49.** a, b, and c

51. (a)   (b) $y = 3 \arctan x$

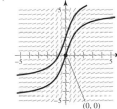

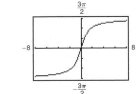

53.   **55.** $\dfrac{\pi}{8}$ **57.** c

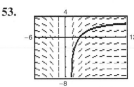

59. (a) $\displaystyle\int_0^1 \dfrac{4}{1 + x^2}\, dx = \left[4 \arctan x \right]_0^1 = 4 \arctan 1 - 4 \arctan 0 = \pi$

  (b) 3.1415918   (c) 3.1415927

61. (a)–(c) Proof

63. (a) $v(t) = -32t + 500$

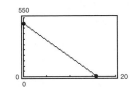

  (b) $s(t) = -16t^2 + 500t$; 3906.25 feet

  (c) $v(t) = \sqrt{\dfrac{32}{k}} \tan \left[\arctan \left(500 \sqrt{\dfrac{k}{32}} \right) - \sqrt{32k}\, t \right]$

  (d)

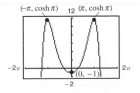

  6.86 seconds

  (e) 1088 feet

  (f) When air resistance is taken into account, the maximum height of the object is not as great.

Section 5.10 (page 403)

1. (a) 10.018   (b) -0.964 **3.** (a) $\dfrac{4}{3}$ (b) $\dfrac{13}{12}$

5. (a) 1.317   (b) 0.962

7. Proof   **9.** Proof   **11.** Proof

13. $\cosh x = \dfrac{\sqrt{13}}{2}$ **15.** $-2x \cosh(1 - x^2)$ **17.** $\coth x$

  $\tanh x = \dfrac{3\sqrt{13}}{13}$

  $\text{csch } x = \dfrac{2}{3}$

  $\text{sech } x = \dfrac{2\sqrt{13}}{13}$

  $\coth x = \dfrac{\sqrt{13}}{3}$

19. $\text{csch } x$ **21.** $\sinh^2 x$ **23.** $\text{sech } t$

25. $\dfrac{y}{x}[\cosh x + x(\sinh x) \ln x] = \dfrac{x^{\cosh x}}{x}[\cosh x + x(\sinh x) \ln x]$

27. $-2(\cosh x - \sinh x)^2 = -2e^{-2x}$

29. Relative maxima: $(\pm \pi, \cosh \pi)$

  Relative minimum: $(0, -1)$

31. Relative maximum: $(1.20, 0.66)$

Relative minimum: $(-1.20, -0.66)$

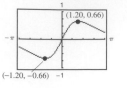

33. $y = a \sinh x$

$y' = a \cosh x$

$y'' = a \sinh x$

$y''' = a \cosh x$

Therefore, $y''' - y' = 0$.

35. $P_1(x) = 0.76 + 0.42(x - 1)$

$P_2(x) = 0.76 + 0.42(x - 1) - 0.32(x - 1)^2$

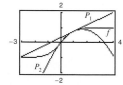

37. (a)

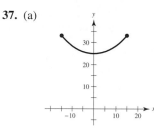

(b) 33.146 units; 25 units   (c) $m = \sinh(1) \approx 1.175$

39. $-\frac{1}{2}\cosh(1 - 2x) + C$ **41.** $\frac{1}{3}\cosh^3(x - 1) + C$

43. $\ln|\sinh x| + C$ **45.** $-\coth\dfrac{x^2}{2} + C$ **47.** $\operatorname{csch}\dfrac{1}{x} + C$

49. $\dfrac{1}{5}\ln 3$ **51.** $\dfrac{\pi}{4}$ **53.** $\dfrac{1}{2}\arctan x^2 + C$ **55.** $\dfrac{3}{\sqrt{9x^2 - 1}}$

57. $|\sec x|$ **59.** $2\sec 2x$ **61.** $2\sinh^{-1}(2x)$

63. See "Definition of the Hyperbolic Functions" on page 395.

65. $-\dfrac{\sqrt{a^2 - x^2}}{x}$

67. $-\operatorname{csch}^{-1}(e^x) + C = -\ln\left(\dfrac{1 + \sqrt{1 + e^{2x}}}{e^x}\right) + C$

69. $2\sinh^{-1}\sqrt{x} + C = 2\ln\left(\sqrt{x} + \sqrt{1 + x}\right) + C$

71. $\dfrac{1}{4}\ln\left|\dfrac{x - 4}{x}\right| + C$ **73.** $\dfrac{1}{2\sqrt{6}}\ln\left|\dfrac{\sqrt{2}(x + 1) + \sqrt{3}}{\sqrt{2}(x + 1) - \sqrt{3}}\right| + C$

75. $\dfrac{1}{4}\arcsin\left(\dfrac{4x - 1}{9}\right) + C$

77. $-\dfrac{x^2}{2} - 4x - \dfrac{10}{3}\ln\left|\dfrac{x - 5}{x + 1}\right| + C$

79. $8\arctan(e^2) - 2\pi \approx 5.207$ **81.** $\dfrac{5}{2}\ln\left(\sqrt{17} + 4\right) \approx 5.237$

83. $\dfrac{52}{31}$ kilograms

85. If k were increased, the time of descent would increase.

87. Proof   **89.** Proof   **91.** Proof

Review Exercises for Chapter 5 (page 405)

1. Vertical asymptote: $x = 0$

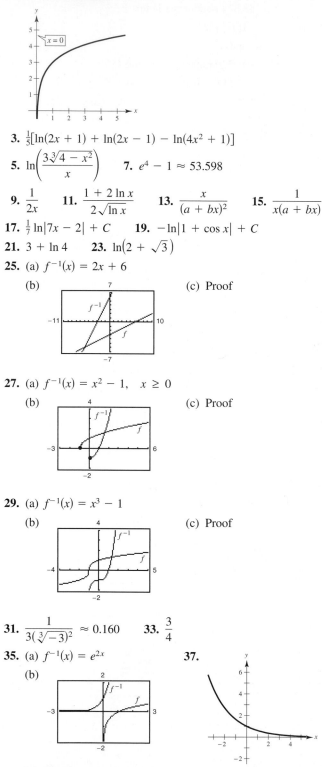

3. $\frac{1}{5}[\ln(2x + 1) + \ln(2x - 1) - \ln(4x^2 + 1)]$

5. $\ln\left(\dfrac{3\sqrt[3]{4 - x^2}}{x}\right)$ **7.** $e^4 - 1 \approx 53.598$

9. $\dfrac{1}{2x}$ **11.** $\dfrac{1 + 2\ln x}{2\sqrt{\ln x}}$ **13.** $\dfrac{x}{(a + bx)^2}$ **15.** $\dfrac{1}{x(a + bx)}$

17. $\frac{1}{7}\ln|7x - 2| + C$ **19.** $-\ln|1 + \cos x| + C$

21. $3 + \ln 4$ **23.** $\ln(2 + \sqrt{3})$

25. (a) $f^{-1}(x) = 2x + 6$

(b)   (c) Proof

27. (a) $f^{-1}(x) = x^2 - 1, \quad x \geq 0$

(b)   (c) Proof

29. (a) $f^{-1}(x) = x^3 - 1$

(b)   (c) Proof

31. $\dfrac{1}{3(\sqrt[3]{-3})^2} \approx 0.160$ **33.** $\dfrac{3}{4}$

35. (a) $f^{-1}(x) = e^{2x}$ **37.**

(b)

(c) Proof

39. $-2x$ **41.** $te^t(t + 2)$ **43.** $\dfrac{e^{2x} - e^{-2x}}{\sqrt{e^{2x} + e^{-2x}}}$

45. $\dfrac{x(2 - x)}{e^x}$ **47.** $-\dfrac{y}{x(2y + \ln x)}$

49. $-\dfrac{1}{6}e^{-3x^2} + C$ **51.** $\dfrac{e^{4x} - 3e^{2x} - 3}{3e^x} + C$

53. $-\dfrac{1}{2}e^{1-x^2} + C$ **55.** $\ln|e^x - 1| + C$

57. $y = e^x(a \cos 3x + b \sin 3x)$

$\quad y' = e^x[(-3a + b) \sin 3x + (a + 3b) \cos 3x]$

$\quad y'' = e^x[(-6a - 8b) \sin 3x + (-8a + 6b) \cos 3x]$

$\quad y'' - 2y' + 10y$

$\qquad = e^x\{[(-6a - 8b) - 2(-3a + b) + 10b] \sin 3x +$

$\qquad\quad [(-8a + 6b) - 2(a + 3b) + 10a] \cos 3x\} = 0$

59. $-\dfrac{1}{2}(e^{-16} - 1) \approx 0.500$

61.

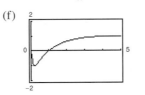

63.

65. $3^{x-1} \ln 3$ **67.** $x^{2x+1}\left(\dfrac{2x + 1}{x} + 2 \ln x\right)$

69. $-\dfrac{1}{\ln 3(2 - 2x)}$ **71.** $\dfrac{5^{(x+1)^2}}{2 \ln 5} + C$

73. (a) ax^{a-1} (b) $(\ln a)a^x$ (c) $x^x(1 + \ln x)$ (d) 0

75. $\approx \$3499.38$ **77.** ≈ 7.79 inches   **79.** About 46.2 years

81. $y = \dfrac{x^2}{2} + 3 \ln|x| + C$ **83.** $y = Ce^{x^2}$

85. $\dfrac{x}{x^2 - y^2} = C$ **87.** Proof; $y = -2x + \dfrac{1}{2}x^3$

89.

91. (a) $\dfrac{1}{2}$ (b) $\dfrac{\sqrt{3}}{2}$

93. $(1 - x^2)^{-3/2}$ **95.** $\dfrac{x}{|x|\sqrt{x^2 - 1}} + \operatorname{arcsec} x$

97. $(\arcsin x)^2$ **99.** $\dfrac{1}{2}\arctan(e^{2x}) + C$

101. $\dfrac{1}{2}\arcsin x^2 + C$ **103.** $\ln\sqrt{16 + x^2} + C$

105. $\dfrac{1}{4}\left(\arctan \dfrac{x}{2}\right)^2 + C$ **107.** $y = A \sin\left(\sqrt{\dfrac{k}{m}}\,t\right)$

109. $2 - \dfrac{\sinh\sqrt{x}}{2\sqrt{x}}$ **111.** $\dfrac{1}{2}\ln\left(\sqrt{x^4 - 1} + x^2\right) + C$

P.S. Problem Solving (page 408)

1. $\theta \approx 1.7263$ or $98.9°$

3. (a) $(0, \infty)$

(b) Answers will vary.

Example: $e^{\pi/2} \approx 4.8105$ and $e^{5\pi/2} \approx 2575.9705$

(c) Answers will vary.

Example: $e^{-\pi/2} \approx 0.2079$ and $e^{3\pi/2} \approx 111.3178$

(d) $[-1, 1]$

(e) $f'(x) = \dfrac{\cos(\ln x)}{x}$; Maximum value is 1.

(f)

Limit does not exist.

(g) Limit does not exist.

5. (a) Area of sector $-\dfrac{t}{2}$

(b) $A(t) = \dfrac{1}{2}$ base $\cdot$ height $- \displaystyle\int_1^{\cosh t} \sqrt{x^2 - 1}\, dx$

$\quad A(t) = \dfrac{1}{2} \cosh t \cdot \sinh t - \displaystyle\int_1^{\cosh t} \sqrt{x^2 - 1}\, dx$

$\quad A(t) = \dfrac{1}{2}t$

7. Tangent line: $y = \dfrac{1}{a}x + (b - 1)$

Passes through $(0, c)$, therefore $c - b - 1$.

Distance between b and c is $b - c = 1$.

9. $2 \ln\left(\dfrac{3}{2}\right) \approx 0.8109$

11. (a) $y = \dfrac{1}{(1 - 0.01t)^{100}}$; $T = 100$

(b) $y = \dfrac{1}{\left[\left(\dfrac{1}{y_0}\right)^e - ket\right]^{1/e}}$; Answers will vary.

13. $22.35°$F

15. (a) $\dfrac{dS}{dt} = kS(L - S)$; $S = \dfrac{100}{1 + 9e^{-0.8109t}}$

(b) 2.7 months

(c)

(d)

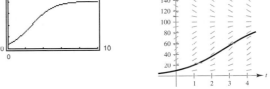

(e) Sales will decrease toward the line $S = L$.

Chapter 6

Section 6.1 (page 418)

1. $-\int_{0}^{6} (x^2 - 6x)\, dx$ **3.** $\int_{0}^{3} (-2x^2 + 6x)\, dx$

5. $-6\int_{0}^{1} (x^3 - x)\, dx$

7.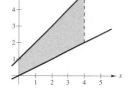

9.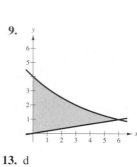

11.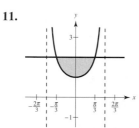

13. d

15. 2

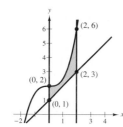

17. $\frac{32}{3}$

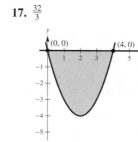

19. $\frac{9}{2}$

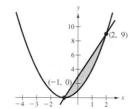

21. 1

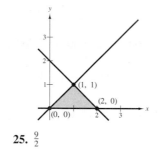

23. $\frac{3}{2}$

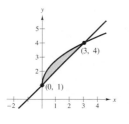

25. $\frac{9}{2}$

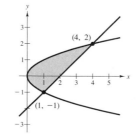

27. 6

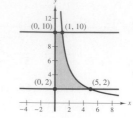

29. 16.094

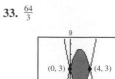

31. $\frac{37}{12}$

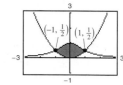

33. $\frac{64}{3}$

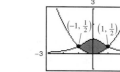

35. 8

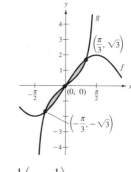

37. $\frac{\pi}{2} - \frac{1}{3} \approx 1.237$

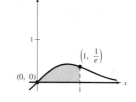

39. ≈ 1.759

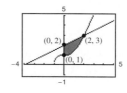

41. $2(1 - \ln 2) \approx 0.614$

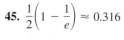

43. $4\pi \approx 12.566$

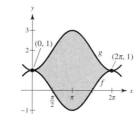

45. $\frac{1}{2}\left(1 - \frac{1}{e}\right) \approx 0.316$

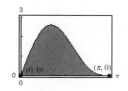

47. 4

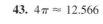

49. ≈ 1.323

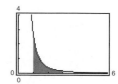

51. (a)

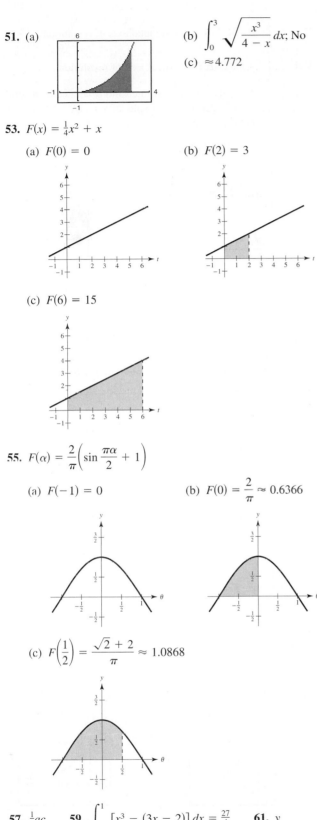

(b) $\displaystyle\int_0^3 \sqrt{\frac{x^3}{4-x}}\,dx$; No

(c) ≈ 4.772

53. $F(x) = \frac{1}{4}x^2 + x$

(a) $F(0) = 0$

(b) $F(2) = 3$

(c) $F(6) = 15$

55. $F(\alpha) = \dfrac{2}{\pi}\left(\sin\dfrac{\pi\alpha}{2} + 1\right)$

(a) $F(-1) = 0$

(b) $F(0) = \dfrac{2}{\pi} \approx 0.6366$

(c) $F\left(\dfrac{1}{2}\right) = \dfrac{\sqrt{2}+2}{\pi} \approx 1.0868$

57. $\frac{1}{2}ac$ **59.** $\displaystyle\int_{-2}^1 [x^3 - (3x - 2)]\,dx = \frac{27}{4}$ **61.** y

63. Answers will vary. Example: $x^4 - 2x^2 + 1 \le 1 - x^2$ on $[-1, 1]$

$\displaystyle\int_{-1}^1 [(1 - x^2) - (x^4 - 2x^2 + 1)]\,dx = \frac{4}{15}$

65. Offer 2 is better because the cumulative salary (area under the curve) is greater.

67. $b = 9\left(1 - \dfrac{1}{\sqrt[3]{4}}\right) \approx 3.330$

69. $\frac{1}{6}$

71. \$1.625 billion

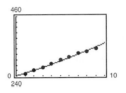

$f(x) = x - x^2$

(1, 0)

(0, 0)

73. (a) $y = 275.0675(1.0537)^t$ or $y = 275.0675e^{0.0523t}$

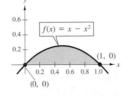

(b) $y = 239.9407(1.0417)^t$ or $y = 239.9407e^{0.0408t}$

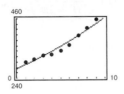

(c) \$649.5 billion

(d) No. The model for total receipts is increasing at a faster rate than the model for total expenditures. No.

75. $\frac{16}{3}(4\sqrt{2} - 5) \approx 3.5$

77. (a) 6.031 square meters   (b) 12.062 cubic meters

(c) 60,310 pounds

79. True

81. False. Let $f(x) = x$ and $g(x) = 2x - x^2$ on the interval $[0, 2]$.

Section 6.2 *(page 428)*

1. $\displaystyle\pi\int_0^1 (-x + 1)^2\,dx = \frac{\pi}{3}$ **3.** $\displaystyle\pi\int_1^4 \left(\sqrt{x}\right)^2\,dx = \frac{15\pi}{2}$

5. $\displaystyle\pi\int_0^1 [(x^2)^2 - (x^3)^2]\,dx = \frac{2\pi}{35}$ **7.** $\displaystyle\pi\int_0^4 \left(\sqrt{y}\right)^2\,dy = 8\pi$

9. $\displaystyle\pi\int_0^1 (y^{3/2})^2\,dy = \frac{\pi}{4}$

11. (a) 8π (b) $\dfrac{128\pi}{5}$ (c) $\dfrac{256\pi}{15}$ (d) $\dfrac{192\pi}{5}$

13. (a) $\dfrac{32\pi}{3}$ (b) $\dfrac{64\pi}{3}$

15. 18π **17.** $\pi\left(16\ln 2 - \frac{3}{4}\right) \approx 32.485$

19. $\dfrac{208\pi}{3}$ **21.** $\dfrac{384\pi}{5}$ **23.** $\pi\ln 4$ **25.** $\dfrac{3\pi}{4}$

27. $\dfrac{\pi}{2}\left(1 - \dfrac{1}{e^2}\right) \approx 1.358$ **29.** $\dfrac{277\pi}{3}$ **31.** 8π

33. $\dfrac{\pi^2}{2} \approx 4.935$ **35.** 1.969   **37.** 49.022   **39.** a

41. (a) See page 422 for the disk method.

(b) Horizontal axis of revolution:

$$V = \pi \int_a^b \left([R(x)]^2 - [r(x)]^2\right) dx$$

Vertical axis of revolution:

$$V = \pi \int_c^d \left([R(y)]^2 - [r(y)]^2\right) dy$$

43. The parabola $y = 4x - x^2$ is a horizontal translation of the parabola $y = 4 - x^2$. Therefore, their volumes are equal.

45. 18π **47.** Proof   **49.** $\pi r^2 h\left(1 - \dfrac{h}{H} + \dfrac{h^2}{3H^2}\right)$ **51.** $\dfrac{\pi}{30}$

53. (a) 60π (b) 50π

55. One-fourth: 32.64 feet; Three-fourths: 67.36 feet

57. (a) ii; right circular cylinder of radius r and height h

(b) iv; ellipsoid whose underlying ellipse has the equation

$$\left(\dfrac{x}{b}\right)^2 + \left(\dfrac{y}{a}\right)^2 = 1$$

(c) iii; sphere of radius r

(d) i; right circular cone of radius r and height h

(e) v; torus of cross-sectional radius r and other radius R

59. (a) $\dfrac{81}{10}$ (b) $\dfrac{9}{2}$

61. (a) $\dfrac{1}{10}$ (b) $\dfrac{\pi}{80}$ (c) $\dfrac{\sqrt{3}}{40}$ (d) $\dfrac{\pi}{20}$

63. Proof   **65.** $5\sqrt{1 - 2^{-2/3}} \approx 3.0415$

67. (a) $\dfrac{2r^3}{3}$ (b) $\dfrac{2r^3 \tan\theta}{3}$, $\displaystyle\lim_{\theta \to 90^\circ} V = \infty$

Section 6.3 (page 437)

1. $2\pi \displaystyle\int_0^2 x^2 \, dx = \dfrac{16\pi}{3}$ **3.** $2\pi \displaystyle\int_0^4 x\sqrt{x} \, dx = \dfrac{128\pi}{5}$

5. $2\pi \displaystyle\int_0^2 x^3 \, dx = 8\pi$ **7.** $2\pi \displaystyle\int_0^2 x(4x - 2x^2) \, dx = \dfrac{16\pi}{3}$

9. $2\pi \displaystyle\int_0^2 x(x^2 - 4x + 4) \, dx = \dfrac{8\pi}{3}$

11. $2\pi \displaystyle\int_0^1 x\left(\dfrac{1}{\sqrt{2\pi}} e^{-x^2/2}\right) dx = \sqrt{2\pi}\left(1 - \dfrac{1}{\sqrt{e}}\right) \approx 0.986$

13. $2\pi \displaystyle\int_0^2 y(2 - y) \, dy = \dfrac{8\pi}{3}$

15. $2\pi\left[\displaystyle\int_0^{1/2} y \, dy + \int_{1/2}^1 y\left(\dfrac{1}{y} - 1\right) dy\right] = \dfrac{\pi}{2}$

17. 16π **19.** 64π

21. (a) $\dfrac{128\pi}{7}$ (b) $\dfrac{64\pi}{5}$ (c) $\dfrac{96\pi}{5}$

23. (a) $\dfrac{\pi a^3}{15}$ (b) $\dfrac{\pi a^3}{15}$ (c) $\dfrac{4\pi a^3}{15}$

25. $V = 2\pi \displaystyle\int_c^d p(y)h(y) \, dy$ for horizontal axis of revolution

$V = 2\pi \displaystyle\int_a^b p(x)h(x) \, dx$ for vertical axis of revolution

27. Both integrals yield the volume of the solid generated by revolving the region bounded by the graphs of $y = \sqrt{x - 1}$, $y = 0$, and $x = 5$ about the x-axis.

29. (a) 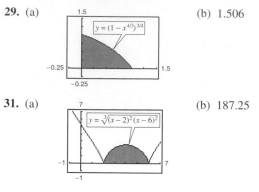   (b) 1.506

$y = (1 - x^{4/3})^{3/4}$

31. (a) 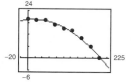   (b) 187.25

$y = \sqrt[3]{(x - 2)^2 (x - 6)^2}$

33. d   **35.** Diameter $= 2\sqrt{4 - 2\sqrt{3}} \approx 1.464$

37. $4\pi^2$ **39.** Proof

41. (a) ii; right circular cone of radius r and height h

(b) v; torus of cross-sectional radius r and other radius R

(c) iii; sphere of radius r

(d) i; right circular cylinder of radius r and height h

(e) iv; ellipsoid whose underlying ellipse has the equation

$$\left(\dfrac{x}{b}\right)^2 + \left(\dfrac{y}{a}\right)^2 = 1$$

43. (a) 1,366,593 cubic feet

(b) $d = -0.000561x^2 + 0.0189x + 19.39$

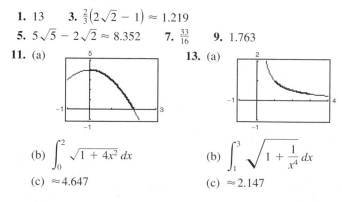

(c) 1,343,345 cubic feet

(d) 10,048,221 gallons

Section 6.4 (page 447)

1. 13   **3.** $\dfrac{2}{3}\left(2\sqrt{2} - 1\right) \approx 1.219$

5. $5\sqrt{5} - 2\sqrt{2} \approx 8.352$ **7.** $\dfrac{33}{16}$ **9.** 1.763

11. (a) [graph]   **13.** (a) [graph]

(b) $\displaystyle\int_0^2 \sqrt{1 + 4x^2} \, dx$ (b) $\displaystyle\int_1^3 \sqrt{1 + \dfrac{1}{x^4}} \, dx$

(c) ≈ 4.647 (c) ≈ 2.147

15. (a)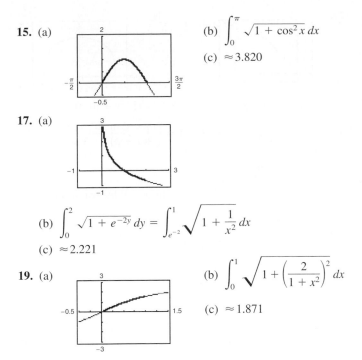

(b) $\displaystyle\int_0^{\pi} \sqrt{1 + \cos^2 x}\, dx$

(c) ≈ 3.820

17. (a)

(b) $\displaystyle\int_0^2 \sqrt{1 + e^{-2y}}\, dy = \int_{e^{-2}}^1 \sqrt{1 + \frac{1}{x^2}}\, dx$

(c) ≈ 2.221

19. (a)

(b) $\displaystyle\int_0^1 \sqrt{1 + \left(\frac{2}{1 + x^2}\right)^2}\, dx$

(c) ≈ 1.871

21. b

23. (a) 64.125   **(b)** 64.525   **(c)** 64.666   **(d)** 64.672

25. (a)

(b) y_1, y_2, y_3, y_4

(c) $s_1 \approx 5.657$; $s_2 \approx 5.759$; $s_3 \approx 5.916$; $s_4 \approx 6.063$

27. Fleeing object: $\frac{2}{3}$ unit

Pursuer: $\displaystyle\frac{1}{2}\int_0^1 \frac{x + 1}{\sqrt{x}}\, dx = \frac{1}{2}\left[\frac{2}{3}x^{3/2} + 2x^{1/2}\right]_0^1$

$\displaystyle = \frac{4}{3} = 2\left(\frac{2}{3}\right)$

29. $20[\sinh 1 - \sinh(-1)] \approx 47.0$ meters

31. $3 \arcsin \frac{2}{3} \approx 2.1892$

33. $\displaystyle 2\pi\int_0^3 \frac{1}{3}x^3 \sqrt{1 + x^4}\, dx = \frac{\pi}{9}\left(82\sqrt{82} - 1\right) \approx 258.85$

35. $\displaystyle 2\pi\int_1^2 \left(\frac{x^3}{6} + \frac{1}{2x}\right)\left(\frac{x^2}{2} + \frac{1}{2x^2}\right) dx = \frac{47\pi}{16}$

37. $\displaystyle 2\pi\int_1^8 x\sqrt{1 + \frac{1}{9x^{4/3}}}\, dx = \frac{\pi}{27}\left(145\sqrt{145} - 10\sqrt{10}\right) \approx 199.48$

39. 14.424

41. A rectifiable curve is a curve with a finite arc length.

43. The integral formula for the area of a surface of revolution is derived from the formula for the lateral surface area of the frustum of a right circular cone. The formula is $S = 2\pi rL$ where $r = \frac{1}{2}(r_1 + r_2)$, which is the average radius of the frustum, and L is the length of a line segment on the frustum.

45. Proof   **47.** $6\pi\left(3 - \sqrt{5}\right) \approx 14.40$

49. Surface area $= \dfrac{\pi}{27}$ square feet ≈ 16.8 square inches

Amount of glass $= \dfrac{\pi}{27}\left(\dfrac{0.015}{12}\right)$ cubic foot

≈ 0.00015 cubic foot

≈ 0.25 cubic inch

51. (a) $y = (1.953 \times 10^{-7})x^4 - (1.804 \times 10^{-4})x^3$
$+ 0.0496x^2 - 4.8323x + 536.927$

(b) 131,734.5 square feet ≈ 3 acres

(c) 794.9 feet

53. (a) $\pi\left(1 - \dfrac{1}{b}\right)$ **(b)** $\displaystyle 2\pi\int_1^b \frac{\sqrt{x^4 + 1}}{x^3}\, dx$

(c) $\displaystyle\lim_{b\to\infty} V = \lim_{b\to\infty} \pi\left(1 - \frac{1}{b}\right) = \pi$

(d) Since $\dfrac{\sqrt{x^4 + 1}}{x^3} > \dfrac{\sqrt{x^4}}{x^3} = \dfrac{1}{x} > 0$ on $[1, b]$,

we have $\displaystyle\int_1^b \frac{\sqrt{x^4 + 1}}{x^3}\, dx > \int_1^b \frac{1}{x}\, dx = \left[\ln x\right]_1^b = \ln b$

and $\displaystyle\lim_{b\to\infty} \ln b \to \infty$. Thus, $\displaystyle\lim_{b\to\infty} 2\pi\int_1^b \frac{\sqrt{x^4 + 1}}{x^3}\, dx = \infty$.

55. (a) Area of circle with radius L: $A = \pi L^2$

Area of sector with central angle θ (in radians):

$$S = \frac{\theta}{2\pi}A = \frac{\theta}{2\pi}(\pi L^2) = \frac{1}{2}L^2\theta$$

(b) Let s be the arc length of the sector, which is the circumference of the base of the cone. Here, $s = L\theta = 2\pi r$, and you have

$$S = \frac{1}{2}L^2\theta = \frac{1}{2}L^2\left(\frac{s}{L}\right) = \frac{1}{2}Ls = \frac{1}{2}L(2\pi r) = \pi rL.$$

(c) The lateral surface area of the frustum is the difference between the large cone and the small one.

$S = \pi r_2(L + L_1) - \pi r_1 L_1$
$= \pi r_2 L + \pi L_1(r_2 - r_1)$

By similar triangles, $\dfrac{L + L_1}{r_2} = \dfrac{L_1}{r_1} \Rightarrow Lr_1 = L_1(r_2 - r_1)$.

Hence,

$S = \pi r_2 L + \pi L_1(r_2 - r_1) = \pi r_2 L + \pi Lr_1$
$= \pi L(r_1 + r_2).$

Section 6.5 (page 456)

1. 1000 foot-pounds **3.** 448 newton-meters

5. If an object is moved a distance D in the direction of an applied constant force F, then the work W done by the force is defined as $W = FD$.

7. c, d, a, b **9.** 30.625 inch-pounds $\approx$ 2.55 foot-pounds

11. 8750 newton-centimeters = 87.5 newton-meters

13. 160 inch-pounds $\approx$ 13.3 foot-pounds

15. 37.125 foot-pounds

17. (a) 487.805 mile-tons $\approx$ 5.151(10^9) foot-pounds

(b) 1395.349 mile-tons $\approx$ 1.473(10^{10}) foot-pounds

19. (a) 2.93×10^4 mile-tons $\approx 3.10 \times 10^{11}$ foot-pounds

(b) 3.38×10^4 mile-tons $\approx 3.57 \times 10^{11}$ foot-pounds

21. (a) 2496 foot-pounds (b) 9984 foot-pounds

23. $470,400\pi$ newton-meters **25.** 2995.2π foot-pounds

27. $20,217.6\pi$ foot-pounds **29.** 2457π foot-pounds

31. 337.5 foot-pounds **33.** 300 foot-pounds

35. 168.75 foot-pounds **37.** 7987.5 foot-pounds

39. $2000 \ln \dfrac{3}{2} \approx 810.93$ foot-pounds **41.** $\dfrac{3k}{4}$

43. 3249.4 foot-pounds **45.** 10,330.3 foot-pounds

Section 6.6 (page 467)

1. $\bar{x} = -\dfrac{6}{7}$ **3.** $\bar{x} = 12$ **5.** (a) $\bar{x} = 17$ (b) $\bar{x} = -3$

7. $x = 6$ feet **9.** $(\bar{x}, \bar{y}) = \left(\dfrac{10}{9}, -\dfrac{1}{9}\right)$ **11.** $(\bar{x}, \bar{y}) = \left(-\dfrac{7}{8}, -\dfrac{7}{16}\right)$

13. $M_x = 4\rho, M_y = \dfrac{64\rho}{5}, \quad (\bar{x}, \bar{y}) = \left(\dfrac{12}{5}, \dfrac{3}{4}\right)$

15. $M_x = \dfrac{\rho}{35}, M_y = \dfrac{\rho}{20}, \quad (\bar{x}, \bar{y}) = \left(\dfrac{3}{5}, \dfrac{12}{35}\right)$

17. $M_x = \dfrac{99\rho}{5}, M_y = \dfrac{27\rho}{4}, \quad (\bar{x}, \bar{y}) = \left(\dfrac{3}{2}, \dfrac{22}{5}\right)$

19. $M_x = \dfrac{192\rho}{7}, M_y = 96\rho, \quad (\bar{x}, \bar{y}) = \left(5, \dfrac{10}{7}\right)$

21. $M_x = 0, M_y = \dfrac{256\rho}{15}, \quad (\bar{x}, \bar{y}) = \left(\dfrac{8}{5}, 0\right)$

23. $M_x = \dfrac{27\rho}{4}, M_y = -\dfrac{27\rho}{10}, \quad (\bar{x}, \bar{y}) = \left(-\dfrac{3}{5}, \dfrac{3}{2}\right)$

25. $A = \displaystyle\int_0^1 (x - x^2)\, dx = \dfrac{1}{6}$

$M_x = \displaystyle\int_0^1 \left(\dfrac{x + x^2}{2}\right)(x - x^2)\, dx = \dfrac{1}{15}$

$M_y = \displaystyle\int_0^1 x(x - x^2)\, dx = \dfrac{1}{12}$

27. $A = \displaystyle\int_0^3 (2x + 4)\, dx = 21$

$M_x = \displaystyle\int_0^3 \left(\dfrac{2x + 4}{2}\right)(2x + 4)\, dx = 78$

$M_y = \displaystyle\int_0^3 x(2x + 4)\, dx = 36$

29.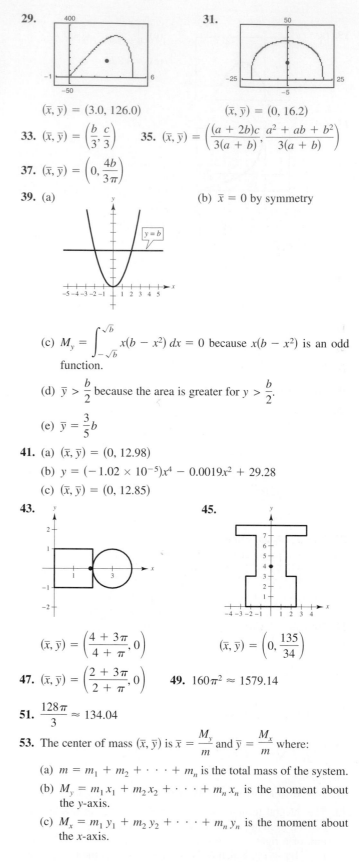

$(\bar{x}, \bar{y}) = (3.0, 126.0)$

31.

$(\bar{x}, \bar{y}) = (0, 16.2)$

33. $(\bar{x}, \bar{y}) = \left(\dfrac{b}{3}, \dfrac{c}{3}\right)$ **35.** $(\bar{x}, \bar{y}) = \left(\dfrac{(a + 2b)c}{3(a + b)}, \dfrac{a^2 + ab + b^2}{3(a + b)}\right)$

37. $(\bar{x}, \bar{y}) = \left(0, \dfrac{4b}{3\pi}\right)$

39. (a) (b) $\bar{x} = 0$ by symmetry

(c) $M_y = \displaystyle\int_{-\sqrt{b}}^{\sqrt{b}} x(b - x^2)\, dx = 0$ because $x(b - x^2)$ is an odd function.

(d) $\bar{y} > \dfrac{b}{2}$ because the area is greater for $y > \dfrac{b}{2}$.

(e) $\bar{y} = \dfrac{3}{5}b$

41. (a) $(\bar{x}, \bar{y}) = (0, 12.98)$

(b) $y = (-1.02 \times 10^{-5})x^4 - 0.0019x^2 + 29.28$

(c) $(\bar{x}, \bar{y}) = (0, 12.85)$

43.

$(\bar{x}, \bar{y}) = \left(\dfrac{4 + 3\pi}{4 + \pi}, 0\right)$

45.

$(\bar{x}, \bar{y}) = \left(0, \dfrac{135}{34}\right)$

47. $(\bar{x}, \bar{y}) = \left(\dfrac{2 + 3\pi}{2 + \pi}, 0\right)$ **49.** $160\pi^2 \approx 1579.14$

51. $\dfrac{128\pi}{3} \approx 134.04$

53. The center of mass $(\bar{x}, \bar{y})$ is $\bar{x} = \dfrac{M_y}{m}$ and $\bar{y} = \dfrac{M_x}{m}$ where:

(a) $m = m_1 + m_2 + \cdots + m_n$ is the total mass of the system.

(b) $M_y = m_1 x_1 + m_2 x_2 + \cdots + m_n x_n$ is the moment about the y-axis.

(c) $M_x = m_1 y_1 + m_2 y_2 + \cdots + m_n y_n$ is the moment about the x-axis.

55. (a) $\left(\frac{5}{6}, 2\frac{5}{18}\right)$; The plane region has been translated 2 units up.

(b) $\left(2\frac{5}{6}, \frac{5}{18}\right)$; The plane region has been translated 2 units to the right.

(c) $\left(\frac{5}{6}, -\frac{5}{18}\right)$; The plane region has been reflected across the x-axis.

(d) Not possible

57. $(\bar{x}, \bar{y}) = \left(0, \dfrac{2r}{\pi}\right)$

59. $(\bar{x}, \bar{y}) = \left(\dfrac{n+1}{n+2}, \dfrac{n+1}{4n+2}\right)$; As $n \to \infty$, the region shrinks towards the line segments $y = 0$ for $0 \le x \le 1$ and $x = 1$ for $0 \le y \le 1$; $(\bar{x}, \bar{y}) \to \left(1, \dfrac{1}{4}\right)$.

Section 6.7 (page 474)

1. 936 pounds   **3.** 748.8 pounds   **5.** 1123.2 pounds

7. 748.8 pounds   **9.** 1064.96 pounds

11. 117,600 newtons   **13.** 2,381,400 newtons

15. 2814 pounds   **17.** 6753.6 pounds   **19.** 94.5 pounds

21. $h(y) = k - y$

$L(y) = 2\sqrt{r^2 - y^2}$

$F = w \displaystyle\int_{-r}^{r} (k - y)\sqrt{r^2 - y^2}\,(2)\,dy$

$= w\left[2k\displaystyle\int_{-r}^{r} \sqrt{r^2 - y^2}\,dy + \displaystyle\int_{-r}^{r} \sqrt{r^2 - y^2}\,(-2y)\,dy\right]$

The second integral is zero since its integrand is odd and the limits of integration are symmetric with respect to the origin. The first integral is the area of a semicircle of radius r.

$F = w\left[(2k)\dfrac{\pi r^2}{2} + 0\right] = wk\pi r^2$

23. $h(y) = k - y$

$L(y) = b$

$F = w \displaystyle\int_{-h/2}^{h/2} (k - y)b\,dy$

$= wb\left[ky - \dfrac{y^2}{2}\right]_{-h/2}^{h/2} = wb(hk) = wkhb$

25. 960 pounds   **27.** 3010.8 pounds   **29.** 6448.7 pounds

31. (a) $\dfrac{3\sqrt{2}}{2} \approx 2.12$ feet

(b) The pressure increases with increasing depth.

33. The fluid force F of constant weight-density w (per unit of volume) against a submerged vertical plane region from $y = c$ to $y = d$ is

$$F = w \lim_{\|\Delta\| \to 0} \sum_{i=1}^{n} h(y_i)L(y_i)\,\Delta y = w\displaystyle\int_{c}^{d} h(y)L(y)\,dy$$

where $h(y)$ is the depth of the fluid at y and $L(y)$ is the horizontal length of the region at y.

Review Exercises for Chapter 6 (page 476)

1. $\dfrac{4}{5}$ **3.** $\dfrac{\pi}{2}$

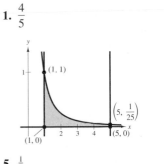

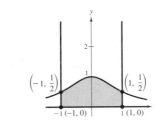

5. $\dfrac{1}{2}$ **7.** $e^2 + 1$

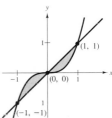

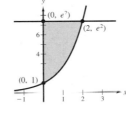

9. $2\sqrt{2}$ **11.** $\dfrac{512}{3}$

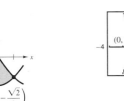

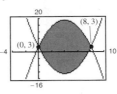

13. $\dfrac{1}{6}$

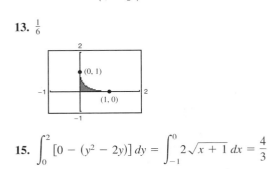

15. $\displaystyle\int_{0}^{2}\left[0 - (y^2 - 2y)\right]dy = \displaystyle\int_{-1}^{0} 2\sqrt{x+1}\,dx = \dfrac{4}{3}$

17. $\displaystyle\int_0^2 \left[1 - \left(1 - \frac{x}{2}\right)\right] dx + \int_2^3 [1 - (x - 2)]\, dx$

$\qquad = \displaystyle\int_0^1 [(y + 2) - (2 - 2y)]\, dy = \frac{3}{2}$

19. Job 1. The salary for job 1 is greater than the salary for job 2 for all the years except the first and tenth years.

21. (a) $\dfrac{64\pi}{3}$ (b) $\dfrac{128\pi}{3}$ (c) $\dfrac{64\pi}{3}$ (d) $\dfrac{160\pi}{3}$

23. (a) 64π (b) 48π **25.** $\dfrac{\pi^2}{4}$

27. $\dfrac{4\pi}{3}(20 - 9 \ln 3) \approx 42.359$

29. $\frac{4}{15}$ **31.** 1.958 feet

33. $\frac{8}{15}(1 + 6\sqrt{3}) \approx 6.076$ **35.** 4018.2 feet **37.** 15π

39. 50 inch-pounds ≈ 4.167 foot-pounds

41. $104{,}000\pi$ foot-pounds ≈ 163.4 foot-tons

43. 250 foot-pounds

45. $a = \dfrac{15}{4}$ **47.** $(\bar{x}, \bar{y}) = \left(\dfrac{a}{5}, \dfrac{a}{5}\right)$

49. $(\bar{x}, \bar{y}) = \left(0, \dfrac{2a^2}{5}\right)$

51. $(\bar{x}, \bar{y}) = \left(\dfrac{2(9\pi + 49)}{3(\pi + 9)}, 0\right)$

53. Let $D =$ surface of liquid; $\rho =$ weight per cubic volume.

$F = \rho \displaystyle\int_c^d (D - y)[f(y) - g(y)]\, dy$

$\quad = \rho\left[\displaystyle\int_c^d D[f(y) - g(y)]\, dy - \int_c^d y[f(y) - g(y)]\, dy\right]$

$\quad = \rho\left[\displaystyle\int_c^d [f(y) - g(y)]\, dy\right]\left[D - \dfrac{\displaystyle\int_c^d y[f(y) - g(y)]\, dy}{\displaystyle\int_c^d [f(y) - g(y)]\, dy}\right]$

$\quad = \rho(\text{area})(D - \bar{y})$

$\quad = \rho(\text{area})(\text{depth of centroid})$

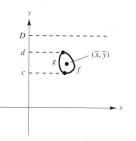

P.S. Problem Solving (page 478)

1. 3 **3.** (a) $4\pi^2$ (b) $2\pi^2 r^2 R$ **5.** $\dfrac{\pi h^3}{6}$

7. (a) Area S is 16 times area R.

(b) Let point A be (a, a^3). The equation of the tangent line to the curve $y = x^3$ at A is $y = 3a^2 x - 2a^3$, and point B is $(-2a, -8a^3)$. Area R is

$\displaystyle\int_{-2a}^{a} (x^3 - 3a^2 x + 2a^3)\, dx = \dfrac{27a^4}{4}.$

Then, the equation of the tangent line to the curve $y = x^3$ at B is $y = 12a^2 x + 16a^3$, and point C is $(4a, 64a^3)$. Area S is

$\displaystyle\int_{-2a}^{4a} (12a^2 x + 16a^3 - x^3)\, dx = 108a^4.$

Therefore, area S is 16 times area R.

9. (a) $\dfrac{ds}{dx} = \sqrt{1 + [f'(x)]^2}$

(b) $ds = \sqrt{1 + [f'(x)]^2}\, dx$; $(ds)^2 = (dx)^2 + (dy)^2$

(c) $\displaystyle\int_1^x \sqrt{1 + \tfrac{9}{4}t}\, dt$

(d) $s(2) \approx 2.0858$. This is the arc length of the curve.

11.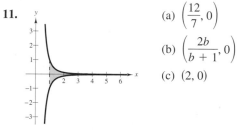

(a) $\left(\dfrac{12}{7}, 0\right)$

(b) $\left(\dfrac{2b}{b + 1}, 0\right)$

(c) $(2, 0)$

13. (a) 12 (b) 7.5

15. Consumer surplus: 1600; Producer surplus: 400

17. Wall at shallow end: 9984 pounds

Wall at deep end: 39,936 pounds

Side wall: $19{,}968 + 26{,}624 = 46{,}592$ pounds

Chapter 7

Section 7.1 (page 486)

1. b **3.** c

5. $\displaystyle\int u^n\, du$ **7.** $\displaystyle\int \dfrac{du}{u}$

$u = 3x - 2, n = 4$ $\qquad u = 1 - 2\sqrt{x}$

9. $\displaystyle\int \dfrac{du}{\sqrt{a^2 - u^2}}$ **11.** $\displaystyle\int \sin u\, du$

$u = t, a = 1$ $\qquad u = t^2$

13. $\displaystyle\int e^u\, du$ **15.** $-\frac{1}{5}(-2x + 5)^{5/2} + C$

$u = \sin x$

17. $-\dfrac{5}{4(z - 4)^4} + C$ **19.** $\dfrac{1}{4}(t^3 - 1)^{4/3} + C$

21. $\dfrac{1}{2}v^2 - \dfrac{1}{6(3v - 1)^2} + C$ **23.** $-\dfrac{1}{3}\ln|-t^3 + 9t + 1| + C$

25. $\frac{1}{2}x^2 + x + \ln|x - 1| + C$ **27.** $\ln(1 + e^x) + C$

29. $\dfrac{x}{15}(12x^4 + 20x^2 + 15) + C$　　**31.** $\dfrac{1}{4\pi}\sin 2\pi x^2 + C$

33. $-\dfrac{1}{\pi}\csc \pi x + C$　　**35.** $\dfrac{1}{5}e^{5x} + C$　　**37.** $2\ln(1 + e^x) + C$

39. $(\ln x)^2 + C$　　**41.** $\ln|\sec x(\sec x + \tan x)| + C$

43. $\csc \theta + \cot \theta + C$　　**45.** $\dfrac{3}{2}\ln(z^2 + 9) + \dfrac{2}{3}\arctan\dfrac{z}{3} + C$

47. $-\dfrac{1}{2}\arcsin(2t - 1) + C$　　**49.** $\dfrac{1}{2}\ln\left|\cos\dfrac{2}{t}\right| + C$

51. $3\arcsin\dfrac{x - 3}{3} + C$　　**53.** $\dfrac{1}{4}\arctan\dfrac{2x + 1}{8} + C$

55. (a)

(b) $\dfrac{1}{2}\arcsin t^2 - \dfrac{1}{2}$

57. $\dfrac{dy}{dx} = 0.2y$

59. $y = \dfrac{1}{2}e^{2x} + 2e^x + x + C$　　**61.** $y = \dfrac{1}{2}\arctan\dfrac{\tan x}{2} + C$

63. $\dfrac{1}{2}$　　**65.** $\dfrac{1}{2}(1 - e^{-1}) \approx 0.316$　　**67.** 4　　**69.** $\dfrac{\pi}{18}$

71. $\dfrac{1}{3}\arctan\left(\dfrac{x + 2}{3}\right) + C$　　**73.** $\tan \theta - \sec \theta + C$

One graph is a vertical　　One graph is a vertical
translation of the other.　　translation of the other.

75. Power rule: $\displaystyle\int u^n \, du = \dfrac{u^{n+1}}{n + 1} + C$; $u = x^2 + 1$, $du = 2x$, $n = 3$

77. Log rule: $\displaystyle\int \dfrac{du}{u} = \ln|u| + C$; $u = x^2 + 1$, $du = 2x$

79. Using laws of logarithms, $y_1 = e^{x+C_1} = e^x \cdot e^{C_1}$ where e^{C_1} is a constant. Therefore, e^{C_1} can be replaced by C resulting in $y_2 = Ce^x$.

81. $a = \sqrt{2}, b = \dfrac{\pi}{4}$

$-\dfrac{1}{\sqrt{2}}\ln\left|\csc\left(x + \dfrac{\pi}{4}\right) + \cot\left(x + \dfrac{\pi}{4}\right)\right| + C$

83. a　　**85.** $\dfrac{4}{3}$　　**87.** $a = \dfrac{1}{2}$

89. (a) $\pi(1 - e^{-1}) \approx 1.986$

(b) $b = \sqrt{\ln\left(\dfrac{3\pi}{3\pi - 4}\right)} \approx 0.743$

91. $\dfrac{2}{\arcsin(4/5)} \approx 2.157$　　**93.** 1.0320

Section 7.2 (page 494)

1. b　　**2.** d　　**3.** c　　**4.** a

5. $u = x, dv = e^{2x}\,dx$　　**7.** $u = (\ln x)^2, dv = dx$

9. $u = x, dv = \sec^2 x\,dx$　　**11.** $-\dfrac{1}{4e^{2x}}(2x + 1) + C$

13. $e^x(x^3 - 3x^2 + 6x - 6) + C$　　**15.** $\dfrac{1}{3}e^{x^3} + C$

17. $\dfrac{1}{4}[2(t^2 - 1)\ln|t + 1| - t^2 + 2t] + C$

19. $\dfrac{(\ln x)^3}{3} + C$　　**21.** $\dfrac{e^{2x}}{4(2x + 1)} + C$　　**23.** $(x - 1)^2e^x + C$

25. $\dfrac{2(x - 1)^{3/2}}{15}(3x + 2) + C$　　**27.** $x\sin x + \cos x + C$

29. $(6x - x^3)\cos x + (3x^2 - 6)\sin x + C$

31. $-t\csc t - \ln|\csc t + \cot t| + C$

33. $x\arctan x - \dfrac{1}{2}\ln(1 + x^2) + C$

35. $\dfrac{1}{5}e^{2x}(2\sin x - \cos x) + C$　　**37.** $y = \dfrac{1}{2}e^{x^2} + C$

39. $y = \dfrac{2}{405}(27t^2 - 24t + 32)\sqrt{2 + 3t} + C$

41. $\sin y = x^2 + C$

43. (a)

(b) $2\sqrt{y} - \cos x - x\sin x = 3$

45.

47. $4 - \dfrac{12}{e^2}$　　**49.** $\dfrac{\pi}{2} - 1$　　**51.** $\dfrac{\pi - 3\sqrt{3} + 6}{6} \approx 0.658$

53. $\dfrac{e[\sin(1) - \cos(1)] + 1}{2} \approx 0.909$　　**55.** $\dfrac{24\ln 2 - 7}{9} \approx 1.071$

57. $8\arcsec 4 + \dfrac{\sqrt{3}}{2} - \dfrac{\sqrt{15}}{2} - \dfrac{2\pi}{3} \approx 7.380$

59. $\dfrac{e^{2x}}{4}(2x^2 - 2x + 1) + C$

61. $(3x^2 - 6)\sin x - (x^3 - 6x)\cos x + C$

63. $x\tan x + \ln|\cos x| + C$　　**65.** Product Rule

67. No　　**69.** Yes. Let $u = x^2$ and $dv = e^{2x}\,dx$.

71. Yes. Let $u = x$, $dv = \dfrac{1}{\sqrt{x+1}}\,dx$. (Substitution also works. Let $u = \sqrt{x+1}$).

73. $-\dfrac{e^{-4t}}{128}(32t^3 + 24t^2 + 12t + 3) + C$

75. $\frac{1}{13}(2e^{-\pi} + 3) \approx 0.2374$　**77.** $\frac{2}{5}(2x - 3)^{3/2}(x + 1) + C$

79. $\frac{1}{3}\sqrt{4 + x^2}(x^2 - 8) + C$

81. $n = 0$: $x(\ln x - 1) + C$

$n = 1$: $\dfrac{x^2}{4}(2 \ln x - 1) + C$

$n = 2$: $\dfrac{x^3}{9}(3 \ln x - 1) + C$

$n = 3$: $\dfrac{x^4}{16}(4 \ln x - 1) + C$

$n = 4$: $\dfrac{x^5}{25}(5 \ln x - 1) + C$

$\displaystyle\int x^n \ln x\, dx = \dfrac{x^{n+1}}{(n+1)^2}\left[(n+1)\ln x - 1\right] + C$

83. Proof　**85.** Proof　**87.** Proof

89. $\dfrac{x^4}{16}(4 \ln x - 1) + C$　**91.** $\dfrac{e^{2x}}{13}(2 \cos 3x + 3 \sin 3x) + C$

93.

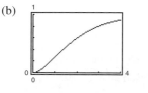

$1 - \dfrac{5}{e^4} \approx 0.908$

95.

$\dfrac{\pi}{1 + \pi^2}\left(\dfrac{1}{e} + 1\right) \approx 0.395$

97. (a) 1　(b) $\pi(e - 2) \approx 2.257$

(c) $\dfrac{(e^2 + 1)\pi}{2} \approx 13.177$

(d) $\left(\dfrac{e^2 + 1}{4}, \dfrac{e - 2}{2}\right) \approx (2.097, 0.359)$

99. $\dfrac{7}{10\pi}(1 - e^{-4\pi}) \approx 0.223$　**101.** \$931,265

103. Proof　**105.** $b_n = \dfrac{8h}{(n\pi)^2}\sin\left(\dfrac{n\pi}{2}\right)$

107. Shell: $V = \pi\left[b^2 f(b) - a^2 f(a) - \displaystyle\int_a^b x^2 f'(x)\, dx\right]$

Disk: $V = \pi\left[b^2 f(b) - a^2 f(a) - \displaystyle\int_{f(a)}^{f(b)} [f^{-1}(y)]^2\, dy\right]$

Both methods yield the same volume because $x = f^{-1}(y)$, $f'(x)\, dx = dy$, if $y = f(a)$ then $x = a$, and if $y = f(b)$ then $x = b$.

109. (a) $f(x) = -xe^{-x} - e^{-x} + 1$

(b)

(c) You obtain the following points.

n	x_n	y_n
0	0	0
1	0.05	0
2	0.10	2.378×10^{-3}
3	0.15	0.0069
4	0.20	0.0134
⋮	⋮	⋮
80	4.0	0.9064

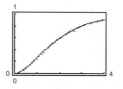

(d) You obtain the following points.

n	x_n	y_n
0	0	0
1	0.1	0
2	0.2	0.0090484
3	0.3	0.025423
4	0.4	0.047648
⋮	⋮	⋮
40	4.0	0.9039

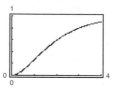

(e) $f(4) = 0.9084$

The approximations are tangent line approximations. The result in (c) is better because Δx is smaller.

Section 7.3 (page 503)

1. (a) $\frac{1}{4}(3 + \cos 4x)$

(b) $2 \cos^4 x - 2 \cos^2 x + 1$

(c) $1 - 2 \sin^2 x \cos^2 x$　(d) $1 - \frac{1}{2} \sin^2 2x$

(e) Four. No; there is often more than one way to rewrite a trigonometric expression.

3. $-\frac{1}{4} \cos^4 x + C$　**5.** $\frac{1}{12} \sin^6 2x + C$

7. $-\frac{1}{3} \cos^3 x + \frac{2}{5} \cos^5 x - \frac{1}{7} \cos^7 x + C$

9. $\frac{2}{3} \sin^{3/2} \theta - \frac{2}{7} \sin^{7/2} \theta + C$　**11.** $\frac{1}{12}(6x + \sin 6x) + C$

13. $\dfrac{\alpha - (1/4)\sin 4\alpha}{8} + C$ or $\dfrac{\alpha}{8} - \dfrac{1}{32}\sin 4\alpha + C$

15. $\frac{1}{8}(2x^2 - 2x\sin 2x - \cos 2x) + C$

17. Proof   **19.** Proof   **21.** $\frac{1}{3}\ln|\sec 3x + \tan 3x| + C$

23. $\frac{1}{15}\tan 5x(3 + \tan^2 5x) + C$

25. $\dfrac{1}{2\pi}\bigl(\sec \pi x \tan \pi x + \ln|\sec \pi x + \tan \pi x|\bigr) + C$

27. $\tan^4\!\left(\dfrac{x}{4}\right) - 2\tan^2\!\left(\dfrac{x}{4}\right) - 4\ln\left|\cos\dfrac{x}{4}\right| + C$

29. $\dfrac{1}{2}\tan^2 x + C$ **31.** $\dfrac{\tan^3 x}{3} + C$ **33.** $\dfrac{\sec^6 4x}{24} + C$

35. $\frac{1}{3}\sec^3 x + C$ **37.** $\ln|\sec x + \tan x| - \sin x + C$

39. $r = \dfrac{1}{32\pi}(12\pi\theta - 8\sin 2\pi\theta + \sin 4\pi\theta) + C$

41. $y = \frac{1}{9}\sec^3 3x - \frac{1}{3}\sec 3x + C$

43. (a)   (b) $y = \frac{1}{2}x - \frac{1}{4}\sin 2x$

45.

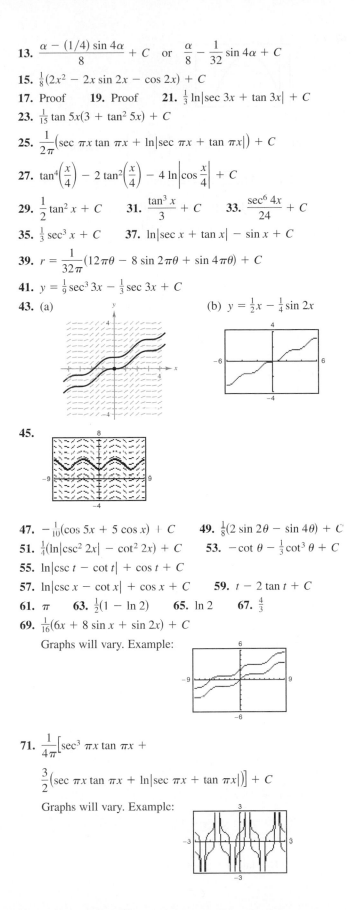

47. $-\frac{1}{10}(\cos 5x + 5\cos x) + C$ **49.** $\frac{1}{8}(2\sin 2\theta - \sin 4\theta) + C$

51. $\frac{1}{4}(\ln|\csc^2 2x| - \cot^2 2x) + C$ **53.** $-\cot\theta - \frac{1}{3}\cot^3\theta + C$

55. $\ln|\csc t - \cot t| + \cos t + C$

57. $\ln|\csc x - \cot x| + \cos x + C$ **59.** $t - 2\tan t + C$

61. π **63.** $\frac{1}{2}(1 - \ln 2)$ **65.** $\ln 2$ **67.** $\frac{4}{3}$

69. $\frac{1}{16}(6x + 8\sin x + \sin 2x) + C$

Graphs will vary. Example:

71. $\dfrac{1}{4\pi}\Bigl[\sec^3 \pi x \tan \pi x +$

$\dfrac{3}{2}\bigl(\sec \pi x \tan \pi x + \ln|\sec \pi x + \tan \pi x|\bigr)\Bigr] + C$

Graphs will vary. Example:

73. $\dfrac{1}{5\pi}\sec^5 \pi x + C$ **75.** $\dfrac{3\sqrt{2}}{10}$ **77.** $\dfrac{3\pi}{16}$

Graphs will vary.
Example:

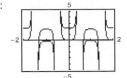

79. (a) Save one sine factor and convert the remaining factors to cosine. Then, expand and integrate.

     (b) Save one cosine factor and convert the remaining factors to sine. Then, expand and integrate.

     (c) Make repeated use of the power-reducing formulas to convert the integrand to odd powers of the cosine. Then, proceed as in part (b).

81. (a) $\dfrac{\tan^6 3x}{18} + \dfrac{\tan^4 3x}{12} + C_1,\ \dfrac{\sec^6 3x}{18} - \dfrac{\sec^4 3x}{12} + C_2$

     (b)   (c) Proof

83. $\dfrac{1}{2}$ **85.** (a) $\dfrac{\pi^2}{2}$ (b) $(\bar{x}, \bar{y}) = \left(\dfrac{\pi}{2}, \dfrac{\pi}{8}\right)$

87. Proof   **89.** Proof

91. $-\frac{1}{15}\cos x(3\sin^4 x + 4\sin^2 x + 8) + C$

93. $\dfrac{5}{6\pi}\tan\dfrac{2\pi x}{5}\left(\sec^2\dfrac{2\pi x}{5} + 2\right) + C$

95. (a) $H(t) = 55.46 - 23.88\cos\dfrac{\pi t}{6} - 3.34\sin\dfrac{\pi t}{6}$

     (b) $L(t) = 39.34 - 20.78\cos\dfrac{\pi t}{6} - 4.33\sin\dfrac{\pi t}{6}$

     (c) Summer

97. Proof

Section 7.4 *(page 512)*

1. b   **2.** d   **3.** a   **4.** c   **5.** $\dfrac{x}{25\sqrt{25 - x^2}} + C$

7. $5\ln\left|\dfrac{5 - \sqrt{25 - x^2}}{x}\right| + \sqrt{25 - x^2} + C$

9. $\ln\left|x + \sqrt{x^2 - 4}\right| + C$ **11.** $\frac{1}{15}(x^2 - 4)^{3/2}(3x^2 + 8) + C$

13. $\frac{1}{3}(1 + x^2)^{3/2} + C$ **15.** $\dfrac{1}{2}\left(\arctan x + \dfrac{x}{1 + x^2}\right) + C$

17. $\frac{1}{2}x\sqrt{4 + 9x^2} + \frac{2}{3}\ln\left|3x + \sqrt{4 + 9x^2}\right| + C$

19. $\sqrt{x^2 + 9} + C$ **21.** $\arcsin\!\left(\dfrac{x}{4}\right) + C$

23. $4 \arcsin\left(\dfrac{x}{2}\right) + x\sqrt{4 - x^2} + C$ **25.** $\ln\left|x + \sqrt{x^2 - 9}\right| + C$

27. $-\dfrac{(1 - x^2)^{3/2}}{3x^3} + C$ **29.** $-\dfrac{1}{3}\ln\left|\dfrac{\sqrt{4x^2 + 9} + 3}{2x}\right| + C$

31. $\dfrac{5\sqrt{x^2 + 5}}{x^2 + 5} + C$ **33.** $\dfrac{1}{3}(1 + e^{2x})^{3/2} + C$

35. $\dfrac{1}{2}\left(\arcsin e^x + e^x\sqrt{1 - e^{2x}}\right) + C$

37. $\dfrac{1}{4}\left(\dfrac{x}{x^2 + 2} + \dfrac{1}{\sqrt{2}}\arctan\dfrac{x}{\sqrt{2}}\right) + C$

39. $x\operatorname{arcsec} 2x - \dfrac{1}{2}\ln\left|2x + \sqrt{4x^2 - 1}\right| + C$

41. $\arcsin\left(\dfrac{x - 2}{2}\right) + C$

43. $\sqrt{x^2 + 4x + 8} - 2\ln\left|\sqrt{x^2 + 4x + 8} + (x + 2)\right| + C$

45. (a) and (b) $\sqrt{3} - \dfrac{\pi}{3} \approx 0.685$

47. (a) and (b) $9\left(2 - \sqrt{2}\right) \approx 5.272$

49. (a) and (b) $-\dfrac{9}{2}\ln\left(\dfrac{2\sqrt{7}}{3} - \dfrac{4\sqrt{3}}{3} - \dfrac{\sqrt{21}}{3} + \dfrac{8}{3}\right)$
$+ 9\sqrt{3} - 2\sqrt{7} \approx 12.644$

51. $\dfrac{1}{2}(x - 15)\sqrt{x^2 + 10x + 9}$
$+ 33\ln\left|\sqrt{x^2 + 10x + 9} + (x + 5)\right| + C$

53. $\dfrac{1}{2}\left(x\sqrt{x^2 - 1} + \ln\left|x + \sqrt{x^2 - 1}\right|\right) + C$

55. (a) Let $u = a\sin\theta$, $\sqrt{a^2 - u^2} = a\cos\theta$, where
$$-\dfrac{\pi}{2} \le \theta \le \dfrac{\pi}{2}.$$
(b) Let $u = a\tan\theta$, $\sqrt{a^2 + u^2} = a\sec\theta$, where
$$-\dfrac{\pi}{2} < \theta < \dfrac{\pi}{2}.$$
(c) Let $u = a\sec\theta$, $\sqrt{u^2 - a^2} = \tan\theta$ if $u > a$ and
$\sqrt{u^2 - a^2} = -\tan\theta$ if $u < -a$, where $0 \le \theta < \dfrac{\pi}{2}$
or $\dfrac{\pi}{2} < \theta \le \pi$.

57. πab **59.** $\dfrac{a^2\pi}{2} - a^2\arcsin\dfrac{h}{a} - h\sqrt{a^2 - h^2}$ **61.** $6\pi^2$

63. $\ln\left[\dfrac{5(\sqrt{2} + 1)}{\sqrt{26} + 1}\right] + \sqrt{26} - \sqrt{2} \approx 4.367$

65. Length of one arch of sine curve: $y = \sin x$, $y' = \cos x$
$$L_1 = \int_0^\pi \sqrt{1 + \cos^2 x}\, dx$$
Length of one arch of cosine curve: $y = \cos x$, $y' = -\sin x$
$$L_2 = \int_{-\pi/2}^{\pi/2} \sqrt{1 + \sin^2 x}\, dx$$
$$= \int_{-\pi/2}^{\pi/2} \sqrt{1 + \cos^2\left(x - \dfrac{\pi}{2}\right)}\, dx, \quad u = x - \dfrac{\pi}{2},\ du = dx$$
$$= \int_{-\pi}^0 \sqrt{1 + \cos^2 u}\, du$$
$$= \int_0^\pi \sqrt{1 + \cos^2 u}\, du = L_1$$

67. (a)

(b) 200 (c) $100\sqrt{2} + 50\ln\left(\dfrac{\sqrt{2} + 1}{\sqrt{2} - 1}\right) \approx 229.559$

69. $(0, 0.422)$ **71.** $\dfrac{\pi}{32}\left[102\sqrt{2} - \ln(3 + 2\sqrt{2})\right] \approx 13.989$

73. (a) 187.2π pounds (b) $62.4\pi d$ pounds

75. (a) $m = \dfrac{dy}{dx} = \dfrac{y - \left(y + \sqrt{144 - x^2}\right)}{x - 0}$
$$= -\dfrac{\sqrt{144 - x^2}}{x}$$
(b) $y = -12\ln\left(\dfrac{12 - \sqrt{144 - x^2}}{x}\right) - \sqrt{144 - x^2}$

(c) $x = 0$ (d) 5.2 meters

77. True

79. False: $\displaystyle\int_0^{\sqrt{3}} \dfrac{dx}{(1 + x^2)^{3/2}} = \int_0^{\pi/3} \cos\theta\, d\theta$ **81.** Proof

Section 7.5 (page 522)

1. $\dfrac{A}{x} + \dfrac{B}{x - 10}$ **3.** $\dfrac{A}{x} + \dfrac{Bx + C}{x^2 + 10}$ **5.** $\dfrac{A}{x} + \dfrac{B}{x^2} + \dfrac{C}{x - 10}$

7. $\dfrac{1}{2}\ln\left|\dfrac{x - 1}{x + 1}\right| + C$ **9.** $\ln\left|\dfrac{x - 1}{x + 2}\right| + C$

11. $\dfrac{3}{2}\ln|2x - 1| - 2\ln|x + 1| + C$

13. $5\ln|x - 2| - \ln|x + 2| - 3\ln|x| + C$

15. $x^2 + \dfrac{3}{2}\ln|x - 4| - \dfrac{1}{2}\ln|x + 2| + C$

17. $\dfrac{1}{x} + \ln|x^4 + x^3| + C$

19. $2\ln|x - 2| - \ln|x| - \dfrac{3}{x - 2} + C$

21. $\ln\left|\dfrac{x^2 + 1}{x}\right| + C$

23. $\dfrac{1}{6}\left[\ln\left|\dfrac{x - 2}{x + 2}\right| + \sqrt{2}\arctan\left(\dfrac{x}{\sqrt{2}}\right)\right] + C$

25. $\dfrac{1}{16}\ln\left|\dfrac{4x^2 - 1}{4x^2 + 1}\right| + C$

27. $\ln|x + 1| + \sqrt{2}\arctan\left(\dfrac{x - 1}{\sqrt{2}}\right) + C$

29. $\ln 2$ **31.** $\dfrac{1}{2}\ln\left(\dfrac{8}{5}\right) - \dfrac{\pi}{4} + \arctan 2 \approx 0.557$

33. $y = 3 \ln|x - 3| - \dfrac{9}{x - 3} + 9$

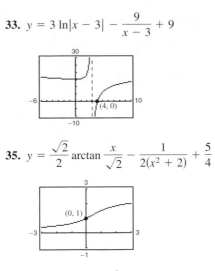

35. $y = \dfrac{\sqrt{2}}{2} \arctan \dfrac{x}{\sqrt{2}} - \dfrac{1}{2(x^2 + 2)} + \dfrac{5}{4}$

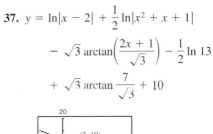

37. $y = \ln|x - 2| + \dfrac{1}{2} \ln|x^2 + x + 1|$

$\qquad - \sqrt{3} \arctan\left(\dfrac{2x + 1}{\sqrt{3}}\right) - \dfrac{1}{2} \ln 13$

$\qquad + \sqrt{3} \arctan \dfrac{7}{\sqrt{3}} + 10$

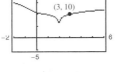

39. $y = \dfrac{1}{4} \ln\left|\dfrac{x - 2}{x + 2}\right| + \dfrac{1}{4} \ln 2 + 4$

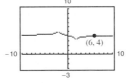

41. $\ln\left|\dfrac{\cos x}{\cos x - 1}\right| + C$ **43.** $\ln\left|\dfrac{-1 + \sin x}{2 + \sin x}\right| + C$

45. $\dfrac{1}{5} \ln\left|\dfrac{e^x - 1}{e^x + 4}\right| + C$ **47.** Proof   **49.** Proof

51. $y = \dfrac{3}{2} \ln\left|\dfrac{2 + x}{2 - x}\right| + 3$ **53.** First divide x^3 by $(x - 5)$.

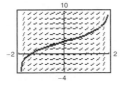

55. (a) Log Rule   (b) Partial fractions

(c) Inverse Tangent Rule

57. 4.90 or $490,000   **59.** c

61. $x = \dfrac{n[e^{(n+1)kt} - 1]}{n + e^{(n+1)kt}}$ **63.** $\dfrac{\pi}{8}$

Section 7.6 (page 528)

1. $-\dfrac{1}{2}x(2 - x) + \ln|1 + x| + C$

3. $\dfrac{1}{2}\left[e^x \sqrt{e^{2x} + 1} + \ln\left(e^x + \sqrt{e^{2x} + 1}\right)\right] + C$

5. $-\dfrac{\sqrt{1 - x^2}}{x} + C$

7. $\dfrac{1}{16}(6x - 3 \sin 2x \cos 2x - 2 \sin^3 2x \cos 2x) + C$

9. $-2\left(\cot\sqrt{x} + \csc\sqrt{x}\right) + C$

11. $x - \dfrac{1}{2}\ln(1 + e^{2x}) + C$ **13.** $\dfrac{1}{16}x^4(4 \ln x - 1) + C$

15. (a) and (b) $e^x(x^2 - 2x + 2) + C$

17. (a) and (b) $\ln\left|\dfrac{x + 1}{x}\right| - \dfrac{1}{x} + C$ **19.** $\dfrac{1}{2}e^{x^2} + C$

21. $\dfrac{1}{2}\left\{(x^2 + 1)\operatorname{arcsec}(x^2 + 1) - \ln\left[(x^2 + 1) + \sqrt{x^4 + 2x^2}\right]\right\} + C$

23. $\dfrac{1}{9}x^3(-1 + 3 \ln x) + C$ **25.** $\dfrac{\sqrt{x^2 - 4}}{4x} + C$

27. $\dfrac{2}{9}\left(\ln|1 - 3x| + \dfrac{1}{1 - 3x}\right) + C$

29. $e^x \arccos(e^x) - \sqrt{1 - e^{2x}} + C$

31. $\dfrac{1}{2}(x^2 + \cot x^2 + \csc x^2) + C$ **33.** $\arctan(\sin x) + C$

35. $\dfrac{\sqrt{2}}{2} \arctan\left(\dfrac{1 + \sin \theta}{\sqrt{2}}\right) + C$ **37.** $-\dfrac{\sqrt{2 + 9x^2}}{2x} + C$

39. $(t^3 - 6t) \sin t + 3(t^2 - 2) \cos t + C$

41. $\dfrac{1}{4}\left(2 \ln|x| - 3 \ln 3 + 2 \ln|x|\right) + C$

43. $\dfrac{3x - 10}{2(x^2 - 6x + 10)} + \dfrac{3}{2} \arctan(x - 3) + C$

45. $\dfrac{1}{2} \ln\left|x^2 - 3 + \sqrt{x^4 - 6x^2 + 5}\right| + C$

47. $-\dfrac{1}{3}\sqrt{4 - x^2}(x^2 + 8) + C$

49. $\dfrac{2}{1 + e^x} - \dfrac{1}{2(1 + e^x)^2} + \ln(1 + e^x) + C$

51. Proof   **53.** Proof   **55.** Proof

57. $y = -\dfrac{2\sqrt{1 - x}}{\sqrt{x}} + 7$

59. $y = \dfrac{1}{2}\left[\dfrac{x - 3}{x^2 - 6x + 10} + \arctan(x - 3)\right]$

61. $y = -\csc\theta + \sqrt{2} + 2$

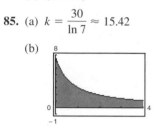

63. $\dfrac{1}{\sqrt{5}} \ln\left|\dfrac{2\tan(\theta/2) - 3 - \sqrt{5}}{2\tan(\theta/2) - 3 + \sqrt{5}}\right| + C$ **65.** $\ln 2$

67. $\frac{1}{2}\ln(3 - 2\cos\theta) + C$ **69.** $2\sin\sqrt{\theta} + C$ **71.** $\frac{40}{3}$

73. Use Formula 23 and let $a = 1$, $u = e^x$, and $du = e^x\,dx$.

75. Use Formula 81 and let $u = x^2$ and $du = 2x\,dx$.

77. Impossible

79. Answers will vary. For example: $\int x^3 \cos x\,dx$ can be integrated using Formula 55 where $u = x$, $du = dx$, and $n = 3$.

81. 1919.145 foot-pounds

83. (a) $V = 80\ln\left(\sqrt{10} + 3\right) \approx 145.5$ cubic feet

       $W = 11{,}840\ln\left(\sqrt{10} + 3\right) \approx 21{,}530.4$ pounds

     (b) $(0, 1.19)$

85. (a) $k = \dfrac{30}{\ln 7} \approx 15.42$

     (b)

87. False. Substitutions may first have to be made to rewrite the integral in a form that appears in the table.

Section 7.7 (page 537)

1.

x	-0.1	-0.01	-0.001
$f(x)$	2.4132	2.4991	2.500

x	0.001	0.01	0.1
$f(x)$	2.500	2.4991	2.4132

2.5

3.

x	1	10	10^2
$f(x)$	0.9900	90,483.7	3.7×10^9

x	10^3	10^4	10^5
$f(x)$	4.5×10^{10}	0	0

0

5. $\frac{1}{3}$ **7.** $\frac{1}{4}$ **9.** $\frac{5}{3}$ **11.** 3   **13.** 0   **15.** 2

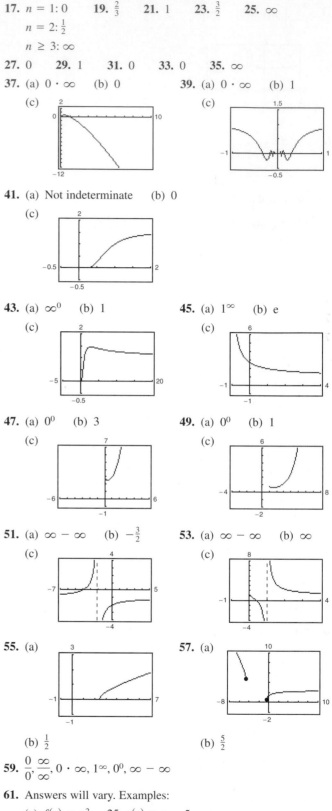

17. $n = 1: 0$ **19.** $\frac{2}{3}$ **21.** 1   **23.** $\frac{3}{2}$ **25.** ∞

     $n = 2: \frac{1}{2}$

     $n \geq 3: \infty$

27. 0   **29.** 1   **31.** 0   **33.** 0   **35.** ∞

37. (a) $0 \cdot \infty$ (b) 0   **39.** (a) $0 \cdot \infty$ (b) 1

     (c)   (c)

41. (a) Not indeterminate   (b) 0

     (c)

43. (a) ∞^0 (b) 1   **45.** (a) 1^∞ (b) e

     (c)   (c)

47. (a) 0^0 (b) 3   **49.** (a) 0^0 (b) 1

     (c)   (c)

51. (a) $\infty - \infty$ (b) $-\frac{3}{2}$ **53.** (a) $\infty - \infty$ (b) ∞

     (c)   (c)

55. (a)   **57.** (a)

     (b) $\frac{1}{2}$ (b) $\frac{5}{2}$

59. $\dfrac{0}{0}, \dfrac{\infty}{\infty}, 0 \cdot \infty, 1^\infty, 0^0, \infty - \infty$

61. Answers will vary. Examples:

     (a) $f(x) = x^2 - 25$, $g(x) = x - 5$

     (b) $f(x) = (x - 5)^2$, $g(x) = x^2 - 25$

     (c) $f(x) = x^2 - 25$, $g(x) = (x - 5)^3$

63. 0   **65.** 0   **67.** 0

69.

x	10	10^2	10^4	10^6	10^8	10^{10}
$\dfrac{(\ln x)^4}{x}$	2.811	4.498	0.720	0.036	0.001	0.000

71. Horizontal asymptote: $y = 1$
Relative maximum: $(e, e^{1/e})$

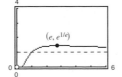

73. Horizontal asymptote: $y = 0$
Relative minimum: $\left(1, \dfrac{2}{e}\right)$

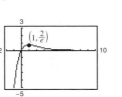

75. Limit is not of the form $0/0$ or ∞/∞.

77. Limit is not of the form $0/0$ or ∞/∞.

79. (a) 1

(b) $\displaystyle\lim_{x \to \infty} \frac{x}{\sqrt{x^2 + 1}} = \lim_{x \to \infty} \frac{\sqrt{x^2 + 1}}{x} = \lim_{x \to \infty} \frac{x}{\sqrt{x^2 + 1}}$

Applying L'Hôpital's Rule twice results in the original limit, so L'Hôpital's Rule fails.

(c)

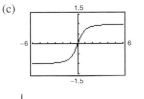

1

81. $v = 32t + v_0$ **83.** $\dfrac{3}{4}$ **85.** $c = \dfrac{2}{3}$ **87.** $c = \dfrac{\pi}{4}$

89. False: L'Hôpital's Rule does not apply, because $\displaystyle\lim_{x \to 0}(x^2 + x + 1) \neq 0$.

91. True

93. (a) $\dfrac{1}{2}\sin\theta - \dfrac{1}{2}\sin\theta\cos\theta$ (b) $\dfrac{1}{2}\theta - \dfrac{1}{2}\sin\theta\cos\theta$

(c) $\dfrac{\sin\theta - \sin\theta\cos\theta}{\theta - \sin\theta\cos\theta}$ (d) $\dfrac{3}{4}$

95. Proof   **97.** Proof

Section 7.8 (page 547)

1. Infinite discontinuity at $x = 0$; 4

3. Infinite discontinuity at $x = 1$; diverges

5. Infinite limit of integration; 1

7. Infinite discontinuity at $x = 0$; diverges

9. 1   **11.** Diverges

13. Diverges   **15.** 2   **17.** $\dfrac{1}{2}$ **19.** $\dfrac{1}{2(\ln 4)^2}$

21. π **23.** $\dfrac{\pi}{4}$ **25.** Diverges   **27.** Diverges

29. 6   **31.** $-\dfrac{1}{4}$ **33.** Diverges   **35.** $\dfrac{\pi}{3}$

37. $\ln(2 + \sqrt{3})$ **39.** 0   **41.** $\dfrac{2\pi\sqrt{6}}{3}$ **43.** $p > 1$

45. Proof   **47.** Diverges   **49.** Converges

51. Converges   **53.** Diverges   **55.** Converges

57. An integral with infinite integration limits, an integral with an infinite discontinuity at or between the integration limits

59. The improper integral diverges.

61. $\dfrac{1}{s}$, $s > 0$ **63.** $\dfrac{2}{s^3}$, $s > 0$ **65.** $\dfrac{s}{s^2 + a^2}$, $s > 0$

67. $\dfrac{s}{s^2 - a^2}$, $s > |a|$ **69.** (a) 1   (b) $\dfrac{\pi}{2}$ (c) 2π

71.

Perimeter $= 48$

73. (a) $\Gamma(1) = 1$, $\Gamma(2) = 1$, $\Gamma(3) = 2$ (b) Proof

(c) $\Gamma(n) = (n-1)!$

75. (a) Proof   (b) $P = 43.53\%$ (c) $E(x) = 7$

77. (a) \$757,992.41   (b) \$837,995.15   (c) \$1,066,666.67

79. $\dfrac{k\left(\sqrt{a^2 + 1} - 1\right)}{a^2\sqrt{a^2 + 1}}$ **81.** Three. All three must converge.

83. (a) $\dfrac{1}{6}$ (b) $\dfrac{1}{24}$ (c) $\dfrac{1}{60}$

85. False. Let $f(x) = \dfrac{1}{x + 1}$.   **87.** True

Review Exercises for Chapter 7 (page 550)

1. $\dfrac{(x^2 - 1)^{3/2}}{3} + C$ **3.** $\dfrac{1}{2}\ln|x^2 - 1| + C$

5. $\dfrac{(\ln(2x))^2}{2} + C$ **7.** $16\arcsin\dfrac{x}{4} + C$

9. $\dfrac{e^{2x}}{13}(2\sin 3x - 3\cos 3x) + C$

11. $\dfrac{2}{15}(x - 5)^{3/2}(3x + 10) + C$

13. $-\dfrac{1}{2}x^2\cos 2x + \dfrac{x}{2}\sin 2x + \dfrac{1}{4}\cos 2x + C$

15. $\dfrac{1}{16}\left[(8x^2 - 1)\arcsin 2x + 2x\sqrt{1 - 4x^2}\right] + C$

17. $\dfrac{1}{3\pi}\sin(\pi x - 1)[\cos^2(\pi x - 1) + 2] + C$

19. $\dfrac{2}{3}\left[\tan^3\left(\dfrac{x}{2}\right) + 3\tan\left(\dfrac{x}{2}\right)\right] + C$ **21.** $\tan\theta + \sec\theta + C$

23. $\dfrac{3\sqrt{4 - x^2}}{x} + C$ **25.** $\dfrac{1}{3}(x^2 + 4)^{1/2}(x^2 - 8) + C$

27. $\dfrac{1}{2}\left(4 \arcsin \dfrac{x}{2} + x\sqrt{4 - x^2}\right) + C$

29. (a), (b), and (c) $\frac{1}{3}\sqrt{4 + x^2}(x^2 - 8) + C$

31. $6 \ln|x + 2| - 5 \ln|x - 3| + C$

33. $\frac{1}{4}[6 \ln|x - 1| - \ln(x^2 + 1) + 6 \arctan x] + C$

35. $x + \frac{9}{8} \ln|x - 3| - \frac{25}{8} \ln|x + 5| + C$

37. $\dfrac{1}{9}\left(\dfrac{2}{2 + 3x} + \ln|2 + 3x|\right) + C$ **39.** $\frac{1}{2}[\tan x^2 - \sec x^2] + C$

41. $\dfrac{1}{2} \ln|x^2 + 4x + 8| - \arctan\left(\dfrac{x + 2}{x}\right) + C$

43. $\left(\dfrac{1}{\pi}\right) \ln|\tan \pi x| + C$

45. Proof **47.** $\frac{1}{8}(\sin 2\theta - 2\theta \cos 2\theta) + C$

49. $\frac{4}{3}[x^{3/4} - 3x^{1/4} + 3 \arctan(x^{1/4})] + C$

51. $2\sqrt{1 - \cos x} + C$ **53.** $\sin x \ln(\sin x) - \sin x + C$

55. $y = \dfrac{3}{2} \ln\left|\dfrac{x - 3}{x + 3}\right| + C$

57. $y = x \ln|x^2 + x| - 2x + \ln|x + 1| + C$ **59.** $\frac{1}{5}$

61. $\frac{1}{2}(\ln 4)^2 \approx 0.961$ **63.** π **65.** $\frac{128}{15}$

67. $(\bar{x}, \bar{y}) = \left(0, \dfrac{4}{3\pi}\right)$ **69.** 3.82 **71.** 0 **73.** ∞ **75.** 1

77. $1000e^{0.09} \approx 1094.17$ **79.** Converges; $\frac{32}{3}$ **81.** Diverges

83. (a) \$6,321,205.59 (b) \$10,000,000

85. (a) 0.4581 (b) 0.0135

P.S. Problem Solving (page 552)

1. (a) $\frac{4}{3}, \frac{16}{15}$ (b) Proof **3.** $\ln 3$

5. Let P be represented by $\left(c, \sqrt{1 - c^2}\right)$. Then

$$S = \int_c^1 \sqrt{1 + \left(\dfrac{-x}{\sqrt{1 - x^2}}\right)^2} \, dx = \dfrac{\pi}{2} - \arcsin c.$$ Then Q is

represented by $\left(1, \dfrac{\pi}{2} - \arcsin c\right)$ and line PQ is represented

by $y - \sqrt{1 - c^2} = \left(\dfrac{\dfrac{\pi}{2} - \arcsin c - \sqrt{1 - c^2}}{1 - c}\right)(x - c).$

Since R is on the x-axis, set $y = 0$. Then simplify and find

$\lim\limits_{c \to -1}\left(c - \dfrac{(1 - c)\sqrt{1 - c^2}}{\dfrac{\pi}{2} - \arcsin c - \sqrt{1 - c^2}}\right).$ This limit is -2 and

therefore the length of segment OR is 2.

7. (a) Area ≈ 0.2986 (b) $\ln 3 - \frac{4}{5}$

(c) $\ln 3 - \frac{4}{5}$

9. $\ln 3 - \frac{1}{2} \approx 0.5986$ **11.** Proof

13. $x^4 + 1 = \left(x^2 + \sqrt{2}x + 1\right)\left(x^2 - \sqrt{2}x + 1\right)$

$$A = \dfrac{\sqrt{2}}{4}\left[\arctan\left(\sqrt{2} + 1\right) + \arctan\left(\sqrt{2} - 1\right)\right]$$

$$+ \dfrac{\sqrt{2}}{8}\left[\ln\left(2 + \sqrt{2}\right) - \ln\left(2 - \sqrt{2}\right)\right]$$

$A \approx 0.8670$

15. (a) ∞ (b) 0 (c) $-\frac{2}{3}$

The indeterminate form $0 \cdot \infty$ does not determine the value of the limit or even whether the limit exists.

17. $\dfrac{1/12}{x} + \dfrac{111/140}{x + 4} + \dfrac{1/42}{x - 3} + \dfrac{1/10}{x - 1}$ **19.** Proof

Chapter 8

Section 8.1 (page 564)

1. 2, 4, 8, 16, 32 **3.** $-\frac{1}{2}, \frac{1}{4}, -\frac{1}{8}, \frac{1}{16}, -\frac{1}{32}$ **5.** 1, 0, -1, 0, 1

7. $-1, -\frac{1}{4}, \frac{1}{9}, \frac{1}{16}, -\frac{1}{25}$ **9.** 5, $\frac{19}{4}, \frac{43}{9}, \frac{77}{16}, \frac{121}{25}$

11. 3, $\frac{9}{2}, \frac{27}{6}, \frac{81}{24}, \frac{243}{120}$ **13.** 3, 4, 6, 10, 18 **15.** 32, 16, 8, 4, 2

17. d **18.** a **19.** c **20.** b

21.

23.

25.

27. 14, 17; add 3 to preceding term

29. $\frac{3}{16}, -\frac{3}{32}$; multiply preceding term by $-\frac{1}{2}$

31. $10 \cdot 9 = 90$ **33.** $n + 1$ **35.** $\dfrac{1}{(2n + 1)(2n)}$

37. 5 **39.** 2 **41.** 0

43.

Converges to 1

45.

Diverges

47. Diverges **49.** Converges to $\frac{3}{2}$

51. Converges to 0 **53.** Converges to 0

55. Converges to 0 **57.** Diverges

59. Converges to 0 **61.** Converges to 0

63. Converges to e^k **65.** Converges to 0

Answers may vary in 67–79.

67. $3n - 2$ **69.** $n^2 - 2$ **71.** $\dfrac{n+1}{n+2}$ **73.** $\dfrac{(-1)^{n-1}}{2^{n-2}}$

75. $\dfrac{n+1}{n}$ **77.** $\dfrac{n}{(n+1)(n+2)}$

79. $\dfrac{(-1)^{n-1}}{1\cdot 3\cdot 5\cdots(2n-1)} = \dfrac{(-1)^{n-1}2^n n!}{(2n)!}$

81. Monotonic, bounded   **83.** Monotonic, bounded

85. Not monotonic, bounded   **87.** Monotonic, bounded

89. Not monotonic, bounded

91. (a) $\left|5 + \dfrac{1}{n}\right| \le 6 \implies$ bounded   (b)

$a_n > a_{n+1} \implies$ monotonic

So, $\{a_n\}$ converges.

Limit $= 5$

93. (a) $\left|\dfrac{1}{3}\left(1 - \dfrac{1}{3^n}\right)\right| < \dfrac{1}{3} \implies$ bounded

$a_n < a_{n+1} \implies$ monotonic

So, $\{a_n\}$ converges.

(b)

Limit $= \frac{1}{3}$

95. (a) No; $\lim\limits_{n\to\infty} a_n$ does not exist.

(b)

n	1	2	3	4
A_n	\$9086.25	\$9173.33	\$9261.24	\$9349.99

n	5	6	7	8
A_n	\$9439.60	\$9530.06	\$9621.39	\$9713.59

n	9	10
A_n	\$9806.68	\$9900.66

97. (a) A sequence is a function whose domain is the set of positive integers.

(b) A sequence converges if it has a limit.

(c) A bounded monotonic sequence is a sequence that has nondecreasing or nonincreasing terms and an upper and lower bound.

99. Answers will vary. Example: $a_n = \dfrac{10n}{n+1}$

101. Answers will vary. Example: $a_n = \dfrac{3n^2 - n}{4n^2 + 1}$

103. (a) \$2,500,000,000(0.8)^n

(b)

Year	1	2
Budget	\$2,000,000,000	\$1,600,000,000

Year	3	4
Budget	\$1,280,000,000	\$1,024,000,000

(c) Converges to 0

105. (a) $a_n = -3.73n^2 + 75.9n + 684$

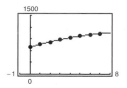

(b) \$1016

107. (a) $a_9 = a_{10} = \dfrac{1,562,500}{567}$

(b) Decreasing

(c) Factorials increase more rapidly than exponentials.

109. 1, 1.4142, 1.4422, 1.4142, 1.3797, 1.3480; Converges to 1

111. (a) 1, 1, 2, 3, 5, 8, 13, 21, 34, 55, 89, 144

(b) 1, 2, 1.5, 1.6667, 1.6, 1.6250, 1.6154, 1.6190, 1.6176, 1.6182

(c) Proof

(d) $\rho = \dfrac{1 + \sqrt{5}}{2} \approx 1.6180$

113. True   **115.** True

117. 1.4142, 1.8478, 1.9616, 1.9904, 1.9976

$\lim\limits_{n\to\infty} a_n = 2$

Section 8.2 (page 573)

1. 1, 1.25, 1.361, 1.424, 1.464

3. 3, -1.5, 5.25, -4.875, 10.3125

5. 3, 4.5, 5.25, 5.625, 5.8125

7. Geometric series: $r = \frac{3}{2} > 1$

9. Geometric series: $r = 1.055 > 1$ **11.** $\lim\limits_{n\to\infty} a_n = 1 \ne 0$

13. $\lim\limits_{n\to\infty} a_n = 1 \ne 0$ **15.** $\lim\limits_{n\to\infty} a_n = \frac{1}{2} \ne 0$

17. c; 3   **18.** b; 3   **19.** a; 3   **20.** d; 3

21. Telescoping series: $a_n = \dfrac{1}{n} - \dfrac{1}{n+1}$; Converges to 1.

23. Geometric series: $r = \frac{3}{4} < 1$

25. Geometric series: $r = 0.9 < 1$

27. (a) $\frac{11}{3}$

(b)

n	5	10	20	50	100
S_n	2.7976	3.1643	3.3936	3.5513	3.6078

(c)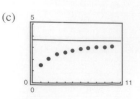

(d) The terms of the series decrease in magnitude relatively slowly, and the sequence of partial sums approaches the sum of the series relatively slowly.

29. (a) 20

(b)

n	5	10	20	50	100
S_n	8.1902	13.0264	17.5685	19.8969	19.9995

(c)

(d) The terms of the series decrease in magnitude relatively slowly, and the sequence of partial sums approaches the sum of the series relatively slowly.

31. (a) $\frac{40}{3}$

(b)

n	5	10	20	50	100
S_n	13.3203	13.3333	13.3333	13.3333	13.3333

(c)

(d) The terms of the series decrease in magnitude relatively rapidly, and the sequence of partial sums approaches the sum of the series relatively rapidly.

33. $\frac{3}{4}$ **35.** 4 **37.** 2 **39.** $\frac{2}{3}$ **41.** $\frac{10}{9}$ **43.** $\frac{9}{4}$

45. $\frac{1}{2}$ **47.** $\sum_{n=0}^{\infty} \frac{4}{10}(0.1)^n = \frac{4}{9}$ **49.** $\sum_{n=0}^{\infty} \frac{3}{40}(0.01)^n = \frac{5}{66}$

51. Diverges **53.** Converges **55.** Diverges

57. Converges **59.** Diverges **61.** Diverges

63. See definition on page 567.

65. The series given by

$$\sum_{n=0}^{\infty} ar^n = a + ar + ar^2 + \cdots + ar^n + \cdots, a \neq 0$$

is a geometric series with ratio r. When $0 < |r| < 1$, the series

converges to the sum $\sum_{n=0}^{\infty} ar^n = \frac{a}{1 - r}$.

67. (a) x

(c)

(b) $f(x) = \frac{1}{1 - x}$, $|x| < 1$

69.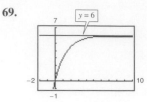

Horizontal asymptote: $y = 6$
The horizontal asymptote is the sum of the series.

71. The required terms for the two series are $n = 100$ and $n = 5$, respectively. The second series converges at a faster rate.

73. $80,000(1 - 0.9^n)$ units

75. $400(1 - 0.75^n)$ million dollars; Sum = $400 million

77. 152.42 feet

79. $\frac{1}{8}$; $\sum_{n=0}^{\infty} \frac{1}{2}\left(\frac{1}{2}\right)^n = \frac{1/2}{1 - 1/2} = 1$

81. (a) $-1 + \sum_{n=0}^{\infty} \left(\frac{1}{2}\right)^n = -1 + \frac{a}{1 - r} = -1 + \frac{1}{1 - 1/2} = 1$

(b) No (c) 2

83. $557,905.82; The $1,000,000 sweepstakes has a present value of $557,905.82. After accruing interest over the 20-year period, it attains its full value.

85. (a) $5,368,709.11 (b) $10,737,418.23

(c) $21,474,836.47

87. (a) $16,415.10 (b) $16,421.83

89. (a) $118,196.13 (b) $118,393.43

91. (a) $a_n = 6110.1832e^{0.0530x}$ (b) $78,530 million

(c) $78,461 million

93. Proof **95.** Proof

97. Answers will vary. Example: $\sum_{n=0}^{\infty} 1$, $\sum_{n=0}^{\infty} (-1)$

99. False. $\lim_{n \to \infty} \frac{1}{n} = 0$, but $\sum_{n=1}^{\infty} \frac{1}{n}$ diverges.

101. False

$$\sum_{n=1}^{\infty} ar^n = \left(\frac{a}{1 - r}\right) - a$$

The formula requires that the geometric series begins with $n = 0$.

103. H = half-life of the drug

n = number of equal doses

P = number of units of the drug

t = equal time intervals

The total amount of the drug in the patient's system at the time the last dose is given is

$$T_n = P + Pe^{kt} + Pe^{2kt} + \cdots + Pe^{(n-1)kt}$$

where $k = -(\ln 2)/H$. One time interval after the last dose is given is

$$T_{n+1} = Pe^{kt} + Pe^{2kt} + Pe^{3kt} + \cdots + Pe^{nkt}$$

and so on. Because $k < 0$, $T_{n+s} \to 0$ as $s \to \infty$.

Section 8.3 (page 580)

1. Diverges   3. Converges   5. Converges

7. Diverges   9. Diverges   11. Converges

13. Diverges   15. Diverges   17. Converges

19. Converges   21. a; Diverges   22. d; Diverges

23. b; Converges   24. c; Converges

25. No. For some series the terms decrease toward 0 too slowly for the series to converge.

27. (a)

M	2	4	6	8
N	4	31	227	1674

(b) No. Because the magnitude of the terms of the series is approaching zero, it requires more and more terms to increase the partial sum by 2.

29. $p > 1$

31. See Theorem 8.10 on page 577. Answers will vary. For example, convergence or divergence can be determined for the series

$$\sum_{n=1}^{\infty} \frac{1}{n^2 + 1}.$$

33. No. Because $\sum_{n=1}^{\infty} \frac{1}{n}$ diverges, $\sum_{n=10,000}^{\infty} \frac{1}{n}$ also diverges. The convergence or divergence of a series is not determined by the first finite number of terms of the series.

35. Proof

37. $S_6 \approx 1.0811$ 39. $S_{10} \approx 0.9818$ 41. $S_4 \approx 0.4049$

   $R_6 \approx 0.0015$ $R_{10} \approx 0.0997$ $R_4 \approx 5.6 \times 10^{-8}$

43. $N \geq 7$ 45. $N \geq 2$ 47. $N \geq 1004$

49. (a) $\sum_{n=2}^{\infty} \frac{1}{n^{1.1}}$ converges by the p-Series Test since $1.1 > 1$.

$\sum_{n=2}^{\infty} \frac{1}{n \ln n}$ diverges by the Integral Test since $\int_{2}^{\infty} \frac{1}{x \ln x}\, dx$ diverges.

(b) $\sum_{n=2}^{\infty} \frac{1}{n^{1.1}} = 0.4665 + 0.2987 + 0.2176 + 0.1703$

$+ 0.1393 + \cdots$

$\sum_{n=2}^{\infty} \frac{1}{n \ln n} = 0.7213 + 0.3034 + 0.1803 + 0.1243$

$+ 0.0930 + \cdots$

(c) $n \geq 3.431 \times 10^{15}$

51. (a) Let $f(x) = 1/x$. f is positive, continuous, and decreasing on $[1, \infty)$.

$$S_n - 1 \leq \int_{1}^{n} \frac{1}{x}\, dx = \ln n$$

$$S_n \geq \int_{1}^{n+1} \frac{1}{x}\, dx = \ln(n + 1)$$

So, $\ln(n + 1) \leq S_n \leq 1 + \ln n$.

(b) $\ln(n + 1) - \ln n \leq S_n - \ln n \leq 1$.

Also, $\ln(n + 1) - \ln n > 0$ for $n \geq 1$. So, $0 \leq S_n - \ln n \leq 1$ and the sequence $\{a_n\}$ is bounded.

(c) $a_n - a_{n+1} = [S_n - \ln n] - [S_{n+1} - \ln(n + 1)]$

$= \int_{n}^{n+1} \frac{1}{x}\, dx - \frac{1}{n + 1} \geq 0$

So, $a_n \geq a_{n+1}$.

(d) Because the sequence is bounded and monotonic, it converges to a limit, γ.

(e) 0.5822

53. Diverges   55. Converges   57. Converges

59. Diverges   61. Diverges   63. Converges

Section 8.4 (page 587)

1. (a)

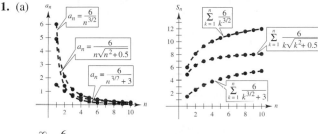

(b) $\sum_{n=1}^{\infty} \frac{6}{n^{3/2}}$; Converges

(c) Magnitudes of terms are less than magnitudes of terms of p-series. Therefore, series converges.

(d) The smaller the magnitudes of the terms, the smaller the magnitudes of the terms of the sequence of partial sums.

3. Converges   5. Diverges   7. Converges

9. Diverges   11. Converges   13. Converges

15. Diverges   17. Diverges   19. Converges

21. Diverges   23. Converges   25. Diverges

27. Diverges   29. Diverges; p-Series Test

31. Converges; Direct Comparison Test with $\sum_{n=1}^{\infty} \left(\frac{1}{3}\right)^n$

33. Diverges; nth-Term Test   35. Converges; Integral Test

37. $\lim_{n \to \infty} \frac{a_n}{1/n} = \lim_{n \to \infty} na_n$

$\lim_{n \to \infty} na_n \neq 0$, but is finite.

The series diverges by the Limit Comparison Test.

39. Diverges   41. Converges

43. $\lim_{n \to \infty} n\left(\frac{n^3}{5n^4 + 3}\right) = \frac{1}{5} \neq 0$

So, $\sum_{n=1}^{\infty} \frac{n^3}{5n^4 + 3}$ diverges.

45. See Theorem 8.12 on page 583. Answers will vary. For example, convergence or divergence of the series

$\sum_{n=1}^{\infty} \frac{1}{3n^2 + 4}$ can be determined by comparing it to the series

$\frac{1}{3}\sum_{n=1}^{\infty} \frac{1}{n^2}.$

47.

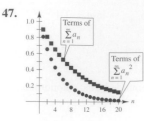

Because $0 < a_n < 1$, $0 < a_n^2 < a_n < 1$.

49. Diverges   **51.** Converges

53. Convergence or divergence is dependent on the form of the general term for the series and not necessarily the magnitude of the terms.

55. False. Let $a_n = \dfrac{1}{n^3}$ and $b_n = \dfrac{1}{n^2}$.   **57.** True   **59.** Proof

61. $\displaystyle\sum_{n=1}^{\infty} \frac{1}{n^2}, \sum_{n=1}^{\infty} \frac{1}{n^3}$ **63.** (a) Proof   (b) Proof

65. Area $= \dfrac{18\sqrt{3}}{5}$; Perimeter is infinite.

Section 8.5 (page 595)

1. b   **2.** d   **3.** c   **4.** a

5. (a)

n	1	2	3	4	5
S_n	1.0000	0.6667	0.8667	0.7238	0.8349

n	6	7	8	9	10
S_n	0.7440	0.8209	0.7543	0.8131	0.7605

(b)

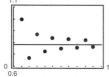

(c) The points alternate sides of the horizontal line $y = \pi/4$ that represents the sum of the series. The distances between the successive points and the line decrease.

(d) The distance in part (c) is always less than the magnitude of the next term of the series.

7. (a)

n	1	2	3	4	5
S_n	1.0000	0.7500	0.8611	0.7986	0.8386

n	6	7	8	9	10
S_n	0.8108	0.8312	0.8156	0.8280	0.8180

(b)

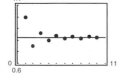

(c) The points alternate sides of the horizontal line $y = \pi^2/12$ that represents the sum of the series. The distances between the successive points and the line decrease.

(d) The distance in part (c) is always less than the magnitude of the next term of the series.

9. Converges   **11.** Converges   **13.** Diverges

15. Converges   **17.** Diverges   **19.** Diverges

21. Diverges   **23.** Converges   **25.** Converges

27. Converges

29. $2.3713 \le S \le 2.4937$ **31.** $0.7305 \le S \le 0.7361$

33. (a) 7 terms (Note that the sum begins with $N = 0$.)
   (b) 0.368

35. (a) 3 terms (Note that the sum begins with $N = 0$.)
   (b) 0.842

37. (a) 1000 terms   (b) 0.693

39. 7   **41.** Converges absolutely

43. Converges conditionally   **45.** Diverges

47. Converges conditionally   **49.** Converges absolutely

51. Converges absolutely   **53.** Converges conditionally

55. Converges absolutely

57. An alternating series is a series whose terms alternate in sign. See Theorem 8.14 on page 590 for the Alternating Series Test.

59. A series $\Sigma\, a_n$ is absolutely convergent if $\Sigma\, |a_n|$ converges. A series $\Sigma\, a_n$ is conditionally convergent if $\Sigma\, a_n$ converges and $\Sigma\, |a_n|$ diverges.

61. Graph (b) represents the partial sums of an alternating series because, by definition of an alternating series, either the even or the odd terms are negative. In this example, the even terms are negative.

63. (a) Proof
   (b) The converse is false. For example: Let $a_n = 1/n$.

65. $\displaystyle\sum_{n=1}^{\infty} \frac{1}{n^2}$ converges, hence so does $\displaystyle\sum_{n=1}^{\infty} \frac{1}{n^4}$.

67. False. Let $a_n = \dfrac{(-1)^n}{n}$.

69. Converges; p-Series Test   **71.** Diverges; nth-Term Test

73. Converges; Geometric Series   **75.** Converges; Integral Test

77. Converges; Alternating Series Test

79. The first term of the series is zero, not one. You cannot regroup series terms arbitrarily.

Section 8.6 (page 603)

1. Proof   **3.** Proof

5. d   **6.** c   **7.** f   **8.** b   **9.** a   **10.** e

11. (a) Proof

   (b)

n	5	10	15	20	25
S_n	9.2104	16.7598	18.8016	19.1878	19.2491

(c)

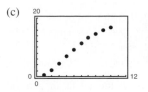

(d) 19.26

(e) The more rapidly the terms of the series approach 0, the more rapidly the sequence of partial sums approaches the sum of the series.

13. Diverges **15.** Converges **17.** Converges

19. Diverges **21.** Converges **23.** Diverges

25. Converges **27.** Converges **29.** Diverges

31. Converges **33.** Proof **35.** Converges

37. Converges **39.** Diverges **41.** Converges

43. Converges; Alternating Series Test

45. Converges; p-Series Test **47.** Diverges; nth-Term Test

49. Diverges; Ratio Test

51. Converges; Limit Comparison Test with $b_n = 1/2^n$

53. Converges; Direct Comparison Test with $b_n = 1/2^n$

55. Converges; Ratio Test **57.** Converges; Ratio Test

59. Converges; Ratio Test **61.** a and c **63.** a and b

65. $\displaystyle\sum_{n=0}^{\infty} \frac{n+1}{4^{n+1}}$ **67.** (a) 9 (b) -0.7769

69. See Theorem 8.17 on page 597.

71. No; the series $\displaystyle\sum_{n=1}^{\infty} \frac{1}{n+10{,}000}$ diverges.

73. Absolutely **75.** Proof

Section 8.7 (page 613)

1. d **2.** c **3.** a **4.** b

5. $P_1 = 6 - 2x$ **7.** $P_1 = \sqrt{2}x + \dfrac{\sqrt{2}(4 - \pi)}{4}$

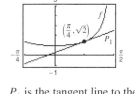

P_1 is the tangent line to the curve $f(x) = 4/\sqrt{x}$ at the point $(1, 4)$.

P_1 is the tangent line to the curve $f(x) = \sec x$ at the point $\left(\dfrac{\pi}{4}, \sqrt{2}\right)$.

9. $P_2 = 4 - 2(x - 1) + \frac{3}{2}(x - 1)^2$

Table for 9

x	0	0.8	0.9	1	1.1
$f(x)$	Error	4.4721	4.2164	4.0000	3.8139
$P_2(x)$	7.5000	4.4600	4.2150	4.0000	3.8150

x	1.2	2
$f(x)$	3.6515	2.8284
$P_2(x)$	3.6600	3.5000

11. (a)

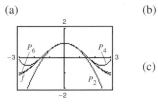

(b) $f^{(2)}(0) = -1$ $P_2^{(2)}(0) = -1$
$f^{(4)}(0) = 1$ $P_4^{(4)}(0) = 1$
$f^{(6)}(0) = -1$ $P_6^{(6)}(0) = -1$

(c) $f^{(n)}(0) = P_n^{(n)}(0)$

13. $1 - x + \frac{1}{2}x^2 - \frac{1}{6}x^3$ **15.** $1 + 2x + 2x^2 + \frac{4}{3}x^3 + \frac{2}{3}x^4$

17. $x - \frac{1}{6}x^3 + \frac{1}{120}x^5$ **19.** $x + x^2 + \frac{1}{2}x^3 + \frac{1}{6}x^4$

21. $1 - x + x^2 - x^3 + x^4$ **23.** $1 + \frac{1}{2}x^2$

25. $1 - (x - 1) + (x - 1)^2 - (x - 1)^3 + (x - 1)^4$

27. $1 + \frac{1}{2}(x - 1) - \frac{1}{8}(x - 1)^2 + \frac{1}{16}(x - 1)^3 - \frac{5}{128}(x - 1)^4$

29. $(x - 1) - \frac{1}{2}(x - 1)^2 + \frac{1}{3}(x - 1)^3 - \frac{1}{4}(x - 1)^4$

31. (a) $P_3(x) = x + \dfrac{1}{3}x^3$ (b) $P_5(x) = x + \dfrac{1}{3}x^3 + \dfrac{2}{15}x^5$

(c) $Q_3(x) = 1 + 2\left(x - \dfrac{\pi}{4}\right) + 2\left(x - \dfrac{\pi}{4}\right)^2 + \dfrac{8}{3}\left(x - \dfrac{\pi}{4}\right)^3$

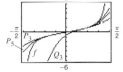

33. (a)

x	0	0.25	0.50	0.75	1.00
$\sin x$	0	0.2474	0.4794	0.6816	0.8415
$P_1(x)$	0	0.25	0.50	0.75	1.00
$P_3(x)$	0	0.2474	0.4792	0.6797	0.8333
$P_5(x)$	0	0.2474	0.4794	0.6817	0.8417
$P_7(x)$	0	0.2474	0.4794	0.6816	0.8415

(b)

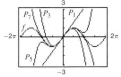

(c) As the distance increases, the polynomial approximation becomes less accurate.

35. (a) $P_3(x) = x + \frac{1}{6}x^3$

(b)

x	-0.75	-0.50	-0.25	0	0.25
$f(x)$	-0.848	-0.524	-0.253	0	0.253
$P_3(x)$	-0.820	-0.521	-0.253	0	0.253

x	0.50	0.75
$f(x)$	0.524	0.848
$P_3(x)$	0.521	0.820

(c)

37.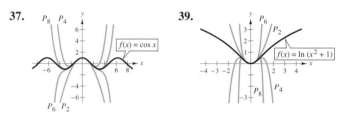

39.

41. 0.6042   **43.** 0.1823   **45.** $R_4 \le 2.03 \times 10^{-5}$

47. $R_3 \le 7.82 \times 10^{-3}$ **49.** 3   **51.** 9; 0.4055

53. $-0.3936 < x < 0$

55. The graph of the approximating polynomial P and the elementary function f both pass through the point $(c, f(c))$, and the slope of P is the same as the slope of the graph of f at the point $(c, f(c))$. If P is of degree n, then the first n derivatives of f and p agree at c. This allows for the graph of P to resemble the graph of f near the point $(c, f(c))$.

57. See "Definition of nth Taylor Polynomial and nth Maclaurin Polynomial" on page 607.

59. As the degree of the polynomial increases, the graph of the Taylor polynomial becomes a better and better approximation of the function within the interval of convergence. Therefore, the accuracy is increased.

61. (a) $f(x) \approx P_4(x) = 1 + x + \frac{1}{2}x^2 + \frac{1}{6}x^3 + \frac{1}{24}x^4$

$g(x) \approx Q_5(x) = x + x^2 + \frac{1}{2}x^3 + \frac{1}{6}x^4 + \frac{1}{24}x^5$

$Q_5(x) = xP_4(x)$

(b) $g(x) \approx P_6(x) = x^2 - \frac{x^4}{3!} + \frac{x^6}{5!}$

(c) $g(x) \approx P_4(x) = 1 - \frac{x^2}{3!} + \frac{x^4}{5!}$

63. (a) $Q_2(x) = -1 + \frac{\pi^2}{32}(x + 2)^2$

(b) $R_2(x) = -1 + \frac{\pi^2}{32}(x - 6)^2$

(c) No. Horizontal translations of the result in part (a) are possible only at $x = -2 + 8n$ (where n is an integer) because the period of f is 8.

65. Proof

67. As you move away from $x = c$, the Taylor polynomial becomes less and less accurate.

Section 8.8 (page 623)

1. 0   **3.** 2   **5.** $R = 1$ **7.** $R = \frac{1}{2}$ **9.** $R = \infty$

11. $(-2, 2)$ **13.** $(-1, 1]$ **15.** $(-\infty, \infty)$ **17.** $x = 0$

19. $(-4, 4)$ **21.** $(0, 10]$ **23.** $(0, 2]$ **25.** $(0, 2c)$

27. $\left(-\frac{1}{2}, \frac{1}{2}\right)$ **29.** $(-\infty, \infty)$ **31.** $(-1, 1)$ **33.** $x = 3$

35. (a) $(-2, 2)$ (b) $(-2, 2)$ (c) $(-2, 2)$ (d) $[-2, 2)$

37. (a) $(0, 2]$ (b) $(0, 2)$ (c) $(0, 2)$ (d) $[0, 2]$

39. c; $S_1 = 1$, $S_2 = 1.33$ **40.** a; $S_1 = 1$, $S_2 = 1.67$

41. b; diverges   **42.** d; alternating

43. A series of the form

$$\sum_{n=0}^{\infty} a_n(x - c)^n = a_0 + a_1(x - c) + a_2(x - c)^2 + \cdots$$
$$+ a_n(x - c)^n + \cdots$$

is called a power series centered at c, where c is a constant.

45. 1. A single point

2. An interval centered at c

3. The entire real line

47. (a) For $f(x)$: $(-\infty, \infty)$; For $g(x)$: $(-\infty, \infty)$

(b) $f'(x) = \sum_{n=0}^{\infty} \frac{(-1)^n(2n + 1)x^{2n}}{(2n + 1)!} = \sum_{n=0}^{\infty} \frac{(-1)^n x^{2n}}{(2n)!} = g(x)$

(c) $g'(x) = \sum_{n=1}^{\infty} \frac{(-1)^n 2n x^{2n-1}}{2n!} = -\sum_{n=0}^{\infty} \frac{(-1)^n x^{2n+1}}{(2n + 1)!} = -f(x)$

(d) $f(x) = \sin x$; $g(x) = \cos x$

49. $y'' - xy' - y = \sum_{n=1}^{\infty} \frac{2n(2n - 1)x^{2n-2}}{2^n n!} - \sum_{n=1}^{\infty} \frac{2nx^{2n}}{2^n n!} - \sum_{n=0}^{\infty} \frac{x^{2n}}{2^n n!}$

$= \sum_{n=0}^{\infty} \frac{2(n + 1)x^{2n}[(2n + 1) - (2n + 1)]}{2^{n+1}(n + 1)!} = 0$

51. (a) $\lim_{k \to \infty} \left| \frac{(-1)^{k+1}x^{2k+2}}{2^{2k+2}[(k + 1)!]^2} \cdot \frac{2^{2k}(k!)^2}{(-1)^k x^{2k}} \right| = \lim_{k \to \infty} \left| \frac{(-1)x^2}{4(k + 1)^2} \right| = 0$

The interval of convergence is $(-\infty, \infty)$.

(b) $x^2 J_0'' + x J_0' + x^2 J_0$

$= \sum_{k=0}^{\infty} (-1)^{k+1} \frac{2(2k + 1)x^{2k+2}}{4^{k+1}(k + 1)!k!} +$

$\sum_{k=0}^{\infty} (-1)^{k+1} \frac{2x^{2k+2}}{4^{k+1}(k + 1)!k!} + \sum_{k=0}^{\infty} (-1)^k \frac{x^{2k+2}}{4^k(k!)^2}$

$= \sum_{k=0}^{\infty} \frac{(-1)^k x^{2k+2}}{4^k(k!)^2} \left[\frac{-4k - 2}{4k + 4} - \frac{2}{4k + 4} + \frac{4k + 4}{4k + 4} \right] = 0$

(c)

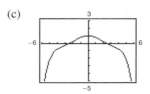

(d) 0.92

53. $f(x) = \cos x$

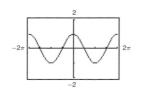

55. $f(x) = \dfrac{1}{1 + x}$

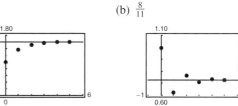

57. (a) $\frac{8}{5}$ (b) $\frac{8}{11}$

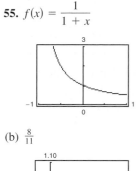

(c) The alternating series converges more rapidly. The partial sums of the series of positive terms approach the sum from below. The partial sums of the alternating series alternate sides of the horizontal line representing the sum.

(d)

M	10	100	1000	10,000
N	4	9	15	21

59. False. Let $a_n = \dfrac{(-1)^n}{n2^n}$.   **61.** True

Section 8.9 (page 630)

1. $\displaystyle\sum_{n=0}^{\infty} \frac{x^n}{2^{n+1}}$ **3.** $\displaystyle\sum_{n=0}^{\infty} \frac{(-1)^n x^n}{2^{n+1}}$

5. $\displaystyle\sum_{n=0}^{\infty} \frac{(x-5)^n}{(-3)^{n+1}}$ **7.** $-3\displaystyle\sum_{n=0}^{\infty} (2x)^n$

$(2, 8)$ $\left(-\frac{1}{2}, \frac{1}{2}\right)$

9. $-\dfrac{1}{11}\displaystyle\sum_{n=0}^{\infty} \left[\frac{2}{11}(x+3)\right]^n$ **11.** $\dfrac{3}{2}\displaystyle\sum_{n=0}^{\infty} \left(-\frac{x}{2}\right)^n$

$\left(-\frac{17}{2}, \frac{5}{2}\right)$ $(-2, 2)$

13. $\displaystyle\sum_{n=0}^{\infty} \left[\left(-\frac{1}{2}\right)^n - 1\right]x^n$ **15.** $\displaystyle\sum_{n=0}^{\infty} x^n[1 + (-1)^n] = 2\displaystyle\sum_{n=0}^{\infty} x^{2n}$

$(-1, 1)$ $(-1, 1)$

17. $2\displaystyle\sum_{n=0}^{\infty} x^{2n}$ **19.** $\displaystyle\sum_{n=1}^{\infty} n(-1)^n x^{n-1}$ **21.** $\displaystyle\sum_{n=0}^{\infty} \frac{(-1)^n x^{n+1}}{n+1}$

$(-1, 1)$ $(-1, 1)$ $(-1, 1]$

23. $\displaystyle\sum_{n=0}^{\infty} (-1)^n x^{2n}$ **25.** $\displaystyle\sum_{n=0}^{\infty} (-1)^n (2x)^{2n}$

$(-1, 1)$ $\left(-\frac{1}{2}, \frac{1}{2}\right)$

27.

x	0.0	0.2	0.4	0.6	0.8	1.0
S_2	0.000	0.180	0.320	0.420	0.480	0.500
$\ln(x+1)$	0.000	0.182	0.336	0.470	0.588	0.693
S_3	0.000	0.183	0.341	0.492	0.651	0.833

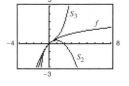

29. c   **30.** d   **31.** a   **32.** b

33. $f(x) = \arctan x$ is an odd function (symmetric to the origin).

35. 0.245   **37.** 0.125

39. (a) $\displaystyle\sum_{n=1}^{\infty} nx^{n-1}, \; -1 < x < 1$

(b) $\displaystyle\sum_{n=0}^{\infty} nx^n, \; -1 < x < 1$

(c) $\displaystyle\sum_{n=0}^{\infty} (2n+1)x^n, \; -1 < x < 1$

(d) $\displaystyle\sum_{n=0}^{\infty} (2n+1)x^{n+1}, \; -1 < x < 1$

41. $E(n) = 2$. Because the probability of obtaining a head on a single toss is $\frac{1}{2}$, it is expected that, on average, a head will be obtained in two tosses.

43. Since $\dfrac{1}{1+x} = \dfrac{1}{1-(-x)}$, substitute $(-x)$ into the geometric series.

45. Since $\dfrac{5}{1+x} = 5\left(\dfrac{1}{1-(-x)}\right)$, substitute $(-x)$ into the geometric series and then multiply the series by 5.

47. Proof   **49.** (a) Proof   (b) 3.14

51. $\ln\frac{3}{2} \approx 0.4055$; See Exercise 21.

53. $\ln\frac{7}{5} \approx 0.3365$; See Exercise 51.

55. $\arctan\frac{1}{2} \approx 0.4636$; See Exercise 54.

57. The series in Exercise 54 converges to its sum at a slower rate because its terms approach 0 at a much slower rate.

59. -0.6931

Section 8.10 (page 641)

1. $\displaystyle\sum_{n=0}^{\infty} \frac{(2x)^n}{n!}$ **3.** $\dfrac{\sqrt{2}}{2}\displaystyle\sum_{n=0}^{\infty} \frac{(-1)^{n(n+1)/2}}{n!}\left(x - \frac{\pi}{4}\right)^n$

5. $\displaystyle\sum_{n=0}^{\infty} \frac{(-1)^n (x-1)^{n+1}}{n+1}$ **7.** $\displaystyle\sum_{n=0}^{\infty} \frac{(-1)^n (2x)^{2n+1}}{(2n+1)!}$

9. $1 + \dfrac{x^2}{2!} + \dfrac{5x^4}{4!} + \cdots$ **11.** Proof

13. $\displaystyle\sum_{n=0}^{\infty} (-1)^n (n+1)x^n$

15. $\dfrac{1}{2}\left[1 + \displaystyle\sum_{n=1}^{\infty} \frac{(-1)^n 1 \cdot 3 \cdot 5 \cdots (2n-1)x^{2n}}{2^{3n}n!}\right]$

17. $1 + \dfrac{x^2}{2} + \displaystyle\sum_{n=2}^{\infty} \dfrac{(-1)^{n+1}1 \cdot 3 \cdot 5 \cdots (2n-3)x^{2n}}{2^n n!}$

19. $1 + \dfrac{x^2}{2} + \dfrac{x^4}{2^2 2!} + \dfrac{x^6}{2^3 3!} + \cdots$ **21.** $\displaystyle\sum_{n=0}^{\infty} \dfrac{(-1)^n(2x)^{2n+1}}{(2n+1)!}$

23. $\displaystyle\sum_{n=0}^{\infty} \dfrac{(-1)^n x^{3n}}{(2n)!}$ **25.** $\displaystyle\sum_{n=0}^{\infty} \dfrac{x^{2n+1}}{(2n+1)!}$

27. $\dfrac{1}{2}\left[1 + \displaystyle\sum_{n=0}^{\infty} \dfrac{(-1)^n(2x)^{2n}}{(2n)!}\right]$ **29.** $\displaystyle\sum_{n=0}^{\infty} \dfrac{(-1)^n x^{2n+2}}{(2n+1)!}$

31. $\begin{cases} \displaystyle\sum_{n=0}^{\infty} \dfrac{(-1)^n x^{2n}}{(2n+1)!}, & x \neq 0 \\ 1, & x = 0 \end{cases}$

33. Proof

35. $P_5(x) = x + x^2 + \frac{1}{3}x^3 - \frac{1}{30}x^5 + \cdots$

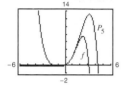

37. $P_5(x) = x - \frac{1}{2}x^2 - \frac{1}{6}x^3 + \frac{3}{40}x^5 + \cdots$

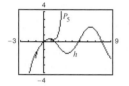

39. $P_4(x) = x - x^2 + \frac{5}{6}x^3 - \frac{5}{6}x^4 + \cdots$

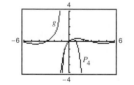

41. a; $y \approx x \sin x$ **42.** b; $y \approx x \cos x$ **43.** c; $y \approx xe^x$

44. d; $y \approx x^2\left(\dfrac{1}{x-1}\right)$ **45.** $\displaystyle\sum_{n=0}^{\infty} \dfrac{(-1)^{(n+1)}x^{2n+3}}{(2n+3)(n+1)!}$

47. 0.6931   **49.** 7.3891   **51.** 0   **53.** 0.9461

55. 0.7040   **57.** 0.2010   **59.** 0.3413

61. $P_5(x) = x - 2x^3 + \frac{2}{3}x^5$

$\left[-\frac{3}{4}, \frac{3}{4}\right]$

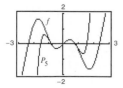

63. $P_5(x) = (x-1) - \frac{1}{24}(x-1)^3 + \frac{1}{24}(x-1)^4 - \frac{71}{1920}(x-1)^5$

$\left[\frac{1}{4}, 2\right]$

65. See "Guidelines for Finding a Taylor Series" on page 636.

67. (a) Replace x with $-x$ in the series for e^x.

   (b) Replace x with $3x$ in the series for e^x.

   (c) Multiply the series for e^x by x.

   (d) Replace x with $2x$ in the series for e^x. Then replace x with $-2x$ in the series for e^x. Then add the two together.

69. Proof

71. (a)   (b) Proof

         (c) $\displaystyle\sum_{n=0}^{\infty} 0x^n = 0 \neq f(x)$

73. Proof

Review Exercises for Chapter 8 (page 643)

1. $a_n = \dfrac{1}{n!}$ **3.** a   **4.** c   **5.** d   **6.** b

7.

Converges to 5

9. Converges to 0   **11.** Diverges

13. Converges to 0   **15.** Converges to 0

17. (a)

n	1	2	3	4
A_n	\$5062.50	\$5125.78	\$5189.85	\$5254.73

n	5	6	7	8
A_n	\$5320.41	\$5386.92	\$5454.25	\$5522.43

(b) \$8218.10

19. (a)

k	5	10	15	20	25
S_k	13.2	113.3	873.8	6648.5	50,500.3

(b)

21. (a)

k	5	10	15	20	25
S_k	0.4597	0.4597	0.4597	0.4597	0.4597

(b)

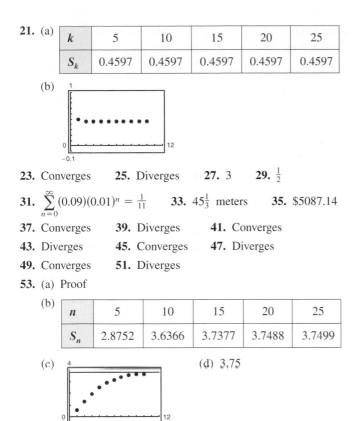

23. Converges   **25.** Diverges   **27.** 3   **29.** $\frac{1}{2}$

31. $\displaystyle\sum_{n=0}^{\infty}(0.09)(0.01)^n = \frac{1}{11}$ **33.** $45\frac{1}{3}$ meters   **35.** \$5087.14

37. Converges   **39.** Diverges   **41.** Converges

43. Diverges   **45.** Converges   **47.** Diverges

49. Converges   **51.** Diverges

53. (a) Proof

(b)

n	5	10	15	20	25
S_n	2.8752	3.6366	3.7377	3.7488	3.7499

(c)   (d) 3.75

55. (a)

N	5	10	20	30	40
$\displaystyle\sum_{n=1}^{N}\frac{1}{n^p}$	1.4636	1.5498	1.5962	1.6122	1.6202
$\displaystyle\int_{N}^{\infty}\frac{1}{x^p}\,dx$	0.2000	0.1000	0.0500	0.0333	0.0250

(b)

N	5	10	20	30	40
$\displaystyle\sum_{n=1}^{N}\frac{1}{n^p}$	1.0367	1.0369	1.0369	1.0369	1.0369
$\displaystyle\int_{N}^{\infty}\frac{1}{x^p}\,dx$	0.0004	0.0000	0.0000	0.0000	0.0000

The series in part (b) converges more rapidly. This is evident from the integrals that give the remainders of the partial sums.

57. $P_3(x) = 1 - \dfrac{x}{2} + \dfrac{x^2}{8} - \dfrac{x^3}{48}$ **59.** 0.996   **61.** 0.560

63. (a) 4   (b) 6   (c) 5   (d) 10

65. $(-10, 10)$ **67.** $[1, 3]$ **69.** converges only at $x = 2$

71. $x^2 y'' + xy' + x^2 y$

$\displaystyle = \sum_{n=0}^{\infty}\frac{(-1)^{n+1}(2n+2)(2n+1)x^{2n+2}}{4^{n+1}[(n+1)!]^2}$

$\displaystyle + \sum_{n=0}^{\infty}\frac{(-1)^{n+1}(2n+2)x^{2n+2}}{4^{n+1}[(n+1)!]^2} + \sum_{n=0}^{\infty}(-1)^n\frac{x^{2n+1}}{4^n(n!)^2} = 0$

73. $\displaystyle\sum_{n=0}^{\infty}\frac{2}{3}\left(\frac{x}{3}\right)^n$ **75.** $\displaystyle\sum_{n=0}^{\infty}\frac{2}{9}(n+1)\left(\frac{x}{3}\right)^n,\ -1 < x < 1$

77. $f(x) = \dfrac{3}{3 - 2x},\ \left(-\dfrac{3}{2}, \dfrac{3}{2}\right)$

79. $\dfrac{\sqrt{2}}{2}\displaystyle\sum_{n=0}^{\infty}\frac{(-1)^{n(n+1)/2}}{n!}\left(x - \dfrac{3\pi}{4}\right)^n$

81. $\displaystyle\sum_{n=0}^{\infty}\frac{(x\ln 3)^n}{n!}$ **83.** $-\displaystyle\sum_{n=0}^{\infty}(x+1)^n$

85. $1 + \dfrac{x}{5} - \dfrac{2x^2}{25} + \dfrac{6x^3}{125} - \dfrac{21x^4}{625} + \cdots$

87. $\ln\dfrac{5}{4} \approx 0.2231$ **89.** $e^{1/2} \approx 1.6487$

91. $\cos\dfrac{2}{3} \approx 0.7859$

93. The series for Exercise 41 converges to its sum at a slower rate because its terms approach 0 at a slower rate.

95. $1 + 2x + 2x^2 + \dfrac{4}{3}x^3$ **97.** $\displaystyle\sum_{n=0}^{\infty}\frac{(-1)^n x^{2n+1}}{(2n+1)(2n+1)!}$

99. $\displaystyle\sum_{n=0}^{\infty}\frac{(-1)^n x^{n+1}}{(n+1)^2}$ **101.** 0

P.S. Problem Solving (page 646)

1. (a) 1   (b) Answers will vary. Example: $0, \frac{1}{3}, \frac{2}{3}$ (c) 0

3. $\dfrac{\pi}{8}$

5. (a) $R = 1$; Sum $= \dfrac{3x^2 + 2x + 1}{1 - x^3}$

(b) $R = 1$; Sum $= \dfrac{a_{p-1}x^{p-1} + a_{p-2}x^{p-2} + \cdots + a_1 x + a_0}{1 - x^p}$

7. $\displaystyle\sum_{n=0}^{\infty}\frac{x^{n+1}}{n!}$; $\displaystyle\sum_{n=1}^{\infty}\frac{1}{n!(n+2)} - \dfrac{1}{2}$

9. Let $a_1 = \displaystyle\int_{0}^{\pi}\frac{\sin x}{x}\,dx = 1.8519$

$a_2 = \displaystyle\int_{\pi}^{2\pi}\frac{\sin x}{x}\,dx = -0.4338$

$a_3 = \displaystyle\int_{2\pi}^{3\pi}\frac{\sin x}{x}\,dx = 0.2566$

$a_4 = \displaystyle\int_{3\pi}^{4\pi}\frac{\sin x}{x}\,dx = -0.1826.$

It follows that the total area is

$\displaystyle\int_{0}^{\infty}\frac{\sin x}{x}\,dx = a_1 - a_2 + a_3 - a_4 + \cdots.$

Also, $\displaystyle\lim_{n\to\infty} a_n = 0$ and $0 < a_{n+1} \le a_n$. Therefore, it follows by the Alternating Series Test that $\int_{0}^{\infty} f(x)\,dx$ converges.

11. (a) $a_1 = 3,\ a_2 = 1.7321,\ a_3 = 2.1753,\ a_4 = 2.2749,$

$a_5 = 2.2967,\ a_6 = 2.3015$

Proof; $L = \dfrac{1 + \sqrt{13}}{2}$

(b) Proof; $L = \dfrac{1 + \sqrt{1 + 4a}}{2}$

13. (a) $1, \frac{9}{8}, \frac{11}{8}, \frac{45}{32}, \frac{47}{32}$

(b) $\displaystyle\lim_{n\to\infty}\left|\frac{a_{n+1}}{a_n}\right| = \lim_{n\to\infty}\left|\frac{\dfrac{1}{2^{(n+1)+(-1)^{n+1}}}}{\dfrac{1}{2^{n+(-1)^n}}}\right|$

$\displaystyle = \lim_{n\to\infty}\left|\frac{2^{n+(-1)^n}}{2^{(n+1)+(-1)^{n+1}}}\right|$; Does not exist

Therefore, the Ratio Test is inconclusive.

(c) $\displaystyle\lim_{n\to\infty}\sqrt[n]{|a_n|} = \lim_{n\to\infty}\sqrt[n]{\frac{1}{2^{n+(-1)^n}}} = \frac{1}{2}$. Therefore, by the Root Test, this series converges.

15. $S_6 = 240$; $S_7 = 440$; $S_8 = 810$; $S_9 = 1490$; $S_{10} = 2740$

Chapter 9

Section 9.1 (page 660)

1. h   **2.** a   **3.** e   **4.** b

5. f   **6.** g   **7.** c   **8.** d

9. Vertex: $(0, 0)$

Focus: $\left(-\frac{3}{2}, 0\right)$

Directrix: $x = \frac{3}{2}$

11. Vertex: $(-3, 2)$

Focus: $\left(-\frac{13}{4}, 2\right)$

Directrix: $x = -\frac{11}{4}$

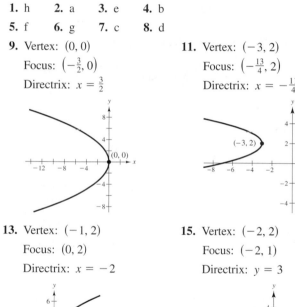

13. Vertex: $(-1, 2)$

Focus: $(0, 2)$

Directrix: $x = -2$

15. Vertex: $(-2, 2)$

Focus: $(-2, 1)$

Directrix: $y = 3$

17. Vertex; $\left(\frac{1}{4}, -\frac{1}{2}\right)$

Focus: $\left(0, -\frac{1}{2}\right)$

Directrix: $x = \frac{1}{2}$

19. Vertex: $(-1, 0)$

Focus: $(0, 0)$

Directrix: $x = -2$

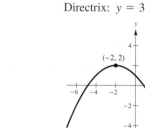

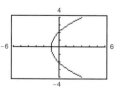

21. $y^2 - 4y + 8x - 20 = 0$ **23.** $x^2 - 24y + 96 = 0$

25. $x^2 + y - 4 = 0$ **27.** $5x^2 - 14x - 3y + 9 = 0$

29. Center: $(0, 0)$

Foci: $(\pm\sqrt{3}, 0)$

Vertices: $(\pm 2, 0)$

$e = \dfrac{\sqrt{3}}{2}$

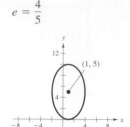

31. Center: $(1, 5)$

Foci: $(1, 9), (1, 1)$

Vertices: $(1, 10), (1, 0)$

$e = \dfrac{4}{5}$

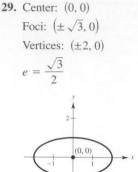

33. Center: $(-2, 3)$

Foci: $\left(-2, 3 \pm \sqrt{5}\right)$

Vertices: $(-2, 6), (-2, 0)$

$e = \dfrac{\sqrt{5}}{3}$

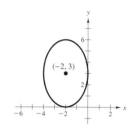

35. Center: $\left(\frac{1}{2}, -1\right)$

Foci: $\left(\frac{1}{2} \pm \sqrt{2}, -1\right)$

Vertices: $\left(\frac{1}{2} \pm \sqrt{5}, -1\right)$

To obtain the graph, solve for y and get

$y_1 = -1 + \sqrt{\dfrac{57 + 12x - 12x^2}{20}}$ and

$y_2 = -1 - \sqrt{\dfrac{57 + 12x - 12x^2}{20}}.$

Graph these equations in the same viewing window.

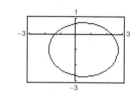

37. Center: $\left(\frac{3}{2}, -1\right)$

Foci: $\left(\frac{3}{2} - \sqrt{2}, -1\right), \left(\frac{3}{2} + \sqrt{2}, -1\right)$

Vertices: $\left(-\frac{1}{2}, -1\right), \left(\frac{7}{2}, -1\right)$

To obtain the graph, solve for y and get

$y_1 = -1 + \sqrt{\dfrac{7 + 12x - 4x^2}{8}}$ and

$y_2 = -1 - \sqrt{\dfrac{7 + 12x - 4x^2}{8}}.$

Graph these equations in the same viewing window.

39. $\dfrac{x^2}{9} + \dfrac{y^2}{5} = 1$ **41.** $\dfrac{(x-3)^2}{9} + \dfrac{(y-5)^2}{16} = 1$

43. $\dfrac{x^2}{16} + \dfrac{7y^2}{16} = 1$

45. Center: $(0, 0)$

Foci: $\left(0, \pm\sqrt{5}\right)$

Vertices: $(0, \pm 1)$

47. Center: $(1, -2)$

Foci: $\left(1 \pm \sqrt{5}, -2\right)$

Vertices: $(-1, -2), (3, -2)$

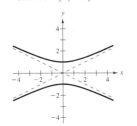

49. Center: $(2, -3)$

Foci: $\left(2 \pm \sqrt{10}, -3\right)$

Vertices: $(1, -3), (3, -3)$

51. Degenerate hyperbola

Graph is two lines

$y = -3 \pm \frac{1}{3}(x + 1)$

intersecting at $(-1, -3)$.

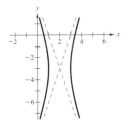

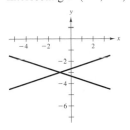

53. Center: $(1, -3)$

Foci: $\left(1, -3 \pm 2\sqrt{5}\right)$

Vertices: $\left(1, -3 \pm \sqrt{2}\right)$

55. Center: $(1, -3)$

Foci: $\left(1 \pm \sqrt{10}, -3\right)$

Vertices: $(-1, -3), (3, -3)$

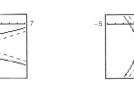

57. $\dfrac{x^2}{1} - \dfrac{y^2}{9} = 1$ **59.** $\dfrac{y^2}{9} - \dfrac{(x-2)^2}{9/4} = 1$

61. $\dfrac{y^2}{4} - \dfrac{x^2}{12} = 1$ **63.** $\dfrac{(x-3)^2}{9} - \dfrac{(y-2)^2}{4} = 1$

65. (a) $\left(6, \sqrt{3}\right)$: $2x - 3\sqrt{3}y - 3 = 0$

$\left(6, -\sqrt{3}\right)$: $2x + 3\sqrt{3}y - 3 = 0$

(b) $\left(6, \sqrt{3}\right)$: $9x + 2\sqrt{3}y - 60 = 0$

$\left(6, -\sqrt{3}\right)$: $9x - 2\sqrt{3}y - 60 = 0$

67. Ellipse **69.** Parabola **71.** Circle

73. Circle **75.** Hyperbola

77. (a) A parabola is the set of all points (x, y) that are equidistant from a fixed line and a fixed point not on the line.

(b) For directrix $y = k - p$: $(x - h)^2 = 4p(y - k)$

For directrix $x = h - p$: $(y - k)^2 = 4p(x - h)$

(c) If P is a point on a parabola, then the tangent line to the parabola at P makes equal angles with the line passing through P and the focus, and with the line passing through P parallel to the axis of the parabola.

79. (a) A hyperbola is the set of all points (x, y) for which the absolute value of the difference between the distances from two distinct fixed points is constant.

(b) Transverse axis is horizontal: $\dfrac{(x - h)^2}{a^2} - \dfrac{(y - k)^2}{b^2} = 1$

Transverse axis is vertical: $\dfrac{(y - k)^2}{a^2} - \dfrac{(x - h)^2}{b^2} = 1$

(c) Transverse axis is horizontal:

$y = k + \dfrac{b}{a}(x - h)$ and $y = k - \dfrac{b}{a}(x - h)$

Transverse axis is vertical:

$y = k + \dfrac{a}{b}(x - h)$ and $y = k - \dfrac{a}{b}(x - h)$

81. $\frac{9}{4}$ meters **83.** $y = 2ax_0x - ax_0^2$

85. (a) Proof (b) Proof

87. $x_0 = \dfrac{2\sqrt{3}}{3}$; Distance from hill: $\dfrac{2\sqrt{3}}{3} - 1$

89. $\dfrac{16\left(4 + 3\sqrt{3} - 2\pi\right)}{3} \approx 15.536$ square feet

91. (a) $y = \dfrac{1}{180}x^2$

(b) $10\left[2\sqrt{13} + 9\ln\left(\dfrac{2 + \sqrt{13}}{3}\right)\right] \approx 128.4$ meters

93.

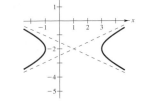

As p increases, the graph of $x^2 = 4py$ gets wider.

95.

97. The tacks should be placed 1.5 feet from the center. The string should be $2a = 5$ feet long.

99. $e = \dfrac{c}{a}$

$A + P = 2a$

$a = \dfrac{A + P}{2}$

$c = a - P = \dfrac{A + P}{2} - P = \dfrac{A - P}{2}$

$e = \dfrac{c}{a} = \dfrac{\dfrac{(A - P)}{2}}{\dfrac{(A + P)}{2}} = \dfrac{A - P}{A + P}$

101. $e \approx 0.9672$ **103.** $\left(0, \frac{25}{3}\right)$

105. Minor-axis endpoints: $(-6, -2), (0, -2)$

Major-axis endpoints: $(-3, -6), (-3, 2)$

107. (a) Area $= 2\pi$

     (b) Volume $= \dfrac{8\pi}{3}$

        Surface area $= \dfrac{2\pi\left(9 + 4\sqrt{3}\pi\right)}{9} \approx 21.48$

     (c) Volume $= \dfrac{16\pi}{3}$

        Surface area $= \dfrac{4\pi\left[6 + \sqrt{3}\ln\left(2 + \sqrt{3}\right)\right]}{3} \approx 34.69$

109. 37.96   **111.** 40   **113.** $\dfrac{(x-6)^2}{9} - \dfrac{(y-2)^2}{7} = 1$

115.

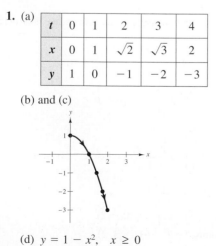

117. Proof

119. $x = \dfrac{-90 + 96\sqrt{2}}{7} \approx 6.538$

      $y = \dfrac{160 - 96\sqrt{2}}{7} \approx 3.462$

121. There are four points of intersection.

     At $\left(\dfrac{\sqrt{2}\,ac}{\sqrt{2a^2 - b^2}}, \dfrac{b^2}{\sqrt{2}\sqrt{2a^2 - b^2}}\right)$, the slopes of the tangent

     lines are $y'_e = -\dfrac{c}{a}$ and $y'_h = \dfrac{a}{c}$.

     Since the slopes are negative reciprocals, the tangent lines are perpendicular. Similarly, the curves are perpendicular at the other three points of intersection.

123. False. See the definition of a parabola.   **125.** True

127. False. $y^2 - x^2 + 2x + 2y = 0$ yields two intersecting lines.

129. True

Section 9.2 (page 672)

1. (a)

t	0	1	2	3	4
x	0	1	$\sqrt{2}$	$\sqrt{3}$	2
y	1	0	-1	-2	-3

     (b) and (c)

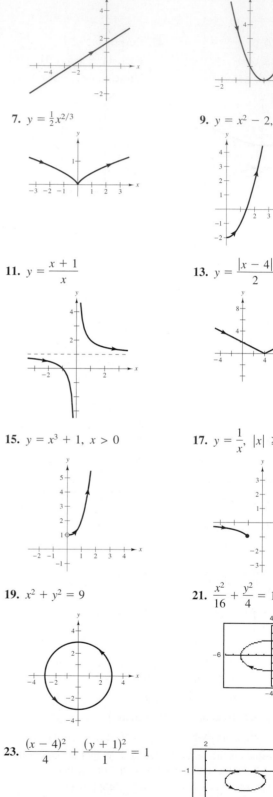

     (d) $y = 1 - x^2, \quad x \geq 0$

3. $2x - 3y + 5 = 0$

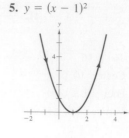

5. $y = (x - 1)^2$

7. $y = \frac{1}{2}x^{2/3}$

9. $y = x^2 - 2, \ x \geq 0$

11. $y = \dfrac{x + 1}{x}$

13. $y = \dfrac{|x - 4|}{2}$

15. $y = x^3 + 1, \ x > 0$

17. $y = \dfrac{1}{x}, \ |x| \geq 1$

19. $x^2 + y^2 = 9$

21. $\dfrac{x^2}{16} + \dfrac{y^2}{4} = 1$

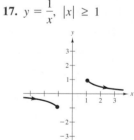

23. $\dfrac{(x - 4)^2}{4} + \dfrac{(y + 1)^2}{1} = 1$

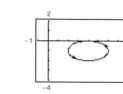

25. $\dfrac{(x-4)^2}{4} + \dfrac{(y+1)^2}{16} = 1$ **27.** $\dfrac{x^2}{16} - \dfrac{y^2}{9} = 1$

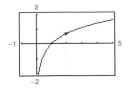   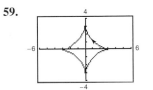

29. $y = \ln x$ **31.** $y = \dfrac{1}{x^3}, \quad x > 0$

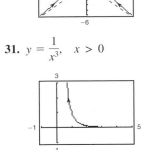

33. Each curve represents a portion of the line $y = 2x + 1$.

	Domain	Orientation	Smooth
(a)	$-\infty < x < \infty$	Up	Yes
(b)	$-1 \le x \le 1$	Oscillates	No, $\dfrac{dx}{d\theta} = \dfrac{dy}{d\theta} = 0$ when $\theta = 0, \pm\pi, \pm 2\pi, \ldots$
(c)	$0 < x < \infty$	Down	Yes
(d)	$0 < x < \infty$	Up	Yes

35. (a) and (b) represent the parabola $y = 2(1 - x^2)$ for $-1 \le x \le 1$. The curve is smooth. The orientation is from right to left in part (a) and in part (b).

37. (a)

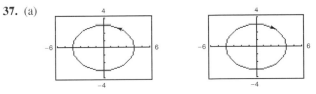

(b) The orientation is reversed.

(c) The orientation is reversed.

(d) Answers will vary. For example,

$x = 2 \sec t \qquad x = 2 \sec(-t)$

$y = 5 \sin t \qquad y = 5 \sin(-t)$

have the same graphs, but their orientation is reversed.

39. $y - y_1 = \dfrac{y_2 - y_1}{x_2 - x_1}(x - x_1)$ **41.** $\dfrac{(x-h)^2}{a^2} + \dfrac{(y-k)^2}{b^2} = 1$

43. $x = 5t$
$y = -2t$
(Solution is not unique.)

45. $x = 2 + 4\cos\theta$
$y = 1 + 4\sin\theta$
(Solution is not unique.)

47. $x = 5\cos\theta$
$y = 3\sin\theta$
(Solution is not unique.)

49. $x = 4\sec\theta$
$y = 3\tan\theta$
(Solution is not unique.)

51. $x = t$
$y = 3t - 2;$
$x = t - 3$
$y = 3t - 11$
(Solution is not unique.)

53. $x = t$
$y = t^3;$
$x = \tan t$
$y = \tan^3 t$
(Solution is not unique.)

55. 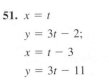   **57.**

Not smooth when $\theta = 2n\pi$

59. 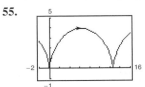   **61.**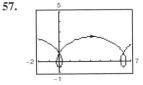

Not smooth when $\theta = \frac{1}{2}n\pi$

63. See page 665.   **65.** See page 670.

67. d   **68.** a   **69.** b   **70.** c

71. $x = a\theta - b\sin\theta; \; y = a - b\cos\theta$

73. False. The graph of the parametric equations is the portion of the line $y = x$ when $x \ge 0$.

75. (a) $x = \left(\dfrac{440}{3}\cos\theta\right)t; \; y = 3 + \left(\dfrac{440}{3}\sin\theta\right)t - 16t^2$

(b)   (c)

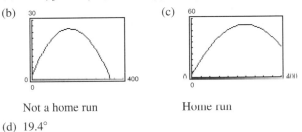

Not a home run   Home run

(d) $19.4°$

Section 9.3 (page 681)

1. $-\dfrac{2}{t}$ **3.** -1

5. $\dfrac{dy}{dx} = \dfrac{3}{2}, \dfrac{d^2y}{dx^2} = 0$; neither concave upward nor concave downward

7. $\dfrac{dy}{dx} = 2t + 3, \dfrac{d^2y}{dx^2} = 2$

At $t = -1, \dfrac{dy}{dx} = 1, \dfrac{d^2y}{dx^2} = 2$; concave upward

9. $\dfrac{dy}{dx} = -\cot\theta, \dfrac{d^2y}{dx^2} = -\dfrac{\csc^3\theta}{2}$

At $\theta = \dfrac{\pi}{4}, \dfrac{dy}{dx} = -1, \dfrac{d^2y}{dx^2} = -\sqrt{2}$; concave downward

11. $\dfrac{dy}{dx} = 2\csc\theta, \dfrac{d^2y}{dx^2} = -2\cot^3\theta$

At $\theta = \dfrac{\pi}{6}, \dfrac{dy}{dx} = 4, \dfrac{d^2y}{dx^2} = -6\sqrt{3}$; concave downward

13. $\dfrac{dy}{dx} = -\tan\theta, \quad \dfrac{d^2y}{dx^2} = \dfrac{\sec^4\theta\csc\theta}{3}$

At $\theta = \dfrac{\pi}{4}, \dfrac{dy}{dx} = -1, \dfrac{d^2y}{dx^2} = \dfrac{4\sqrt{2}}{3}$; concave upward

15. $\left(-\dfrac{2}{\sqrt{3}}, \dfrac{3}{2}\right)$: $3\sqrt{3}x - 8y + 18 = 0$

$(0, 2)$: $y - 2 = 0$

$\left(2\sqrt{3}, \dfrac{1}{2}\right)$: $\sqrt{3}x + 8y - 10 = 0$

17. (a) and (d)

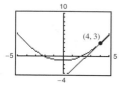

(b) At $t = 2$, $\dfrac{dx}{dt} = 2$, $\dfrac{dy}{dt} = 4$, and $\dfrac{dy}{dx} = 2$.

(c) $y = 2x - 5$

19. (a) and (d)

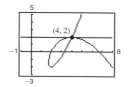

(b) At $t = -1$, $\dfrac{dx}{dt} = -3$, $\dfrac{dy}{dt} = 0$, and $\dfrac{dy}{dx} = 0$.

(c) $y = 2$

21. $y = \pm\dfrac{3}{4}x$

23. Horizontal: $(1, 0), (-1, \pi), (1, -2\pi)$

Vertical: $\left(\dfrac{\pi}{2}, 1\right), \left(-\dfrac{3\pi}{2}, -1\right), \left(\dfrac{5\pi}{2}, 1\right)$

25. Horizontal: $(1, 0)$

Vertical: none

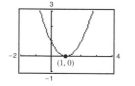

27. Horizontal: $(0, -2), (2, 2)$

Vertical: none

29. Horizontal: $(0, 3), (0, -3)$

Vertical: $(3, 0), (-3, 0)$

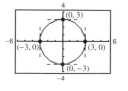

31. Horizontal: $(4, 0), (4, -2)$

Vertical: $(2, -1), (6, -1)$

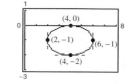

33. Horizontal: none

Vertical: $(1, 0), (-1, 0)$

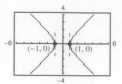

35. $2\sqrt{5} + \ln(2 + \sqrt{5}) \approx 5.916$ **37.** $\sqrt{2}(1 - e^{-\pi/2}) \approx 1.12$

39. $\dfrac{1}{12}\left[\ln(\sqrt{37} + 6) + 6\sqrt{37}\right] \approx 3.249$ **41.** $6a$ **43.** $8a$

45. (a)

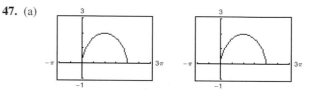

(b) 219.2 feet

(c) 230.8 feet

(d) The range is maximized when $\theta = 45°$; the arc length is maximized when $\theta = 90°$.

47. (a)

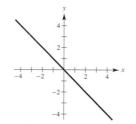

(b) The average speed of the particle on the second path is twice the average speed of the particle on the first path.

(c) 4π

49. (a) $32\pi\sqrt{5}$ (b) $16\pi\sqrt{5}$ **51.** 32π **53.** $\dfrac{12\pi a^2}{5}$

55. See Theorem 9.7, Parametric Form of the Derivative, on page 675.

57. Answers will vary. Example:

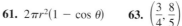

59. See Theorem 9.8, Arc Length in Parametric Form, on page 678.

61. $2\pi r^2(1 - \cos\theta)$ **63.** $\left(\dfrac{3}{4}, \dfrac{8}{5}\right)$ **65.** 36π **67.** $\dfrac{3\pi}{2}$

69. d   **70.** b   **71.** f   **72.** c   **73.** a   **74.** e

75.

(a) Circle of radius 1 and center at $(0, 0)$ except the point $(-1, 0)$

(b) As t increases from -20 to 0, the speed increases, and as t increases from 0 to 20, the speed decreases.

77. False: $\dfrac{d^2y}{dx^2} = \dfrac{\dfrac{d}{dt}\left[\dfrac{g'(t)}{f'(t)}\right]}{f'(t)} = \dfrac{f'(t)g''(t) - g'(t)f''(t)}{[f'(t)]^3}$.

Section 9.4 (page 691)

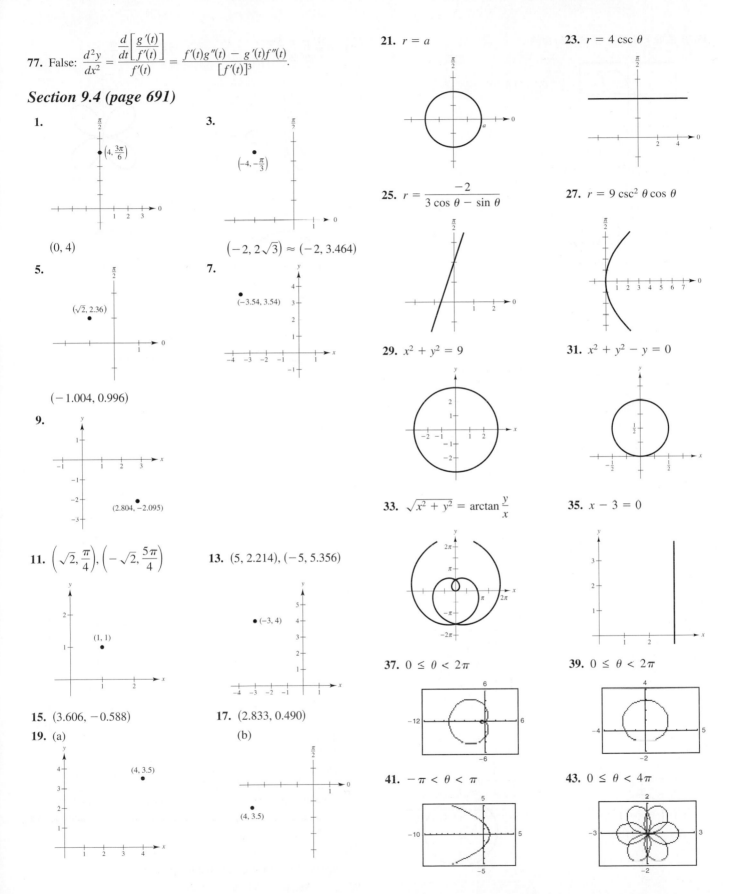

1.

$(0, 4)$

3.

$(-2, 2\sqrt{3}) \approx (-2, 3.464)$

5.

$(-1.004, 0.996)$

7.

9.

11. $\left(\sqrt{2}, \dfrac{\pi}{4}\right), \left(-\sqrt{2}, \dfrac{5\pi}{4}\right)$

13. $(5, 2.214), (-5, 5.356)$

15. $(3.606, -0.588)$

17. $(2.833, 0.490)$

19. (a)   (b)

21. $r = a$

23. $r = 4 \csc \theta$

25. $r = \dfrac{-2}{3 \cos \theta - \sin \theta}$

27. $r = 9 \csc^2 \theta \cos \theta$

29. $x^2 + y^2 = 9$

31. $x^2 + y^2 - y = 0$

33. $\sqrt{x^2 + y^2} = \arctan \dfrac{y}{x}$

35. $x - 3 = 0$

37. $0 \le \theta < 2\pi$

39. $0 \le \theta < 2\pi$

41. $-\pi < \theta < \pi$

43. $0 \le \theta < 4\pi$

45. $0 \le \theta < \pi/2$

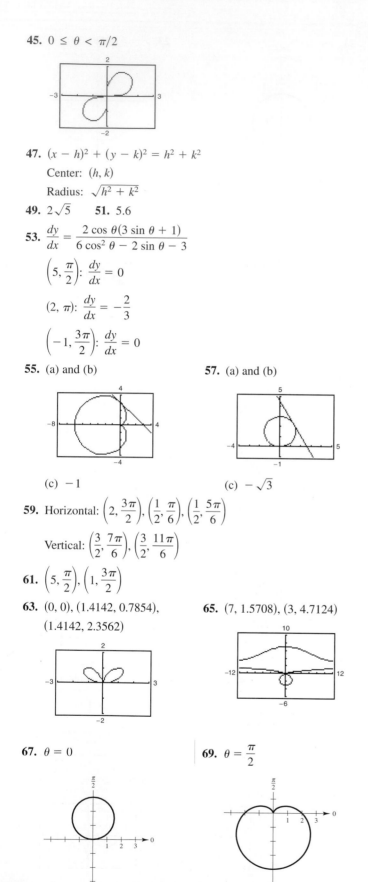

47. $(x - h)^2 + (y - k)^2 = h^2 + k^2$

Center: (h, k)

Radius: $\sqrt{h^2 + k^2}$

49. $2\sqrt{5}$ **51.** 5.6

53. $\dfrac{dy}{dx} = \dfrac{2 \cos \theta(3 \sin \theta + 1)}{6 \cos^2 \theta - 2 \sin \theta - 3}$

$\left(5, \dfrac{\pi}{2}\right)$: $\dfrac{dy}{dx} = 0$

$(2, \pi)$: $\dfrac{dy}{dx} = -\dfrac{2}{3}$

$\left(-1, \dfrac{3\pi}{2}\right)$: $\dfrac{dy}{dx} = 0$

55. (a) and (b)   **57.** (a) and (b)

(c) -1 (c) $-\sqrt{3}$

59. Horizontal: $\left(2, \dfrac{3\pi}{2}\right), \left(\dfrac{1}{2}, \dfrac{\pi}{6}\right), \left(\dfrac{1}{2}, \dfrac{5\pi}{6}\right)$

Vertical: $\left(\dfrac{3}{2}, \dfrac{7\pi}{6}\right), \left(\dfrac{3}{2}, \dfrac{11\pi}{6}\right)$

61. $\left(5, \dfrac{\pi}{2}\right), \left(1, \dfrac{3\pi}{2}\right)$

63. $(0, 0), (1.4142, 0.7854),$
$(1.4142, 2.3562)$

65. $(7, 1.5708), (3, 4.7124)$

67. $\theta = 0$ **69.** $\theta = \dfrac{\pi}{2}$

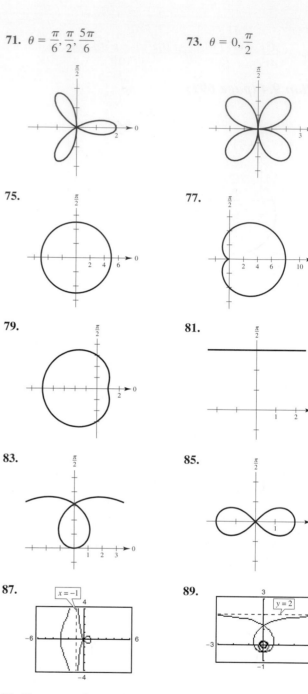

71. $\theta = \dfrac{\pi}{6}, \dfrac{\pi}{2}, \dfrac{5\pi}{6}$ **73.** $\theta = 0, \dfrac{\pi}{2}$

75.   **77.**

79.   **81.**

83.   **85.**

87.   **89.**

91. The rectangular coordinate system is a collection of points of the form (x, y), where x is the directed distance from the y-axis to the point and y is the directed distance from the x-axis to the point. Every point has a unique representation.

The polar coordinate system is a collection of points of the form (r, θ), where r is the directed distance from the origin O to a point P and θ is the directed angle, measured counterclockwise, from the polar axis to the segment $\overline{OP}$. Polar coordinates do not have unique representations.

93. $r = a$: Circle of radius a centered at the pole

$\theta = b$: Line passing through the pole

95. c   **96.** b   **97.** a   **98.** d

99. (a) 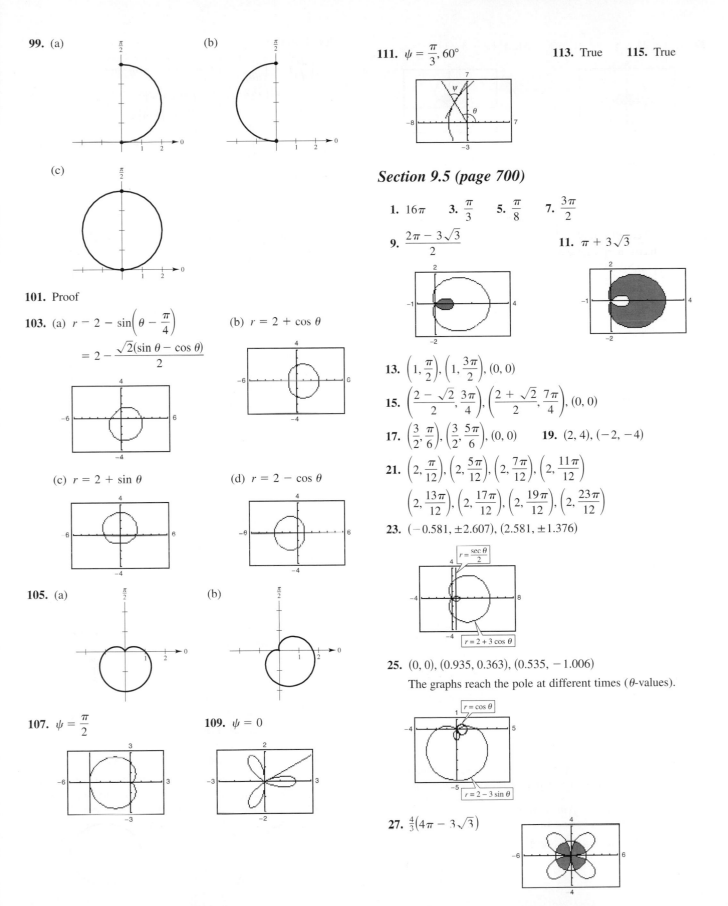 (b)

(c)

101. Proof

103. (a) $r - 2 - \sin\left(\theta - \dfrac{\pi}{4}\right)$ (b) $r = 2 + \cos\theta$

$\qquad = 2 - \dfrac{\sqrt{2}(\sin\theta - \cos\theta)}{2}$

(c) $r = 2 + \sin\theta$ (d) $r = 2 - \cos\theta$

105. (a) (b)

107. $\psi = \dfrac{\pi}{2}$ **109.** $\psi = 0$

111. $\psi = \dfrac{\pi}{3}, 60°$ **113.** True **115.** True

Section 9.5 (page 700)

1. 16π **3.** $\dfrac{\pi}{3}$ **5.** $\dfrac{\pi}{8}$ **7.** $\dfrac{3\pi}{2}$

9. $\dfrac{2\pi - 3\sqrt{3}}{2}$ **11.** $\pi + 3\sqrt{3}$

13. $\left(1, \dfrac{\pi}{2}\right), \left(1, \dfrac{3\pi}{2}\right), (0, 0)$

15. $\left(\dfrac{2 - \sqrt{2}}{2}, \dfrac{3\pi}{4}\right), \left(\dfrac{2 + \sqrt{2}}{2}, \dfrac{7\pi}{4}\right), (0, 0)$

17. $\left(\dfrac{3}{2}, \dfrac{\pi}{6}\right), \left(\dfrac{3}{2}, \dfrac{5\pi}{6}\right), (0, 0)$ **19.** $(2, 4), (-2, -4)$

21. $\left(2, \dfrac{\pi}{12}\right), \left(2, \dfrac{5\pi}{12}\right), \left(2, \dfrac{7\pi}{12}\right), \left(2, \dfrac{11\pi}{12}\right)$

$\left(2, \dfrac{13\pi}{12}\right), \left(2, \dfrac{17\pi}{12}\right), \left(2, \dfrac{19\pi}{12}\right), \left(2, \dfrac{23\pi}{12}\right)$

23. $(-0.581, \pm 2.607), (2.581, \pm 1.376)$

25. $(0, 0), (0.935, 0.363), (0.535, -1.006)$

The graphs reach the pole at different times (θ-values).

27. $\frac{4}{3}\left(4\pi - 3\sqrt{3}\right)$

29. $11\pi - 24$ **31.** $\frac{2}{3}\left(4\pi - 3\sqrt{3}\right)$

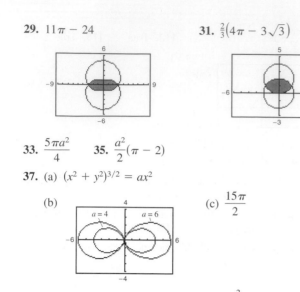

33. $\dfrac{5\pi a^2}{4}$ **35.** $\dfrac{a^2}{2}(\pi - 2)$

37. (a) $(x^2 + y^2)^{3/2} = ax^2$

(b) 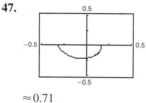   (c) $\dfrac{15\pi}{2}$

39. The area enclosed by the function is $\dfrac{\pi a^2}{4}$ if n is odd and is $\dfrac{\pi a^2}{2}$ if n is even.

41. $2\pi a$ **43.** 8

45.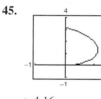

≈ 4.16

47.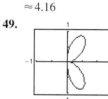

≈ 0.71

49.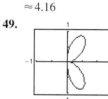

≈ 4.39

51. 36π **53.** $\dfrac{2\pi\sqrt{1 + a^2}}{1 + 4a^2}(e^{\pi a} - 2a)$ **55.** 21.87

57. Area $= \dfrac{1}{2}\displaystyle\int_\alpha^\beta r^2\, d\theta$; Arc length $= \displaystyle\int_\alpha^\beta \sqrt{r^2 + \left(\dfrac{dr}{d\theta}\right)^2}\, d\theta$

59. The integral (a) yields the correct arc length.

61. $4\pi^2 ab$

63. False. The graphs of $f(\theta) = 1$ and $g(\theta) = -1$ coincide.

65. In parametric form,
$$s = \int_a^b \sqrt{\left(\frac{dx}{dt}\right)^2 + \left(\frac{dy}{dt}\right)^2}\, dt.$$
Using θ instead of t gives $x = r\cos\theta$ and $y = r\sin\theta$. Let $r = f(\theta)$. Now we have $x = f(\theta)\cos\theta$ and $y = f(\theta)\sin\theta$.

So, $\dfrac{dx}{d\theta} = f'(\theta)\cos\theta - f(\theta)\sin\theta$ and

$\dfrac{dy}{d\theta} = f'(\theta)\sin\theta + f(\theta)\cos\theta.$

(continued)

It follows that
$$\left(\frac{dx}{d\theta}\right)^2 + \left(\frac{dy}{d\theta}\right)^2 = [f'(\theta)\cos\theta - f(\theta)\sin\theta]^2$$
$$+ [f'(\theta)\sin\theta + f(\theta)\cos\theta]^2$$
$$= [f(\theta)]^2 + [f'(\theta)]^2.$$
Therefore, $s = \displaystyle\int_\alpha^\beta \sqrt{[f(\theta)]^2 + [f'(\theta)]^2}\, d\theta.$

Section 9.6 (page 707)

1.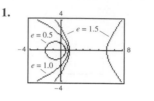

(a) Parabola

(b) Ellipse

(c) Hyperbola

3.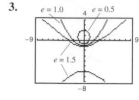

(a) Parabola

(b) Ellipse

(c) Hyperbola

5. (a) Ellipse

As $e \to 1^-$, the ellipse becomes more elliptical, and as $e \to 0^+$, it becomes more circular.

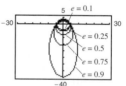

(b) Parabola

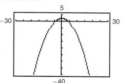

(c) Hyperbola

As $e \to 1^+$, the hyperbola opens more slowly, and as $e \to \infty$, it opens more rapidly.

7. c   **8.** f   **9.** a   **10.** e   **11.** b   **12.** d

13. Parabola   **15.** Ellipse

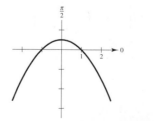   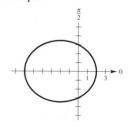

17. Ellipse

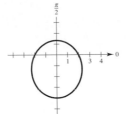

19. Hyperbola

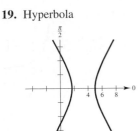

21. Hyperbola

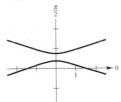

23. Ellipse

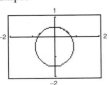

25. Parabola

27. Rotated $\pi/4$ radians
counterclockwise

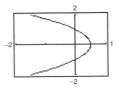

29. Rotated $\pi/6$ radians
clockwise

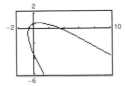

31. $r = \dfrac{5}{5 + 3\cos\left(\theta + \dfrac{\pi}{4}\right)}$

33. $r = \dfrac{1}{1 - \cos\theta}$ **35.** $r = \dfrac{1}{2 + \sin\theta}$

37. $r = \dfrac{2}{1 + 2\cos\theta}$ **39.** $r = \dfrac{2}{1 - \sin\theta}$

41. $r = \dfrac{16}{5 + 3\cos\theta}$ **43.** $r = \dfrac{9}{4 - 5\sin\theta}$

45. If $0 < e < 1$, the conic is an ellipse.
If $e = 1$, the conic is a parabola.
If $e > 1$, the conic is a hyperbola.

47. (a) Hyperbola   (b) Ellipse
(c) Parabola   (d) Hyperbola

49. $r^2 = \dfrac{9}{1 - (16/25)\cos^2\theta}$ **51.** $r^2 = \dfrac{-16}{1 - (25/9)\cos^2\theta}$

53. 10.88   **55.** $r = \dfrac{345,996,000}{43,373 - 40,627\cos\theta}$; 11,004 miles

57. $r = \dfrac{92,931,075.2223}{1 - 0.0167\cos\theta}$

Perihelion: 91,404,618 miles
Aphelion: 94,509,382 miles

59. $r = \dfrac{5.537 \times 10^9}{1 - 0.2481\cos\theta}$

Perihelion: 4.436×10^9 kilometers
Aphelion: 7.364×10^9 kilometers

61. (a) 9.341×10^{18} square kilometers; 21.867 years
(b) 0.8995 radians; Larger angle with the smaller ray to generate an equal area
(c) Part (a): 2.559×10^9 kilometers; 1.17×10^8 kilometers per year
Part (b): 4.119×10^9 kilometers; 1.88×10^8 kilometers per year

63. Let $r_1 = \dfrac{ed}{1 + \sin\theta}$ and $r_2 = \dfrac{ed}{1 - \sin\theta}$.

The points of intersection of r_1 and r_2 are $(ed, 0)$ and (ed, π).
The slope of the tangent line to r_1 at $(ed, 0)$ is -1 and at (ed, π)
is 1. The slope of the tangent line to r_2 at $(ed, 0)$ is 1 and at
(ed, π) is -1. Therefore, at $(ed, 0)$, $m_1 m_2 = -1$ and at (ed, π),
$m_1 m_2 = -1$ and the curves intersect at right angles.

Review Exercises for Chapter 9 (page 709)

1. d   **2.** b   **3.** a   **4.** c

5. Circle
Center: $\left(\frac{1}{2}, -\frac{3}{4}\right)$
Radius: 1

7. Hyperbola
Center: $(-4, 3)$
Vertices: $\left(-4 \pm \sqrt{2}, 3\right)$

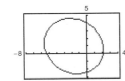

9. Ellipse
Center: $(2, -3)$
Vertices: $\left(2, -3 \pm \dfrac{\sqrt{2}}{2}\right)$

11. $y^2 - 4y - 12x + 4 = 0$

13. $\dfrac{(x-2)^2}{25} + \dfrac{y^2}{21} = 1$ **15.** $\dfrac{x^2}{16} - \dfrac{y^2}{20} = 1$

17. 15.87   **19.** $4x + 4y - 7 = 0$

21. (a) 192π cubic feet   (b) 7057.3 pounds
(c) 4.212 feet   (d) 429.105 square feet

23. $4y + 3x - 11 = 0$

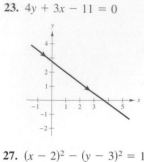

25. $x^2 + y^2 = 36$

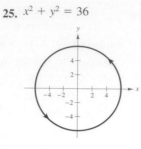

27. $(x - 2)^2 - (y - 3)^2 = 1$

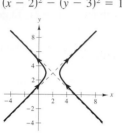

29. $x = 5t - 2$
    $y = 6 - 4t$

31. $x = 4\cos\theta - 3$
    $y = 4 + 3\sin\theta$

33.

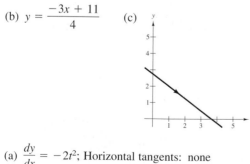

35. (a)

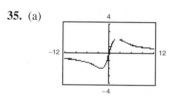

(b) From $x = 2\cot\theta$, it follows that $\cot\theta = \dfrac{x}{2}$.

Substituting into $y = 4\sin\theta\cos\theta$ results in

$$y = 4\left(\frac{x}{\sqrt{x^2 + 4}}\right)\left(\frac{2}{\sqrt{x^2 + 4}}\right).$$

This simplifies to $y = \dfrac{8x}{x^2 + 4}$ or $8x = (4 + x^2)y$.

37. (a) $\dfrac{dy}{dx} = -\dfrac{3}{4}$; Horizontal tangents: none

(b) $y = \dfrac{-3x + 11}{4}$ (c)

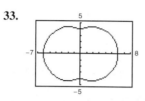

39. (a) $\dfrac{dy}{dx} = -2t^2$; Horizontal tangents: none

(b) $y = 3 + \dfrac{2}{x}$

(c)

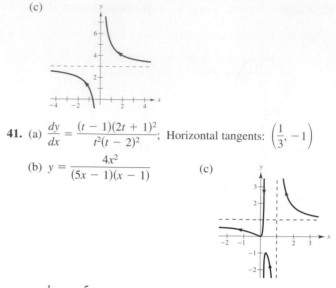

41. (a) $\dfrac{dy}{dx} = \dfrac{(t - 1)(2t + 1)^2}{t^2(t - 2)^2}$; Horizontal tangents: $\left(\dfrac{1}{3}, -1\right)$

(b) $y = \dfrac{4x^2}{(5x - 1)(x - 1)}$ (c)

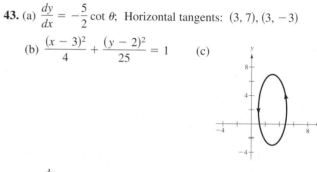

43. (a) $\dfrac{dy}{dx} = -\dfrac{5}{2}\cot\theta$; Horizontal tangents: $(3, 7), (3, -3)$

(b) $\dfrac{(x - 3)^2}{4} + \dfrac{(y - 2)^2}{25} = 1$ (c)

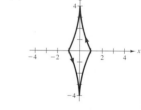

45. (a) $\dfrac{dy}{dx} = -4\tan\theta$; Horizontal tangents: none

(b) $x^{2/3} + (y/4)^{2/3} = 1$ (c)

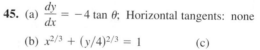

47. (a) and (c)

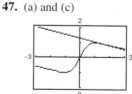

(b) $\dfrac{dx}{d\theta} = -4, \dfrac{dy}{d\theta} = 1, \dfrac{dy}{dx} = -\dfrac{1}{4}$

49. $\dfrac{\pi^2 r}{2}$

51.

$\left(4\sqrt{2}, \dfrac{7\pi}{4}\right), \left(-4\sqrt{2}, \dfrac{3\pi}{4}\right)$

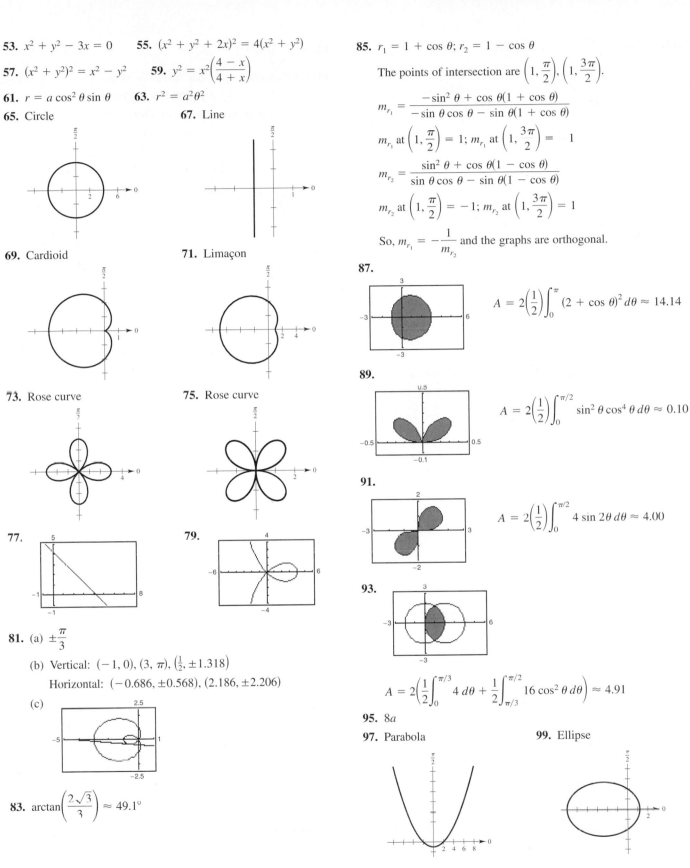

53. $x^2 + y^2 - 3x = 0$ **55.** $(x^2 + y^2 + 2x)^2 = 4(x^2 + y^2)$

57. $(x^2 + y^2)^2 = x^2 - y^2$ **59.** $y^2 = x^2\left(\dfrac{4 - x}{4 + x}\right)$

61. $r = a\cos^2\theta\sin\theta$ **63.** $r^2 = a^2\theta^2$

65. Circle   **67.** Line

69. Cardioid   **71.** Limaçon

73. Rose curve   **75.** Rose curve

77.   **79.**

81. (a) $\pm\dfrac{\pi}{3}$

(b) Vertical: $(-1, 0), (3, \pi), \left(\frac{1}{2}, \pm 1.318\right)$

Horizontal: $(-0.686, \pm 0.568), (2.186, \pm 2.206)$

(c)

83. $\arctan\left(\dfrac{2\sqrt{3}}{3}\right) \approx 49.1°$

85. $r_1 = 1 + \cos\theta; r_2 = 1 - \cos\theta$

The points of intersection are $\left(1, \dfrac{\pi}{2}\right), \left(1, \dfrac{3\pi}{2}\right)$.

$m_{r_1} = \dfrac{-\sin^2\theta + \cos\theta(1 + \cos\theta)}{-\sin\theta\cos\theta - \sin\theta(1 + \cos\theta)}$

m_{r_1} at $\left(1, \dfrac{\pi}{2}\right) = 1$; m_{r_1} at $\left(1, \dfrac{3\pi}{2}\right) = 1$

$m_{r_2} = \dfrac{\sin^2\theta + \cos\theta(1 - \cos\theta)}{\sin\theta\cos\theta - \sin\theta(1 - \cos\theta)}$

m_{r_2} at $\left(1, \dfrac{\pi}{2}\right) = -1$; m_{r_2} at $\left(1, \dfrac{3\pi}{2}\right) = 1$

So, $m_{r_1} = -\dfrac{1}{m_{r_2}}$ and the graphs are orthogonal.

87.

$A = 2\left(\dfrac{1}{2}\right)\displaystyle\int_0^\pi (2 + \cos\theta)^2\, d\theta \approx 14.14$

89.

$A = 2\left(\dfrac{1}{2}\right)\displaystyle\int_0^{\pi/2} \sin^2\theta\cos^4\theta\, d\theta \approx 0.10$

91.

$A = 2\left(\dfrac{1}{2}\right)\displaystyle\int_0^{\pi/2} 4\sin 2\theta\, d\theta \approx 4.00$

93.

$A = 2\left(\dfrac{1}{2}\displaystyle\int_0^{\pi/3} 4\, d\theta + \dfrac{1}{2}\displaystyle\int_{\pi/3}^{\pi/2} 16\cos^2\theta\, d\theta\right) \approx 4.91$

95. $8a$

97. Parabola   **99.** Ellipse

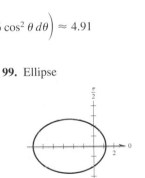

101. Hyperbola   **103.** $r = 10 \sin \theta$

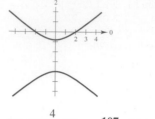

105. $r = \dfrac{4}{1 - \cos \theta}$ **107.** $r = \dfrac{5}{3 - 2 \cos \theta}$

P.S. Problem Solving (page 712)

1. (a)

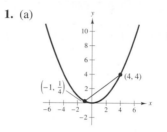

   (b) The slope of the tangent line to the parabola at $\left(-1, \frac{1}{4}\right)$ is $-\frac{1}{2}$. The slope of the tangent line to the parabola at $(4, 4)$ is 2. The product of the two slopes is -1 and therefore the tangent lines are perpendicular.

   (c) The directrix of the parabola is $y = -1$. The equations of the two tangent lines are $y = -\frac{1}{2}x - \frac{1}{4}$ and $y = 2x - 4$. They intersect at the point $\left(\frac{3}{2}, -1\right)$, which lies on the directrix.

3. Proof

5. (a) $r = 2a \tan \theta \sin \theta$

   (b) $x = \dfrac{2at^2}{1 + t^2}$

      $y = \dfrac{2at^3}{1 + t^2}$

   (c) $y^2 = \dfrac{x^3}{2a - x}$

7. $x = a \arccos\left(\dfrac{a - y}{a}\right) - \sqrt{2ay - y^2},\ 0 \le y \le 2a$

9. ∞

11. (a) Area of triangle $= \frac{1}{2} \times$ base $\times$ height

$$= \tfrac{1}{2}(1)(\tan \alpha)$$
$$= \tfrac{1}{2} \tan \alpha$$

and $A(\alpha) = \dfrac{1}{2} \displaystyle\int_0^\alpha \sec^2 \theta\, d\theta$

$$= \frac{1}{2}\Big[\tan \theta\Big]_0^\alpha$$
$$= \tfrac{1}{2} \tan \alpha$$

   (b) $\displaystyle\int_0^\alpha \sec^2 \theta\, d\theta = \Big[\tan \theta\Big]_0^\alpha$

$$= \tan \alpha$$

   (c) $\dfrac{d}{d\alpha}(\tan \alpha) = \sec^2 \alpha$

13. $r = \dfrac{1}{\sqrt{2}}\, de^{(\pi/4 - \theta)}$

15. (a) First plane: $x_1 = \cos 70(150 - 375t)$
$$y_1 = \sin 70(150 - 375t)$$
Second plane: $x_2 = \cos 45(450t - 190)$
$$y_2 = \sin 45(190 - 450t)$$

   (b) $\{[\cos 45(450t - 190) - \cos 70(150 - 375t)]^2$
$$+\ [\sin 45(190 - 450t) - \sin 70(150 - 375t)]^2\}^{1/2}$$

   (c)

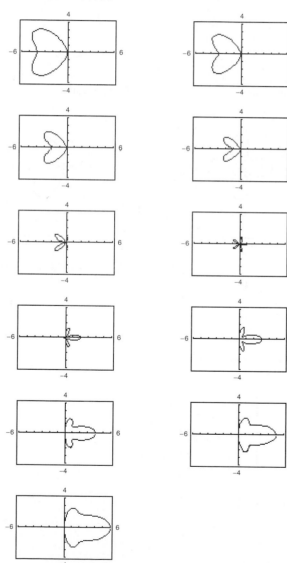

0.4145 hours; Yes

17. $r = \cos 5\theta + n \cos \theta$

$n = 1, 2, 3, 4, 5$ produce "bells"; $n = -1, -2, -3, -4, -5$ produce "hearts."

Appendix

Appendix A (page A6)

1.

x	−4	−2	0	2	4	8
y	2	0	4	4	6	8
dy/dx	−2	Undef.	0	$\frac{1}{2}$	$\frac{2}{3}$	1

3. (a) Answers will vary.   (b) $y = \frac{1}{2}(e^x + e^{-x})$

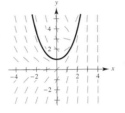

(c)

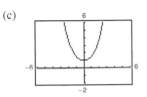

5. (a) Answers will vary.

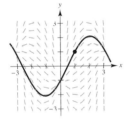

(b) $y = -\cos x + 1.8305 \sin x$

(c)

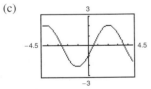

7.

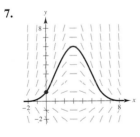

9.

n	0	1	2	3	4	5
x_n	0	0.1	0.2	0.3	0.4	0.5
y_n	2	2.2	2.43	2.693	2.9923	3.3315

n	6	7	8	9	10
x_n	0.6	0.7	0.8	0.9	1.0
y_n	3.7147	4.1462	4.6308	5.1738	5.7812

11.

n	0	1	2	3	4	5
x_n	0	0.05	0.10	0.15	0.20	0.25
y_n	3	2.7	2.4375	2.2088	2.0104	1.8393

n	6	7	8	9	10
x_n	0.30	0.35	0.40	0.45	0.50
y_n	1.6929	1.5686	1.4643	1.3778	1.3075

13.

n	0	1	2	3	4	5
x_n	0	0.1	0.2	0.3	0.4	0.5
y_n	1	1.1	1.2116	1.3390	1.4885	1.6699

n	6	7	8	9	10
x_n	0.6	0.7	0.8	0.9	1.0
y_n	1.9003	2.2131	2.6838	3.5398	5.9584

15. False. $y' + xy = x^2$ is linear.

17. $y = x^2 + 2x + \dfrac{C}{x}$

19. $y = e^{x^3}(x + C)$

21. $y = \frac{1}{2}(\sin x - \cos x) + Ce^x$

23. $y = x(\ln|x| + C)$

25. $y = -\frac{1}{13}(3 \sin 2x + 2 \cos 2x) + Ce^{3x}$

27. $y = \dfrac{x^3 - 3x + C}{3(x - 1)}$

29. $y = e^x(1 + \tan x) + C \sec x$

31. $y = \dfrac{bx^4}{4 - a} + Cx^a$

33. $y = 1 + 4e^{\tan x}$

35. $y - \sin x + (x + 1) \cos x$

37. $y = \dfrac{4}{x}$

39. $y = x \ln|x| + 12x - 2$

41. (a)

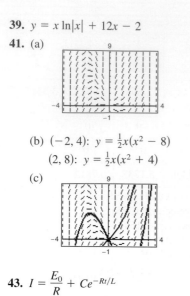

(b) $(-2, 4)$: $y = \frac{1}{2}x(x^2 - 8)$

   $(2, 8)$: $y = \frac{1}{2}x(x^2 + 4)$

(c)

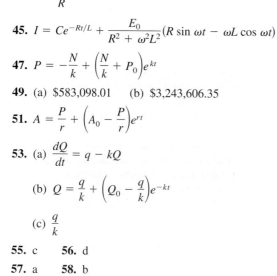

43. $I = \dfrac{E_0}{R} + Ce^{-Rt/L}$

45. $I = Ce^{-Rt/L} + \dfrac{E_0}{R^2 + \omega^2 L^2}(R \sin \omega t - \omega L \cos \omega t)$

47. $P = -\dfrac{N}{k} + \left(\dfrac{N}{k} + P_0\right)e^{kt}$

49. (a) \$583,098.01   (b) \$3,243,606.35

51. $A = \dfrac{P}{r} + \left(A_0 - \dfrac{P}{r}\right)e^{rt}$

53. (a) $\dfrac{dQ}{dt} = q - kQ$

(b) $Q = \dfrac{q}{k} + \left(Q_0 - \dfrac{q}{k}\right)e^{-kt}$

(c) $\dfrac{q}{k}$

55. c   **56.** d

57. a   **58.** b

Index of Applications

Index

ALGEBRA

Factors and Zeros of Polynomials

Let $p(x) = a_n x^n + a_{n-1} x^{n-1} + \cdots + a_1 x + a_0$ be a polynomial. If $p(a) = 0$, then a is a *zero* of the polynomial and a solution of the equation $p(x) = 0$. Furthermore, $(x - a)$ is a *factor* of the polynomial.

Fundamental Theorem of Algebra

An nth degree polynomial has n (not necessarily distinct) zeros. Although all of these zeros may be imaginary, a real polynomial of odd degree must have at least one real zero.

Quadratic Formula

If $p(x) = ax^2 + bx + c$, and $0 \le b^2 - 4ac$, then the real zeros of p are $x = \left(-b \pm \sqrt{b^2 - 4ac}\right)/2a$.

Special Factors

$$x^2 - a^2 = (x - a)(x + a) \qquad\qquad x^3 - a^3 = (x - a)(x^2 + ax + a^2)$$

$$x^3 + a^3 = (x + a)(x^2 - ax + a^2) \qquad\qquad x^4 - a^4 = (x^2 - a^2)(x^2 + a^2)$$

Binomial Theorem

$$(x + y)^2 = x^2 + 2xy + y^2 \qquad\qquad (x - y)^2 = x^2 - 2xy + y^2$$

$$(x + y)^3 = x^3 + 3x^2 y + 3xy^2 + y^3 \qquad\qquad (x - y)^3 = x^3 - 3x^2 y + 3xy^2 - y^3$$

$$(x + y)^4 = x^4 + 4x^3 y + 6x^2 y^2 + 4xy^3 + y^4 \qquad\qquad (x - y)^4 = x^4 - 4x^3 y + 6x^2 y^2 - 4xy^3 + y^4$$

$$(x + y)^n = x^n + nx^{n-1} y + \frac{n(n-1)}{2!} x^{n-2} y^2 + \cdots + nxy^{n-1} + y^n$$

$$(x - y)^n = x^n - nx^{n-1} y + \frac{n(n-1)}{2!} x^{n-2} y^2 - \cdots \pm nxy^{n-1} \mp y^n$$

Rational Zero Theorem

If $p(x) = a_n x^n + a_{n-1} x^{n-1} + \cdots + a_1 x + a_0$ has integer coefficients, then every *rational zero* of p is of the form $x = r/s$, where r is a factor of a_0 and s is a factor of a_n.

Factoring by Grouping

$$acx^3 + adx^2 + bcx + bd = ax^2(cx + d) + b(cx + d) = (ax^2 + b)(cx + d)$$

Arithmetic Operations

$$ab + ac = a(b + c) \qquad \frac{a}{b} + \frac{c}{d} = \frac{ad + bc}{bd} \qquad \frac{a + b}{c} = \frac{a}{c} + \frac{b}{c}$$

$$\frac{\left(\dfrac{a}{b}\right)}{\left(\dfrac{c}{d}\right)} = \left(\frac{a}{b}\right)\left(\frac{d}{c}\right) = \frac{ad}{bc} \qquad \frac{\left(\dfrac{a}{b}\right)}{c} = \frac{a}{bc} \qquad \frac{a}{\left(\dfrac{b}{c}\right)} = \frac{ac}{b}$$

$$a\left(\frac{b}{c}\right) = \frac{ab}{c} \qquad \frac{a - b}{c - d} = \frac{b - a}{d - c} \qquad \frac{ab + ac}{a} = b + c$$

Exponents and Radicals

$$a^0 = 1, \quad a \ne 0 \qquad (ab)^x = a^x b^x \qquad a^x a^y = a^{x+y} \qquad \sqrt{a} = a^{1/2} \qquad \frac{a^x}{a^y} = a^{x-y} \qquad \sqrt[n]{a} = a^{1/n}$$

$$\left(\frac{a}{b}\right)^x = \frac{a^x}{b^x} \qquad \sqrt[n]{a^m} = a^{m/n} \qquad a^{-x} = \frac{1}{a^x} \qquad \sqrt[n]{ab} = \sqrt[n]{a}\,\sqrt[n]{b} \qquad (a^x)^y = a^{xy} \qquad \sqrt[n]{\frac{a}{b}} = \frac{\sqrt[n]{a}}{\sqrt[n]{b}}$$